ADVANCES IN CARBONATE SEQUENCE STRATIGRAPHY: APPLICATION TO RESERVOIRS, OUTCROPS AND MODELS

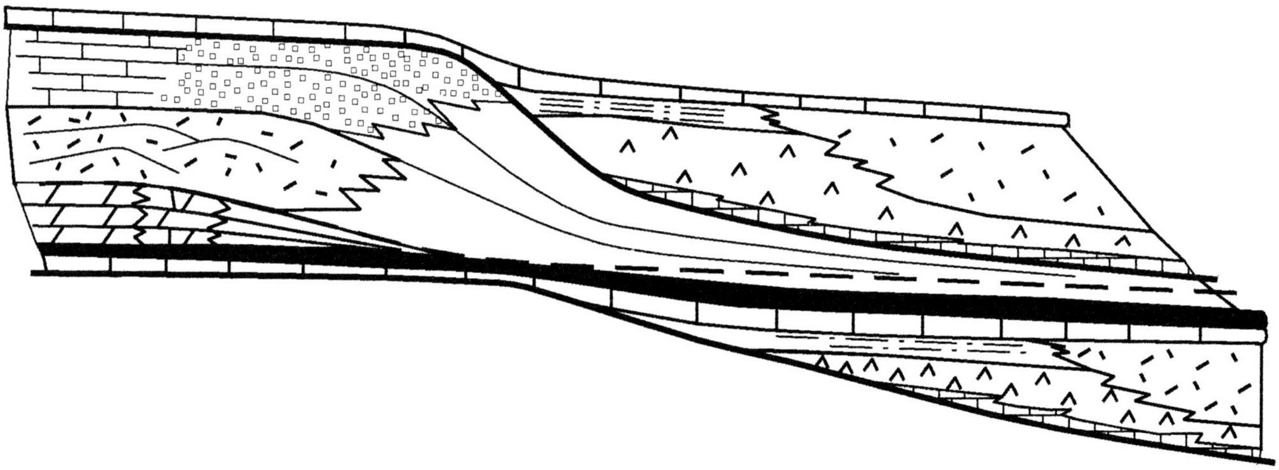

Edited by

Paul M. (Mitch) Harris, Chevron Petroleum Technology Company

Art H. Saller, Unocal

and

J.A. (Toni) Simo, University of Wisconsin

Robert W. Dalrymple, Editor of Special Publications

Special Publication No. 63

Tulsa, Oklahoma, U.S.A. *November, 1999*

SEPM thanks the following for their generous contributions to
Advances in Carbonate Sequence Stratigraphy: Application to Reservoirs, Outcrops and Models:

Mobil Oil Italiana S.p.A.
Exxon Production Research Company, Houston, Texas

ISBN: 1-56576-060-3

SEPM (Society for Sedimentary Geology)
1731 E. 71st Street
Tulsa, OK 74136-5108
1-800-865-9765

Printed in the United States

CONTENTS

SEQUENCE STRATIGRAPHY AND FACIES OF SLOPE AND BASIN CARBONATES

PALEO-OCEANOGRAPHIC FACTORS THAT CAUSE VARIATION IN CARBONATE DEPOSITION

INTRODUCTION

PAUL M. (Mitch) HARRIS
Chevron Petroleum Technology Company, 2811 Hayes Road, Houston, TX 77082
ART H. SALLER
Unocal, 14141 Southwest Freeway, Sugar Land, Texas 77478
AND
J. A. (Toni) SIMO
Department of Geology and Geophysics, University of Wisconsin, 1215 West Dayton Street, Madison, Wisconsin 53706

The SEPM Research Symposium for the 1997 AAPG–SEPM Annual Meeting in Dallas, Texas, was organized around the theme "Advances in Carbonate Sequence Stratigraphy—Application to Reservoirs, Outcrops, and Models." The symposium consisted of both oral and poster presentations. The sessions proved to be thought-provoking and forward-looking and showed how sequence stratigraphy is influencing many aspects in our understanding of carbonates. The very favorable response to the symposium encouraged many of the authors to publish their material and led to this SEPM Special Publication.

REVIEW OF CARBONATE SEQUENCE STRATIGRAPHY

Comparing stratigraphic patterns in seismic data to large-scale outcrop geometries started "sequence stratigraphy." Currently, sequence stratigraphy also encompasses the analysis of small-scale stratigraphic patterns from cores, logs, and outcrop studies and attempts to integrate those into larger-scale stratal geometries. As such, sequence stratigraphy is a multidisciplinary approach to stratigraphy in which genetically related facies are correlated between chronostratigraphic surfaces. Modern sequence stratigraphic studies commonly integrate seismic, well-log, lithostratigraphic, and biostratigraphic data. General advantages of the sequence stratigraphic approach are (1) more accurate correlations, (2) improved prediction of lithofacies, biofacies, and reservoir quality, (3) a more logical framework for depositional and diagenetic history, and (4) a temporal framework for integrating many types of stratigraphic, geochemical, and faunal data.

Roots in Seismic Stratigraphy

Sloss's (1963) recognition of craton-wide, unconformity-bounded sequences may be considered the earliest work on sequence stratigraphy. The unconformity-bounded sequence concept greatly expanded when sedimentary sequences were recognized in seismic lines. Seismic stratigraphy as advanced by AAPG Memoir 26 (Payton, 1977) is the root of sequence stratigraphic concepts. Increased seismic resolution developed during the late 1960s and 1970s allowed seismic data to portray large-scale stratal geometries in structurally undeformed areas. Seismic data made large stratal geometries visible on a scale never before possible, hundreds of meters thick and tens of kilometers long. Seismic stratigraphy of the 1970s (Vail and Exxon colleagues) emphasized large-scale stratal geometries and the repeated occurrence of certain seismic stratal patterns. An important assumption, based on observation, was that seismic reflectors followed stratigraphic time lines. Thus, seismic geometries could be used for time-stratigraphic correlation, to reconstruct paleobathymetry, and to interpret depositional environments and shifts in facies.

Sequence boundaries were the main surfaces for interregional stratigraphic correlation of depositional packages (Vail et al., 1977; Vail, 1987). Onlap, toplap, and erosional truncation were the main seismic relationships used to define sequence boundaries. Sequence boundaries were generally interpreted to be surfaces of subaerial exposure, even though they were identified by seismic relationships. Bubb and Hatlelid (1977) illustrated the seismic expression of carbonate buildups.

Sequence Stratigraphy

In the 1980s, as seismic reflection data continued to improve, seismic stratigraphic patterns within sequences were better defined. Sequence stratigraphy, an outgrowth of seismic stratigraphy, was also applied to outcrop and subsurface well data. Advances made in sequence stratigraphy during much of the 1980s were documented in SEPM Special Publication 42 (Wilgus et al., 1988). The "slug diagram" for a typical depositional sequence (Van Wagoner et al., 1988) was developed by using outcrop geometries, more refined seismic data, and computer modeling of subsidence and sedimentation. General computer models were built; they assumed a constant, basinward-increasing tectonic subsidence and a sinusoidal eustatic sea-level curve that resulted in a relative subsidence curve (Jervey, 1988). When a uniform sediment supply was superimposed on these relative sea-level curves, computer modeling produced the stratigraphic patterns shown in the slug diagram. Sequences were subdivided into lowstand, transgressive, and highstand systems tracts (Van Wagoner et al., 1988).

The Exxon sequence stratigraphic group of the 1980s generally emphasized that seismic sequences and other medium-scale stratal geometries were controlled by fluctuations of eustatic sea

Advances in Carbonate Sequence Stratigraphy: Application to Reservoirs, Outcrops and Models, SEPM Special Publication No. 63
 ISBN 1-56576-060-3, p. 1–10

level. Well logs, cores, and outcrops were increasingly being incorporated into the development of stratigraphic frameworks (Van Wagoner et al., 1990). A relatively detailed eustatic sea-level curve was published for the Mesozoic and Cenozoic by Haq et al. (1987, 1988), and a similar curve was published for the Paleozoic by Ross and Ross (1987). Although many of the general sequence stratigraphic patterns and principles were widely accepted, debate raged as to whether eustatic sea-level fluctuations were the main driving force in producing stratigraphic patterns on a time frame of 0.5–10 m.y. Stratigraphic modeling by Lawrence et al. (1990), Bosence (1990), and Aigner et al. (1991), and insightful geological discussion by Schlager (1991, 1992) better illustrated the interplay between sea-level change, subsidence variation, and other parameters as fundamental controls over stratigraphy.

The Haq et al. (1987, 1988) eustatic curve and most other sea-level curves have been intensely debated. Causes for the proposed large-amplitude sea-level fluctuations are unclear for most of geologic time. During icehouse times (late Paleozoic, Oligocene–Holocene), large-amplitude sea-level fluctuations were clearly due to continental glaciation. For instance, the Oligocene to middle Miocene Atlantic passive-margin succession shows a close relationship between the timing of the formation of sequence boundaries and global $\delta^{18}O$ increases (inferred ice buildup in the poles), thus suggesting that both occurred during global sea-level lowering (Miller et al., 1996). However, large continental glaciers did not exist during most of geologic history, and the lack of a viable mechanism to cause significant sea-level fluctuations has been a problem for advocates of large-scale eustatic fluctuations and generalized eustatic curves.

Depositional-facies models for carbonates and siliciclastics were superimposed on stratal geometries to create large-scale stratigraphic and facies models for generalized sequences; for example, Sarg (1988) showed how carbonate depositional systems might fit into a generalized sequence stratigraphic model. Landward- and basinward-stepping of depositional facies were shown within sequences in which smaller-scale cycles were identified. However, some unique aspects of carbonates were not fully captured in these models. Carbonates are characterized by *in situ* growth, high sedimentation and aggradation rates, rapid lithification, and deposition that is out of phase relative to siliciclastics when related to sea level. Schlager (1992) and Fitchen (1997) summarized additional important differences between siliciclastics and carbonates. As a result of these many differences, carbonate sequences and systems tracts can differ from those of siliciclastics.

Carbonate Sequence Stratigraphy

Although written before the term "sequence stratigraphy" was introduced, the reciprocal sedimentation model of Wilson (1967) is a classic work in sequence stratigraphy, especially carbonate sequence stratigraphy. Many later sequence stratigraphic "ideas" follow Wilson's reciprocal sedimentation model. Meissner (1972) immediately applied the concept in relating shelf deposits to basin deposits in the Permian basin. Changes in relative sea level and siliciclastic bypass were the cornerstone of Wilson's (1967) model. Wilson proposed that the locus of clastic deposition was pushed landward when sea level rose over the shelf, allowing prolific carbonate sedimentation in clear water on basinward parts of the shelf when the shelf was flooded. During sea-level lowstands, sea level was below the shelf top, and terrigenous sediments moved across the shelf and were deposited as thick wedges on the slope and in the basin.

Kendall and Schlager (1981) summarized the interplay between carbonate deposition and relative sea-level change. But carbonate sequence stratigraphic research exploded in the middle and late 1980s, and the nature of that interplay became scrutinized in more detail. Sarg and Lehmann's (1986) pioneering work on outcrops and subsurface data in west Texas and southeast New Mexico showed how sequence stratigraphy could be used to correlate from the Permian carbonate shelf into the Delaware basin. They used outcrops and seismic data to show how carbonate shelves and ramps retrograded, aggraded, and prograded apparently in response to variations in sea level. Detailed, high-resolution sequence stratigraphic studies of the same outcrops by Sonnenfeld (1991), Sonnenfeld and Cross (1993), Kerans et al. (1994), Kerans and Fitchen (1995), and others showed how depositional cycles and facies change systematically within the larger-scale sequences. These studies illustrate the need to continue refining stratigraphic correlations and interpretations.

Carbonate sequence stratigraphic models (e.g., Sarg, 1988; Handford and Loucks, 1993; Greenlee and Lehmann, 1993; Fitchen, 1997) propose generalized predictive successions applicable to carbonate-dominated basins. However, detailed work in shallow- and deep-water carbonates from the Precambrian to the Holocene has demonstrated that carbonate sequence stratigraphic systems vary owing to styles of tectonic subsidence, basin or oceanographic configuration, carbonate organisms living at a particular time, climate, and characteristics of eustatic sea-level fluctuation (e.g., papers in Crevello et al., 1989; Tucker et al., 1990; Biddle and Schlager, 1991). As a result, empirical studies of carbonate systems of different ages and in different basins still have great value as important analogues. As resolution of reflection seismic data improved, matching of seismic geometries with stratigraphy and prediction of carbonate facies also improved (e.g., Eberli and Ginsburg, 1989; Wendte et al., 1992; Pomar, 1993; Fitchen, 1997). Many examples of improved sequence stratigraphic description and interpretation are found in AAPG Memoir 57 (Loucks and Sarg, 1993). Tucker and Wright (1990) and Kerans and Tinker (1997) have provided textbooks illustrating modern concepts in carbonate sequence stratigraphy.

Because carbonate depositional systems differ from those of siliciclastics (Kendall and Schlager, 1981), the sequences and systems tracts that they build are often different (Handford and Loucks, 1993; Pomar, 1993; P. M. Harris, unpublished notes for short course; Kerans and Fitchen, 1995; Fitchen, 1997). Facies-tracts distribution, stratal geometries, and therefore reservoir quality are highly variable within stratigraphic units and are unique to carbonate systems. This variability depends on the depositional system, platform-margin profile, and accommodation history (Fitchen, 1997), but is predictable with sequence models that are tailored to the intrinsic attributes of the depositional system. There is no single generalized sequence stratigraphic model that can be applied to all carbonate systems. As a result, models applied to any carbonate area should contain as many of the following features as possible: (1) stratal patterns from seismic data and appropriate outcrop analogues; (2) cycle definition and stacking patterns from core and log data, outcrop analogues, and perhaps analogous modern examples; and (3) a correlation framework from the integration of seismic, core, log, and outcrop data. Comparing the interpretations of Jacquin et al. (1991) and Everts et al. (1995) shows how different approaches to developing a sequence framework led to different results for the same area.

Carbonate Cyclostratigraphy

Beginning in the late 1980s, carbonate stratigraphers systematically analyzed meter-scale depositional cycles (i.e., upward-shallowing cycles of James, 1979; parasequences of Van Wagoner et al., 1988; or fundamental depositional sequences of Wanless, 1991) on ancient carbonate shelves and ramps. Fischer plots were constructed to show variations in cycle thickness versus time for shallow-marine carbonates (e.g., Goldhammer et al., 1987; Read and Goldhammer, 1988). Sadler et al. (1993) pointed out limitations of Fischer plots and showed that they must be used with care. Theoretical modeling of small-scale cyclic sedimentation demonstrated that composite sea-level curves containing short-term sea-level change within long-term sea-level oscillations could cause variations in cycle thickness that are observed in shelf carbonates (Read et al., 1986; Goldhammer et al., 1987, 1990).

Reconciling cycle stacking patterns with previous sequence stratigraphic concepts was not always obvious. Seismic stratigraphy had emphasized single surfaces (e.g., sequence boundaries and maximum flooding surfaces). However, cycle stacking patterns suggested that sequence boundaries and maximum flooding surfaces could actually be represented by multiple cycles. Several studies helped seismic-based sequence stratigraphers better realize that seismic-scale sequences are composed of numerous depositional cycles that change in thickness and character during lowstands, transgression, and progradation (e.g., Goldhammer et al., 1990, 1991; Montanez and Osleger, 1993; Kerans and Fitchen, 1995; Read, 1995; Fitchen, 1997). Small-scale stratigraphic units, e.g., upward-shoaling cycles and high-frequency sequences, are the fundamental building blocks of carbonate reservoirs, carbonate platforms, and shelf to basin relationships. The importance of correlating these small-scale stratigraphic units in hydrocarbon reservoir studies is widely accepted. During regional studies and hydrocarbon exploration, however, also critical is an understanding that platforms and basins interact together as part of the same small-scale units.

A corollary of depositional cycles is the relationship of carbonate cycle stratigraphy to Milankovitch orbital patterns, which has been the subject of many papers. Heckel (1986) suggested that orbital variations might have been critical to the Kansas cyclothems during the Pennsylvanian. Systematic variations in cycle thickness in some Triassic carbonates (as shown by Fischer plots: Goldhammer et al., 1987, 1990) suggest that those cycles are related to orbital cycles. More recently, however, the interpreted duration of these Triassic cycles and hence their relationship to orbital parameters has been challenged (Brack et al., 1996). Papers in Fischer and Bottjer (1991) and Schwarzacher (1993) summarize much of our understanding of orbital forcing, sedimentation, and stratigraphy. Although some (e.g., Osleger and Read, 1991; Borer and Harris, 1991; Read and Horbury, 1993; Goldhammer et al., 1993; Read, 1995) have considered Milankovitch-related eustasy as the main control for cyclic carbonates, others (e.g., Drummond and Wilkinson, 1993) have argued that much of the cyclicity is random. In addition, Schlager (1991, 1992) has identified factors other than sea level that control carbonate facies development and stratigraphy.

Carbonate Diagenesis

Early diagenesis is frequently pivotal in the porosity evolution of carbonates; therefore, by putting early diagenesis into a sequence stratigraphic framework, porosity may become more predictable (Sarg, 1988; Read and Horbury, 1993; Tucker, 1993; Sun and Wright, 1998). This includes dolomitization and subaerial exposure. Dolomitization is critical to porosity preservation in many carbonates (Schmoker and Halley, 1982). Montanez and Read (1992) showed how thin cycles deposited during late highstands were preferentially dolomitized; this finding suggested dolomitization by restricted seawater or tidal-flat circulation during late highstands. Similarly, Oswald (1992) and Pomar et al. (1996) showed dolomitization by hypersaline waters during basin-wide transgressions. Porosity distribution in carbonates is altered systematically during subaerial exposure and meteoric diagenesis (James and Choquette, 1988; Esteban and Wilson, 1993; Read and Horbury, 1993; Saller et al., 1994a). Since duration of subaerial exposure is one of the related factors controlling carbonate stratigraphy as well as diagenesis, sequence stratigraphy should help predict duration of subaerial exposure and hence final products of subaerial exposure (Read and Horbury, 1993). Saller et al. (1994b) showed that cyclic late-highstand limestones subjected to prolonged subaerial exposure have less porosity than early-highstand equivalents subjected to brief subaerial exposure. Sonnenfeld (1996) and Tinker et al. (1995) related karst events to a sequence stratigraphic framework. Tinker et al. (1995) showed that the three-dimensional distribution of caves in the giant Yates field of the Permian basin relates to four major cycles, each of which shoaled upward and was capped by a subaerially exposed island complex.

Carbonate Modeling

A variety of modeling approaches is advancing our understanding of carbonate seismic response, facies patterns, stratigraphy, and diagenesis. Seismic modeling of outcrops has shown interesting results that can be applied to the interpretation of seismic data. By building seismic models of key outcrop localities, Biddle et al. (1992), Campbell and Stafleu (1992), Stafleu and Schlager (1993), Stafleu and Sonnenfeld (1994), Anselmetti et al. (1997), and others have shown the sedimentological and stratigraphic details that potentially exist in apparent seismic downlap, onlap, and toplap relationships. In addition to studying "lap-out" patterns, these studies have also examined amplitude variations, seismic resolution, and seismic facies. Modeling of carbonate sedimentation patterns (e.g., Bice, 1988; Bosence, 1990; Bosence et al., 1994), the development of carbonate cycles (Read et al., 1986; Hardie et al., 1991), and larger-scale stratigraphic relationships (Kendall et al., 1989; Lawrence et al., 1990; Elrick and Read, 1991; Franseen et al., 1991; Borer and Harris, 1995) has provided insights into carbonate depositional systems and sequence stratigraphy that are leading to better interpretations of both rock and seismic data. Modeling of carbonate diagenesis within the context of a sequence stratigraphic framework is only now beginning. The dolomitization of carbonate platforms was modeled by Kaufman (1994). Other examples of basin-wide fluid migration and modeling can be found in Montanez et al. (1997).

Carbonate Reservoirs

In recent years, special attention has been given to applications of carbonate sequence stratigraphy to description of carbonate reservoirs (e.g., Read, 1995; Fitchen, 1997; Kerans and Tinker, 1997). Sequence stratigraphic boundaries are also commonly boundaries for major reservoir units, and cycles or sets of cycles frequently equate with flow units in a reservoir. The upward change from fine- to coarse-grained textural fabrics that commonly occurs within a depositional cycle can be translated into predictable, petrophysically significant relationships that aid log interpretation of permeability (Lucia, 1983). The cycle framework thus has direct applications for reservoir characterization and flow-modeling studies in carbonate reservoirs. By analyzing the distribution of reservoir-quality facies within a cyclic framework, several studies have shown cycles to be the basic architectural element of carbonate-platform reservoirs (Goldhammer et al., 1991; Kerans and Nance, 1991; Lindsay, 1991; Kendall et al., 1991; Borer and Harris, 1991; Cross et al., 1993; Harris et al., 1993; Kerans and Tinker, 1997). Dividing a reservoir interval into smaller time-bounded units facilitates the accurate assessment of depositional and diagenetic facies and can form the basis of stochastic reservoir modeling (Grant et al., 1994; Kerans et al., 1994; Eisenberg et al., 1994).

Carbonate Sequence Stratigraphy and Paleoceanography

Sequence stratigraphy provides a detailed temporal framework that can be correlated to global and regional paleoceanographic changes. Carbonate systems record biological and geochemical changes and have the potential to provide a record of events in Earth history. For instance, fluctuations in the stable isotopes of marine carbonate reflect changes in the stable isotope compositions of the seawater in which they formed. Stable isotope data from carbonates deposited on the ocean floor (DSDP and ODP results) have been essential for interpreting changes in Cenozoic and Mesozoic climates and organic carbon burial (e.g., Shackleton, 1987; Miller et al., 1987). The isotopic anomalies are used as chemostratigraphic correlation lines between deep- and shallow-water facies and global events (e.g., Drzewiecki and Simo, 1997). The powerful combination of measuring variations in isotopes from time-equivalent pelagic and benthic fauna in different latitudes makes it possible to map global isotope variability and trace water-mass movements. The correlation of these changes to shallow-water carbonates provides new avenues to interpret environmental variation.

Ideally, interpretation of stable isotope fluctuations includes carbon and oxygen isotope values and well-constrained stratigraphic, paleontologic, and sedimentologic data, as well as minor and major element compositions. Since diagenesis modifies the original stable isotope signature, especially that of oxygen (Land, 1995), detailed sampling, petrography, and sedimentology and reproducibility of the results in more than one section on the shelf and basin are needed to build solid interpretations. The greater diagenetic stability of carbon isotopes has made them the preferred isotopic tool for studies of carbonates. Generally, positive $\delta^{13}C$ excursions correlate with high sea levels (Arthur et al., 1987) and may be enhanced by increased warm, saline, anoxic bottom waters or during periods of enhanced ocean circulation. As is evidenced by the late Cenozoic record, however, these ocean- and organic-carbon–driven processes can be overwhelmed by tectonics (Beck et al., 1995). On the basis of oxygen isotopes from the Great Barrier Reef, Isern et al. (1996) concluded that fluctuations in sea-surface temperature could account for much of the observed spatial and temporal variations of reef growth and carbonate facies change. This method becomes progressively less effective in older rocks because of the more problematic calculations of ice volumes and more complicated diagenesis.

THIS SEPM SPECIAL PUBLICATION

This SEPM Special Publication contains studies that represent advances in our understanding of stratigraphy and the distribution of carbonate facies and diagenesis within a sequence stratigraphic framework. The case studies in this volume reinforce the observation that the most important factor controlling stratigraphic relationships is accommodation change, which is generally a product of variations in eustatic sea level and tectonic subsidence. Variations in input of terrigenous sediment, climate, hydrologic factors, and other more subtle variables like nutrient influx are also important. Although many earlier sequence stratigraphic studies tended to emphasize one of the above variables, most of the studies in this volume show that there is an interplay of several variables and their relative importance varies by time (icehouse versus greenhouse), tectonic setting, and location in the depositional system.

The sequence stratigraphic studies included in this SEPM Special Publication develop a better framework for interpreting depositional and diagenetic processes in carbonate strata. The subject matter falls generally into five main categories: (1) platform to basin stratigraphic correlation and facies distribution, (2) high-resolution sequence framework bridging the gap between seismic-scale and reservoir-scale stratigraphy, (3) cycle-stacking patterns and their relationship to longer-term sequences, (4) sequence stratigraphy and facies of slope and basin carbonates, and (5) paleoceanographic factors that cause variation in carbonate deposition.

Platform to Basin Correlation and Facies Distribution

Large-scale platform to basin correlation and facies distribution have been the "heart and soul" of sequence stratigraphy since its advent. Sequence stratigraphic models continue to improve, making correlation more accurate and prediction of facies more reliable from seismic, well, and outcrop data. Hence, better-constrained type examples aid in the interpretation of other systems.

Sarg, Markello, and Weber have developed a second-order (10–20 m.y.) sequence stratigraphic model for basin evolution as a framework to predict the distribution of reservoir rocks, source rocks, and seals in late Paleozoic rocks in the southwestern United States. Their work indicates that source-rock-quality (source-prone), organic-rich calcareous mudstones are deposited on shelves and in basins during the transgressive phase of second-order cycles. In the view of these authors, carbonate reservoirs are best developed during third-order highstands of second-order transgressions and early highstands. The authors also discuss the relationship between carbonate shelves and evaporite- or sandstone-filled basins. In the Pennsylvanian Paradox basin, basin restriction was insufficient for evaporites to accumulate during fourth-order lowstands, but thick evaporitic successions accumulated during third-order composite-sequence lowstands. Siliciclastic bypass filled the Delaware basin during composite high-frequency lowstands, and the major time of sandstone deposition was during the second-order transition from transgression to progradation.

Schlager convincingly argues that the boundary between a highstand systems tract and overlying transgressive systems tract should be recognized as a specific type of sequence boundary if a subaerial unconformity is not present. Such a boundary is commonly associated with long submarine hiatuses, drowned carbonate platforms, and pronounced seismic reflectors and results from a rate of subsidence that exceeds sea-level fall during a particular depositional cycle. A natural consequence of correctly identifying this type of boundary is better correlation between platforms lacking subaerial exposure and basins lacking lowstand systems tracts.

Ward studied the platform to basin transition in three aggradational to landward-stepping sequences in the Devonian of the Canning basin in Australia. Many of the platform margins underwent massive erosional truncation (collapse), resulting in near-vertical escarpments. The platform margins contain massive reef limestones laterally adjacent to stratified platform strata; however, detailed mapping of stratigraphic surfaces shows reef strata to be distinctly younger than the adjacent platform strata. The main cause of landward-stepping platform margins is interpreted as tectonic subsidence and faulting.

Hunt and Fitchen found that differential compaction can substantially alter stratal geometries and distribution of facies during and shortly after deposition. They studied upper Paleozoic shelf to basin transitions in outcrops and the subsurface of west Texas and southeast New Mexico. Modification of stratal geometries is especially dramatic over and basinward of previous carbonate-shelf margins. Stratal inflections associated with differential compaction could easily be misinterpreted as depositional shelf margins. Compaction over underlying shelf margins can also cause fracturing and localization of karst processes in overlying carbonates.

High-Resolution Sequence Framework Bridging the Gap Between Seismic-Scale and Reservoir-Scale Stratigraphy

Several articles in this volume continue to bridge the gap between understanding the seismic-scale sequence and small or reservoir-scale cycles. In general, the relationships are more complex than would have been predicted by models in AAPG Memoir 26 or by Sarg (1988).

Smith and Read show how limestone and sandstone facies vary in the Mississippian of the Illinois basin. Their stratigraphic interval straddles the greenhouse to icehouse transition. The stratigraphic patterns show distinct differences as a result of the variations in eustatic sea-level fluctuations and rates of subsidence. The ramp system changes from one with broad facies belts to one with abrupt facies changes. Deep erosional incision is associated with major sequence boundaries.

Rankey, Bachtel, and Kaufman show the stratigraphic complexities associated with high-frequency sequences in a mixed carbonate-siliciclastic system in the Upper Pennsylvanian of the Sacramento Mountains of New Mexico. Their stratigraphic section extends from the shelf to the basin, and they used sequence boundaries and subaerial exposure surfaces formed during lowstands of sea level to make a high-resolution correlation. Depositional facies in this setting show dramatic lateral variations. The shelf margin is dominated by marine carbonates whereas landward and basinward strata are characterized by marine carbonates interbedded with alluvial and marine siliciclastics.

Mazzullo presents a stratigraphic framework based on a large number of measured sections for the Lower Permian ramp succession in the U.S. Midcontinent. The complex internal architecture of systems tracts within most of his sequences reflects an interplay between forcing parameters. Syndepositional tectonism—specifically regional uplift or dramatically reduced rates of subsidence—overprinted thickness variations on the fundamental glacio-eustatically forced stratigraphy and resulted in the geometrically complex internal architecture of the sequences and cycles.

To investigate the three-dimensional architecture of a part of the Capitan margin, Osleger and Tinker compare widely spaced, dip-oriented profiles and evaluate sequence boundaries, facies tracts, stacking patterns, and offlap angles. Systematic regional trends are recognized on plots of key depositional variables that were measured on individual high-frequency sequences. The trends illustrate the strike-parallel evolution of the margin and suggest control by regional sea level.

Cycle-Stacking Patterns and Their Relationship to Longer-Term Sequences

The recognition of differences in styles of cycle-stacking patterns is critical for better defining sequences and potentially serves as a guide for constructing higher-resolution zonation in reservoir studies.

Lehrmann and Goldhammer describe icehouse platforms containing stacks of high-frequency sequences that are capped with subaerial-exposure surfaces. The sequences have variable thicknesses but high lateral continuity. In contrast, greenhouse platforms contain broad transgressive-regressive sequences showing systematic change in cycle thickness, facies proportions, deepening and shallowing trends, and sometimes a capping karst surface. The authors evaluate levels of stratigraphic order and secular variations in stratigraphic style for platform carbonates by analyzing a digital database from diverse settings and ages according to a variety of statistical tests. Their results of different classes of stratigraphic style based on age will help to adjust stacking-pattern analysis techniques in order to construct accurate chronostratigraphic zonation between platforms in which a broad spectrum of stratigraphic behavior exists.

Saller, Dickson, Rasbury, and Ebato show that facies, diagenesis, and porosity in subsurface upper Paleozoic limestone of west Texas are related to cycle thickness and duration of subaerial exposure. Thinner cycles show more intense meteoric alteration and cementation and, thus, reduced porosity. They describe temporal changes in cycle characteristics and thickness and interpret these changes in relationship to changes in accommodation space. Thinner cycles are interpreted to represent the same duration as the thicker cycles, but thinner cycles have a longer duration of subaerial exposure and meteoric diagenesis. Their observation can be used in hydrocarbon exploration and general examination of other upper Paleozoic carbonates as well as strata deposited during other icehouse periods.

Sequence Stratigraphy and Facies of Slope and Basin Carbonates

Carbonate in slope and basin environments is either transported downslope from shallow-marine environments or settles from the water column following chemical or biologic precipitation. As a result, slope and basin strata can record sequence stratigraphic events affecting platforms and platform margins.

Westphal, Reijmer, and Head describe compositional changes and diagenetic trends for Cenozoic sediments on the western slope of the Great Bahama Bank and relate them to sea-level history and platform evolution. Their work shows that compositional changes in slope sediment indicate whether the platform was flat topped or sloping basinward (a ramp). Ramps continue to produce carbonate during changes in sea level, and changes in carbonate composition are minor. In contrast, sediment formation shuts down on flat-topped platforms during falls of sea level, and the composition of slope sediments also changes dramatically. The cementation and compaction histories of slope sediments respond to sediment composition and grain size, supply, and sea-level changes. Lowstand, coarse-grained layers are tightly cemented and uncompacted. Both lowstand and highstand fine-grained sediments show an alternation between cemented and uncemented layers. Early cementation history is an important factor for acoustic impedance, leading to the formation of reflectors on seismic profiles.

Choi, Simo, and Saylor discuss the sequence stratigraphy of Upper Ordovician carbonate and siliciclastic strata of Wisconsin. Their focus is on distinguishing depositional sequences and evaluating controls on strata composition and geometry in a ramp system developed within an epeiric sea. Both siliciclastic content and proportion of skeletal grains increase in shallow, more proximal facies so that fining-upward and coarsening-upward successions are interpreted in terms of retrogradation and progradation. Many cycles and cycle sets are asymmetric, sharp-based units that gradationally fine and become cleaner upward. Although seemingly indicative of deepening-upward cycles, the authors interpret the asymmetry in terms of reciprocal sedimentation, with breaks in sedimentation that may be unrelated to sea level. Additional evidence for frequent breaks in sedimentation is found in the numerous closely spaced omission surfaces distributed through much of the succession.

Van Konijnenburg, Bernoulli, and Mutti compare the stratigraphic architecture of a Lower Cretaceous–lower Tertiary base-of-slope succession with a nearby carbonate margin in the central Apennines of Italy. Their analysis, based in part on a comparison of "geohistory" curves (i.e., depth vs. time), suggests that changes in facies associations in the base-of-slope deposits are controlled by cyclic changes in sedimentary regime on the platforms.

Mettraux, Homewood, Schwab, and Guillocheau describe a sedimentologic model for the Cretaceous chalk in the Paris basin that relates facies, hardgrounds, and chert abundance to times of progradation and retrogradation. Cherts are more abundant during times of progradation at the cycle scale, and hardgrounds indicate sediment bypass at the change from longer-term decreasing to increasing accommodation. Progradational intervals with high chert abundance capped by hardgrounds are shown on synthetic seismograms to form reflectors that allow for basin-wide correlation and separation of different rock units. Their model suggests that facies changes and porosity in the chalk are an indication of changes in both accommodation space and sediment supply. They recognize that minor changes in bathymetry were enough to change sediment supply and increase or decrease progradation rates.

Paleoceanographic Factors That Cause Variation in Carbonate Deposition

Previous work on carbonate systems has shown that oceanographic factors like nutrient supply can affect carbonate systems and their sequence stratigraphy (Hallock and Schlager, 1986). These factors are often difficult to evaluate directly in the rock record; however, indirect stratigraphic or geochemical evidence may suggest their importance.

Zempolich and Erba, in describing the Middle Jurassic Vajont Limestone slope sequences in northern Italy, emphasize the growth of carbonate slopes during times of high sea level and carbonate production and demonstrate the use of chemostratigraphy in correlating slope strata to changes in sea level and ocean chemistry. Prolific oolitic sedimentation was the result of a short-term eustatic rise. Resedimentation occurred by wind-driven offshore transport enhanced by the lack of a shelf margin. Radial oolites in the Vajont provide a proxy for Middle Jurassic marine calcite composition. Higher $\delta^{18}O$ values correlate with the skeletal-grain–poor part of slope sequences and may reflect unrestricted circulation of open-ocean water across the platform during the transgression. Transgressive slope carbonates are depleted in ^{13}C, and highstand slope carbonates are enriched in ^{13}C. The $\delta^{13}C$ shifts correlate to secular variations in basinal strata and probably represent erosion of organic carbon stored in coastal environments during regressions and reestablishment of the oceanic carbon reservoirs during highstands.

Detailed isotopic and biostratigraphic data presented by Mutti, Bernoulli, Spezzaferri, and Stille test the relationship between facies changes, global cooling trends, and the paleoceanographic conditions existing in the Mediterranean area during the Neogene. Their data show that a change in carbonate facies from tropical to temperate predates a major cooling event and is related to a regional change in water circulation. Those changes affected rates of carbonate production during the deposition of a second-order sequence and accentuated the transgressive trend within sequences. Their results argue for more studies to document the response of carbonate facies to paleoceanographic changes in order to better estimate the reliability of the sea-level signature preserved in carbonate sequences.

Seismic data interpreted by Ferro, Droxler, Anderson, and Mucciarone show the recent geologic evolution of part of the Belize lagoon, which is intimately tied to shifts between siliciclastic and carbonate sediments induced by sea-level fluctuations. During a very restricted highstand interval, carbonate deposition occurred on a substratum of lowstand and transgressive siliciclastic deposits. The long-term colonization of patch reefs over a lowstand shelf-edge delta, beach ridges, and flooded levees is a possible mechanism by which the present barrier reef established itself.

Summary

A sequence stratigraphic study should vary depending on the type of data available, as shown by the papers in this volume. The approach depends on the resolution and dimensionality of the data. Different types of data have different advantages, and the most effective studies will combine all data types. Seismic data should be used to determine the geometry of large-scale sequences and their internal boundaries. Log and core data can establish a stratigraphic hierarchy by defining depositional cycles and bounding surfaces that are correlated within that framework. Outcrop data can combine the advantageous features of the above data types, depending on outcrop scale. Furthermore, outcrops can also be used to assess lateral facies relationships within cycles and relate them to geochemistry. It is our impression that the papers in this SEPM Special Publication introduce new ideas and document stratigraphic frameworks for several key areas. These new studies develop a better basis for understanding depositional and diagenetic processes in carbonate strata. Such a basis will allow (1) better prediction of reservoir rocks and seals, (2) improved understanding of flow units and reservoir architecture, (3) more accurate correlation between platform and basin, (4) temporal frameworks to compare modern and ancient systems, (5) quantification of lithologic changes and geometries in outcrops, and (6) linkages between shallow- and deep-water carbonates, oceans, and causal mechanisms.

Much remains to be learned about carbonate systems and their sequence stratigraphy. We think that the papers in this volume significantly advance our knowledge of this topic, but realize that carbonate sequence stratigraphic concepts are still evolving and new ideas will continue to be tested. The future will continue to show that sequence stratigraphy is most effective when all types of data and approaches are integrated. Seismic stratigraphy when combined with well data, depositional and diagenetic models, and basin modeling should make carbonate facies and porosity patterns even more predictable in the future. High-resolution seismic data from outcrops and quantification of parameters that can be measured in the field and laboratory (e.g., volume of facies, trends in fauna, diagenetic trends, and permeability) will form the basis for calibration of subsurface data and models.

ACKNOWLEDGMENTS

We thank the authors of the papers in this volume for their interesting contributions and for adhering to our rigorous schedule for submitting and editing the manuscripts. We thank SEPM for their support of our Research Symposium and publishing this volume. Finally, we very much thank the following for their assistance in editing the manuscripts: Steve Bachtel, Darwin Boardman, Clint Cowan, Steve Dorobek, Bill Fitchen, Bob Goldstein, Mark Harris, Brian Keith, Brenda Kirkland-George, Dennis Kolata, Zak Lasemi, Dan Lehrmann, Bob Loucks, Rick Major, Jim Markello, Jim Mazzullo, Sal Mazzullo, Tim McHargue, Leslie Melim, Bill Morgan, Dave Osleger, Mike Pope, Fred Read, Rick Sarg, Beverly Saylor, Scott Tinker, Jim Wilson, Bob Wright.

REFERENCES

AIGNER, T., BRANDENBURG, A., VAN VLIET, A., DOYLE, M., LAWRENCE, D., AND WESTRICH, J., 1991, Stratigraphic modeling of epicontinental basins: Two applications: Sedimentary Geology, v. 69, p. 167–190.

ANSELMETTI, F. S., EBERLI, G. P., AND BERNOULLI, D., 1997, Seismic modeling of a carbonate platform margin (Montagna della Maiella, Italy): Variations in seismic facies and implications for sequence stratigraphy, *in* Palaz, I. and Marfurt, K. J., eds., Carbonate Seismology: Tulsa, Society of Exploration Geophysicists, Geophysical Developments Series No. 6, p. 121–178.

ARTHUR, M. A., SCHLANGER, S. O., AND JENKYNS, H. C., 1987, The Cenomanian-Turonian oceanic anoxic event: II. Palaeoceanographic controls on organic-matter production and preservation, *in* Brooks, J. and Fleet, A. J., eds., Marine Petroleum Source Rocks: London, Geological Society of London Special Publication No. 26, p. 401–420.

BECK, R. A., BURBANK, D. W., SERCOMBE, W. J., OLSON, T. L., AND KHAN, A. M., 1995, Organic carbon exhumation and global warming during the early Himalayan collision: Geology, v. 23, p. 387–390.

BICE, D., 1988, Synthetic stratigraphy of carbonate platform and basin systems: Geology, v. 16, p. 703–706.

BIDDLE, K. T., AND SCHLAGER, W., eds., 1991, The record of sea level fluctuations: Sedimentary Geology, v. 70, p. 85–270.

BIDDLE, K. T., SCHLAGER, W., RUDOLPH, K. W., AND BUSH, T. L., 1992, Seismic model of a progradational carbonate platform, Picco di Vallandro, the Dolomites, northern Italy: American Association of Petroleum Geologists Bulletin, v. 76, p. 14–30.

BORER, J. M., AND HARRIS, P. M., 1991, Lithofacies and cyclicity of the Yates Formation, Permian basin: Implications for reservoir heterogeneity: American Association of Petroleum Geologists Bulletin, v. 75, p. 726–779.

BORER, J. M., AND HARRIS, P. M., 1995, Computer simulation of the Yates Formation (Permian, Delaware basin)—Sequence stratigraphy and shelf-to-basin correlation implications, *in* Martin, R. L., ed., In Search of New Permian Basin Oil and Gas Fields: Using Today's Technologies and Tomorrow's Ideas for Exploration, Development, and 3D Seismic in a Mature Basin: Midland, West Texas Geological Society Symposium Publication No. 95-98, p. 111–132.

Bosence, D. W. J., 1990, Computer modeling and internal architecture of carbonate platforms: Geology, v. 18, p. 26–30.

BOSENCE, D. W. J., POMAR, L., WALTHAM, D. A., AND LANKESTER, T. H. G., 1994, Computer modeling of a Miocene carbonate platform, Mallorca, Spain: American Association of Petroleum Geologists Bulletin, v. 78, p. 247–266.

BRACK, P., MUNDIL, R., OBERLI, F., MEIER, M., AND RIEBER, H., 1996, Biostratigraphic and radiometric age data question the Milankovitch characteristic of the Latemar cycles (Southern Alps, Italy): Geology, v. 24, p. 371–375.

BUBB, J. N., AND HATLELID, W. G., 1977, Seismic stratigraphy and global changes of sea level: Part 10. Seismic recognition of carbonate buildups, *in* Payton, C. E., ed., Seismic Stratigraphy—Applications to Hydrocarbon Exploration: Tulsa, American Association of Petroleum Geologists Memoir 26, p. 185–204.

CAMPBELL, A. E., AND STAFLEU, J., 1992, Seismic modeling of an Early Jurassic, drowned carbonate platform: Djebel Bou Dahar, High Atlas, Morocco: American Association of Petroleum Geologists Bulletin, v. 76, p. 1760–1777.

CREVELLO, P. D., SARG, J. F., WILSON, J. L., AND READ, J. F., eds., 1989, Controls on Carbonate Platform and Basin Evolution: Tulsa, Society of Economic Paleontologists and Mineralogists Special Publication No. 44, 399 p.

CROSS, T. A., BAKER, M. R., CHAPIN, M. A., CLARK, M. S., GARDNER, M. H., HANSON, M. S., LESSENGER, M. A., LITTLE, L. D., MCDONOUGH, K-J., SONNENFELD, M. D., VALASEK, D. W., WILLIAMS, M. R., AND WITTER, D. N., 1993, Applications of high-resolution sequence stratigraphy to reservoir analysis, *in* Eschard, R. and Doligez, B., eds., Subsurface Reservoir Characterization from Outcrop Observations: Paris, Editions Technip, Proceedings of the 7th IFP Exploration and Production Research Conference, p. 11–33.

DRZEWIECKI, P. A., AND SIMO, J. A., 1997, Carbonate platform drowning and oceanic anoxic events on a mid-Cretaceous carbonate platform, south-central Pyrenees, Spain: Journal of Sedimentary Research, v. 67, p. 698–714.

DRUMMOND, C. N., AND WILKINSON, B. H., 1993, Carbonate cycle stacking patterns and hierarchies of orbitally forced eustatic sea-level change: Journal of Sedimentary Petrology, v. 63, p. 369–377.

EBERLI, G. P,. AND GINSBURG, R. N., 1989, Cenozoic progradation of northwestern Great Bahama Bank, a record of lateral platform growth and sea level fluctuations, *in* Crevello, P. D., Sarg, J. F., Wilson, J. L., and Read, J. F., eds., Controls on Carbonate Platform and Basin Evolution: Tulsa, Society of Economic Paleontologists and Mineralogists Special Publication No. 44, p. 339–355.

EISENBERG, R. A., HARRIS, P. M., GRANT, C. W., GOGGIN, D. J., AND CONNER, F. J., 1994, Modeling reservoir heterogeneity within outer ramp carbonate facies using an outcrop analog, San Andres Formation of the Permian basin: American Association of Petroleum Geologists Bulletin, v. 78, p. 1337–1359.

ELRICK, M., AND READ, J. F., 1991, Cyclic ramp-to-basin carbonate deposits, Lower Mississippian, Wyoming and Montana: A combined field and computer modeling study: Journal of Sedimentary Petrology, v. 61, p. 1194–1224.

ESTEBAN, M., AND WILSON, J. L., 1993, Introduction to karst systems and paleokarst reservoirs, *in* Fritz, R. D., Wilson, J. L., and Yurewicz, D. A., eds., Paleokarst-Related Hydrocarbon Reservoirs: Tulsa, SEPM (Society for Sedimentary Geology) Core Workshop No. 18, p. 1–9.

EVERTS, A. J. W., STAFLEU, J., SCHLAGER, W., FOUCKE, B. W., AND SWART, E. W., 1995, Stratal patterns, sediment composition and sequence stratigraphy at the margin of the Vercors platform: Journal of Sedimentary Research, v. B65, p. 119–131.

FISCHER, A. G., AND BOTTJER, D. J., eds., 1991, Orbital forcing and sedimentary sequences: Journal of Sedimentary Petrology, Special Issue, v. 61, p. 1063–1252.

FITCHEN, W. M., 1997, Carbonate sequence stratigraphy and its application to hydrocarbon exploration and reservoir development, *in* Palaz, I. and Marfurt, K. J., eds., Carbonate Seismology: Tulsa, Society of Exploration Geophysicists, Geophysical Developments Series No. 6, p. 121–178.

FRANSEEN, E. K., WATNEY, W. L., KENDALL, C. G. ST. C., AND ROSS, W., eds., 1991, Sedimentary modeling: Computer simulations and methods for improved parameter definition: Lawrence, Kansas Geological Survey Bulletin 233, 524 p.

GOLDHAMMER, R. K., DUNN, P. A., AND HARDIE, L. A., 1987, High frequency glacio-eustatic sea level oscillations with Milankovitch characteristics recorded in Middle Triassic platform carbonates in northern Italy: American Journal of Science, v. 287, p. 853–892.

GOLDHAMMER, R. K., DUNN, P. A., AND HARDIE, L. A., 1990, Depositional cycles, composite sea-level changes, cycle stacking patterns, and the hierarchy of stratigraphic forcing—Examples from platform carbonates of the Alpine Triassic: Geological Society of America Bulletin, v. 102, p. 535–562.

GOLDHAMMER, R. K., OSWALD, E. J., AND DUNN, P. A., 1991, Hierarchy of stratigraphic forcing: Example from Middle Pennsylvanian shelf carbonates of the Paradox basin, *in* Franseen, E. K., Watney, W. L., Kendall, C. G. St. C., and Ross, W., eds., Sedimentary Modeling: Computer Simulations and Methods for Improved Parameter Definition: Lawrence, Kansas Geological Survey Bulletin 233, p. 361–413.

GOLDHAMMER, R. K., LEHMANN, P. J., AND DUNN, P. A., 1993, The origin of high-frequency platform carbonate cycles and third-order sequences (Lower Ordovician El Paso Group, west Texas): Constraints from outcrop data and stratigraphic modeling: Journal of Sedimentary Petrology, v. 63, p. 318–359.

GRANT, C. W., GOGGIN, D. J., AND HARRIS, P. M., 1994, Outcrop analog for cyclic-shelf reservoirs, San Andres Formation of Permian basin: Stratigraphic framework, permeability distribution, geostatistics, and fluid-flow modeling: American Association of Petroleum Geologists Bulletin, v. 78, p. 23–54.

GREENLEE, S. M., AND LEHMANN, P. J., 1993, Stratigraphic framework of productive carbonate buildups, *in* Loucks, R. G. and Sarg, J. F., eds., Carbonate Sequence Stratigraphy—Recent Developments and Applications: Tulsa, American Association of Petroleum Geologists Memoir 57, p. 43–62.

HALLOCK, P., AND SCHLAGER, W., 1986, Nutrient excess and the demise of coral reefs and carbonate platforms: Palaios, v. 1, p. 389–398.

HANDFORD, C. R., AND LOUCKS, R. G., 1993, Carbonate depositional sequences and systems tracts—Responses of carbonate platforms to relative sea-level changes, *in* Loucks, R. G. and Sarg, J. F., eds., Carbonate Sequence Stratigraphy—Recent Developments and Applications: Tulsa, American Association of Petroleum Geologists Memoir 57, p. 3–41.

HAQ, B. U., HARDENBOL, J., AND VAIL, P. R., 1987, Chronology of fluctuating sea levels since the Triassic: Science, v. 235, p. 1156–1167.

HAQ, B. U., HARDENBOL, J., AND VAIL, P. R., 1988, Mesozoic and Cenozoic chronostratigraphy and eustatic cycles of sea-level change, *in* Wilgus, C. K., Hastings, B. S., Kendall, C. G. St. C., Posamentier, H. W., Ross, C. A., and Van Wagoner, J. C., eds., Sea-Level Changes—An Integrated Approach: Tulsa, Society of Economic Paleontologists and Mineralogists Special

Publication No. 42, p. 71–108.

Hardie, L. A., Dunn, P. D., and Goldhammer, R. K., 1991, Field and modeling studies of Cambrian carbonate cycles, Virginia Appalachians—Discussion: Journal of Sedimentary Petrology, v. 61, p. 636–646.

Harris, P. M., Kerans, C., and Bebout, D. G., 1993, Ancient outcrop and modern examples of platform carbonate cycles—Implications for subsurface correlation and understanding reservoir heterogeneity, *in* Loucks, R. G. and Sarg, J. F., eds., Carbonate Sequence Stratigraphy—Recent Developments and Applications: Tulsa, American Association of Petroleum Geologists Memoir 57, p. 475–492.

Heckel, P. H., 1986, Sea-level curve for Pennsylvanian eustatic marine transgressive-regressive depositional cycles along the Midcontinent outcrop belt, North America: Geology, v. 14, p. 330–334.

Isern, A. R., McKenzie, J. A., and Feary, D. A., 1996, The role of sea-surface temperature as a control on carbonate platform development in the western Coral Sea: Palaeogeography, Palaeoclimatology, Palaeoecology, v. 124, p. 247–272.

Jacquin, T., Arnaud-Vanneau, A., Arnaud, H., Ravenne, C., and Vail, P. R., 1991, Systems tracts and depositional sequences in a carbonate setting: A study of continuous outcrops from platform to basin at the scale of seismic lines: Marine and Petroleum Geology, v. 8, p. 122–139.

James, N. P., 1979, Shallowing-upward sequences in carbonates, Chapter 10, *in* Walker, R. G., ed., Facies Models: St. John's, Geological Association of Canada, Geoscience Canada Reprint Series 1, p. 109–119.

James, N. P., and Choquette, P. W., eds., 1988, Paleokarst: New York, Springer-Verlag, 416 p.

Jervey, M. T., 1988, Quantitative geological modeling of siliciclastic rock sequences and their seismic expression, *in* Wilgus, C. K., Hastings, B. S., Kendall, C. G. St. C., Posamentier, H. W., Ross, C. A., and Van Wagoner, J. C., eds., Sea-Level Changes—An Integrated Approach: Tulsa, Society of Economic Paleontologists and Mineralogists Special Publication No. 42, p. 47–69.

Kaufman, J., 1994, Numerical models of fluid flow in carbonate platforms: Journal of Sedimentary Research, v. A64, p. 128–139.

Kendall, C. G. St. C., and Schlager, W., 1981, Carbonates and relative changes in sea level: Marine Geology, v. 44, p. 181–212.

Kendall, C. G. St. C., Strobel, J., Tang, J., Moore, P., Cannon, R., Bezdek, J., and Giswas, G., 1989, Simulation of the West Texas, New Mexican Permian Guadalupian basin margin—A response to eustatic change, an example of SEDPAK, *in* Harris, P. M. and Grover, G. A., eds., Subsurface and Outcrop Examination of the Capitan Shelf Margin: Tulsa, Society of Economic Paleontologists and Mineralogists Core Workshop No. 13, p. 423–426.

Kendall, C. G. St. C., Bowen, B., Alsharhan, A., Cheong, D.-K., and Stoudt, D. L., 1991, Eustatic controls on carbonate facies in reservoirs and seals associated with Mesozoic hydrocarbon fields of the Arabian Gulf and Gulf of Mexico: Marine Geology, v. 102, p. 215–238.

Kerans, C., and Fitchen, W. M., 1995, Sequence hierarchy and facies architecture of a carbonate-ramp system: San Andres Formation of Algerita Escarpment and western Guadalupe Mountains, West Texas and New Mexico: Austin, University of Texas, Bureau of Economic Geology, Report of Investigations 235, 86 p.

Kerans, C., and Nance, H. S., 1991, High frequency cyclicity and regional depositional patterns of the Grayburg Formation, Guadalupe Mountains, New Mexico, *in* Meader-Roberts, S., Candelaria, M. P., and Moore, G. E., eds., Sequence Stratigraphy, Facies, and Reservoir Geometries of the San Andres, Grayburg, and Queen Formations, Guadalupe Mountains, New Mexico and Texas: Midland, Permian Basin Section–SEPM (Society for Sedimentary Geology) Publication No. 91-32, p. 53–69.

Kerans, C., and Tinker, S. W., 1997, Sequence Stratigraphy and Characterization of Carbonate Reservoirs: Tulsa, SEPM (Society for Sedimentary Geology) Short Course Notes No. 40, 130 p.

Kerans, C., Lucia, F. J., and Senger, R. K., 1994, Integrated characterization of carbonate ramp reservoirs using Permian San Andres Formation outcrop analogs: American Association of Petroleum Geologists Bulletin, v. 78, p. 181–216.

Land, L. S., 1995, Comment on "Oxygen and carbon isotopic composition of Ordovician brachiopods: Implications for coeval seawater" by H. Qing and J. Veizer: Geochimica et Cosmochimica Acta, v. 59, p. 2843–2844.

Lawrence, D. T., Doyle, M., and Aigner, T., 1990, Stratigraphic simulation of sedimentary basins—Concepts and calibrations: American Association of Petroleum Geologists Bulletin, v. 74, p. 273–295.

Lindsay, R. F., 1991, Grayburg Formation (Permian–Guadalupian): Comparison of reservoir characteristics and sequence stratigraphy in the Northwest Central Basin platform with outcrops in the Guadalupe Mountains, New Mexico, *in* Meader-Roberts, S., Candelaria, M. P., and Moore, G. E., eds., Sequence Stratigraphy, Facies, and Reservoir Geometries of the San Andres, Grayburg, and Queen Formations, Guadalupe Mountains, New Mexico and Texas: Midland, Permian Basin Section–SEPM (Society for Sedimentary Geology) Publication 91-32, p. 111–118.

Lucia, F. J., 1983, Petrophysical parameters estimated from visual descriptions of carbonate rocks: A new field classification of carbonate pore space: Journal of Petroleum Technology, v. 35, p. 629–637.

Loucks, R. G., and Sarg, J. F., eds., 1993, Carbonate Sequence Stratigraphy—Recent Developments and Applications: Tulsa, American Association of Petroleum Geologists Memoir 57, 545 p.

Meissner, F. F., 1972, Cyclic sedimentation in middle Permian strata of the Permian basin, *in* Elam, J. G. and Chuber, S., eds., Cyclic Sedimentation in the Permian Basin (2nd ed.): Midland, West Texas Geological Society Publication No. 72-60, p. 203–232.

Miller, K. G., Fairbanks, R. G., and Mountains, G. S., 1987, Tertiary oxygen isotope synthesis, sea level history, and continental margin erosion: Paleoceanography, v. 2, p. 1–19.

Miller, K. G., Mountain, G. S., the Leg 150 Shipboard Party, and Members of the New Jersey Coastal Plain Drilling Project, 1996, Drilling and dating New Jersey Oligocene-Miocene sequences: Ice volume, global sea level, and Exxon records: Science, v. 271, p. 1092–1095.

Montanez, I. P., and Osleger, D. A., 1993, Parasequence stacking patterns, third-order accommodation events, and sequence stratigraphy of Middle to Upper Cambrian platform carbonates, Bonanza King Formation, southern Great Basin, *in* Loucks, R. G. and Sarg, J. F., eds., Carbonate Sequence Stratigraphy—Recent Developments and Applications: Tulsa, American Association of Petroleum Geologists Memoir 57, p. 305–326.

Montanez, I. P., and Read, J. F., 1992, Eustatic control on dolomitization of peritidal carbonates: Evidence from the Early Ordovician Knox Group, Appalachians: Geological Society of America Bulletin, v. 104, p. 872–886.

Montanez, I. P., Gregg, J. M., and Shelton, K. L., 1997, Basin-Wide Diagenetic Patterns: Integrated Petrologic, Geochemical, and Hydrologic Considerations: Tulsa, SEPM (Society for Sedimentary Geology) Special Publication No. 57, 302 p.

Osleger, D. A., and Read, J. F., 1991, Relation of eustasy to stacking patterns of meter-scale carbonate cycles, Late Cambrian, U.S.A.: Journal of Sedimentary Petrology, v. 61, p. 1225–1252.

Oswald, E. J., 1992, Dolomitization of a Miocene reef complex, Mallorca, Spain: Unpublished Ph.D. Thesis, State University of New York, Stony Brook, 424 p.

Payton, C. E., ed., 1977, Seismic Stratigraphy—Applications to Hydrocarbon Exploration: Tulsa, American Association of Petroleum Geologists Memoir 26, 516 p.

Pomar, L., 1993, High resolution sequence stratigraphy in prograding carbonates: Application to seismic interpretation, *in* Loucks, R. G. and Sarg, J. F., eds., Carbonate Sequence Stratigraphy—Recent Developments and Applications: Tulsa, American Association of Petroleum Geologists Memoir 57, p. 389–407.

Pomar, L., Ward, W. C., and Green, D. G., 1996, Upper Miocene reef complex of the Llucmajor area, Mallorca, Spain, *in* Franseen, E. K., Esteban, M., Ward, W. C., and Rouchy, J-M., eds., Models for Carbonate Stratigraphy from Miocene Reef Complexes of Mediterranean Regions: Tulsa, SEPM (Society for Sedimentary Geology) Concepts in Sedimentology and Paleontology No. 5, p. 191–225.

Read, J. F., 1995, Overview of carbonate platform sequences, cycle stratigraphy and reservoirs in greenhouse and ice-house worlds, Part 1, *in* Read, J. F., Kerans, C., and Weber, L. J., eds., Milankovitch Sea Level Changes, Cycles and Reservoirs on Carbonate Platforms in Greenhouse and Ice-House Worlds: Tulsa, SEPM (Society for Sedimentary Geology) Short Course Notes No. 35, p. 1–102.

READ, J. F., AND GOLDHAMMER, R. K., 1988, Use of Fischer plots to define third-order sea-level curves in peritidal cyclic carbonates, Ordovician, Appalachians: Geology, v. 16, p. 895–899.

READ, J. F., AND HORBURY, A. D., 1993, Eustatic and tectonic controls on porosity evolution beneath sequence-bounding unconformities and parasequence disconformities on carbonate platforms, *in* Horbury, A. D. and Robinson, A. G., eds., Diagenesis and Basin Development: Tulsa, American Association of Petroleum Geologists Studies in Geology No. 36, p. 155–197.

READ, J. F., GROTZINGER, J. P., BONE, J. A., AND KOERSCHNER, W. F., 1986, Models for generation of carbonate cycles: Geology, v. 14, p. 107–110.

ROSS, C. A., AND ROSS, J. R. P., 1987, Late Paleozoic sea levels and depositional sequences: Ithaca, Cushman Foundation for Foraminiferal Research Special Publication No. 24, p. 137–149.

SADLER, P. M., OSLEGER, D. A., AND MONTANEZ, I. P., 1993, On the labeling, length, and objective basis of Fischer plots: Journal of Sedimentary Petrology, v. 63, p. 360–368.

SALLER, A. H., BUDD, D. A., AND HARRIS, P. M., 1994a, Unconformities and porosity development in carbonate strata: Ideas from a Hedberg Conference: American Association of Petroleum Geologists Bulletin, v. 78, p. 857–872.

SALLER, A. H., DICKSON, J. A. D., AND BOYD, S. A., 1994b, Cycle stratigraphy and porosity in Pennsylvanian and Lower Permian shelf limestones, eastern Central Basin platform: American Association of Petroleum Geologists Bulletin, v. 78, p. 1820–1842.

SARG, J. F., 1988, Carbonate sequence stratigraphy, *in* Wilgus, C. K., Hastings, B. S., Kendall, C. G. St. C., Posamentier, H. W., Ross, C. A., and Van Wagoner, J. C., eds., Sea-Level Changes—An Integrated Approach: Tulsa, Society of Economic Paleontologists and Mineralogists Special Publication No. 42, p. 155–181.

SARG, J. F., AND LEHMANN, P. J., 1986, Lower-middle Guadalupian facies and stratigraphy, San Andres–Grayburg Formations, Permian basin, Guadalupe Mountains, New Mexico, *in* Moore, G. E. and Wilde, G. L., eds., Lower and middle Guadalupian facies, stratigraphy, and reservoir geometries, San Andres–Grayburg Formations, Guadalupe Mountains, New Mexico and Texas: Midland, Permian Basin Section, Society of Economic Paleontologists and Mineralogists Publication No. 86-25, p. 1–36.

SCHLAGER, W., 1991, Depositional bias and environmental change—Important factors in sequence stratigraphy: Sedimentary Geology, v. 70, p. 109–130.

SCHLAGER, W., 1992, Sedimentology and Sequence Stratigraphy of Reefs and Carbonate Platforms, A Short Course: Tulsa, American Association of Petroleum Geologists Continuing Education Course Notes Series No. 34, 71 p.

SCHMOKER, J. W., AND HALLEY, R. B., 1982, Carbonate porosity versus depth—A predictable relation for South Florida: American Association of Petroleum Geologists Bulletin, v. 66, p. 2561–2570.

SCHWARZACHER, W., 1993, Cyclostratigraphy and the Milankovitch Theory: Amsterdam, Elsevier, 225 p.

SHACKLETON, N. J., 1987, The carbon isotope record of the Cenozoic: History of organic carbon burial and of oxygen in the oceans and atmosphere, *in* Brooks, J. and Flaeet, A. J., eds., Marine Petroleum Source Rocks: London, Geological Society of London Special Publication No. 26, p. 423–434.

SLOSS, L. L., 1963, Sequences in the cratonic interior of North America: Geological Society of America Bulletin, v. 74, p. 93–113.

SONNENFELD, M. D., 1991, High-frequency cyclicity within shelf-margin and slope strata of the upper San Andres sequence, Last Chance Canyon, *in* Meador-Roberts, S., Candelaria, M. P., and Moore, G. E., eds., Sequence Stratigraphy, Facies, and Reservoir Geometries of the San Andres, Grayburg, and Queen Formations, Guadalupe Mountains, New Mexico and Texas: Midland, Permian Basin Section–SEPM (Society for Sedimentary Geology) Publication No. 91-32, p. 11–51.

SONNENFELD, M. D., 1996, Sequence evolution and hierarchy within the Lower Mississippian Madison Limestone of Wyoming, *in* Longman, M. W. and Sonnenfeld, M. D., eds., Paleozoic Systems of the Rocky Mountain Region: Rocky Mountain Section–SEPM (Society for Sedimentary Geology), p. 165–192.

SONNENFELD, M. D., AND CROSS, T. A., 1993, Volumetric partitioning and facies differentiation within the Permian Upper San Andres Formation of Last Chance Canyon, Guadalupe Mountains, New Mexico, *in* Loucks, R. G. and Sarg, J. F., eds., Carbonate Sequence Stratigraphy—Recent Developments and Applications: Tulsa, American Association of Petroleum Geologists Memoir 57, p. 435–474.

STAFLEU, J., AND SCHLAGER, W., 1993, Pseudo-toplap in seismic models of the Schlern-Raibl contact (Stella platform, northern Italy): Basin Research, v. 5, p. 55–65.

STAFLEU, J., AND SONNENFELD, M. D., 1994, Seismic models of a shelf-margin depositional sequence: Upper San Andres Formation, Last Chance Canyon, New Mexico: Journal of Sedimentary Research, v. B64, p. 481–499.

SUN, S. Q., AND WRIGHT, V. P., 1998, Controls on reservoir quality of an Upper Jurassic reef mound in the Palmers Wood Field area, Weald basin, southern England: American Association of Petroleum Geologists Bulletin, v. 82, p. 497–515.

TINKER, S. W., EHRETS, J. R., AND BRONDOS, M. D., 1995, Multiple karst events related to stratigraphic cyclicity: San Andres Formation, Yates Field, West Texas, *in* Budd, D. A., Saller, A. H., and Harris, P. M., eds., Unconformities and Porosity in Carbonate Strata: Tulsa, American Association of Petroleum Geologists Memoir 63, p. 213–237.

TUCKER, M. E., 1993, Carbonate diagenesis and sequence stratigraphy, *in* Wright, V. P., ed., Sedimentology Review, v. 1: Oxford, Blackwell Scientific Publications, p. 51–72.

TUCKER, M. E., AND WRIGHT, V. P., 1990, Carbonate Sedimentology: Oxford, Blackwell Scientific Publications, 482 p.

TUCKER, M. E., WILSON, J. L., CREVELLO, P. D., SARG, J. F., AND READ, J. F., eds., 1990, Carbonate Platforms, Facies, Sequences, and Evolution: London, International Association of Sedimentologists Special Publication No. 9, 328 p.

VAIL, P. R., 1987, Seismic stratigraphy interpretation using sequence stratigraphy: Part 1 . Seismic stratigraphy interpretation procedure, *in* Bally, A. W., ed., Atlas of Seismic Stratigraphy, v. 1: Tulsa, American Association of Petroleum Geologists Studies in Geology No. 27, p. 1–10.

VAIL, P. R., MITCHUM, R. M., JR., TODD, R. G., WIDMIER, J. M., THOMPSON, S., III, SANGREE, J. B., BUBB, J. N., AND HATLELID, W. G., 1977, Seismic stratigraphy and global changes of sea level, *in* Payton, C. E., ed., Seismic Stratigraphy—Applications to Hydrocarbon Exploration: Tulsa, American Association of Petroleum Geologists Memoir 26, p. 49–212.

VAN WAGONER, J. C., POSAMENTIER, H. W., MITCHUM, R. M., JR., VAIL, P. R., SARG, J. F., LOUTIT, T. S., AND HARDENBOL, J., 1988, An overview of the fundamentals of sequence stratigraphy and key definitions, *in* Wilgus, C. K., Hastings, B. S., Kendall, C. G. St. C., Posamentier, H. W., Ross, C. A., and Van Wagoner, J. C., eds., Sea-Level Changes: An Integrated Approach: Tulsa, Society of Economic Paleontologists and Mineralogists Special Publication No. 42, p. 39–46.

VAN WAGONER, J. C., MITCHUM, R. M., CAMPION, K. M., AND RAHMANIAN, V. D., 1990, Siliciclastic Sequence Stratigraphy in Well Logs, Cores, and Outcrops: Tulsa, American Association of Petroleum Geologists Methods in Exploration Series No. 7, 55 p.

WANLESS, H. R., 1991, Observational foundation for sequence modeling, *in* Franseen, E. K., Watney, W. L., Kendall, C. G. St. C., and Ross W., eds., Sedimentary Modeling: Computer simulations and methods for improved parameter definition: Lawrence, Kansas Geological Survey Bulletin 233, p. 43–62.

WENDTE, J., STOAKES, F. A., AND CAMPBELL, C. V., 1992, Devonian–Early Mississippian Carbonates of the Western Canada Sedimentary Basin: A Sequence Stratigraphic Framework: Tulsa, SEPM (Society for Sedimentary Geology) Short Course No. 28, 255 p.

WILGUS, C. K., HASTINGS, B. S., KENDALL, C. G. ST. C., POSAMENTIER, H. W., ROSS, C. A., AND VAN WAGONER, J. C., eds., 1988, Sea-Level Changes—An Integrated Approach: Tulsa, Society of Economic Paleontologists and Mineralogists Special Publication No. 42, 407 p.

WILSON, J. L., 1967, Cyclic and reciprocal sedimentation in Virgilian strata of southern New Mexico: Geological Society of America Bulletin, v. 78, p. 805–818.

THE SECOND-ORDER CYCLE, CARBONATE-PLATFORM GROWTH, AND RESERVOIR, SOURCE, AND TRAP PREDICTION

J. F. SARG, J. R. MARKELLO, AND L. J. WEBER*

Mobil Exploration & Production Technical Center, P.O. Box 650232, Dallas, Texas 75265-0232

ABSTRACT: A distinctive second-order (5–20 m.y.) stratigraphic signature is present within upper Paleozoic, evaporite-related, carbonate-platform successions of the Paradox and Permian basins in the southwestern United States. This supersequence scale of cyclicity occurs under icehouse and transitional conditions and in synorogenic and postorogenic phases of basin development and appears to control the depositional and stratigraphic-trapping components of petroleum systems in these basins. Primary controls on the development of second-order architecture include basin geometry and subsidence patterns, sediment supply, and long-term sea-level changes. Climate and paleoceanography are important controls on facies development within sequences. *Supersequences* are defined as long-term transgressive-regressive cycles that may be symmetric or asymmetric and that are composed of third-order sequences.

The most normal, open-marine depositional conditions occur during the transgressive phase of second-order cycles. Organic-rich, source-rock-quality (source-prone) lime mudstones are deposited across shelves and in the basin. Carbonate reservoirs are best developed during second-order transgression and turnaround to highstand. Open-marine conditions and diverse skeletal-grain bank deposition occur during third-order highstands of second-order transgressions. Early marine cement is relatively minor in occurrence, and reservoir quality is excellent. During the second-order turnaround and early regression, shelf areas become more restricted, and third-order highstands are dominated by nonskeletal-grain (ooid and peloid) bank-margin deposition. During the later stages of second-order regression, if platform progradation accesses a deeper basin or encounters elevated salinities, or if siliciclastic input increases, the biota become more specialized, marine cementation increases, and reservoir quality is reduced.

The Lower Pennsylvanian (Morrowan–lower Missourian) supersequence, in the Paradox basin of the Four Corners region, United States, is a symmetrical cycle of approximately 20–22 m.y. duration and is composed of third-order composite depositional sequences, six of which are present in the Paradox basin. Third-order transgressive systems tracts contain organic-rich, black, laminated lime mudstones. Third-order highstand systems tracts of the second-order transgression are characterized by thick mound facies containing a diverse biota and skeletal grainstone banks. Third-order highstands of the second-order turnaround are composed of thick aggradational to progradational stacks of cyclically deposited, relatively thin, phylloid-algal buildups and areally extensive, porous nonskeletal grainstone banks. Third-order highstands of the second-order late-regressive phase are characterized by abundant siliciclastics and nonporous crinoid-rich, skeletal grainstone banks.

The Upper Permian (Guadalupian) supersequence in the Permian basin, west Texas, United States, is an asymmetrical cycle, consisting of 11 third-order depositional sequences; it was deposited over a period of about 10 to 12 m.y. Normal-marine conditions existed over a wide area during second-order transgression, and skeletal-grain bank margins exhibit the most diverse biota. Deep-water lime mudstones are source prone. The third-order highstands of the initial second-order regressive phase are characterized by prograding ooid and peloid ramp margins with excellent reservoir quality. Cemented, upright margins of organic-cement boundstone with steep (35°–40°) foreslopes and poorer reservoir quality developed during the late stages of second-order regression.

Postdepositional compactional drape of turnaround third-order highstand bank margins over underlying structural blocks, combined with shelfward facies changes to nonporous inner-shelf strata, provides a combined structural and stratigraphic trapping configuration for large petroleum accumulations. Stratigraphic traps form (1) during second-order transgressions where isolated platforms and buildups develop and are draped by late transgressive mudstones and (2) in shelf settings during the transgression and turnaround at facies changes from porous carbonate sandstones to nonporous lagoonal dolomites and evaporites.

INTRODUCTION

The seismic-scale, high-quality outcrops of Permian-Pennsylvanian age in the southwestern United States, and their adjacent mature hydrocarbon provinces (Permian and Paradox basins), provide the opportunity to study the large-scale architecture and stratigraphic controls on the deposition of mixed carbonate, evaporite, and siliciclastic sequences and to examine their relationships to the development of major hydrocarbon systems. The widely spaced mountain ranges and the subsurface oil and gas fields of west Texas, New Mexico, and southeastern Utah have received extensive study during the past 75 years. Work includes regional and local geologic mapping, stratigraphic, sedimentologic, and paleontologic descriptions, diagenetic interpretations, and tectonic and resource studies (e.g., King, 1948; Galley, 1958; Wengerd and Matheny, 1958; Peterson and Hite, 1969; Ross and Ross, 1987b; Wardlaw and Rohr, 1996; see Hill, 1996, for summary). Much is known and

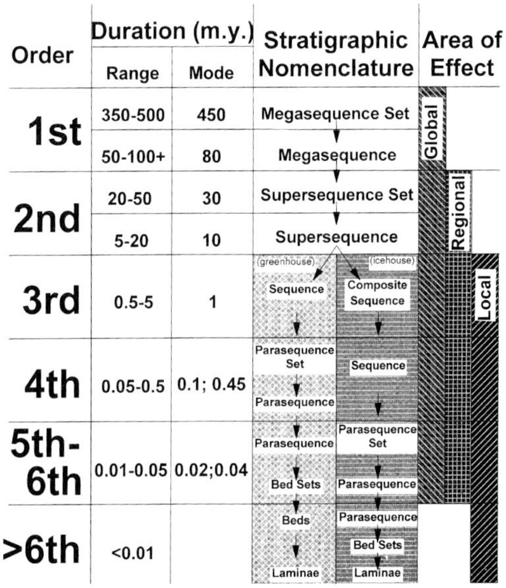

FIG. 1.—Sequence-based hierarchy of stratigraphy (modified after Vail et al., 1991).

*Present address: Exxon Production Research Company, P.O. Box 2189, Houston, TX 77252-2189

Advances in Carbonate Sequence Stratigraphy: Application to Reservoirs, Outcrops and Models, SEPM Special Publication No. 63
 ISBN 1-56576-060-3, p. 11–34

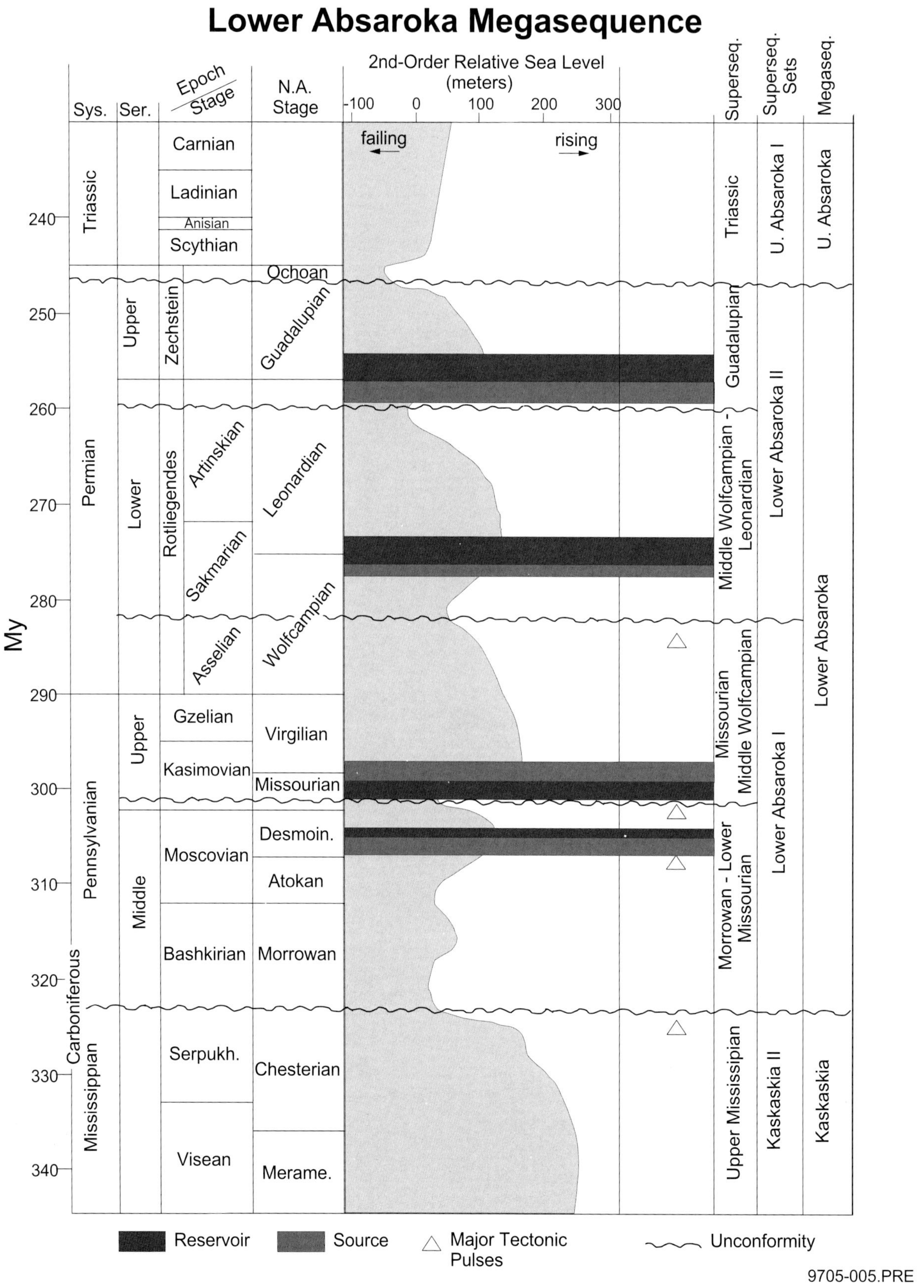

FIG. 2.—Summary of lower Absaroka megasequence of Sloss (1963) showing supersequence subdivisions as defined in North American basins and tied to major tectonic pulses during the Marathon-Ouachita-Appalachian orogeny. Bounding unconformities occur in both synorogenic and postorogenic intervals and are defined by major reorganization in basin deposition patterns. Ages of major hydrocarbon source and reservoir units correspond to transgressive and turnaround units within supersequences, respectively. Time scale modified from Harland et al. (1989).

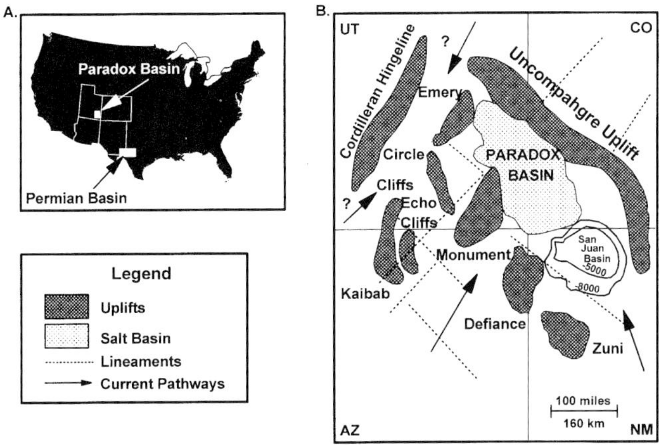

FIG. 3.—(A) Location of Paradox and Permian basins. (B) Generalized tectonic map showing major uplifts relative to Paradox and San Juan basins in Four Corners area of the United States. UT—Utah, CO—Colorado, AZ—Arizona, NM—New Mexico.

published about the geology of this area, and geologic understanding ranges from overall geologic history to details of specific formations and members. The interpretations of most studies are geographically and/or stratigraphically restricted. A broader synthesis is attempted here to provide the basis for better understanding the depositional history of the region and to explain the known hydrocarbon occurrences.

Large-scale, long-term transgressive-regressive cycles are evident in the upper Paleozoic of North America (Sloss, 1963; Ross and Ross, 1987a, b) (Figs. 1, 2). They developed over periods of approximately 5–20 m.y., on a scale of 500 to more than 1,000 m of stratigraphic section, and represent second-order supersequences (Haq et al., 1987; Vail et al., 1991). Supersequences may be symmetric or asymmetric, and their stratigraphic signature results from the interaction of changing depositional systems, basin geometry, subsidence, sea level, climate, and paleoceanographic conditions. They appear to be driven by global-scale tectonic processes related to changing rates

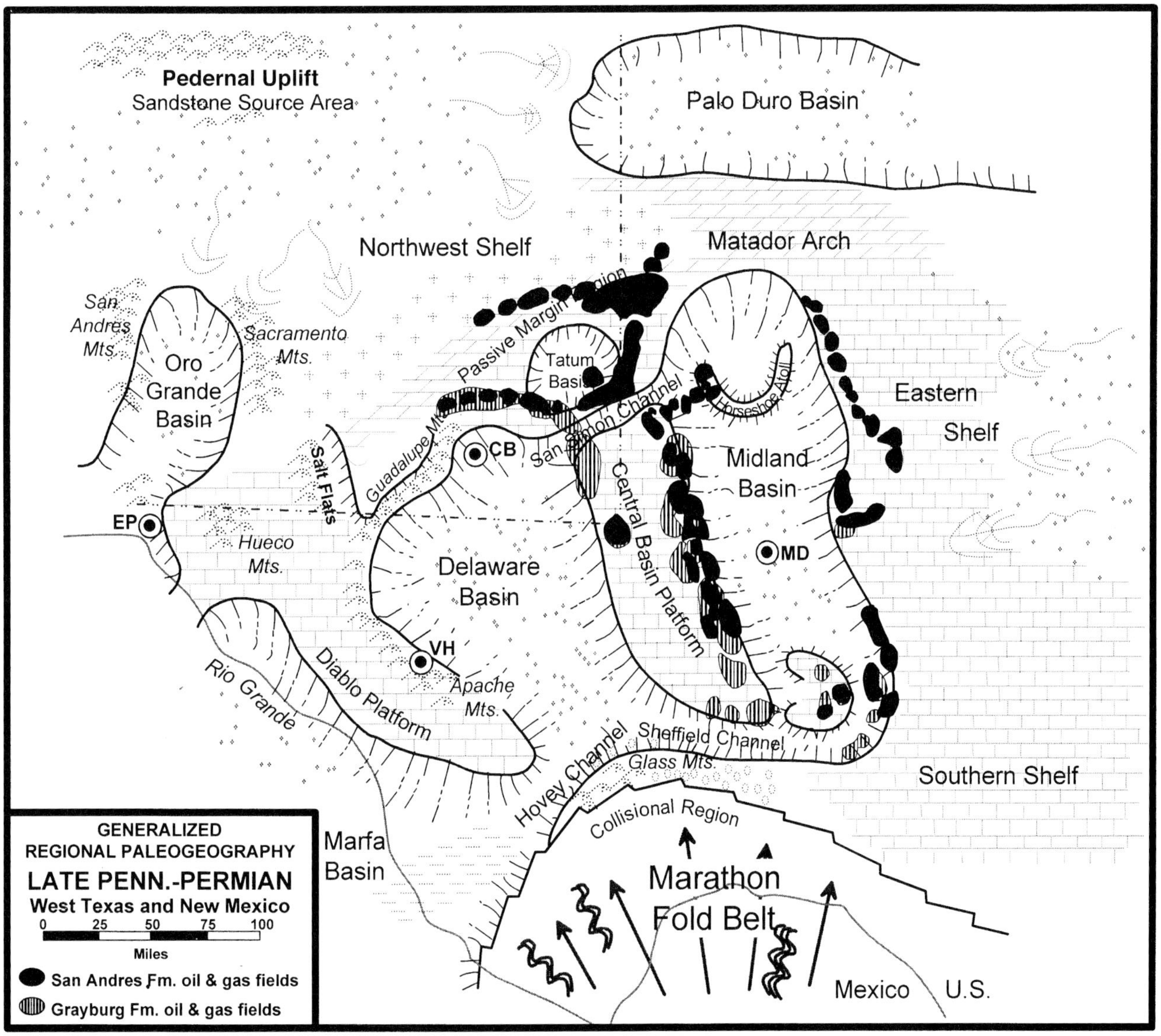

FIG. 4.—Generalized tectonic map showing major basin and shelf elements of west Texas and southeast New Mexico. Oil and gas fields that have San Andres (Ward et al., 1986) and Grayburg Formations (Guadalupian, Permian) as reservoirs are superimposed on basin elements. (CB—Carlsbad, EP—El Paso, MD—Midland, VH—Van Horn).

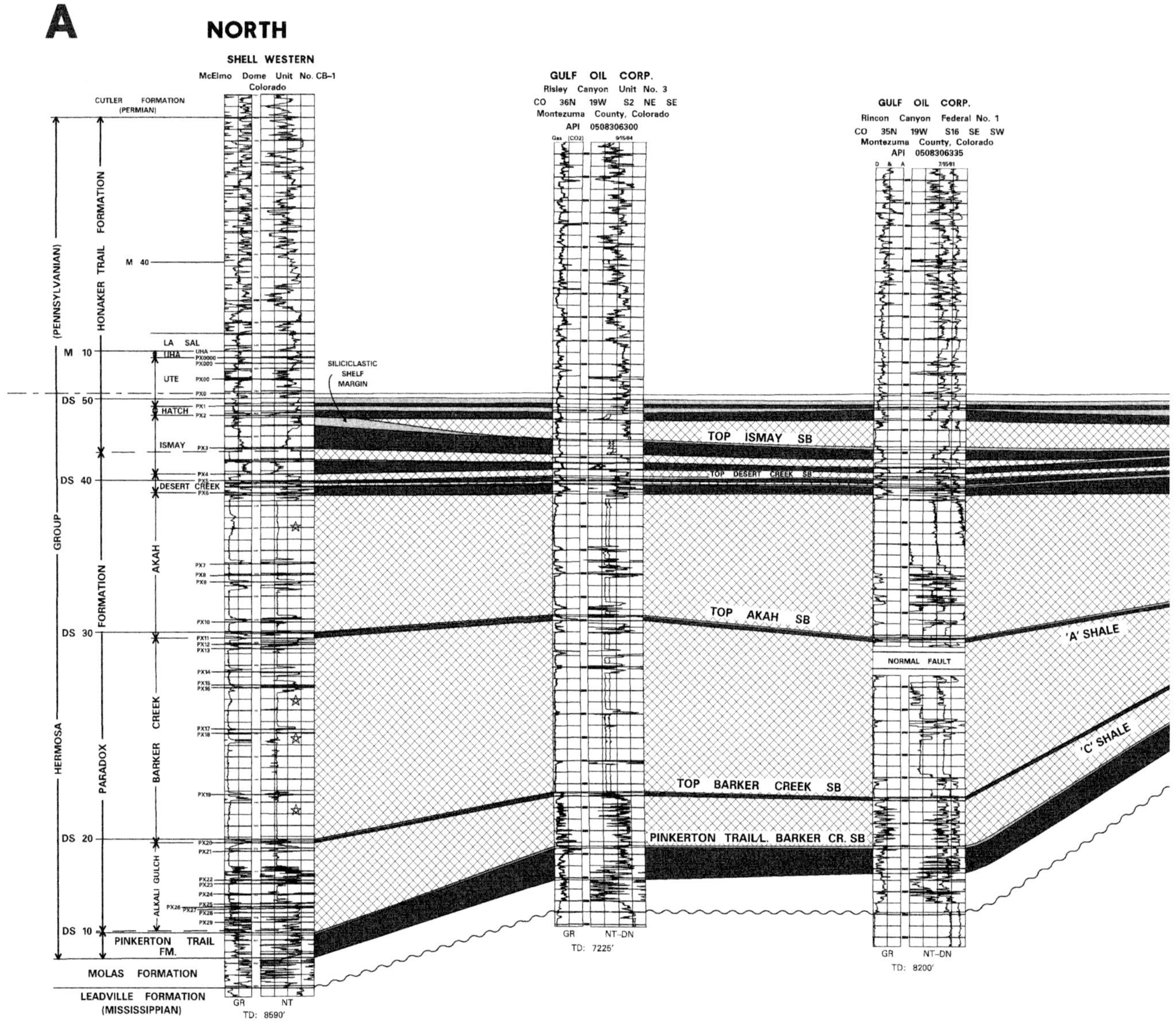

FIG. 5.—Paradox basin regional outcrop and well-log cross section showing sequence correlation from surface outcrops of the Goosenecks region of the San Juan River to the subsurface of the Aneth oil field and central Paradox basin areas. Recognition of third-order sequence boundaries (DS10, DS20, etc.) at the tops of major highstand carbonate-platform units, and the high gamma-ray readings of the transgressive and highstand black lime mudstone condensed sections provide excellent markers for regional correlation. (Log curves: GR—gamma ray, SP—spontaneous potential, NT or NT-DN—neutron-density.)

of sea-floor spreading and tend to be composed of groups of 5 to 12 third-order cycles that developed on a time scale of 0.5–5 m.y. (Figs. 1, 2). Supersequences are bounded by sequence boundaries that document major relative falls in sea level, greater than 50–100 m, and large-scale reorganizations of basin depositional systems. They result in distinctive, recognizable stratigraphic architectures.

The late Paleozoic, Permian-Pennsylvanian, time period of western North America contains four supersequences, here designated: (1) Morrowan–lower Missourian, (2) Missourian–middle Wolfcampian, (3) middle Wolfcampian–Leonardian, and (4) Guadalupian (Fig. 2). Each supersequence contains a well-developed petroleum system. All four supersequences are within the lower Absaroka megasequence of Sloss (1963) that is bounded by cratonwide unconformities. This megasequence is bounded below by an unconformity in the uppermost Chester and near the basal Pennsylvanian, and above by the basal Ochoan (Fig. 2). In this paper and for the purpose of illustration of concepts, we synthesize and summarize the history of two of the supersequences within the lower Absaroka megasequence, the Morrowan–lower Missourian supersequence, and the Guadalupian supersequence. These supersequences will be described in terms of their large-scale transgressive-regressive stratigraphic architecture and cyclicity. We believe, as was first delineated by Greenlee and Lehmann (1993), that this scale of stratigraphy has had a profound effect on the development of major hydrocarbon systems.

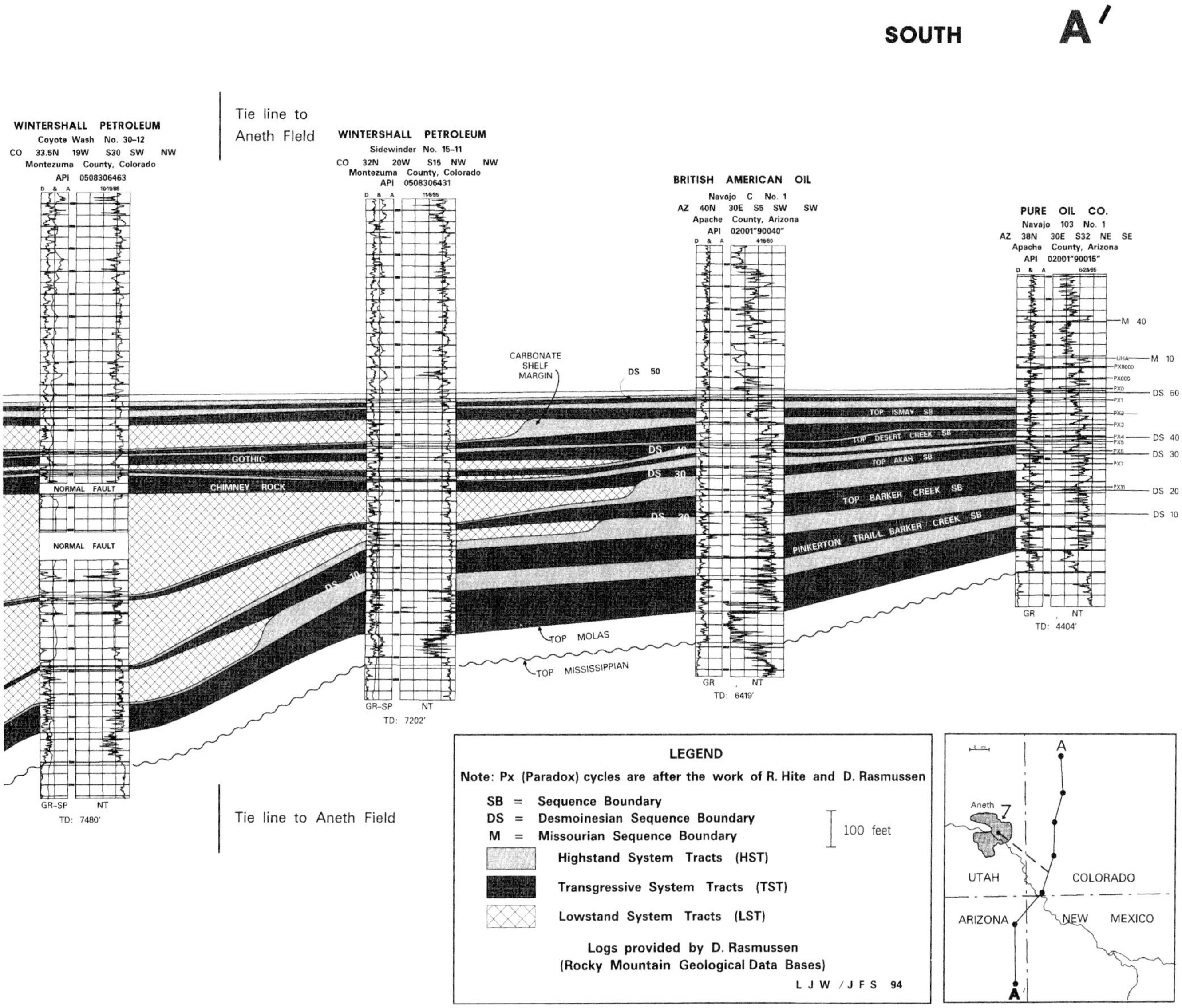

FIG. 5.—Continued.

BASIN HISTORY

The depositional systems and sequence stratigraphy of the Permian and Paradox basins must be understood within the context of the Paleozoic Pangean supercontinent cycle and its stages of basin evolution. The sedimentary fill of these two basins is divided into three major tectono-stratigraphic packages (Clair, 1952; Wengerd and Matheny, 1958; Hills, 1984; Ross, 1986; Frenzel et al., 1988; Hill, 1996; Markello and Sarg, 1996): (1) a lower Paleozoic (Upper Cambrian to Mississippian) passive-margin succession; (2) an Upper Mississippian–Lower Permian synorogenic foreland succession; and (3) an Upper Permian postorogenic basin-fill succession.

The lower Paleozoic passive-margin succession is present throughout the southwestern region of the United States and is as much as 1.5 km thick. Shallow-marine cratonic deposition dominated during the early Paleozoic postrift and drift phases of Laurentia following the breakup of the late Precambrian supercontinent (Hills, 1984; Frenzel et al., 1988). The Pennsylvanian collision of North and South America had profound effects on this passive margin and the adjacent craton. Geometry of the Marathon thrust belt demonstrates southeast-northwest compression during collision, a direction normal to the depositional strike of the continental margin (Ross, 1986; Frenzel et al., 1988). The results of collision were to structurally differentiate the area into several northwest-aligned foreland basins (e.g., Delaware, Midland, Orogrande, Paradox) with intervening high areas (e.g., Central Basin and Diablo platforms and Pedernal, Uncompaghre, and Zuni uplifts (Hills, 1984; Ross, 1986; Frenzel et al., 1988; Hill, 1996) (Figs. 3, 4). This event created the present-day configuration of the Permian and Paradox basins. The synorogenic sedimentary fill deposited in the Permian and Paradox basins during Late Mississippian through Early Permian time has highly variable thickness, from 100's to

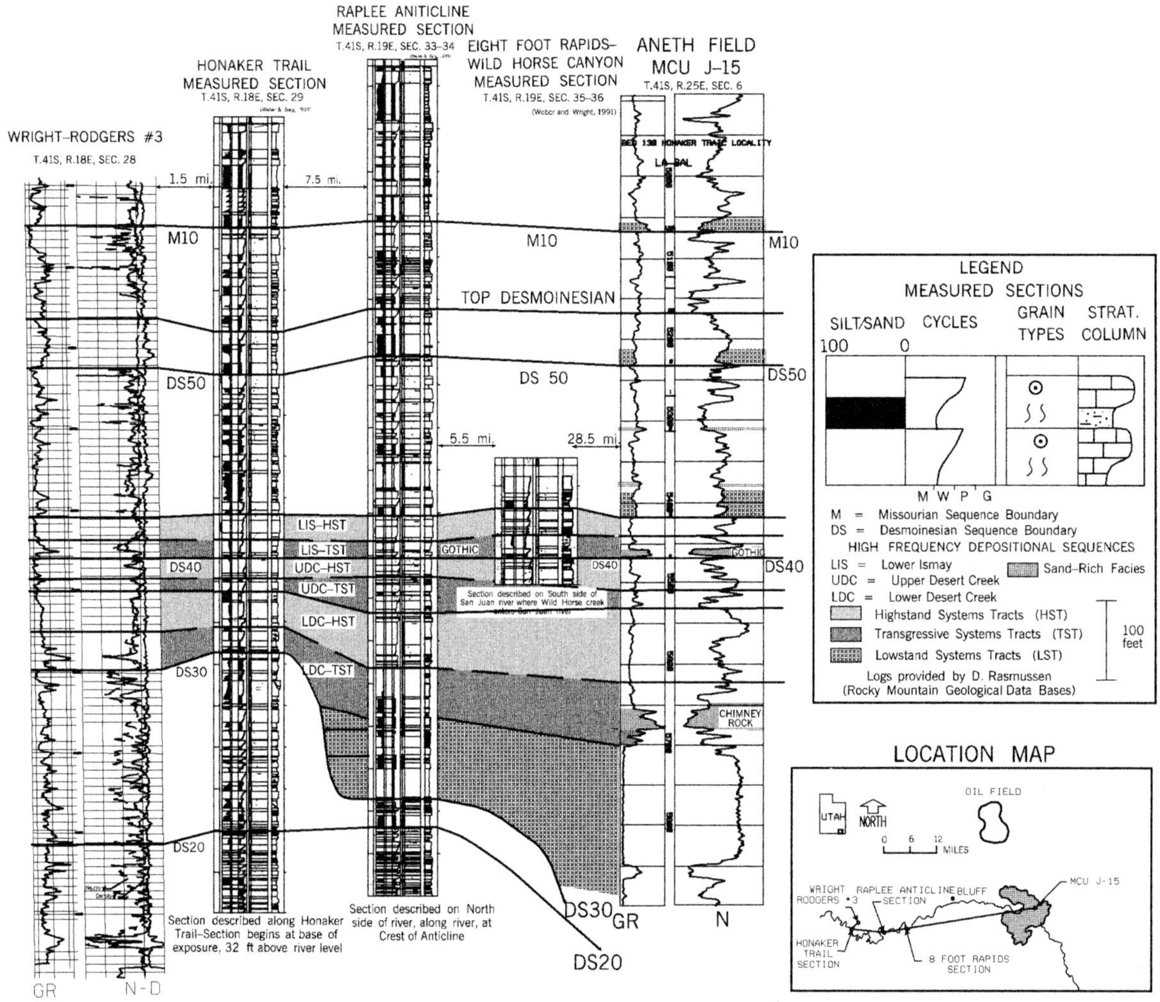

FIG. 6.—Cross section from outcrops of the Goosenecks area of the San Juan River to the Aneth oil field. Highlighted are high-frequency fourth-order depositional sequences defined within the reservoir interval of the Desert Creek and Ismay Members of the Paradox Formation. These fourth-order sequences compose the transgressive and highstand systems tracts of the DS30 and the transgressive systems tract of the DS40 third-order composite sequences. Detail of high-frequency fourth- and fifth-order cycles, as defined in outcrop and core, is described in Weber et al. (1995b).

more than 1,500 m (Ross, 1986). Shallow-marine carbonate, evaporitic, and fluvial-deltaic siliciclastic sediments filled the basin areas. This synorogenic succession contains three angular unconformities that bracket multiple collisional pulses and is capped by the regionally extensive middle Wolfcampian angular unconformity (Ross, 1986) (Fig. 2). The first two upper Paleozoic supersequences (lower Absaroka I supersequence set) are bounded by these unconformities.

Following the Pennsylvanian–Early Permian orogenesis, the Permian and Paradox basins filled with a postorogenic sedimentary succession that is thicker than 4.5 km and comprises the last two supersequences of the late Paleozoic (lower Absaroka II supersequence set) (Fig. 2). This succession, in the Permian basin, is composed of middle and upper Permian, basin-centered sandstones, shelf and shelf-margin carbonates, and basinal and shelf evaporites (Matchus and Jones, 1984; Hills, 1984; Frenzel et al., 1988; Hill, 1996). To the north, continental sandstones and shales spread across the Paradox basin.

THE MORROWAN–LOWER MISSOURIAN SUPERSEQUENCE: PARADOX BASIN

Geologic Setting

The Paradox basin is a Pennsylvanian to Early Permian asymmetric trough that covers parts of Colorado, Utah, Arizona, and New Mexico (Fig. 3). Episodic uplift of the Uncompahgre basement block occurred during the Pennsylvanian, and 4 to 6 km of relief may have existed between the crestal portion of the Uncompahgre uplift and the Paradox basin. The Molas Formation and Hermosa Group (Fig. 5) compose the initial synorogenic basin fill and represent the oldest second-order supersequence of the lower Absaroka I supersequence set within the first-order lower Absaroka megasequence of Sloss (1963) (Fig. 2). Detailed descriptions of the local and regional stratigraphy are found in Clair (1952), Wengerd and Strickland (1954), Herman and Sharps (1956), Wengerd and Matheny (1958), Baars et al. (1967), Peterson and Hite (1969), Molenaar (1981), and Peterson (1992).

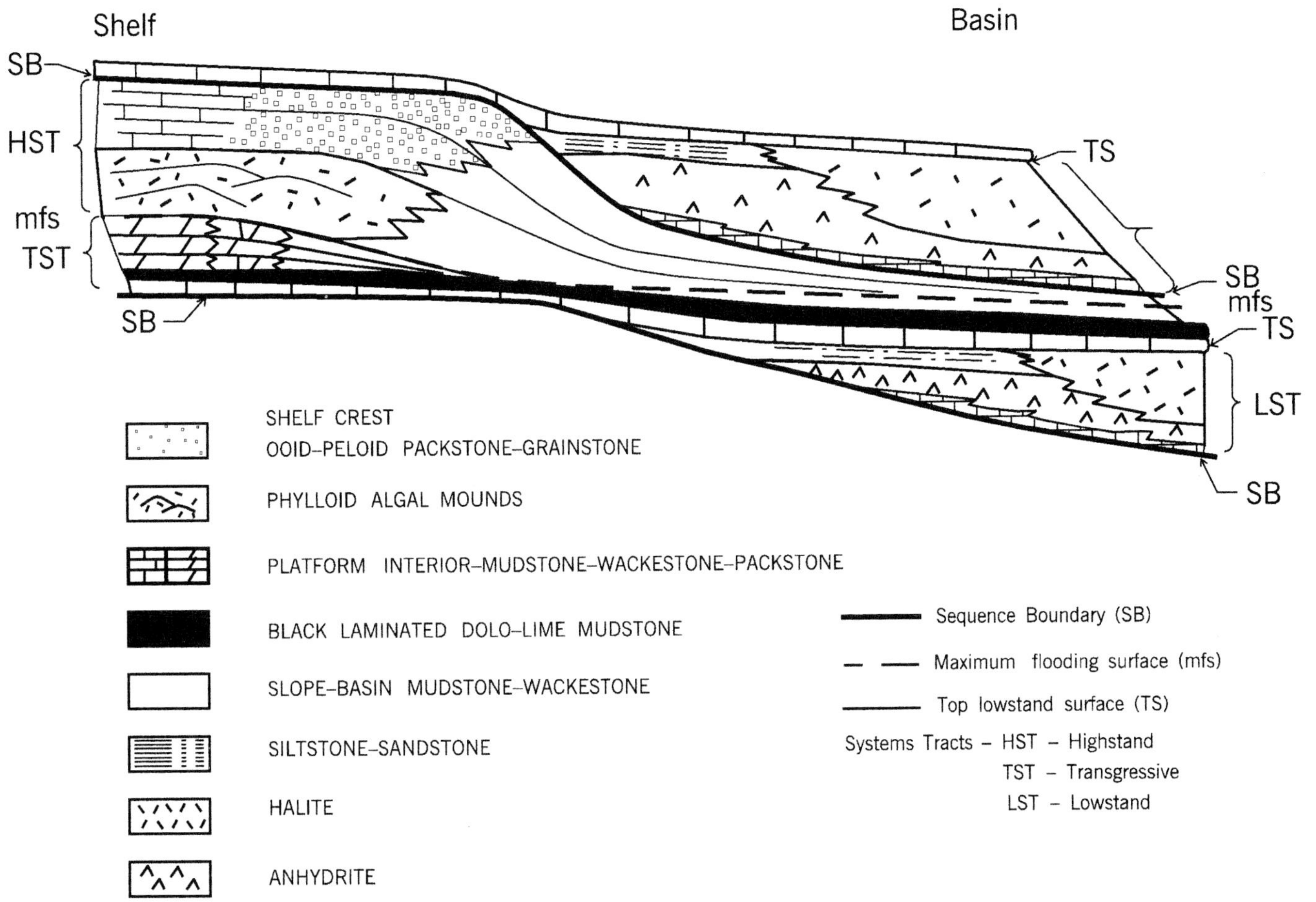

FIG. 7.—Depositional sequence model for the Desmoinesian strata of the Paradox basin. Sequences are developed on both third- and fourth-order scales and range in thickness from 10s to 100s of meters.

The basal Molas Formation is a red-bed section that overlies a paleokarst surface on the Mississippian limestones. The Hermosa Group has a gradational lower contact with the Molas Formation and is divided, from oldest to youngest, into the Pinkerton Trail, Paradox, and Honaker Trail formations (Fig. 5). The Pinkerton Trail Formation is time transgressive and ranges in age from Atokan to earliest Desmoinesian (Wengerd and Matheny, 1958). It is 0 to 60 m thick and is composed of fossiliferous crinoid- and fusulinid-rich limestone and light to dark gray shale, siltstone, and dolostone.

The Paradox Formation is separated areally into three facies zones distributed over the basin. Proximal to the Uncompahgre uplift, and extending to the south and west toward the central part of the basin, is a thick siliciclastic wedge of quartzarenite and arkose. In the central basin area, these siliciclastics interfinger with thick evaporites that are punctuated by thin, organic-rich dolostone and shale. Estimates of original depositional thickness of this facies tract range from 1.2 to 3.1 km (Hite, 1960; Peterson and Hite, 1969). A carbonate-shelf facies is laterally equivalent to the siliciclastic and evaporitic units in the southern and southwestern part of the Paradox basin and is about 300 m thick (Wengerd and Strickland, 1954; Wengerd and Matheny, 1958; Peterson and Hite, 1969). More detailed descriptions of the Paradox Formation are given by Wengerd (1951), Choquette and Traut (1963), Peterson and Hite (1969), Hite (1970), Wilson (1975), Choquette (1983), Goldhammer et al. (1991), Weber et al. (1995a, b), and Gianniny and Simo (1996).

The Honaker Trail Formation overlies the Paradox Formation and reaches thicknesses of 500 to 600 m (Fetzner, 1960; Wengerd and Matheny, 1958). Fossiliferous limestone is interbedded with sandstone, siltstone, and shale and grades into arkosic clastics of the undifferentiated Cutler Formation to the northeast, toward the Uncompahgre uplift.

Supersequence Components and Sequence Hierarchy

The Morrowan–lower Missourian supersequence was deposited during icehouse conditions (Fisher, 1982) and is composed of third-order composite sequences (Mitchum and Van Wagoner, 1991), each of which includes lowstand, transgressive, and highstand systems tracts and which are composed of fourth-order sequences (Figs. 5, 6). All of these sequences and composite sequences are Type 1 sequences (Van Wagoner et al., 1988), and a depositional model for them is summarized in Figure 7 (Weber, 1992; Weber et al., 1995a, b). Sequences are tied to the North American fusulinid (Ross and Ross, 1988) and conodont (B. R. Wardlaw, pers. commun., 1995) zonation. This

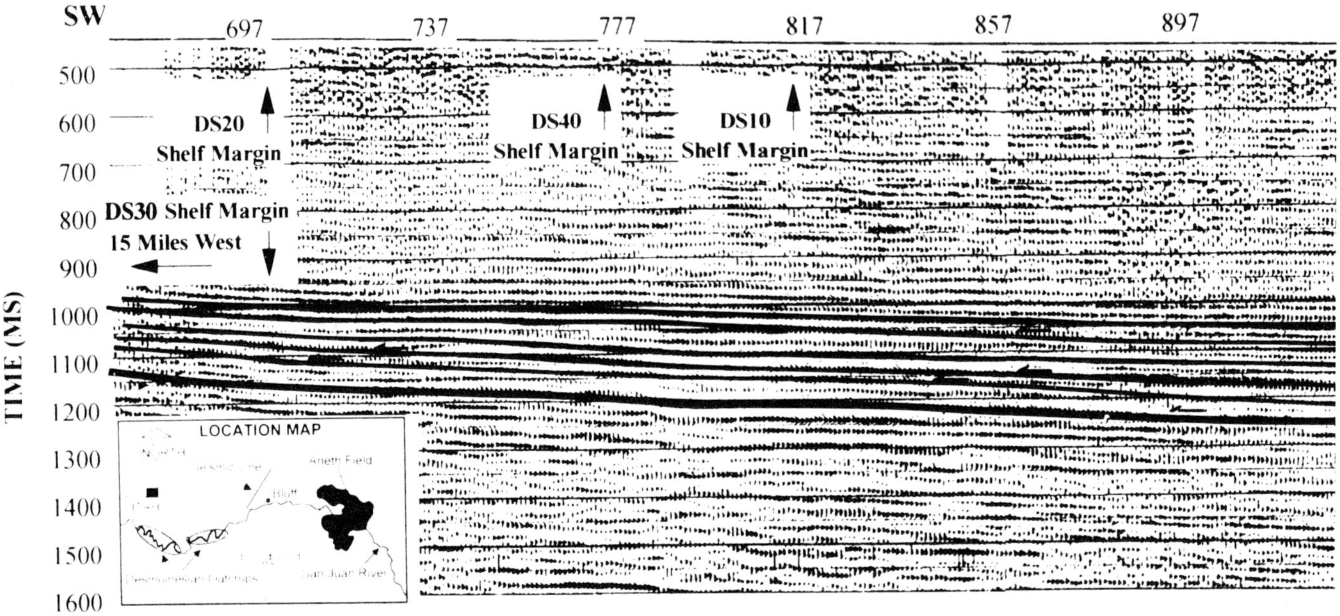

FIG. 8.—Paradox basin seismic line showing interpreted stratigraphy tied through nearby well control. Major onlapping salt wedges are interpreted to be third-order lowstand units for DS10, DS20, and DS30 composite sequences. The positions of low-angle breaks in slope at salt onlap are interpreted to represent third-order highstand platform margins and show a progressive backstepping (i.e., DS10, DS20, and DS30 shelf margins were established in progressively shelfward position to the southwest) followed by forestepping (i.e., DS40 shelf margin is basinward of DS30) controlled by the development of the second-order supersequence.

sequence stratigraphic framework is based on the recognition of regionally significant sequence boundaries that contain evidence for subaerial exposure (Weber et al., 1995a, b).

The icehouse conditions prevalent during the Pennsylvanian and the resultant high-frequency, high-amplitude sea-level changes caused both third- and fourth-order sequences to have similar development of depositional systems. They are difficult to differentiate in outcrop, and we believe that the following criteria can serve to differentiate third-order composite sequences from higher frequency sequences.

1. Major episodes of basin filling by thick accumulations of anhydrite, halite, and potassium-bearing salts (Alkali Gulch, Barker Creek, Akah, Desert Creek, and Ismay evaporites) (Figs. 5, 6). These thick evaporite successions represent major periods of basin restriction and evaporative drawdown, and they onlap carbonate highstand systems tracts. The Alkali Gulch and Barker Creek evaporites are thick enough to be resolved on seismic data (Fig. 8).
2. Third-order highstand systems develop major platform successions that are composed of thick stacks of carbonate-rich, shallowing-upward cycles (Figs. 6, 9). Highstand systems tracts are aggradational to progradational and show a shallowing-upward character at platform margins. Parasequences in early highstands commonly exhibit biologic communities that coalesce and stack (e.g., phylloid-algal–mound buildups). During late highstands, the dominance of biologic communities waned, and stacked ooidpeloid carbonate shoals formed.
3. Caliche, silcrete, trapped eolian silt, microkarst, and silt-filled fissures and cracks are observed at the tops of all third- and fourth-order sequences on the carbonate shelf. Strata beneath third-order composite sequence boundaries show evidence of prolonged exposure when compared to similar rocks deposited beneath higher-frequency sequence boundaries. Platform-crest grainstone facies show considerable evidence of subaerial exposure such as leached allochems (Fig. 10) and recrystallization fabrics, and these features extend downward 8–10 m below sequence boundaries. Meteoric diagenesis below higher-order sequence boundaries is generally restricted to less than 3 m below sequence boundaries (Weber, 1992; Weber et al., 1995a, b).
4. Regionally extensive black, laminated mudstones (e.g., C shale, A shale, Chimney Rock, Gothic, and Ute members) (Fig. 5) are observed within transgressive systems tracts overlying third-order boundaries. These mudstones occur over platform highs, are from 1 to 23 m thick, and grade up into clean, light brown dolomudstone-wackestone facies of peritidal to shallow-subtidal origin (Weber et al., 1995a, b). On the basis of reconstructions immediately around the Aneth platform, water depths for this facies ranged from near wave base to 60 m. Restriction of biota, lack of burrows, preservation of flat lamination, and presence of pyrite suggest disaerobic (i.e., not completely anoxic) to anoxic conditions. This facies generally follows a period of basin restriction and evaporite deposition and is interpreted to have been deposited during major rises in sea level. Initial invasion of marine water into the shallow hypersaline sea of the central Paradox basin probably caused dissolution of halite, and an increase in chloride concentration of the basin waters.

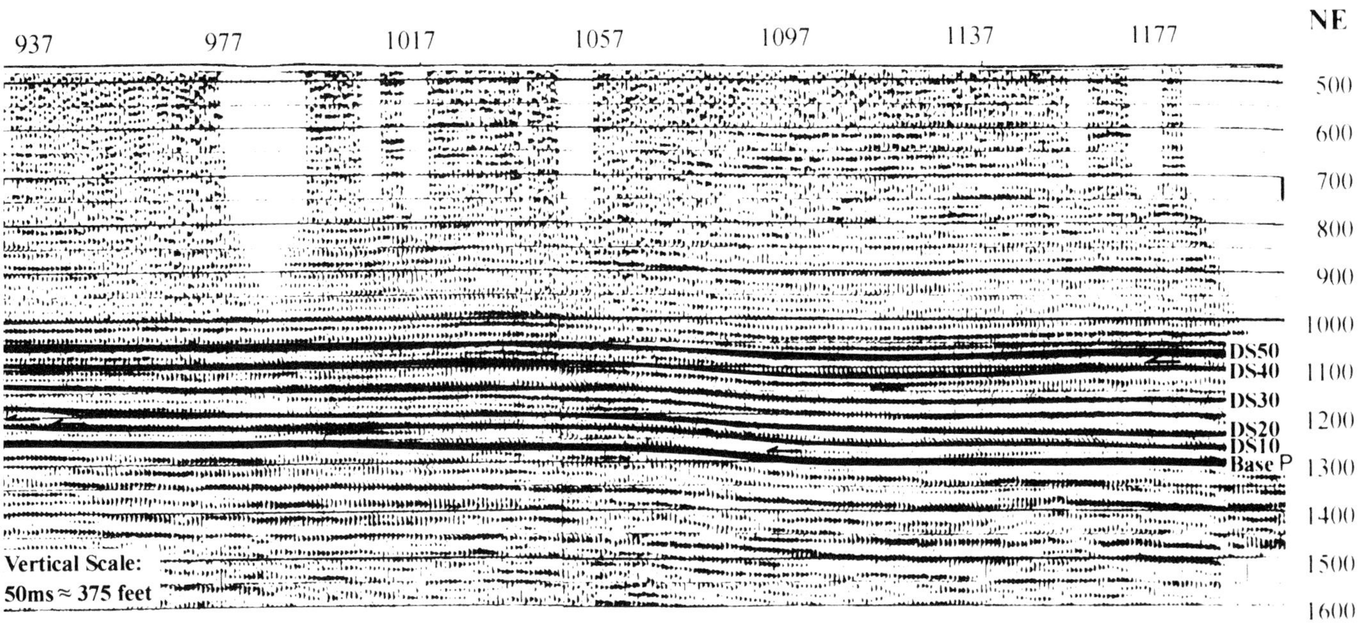

FIG. 8.—Continued.

Continued transgression and increased circulation eventually returned the basin to normal-marine conditions.

5. Significant aggradational growth of phylloid-algal banks occurred in the lower parts of third-order highstand systems tracts. Examples include the lower Desert Creek and lower Ismay phylloid-algal banks. These banks formed during increased accommodation, resulting from third-order rises in sea level. Commonly, highstand systems tracts above fourth-order sequence boundaries are characterized by relatively shallow water and less accommodation, and algal buildups are thin and less areally extensive.

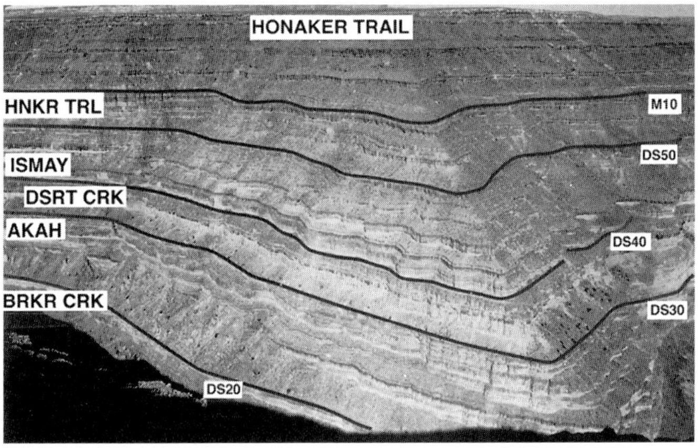

FIG. 9.—View from Honaker Trail (see Fig. 6 for location) across San Juan River canyon to outcrop of Desmoinesian strata that shows upward change in lithofacies contained within the third-order composite sequences of the middle and upper parts of the Morrowan–lower Missourian supersequence. Prominent cliff formers below DS20, DS30, and DS40 third-order sequence boundaries are carbonate highstand units containing shallow-water platform-margin buildups and grainstones. An increase in siliciclastic-rich outcrops that show recessive weathering is evident in DS40 and DS50 sequences that form the late-regressive part of this supersequence. Interbedded carbonate ledges are composed of brachiopod-crinoid grainstone bank deposits. Nonmarine siliciclastics that cap the hill are Wolfcampian and younger Permian strata.

Third-order sequence boundaries are correlated over approximately 10,000 km^2 in the Four Corners region and are inferred to correlate over the entire Paradox basin. These sequence boundaries divide the upper Atokan through the lower Missourian into seven composite sequences (upper Atokan, DS10, DS20, DS30, DS40, DS50, and M10) (Figs. 5, 8, 9). The Desmoinesian, which makes up the greatest portion of this time period, has been assigned durations ranging from 4 m.y. (Odin and Gale, 1982) to 10 m.y. (Van Eysinga, 1975). Harland et al. (1989) consider the Desmoinesian time span to have been approximately 8 m.y. These estimates provide an age range for each of the six Desmoinesian composite sequences of between 0.667 and 1.7 m.y., with a mean of 1.3 m.y. The older, Morrowan and Atokan parts of the supersequence are represented in the Paradox basin by the red beds of the Molas Formation and the transgressive limestones of the Pinkerton Trail Formation (Fig. 5).

The Morrowan–lower Missourian supersequence has a symmetrical architecture (Fig. 5). The lower two composite sequences (upper Atokan and DS10) are characterized by a progressive shift of the shelf margins toward the basin edges ("backstepping"; Figs. 5, 8). The middle two composite sequences (DS20 and DS30) are aggradational. The three youngest composite sequences (DS40, DS50, and M10) prograded and ultimately filled the basin.

The symmetry of the supersequence is shown in the outcrop of the Honaker Trail area of the San Juan River canyon (Fig. 9). Thick carbonate outcrops that are resistant to weathering compose the lower half of the canyon wall and include the upper part of the Atokan composite sequence and the DS10, DS20, and

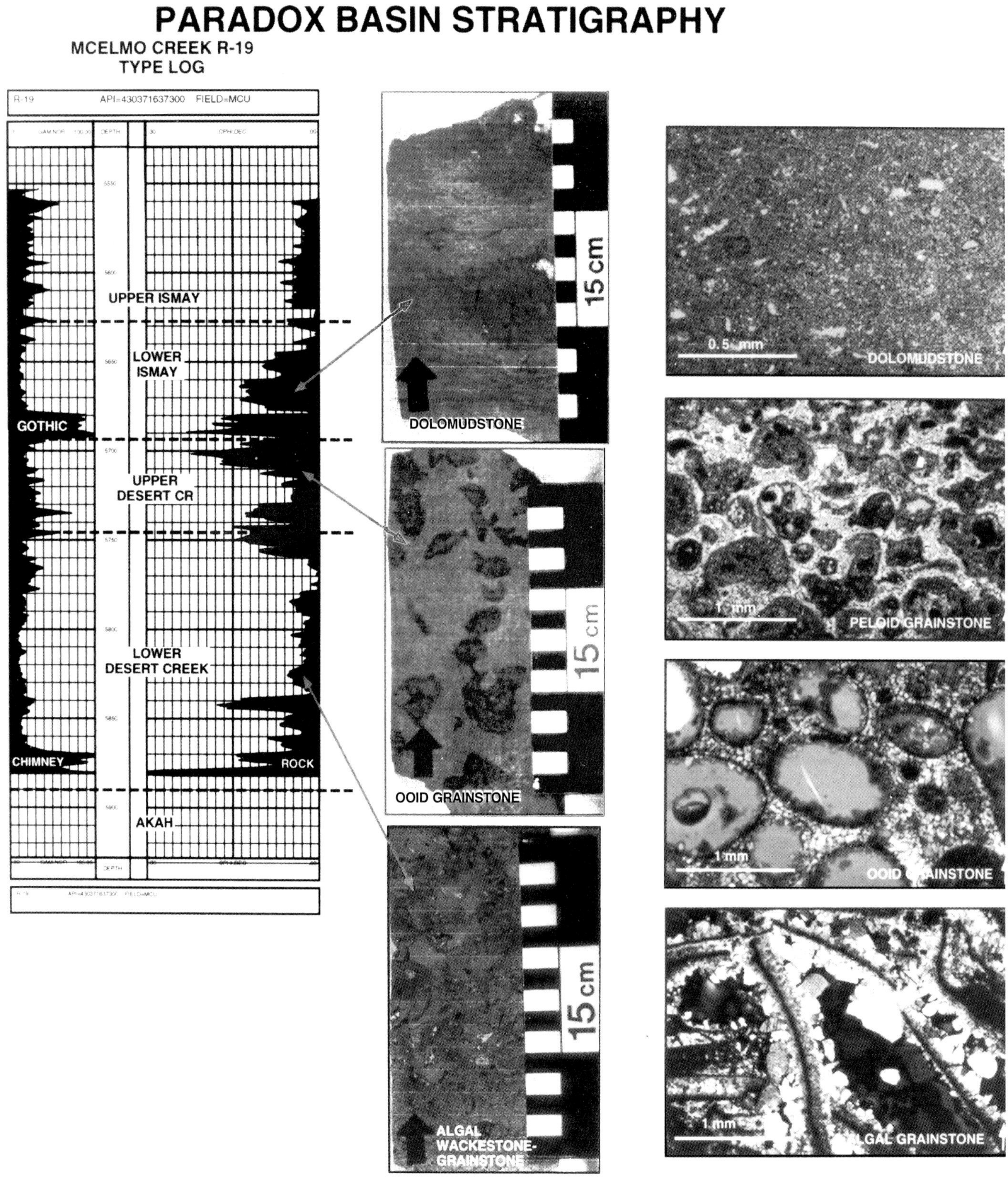

FIG. 10.—Type log, core, and photomicrographs from the McElmo Creek unit of the Aneth oil field showing major reservoir units of the Desert Creek and lower Ismay Members of the Paradox Formation. Pore types include shelter and interparticle porosity within phylloid-algal grainstones, interparticle porosity and moldic porosity beneath sequence boundaries in peloid and ooid grainstones, and microvug porosity in dolomudstone units.

DS30 composite sequences. The second-order maximum flooding surface and turnaround are interpreted to be within the Akah Member. The thickest carbonate buildups occur within this composite sequence and represent the maximum accommodation over the platform areas. The regressive part of the supersequence, in the upper half of the outcrop (Fig. 9), is composed of interbedded siliciclastics and limestones of the DS40 and DS50 composite sequences and the lower part of the lowest Missourian M10 composite sequence.

Late Atokan composite sequence.—
The Pinkerton Trail Formation represents the initial transgressive systems tract of the oldest composite sequence identified in the Paradox basin and represents the first marine transgression over the nonmarine red beds of the Molas Formation. Shallow-water carbonate-platform facies are present over the basin area. The lower part of the Barker Creek Formation is interpreted to be the highstand systems tract for this composite sequence. In shelf areas, the top of the highstand and the first Desmoinesian third-order sequence boundary lie just below the C shale. This sequence boundary correlates to the base of the Alkali Gulch evaporites in the basin (Fig. 5) (base of cycle 30 of Hite and Buckner, 1981).

DS10 composite sequence. —
The Alkali Gulch evaporite cycles comprise a single third-order lowstand wedge deposited during a long-term drawdown of the basin and represent the basal systems tract for the DS10 composite sequence (Figs. 5, 8). Onlap of this wedge suggests that the lower Barker Creek highstand was subaerially exposed during the entire time of deposition of this evaporite succession. The Alkali Gulch lowstand evaporite wedge is composed of a series of high-frequency fourth-order cycles (Fig. 5) (cycles 20–30, Hite and Buckner, 1981). These are composed of anhydrite- and halite-bearing silty dolomite (units B–D of Hite and Buckner, 1981) and black shale to black, laminated lime mudstone (unit A of Hite and Buckner, 1981). The evaporite and silty dolomite units are interpreted to be lowstand units, and the black shale units represent times of mesosalinity. The transgressive and highstand systems tracts for the DS10 composite sequence are composed of the C shale and upper Barker Creek carbonate platform. The upper sequence boundary of this composite sequence occurs at the base of the A shale on the shelf and correlates to the base of the Barker Creek evaporites in the basin (Fig. 5).

DS20 composite sequence. —
The Barker Creek evaporites are also interpreted to represent a single third-order, lowstand evaporite wedge and represent cycles 11 through 19 of Hite and Buckner (1981). These strata are interpreted to form the basal lowstand unit of the DS20 composite sequence (Fig. 5). The transgressive and highstand systems tracts of this composite sequence comprise the A shale and the Akah Member carbonate platform (Figs. 5, 9).

DS30 and DS40 composite sequences.—
The next two composite sequences contain economically significant reservoir facies (i.e., algal-mound buildups and carbonate-sand shoals) and represent the upper aggradational and first progradational composite sequences of the supersequence (Fig. 5). The DS30 composite sequence consists of a basal lowstand systems tract composed of the Akah evaporite cycles (cycles 6–10). The overlying transgressive and highstand systems tracts are, respectively, the Chimney Rock black, laminated lime mudstone and the platform carbonates of the Desert Creek Member (Figs. 5, 9). Outcrop and core data suggest that the maximum flooding surface of this composite sequence is at the top of a silty dolomudstone-wackestone and at the base of the algal-buildup phase of the Desert Creek. Phylloid-algal buildups have been identified in the subsurface at Aneth field and formed as the initial aggradational part of this highstand systems tract. They are overlain at the platform margin or shelf crest by ooid-peloid packstones and grainstones (Figs. 5, 9, 10). DS40, the first progradational composite sequence of the supersequence, is interpreted to begin with deposition of evaporite cycle 5. This lowstand system is succeeded by the transgressive and highstand systems tracts of the Gothic black, laminated lime mudstone and the carbonate-platform facies of the Ismay Member and lower Honaker Trail Formation (Figs. 5, 9).

DS50 composite sequence.—
The uppermost Desmoinesian composite sequence of this supersequence, DS50, includes the transgressive Ute shale and the cyclic marine limestones and siliciclastics of the highstand upper Honaker Trail Formation (Figs. 5, 9). Composite sequences DS40 and DS50 are strongly dominated by cyclic fourth-order sequences, each with a relatively thin evaporite lowstand unit. These fourth-order lowstand evaporite units are represented, from oldest to youngest, by the basinal cycles 4, 3, 2, 1, 0, 00, 000, 0000, and UHA (naming convention after Hite, 1960; D. L. Rasmussen, pers. commun., 1994).

Discussion

During the transgressive phase of the second-order cycle, third-order composite sequences initially backstep and then aggrade (Fig. 5). Basin restriction and evaporitic sediment accumulation occurred in the basin during third-order lowstands and filled much of the basin topography. Basin restriction was insufficient for evaporites to accumulate during higher-frequency fourth-order lowstands that occurred during deposition of the transgressive and highstand systems of these third-order composite sequences. During transgressions and highstands, carbonate production in shallow-water shelf settings was sufficient to allow the shelf deposition to keep pace with accommodation increase. The basin remained sediment starved. During basin drawdown, the shelf areas were subaerially exposed. Icehouse conditions suggest glacio-eustatic sea-level drops of more than 100 m. Maximum relief between the shelf margin and the basin floor exceeded 50–60 m and may have been as much as 180–250 m in the southwestern part of the basin (Fig. 5).

During the regressive phase of the supersequence, the shelf margin aggraded and subsequently prograded. During this time, fourth-order lowstands within all the third-order systems tracts were sufficiently restrictive to allow evaporites to accumulate in the basin (Fig. 5). A stratified basin setting and deposition of black, laminated dolomudstones as condensed sections overlying the evaporites probably allowed preservation of evaporites during marine highstand conditions. Reciprocal sedimentation kept the basin filled, and no steepened shelf-to-basin profile existed during deposition of this regressive part of the supersequence. Depositional relief rarely exceeded 30–50 m.

The composition of carbonate buildups and associated biotic communities within the supersequence changed through time. The Barker Creek and Akah transgressive composite sequences (i.e., Atokan, DS10, and DS20) contain steep-sided buildups that are 3 to 20 m thick and have a diverse biotic assemblage that is distinctive for the abundance of *Chaetetes*, encrusting calcitornellid and opthalmid foraminifera, and cyanobacteria. Other common biotic constituents include brachiopods, echinoderms, fenestrate bryozoans, *Syringopora* and rugose corals, fusulinids, and minor phylloid algae (Kelly, 1994; Gianniny, 1995). The buildups of the regressive Desert Creek and Ismay sequences (i.e., DS30 and DS40) are lower-relief biostromes to bioherms; they range from 3 to 15 m thick and have an internal fabric that varies from algal bafflestones to algal wackestones and packstones (Weber et al., 1995a, b; Grammer et al., 1996). They are composed dominantly of phylloid-algal fragments and subordinate amounts of brachiopods, fusulinids and other foraminifers, bryozoans, crinoids, and *Chaetetes*. The late highstands of the DS40, DS50, and M10 composite sequences show an increase in siliciclastic component and a marked change in biotic composition and depositional geometry (Fig. 9). Crinoid-brachiopod grainstone and packstone banks compose the platform-margin facies and occur seaward of broad siliciclastic-rich shelf interiors.

All the mound facies throughout the supersequence are associated with well-washed grainstones that were deposited in high-energy shoals. The late-regressive crinoid banks display planar and trough cross-bedding, indicating that these skeletal banks, as well as the older mounds, were deposited in relatively shallow water. Water depths were probably less than 15 m (Weber, 1992; Gianniny, 1995; Grammer et al., 1996). Thus, it is unlikely that differences in depositional environment can account for this temporal change in the biotic composition of the supersequence. It is more likely that this change is the result of changing paleoenvironmental conditions in the basin over the duration of the supersequence, perhaps combined with long-term shifts that spanned the entire Pennsylvanian (West, 1988). Long-term increasing basin restriction, progressive elevation of salinity to levels at which fourth-order sequences developed evaporitic lowstand deposits, and an increased influx of siliciclastic material are characteristic of the regressive part of the supersequence. This paleoenvironmental evolution may have caused the change from an initially diverse cyanobacteria-foraminifera-*Chaetetes* community (Atokan, DS10, and DS20 sequences) to a more monospecific phylloid-algal community (DS30 and DS40 sequences), and, finally, to a turbidity-resistant crinoid-brachiopod community (upper parts of DS40, DS50, and M10 composite sequences).

Economic Significance of Supersequence Architecture

The development of the commercially productive hydrocarbon system in the Paradox basin is closely related to the development of the Morrowan–lower Missourian supersequence and its component third-order composite sequences (Fig. 2). Organic-rich, black, laminated lime mudstones, interpreted to be the source of hydrocarbons for the oil fields producing from the supersequence, are contained within the third-order transgressive systems tracts that occupy the transgressive and turnaround parts of the supersequence (i.e., A shale, C shale, Chimney Rock unit, and Gothic unit) (Figs. 5, 6). They extend across the basin and over paleotopographic highs. They are sapropelic and have total organic carbon (TOC) contents that can exceed 15% (Hite, 1970; Roylance, 1984). High organic production in waters of elevated salinity is probably due to the proliferation of phytoplankton (Kirkland and Evans, 1981). High salinities decrease the solubility of oxygen, promote anoxic bottom conditions, and allow preservation of organic material.

The most prospective reservoirs in the supersequence are composed of carbonate-sand shoals and algal-mound buildups. Carbonate-sand and grain-supported algal-mound facies are best developed along the inner shelf and shelf margin and in isolated carbonate platforms. The Aneth platform, for example, contains in excess of 1.3 billion barrels of oil and represents the largest known accumulation in the basin. About 40% of the total pore volume in this field is in grain-dominated facies composed of peloids, ooids, and skeletal debris (Weber, 1992). Primary porosity is volumetrically the most important and includes both interparticle and shelter-pore types in packstones, grainstones, and bafflestones (Fig. 10). Some early marine cement has partly filled pores and provided a rigid framework that resisted porosity reduction during compaction. Secondary moldic and dissolution-enhanced interparticle porosity is high beneath third-order sequence boundaries (Fig. 10). Peritidal dolomudstones and dolowackestones are secondary reservoirs and have intercrystalline and dissolution-enhanced secondary porosity.

The best potential for thick reservoir development in this supersequence is associated with highstands developed in the second-order late-transgressive and turnaround to regressive phases, i.e., the DS20 and DS30 composite sequences. At the second-order turnaround, depositional accommodation space is sufficient to allow mound-buildup and grain-supported shoal facies to coalesce laterally and stack vertically. The abundance of phylloid-algal–dominated buildups at the second-order turnaround in the DS30 and lower part of the DS40 composite sequences adds a significant reservoir type that is not widespread in earlier sequences and is absent in the late-regressive part of the supersequence. Reservoir performance is improved where reservoir facies stack vertically and are connected laterally. Reservoir quality may be enhanced by meteoric leaching associated with third-order sequence boundaries. During the late stage of the regression (upper part of DS40 and DS50 compos-

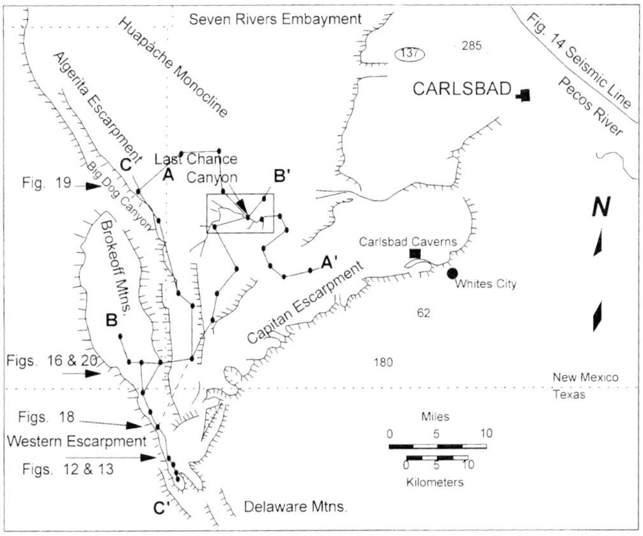

FIG. 11.—Map of region around Guadalupe Mountains showing locations of well-log cross-sections, outcrop profiles, and nearby seismic line (Fig. 14) (Sarg et al., 1997) that have served to provide the stratigraphic framework summarized here. Key outcrop localities include Western Escarpment below Guadalupe Peak (Fig. 13), Brokeoff Mountains (Figs. 16, 18, and 20), Algerita Escarpment (Fig. 19), and Last Chance Canyon.

ite sequences), the grain-supported strata are dominantly echinoderm-brachiopod banks that have little preserved primary interparticle porosity and poor reservoir quality.

GUADALUPIAN SUPERSEQUENCE: PERMIAN BASIN

Geologic Setting and Sequence Hierarchy

The Guadalupian supersequence is the youngest second-order supersequence of the lower Absaroka II supersequence set within the first-order lower Absaroka megasequence (Sloss, 1963). Deposition of this supersequence lasted for 10–12 m.y. (Fig. 2) during the transition from Permian-Pennsylvanian ice-house conditions to Mesozoic greenhouse conditions (Fisher, 1982). In the Permian basin region (Figs. 4, 11), the strata of this supersequence form a series of ramps and rimmed platforms that separate broad shelf provinces from deeper continental-margin basins and compose the postorogenic fill of the basin (Fig. 12). During the Late Permian, the rate of basin subsidence was low enough to permit the shelf to remain shallow and to permit the deposition of a series of prograding carbonate shelves, grain banks, and reef margins. Evaporites, composed dominantly of anhydrite (crystallized in the subsurface) and gypsum (crystallized at the surface), are present shelfward of these carbonate platforms. Siliciclastics form thick, wedge-shaped deposits that are the primary fill in the basin areas. At the transition to shelf-edge carbonates, these siliciclastics onlap the toe-of-slope carbonates and are commonly intercalated with beds of carbonate debris.

The outcrops in the Guadalupe Mountains (Fig. 11) represent a dip-oriented cross section of the Guadalupian supersequence (Figs. 12, 13). The correlation scheme developed in this study builds on the great volume of prior work in this clas-

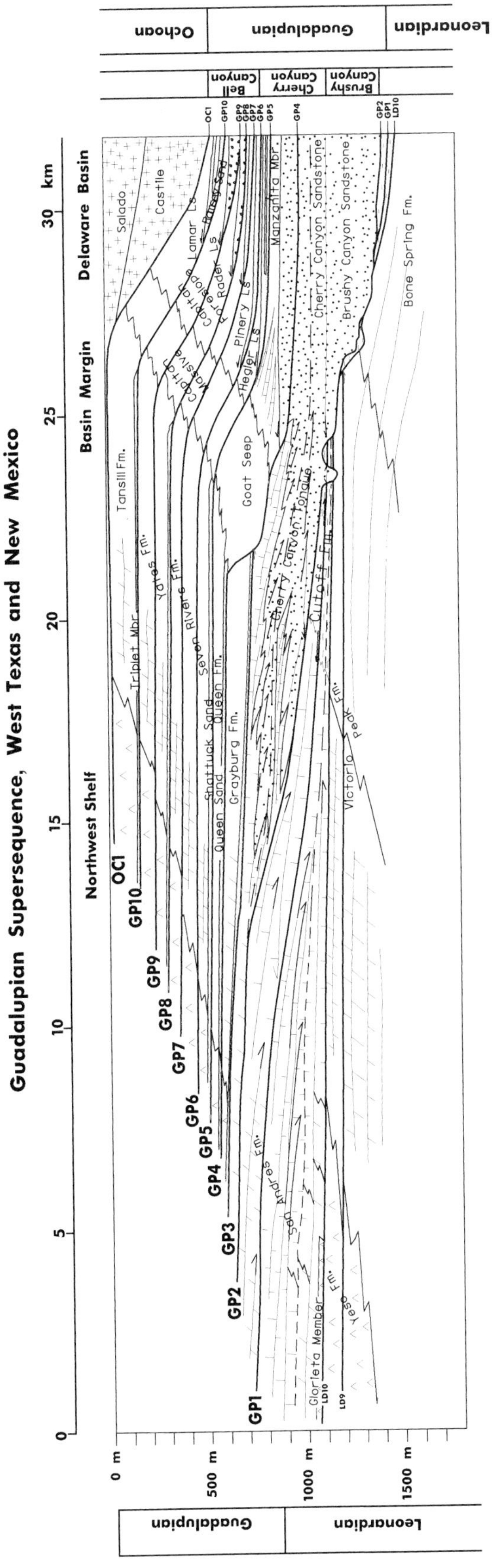

FIG. 12.—Schematic summary of Guadalupian supersequence architecture, west Texas and New Mexico, and enclosed third-order depositional sequences (LD10, GP1, GP2, etc., are sequence boundaries). Location shown in Figure 11.

FIG. 13.—Western Escarpment of Guadalupe Mountains. Outcrop profile shows large-scale second-order supersequence architecture. LD10 is the basal sequence boundary of the Guadalupian supersequence. Underlying Victorio Peak shelf edge is the youngest and most basinward Leonardian margin. Second-order transgression resulted in deposition of deep-water limestones of the Cutoff Formation that drape the underlying Victorio Peak bank. Onlapping siliciclastics (Brushy Canyon and Cherry Canyon Formations) are interpreted to represent third-order lowstand systems tracts. The late-regressive phase of the supersequence is manifested in the capping cliffs that show high-angle clinothems of the Goat Seep and Capitan platform-margin "reef complexes." Location shown in Figure 11.

TABLE 1.—THIRD-ORDER DEPOSITIONAL SEQUENCES COMPOSING SECOND-ORDER GUADALUPIAN SUPERSEQUENCE

Third-Order Sequence	Third-Order Sequence Boundary	Stratigraphic Position or Location of Sequence Boundary
Castile	OC1	base of the Castile Formation
Tansill and Lamar	GP10	base of the Triplett Member, Yates Formation†
Upper Yates	GP9	base of the upper Rader Member, Bell Canyon Formation‡
Lower Yates	GP8	base of the lower Rader Member and Bell Canyon Formation‡
Upper Seven Rivers	GP7	top of the Pinery Limestone Member, Bell Canyon Formation
Middle Seven Rivers	GP6	base of the Azotea Tongue and top of the Hegler Limestone Member, Bell Canyon Formation
Lower Seven Rivers	GP5	base of the Shattuck Sandstone Member, base of the Capitan Formation, and base of the Manzanita Member, Cherry Canyon Formation‡
Queen and Goat Seep	GP4	base of the Queen and Goat Seep Formations
Grayburg	GP3	base of the Grayburg Formation and intra–Cherry Canyon Formation
San Andres 3 and Brushy Canyon	GP2	intra–San Andres Formation and base of the Brushy Canyon Formation
San Andres 2	GP1	intra–San Andres Formation, upper Cutoff Formation, and base of the intra Cutoff Sandstone
San Andres 1	LD10	top of the Victorio Peak and Bone Spring Formations and base of the San Andres and Cutoff Formations

* Abbreviations for sequence boundary names: OC = Ochoan, GP = Guadalupian, LD = Leonardian.

† Based on stratigraphic nomenclature of Esteban and Pray (1977).

‡ Based on stratigraphic nomenclature of King (1948).

sic outcrop region, including: (1) the pioneering mapping work by King (1948), Boyd (1958), and Hayes (1964); (2) a series of thesis investigations under the supervision of Lloyd C. Pray, University of Wisconsin–Madison, on various aspects of the Western Escarpment of the Guadalupe Mountains (Crawford, 1981; Harris, 1982; Kirkby, 1982; Rossen, 1985; Franseen, 1985; Fekete, 1986; New, 1988; Franseen et al., 1989); (3) the initial and current sequence stratigraphic work by Sarg and Lehmann (1986) and Sarg et al. (1997); and (4) the more recent detailed work by Kerans and Fitchen (1995) on the high-frequency cyclicity of the San Andres and Grayburg Formations. The stratigraphic framework presented in this paper is consistent with the fusulinid paleontology of Wilde (1986) and the now-established conodont biostratigraphy of Wardlaw and Rohr (1996).

The Guadalupian supersequence is composed of 11 third-order sequences (Fig. 12). Each sequence ranged in duration from about 0.9 to 1.2 m.y. By convention, the name of the third-order sequence is associated with the name of the formation that forms the bulk of the sequence, and the defining sequence boundary is always the basal sequence boundary (Table 1). The Guadalupian supersequence is bounded by the LD10 and OC1 sequence boundaries. These are recognized as major unconformities that separate thick packages of strata and are classified as both second- and third-order sequence boundaries. Chronostratigraphically, the Guadalupian supersequence is not entirely contained within the Guadalupian Series. Recent fusulinid and conodont biostratigraphic studies of the San Andres and Cutoff formations (Fig. 12, Table 1) that compose the lower parts of the supersequence document that the Leonardian/Guadalupian boundary occurs within these formations (Wilde, 1986; Wilkinson et al., 1991; Wardlaw and Rohr, 1996).

Summary of Third-Order Systems Tracts

The third-order sequences contain well-developed, and seismically resolvable, lowstand and highstand systems tracts (Fig. 14). Figure 15 summarizes the different facies present in these strata. Lowstand systems tracts are dominantly siliciclastic in character (Fig. 13), but have components of both allochthonous and *in situ* carbonate material. Lowstand systems tracts can include both a basal, onlapping siliciclastic wedge composed of sandstone and siltstone and an upper, prograding complex. In places, the siliciclastic wedges contain beds of carbonate material that originated by erosion of the contiguous bank margins and are interpreted to be debris flows (Sarg, 1988; Sarg et al., 1988; Franseen et al., 1989). The prograding complexes onlap at or below the underlying highstand shelf-margin breaks (Figs. 12, 14) and are oblique progradational units composed of mixed carbonate and siliciclastic sediments (Fig. 16). The sandstones of these clinothem packages result from bypass of sand and silt across the shelf and onto the slope and the toe of the slope, where they were deposited as onlapping wedges. The carbonates represent progradation of shallow-water bank sediments.

The third-order transgressive systems tracts are generally thin. They are composed of mixed sandstone and carbonate sediments that, in outcrop and core, are characterized by shallow-water, shelf-crest grain-supported shoal lithofacies to mud-dominated, platform-interior, subtidal and peritidal lithofacies (Fig. 15). Highstand platforms are flat topped and have gentle to abrupt, relatively steep platform edges. We interpret bank tops to be flooded rapidly during third-order rises in sea level, causing quick turnarounds to highstand conditions. The exception to this scenario is the initial transgressive systems tract (LD10 sequence) of the Guadalupian supersequence in which carbonate sedimentation kept up with a long-term rise in sea level.

Highstand systems tracts are either dominated by carbonate-bank production or a mix of carbonate and siliciclastic deposition. They are aggradational in the first third-order sequence of the second-order supersequence (LD10 sequence) and sigmoidal to obliquely progradational in the ensuing third-order sequences of the second-order regression (Fig. 14). Highstand shelf margins change in character through the supersequence (Fig. 17). During the second-order turnaround and early-regressive periods, the shelf-crest facies is dominantly nonskeletal (ooid-peloid) grainstone lithofacies (LD10, GP1, GP2, and GP3 highstand platforms of the San Andres and Grayburg Formations, Fig. 12). The remaining carbonate platforms of the second-order regression (Goat Seep and Capitan formations, Fig. 12) face the deep Delaware basin and are characterized by organic-cement boundstones at the platform margin (Fig. 17).

Component Third-Order Sequences

LD10 third-order depositional sequence—Initiation of the Guadalupian supersequence.—
The Guadalupian supersequence rests immediately above the LD10 second-order sequence boundary. The final Leonardian highstand platform-margin position, on the Western Escarpment of the Guadalupe Mountains, is at Shumard Canyon at the top of the Victorio Peak Formation (Fig. 13) Skeletal dolograinstones of the shallow-water, high-energy bank-margin facies pass basinward into skeletal lime wackestones of the upper slope and, then, into the dark gray, lime mudstones of the lower slope and basin (Bone Spring Formation). The bank margin here is oversteepened at the LD10 boundary, owing to approximately 300 m of erosion over a lateral distance of 1 to 2 km (Pray, 1971). Debris-flow megabreccias fill 10–15-m-deep channels cut into the Victorio Peak slope (Pray and Stehli, 1962). This slope erosion is interpreted to be largely submarine in origin (Pray, 1971; Kirkby, 1982). LD10 passes shelfward to the northwest to a disconformity where dissolution-enlarged cavities form in the bank-margin facies (K. W. Rudolph and J. F. Sarg, pers. observation, 1984; M. D. Sonnenfeld, pers. commun., 1995). Truncation of approximately 10 to 15 m of bank-crest skeletal dolograinstones occurs in the Cutoff Mountain locality, immediately north of the Western Escarpment of the Guadalupe Mountains (Fig. 18). In the West Dog and Panther Canyon areas of the Brokeoff Mountains to the north (Fig. 11), a thin quartz sandstone unit directly overlies the LD10 sequence

boundary and is interpreted to be a transgressive lag deposit (Sarg et al., 1997).

Siliciclastic bypass occurred at the LD10 sequence boundary in other parts of the Delaware basin (Fig. 14) and also in the Midland basin (Sarg et al., 1997). Regional onlap of a basin-restricted siliciclastic wedge occurs on this sequence boundary in these basin areas. This siliciclastic unit has recently been identified in the region around the southern Delaware Mountains (M. H. Gardner, pers. commun., 1997). Bypass of the Victorio Peak bank coincided with a major reorganization of basin sediment patterns, and the LD10 unconformity became the basal boundary for the ensuing Guadalupian supersequence.

The transgression of the LD10 third-order sequence encompasses the entire second-order transgression. This transgression caused a major deepening of the shelf, and the position of the first shelf margin is from 2 to 25 km shelfward to the northwest of the underlying Leonardian shelf edge. Deep-water, dark gray to black lime mudstones of the Cutoff Formation drape the older Victorio Peak bank-margin topography (Harris, 1982, 1988) (Figs. 12, 13). This deep-water lithofacies passes shelfward into slope-forming skeletal dolowackestones. The bank-margin lithofacies consists of skeletal dolopackstones and dolograinstones and is exposed on the Algerita Escarpment (Sarg and Lehmann, 1986; Kerans and Fitchen, 1995) (Figs. 11, 19). For much of the transgression, the carbonate productivity kept up with sea level and resulted in a thick (>130 m), aggradational carbonate bank and platform buildup. The maximum flooding surface crops out about half way up the Algerita Escarpment (Fig. 19) and is overlain by a 150-m-thick, aggradational highstand systems tract that has a ramp geometry (middle San Andres Formation). Here, the highstand ramp margin consists of shelf-crest ooid-peloid dolograinstone and outer-shelf fusulinid skeletal dolopackstones to dolograinstones (Sarg and Lehmann, 1986; Kerans and Fitchen, 1995).

The transgressive and highstand systems tracts of the LD10 third-order sequence are also well displayed in the subsurface. On the Delaware basin seismic line in Figure 14, the lateral shift is less than in the outcrop, covering a distance of only about 2 km versus about 25 km for the outcrop. This difference is attributed to different subsidence rates in the two areas. Subsidence was more rapid in the outcrop area and over much of the Delaware and Midland basin areas, causing a greater amount of backstep. The stratigraphically next-higher three third-order depositional sequences (GP1, GP2, and GP3) represent the turnaround and beginning of the second-order Guadalupian regression.

GP1 third-order depositional sequence.—

In the northern part of the Guadalupe Mountains, the first third-order sequence boundary of the Guadalupian supersequence (GP1) is interpreted to be situated at the top of the middle San Andres Formation as defined by Sarg and Lehmann (1986). An abrupt basinward or down-ramp shift to the southeast, of shelf-crest ooid and skeletal grainstone lithofacies defines this sequence boundary. In places, quartz sandstone is present on the shelf and in down-ramp positions overlying GP1

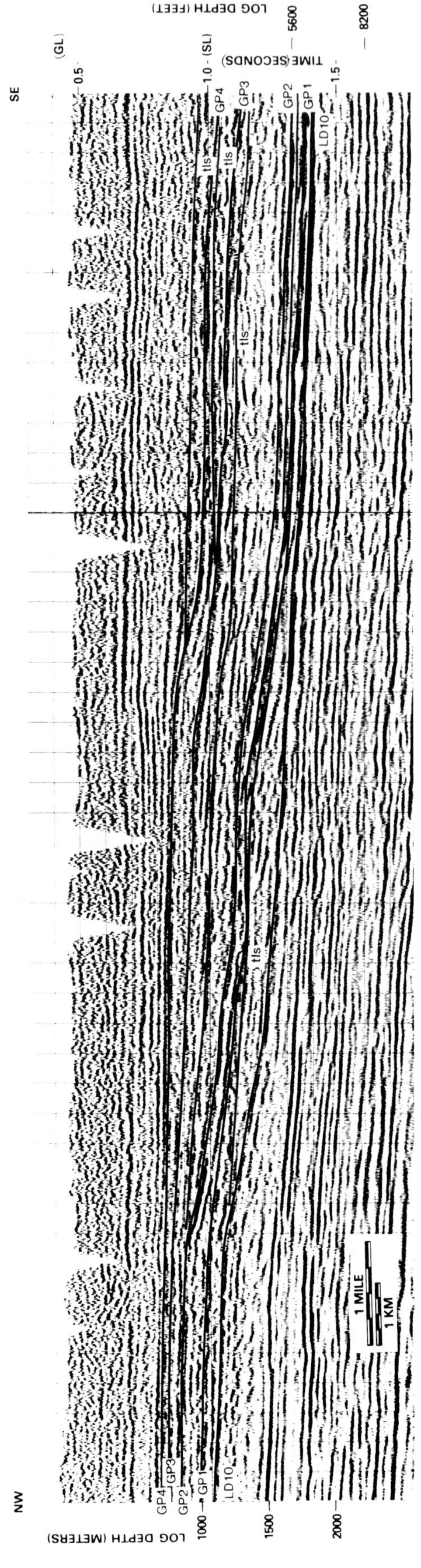

FIG. 14.—Seismic line located northeast of Carlsbad, New Mexico, on edge of Delaware basin showing aggradation and progradation of early-middle Guadalupian third-order depositional sequences similar to observed outcrop geometries. Well to seismic ties indicate that lowstand systems tracts are composed of onlapping siliciclastic wedges and mixed carbonate and siliciclastic prograding complexes (e.g., GP2 sequence). GL—ground level; SL—sea level; tls—top lowstand system; GP1, etc.—sequence boundaries. Location shown in Figure 11.

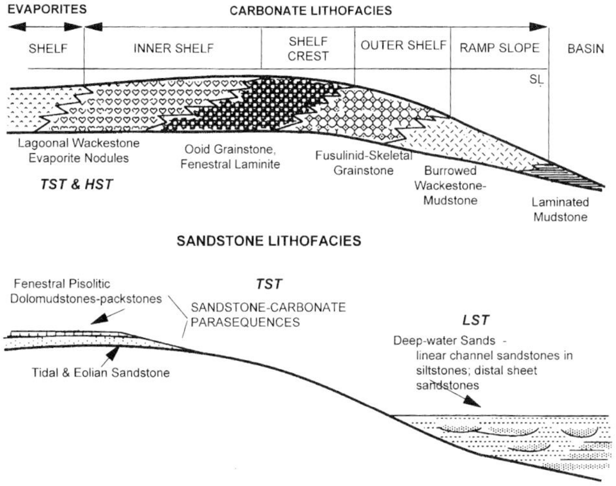

FIG. 15.—Lithofacies summary of components of Guadalupian systems tracts, Permian mixed siliciclastic and carbonate depositional system, Permian basin. TST—transgressive systems tract, HST—highstand systems tract, LST—lowstand systems tract, SL—sea level.

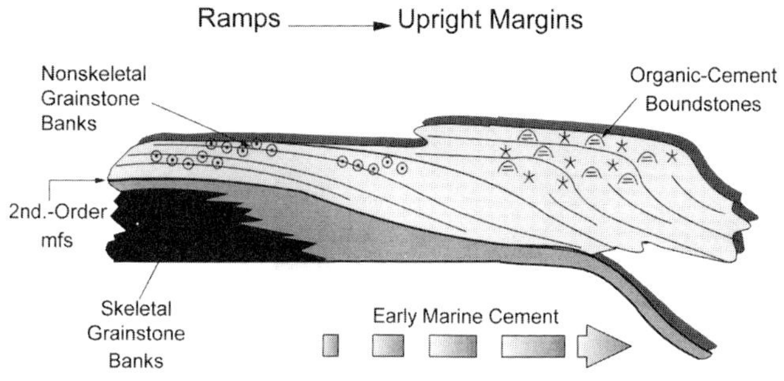

FIG. 17.—Generalized schematic summary of asymmetric Guadalupian supersequence showing evolution of highstand carbonate-platform margins from skeletal to nonskeletal grainstone banks during the second-order transgression, to upright margins of organic-cement boundstones in the late-regressive phases of the supersequence (mfs = maximum-flooding surface). Upright margins with steep foreslopes evolved as the depositional systems reached the deeper and underfilled Delaware basin; the amount of early marine cement increased, filling most of the original depositional porosity in the shelf-edge complexes.

FIG. 16.—Cherry Canyon Sandstone Tongue prograding complex, eastern West Dog Canyon, Brokeoff Mountains. View to northeast showing progradation to the southeast into the Delaware basin. Toplap surface occurs within the complex and is overlain by an aggradational stack of peritidal dolomites and sandstones that cap the lowstand system tract (LST). These peritidal units are interpreted to pass farther basinward to the south into prograding clinothems.

FIG. 18.—Erosion of Victorio Peak Formation carbonate-platform top at the LD10 sequence boundary, at Cutoff Mountain in the southern Brokeoff Mountains (location in Fig. 11). Here, erosion at the sequence boundary has removed 10 to 12 m of section, as measured on the large-scale truncated bed forms evident below the sequence boundary. These strata are fusulinid-rich skeletal dolograinstones of the platform-margin facies and are interpreted to be tidally influenced spillover lobes oriented obliquely to the trend of the margin.

(Fig. 14). The GP1 sequence boundary is the least well expressed discontinuity in the region around the Guadalupe Mountains, and there is debate over its exact position in outcrops of shelf lithofacies. Kerans et al. (1993) and Kerans and Fitchen (1995) have interpreted a sequence boundary to be on top of the first significant ooid-peloid grainstone unit on the northern Algerita Escarpment, a location that is somewhat higher stratigraphically than where we would place the boundary. The uncertainty results from the difficulty of tracing the boundary from outcrop to outcrop. However, regardless of its position in any one outcrop, it is well defined in the Midland and Delaware basin areas, where siliciclastic lowstand wedges onlap the slope (Fig. 14).

The transgressive and highstand systems tracts of the GP1 sequence reestablished carbonate production over the shelf areas. Highstand progradation in the outcrop brought the shelf-crest ooid grainstone lithofacies to a position approximately 5 km shelfward of the underlying LD10 highstand margin. To the northeast, and on the Delaware basin seismic line (Fig. 14), the GP1 highstand was aggradational and developed a well-defined platform margin. Significant shelf-margin and upper-slope erosion occurred at the overlying GP2 sequence boundary. This erosion is observed both in the outcrop area of the Western Escarpment of the Guadalupe Mountains (Fig. 13) and on the seismic line (see prominent reflection terminations at platform edge, Fig. 14). On the shelf, interbedded peritidal sandstones and carbonates are present directly above a well-developed GP2 karst surface (Kerans and Fitchen, 1995).

FIG. 19.—The second-order transgressive systems tract (TST), lower San Andres Formation, Algerita Escarpment, Guadalupe Mountains (location in Fig. 11). An approximately 130-m-thick, subtidal skeletal dolograinstone bank is the resistant, light-colored unit in the middle of the escarpment outcrop. Resistant outcrops are bank-margin skeletal dolograinstone facies that pass basinward to the south (i.e., to the right) into laterally equivalent and recessive-weathering dolowackestones and mudstones of the ramp or bank slope. This bank unconformably overlies the LD10 sequence boundary and the platform-interior lithofacies of the older Victorio Peak Formation highstand systems tract. Two prominent flooding surfaces (fs) are present within this transgressive systems tract and serve to subdivide it into three backstepped parasequence sets. The TST is capped by a maximum-flooding surface (mfs) that separates deep-water cherty dolomudstones from the underlying grainstones.

GP2 third-order depositional sequence.—

Major siliciclastic bypass occurred at the GP2 sequence boundary and resulted in deposition in the basin of the thick Brushy Canyon Formation and Cherry Canyon Tongue Member. These strata represent the GP2 lowstand systems tract. This lowstand systems tract is particularly well exposed and well studied on the Western Escarpment of the Guadalupe Mountains and farther basinward to the south in the adjacent Delaware Mountains (Harms, 1974; Rossen, 1985; Rossen and Sarg, 1988; Zelt and Rossen, 1995; Gardner and Sonnenfeld, 1996). Shelfward, to the north-northwest, the upper part of the GP2 lowstand systems tract is interpreted to consist of a prograding complex (Fig. 16) that onlaps the underlying highstand shelf margin (Fig. 14).

The updip onlap and pinch out of the Cherry Canyon Tongue crops out in the west-central part of the Brokeoff Mountains (Figs. 11, 20). Here, sandstone beds onlap the underlying GP1 highstand of the San Andres Formation. The GP1 highstand is composed of cherty, fusulinid-rich, skeletal dolowackestones to dolomudstones and skeletal and peloidal dolopackstones to dolograinstones deposited on the slope and outer shelf, respectively (Fitchen, 1993). The onlapping sandstones are thick to massive bedded, fine- to medium-grained quartz sandstones. At the pinch out (Fig. 20), they exhibit parallel to low-angle, tangential cross-lamination, have burrowed tops, and are interpreted to have been deposited in shoreface to offshore environments (Fitchen, 1993).

A mixed carbonate-siliciclastic lowstand prograding complex directly overlies the sandstones (Fig. 20) and is interpreted to thin and pinch out farther shelfward, to the north, at the GP1 highstand bank margin. This prograding complex also is evident on the seismic line (Fig. 14). Carbonate clinothems of the prograding complex are composed of fusulinid-peloid dolowackestones to dolopackstones that change facies up slope to shelf-crest skeletal and peloidal dolograinstones (Fitchen, 1993). Along strike, intercalated sandstone beds within the prograding complex onlap carbonate clinothems (Fig. 16).

FIG. 20.—Onlap edge of Cherry Canyon Tongue lowstand sandstone and overlying carbonate prograding complex, western West Dog Canyon, Brokeoff Mountains. Sandstones occur below the downlap surface (dls) at the base of the prograding carbonate and above the GP2 sequence boundary.

The geometry of the prograding complex changes upward from oblique progradation to an aggradational stack of meter-scale sandstone and carbonate parasequences (Fig. 16). Basal brown sandstones grade upward into thin-bedded, light gray, peloidal dolomudstones to pisolitic dolopackstones that contain fenestral fabrics. Small-scale trough and low-angle tangential cross-beds and crinkly, wavy parallel, and ripple laminations with reactivation surfaces occur within the sandstones. Both sandstones and carbonates are interpreted to have a peritidal origin (Sarg et al., 1997). In the area of the Brokeoff Mountains, the lowstand systems tract of the GP2 sequence is overlain by a relatively thin, 10–20-m-thick, transgressive systems tract that is composed of a series of shoaling-upward ooid-rich and peritidal parasequences (Sarg et al., 1997).

Along strike in Last Chance Canyon (Fig. 11), where the dark gray lime mudstones of the GP1 highstand toe-of-slope facies crop out in the canyon bottom, the GP2 sequence boundary surface was differentially eroded into these mudstones to form a gullied slope. Lime mud–rich sponge bioherms grew on intergully highs. These bioherms attained thicknesses of more than 20 m and are interpreted to be part of the late GP2 lowstand and GP2 transgressive systems tracts (Sarg et al., 1997). In the Last Chance Canyon slope area, the bioherms are overlain by later transgressive siliceous dolomudstones and quartz sandstones. A prominent downlap surface and maximum flooding surface records the turnaround to highstand conditions (Sarg and Lehmann, 1986). The GP2 highstand systems tract is 40–70 m thick and prograded into the basin for 2–5 km as a series of sigmoidal to oblique clinothem units (Sarg and Lehmann, 1986; Sonnenfeld and Cross, 1993) (Fig. 14).

The GP2 lowstand wedge, as well as other Guadalupian lowstand siliciclastics, may have had point sources. An isopach of the GP2 lowstand wedge in the Guadalupe Mountains (Fig. 21) shows that, in this area of the Delaware basin, the basin margin has an irregular trend. A locally thickened part of the wedge extends back into the Northwest shelf in the Last Chance Canyon area (northeast part of map in Fig. 21). In the Western Escarpment of the Guadalupe Mountains and in the northern Delaware Mountains (southwest part of map in Fig. 21), the isopach contours extend in a shelfward direction to the north-

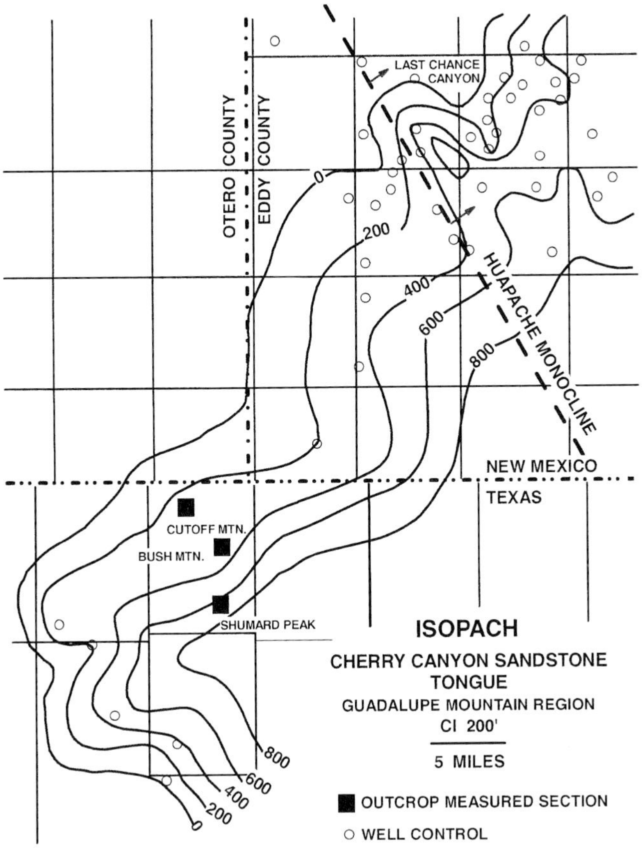

FIG. 21.—Brushy Canyon–Cherry Canyon Tongue isopach map incorporates data from outcrops and wells (circles) (contours in feet; CI = contour interval). Siliciclastic conduits into the basin are suggested by a locally thickened part of the wedge that extends into the Northwest shelf in the Last Chance Canyon area (northeast part of map) and by a shelfward (i.e., northwest) extension of isopach contours in the southwest, at the turn in the basin-margin trend from the northeast to the south.

west where the basin margin turns from a southwesterly to a southerly trend. Both of these locations are interpreted to be sites of major conduits that fed sand and silt into this part of the Delaware basin. They also appear to correspond to abrupt changes in underlying structural configurations. The Last Chance Canyon conduit extends shelfward to the northwest from a dip increase in the Huapache monocline that formed in the Pennsylvanian and plunges northeastward into the Delaware basin. The Western Escarpment conduit coincides with an abrupt change in the basin-margin trend at the northwest corner of the Delaware basin. This structural control on Permian deposition is common in other parts of the Delaware basin (e.g., Central Basin platform) and has resulted from the Permian-Pennsylvanian Marathon orogeny.

GP3 third-order depositional sequence.—
The GP3 sequence boundary that caps the GP2 sequence in the Last Chance Canyon and Algerita Escarpment outcrops has been recognized and mapped as an erosional unconformity. It corresponds to the top of the San Andres Formation over much of the Guadalupe Mountains (Hayes, 1964; Sarg and Lehmann, 1986). This sequence boundary displays decimeter-scale paleokarst solution cavities now filled with quartz sandstone (Sarg and Lehmann, 1986; Kerans and Fitchen, 1995). The updip, proximal part of the GP3 lowstand wedge is exposed at the mouth of Last Chance Canyon, where it comprises a series of peritidal sandstone and carbonate cycles in the lower part of the Grayburg Formation (Sarg and Lehmann, 1986; Kerans et al., 1991). This peritidal facies pinches out by onlap to the northwest, in a shelfward direction and grades into shelf-crest, ooid grainstone lithofacies toward southeast into the basin. Over the Guadalupe Mountains region from the Algerita Escarpment eastward, and in the subsurface, the transgressive and highstand systems tracts of this sequence are contained within the carbonate-platform lithofacies of the Grayburg Formation (Fig. 12). The capping boundary of this sequence (GP4) records subaerial exposure and major bank-margin erosion (Franseen et al., 1989) (Fig. 13).

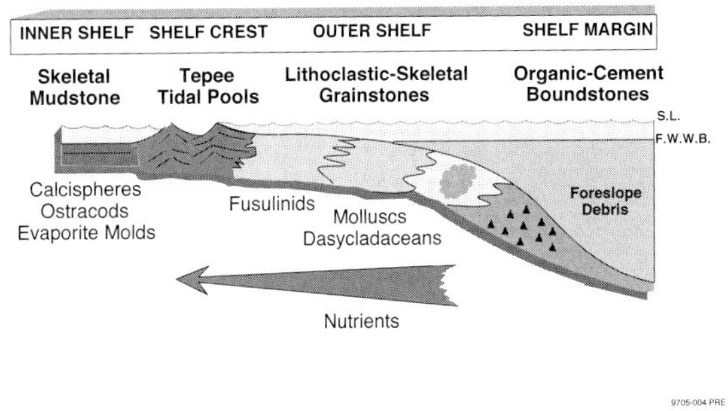

FIG. 22.—Schematic summary of Capitan shelf-margin transition showing seaward progression of outer-shelf, grain-supported lithofacies from fusulinid grainstones to mollusk and dasycladacean (green algal) grainstones to shelf-edge, sponge-algal-cement boundstones.

Remaining Guadalupian third-order depositional sequences.—
The remaining third-order sequences of the Guadalupian complete the regressive highstand part of the supersequence (Table 1, Fig. 12). These sequences comprise the successively younger Queen, Seven Rivers, Yates, and Tansill Formations on the shelf and include the shelf-edge reef complexes of the Goat Seep and Capitan formations (Figs. 12, 13). As progradation continued through the second-order highstand, the shelf-to-slope transitions encountered a deeper-basin environment and changed from ramp geometries to submerged, rimmed shelves with foreslopes that attained angles of 35°–40°. Sand and silt bypassed shelf areas during interpreted lowstands in sea level and compose the basin-restricted siliciclastic wedges of the upper part of the Cherry Canyon and Bell Canyon formations (Fig. 12).

The platform or shelf crest is dominated by tepee-fenestral grainstone complexes (Esteban and Pray, 1977; Mazzullo et al., 1989). Seaward, to the southeast, lithoclastic-skeletal grainstones were deposited in shoal-water environments. Organic-cement boundstones formed in deeper (40–50 m) water at the

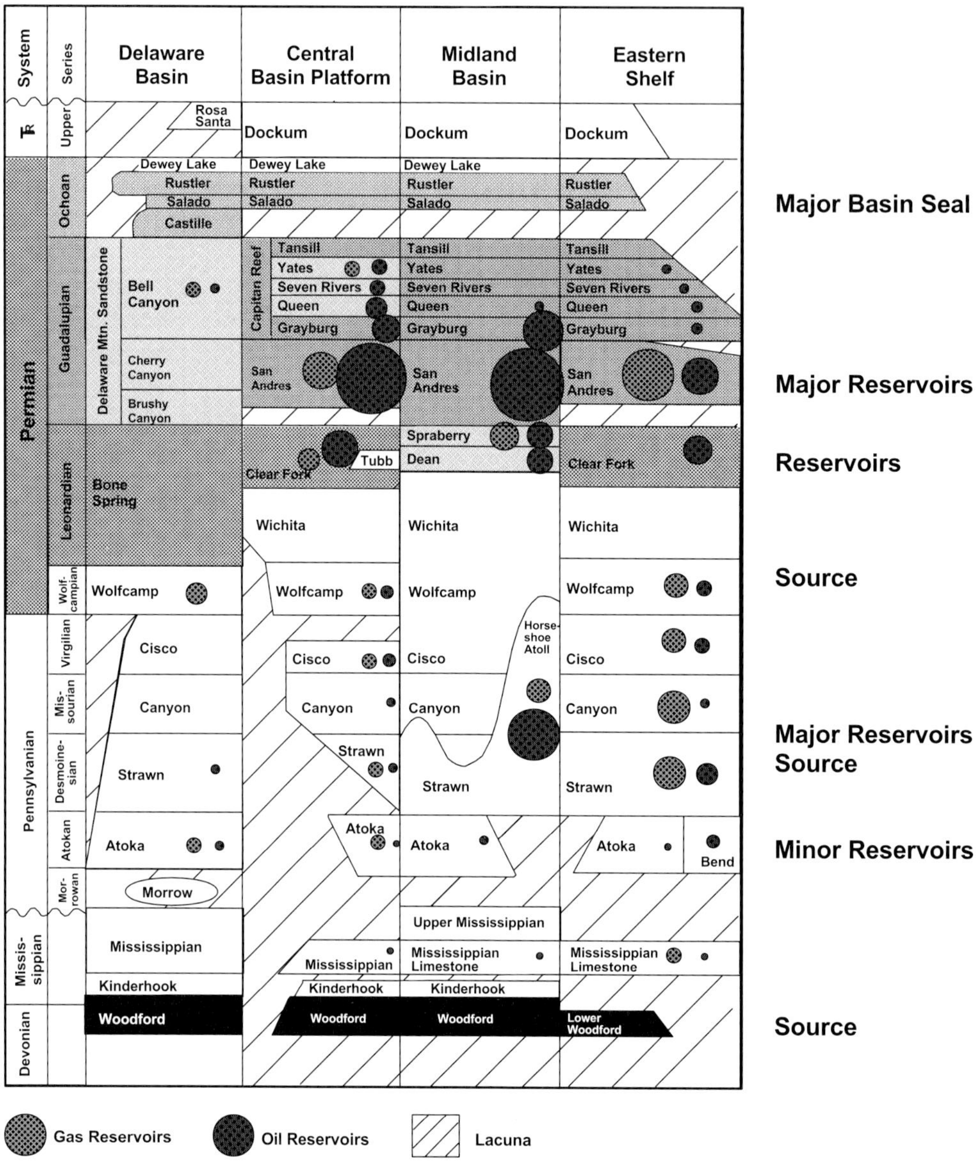

FIG. 23.—Lithostratigraphic correlation chart for the Permian basin with the stratigraphic components of the basin's petroleum systems (modified from Galloway et al., 1983, and Kosters et al., 1989).

transition from the shelf to the slope (Babcock, 1977; Yurewicz, 1977; Crawford, 1981). Early marine cement fills most of the pore space in all three of these facies (Babcock, 1977; Esteban and Pray, 1977; Hurley, 1978; Mazzullo et al., 1989). This changeover from nonskeletal-dominated ramp margins to upright margins supported by organic-cement boundstones is interpreted to be the result, at least in part, of increased access to nutrient-rich waters of the deeper Delaware basin.

There is a marked biotic segregation in the facies transition across the outer shelf to the shelf edge of these upright margins (Fig. 22). Fusulinid-rich grainstones, composed of large and internally complex schwagerinid forms, are immediately seaward of the shelf-crest tepee-fenestral complex. These pass farther seaward, to the southeast, into a mollusk- and dasycladacean-rich grainstone and packstone facies. This mollusk-algal outer-shelf belt then changes facies laterally into the shelf-margin sponge-algal-cement boundstones. The biotic distribution across the margin is due, perhaps, to a marked nutrient gradient across the outer shelf. The sponge-algal community, with its high nutrient requirements, dominated the shelf margin but gave way in the outer shelf to the large, complexly chambered fusulinids that were efficient and had low-nutrient requirements (Brasier, 1995).

The Guadalupian supersequence ended with major evaporative drawdown, subaerial exposure of the shelf, and cessation of carbonate deposition. The thick evaporites of the Castile and Salado formations filled the basin and overlapped the final Capitan shelf.

Economic Significance of Supersequence Architecture

The rocks of the lower part of the Guadalupian supersequence are of preeminent importance in the oil and gas economics of the Permian basin (Figs. 4, 23). The porous and permeable dolopackstones and dolograinstones of the bank-margin lithofacies of the first four third-order sequences of the supersequence (LD10, GP1, GP2, and GP3) are the major reservoirs of the Permian basin petroleum province (Bebout and Harris, 1986; Ward et al., 1986). Like the Paradox basin example, the turnaround sequences have the highest reservoir quality and trapping potential. These reservoirs have produced 7.7 billion barrels of oil and contain approximately 65% of all the hydrocarbons trapped in the Permian basin. Pore types are dominantly interparticle primary, moldic after fusulinids, and intercrystalline in dolomites. Porosity pinchouts occur at the shelfward facies change to evaporites and evaporitic dolomites that provide lateral seals for hydrocarbon accumulations. Wherever these porosity pinchouts coincide with structural noses, major hydrocarbon traps occur over the Northwest shelf and Central Basin platform of the Permian basin (Kosters et al., 1989) (Fig. 4). The platform-margin lithofacies of the late-regressive rimmed shelves (i.e., Goat Seep and Capitan formations) have lower reservoir quality owing to the large amounts of pore-occluding marine cement.

Source-rock strata that have probably contributed hydrocarbons to the Guadalupian reservoirs include the Leonardian basinal lime mudstones of the Bone Spring Formation and the Virgilian-Wolfcampian basinal shales (Figs. 2, 23) (Williams, 1977; Ramondetta, 1982; Hills, 1984; Robinson, 1988; see Hill, 1996, for recent summary). New TOC data from the Cutoff Formation suggest that this upper Leonardian–lower Guadalupian basin facies may also be, where mature, a potential source interval for the hydrocarbons that fill Guadalupian reservoirs (B. B. Sageman, pers. commun., 1997). Each of the above organic-rich units was deposited during late Paleozoic second-order transgressions (Fig. 2). Migration pathways are likely to be faults and fractures of Early Permian age and the sandstone-rich, prograding clinothems of highstand and lowstand systems tract units that are present basinward of the shelf-margin reservoirs.

CONCLUSIONS

1. A distinctive second-order (10–20 m.y.) stratigraphic signature is present within ancient, evaporite-related, carbonate-platform successions in both icehouse and transitional to greenhouse conditions of the Permian-Pennsylvanian of the southwestern United States. Supersequences can be symmetric or asymmetric and result from the interaction of long-term changes in depositional systems, basin geometry and subsidence, sea level, and paleoceanographic conditions.
2. The most normal, open-marine conditions occurred during the transgressive component of supersequences, and skeletal-grain banks exhibit the most diverse biota. Source-prone, organic-rich lime mudstones tend to be deposited across shelves and in the basin during second-order transgressions.
3. As shelf areas became more restricted during the turnaround and early-regressive components of supersequences, highstand platforms were dominated by nonskeletal bank-margin deposition. Early marine cement is a minor component, and carbonate reservoirs are best developed during the transgression and turnaround to regression.
4. During late regression, when platform progradation directly accesses a deeper basin (i.e., the Permian basin) or conditions change to elevated salinity and increased siliciclastic deposition (i.e., the Paradox basin), the biota become more specialized. Marine and early burial cementation increases, and reservoir quality is reduced.
5. Large petroleum accumulations occur at: (1) postdepositional drape of turnaround highstands over underlying structural blocks, combined with shelfward facies changes to nonporous inner-shelf strata (e.g., San Andres–Grayburg sequences); (2) isolated platforms and buildups that developed during second-order transgressions and turnaround phases and that are draped by sealing late-transgressive mudstone (e.g., Aneth platform); and (3) facies changes from carbonate sand deposition to evaporitic lagoons that are developed on the shelf during second-order transgression and turnaround (e.g., lower San Andres transgressive system).

ACKNOWLEDGMENTS

We gratefully acknowledge the support of Mobil Technology Company for this research and thank them for permission to publish this work. The concepts and interpretations reported herein depend a great deal on the prior work of the many geologists, both within and outside Mobil, who have studied these rocks, been fascinated with the stratigraphy, and shared their knowledge and perspectives with us. This manuscript has been greatly improved by the reviews of Toni Simo, Sal Mazzullo, and Mike Pope, and we appreciate their efforts. We would also like to single out for special thanks the mentoring and insight of Lloyd Pray, Fred Read, and Peter Vail. Their encouragement, support, and enthusiasm aided greatly in our work. The interpretations presented here, however, are our responsibility alone.

REFERENCES

Baars, D. L., Parker, J. W., and Chronic, J., 1967, Revised stratigraphic nomenclature of Pennsylvanian System, Paradox basin: American Association of Petroleum Geologists Bulletin, v. 51, p. 292–403.

Babcock, J. A., 1977, Calcareous algae, organic boundstones, and the genesis of the upper Capitan Limestone (Permian, Guadalupian), Guadalupe Mountains, west Texas and New Mexico, *in* Hileman, M. E. and Mazzullo, S. J., eds., Upper Guadalupian Facies, Permian Reef Complex, Guadalupe Mountains, New Mexico and West Texas: Midland, Permian Basin Section, Society of Economic Paleontologists and Mineralogists Publication No. 77-16, p. 3–44.

Bebout, D. G., and Harris, P. M., eds., 1986, Hydrocarbon Reservoir Studies, San Andres/Grayburg Formations, Permian Basin: Proceedings, Research Conference: Midland, Permian Basin Section, Society of Economic Paleontologists and Mineralogists Publication No. 86-26, 143 p.

Boyd, D. W., 1958, Permian sedimentary facies, central Guadalupe Mountains, New Mexico: Socorro, New Mexico Bureau of Mines and Mineral Resources Bulletin 49, 100 p.

Brasier, M. D., 1995, Fossil indicators of nutrient levels: 2. Evolution and extinction in relation to oligotrophy, *in* Bosence, D. W. J. and Allison, P. A., eds., Marine Palaeoenvironmental Analysis from Fossils: London, Geological

Society of London Special Publication No. 83, p. 133–150.

CHOQUETTE, P. W., 1983, Platy algal reef mounds, Paradox basin, *in* Scholle, P. A., Bebout, D. G., and Moore, C. H., eds., Carbonate Depositional Environments: Tulsa, American Association of Petroleum Geologists Memoir 33, p. 454–462.

CHOQUETTE, P. W., AND TRAUT, J. D., 1963, Pennsylvanian carbonate reservoirs, Ismay Field, Utah and Colorado, *in* Bass, R. O., ed., Shelf Carbonates of the Paradox Basin: Durango, Four Corners Geological Society Fourth Field Conference Guidebook, p. 157–184.

CLAIR, J. R., 1952, Paleozoic rocks of the southern Paradox basin, *in* Geological Symposium of the Four Corners Region: Durango, Four Corners Geological Society, p. 36–39.

CRAWFORD, G. A., 1981, Depositional history and diagenesis of the Goat Seep Dolomite (Guadalupian, Permian) Guadalupe Mountains, west Texas–New Mexico: Unpublished Ph.D. Dissertation, University of Wisconsin, Madison, 300 p.

ESTEBAN, M., AND PRAY, L. C., 1977, eds., Upper Guadalupian facies, Permian reef complex, Guadalupe Mountains, New Mexico and west Texas: 1977 Field Conference Guidebook: Midland, Permian Basin Section, Society of Economic Paleontologists and Mineralogists Publication No. 77-16, v. 2, 194 p.

FEKETE, T. E., 1986, The sedimentology and stratigraphy of the Grayburg Formation and its associated erosion surface along the high Western Escarpment of the Guadalupe Mountains, Texas: Unpublished M.S. Thesis, University of Wisconsin, Madison, 174 p.

FETZNER, R. W., 1960, Pennsylvanian paleotectonics of Colorado Plateau: American Association of Petroleum Geologists Bulletin, v. 44, p. 1371–1413.

FISHER, A. G., 1982, Long-term climatic oscillations recorded in stratigraphy, *in* Climate in Earth History: Washington, D.C., National Academy Press, p. 97–104.

FITCHEN, W. M., 1993, Sequence stratigraphic framework of the upper San Andres Formation and equivalent basinal strata in the Brokeoff Mountains, Otero County, New Mexico, *in* Love, D. W., et al., eds., Carlsbad Region, New Mexico and West Texas: Roswell, New Mexico Geological Society Fourth Annual Field Conference, p. 185–194.

FRANSEEN, E. K., 1985, Sedimentology of the Grayburg and Queen Formations (Guadalupian) and the shelf margin erosion surface, Western Escarpment of the Guadalupe Mountains, west Texas: Unpublished M.S. Thesis, University of Wisconsin, Madison, 189 p.

FRANSEEN, E. K., FEKETE, T. E., AND PRAY, L. C., 1989, Evolution and destruction of a carbonate bank at the shelf margin: Grayburg Formation (Permian), Western Escarpment, Guadalupe Mountains, Texas, *in* Crevello, P. D., Wilson, J. L., Sarg, J. F., and Read, J. F., eds., Controls on Carbonate Platform and Basin Development: Tulsa, Society of Economic Paleontologists and Mineralogists Special Publication No. 44, p. 289–304.

FRENZEL, H. N., ET AL., 1988, The Permian basin region, Chapter 11, *in* Sloss, L. L., ed., Sedimentary Cover—North American Craton: Boulder, Geological Society of America, The Geology of North America, v. D-2, p. 261–306.

GALLEY, J. E., 1958, Oil and geology in the Permian basin of Texas and New Mexico, *in* Weeks, L. G., ed., Habitat of Oil: Tulsa, American Association of Petroleum Geologists, p. 395–446.

GALLOWAY, W. E., EWING, T. E., GARRETT, C. M., TYLER, N., AND BEBOUT, D. G., 1983, Atlas of major Texas oil reservoirs: Austin, University of Texas, Bureau of Economic Geology Special Publication, 139 p.

GARDNER, M. H., AND SONNENFELD, M. D., 1996, Stratigraphic changes in facies architecture of the Permian Brushy Canyon Formation in Guadalupe Mountains National Park, west Texas, *in* DeMis, W. D. and Cole, A. G., eds., The Brushy Canyon Play in Outcrop and Subsurface: Concepts and Examples: Midland, Permian Basin Section–SEPM (Society for Sedimentary Geology) Publication 96-38, p. 17–40.

GIANNINY, G. L., 1995, Facies and sequence stratigraphic evolution of the mixed carbonate-siliciclastic strata, lower Desmoinesian, southwest Paradox basin, Utah: Unpublished Ph.D. Dissertation, University of Wisconsin, Madison, 331 p.

GIANNINY, G. L., AND SIMO, J. A., 1996, Implications of unfilled accommodation space for sequence stratigraphy on mixed carbonate-siliciclastic platforms: An example from the lower Desmoinesian (Middle Pennsylvanian), southwestern Paradox basin, Utah, *in* Longman, M. W. and Sonnenfeld, M. D., eds., Paleozoic Systems of the Rocky Mountain Region: Rocky Mountain Section–SEPM (Society for Sedimentary Geology), p. 213–234.

GOLDHAMMER, R. K., OSWALD, E. J., AND DUNN, P. A., 1991, The hierarchy of stratigraphic forcing: An example from Middle Pennsylvanian shelf carbonates of the Paradox basin, *in* Franseen, E. K., et al., eds., Sedimentary modeling: Computer simulations and methods for improved parameter definition: Lawrence, Kansas Geological Survey Bulletin 233, p. 361–413.

GRAMMER, G. M., EBERLI, G. P., VAN BUCHEM, F. S. P., STEVENSON, G. E., AND HOMEWOOD, P., 1996, Application of high-resolution sequence stratigraphy to evaluate lateral variability in outcrop and subsurface—Desert Creek and Ismay intervals, Paradox basin, *in* Longman, M. W. and Sonnenfeld, M. D., eds., Paleozoic Systems of the Rocky Mountain Region: Rocky Mountain Section–SEPM (Society for Sedimentary Geology), p. 235–266.

GREENLEE, S. M., AND LEHMANN, P. J., 1993, Stratigraphic framework of productive carbonate buildups, *in* Loucks, R. G. and Sarg, J. F., eds., Carbonate Sequence Stratigraphy, Recent Developments and Applications: Tulsa, American Association of Petroleum Geologists Memoir 57, p. 43–62.

HAQ, B. U., HARDENBOL, J., AND VAIL, P. R., 1987, Chronology of fluctuating sea levels since the Triassic: Science, v. 235, p. 1156–1166.

HARLAND, W. B., ARMSTRONG, R. L., COX, A. V., CRAIG, L. E., SMITH, A. G., AND SMITH, D. G., 1989, A Geologic Time Scale: New York, Cambridge University Press, 128 p.

HARMS, J. C., 1974, Brushy Canyon Formation, Texas: Deep-water density current deposit: Geological Society of America Bulletin, v. 85, p. 1763–1784.

HARRIS, M. T., 1982, Sedimentology of the Cutoff Formation, western Guadalupe Mountains, west Texas and New Mexico: Unpublished M.S. Thesis, University of Wisconsin, Madison, 176 p.

HARRIS, M. T., 1988, Sedimentology of the Cutoff Formation (Permian), western Guadalupe Mountains, west Texas, *in* Reid, S. T., Bass, R. O., and Welch, P., eds., Guadalupe Mountains Revisited, Texas and New Mexico: Midland, West Texas Geological Society Publication 88-84, p. 133–142.

HAYES, P. T., 1964, Geology of the Guadalupe Mountains, New Mexico: Washington, D.C., United States Geological Survey Professional Paper 446, 69 p.

HERMAN, G., AND SHARPS, S. L., 1956, Pennsylvanian and Permian stratigraphy of the Paradox salt embayment, *in* Peterson, J. A., ed., Geology and Economic Deposits of East Central Utah: Salt Lake City, Intermountain Association of Petroleum Geologists, Seventh Annual Field Conference, p. 77–84.

HILL, C. A., 1996, Geology of the Delaware Basin, Guadalupe, Apache, and Glass Mountains, New Mexico and West Texas: Midland, Permian Basin Section–SEPM (Society for Sedimentary Geology) Publication No. 96-39, 480 p.

HILLS, J. M., 1984, Sedimentation, tectonism, and hydrocarbon generation in Delaware basin, west Texas and southeastern New Mexico: American Association of Petroleum Geologists Bulletin, v. 68, p. 250–267.

HITE, R. J., 1960, Stratigraphy of the saline facies of the Paradox Member of the Hermosa Formation of southeastern Utah and southwestern Colorado, *in* Geology of the Paradox Basin Fold and Fault Belt: Durango, Four Corners Geological Society, Third Annual Field Conference, p. 86–89.

HITE, R. J., 1970, Shelf carbonate sedimentation controlled by salinity in the Paradox basin, southeast Utah, *in* Ron, J. L. and Dellwig, L. F., eds., Third Symposium on Salt: Cleveland, Northern Ohio Geologic Society, v. 1, p. 48–66.

HITE, R. J., AND BUCKNER, D. H., 1981, Stratigraphic correlations, facies concepts, and cyclicity in Pennsylvanian rocks of the Paradox basin, *in* Wiegand, D. L., ed., Geology of the Paradox Basin: Denver, Rocky Mountain Association of Geologists 1981 Field Conference, p. 147–159.

HURLEY, N. F., 1978, Facies mosaic of the lower Seven Rivers Formation (Permian), North McKittrick Canyon, Guadalupe Mountains, New Mexico: Unpublished M.S. Thesis, University of Wisconsin, Madison, 198 p.

KELLY, M. A., 1994, Carbonate buildups of the lower Desmoinesian Barker Creek interval, southeast Utah: Unpublished M.S. Thesis, University of Wisconsin, Madison, 189 p.

KERANS, C., AND FITCHEN, W. M., 1995, Sequence hierarchy and facies architecture of a carbonate-ramp system: San Andres Formation of Algerita Escarpment and western Guadalupe Mountains, west Texas and New Mexico: Austin, University of Texas, Bureau of Economic Geology, Report of Investigations No. 235, 86 p.

KERANS, C., LUCIA, F. J., SENGER, R. K., FOGG, G. E., NANCE, H. S., KASAP, E.,

AND HOVORKA, S. D., 1991, Characterization of reservoir heterogeneity in carbonate ramp systems, San Andres–Grayburg Formations, Permian basin: Austin, University of Texas, Bureau of Economic Geology, Reservoir Characterization Laboratory Final Report, 245 p.

KERANS, C., FITCHEN, W. M., GARDNER, M. H., AND WARDLAW, B. R., 1993, A contribution to the evolving stratigraphic framework of Middle Permian strata of the Delaware basin, Texas and New Mexico, *in* Love, D. W., et al., eds., Carlsbad Region, New Mexico and West Texas: Roswell, New Mexico Geological Society Fourth Annual Field Conference, p. 175–184.

KING, P. B., 1948, Geology of the Southern Guadalupe Mountains, Texas: Washington, D.C., United States Geological Survey Professional Paper 215, 183 p.

KIRKBY, K. C., 1982, Deposition, erosion, and diagenesis of the upper Victorio Peak Formation (Leonardian), southern Guadalupe Mountains, west Texas: Unpublished M.S. Thesis, University of Wisconsin, Madison, 165 p.

KIRKLAND, D. W., AND EVANS, R., 1981, Source-rock potential of evaporite environment: American Association of Petroleum Geologists Bulletin, v. 65, p. 181–189.

KOSTERS, E. C., BEBOUT, D. G., SENI, S. J., GARRETT, C. M., JR., BROWN, L. F., JR., HAMLIN, H. S., DUTTON, S. P., RUPPEL, S. C., FINLEY, R. J., AND TYLER, N., 1989, Atlas of major Texas gas reservoirs: Chicago, Gas Research Institute, and Austin, University of Texas, Bureau of Economic Geology Publication, 161 p.

MARKELLO, J. R., AND SARG, J. F., 1996, Phanerozoic tectono-stratigraphic evolution of the Trans-Pecos and Permian basin regions (Mexico, Texas, New Mexico) using Landsat imagery, subsurface, and outcrop data, *in* Proceedings, 11th Thematic Conference on Applied Geologic Remote Sensing, Las Vegas, Nevada: Ann Arbor, Environmental Research Institute of Michigan, v. II, p. 651–664.

MATCHUS, E. J., AND JONES, T. S., 1984, East-west cross section through the Permian basin of west Texas: Midland, West Texas Geological Society Publication No. 84-79, 1 p.

MAZZULLO, S. J., BISCHOFF, W. D., AND HEDRICK, C. L., 1989, Stacked island facies in Tansill outer-shelf platform, Dark Canyon, Guadalupe Mountains, New Mexico, *in* Harris, P. M. and Grover, G. A., eds., Subsurface and Outcrop Examination of the Capitan Shelf Margin, Northern Delaware Basin: Tulsa, Society of Economic Paleontologists and Mineralogists Core Workshop No. 13, p. 287–293.

MITCHUM, R. M., AND VAN WAGONER, J. C., 1991, High-frequency sequences and their stacking patterns: Sequence-stratigraphic evidence of high-frequency eustatic cycles: Sedimentary Geology, v. 70, p. 131–160.

MOLENAAR, C. M., 1981, Mesozoic stratigraphy of the Paradox basin—An overview, *in* Weigand, D. L., ed., Geology of the Paradox Basin: Denver, Rocky Mountain Association of Geologists, 1981 Field Conference, p. 119–127.

NEW, M. E., 1988, Sedimentology of the Cherry Canyon Tongue (Cherry Canyon Formation, Permian), western Guadalupe Mountains, Texas and New Mexico: Unpublished M.S. Thesis, University of Wisconsin, Madison, 287 p.

ODIN, G. S., AND GALE, N. H., 1982, Mise a jour d'échelles des temps caledoins et hercyniens: Academie des Sciences, Comptes Rendus, v. 294, ser. 2, p. 453–456.

PETERSON, J. A., 1992, Aneth Field, USA, Paradox basin, Utah, *in* Foster, N. H. and Beaumont, E. A., eds., Stratigraphic Traps III, Treatise of Petroleum Geology, Atlas of Oil and Gas Fields: Tulsa, American Association of Petroleum Geologists, p. 41–82.

PETERSON, J. A., AND HITE, R. J., 1969, Pennsylvanian evaporite carbonate cycles and their relation to petroleum occurrence: American Association of Petroleum Geologists Bulletin, v. 53, p. 884–908.

PRAY, L. C., 1971, Submarine slope erosion along Permian bank margin, west Texas (abs.): American Association of Petroleum Geologists Bulletin, v. 55, p. 358.

PRAY, L. C., AND STEHLI, F. G., 1962, Allochthonous origin, Bone Spring "patch reefs" (abs.): Boulder, Geological Society of America Special Paper 73, p. 219.

RAMONDETTA, P. J., 1982, Genesis and emplacement of oil in the San Andres Formation, northern shelf of the Midland basin, Texas: Austin, University of Texas, Bureau of Economic Geology, Report of Investigations 116, 39 p.

ROBINSON, K., 1988, Petroleum geology and hydrocarbon plays of the Permian basin petroleum province, west Texas and southeast New Mexico: United States Geological Survey Open-File Report 88-4502, 53 p.

ROSS, C. A., 1986, Paleozoic evolution of southern margin of Permian basin: Geological Society of America Bulletin, v. 97, p. 536–554.

ROSS, C. A., AND ROSS, J. R. P., 1987a, Late Paleozoic sea levels and depositional sequences, *in* Ross, C. A. and Haman, D., eds., Timing and Depositional History of Eustatic Sequences: Constraints on Seismic Stratigraphy: Ithaca, Cushman Foundation for Foraminiferal Research Special Publication 24, p. 137–149.

ROSS, C. A., AND ROSS, J. R. P., 1987b, Biostratigraphic zonation of late Paleozoic depositional sequences, *in* Ross, C. A. and Haman, D., eds., Timing and Depositional History of Eustatic Sequences: Constraints on Seismic Stratigraphy: Ithaca, Cushman Foundation for Foraminiferal Research Special Publication 24, p. 151–168.

ROSS, C. A., AND ROSS, J. R. P., 1988, Late Paleozoic transgressive-regressive deposition, *in* Wilgus, C. K., Hastings, B. S., Kendall, C. G. St. C., Posamentier, H. W., Ross, C. A., and Van Wagoner, J. C., eds., Sea-Level Changes: An Integrated Approach: Tulsa, Society of Economic Paleontologists and Mineralogists Special Publication No. 42, p. 227–247.

ROSSEN, C., 1985, Sedimentology of the Brushy Canyon Formation (Permian, early Guadalupian) in the onlap area, Guadalupe Mountains, west Texas: Unpublished M.S. Thesis, University of Wisconsin, Madison, 314 p.

ROSSEN, C., AND SARG, J.F., 1988, Sedimentology and regional correlation of a basinally restricted deep-water siliciclastic wedge: Brushy Canyon Formation–Cherry Canyon Tongue (lower Guadalupian), Delaware basin, *in* Reid, S. T., Bass, R. O., and Welch, P., eds., Guadalupe Mountains Revisited, Texas and New Mexico: Midland, West Texas Geological Society Publication 88-84, p. 127–132.

ROYLANCE, M. H., 1984, Depositional and diagenetic control of petroleum entrapment in the Desert Creek interval, Paradox Formation, southeastern Utah and Colorado: Unpublished M.S. Thesis, University of Kansas, Lawrence, 178 p.

SARG, J. F., 1988, Carbonate sequence stratigraphy, *in* Wilgus, C. K., Hastings, B. S., Kendall, C. G. St. C., Posamentier, H. W., Ross, C. A., and Van Wagoner, J. C., eds., Sea-Level Changes: An Integrated Approach: Tulsa, Society of Economic Paleontologists and Mineralogists Special Publication No. 42, p. 155–181.

SARG, J. F., AND LEHMANN, P. J., 1986, Lower and middle Guadalupian facies, stratigraphy, and reservoir geometries, San Andres–Grayburg Formations, Guadalupe Mountains, New Mexico and Texas, *in* Moore, G. E. and Wilde, G. L., eds., Lower and Middle Guadalupian Facies, Stratigraphy, and Reservoir Geometries, San Andres–Grayburg Formations, Guadalupe Mountains, New Mexico and Texas: Midland, Permian Basin Section, Society of Economic Paleontologists and Mineralogists Publication No. 86-25, p. 1–35.

SARG, J. F., ROSSEN, C., LEHMANN, P. J., AND PRAY, L. C., 1988, Geologic Guide to the Western Escarpment, Guadalupe Mountains, Texas: Midland, Permian Basin Section, Society of Economic Paleontologists and Mineralogists Publication 88-30, 60 p.

SARG, J. F., MARKELLO, J. R., BLOCH, R. B., AND CLARKE, R. T., 1997, Stratigraphic Architecture of the Transgressive and Turnaround Phases, Guadalupian Supersequence, Cherry Canyon and San Andres Formations, Brokeoff and Guadalupe Mountains, New Mexico: Tulsa, American Association of Petroleum Geologists, 1997 Annual Convention, Dallas, Guidebook Field Trip No. 16, 213 p.

SLOSS, L. L., 1963, Sequences in the cratonic interior of North America: Geological Society of America Bulletin, v. 74, p. 93–114.

SONNENFELD, M. D., AND CROSS, T. A., 1993, Volumetric partitioning and facies differentiation within the Permian upper San Andres Formation of Last Chance Canyon, Guadalupe Mountains, New Mexico, *in* Loucks, R. G. and Sarg, J. F., eds., Carbonate Sequence Stratigraphy, Recent Developments and Applications: Tulsa, American Association of Petroleum Geologists Memoir 57, p. 435–474.

VAIL, P. R., Audemard, F., Bowman, S. A., Eisner, P. N., and Perez-Cruz, C., 1991, The stratigraphic signatures of tectonics, eustasy and sedimentology—An overview, Chapter 6.1, *in* Einsele, G., et al., eds., Cycles and Events in Stratigraphy: Berlin, Springer-Verlag, p. 617–659.

VAN EYSINGA, F. W. B., 1975, Geologic Time Table: Amsterdam, Elsevier, 1 p.

VAN WAGONER, J. C., POSAMENTIER, H. W., MITCHUM, R. M., JR., VAIL, P. R., SARG, J. F., LOUTIT, T. S., AND HARDENBOL, J., 1988, An overview of the fundamentals of sequence stratigraphy and key definitions, *in* Wilgus, C. K., Hastings, B. S., Kendall, C. G. St. C., Posamentier, H. W., Ross, C. A., and Van Wagoner, J. C., eds., Sea-Level Changes: An Integrated Approach: Tulsa, Society of Economic

Paleontologists and Mineralogists Special Publication No. 42, p. 39–46.

Ward, R. F., Kendall, C. G. C., and Harris, P. M., 1986, Upper Permian (Guadalupian) facies and their association with hydrocarbons—Permian basin, west Texas and New Mexico: American Association of Petroleum Geologists Bulletin, v. 70, p. 239–262.

Wardlaw, B. R., and Rohr, D. M., 1996, Guadalupian II: Abstracts and Proceedings of the Second International Guadalupian Symposium, April 10–13, 1996: Alpine, Sul Ross State University, 80 p.

Weber, L. J., Wright, F. M., Sarg, J. F., Shaw, E., Harman, L. P., Vanderhill, J. B., and Best, D. A., 1995a, Reservoir delineation and performance: Application of sequence stratigraphy and integration of petrophysics and engineering data, Aneth field, southeast Utah, USA, *in* Stoudt, E. L. and Harris, P. M., eds., Hydrocarbon Reservoir Characterization: Geologic Framework and Flow Modeling: Tulsa, SEPM (Society for Sedimentary Geology) Short Course No. 34, p. 31–51.

Weber, L. J., Sarg, J. F., and Wright, F. M., 1995b, Part 3. Sequence stratigraphy and reservoir delineation of the Middle Pennsylvanian (Desmoinesian) Paradox basin and Aneth field, southwestern USA, *in* Read, J. F., et al., eds., Milankovitch Sea-Level Changes, Cycles, and Reservoirs on Carbonate Platforms in Greenhouse and Ice-House Worlds: Tulsa, SEPM (Society for Sedimentary Geology) Short Course No. 35, p. 1–81.

Wengerd, S. A., 1951, Reef limestones of Hermosa Formation, San Juan Canyon, Utah: American Association of Petroleum Geologists Bulletin, v. 35, p. 1038–1051.

Wengerd, S. A., and Matheny, M. L., 1958, Pennsylvanian System of the Four Corners region: American Association of Petroleum Geologists Bulletin, v. 42, p. 2048–2106.

Wengerd, S. A. and Strickland, J. W., 1954, Pennsylvanian stratigraphy of the Paradox salt basin, Four Corners region, Colorado and Utah: American Association of Petroleum Geologists Bulletin, v. 38, p. 2157–2199.

West, R. R., 1988, Temporal changes in Carboniferous reef mound communities: Palaios, v. 3, p. 152–169.

Wilde, G. L., 1986, Stratigraphic relationship of the San Andres and Cutoff Formations, northern Guadalupe Mountains, New Mexico and Texas, *in* Moore, G. E. and Wilde, G. L., eds., Lower and Middle Guadalupian Facies, Stratigraphy, and Reservoir Geometries, San Andres–Grayburg Formations, Guadalupe Mountains, New Mexico and Texas: Midland, Permian Basin Section, Society of Economic Paleontologists and Mineralogists Publication No. 86-25, p. 49–63.

Wilkinson, C. E., Jr., Ritter, S. M., Lambert, L. L., and Wilde, G. L., 1991, Stratigraphy and biostratigraphy of the lower and middle San Andres Formation in Last Chance Canyon, Guadalupe Mountains, New Mexico, *in* Meador-Roberts, S., et al., eds., Sequence Stratigraphy, Facies, and Reservoir Geometries of the San Andres, Grayburg, and Queen Formations, Guadalupe Mountains, New Mexico and Texas: Midland, Permian Basin Section–SEPM (Society for Sedimentary Geology) Publication 91-32, p. 99–110.

Williams, J. A., 1977, Characterization of oil types in the Permian basin: Abilene, Texas, American Association of Petroleum Geologists, Southwest Section, Proceedings, 42 p.

Wilson, J. A., 1975, Carbonate Facies in Geologic History: New York, Springer-Verlag, 471 p.

Yurewicz, D. A., 1977, The origin of the massive facies of the lower and middle Capitan Limestone (Permian), Guadalupe Mountains, New Mexico and Texas, *in* Hileman, M. E. and Mazzullo, S. J., eds., Upper Guadalupian Facies, Permian Reef Complex, Guadalupe Mountains, New Mexico and West Texas: Midland, Permian Basin Section, Society of Economic Paleontologists and Mineralogists Publication No. 77-16, p. 45–92.

Zelt, F. B., and C. Rossen, 1995, Geometry and continuity of deep-water sandstones and siltstones, Brushy Canyon Formation (Permian), Delaware Mountains, Texas, *in* Pickering, K. T., et al., eds., Atlas of Deep Water Environments—Architectural Style in Turbidite Systems: London, Chapman and Hall, p. 167–183.

TYPE 3 SEQUENCE BOUNDARIES

WOLFGANG SCHLAGER

Vrije Universiteit, Earth Sciences, De Boelelaan 1085, 1081 HV Amsterdam, Netherlands

ABSTRACT: Sequence boundaries were originally defined as bounding surfaces of conformably stratified units. Within this descriptive definition, two types of boundaries were specifically recognized and widely used—the boundary between highstand and lowstand systems tracts and that between highstand systems tract and shelf-margin wedge. A third type of boundary, the flooding surface between a highstand tract and an overlying transgressive tract, is increasingly used as a sequence boundary, particularly in carbonates. It should be formally recognized as a specific type of sequence boundary because it is prone to developing long submarine hiatuses, occurs commonly in the geologic record, particularly on drowned carbonate platforms, tends to be a pronounced reflector and unconformity in seismic images, and represents a logical third combination in the interplay of subsidence and eustasy—one in which the rate of subsidence exceeds even the most rapid fall of sea level during a particular eustatic cycle. Including type 3 unconformities as sequence boundaries has two advantages: (1) the general definition of sequence boundary remains descriptive and thus workable even where the boundary must be drawn solely on geometric grounds without supporting evidence for sea-level fall and exposure, and (2) the designation of boundary types 1 and 2 can be restricted to surfaces with reliable evidence for relative fall in sea level.

On drowned reefs and carbonate platforms, the hiatuses associated with type 3 unconformities frequently exceed 10 m.y., occasionally even 100 m.y. Relicts of pelagic cover in hardgrounds, along with marine diagenesis, attest to the submarine, erosional nature of these hiatuses. On rapidly flooded siliciclastic shelves, the effects are similar but less pronounced. The mechanism behind this persistent and powerful marine erosion is amplification of sluggish oceanic tides by the interaction with sharp topography. This process continues and expands the effects of marine shoreface erosion at greater water depth.

INTRODUCTION

The standard model of sequence stratigraphy currently recognizes two types of sequence boundaries: type 1 between a highstand systems tract (HST) and an overlying lowstand systems tract (LST) and type 2 between a highstand tract and a shelf-margin wedge (Vail et al., 1984; Sarg, 1988, for carbonates). I propose to acknowledge a third boundary, namely, the combination of a highstand systems tract immediately followed by a transgressive systems tract (TST) without a recognizable sea-level fall or lowstand deposits in between. Vail et al. (1984, p. 184) have already indicated this configuration as a way to generate condensed sections but not as a potential locus for erosional unconformities. Reviewing examples from carbonate platforms and seamounts, Schlager (1998) has reported abundant evidence of extensive marine erosion and long hiatuses at the HST-TST boundaries. This review presents examples of such boundaries in the rock record and in seismic data, examines the processes that cause the submarine hiatuses, and finally discusses the pros and cons of HST-TST boundaries as limits of stratigraphic sequences.

EXAMPLES OF HIGHSTAND TRACTS FOLLOWED BY TRANSGRESSIVE TRACTS

The best examples for transgressive tracts overlying highstand tracts without significant exposure are found on drowned reefs and carbonate platforms. Shoal-water carbonates lithify quickly when sedimentation is interrupted by subaerial exposure or by submarine erosion (e.g., Bathurst, 1971). The diagenetic overprint resulting from exposure is distinctly different from that of marine lithification, and the record is commonly preserved because of the early hardening of the deposits (James and Choquette, 1984). In siliciclastics, cessation of sedimentation normally leads to extensive reworking whereby the record of the late highstand and its subsequent alteration, on land or under water, is lost.

Figures 1–3 illustrate examples of highstand tracts followed by transgressive tracts in which the boundaries have been used as sequence boundaries. The Liuhua platform of the South China Sea has been well documented in the literature by using seismic and core borings (Erlich et al., 1990; Moldovanyi et al., 1995; Wagner et al., 1995). This early Miocene platform shows the common pattern of a first, extensive, but low-relief platform that shrinks and disintegrates into an assembly of smaller platforms that finally drown and are buried by shale. The growth history shows several exposure surfaces, complicated by syndepositional block faulting (Moldovanyi et al., 1995). The exposure events are recognizable in stable isotope data, leaching porosity, and other indicators of fresh-water diagenesis. However, the dominant unconformity at the top of the drowned platform lacks evidence of exposure; it simply records rapid deepening of the environment with indication of winnowing and starvation prior to the onset of shale deposition (Fig. 1; Erlich et al., 1990; Moldovanyi et al., 1995, p. 131). J. F. Sarg (written commun.) suggested a sequence stratigraphic interpretation that conforms to the standard model by designating the drowning unconformity as a maximum flooding surface in a sequence whose lower boundary is put at the nearest prominent exposure surface below, for instance, the one at 1.36 s in well LH4-1-1 (Figs. 1, 2; see discussion section).

The Oligocene platform of Kalimantan (Fig. 3) was described by Saller et al. (1993) from seismic sections, boreholes, and outcrop sections. Four sequences were recognized, but exposure could be demonstrated for only one of five sequence boundaries. The others are typical type 3 boundaries where highstand tracts, recognizable by shoaling and prograding successions, are capped by flooding surfaces and backstepping transgressive tracts. Here again, the boundaries between highstand and overlying transgressive tracts provide good seismic unconformities and reflectors.

Detailed descriptions of other drowning unconformities without prior exposure have been contributed by, among others,

Advances in Carbonate Sequence Stratigraphy: Application to Reservoirs, Outcrops and Models, SEPM Special Publication No. 63
 ISBN 1-56576-060-3, p. 35–45

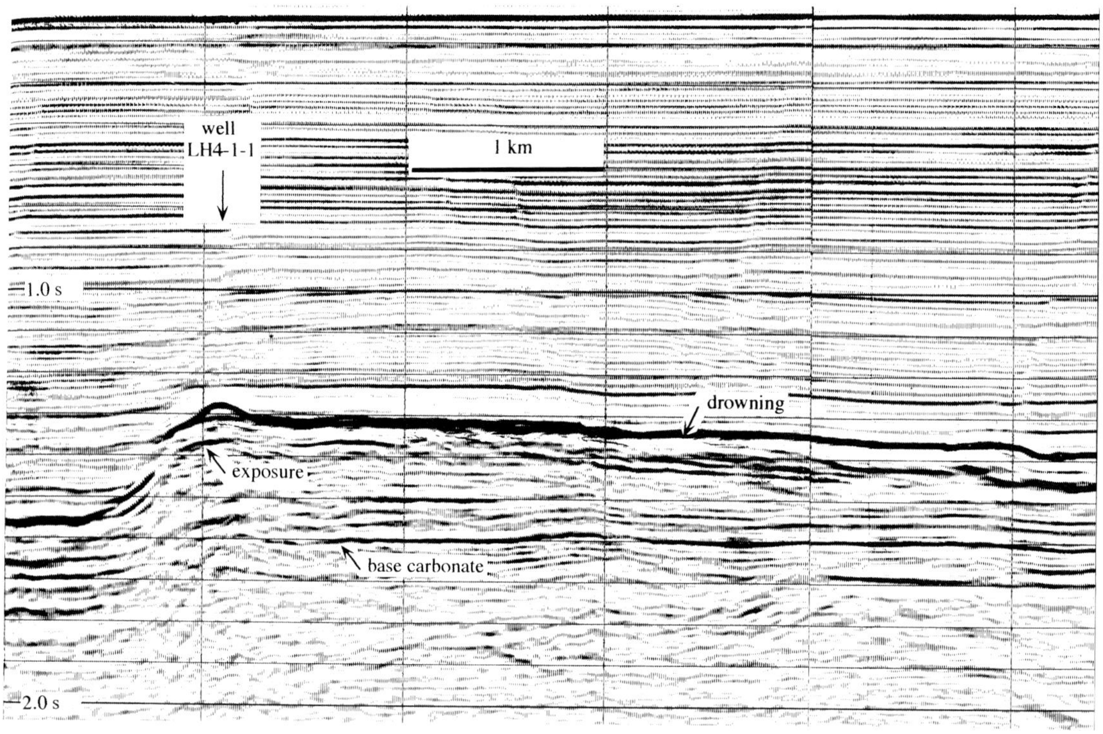

FIG. 1.—Seismic profile of the Miocene Liuhua platform, South China Sea. The platform is overlain by shale. The platform was exposed repeatedly during growth. However, even the most prominent exposure unconformity (at about 1.36 s at site LH4-1-1) is difficult to trace seismically. The dominant seismic reflector is a drowning unconformity where the platform repeatedly backstepped and finally drowned; after a period of condensed sedimentation, it was covered by shale. After Erlich et al. (1990), reprinted by permission of the American Association of Petroleum Geologists.

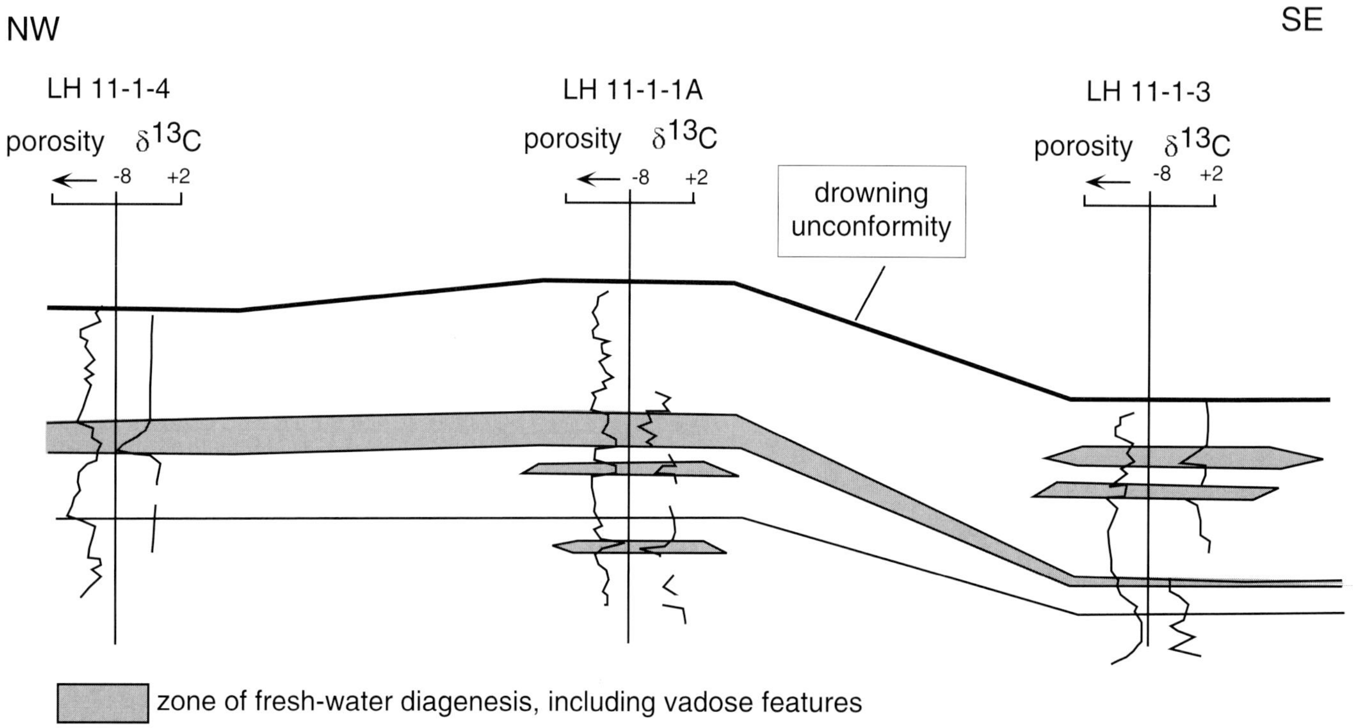

FIG. 2.—Core data from the Miocene Liuhua platform indicate several exposure events, recorded in carbon isotope compositions, leaching porosity, and calcite cements. Irregular distribution of these events suggests control by movement of fault blocks. No exposure was found at the drowning surface, a sharp boundary between limestone and shale in a regionally rather uniform, deepening-upward succession. Modified after Moldovanyi et al. (1995), reprinted by permission of the American Association of Petroleum Geologists.

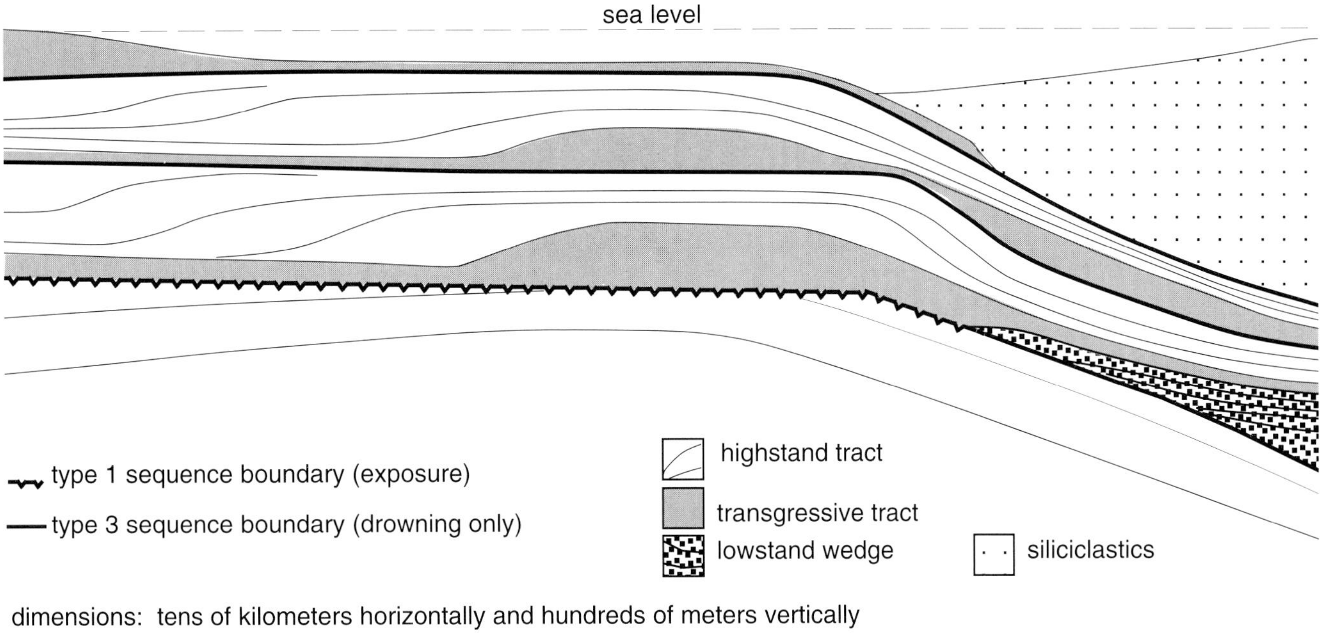

FIG. 3.—Growth history and sequence stratigraphy of Oligocene platform in Kalimantan. Platform oscillated between a highstand mode of strong progradation and a transgressive mode of deep flooding during which only the margin maintained a flat top at sea level. Evidence for exposure and basin-restricted lowstand wedge has been found only at lowermost sequence boundary. Other boundaries are of type 3—transgressive tract overlying highstand tract after deep flooding. Note bidirectional progradation of highstand tract, a characteristic of carbonate systems after deep flooding. Modified after Saller et al. (1993).

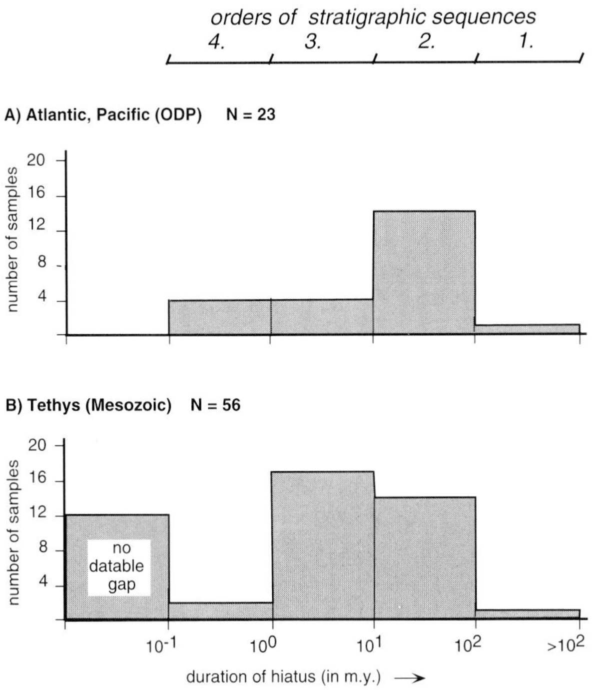

FIG. 4.—Duration of marine hiatuses on drowned platforms, based on numerous sources. (A) Pacific seamounts. (B) Platforms on the margins of the Jurassic–Cretaceous Tethys ocean.

Dravis (1989), Föllmi et al. (1994), and Wendte et al. (1992) (see Schlager, 1998, for review). A recurring message of many case studies is that rock samples are needed to determine the presence or absence of subaerial exposure; geometry alone is but a very ambiguous guide to exposure history (e.g., Fouke et al., 1995).

A common feature of the drowning successions on carbonate platforms and starved clastic shelves is the appearance of authigenic minerals such as iron and manganese oxides, phosphates, or glauconite. These minerals form at the sea floor or very close to it, their growth is slow compared to average rates of pelagic or hemipelagic sedimentation, and their abundance indicates very slow sedimentation. Hardgrounds with truncation of underlying beds and pebble-size lag deposits on top indicate that sedimentation alternated with pulses of erosion. Again, the record of internal sediments in rock cavities clearly indicates that these erosive pulses occurred over a long time and in the marine environment (Sliter, 1995; Camoin et al., 1998).

SHARP TOPOGRAPHY AND MARINE EROSION

Siliciclastic sequence stratigraphy has emphasized that transgressive systems tracts are often sediment starved and rich in condensed intervals as sediment capture shifts shoreward. Occasionally, the entire transgressive tract may be replaced by a single flooding surface that forms the basis of the next prograding and downlapping highstand tract (Vail and Todd, 1981; Loutit et al., 1988).

The stratigraphic ages reported from the drowning successions of reefs and carbonate platforms indicate that hiatuses of

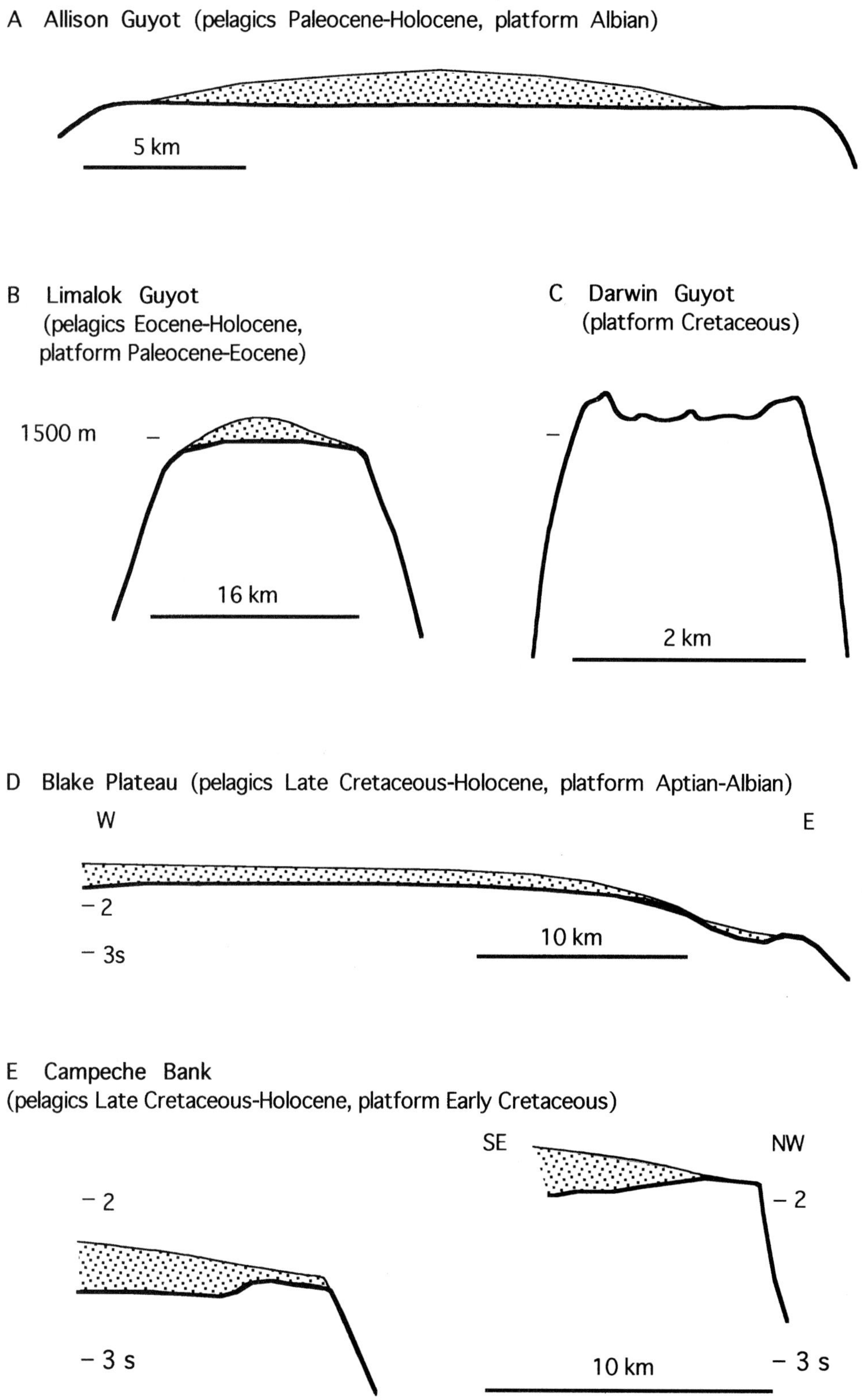

FIG. 5.—Pelagic cover on drowned platforms is lens-shaped and absent on the margins. This geometry and the long duration of the gaps attest to submarine erosion. Modified after Schlager (1998).

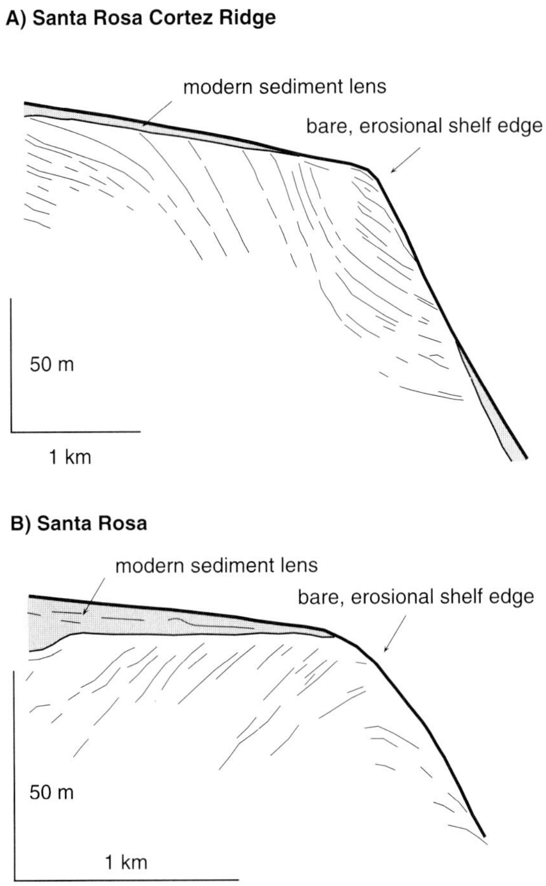

FIG. 6.—Sediment cover on starved siliciclastic shelves is thin and lens-shaped, similar to the cover of drowned platforms. Oceanographic studies indicate similar current amplification by interaction with sharp topography as on seamounts. Interpretations of seismic lines from the California Borderland. Modified after Field et al. (1983).

tens of millions of years are common in these deposits and that occasionally more than 100 m.y. of record may be missing (Fig. 4). Seismic profiles show that the pelagic or hemipelagic cover of drowned platforms is rarely ever an even pelagic drape. Rather, it appears as a lens-shaped body that is thickest in the platform interior and thin or absent at the margins. Both detached, atoll-like platforms as well as land-attached platforms show this pattern (Fig. 5). It indicates that erosion and winnowing prevent the pelagic or hemipelagic sediments from evenly draping over the platform. The geometry of the pelagic caps indicates that erosion and winnowing are strongest at the platform margin, and study of the vertical succession frequently shows the most pronounced gaps at the very base or in the oldest parts of the pelagic cover. Because of the combined effects of extensive erosion and change in sediment supply, the boundary between a drowned carbonate platform and its cover of deep-water sediments tends to be a prominent unconformity and seismic reflector that marks both a basic change in sediment input as well as a change in sediment composition (Schlager, 1989).

Many of these drowning unconformities are preceded by exposure and thus qualify as type 1 sequence boundaries in the sense of Vail and Todd (1981) and Sarg (1988). However, the growing discoveries of drowning unconformities without prior exposure shows that exposure is not required to produce the unconformity. A basic change in depositional regime coupled with marine erosion will suffice. The intensity and persistence of marine erosion are borne out by the extremely low sedimentation rates in the pelagic cover of drowned platforms and the abundance of erosional hiatuses at the base of the pelagic cover and within it. The hiatuses within the pelagic section are particularly well constrained as bathyal features, far beyond the range of subaerial exposure (e.g., Sliter, 1995, and Camoin et al., 1998, for extant seamounts; Clari et al., 1995, for Tethyan examples). These intrapelagic hiatuses attest to the vigor of submarine erosion on top of drowned platforms.

Siliciclastic shelves with well-defined shelf edges show a similar, albeit more subdued trend toward starvation and erosion on the outer shelf. Much of the shelves off the U.S. Pacific coast are sediment starved with sediment geometries akin to those of drowned platforms (Fig. 6; Field et al., 1983). Stanley et al. (1983) emphasized the general tendency of shelf edges to be winnowed and sediment starved. They attributed these characteristics to the combined effects of (1) low sediment supply because of the distance from onshore sources and (2) high energy delivered by the adjacent ocean. Repeatedly, unconformities have been observed where shallow shelves were rapidly flooded and subsequently covered by highstand sediment. A very well constrained example by Coleman and Roberts (1988) is shown in Figure 7. This example differs from the drowned platforms cited above because the flooded delta lobe beneath the transgressive tract is a genuine lowstand systems tract. The common feature of the siliciclastic shelves in Figures 6 and 7 and the carbonate platforms in Figures 1 and 5 is the unconformable boundary that forms where a shelf break is rapidly flooded and abandoned.

The regional distribution of winnowed shelves and drowned carbonate platforms does not show a preference for areas with strong geostrophic currents (such as the Gulf Stream or the Circum-Antarctic Current). The common denominator seems to be the presence of "sharp topography," i.e., a flat shelf with a well-defined shelf edge on top of a slope rising from bathyal depths. I submit that the cause of the frequent and long-lasting erosion and winnowing of outer shelves and drowned platforms lies in the interaction between this sharp topography and the (relatively sluggish) oceanic tides and currents. A particularly successful physical model for such current amplification is the notion of internal waves "trapped" by topography such that they travel with the shallow topography to the right in the Northern Hemisphere and to the left in the Southern Hemisphere (see Brink, 1987, for review of "coastal-trapped waves" near shelf breaks; Brink, 1990, and Eriksen, 1991, for discussions of "seamount-trapped waves"; and Roden, 1987, for general review of processes related to seamount topography).

The current amplification is particularly pronounced on detached platforms and volcanic seamounts. Figure 8 shows the resonance effect between a flat East Pacific guyot and the diurnal tide. The tidal wave, normally associated with velocities of less than 1 cm/s in the open ocean, reaches peak velocities of nearly 50 cm/s over the seamount (Genin et al., 1989).

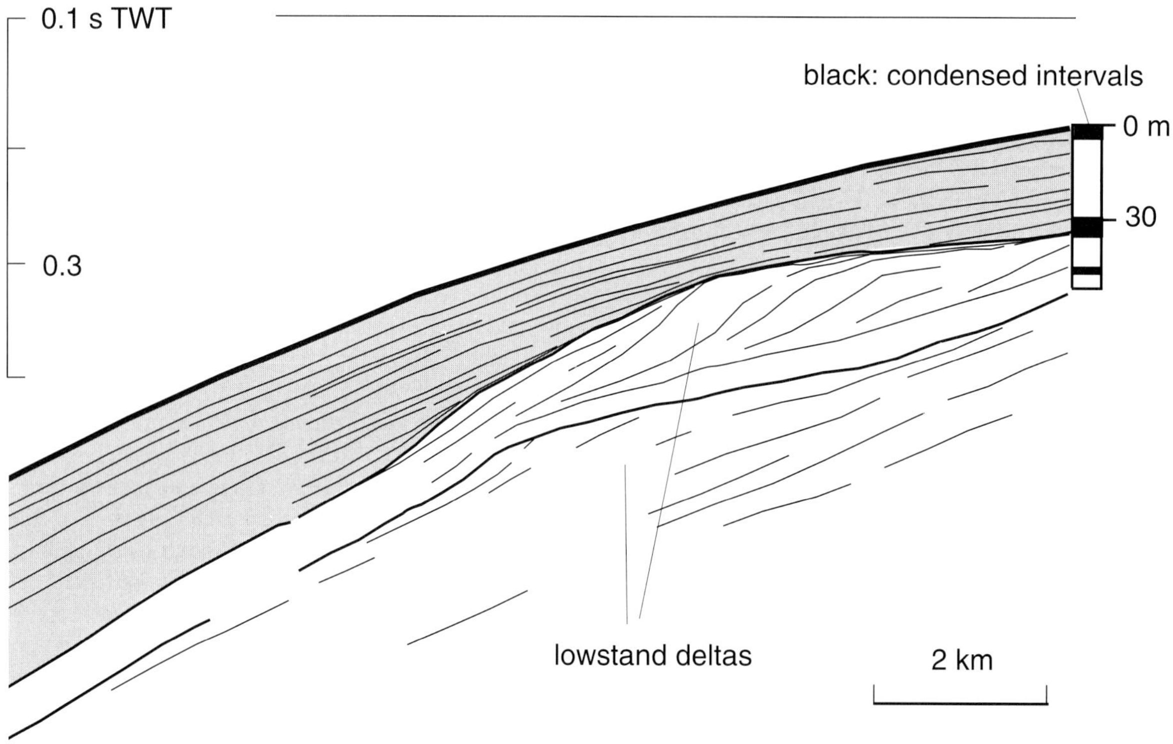

FIG. 7.—Glacial delta lobes of Mississippi, unconformably covered by the Holocene transgressive and highstand tracts. Boreholes indicate that the delta lobes formed during isotope stages 2 and 4, at glacial sea-level minima. Thus, the unconformity on top of the younger lobe must have formed by marine erosion during the rapid Holocene eustatic rise. Its mode of formation is analogous to that of a drowning unconformity on a carbonate platform. Modified after Coleman and Roberts (1988). TWT = two-way traveltime.

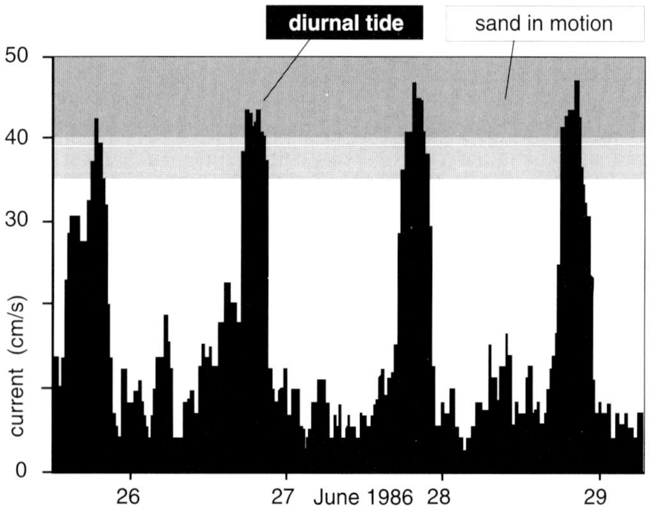

FIG. 8.—Current-meter record from the flat top of Fieberling Guyot at 500-m water depth in the East Pacific. Sluggish currents of a few centimeters per second rise every day to short peaks of more than 40 cm/s. These peaks, capable of moving sand and silt, are interpreted as a resonance effect between the diurnal tide (generally less than 1 cm/s) and the sharp topography of the seamount. Current amplification of this kind may explain the extensive and long hiatuses on drowned carbonate strata. Modified after Genin et al. (1989). Sand threshold is estimated from Rubin and McCulloch (1980).

Numerous authors have observed similar current intensification on seamounts or deduced it from the sediment cover or benthic organisms (e.g., Lonsdale et al., 1972; Hein et al., 1985; Cacchione et al., 1988).

Analogous, albeit less extreme effects have been observed on siliciclastic shelves. McGrail and Carnes (1983) illustrated the acceleration of currents over the shelf break in the Gulf of Mexico by what those authors called "cross-shelf oscillations." Stanley et al. (1983) documented the nearly worldwide tendency of siliciclastic systems to be sediment starved or at least winnowed and sand rich on the outer shelf and uppermost slope with a transition into muddy sediments below (the "mud-line concept"). The observations on currents and sediment textures agree with the sediment geometries observed during rapid flooding of shelf edges (Fig. 7).

DISCUSSION

The Type 3 Sequence Boundary—A Fingerprint

As the boundary between a highstand tract and an overlying transgressive tract, the type 3 sequence boundary is a flooding surface that is not preceded by forced regression in the sense of Posamentier et al. (1992). The unconformity is generated by the combination of marine erosion and change in sediment input. Consequently, the hallmark of the boundary is a marine hiatus and a change in sedimentation associated with rapid deepening of the marine depositional environment at a given place or a rapid landward shift of depositional environments when viewed in cross section. The boundaries between drowned reefs and carbonate platforms and their pelagic or hemipelagic sediment cover are particularly impressive examples of type 3 sequence boundaries. However, rapidly flooded siliciclastic shelves, too, may develop marine unconformities by a similar interplay of marine erosion and abrupt change in sediment input and dispersal (e.g., Fig. 7).

The type 3 sequence boundary proposed here may be further characterized by comparing it with other classifications. The type 3 sequence boundary will not normally fall on the maximum flooding surface in the sense of Vail et al. (1977). It represents a transgressive surface in the sense of Van Wagoner et al. (1988) but with the distinction that there is no evidence for a sea-level lowstand between this surface and the underlying highstand tract. Highstand deposition is terminated by deep flooding of siliciclastic shelves and drowning of the shoal-water carbonate factory on reefs and platforms. When characterized in terms of shoreline trajectories (Helland-Hansen and Gjelberg, 1994), the type 3 boundary is very similar to the case of "non-accretionary transgression." This analogy holds with the proviso that marine erosion largely results from current amplification by sharp topography, not via ravinement, i.e., erosion by a transgressing shoreface.

Vail and Todd (1981) already proposed to distinguish three types of sequence boundaries. A comparison with the unconformity types of Vail and Todd (1981) is hampered by the fact that they consider only systems where sediment supply is approximately constant and tectonic movements are reduced to steady subsidence; consequently, the three types of sequence boundaries are classified in terms of eustatic sea-level movements. The classification proposed here is based on the geometry and facies patterns of systems tracts and proceeds on the assumption that these patterns reflect the balance of changes in relative sea level (the sum of regional tectonics and eustasy) and changes in sediment supply. If one bases a comparison with Vail and Todd (1981) on the sketch of sediment anatomy in Figure 3 (rather than on the inferred eustatic movements), the type 3 boundary proposed here would certainly include the situation in Figure 3C, marine hiatus with marine transgression. It would also include situation 3e (termed type 2 unconformity by Vail and Todd, 1981) if the wavy unconformity line represents a marine hiatus not immediately preceded by exposure.

Where the transgressive tract is missing such that transgressive surface and maximum flooding surface coincide, type 3 boundaries coincide with the genetic sequence boundary of Galloway (1989a, b).

Consequent application of type 3 sequence boundaries may lead to a pattern of alternating highstand tracts and transgressive tracts on carbonate platforms, as illustrated in exemplary fashion by Saller et al. (1993). This pattern also follows from many studies that used cycle-stacking patterns for delineating systems tracts and sequence boundaries on carbonate platforms (e.g., Goldhammer et al., 1990; Montanez and Osleger, 1993). The underlying assumption of all these studies is that the topset beds of carbonate platforms are, by definition, composed of highstand and transgressive systems tracts only. Lowstand tracts are confined to the areas seaward of the platform margin.

The sequence stratigraphy of Embry (1993), which recognizes only transgressive and regressive systems tracts, puts the sequence boundary between the regressive tract below and the transgressive tract above. If regression is caused by progradation rather than by sea-level fall (i.e., in the absence of forced regression), this boundary is identical with the type 3 sequence boundary proposed here.

Type 3 Boundary—Pros and Cons

The proposal for a third type of sequence boundary rests on three principal arguments:

1. HST-TST successions are common in the geologic record and may contain marine hiatuses of tens of millions of years because of the current amplification over the "sharp topography" of the underlying highstand shelf edge. Under these conditions, the standard approach in sequence stratigraphy—classifying the marine hiatus as a transgressive surface and picking the closest underlying exposure surface as sequence boundary—may lead to unreasonable results because the marine hiatus within a sequence may be as long as a third-order or even second-order sequence cycle (see Fig. 4).
2. HST-TST boundaries commonly represent major stratigraphic turning points and levels of reorganization of the sedimentation pattern. This is so because marine hiatuses in this succession are so frequent and last so long. Furthermore, many HST-TST successions represent the drowning of carbonate platforms and the subsequent burial by siliciclastics. Particularly in the last configuration, the HST-TST boundary is likely to be a major seismic reflector as relatively soft, clay-rich pelagic or hemipelagic sediments rest on hard shoal-water carbonates.
3. The succession of TST over HST is a logical third possibility in the superposition of a eustatic cycle and long-term subsidence (Fig. 9). Subsidence on passive margins and normal ocean crust is driven by cooling and associated sediment loading. Both processes are rather steady, and their rates are well constrained by direct observation and crustal modeling. As loading can only partly compensate for the decrease in cooling subsidence, the maximum rates of subsidence of ocean crust and passive margins are largely determined by the thermal component. According to the empirical equation of Parsons and Sclater (1977), thermal subsidence starts at 350 m/m.y. for the first million years and quickly decreases to rates below 100 m/m.y. Typical rates of subsidence on passive margins are between 10 and 100 m/m.y. The rates of sea-level fluctuations for cycles in the 1–10-m.y. domain depicted by Haq et al. (1987) fall in the same range. With the range of rates so similar, the balance between eustasy and subsidence must have shifted frequently in the geologic record: depending on location and time, eustatic falls of sea level must have sometimes exceeded subsidence, sometimes equaled it, and sometimes remained below it. The same

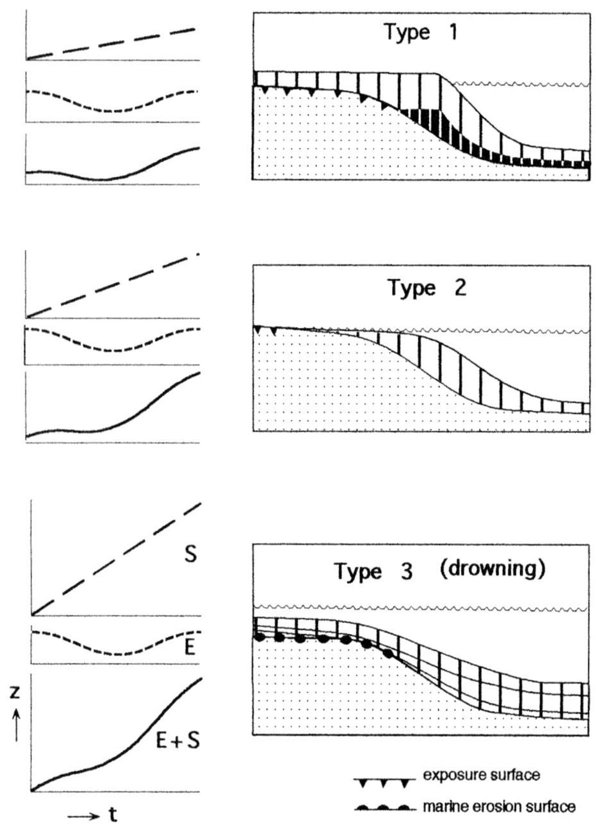

FIG. 9.—Record of an imaginary eustatic cycle in three areas with different rates of subsidence. (Top) Maximum rate of eustatic fall comfortably exceeds the rate of subsidence, resulting in a type 1 sequence boundary. (Middle) Maximum rate of eustatic fall is approximately equal to the rate of subsidence, resulting in a type 2 sequence boundary. (Bottom) Maximum rate of eustatic fall (*E*) is distinctly less than the rate of subsidence (*S*). Rate of accommodation creation decreases (but never ceases) during eustatic fall, and only subtle changes in sedimentation ensue from this slowdown. Major change in sedimentation may occur during subsequent eustatic rise when eustasy and subsidence are in phase and drown platforms or deeply flood siliciclastic shelves, resulting in type 3 sequence boundaries.

eustatic fall may be recorded as a type 1 sequence boundary (eustatic fall exceeding subsidence) in one area and as a type 2 boundary (eustatic fall and subsidence approximately equal) in another. In rapidly subsiding areas where the rate of eustatic fall always remained lower than the rate of subsidence, the falling part of the sea-level cycle may not produce a significant sedimentologic event at all. However, a significant sedimentologic event may occur during the subsequent rise in the eustatic cycle when the superposition of rapid subsidence and eustatic rise causes drowning of carbonate platforms and deep flooding of siliciclastic shelves. The resulting surface is the type 3 boundary proposed here.

The rather unsatisfactory performance of the standard model in the presence of type 3 unconformities may be illustrated by the Liuhua platform (Figs. 1, 2). J. F. Sarg's interpretation in terms of the standard model (written commun. referred to above) designates the drowning unconformity as a maximum flooding surface (or a transgressive surface) in a sequence whose lower boundary is put at the nearest prominent exposure surface below. In this procedure, the most prominent unconformity and the most drastic lithologic change in the succession falls within a sequence; as a consequence, this sequence is extremely asymmetric with the highstand tract replaced by a nearly layer-cake package of hemipelagic shale that extends to the Quaternary sea floor. The approach recommended here is to recognize two sequence boundaries: a type 1 boundary at the exposure surface of 1.36 s in well LH4-1-1 and a type 3 boundary at the platform top.

One drawback of the type 3 sequence boundary is the lack of forced regression, expressed as exposure and karst phenomena in shoal-water carbonates and as down-stepping shorelines, river rejuvenation, etc., in clastics. It remains uncertain if a type 3 boundary is caused by a rapid relative rise of sea level or by abrupt reduction in sediment supply coupled with steady subsidence. However, this uncertainty is a fundamental one, and it is only advantageous to recognize it. If relative sea-level movements are defined as changes in the distance between the sea surface and a deep-crustal reference level (e.g., the top of basement), then most features in the systems tract model are not reliable clues to sea level but reflect the balance between the rate of change of relative sea level and the rate of sediment supply (Jervey, 1988; Schlager, 1993). The only reliable evidence of relative sea-level cycles is proof of forced regression because it requires a fall of relative sea level irrespective of changes in supply (Posamentier et al., 1992). Admitting type 3 sequence boundaries has two beneficial effects. First, the sequence definition remains sufficiently broad to be workable. In everyday applications, many sequence boundaries have to be drawn on geometry alone, without diagnostic evidence for exposure and sea-level fall; they should simply be termed sequence boundaries without further subdivision. Second, the three types of sequence boundaries allow one to separate the grain from the chaff when searching for the relative sea-level signature: type 1 provides evidence for a sea-level fall to beyond the shelf break, type 2 provides evidence for a sea-level fall to the inner shelf, and type 3 implies that there is no evidence for a fall in relative sea level.

The type 3 boundary is compatible with the original definition of sequence boundary because this definition is descriptive and says nothing about the cause(s): Vail et al. (1977, p. 53) defined the term *sequence boundary* as an unconformity that separates conformably stratified units, the sequences. Sedimentologically speaking, this definition implies that the sequence boundary represents a geometrically manifest change in the pattern of sediment input and dispersal (Schlager, 1992). Sea level is but one of several processes that can induce a rapid change in the sedimentation pattern.

Condensed versus Erosional Intervals

The standard model of sequence stratigraphy recognizes condensed intervals in the transgressive systems tract and also the highstand systems tract. Condensed intervals are largely attributed to sediment starvation during high rising sea level when sediment sources are far away (Loutit et al., 1988). Hiatuses by marine erosion, commonly attributed to shifting ocean currents, are included in the condensed sections (e.g., Haq, 1991).

I believe that genuine marine hiatuses should be identified and, wherever possible, separated from what sequence stratigraphers call "condensed sections." Haq (1991) has rightly pointed out that where nearshore sediment supply ceases, pelagic sedimentation takes over; as a consequence, many condensed intervals simply represent periods of pelagic sedimentation with minimal input from shore. Pelagic sedimentation is generally lower than nearshore sedimentation, but it is not zero. Even where all production by siliceous and calcareous plankton is redissolved, there remains a background sedimentation of terrigenous and volcanogenic fines, partly airborne, that accounts for accumulation at rates of 4–6 m/m.y. in the middle of vast oceans and significantly more in small ocean basins. A genuine hiatus will form only where currents remove this background sediment. Most often, study of seismic sections cannot distinguish between genuine (erosional) gaps and condensed deposits (Van Hinte, 1982; Schlager, 1996); conceptually, however, these two situations are different.

Darwin Guyot (Fig. 5) may serve as a case in point. This carbonate platform was drowned 100 m.y. ago and has since subsided to 1500-m water depth; its pelagic cover is negligible (Ladd et al., 1974; E. L. Winterer, written commun.). Even if all nearby sediment sources had been shut off, the global background sedimentation should have produced a cap of 400–600 m of pelagic clay on the drowned atoll. The absence of such a cover attests to erosion and winnowing. These processes must have operated in the deep-marine environment because the guyot has spent much of the past 100 m.y. far below wave base. The same argument can be made for most of the guyots in the central Pacific (Winterer and Sager, 1995, p. 526).

Two agents capable of extensive marine erosion are wave abrasion in a transgressing shoreface (the ravinement zone) and deep-sea currents sweeping continental rises and abyssal plains. None of them can be invoked for the erosion of Darwin Guyot and similar drowned platforms. This review emphasizes a third process—interaction of sharp topography with sluggish ocean currents, such as the astronomic tides. Particularly the phenomenon of "trapped waves" may go a ways toward explaining the distribution of large hiatuses on shelves and oceanic plateaus. An important characteristic of this mechanism is its independence of major contour-hugging currents in the ocean. Currents may be amplified and marine erosion may set in wherever sharp topography (such as a shelf margin) faces the open ocean.

Procedure for Establishing Sequence Boundaries

Accepting the marine erosion surface between a highstand tract and a transgressive tract as a third type of sequence boundary simplifies the picking of sequence boundaries in case studies with mainly seismic data and little ground truth. The addition of flooding events as possible sequence boundaries allows one to start from the descriptive definition of a sequence boundary by Vail et al. (1977) that was discussed above. This broad characterization of sequence boundaries provides the basis for a three-step approach:

1. Consider the coast-to-basin system (or as much of it as possible) and identify all pronounced unconformities in the sediment pile under consideration.
2. Characterize these surfaces in terms of (a) presence or absence of exposure surfaces and subaerial hiatuses, (b) presence or absence of flooding surfaces and marine hiatuses, and (c) with regard to geometry by using the systems-tract concept (Posamentier and Vail, 1988) or the concept of shoreline trajectories (Helland-Hansen and Gjelberg, 1994).
3. Assign types of sequence boundaries where information is sufficient to do so. Otherwise, arrest the process at step 1 and use "sequence boundary" in the descriptive definition without further subdivision into boundary types.

CONCLUSIONS

The boundary between a highstand systems tract and an overlying transgressive systems tract is a good choice for a sequence boundary even if there is no evidence for exposure. This "type 3 sequence boundary" is particularly well developed on drowned reefs and carbonate platforms, but rapidly flooded siliciclastic shelves may also develop distinct unconformities between highstand and overlying transgressive tracts.

Type 3 sequence boundaries tend to be good seismic reflectors because of a pronounced change in sediment composition. They are unconformities because the flooding event changes the sediment-dispersal pattern, triggers submarine erosion, and creates submarine hiatuses.

The efficiency of submarine erosion on rapidly flooded shelves and drowned platforms deserves more attention in sequence stratigraphy. Fueled by the interaction of (sluggish) oceanic tides and sharp bottom topography, the erosion on seamounts and shelf edges is different from the erosion by a transgressing shoreface (ravinement) and from the erosion by strong geostrophic currents. On drowned platforms, this process generates marine hiatuses that easily span 10 m.y. and sometimes exceed 100 m.y. They are among the longest nontectonic hiatuses in the geologic record. On siliciclastic shelves, patterns are similar, but hiatuses are shorter because of slower lithification and more intensive reworking.

ACKNOWLEDGMENTS

Much of the research that led to this paper was supported by the VU Industrial Associates in Sedimentology. Wil P. M. de Ruijter (Utrecht) pointed me in the right direction with the oceanographic aspects of the topic. To J. F. Sarg (Mobil, Dallas), R. Stephenson (Amsterdam), R. P. Wright (Chevron, Houston), and P. M. Harris (Chevron, La Habra) I owe thanks for thoughtful reviews.

REFERENCES

Bathurst, R. G. C., 1971, Carbonate Sediments and Their Diagenesis: Amsterdam, Elsevier, 620 p.

Brink, K. H., 1987, Coastal ocean physical processes: Reviews of Geophysics and Space Physics, v. 25, p. 204–216.

Brink, K. H., 1990, On the generation of seamount-trapped waves: Deep-Sea Research, v. 37, p. 1569–1582.

Cacchione, D. A., Schwab, A. W. C., Noble, M., and Tate, G., 1988, Internal tides and sediment movement on Horizon Guyot, Mid-Pacific mountains: Geo-Marine Letters, v. 8, p. 11–17.

Camoin, G. F., Arnaud-Vannau, A., Bergersen, D. D., Enos, P., and Ebren, P., 1998, Development and demise of mid-oceanic carbonate platforms, Wodejebato Guyot (NW Pacific), *in* Camoin, G. F. and Davies, P., eds., Reefs and Carbonate Platforms of the SW Pacific and Indian Oceans: London, International Association of Sedimentologists Special Publication 25, p. 39–67.

Clari, P. A., Dela Pierre, F., and Martire, L., 1995, Discontinuities in carbonate successions: Identification, interpretation and classification of some Italian examples: Sedimentary Geology, v. 100, p. 97–121.

Coleman, J. M., and Roberts, H. H., 1988, Sedimentary development of the Louisiana continental shelf related to sea level cycles, Part I, Sedimentary sequences: Geo-Marine Letters, v. 8, p. 63–108.

Dravis, J. J. , 1989, Deep-burial microporosity in Upper Jurassic Haynesville oolitic grainstones, east Texas: Sedimentary Geology, v. 63, p. 325–341.

Embry, A. F., 1993, Transgressive-regressive (T-R) sequence analysis of the Jurassic succession of the Sverdrup Basin, Canadian Arctic Archipelago: Canadian Journal of Earth Sciences, v. 30, p. 301–320.

Eriksen, C. C., 1991, Observations of amplified flows atop a large seamount: Journal of Geophysical Research, v. 96, p. 15227–15236.

Erlich, R. N., Barrett, S. F., and Guo, B. J., 1990, Seismic and geologic characteristics of drowning events on carbonate platforms: American Association of Petroleum Geologists Bulletin, v. 74, p. 1523–1537.

Field, E. M., Carlson, P. R., and Hall, R. K., 1983, Seismic facies of shelfedge deposits, U.S. Pacific continental margin, *in* Stanley, D. J. and Moore, G. T., eds., The Shelfbreak: Critical Interface on Continental Margins: Tulsa, Society of Economic Paleontologists and Mineralogists Special Publication 33, p. 299–313.

Föllmi, K. B., Weissert, H., Bisping, M., and Funk, H. P., 1994, Phosphogenesis, carbon-isotope stratigraphy, and carbonate-platform evolution along the Lower Cretaceous northern Tethyan margin: Geological Society of America Bulletin, v. 106, p. 729–746.

Fouke, B. W., Everts, A.-J. W., Zwart, E., Schlager, W., Smalley, P. C., and Weissert, H., 1995, Subaerial exposure unconformities on the Vercors carbonate platform (SE France) and their sequence stratigraphic significance, *in* Howell, J. A. and Aitken, J. F., eds., High Resolution Sequence Stratigraphy: Innovations and Applications: London, Geological Society of London Special Publication 104, p. 295–320.

Galloway, W. E., 1989a, Genetic stratigraphic sequences in basin analysis: I. Architecture and genesis of flooding-surface bounded depositional units: American Association of Petroleum Geologists Bulletin, v. 73, p. 125–142.

Galloway, W. E., 1989b, Genetic stratigraphic sequences in basin analysis: II. Application to northwest Gulf of Mexico Cenozoic basin: American Association of Petroleum Geologists Bulletin, v. 73, p. 143–154.

Genin, A., Noble, M., and Lonsdale, P. F., 1989, Tidal currents and anticyclonic motions on two North Pacific seamounts: Deep-Sea Research, v. 36, p. 1803–1815.

Goldhammer, R. K., Dunn, P. A., and Hardie, L. A., 1990, Depositional cycles, composite sea-level changes, cycle stacking patterns, and the hierarchy of stratigraphic forcing: Examples from Alpine Triassic platform carbonates: Geological Society of America Bulletin, v. 102, p. 535–562.

Haq, B. U., 1991, Sequence stratigraphy, sea-level change, and significance for the deep sea, *in* Macdonald, D. I. M., eds., Sedimentation, Tectonics and Eustasy: London, International Association of Sedimentologists Special Publication 12, p. 3–40.

Haq, B. U., Hardenbol, J., and Vail, P. R., 1987, Chronology of fluctuating sea levels since the Triassic (250 million years ago to present): Science, v. 235, p. 1156–1167.

Hein, J. R., Manheim, F. T., Schwab, W. C., and Davis, A. W., 1985, Ferromanganese crusts from Necker Ridge, Horizon Guyot and S. P. Lee Guyot: Geomorphological considerations: Marine Geology, v. 69, p. 5–54.

Helland-Hansen, W,. and Gjelberg, J. G., 1994, Conceptual basis and variability in sequence stratigraphy: A different perspective: Sedimentary Geology, v. 92, p. 31–52.

James, N. P., and Choquette, P. W., 1984, Diagenesis 9: Limestones—The meteoric diagenetic environments: Geoscience Canada, v. 11, p. 161–194.

Jervey, M. T., 1988, Quantitative geological modeling of siliciclastic rock sequences and their seismic expression, *in* Wilgus, C. K., Hastings, B. S., Kendall, C. G. St. C., Posamentier, H. W., Ross, C. A., and Van Wagoner, J. C., eds., Sea-Level Changes: An Integrated Approach: Tulsa, Society of Economic Paleontologists and Mineralogists Special Publication 42, p. 47–69

Ladd, H. S., Newman, W. A., and Sohl, N. F., 1974, Darwin Guyot, the Pacific's oldest atoll, *in* Second International Coral Reef Symposium Proceedings: Brisbane, Great Barrier Reef Committee, part 2, p. 513–522.

Lonsdale, P. F., Normark, W. R., and Newman, W. A., 1972, Sedimentation and erosion on Horizon Guyot: Geological Society of America Bulletin, v. 83, p. 289–315.

Loutit, T. S., Hardenbol, J., Vail, P. R., and Baum, G. R., 1988, Condensed sections: The key to age determination and correlation of continental margin sequences, *in* Wilgus, C. K., Hastings, B. S., Kendall, C. G. St. C., Posamentier, H. W., Ross, C. A., and Van Wagoner, J. C., eds., Sea-Level Changes: An Integrated Approach: Tulsa, Society of Economic Paleontologists and Mineralogists Special Publication 42, p. 183–213.

McGrail, D. W., and Carnes, M., 1983, Shelfedge dynamics and the nepheloid layer in the northwestern Gulf of Mexico, *in* Stanley, D. J. and Moore, G. T., eds., The Shelfbreak: Critical Interface on Continental Margins: Tulsa, Society of Economic Paleontologists and Mineralogists Special Publication 33, p. 251–264.

Moldovanyi, E. P., Waal, F. M., and Zhang Jun Yan, 1995, Regional exposure events and platform evolution of Zhujiang Formation carbonates. Pearl River Mouth Basin: Evidence from primary and diagenetic seismic facies, *in* Budd, D. A., Saller, A. H., and Harris, P. M., eds., Unconformities and Porosity in Carbonate Strata: Tulsa, American Association of Petroleum Geologists Memoir 63, p. 125–140.

Montanez, I. P., and Osleger, D. A., 1993, Parasequence stacking patterns, third-order accommodation events, and sequence stratigraphy of Middle to Upper Cambrian platform carbonates, Bonanza King Formation, southern Great Basin, *in* Loucks, R. G. and Sarg, J. F., eds., Carbonate Sequence Stratigraphy: Tulsa, American Association of Petroleum Geologists Memoir 57, p. 305–326.

Parsons, B., and Sclater, J. G., 1977, An analysis of the variation of ocean floor bathymetry and heat flow with age: Journal of Geophysical Research, v. 82, p. 803–827.

Posamentier, H. W., and Vail, P. R., 1988, Eustatic controls on clastic deposition II—Sequence and systems tract models, *in* Wilgus, C. K., Hastings, B. S., Kendall, C. G. St. C., Posamentier, H. W., Ross, C. A., and Van Wagoner, J. C., eds., Sea-Level Changes: an Integrated Approach: Tulsa, Society of Economic Paleontologists and Mineralogists Special Publication 42, p. 125–154.

Posamentier, H. W., Allen, G. P., James, D. P., and Tesson, M., 1992, Forced regressions in a sequence stratigraphic framework: Concepts, examples, and exploration significance: American Association of Petroleum Geologists Bulletin, v. 76, p. 1687–1709.

Roden, G. I., 1987, Effect of seamounts and seamount chains on ocean circulation and thermohaline structure, *in* Keating, B. H., Fryer, P., Batiza, R., and Boehlert, G. W., eds., Seamounts, Islands and Atolls: Washington, D.C., American Geophysical Union Geophysical Monograph 43, p. 335–354.

Rubin, D. M., and McCulloch, D. W., 1980, Single and superimposed bedforms: Synthesis of San Francisco Bay and flume observations: Sedimentary Geology, v. 26, p. 207–231.

Saller, A., Armin, R., Ichram, L. O., and Glenn-Sullivan, C., 1993, Sequence stratigraphy of aggrading and backstepping carbonate shelves, Oligocene, Central Kalimantan, Indonesia, *in* Loucks, R. G. and Sarg, J. F., eds., Carbonate Sequence Stratigraphy: Tulsa, American Association of Petroleum Geologists Memoir 57, p. 267–290.

Sarg, J. F., 1988, Carbonate sequence stratigraphy, *in* Wilgus, C. K., Hastings,

B. S., Kendall, C. G. St. C., Posamentier, H. W., Ross, C. A., and Van Wagoner, J. C., eds., Sea-Level Changes: An Integrated Approach: Tulsa, Society of Economic Paleontologists and Mineralogists Special Publication 42, p. 155–181.

SCHLAGER, W., 1989, Drowning unconformities on carbonate platforms, *in* Crevello, P. D., Wilson, J. L., Sarg, J. F., and Read, J. F., eds., Controls on Carbonate Platform and Basin Development: Tulsa, Society of Economic Paleontologists and Mineralogists Special Publication 44, p. 15–25.

SCHLAGER, W., 1992, Sedimentology and Sequence Stratigraphy of Reefs and Carbonate Platforms: Tulsa, American Association of Petroleum Geologists Continuing Education Course Note Series, v. 34, 71 p.

SCHLAGER, W., 1996, The mismatch between outcrop unconformity and seismic unconformity, *in* Caughey, C. A., Carter, D. C., Clure, J., and Gresko, M. J., eds., Proceedings, International Symposium on Sequence Stratigraphy in Southeast Asia: Jakarta, Indonesian Petroleum Association, p. 4–18.

SCHLAGER, W., 1998, Exposure, drowning and sequence boundaries on carbonate platforms, *in* Camoin, G. and Davies, P., eds., Reefs and Carbonate Platforms of the W Pacific and Indian Oceans: London, International Association of Sedimentologists Special Publication 25, p. 3–21.

SLITER, W., 1995, Cretaceous planktonic foraminifers from sites 865, 866 and 869: A synthesis of Cretaceous pelagic sedimentation in the central Pacific Ocean Basin, *in* Sager, W. W., Winterer, E. L., and Firth, J. V., Proceedings of the Ocean Drilling Program, Scientific Results: College Station, Ocean Drilling Program, v. 143, p. 15–30.

STANLEY, D. J., ADDY, S. K., AND BEHRENS, E. W., 1983, The mudline: Variability of its position relative to shelfbreak, *in* Stanley, D. J. and Moore, G. T., eds., The Shelfbreak: Critical Interface on Continental Margins: Tulsa, Society of Economic Paleontologists and Mineralogists Special Publication 33, p. 279–298.

VAIL, P. R., AND TODD, R. G., 1981, Northern North Sea Jurassic unconformities, chronostratigraphy and sea-level changes from seismic stratigraphy, *in* Illing, L. V. and Hobson, G. D., eds., Petroleum Geology of the NW Continental Shelf: London, Institute for Petroleum, Proceedings, Petroleum Geology of the Continental Shelf of Northwest Europe, p. 216–233.

VAIL, P. R., MITCHUM, R. M., TODD, R. G., WIDMIER, J. M., THOMPSON, S., SANGREE, J. B., BUBB, J. N., AND HATLELID, W. G., 1977, Seismic stratigraphy and global changes of sea level, *in* Payton, C. E., ed., Seismic Stratigraphy—Applications to Hydrocarbon Exploration: Tulsa, American Association of Petroleum Geologists Memoir 26, p. 49–212.

VAIL, P. R., HARDENBOL, J., AND TODD, R. G., 1984, Jurassic unconformities, chronostratigraphy, and sea-level changes from seismic stratigraphy and biostratigraphy, *in* Schlee, J. S., ed., Interregional Unconformities and Hydrocarbon Accumulation: Tulsa, American Association of Petroleum Geologists Memoir 36, p. 129–144.

VAN HINTE, J. E., 1982, Synthetic seismic sections from biostratigraphy, *in* Watkins, J. S. and Drake, C. L., eds., Studies in Continental Margin Geology: Tulsa, American Association of Petroleum Geologists Memoir 34, p. 675–685.

VAN WAGONER, J. C., POSAMENTIER, H. W., MITCHUM, R. M., VAIL, P. R., SARG, J. F., LOUTIT, T. S., AND HARDENBOL, J., 1988, An overview of the fundamentals of sequence stratigraphy and key definitions, *in* Wilgus, C. K., Hastings, B. S., Kendall, C. G. St. C., Posamentier, H. W., Ross, C. A., and Van Wagoner, J. C., eds., Sea-Level Changes: An Integrated Approach: Tulsa, Society of Economic Paleontologists and Mineralogists Special Publication 42, p. 39–45.

WAGNER, P. D., TASKER, D. R., AND WAHLMAN, G. P., 1995, Reservoir degradation and compartmentalization below subaerial unconformities: Limestone examples from West Texas, China and Oman, *in* Budd, D. A., Saller, A. H., and Harris, P. M., eds., Unconformities and Porosity in Carbonate Strata: Tulsa, American Association of Petroleum Geologists Memoir 63, p. 177–196.

WENDTE, J. C., STOAKES, F. A., AND CAMPBELL, C. V., 1992, Devonian–Early Mississippian carbonates of the Western Canada sedimentary basin: A sequence stratigraphic framework: Tulsa, SEPM (Society for Sedimentary Geology) Short Course Notes No. 28, 255 p.

WINTERER, E. L., AND SAGER, W. W., 1995, Synthesis of drilling results from the Mid-Pacific mountains: Regional context and implications, *in* Sager, W. W., Winterer, E. L., and Firth, J. V., Proceedings of the Ocean Drilling Program: Scientific Results: College Station, Ocean Drilling Program, v. 143, p. 497–535.

TECTONIC CONTROL ON BACKSTEPPING SEQUENCES REVEALED BY MAPPING OF FRASNIAN BACKSTEPPED PLATFORMS, DEVONIAN REEF COMPLEXES, NAPIER RANGE, CANNING BASIN, WESTERN AUSTRALIA

W. Bruce Ward*

Department of Earth and Space Sciences, State University of New York at Stony Brook, Stony Brook, New York 11794-2100

ABSTRACT: Mapping Upper Devonian reef complexes along 80 km of the southern Napier Range in the Canning basin, Western Australia, has documented three generations (A, B, C) of Frasnian reef-rimmed carbonate-siliciclastic platforms (B and C are backstepped). The tops of these platforms are sequence boundaries. The seaward margins of the platforms are typically upright escarpments developed by both reef growth and collapse. Backstepping of platform B over platform A is most consistent with backstepping during a sea-level rise following a sea-level fall. The internal stratigraphy within platform A predicts this fall and suggests that the fall was predominantly eustatic. This finding is also consistent with the record of the sea-level rise recorded by the lower strata of platform B.

Platform B, which lacks subaerial exposure features, marks a major episode of margin reconfiguration by collapse along neptunian fractures (oblique, parallel, and normal to platform trends). Overlying strata record a prolonged postcollapse submergence prior to and during the development of backstepped platform C. During submergence, the near-vertical escarpments were coated by calcareous cyanobacteria. The stratigraphic sequence shows no evidence of eustatic control on this backstepping. The record of the period between the termination of platform B deposition and backstepping to form platform C is most consistent with rift-related tectonic movement that submerged the platform, produced an extensive submarine erosional escarpment, and forced backstepping to occur. Tectonic movement effecting a change in relative sea level and backstepping is consistent with (1) the regional extent of a mixed carbonate-siliciclastic conglomerate that indicates erosion of the interior of several platforms, (2) the structural control on neptunian-fracture and platform-margin orientations, (3) siliciclastic distribution, and (4) large (meter-scale) Precambrian boulders. The above interpretation implies that basinal wedges with "lowstand geometries" in the subsurface actually record tectonically induced sedimentation during a sea-level rise.

INTRODUCTION

Recognizing different hierarchies of depositional packages that record transgressive to progradational and aggradational relationships has been important for providing frameworks for correlations and for interpreting and predicting facies distributions. The controls on the changes in relative sea level recorded by these depositional packages can affect the distribution of facies and their predictive relationships. For purely eustatic-controlled sea-level changes, not only is there a potential for worldwide correlation, but also for predicting facies distributions and proportions. This is because facies may change systematically with sea-level fluctuations from transgressive to highstand and then to lowstand deposits. Tectonic influences on relative sea level have the potential to punctuate systematic changes and add complications to predicting facies distribution within a stratigraphic framework. Many carbonate platforms through geologic history have developed in tectonically active areas, especially in rift basins. The stratigraphic geometry of these platforms may record relative sea-level changes that could be either mainly eustatic or mainly tectonic. Separating these influences remains a challenge.

The exposures of Devonian reef complexes in the Canning basin provide an opportunity to evaluate the relative roles of eustasy and tectonism. These reef complexes record multiple hierarchies of relative-sea-level changes that range from meter-scale cycles, possibly recording high-frequency eustasy (e.g., Read, 1973a, b), to the entire Devonian reef package, recording the long-term tectonic evolution of the Lennard shelf (e.g., Playford et al., 1989). The entire succession includes the predominantly backstepping Pillara cycle followed by the predominantly prograding Nullara cycle (Playford, 1980; Playford et al., 1989). Within this framework, controls on stratigraphic successions are less clear. Depositional sequences (interpreted to be third-order) within the reef complexes have been defined on (1) cross-sectional platform relationships in the outcrops and in the subsurface (e.g., Playford, 1991; Kennard et al., 1992), (2) facies changes within slope deposits (e.g., George et al., 1994, 1997; George and Powell, 1997), and (3) seismic relationships showing apparent siliciclastic lowstand wedges in subsurface strata (e.g., Southgate et al., 1991, 1993; Kennard et al., 1992). Holmes and Christie-Blick (1993) pointed out, as do George et al. (1997), Ward (1996), and this paper, that many of the sequences may be the result of extension-related movement along fault blocks. Syndepositional faults occur throughout the outcrop area (e.g., Playford, 1980; Playford et al., 1989).

Through a series of maps and cross sections, this paper documents the stratigraphy of the Upper Devonian reef complexes exposed along 80 km of the southern Napier Range on the northeastern edge of the Lennard shelf (Fig. 1). The focus is on the three-dimensional relationships of Frasnian reef-rimmed carbonate-siliciclastic platforms of the Pillara Limestone. The platforms evolved through periods of backstepping, margin collapse, and progradation. There are three generations of Frasnian platforms: A, B, and C (oldest to youngest), of which B and C are backstepped (Fig. 2). The upright escarpment margins of these platforms are flanked by aprons of steeply dipping slope strata of the Napier Formation.

These depositional packages are referred to as platforms A, B, and C and their equivalents. These are chronostratigraphic

Present address: Earthworks LLC, and Theoretical & Applied Geology, 50 Day Street, Norwalk, CT 06854

Advances in Carbonate Sequence Stratigraphy: Application to Reservoirs, Outcrops and Models, SEPM Special Publication No. 63
 ISBN 1-56576-060-3, p. 47–74

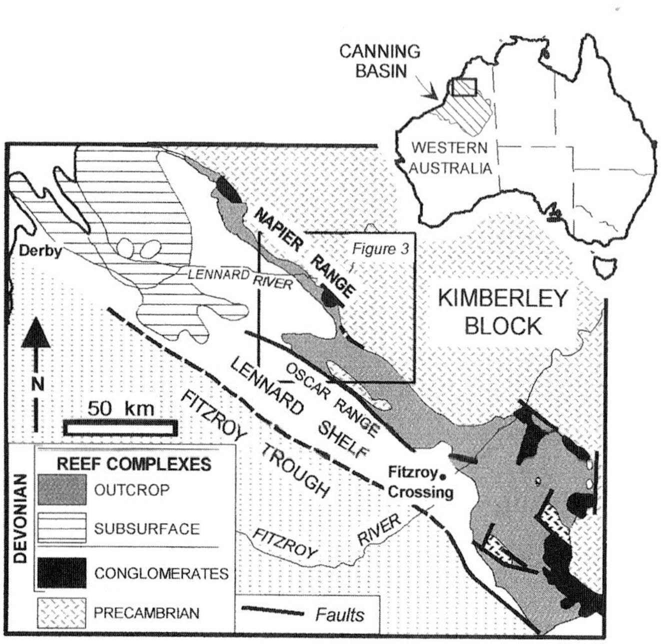

FIG. 1.— Index map of Australia (upper right corner) and location map of the Devonian reef complexes on the Lennard shelf in the Canning basin. Rectangle indicates area shown in Figure 3.

units bounded by surfaces of lateral lithofacies shift and/or unconformities. In the Napier Range, two of these boundaries are defined by backstepped lithofacies shifts (top of A, top of B) and one by a karst unconformity (top of C). Platforms A, B, and C appear to be equivalent to the depositional sequences Frasnian 2, 3, and 4, respectively, as defined by Kennard et al. (1992) and Southgate et al. (1993). The usage of *platform* is preferred over the usage of *sequence* because the nature of the boundaries are debatable. *Platform* as used here is in the spirit of earlier field studies (Playford, 1980; Kerans, 1985; Hurley, 1986; Wallace, 1987).

The transition from platform A to B is foreshadowed by systematic changes in the internal stratigraphy of platform A. The transition from platform B to C is problematic. The top surface of platform B, which lacks subaerial exposure features, marks a major episode of margin reconfiguration by collapse along neptunian fractures (oblique, parallel, and normal to platform trends). Oblique fracture sets are interpreted as opened structural joints. Features that are consistent with the backstepping being fault controlled include (1) a regional margin collapse, (2) carbonate conglomerate indicating erosion of the interior of several platforms, (3) the structural control on neptunian-fracture and platform-margin orientations, (4) the siliciclastic distribution, and (5) deposits with large Precambrian boulders.

The Devonian reef complexes in the Canning basin of Western Australia are classic examples of ancient reef-rimmed carbonate platforms. The documentation of these has been made possible by the extensive outcrop exposures of relatively undeformed carbonate-platform complexes that make up ranges extending along 350 km of the Lennard shelf on the northeastern edge of the Canning basin (Fig. 1). The outcrop topography of these Devonian carbonate rocks approaches being exhumed Devonian depositional bathymetry of the reef complexes (Playford, 1980, 1984). A few gorges through the ranges present well-exposed and often-photographed cross sections across platform margins and steeply dipping slope strata. *The* example of such cross sections is "the classic face" in Windjana Gorge (e.g., Playford and Lowry, 1966; Playford, 1981, 1984; Playford et al., 1989). Cross-sectional outcrop exposures, however spectacular, are somewhat limited; their coverage is relatively narrow and does not extend through the entire stratigraphic section. This condition exists partially because the ranges are relatively low and narrow and the basin facies are buried or eroded. The truly exceptional aspect of the Devonian reef complexes is that their lateral exposures permit regionally extensive three-dimensional views of similar-aged platforms. Depositional packages such as cycles and sequences are three-dimensional bodies that require at least two-dimensional cross sec-

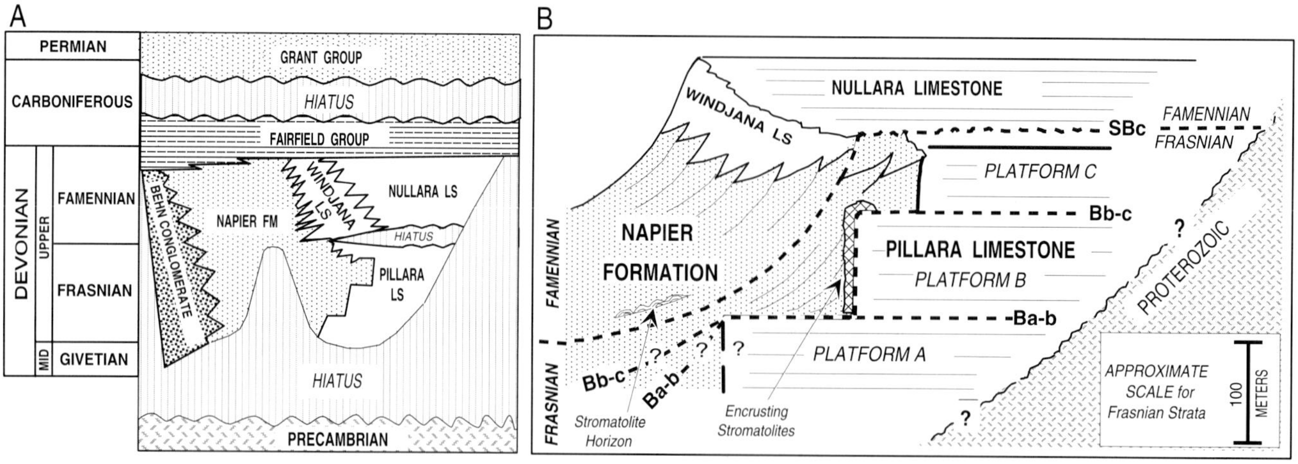

FIG. 2.—(A) Stratigraphic relationships of the Devonian reef complexes in the Napier Range. The stratigraphic "thickness" represented by the hiatus at the unconformity between the Frasnian Pillara Limestone and Famennian platform strata is unknown. The extent and timing of other periods of exposure of the platform facies continue to be debated. The Behn Conglomerate also interfingers with platform facies. (B) Stratigraphy of the Devonian reef complexes in the Napier Range. Three Frasnian platforms (A, B, and C) were mapped throughout the study area. The sequence-boundary contacts between platforms A and B and between platforms B and C are labeled Ba-b and Bb-c. The sequence boundary between platform C and the Nullara Limestone (SB-c) is a karst surface. The Nullara Limestone is not exposed in the southern Napier Range, and the other Famennian strata were not studied in detail. Therefore Famennian sequences are not shown.

tions, typically dip sections for carbonate platforms and ramps, to define the geometries of transgressive and progradational bodies (e.g., Kerans and Tinker, 1997). The more lateral variability there is within a platform and the more limited the platform-to-basin section is, the more three-dimensional control is required to define these packages. The Frasnian platform B is exposed throughout the southern Napier Range and, thus, provides three-dimensional control on internal facies and on platform B's relationship to platforms A and C.

The events that contributed to backstepping of platform C and formation of the sequence boundary between B and C remains unknown. The interrelationships between siliciclastic sandstones and conglomerates, the reef margins, tectonics, and sea level are complex. This paper suggests that the platform and slope depositional record that defines this backstepping event is most consistent with being the result of faulting and movement of rift blocks.

REGIONAL SETTING

The Givetian through Famennian reef complexes occur on the surface and in the subsurface of the Lennard shelf (Fig. 1). The Canning basin initially developed during the Ordovician, while significant rifting started during the Early Devonian. The Lennard shelf is bounded on the southwest by the Fitzroy trough, which is filled with at least 10 km of Paleozoic sedimentary rock. The Devonian reef-rimmed carbonate platforms developed on fault blocks of Precambrian and/or Ordovician strata and along the Devonian mainland of the Kimberley block. In the Napier Range, the Devonian strata directly overlie the Proterozoic terrane of the Kimberley block (Fig. 1).

The Napier Range is dominated by the Frasnian-Famennian Napier Formation and the Frasnian Pillara Limestone (Figs. 2B and 3). The Napier marginal-slope strata are exposed throughout most of its western edge, whereas discontinuous regions of

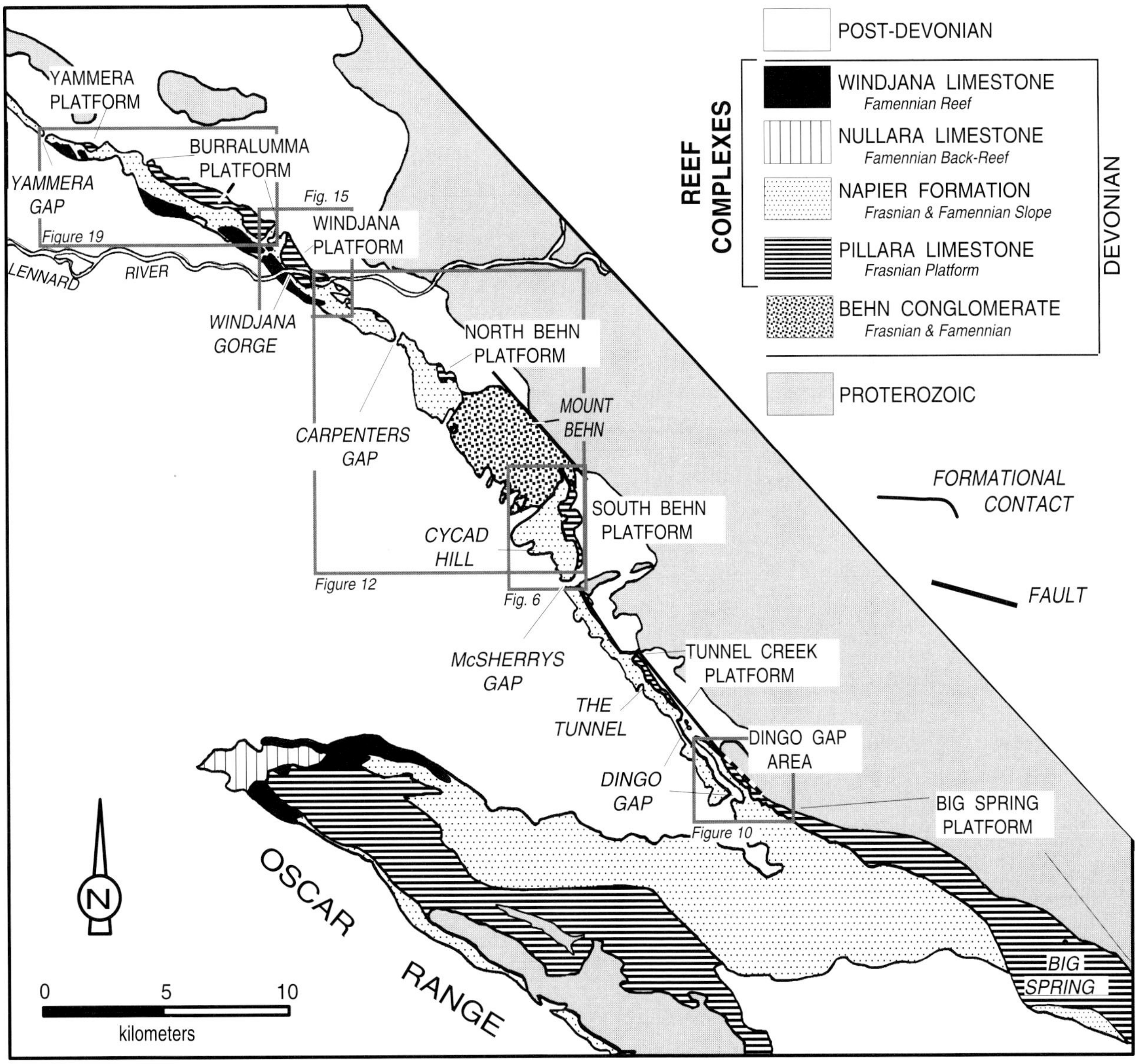

FIG. 3.—Outcrop geologic map of the southern Napier Range and northern Oscar Range. The Oscar Range geology is after Hurley (1986), and the Napier Range geology is based on Ward (1996) and the maps presented in this paper. Rectangles outline map areas shown in the figures indicated.

Pillara platform strata crop out along the eastern edge of the range. Famennian reef strata of the Windjana Limestone crop out at and north of Windjana Gorge. Famennian back-reef strata are not exposed in the southern part of the Napier Range, but are exposed in the northernmost Napier Range near Chedda Cliffs and also in the Oscar Range. The basin facies are not exposed in the southern range, but presumably occur in the subsurface just west of the Napier Range. The nearest exposures of basin facies are in the Oscar Plateau just southwest of the field area (Hurley, 1986) and farther to the south in Copley Valley near Geikie Gorge (Wallace, 1987). Precambrian metamorphic rocks of the Halls Creek Province border the eastern edge of the range.

During the development of the Devonian reef complexes, siliciclastics derived from the Kimberley block formed both well-bedded deposits within the carbonate platform and slope strata and massive fans that interfingered with the platform and slope carbonates. Many of the fanglomerates, including the Behn Conglomerate in the study area, apparently formed at fault escarpments (Figs. 1 and 3). The Behn Conglomerate has been interpreted by a few workers (e.g., Southgate et al., 1993) as lowstand deposits incised into the platform; however, the relationships described in this paper show this origin is not the case. The Behn Conglomerate interfingers with Pillara platform strata.

MAPPING IN THE NAPIER RANGE

In addition to providing the basic stratigraphic data for this paper, mapping has established the regional extent and correlation of the Frasnian deposits. Playford and Lowry (1966) published maps of the Devonian reef complexes exposed throughout the Lennard shelf showing the platform-, slope-, and basin-facies relationships. With the exception of the Windjana Gorge area, their maps of the Napier Range persisted as the most up-to-date maps up to the time of this study. Read (1973a, b) measured stratigraphic sections and conducted detailed studies of cycles in the Pillara Limestone back-reef strata, including that in the Napier Range. His interpretations of back-reef strata, however, were divorced from discussion of the platform-margin and slope facies.

Playford and Cockbain (1976) refined the stratigraphic nomenclature to that used at the present time (Fig. 2A). Most important, they distinguished between the Givetian and Frasnian Pillara Limestone (platform facies) and the Famennian Nullara Limestone (back-reef subfacies) and the Famennian Windjana Limestone (reef-massive subfacies). Playford (1980, 1981) set the groundwork for the present "second" generation of maps and studies by updating the concepts of Playford and Lowry (1966). In addition, Playford (1976, 1981) updated maps of individual areas, including the Windjana Gorge area in the Napier Range. Kerans (1985) discussed updated interpretations of several localities within the Napier Range. In areas adjacent to the Napier Range, Hurley (1986) in the Oscar Range (Fig. 3) and Wallace (1987) in Geikie Gorge (location not shown) greatly refined Playford and Lowry's maps by using modern concepts.

This paper presents "second-generation" maps of the southern Napier Range that emphasize the Pillara platforms and their margins (Ward, 1996; Ward and Meyers, 1987a, b; Ward, in prep). Griffin et al. (1993) incorporated the Devonian mapping from Ward (1996) and prior studies into their regional geology map. Subsequently, George (George et al., 1994, 1995, 1997; George and Powell, 1997) mapped in detail the marginal-slope strata of the Napier Formation in a part of the southern Napier Range.

Mapping Methodology

The Devonian outcrops from Yammera Gap to Big Spring in the southern half of the Napier Range (Fig. 3) were mapped on color aerial photographs at 1:20,000 (Ward, 1996). Mapping was based on the concepts of Playford (1980) and focused on the Frasnian platform strata of the Pillara Limestone and the adjacent Frasnian marginal-slope strata of the Napier Formation (Figs. 2B, 4). Specific attention was paid to stratigraphic relationships of the Pillara Limestone platform margin. The mappable units are based on depositional facies, original and modified dips of beds, occurrences of macrofossils, occurrences of marine cements, and relative stratigraphic positions (Table 1). "Platform" and "marginal-slope" facies and "reefal-slope," "reef-flat," and "back-reef" subfacies are mapping terms of Playford and Cockbain (1976) and Playford (1980). The general distribution of these units and facies is schematically shown in Figures 2B and 4.

Localities of contiguous exposures of platform strata are informally named as "platforms." These include Big Spring–Dingo Gap, Tunnel Creek, South Behn, North Behn, Horizontal Wandjina Man, Wandjina Man, Windjana, South Burralumma, Burralumma, and Yammera platforms (Fig. 3). The Behn conglomerate fan exposed at Mount Behn separated the South and North Behn platforms during deposition.

The reliability of mapped features varies according to the

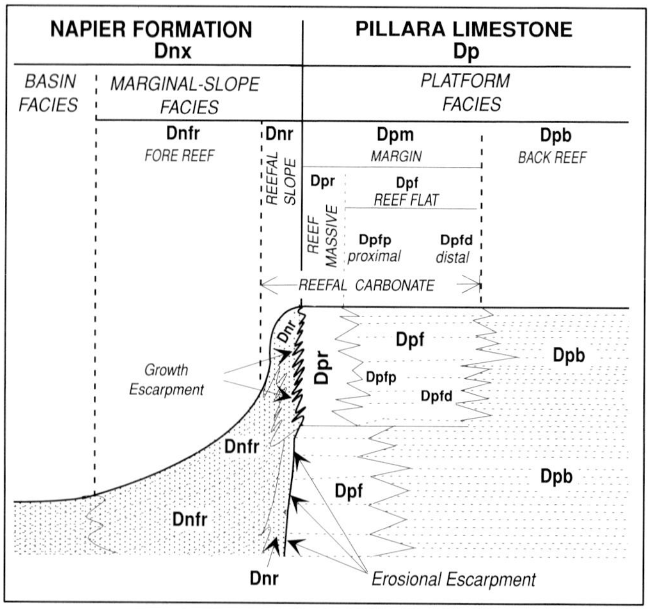

FIG. 4.—General distribution of facies and subfacies within the Frasnian Pillara Limestone and the Frasnian part of the Napier Formation. Map unit symbols are explained in Table 1.

TABLE 1.—STRATIGRAPHIC UNITS OF DEVONIAN REEF COMPLEXES MAPPED IN SOUTHERN HALF OF NAPIER RANGE

Mapping Unit	Symbol	Description
Pillara Limestone (Frasnian)		
Platform facies	Dp	flat-lying, well-bedded carbonate-platform and siliciclastic strata *common rock types:* stromatoporoid-bearing limestones and dolostones, fenestral peloidal and ooid limestones, poorly consolidated siliciclastic sandstones, siliciclastic pebble conglomerates
Platform-margin subfacies	Dpm	well-bedded to massive reef facies; key indicators include presence of *Renalcis* and fibrous marine cement (Kerans, 1985) *common rock types:* framestones, boundstones, grainstones
Reef-massive strata	Dpr	massive reef-margin rock (>2 m thick); uncommon in field area; depositional dip width <10 m; common to abundant neptunian dikes *common rock types: Renalcis-Stachyodes*-form boundstones, massive-stromatoporoid framestones, fenestral-stromatolite boundstones
Reef-flat strata (note: mapping term; not necessarily equivalent to reef flats of modern reefs [Kerans 1985])	Dpf	reef-margin rock with originally flat-lying, well-developed beds (<2 m thick); formed along entire extent of all platforms; depositional dip width <10 to 100 m; distinguished from back reef by its fibrous cements and *Renalcis;* it also tends to have thicker bedding and fewer siliciclastic beds than back-reef strata; locally siliciclastic-rich layers define bedding in what otherwise would be massive reef limestones; neptunian dikes are common to abundant near the margin and decrease in abundance toward the back reef *common rock types: Renalcis-Stachyodes*-form boundstones, massive stromatoporoid rudstones to framestones, fenestral grainstones, siliciclastic-rich grainstones and oncolitic siliciclastic and carbonate granular and pebble conglomerates
Proximal reef-flat strata	Dpfp	reef flat adjacent to platform margin or gradational with reef-massive strata; postdepositional dips are toward back reef because of preferential pressure solution of distal reef-flat and back-reef strata (Playford, 1980; Hurley, 1986); distinguished by abundance of boundstone and marine cement; distinguished from reef-massive subfacies by its well-developed bedding
Distal reef-flat strata	Dpfd	reef flat adjacent to back-reef; gradational with proximal-reef-flat and back-reef strata; postdepositional dips parallel that of back reef; contains only about 1–10% fibrous marine calcite
Back-reef subfacies	Dpb	well-bedded platform interior strata (<2 m, typically <1 m thick) with horizontal depositional dips composed of nonreef facies; repetitive shoaling-upward vertical facies successions common; laterally persistent beds that can be traced into reef flat; neptunian fractures uncommon *common rock types*: stromatoporoid beds (<1 m thick) that range laterally from rudstones to framestones and typically have skeletal-peloidal packstone matrices; oncolite rudstones, peloidal-skeletal packstones and grainstones, peloidal oolitic fenestral grainstones, packstones, and calcareous fenestral sandstones, poorly consolidated siliciclastic sandstones, siliciclastic pebble conglomerates
Nullara Limestone (Famennian)		
Back-reef facies	Du	flat-lying platform-interior strata; crops out in northern Napier Range and northern Oscar Range (Playford, 1981); no stromatoporoids *common rock types*: fenestral intraclast-oolite grainstones and rudstones, and fenestral oolite-peloid packstones and rudstones (Hurley, 1986); includes exposure and vadose features, pisoids, and tepees
Windjana Limestone (Famennian)		
Reef-margin subfacies	Dw	Famennian counterpart to Frasnian reef-massive strata and some of Frasnian reefal-slope strata; predominantly prograding with upright scarp margins; marine calcite cements are common to abundant in shelter cavities, neptunian fractures, and secondary stromatactis cavities; no stromatoporoids *common rock types:* massive *Renalcis*-cyanobacterial boundstones
Napier Formation (Frasnian and Famennian)		
Marginal-slope facies (Frasnian and Famennian)	Dn	predominantly basinward-dipping slope strata basinward of Pillara Limestone; main subdivisions (Dnx, Dny, Dnz) based on age and/or relative position, macrofossils, and lithology *common rock stypes:* in Napier Range, includes reefal-slope limestones and dolostones, fore-reef breccias, allochthonous reef blocks, siliciclastic sandstones and conglomerates, deep-water nonfenestral stromatolitic limestones, and dolostones

TABLE 1.—CONTINUED

Mapping Unit	Symbol	Description
Napier Formation (Frasnian and Famennian), continued		
Marginal-slope facies (Frasnian) (note: much of Frasnian marginal-slope facies has not been differentiated into subfacies)	Dnx	lowermost unit of Napier Formation; distinguished in the field from Famennian slope facies by presence of stromatoporoids, allochthonous Pillara Limestone clasts, and Frasnian megalodont bivavles; upper-slope strata interfinger with Frasnian Pillara platform strata or encrust or overlie platform-margin escarpment; lowermost slope is not exposed but presumably interfingers basinward with fine-grained basin sediments; rich in siliciclastic sandstones and conglomerates mixed with carbonate debris; neptunian fractures locally common *common rock types*: skeletal, peloidal, and/or oolitic packstones and grainstones, carbonate siltstones, cyanobacterial limestones, and siliciclastic sandstones to gravel conglomerates, megabreccias, and allochthonous reef blocks
Reefal-slope subfacies (Frasnian)	Dnr	steeply dipping reefal-slope facies; typically deeper-water reef rocks than those of shallow-water platform margin (Fig. 4); map width <10 m; common to abundant stromatoporoids, *Renalcis*, sponges, and marine cements and internal sediments in primary cavities; *Stachyodes*-form, crinoids, brachiopods, ammonoids, and nautiloids locally common; 20° to near-vertically dipping reefal-slope strata are accreted-margin escarpments; these encrustations are typically dominated by calcareous stromatolitic cyanobacterial limestones and dolostones or by platy stromatoporoid and *Renalcis* limestones or dolostones *common rock types:* framestones and boundstones of *Renalcis;* stromatolites, cyanobacteria, laminar stromatoporoids, and *Stachyodes-forms*
Upper-fore-reef subfacies (Frasnian)	Dnfr	slope facies downslope from and/or interbedded with reefal-slope strata; dominated by allochthonous debris breccias derived from platform margin and slope; common basinward dips 12° to 35°; back-reef clasts are rare to nonexistent except for Lower Napier Conglomerate; common marine internal sediment and marine fibrous calcite cement *common rock types:* skeletal packstones and grainstones, megabreccias (clasts to several meters), adnet and scheck breccias (Kerans, 1984), micritic limestones, cyanobacterial encrustations
Lower Napier Conglomerate	Dnlnc	matrix-poor "limestone clast conglomerate" in lower Napier Formation (G. Beere, pers. commun., 1985); consists of siliciclastic and limestone clasts derived from platform and includes material from back reef; locally dolomitized; clasts include back-reef stromatoporoid limestones, oncolitic rudstones, and calcareous siliciclastic sandstones and platform-margin *Renalcis*-stromatoporoid boundstones; clasts range up to several meters across; matrix is red carbonate and siliciclastic silt or, less commonly, siliciclastic sand
Nonfenestral stromatolite (Frasnian and Famennian)	Dnrs	micritic stromatolitic and cyanobacterial limestones; common throughout slope; commonly encrusts tops of allochthonous blocks and debris flows
	sss	a nonfenestral-stromatolite horizon or series of horizons within lower Famennian slope section and locally in uppermost Frasnian strata; locally this horizon forms elongate antiformal ridges extending downslope (Kerans, 1985)
Marginal-slope facies (Famennian)	Dny	steeply dipping well-bedded fore-reef strata below, adjacent to, and interfingering with prograding Windjana Limestone; similar to the Frasnian, except that stromatoporoids are not present; steeply dipping strata underlie reef-massive facies (Windjana Limestone) distinguished in the field by their stratigraphic position and absence of stromatoporoids and Frasnian megalodont fossils; commonly true Frasnian/Famennian boundary is difficult to place without conodont data (e.g., George et al., 1997); subdivisions are considered to be predominantly Frasnian (Dnx) and predominantly Famennian (Dny); in most places the contact between them indicated on the maps is approximate; it is placed at uppermost occurrence of stromatoporoids (probably still in Frasnian strata) or, where macrofossils are scarce, at base of nonfenestral-stromatolite (usually in Famennian strata) *common rock types:* skeletal, oolitic, and peloidal grainstones and packstones
Crinoidal limestone	Dnz	uppermost unit of Napier Formation; locally may be equivalent to basal Fairfield Group (P. E. Playford, pers. commun.) *common rock types:* cross-bedded crinoidal-skeletal packstones and grainstones
Behn Conglomerate		
Behn Conglomerate	Db	siliciclastic bimodal conglomerate cropping out at Mount Behn (Playford and Lowry, 1966; Botten, 1984); well-rounded, cobble- to boulder-sized clasts of Precambrian rocks (quartzites, granites, metavolcanic rocks) floating in a poorly cemented, coarse-grained arkosic sandstone matrix; estimated thickness of 500 m (G. Beere, pers. commun., 1986)
Behn-like conglomerate	Db	bimodal Precambrian-clast conglomerate occurring between Mount Behn and Windjana Gorge

amount of time spent on the outcrop walking out features versus depending on aerial-photograph interpretations. The platform interior strata in the South Behn, North Behn, Horizontal Wandjina Man, and Windjana platforms were walked out extensively. Pillara strata to the south and north of these were mapped on traverses. Stratigraphic sections were measured through Pillara Limestone in the Dingo Gap, South Behn, North Behn, and Windjana platforms. Field studies were complemented with hand-sample and thin-section petrography.

Traces of bedding surfaces in both the Pillara and Napier shown on the map are generally considered chronostratigraphic horizons that help define depositional strike and dip and aid in correlation. For the Pillara Limestone, most of the bedding traces shown on the maps trace stromatoporoid limestones that define the bases of repetitive, dominantly meter-scale shallowing-upward, vertical facies successions (Fig. 5) (cycles of Read, 1973a, b; Kerans, 1985). Locally these stromatoporoid limestones are overlain by oncolitic rudstones. The upper parts of the repetitive vertical facies successions include significant amounts of siliciclastic sandstones and siltstones and fenestral packstones, grainstones, and sandstones. Desiccation cracks occur within the upper beds of some repetitive vertical facies successions. Repetitive "vertical facies successions" are defined with individual measured sections and thus may be only local. Many, and possibly, most of these repetitive vertical facies successions, however, can be traced throughout the back reef and into the reef flat (Kerans, 1985; Ward, 1996). A vertical facies succession that can be traced or confidently correlated across lateral facies is defined as a high-frequency cycle (Kerans and Tinker, 1997). The generally shallowing-upward succession of facies that occurs throughout the platform in different lateral facies indicates that the high-frequency cycle records relative sea-level fluctuations affecting an area on the scale of the correlations. The mechanism controlling sea-level fluctuations is most likely external to the Lennard shelf (Ward, 1996). Several of these high-frequency cycles have been walked out and traced on photographs to establish the stratigraphic relationships throughout the mapped platforms. Most of the potential high-frequency cycles, however, were not walked out systematically or mapped in detail.

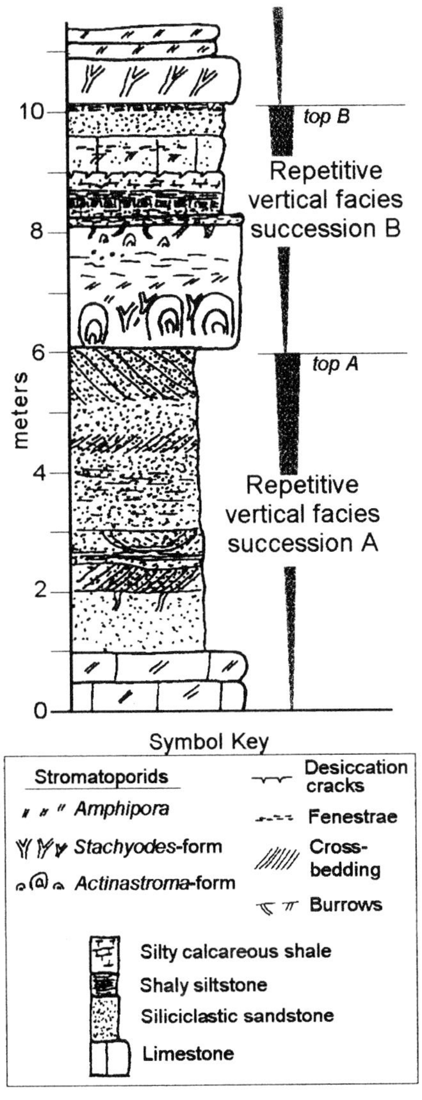

FIG. 5.—Example of repetitive vertical facies successions (A and B) in part of a measured section through back-reef strata of the Pillara Limestone exposed in the Windjana platform. The upper part of RSa is composed of siliciclastic sandstones, whereas the top of RSb is composed of desiccation-cracked silty mudstones and sandstones. Such successions that can be traced throughout the platform are considered high-frequency cycles. Cycles generally cannot be traced through reef-massive and upper-slope facies. In these facies, cycles are recorded in primary cavities by vertical facies successions of siliciclastic and carbonate internal sediments and marine cements (Ward, 1996).

Although local, detailed correlation within platforms is possible, precise bed-to-bed correlation of strata throughout the Napier Range is problematic. Bed-to-bed correlation between separated platform exposures is difficult because of lateral facies changes and lack of marker beds. Detailed biostratigraphy is lacking; conodonts have yet to be found within the Pillara Limestone, and ammonites are rare. Thus, correlation between platforms is dependent on vertical successions of Pillara Limestone strata, backstepping relationships, and successions of marginal-slope strata.

Even though the Napier Formation crops out almost continuously through the Napier Range, precise correlation of strata within the Napier Formation is complicated by the discontinuity of beds. Precise correlation is especially difficult for the upper-slope sediments, including the reefal-slope subfacies. Correlation is even further complicated because the depositional depth and thus facies (e.g., upper slope vs. lower slope) of the slope deposits varied along strike.

Descriptions of key stratigraphic relationships for individual platform localities are described in the following sections.

STRATIGRAPHIC RELATIONSHIPS

South Behn Platform

All of the key Frasnian stratigraphic units and relationships are exposed at the South Behn platform, and these are the basis for interpreting the Frasnian stratigraphy of the Napier Range. Two major subdivisions of the Pillara Limestone occur at the South Behn platform: a lower unit of sandstone-dominated back-reef strata (platform A) and an upper unit of stromatoporoid-limestone–dominated back-reef to reef-margin strata (platform B; Figs. 6, 7). The reef-flat facies of platform B overlies the back-reef facies of platform A and thus forms a backstepping geometry. Overlying the upper unit are fore-reef and reefal-slope strata (Fig. 7). These indicate that a reef margin occurred landward (eastward) and thus prove the existence of a second backstepped margin (platform C), now removed by erosion. The preserved morphology of the platform B margin is that which existed during the submergence that led to the second backstepping episode (Ward and Meyers, 1987a; Ward, 1996). The 100-m-thick platform B has a nearly vertical escarpment margin that truncates reef-massive to distal reef-flat strata. The upper 50 m of the vertical escarpment is encrusted by a *Frutexites* cyanobacterial limestone layer (~10–20 cm thick) that has internal laminae paralleling the escarpment. This is an approximate time line indicating that the entire escarpment was exposed as a submarine cliff (Ward, 1996, and in prep.). The *Frutexites* layer and the rest of the escarpment is encrusted with a nearly vertical nonfenestral-stromatolite layer as much as 2 m wide that has horizontal internal laminations perpendicular to the escarpment. This stromatolite also encrusts nonvertical parts of the erosional escarpment. Figure 8 shows the relative timing of depositional and erosional events. The final escarpment geometry was formed after (1) deposition of platform B, (2) fracturing of the margin, and (3) filling of neptunian fractures by marine encrustations, sediment, and fossils.

Locally, at or near the top of platform B there are two types of features that potentially could be have been caused by exposure and thus warrant further investigation: *in situ* breccias and dolomite-sediment–filled fractures. At two localities, an *in situ* breccia is preserved on the upper surface of platform B. This breccia formed after or at the end of platform B deposition and before the deposition of fore-reef sediments equivalent to platform C. One of the breccias overlies an irregular erosional surface with less than 2 cm relief. This surface appears similar to a microkarstic surface; however, petrographically there is no evidence of dissolution- or exposure-related diagenesis (Ward, 1996). Limited outcrop exposures of these 3- to 5-cm-thick breccias prevent tighter timing constraints. Intersecting fractures filled with dolomitized sediment occur in the uppermost several meters of platform B strata. Their crosscutting relationships provide poor timing constraints.

The Lower Napier Conglomerate is an important stratigraphic horizon marking the backstepping of platform C. It is the lowest part of the Napier Formation exposed throughout most of the southern Napier Range. This unit is recognized as far north in the Napier Range as Hawkstone Creek near the Van Emmerick Conglomerate (V. A. Pedone and W. J. Meyers, oral commun., 1986). In the field area it extends at least 10 km south of Dingo Gap. It is varied laterally and vertically and represents deposition over a period of time, involving several depositional events. The relationship of the lower boundary of the Lower Napier Conglomerate is variable (Figs. 7B, 8). It ranges from being conformable to unconformable with Precambrian-clast conglomerates (including the Behn Conglomerate) and is unconformable with the Pillara platform A and platform B

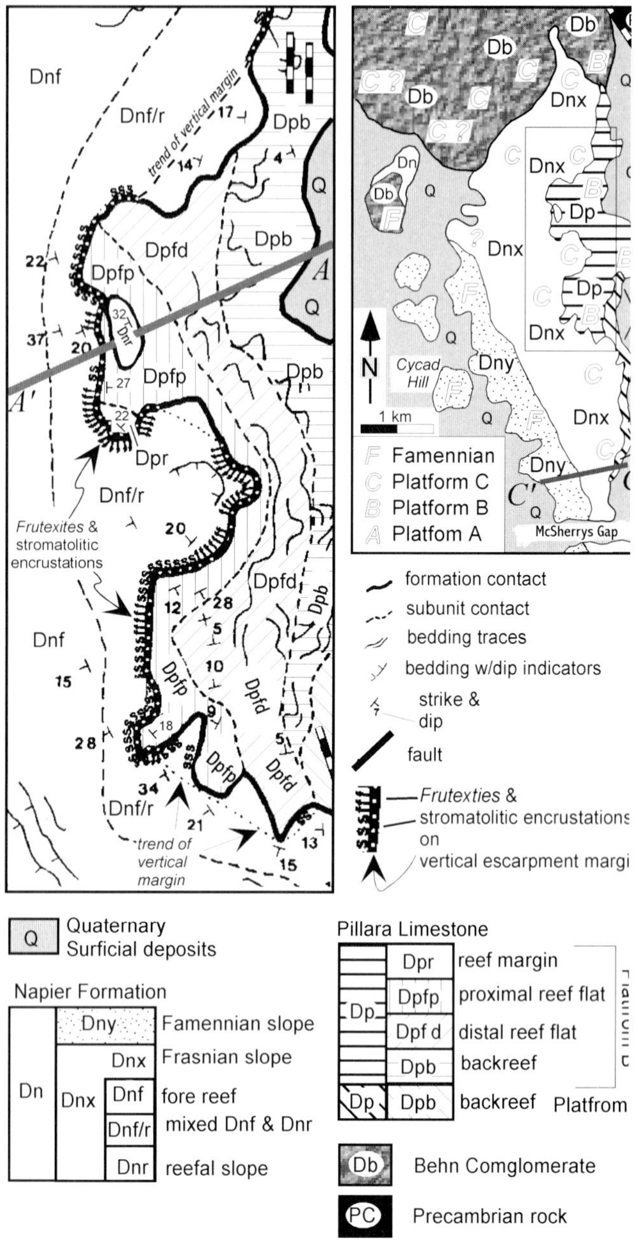

FIG. 6.—Geologic map of the South Behn platform showing the Frasnian platform chronostratigraphic units (A, B, and C). The map enlargement shows the details along the margin of platform B of South Behn. The exposed Frasnian Napier Formation (Dn symbols) is interpreted to be time equivalent to platform C. A–A′ is the line of the cross section in Figure 7A. C–C′ is the line of the schematic section in Figure 7C.

strata. The lowermost Lower Napier Conglomerate is typically siliciclastic rich and exclusively composed of clasts from the interior of platforms A and B. These clasts do not include clasts of reefal rock. Reef-margin clasts appear in the upper part of the Lower Napier Conglomerate. The uppermost Lower Napier Conglomerate sediments grade upward into the fore-reef equivalent to platform C that contains large reef blocks and lacks clasts derived from the interior of the platform (Figs. 7B–C). The upper part of the Lower Napier Conglomerate unconformably overlies the stromatolite layers encrusting the platform

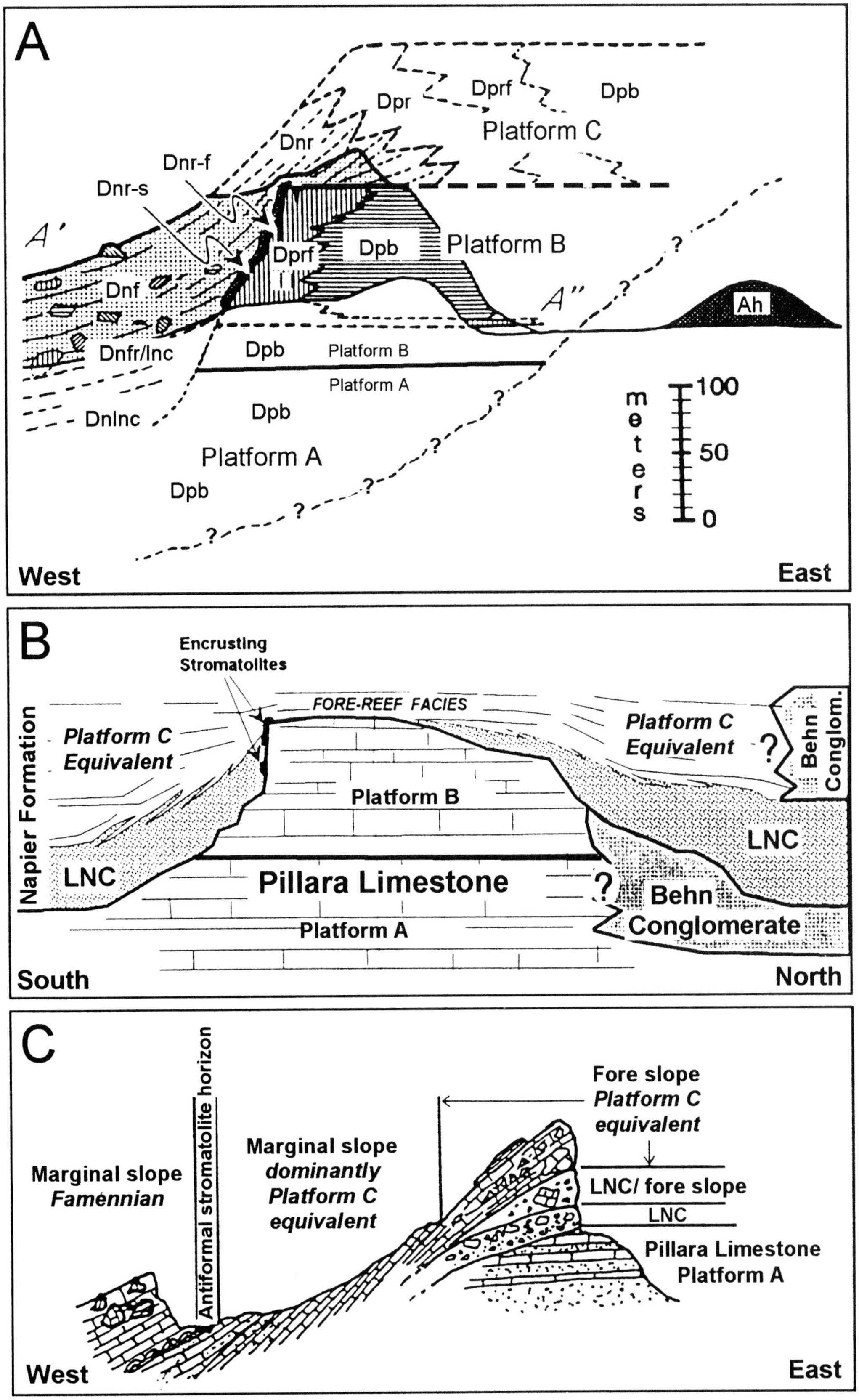

FIG. 7.—(A) Generalized interpretive east-west cross section through the South Behn platform. Symbol-filled parts indicate exposed strata. The upper bounding line of the symbol-filled area is the present-day ground surface along section line A–A′ in Figure 6. Horizontal dimensions are compressed and not to scale. (B) Interpretive relationships between the Lower Napier Conglomerate (LNC) and other Frasnian strata shown along an approximately north-south cross section through the South Behn platform. Horizontal dimensions are compressed and are not to scale. (C) Schematic east-west cross section, drawn through the southernmost end of the South Behn platform, shows the Lower Napier Conglomerate overlying the oldest exposed part of platform A. Map unit symbols are explained in Table 1.

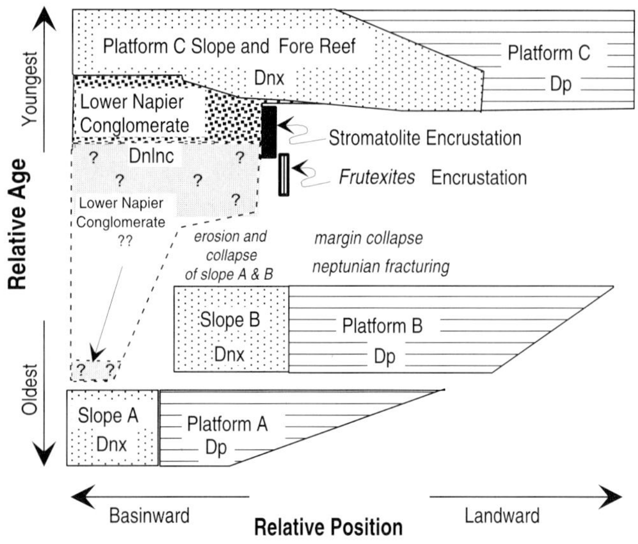

FIG. 8.—Timing of depositional events at the South Behn platform. The time versus space diagram above shows the interpreted relative timing of the deposition of Frasnian sediments in the South Behn area and their relative position along depositional dip. The timing of deposition of the basal strata of the Lower Napier Conglomerate is poorly constrained. Platform C is not exposed. Map unit symbols are explained in Table 1.

B escarpment. This relationship is consistent with grading of the upper conglomerate into platform C slope deposits and suggests that the youngest Lower Napier Conglomerate was deposited early in the development of platform C.

Stratigraphic relationships of the oldest strata of the Lower Napier exposed at South Behn are problematic. The erosional escarpment surface on platform B extends down to the unconformity between the Lower Napier Conglomerate and platform A. Stratigraphic relationships, therefore, allow for multiple scenarios of erosion, platform encrustation by the stromatolite layers, and deposition of the Lower Napier Conglomerate. Possible scenarios include the following:

1A. The platform B escarpment and the unconformity overlying platform A formed together, after which the *Frutexites* and stromatolite layers encrusted platform B. These events were followed by deposition of the Lower Napier Conglomerate.

1B. The escarpment was formed and encrusted before platform A was eroded and the Lower Napier Conglomerate was deposited.

2A. The escarpment and unconformity formed together, and the lowermost part of the Lower Napier Conglomerate was deposited before platform B was encrusted whereas the uppermost part of the Lower Napier Conglomerate was deposited after platform B encrustation.

2B. The escarpment and unconformity formed together, and deposition of the lowermost Lower Napier Conglomerate started before platform B encrustation, continued through encrustation, and ended with the progradation of platform C fore-reef sediments over the platform B margin.

The latter seems most reasonable, as explained in the following, and is the basis for the reconstruction shown in Figure 9. The significant erosion of platform A is along an escarpment (Fig. 9), not an incised channel cutting through a platform. The resultant unconformity extends to the toe of the platform B escarpment and represents spalling off of large blocks of reef rock that were transported beyond the base of the escarpment and thus were not deposited as part of the Lower Napier Conglomerate.

The relationships of the Behn Conglomerate indicate that although its deposition may have been punctuated, it was deposited during deposition of the Frasnian and Famennian carbonate platform and slope facies. The deposition of the Behn Conglomerate was interrupted by that of the Lower Napier Conglomerate. The oldest exposed strata of the Behn Conglomerate appear to interfinger with the Pillara Limestone (platform B) at the South Behn platform and clearly interfinger with the Pillara Limestone at the North Behn platform. Most of the exposures of the Behn Conglomerate, however, are younger than Pillara platform B. The Behn Conglomerate interfingers with the platform C–equivalent Napier Formation *and* the Famennian Napier Formation. The vertical contacts with the Napier Formation are erosional where the Behn Conglomerate overlies the Napier Formation. The contacts where the Napier Formation overlies the Behn Conglomerate are vertically gradational on a scale of decimeters.

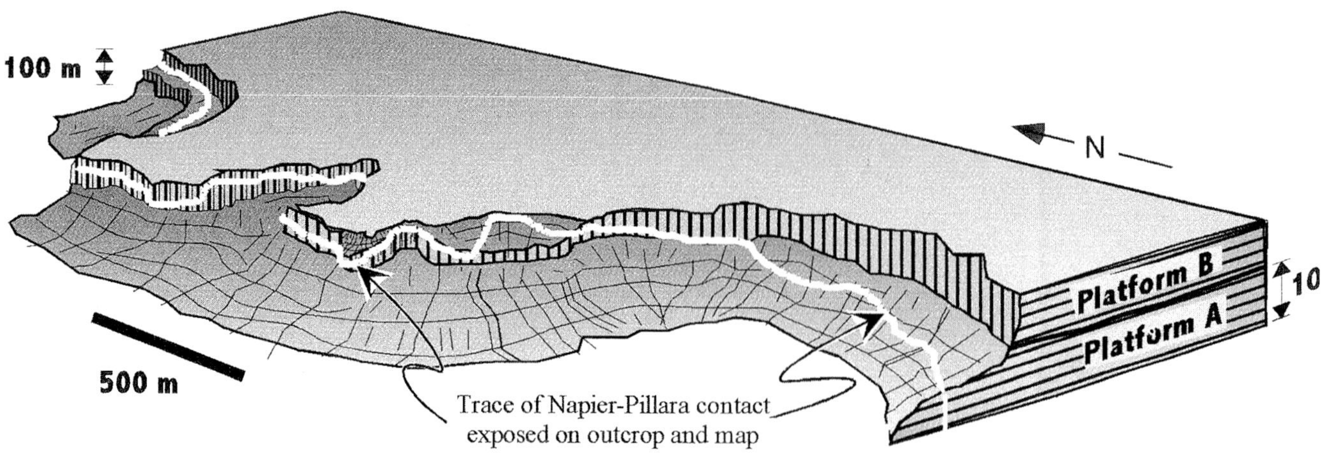

FIG. 9.—Reconstruction of the South Behn platform after platform B margin collapse and before deposition of the Lower Napier Conglomerate. Line indicates position of the Pillara-Napier formational contact shown in Figure 6.

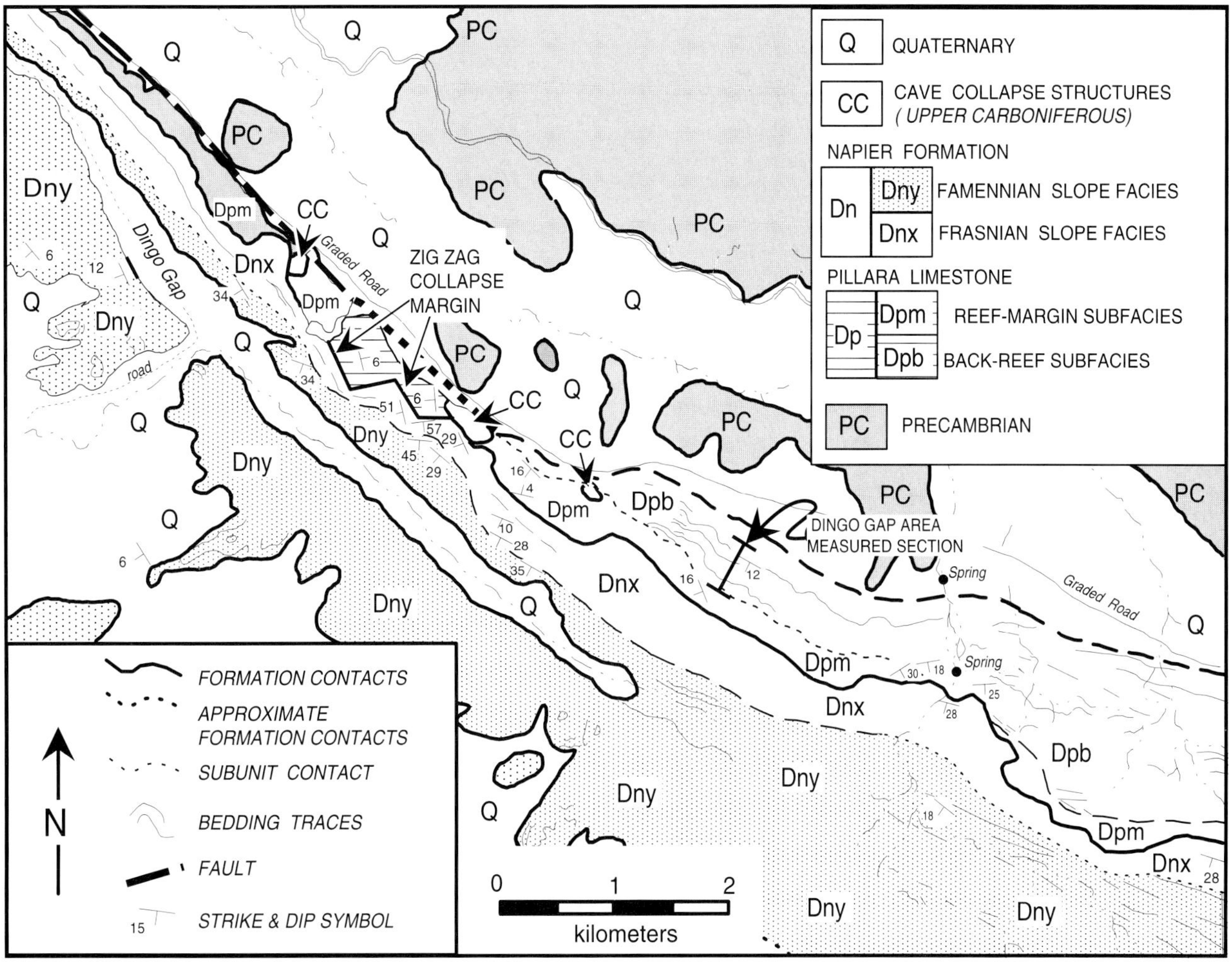

FIG. 10.—Geologic map of the Dingo Gap area.

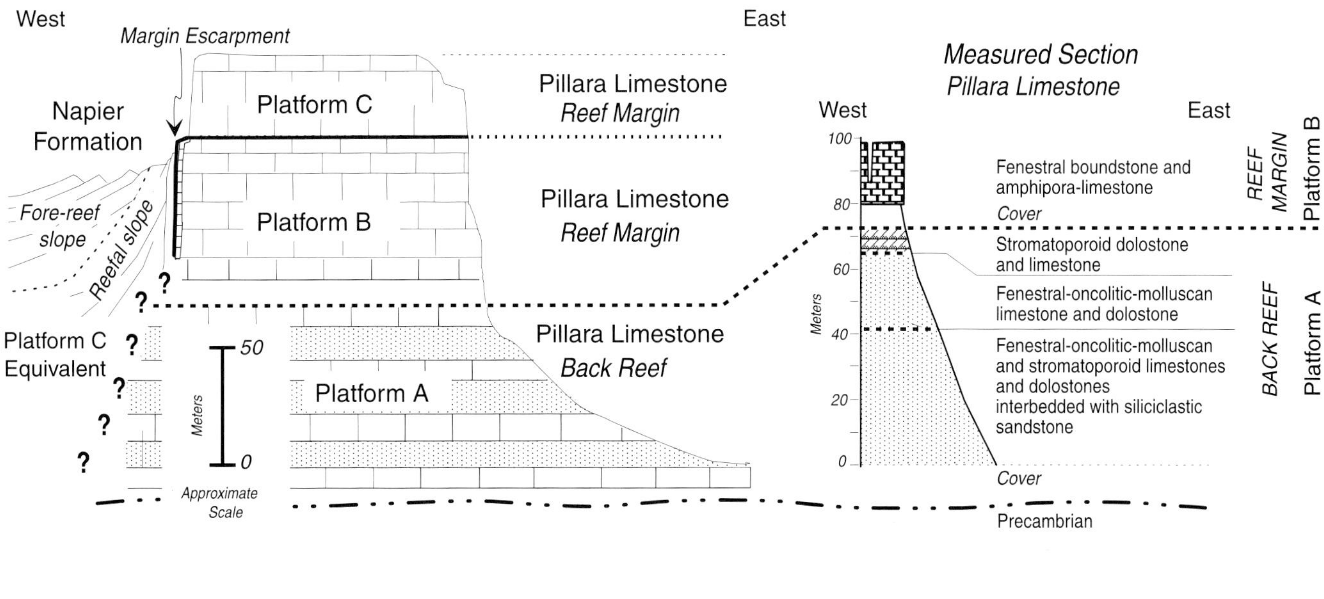

FIG. 11.—(Left) A schematic east-west cross section through the Frasnian platform for the entire Dingo Gap area. (Right) The generalized section measured along the line in the map in Figure 10.

Predominantly Frasnian upper-slope to lower-slope strata overlie the oldest fore-reef strata that are equivalent to platform C. These Frasnian slope strata extend upward through the section to the deep-water nonfenestral-stromatolite unit above the Frasnian-Famennian boundary. The stromatolite unit occurs from the Dingo Gap area to Carpenters Gap and from the North Burralumma area to north of Yammera Gap. It also occurs at the northernmost outcrops in the Van Emmerick Range (at the north end of the Napier Range; Kerans, 1985). A similar horizon occurs in the Oscar Range (Hurley, 1986) and encrusts the platform at Geikie Gorge (Wallace, 1987) and also occurs in the McWhae Ridge area (Playford, 1981; Kerans, 1985). In the Oscar Range, the stromatolitic horizon expands into the Frasnian and Famennian Elimberrie stromatolite bioherms (Playford, 1981; Hurley, 1986). In the northern Napier Range, this horizon extends into stromatolitic bioherms, whose exposures in map view are U or V shaped and whose open ends point landward (Kerans, 1985).

Dingo Gap

The Pillara platform margin exposed in the Dingo Gap area (Fig. 10) is the best evidence for a structural control on neptunian fractures and margin-collapse scars (Ward and Meyers, 1987b; Ward, 1996, and in prep.). At the Dingo Gap measured section (Figs. 10 and 11), two main subfacies occur in the Pillara Limestone: an older back-reef subfacies (DG-1) and a younger reef-margin subfacies (DG-2). The vertical succession of the reef-flat strata overlying back-reef strata indicates a transgression and suggests a backstepping relationship. Backstepping of

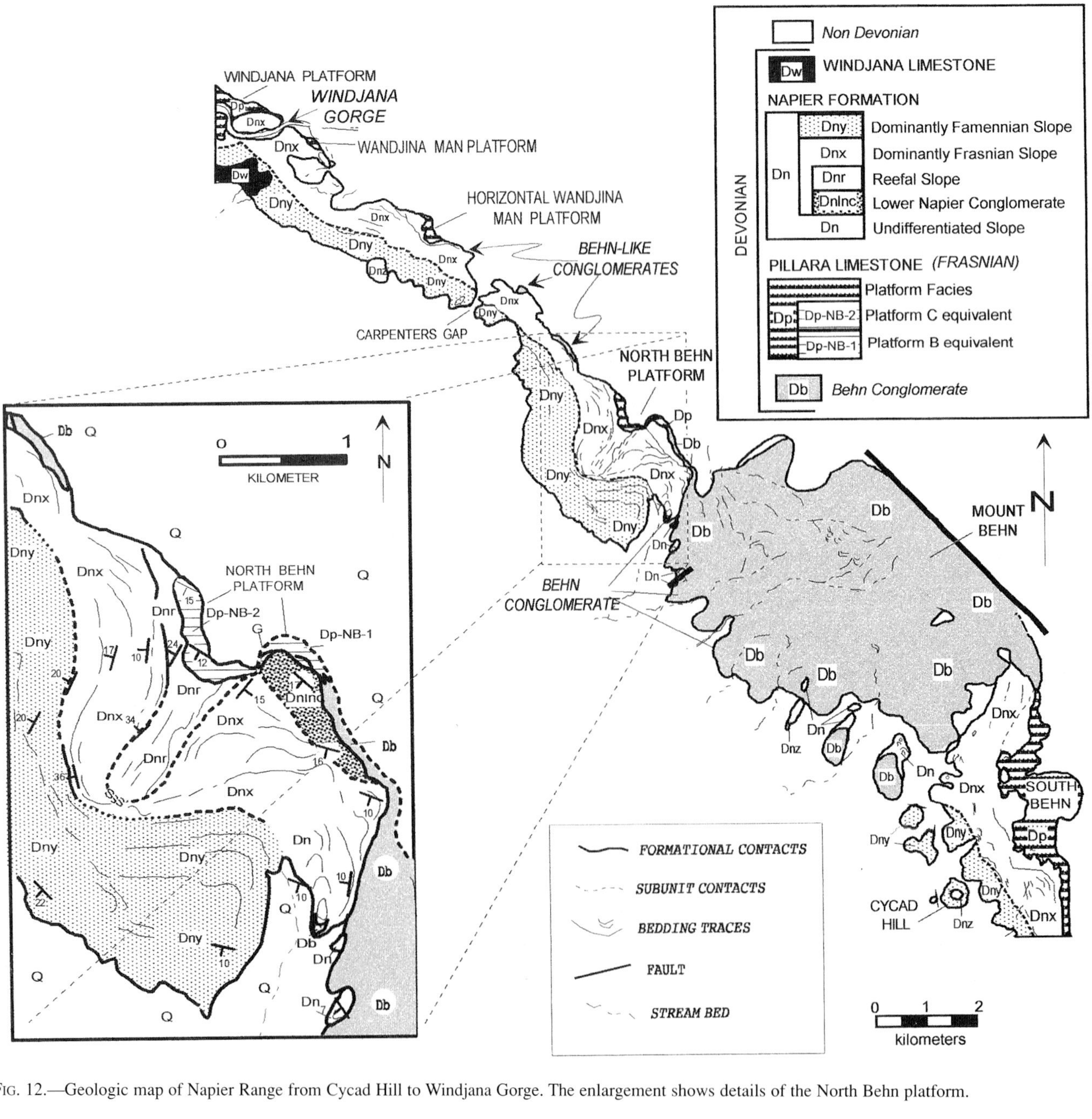

FIG. 12.—Geologic map of Napier Range from Cycad Hill to Windjana Gorge. The enlargement shows details of the North Behn platform.

the margin cannot be confirmed in this area because the margin of platform DG-1 is not exposed. Several kilometers to the south, a backstepping relationship occurs where the Lower Napier Conglomerate and other marginal-slope strata unconformably overlie the Pillara back-reef strata. North of the measured section at Dingo Gap is a small backstep in the platform where reefal-slope rocks overlie the platform margin. From that point north to the Tunnel Creek Cave area (Fig. 3), the platform margin is a rollover margin with reef-massive grading and interfingering into reefal-slope strata. Thus, in this region, there are at least three platform units, which are labeled DG-1, DG-2, and DG-3 (Fig. 11). These units are interpreted to be equivalent to platforms A, B, and C, respectively.

Locally, the escarpment margin of platform DG-2 is encrusted with a vertical calcareous cyanobacterial layer and, at one location, a nonfenestral-stromatolite layer. To the south of this area, the Lower Napier Conglomerate overlies platform DG-1 strata. In the Dingo Gap area, the marginal-slope strata postdate platform DG-2. Basinward (west) of the margin are mainly strata that are interpreted to be upper-slope to middle-slope strata; they extend up to the nonfenestral-stromatolite unit near the Frasnian-Famennian boundary. Below and above this unit are several neptunian sills with marine cements encasing three generations of thin *Frutexites* layers (Ward, 1996; Ward et al., 1994; Kwong, 1995). The Napier Formation has been subsequently mapped in detail by George (George et al., 1995, 1997; George and Powell, 1997). She recognizes two sequences in the

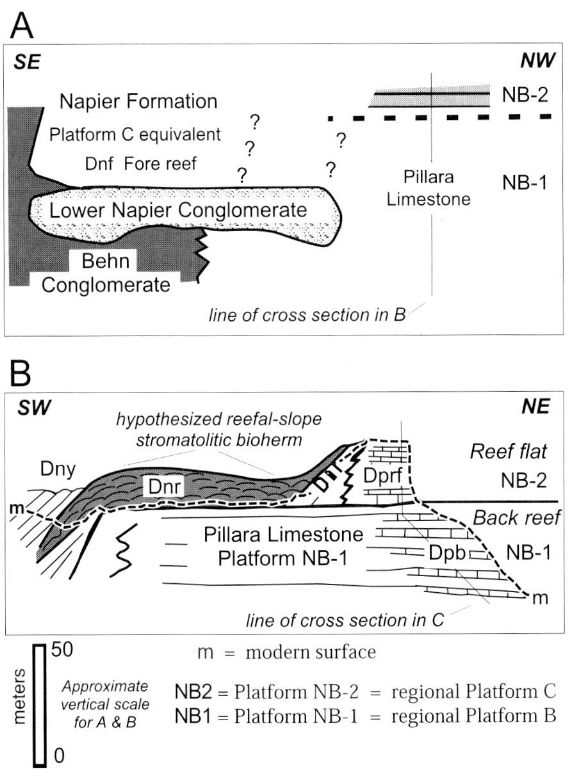

FIG. 13.—(A) A southeast-northwest schematic cross section through the North Behn platform. (B) A southwest-northeast schematic interpretive cross section through the North Behn platform and through the axis of a structure with a U-shaped plan view in the reefal-slope strata. Platforms NB-1 and NB-2 appear to correlate with the regional Pillara platforms B and C, respectively.

Frasnian slope that correspond to the lower and upper marginal-slope strata equivalent to platform C in this paper.

North Behn Platform

The main Frasnian stratigraphic units are exposed at the North Behn platform and suggest local backstepping of 1 km. There are two Pillara Limestone units at the North Behn platform, which are labeled NB-1 and NB-2 (Figs. 12 and 13). The lower unit, NB-1, is dominated by siliciclastic sandstones interlayered with thin back-reef stromatoporoid limestone beds. This unit appears to grade laterally south into the Behn Conglomerate. Overlying the back-reef strata of NB-1 are the reef-flat strata of unit NB-2. These include oncolitic limestone, pebble conglomerate, and stromatoporoid limestone. This upper unit grades basinward (southwestward) into reef-massive facies. The margin of platform NB-2 is an upright rollover margin, in which reef-massive grades into reefal-slope strata without a major escarpment. In this area the reef-massive facies is difficult to distinguish from reefal-slope facies.

On the southern side of the North Behn platform, the Lower Napier Conglomerate unconformably overlies both the lowermost strata of the Behn Conglomerate and the platform NB-1 (Fig. 13A). The relationship between platform NB-2 and the Lower Napier Conglomerate is uncertain.

In front (southwest) of the margin of platform NB-2 is a structure, U shaped in plan view, that consists of a broad expanse of shallow-dipping reefal-slope material, which includes cyanobacterial and sponge boundstones (Fig. 12). To the southeast, southwest, and northwest of this structure, the slope beds dramatically steepen. The southwestern nose of the structure is encrusted with a nearly vertical 10-cm-thick nonfenestral-stromatolite layer. This is overlain by a succession of steeply dipping reefal-slope and fore-reef beds, which, in turn, are overlain by the nonfenestral-stromatolite unit.

The reefal-slope facies of platform NB-2 overlying the back-reef facies of platform NB-1 indicates transgression and a backstepping geometry between the two platform units. The margin of platform NB-1 is not exposed; therefore, the backstepping geometry of the margin cannot be observed. The reefal-slope beds, however, hint at the geometry of platform NB-1 (Fig. 12). The dips of the reefal-slope strata making the U-shaped structure suggest that they overlie a bathymetric high. These reefal-slope beds are interpreted to be the central core of a "bioherm" that has overgrown platform NB-1 (Fig. 13B). This bioherm may be similar to the Elimberrie bioherms in the Oscar Range (Playford, 1981; Hurley, 1986) and the bioherms with U-shaped plan views in the northern Napier Range (Kerans, 1985). If the interpretation shown in Figure 13B is correct, it means that platform NB-1 protrudes almost 1 km basinward past the margin of platform NB-2 and, thus, there was a major backstep from platform NB-1 to platform NB-2 (Figs. 12, 13B).

Horizontal Wandjina Man Platform

The Horizontal Wandjina Man platform, which is the first

Pillara Limestone exposure north of the North Behn platform, shows the consistent relationships among the escarpment margin of platform B, the overlying lower Napier Conglomerate, and the slope strata equivalent to platform C (Figs. 12, 14). Between the North Behn platform and Carpenters Gap, the Lower Napier Conglomerate locally overlies a Behn-like conglomerate (Fig. 12).

The relatively small exposure of Pillara Limestone at the Horizontal Wandjina Man platform consists of only one Pillara unit (reef-flat facies). Stromatoporoid limestones and dolostones (Fig. 14) and *Sphaerocodium* boundstones are interlayered with siliciclastic-rich sandstones and gravels. This platform has a vertical escarpment margin that, locally, is encrusted with a thin (10-cm-thick) nonfenestral-stromatolite layer (Fig. 14). Basinward of this escarpment is the Lower Napier Conglomerate, which also unconformably overlies the platform strata and grades stratigraphically upward into fore-reef strata (Fig. 14). The fore-reef strata indicate that a backstepped platform margin occurred landward (northeast) of this margin. These relationships are consistent with an interpretation that the strata in the Horizontal Wandjina Man platform is equivalent to platform B and the fore-reef strata is equivalent to platform C.

Wandjina Man Platform to the Eastern Entrance of Windjana Gorge

The outcrops from the Wandjina Man platform to the eastern entrance of Windjana Gorge reveal important relationships of the platform B escarpment margin and the overlying Napier strata (Figs. 15, 16). The small exposure of Pillara Limestone at the Wandjina Man platform has a rollover margin that grades into reefal-slope strata (Fig. 15). In the cliff exposure north of the Wandjina Man platform and south of Windjana Gorge is a stratal succession of slope facies that is younger than the Wandjina Man platform (Fig. 16A). The base of this succession consists of limestone blocks and sandstone-filled channels. Above this are Behn-like Precambrian-clast boulder conglomerate (granitic and metamorphic boulders) and limestone-clast conglomerate. Parts of this mixed conglomerate unit appear similar to the Lower Napier Conglomerate. Above this are scheck breccias overlain by a tongue of reefal-slope strata. The same sequence occurs to the south in a cliff exposure near Little Spring. Southgate et al. (1993) interpreted the Behn-like conglomerate at this locality as a lowstand deposit.

A similar succession also occurs at the base of the platform-margin escarpment exposed on the northern side of the eastern entrance to Windjana Gorge (Fig. 16B), with the exception that below this succession is a spectacular breccia unit that consists of large clasts of Pillara Limestone, lacks a fine-grained matrix, and is cemented by white calcite cement (both marine and nonmarine; Ward, 1996). Most of the clasts could have been derived from the platform margin. This is interpreted to be a talus deposit. The platform-margin escarpment at this locality is encrusted with cyanobacterial limestone layers (Fig. 16B).

On the eastern side of the range, just north of Windjana Gorge's eastern entrance, there are siliciclastic sandstone channels as much as 5 m across that occur in the Napier Formation ("5" in Fig. 16B). The sandstone consists predominantly of quartz grains and/or schist fragments. The landward projection of these channels indicates that they could have cut through the Pillara Limestone margin. The details at this outcrop, unfortunately, are obscured by cover.

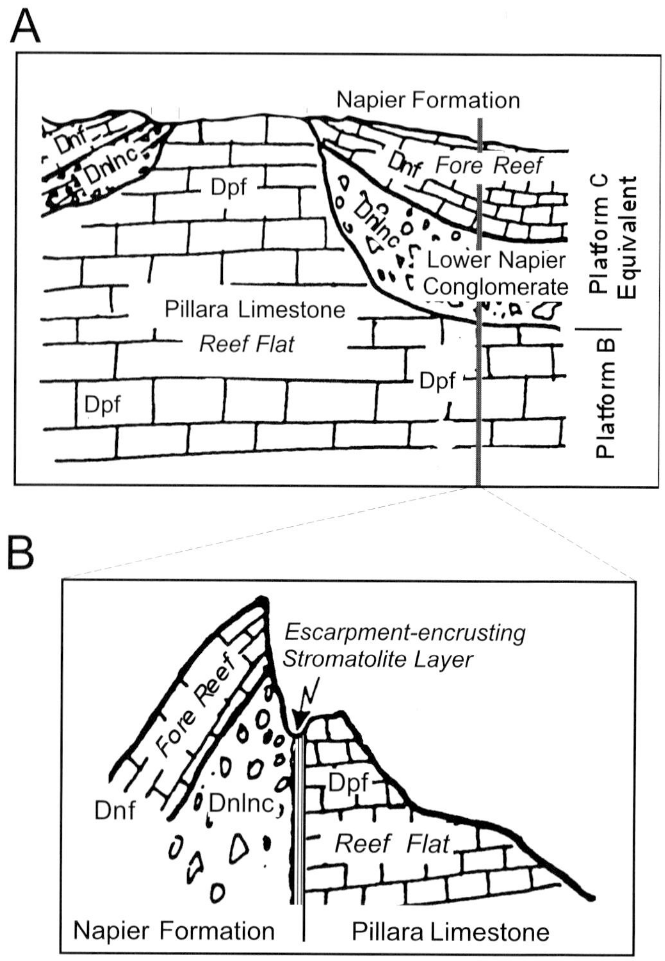

FIG. 14.—Schematic cross sections through Horizontal Wandjina Man platform. (A) This north-south cross section is essentially the view looking west toward the outcrop. The vertical dashed line indicates the line of section for B. (B) Interpretive east-west cross section through the north edge of the Horizontal Wandjina platform along the section indicated by the dashed line in Figure 14A. The vertical encrusting stromatolite layer is as thick as 10 cm.

Windjana and South Burralumma Platforms

The Pillara Limestone exposed in and around Windjana Gorge (the Windjana platform, Fig. 15) not only forms the basis for early stratigraphic concepts of the Devonian reef complexes (e.g., Playford and Lowry, 1966; Playford, 1980), but it also is one of a few outcrops cited in many of the more recent stratigraphic discussions (e.g., Southgate, 1993, and Playford et al., 1989). The above is especially true for the cliff exposure of the "classic face" (Fig. 15). The stratigraphic succession of the Windjana platform is divided into five subunits that define three

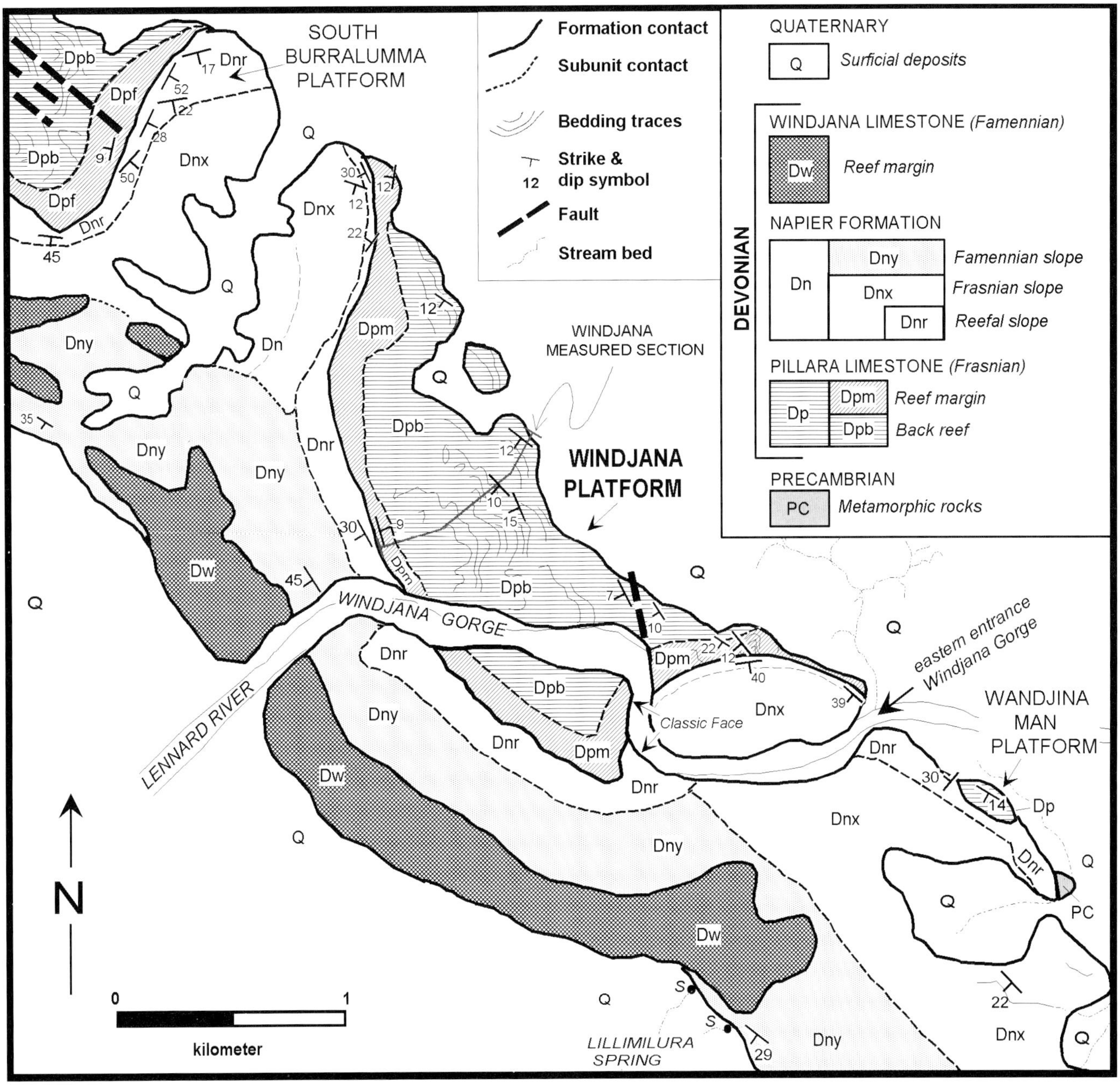

FIG. 15.—Geologic map of the Windjana Gorge area.

main platform divisions or sequences (WG-A , WG-B, and WG-C; Fig. 17). These are interpreted to be the equivalent of platforms A, B, and C. These basic platform divisions also occur just to the north in the South Burralumma platform (Fig. 15). For all three units, back-reef to reef-flat lithofacies crop out. The lower section is a stromatoporoid-limestone and sandstone unit (WG-A). The middle and upper sections are stromatoporoid limestone and fenestral limestone units (WG-B and WG-C). Repetitive, meter-scale, shoaling-upward, vertical facies successions are well developed throughout the platform. In platform WG-B, the vertical facies successions commonly end with desiccation-cracked fenestral grainstones or mudstones.

In the Windjana platform, at approximately the division between WG-A and WG-B, there is an interval that includes (1) siliciclastic gravel, (2) originally loose stromatoporoid fragments, (3) a limestone-clast channel fill, and (4) a "chaotic" limestone-clast unit (Fig. 17). The chaotic unit consists of a few limestone clasts, stromatoporoid and other fossil fragments, and fine-grained carbonate with no apparent internal stratification or other depositional structures.

Two generations of margin development are revealed in the classic face within Windjana Gorge (Ward, 1996, and in prep.). The lower unit has an upright escarpment margin and is encrusted by steeply dipping reefal-slope strata. The upper unit has a backstepped margin. Although correlation across the gorge is tenuous, these units approximately correspond to WG-B and WG-C. The rest of the margin at the Windjana platform and along the South Burralumma platform is a variation of that exposed in the

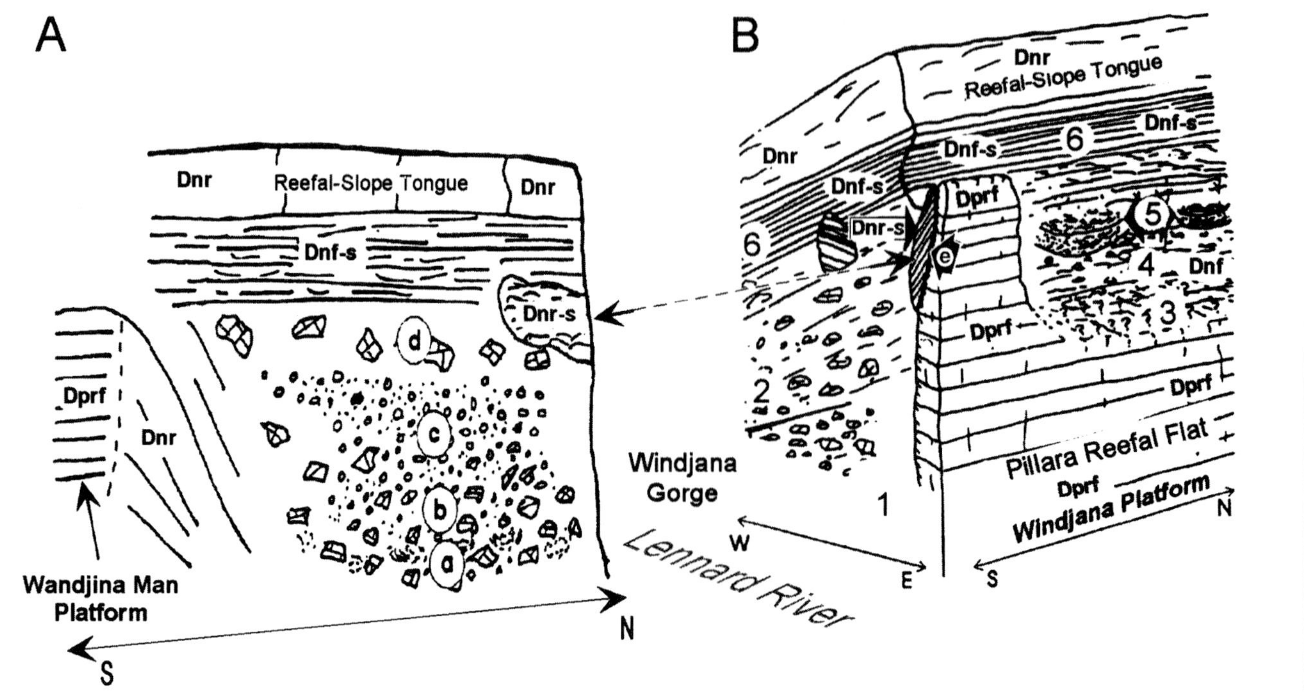

Dpf	Pillara reef-flat strata
Dnr	Reefal-slope strata
Dnr-s	Escarpment encrusting reefal-slope strata
Dnf-s	Scheck breccia and other fore-reef strata
d	Breccia of limestone blocks
c	Precambrian-clast boulders (Behn-like Conglomerate)
b	Precambrian-clast boulders mixed with limestone clasts
a	Breccia with blocks of Pillara Limestone and siliciclastic-sandstone-filled channels

6	Scheck and adnet breccias mixed with siliciclastics.
5	Siliciclastic-sandstone-filled channels (with gravels). These appear to cross cut limestone breccia beds.
4	Limestone-clast-breccia that grades into mixed limestone-clast mega-breccias and siliciclastic sandstones and conglomerates (sand-to-boulder size).
3	Poor exposure on step back: probable mega scheck breccias.
2	Limestone mega-breccia and mega-schecks mixed with siliciclastic sandstone and boulder conglomerate.
1	Mega-scheck breccia (talus) exposed at foot of cliff in Windjana Gorge.

FIG. 16.—(A) Schematic drawing of Frasnian strata exposed in the north-trending cliff between the Wandjina Man platform and the eastern entrance to Windjana Gorge. Strata dip basinward into the cliff (into the paper away from the viewer). Arrows correlate an escarpment-encrusting reefal-slope limestone (Dnr-s) across the gorge. (B) Schematic drawing of the northern side of the eastern entrance to Windjana Gorge and north-trending eastern side of the range. A remnant of a cyanobacterial reefal-slope limestone (Dnr-s) that encrusts the platform escarpment (e) is preserved. Siliciclastic sandstone- and gravel-filled channels (label number 5) cut through Napier Formation limestone breccia.

classic face and that at the eastern entrance to Windjana Gorge. The interpreted composite cross section in Figure 18 summarizes these relationships.

North Burralumma Platform

There is nearly a continuous exposure of Pillara Limestone in the composite Burralumma platform, from the South Burralumma platform to the North Burralumma platform (Fig. 19). At the northern end, there are two divisions of the Pillara Limestone strata: a back-reef bulbous-stromatoporoid limestone and sandstone unit (platform B-B) overlain by a reef-massive to reef-flat unit (platform B-C) (Figs. 19, 20). These correlate with platforms B-B and B-C at the South Burralumma platform and, thus, with the regional Pillara platforms B and C. Platform B-C has backstepped onto the back-reef strata of platform B-B (Fig. 20). The Lower Napier Conglomerate lies unconformably on and west of platform B-B. Basinward (west to southwest) of platform B-C's margin and overlying platform B-B are reefal-slope and fore-reef strata associated with platform B-C. These fore-reef strata also overlie the Lower Napier Conglomerate. Overlying the fore-reef strata is a nonfenestral-stromatolite horizon that "wraps" around the corner of the platform (Figs. 19, 20). This appears to be the remnant of a larger "bioherm" encrusting the corner of this platform. Basinward of this stromatolite unit are Famennian fore-reef beds.

From the northwestern end of North Burralumma platform to Yammera platform, only the Napier Formation is exposed (Fig. 19). At the base of this Napier Formation exposure is a mixture of megascheck conglomerates, Precambrian-rock clasts of all sizes and Lower Napier Conglomerate. At one locality within the basal mixed conglomerate lies a large Precambrian granitic block whose exposed cross section is 4.6 m × 2.6 m. The basal mixed conglomerate is overlain by a nonfenestral-stromatolite horizon.

Yammera Platform and Northward

The relatively small exposure of Pillara Limestone in the Yammera platform (Fig. 19), the last platform exposure toward the north for several kilometers, shows evidence of extensive syndepositional fracturing. Much of the platform strata that crop out is reef-flat facies. These platform-margin strata are intensely crosscut by multiple generations and multiple directions of marine-calcite-cement–filled neptunian fractures. Basinward

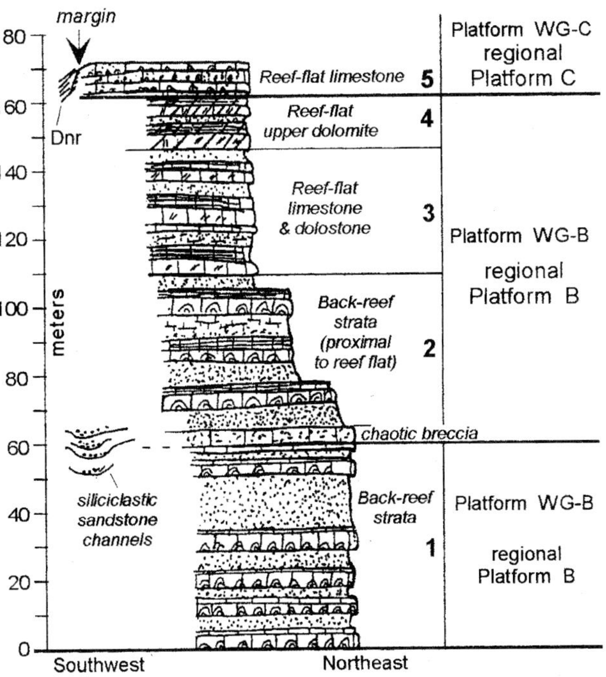

FIG. 17.—Stratigraphy of Windjana platform based on a measure section. The right-hand side shows interpreted platform units. The section was measured on a hill slope dipping to the east away from the platform margin. Therefore, the top of the section is within 50 m of the margin, whereas the bottom part is approximately 1 km from the margin. From the bottom to the top of the section there are fewer siliciclastic sandstones, more fenestral limestones, more *Amphipora* versus *Actinostroma*-forms, more coated grains, and apparently more gravel conglomerates. Siliciclastic sandstone-filled and fossil-filled channels occur at the same the horizon of a chaotic breccia limestone bed. The Pillara reef-flat strata of WG-C can be traced out to the upright margin (m) and the associated reefal-slope strata (Dnr). The back-reef strata (unit 1) consists of repetitive facies successions of siliciclastic sandstone and stromatoporoid limestones and dolostones. Back-reef strata proximal to reef-flat strata (unit 2) consist of repetitive facies successions of fenestral limestones, stromatoporoid limestones and dolostones, and siliciclastic sandstone. The reef-flat limestone and dolostone (unit 3) consists of repetitive facies successions of fenestral limestones, *Amphipora* limestones and dolostones, and siliciclastic sandstone. The reef-flat upper dolomite (unit 4) consists of repetitive facies successions of fenestral dolostone, *Amphipora* dolostone, and siliciclastic sandstone. Reef-flat limestone (unit 5) consists of stromatoporoid limestones, fenestral boundstones, skeletal-oncolitic and siliciclastic-sand grainstone and gravels.

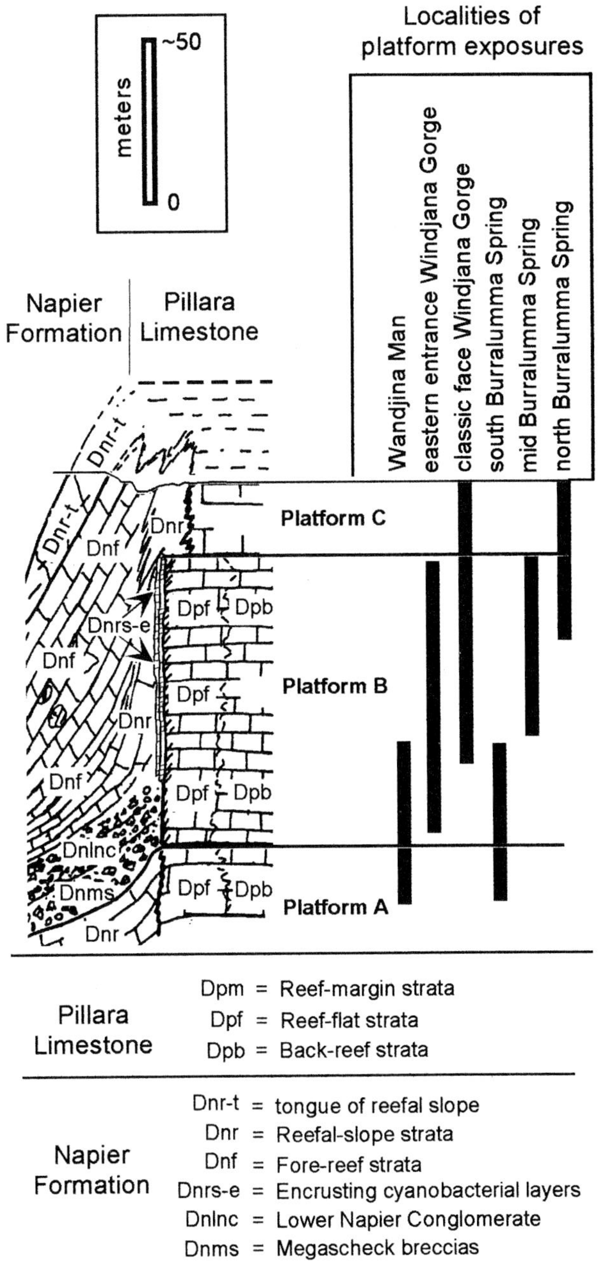

FIG. 18.—Interpretive composite cross section of the platform margin of the Wandjina Man, Windjana, and Burralumma platforms. Vertical lines indicate interpreted range exposed at various localities.

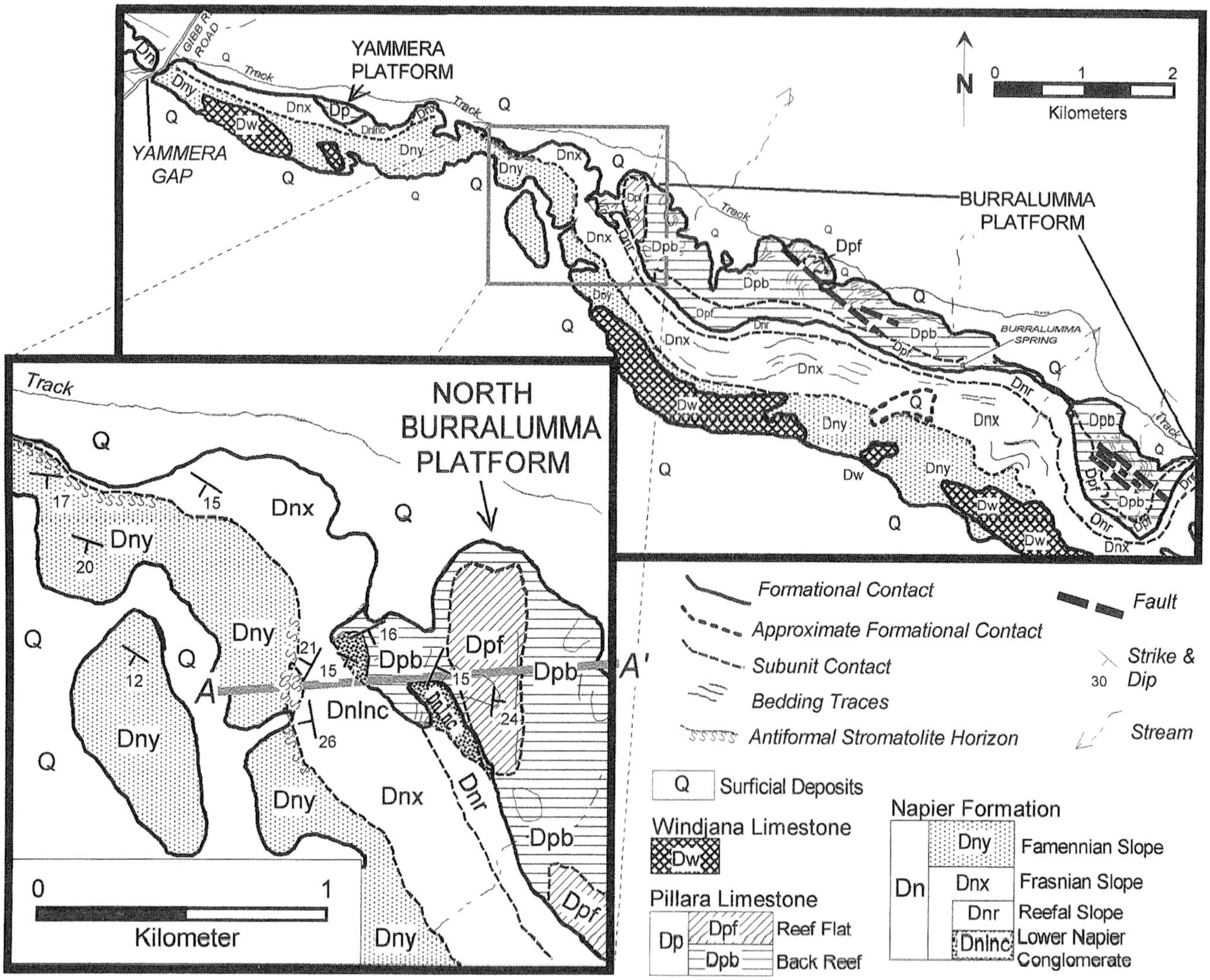

FIG. 19.—Geologic map from South Burralumma platform to the Yammera Gap. Inset map shows detail of the northern end of the Burralumma platform. A-A′ is the line of the cross section shown in Figure 20.

(westward), in the stratigraphically highest platform strata exposed in outcrop, the reef-flat subfacies grades into reef-massive subfacies that in turn, at least locally, grades into reefal-slope subfacies forming a rollover margin. The walls and roof of Emu Cave at the southern end of the Yammera platform are composed of the Lower Napier Conglomerate that unconformably overlies the Pillara. The detailed stratigraphic and marginal relationships are largely unknown because this platform was not studied thoroughly.

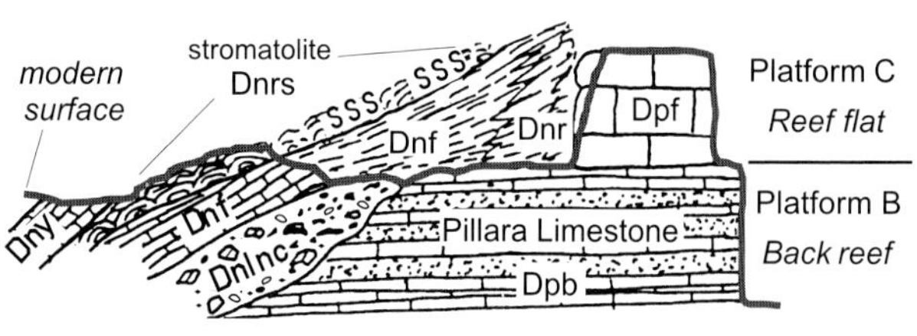

FIG. 20.—Schematic east-west section through the northern Burralumma platform.

The stratigraphy exposed north of Yammera Gap and the Napier Downs Station suggests that platform C and its slope deposits backstepped onto essentially bare Precambrian rock. In this area the Napier Formation lies over a several-meter-thick layer of poorly consolidated Precambrian-rock-clast conglomerate, which directly overlies Precambrian garnet schist. Reef debris, including reef-margin blocks, overlies this conglomerate. The reef blocks are encrusted by nonfenestral-stromatolite layers. The stratigraphy here suggests that the Napier Formation directly over the Precambrian is equivalent to platform C.

INTERPLATFORM CORRELATIONS

The relationships of platform units described above indicate that there are three different platforms (A, B, and C, following the division at the South Behn platform) exposed throughout the southern Napier Range and that the upper two platforms have backstepped geometries (Fig. 2B). Figure 21 shows interplatform correlations for the Pillara Limestone and associated slope

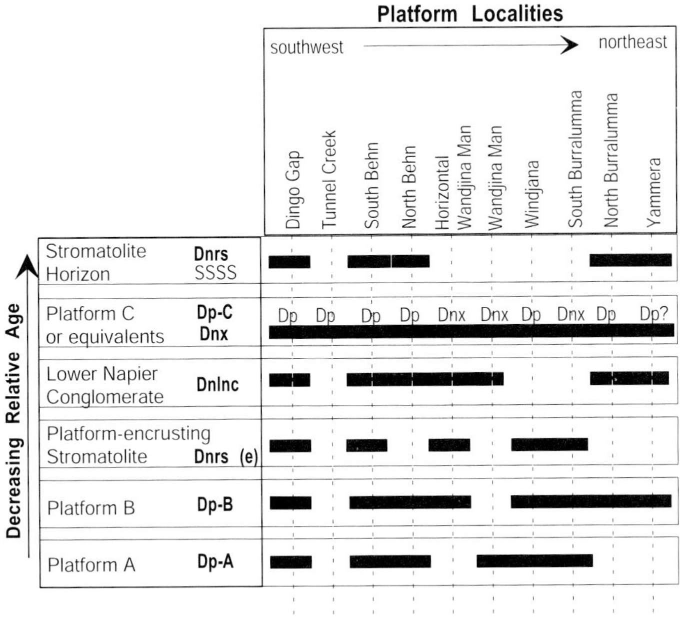

FIG. 21.—Interplatform correlation of subunits of the Frasnian Pillara Limestone and the Napier Formation in the southern Napier Range. The units indicated for each area are those local divisions shown on the respective maps and cross sections of each area.

strata. This correlation generally corresponds to that of Read (1973a, b) for the Pillara in the Napier Range based on the vertical succession of back-reef strata.

The platform A to B transition was previously undocumented in the Napier Range, whereas Playford (1981) and Playford et al. (1989) recognized the backstepping of platform C over platform B in Windjana Gorge. In the Oscar Range, Hurley (1986) recognized two generations of backstepped Frasnian platforms. In outcrops in the southwestern part of the Lennard shelf, four generations of Pillara platforms are recognized (Playford, 1980; Playford et al., 1989). In the subsurface there are as many as four generations of backstepped Frasnian platform that are interpreted as Frasnian depositional sequences Frasnian 1–4 (Kennard et al., 1992; Southgate et al., 1993).

The correlations within the Frasnian part of the Napier Formation are based on relationships with backstepped platforms and similarities in stratigraphy. Throughout the Napier Range, the Lower Napier Conglomerate has a similar occurrence in the stratigraphic succession. This succession is framed by the same stratigraphy from the backstepped margin of platform B to the antiformal nonfenestral-stromatolite unit. The Lower Napier Conglomerate records submarine deposition that occurred between the cessation of platform B growth and early platform C development (Fig. 8). The lower part of the Lower Napier Conglomerate is derived mainly from erosion of the inactive platform B and includes a minor contribution from the underlying platform A. It is unclear whether the lowest and oldest part of the Lower Napier Conglomerate is coeval with, or older than, the submergence of platform B and the backstepping event. The upper part of the Lower Napier Conglomerate records the early deposition of the fore-reef deposits that flank the backstepped margin of platform C.

Identification of the platform depositional units on the outcrop may not lead to defining classic depositional sequences. The top of C is the only bounding surface with a demonstrable subaerial unconformity and, thus, fits the definition of a sequence boundary. The top of platform B marks the only Frasnian bounding surface in the Napier Range that confidently can be extended beyond the platform. It is the most significant reconfiguration of the Frasnian platform. The unconformity, however, is vertical (i.e., the escarpment margin). This relationship also results in adjacent platform and slope strata that are not time equivalent. The top of platform B as expressed in the classic face of Windjana Gorge commonly is cited as a subaerial-exposure surface. The geometry of this contact, however, demonstrates that the boundary is a near-vertical escarpment-margin unconformity and does not show unconformable relief on a horizontal surface (Ward, 1996, and in prep.). At the South Behn platform, the thin breccia and dolomite-filled fractures are suggestive of minor subaerial exposure of the top of B, but are inconclusive. The top of platform A is poorly exposed, but at the South Behn platform, the top is marked by siliciclastic sandstone and, at the Windjana platform, by chaotic breccia. It is unclear, however, whether the surface has been subaerially exposed.

Holmes and Christie-Blick (1993) suggested that periods of extension-related fault-block adjustments lasting into the Famennian could produce sequencelike boundaries without having subaerial exposure of the platform. They noted that such "rift-onset unconformities" fit the definition of the classic sequence boundary in that the surfaces are equivalent to subaerial unconformities on the hanging wall of fault blocks. This model provides exposure of the fault scarp as a mechanism other than lowstand exposure for depositing siliciclastic conglomerates into the basin during backstepping (Holmes and Christie-Blick, 1993; Ward, 1996).

Such movement of rift blocks potentially could affect much of the Lennard shelf. For the Napier Range, one series of faults along which the Devonian sedimentary rocks were potentially down dropped is that which currently bounds much of the northeastern edge of the range. This series includes the fault extending from McSherrys Gap to Dingo Gap (the Tunnel Creek fault; Ward, 1996) and the fault landward of the Behn Conglomerate now bounding the eastern edge of the Napier Range (Holmes and Christie-Blick, 1993). Post-Devonian movement as shown in present-day bedding attitudes is consistent with these being listric faults (Ward, 1996, and in prep.).

GENERIC BACKSTEPPING MODELS

Backstepping occurs during relative submergence of the platform, when a platform margin develops landward of the preexisting platform margin. "What was the cause of backstepping?" is a question that has been asked since the backstepped geometries were recognized (e.g., Playford, 1980). Critical to any interpretation of underlying causes of backstepping is determining the controls on the initial steps of backstepping, especially on what slowed or shut down carbonate production on the platform prior to or during a relative-sea-level rise (e.g.,

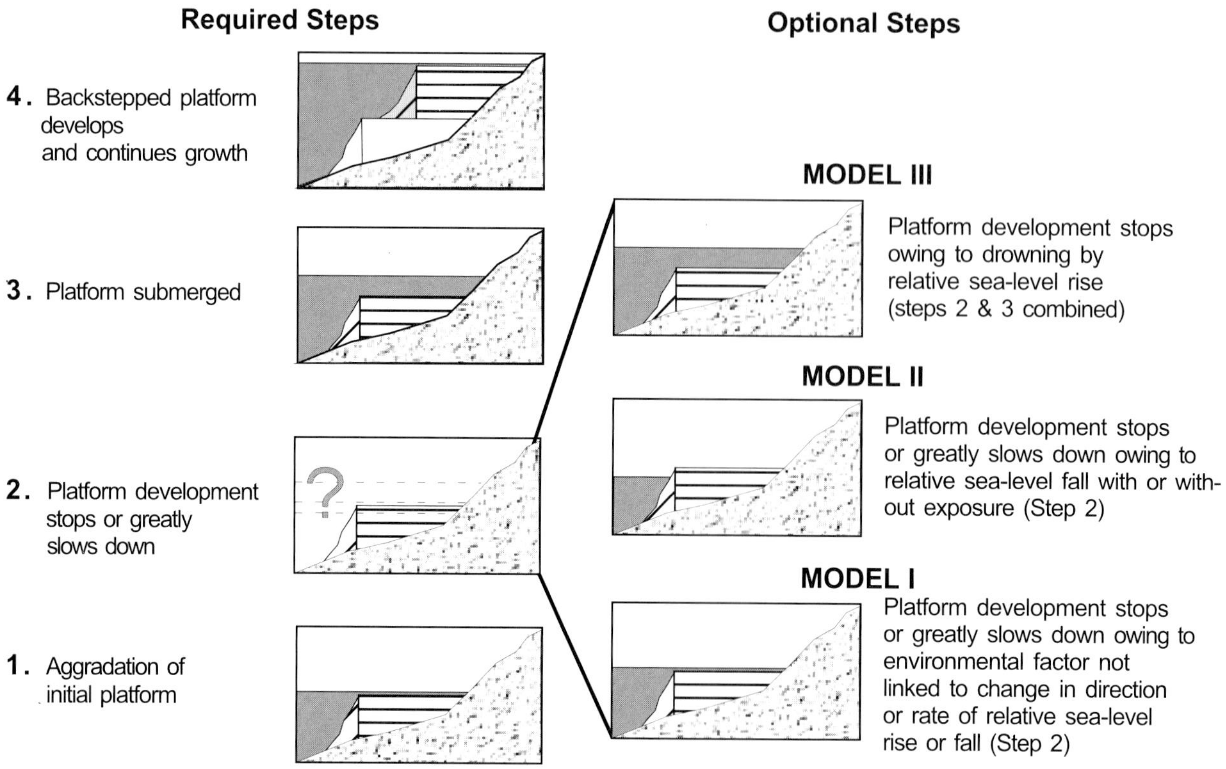

FIG. 22.—Steps required for backstepping are shown on the left. There are three basic models for the termination of platform deposition prior to backstepping. Model I has two parts: IA, purely environmental change stopping platform growth; IB, removal of the margin through collapse. In both models II and III, the change in sea level can be driven by eustasy (IIA and IIIA) or tectonism (IIB and IIIB).

Schlager, 1981). Were the platforms simply drowned during backstepping? Were the platforms exposed prior to submergence and backstepping?

Figure 22 shows the steps for possible scenarios of platform backstepping. In a generic backstepping scenario, carbonate production on the initial platform must slow down or stop prior to or during backstepping. At least three end-member models can account for cessation of platform development followed by backstepping (Fig. 22). Two (models II and III) require a change in direction or the rate of relative-sea-level rise or fall, and the other (model I) does not. Both eustatic and tectonic controls on relative sea level are compatible with either model II or III. Table 2 lists stratigraphic relationships predicted for each model.

Model I

In model I (Fig. 22), platform growth ceases prior to backstepping that is not tied to a change in relative sea level. This model can involve platform termination due to either (1) an environmental change (model IA) or (2) wholesale removal of the margin through collapse or faulting (model IIA). Evidence of change in sea level, such as platform exposure, rules out this model.

Model IA is considered the least likely scenario. Inhospitable waters, a change in salinity, and smothering by siliciclastic sediments are all possible causes for shutting down carbonate production (e.g., Schlager, 1981). Submergence would result from continuation of subsidence or a continuation of a sea-level rise. Siliciclastic deposition should be continuous through backstepping. This model does not easily account for regional correlation of backstepping. Such a model is most likely to have occurred on a limited scale. The only possibility of a non-sea-level effect would have to be a basin-wide change in ocean conditions. This model does not predict basinal siliciclastic wedges and margin collapse.

In model IB, abruptly removing the margin at least momentarily stops carbonate production at the margin. This process does not refer to the partial or local collapse that appears to be common in the development of the Devonian reef complexes (e.g., Playford, 1980). Partial removal of margin facies only produces marginal escarpments without backstepping relationships. Removal of the entire margin facies requires backstepping relationships to occur if platform growth continues. Such removal can occur through erosional collapse of the margin in which a backstepping relationship would form when a new margin develops on or landward of the collapse scar. For this scenario to produce a regional backstepped margin, it seems most likely that it would have to be related to tectonism in the form of earthquakes or faulting undermining the slope and margin.

An alternative way to remove the margin is to down drop it by faulting. A reef margin developing on a fault scarp that cuts back-reef facies results in a backstepping geometry. The faulted margin, however, would record drowning and thus falls into model III. In any case, this type of backstepped geometry would only extend the length of the fault.

Table 2.—STRATIGRAPHIC RELATIONSHIPS PREDICTED BY THE THREE MODELS FOR BACKSTEPPING

Model I: Platform Termination Not Linked to Sea Level		Model II: Platform Termination Linked to Sea-Level Fall before Rise		Model III: Platform Termination Linked to Sea-Level Rise Only	
IA: Environmental Crisis Only	IB: Margin Collapse Only	IIA: Eustatic Sea-Level Fall	IIB: Tectonic Sea-Level Fall	IIIA: Eustatic Sea-Level Rise	IIIB: Tectonic Sea-Level Rise
Predicted Features Possible in Prebackstepped Platform Strata					
Internal stratigraphy not linked	internal stratigraphy not linked	internal stratigraphy linked = record of decreasing acc.	internal stratigraphy not linked	internal strigraphy linked = record of increasing acc.	internal stratigraphy not linked
change in facies at backstepping	local uninterrupted carbonate production possible				
no exposure features	no exposure features	exposure features: karst and incised valleys	exposure features: karst and incised valleys	no exposure features	no exposure features
	collapse escarpment; fracturing and faulting possible		related faulting, fracturing, and collapse possible		related faulting, fracturing, and collapse possible
Predicted Features Possible in Backstepped Platform Strata					
internal stratigraphy not linked	internal stratigraphy not linked	internal stratigraphy linked = TST to HST	internal stratigraphy not strongly linked = maybe TST to HST	internal stratigraphy linked = continued TST to HST and LST*	internal stratigrahy not strongly linked = maybe TST to HST and LST*
	new margin probable at edge of collapse scar				
Predicted Features Possible in Slope Strata					
uninterrupted carbonate sedimentation on slope	uninterrupted carbonate sedimentation possible on preserved slope	platform-derived carbonate deposition ends with demise of old platform and restarts after backstepping	platform-derived carbonate deposition ends with demise of old platform and restarts after backstepping	slowed carbonate deposition = deep-water sedimentation; platform-derived carbonate deposition ends with demise of old platform and restarts after backstepping	slowed carbonate deposition = deep-water sedimentation; platform-derived carbonate deposition ends with demise of old platform and restarts after backstepping
platform-derived sediments change with facies change at backstepping	upper slope of older platform removed				
collapse blocks possible	collapse debris in lower slope and basin	collapse blocks possible	collapse blocks possible	collapse blocks possible	collapse blocks possible
		lowstand buildups	lowstand buildups	no lowstand buildups	no lowstand buildups
exposure not possible	exposure not possible	exposure not possible	exposure not possible	exposure not possible	exposure not possible
increase in siliciclastics may be cause of platform demise	no mechanism to significantly change siliciclastics	increase in total siliciclastics = point-source deposits that possibly bypass slope to basin = lowstand wedge	increase in total siliciclastics = point-source dposites that possibly bypass slope to basin = lowstand wedge	no mechanism to increase total coarse siliciclastics, they may decrease or possibly bypass slope	increase in siliciclastics = point-source deposits that possibly bypass slope to basin = TST wedges

NOTE: Environmental changes induced by sea-level changes play role in models II and III. Margin collapse may be especially effective when coupled with sea-level rise (model III). If sea-level drop is relatively slow, then it may not be possible to distinguish between eustasy and tectonism by only local stratigraphy and facies successions. Acc. = accommodation.

*Sequence boundary at base of backstepped platform may form in the middle of a eustatic rise, especially if a coupled tectonic and eustatic rise.

Model IA requires an escarpment margin or fault. Of the different models, IA is considered most unlikely to be of regional extent. Removing the margin could result in increased siliciclastic deposition off of the platform. With this model, there is no reason for carbonate production throughout the platform to stop. Thus, backstepped marginal relationships may not be easily tied to a surface within the platform interior. A facies change, however, could occur throughout the platform. Removing a rimmed margin may open up circulation within the platform and change environmental conditions and facies distributions. This model does not change accommodation on the surviving platform. Thus, although facies changes may occur in the backstepped platform, the trends of accommodation should not change from the underlying platform to the backstepped platform. Collapse removal of margins has a greater chance of being related to a regional backstepping event if it is coupled with a change in sea level (model II or III).

Model II

In model II (Fig. 22), carbonate production is dramatically reduced or stopped by a drop in relative sea level followed by a rise and backstepping. This model allows for one of the most effective ways to stop carbonate production both throughout an entire platform and regionally: subaerial exposure of the platform. Model II follows that of a classic sequence-stratigraphy interpretation of backstepping. This is the common interpretation for all the backstepped platforms of the Devonian reef complexes (Playford, 1991; Southgate et al., 1991; Holmes, 1991; Kennard et al., 1992). In addition to exposure features, the model predicts a reciprocal relationship with carbonate-platform growth and siliciclastic deposition in that siliciclastics would be deposited as basinal lowstand wedges (e.g., Southgate et al., 1993). Carbonate autochthonous and line-source allochthonous deposition on the slope would be replaced with siliciclastic sediments being channelized through the platform and slope. A large fall in sea level would expose slope deposits. Additionally, lowstand shallow-water reefal buildups potentially could develop on the slope below the initial shelf margin. Margin collapse can occur during lowstands through subaerial exposure or during early transgression (Southgate et al., 1993; George and Powell, 1997).

In this model a fall in sea level may be controlled by eustasy (model IIA) or tectonic movement (model IIB). A eustatic rise and fall has a higher potential of effecting a characteristic sequence of deposition than a tectonically induced fall. Ideally, prior to approaching a fall, platform facies and strata would reflect progressively decreasing accommodation. For such a succession to be deposited during a fall purely controlled by tectonic movement, the movement would have to be drawn out or occur incrementally over time. In such a case it may be impossible to distinguish the effects of tectonism from eustasy through study of the intraplatform strata. Comparisons with sedimentary units outside the tectonic influence would be required (e.g., worldwide correlation equates to eustasy). Relatively rapid tectonic movements would not be linked to the progressive stratal successions within the platform.

A eustatic fall is predicted to be followed by a transgression with an increasing rate of rise that may be recorded in the backstepped platform. Tectonic movement may interrupt a sea-level fluctuation at any stage, thus precluding a predictive sedimentary succession for backstepped platforms. A tectonically induced fall in conjunction with a eustatic fall would accentuate exposure and lowstand sedimentation, and the following rise would be eustasy induced. A tectonically induced fall occurring during a eustatic rise may result in a brief lowstand followed by a shortened transgression.

Model III

The third model (model III, Fig. 22) involves rapid submergence of the platform in marine waters, i.e., drowning. The rise could result from either eustasy (model IIA) or from tectonic down dropping (model IIIB). In this scenario, only transgressive and highstand sedimentation would occur during and following backstepping; lowstand sedimentation and platform exposure are not predicted. Siliciclastic deposition, however, may be redirected by removing the effects of a protective reef-rimmed platform. Although the shoreline and source of siliciclastic sediment would be moved landward, longshore currents might be more effective in transporting sand to channels, which then have the potential to move more sand to and through existing fans. With a eustatic-controlled rise, there is no mechanism to greatly increase siliciclastic deposition. Although sand may be shed into the basin, the coarsest sediments would be trapped on the platform. Tectonic movement that exposes a fault scarp, in addition to raising relative sea level over the platform, would create a new depositional profile and a newly active source for siliciclastics, sand sized and coarser. In such a case, siliciclastics could be shed into the basin during a rise, producing deposits whose geometries resemble lowstand wedges.

Margin collapse has the potential of being triggered during submergence by increasing overlying water weight and change in pore pressures (Ward, 1996, and in prep.). Such a collapse scenario seems more likely with the preconditioning of the margin through fracturing or karst (e.g., George and Powell, 1997). Faulting, fracturing, and earthquakes associated with tectonic movement would greatly increase the likelihood of margin collapse.

As in model II, the internal stratigraphy of a platform prior to backstepping potentially can foreshadow a eustatic rise. A tectonic down drop and associated relative-sea-level rise is unlikely to be foreshadowed by preceding deposition. For a tectonically induced rise, the stratigraphic succession in the backstepped platform would be dependent on when the tectonism occurred during a eustatic sea-level fluctuation. Depending on the rate of submergence, a backstepped platform may not begin forming until the sea-level rise slows or actually falls. In such a scenario, the time between platform drowning and the development of a backstepped platform would depend on when a tectonically induced rise occurred during a eustatic fluctuation. Thus, even if a drowning results from tectonism, the depositional succession overlying the backstepped platform may record eustasy.

SEQUENCE OF DEPOSITIONAL EVENTS AND INTERPRETATION OF BACKSTEPPING

Table 3 summarizes characteristics of platform A, B, and C, and their backstepping relationships. Understanding the events leading to backstepping and ruling out the inappropriate backstepping models are hampered by the lack of exposed slope deposits for platforms A and B. However, the lateral three-dimensional control on these platforms greatly aids the interpretation of their evolution.

Model IA (Fig. 22) is considered unlikely for either of the backstepping events. Such causes are most likely to have occurred on a limited scale and not have affected platforms on the scale of the Napier Range. For example, platform A at South Behn platform could have been overwhelmed by the prograding siliciclastic sands that overlie the platform. The apparently equivalent transition at Dingo Gap in a more seaward position, however, lacks the thick sandstone. Thus this model does not easily account for the regional correlation of backstepping. The only possibility of a non-sea-level event affecting platform development along at least 80 km would be a Lennard shelf–wide or greater change in ocean conditions. More important, such a change does not explain the progressive change from platform A to platform B, margin collapse (Ward, 1996), the deposition of the Lower Napier Conglomerate, and the basinal siliciclastic wedges.

The other models are more difficult to dismiss and are discussed below for each backstepping event.

Platform A to Platform B Backstepping

The internal stratigraphy of platform A at South Behn records a transgression from siliciclastic sandstone to stromatoporoid-limestone–based sand-rich cycles. This was followed by shallowing, recorded by the deposition of a fenestral grainstone cap at the top of platform A. These strata are interpreted as recording a predominantly eustatic sea-level fluctuation, as does the base of the overlying platform B. Siliciclastic sands were deposited over the platform, consistent with a sea-level fall. The resulting sandstone is overlain by stromatoporoid reefal limestones that extend well onto the platform, recording the initial reflooding of the platform during early transgression. These reefal subfacies are overlain by a succession of interbedded shales and grainstones that is interpreted to record continued early flooding. These are the only shales occurring in the Pillara platform successions of the Napier Range. Overlying this is another reefal limestone that extends onto the platform. Above this, a platform margin and its associated back reef is developed. Reef facies occurring beyond the platform margin may record maximum flooding or may record the initiation of a reef margin that localized reef growth at the edge of the platform.

In light of the above, platform A and its transition to platform B appear consistent with a long-term sea-level fluctuation that led to a fall and stoppage of carbonate production followed by a rise and, after initial flooding, backstepping. A eustatic fluctuation is the simplest explanation. For the Windjana platform, the high-frequency cycles decrease in thickness approaching the platform A upper boundary (Fischer plot [Fischer, 1964] in Ward, 1996). A straightforward interpretation of these data is that accommodation is decreasing prior to a sea-level lowstand. There are no indications that a rapid tectonic movement induced a sea-level rise and backstepping. It should be kept in mind that there is limited exposure of platform A, limited preservation of its margin, and essentially no exposure of its slope strata. At South Behn, the exact top of A is covered. In the Windjana platform, the top is marked locally by either a limestone breccia or an *in situ* breccia with rotated clasts that is a possible karst-collapse feature. This feature is consistent with the interpreted sea-level fall. Model IIA applied to the platform A to platform B transition predicts that there was increased siliciclastic deposition into the basin and a high potential for a lowstand-wedge geometry to have developed.

Platform B to Platform C Backstepping

The transition from platform B to platform C is more problematic. After initial transgression, there is not an internal stratigraphic succession clearly foreshadowing either a sea-level drop or rise. After platform B stopped growing, the margin collapsed along neptunian fractures (Ward, 1996) for the final time, and the resultant escarpment was encrusted with cyanobacterial limestone layers. These clearly indicate submergence of the platform in marine waters. Before the encrustation, the top of the margin, at least locally at South Behn, was brecciated, and polygonal fractures formed. These features might suggest minor subaerial exposure at the top of platform B. However, they could be submarine exposure features (Ward, 1996), because there is no clear evidence of extensive subaerial exposure in outcrop or thin section.

Playford (1991) and Southgate et al. (1991) suggested that the contact between platform B and platform C at Windjana Gorge is a sequence boundary representing a lowstand event with subaerial exposure of the platform. Playford (1981, 1991) and Playford et al. (1989) interpreted apparent topographical relief on top of platform B in the classic face in Windjana Gorge to indicate subaerial exposure. Ward (1996, and in prep.) showed this "relief" to be the combined result of the geometry of a collapse escarpment and the geometry of the present-day outcrop. Southgate et al. (1991) and Kennard et al. (1992) partly defined their sequences on seismic interpretation of subsurface basinal siliciclastic deposits with lowstand-wedge geometries. One wedge approximately correlates with the platform B to platform C transition. The lowstand geometries also could be a result of backstepping regardless of the position of sea level. The Behn Conglomerate and siliciclastic deposits throughout the Napier Formation indicate that a substantial amount of siliciclastics were deposited in

TABLE 3.—SUMMARY OF CHARACTERISTICS OF PLATFORMS A, B, AND C AND THEIR BOUNDING SURFACES

	Platform A Sequence	A to B Transition	Platform B Sequence	B to C Transition	Platform C Sequence	Transition from C
Platform Relationships						
Extent of Outcrop	exposed locally	exposed locally	exposed throughout	exposed	least exposed	not exposed
Internal stratigraphy	transgression from channeled siliciclastic sandstones to stromatoporoid limestone-based sandstone-rich cycles shallowing to fenestral grainstone top	sea-level drop and turn-around that is foreshadowed by strata in A and continued in B	transgression from sands to reef facies onto platform at initiation of backstepped platform B; then aggradation	no obvious foreshadow-ing in B; only submergence after collapse recorded	not well documented	sea-level drop and exposure fore-shadowed by margin relationship of C
Margin relationships	upright margin where exposed; locally eroded by collapse	backstepping: B margin overlying A back reef	upright with escarpment margin	backstepping: C margin overlying B reef flat to back reef	upright margin to prograding margin	prograding: Famennian back reef overlying C margin
Top bounding surface (sequence boundary)	locally at base of silici-clastic unit; precise contact poorly ex-posed; no evidence of extensive exposure		no evidence of extensive exposure		outside of mapped area, surface shows karst topography with at least 1.5 m relief	in northern Napier Range, Famennian back-reef fenestral limestones overlie Frasnian reef-massive dolostones that have a karst surface
Slope Relationships						
Extent of outcrop	reefal slope rarely exposed	not exposed	reefal slope rarely exposed	postcollapse slope exposed	slope facies exposed throughout	exposed
Stratigraphic succession	mostly eroded prior to C	eroded prior to C	eroded prior to C	platform C slope overlying unconformity on platforms B and A and on Precambrian surface	transgression recorded by cyanobacterial en-crustrations and Lower Napier Conglomerate followed by progra-dation of fore reef; reefal tongue may indicate sea-level fall	difficult to define without biostrati-graphy (e.g., George et al., 1997)
Top bounding surface (sequence boundary)	mostly eroded prior to C	eroded prior to C	eroded prior to C	at base, encrusting *Frutexites;* locally at base of stromatolite or Lower Napier Conglomerate	poorly defined by facies (also see George et al., 1997)	poorly defined by facies or stratal surfaces (also see George et al., 1997)

marine waters during apparent highstands when platforms were submerged and actively growing. If the platforms were drowned and the carbonate production shut down, the deep-water deposition of siliciclastic sediments would not be required to stop. If backstepping were the result of faulting landward of the platforms, then there could have been a pulse of siliciclastic sediments from increased erosion along a fault escarpment.

The unconformity below the Lower Napier Conglomerate is not channelized, but appears to be along an escarpment. In a fallen-sea-level scenario, this escarpment could form from large-scale margin collapse or from a major drop (>150 m) in sea level and associated sea-cliff erosion. However, the encrusting stromatolite layers record transgression. Both the Lower Napier Conglomerate and encrusting stromatolite layers appear to represent prolonged submergence and transgression during which carbonate production was stopped or slowed down. In addition, the backstepped platform C records continued transgression to highstand deposition (followed by eventual exposure and karst development).

If platform B to platform C backstepping records only a rise in sea level, the basinal siliciclastic wedge must record periods of low carbonate sedimentation during the submergence and transgression. During periods of high carbonate production, siliciclastic sediments are mixed with the carbonates and develop nonwedge geometries. In such a case, coarse siliciclastics that were transported to the basin, past the slopes, would be most likely to travel through channels associated with siliciclastic fans such as the Behn Conglomerate. Active carbonate sedimentation, however, would limit the amount of siliciclastics delivered through such point sources by spreading sediments out along marginal-slope aprons. Platform development could trap more siliciclastics on the platform. Additionally, siliciclastics that were transported across the platform would mix with carbonate sediments and spread out along marginal-slope deposits. Removing nearshore siliciclastic sedimentation from the protection of a reef-rimmed platform by dramatically flooding the platform could have the effect of delivering more siliciclastics to channels and fans. For example, removing the protective reef rim could permit longshore currents to be more effective in transporting sediments to major channels cutting through the platforms. Thus, the siliciclastic-fan deposition would increase and result in the spilling of more sediments into the basin. The addition of a tectonic pulse, such as major faulting along rift blocks, could uplift Precambrian terrane and increase erosion and thus siliciclastic deposition. In this scenario, the erosional unconformity beneath the Lower Napier Conglomerate could only have resulted from large-scale submarine-margin collapse. Faulting would have facilitated such a collapse. The neptunian fractures suggest that faulting and related stresses did effect margin collapse of platform B (Ward, 1996, and in prep).

The Lennard shelf is considered to have undergone relatively rapid tectonic subsidence during the Frasnian (Playford et al., 1989). The upright margins are considered to have resulted from the carbonate platforms keeping up with a fast rise in relative sea level that, in turn, was due to a high subsidence rate (Playford et al., 1989). If this is true, a rapid eustatic rise superimposed on a tectonic subsidence might be able to outpace carbonate production and effectively drown a platform.

Alternatively, faulting-induced down drop, regardless of eustatic sea-level fluctuations, also could lead to backstepping (model IIIB, Fig. 22). Penecontemporaneous faulting within Frasnian strata has been documented in the Devonian reef complexes (Playford, 1980; Playford et al., 1989; Hurley, 1986; Wallace, 1987). Hurley (1986) documented Frasnian reef-margin facies forming on a fault scarp cutting the backreef strata in the southern Oscar Range. This fault appears to be local and not correlated to the regional backstepped margins. In the Napier Range, the backstepped platform C did not develop on a fault scarp. In the Napier Range, tectonic activity during the Frasnian is indicated by neptunian fractures related to margin collapses (Ward, 1996, and in prep.) and large siliciclastic boulders predating the Lower Napier Conglomerate. The backstepping platforms appear to correlate throughout the Napier Range and probably throughout the Lennard shelf (Playford, 1980). Therefore, any tectonic control would have to affect the entire Lennard shelf. This scenario would be consistent with rift-related faulting, as also suggested by Holmes and Christie-Blick (1993).

Platform B does have a regional erosional escarpment that resulted from collapse and removal of a part of the platform margin. The incomplete removal of the margin reduces the chances that collapse alone caused backstepping (model IA). It is likely that the intact reef facies would continue to perpetuate and redevelop the margin. This process would cause a backstep geometry at the margin, but would not account for the submergence or the prolong hiatus in platform growth recorded in the deposition of the Lower Napier Conglomerate. Collapse alone also does not account for the siliciclastic conglomerates deposited beneath and within the Lower Napier Conglomerate. Collapse on a smaller scale appears to be an aspect of the development of these upright margins. This possibility is indicated by large allochthonous reef blocks in the marginal-slope strata (e.g., Playford, 1980). These smaller collapses do not appear to have adversely affected the reef-margin growth.

The amount platform C backstepped over platform B varies along the range. At Windjana Gorge, backstepping was only a few meters. At the South Behn and North Burralumma platforms, margin C backstepped 50 to 100 m landward. If the interpretation at the North Behn platform is correct, then the margin backstepped approximately 1 km. At several localities in the Napier Range, most notably north of Napier Downs, the Pillara platform C appears to have developed over bare Precambrian terrane or on a thin veneer of Precambrian-clast conglomerate. This situation suggests that either (1) the margin backstepped a significant amount, (2) the platform at these localities formed for the first time during the late Frasnian, (3) erosion removed older Frasnian platform strata, or (4) a combination of the above occurred in connection with faulting and reconfiguration of the shoreline.

Wallace et al. (1991) suggested that the regional lower Famennian ferruginous deep-water stromatolite horizon in the Devonian reef complexes represents the initial transgression and marks a sequence boundary resulting from rapidly drowning platforms. This stromatolite unit records a time of slow sedimentation. The base of the older Frasnian stromatolite encrustation on the South Behn platform also suggests initial transgression over a sequence boundary. This encrusting cyanobacterial layer could have been the result of rapid drowning or encrustation on a submarine cliff that was bypassed by sediment. It cannot represent, however, a time of slow sedimentation during subaerial exposure of platform B, because the entire platform is encrusted.

The transition from platform B to platform C is interpreted to record a drowning-induced backstepping (model III, Fig. 22) because there is a record of a prolonged submergence of platform B and no record of a fall. This finding is consistent with the absence of subaerial-exposure features and incised valleys in the top of platform B and the absence of an internal stratigraphy of platform B predicting a fall. Submergence is indicated by the cyanobacterial encrustations of platform B and the Lower Napier Conglomerate. Platform B does have a regional erosional escarpment that resulted from collapse and removal of the platform margin. The regional collapse of platform B's margin may have facilitated the drowning; however, collapse alone is not considered to be the cause of backstepping (model IA) because (1) there is a record of prolonged submergence, (2) the entire margin facies was not removed, and (3) the Lower Napier Conglomerate and its associated siliciclastic boulders indicate that deposition was effected landward of the margins.

The rise in sea level that led to drowning is interpreted to result from tectonism (model IIB) because of (1) the extensive margin collapse that occurred along neptunian fractures affecting both platforms A and B, (2) the extensive fracturing in the margin of platform B, (3) the internal stratigraphy of platform B not clearly linked to backstepping, and (4) the siliciclastic sandstones and conglomerates with boulders as much as 4 m across below the Lower Napier Conglomerate. A tectonically induced backstepping would be consistent with the differential amounts of backstepping and backstepping onto bare Precambrian terrane. The above interpretation implies that the basinal wedges with "lowstand geometries" actually record tectonically induced sedimentation during a sea-level rise.

Termination of Platform C

The rise and demise of platform C contrasts with the rise and demise of platform B. The change in platform C's margin geometry from upright to prograding indicates a long-term change in sea level that is consistent with the subaerial exposure of platform C, which is indicated by karst with about 2 m of relief in reef-massive subfacies. Within the marginal-slope strata of the Napier Range, there is no clear evidence of this exposure event (e.g., Playford et al., 1989). In contrast, the event or events resulting in platform C's backstepping over platform B, which did not undergo karst development, had a prolonged impact on marginal-slope deposition. The B-to-C transition not only effected a major margin reconfiguration by collapse of platform B's margin and associated slope, it also resulted in postcollapse deposition of siliciclastic conglomerates and the mixed carbonate-siliciclastic Lower Napier Conglomerate.

CONCLUSIONS

The Frasnian carbonate platforms exposed in the Napier Range developed along the Devonian mainland underlain by the Proterozoic Kimberley block. Laterally, platforms formed of the Frasnian Pillara Limestone are separated by siliciclastic fans of sandstones and conglomerates such as the Behn Conglomerate. The seaward margins of the platforms are typically upright escarpments developed by both reef growth and collapse.

Three generations of backstepped Frasnian platforms define an overall transgression of the Pillara cycle. The tops of these platforms are sequence boundaries. Two submergence and backstepping events that created these boundaries are recorded throughout the southern Napier Range. The first one, the backstepping of platform B over platform A, is most consistent with backstepping during a sea-level rise following a sea-level fall. The internal stratigraphy of platform A predicts this fall and suggests that the fall was predominantly eustatic. This interpretation is also consistent with the record of the sea-level rise recorded by the lower strata of platform B.

The A to B transition contrasts with the platform B to C transition. Submergence of the South Behn platform is well documented. Initial submergence is recorded by fractures, fracture fills of marine sediments, cements and fossils, and marine diagenetic features in the platform rock. During submergence the near-vertical escarpments were coated by calcareous cyanobacteria. After a period of drowning, the platform was covered by marginal-slope deposits of the younger backstepped platform. All the features and events during the period between the termination of platform B deposition and the backstepping to form platform C may be explained by rift-related tectonic movement that submerged the platform and produced an extensive submarine erosional escarpment. There is no evidence that platform B was significantly subaerially exposed. The basinal siliciclastic wedges could have formed whether or not the platform was exposed. Tectonic movement effecting a change in relative sea level and backstepping is consistent with (1) the regional extent of the Lower Napier Conglomerate that indicates erosion of the interior of several platforms, (2) the structural control on neptunian-fracture and platform-margin orientations, (3) siliciclastic distribution, and (4) large Precambrian boulders.

ACKNOWLEDGMENTS

This paper is modified from Chapter 1 of the author's dissertation at SUNY at Stony Brook under the principal advisorship of W. J. Meyers and coadvisorship of P. E. Playford (Geological Survey of Western Australia). This project was funded by the Geological Survey of Western Australia (GSWA), contributors to the Petroleum Research Foundation of the American Chemical Society (grant to W. J. Meyers), and Texaco. GSWA assistance, which was essential for the field work, was extensive and included use of a field vehicle and field equipment, a field assistant, and abundant scientific and logistical support. This study was initially coordinated with the 1985 field work of Glenn Beere (formerly of the GSWA) concentrating on the stratigraphy and sedimentology of the Devonian siliciclastic conglomerates on the Lennard shelf.

Ideas presented in this paper have been influenced by discussions with P. E. Playford, C. Kerans, W. J. Meyers, N. Hurley, V. Pedone, G. Beere, M. Wallace, W. C. Ward, K. A. Ward, E. Oswald, B. Fouke, J. L. Banner, A. George, and A. E. Holmes. Peter Boner, Richard Hulyko, and Garrie Meddings deserve special thanks for assistance in the field that made life in the bush enjoyable and made this study possible. Lauren Shepherd's assistance with drafting is greatly appreciated. This paper was greatly improved by reviews of the original chapter by Meyers, Playford, Kerans, S. McLellan, M. Schoonen, and R. J. Reeder and by reviews of the present manuscript by W. C. Ward, W. Fitchen, J. L. Wilson, and A. Saller.

REFERENCES

BOTTEN, P., 1984, Uranium exploration in the Canning basin: A case study, *in* Purcell, P. G., ed., The Canning basin, Western Australia (Proceedings: Canning Basin Symposium): Perth, Geological Society of Australia and Petroleum Exploration Society of Australia, p. 485–501.

FISCHER, A. G., 1964, The Lofer cyclothems of the Alpine Triassic: Lawrence, Kansas Geological Survey Bulletin, v. 169, p. 107–149.

GEORGE, A. D., AND POWELL, C. MCA., 1997, Paleokarst in an Upper Devonian reef complex of the Canning basin, Western Australia: Journal of Sedimentary Research, v. 67, p. 935–944.

GEORGE, A. D., PLAYFORD, P. E., AND POWELL, C. McA., 1994, Carbonate breccias and quartzo-feldspathic sandstones of the marginal slope, Devonian Reef Complex, Canning basin, Western Australia, *in* Purcell, P. G., and Purcell, R. R., eds. Proceedings of the West Australian Basin Symposium: Perth, Petroleum Exploration Society of Australia, p. 727–738.

GEORGE, A. D., PLAYFORD, P. E., AND POWELL, C. McA., 1995, Platform-margin collapse during Famennian reef evolution, Canning basin, Western Australia: Geology, v. 23, p. 691–694.

GEORGE, A. D., PLAYFORD, P. E., POWELL, C. MCA., AND TORNATORA, P. M., 1997, Lithofacies and sequence development on an Upper Devonian mixed carbonate-siliciclastic fore-reef slope, Canning basin, Western Australia: Sedimentology, v. 44, p. 843–867.

GRIFFIN, T. J., TYLER, I. M., AND PLAYFORD, P. E., 1993, Lennard River: Perth, Geological Survey of Western Australia, 1:250,000 Geological Map Series and Explanatory Notes.

HOLMES, A. E., 1991, Late Devonian carbonate-conglomerate association in the Canning basin, Western Australia: American Association of Petroleum Geologists Bulletin, v. 75, p. 596.

HOLMES, A. E., AND CHRISTIE-BLICK, N., 1993, Origin of sedimentary cycles in mixed carbonate-siliciclastic systems: An example from the Canning basin, Western Australia, *in* Loucks, R. G., and Sarg, J. F., eds., Recent Advances and Applications of Carbonate Sequence Stratigraphy: American Association of Petroleum Geologists Memoir 57, p. 181–212.

HURLEY, N. F., 1986, Geology of the Oscar Range Devonian reef complex, Canning basin, Western Australia: Unpublished Ph.D. dissertation, University of Michigan, Ann Arbor, 269 p.

KENNARD, J. M., SOUTHGATE, P. N., JACKSON, M. J., O'BRIEN, P. E., CHRISTIE-BLICK, NICHOLAS, HOLMES, A. E., AND SARG, J. F., 1992, New sequence perspective on the Devonian reef complex and the Frasnian-Famennian boundary, Canning basin, Australia: Geology, v. 20, p. 1135–1138.

KERANS, C., 1985, Petrology of Devonian and Carboniferous Carbonates of the Canning and Bonaparte Basins: Perth, Western Australian Mining and Petroleum Research Institute, Report 12, 203 p.

KERANS, C., AND TINKER, S. W., 1997, Sequence Stratigraphy and Characterization of Carbonate Reservoirs: Tulsa, SEPM (Society for Sedimentary Geology) Short Course Notes No. 40, 140 p.

KWONG, M. H., 1995, Salinity and stable isotope constraints for Late Devonian seawater from calcite cement sequences within Upper Devonian reef complexes, Canning basin, Western Australia: Unpublished M.S. Thesis, University of Texas, Austin, 140 p.

PLAYFORD, P. E., 1976, Devonian reef complexes of the Canning basin, Western Australia: 25th International Geological Congress Excursion Guide 38A, 39 p.

PLAYFORD, P. E., 1980, Devonian "Great Barrier Reef" of the Canning basin, Western Australia: American Association of Petroleum Geologists Bulletin, v. 64, p. 814–840.

PLAYFORD, P. E., 1981, Devonian reef complexes of the Canning basin, Western Australia: Geological Society of Australia, Fifth Australian Geological Convention, Guidebook, 64 p.

PLAYFORD, P. E., 1984, Platform-margin and marginal-slope relationships in Devonian reef complexes of the Canning basin, *in* Purcell, P. G., ed., The Canning Basin, Western Australia (Proceedings: Canning Basin Symposium): Perth, Geological Society of Australia and Petroleum Exploration Society of Australia, p. 189–234.

PLAYFORD, P. E., 1991, Reef development and extinction in response to sea level change in the Devonian of the Canning basin: American Association of Petroleum Geologists Bulletin, v. 75, p. 654.

PLAYFORD, P. E., AND COCKBAIN, A. E., 1976, Revised stratigraphy and facies nomenclature in Devonian reef complexes of the Canning Basin: Perth, Western Australia Geological Survey Annual Report for 1975.

PLAYFORD, P. E., AND LOWRY, D. C., 1966, Devonian reef complexes of the Canning Basin, Western Australia: Western Australia Geological Survey Bulletin 118, 150 p.

PLAYFORD, P. E., HURLEY, N. F. KERANS, C., AND MIDDLETON, M. F., 1989, Reefal platform development, Devonian of the Canning basin, Western Australia, *in* Crevello, P. D., Wilson, J. L., Sarg, J. F., and Read, J. F., eds., Controls on Carbonate Platform and Basin Development: Tulsa, Society of Economic Paleontologists and Mineralogists Special Publication No. 44, p. 187–202.

READ, J. F., 1973a, Carbonate cycles, Pillara Fm. (Devonian), Canning basin, Western Australia: Canadian Petroleum Geology Bulletin, v. 21, 38–51.

READ, J. F., 1973b, Paleo-environments and paleogeography, Pillara Fm. (Devonian), Western Australia: Canadian Petroleum Geology Bulletin, v. 21, p. 344–394.

SCHLAGER, W., 1981, The paradox of drowned reefs and carbonate platforms: Geological Society of America Bulletin, v. 92, p. 197–211.

SOUTHGATE, P. N., JACKSON, J., KENNARD, J. M., O'BRIEN, P. E., PASSMORE, W. L., LINDSAY, J. F., HOLMES, A. E., AND CHRISTIE-BLICK, N., 1991, Subsurface sequence stratigraphy of Devonian carbonates, Canning basin, Western Australia (abs.): American Association of Petroleum Geologists Bulletin, v. 75, p. 675.

SOUTHGATE, P. N., KENNARD, J. M., JACKSON, M. J., O'BRIEN, P. E., AND SEXTON, M. J., 1993, Reciprocal lowstand clastic and highstand carbonate sedimentation, subsurface Devonian reef complex, Canning basin, Western Australia, *in* Loucks, R. G., and Sarg, J. F., eds., Carbonate Sequence Stratigraphy: Tulsa, American Association of Petroleum Geologists Memoir 57, p. 157–179.

WALLACE, M. W., 1987, Sedimentology and diagenesis of Upper Devonian carbonates, Canning basin, Western Australia: Unpublished Ph.D. dissertation,

University of Tasmania, Hobart, 184 p.

WALLACE, M. W., KEAYS, R. R., AND GOSTIN, V. A., 1991, Stromatolitic iron oxides: Evidence that sea level changes can cause sedimentary iridium anomalies: Geology, v. 19, p. 551–554.

Ward, W. B., 1996, Evolution and diagenesis of Frasnian carbonate platforms, Devonian reef complexes, Napier Range, Canning basin, Western Australia: Unpublished Ph.D. Dissertation, State University of New York, Stony Brook, 540 p.

WARD, W. B., AND MEYERS, W. J., 1987a, Morphology and Evolution of an Upper Devonian reef-rimmed platform, Napier Range, Canning basin, Western Australia (abs.): Tulsa, Society of Economic Paleontologists and Mineralogists Annual Midyear Meeting, Abstracts, v. IV, p. 89.

WARD, W. B., AND MEYERS, W. J., 1987b, Neptunian-fracture control on platform-margin geometry, Upper Devonian reef complexes, Napier Range, Canning basin, Western Australia (abs.): Tulsa, Society of Economic Paleontologists and Mineralogists Annual Midyear Meeting, Abstracts, v. IV, p. 89.

WARD, W. B., KWONG, M. H., AND BANNER, J. L., 1994, Marine cement stratigraphy of neptunian fractures, Devonian reef complexes, Canning basin, Western Australia (abs.): Tulsa, American Association of Petroleum Geologists Annual Convention Program, p. 278.

COMPACTION AND THE DYNAMICS OF CARBONATE-PLATFORM DEVELOPMENT: INSIGHTS FROM THE PERMIAN DELAWARE AND MIDLAND BASINS, SOUTHEAST NEW MEXICO AND WEST TEXAS, U.S.A.

DAVID HUNT

Department of Earth Sciences, The University of Manchester, Oxford Road, Manchester, M13 9PL, United Kingdom

AND

WILLIAM M. FITCHEN

Exxon Production Research Company, P.O. Box 2189, Houston, Texas 77252

ABSTRACT: This study evaluates the role of compaction-induced differential subsidence as a control on carbonate-platform architecture, sequence development, and early diagenesis in Permian strata of the Delaware and Midland basins, west Texas and southeast New Mexico. The geometry and stratal patterns of compaction-modified sequences identified from outcrops in the Sierra Diablo, Guadalupe, and Brokeoff Mountains are comparable to those observed in seismic lines from the Northwest and Eastern shelf areas. The modification of platform geometries by differential compaction varies spatially and temporally in the studied area and is most important where Leonardian and Guadalupian highstand strata (sequences and sequence sets) have prograded over antecedent aggradational and/or erosional platform margins and wedge-shaped packages of compactible basin facies.

A "compaction hinge" delineates two mechanically distinct parts of a carbonate platform with contrasting stratal patterns, facies, and accommodation histories. The hinge normally develops in highstand strata deformed by compaction-induced differential subsidence over a preexisting platform margin. Updip of the compaction hinge, platform topsets are generally parallel and of constant thickness, whereas downdip, the same topsets thicken, dip, and diverge basinward toward a downward-deflected depositional platform margin. The alteration of platform architecture by differential compaction during a sea-level lowstand forms a compaction-modified sequence boundary. During a lowstand, flexural deformation driven by compaction-induced differential subsidence (at an outcrop-constrained average rate of 0.19 m/k.y.) modifies platform-top physiography and can generate relief and fractures to control facies patterns, stacking patterns, and the distribution of penetrative paleokarst. The resulting compaction-modified platform architectures and sequence boundaries show a characteristic suite of features that allows their identification using the criteria presented herein. The progressive further steepening of the dips formed across compaction-modified sequence boundaries during burial aids their recognition in outcrops and especially on regional seismic lines.

The control of compaction-induced differential subsidence on carbonate-platform development has previously been underestimated. Differential compaction can significantly modify the development and dynamics of carbonate platforms and holds important implications for the interpretation of stacking patterns, platform architecture, paleoecology, diagenesis, and accommodation history.

INTRODUCTION

Carbonate sequence stratigraphic models emphasize the role of inherited physiography as an important control on platform development (e.g., Crevello et al., 1989; Burchette and Wright, 1992; Schlager, 1993; Hunt and Tucker, 1993; Handford and Loucks, 1993; Loucks and Sarg, 1993; Sarg et al., 1995; Fitchen, 1996a). Antecedent physiography can control the area available for *in situ* carbonate production and the juxtaposition, volume, facies, and stacking patterns of stratal units formed in response to changes in relative sea level. The control of compaction on carbonate-platform development warrants full and careful consideration because flexural deformation induced by differential compaction can alter platform physiography, dynamics, stratal patterns, and diagenesis (Hunt et al., 1995, 1996).

This paper describes the attributes of compaction-modified platforms and sequence boundaries from outcrops in the Sierra Diablo, Guadalupe, and Brokeoff Mountains and compares these attributes to subsurface relationships from the Eastern and Northwest shelf areas that flank the Midland and Delaware basins. The principles that explain the development of compaction-modified sequence boundaries provide new insights into carbonate-platform dynamics in the Permian basin. We discuss the implications for the interpretation of stratal patterns, accommodation history, diagenesis, and the paleoecology of carbonate platforms in the study area. Although our discussion focuses on platforms in the Permian basin, the principles developed and conclusions reached apply to carbonate platforms in general.

Compaction

Compaction-driven subsidence can contribute as much as one-fifth of the total subsidence in sedimentary basins (Steckler and Watts, 1978; Watts, 1981; Bond and Kominz, 1984). However, the compaction process is often underestimated as a control on structural deformation and stratigraphy (Suppe, 1985; Gay, 1989; Hunt et al., 1995, 1996). For example, Jervey (1988) argued that compaction does not directly control sequence development. Such a view reflects the often-held perception of compaction as a background process that takes place at a low and gradually decreasing rate over a time interval comparable to thermal subsidence (e.g., 10^6 to 10^7 yr). On the contrary, this paper, which builds from the studies of Hunt et al. (1995, 1996) and Saller (1996), presents evidence to the contrary that compaction-induced subsidence can

Advances in Carbonate Sequence Stratigraphy: Application to Reservoirs, Outcrops and Models, SEPM Special Publication No. 63

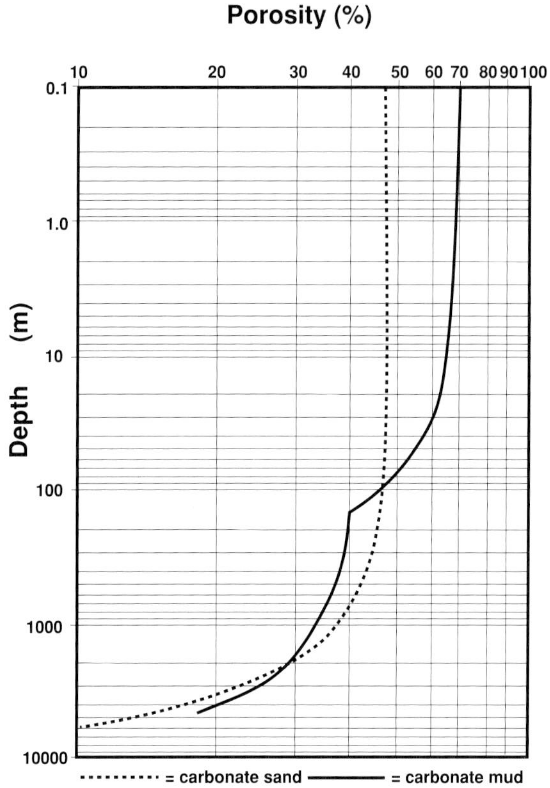

FIG. 1.—Contrasting porosity vs. depth curves for carbonate muds and sands. In the absence of early marine cementation, carbonate sands compact more slowly than muds (redrawn from Goldhammmer, 1997).

directly modify sequence development and control the architecture and diagenesis of carbonate platforms.

Experimental and empirical data established in the past 20 years show that in the absence of early cementation, carbonate lime muds with initial porosities of 70%–80% can undergo a thickness reduction of 50% within 100–200 m of burial (e.g., Fig. 1; Enos and Sawatsky, 1981; Shinn and Robbin, 1983; Goldhammer, 1997). Significant potential exists to derive accommodation space from the compaction of carbonate muds in the shallow subsurface (<1–200 m). In comparison, carbonate sands compact more slowly than muds, even in the absence of early marine cements, as indicated by the quite different trends of porosity vs. depth curves for carbonate sands and muds for 1–200 m of burial (Fig. 1; Goldhammer, 1997). Such lithological contrasts can promote differential compaction over platform margins (e.g., Fig. 2; Hunt et al., 1995, 1996; Saller, 1996). Differential subsidence results in the tilting of strata and can generate relief, cause thickness and facies changes, and fracture lithified sediments. Such features allow identification of compaction-induced subsidence as a control on stratigraphy and structure (Gay, 1989; Hunt et al., 1995, 1996). In contrast, in the case of uniform compaction (Fig. 3), there is little chance of being able to isolate the contribution to accommodation development of compaction-driven subsidence compared to the contribution from tectonic subsidence and/or a eustatic sea-level rise.

Differential Compaction

Compaction-induced differential subsidence can occur over areas of lateral facies change, thickness change, and/or the antecedent relief of a relatively incompactible substrate (Fig. 2; Gay, 1989; Hunt et al., 1995, 1996). Some important prerequisites that promote differential compaction across carbonate-platform margins include (1) deposition of wedge-shaped stratal units of highly compactible strata, (2) abrupt lateral facies transitions, (3) juxtaposition of strata with markedly different compaction potentials (e.g., cemented platform margins adjacent to compactible uncemented basinal mudstones), (4) deposition of basin and foreslope facies with high depositional porosities (<70%–80%, Fig. 1; e.g., mudstones, wackestones, and/or fine siliciclastics), (5) burial of relatively incompactible antecedent topography, such as cemented platform margins, foreslopes, reefs, and buildups, and (6) the ability to build high-relief steep slopes (that may be onlapped by a thick wedge of compactible facies). Figure 2 illustrates some common platform-basin profiles in which thinning of compactible sedimentary units toward the platform may induce differential compaction and result in subsidence of the terminal shelf-slope break, the basinward subsidence and tilting of topset strata, and the formation of a "compaction hinge" marked by a secondary slope break located over an antecedent incompactible platform margin (Hunt et al., 1995).

Outcrop studies and theoretical considerations independently led Hunt et al. (1995) and Saller (1996) to suggest that differential compaction can also occur in response to the spatial variability of an applied load, even where the compacting unit has constant thickness. Figure 4 illustrates how the load distribution within each sigmoidal clinoform package causes the resultant compaction-induced differential subsidence to decrease both platformward and basinward. In this example, the locus of compaction-induced subsidence is coincident with the greatest load—the position of the terminal shelf-slope break in each sigmoidal progradational package (Fig. 4; Hunt et al., 1995; Saller, 1996). However, the load distribution during progradation is a function of slope height, slope angle, and the lateral accretion of each sigmoidal accretionary package. Therefore, the locus of loading-induced differential compaction is not always coincident with the platform margin (Fig. 5).

Timing of Differential Compaction Versus Deposition and Recognition Criteria

Quite different platform physiography, stratal patterns, and architecture can form depending on whether compaction-induced subsidence occurs during and/or after deposition (Fig. 6). When differential compaction takes place during progradation, a basinward increase of compaction-driven accommodation will result in the divergence and thickening of topsets toward the platform margin to form a "rollover" type geometry in which younger topsets dip less steeply than their precursors (Fig. 6A). It is significant that there is normally a temporal decrease in rollover within a highstand sequence or sequence set because the basinward dips of the oldest prograd-

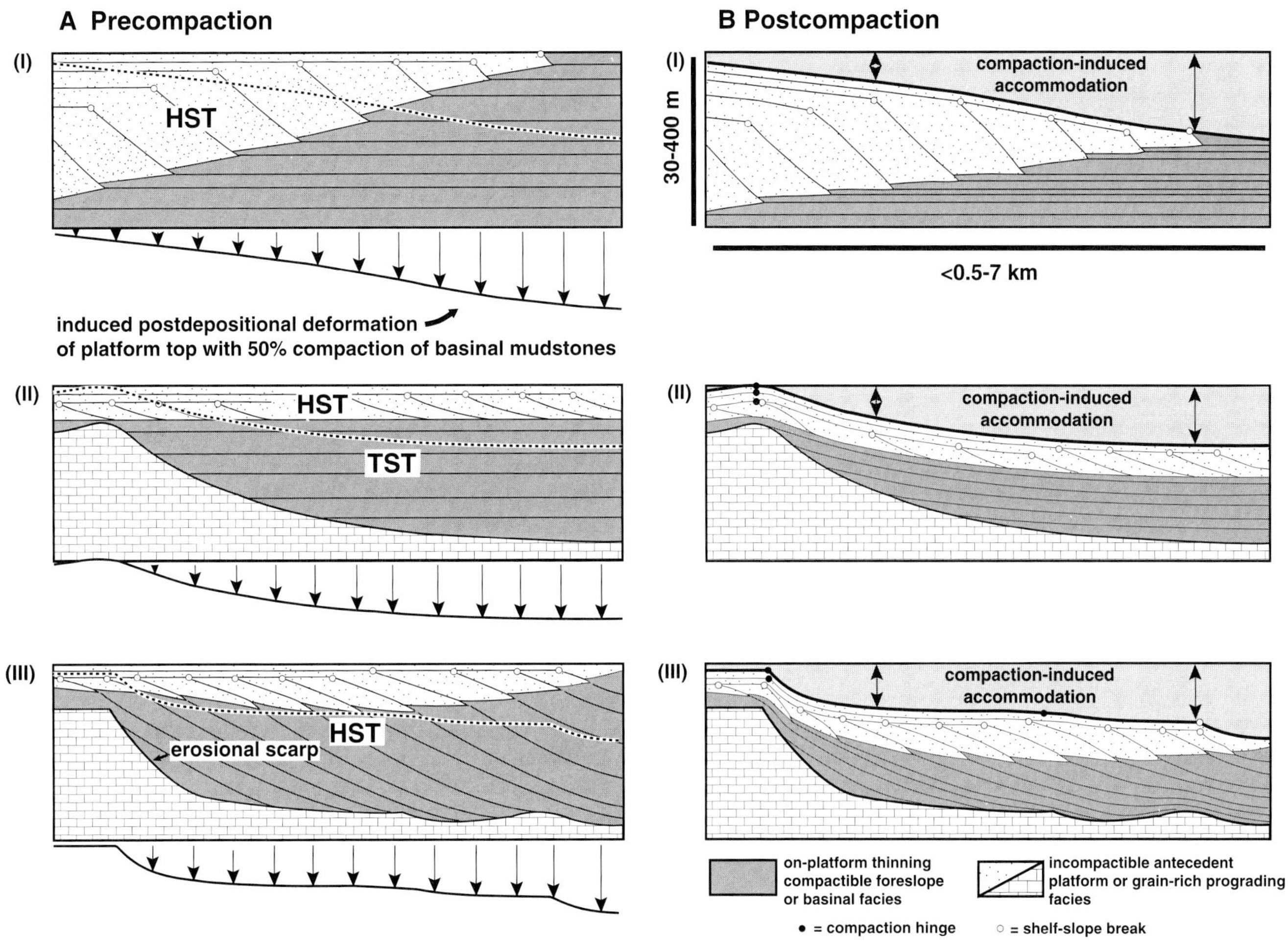

FIG. 2.—Three common situations (I–III) in which differential compaction can occur over carbonate-platform margins. Parallel subhorizontal platform topsets indicate that no differential subsidence occurred during highstand progradation. On-platform thinning of compactible basin facies occurs through (I) basinward climbing progradation, (II) onlap against a drowned platform margin, and (III) basinward thickening of foreslope facies to fill antecedent erosional relief. (A) In the precompaction state, the dark solid line below each of the profiles in the column indicates a 50% thickness reduction of the compactible lithology. The compaction induces subsidence and deformation of the platform top as marked by the dashed line. (B) In the postcompaction state, a secondary slope break forms over a compaction hinge. These idealized diagrams make several important assumptions, namely (1) that only the unit with the gray-toned pattern compacts, by 50%, (2) that other lithologies are incompactible (e.g., contain early marine cements), and (3) that none of the compaction takes place during deposition of the rocks with the random-dot pattern that overlie the rocks with the gray-toned pattern (modified from Hunt et al., 1995, 1996).

ing strata are further steepened during subsequent platform aggradation and progradation. Our outcrop observations and computer modeling (D. Waltham and D. Bosence, pers. communs., 1997–1998) show this relationship to hold true even when there is a concurrent increase of platform-to-basin relief though a highstand.

In contrast, when there is a delay of compaction so that the induced differential subsidence occurs after progradation, then the topsets *within* sequences are parallel and maintain a uniform thickness. In this situation, flexure induced by compaction-induced differential subsidence following progradation (e.g., during subaerial hiatus; Fig. 6B) will tilt platform strata basinward to form postdepositional relief and a "compaction-modified" sequence boundary (Hunt et al., 1995, 1996). The recognition of compaction-modified sequence boundaries relies on the identification of more than one of the following key attributes: (1) regional marker surfaces from which the contribution of tectonic subsidence can be evaluated, (2) markers of the paleohorizontal (geopetals), (3) formation of a compaction hinge over an antecedent platform margin, (4) topsets that dip and diverge basinward from a compaction hinge toward a subsided depositional platform margin, and (5) onlap and landward thinning of younger sequences onto basinward-tilted topsets. These and other features are shown in Figure 7.

THE PERMIAN BASIN

In the following sections we evaluate compaction-induced differential subsidence as a control on carbonate-platform architecture and development in the Permian basin of west Texas and southeast New Mexico (Fig. 8). Numerous publications describe the geology and sequence stratigraphy of the Permian basin, where a long tradition of comparison between outcrop and the subsurface reflects the proximity of important hydrocarbon reservoirs to spectacular outcrop analogues (e.g., King, 1942,

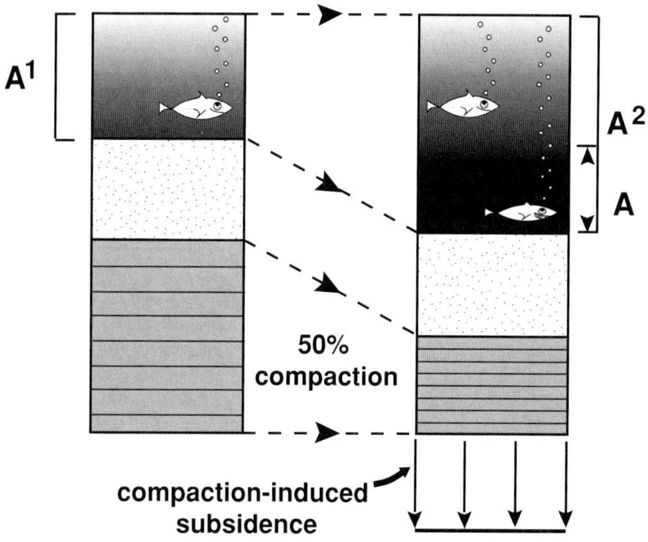

FIG. 3.—Accommodation developed through uniform 50% compaction of lime mudstones loaded by a grain-rich package. In this example, compaction is uniform, and there is little chance of isolating the contribution of compaction-induced subsidence to accommodation development versus that of a eustatic sea-level rise and/or tectonic subsidence. A^1 = precompaction accommodation space; A^2 = postcompaction accommodation space; A = compaction-induced accommodation (modified from Hunt et al., 1995, 1996).

1948, 1965; Adams and Freznel, 1950; Newell et al., 1953; Hayes, 1964; Hileman and Mazzullo, 1977; Hills, 1985; Sarg and Lehmann, 1986; Reid et al., 1988; Borer and Harris, 1991, 1995; Kerans et al., 1992, 1994; Sonnenfeld and Cross, 1993; Stafleu and Sonnenfeld, 1994; Grant et al., 1994; Tinker, 1996; among many references).

Sarg and Lehmann (1986), following the principles of sequence stratigraphy outlined in Payton (1977), led the application of sequence stratigraphy in the Permian basin with a direct comparison of stratal relationships at outcrop and in seismic data. Subsequently, the work of Stafleu and Sonnenfeld (1994) highlighted the limitations of such analogies due to the resolution of regional 23–35 Hz seismic data. However, we further pursue the original course set by Sarg and Lehmann (1986) with a comparison of the architecture of compaction-modified platforms interpreted from outcrops to carbonate-platform geometries commonly seen in seismic sections from the Permian basin. First, we briefly introduce the basin setting.

Tectonic Framework

The north-northwest–trending Delaware basin, Midland basin, and surrounding platforms represent parts of a segmented Pennsylvanian-Permian foreland basin formed inboard of the

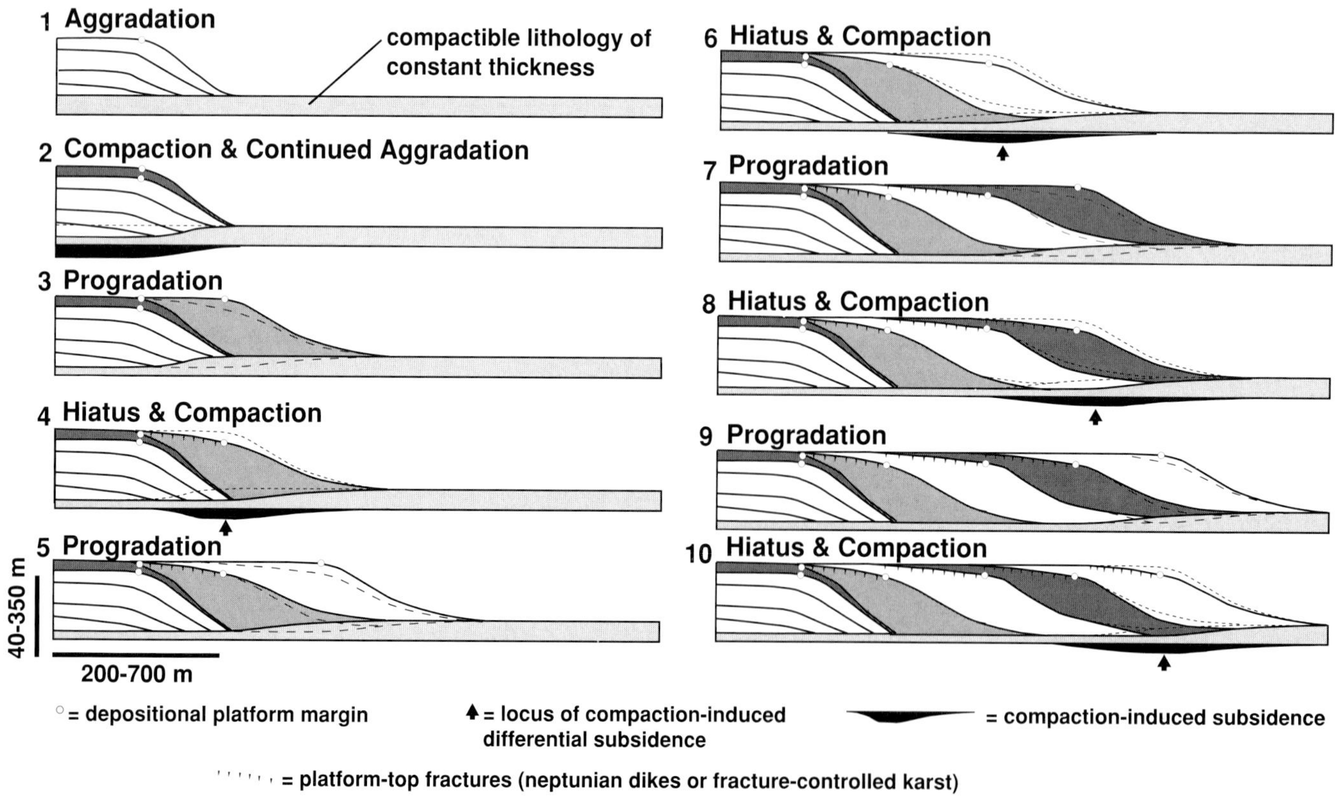

FIG. 4.—Compaction-induced differential subsidence resulting from spatial variation in loading. In this model, compaction-induced subsidence follows deposition of each unit during a hiatus. The maximum reduction in the thickness of the compactible lithology is 60%, induced by a sedimentary column equivalent in height to the distance between the depositional platform margin and the toe of slope. The locus of the applied load and compaction-induced subsidence decreases both landward and basinward of the platform margin. The resulting differential compaction-induced subsidence forms rollover-type platform geometries, where outer-platform topsets steepen toward a depressed shelf-slope break, a common platform geometry in the Permian basin (e.g., Dunham, 1972; Fig. 9). The result of this flexure is extension and fracturing of the platform top, forming neptunian dikes and/or fracture-controlled karst (modified from Hunt et al., 1995).

Ouachita and Marathon fold-and-thrust belts. Yang and Dorobek (1992) related the configuration of basement blocks and synorogenic sediment wedges to intraforeland loading induced by uplift of the Central Basin platform. Flexural subsidence and the intraforeland uplift of basement blocks culminated in the early to middle Wolfcampian to form a regional unconformity (Kluth and Coney, 1981; Thomas, 1983; Hills, 1985; Ross, 1986; Wuellner et al., 1986; Ewing, 1991; Yang and Dorobek, 1992). A period of isostatic adjustment followed.

Beginning in the middle Wolfcampian, subsidence rates in the Delaware and Midland basins exceeded those on adjacent platforms (Horak, 1985; Yang and Dorobek, 1992). Transgression and onlap of the uplifted blocks by marine carbonates occurred as the platform areas subsided (King, 1965; Hills, 1985). Following transgression, mixed carbonate-siliciclastic sedimentation generally kept pace with or exceeded rates of relative rise in sea level, resulting in significant platform aggradation and progradation. The relatively constant thickness of late Wolfcampian–Guadalupian age carbonate platforms attests to their development after differential uplift and active faulting.

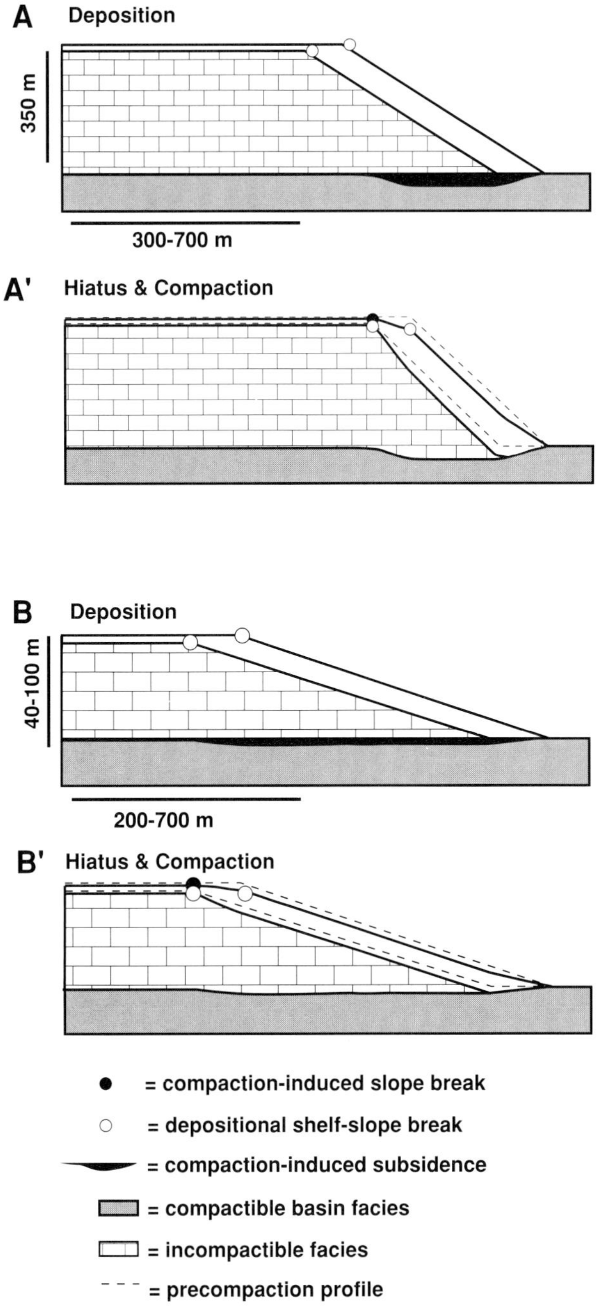

FIG. 5.—The control of slope height and slope angle on load-induced differential compaction. In the situations shown, there is a delay between loading and the resulting compaction-induced differential subsidence (A to A′ and B to B′, respectively). Note the secondary basinward dip induced in toplap strata to form rollover-type geometries.

Stratigraphy

The Leonardian stratigraphy of the Permian basin region is best known from outcrop studies in the Glass, Sierra Diablo, and Guadalupe Mountains of west Texas (King, 1931, 1942, 1948, 1965; Ross, 1963, 1986; McDaniel and Pray, 1967; Rogers, 1972; Pray et al., 1980; Fitchen et al., 1995; Wilde, 1995;

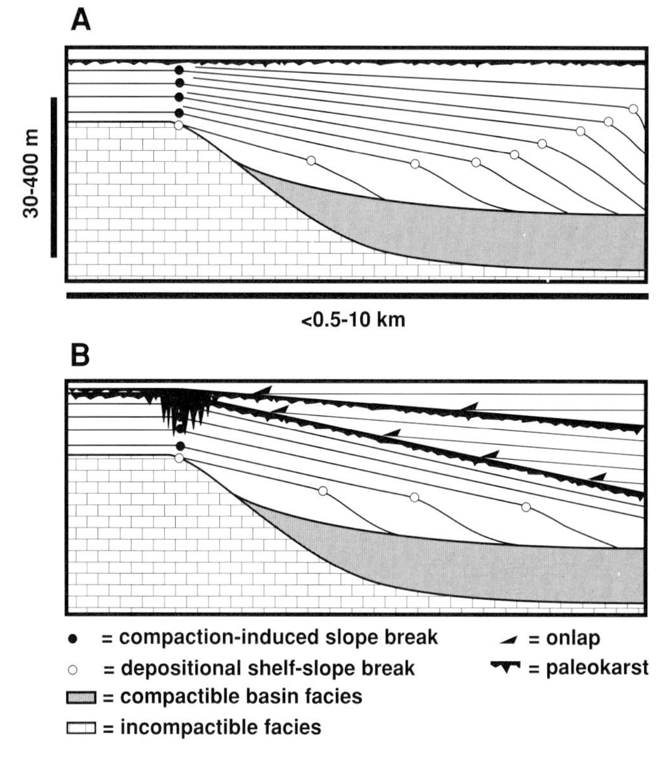

FIG. 6.— Differential compaction over carbonate-platform margins results in contrasting stratal patterns that depend on whether the induced subsidence occurred during or after progradation. A compaction hinge separates parallel topsets from toplap surfaces that diverge basinward in both situations. Note steeper dip of older topsets. (A) Syndepositional differential subsidence; toplap surfaces diverge and split basinward. (B) Postdepositional differential subsidence; toplap surfaces are parallel within sequences but abruptly change dip across compaction-modified sequence boundaries, against which less-steep surfaces of the succeeding sequence onlap. In these quite different situations (A, B), the external geometry is similar; the distinction relies on examination of internal stratigraphic relationships. Significant contrast exists between the relief inherited by successor platforms across sequence boundaries in these two cases. In A, the succeeding platform inherits the topography of a flat-topped shelf, whereas in B, the physiography is akin to a distally steepened ramp.

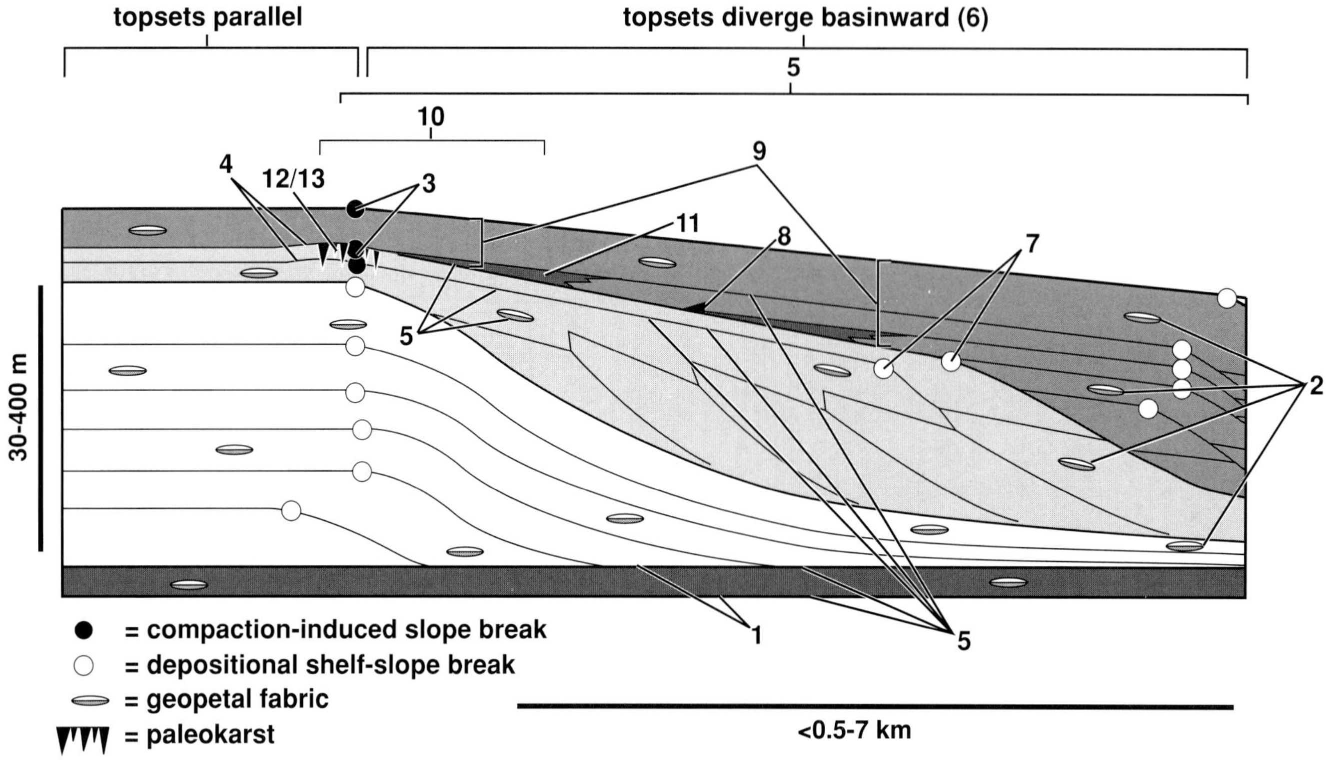

FIG. 7.—Criteria for recognition of compaction-modified stratal patterns at outcrop and in the subsurface: (1) the presence of regional surfaces from which the contribution of differential tectonic subsidence can be evaluated, (2) recognition of markers of the paleohorizontal (geopetals), (3) development of a compaction hinge, (4) elevated topsets across the compaction hinge, (5) the steeper attitude of tilted topsets compared to regional dips, (6) divergence of topsets basinward from the compaction hinge, (7) the presence of depressed depositional platform margin, (8) onlap of basinward-tilted topsets and/or toplap surfaces toward the compaction hinge by less steeply dipping younger strata, (9) the thinning of sequences across tilted topsets toward a compaction hinge, (10) the development of basin-facing monoclines localized above antecedent platform margins, (11) facies changes basinward across the compaction hinge, and (12) the localization of fractures across the compaction hinge that (13) can localize karst penetration or meteoric influence.

Fitchen, 1997; among other references; Fig. 8) and from subsurface studies in the Delaware and Midland basins and surrounding platforms (e.g., Silver and Todd, 1969; Mazzullo et al., 1987; Mazzullo and Reid, 1987, 1989; Saller et al., 1989; Sarg, 1989). Leonardian platform-to-basin systems are characterized by high-angle carbonate-ramp (3°–5° slopes) to rimmed-shelf (as steep as 35° slopes) platform sequences and mixed carbonate-siliciclastic basinal sequences (e.g., Fig. 9). The basinal siliciclastic wedges onlap the toe of slope of older platform sequences and correlate to subaerial unconformities on the platforms. The basinal carbonate units consist of organic-rich argillaceous mudstones whereas the platform carbonate units generally consist of carbonate grain–rich, siliciclastic-poor facies. The Leonardian is subdivided into eight high-frequency sequences (L1–L8) on the basis of outcrop studies in the Sierra Diablo (Fitchen et al., 1995; Wilde, 1995; Fitchen, 1997) and Guadalupe Mountains (Kerans et al., 1993; Kerans and Fitchen, 1995; Figs. 8–10). Correlation of the L1–L8 sequences to equivalent subsurface lithostratigraphic and chronostratigraphic units is based on fusulinid and conodont biostratigraphic studies and lithostratigraphic comparison (Fitchen et al., 1995; Wilde, 1995). Leonardian sequences exhibit a fairly consistent stacking pattern: the L1 sequence (Wichita-Albany Formation or Abo Formation) is relatively progradational, the L2–L3 sequences (lower and middle Clear Fork Formations) are relatively aggradational, the L4–L6 sequences (upper Clear Fork and Glorieta Formations) are highly progradational, and the L7–L8 sequences (lower part of the San Andres Formation or the Holt Formation) are retrogradational.

The Guadalupian stratigraphy of the Permian basin region is best known from outcrop studies in the Guadalupe Mountains (King, 1942, 1948; Newell et al., 1953; Hayes, 1964; Tyrrell, 1969; Hileman and Mazzullo, 1977; Kerans et al., 1992; among other references) and from nearby subsurface studies (e.g., Meissner, 1972; Ward et al., 1986). Guadalupian platform-to-basin systems developed from pure carbonate ramps (G1–G4 sequences; middle part of the San Andres Formation) to low-angle, mixed carbonate-siliciclastic rimmed shelves (G12–G14 sequences; upper part of the San Andres Formation–lower part of the Grayburg Formation) to high-angle, carbonate-rimmed shelves (G15–G29 sequences; upper Grayburg, Queen, Seven Rivers, Yates, and Tansill Formations; Figs. 9, 10). Basinal lowstand to transgressive systems tract siliciclastic wedges are developed in most of the post-G4 sequences (Brushy Canyon, Cherry Canyon, and Bell Canyon Formations; Figs. 9, 10). Most of the Guadalupian sequences exhibit progradation of ramp–shelf-margin and distal-ramp–slope facies over basinal siliciclastic wedges; the most pronounced progradational phases

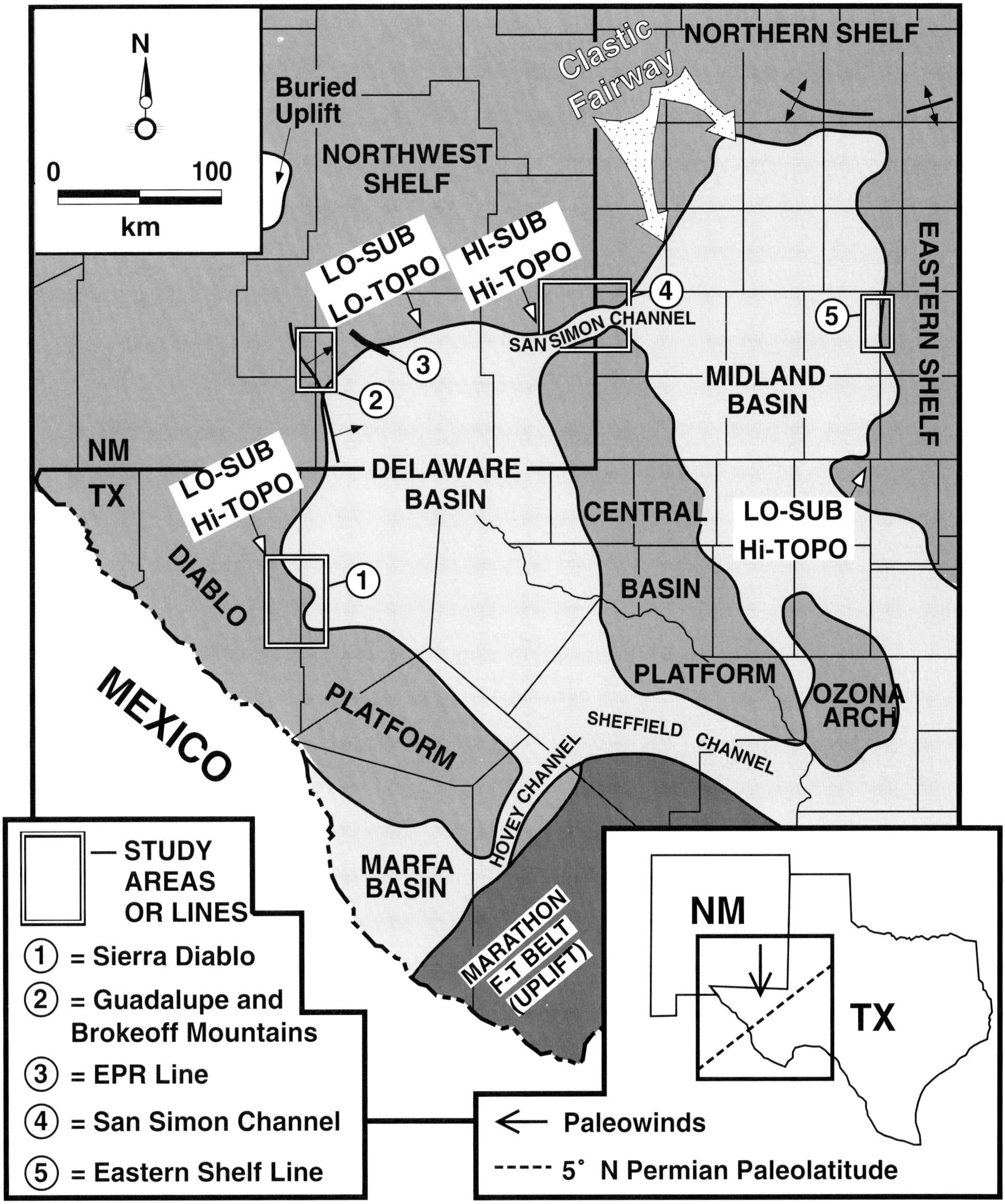

FIG. 8.—Location map showing the early Leonardian paleogeography of the Permian Midland and Delaware basins and adjacent uplifted platform areas, west Texas (TX) and southeast New Mexico (NM) (modified after Wright, 1962). Bottom right inset shows location of main map area and orientations of the 5° north paleolatitude and the dominant trade winds (from Fischer and Sarnthein, 1988). On the main map, labels in the unshaded boxes refer to the prevailing platform-basin relief and subsidence rates across the Permian basin: Lo- and Hi-Sub = low and high subsidence rates, respectively, Lo- and Hi-Topo = low and high topographic relief platforms, respectively (Fitchen, 1996b). F-T belt = fold-and-thrust belt.

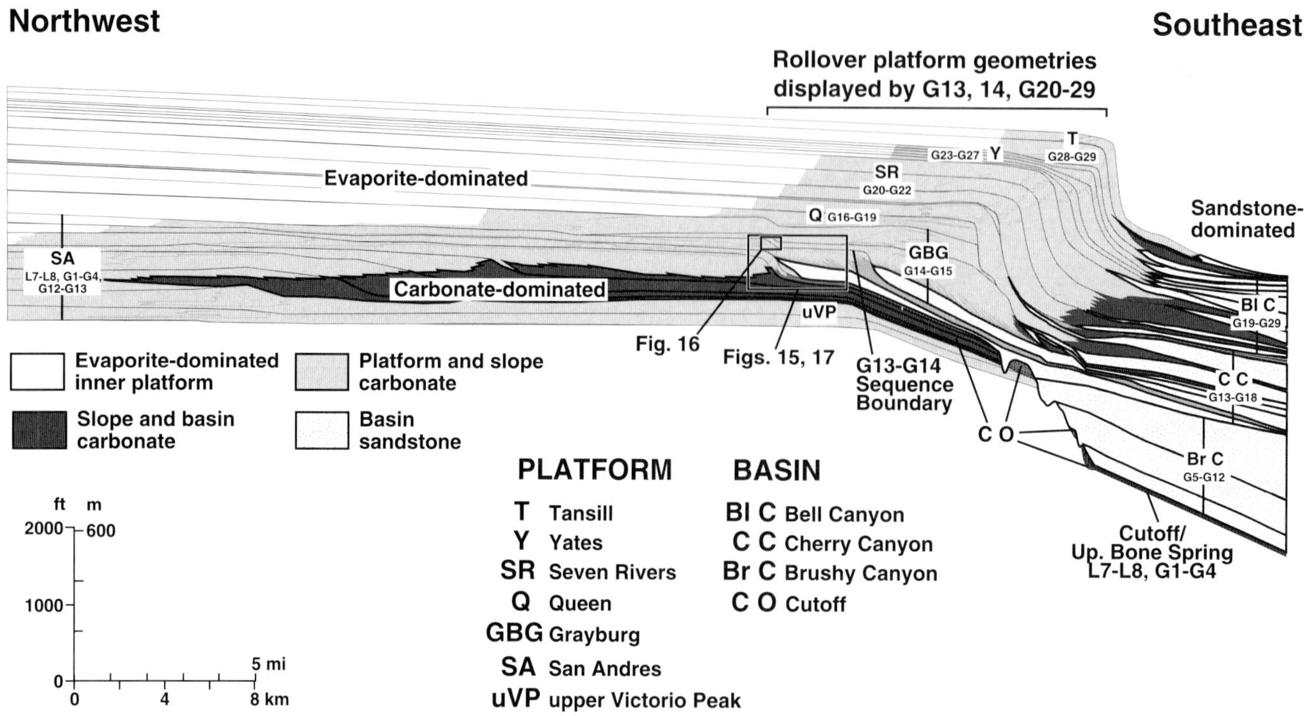

FIG. 9.—Sequence stratigraphic framework of the Guadalupe and Brokeoff Mountains showing nomenclature of sequences used in comparison to classic formation boundaries. Note the development of rollover geometries in sequences G13, G14, and G20–G29 that prograde over on-platform thinning wedges of basinal sandstone and carbonate (modified from Kerans et al., 1992).

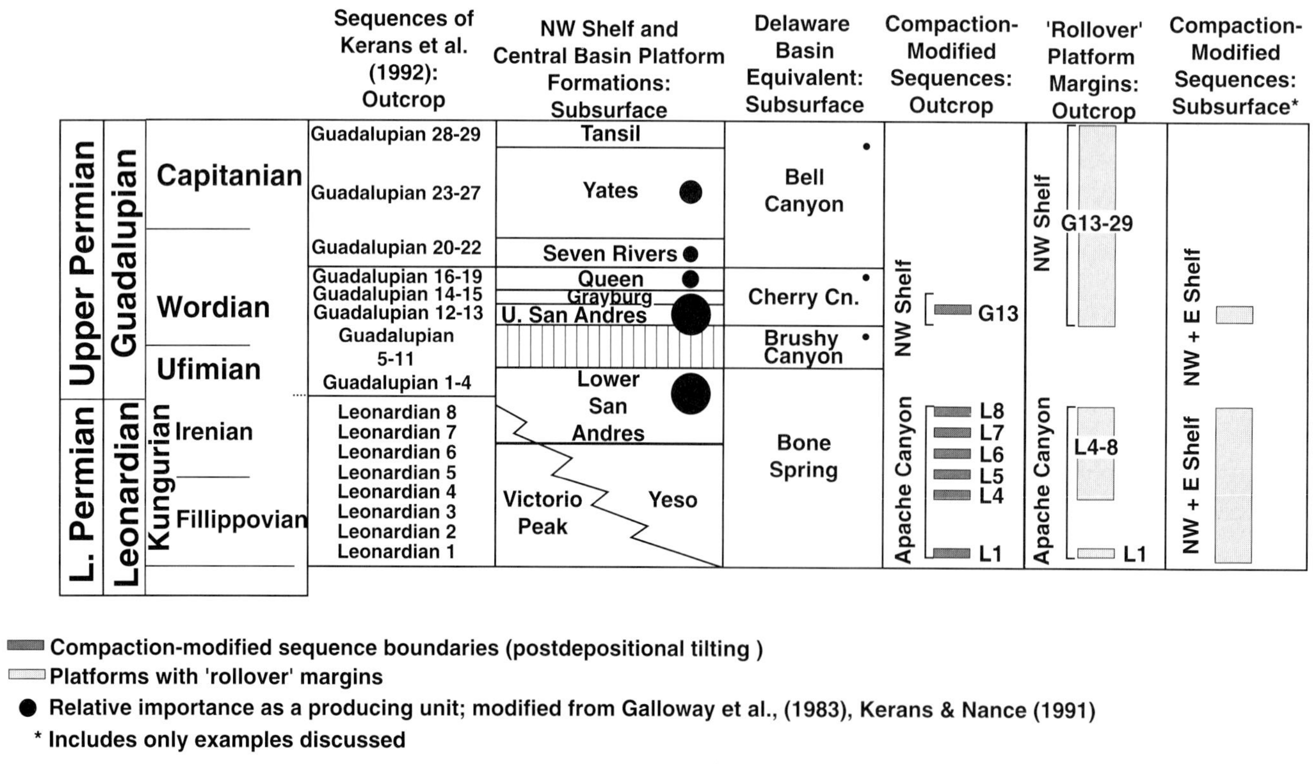

FIG. 10.—Nomenclature of Leonardian–Guadalupian stratigraphy on the Northwest shelf, Central Basin platform, and Delaware basin showing the distribution and relative volumetric importance of hydrocarbon-producing units in the Permian basin (following Galloway et al., 1983; Ward et al., 1986; Kerans and Nance, 1991; Kerans and Fitchen, 1995). The stratigraphic distribution of compaction-modified sequences discussed herein is shown in the right and third-from-right columns. However, this figure is not meant to imply that the compaction-modified sequences are time specific and therefore developed in platforms of a certain age across the whole Permian basin. Along-strike variations in sediment supply, accommodation, and shelf-to-basin relief mean that exact timing and importance of compaction-related modification of platforms vary along strike and though time. Modified from Kerans and Fitchen (1995).

include the G13–G15 sequences (upper San Andres–lower Grayburg) and the G16–G27 sequences (Queen, Seven Rivers and Yates Formations equivalent to the Goat Seep Formation and Capitan Formation shelf-margin facies).

COMPACTION-MODIFIED SEQUENCES SEEN AT OUTCROP

We here examine the stratal geometry and attributes of compaction-modified sequences interpreted from the Sierra Diablo, Brokeoff, and Guadalupe Mountains (Fig. 8). These examples of compaction-modified geometries and relationships are at a scale comparable to seismic sections, allowing evaluation of the control of compaction-induced differential subsidence on the platform-to-basin systems system as a whole. The dips associated with compaction-induced deformation are commonly low (0.5°–5°). Such low dips are obvious to the eye, but difficult to measure without significant error when using a clinometer alone. It is important to point out that detailed mapping onto photographic panoramas in combination with multiple measured sections along kilometer-scale outcrops accurately constrains the dips discussed in the following section.

Apache Canyon, Sierra Diablo Mountains

Fitchen (1997) developed a sequence stratigraphic framework for Apache Canyon by building on the classic study of King (1965). Fitchen (1997) divided the Leonardian succession into six sequences (L1–L6), each of approximately 1.6–2.0 m.y. duration (Fig. 11). These stack to form two larger composite sequences, the lower composed of sequences L1–L4 and the upper of L5–L6 (Fig. 12; Fitchen et al., 1995; Fitchen, 1997). Figure 11 shows two cross sections constructed by Fitchen (1997) with a 1.3–2 km along-strike spacing through the Wolfcampian–Leonardian platform-to-basin transition exposed in Apache Canyon. These cross sections are constrained by a combination of detailed mapping and 23 complimentary measured sections. Fitchen (1997) correlated sequences L5–L8 from the Sierra Diablo to sequences L5–L8 identified by Kerans et al. (1992) on the western escarpment of the Guadalupe Mountains (e.g., Fig. 9) on the basis of sequence stratigraphic, biostratigraphic, and lithostratigraphic considerations. Of particular interest in Apache Canyon are the relationships of sequences L1 and L4–L6, developed over the antecedent Wolfcampian and L2–L3 platform margins, respectively

Sequence L1.—

The L1 sequence nucleated on the remnant topography of an eroded middle to late Wolfcampian precursor platform cut by a 170-m-relief, stepped erosional listric scarp (Fig. 11; Fitchen et al., 1995; Fitchen, 1997). Aggradational peritidal cycles characterize middle Wolfcampian deposition (mW1–mW2, Fig. 11). Strata of late Wolfcampian age exhibit predominantly subtidal cyclic deposition, landward stepping (i.e., backstepping) of the platform margin, and the onset of platform-margin collapse in response to a regional relative rise in sea level (Fitchen, 1997). The subsequent relative fall in sea level (>43 m) at the end of the Wolfcampian produced karstic breccias across the eroded platform margin. The erosional topography cut into Wolfcampian strata is thus interpreted to be a product of both submarine and subaerial processes (Fitchen et al., 1995; Fitchen, 1997; Fig. 11).

The L1 sequence displays 3–4 km of basinward progradation and 10–15 m of platform aggradation; the sequence is 11.5–220 m thick (Figs. 11, 12). A subaerial-exposure profile characterized by paleokarst, calcrete, and limestone conglomerate marks the upper surface of L1 (Fitchen et al., 1995; Fitchen, 1997). Detailed mapping by Fitchen (1997) showed the most spectacular paleokarst across the L1 sequence boundary to be localized to a 200–500-m-wide zone located over and basinward of the underlying Wolfcampian scarp (Fig. 13A). The penetration depth of this paleokarst in places is as much as 25–30 m but decreases basinward of the antecedent Wolfcampian scarp and is notably less pronounced across the outermost 0.5 km of the L1 platform (Fig. 13; Fitchen, 1997).

Thickness of the L1 sequence is generally proportional to that of strata removed at the Wolfcampian–Leonardian unconformity (Fig. 11). The L1 foreslope facies (mudstones-wackestones of the Bone Spring Formation) fill the erosional relief cut into the Wolfcampian (mW1–mW2, uW1; Figs. 11, 13). On the northern wall of Apache Canyon, the L1 toplap surface rises basinward to a crest above the underlying Wolfcampian erosional scarp, landward of which it has a concave-up (synclinal) morphology (Figs. 11, 13). Basinward of the underlying Wolfcampian scarp, the L1 toplap surface dips down to the basin at an angle of 2.6°. Consequently, the terminal L1 shelf-slope break is 67 m below its oldest topsets and forms a significantly less pronounced feature in cross section than the slope break developed over the Wolfcampian escarpment (Figs. 12, 13); it is interesting to speculate which feature would be interpreted as the shelf-slope break on regional seismic data. The relief of the L1 sequence boundary must have developed prior to L2 deposition since it is onlapped by the succeeding L2 lowstand–transgressive systems tract (Fig. 11). On the southern side of the canyon, the L1 sequence displays less relief (58 m), with a gentle basinward dip of 0.6 (abruptly increasing to 5.2° basinward of the underlying Wolfcampian scarp (Figs. 11, 13).

Interpretation.—Formation of the basinward-descending L1 toplap surface could result from (1) compaction-induced differential subsidence, (2) a forced regression, (3) differential tectonic subsidence, and/or (4) erosional ravinement during transgression (Fitchen et al., 1995; Hunt et al., 1995, 1996; Fitchen, 1997). We are able to rule out the latter two factors because the thickness and preservation of L1 shelf-margin facies are relatively constant and the dip of Wolfcampian peritidal cycles constrains the contribution of tectonic deformation and tilting to the architecture of L1 (Figs. 11, 13). Therefore, the relief displayed by the L1/L2 sequence boundary basinward of the underlying Wolfcampian scarp resulted from the effects of differential compaction and/or relative fall in sea level (Fitchen et al., 1995; Hunt et al., 1995; Fitchen, 1997).

Distinction between the effects of relative fall in sea level and compaction-induced differential subsidence is difficult in the absence of aggradational topsets. Figure 13 illustrates an attempt to isolate these controls; the P1 profile shows an attempt

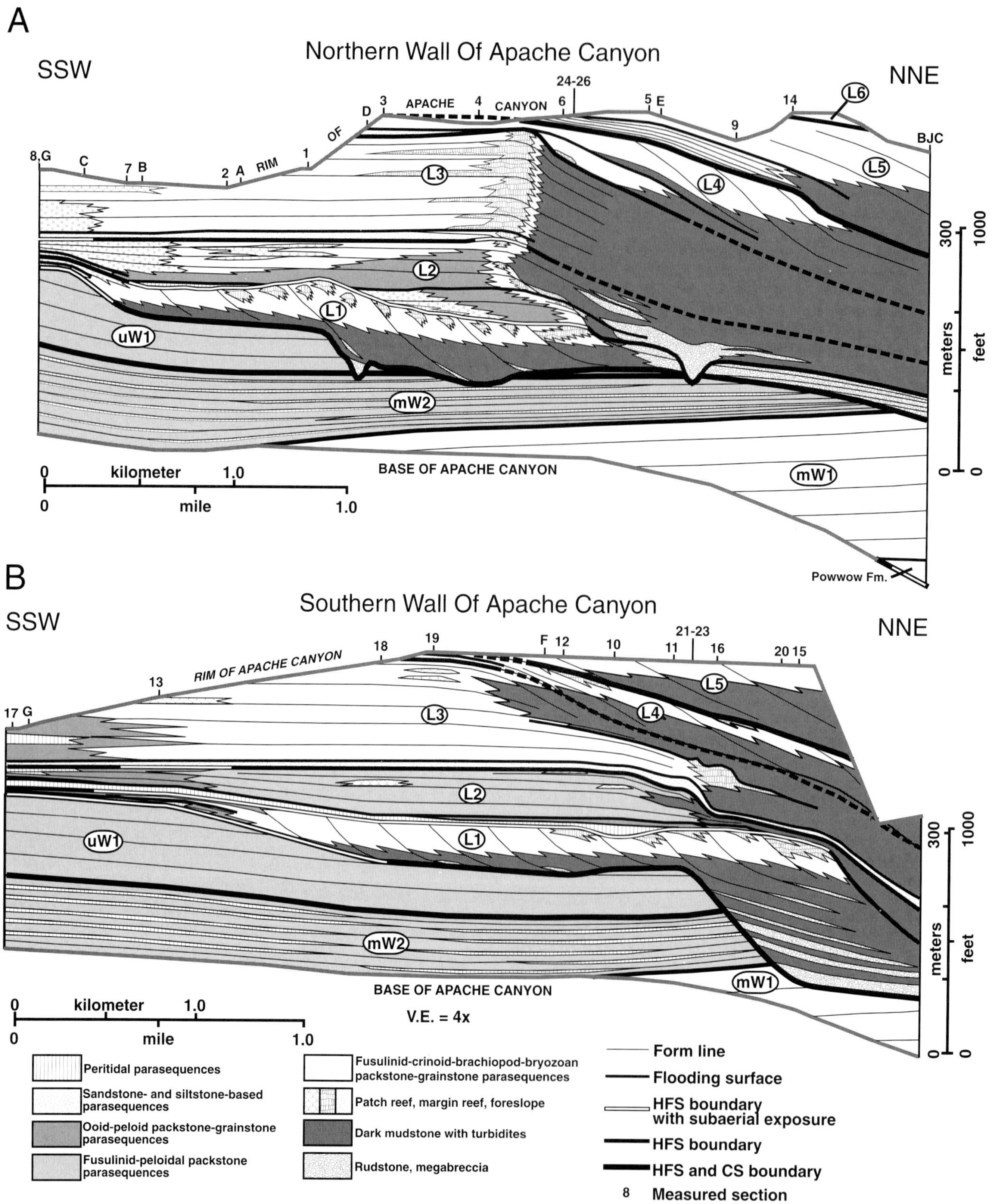

FIG. 11.—Sequence stratigraphic cross sections through Wolfcampian (mW1, mW2, and uW1) and Leonardian (L1–L6) sequences on the northern (A) and southern (B) sides of Apache Canyon, Sierra Diablo Mountains, west Texas (from Fitchen, 1997). The Wolfcampian and Leonardian sequences are separated by a prominent stepped unconformity that cuts 170 m through inner-platform Wolfcampian peritidal cycles. Deformation of these peritidal cycles allows us to constrain the contribution of tectonism to the geometry of Leonardian platforms (L1–L6; see Fig. 13). The Leonardian strata demonstrate the modification of carbonate-platform architecture by compaction-induced differential subsidence over antecedent erosional (L1, over the Wolfcampian) and aggradational platform margins (L4–L6, over L2–L3). Progradational sequences are tilted basinward over antecedent platform margins, interpreted to be compaction hinges, forming relationships directly comparable to time equivalents observed in seismic section (e.g., Figs. 18–21). See text for detailed discussion. HFS = high-frequency sequence; CS = composite sequence; mW and uW = Wolfcampian Hueco Group; L1–L6 = Leonardian. Numbers at the top of the section refer to measured sections of Fitchen (1997).

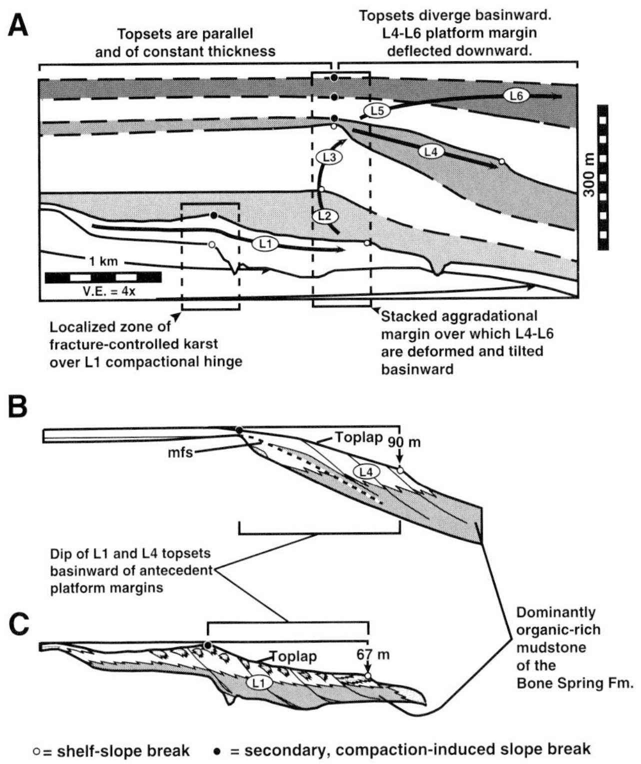

FIG. 12.—(A) Stacking of high-frequency sequences L1–L6 to form two lower-order composite sequences; the lower one is composed of lowstand (L1), transgressive (L2–L3), and highstand (L4) sequence sets, and the upper one is composed of transgressive (L5) and highstand (L6) sequences. L1 and L4–L6 topsets dip basinward over erosional and aggradational platform margins, which form compaction hinges. The compaction-modified relationships of sequences L4–L6 were exaggerated through burial and are remarkably similar to those of Leonardian sequences seen in the subsurface (Figs. 18–21). Viewing this section as if it were a seismic line, the elevated topsets in L1 over the compaction hinge would probably be mistaken as a terminal reefal shelf margin. The terminal L1 shelf-slope break is subdued owing to basinward dip (steepening) of topsets by differential compaction. In this case, there is potential to misinterpret L1 as aggradational; over the Wolfcampian scarp, the basinward-dipping subsided L1 topsets would be interpreted as slope facies, and therefore L2 would be interpreted as a progradational sequence. A similar relationship is developed in L4. (B and C) Detail of the architecture of sequences L1 and L4, both characterized by parallel platform strata landward of an antecedent margin, basinward of which topset and toplap strata dip basinward toward a subsided platform margin.

to eliminate the contribution of tectonic deformation to the dips observed in Apache Canyon and specifically the L1 sequence architecture. This result is based on the flattening of dips in Wolfcampian peritidal deposits. The P1 profile represents an initially horizontal platform top (i.e., no compaction-induced subsidence, no sea-level fall) and is constrained by the dip of Wolfcampian peritidal sediments (as defined by P1, Fig. 13). Surface P2 illustrates the effects of 50% compaction of foreslope and basinal facies on an originally flat-topped platform top (e.g., as defined by P1; Fig. 13). Despite the simplicity of this approach, comparison of the P2 profile and L1 toplap surface reveals a relatively good match. On either side of the canyon there is excellent fit of the P1 profile with L1 updip of the antecedent Wolfcampian scarp. Accordingly, across this area, we interpret the gentle basinward slope of the L1 platform to result from the differential compaction of foreslope facies.

Two areas of poor fit between the P1 profile and L1 toplap surface occur downdip of the Wolfcampian erosional scarp (Fig. 13). On the north side of the canyon, the first disparity between the L1 toplap surface and P2 profile is over the crest of the Wolfcampian erosional scarp. Here the P2 profile is as much as 20 m lower than that of L1 (Fig. 13A; area labeled "disparity i"). We relate this overestimate of compaction-induced subsidence to the more grain-rich nature of foreslope strata and an increase of early marine cementation adjacent to the Wolfcampian scarp. In turn, we attribute these factors to sediment bypass and a reduced rate of progradation as the L1 sequence prograded over the Wolfcampian escarpment (Fitchen, 1997; see Fig. 13A). The second area of disagreement between the P2 profile and the L1 toplap surface occurs across the outer 0.55 km of the platform (Fig. 13A; area labeled "disparity ii"). The underestimate of the basinward deflection of the L1 toplap surface increases toward its terminal shelf-slope break (<42 m). We attribute this discrepancy to the effects of a 40-m sea-level fall during progradation of the outer 0.55 km of the platform. Incorporating the effects of a relative fall in sea level accounts for the foreshortening of stratigraphy and thinning of foreslope strata observed in the most basinward part of the L1 platform (Fig. 13A). Figure 13B shows the opposite L1-P2 relationship across the outermost 0.7 km of the L1 platform on the south side of the canyon. Here, the P2 profile increasingly overestimates the position of the L1 sequence boundary toward the shelf-slope break (<40 m). This discrepancy is opposite to that on the north side of Apache Canyon and reflects the more grain-rich nature of the foreslope facies in the southern transect (Fitchen, 1997; e.g., see Fig. 13B).

Profile P4 provides the best fit to the observed L1 profile and incorporates the effects of 50% compaction of foreslope facies and a 40-m relative fall in sea level on the north side of the canyon (Fig. 13A). We interpret that the major slope break in sequence L1 formed over the Wolfcampian scarp as a compaction hinge (Figs. 12, 13). On the southern side of Apache Canyon, the best fit to the observed profile that incorporates a 40-m relative fall in sea level shows a basinward decrease of compaction-induced subsidence across the Wolfcampian scarp that varies between 25% and 40%. We attribute along-strike variation in the total amount of compaction to the more grain-rich nature of foreslope strata on the southern side of Apache Canyon compared to the north, as reported by Fitchen (1997; e.g., see Figs. 11, 13).

The compaction hinge separates relatively undeformed topsets on the shelf from more steeply basinward-dipping topsets, which were deformed by compaction-induced differential subsidence and onlapped by the succeeding L2 lowstand and transgressive systems tract (sensu Figs. 2, 6, and 7). It is significant that compaction-induced deformation apparently controlled the distribution of penetrative paleokarst over and downdip of the compaction hinge across the L1 sequence boundary (e.g., Fig. 13A). We interpret the major karstic fissures as solution-enlarged extensional fractures that formed by compaction-induced deformation of the platform top over the compaction hinge (Fitchen et al., 1995; Hunt et al., 1995).

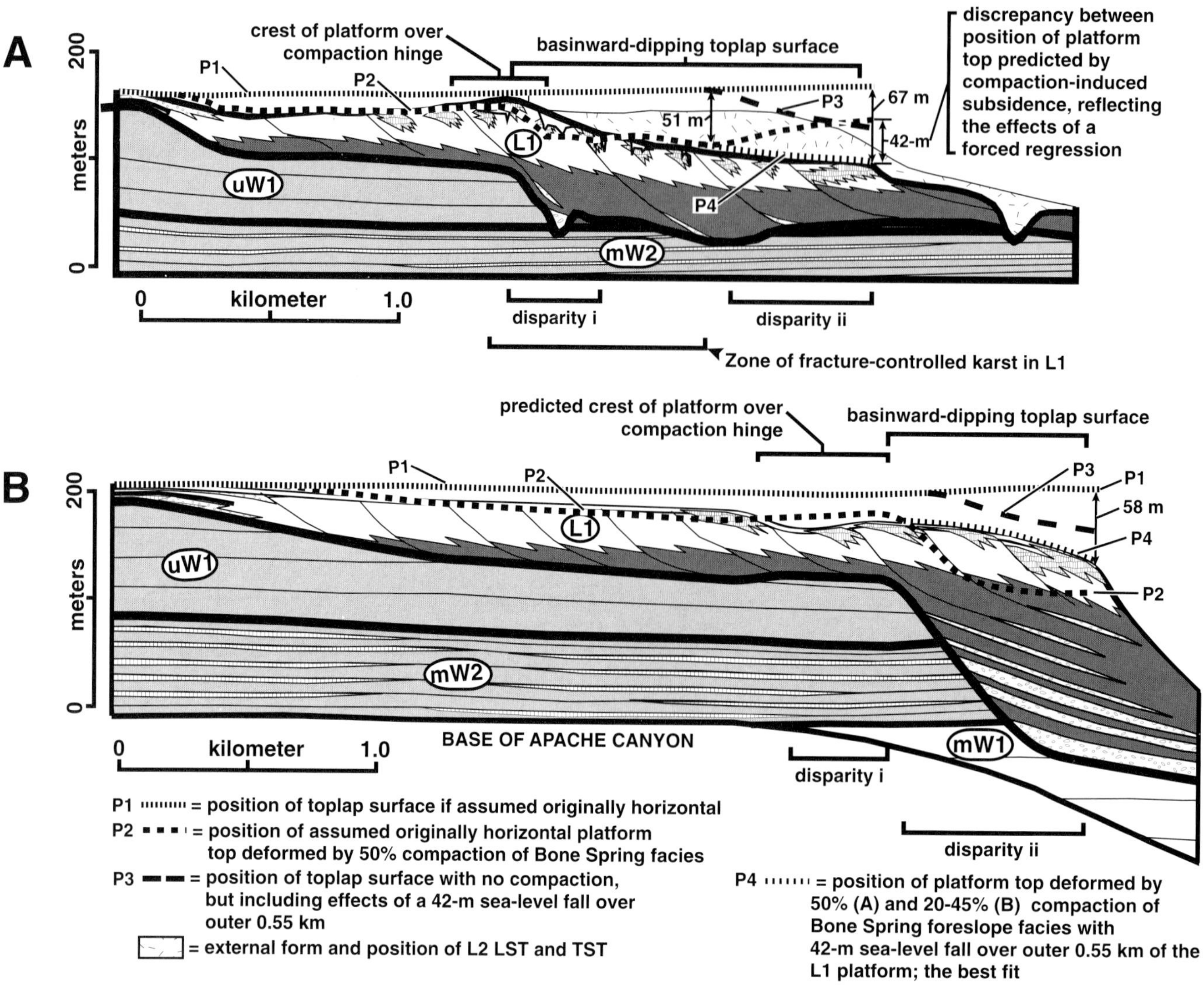

FIG. 13.—Comparison of the stratal architecture of L1 on the north and south sides of Apache Canyon. In both cases, the dips of peritidal cycles are flattened. (A) Relationship of the Wolfcampian erosional relief and L1 sequence on the north wall of the canyon showing comparison of modeled platform profiles (P1–P4) with the observed L1 platform relief. Important features to note are (1) the syncline shelfward of the compaction hinge, (2) rise of the L1 surface to a crest over the compaction hinge (a feature apparent on regional seismic lines shown in Figs. 20 and 21), and (3) the basinward-descending geometry of the L1 toplap surface basinward of the interpreted compaction hinge. The P4 profile provides the best fit to the observed L1 toplap surface, incorporating the effects of 50% compaction of foreslope and basinal facies, and a relative fall in sea level of 40 m over the outer 0.55 km of platform progradation. (B) Comparison of the L1 platform profile on the southern side of the canyon to that of a horizontal platform top (P1) and with 50% compaction of foreslope strata (P2). Important features in this profile are the lower relief displayed on the north side and the basinward-descending geometry of the L1 toplap surface basinward of the Wolfcampian scarp. Here, the estimate of 50% compaction of foreslope facies yields a good fit except basinward of the compaction hinge, where the P2 profile is lower than observed; this is a reflection of the more grain-rich nature of foreslope sediments, retarding compaction. The best fit is given by P4 that incorporates the effects of a relative fall in sea level of 40 m, constrained from A, with 20%–45% compaction of the Bone Spring lithology foreslope facies. We have assumed the foreslope breccias to be relatively incompactible in calculating the P4 profile. For key to facies and stratigraphic surfaces, see Figure 11.

The contrasting morphology and relief developed over the L1 toplap surface between the north and south sides of Apache Canyon had a significant impact on the distribution, external form, and stacking patterns of the succeeding L2 and L3 sequences. On the north side of Apache Canyon, the L2 lowstand and transgressive systems tract is an onlapping wedge <38 m thick, succeeded by the aggradational stacking of the L2–L3 shelf margin close to the terminal L1 margin (Figs. 11, 13, 14). In contrast, on the southern side of the canyon, the L2 lowstand and transgressive systems tract is a more sheetlike deposit generally 10 m thick (but <26 m; Fig. 11). Here, the L2–L3 shelf margin backstepped more than 600 m from the terminal L1 shelf margin. Fitchen (1997) argued that these along-strike changes in stacking pattern do not relate to variations in accommodation development during deposition of L2 and L3. Instead, he related the variability of L2 and L3 to the physiography of the antecedent L1 platform. This physiography, which is related here to the effects of compaction-induced differential subsidence and relative fall in sea level, acted to control the area suitable for shallow-water sedimentation as well as rates of transgression. These factors in turn controlled the facies and stacking patterns of L2 and L3.

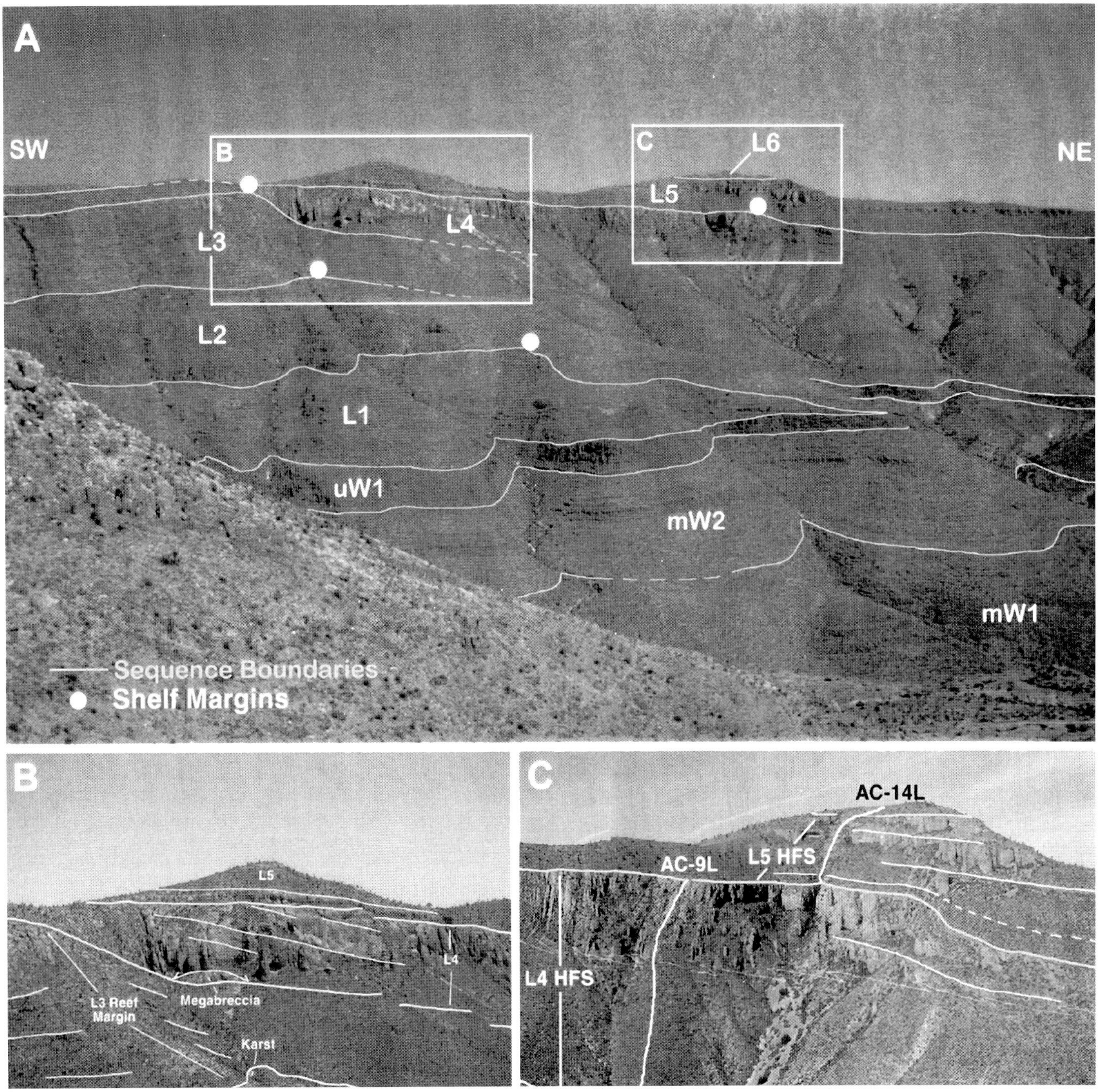

FIG. 14.—Interpreted photographs of the northern wall of Apache Canyon, Sierra Diablo. (A) View of much of the northern wall showing the erosional truncation at the top of the Wolfcampian, over which the L1 topsets dip basinward as a result of compaction-induced differential subsidence and relative fall in sea level. The 3°–5° dip of sequence L4 basinward of the L2–L3 shelf margin is clear (cf. Fig. 11A). Circles mark the position of the terminal shelf-slope breaks of Leonardian sequences. (B) View showing the uppermost L2 shelf margin, aggradational L3 reef-dominated shelf margin and upper slope, and initial shelf-margin and upper-slope stratigraphy of L4. Aggradational L4 topsets are deformed basinward over the aggradational L3 margin at top left of photograph. (C) View to the northeast of L4–L5 on the northwest wall of the canyon showing the clear basinward slope on the L4 sequence and onlap of overlying outer shelf and slope sandstones. From Fitchen (1997).

Consideration of the various platform profiles (P1–P4; Fig. 3) illustrates how changing the contribution of compaction-induced differential subsidence and/or sea-level fall would affect the relief developed across L1 and, in turn, the likely position, volume, and stratal relationships of the subsequent L2 platform (Fig. 13A). Without the contribution of compaction-induced differential subsidence (i.e., profiles P1, P3; Fig. 13), the L2 lowstand and transgressive systems tract would (1) be situated in a more basinward position; (2) be situated below the terminal L1 shelf-slope break; (3) onlap foreslope strata instead of subsided basinward-dipping topset facies; and (4) be absent over the L1 platform top. In addition, the L2 lowstand and transgressive systems tract would likely be narrower and less volumetrically important, owing to steeper foreslope angles that would limit the area available for shallow-water carbonate production (e.g., profile P3; Fig. 13). Comparison of profiles P2 and P4 with P1 and P3 demonstrates that the contribution of compaction-induced subsidence was the most important control on the position, size, and stacking patterns of the L2 lowstand and transgressive systems tract.

Sequences L4–L6.—Architecture of the L4–L6 highstand sequence set is comparable to that displayed by L1 (e.g., Fig. 12). Like L1, shelfward of an antecedent platform margin, the L4–L6 shelf strata are parallel and of constant thickness, whereas downdip the platform tops of these sequences diverge from a common point—the aggradational L2–L3 platform margin (Figs. 11, 14). The L4–L6 sequence boundaries dip basinward at a progressively lower angle (e.g., 3°–5° for L4; 1°–2° for L5), and the terminal shelf-slope break of a sequence is topographically lower than aggradational shelf strata of the same sequence. Sequence L4 shows only minor lateral variability in attributes such as progradation:aggradation ratio, facies tracts, and stratal patterns (Fig. 11; Fitchen, 1997). The upper surface of L4 descends 90 m across the 1.4–1.6 km over which it prograded basinward of L3 (Figs. 12, 14).

Interpretation.—Fitchen (1997) interpreted a 20–25-m downdip increase of water depth across the basinward-dipping L4 sequence boundary on the basis of facies changes in the basal L5 deposits, suggesting formation of 20–25 m of relief across the L4 platform top prior to L5 deposition. The deformation of aggradational L4 topsets across the aggradational L2–L3 platform margin, at a comparable dip to the overlying L4 sequence boundary (Figs. 12, 14), indicates that the relief displayed by L4 resulted from compaction-induced differential subsidence, not from sea-level fall. Accordingly, we interpret most of the present 90 m of relief displayed by L4 basinward of the L2–L3 platform margin resulted from compaction-induced differential subsidence following the L5 lowstand. The analogous basinward divergence and thickening of sequences L5 and L6 from the antecedent L2–L3 platform margin suggests that compaction-induced differential subsidence also acted as an important control on their development. It is likely that much of the basinward tilting and thickening in L5 and L6 observed across the L4 sequence-boundary toplap surface occurred in response to the depositional loading induced by sequences L5 and L6. However, the exact timing of the resulting subsidence (syndepositional or postdepositional) is unclear.

West Dog Canyon, Brokeoff Mountains, New Mexico

The Brokeoff Mountains in Otero County, New Mexico, are the northwestern extension of the Guadalupe Mountains (Fig. 8). Fitchen (1992, 1993) developed a new sequence stratigraphic framework for the lower and upper parts of the San Andres Formation and basinal equivalents on the basis of detailed mapping and measured sections and comparison to the work by Kerans et al. (1993) on the Algerita Escarpment (Fig. 5; see also Kerans and Fitchen, 1995). Interest here focuses on the relationships formed between the terminal middle San Andres platform margin (G4), the overlying aggradational G12 platform margin, and the sequence boundary separating the G13 (upper San Andres) and G14 (Grayburg) fourth-order sequences (Figs. 9, 15).

Stratigraphy.—The lower and middle parts of the San Andres Formation include a composite sequence that consists of transgressive (L7, L8) and highstand (G1–G4) sequence sets. The composite G1–G4 highstand bank-ramp complex prograded approximately 15 km (Fig. 9; Kerans et al., 1992; Kerans and Fitchen, 1995). A subaerial-exposure surface terminates this highstand sequence set and passes downdip to a major slope collapse and bypass surface, which is onlapped by a basin-restricted wedge of basin-floor and slope siliciclastic-dominated sediments that now make up the Brushy Canyon Formation (Fig. 9). Sarg and Lehmann (1986) and Kerans et al. (1992; among other references) interpreted the upper surface of the lower and middle parts of the San Andres Formation as a major sequence boundary that formed in response to relative fall in sea level. The subsequent relative rise in sea level reestablished the aggradational (transgressive) G12 platform margin over that of G4 and mixed carbonate-siliciclastic sedimentation across the platform (Figs. 9, 15; Fitchen, 1992, 1993). The overlying lower Cherry Canyon Tongue is a basin-restricted wedge of sandstones that separates sequences G12 and G13 (Fig. 9, 15). Fitchen (1992, 1993) interpreted these basinal sandstones as a lowstand to transgressive deposit, with the succeeding highstand represented by progradational carbonate-siliciclastic cycles of G13 (Figs. 9, 15).

Early-highstand carbonate-dominated sigmoidal cycles prograded the G13 platform margin basinward by approximately 1 km and aggraded it by 25 m (Fig. 15). Sigmoid- to sigmoid-oblique mixed carbonate-siliciclastic cycles of the late G13 highstand phase prograded the platform an additional 5–6 km basinward, but display little aggradation (Fitchen, 1992, 1993). Landward (i.e., northwest) of the vertically stacked G4 and G12 platform margins, the G13/G14 sequence boundary is parallel to underlying sequence boundaries (Figs. 9, 15). However, the topsets of the G13 sigmoidal clinoforms and the G13/G14 sequence boundary steepen to 1.2° basinward dips over the antecedent aggradational G4 and G12 platform margins (Fitchen, 1992, 1993). In Figure 15, the G13/G14 sequence boundary displays 14 m of relief basinward of the antecedent G4 and G12 platform margins. This relief controlled the updip pinch-out of the basal Grayburg sandstones immediately basinward of the antecedent aggradational G4 and G12 platform margin. It is significant that Fitchen (1992, 1993) mapped a 200–300-m-wide strike-parallel zone of unusually penetrative paleokarst (as deep as 30 m) formed in G13 over and basinward of the slope break in the G13/G14 sequence boundary (Fig. 15). The interpreted paleosinkholes and grikes (i.e., solution fissures) have a fine- to medium-grained sandstone fill derived from the succeeding Grayburg, along with tilted angular breccia blocks of host dolomite (Fig. 16; Fitchen, 1992, 1993). Fitchen (1992, 1993) noted that paleokarst is not as pervasive or penetrative elsewhere along the G13/G14 sequence boundary,

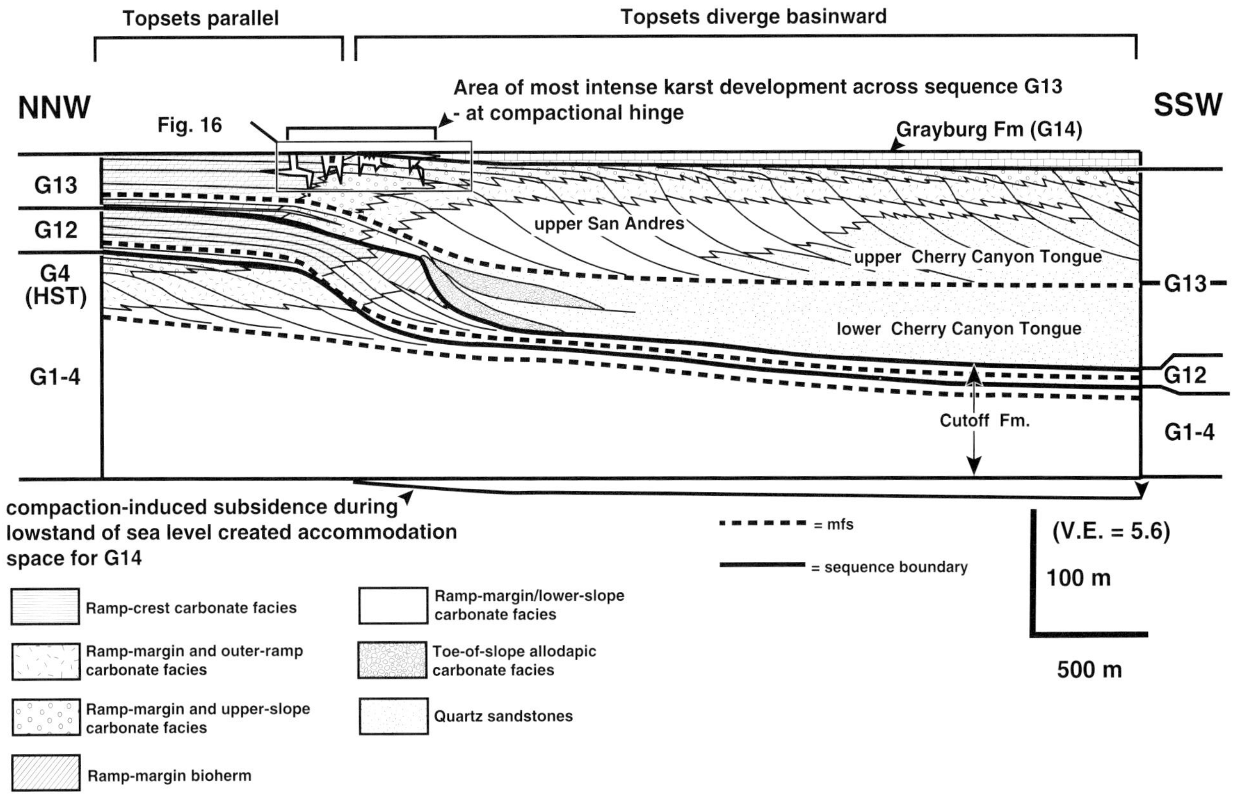

FIG. 15.—Stratal geometries developed within the lower and middle San Andres (G1–G4), upper San Andres (G12–G13), and Grayburg (G14) Formations in West Dog Canyon, Brokeoff Mountains, New Mexico (approximately 12 km northwest of the western escarpment of the Guadalupe Mountains). Here, the early phase of the G13 composite highstand is characterized by carbonate-dominated sigmoidal clinoforms, whereas sigmoid-oblique mixed carbonate-siliciclastic clinoforms with little aggradation typify the late phase (compare to Last Chance Canyon, Fig. 17). Relief developed across the G13/G14 sequence boundary (14 m in this figure) by compaction-induced differential subsidence is interpreted to have controlled the lap-out of sandstones at the base of the overlying Grayburg, a relationship also observed on regional seismic lines (Figs. 18, 19). The compaction hinge developed in G13 strata over the G4 terminal shelf-slope break and the aggradational G12 margin. The compaction hinge in G13 is associated with the development of a 200–300-m-wide strike-parallel zone of penetrative paleokarst interpreted as solution-enlarged extensional fractures, as also observed in the subsurface.

especially updip of the antecedent aggradational G4 and G12 platform margins.

Interpretation.—The increase of dip and paleotopography formed across the G13 platform and G13/G14 sequence boundary is coincident with (1) the updip pinch-out of a wedge of very fine grained toe-of-slope to basinal sandstones of the lower Cherry Canyon Tongue, (2) the updip pinch-out of a wedge of G13 foreslope carbonate mudstones, wackestones, and very fine sandstones, (3) the antecedent aggradational G4 and G12 platform margins, and (4) the updip pinch-out of basal Grayburg sandstones (Fitchen, 1992, 1993; Figs. 15, 16). The increase of dip displayed by G13 is not evident in the underlying sequences (Fitchen, 1992, 1993; Kerans et al., 1992; Figs. 9, 15), ruling out the role of tectonic subsidence. Deformation of the aggradational G13 sigmoids also rules out relative fall in sea level as a contributing factor to the relief of the G13/G14 sequence boundary in this updip part of G13. For these reasons, and the coincidence of features outlined above, we interpret the upper G13 surface as a compaction-modified sequence boundary deformed during a relative lowstand of sea level by compaction-induced differential subsidence over and basinward of the antecedent G4 and G12 platform margins. The aggradational G4 and G12 platform margins localized a compaction hinge that separates an updip portion of the G13 platform in which topsets to the G4, G12, and G13 platforms are parallel from a basinward portion in which G13 and G14 topsets diverge basinward.

The postdepositional relief developed across the G13 platform (<14 m; Fig. 15) acted to control the architecture, facies patterns, thickness, and lap-out within the overlying G14 (Grayburg) sequence (Figs. 15, 16; Barnaby and Ward, 1995), a relationship observed in the subsurface and discussed later (cf. Figs. 18, 19). The relief formed across the G13 platform controlled (1) landward tapering of the Grayburg across the tilted G13 platform top toward the antecedent aggradational G4 and G12 platform margins and (2) the pinch-out of its basal sandstones against the G13 compaction hinge. The positions of the compaction hinge and Grayburg pinch-out are coincident with the linear 200–300-m-wide strike-parallel zone of penetrative paleokarst in G13 (Fitchen, 1992, 1993; Figs. 15, 16). We consider this paleokarst to represent a zone of solution-enlarged fractures, localized over the G4 and G12 platform margins by compaction-induced flexure and extension of the lithified G13 platform top.

FIG. 16.—(A) Photograph looking southward obliquely downdip at the deformed G13/G14 sequence boundary over the compaction hinge. Here, the onlap of dark transgressive sandstones is observed against the G13/G14 sequence boundary over the compaction hinge where (B) paleokarst features filled by fine- to medium-grained sandstones (dark and laminated in photo) penetrate as far as 30 m below the sequence boundary. This paleokarst is restricted to a 200–300-m-wide zone and is characterized by solution-enlarged extensional fractures (grikes) and paleocaverns. Scale in B is a geologic hammer, circled.

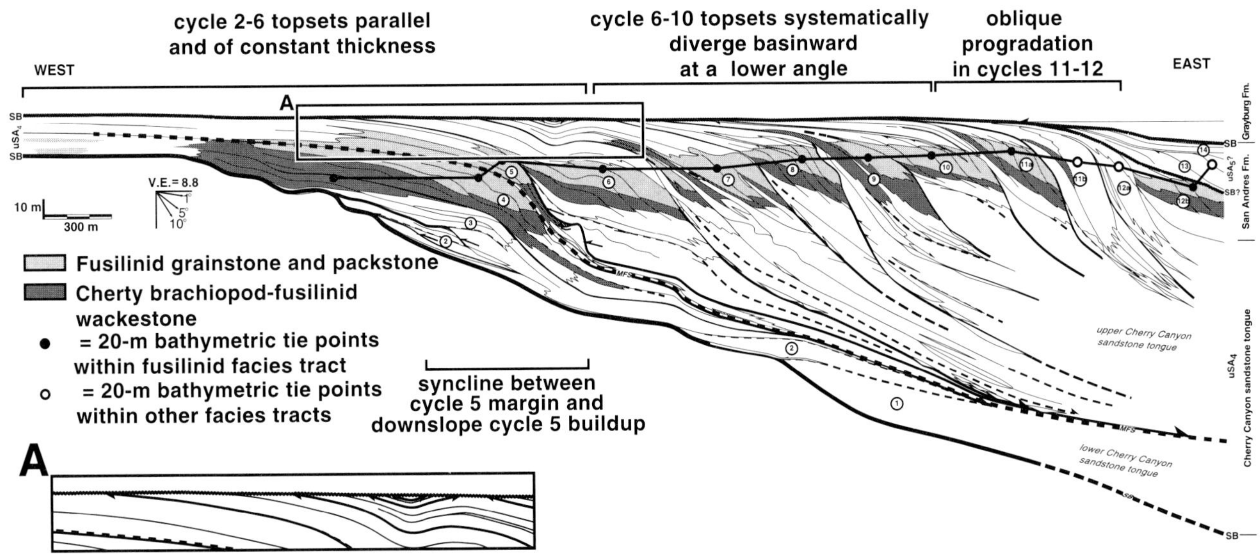

FIG. 17.—Stratigraphic cross section through the upper San Andres 4 sequence (uSA_4) mixed carbonate-siliciclastic succession in Last Chance Canyon, New Mexico. Here the effects of compaction-induced differential subsidence are interpreted to have occurred both during highstand progradation, resulting in the basinward "fanning" of cycles 6–10, and during lowstand to form an erosional angular unconformity between the upper San Andres and Grayburg Formations. In Last Chance Canyon, the relief developed across the uSA/Grayburg sequence boundary is not preserved but was eroded, a situation that contrasts with the same compaction-modified sequence boundary some 30 km along strike (Figs. 15, 16). The differential compaction of lower-slope strata during the lowstand deformed cycle 6–9 strata and formed a syncline (A) between the relatively incompactible antecedent aggradational platform margin to cycles 4–5 and a lower-slope buildup (M. D. Sonnenfeld, pers. commun., 1995). Modified from Sonnenfeld and Cross (1993).

Last Chance Canyon, Guadalupe Mountains, New Mexico

The mapping of Hayes (1964), Sonnenfeld (1991, 1993), and Fitchen (1992, 1993) showed the upper San Andres Formation in the Brokeoff Mountains and Last Chance Canyon to be lateral equivalents. In Last Chance Canyon, as in the Brokeoff Mountains, the upper San Andres and Grayburg Formations are separated by a sequence boundary interpreted to have formed during a sea-level lowstand (Sarg and Lehmann, 1986; Sarg, 1988, 1989; Sonnenfeld, 1991, 1993; Fitchen, 1992, 1993; Kerans et al., 1992; Sonnenfeld and Cross, 1993; Figs. 9, 15, 17). Sonnenfeld (1991, 1993) and Sonnenfeld and Cross (1993) described the sedimentology, stratal patterns, and sequence stratigraphy of the upper San Andres Formation (uSA) in Last Chance Canyon. We follow the stratigraphic nomenclature of Sonnenfeld and Cross (1993) whose upper San Andres 3 and 4 sequences (uSA_3, uSA_4) are correlative to the G12 and G13 sequences of Kerans et al. (1992), Fitchen (1993), and Kerans and Fitchen (1995) (e.g., compare Figs. 9, 15, and 17).

Stratigraphy.—

Sonnenfeld and Cross (1993) recognized 14 fifth-order cycles, which build the composite fourth-order uSA_4 sequence in Last Chance Canyon (Fig. 17). The uSA_4 sequence contains a clastic-dominated lowstand systems tract (the lower Cherry Canyon Tongue; cycles 1–2), a carbonate-dominated transgressive systems tract (cycles 3–4) and mixed carbonate-siliciclastic progradational highstand (cycles 5–10) and forced regressive systems tracts (cycles 11a–12b; Sonnenfeld and Cross, 1993; M. D. Sonnenfeld, pers. commun., 1995). Sonnenfeld and Cross (1993) regarded the angular truncation of cycles 5–12a across the upper bounding surface of the uSA4 to have formed through a combination of subaerial and marine processes (Fig. 17).

Highstand cycles of the uSA_4 sequence display sigmoidal progradation; cycles 5, 6, 7, 8, and 10 show the most aggradation. Oblique cycles of the forced regressive systems tract (cycles 11a–12b) show little aggradation (Fig. 17). The bounding surfaces of cycles 5–6 are subparallel updip of their equivalent platform margins and show little landward convergence, whereas those of cycles 7–13 diverge basinward (Sonnenfeld and Cross, 1993). A pre-Grayburg syncline is developed in cycles 6–8, with dip reversal occurring between the aggradational margin to cycles 2–5 and a downdip carbonate buildup of cycle 5 (Fig. 17). However, surfaces bounding the highstand cycles (5–10) display an overall basinward dip toward their depositional platform margin, defining a rollover-type architecture (Sonnenfeld and Cross, 1993). The dip and rollover in the highstand cycles systematically decreases from 1.5° in cycle 6 to 0.6° in cycle 10 (relative to the uSA_4/Grayburg sequence boundary). Cycle 10 is nearly parallel to the bounding uSA_4/Grayburg sequence boundary and displays little rollover, although cycles 11a–12a reverse this trend.

Interpretation.—Fundamental to understanding stratal geometries within the uSA_4 highstand and forced regressive systems tracts (cycles 6–12) are the following stratal relationships: (1) the syncline immediately basinward of the aggradational margin to cycles 2–5, (2) the systematic rollover and changes in dip of cycle 6–10 topsets, and (3) the angular truncation of cycle 5 to cycle 12 topsets and toplap strata (Fig. 17).

The syncline deforms strata of several cycles and is thus unlike depositional synforms formed within uSA_4 cycles through the accretion of depositional lobes observed in strike section (Sonnenfeld and Cross, 1993). The syncline's updip limb is coincident with the landward pinch-out of a thick tract of uSA_4 lower-slope and basinal facies (e.g., downdip of the shaded facies in Fig. 17). The syncline axis occurs between the relatively incompactible aggradational platform margin to cycles 3–5 and a uSA_4 downslope carbonate buildup in cycle 5 (Fig. 17). Erosional truncation of the syncline at the uSA_4/Grayburg sequence boundary shows that deformation occurred prior to Grayburg transgression (Fig. 17). Preservation of topset strata in cycle 8 within the syncline indicates that a minimum 7 m of subsidence, structural deformation, and erosion of the uSA_4 sequence occurred prior to Grayburg deposition (constrained from Fig. 17). Sonnenfeld and Cross (1993) estimated that prior to erosion, highstand cycles 5–8 extended 0.5–1.5 km farther updip. Within the uSA_4 highstand, there is also a decrease in rollover and basinward dips of cycle 6–10 topsets that apparently records a temporal change in platform morphology. The systematic nature of these changes suggests an external forcing mechanism. Elsewhere in the Permian basin, temporal changes of platform architecture may relate to changes in the rate of accommodation development, the physiography of precursor platforms, and/or platform-basin relief (e.g., Kerans et al., 1992; Kerans and Fitchen, 1995; Fig. 9). However, none of these factors applies to cycles 6–10 in which Sonnenfeld and Cross (1993) interpreted platform-to-basin relief, antecedent physiography, and the rate of accommodation development to have been relatively constant.

We think that syndepositional and postdepositional compaction-induced differential subsidence, with the effects of sea-level fall also contributing to the stratal geometry of cycles 11a–12, best explains the architecture, basinward dips, and deformation of the progradational uSA_4 cycles. The subparallel nature of cycles 2–7 topsets indicates that little differential subsidence took place during their deposition. Progressive tilting due to compaction-induced differential subsidence is interpreted to have occurred during the late-uSA_4, late-highstand, and forced regressive systems tracts; this tilting is thought to have formed the rollover, the basinward dips, and the syncline in cycles 6–12. Further deformation and tilting of the uSA_4 platform likely occurred during sea-level lowstand (as in the Brokeoff Mountains). However, rather different from the situation in West Dog Canyon (Figs. 15, 16), erosion prior to and/or during Grayburg transgression removed most of the postdepositional relief formed in the uSA_4 platform (Fig. 17). By way of analogy, the geometry of cycles 6–10 is comparable to that seen in Figures 4–6, in which the tilting of topsets is greatest in the oldest cycles and decreases during progradation. This is a natural stratal pattern of highstand sequence sets in the Permian basin, which we think develops during deposition, and/or lowstand, and is exaggerated during burial.

COMPACTION-MODIFIED SEQUENCE GEOMETRIES FROM THE SUBSURFACE PERMIAN BASIN

In this section we present examples of platform geometries and relationships in seismic section from the Northwest and Eastern shelf areas of the Permian basin that are directly comparable, and in some cases time equivalent, to the architecture and relationships of compaction-modified sequences observed in outcrops. The data used to identify compaction-modified sequences require one to distinguish between the effects of differential subsidence induced by tectonism and compaction and the effects of relative fall in sea level (e.g., Fig. 7). In addition, it is necessary to account for the exaggeration of relief in seismic time sections across the platform-to-basin transition that is caused by the contrast between high-velocity carbonates and relatively low-velocity mixed carbonate-siliciclastic basin sedimentary units. This contrast often leads to velocity pull-up over platform areas and pull-down across basinal sedimentary units. The relief across a compaction-modified platform top often controls the thickness and updip limit of relatively low-velocity transgressive sandstones (e.g., Figs. 15, 16) and so is often exaggerated in seismic time sections, a fact that aids recognition.

The Northwest Shelf

Sarg and Lehmann (1986) and Sarg (1988, 1989) showed that the seismic sequence stratigraphy of Permian mixed carbonate-siliciclastic-evaporite platforms across the Northwest shelf is directly comparable to nearby outcrops, albeit with limitations of seismic (Stafleu and Sonnenfeld, 1994) and biostratigraphic resolution. Sarg and Lehmann (1986) and Sarg (1988, 1989) defined seismic sequences in the Exxon Production Research (EPR) Line on the basis of toplap, erosional truncation, and onlap of basin-restricted wedges. The 13 seismic sequences recognized by Sarg (1988, 1989) overlie a regional throughgoing "basal" reflector that displays a gentle basinward dip, attributed to regional tectonism and the basinward thickening of low-velocity sandstones (Fig. 18). Hunt et al. (1995) reported the development of two distinctive slope breaks in highstand sequence sets over antecedent aggradational platform margins (Fig. 18). The older, more landward structure affects multiple sequences (3–5, 7, and 9–12; Fig. 18), whereas the younger, more basinward flexure occurs in a single sequence (11; Figs. 18, 19).

Primary Compaction Hinge.—
The primary structural feature of the platforms imaged in the EPR seismic line is a basin-facing monocline formed in interpreted aggradational lagoonal (back-reef) strata of sequences 3–5, 7, and 9–12 over the aggradational margin to sequences 1 and 2 (Figs. 18, 19; Hunt et al., 1995). Sequences 1–2 and 3–5 form aggradational transgressive and progradational highstand sequence sets, respectively. The topset strata of sequences 3–5 (and younger) are subhorizontal and of constant thickness; the strata display parallel relationships landward of the aggradational margin to sequences 1 and 2 (Figs. 18, 19). In sequences 3–5, 7, and 9–12 over this antecedent margin there is a distinctive slope break, basinward of which the upper surfaces to sequences 3–5 dip and diverge basinward. Each upper surface to sequences 3–5 dips basinward at a lower angle than its precursor but at a steeper angle than the regional basal reflector (Figs. 18, 19). The tops of sequences 3–5 dip toward their ter-

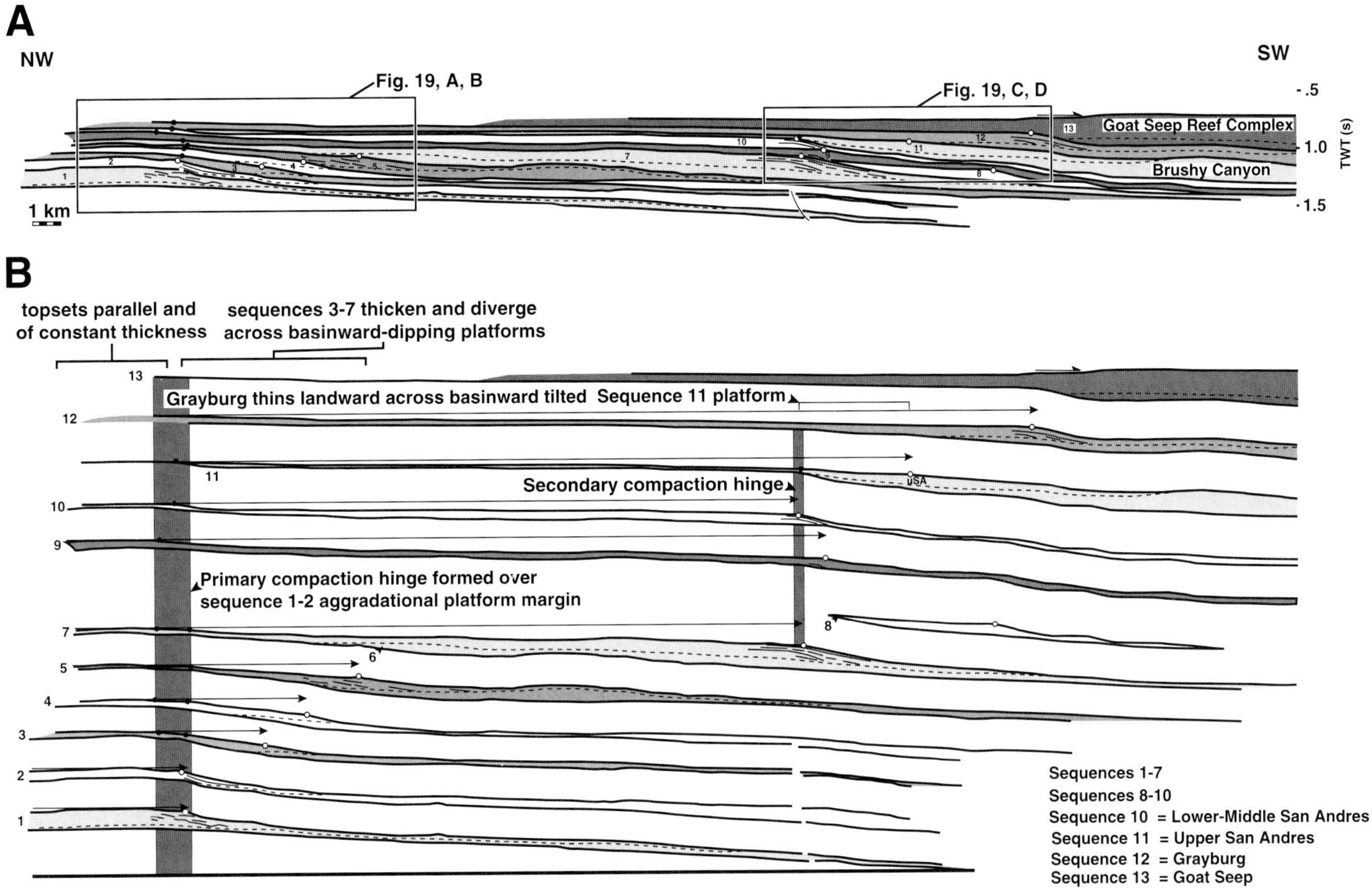

FIG. 18.—Seismic sequence stratigraphic interpretation of the Exxon Production Research Line from across the Northwest shelf into the Delaware basin (line located in Fig. 8). In this line, the regional dip of a throughgoing basal reflector helps to constrain the role of tectonism. Sarg and Lehmann (1986) and Sarg (1988, 1989) divided the section into 13 seismic sequences that we follow here. Two compaction hinges are developed, the positions of which are shown with the vertical gray shading (B). The primary hinge developed over aggradational (Wolfcampian–Leonardian?) platform margins (sequences 1–2) and migrated upward through the overlying 11 sequences. Farther downdip, a subordinate compaction hinge is developed over the aggradational lower and middle San Andres margins (sequence 10) and propagates through the overlying upper San Andres and Grayburg sequences. Depositional shelf-slope breaks are shown with an open circle.

minal depositional platform margin, recognized on the basis of its distinctive physiographic expression (Sarg and Lehmann, 1986; Sarg, 1989; Hunt et al., 1995). Consequently, the depositional platform margins of sequences 3–5 are positioned deeper on the seismic section in two-way traveltime and appear to be lower, in terms of elevation, than their platform equivalents deposited landward of the antecedent platform margin to sequences 1 and 2. The basinward-dipping platform tops of sequences 3–5 are seen in Figures 18 and 19 to be onlapped by the landward-thinning, less steeply dipping reflectors of sequences 4–6, respectively (Figs. 18, 19).

Interpretation.—The stratal pattern displayed by sequences 3–5 over the aggradational platform margin antecedent to sequences 1–2 is analogous to that of sequences L4–L6 formed above the L2–L3 margins in the Sierra Diablo Mountains (e.g., Figs. 11, 12, 14). We interpret sequences 3–5 in Figures 18 and 19 to be compaction-modified platform sequences. The onlap of younger, less steeply dipping strata against the basinward-dipping platform top of sequences 3–5 suggests that differential subsidence took place during lowstands (Figs. 19A–B). This relief formed basinward of the aggradational sequence 1–2 platform margin and appears to have controlled the landward thinning of sequences 4, 5, and 6–7 across sequence boundaries 3/4, 4/5, and 5/6–7, respectively. We interpret the platform margin to sequences 1–2 to have localized a compaction hinge in younger sequences (3–5, 7, 9–12; Figs. 18, 19). Further exaggeration of the early-formed relief across compaction-modified sequences 3–5 occurred during subsequent burial, as indicated by the basinward dip of sequence 7 and 9–12 topset strata downdip of the compaction hinge (Figs. 18, 19).

Subordinate Compaction Hinge.—

This more basinward slope break is a subtle flexure across the top of sequence 11 located over the aggradational platform margin to sequences 7, 9, and 10 (Figs. 18, 19). Sarg and Lehmann (1986) and Sarg (1988, 1989) interpreted sequences 10, 11, and 12 to represent the lower to middle San Andres, the upper San Andres, and the Grayburg sequences, respectively. This interpretation makes the flexure across sequence 11 (Figs. 18, 19) equivalent to that observed in the upper San

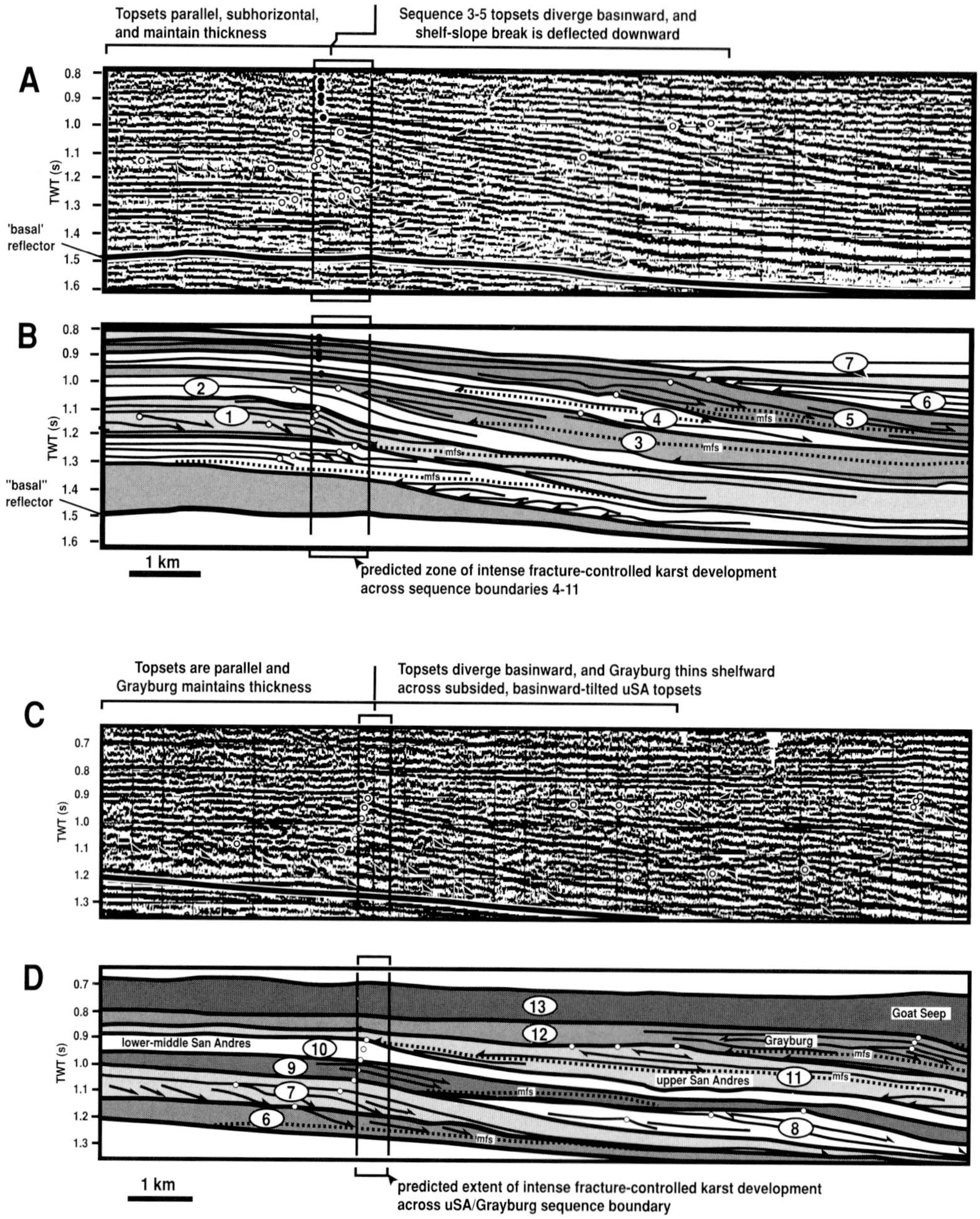

FIG. 19.—Enlargements of parts of the Exxon Production Research Line (see Fig. 18 for locations). (A and B) Detail of the stratigraphic relationships developed over the aggradational platform margin to Leonardian sequences 1–2 of Sarg (1988, 1989). (A) The uninterpreted line. (B) The interpretation. The dips of sequence boundaries 3/4, 4/5, and 5/6 are steeper than regional dips (<3°). Leonardian bank-ramp platforms of many authors in the Permian basin often have low-angle foreslopes (<6°) so that dips induced by differential compaction can completely subdue the expression of the depositional shelf-slope break, making for significant problems in seismic facies interpretation. The relationships displayed by this line are comparable in age and architecture to those observed from Apache Canyon, the Eastern shelf, the Northwest shelf, and along the San Simon channel. Over the compaction hinge we would predict the development of penetrative fracture-controlled diagenesis across compaction-modified sequence boundaries (3–5). (C and D) Detail of the "subordinate" compaction hinge developed in the upper San Andres (uSA) and Grayburg Formations over the lower and middle San Andres Formation and older platform margins (sequences 7, 9, 10). Updip of the compaction hinge, topsets are subhorizontal, maintain constant thickness, and are parallel in the lower and upper San Andres and the Grayburg Formations. Dip of the uSA topsets changes abruptly (by approximately 1°) across the antecedent lower and middle San Andres Formation platform margin. The terminal shelf-slope break of the uSA sequence sits lower than equivalents shelfward of the compaction hinge. This relief is attributed to the affects of postdepositional compaction-induced relief and relative fall in sea level. The relief developed across the uSA sequence controlled the thinning and onlap(?) of the basal Grayburg shelfward across the uSA platform, as is seen in West Dog Canyon (Figs. 15, 16). Deformation and basinward dip of toplap surfaces induced by compaction-related subsidence present difficulties for sequence stratigraphic and sea-level history interpretation (see text for further discussion). TWT = two-way traveltime.

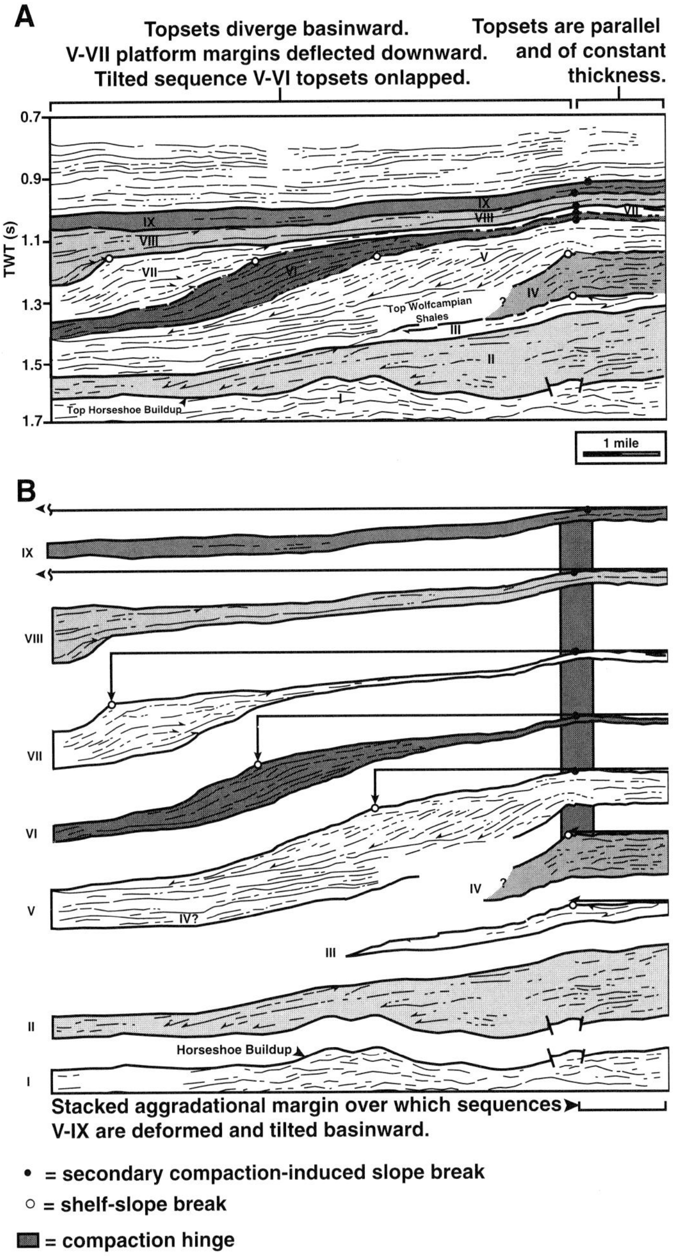

FIG. 20.—(A) Seismic sequence stratigraphic interpretation of Wolfcampian–Leonardian sequences from the Eastern shelf. Highstand sequences prograde westward into the Midland basin and develop relationships that are directly comparable to those observed in the Sierra Diablo, some 400 km to the southwest. (B) This view of the component sequences clearly illustrates the fundamental division of the Leonardian highstand sequence set by the compaction hinge. Shelfward of the compaction hinge, sequence boundaries and topsets are parallel, subhorizontal, and of constant thickness. In contrast, basinward of the compaction hinge, topset facies and surfaces of the highstand sequences diverge and dip basinward. Onlap of less steeply dipping topsets onto basinward-tilted platform tops is clearly seen at the base of sequences VI and VII. The topsets to the highstand sequence set (V–VIII) all diverge from a common point (the compaction hinge) that is developed over the aggradational platform margin of sequences III–IV. The oldest topset surface dips most steeply (approximately 5°), with successive toplap surfaces gradually shallowing. Line drawing and interpretation modified from Hunt et al. (1995) and made from a seismic line originally published by Greenlee and Lehmann (1993). TWT = two-way traveltime.

Andres Formation in the outcrops in West Dog and Last Chance Canyons (Figs. 15–17).

In Figures 18 and 19, the dip of the uSA sequence (11) changes abruptly basinward of the antecedent lower to middle San Andres (sequence 10) platform margin by approximately 1°. Landward of this antecedent platform margin, the topset facies of the lower, middle, and upper San Andres and the Grayburg sequences are parallel and maintain constant thicknesses (Figs. 18, 19). Basinward of the terminal platform margin to sequences 6–7 and 9–10, (1) the uSA toplap surface dips basinward, (2) the terminal uSA platform margin is lower than equivalent strata landward of the lower to middle San Andres sequence margin, (3) the uSA and Grayburg toplap surfaces diverge, and (4) the Grayburg sequence thins landward across the uSA platform top (Figs. 18, 19).

Interpretation.—Comparison of these stratal relationships to those observed in outcrop equivalents (e.g., West Dog and Last Chance Canyons; Figs. 15–17) leads us to interpret the uSA/Grayburg contact as a compaction-modified sequence boundary, likely exaggerated by a relative fall in sea level across its outer 0.4–1 km (e.g., 10–25% of uSA progradation, estimated from the work of Sonnenfeld and Cross, 1993, and Stafleu and Sonnenfeld, 1994). From the seismic data, we interpret the effects of compaction-induced differential subsidence and sea-level fall to have created approximately 70 m of relief across the outer 4 km of the uSA platform. This relief controlled landward thinning at the base of the Grayburg Formation across the uSA platform top, as observed in seismic and outcrop section alike (Figs. 15, 16, 18, 19). Outcrop observations also lead us to predict a narrow zone (200–1000 m) of fracture-controlled paleokarst in the uSA sequence over the compaction hinge formed above the antecedent platform margin to sequences 6–7 and 9–10 (Figs. 18, 19). Such a relationship does occur in the subsurface along-strike in the Maljamar field, Northwest shelf (C. Modica, pers. commun., 1997), and in the Means field on the Central Basin platform.

The Eastern Shelf

Figure 20 shows a line drawing and sequence stratigraphic interpretation of a seismic section through Eastern shelf carbonate platforms that prograded over Horseshoe atoll (approximate line location given in Fig. 8; see also Greenlee and Lehmann, 1993). We have identified nine seismic sequences (I–IX) on the basis of erosional truncation, toplap, and/or slope onlap. Horseshoe atoll is at the base of the section and forms a useful subhorizontal reference horizon that shows there is negligible differential tectonic subsidence across this part of the line.

Seismic Sequence Stratigraphy.—

Sequences III–IV form an aggradational sequence set that is succeeded by a progradational sequence set composed of sequences V–VIII (Fig. 20). Landward of the aggradational margin to sequences III–IV, the topsets of sequences III–IX are subparallel, and maintain thickness. Slight elevation, the thickening of sequence V, and landward dip of sequence V–VIII

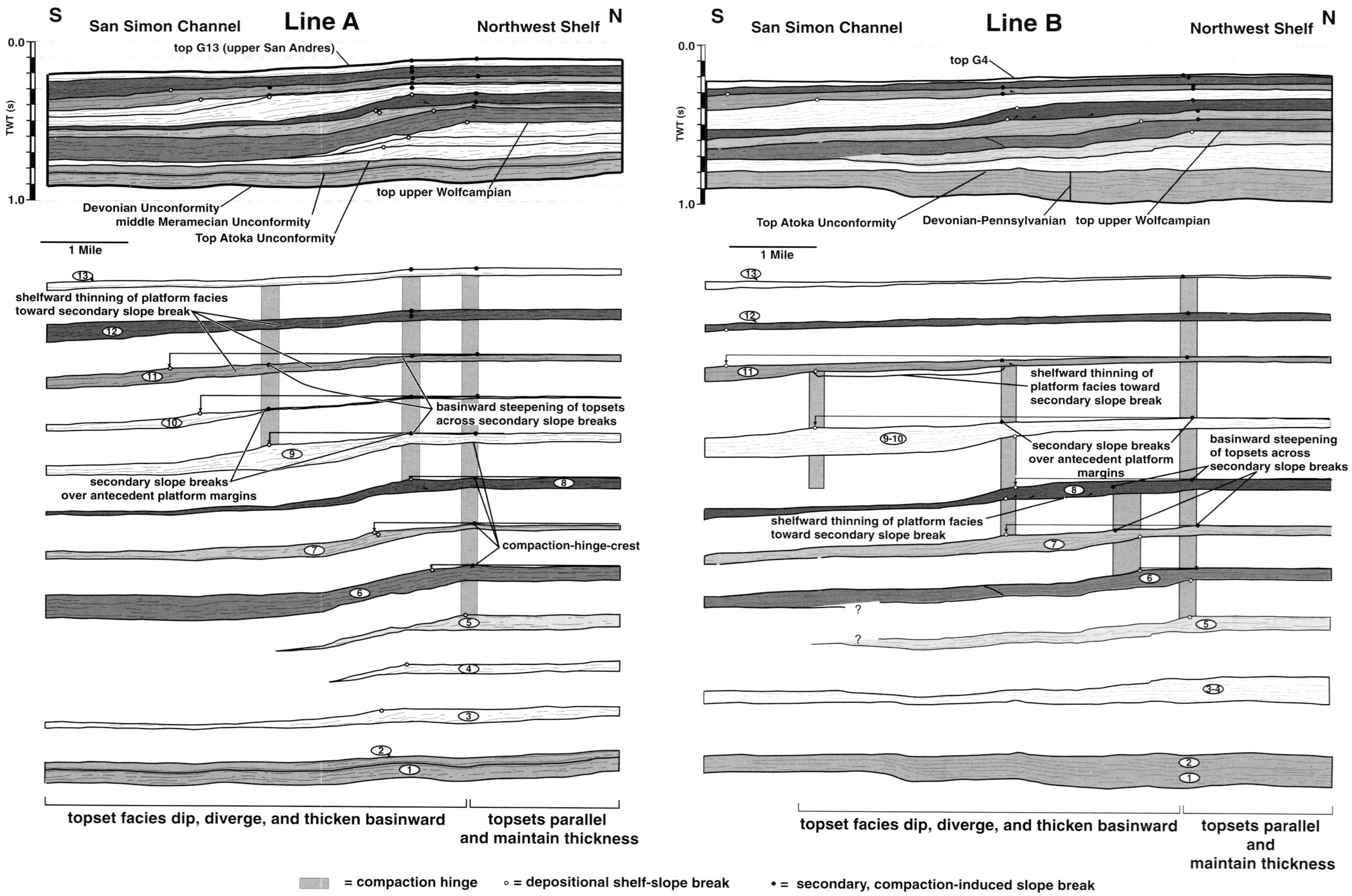

FIG. 21.—Line drawing of seismic sequence stacking patterns of Leonardian platforms with multiple compaction hinges from the Northwest shelf along the San Simon channel. The two lines are approximately 5 km apart and show a similar structure in the Wolfcampian–Leonardian platforms despite the reversal of dips evident in the underlying Devonian–Pennsylvanian. The compaction hinges show variability in their longevity. In both lines, sequences 7 and 10 have been modified by compaction. These sequences are onlapped or downlapped by successor platforms (8 and 11) that thin landward across basinward-tilted and subsided topsets. TWT = two-way traveltime. See text for further discussion.

topsets is apparent over 0.5–1.1 km (0.3–0.7 mi) updip of the antecedent sequence III–IV platform margin. Landward of the sequence III–IV margin, topsets of sequence III–IV dip basinward whereas those of VI–IX display negligible dip (Fig. 20B). In contrast, downdip of the aggradational sequence III–IV platform margin, the topsets to sequences V–IX all dip, diverge, and thicken basinward. The abrupt basinward increase of dip in sequence V–IX topsets is coincident with the sequence III–IV platform margin. In turn, the dip displayed by sequence V–VIII topsets and decreases up section. Consequently, the platform margins to sequences V–VII occur 0.11, 0.14, and 0.16 s TWT (two-way traveltime) below their equivalent platform strata landward of the sequence III–IV margin. Sequences VI and VII onlap the basinward dipping topset reflectors of preceding sequences (Fig. 20). This indicates that relief developed across sequences V and VI during sequence-boundary formation.

Interpretation.—The dip of the Horseshoe atoll succession allows us to rule out tectonic subsidence as a mechanism to account for the dips displayed by sequence V–IX topsets. In addition, the presence of aggradational topset reflectors at the margin to sequences V–VII also allows us to rule out the role of relative fall in sea level during their deposition. Accordingly, we interpret that a compaction hinge formed over the antecedent sequence III–IV platform margin to fundamentally divide and modify the architecture and development of the progradational sequence V–IX set. A distinctive slope break marks the compaction hinge in sequence V–IX topsets. The elevation of sequence V–VII topsets and local reversal of dip over the compaction hinge is analogous to that of L1 topsets across the compaction hinge on the north side of Apache Canyon (e.g., compare Fig. 20 with Figs. 11A, 12, 13A). As in Apache Canyon, we attribute this platform-hinge crest to the presence of cemented and/or grain-rich strata over an antecedent platform margin, most likely a transgressive or early-highstand reef in sequence V (Fig. 20). The development of the compaction hinge began with the subsidence and basinward tilting of the sequence V and VI platforms during times of sequence-boundary formation (sea-level fall and lowstand). The tops of these basinward-dipping platforms were then onlapped by sequences VI and VII, respectively (Fig. 20). The basinward divergence of younger sequences in Figure 20 shows that the structure that formed over the compaction hinge became exaggerated further during burial. The resulting highstand sequence-set platform architecture is directly comparable to that seen in Apache Canyon and on the EPR Line (e.g., Figs. 12, 19A).

San Simon Channel

The San Simon channel represents an ancient seaway that separated the Central Basin platform from the Northwest shelf in the Wolfcampian, Leonardian, and early Guadalupian (Fig. 8). Steep oblique and dip-slip faults cut and deform the underlying basement, which has a relatively complex structure. In contrast, the structure observed in overlying carbonate-siliciclastic Wolfcampian-Guadalupian platforms is relatively simple. Platform topsets are generally subhorizontal and subparallel or dip and diverge basinward at as much as 6.5° (Fig. 21). The integration of well data via synthetic seismograms into a high-quality two-dimensional seismic grid with a variable 3–7-km line spacing constrains the stratigraphy of the Wolfcampian–Leonardian platforms over an area of approximately 3000 km^2 (Fig. 8). One of us (Fitchen) has correlated and mapped 13 sequences through this area on the basis of seismic reflector configuration, well-log ties, and sandstone markers. These data suggest that sequences 3–5 are of Wolfcampian age and sequences 6–11 are of Leonardian age.

Wolfcampian–Leonardian sequences.—

Figure 21 shows representative platform geometries from the Northwest shelf toward the western end of the San Simon channel. In the eastern line (line A), the Devonian–Pennsylvanian "basement" dips gently basinward (south), whereas this dip reverses 5.2 km to the west (line B; Fig. 21). The Wolfcampian platforms (sequences 3–5) developed on and subdued the underlying basement structure. Topsets within these sequences are generally subhorizontal, parallel, and of constant thickness, indicating that they postdate basement faulting, that negligible differential tectonic subsidence occurred during platform growth, and that there has been little postdepositional tectonic tilting (e.g., sequences 3–5, Fig. 21). In terms of stratigraphy, the Wolfcampian platforms form an aggradational-retrogradational sequence set (Fig. 21). The Leonardian sequences form two progradational sequence sets composed of sequences 6–7 and 9–13 (Fig. 21). They are separated by sequence 8, which is retrogradational-aggradational and downlaps onto basinward-dipping sequence 7 topsets (Fig. 21).

In both lines of Figure 21, the aggradational margin of sequence 5 is a feature of fundamental importance to the structure and architecture of younger sequences. Landward of this margin, the topsets to sequences 6–13 generally maintain thickness, are horizontal, and display parallel reflectors (Fig. 21); shingled reflectors are limited to sequence 12. In sequences 6–9, a zone 0.5–0.7 km (0.3–0.45 mi) wide of elevated topsets and a local reversal of dips (to landward) occur over and landward of the antecedent sequence 5 margin (line A; Fig. 21). In contrast, basinward of the aggradational sequence 5 platform margin, the upper surfaces of sequences 6–13 dip downward toward their depositional margins. The topsets at these platform margins (sequences 6–11) are 0.02–0.11 s TWT lower than equivalents landward of the sequence 5 platform margin. In sequences 6–8 of Line A and sequence 6 of Line B, a single slope break occurs landward of the depositional platform margin (Fig. 21). This secondary slope break is located over the sequence 5 margin. In younger sequences, several discrete slope breaks occur in topsets landward of the depositional margins to sequences 10 and 11 in Line A and 7–11 in Line B (Fig. 21). These breaks of slope all occur over the terminal margin of antecedent platforms. However, updip (landward) thinning across basinward-dipping topsets of the preceding platform sequence is limited to sequences 8, 11, and 12. Sequence 8 downlaps onto sequence 7, whereas sequence 11 locally onlaps sequence 10. Basinward thickening of sequence 12 is accompanied by the gradual divergence of reflectors.

Interpretation.—The occurrence of aggradational topset reflectors below the terminal platform margin to sequences 6–9 and 11 indicates that the effects of relative fall in sea level during progradation did not contribute to the basinward-descending stratal geometry characteristic of these sequences. Precursor platform margins clearly localize the increase in basinward dip (to as much as 6.5° basinward as constrained by tying seismic data to well logs) and deformation of sequence 6–9 and 11 platform topsets (Fig. 21). Deformation over antecedent margins formed basin-facing monoclines that are not related to basement structure. For example, the monocline in sequences 6–9 that is localized over the undeformed subhorizontal platform top of sequences 3–4 cannot be attributed to tectonic subsidence or deformation (line A; Fig. 21). We think that the fact that the tilted topsets dip and/or diverge basinward from over antecedent platform margins is no coincidence and, in combination with the above observations, points toward compaction-induced differential subsidence as the mechanism responsible for the architecture and structure of sequences 6–13. We interpret sequences 6–12 to show compaction-modified stratal architecture resulting from differential subsidence over three discrete compaction hinges (Fig. 21). The formation of a narrow zone of elevated topsets in an anticline over the compaction hinge is a common attribute of platforms along the Northwest shelf (Fig. 21). This is not a velocity pull-up because no pull-up occurs in the reflectors below. Instead, we think that these seismic features are analogous to the compaction-hinge crest developed in sequence L1 over the erosional Wolfcampian platform margin in Apache Canyon (e.g., Figs. 11–13). As in Apache Canyon, we interpret formation of these anticlinal structures to localized changes of facies and diagenetic patterns over the compaction hinge that inhibit compaction.

In terms of the timing of compaction-induced differential subsidence, only the top of sequences 7 and 10 meet the criteria of compaction-modified sequence boundaries. The relief formed during sequence-boundary formation across sequences 7 and 10 modified their platform-top physiography to control the updip onlap or downlap and thinning or pinch-out of the succeeding sequences 8 and 11 (Fig. 21). However, the deformation across sequences 7 and 10 and the associated increase in thickness of sequences 8 and 11 can only account for a small proportion of the deformation observed in sequences 6–13. This observation, in addition to the general shallowing of dips in topsets of younger sequences (i.e., 6–13) across the sequence 5 platform margin, indicates progressive monocline development (Fig. 21). It appears that most of the compaction-induced deformation in this area modified platform architecture during the deposition of sequences 11 and younger.

DISCUSSION

This study of Permian strata in west Texas and southeast New Mexico has presented outcrop and subsurface data supporting the contention that the modification of carbonate platforms and sequence boundaries by compaction-induced differential subsidence can be an important control on their architecture, development, and diagenesis. We have shown that compaction-modified sequences displayed in outcrops in the Sierra Diablo, Guadalupe, and Brokeoff Mountains developed characteristic features that are directly comparable to subsurface equivalents on the Northwest and Eastern shelves. In the following section, we discuss the implications of the study for understanding (1) the compaction process, (2) the compaction rates, (3) the distribution and (4) the architecture of compaction-modified sequences, (5) their along-strike variability, (6) the basinward tilting of platform topsets, (7) the distribution and significance of rollover in terms of reef paleoecology, (8) the control of relief formed on lowstand and transgressive systems tracts, (9) the accommodation interpretation, (10) the prediction of fractures and paleokarst development, and (11) the structures formed in compaction-modified platform sequences.

Compaction Process

We have shown that some of the best examples of compaction-modified sequence geometries, such as L1 in Apache Canyon, formed during lowstands (Hunt et al., 1995, 1996). The existence of compaction-modified sequence boundaries has important implications for the timing and continuity of the compaction process and points to several underlying processes; it is difficult, however, to isolate these mechanisms. Since compaction is a load-driven process, it appears paradoxical that compaction-modified sequences boundaries are developed during times of subaerial exposure and, hence, during a hiatus in deposition and loading across the exposed platform top. The formation of compaction-modified sequences boundaries implies the existence of a mechanism that can account for the delay between loading and the resulting compaction.

In previous articles, we argued that the delay between loading and compaction is related to the buildup and bleed-off of pore-fluid overpressure in the shallow subsurface (<250 m) (Hunt et al., 1995, 1996). During the addition of overburden, overpressure and a state of compaction disequilibrium occur in low-permeability basinal facies, and/or in sediments bounded by low-permeability surfaces or facies, from which rates of fluid flow are retarded. The development of a state of compaction disequilibrium can be viewed as a twofold process. First, rapid rates of deposition lead to pressure buildup when rates of pore-fluid escape are exceeded. The buildup of abnormal pressures temporarily retards compaction so that the overpressured strata retain a "stored" capacity to compact. Subsequently, when rates of pressure release from pore-fluid bleed-off exceed rates of pressure increase due to loading (e.g., during hiatus), excess pressure and pore fluids dissipate, and the "delayed" compaction proceeds until a state of pore-fluid and compaction equilibrium is reached. The development of compaction-modified sequence boundaries on carbonate platforms suggests that this twofold process of pressure buildup and bleed-off is associated with highstands and lowstands, respectively (Hunt et al., 1995, 1996).

An alternative mechanism that may induce compaction during sea-level lowstands is the effect of fluid buoyancy. Pore fluids partially support the weight of the enclosing solid so that removal of pore fluids from a saturated (and previously uncompacted) sediment will result in the withdrawal of this support during lowstands and, hence, compaction. Across a carbonate platform, the contribution of this effect will depend on (1) amplitude of the relative fall in sea level and (2) cementation of the compactible lithologies. Buoyancy effects should be most pronounced when high-amplitude sea-level falls drain pore fluids from uncemented foreslope and/or basinal facies underlying partially cemented platform deposits. These two mechanisms are likely to operate in concert.

Compaction Rates

To provide an estimate of compaction rates we follow the biostratigraphy of Wilde (1995), integrated with the time scales of Harland et al. (1989), and we have assumed a sinusoidal-shaped sea-level curve, where the lowstand represents 25% of sequence duration. Kerans et al. (1993) estimated that the average duration of Leonardian sequences is approximately 1.6 m.y. In Apache Canyon, the compaction-induced subsidence of L1 is 10 m on the shelf, 51 m over the Wolfcampian scarp, and 25 m at the terminal shelf-slope break (Fig. 13). Assuming that the lowstand represents 25% of the sequence duration, these subsidence values imply a rate of compaction in the range of 0.03–0.13 m/k.y. By using this approach and following Kerans and Fitchen (1995) in taking 2.0–4.0 m.y. for the duration of the entire San Andres Formation that is composed of 15 high-frequency sequences (L8–G13), we measure rates of 0.21–0.42 m/k.y. and 0.1–0.21 m/k.y. for compaction rates at West Dog Canyon and Last Chance Canyon, respectively (Figs. 15, 17). Within the upper San Andres Formation highstand of Last Chance Canyon, 17 m of relief is displayed by cycle 6 relative to the uSA_4/Grayburg sequence boundary. If we add this relief to the minimum estimate of 7 m of relief developed during the lowstand and assume that the highstand and lowstand represent 66% of the sinusoidal sea-level curve, we obtain an average rate of compaction-induced subsidence of 0.14–0.27 m/k.y. The average compaction-induced differential subsidence rate during a lowstand that we calculated from our examples (based on the stated assumptions) was 0.19 m/k.y.

Distribution of Compaction-Modified Sequences

In the Permian basin, the modification of stratal architecture by compaction-induced differential subsidence appears to be a characteristic of highstand sequence sets (Figs. 11–22). In general, compaction-modified stratal geometries do not occur in aggradational or retrogradational sequences because the foundation of these sequences is normally relatively homogeneous, and their margin is positioned over or landward of antecedent platform margins (e.g., sequences 1–2, Figs. 18, 19A; sequences 8, 9, 10, Fig. 19; sequences 3–5, Fig. 21). We have, however, identified a single example of a compaction-modified sequence associated with platform aggradation-retrogradation (sequence 8, Fig. 21). The modification of this platform occurred because it has a progradational sequence set as a foundation (sequences 6–7; Fig. 21). In the examples we have identified, compaction-induced differential subsidence is most important where platforms have prograded over an antecedent platform margin and wedge-shaped packages of compactible fine-grained basinal carbonate and siliciclastic sediments (e.g., Figs. 11, 12, 14, 15, 17, 18–22).

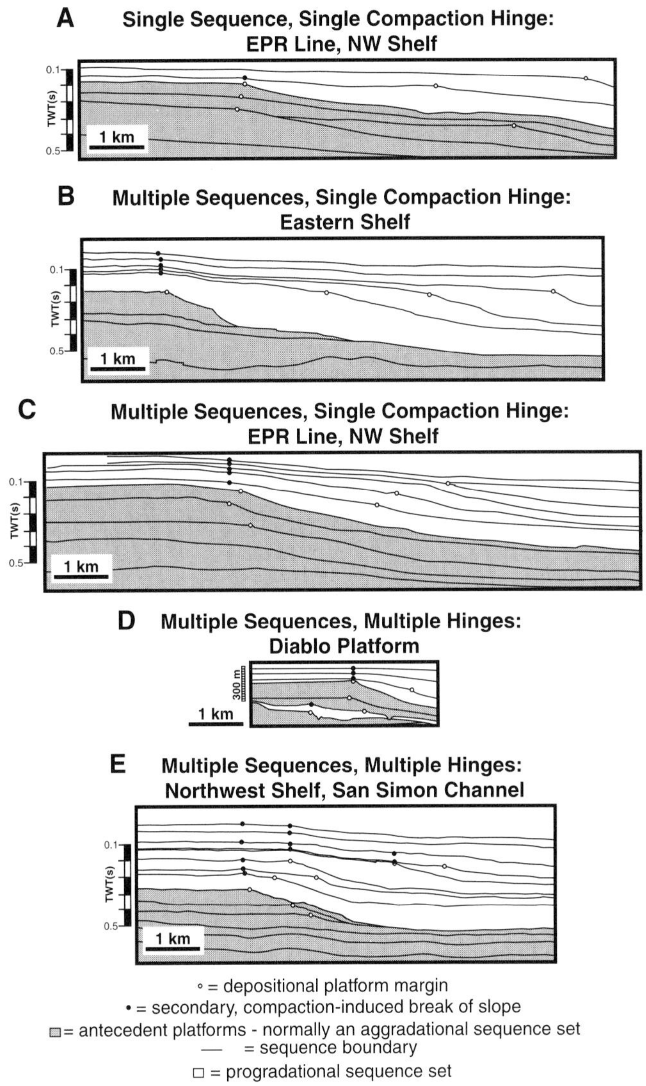

FIG. 22.—Simplified line drawings of compaction-modified platforms and sequence boundaries summarized from examples presented herein. All sections are at the same horizontal and vertical scale in two-way traveltime (TWT), except for the Diablo platform, where the vertical scale is in meters. The modification of platform geometries by differential compaction became more complex, and generally longer lived, from the top to the bottom of the diagram. Note how topsets diverge basinward from a common point: the compaction hinge. Little or no rollover is displayed by antecedent aggradational platforms. Rollover is a characteristic of highstand sequences and sequence sets. The amount of rollover generally decreases upward through a sequence set. The angle of dip across individual compaction-modified sequence boundaries has become exaggerated through further compaction-induced differential subsidence during burial (<6.5°). The stratal patterns of these highstands are analogous to those of Figure 6A, although in detail the individual compaction-modified sequence boundaries show stratal patterns analogous to those in Figure 6B, as discussed in the text and detailed in earlier sections.

Architecture of Compaction-Modified Sequences

The most prominent and characteristic feature of compaction-modified sequences developed over antecedent platform margins is the compaction hinge (Fig. 22). The hinge separates two mechanically distinct parts of a carbonate platform characterized by differing accommodation histories, facies patterns, and stacking patterns as seen in outcrops in the Sierra Diablo, Brokeoff, and Guadalupe Mountains (Figs. 11–17). Shelfward of the compaction hinge, topset facies are parallel and of constant thickness. In contrast, traced basinward, the same topsets and platform sequences are observed to diverge and thicken toward a downward-deflected shelf-slope break; this characteristic is especially evident in regional seismic lines (Figs. 18–22). Comparison of the lines in Figure 22 shows that the stratigraphic importance of an individual compaction hinge can be quite variable. Over antecedent high-relief platform margins (>200–300 m), a compaction hinge is often a long-lived feature and will propagate upward through several sequences as the effects of differential compaction continue from the shallow subsurface through the burial environment. This stratal pattern and style of deformation are exemplified by sequences L4–L6 in Apache Canyon, by sequences 3–7 on the EPR Line, and by sequences in the San Simon channel (Figs. 12, 19, 22). In such cases, the compaction hinge can form the main structural feature and in outcrops or on regional seismic lines should be easily recognized.

However, over low-relief platform margins (<100–150 m), a compaction hinge may only influence one or two succeeding sequences and, in comparison to high-relief situations, may be a short-lived feature with a less pronounced seismic expression, as is observed in the EPR Line over the upper San Andres Formation margin (Fig. 18) and in Apache Canyon across sequence L1 (Fig. 13). In these two examples, most of the basinward dip of topsets occurred in response to both differential compaction and relative fall in sea level; these processes formed space that controlled the onlap and landward extent of the succeeding sequences (Figs. 13, 15, 16, 19C, D). In both cases, the relief across the compaction-modified sequence boundaries took place very early during lowstands, with little discernible compaction during subsequent burial.

The spectrum of scales at which compaction hinges may be developed means that several may simultaneously have an effect on the architecture and stratigraphy of any one sequence, as is well illustrated in the EPR Line by sequence 11 (Fig. 18) and in Lines A and B by sequences 7–11 (Fig. 21). In these cases, the deformed platform top will have several discrete slope breaks and changes of gradient. It is normal in such situations to see a "primary" updip compaction hinge with "subordinate" or "secondary" hinges downdip whose expressions are more subtle (Figs. 18, 22E).

Other important architectural characteristics of compaction-modified platforms include the basinward subsidence of the platform top and margin and the basinward dip of topsets and toplap surfaces. Tilting takes place as the outer platform and platform margin subsides and is normally concurrent with the shallowing of foreslope angles (e.g., sequence L1, Apache Canyon; Figs. 11–13). As a compaction-modified sequence boundary is formed, postdepositional relief is developed across the platform. This relief substantially modifies the platform-top physiography from a flat-topped carbonate shelf to a distally steepened ramplike profile (Figs. 2, 7, 8; terminology of Read, 1982, 1985).

Strike Variability

Two aspects of strike variability have been illustrated in terms of the development of compaction-induced relief and the preservation potential of the compaction-modified sequence boundary. In Apache Canyon, there is a clear contrast in the morphology and relief of the basinward-tilted L1 topsets between the two dip sections (1.3–2 km apart). This morphology and relief later controlled the position and stratal relationships of sequences L2 and L3 (e.g., Figs. 11–13). The examples from West Dog and Last Chance Canyons are rather different and illustrate variability in the preservation of the G13/G14 compaction-modified sequence boundary. In the former example, compaction-induced relief was preserved to control shelfward thinning and lap-out at the base of the G14 (Figs. 15, 16). In Last Chance Canyon, 30 km to the northeast, relief that formed by differential compaction was later removed by subaerial and transgressive erosion to form an angular unconformity.

Tilting of Platform Topsets

We have shown that the deformation of progradational platform strata over a compaction hinge often begins during and/or immediately after deposition (during a lowstand, e.g., L1 and L4–L6 in Apache Canyon, Figs, 11, 13, 14; and the upper San Andres Formation in West Dog Canyon, Fig. 15). However, analysis of seismic data shows that significant deformation continues during burial to further steepen dips formed across compaction-modified sequence boundaries (e.g., sequence boundaries 3–5; Figs. 18, 19; V–VIII, Fig. 20; sequence boundaries 7/8 and 10/11, Fig. 21). The dips formed over compaction-modified sequence boundaries were generally <1° prior to transgression, but the dips of platform topsets have commonly been further tilted and steepened to an angle of <5°–7° during subsequent burial. As a result of the steepening of strata during burial, it is normal to observe a divergence of platform topsets within a highstand sequence set that is basinward of a compaction hinge (Figs. 22B–D). Topsets of the oldest progradational sequence dip most steeply, and dips are reduced upward in younger sequences within a highstand set. This stratal pattern within highstand sequence sets is analogous to that shown in Figure 6A.

Significance of Rollover

We have presented examples from the Permian basin that show rollover or basinward dip of topsets and associated deepening of topset facies toward a downward-deflected, subsided platform margin. We think that these relationships are a natural consequence of platform progradation over antecedent margins (e.g., Fig. 22). In the Permian basin, it is normal to see that (1) the preceding aggradational platform precursor normally shows an elevated margin and a flat top with no rollover; (2) rollover is limited to progradational sequences and/or sequence sets; (4) within the highstand sequence set, the amount of rollover is normally greatest in the oldest strata and decreases through time; (4) rollover margins are limited to platforms where the equivalent back-reef strata thicken basinward over tilted, earlier platform topsets, indicating differential subsidence during progradation where equivalent platform topsets are parallel updip of a compaction hinge; and (5) this stratal pattern is developed across the Permian basin independent of strata age, from the lowermost Leonardian to the uppermost Guadalupian rocks, as illustrated by Figures 18–22. These observations are important to bear in mind when interpreting the original morphology and facies relationships of platforms in the Permian basin and elsewhere. Many rollover-type platform margins have been interpreted to be of purely depositional origin, with platform-margin reef facies interpreted to have formed in relatively deep waters. Wood et al. (1994) and Saller (1996) put forward similar arguments in their interpretation of the Capitan reef, with whom our observations from this study and recent work in Slaughter Canyon agree.

Control on Lowstand and Transgressive Systems Tract Development

We have emphasized the modification of platform geometries by differential compaction during lowstands to develop compaction-modified sequence boundaries. Compaction-induced differential subsidence can significantly alter the morphology of carbonate platforms during relative lowstands of sea level to control the size, extent, and stratal relationships of the succeeding lowstand and transgressive systems tracts (e.g., Figs. 1–21). As a consequence of compaction-induced deformation during lowstands, the tilted platform topsets develop secondary relief with one or more compaction-induced slope breaks (Figs. 2, 6B). If preserved, the succeeding lowstand and/or transgressive deposits inherit and onlap this relief (Figs. 11–21). They inherit "additional" compaction-induced accommodation space across the platform top formed during lowstands (Fig. 23). As a consequence, transgressive systems tracts are both thicker and are located in a more shelfward position than would otherwise be predicted (e.g., Figs. 13, 15); the induced relief across the compaction-modified part of the platform (basinward of the compaction hinge), both reduces rates of transgression and provides a greater suitable area (<20 m water depths) for sediment production.

Accommodation Interpretation

Differentiating the effects on platform morphology caused by relative fall in sea level or forced regression (cf. Plint, 1990; Posamentier et al., 1992; Hunt and Tucker, 1992, 1993, 1995; D. Hunt and R. L. Gawthorpe, in prep.) versus compaction-induced subsidence can be problematic. In much of the literature concerning forced regression, the basinward-descending geometry of prograding systems is portrayed as unique. This study has highlighted that analogous geometries form through compaction-induced differential subsidence. Distinguishing between compaction-modified geometries and geometries formed by a forced regression is aided by the use of several features, including (1) the presence of basinward-dipping aggradational topset facies that (2) are tilted across a compaction hinge positioned over an antecedent platform, the effects of which (3) propagate through several preceding and overlying sequences, in combination with (4) along-strike correlation to areas of lower relief, as shown in Figure 21. However, where no aggradational topset strata (sigmoids) are present, the geometries formed by relative fall in sea level and differential compaction are difficult to distinguish. This is the case in Apache Canyon, for example (Figs. 11–13). In such situations, the presence of the other cited criteria, in combination with simple forward and reverse modeling, can aid differentiation.

A case in point is schematically illustrated in Figure 23, a diagram derived from our experience in interpreting the upper San Andres Formation in outcrops (e.g., Figs. 15–17) and in the subsurface (as in Figs. 18 and 19). In Figure 24, contrasting precompaction and postcompaction stratal patterns and sequence stratigraphic and accommodation interpretations are made from tracing the translation of the shelf-slope break. In the precompaction state (Fig. 23), the platform evolved from aggradational to progradational stacking patterns. Vertical aggradation of the platform margin characterizes times 1–2, as there is a balance between rates of accommodation development (A′) and sediment supply (S′). Sigmoidal clinoform progradation in times 2–5 (S′ > A′) is succeeded by oblique clinoform progradation (times 5–7; S′ >> A′, A′ = 0), and by progradation and downstepping (S′ >> A′, A′ is negative). In this situation, oblique clinoform progradation between times 5 and 7 is best interpreted in terms of a high stillstand of relative sea level (Fig. 23). Downstepping of the platform margin during times 7–9 is interpreted as deposition during relative fall in sea level.

In the postcompaction state, there is ambiguity as to the interpretation of the oblique clinoforms (times 5–7, Fig. 23). Instead of representing a high-stillstand condition, these clinoforms would likely be interpreted to represent deposition during falling relative sea level (S′ >> A′, A′ is negative). The interpretation of the platform's development and response is changed as accommodation is "lost" from tracing the position of the depositional shelf-slope break in the postcompaction state. In the upper San Andres Formation, the affects of compaction-induced differential subsidence can lead to an underestimation of platform aggradation during progradation (e.g., Fig. 17). The basinward tilting of the platform leads to practical problems in differentiating the effects of relative fall in sea level and compaction-induced differential subsidence in the outer parts of the platform both in the subsurface and in outcrops.

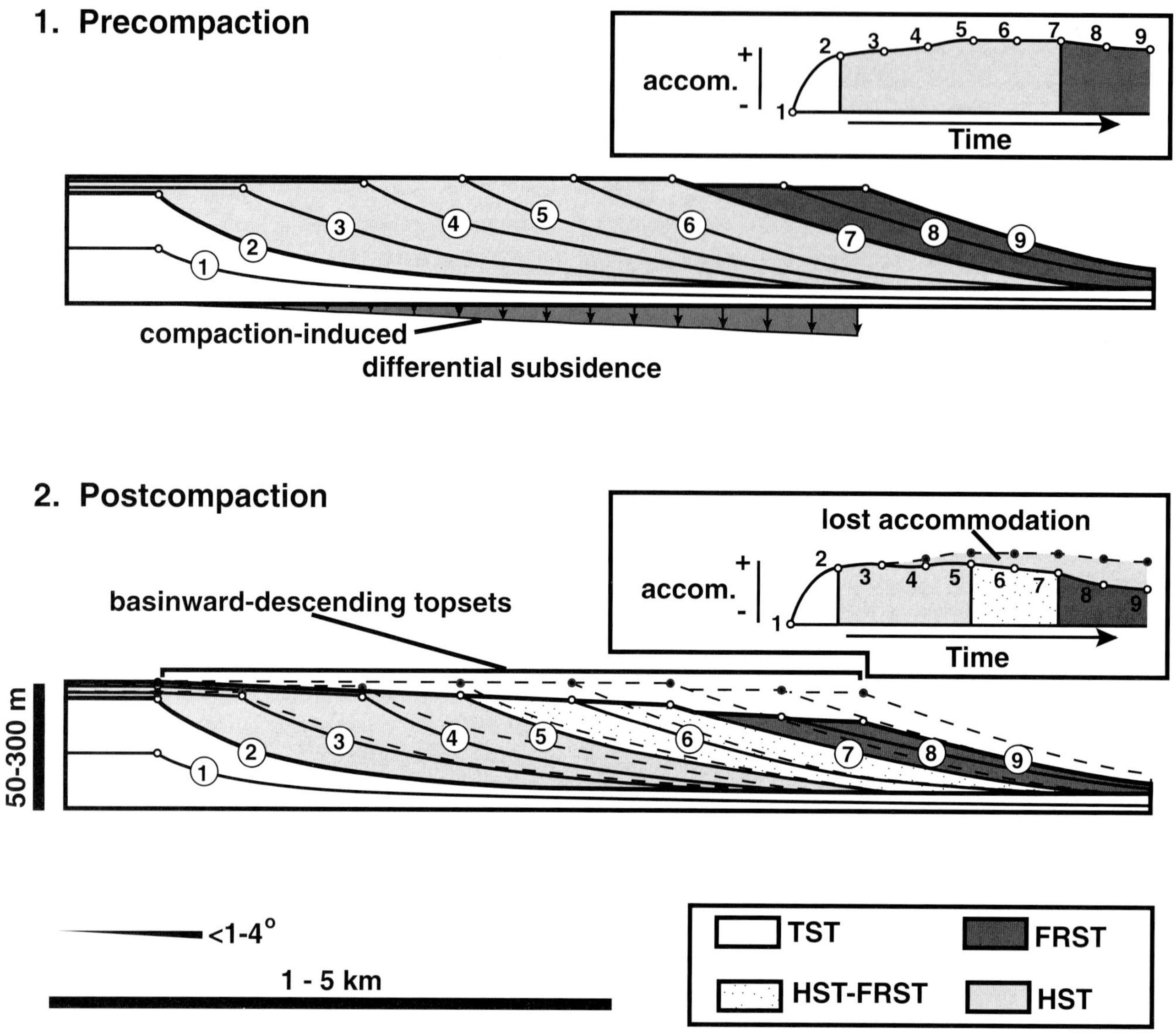

FIG. 23.—Two cross sections illustrating the problems of sequence stratigraphy and sea-level history interpretation of oblique clinoform progradation following the effects of compaction-induced differential subsidence, based on observations of the upper San Andres Formation in the Brokeoff Mountains, Last Chance Canyon, and in the subsurface. In the precompaction state, the oblique clinoform progradation characteristic of times 5–7 would most likely be interpreted to represent a sea-level stillstand during late highstand, with times 7–9 interpreted as progradation during a relative fall in sea level or forced regression (cf. Plint, 1990; Posamentier et al., 1992; Hunt and Tucker, 1992, 1995). In contrast, following compaction, the interpretation of the oblique clinoform progradation becomes ambiguous; it may be interpreted as either part of the highstand or, more likely, part of the forced regressive systems tract. Using the position of the shelf-slope break to interpret accommodation development leads to an underestimation of relative rise in sea level and an overestimation of progradation during sea-level fall in the compacted profile, compared to the precompaction interpretation. TST = transgressive systems tract, HST = highstand systems tract, FRST = forced regressive systems tract, accom. = accommodation.

Prediction of Fractures and Paleokarst Development

Flexural deformation of the platform top by compaction-induced subsidence is focused at the compaction hinge. In the examples from Apache Canyon and West Dog Canyon, solution-widened fractures (grikes) and paleokarst sinkhole fills are best developed within a strike-parallel, 200–500-m-wide zone over and immediately downdip of the compaction hinge. This preferential distribution of penetrative paleokarst is attributed to enhanced fracture development within basinward-tilted, early-lithified topset strata across the compaction hinge (Figs. 13, 15, 16) and perhaps to steeper topography and associated higher rates of meteoric water flow. These observations at outcrop have led us to suggest that the compaction hinge is where intensive fracture-controlled paleokarst might be predicted in the subsurface. This prediction has been supported by studies in the Maljamar field on the Northwest shelf (C. Modioca, pers. commun., 1997) and in the Means field on the Central Basin platform. Recognition of the compaction hinge from simple geometries has obvious potential for the prediction of fracture-controlled karst and associated permeability networks in the subsurface.

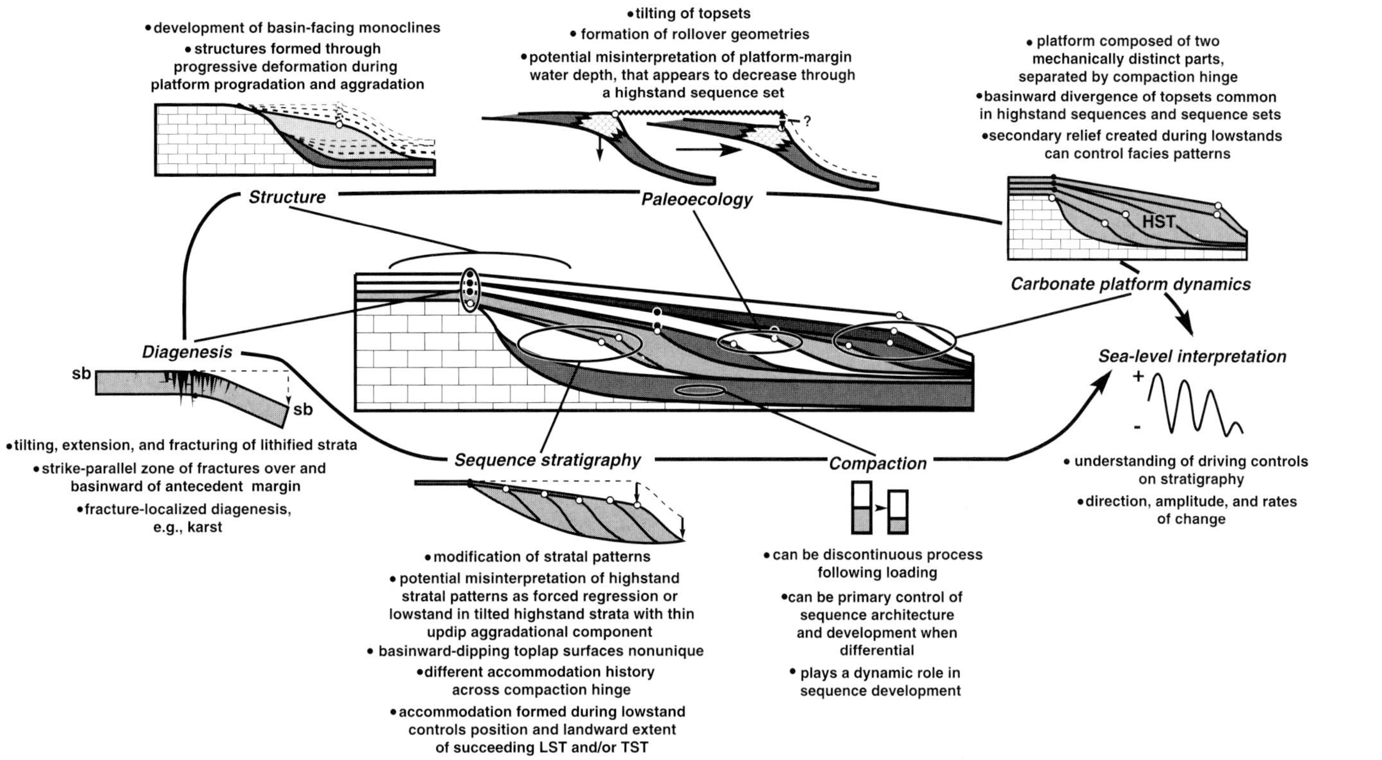

FIG. 24.—Summary diagram showing the characteristic stratal patterns of compaction-modified sequences and the implications for carbonate-platform development and interpretation. See discussion and conclusions in text for details.

Structural Development of Compaction-Modified Carbonate Platforms

Lithified strata that are differentially compacted across a compaction hinge undergo significant extension. Fracturing that occurs as a result of this extension can control the distribution of penetrative paleokarst at sequence boundaries, as already discussed. The deformation of topsets facies across the compaction hinge leads to the development of low-amplitude (<100-m amplitude) monoclines or asymmetric anticlines that began to grow during and/or immediately after deposition (during lowstands). The growth of these structures is interpreted to be progressive and continues with increasing burial (e.g., Figs. 22, 24). The resulting folds are asymmetric anticlines, with a steep limb dipping toward the basin. In the Permian basin, such folds form some of the most important structural traps (e.g., Galloway et al., 1983). Since the location and form of these monoclines and asymmetric anticlines are controlled by underlying structures (i.e., carbonate-platform margins), these are classified as forced folds (sensu Sandford, 1959; Sterns, 1978; Withjack et al., 1990).

CONCLUSIONS

The control of compaction-induced differential subsidence on carbonate-platform development is frequently overlooked. In Permian strata of the Delaware and Midland basins, west Texas and southeast New Mexico, we have shown that compaction can play a dynamic role in the development of carbonate platforms. The modification of platform geometries by differential compaction varies spatially and temporally. It is most important where Leonardian-Guadalupian highstand strata (sequences and sequence sets) prograded over antecedent aggradational or erosional platform margins and/or wedge-shaped packages of compactible basin facies. Figure 24 summarizes implications of the study results (1) for understanding the compaction process as well as the diagenesis, sequence stratigraphy, structure, and paleoecology of carbonate platforms and (2) for the interpretation of accommodation histories.

Compaction-modified sequences develop characteristic attributes, the most important of which are the basinward dip of topsets toward a subsided platform margin and the formation of a compaction hinge. The tilting of topsets during lowstands can control the location, size, and areal extent of the succeeding lowstand and transgressive systems tracts. It can also develop postdepositional stratal patterns in highstand strata that resemble those formed during a relative fall in sea level. Deformation by compaction-induced differential subsidence over an antecedent platform margin forms a compaction hinge and a secondary slope break in platform topsets. The compaction hinge separates two mechanically distinct parts of a platform with contrasting accommodation histories, facies, and stratal patterns. Updip of the compaction hinge, platform topsets are generally parallel and of constant thickness, whereas downdip, the same topsets thicken, dip, and diverge basinward toward a downward-deflected depositional shelf-slope break. The expression and persistence of a compaction hinge may vary both spatially and temporally. The further steepening of dips across compaction-modified sequence boundaries during burial aids recognition in outcrops and especially in the subsurface. Fracturing over the compaction hinge can also localize unusually penetrative paleokarst to a narrow (200–1000 m) strike-parallel zone.

The formation of topset rollover as a result of compaction-induced differential subsidence is a natural consequence of, and is characteristic of, progradation within highstand sequences and/or sequence sets in the Permian basin. Most rollover occurs in the oldest highstand strata and sequences and systematically decreases updip. It is significant that antecedent aggradational platforms and the final sequence of a highstand sequence set normally display little or no rollover. The formation of post-depositional rollover in highstand topsets has important implications for the interpretation of reef paleoecology and the interpretation of accommodation histories from stratal patterns. Asymmetric low-amplitude monoclines (<100-m relief) are the dominant structure of progradational platforms in the Permian basin. These forced folds form some of the most important structural traps in the Permian basin; they developed progressively through differential compaction-induced subsidence during deposition and subsequent burial.

It is clear from this study that an understanding of how differential compaction can modify sequence development, architecture, and structure helps one to better understand and interpret the often complex stratal patterns of carbonate platforms. The results of this study are not restricted to the Permian basin, but apply to carbonate platforms in general.

ACKNOWLEDGMENTS

Initial support for this project was provided by a grant from the Natural Environment Research Council (to D. H.). W. M. F. thanks Exxon, Texaco, Unocal, ARCO, Chevron, Marathon, Mobil, and the University of Texas Geology Foundation for financial support of field studies in the Brokeoff and Sierra Diablo Mountains. We thank Tony Adams, Rob Gawthorpe, and Tim Allsop for their reviews of an earlier draft of the manuscript. Art Saller, Dan Lehrmann, and Scott Tinker suggested many improvements to a later version of the manuscript. We also thank J. Kennard and W. Schlager for making suggestions on earlier papers concerning the subjects that are addressed here; their comments have proved insightful. Finally, we thank Lloyd Pray for introducing us to the Guadalupe Mountains. His teaching, research contribution, and enthusiasm have been invaluable to us and continues to inspire.

REFERENCES

ADAMS, J. E., AND FRENZEL, H. N., 1950, Capitan reef barrier, Texas and New Mexico: Journal of Geology, v. 58, p. 289–312.

BARNABY, R. J., AND WARD, W. B., 1995, Sequence stratigraphic framework, high-frequency cyclicity, and three-dimensional heterogeneity: Grayburg Formation, Brokeoff Mountains, New Mexico, *in* Pausé, P. H. and Candelaria, M. P., eds., Carbonate Facies and Sequence Stratigraphy: Practical Applications of Carbonate Models: Midland, Permian Basin Section–SEPM (Society for Sedimentary Geology) Publication No. 95-36, p. 37–49.

BOND, G. C., AND KOMINZ, M. A., 1984, Construction of tectonic subsidence curves for early Paleozoic miogeocline, southern Canadian Rocky Mountains: Implications for subsidence mechanisms, age of breakup, and crustal thinning: Geological Society of America Bulletin, v. 95, p. 155–173.

BORER, J. M., AND HARRIS, P. M., 1991, Lithofacies and cyclicity of the Yates Formation, Permian basin—Implications for reservoir heterogeneity: American Association of Petroleum Geologists Bulletin, v. 75, p. 726–779.

BORER, J. M., AND HARRIS, P. M., 1995, Computer simulation of the Yates Formation (Permian, Delaware basin)—Sequence stratigraphy and shelf-to-basin correlation implications, *in* Martin, R. L., ed., In Search of New Permian Basin Oil and Gas Fields: Using Today's Technologies and Tomorrow's Ideas for Exploration, Development, and 3D Seismic in a Mature Basin: Midland, West Texas Geological Society Publication No. 95-98, p. 111–132.

BURCHETTE, T. P., AND WRIGHT, V. P., 1992, Carbonate ramp depositional systems: Sedimentary Geology, v. 79, p. 3–57.

CREVELLO, P. D., WILSON, J. L., SARG, J. F., AND READ, J. F., eds., 1989, Controls on Carbonate Platform and Basin Development: Tulsa, Society of Economic Paleontologists and Mineralogists Special Publication 44, 405 p.

DUNHAM, R. J., 1972, Capitan reef, New Mexico and Texas—Facts and questions to aid interpretation and discussion: Midland, Permian Basin Section, Society of Economic Paleontologists and Mineralogists Publication 77-16, p. 479–486.

ENOS, P., AND SAWATSKY, L. H., 1981, Pore networks in Holocene carbonate sediments: Journal of Sedimentary Petrology, v. 51, p. 961–985.

EWING, T. E., 1991, The Tectonic Framework of Texas: Text to Accompany "The Tectonic Map of Texas": Austin, University of Texas, Bureau of Economic Geology, 36 p.

FISCHER, A. G., AND SARNTHEIN, M., 1988, Airborne silts and dune-derived sands in the Permian of the Delaware basin: Journal of Sedimentary Petrology, v. 58, p. 637–643.

FITCHEN, W. M., 1992, Sequence stratigraphy of the upper San Andres Formation and Cherry Canyon Tongue (Permian, Guadalupian), southern Brokeoff Mountains, New Mexico: Unpublished M.A. Thesis, University of Texas, Austin, 155 p.

FITCHEN, W. M., 1993, Sequence stratigraphic framework of the upper San Andres Formation and equivalent basinal strata in the Brokeoff Mountains, Otero County, New Mexico, *in* Love, D. W., et al., eds., Carlsbad Region, New Mexico and West Texas: Socorro, New Mexico Geological Society Guidebook, 44th Field Conference, p. 185–193.

FITCHEN, W. M., 1996a, Carbonate sequence stratigraphy and its application to hydrocarbon exploration and reservoir development, *in* Ibrahim, P. and Marfurt, K. J., eds., Carbonate Seismology: Tulsa, Society of Exploration Geophysicists, Geophysical Developments Series No. 6, p. 121–178.

FITCHEN, W. M., 1996b, Strike variability of carbonate platform margin stratal architecture and cycle stacking patterns: Outcrop and seismic examples from Lower Permian depositional sequences of the Permian basin, U.S.A.: Tulsa, American Association of Petroleum Geologists, Annual Convention Official Program with Abstracts, p. A46.

FITCHEN, W. M., 1997, Lower Permian sequence stratigraphy of the western Delaware basin margin, Sierra Diablo, west Texas: Unpublished Ph.D. Thesis, University of Texas, Austin, 264 p.

FITCHEN, W. M., STARCHER, M. A., BUFFLER, R. T., AND WILDE, G. L., 1995, Sequence stratigraphic framework of early Permian carbonate platform margins, Sierra Diablo, west Texas, *in* Garber, R. and Lindsay, R. F., eds., Wolfcampian–Leonardian Shelf Margin Facies of the Sierra Diablo: Seismic Models for Subsurface Exploration: Midland, West Texas Geological Society Publication No. 95-97, p. 23–66.

GALLOWAY, E. E., EWING, T. E., GARRETT, C. M., TYLER, N., AND BEBOUT, D. G., 1983, Atlas of major Texas oil reservoirs: Austin, University of Texas, Bureau of Economic Geology, 139 p.

GAY, S. P., 1989, Gravitational compaction, a neglected process in structural and stratigraphic studies: New evidence from mid-continent, USA: American Association of Petroleum Geologists Bulletin, v. 73, p. 641–657.

GOLDHAMMER, R. K., 1997, Compaction and decompaction algorithms for sedimentary carbonates: Journal of Sedimentary Research, v. 67, p. 26–35.

GRANT, C. W., GOGGIN, D. J., AND HARRIS, P. M., 1994, Outcrop analog for cyclic-shelf reservoirs, San Andres Formation of Permian basin: Stratigraphic framework, permeability distribution, geostatistics, and fluid-flow modeling: American Association of Petroleum Geologists Bulletin, v. 78, p. 23–54.

GREENLEE, S. M., AND LEHMANN, P. J., 1993, Stratigraphic framework of productive carbonate buildups, *in* Loucks, R. G. and Sarg, J. F., eds., Carbonate Sequence Stratigraphy—Recent Developments and Applications: Tulsa, American Association of Petroleum Geologists Memoir 57, p. 43–62.

HANDFORD, C. R., AND LOUCKS, R. G., 1993, Carbonate depositional sequences and systems tracts—Responses of carbonate platforms to relative sea-level changes, *in* Loucks, R. G. and Sarg, J. F., eds., Carbonate Sequence Stratigraphy—Recent Developments and Applications: Tulsa, American Association of Petroleum Geologists Memoir 57, p. 435–474.

HARLAND, W. B., ARMSTRONG, R. L., COX, A. V., CRAIG, L. E., SMITH, A. G., AND SMITH, D. G., 1989, A Geologic Time Scale 1989: Cambridge, Cambridge University Press, 263 p.

HAYES, P. T., 1964, Geology of the Guadalupe Mountains, New Mexico: Washington, D.C., United States Geological Survey Professional Paper 466, 69 p.

HILEMAN, M. E., AND MAZZULLO, S. J., eds., 1977, Upper Guadalupian Facies, Permian Reef Complex, Guadalupe Mountains, New Mexico and West Texas: Field Conference Guidebook, Volume 1: Midland, Permian Basin Section, Society of Economic Paleontologists and Mineralogists Publication. 77-16, 450 p.

HILLS, J. M., 1985, Structural evolution of the Permian basin of west Texas and New Mexico, in Dickerson, P. W. and Muehlburger, W. R., eds., Structure and Tectonics of Trans-Pecos Texas, Field Conference Guidebook: Midland, West Texas Geological Society Publication No. 85-81, p. 89–100.

HORAK, R. L., 1985, Trans-Pecos tectonism and its effect on the Permian basin, *in* Dickerson, P. W. and Muehlburger, W. R., eds., Structure and Tectonics of Trans-Pecos Texas, Field Conference Guidebook: Midland, West Texas Geological Society Publication No. 85-81, p. 81–87.

HUNT, D., AND TUCKER, M. E., 1992, Stranded parasequences and the forced regressive wedge systems tract: Deposition during base-level fall: Sedimentary Geology, v. 81, p. 1–9.

HUNT, D., AND TUCKER, M. E., 1993, The sequence stratigraphy of carbonate shelves with an example from the mid-Cretaceous of S. E. France, *in* Posamentier, H. W., Summerhayes, C. P., Haq, B. U., and Allen, G. P., eds., Sequence Stratigraphy and Facies Associations: London, International Association of Sedimentologists Special Publication No. 18, p. 307–342.

HUNT, D., AND TUCKER, M. E., 1995, Stranded parasequences and the forced regressive wedge systems tract: Deposition during base-level fall: Reply: Sedimentary Geology, v. 95, p. 147–160.

HUNT, D., FITCHEN., W. M., SWARBRICK, R., AND ALLSOP, T., 1995, Differential compaction as a primary control of sequence architecture and development in the Permian basin: Geological significance and potential as a hydrocarbon exploration model, *in* Garber, R. and Lindsay, R. F., eds., Wolfcampian–Leonardian Shelf Margin Facies of the Sierra Diablo: Seismic Models for Subsurface Exploration: Midland, West Texas Geological Society Publication No. 95-97, p. 83–104.

HUNT, D., ALLSOP, T., AND SWARBRICK, R. E., 1996, Compaction as a primary control on the architecture and development of depositional sequences: Conceptual framework, applications and implications, *in* Howell, J. and Aitken, J. F., eds., High Resolution Sequence Stratigraphy: Innovations and Applications: London, Geological Society of London Special Publication No. 104, p. 321–346.

JERVEY, M. T., 1988, Quantitative geological modeling of siliciclastic rock sequences and their seismic expression, *in* Wilgus, C. K., Hastings, B. S., Kendall, C. G. St. C., Posamentier, H. W., Ross, C. A., and Van Wagoner, J. C., eds., Sea-Level Changes: An Integrated Approach: Tulsa, Society of Economic Paleontologists and Mineralogists Special Publication 42, p. 47–69.

KERANS, C., AND FITCHEN, W. M., 1995, Sequence hierarchy and facies architecture of a carbonate ramp system, San Andres Formation, Algerita Escarpment and western Guadalupe Mountains: Austin, University of Texas, Bureau of Economic Geology, Report of Investigations No. 235, 86 p.

KERANS, C., AND NANCE, H. S., 1991, High-frequency cyclicity and regional depositional patterns of the Grayburg Formation, Guadalupe Mountains, New Mexico, *in* Meador-Roberts, S., Candelaria, M. P., and Moore, G. E., eds., Sequence Stratigraphy, Facies and Reservoir Geometries of the San Andres, Grayburg and Queen Formations, Guadalupe Mountains, New Mexico and Texas: Midland, Permian Basin Section–SEPM (Society for Sedimentary Geology) Publication No. 91-32, p. 53–96.

KERANS, C., FITCHEN, W. M., GARDNER, M. H., SONNENFELD, M. D., TINKER, S. W., AND WARDLAW, B. R., 1992, Styles of sequence development within uppermost Leonardian through Guadalupian strata of the Guadalupe Mountains, Texas and New Mexico, *in* Mruk, D. H. and Curran, B. C., eds., Permian Basin Exploration and Production Strategies: Midland, West Texas Geological Society Publication No. 92-91, p. 1–6.

KERANS, C., FITCHEN, W. M., GARDNER, M. S., AND WARDLAW, B. R., 1993, A contribution to the evolving stratigraphic framework of middle Permian strata of the Delaware basin, Texas and New Mexico, *in* Love, D. W., et al., eds., Carlsbad Region, New Mexico and West Texas: Socorro, New Mexico Geological Society Guidebook, 44th Field Conference, p. 175–184.

KERANS, C., LUCIA, F. J., AND SENGER, R. K., 1994, Integrated characterization of carbonate ramp reservoirs using Permian San Andres Formation outcrop analogs: American Association of Petroleum Geologists Bulletin, v. 78, p. 181–216.

KING, P. B., 1931, Geology of the Glass Mountains, Texas: Part I. Descriptive geology: Austin, University of Texas Bulletin 3038, 167 p.

KING, P. B., 1942, Permian of west Texas and southeastern New Mexico: American Association of Petroleum Geologists Bulletin, v. 26, p. 535–763.

KING, P. B., 1948, Geology of the southern Guadalupe Mountains: Washington, D.C., United States Geological Survey Professional Paper 215, 183 p.

KING, P. B., 1965, Geology of the Sierra Diablo region, Texas: Washington, D.C., United States Geological Survey Professional Paper 480, 185 p.

KLUTH, C. F., AND CONEY, P. J., 1981, Plate tectonics of the Ancestral Rocky Mountains: Geology, v. 9, p. 10–15.

LOUCKS, R. G., AND SARG, J. F., eds., 1993, Recent Advances and Applications of Carbonate Sequence Stratigraphy: Tulsa, American Association of Petroleum Geologists Memoir 57, 545 p.

MAZZULLO, S. J., AND REID, A. M., 1987, Stratigraphy of the Bone Spring Formation (Leonardian) and depositional setting in the Scharb field, Lea County, New Mexico, *in* Cromwell, D. and Mazzullo, L. J., eds., The Leonardian Facies in W. Texas and S. E. New Mexico and Guidebook to the Glass Mountains: Midland, Permian Basin Section, Society of Economic Paleontologists and Mineralogists Publication 87-27, p. 107–111.

MAZZULLO, S. J., AND REID, A. M., 1989, Lower Permian platform and basin depositional systems, northern Midland basin, Texas, *in* Crevello, P. D., Wilson, J. L., Sarg, J. F., and Read, J. F., eds., Controls on Carbonate Platform and Basin Development: Tulsa, Society of Economic Paleontologists and Mineralogists Special Publication 44, p. 305–320.

MAZZULLO, S. J., REID, A. M., AND MAZZULLO, L. J., 1987, Basinal Lower Permian facies, Permian basin: Part I. Stratigraphy of the Wolfcampian–Leonardian boundary: Midland, West Texas Geological Society Bulletin, v. 26, p. 5–9.

MCDANIEL, P. N., AND PRAY, L. C., 1967, Bank-to-basin transition in Permian (Leonardian) carbonates, Guadalupe Mountains, Texas (abs.): American Association of Petroleum Geologists Bulletin, v. 51, p. 474.

MEISSNER, F., 1972, Cyclic sedimentation in middle Permian strata of the Permian basin, *in* Elam, J. G. and Chuber, S., eds., Cyclic Sedimentation in the Permian Basin (2nd ed.): Midland, West Texas Geological Society Publication No. 72-60, p. 203–232.

NEWELL, N. D., RIGBY, J. K., FISCHER, A. G., WHITEMAN, A. J., HICKOX, J. E., AND BRADLEY, J. S., 1953, The Permian Reef Complex of the Guadalupe Mountains Region, Texas and New Mexico: San Francisco, Freeman and Company, 236 p.

PAYTON, C. E., ed., 1977, Seismic Stratigraphy—Applications to Hydrocarbon Exploration: Tulsa, American Association of Petroleum Geologists Memoir 26, 516 p.

PLINT, A. G., 1990, High-frequency relative sea-level oscillations in Upper Cretaceous shelf clastics of the Alberta foreland basin: Possible evidence for a glacio-eustatic control?, *in* Macdonald, D. I. M., ed., Sedimentation, Tectonics and Eustasy: Sea-Level Changes at Active Margins: London, International Association of Sedimentologists Special Publication No. 12, p. 409–428.

POSAMENTIER, H. W., ALLEN, G. P., JAMES, D. P., AND TESSON, M., 1992, Forced regressions in a sequence stratigraphic framework: Concepts, examples, and exploration significance: American Association of Petroleum Geologists Bulletin, v. 76, p. 1687–1709.

PRAY, L. C., CRAWFORD, G. A., HARRIS, M. T., AND KIRKBY, K., 1980, Early

Guadalupian (Permian) bank margin erosion surfaces, Guadalupe Mountains, Texas: American Association Petroleum Geologists Bulletin, Programs with Abstracts, v. 64, p. 78.

READ, J. F., 1982, Carbonate platforms of passive (extensional) continental margins: Types, characteristics and evolution: Tectonophysics, v. 81, p. 195–212.

READ, J. F., 1985, Carbonate platform facies models: American Association of Petroleum Geologists Bulletin, v. 66, p. 860–878.

REID, S. T., BASS, R. O., AND WELCH, P., eds., 1988, Guadalupe Mountains Revisited, Texas and New Mexico: 1988 Field Seminar: Midland, West Texas Geological Society Publication No. 88-84, 202 p.

ROGERS, W. B., 1972, Depositional environments of the Skinner Ranch and Hess Formations (Lower Permian), Glass Mountains, West Texas: Unpublished Ph.D. Dissertation, University of Texas, Austin, 392 p.

ROSS, C. A., 1963, Standard Wolfcampian Series (Permian), Glass Mountains, Texas: Boulder, Geological Society of America Memoir 88, 205 p.

ROSS, C. A., 1986, Paleozoic evolution of southern margin of Permian basin: Geological Society of America Bulletin, v. 97, p. 536–544.

SALLER, A. H., 1996, Differential compaction and basinward tilting of the prograding Capitan reef complex, Permian, west Texas and southeast New Mexico, USA: Sedimentary Geology, v. 101, p. 21–30.

SALLER, A. H., BARTON, J. W., AND BARTON, R. E., 1989, Slope sedimentation associated with a vertically building shelf, Bone Spring Formation, Mescalero Escarpe field, south-central New Mexico, *in* Crevello, P. D., Wilson, J. L., Sarg, J. F., and Read, J. F., eds., Controls on Carbonate Platform and Basin Development: Tulsa, Society of Economic Paleontologists and Mineralogists Special Publication 44, p. 275–288.

SANDFORD, A. R., 1959, Analytical and experimental study of simple geological structures: Geological Society of America Bulletin, v. 70, p. 19–52.

SARG, J. F., 1988, Carbonate sequence stratigraphy, *in* Wilgus, C. K., Hastings, B. S., Kendall, C. G. St. C., Posamentier, H. W., Ross, C. A., and Van Wagoner, J. V., eds., Sea-Level Changes: An Integrated Approach: Tulsa, Society of Economic Paleontologists and Mineralogists Special Publication 42, p. 155–181.

SARG, J. F., 1989, Middle–Late Permian Depositional Sequences, Permian, West Texas and New Mexico, v. 3: Tulsa, American Association of Petroleum Geologists Studies in Geology No. 27, p. 140–154.

SARG, J. F., AND LEHMANN, P. J., 1986, Lower–middle Guadalupian facies and stratigraphy, San Andres/Grayburg Formations, Guadalupe Mountains, New Mexico, *in* Moore, G. E. and Wilde, G. L., eds., San Andres/Grayburg Formations: Lower–Middle Guadalupian Facies, Stratigraphy, and Reservoir Geometries, Guadalupe Mountains, New Mexico: Midland, Permian Basin Section — Society of Economic Paleontologists and Mineralogists Publication No. 86-25, p. 1–8.

SARG, J. F., WEBER, L. J., AND MARKELLO, J. R., 1995, Carbonate sequence stratigraphy—A summary and perspective, *in* Pausé, P. H. and Candelaria, M. P., eds., Carbonate Facies and Sequence Stratigraphy: Practical Applications of Carbonate Models: Midland, Permian Basin Section–SEPM (Society for Sedimentary Geology) Publication No. 95-36, p. 181–200.

SCHLAGER, W., 1993, Accommodation and supply—A dual control on stratigraphic sequences: Sedimentary Geology, v. 86, p. 111–136.

SHINN, E. A., AND ROBBIN, D. M., 1983, Mechanical and chemical compaction in fine-grained shallow-water limestones: Journal of Sedimentary Research, v. 53, p. 595–618.

SILVER, B. A., AND TODD, R. G., 1969, Permian cyclic strata, northern Midland and Delaware basins, West Texas and southeastern New Mexico: American Association of Petroleum Geologists Bulletin, v. 53, p. 2223–2251.

SONNENFELD, M. D., 1991, Anatomy of offlap in a shelf-margin depositional sequence: Upper San Andres Formation (Permian, Guadalupian), Last Chance Canyon, Guadalupe Mountains, New Mexico: Unpublished M.Sc. Thesis, Colorado School of Mines, Golden, 297 p.

SONNENFELD, M. D., 1993, Anatomy of offlap: Upper San Andres Formation (Permian, Guadalupian), Last Chance Canyon, Guadalupe Mountains, New Mexico, *in* Love, D. W., et al., eds., Carlsbad Region, New Mexico and West Texas: Socorro, New Mexico Geological Society Guidebook, 44th Field Conference, p. 195–204.

SONNENFELD, M. D., AND CROSS, T. A., 1993, Volumetric partitioning and facies differentiation within the Permian upper San Andres Formation of Last Chance Canyon, Guadalupe Mountains, New Mexico, *in* Loucks, R. G. and Sarg, J. F., eds., Recent Advances and Applications of Carbonate Sequence Stratigraphy: Tulsa, American Association of Petroleum Geologists Memoir 57, p. 435–474.

STAFLEU, J., AND SONNENFELD, M. D., 1994, Seismic models of a shelf-margin depositional sequence: Upper San Andres Formation, Last Chance Canyon, New Mexico: Journal of Sedimentary Petrology, v. B64, p. 481–499.

STECKLER, M. S., AND WATTS, A. B., 1978, Subsidence of the Atlantic-type continental margin off New York: Earth and Planetary Science Letters, v. 41, p. 1–13.

STERNS, D. W., 1978, Faulting and forced folding in the Rocky Mountains foreland, *in* Matthews, V., ed., Laramide Folding Associated with Basement Block Faulting in the Western United States: Boulder, Geological Society of America Memoir 151, p. 1–37.

SUPPE, J., 1985, Principles of Structural Geology: Englewood Cliffs, Prentice-Hall, 537 p.

THOMAS, W. A., 1983, Continental margins, orogenic belts and intracratonic structures: Geology, v. 11, p. 270–272.

TINKER, S., 1996, Building the 3-d jigsaw puzzle: Applications of sequence stratigraphy to 3-d reservoir characterization, Permian basin: American Association of Petroleum Geologists Bulletin, v. 80, p. 460–485.

TYRRELL, W. W., JR., 1969, Criteria useful in interpreting environments of unlike but time-equivalent carbonate units (Tansill-Capitan-Lamar), Capitan reef complex, west Texas and New Mexico, *in* Friedman, G. M., ed., Depositional Environments in Carbonate Rocks: Tulsa, Society of Economic Paleontologists and Mineralogists Special Publication 14, p. 80–97.

WARD, R. F., KENDALL, C. G. ST. C., AND HARRIS, P. M., 1986, Upper Permian (Guadalupian) facies and their association with hydrocarbons—Permian basin, west Texas and New Mexico: American Association of Petroleum Geologists Bulletin, v. 70, p. 239–262.

WATTS, A. B., 1981, The U. S. Atlantic continental margin: Subsidence history, crustal structure and thermal evolution: Tulsa, American Association of Petroleum Geologists, Education Course Notes Series No. 19, p. 2–75.

WILDE, G. L., 1995, Wolfcampian–Leonardian biostratigraphy, Sierra Diablo: Its relationship to sequence stratigraphic markers in the surface and subsurface, *in* Garber, R. and Lindsay, R. F., eds., Wolfcampian–Leonardian Shelf Margin Facies of the Sierra Diablo: Seismic Models for Subsurface Exploration: Midland, West Texas Geological Society Publication No. 95-97, p. 67–81.

WITHJACK, M. O., OLSON, J., AND PETERSON, E., 1990, Experimental models of extensional forced folds: American Association of Petroleum Geologists Bulletin, v. 74, p. 1038–1054.

WOOD, R., DICKSON, J. A. D., AND KIRKLAND-GEORGE, B., 1994, Turning the Capitan reef upside down: A new appraisal of the ecology of the Permian Capitan reef, Guadalupe Mountains, Texas and New Mexico: Palaios, v. 9, p. 422–427.

WRIGHT, 1962, Abo reef: Prime west Texas target (part 2): Oil and Gas Journal, p. 226–235.

WUELLNER, D. E., LEHTONEN, L. R., AND JAMES, W. C., 1986, Sedimentary-tectonic development of the Marathon and Val Verde basins, west Texas, U.S.A.: A Permo-Carboniferous migrating foredeep, *in* Allen, P. A. and Homewood, P., eds., Foreland Basins: London, International Association of Sedimentologists Special Publication No. 8, p. 347–368.

YANG, K.-M., AND DOROBECK, S. L., 1992, Mechanisms for late Paleozoic synorogenic subsidence of the Midland and Delaware basins, Permian basin, Texas and New Mexico, *in* Mruk, D. H. and Curran, B. C., eds., Permian Basin Exploration and Production Strategies: Applications of Sequence Stratigraphic and Reservoir Characterization Concepts: Midland, West Texas Geological Society Publication No. 92-91, p. 45–60.

APPLICATION OF HIGH-RESOLUTION SEQUENCE STRATIGRAPHY TO TIDALLY INFLUENCED UPPER MISSISSIPPIAN CARBONATES, ILLINOIS BASIN

LANGHORNE B. SMITH, JR.

Comparative Sedimentology Laboratory, University of Miami, 4600 Rickenbacker Causeway, Miami, Florida 33149

AND

J. FRED READ

Department of Geological Sciences, Virginia Tech, Blacksburg, Virginia 24061

ABSTRACT: The Upper Mississippian (Chesterian) Ste. Genevieve and Paoli Formations of the Illinois basin are composed of five carbonate and mixed carbonate-siliciclastic fourth-order (100–500 k.y.) sequences. Interpretive cross sections based on detailed measured sections of 57 closely spaced outcrops and cores show that sequence-bounding carbonate and siliciclastic paleosol horizons and distinctive marker beds can be correlated basin-wide, within the available biostratigraphic framework. Each sequence is composed of as many as nine regionally developed parasequences. Carbonate parasequences have patchy updip carbonate eolianites at the base, overlain by laterally discontinuous ooid tidal ridge and skeletal bank facies and laterally extensive muddy carbonate caps. Mixed carbonate-siliciclastic parasequences are composed of skeletal limestone capped by fossiliferous shale. Carbonate tidal-flat facies are rare in the succession.

The sequences were produced by moderate–amplitude, fourth-order (100–500 k.y.) glacio-eustatic sea-level changes, whereas the parasequences were likely produced by fifth-order (10–100 k.y.) glacio-eustatic sea-level changes during a time of transition from the greenhouse conditions of the Early Mississippian to the icehouse conditions of the late Paleozoic. This transition is reflected in upward change from carbonate-dominated, disconformity-bounded sequences with little lowstand erosion at the base to mixed carbonate-clastic sequences with significant lowstand erosion at the top of the study interval and in overlying sequences. These changes were likely caused by an increase in the magnitude of sea-level fluctuations coupled with increasingly humid wet-dry seasonal climate. The sequences may be stacked into one or two third-order (0.5–5 m.y.) composite sequences that are bounded by relatively well developed unconformities. However, the magnitude of the fourth-order signal makes third-order bundling less clear than in strata deposited during greenhouse times of lower-amplitude fourth- and fifth-order sea-level fluctuations. The dominance of fourth-order disconformity-bounded sequences and the lack of regional tidal-flat facies reflect moderate-amplitude eustasy associated with the transition from global greenhouse climates of the Early Mississippian to times of abundant global ice in the late Paleozoic.

INTRODUCTION

Previous work on the lower Chesterian Ste. Genevieve to Paoli Formations in the Illinois basin includes regional surface and subsurface mapping of limestone, sandstone, and shale formations by Butts (1917), Weller and Sutton (1940), Swann and Atherton (1948), McFarlan et al. (1955), Swann (1963, 1964), Potter (1962), Sable and Dever (1990), and Droste and Carpenter (1990) and detailed, local sedimentologic studies by Kissling (1967), Choquette and Steinen (1980), Cluff and Lineback (1981), Liebold (1982), Merkley (1991), and Dodd et al. (1996), among others. The lateral heterogeneity of the ooid grainstone reservoirs has made the Ste. Genevieve a "correlation nightmare" in the subsurface (Zuppann, 1993). Most paleosols were considered to be local in extent and to have formed as a result of local island shoaling (Droste and Carpenter, 1990). Recent work has shown that some sequences and parasequences bounded by paleosols and eolianite horizons can be correlated both locally and regionally (Smith et al., 1995; Smith and Read, 1995; Smith, 1996; Dodd et al., 1996).

This paper demonstrates the value of combining detailed sedimentology with a basin-wide stratigraphic approach. Only through sedimentologic study of all rock types and exposure surfaces can the sequence stratigraphy be accurately interpreted, and only through basin-wide stratigraphic study can local features be distinguished from laterally extensive marker horizons and genetically significant surfaces. Detailed lithologic logs of 57 outcrops and cores were used to construct regional lithologic cross sections of the Ste. Genevieve Formation and the Paoli Formation (and its time-equivalent units) along more than 600 km of the eastern, southern, and western outcrop belts of the Illinois basin. The cross sections show that, despite the lateral heterogeneity of much of the stratigraphy, several sequence-bounding paleosol horizons can be correlated over much of the basin within a framework of biostratigraphic and lithologic markers (Smith and Read, 1995; Smith et al., 1995). The paleosols provide additional lithologic markers that make it possible to correlate parasequences within the sequences. The cross sections provide an opportunity to better understand the relative effects of eustasy, tectonics, and tides and to test sequence stratigraphic concepts on carbonate and mixed carbonate-siliciclastic sequences deposited updip on a gently dipping ramp. This study also provides insight into the stratigraphic signature of the transition from global greenhouse conditions of the Early Mississippian to times of abundant global ice in the Pennsylvanian and Early Permian.

GEOLOGIC SETTING

Tectonic and Paleogeographic Setting

The Illinois basin was located between 5° and 15° south of the equator during the Late Mississippian (Craig and Connor, 1979; Scotese and McKerrow, 1990) (Fig. 1A). Throughout much of the Paleozoic, the basin was an embayment open to the present-day south, which covered southern Illinois, southwestern Indiana, and western Kentucky. The embayment was part of an extensive carbonate ramp that was continuous from Pennsylvania to western Canada (Craig and Connor, 1979) (Fig. 1A). The ramp was separated from the Antler foredeep of

Advances in Carbonate Sequence Stratigraphy: Application to Reservoirs, Outcrops and Models, SEPM Special Publication No. 63

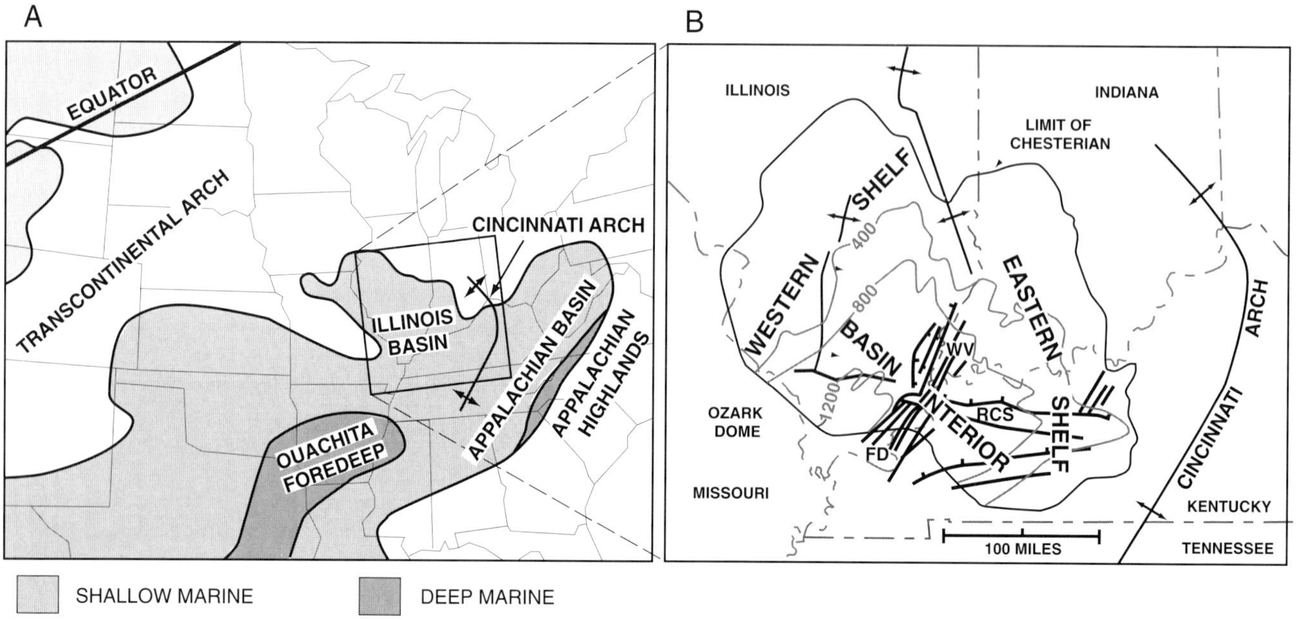

FIG. 1.—(A) Paleogeographic map showing regional setting of the Illinois basin (modified from Craig and Connor, 1979). (B) Tectonic map of Illinois basin. Isopachs in feet for the interval from the Bethel to the Mississippian/Pennsylvanian boundary (from Swann, 1963); 800-ft isopach is approximate boundary of basin interior. Faults: Wabash Valley fault system (WV); Rough Creek–Shawneetown fault system (RCS); Fluorspar District fault complex (FD).

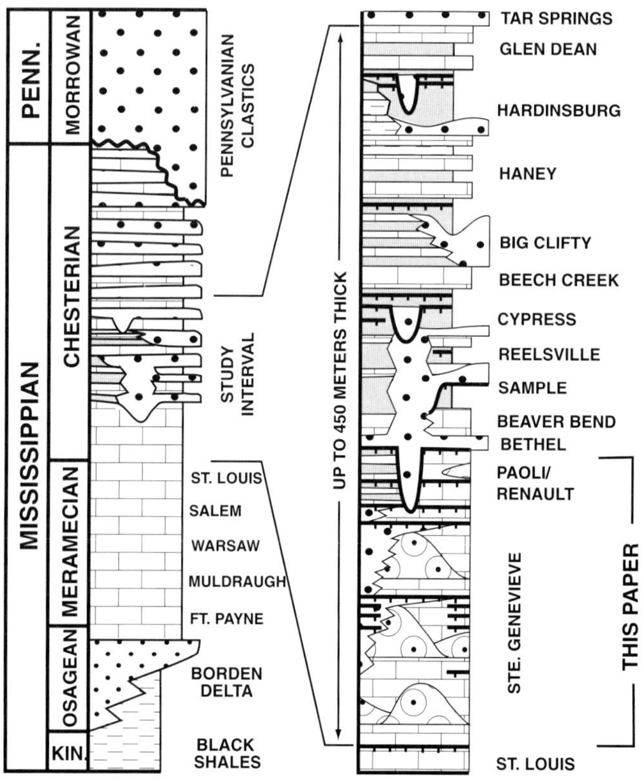

FIG. 2.—Left column is a generalized stratigraphic column for the Mississippian in the Illinois basin. Right column shows more detailed stratigraphy of lower to middle Chesterian units. This paper concentrates on the Ste. Genevieve to Paoli interval. For explanation of patterns of units, see Figure 4.

the western United States by the Transcontinental arch, which was emergent throughout the Chesterian (Craig and Connor, 1979). The Illinois basin lay at least 300 km updip from the ramp margin that may have started to develop along the Ouachita foredeep to the south.

The Illinois basin can be subdivided into three main provinces along the outcrop belt: the eastern "shelf," "basin interior," and the western "shelf" (Fig. 1B). The term shelf is used by local geologists to distinguish areas of lower subsidence rates from the more rapidly subsiding basin interior. The eastern shelf formed along the western flank of the Cincinnati arch, and the western shelf formed over the DuQuoin monocline. The basin interior appears to be centered on the Wabash Valley and Fluorspar district fault systems that trend northeast and may have been active during the Late Mississippian. Despite the varying subsidence rates, facies interpretations suggest that shallow-marine to nonmarine conditions prevailed throughout the Illinois basin during the early and middle Chesterian.

Stratigraphic Setting

The Ste. Genevieve and Paoli Formations lie at the top of a thick Middle Mississippian carbonate succession that includes the Osagean to Meramecian Muldraugh, Warsaw, Salem, and St. Louis Formations (Cluff and Lineback, 1981) (Fig. 2). The St. Louis Formation directly underlies the study interval (Figs. 2, 3) and is primarily composed of shallow-water muddy carbonates and skeletal limestone (Cluff and Lineback, 1981). The base of the Ste. Genevieve is picked at a biostratigraphic zone boundary that coincides with a regional disconformity that commonly occurs within a muddy carbonate unit overlying the Lost River Chert of Elrod (1899). In outcrops and cores, the Lost River

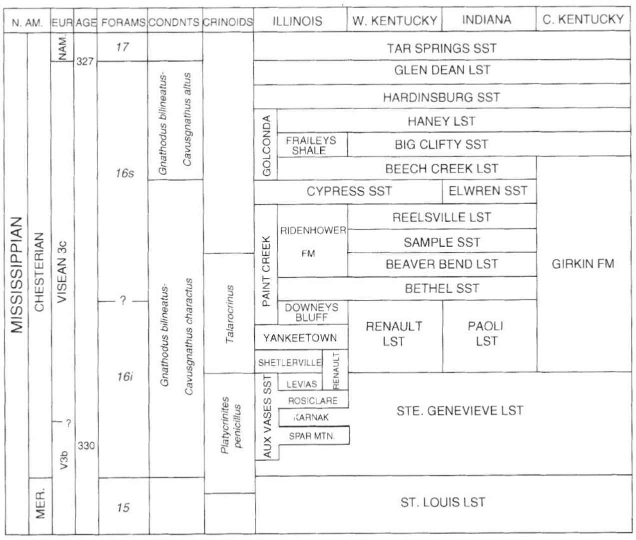

FIG. 3.—Biostratigraphy and correlation chart for Ste. Genevieve to Glen Dean interval (modified from Sable and Dever, 1990, and Maples and Waters, 1987). MER. = Meramecian, SST = sandstone, LST = limestone, Fm = formation.

Chert is a distinctive marker bed that occurs in the upper part of a laterally extensive grainstone unit in which the outlines of fenestrate bryozoans are commonly preserved within chert nodules. Conodont, foraminifer, and coral biostratigraphic zone boundaries occur at this unconformity (Rexroad and Fraunfelter, 1977). Maples and Waters (1987) moved the Meramecian/Chesterian provincial series boundary to this unconformity because the fauna above the boundary more closely resembles the fauna in the middle and upper Chesterian than the fauna in the underlying St. Louis Formation. Thus, age of the Ste. Genevieve to Paoli interval is early Chesterian or late Visean (middle V3b to early V3c).

The Ste. Genevieve Formation (Fig. 3) is primarily composed of ooid grainstone (a common reservoir facies), skeletal grainstone, commonly dolomitized muddy carbonates, and numerous disconformities that can be recognized by the presence of caliche, breccia, and eolianites. The boundary between the middle and upper Ste. Genevieve and the Aux Vases Sandstone in the western part of the basin was thought to be gradational (Cluff and Lineback, 1981) but has recently been interpreted to be primarily erosional (Leetaru, 1997). The source of the Aux Vases Sandstone lay to the northwest; the formation is composed of quartz sandstone, flaser-, wavy-, and lenticular-bedded sandstone and shale and slickensided mudrock and caliche paleosols beneath disconformities. The Rosiclare Tongue of the Aux Vases extends east into western Kentucky and separates the lower Ste. Genevieve from the overlying Levias Member of the Renault Formation. The Levias Member contains the crinoid *Platycrinites penicillus* and is time-equivalent to the uppermost Ste. Genevieve Formation in Kentucky and Indiana (Swann, 1963). The Ste. Genevieve Formation and the Levias Member of the Renault Formation are capped by a regional disconformity called the Bryantsville Breccia Bed in Indiana and Kentucky (Mallot, 1952). This boundary coincides with the top of the Platycrinites penicillus zone and the base of the *Talarocrinus* zone.

Strata above the *Platycrinites penicillus* zone and below the Bethel Sandstone (Renault, Shetlerville, Yankeetown, Downeys Bluff, and Paoli Formations) will be referred to as the Paoli interval in this paper. Along the eastern outcrop belt, the Paoli interval is primarily composed of shallow-water carbonates similar to those in the Ste. Genevieve. In the basin interior, the Paoli interval is composed of skeletal limestone and fossiliferous dark gray shales and, on the western shelf, of shallow-water limestones, fossiliferous shale, and quartz sandstone, which suggests that the clastic source remained to the northwest (Potter, 1962). The Paoli is capped by a regional disconformity and is generally overlain by the Bethel Sandstone. In the southeastern part of the basin, the Bethel Sandstone pinches out, and the Paoli is included in the Girkin Limestone. The Girkin Limestone includes the Paoli, Beaver Bend, Reelsville, and Beech Creek Limestones, each separated by well-developed carbonate paleosol horizons. The Paoli interval is overlain by middle and upper Chesterian mixed carbonate-clastic sequences (Swann, 1963, 1964; Smith, 1996).

CONSTRUCTION OF LITHOFACIES CROSS SECTIONS

Cross sections through the study interval have been constructed on the basis of closely spaced measured sections of core and outcrop (Fig. 4). Each core and outcrop was logged bed-by-bed, and the thickness, color, grain size, type and shape, depositional texture, biota, sedimentary structures, and evidence for subaerial exposure were noted for each unit. In outcrop, units were traced laterally to determine facies relationships where possible. The Ste. Genevieve cross sections, A–A′ and B–B′(Fig. 4), are hung on the disconformity that coincides with the top of the *Platycrinites penicillus* zone and the Bryantsville Breccia Bed in Indiana and western Kentucky. The overlying Paoli cross sections, C–C′ and D–D′ (Fig. 4), have been hung on the disconformity at the top of the interval except where pre-Bethel erosion occurred. In these locations, the cross section was hung on a regional disconformity (I) within the Paoli. Correlations were made by using distinctive marker beds such as the Lost River Chert Bed, numerous paleosol horizons, and laterally extensive muddy carbonate and shale units.

Parasequences have been numbered to help describe the cross sections. We are not suggesting that this numbering system be adopted by others because it is based on interpretive cross sections and subject to change with the addition of new data. For clarity, measured sections have been equally spaced on the cross sections, except in areas of major changes in thickness or larger-than-usual gaps between sections where they are spaced farther apart. Actual distances between sections can be determined from the location maps (Fig. 4).

LITHOFACIES AND DEPOSITIONAL ENVIRONMENTS

Characteristics of major rock types are summarized in Table 1. The relative abundance of facies at each location (Fig. 5), field observations, and subsurface data from the literature (Choquette

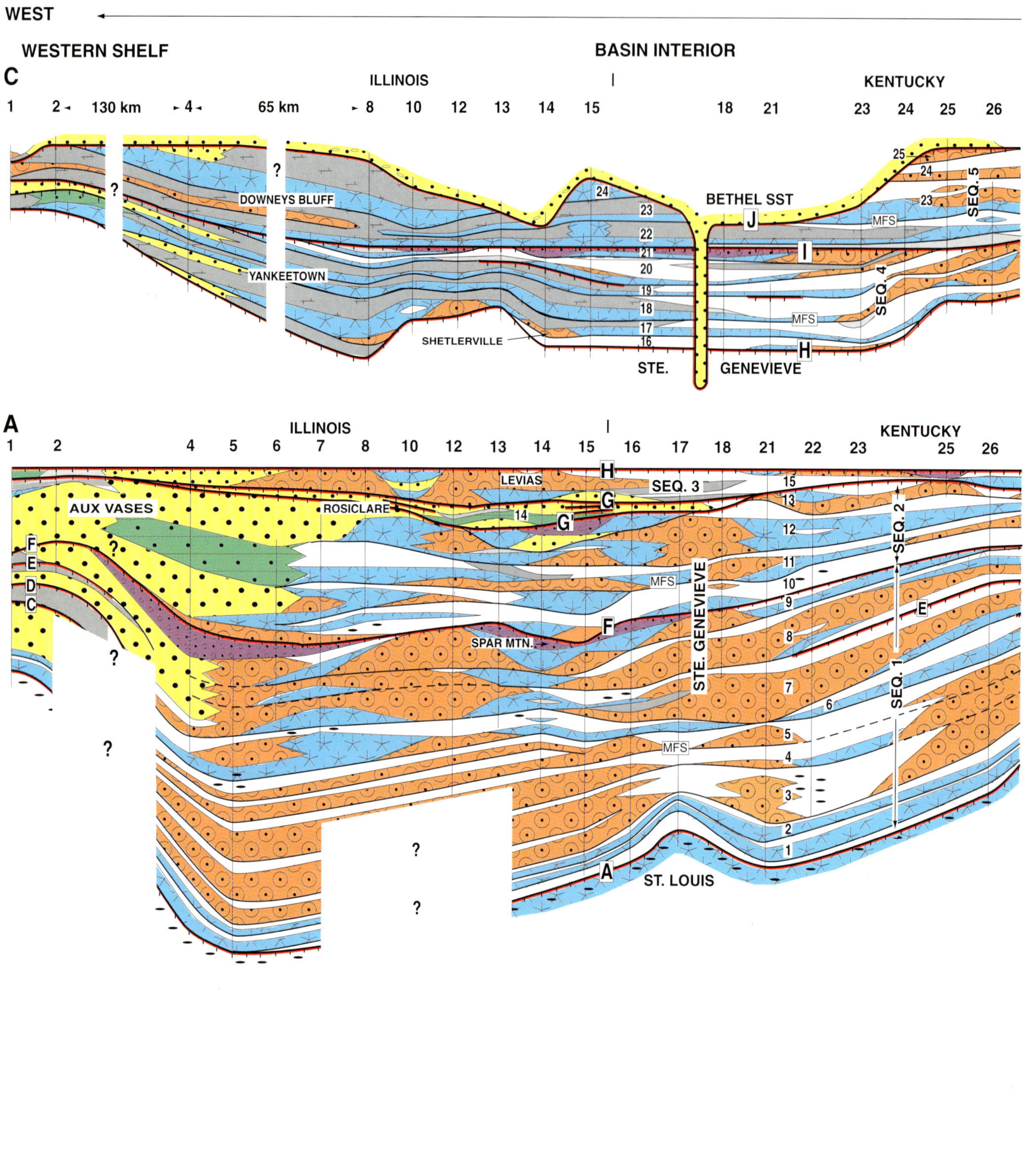

FIG. 4.—Interpretive cross sections of sequences 1 through 5 in the Ste. Genevieve to Paoli interval. Measured sections are equally spaced except where there are major gaps between sections (see location maps for actual horizontal spacing). Disconformities are lettered from A to J, and parasequences are numbered from 1 to 25. The Bethel channel is not drawn to scale; its actual depth is 75 m (Friberg et al., 1969). Cross section A-A' includes the Ste. Genevieve and trends west to east from location 1 to location 35 and south to north from location 35 to location 75. Cross section B-B' shares location 35 and trends south to north parallel and downdip of cross section A-A'. Cross section C-C' trends west to east from location 1 to location 33 and south to north from location 33 to location 74. Cross section D-D' shares location 33 and trends south to north parallel and downdip of cross section C-C'. MFS = maximum-flooding surface, GRST. = grainstone, PKST. = packstone, MDST. = mudstone, WKST. = wackestone, LST. = limestone.

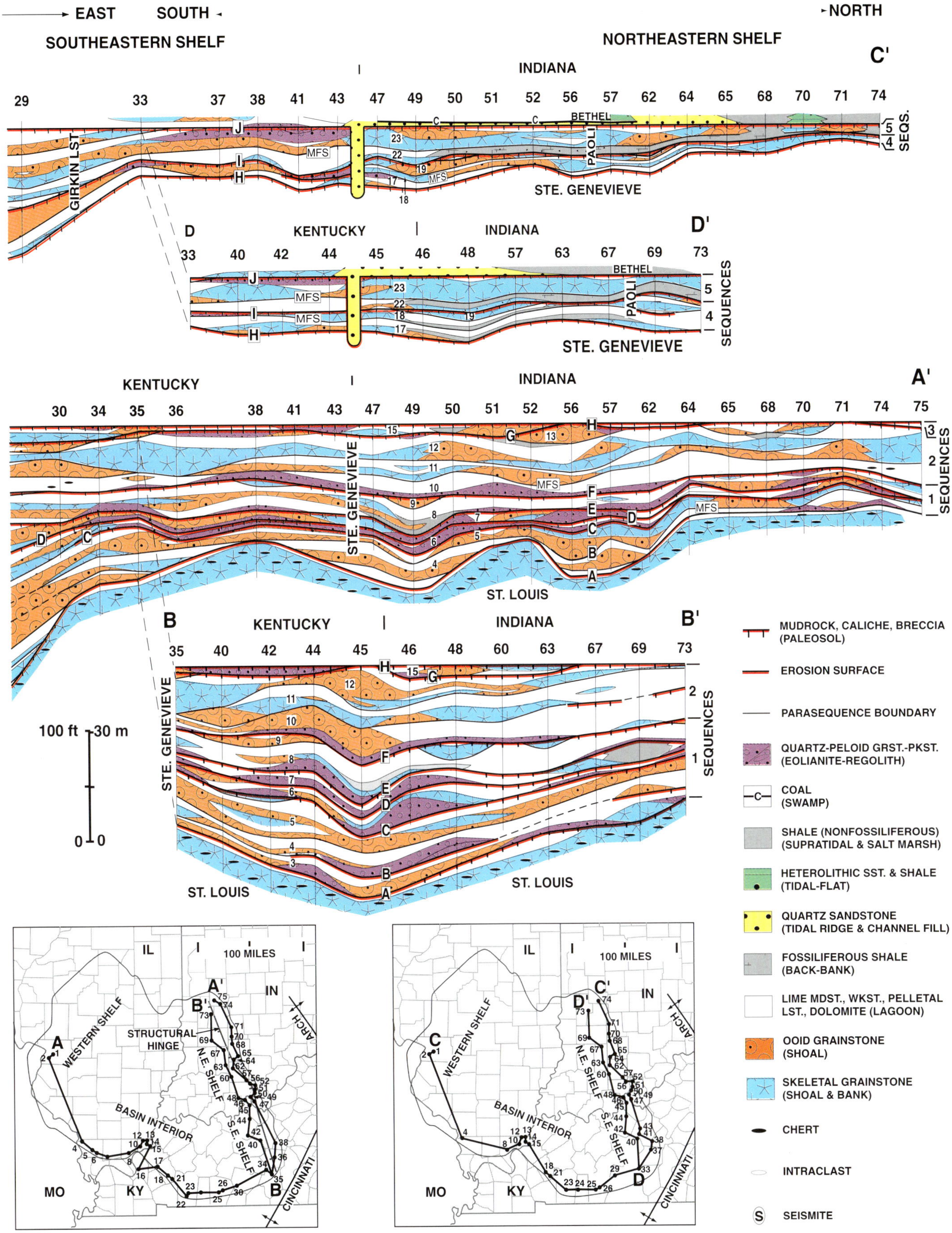
EAST
SOUTH
NORTH
SOUTHEASTERN SHELF
NORTHEASTERN SHELF
C'
INDIANA
KENTUCKY
BETHEL
PAOLI
GIRKIN LST
STE. GENEVIEVE
MFS
SEQS.
D
D'
SEQUENCES
A'
ST. LOUIS
B
B'
100 ft 30 m
0 0
MUDROCK, CALICHE, BRECCIA (PALEOSOL)
EROSION SURFACE
PARASEQUENCE BOUNDARY
QUARTZ-PELOID GRST.-PKST. (EOLIANITE-REGOLITH)
COAL (SWAMP)
SHALE (NONFOSSILIFEROUS) (SUPRATIDAL & SALT MARSH)
HETEROLITHIC SST. & SHALE (TIDAL-FLAT)
QUARTZ SANDSTONE (TIDAL RIDGE & CHANNEL FILL)
FOSSILIFEROUS SHALE (BACK-BANK)
LIME MDST., WKST., PELLETAL LST., DOLOMITE (LAGOON)
OOID GRAINSTONE (SHOAL)
SKELETAL GRAINSTONE (SHOAL & BANK)
CHERT
INTRACLAST
SEISMITE
IL
IN
MO
KY
100 MILES
WESTERN SHELF
STRUCTURAL HINGE
N.E. SHELF
S.E. SHELF
BASIN INTERIOR
CINCINNATI ARCH
A
C

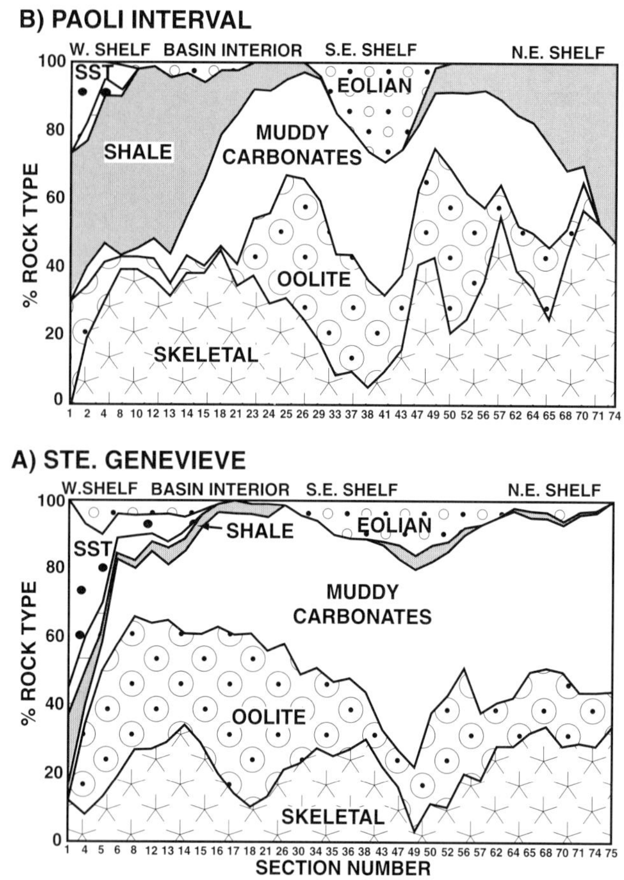

FIG. 5.—Facies distribution on the ramp based on percentage of rock type at each location. Curves smoothed by using three-point average. (A) Ste. Genevieve facies distribution (sequences 1 and 2); (B) Paoli interval facies distribution (sequences 3 and 4).

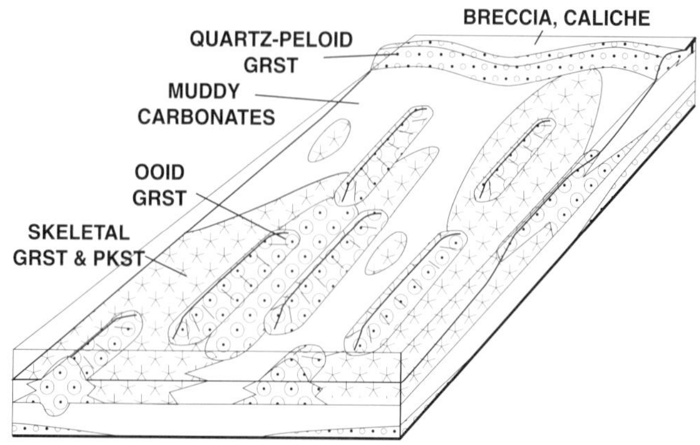

FIG. 6.—Schematic depositional environment for carbonate-dominated ramp in the Ste. Genevieve Formation (not to scale). See Figure 4 for legend and rock types.

and Steinen, 1980; Cluff and Lineback, 1981; Zuppann and Keith, 1988) were used to determine relative facies distribution and depositional environments for carbonate-dominated and mixed carbonate-siliciclastic parts of the ramp (Fig. 6). The ramp slope was very gentle (average 1.5 cm/km [0.001°] for the basin and no more than 5 cm/km [0.003°] along the margins of the shelves [Smith, 1996]), which accounts for the similarity of facies deposited in the central and eastern parts of the basin in the Ste. Genevieve Formation (Smith, 1996). Although the facies on the eastern shelf and the basin interior are different in sequences 4 and 5, this variation is primarily due to the northwestern location of the clastic source rather than a change in ramp slope.

Increasingly humid wet-dry seasonality prevailed throughout the Ste. Genevieve to Paoli interval, as indicated by the presence of caliche, breccia, and red, slickensided mudrock paleosols. Evaporites, which are common in the underlying St. Louis Formation, are absent in the Ste. Genevieve to Paoli interval, which suggests that the climate became more humid in the early Chesterian. The duration of the wet season must have been relatively short during Ste. Genevieve time, on the basis of the abundance of ooids, dolomite, and eolianites, all of which are most common in areas with arid and semiarid climates (Witzke, 1990; Ettensohn et al., 1988; Ambers and Petzold, 1992). The overall upward increase in siliciclastic sedimentation up through the study interval is related to the source area's becoming more humid with time. Increased influx of shale into the basin in sequences 3 and 4 coincides with a decrease in ooid production in the basin, possibly related to increased fresh-water influx from rivers to the northwest. The trend of increasingly humid wet-dry seasonality continued into the overlying sequences, which have more siliciclastics as well as thin coal beds (Smith, 1996).

Carbonate-Dominated Ramp

The Ste. Genevieve Formation and most of the Paoli interval of central Kentucky and southeastern Indiana were deposited on a carbonate-dominated part of the ramp (Fig. 6) (facies summarized in Table 1).

Skeletal grainstone and packstone.—
Skeletal grainstone and packstone (Fig. 7A) formed in moderate- to high-energy conditions on the basis of the lack of mud, the medium to very coarse grain size, and the abundance of echinoderms. In this setting, skeletal grainstone formed landward, in between and seaward of ooid shoals in water depths that rarely exceeded 10 m (Smith, 1996).

Ooid grainstone.—
This reservoir facies (Fig. 7B) formed in a high-energy shallow-marine setting where strong tidal currents shaped tidal ridges oriented subparallel to the paleoslope (Choquette and Steinen, 1980) and locally cut tidal channels into underlying strata. Ooid grainstone is widespread in the Ste. Genevieve, with the thickest accumulations in the basin interior, but largely confined to the eastern shelf in the Paoli interval.

Tidal influence on grainstone units.—
Strong tidal influence on the gently sloping ramp caused much lateral heterogeneity of facies. Ooid grainstone reservoir

TABLE 1.—CARBONATE AND MIXED CARBONATE–CLASTIC RAMP FACIES

Lithology (Depositional Environment)	**Blocky mudrock and caliche/breccia (paleosol)**	**Quartz-peloid grainstone (eolianite)**	**Muddy carbonates (lagoon and intershoal)**	**Ooid grainstone (tidal bar and shoal)**	**Skeletal grainstone and packstone (bank, shoal)**	**Quartz sandstone (tidal sand ridge, channel fill)**	**Fossiliferous Shale (carbonate-clastic transition)**
Occurrence	Common throughout basin, better developed on shelves. Commonly underlies and overlies marginal marine shale with plant fragments.	Dune-forms; units commonly thicken and thin over short distances; underlie and overlie disconformities at tops and bases of sequences, respectively.	Commonly occur as regional, traceable sheets capping parsequences. Also lagoons and local intershoal swale-fills.	Dip-parallel oolitic bar-forms (Zuppann, 1993); commonly occur at base of parasequences in Ste. Genevieve and Paoli of eastern shelf.	Common throughout basin in all limestone units. Forms sheet-like units at base of most regressive parasequences.	Commonly fills incised valleys and forms in linear sand ridges.	Common throughout basin. Transition between siliciclastic and carbonate rock types.
Color	Caliche is light brown; Breccia zones are dark gray; Mudrock is red and maroon.	Dark gray.	Dolomitized units are tan; limestone units light gray brown to dark gray.	White to light gray.	Light to medium to dark gray.	White to light gray/green.	Olive green and dark gray
Depositional texture and grain type	Caliche is stringy to laminated micrite. Breccia has fitted fabric. Mudrock is blocky, brecciated, slickensided clay w/dolomitic nodules.	Grainstone; fine to medium, tightly packed and well sorted within laminae; dominated by peloids, rounded lithoclasts, subangular quartz; lesser ooids (whole and broken), echinoderm grains.	Mudstone and wackestone; predominantly composed dolomitic or lime mud and 20-100 μm pellets; common whole thin-shelled fossils.	Grainstone, rare packstone; radial and concentric, fine- to medium-grained ooids with variable amounts of skeletal grains; ooid nuclei include skeletal grains, lithoclasts, quartz and peloids.	Grainstone and packstone with up to 15% shale. Medium to very coarse grained.	Very-fine to medium-grained, subangular, well-sorted; may have up to 15% shale.	Wackestone/fissile shale; whole brachiopods and bryozoans.
Bedding and sedimentary structures	Mudrock occurs in breccia and calichified zones up to three meters thick. Massive beds up to one meter thick; rooting and reduction halos common.	Millimeter-scale inverse-graded laminations form cross-beds that dip <20%; cross-beds dip multidirectionally but dominantly to the north (Hunter, 1993).	Massive to burrowed, rare fenestrae, laminations, and mud-cracks.	Cross-bedded, less commonly thick bedded and massive; foresets dip as much as 25%.	Thick-bedded, massive and cross-bedded.	Cross-bedded, flat-bedded and massive bedded, flaser-bedding common.	Fissile, flat-bedded.
Terrigenous clastics admixed or interbedded		Average 15% very fine quartz sand on eastern shelf; up to 75% quartz sand in basin interior and western shelf.	Green shale, quartz silt and quartz sand common.	Quartz grains form nuclei of ooids above some quartz sandstone units.	Quartz sand and silt are rare; thin green and gray shale beds and stringers are common.		
Biota		Abraded, rounded skeletal grains common.	Lagoonal facies: ostracodes and gastropods; Intershoal swales: bryozoans, brachiopods and echinoderms.	Echinoderm and brachiopod fragments common.	Abundant echinoderms, brachiopods, bryozoans, common mollusks and foraminifera.	Rare echinoderm fragments, could be transported or eroded from underlying limestone.	Bryozoans, brachiopods and echinoderms.

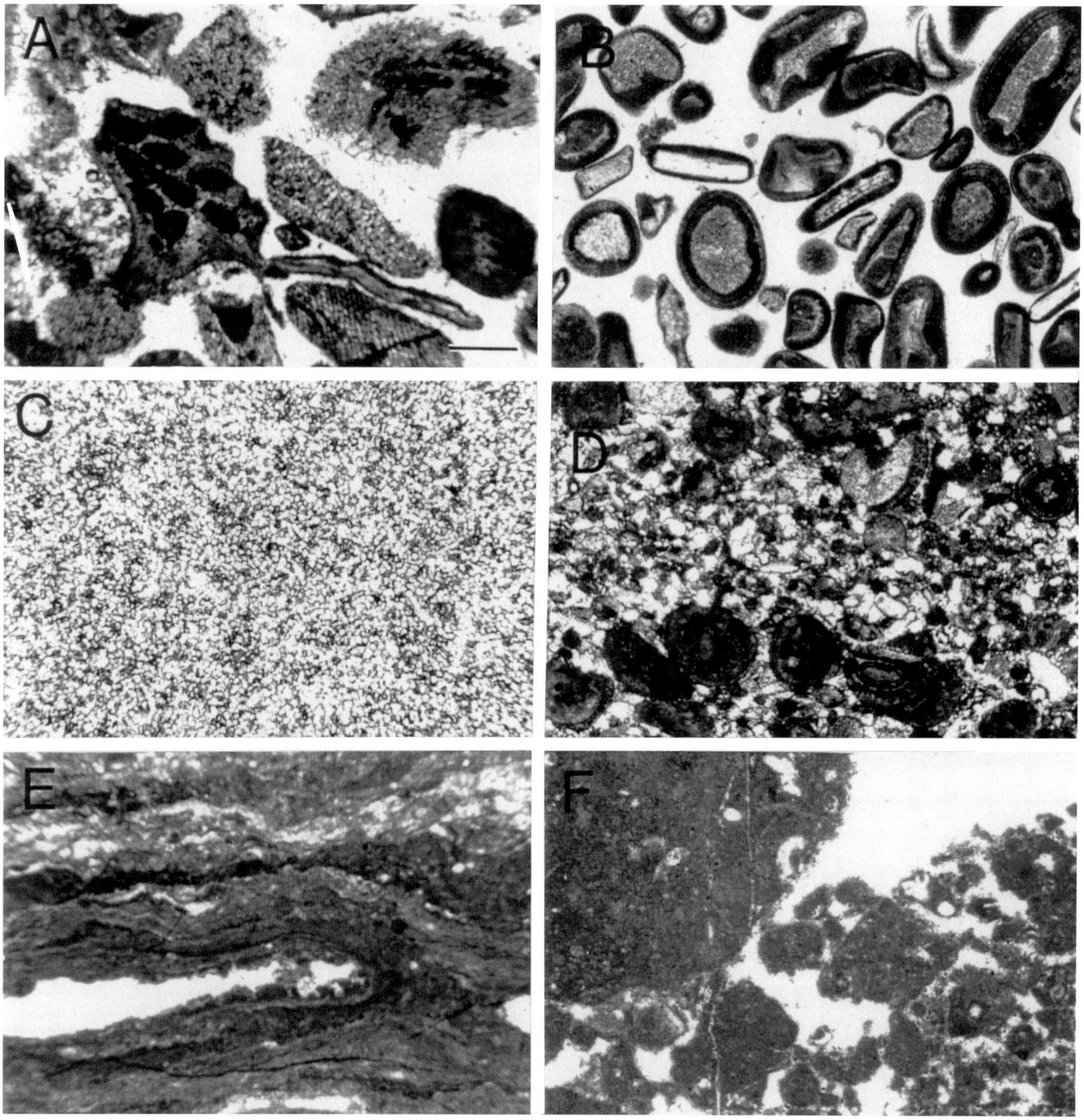

FIG. 7.—Photomicrographs of important rock types: (A) skeletal grainstone, (B) ooid grainstone, (C) microcrystalline dolomite, (D) quartz-peloid grainstone (eolianite), (E) caliche, (F) breccia. Scale on A represents 0.25 mm and is same for all photographs.

facies thicken, thin, and pinch out into skeletal grainstone and muddy carbonates and reappear along the same horizon. Facies differentiation was due to local tidal energy rather than water depth or regional position on the ramp. If the ramp slopes had been steeper, or if the ramp had been wave-dominated, then strike-parallel facies belts would have formed. The dip-parallel orientation of the tidal sand ridges focused and increased the velocity of the tidal currents, which may have allowed synchronous ooid formation over a large area. The lack of a continuous barrier allowed open-marine organisms such as echinoderms to live landward of, and between, ooid shoals. During times of maximum tidal influence, the central and eastern parts of the basin were a mosaic of ooid ridges and skeletal grainstone sheets.

Muddy carbonates.—

These have been interpreted as low-energy subtidal lagoon facies (Choquette and Steinen, 1980), as open-shelf facies formed seaward of ooid and skeletal grainstone (Dodd et al., 1996), and as facies formed both landward and seaward of the ooid shoals (Cluff and Lineback, 1981). On the basis of the fauna, grain size, regional distribution, and stratigraphic relationships, it is likely that most muddy carbonates formed in low-energy environments updip of the ooid shoals, but also filled topography in swales and channels between and in front of shoals. Open-marine biota (especially echinoderms) in some units indicate that some muddy carbonates formed in waters with normal salinity whereas dolomitized ostracode-bearing mudstone and wackestone (Fig. 7C) likely formed in more saline, restricted settings. Close association of these facies suggests that waters in lagoons and intershoal swales fluctuated from open marine to restricted. The lack of laminites suggests that microbial mat–covered carbonate tidal flats were rare despite the arid Late Mississippian climate.

Quartz-peloid grainstone.—
This rock type (Fig. 7D) is most common on the central eastern shelf in the Ste. Genevieve and on the south part of the eastern shelf in the Paoli interval, but also occurs beneath disconformities in the basin interior. The cross-bedded, quartz-peloid grainstone units are eolianites, on the basis of their inverse grading, sorting, rhizoliths, rounded broken ooids, lack of fossils or coarse grains, and close association with carbonate paleosols (Merkley, 1991; Hunter, 1993; Dodd et al., 1993). Some subhorizontally laminated quartz-peloid grainstone with coarse-grained fossiliferous laminae may be beach facies. Quartz-peloid grainstone eolianites have also been reported in time-equivalent strata in Kansas (Abegg, 1994), eastern Kentucky, and West Virginia (Al-Tawil and Read, 1996). Quartz-peloid grainstone commonly overlies or is capped by caliche and breccia and occurs in beds that are as much as 6 m thick and that thicken and thin over short distances.

Carbonate paleosols.—
Quartz-peloid packstone, caliche, and breccia are paleosols that formed during subaerial exposure of carbonate rocks in a seasonally wet-dry climate. In the early stages of Late Mississippian carbonate paleosol development, quartz-peloid packstone formed when eolian peloids, quartz sand, and ooids accumulated along with clay carried by wind or sheetwash. Some lithoclasts eroded from underlying carbonates were transported and deposited in swales between eolian dunes and on exposure surfaces (Hunter, 1993); other lithoclasts may be in situ pedogenic concretions of caliche analogous to calcrete ooids or pisolites (Read, 1974). With continued exposure, caliche-coated grains and interstitial fine-grained carbonate were precipitated to form gray-brown, quartz-peloid packstone paleosols. With time, the sediments were cemented, and laminar caliche (Fig. 7E) lined vertical fractures, joints, and bedding planes. In the most mature profiles, exposed carbonates were brecciated (Fig. 7F), and laminar caliche crusts and subhorizontal layers of caliche formed as much as 4 m below the exposed surface along joints and bedding planes (Walls et al., 1975; Ettensohn, 1975; Harrison and Steinen, 1978; Ettensohn et al., 1988).

Mixed Carbonate-Siliciclastic Ramp

The Aux Vases Formation and the Paoli interval of Illinois, western Kentucky, and western and central Indiana were deposited on a mixed carbonate-siliciclastic part of the ramp (Smith, 1996).

Skeletal grainstone and packstone.—
Where associated with siliciclastics, skeletal grainstone and packstone likely formed in a deeper, more distal environment than in the Ste. Genevieve on the basis of the lack of cross-bedding, lack of ooids, and common presence of graded beds that suggest storm influence. Skeletal grainstone and packstone grade laterally into fossiliferous shale that represents a gradation between carbonate and siliciclastic environments.

Fossiliferous shale.—
This rock type commonly interfingers with muddy carbonates to the east and grades laterally into tidally influenced heterolithic interlaminated sandstone and shale toward the northwestern siliciclastic source. The heterolithic facies formed both on tidal flats and in the transition zone seaward of quartz sandstone.

Quartz sandstone.—
Quartz sandstone was deposited in a tide-dominated shallow-marine environment on the basis of abundant channel-form structures with tidal fills, bidirectional cross-bedding, and sand ridges oriented parallel to the paleoslope (Seyler, 1986; Cole, 1990; Huff, 1993). Although they are somewhat smaller, the sand ridges resemble those found in tide-dominated environments such as the North Sea (Off, 1963; Huff, 1993). Quartz sandstone with conglomerate beds also commonly fills channels scoured by strong tidal currents.

Paleosols.—
On the mixed carbonate-siliciclastic ramp, paleosols are most commonly blocky, slickensided mudrock that formed from clay on exposed shale beds, but caliche and breccia are also common exposure features on quartz sandstone. The red, green, or dark gray mudrock paleosols are brecciated, root disrupted, and commonly have brecciated carbonate nodules and tepee structures (Smith, 1996).

Relative Ranking of Disconformities

Disconformities (lettered from A to J, oldest to youngest, on cross sections) can be crudely ranked in terms of the relative duration of exposure, estimated from the relative abundance of exposure features (Fig. 8). From immature to most mature, the exposure features are (1) eolianite; (2) “dirty” packstone and minor caliche; (3) thick laminar caliche extending deep beneath the surface; and (4) brecciated or karst-weathered limestone. Blocky, slickensided mudrock paleosols and incised valleys are treated separately because they are difficult to rank relative to other features that indicate carbonate exposure.

Disconformities H and J, on which the cross sections are hung, are the best developed and likely represent the longest period of emergence (major disconformities). Disconformity H has the highest occurrence of caliche and breccia, and there is major incision associated with Disconformity J. Although depth of incision is a better measure of sea-level fall than of duration of exposure, it would have taken a relatively long period of time

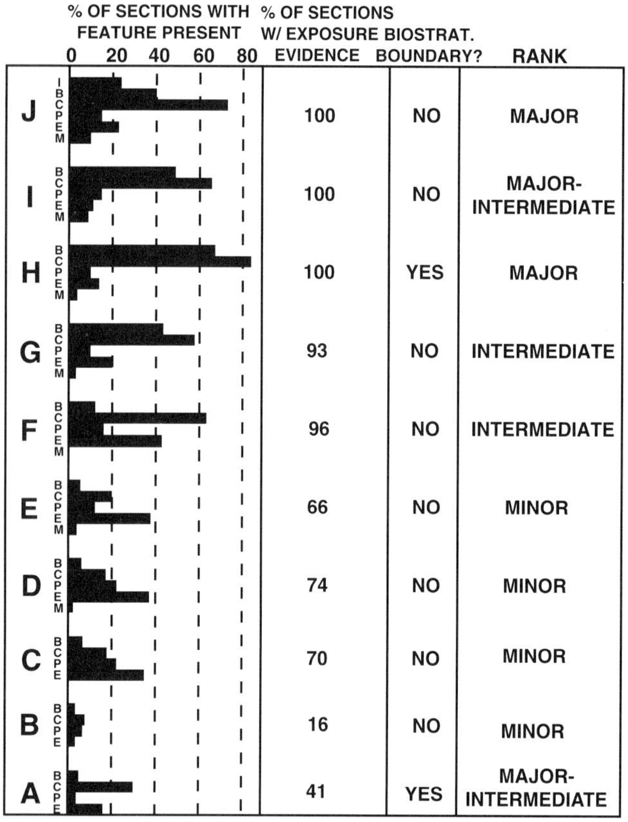

FIG. 8.—Relative ranking of disconformities based on percentage of locations measured with breccia (B), caliche (C), dirty packstone (P), eolianites (E), and slickensided mudrock (M). Incision (section I) only occurs on the uppermost boundary and is a better measure of how far sea level fell rather than the duration of exposure. Disconformities that coincide with biostratigraphic boundaries may represent relatively long-term exposure.

to incise a valley more than 75 m through underlying carbonate strata. Disconformity A may also represent a major hiatus because it shows evidence for a major biostratigraphic break, but paleosols are poorly developed or absent at this horizon. Exposure surfaces F, G, I, and J can be traced into sections in the more rapidly subsiding basin interior and therefore represent periods of widespread emergence, but the lesser development of caliche and breccia suggests that the duration may not have been as long (intermediate disconformities). Disconformities B, C, D, and E can only be traced along the eastern and western shelves and are marked predominantly by eolianites and dirty packstone paleosols and therefore are minor disconformities that represent relatively short periods of exposure.

SEQUENCE STRATIGRAPHY

A *sequence* is defined as a genetically related succession of strata lacking apparent internal unconformities and composed of parasequences and parasequence sets arranged in systems tracts and bounded by unconformities or their correlative conformities (Mitchum and Van Wagoner, 1991). It is not possible to meet all of these criteria with the disconformity-bounded units preserved in the Upper Mississippian strata of the Illinois basin because they were deposited hundreds of kilometers updip from the shelf edge. In this updip setting, in which lowstand systems tracts are absent and are represented only by unconformities, it is difficult to recognize parasequence sets, and transgressive and highstand systems tracts or entire sequences may be composed of as few as one parasequence. Stratigraphic units are called sequences in this study if they have maximum-flooding surfaces and are bounded by regionally extensive disconformities. On the basis of this definition, there are five disconformity-bounded sequences in the Ste. Genevieve to Paoli interval that are each composed of as many as nine parasequences.

A *parasequence* is defined as a relatively conformable succession of genetically related beds and bedsets bounded by marine flooding surfaces and their correlative surfaces (Mitchum and Van Wagoner, 1991). Most parasequences in the study interval can be traced across the eastern shelf and into the basin interior within a framework of disconformities and laterally extensive marker beds. Some parasequences are confined to local topographic lows and form multilateral, discontinuous units, especially beneath sequence boundaries. Parasequences are 1 to 8 m thick on the eastern shelf, where some parasequences are capped by disconformities, and 4 to 12 m thick in the basin interior where they are generally conformable.

Typical parasequences consist of, from base to top, (1) a disconformity and/or marine flooding surface; (2) locally developed muddy carbonates and/or quartz-peloid grainstone (eolianite); (3) laterally discontinuous ooid grainstone ridges and channel fills that overlie, underlie, and pass laterally into skeletal grainstone sheets and banks; (4) laterally extensive muddy carbonate or fossiliferous shale units; (5) locally developed quartz-peloid grainstone (eolianite) units; and (6) a disconformity and/or marine flooding surface. The facies within sequences 1 through 5 vary from the western shelf to the basin interior to the eastern shelf and are discussed individually to illustrate their differences and similarities.

Sequence 1

Sequence 1 is bounded at the base by disconformity A and at the top by disconformity F. It includes the Fredonia Member and the lower McCloskey oolite reservoirs of the Ste. Genevieve and the lower half of the Aux Vases Formation on the western shelf, and it is capped by the Spar Mountain Member of the Aux Vases in the basin interior. Sequence 1 is 27 m thick on the western shelf, thickens to 63 m in the basin interior, thins to 20 m near the Cincinnati arch and 9 m on the northern part of eastern shelf.

On the updip parts of the eastern shelf (A–A′), the lower boundary is picked at an undulose surface in a dolomite unit overlying the Lost River Chert Bed that coincides with a major biostratigraphic zone boundary at the St. Louis/Ste. Genevieve contact (Rexroad and Fraunfelter, 1977; Maples and Waters, 1987; Merkley, 1991; Dodd et al., 1996). Caliche and eolianites mark the boundary in the basin interior and the downdip sections of the eastern shelf (cross section B–B′). Well-developed breccia and conglomerate have been reported at the basal boundary in Missouri and in the Appalachian basin (Weller and

Sutton, 1940; Al-Tawil and Read, 1996). Although it appears that there was topography on the Lost River Chert on cross section A–A′ on the northern part of the eastern shelf, the cross section is composed of locations that are on either side of the hinge of a northwest-trending structural high that caused thinning of as much as 16 m over 35 km in the basal part of sequence 1 (see location map, Fig. 4). Locations 29, 35, 36, and 40–46 were on the structural high and sections 26, 32, 33, 37, 38 and probably 39 were on the more rapidly subsiding basinward side of the hinge. This structure parallels the northwest-trending part of the Cincinnati arch and the Lasalle anticlinal belt (Fig. 1B).

The transgressive systems tract (TST) of sequence 1 is composed of four onlapping parasequences (1–4) (Fig. 4). Parasequences 1 and 2 are thin, regional, skeletal grainstone–muddy carbonate parasequences that onlap the margin of the eastern shelf and are present only in the basin interior. Parasequences 3 and 4 are well-defined ooid grainstone–muddy carbonate units in the downdip sections of the eastern shelf (cross section B–B′) that onlap the structural high in the northeastern part of the basin. In the basin interior, it is difficult to correlate these parasequences because they are commonly amalgamated in thick ooid grainstone and muddy carbonate deposits (locations 17, 22, and 26). The maximum-flooding surface (MFS) is picked at the base of the grainstone unit in parasequence 5 because it is the first laterally extensive grainstone to be deposited on the structural high in the northeastern part of the basin.

The highstand systems tract (HST) is composed of five throughgoing parasequences (5–9). Parasequence 5 can be correlated across the eastern shelf but is difficult to correlate in parts of the basin interior because it is amalgamated with parasequences 3 and 4 in sections 17, 22, and 26 and correlates laterally with two parasequences at locations 5, 12, and 14. Parasequences 5, 6, 7, and 9 are capped by offlapping disconformities on the eastern shelf and are primarily composed of muddy carbonates and eolianites. These HST parasequences are conformable in much of the basin interior and are locally amalgamated into thick ooid grainstone units (locations 5 and 12). Parasequence 9, which underlies the sequence boundary, consists of thin multilateral grainstone units confined to local lows on the eastern shelf and is capped by an eolianite in the basin interior. The eolianite—which is the Spar Mountain Member of the Ste. Genevieve—passes westward into marine siliciclastics that prograded east at the top of the sequence (Swann, 1964). On the western shelf, the HST is composed of siliciclastic parasequences bounded by four paleosols that may correlate with the four regional disconformities on the eastern shelf (disconformities C to F). The sequence is bounded at the top by disconformity F. Leetaru (1997) mapped an erosional contact with as much as 24 m of relief at the contact between the siliciclastics of the Aux Vases and the carbonates of the Ste. Genevieve. However, this erosion is not related to the sequence-bounding disconformity F that can be traced through the middle of the siliciclastic unit. The erosion may be the result of scour caused by a prograding tidal delta in the late HST of the sequence. If this erosion were related to a sea-level lowstand, the sequence-bounding paleosol would be absent in locations where erosion occurred.

Sequence 2

Sequence 2 is bounded at the base by disconformity F and at the top by disconformity H. It is composed of the Karnak and Joppa Members of the Ste. Genevieve and the Rosiclare Member of the Aux Vases. Sequence 2 is 16 m thick on the western shelf, thickens to 30 m in the basin interior, and thins to 13 to 15 m on the eastern shelf. Sequence 2 is composed of tidally influenced siliciclastics on the western shelf that cannot be subdivided without a more extensive data set.

Disconformity F is an intermediate-rank paleosol that commonly overlies eolian facies in the basin interior and underlies eolianites on the eastern shelf (Fig. 8). The TST of sequence 2 consists of parasequence 10 that is composed of patchy eolianites and grainstone capped by a 1–10-m-thick muddy carbonate unit that can be correlated over most of the basin interior and eastern shelf. The MFS is picked at the base of parasequence 11 because it was the first laterally extensive grainstone unit to be deposited on the eastern shelf. The HST is composed of five parasequences (11–15). Parasequences 11 and 12 are generally conformable grainstone–muddy carbonate units that can be correlated across most of the basin interior and eastern shelf. In many sections on the eastern shelf, the absence of the muddy cap to parasequence 11 suggests either nondeposition or erosion at the base of parasequence 12. Parasequence 13, which underlies disconformity G/G′, consists of multilateral grainstone units and is laterally discontinuous across the basin. The HST is capped by parasequence 14, a disconformity-bounded parasequence composed of tidally influenced siliciclastics and multiple local disconformities; the HST is only present in the basin interior and is capped by a well-developed unconformity (disconformity G). The contact between the Aux Vases and Ste. Genevieve is commonly erosional near the top of this sequence and has also been picked as a sequence boundary (Leetaru, 1997). However, the fact that the regional disconformity can again be traced through the siliciclastic unit suggests that any erosion occurred by tidal scour in the late part of the highstand rather than during sea-level lowstand.

Sequence 3

Sequence 3 is bounded at the base by disconformity G and at the top by disconformity H. This sequence is composed of a single disconformity-bounded parasequence at the top of the Ste. Genevieve Formation in Kentucky and Indiana and the Levias Member of the Renault Formation in Illinois. The lower boundary (disconformity G) is a well-developed, basin-wide paleosol that caps siliciclastics of the Rosiclare Tongue of the Aux Vases. Sequence 3 is predominantly composed of grainstone and minor siliciclastics in the basin interior, eolianites on the southern part of the eastern shelf, and muddy carbonates and grainstone on the northern part of the eastern shelf. The MFS is interpreted to be coincident with the lower sequence boundary. The upper sequence boundary (disconformity H) is the best-developed paleosol in the study interval, which is called the Bryantsville Breccia Bed in Indiana and central Kentucky. In earlier work (Smith, 1996), this unit was called a disconformity-bounded parasequence at the top of sequence 2, but is here reinterpreted as a separate sequence. The progradation of siliciclastics at the top of the underlying unit and the intensity of the brecciation at the top of the sequence suggest that it is a thin updip extension of a sequence that is better developed downdip.

Sequence 4

Sequence 4 is bounded at the base by disconformity H and at the top by disconformity I. Sequence 4 includes the lower part of the Paoli interval, the basal disconformity-bounded unit in the Girkin Limestone on the eastern shelf, and the Shetlerville Member of the Renault Formation and the Yankeetown Formation in the central and western parts of the basin. The sequence is 6 m thick on the western shelf, thickens to 20 m in the basin interior, and thins to 4 m near the Cincinnati Arch and 2 m in the northeastern sections.

Disconformity H is a major disconformity characterized by accumulations of breccia and caliche as much as 4 m thick and called the Bryantsville Breccia Bed in Indiana and Kentucky (Mallot, 1952; Liebold, 1982). Disconformity H coincides with the top of the *Platycrinites penicillus* zone and the base of the *Talarocrinus* zone (Swann, 1963). The TST of sequence 4 is composed of parasequences 16 and 17. Parasequence 16 is a thin parasequence, confined to the most rapidly subsiding parts of the basin interior, which onlaps the shelves and a local tectonic high in the basin interior. Parasequence 17 has a patchy grainstone unit at the base that pinches out onto tectonic highs and is capped by a laterally extensive muddy carbonate—fossiliferous shale unit that can be correlated across most of the basin interior and eastern shelf. On the western shelf, the TST is absent except at section 3 where parasequence 16 is composed of quartz sandstone and parasequence 17 is composed of a fossiliferous shale capped by quartz sandstone. The Yankeetown Shale that occurs near the top of the sequence grades updip into quartz sandstone to the northwest (Swann, 1963).

The MFS is picked at the base of parasequence 18 because it is the first laterally extensive grainstone unit to be deposited on the eastern and western shelves. The HST consists of parasequences 18 to 21 in the basin interior, but only parasequences 18 and 19 were deposited on the shelves. On the eastern shelf, parasequence 18 is a grainstone–muddy carbonate parasequence with the exception of a few downdip sections, where it is composed of muddy carbonates and green fossiliferous shale. Parasequence 19 is composed of multilateral grainstone deposits that underlie the sequence boundary on the eastern shelf. In the basin interior, parasequences 18 and 19 have skeletal grainstone or packstone bases with fossiliferous shale caps that grade laterally to muddy carbonates in the eastern basin interior. Ooid grainstone is rare in the basin interior, but does occur at the base of the HST near tectonic highs. Parasequence 19 is capped by a local disconformity at sections 10, 11, and 12. On the western shelf, parasequence 18 is composed of skeletal grainstone overlain by tidally influenced siliciclastics and capped by a paleosol, and parasequence 19 is composed of quartz siltstone. Late-highstand parasequences 20 and 21 progradationally offlap and are only present in the basin interior. Sequence 4 is capped by disconformity I.

Sequence 5

Sequence 5 is bounded at the base by disconformity I and at the top by disconformity J. In the basin interior and western shelf, sequence 5 includes the upper part of the Paoli interval, a second disconformity-bounded unit in the Girkin Limestone, and the Downeys Bluff Formation. Sequence 5 is 6 m thick on the western shelf, thickens to 20 m in the eastern part of the basin interior, and thins to 7 m near the Cincinnati arch and to less than 3 m on the northern part of the eastern shelf. The sequence may have been thicker in the basin interior but was eroded during pre-Bethel incision.

Disconformity I is an intermediate unconformity that overlies a thin eolianite in the basin interior (Fig. 8). The TST is composed of parasequence 22 that consists of a thin basal shale in the eastern part of the basin interior, overlain by a grainstone unit that is very patchy on the eastern shelf and a laterally extensive fossiliferous shale or muddy carbonate cap. The MFS is picked at the base of the grainstone unit in parasequence 23 because it is the first laterally extensive grainstone unit to cover the shelves. The HST is composed of parasequences 23 to 25 in the basin interior, but only parasequence 23 was deposited on the shelves. Parasequence 23 is composed of a thick skeletal-oolitic grainstone unit overlain by a patchy muddy carbonate unit on the eastern shelf, ooid-skeletal grainstone capped by fossiliferous shale and tidally influenced siliciclastics on the western shelf, and skeletal limestone capped by fossiliferous shale in the basin interior. Parasequences 24 and 25 are only preserved basinward of the shelves, suggesting that progradational offlap occurred at the top of sequence 5. These parasequences are composed of thick accumulations of ooid grainstone along the margin of the eastern shelf and skeletal limestone and shale in most of the basin interior. A widespread eolianite overlies the sequence on much of the southern part of the eastern shelf. The sequence is capped by an unconformity (disconformity J) that consists of a well-developed paleosol on the eastern and western shelves that can be traced into an incised valley that trends southwest from south-central Indiana to western Kentucky and is as deep as 75 m (Friberg et al., 1969).

Composite Sequences

A composite sequence is a succession of genetically related sequences in which the individual sequences stack into lowstand, transgressive, and highstand sequence sets (Mitchum and Van Wagoner, 1991). The updip location of the Illinois basin on the ramp makes it difficult to trace sequence boundaries to determine which extends the farthest downdip. In this setting, composite sequence boundaries can only be distinguished from higher-frequency sequence boundaries by the degree of paleosol development, depth of incision, and/or coincidence with biostratigraphic zone boundaries (Smith, 1996). On the basis of these criteria, the major disconformities at the base of sequence

1 (disconformity A), at the top of sequence 3 (disconformity H), and at the top of sequence 5 (disconformity J) may be composite sequence boundaries (Smith, 1996). In Smith (1996), sequences 1 through 5 were picked as a composite sequence on the basis of the development of the bounding unconformities (A and J) and apparent backstepping of ooid grainstone upward through the interval. Disconformity H was picked as a higher-frequency composite sequence boundary. Disconformities A and H have been interpreted to be composite sequence boundaries in the Appalachian basin (Al-Tawil, 1998), but J is not any better developed than the other sequence boundaries. Thus, the sequences are likely bundled into composites sequences, but the updip location of the Illinois basin and the intensity of the sequence development make it difficult to confidently pick composite sequence boundaries.

DISCUSSION

Duration of Sequences and Parasequences

The most recent ages for the duration of the Mississippian Period come from SHRIMP dating of volcanic rocks interbedded with marine sedimentary rocks in Australia (Roberts et al., 1995). Brachiopod biostratigraphy was used to correlate with European and North American conodont- and foraminiferal-zones that tied the SHRIMP ages to radiometric dates from Europe (Roberts et al., 1995). On the basis of foraminiferal and conodont zones, the Ste. Genevieve to Glen Dean interval is uppermost Visean (middle of V3b to V3c) (Baxter and Brenckle, 1982). This interval has a duration of approximately 3 m.y. (Roberts et al., 1995), but is poorly constrained because of the difficulties inherent in correlating between brachiopod, conodont, and foraminiferal zones on different continents and the error bars on the radiometric dates.

There are five disconformity-bounded sequences in the Ste. Genevieve to Paoli interval and four in the overlying Bethel to Glen Dean interval (Smith, 1996), for a total of nine sequences. Dividing ten sequences into 3 m.y. gives an average period of 333 k.y. per sequence—which is suggestive of the long-term Milankovitch eccentricity signal (~414 k.y.) (Berger, 1988) and comparable to the calculated duration for some of the Pennsylvanian cyclothems (Heckel, 1986). Such fourth-order (0.1 to 0.5 m.y.) sequences are compatible with times of abundant global ice (Weber et al., 1995). Calculated average duration of the preserved parasequences would be about 37 k.y. in sequence 1, 55 k.y. in sequences 2 and 4, and 83 k.y. in sequence 5. These results suggest that the parasequences are dominantly fifth-order (10 to 100 k.y.) (cf. Weber et al., 1995). Any composite sequences would be third-order (0.5–5 m.y.) (cf. Weber et al., 1995).

Origin and Development of Sequences and Parasequences

Origin of sequences.—

The sequences and sequence boundaries in the Chesterian interval have been interpreted to have formed by local tectonics (Leetaru, 1997), cyclic climate change (Swann, 1963), and eustatic sea-level change (Smith et al., 1995; Smith, 1996). The conclusion that the sequence boundaries likely were caused by fourth-order (0.1–0.5 m.y.) eustatic sea-level changes is based on four facts: (1) by using distinct marker beds and biostratigraphy, the sequence-bounding unconformities in both siliciclastics and carbonates can be correlated across the Illinois basin and in some cases into the Appalachian basin (Smith et al., 1995); (2) the unconformities coincide with biostratigraphic zone boundaries; (3) disconformity-bounded sequences with similar characteristics occur in time-equivalent strata in Kansas (Abegg, 1994), Great Britain (Walkden, 1987), and Kazakhstan (Lehmann et al., 1996); and (4) the duration of the sequences roughly coincides with the Milankovitch long-term (~400 k.y.) eccentricity signal and the calculated period for the overlying Pennsylvanian cyclothems (Heckel, 1986). If the sequences were made by local tectonic uplift, the sequence boundaries would be confined to the area of uplift, but instead they are basin-wide. If the unconformities were caused by a migrating peripheral bulge, they would be diachronous, but the fact that the unconformities coincide with biostratigraphic zone boundaries suggests that they are relatively synchronous across the basin. Sequence-bounding disconformities formed during periods of high, moderate, and low differential subsidence in the Illinois basin (Smith, 1996). Regional uplift of the Illinois and Appalachian basins could have caused the unconformities, but there is no known mechanism that would cause rhythmic uplift and subsidence of the craton with a period on the order of 400 k.y. If the sequences were made by variations in siliciclastic sediment supply caused by climate change (Swann, 1963), the unconformities would be present only in areas of mixed carbonates and clastics, but they can be traced from mixed carbonate-clastic intervals into carbonate-dominated strata.

Sequence development.—

The magnitude of sea-level fluctuations that produced the boundaries of sequences 1 through 4 can be estimated in the following manner. After deposition of the Lost River Chert Bed, which can be traced from the basin interior to the northern part of the eastern shelf, the sea-level fall was sufficient to expose the basin interior as subsidence continued. The exposed profile can be roughly reconstructed by using the thickness of the TST of sequence 1, which is 27 m thick in the basin interior, thinning to less than 1 m on the northern part of the eastern shelf. These data suggest that a sea-level change of about 30 m formed the sequence boundary at the base of sequence 1, given that the effects of compaction (which would increase the estimate) and differential subsidence during deposition of the TST (which would decrease the estimate) would tend to cancel each other out. The lack of incision on the boundaries of sequences 1 through 4 suggests that sea level may not have fallen far below the platform. The incision within the Aux Vases Formation may have been caused by prograding tidal deltas or local tectonics (Leetaru, 1997), but these erosion surfaces are not sequence boundaries because they do not correlate with paleosols in areas of no incision. Thus, sequences 1 through 4 likely were made by moderate-amplitude sea-level fluctuations of 30 m or less. In

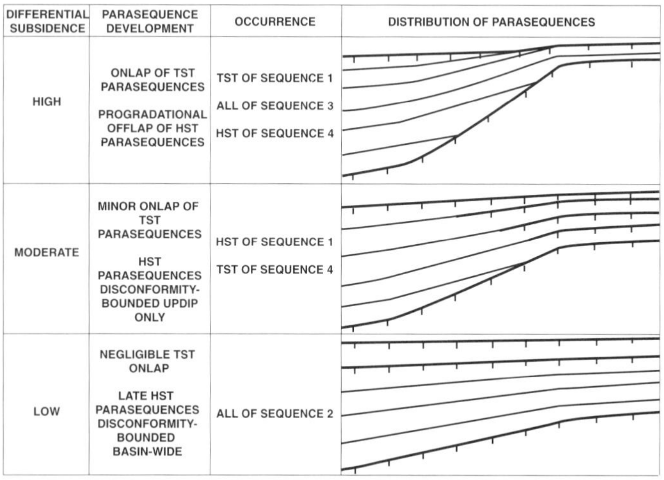

FIG. 9.—Effects of variations in differential subsidence of the eastern shelf and the basin interior on parasequence distribution within the sequences. TST = transgressive systems tract, HST = highstand system tract.

contrast, sequences in the overlying middle Chesterian mixed carbonate-clastic sequences also have incised valleys (indicating as much as 75 m of erosion) as well as deeper-water carbonate facies that together suggest that the magnitude of sea-level changes increased to as much as 30–95 m (Smith, 1996).

Lowstand systems tracts were not preserved in the study area because the Illinois basin was more than 300 km updip from the regional ramp margin. As sea level rose, transgressive systems tracts composed of one to four parasequences onlapped the shelves. Onlap was greatest in the transgressive systems tracts during periods of high differential subsidence between the shelves and the basin interior (Fig. 9) (Smith, 1996). Initial widespread flooding caused the uppermost of these parasequences to extend onto the eastern shelf where they typically consist of laterally extensive muddy carbonate units. Maximum flooding generated extensive grainstone units over the eastern shelf, when water depths increased enough to accommodate the tidal wedge and generate strong tidal currents. The lack of deeper-water facies indicates that water depths never exceeded those of ooid and skeletal grainstone lithotopes. In the HST, three to five aggradational to progradational parasequences were deposited as relative sea level peaked and began to fall. Progradational offlap of parasequences occurred in the HST during periods of high differential subsidence in the basin, and aggradation occurred during periods of relatively low differential subsidence (Fig. 9) (Smith, 1996).

Glacio-eustatic origin of parasequences.—

The lateral extent and frequency of the regionally developed parasequences indicate that these strata were produced by fifth-order (10–100 k.y.) glacio-eustatic sea-level changes. Neither local tectonics nor autocyclic processes could have produced basin-wide parasequences that appear to have formed in tens of thousands of years. The presence of tidal sand ridges 8–10 m high (Carr, 1973; Cluff and Lineback, 1981) suggest that rapid sea-level rises of as much as 10 m would have been required to provide the necessary accommodation for typical parasequences. The lack of deep-water facies extending onto the ramp and the restriction of some parasequence-bounding disconformities to updip areas suggests that the magnitude of high-frequency sea-level fluctuations probably was not much more than 10 m.

Assuming that the sequences are 400 k.y. in average duration, and from 2 to 63 m thick (ignoring the effects of compaction), subsidence rates would have been between 0.5 and 15 cm/k.y. These rates would not have drowned healthy, shallow, warm-water carbonate platforms, many of which are capable of generating carbonate sediment at 20 cm/k.y. or more (Schlager, 1981). Thus, parasequence flooding likely required rapid glacio-eustatic sea-level rises of as much as 10 m to generate water depths necessary for deposition of thick grainstone deposits. If sea-level rises occurred during the first 15% of the parasequence duration (similar to Pleistocene sea-level rises), then rise rates would have been between 60 and 300 cm/k.y. for parasequences with durations of 20 to 100 k.y. These rise rates would have been able to incipiently drown the ramp. Sea-level rises much less than 10 m would have tended to generate meter-scale tidal-flat cycles on the inner ramp similar to those described from greenhouse platforms (Goodwin and Anderson, 1985; Koerschner and Read, 1989; Goldhammer et al., 1990), which are notably absent on the Mississippian platform.

Parasequence development.—

Parasequence deposition was initiated by high-frequency sea-level rises that flooded both disconformable and conformable surfaces on shallow-water mudstone that had shoaled to, or near to, sea level (Fig. 10). High-frequency (20 to 100 k.y.), rapid, eustatic sea-level rise coupled with background subsidence, generated water depths that were sufficient for the tidal wedge to develop and generate widespread high-energy tidal currents and wave reworking. These high-energy conditions allowed deposition of skeletal and ooid grainstone in discontinuous sheets and tidal ridges. Ooid and skeletal grainstone and packstone accumulated to form a broadly undulating submarine topography of ridges and passes. Tidal energy appears to have diminished in updip parts of the ramp where muddy carbonates were deposited in widespread lagoons. As ooid and skeletal grainstone filled much of the accommodation space on the ramp, tidal energy diminished, and muddy lagoonal facies and fossiliferous shale prograded to form widespread sheets that capped and filled swales between carbonate sand ridges. Tidal-flat laminites were not deposited at the tops of most parasequences because sea-level fall caused rapid movement of the shoreline seaward across the shelf, stranding any tidal flats updip. In areas with low accommodation rates, parasequences were capped by patchy eolianites and disconformities. These disconformities pinched out into areas of higher accommodation where relative sea-level fall did not expose sediments for a period of time sufficient to develop disconformities. In these areas, renewed flooding generated conformable parasequences.

Shallowing-upward trends within the individual carbonate parasequences are complicated by the gentle dip of the ramp and the strong tidal currents, which generated the dip-parallel orientation of the ooid shoals. Thus, facies were as much dependent

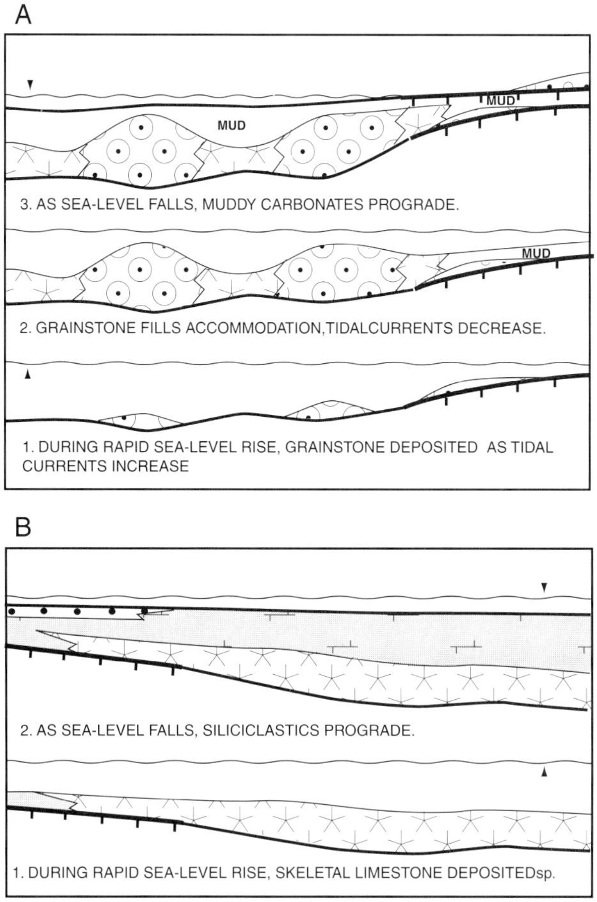

FIG. 10.—(A) Development of carbonate parasequences: (1) Rapid glacioeustatic sea-level rise promotes increase in tidal energy and deposition of ooid ridges and skeletal banks. (2) Ooid and skeletal grainstone fills much of the accommodation space, and tidal energy diminishes. (3) Muddy carbonates prograde, filling topography and forming laterally extensive sheets. (B) Skeletal limestone-shale parasequence development: (1) During sea-level rise, successive siliciclastic layers are deposited farther updip toward the source, and carbonate-producing organisms thrive and make skeletal limestone. (2) As sea level falls, shale progrades from the west-northwest and caps parasequences. In areas close to the siliciclastic source, sandstone progrades over shale.

on local tidal energy as water depth. If ooid shoals had formed wave-dominated strike-parallel barriers (rather than dip-parallel tidal ridges), a more predictable succession of skeletal grainstone to ooid grainstone to muddy carbonate and shale to eolianite would have been generated by upward shallowing within prograding parasequences. Instead, skeletal grainstone occurs above and below and passes laterally into ooid grainstone as a product of dip-parallel grainstone geometries.

Most apparently symmetrical parasequences with muddy carbonates at the base and top (Pryor et al., 1990) can be traced downdip where the basal muddy units commonly form the tops of parasequences with basal, patchy, high-energy grainstone units. Thus many updip basal mudstones are actually the updip, low-energy tongues of downdip parasequences. At the other extreme, some well-developed updip parasequences merge downdip into thick ooid grainstone or mudstone units (e.g., parasequences 3, 4 and 5) that represent amalgamated parasequences. These developed where ooid grainstone and muddy carbonate deposition persisted through time, even though sea level was changing. Other apparently amalgamated parasequences could be due to erosion of muddy carbonate caps during subsequent transgressions, leaving stacked grainstone units.

Autocyclic shallowing associated with local island development formed local parasequences within thicker, regionally traceable parasequences. Such autocycles are clearly evident with the detailed, closely spaced cross sections but could not have been distinguished from regional parasequences on the basis of a single stratigraphic section.

Late Mississippian Sea-Level Curve

Basin-wide paleosols and facies-stacking patterns in the Mississippian succession can be used to construct a detailed relative-sea-level curve (Fig. 11). Ross and Ross (1988) constructed an onlap curve for the late Paleozoic but only correlated large-scale sequences (Fig. 11). They picked a larger-scale sequence boundary at the St. Louis/Ste. Genevieve contact, and the Ste. Genevieve–Paoli interval only makes up part of one of their sequences. Swann (1963) constructed a transgression-regression, or relative-sea-level, curve on the basis of the seaward extent of siliciclastics within the Illinois basin (Fig. 11). Swann's curve roughly parallels this paper's fourth-order sequence curve, because the siliciclastic units commonly prograded toward the tops of the sequences. However, the Swann (1963) curve does not show a regression at the St. Louis/Ste. Genevieve boundary, because there was no siliciclastic influx associated with it.

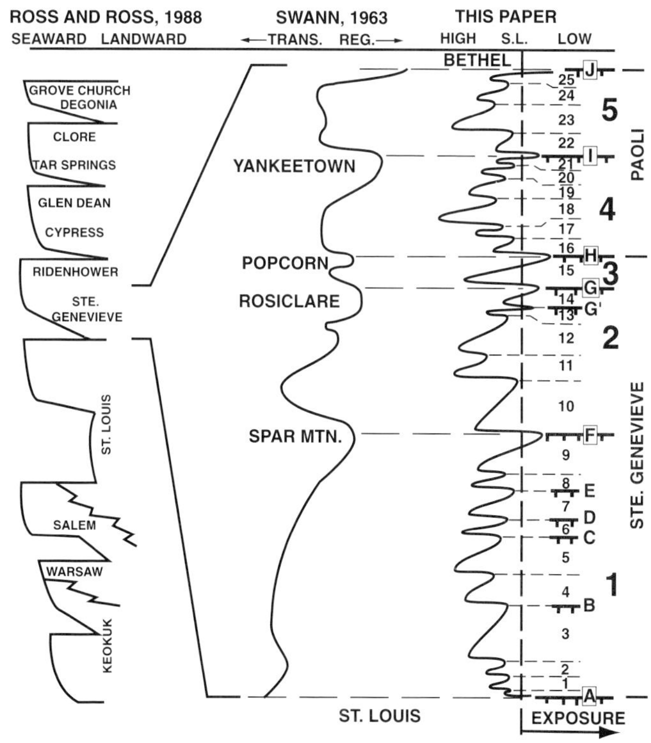

FIG. 11.—Comparison of onlap curve from Ross and Ross (1988), transgression-regression curve of Swann (1963), and relative sea-level curve prepared for this paper. Disconformities are lettered from A to J, sequences are numbered from 1 to 5, and parasequences are numbered from 1 to 25.

Stratigraphic Signature of the Greenhouse to Icehouse Transition

The first ice-rafted deposits of the late Paleozoic glaciation of Gondwana suggest that major continental ice sheets began to grow during the earliest Namurian or middle Chesterian (Fig. 12) (Garrasino, 1981; Hambrey and Harland, 1981; Frakes et al., 1992; Caputo and Crowell, 1985; Roberts et al., 1995). The base of the overlying Glen Dean Formation is equivalent to the base of the Namurian; therefore the age of the Ste. Genevieve to Paoli interval is late Visean (Baxter and Brenckle, 1982). Thus the study interval spans the later part of the transition from a relatively ice-free globe in the Early to middle Mississippian to the Pennsylvanian–Permian glaciation. Ice-rafted deposits time-equivalent to the study interval have not been reported, but such deposits likely would have been destroyed during buildup of progressively larger ice sheets.

The waxing and waning of progressively larger ice sheets may have caused an increase in the amplitude of sea-level fluctuations forming these sequences and those in the overlying middle Chesterian interval (Fig. 13) (Smith, 1996). Sequences 1 through 4 are bounded by surfaces with minimal erosion and were likely formed by sea-level fluctuations of ~30 m or less. In contrast, sequences 5 through 10 have bounding surfaces with as much as 75 m of incision, which may have been produced by sea-level changes of as much as 95 m (Smith, 1996). Time-equivalent strata in Great Britain show a similar upward change from nonerosional to erosional boundaries that has been attributed to increasing amplitude of relative sea-level fluctuations coupled with tectonics (Walkden, 1987). This increasing amplitude is likely related to the onset of continental glaciation on Gondwana.

The Early Mississippian was dominated by precessionally driven, low-amplitude, high-frequency sea-level fluctuations, which suggests that it was relatively ice free (Vaughn and Adams, 1984; Adams et al., 1990; Read and Horbury, 1993). The stratigraphy of the updip parts of large carbonate platforms that developed during greenhouse times of little global ice is dominated by meter-scale peritidal cycles capped by regional tidal-flat facies, especially in highstand systems tracts of third-order sequences (Goodwin and Anderson, 1985; Koerschner and Read, 1989; Goldhammer et al., 1990; Elrick, 1991; Read, 1995). These peritidal cycles may be bundled into obliquity or eccentricity cycle sets (Goldhammer et al., 1990). Disconformities are poorly developed on these greenhouse platforms except at third-order sequence boundaries.

The parasequences formed during the transition period of the Late Mississippian covered by this study generally lack

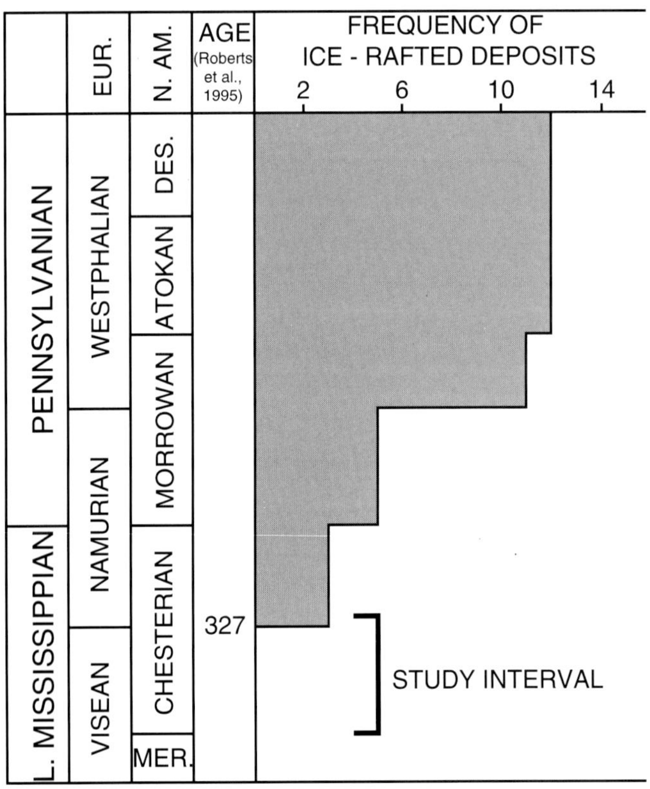

FIG. 12.—Global climate setting during Late Mississippian and Pennsylvanian. Graph shows the number of occurrences of ice-rafted deposits worldwide leading up to late Paleozoic glaciation of Gondwana (modified from Frakes et al., 1992; age from Roberts et al., 1995). The study interval occurs just below the first glacial deposits and during the transition from little global ice to major continental glaciation. MER. = Meramecian, DES. = Desmoinesian.

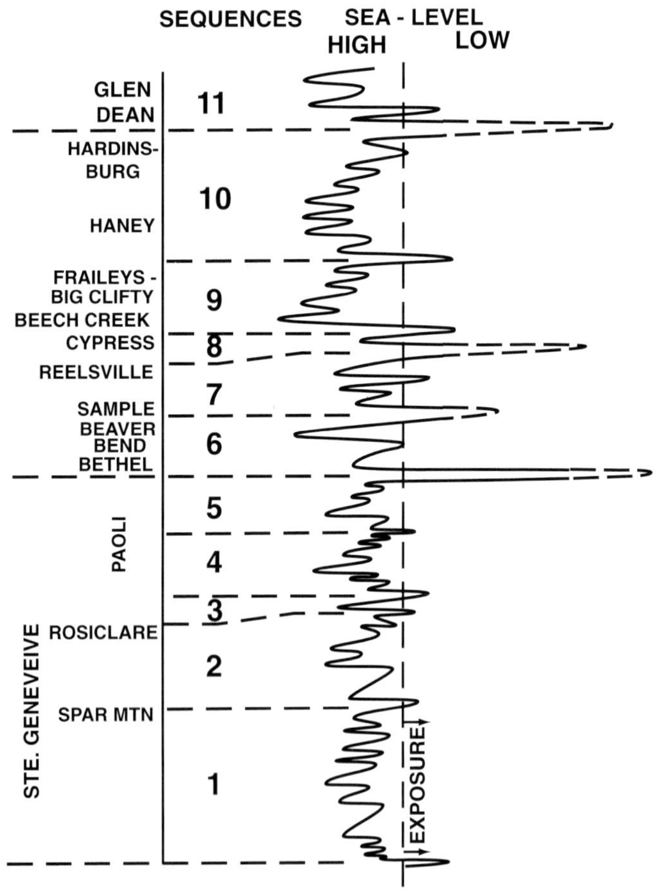

FIG. 13.—Composite sea-level curve for the Ste. Genevieve to Paoli interval and the Bethel through Glen Dean interval (sequences 6 to 10) (Smith, 1996). The amplitude of sea-level changes increased through the Chesterian in response to onset of major continental glaciation on Gondwana.

intertidal-supratidal caps, and disconformities are developed on subtidal lagoonal and shoal facies of fourth-order and some updip fifth-order units. Third-order composite sequences are difficult to recognize because of the frequency of the disconformities and the magnitude of the moderate-amplitude fourth-order sea-level changes.

Global icehouse times such as the Pennsylvanian and Pleistocene were characterized by high-frequency fourth-order sequences (Heckel, 1986; Goldhammer et al., 1991; Weber et al., 1995; Mitchum and Van Wagoner, 1991; Read, 1995) This paper shows that this type of sequence development is evident as early as the Late Mississippian. The oxygen isotope record suggests that the dominant episodes of waxing and waning of Pleistocene ice sheets correlate with the 41-k.y. obliquity and the 100-k.y. eccentricity periods (Ruddiman et al., 1986). Given that there are missing beats associated with the disconformities, the average durations of the parasequences (40 to 100 k.y.) suggest that the parasequences may have been produced by sea-level changes associated with the 41-k.y. obliquity signal in possible combination with the 100-k.y. short-term eccentricity signal. Unlike during greenhouse times, there is no bundling of parasequences (e.g., precessional cycles bundled into obliquity-eccentricity cycles [Goldhammer et al., 1990]) at a scale below that of the sequences.

The increase in siliciclastic content from sequences 1 to 4 continues into the overlying sequences and may be related in part to the increasing amplitude of sea-level fluctuations due to the onset of glaciation. A similar increase in siliciclastic sedimentation occurred in time-equivalent strata in Great Britain that was attributed to increasing-amplitude eustasy and local tectonics (Walkden, 1987). Increasing-amplitude eustasy caused greater base-level falls that in turn caused greater erosion in the headlands and increased siliciclastic sediment supply. At the same time, synchronous climate change and increasingly humid wet-dry seasonality would have increased clastic influx during lowered sea level.

CONCLUSIONS

Regional disconformities were previously thought to be of relatively local extent in the Ste. Genevieve to Paoli interval in the Upper Mississippian (Chesterian) of the Illinois basin. However, these disconformities, which probably are the result of moderate-amplitude fourth-order eustasy, can be mapped regionally to define a high-resolution sequence stratigraphic framework of fourth-order sequences and component parasequences.

Five disconformity-bounded fourth-order sequences (2 to 63 m thick) can be traced across the Illinois basin in the Ste. Genevieve to Paoli interval. The boundaries of the lower four sequences are marked by caliche, breccia, red mudrock paleosols, and carbonate eolianites. The boundary at the top of sequence 5 is marked by a well-developed paleosol that can be traced into an erosional unconformity with as much as 75 m of incision. Lowstand systems tracts are not preserved because of the updip position of the study area on the ramp. Transgressive systems tracts are relatively thin and are composed of one to four onlapping parasequences. Maximum-flooding surfaces are picked at the base of regional grainstones that extend over the eastern shelf. Sequences are mainly composed of highstand systems tract, aggradational to progradational parasequences. The sequences correlate with those in the Appalachian basin, indicating a eustatic origin. The magnitude of sea-level changes was probably 30 m or less for the nonerosional sequence boundaries but increased to as much as 95 m to create the erosional unconformity at the top of the interval. The increasing magnitude of sea-level fluctuations was probably driven by waxing and waning of progressively larger ice sheets during the onset of Gondwana glaciation.

Parasequences contain the individual reservoirs and consist of, from base to top, (1) a basal disconformity or conformable boundary, (2) ooid (reservoir) and skeletal grainstone, (3) muddy carbonate and/or fossiliferous shale caps; and (4) a capping disconformity or conformable boundary. These parasequences differ from typical carbonate cycles developed during global greenhouse conditions in their lack of tidal-flat caps and common disconformities. Strong tidal currents formed laterally discontinuous, dip-parallel ooid grainstone ridges and channel-fills (reservoirs) that pass laterally into skeletal grainstone sheets and banks. Grainstones were capped by muddy carbonates that were deposited in laterally extensive sheets and swale fills. This process generated highly compartmentalized reservoirs. Without the sequence framework defined by regional tracing of disconformities, reservoirs appear to be randomly distributed. The sequence and parasequence framework more clearly shows how individual reservoirs are related. The parasequences can be correlated between areas with vastly different subsidence rates; thus they were likely formed by fifth-order sea-level fluctuations, superimposed on the fourth-order eustatic signal.

Carbonate reservoir facies of Mississippian age worldwide are likely to show similar high-frequency sequence stratigraphy because of the global eustatic control during the time of transition into the late Paleozoic glaciation.

ACKNOWLEDGMENTS

We thank Steve Bachtel, Brian Keith, Art Saller, and Zak Lasemi for their help in revising this manuscript. We also thank the members of the Indiana, Illinois, and Kentucky State Geological Surveys for providing cores and valuable discussion. Aus Al-Tawil, Mike Pope, Garland Dever, and John Nelson provided help and guidance throughout the course of this research. Financial assistance was provided by the Petroleum Research Fund (ACS-PRF#30516-AC8), the Geological Society of America, Sigma Xi, and Virginia Tech.

REFERENCES

Abegg, F. E., 1994, Recognition of carbonate eolianites and sequence boundaries in the Ste. Genevieve and St. Louis Limestones (Upper Mississippian), southwestern Kansas (abs.): Tulsa, American Association of Petroleum Geologists Official Program, v. 3, p. 91.

Adams, A. E., Horbury, A. D., and Abdel Aziz, A. A., 1990, Controls on Dinantian sedimentation in South Cumbria and surrounding areas of northwest England: London, Proceedings of the Geologists' Association, v. 101, p. 19–30.

Al-Tawil, A. A., 1998, High-resolution sequence stratigraphy of Late

Mississippian carbonates in the Appalachian Basin: Unpublished Ph.D. Dissertation, Virginia Tech, Blacksburg, 109 p.

AL-TAWIL, A. A., AND READ, J. F., 1996, High resolution sequence stratigraphy of Late Mississippian carbonates in the Appalachian basin, implications for compartmentalization of reservoir facies (abs.): Tulsa, American Association of Petroleum Geologists Official Program, v. 5, p. A3.

AMBERS, C. P., AND PETZOLD, D. D., 1992, Ephemeral arid exposure during deposition of the Elwren Formation (Chesterian) in Indiana, *in* Horowitz, A. S. and Dodd, J. R., eds., Chesterian sections (Late Mississippian) along Interstate 64 in southern Indiana: Great Lakes Section–SEPM (Society for Sedimentary Geology) Field Guide, p. 98–145.

BAXTER, J. W., AND BRENCKLE, P. L., 1982, Preliminary statement on Mississippian calcareous foraminiferal successions of the Midcontinent (U.S.A.) and their correlation to western Europe: Newsletters on Stratigraphy, v. 11, p. 136–153.

BERGER, A., 1988, Milankovitch theory and climate: Reviews of Geophysics, v. 26, p. 624–657.

BUTTS, C., 1917, Descriptions and correlations of the Mississippian formations of western Kentucky: Lexington, Kentucky Geological Survey, 119 p.

CAPUTO, M. V., AND CROWELL, J. C., 1985, Migration of glacial centers across Gondwana during Paleozoic Era: Geological Society of America Bulletin, v. 96, p. 1020–1036.

CARR, D. D., 1973, Geometry and origin of oolite bodies in the Ste. Genevieve Limestone (Mississippian) in the Illinois basin: Bloomington, Indiana Geological Survey Bulletin 48, 81 p.

CHOQUETTE, P. W., AND STEINEN, R. P., 1980, Mississippian non-supratidal dolomite Ste. Genevieve Limestone, Illinois basin: Evidence for mixed-water dolomitization: Tulsa, SEPM (Society for Sedimentary Geology)Special Publication 28, p. 163–196.

CLUFF, R. M., AND LINEBACK, J. A, 1981, Middle Mississippian carbonates of the Illinois basin: Champaign, Illinois Geological Society and Illinois Geological Survey Joint Publication, 88 p.

CRAIG, L. C., AND CONNOR, C. W., 1979, Paleotectonic Investigations of the Mississippian System in the United States: Washington, D.C., United States Geological Survey Professional Paper 1010, 559 p.

COLE, R. D., 1990, The stratigraphy, petrology and depositional environments of the Mississippian Aux Vases Formation across the southern portion of the Illinois basin: Unpublished Ph.D. Thesis, Southern Illinois University, Carbondale, 260 p.

DODD, J. R., ZUPPANN, C. W., HARRIS, C. D., LEONARD, K. W., AND BROWN, T. W., 1993, Petrologic method for distinguishing eolian and marine grainstones, Ste. Genevieve Limestone (Mississippian) of Indiana, *in* Keith, B. D. and Zuppann, C. W., eds., Mississippian Oolites and Modern Analogs: Tulsa, American Association of Petroleum Geologists Studies in Geology No. 35, p. 49–59.

DODD, J. R., HUNTER, R. E., AND MERKLEY, P. A., 1996, Eolianite-bearing depositional parasequences in the Ste. Genevieve Limestone of Indiana and Kentucky: Evidence for Mississippian Eustasy? (abs.): Tulsa, American Association of Petroleum Geologists Official Program, v. 5, p. A36.

DROSTE, J. B., AND CARPENTER, G. L., 1990, Subsurface Stratigraphy of the Blue River Group (Mississippian) in Indiana: Bloomington, Indiana Geological Survey Bulletin 62, 45 p.

ELRICK, M. E., 1991, Development of cyclic ramp-to-basin carbonate deposits, Lower Mississippian, Wyoming and Montana: Unpublished Ph.D. Thesis, Virginia Polytechnic Institute and State University, Blacksburg, 169 p.

ELROD, M. N., 1899, The geologic relations of some St. Louis group caves and sinkholes: Indianapolis, Indiana Academy of Science Proceedings, v. 8, p. 258–267.

ETTENSOHN, F. R., 1975, Stratigraphic and paleoenvironmental aspects of Upper Mississippian rocks (Upper Newman Group), east-central Kentucky: Unpublished Ph.D. Thesis, University of Illinois, Urbana, 320 p.

ETTENSOHN, F. R., DEVER, G. R., AND GROW, J. S., 1988, A paleosol interpretation for profiles exhibiting subaerial exposure crusts from the Mississippian of the Appalachian basin, *in* Reinhart, J. and Singleo, W. R., eds., Paleosols and Weathering through Geologic Time: Principles and Applications: Boulder, Geological Society of America Special Paper 216, p. 49–79.

FRAKES, L. A., FRANCIS, J. E., AND SYKTU, J. I., 1992, Climate Modes of the Phanerozoic: Cambridge, Cambridge University Press, 274 p.

FRIBERG, J. F., SCHWARTZ, R. K., SIEMERS, C. T., STRAW, P., POTTER, P. E., AND HRABAR, S. V., 1969, Sedimentation Seminar, Indiana University and University of Cincinnati, Bethel Sandstone (Mississippian) of western Kentucky and south-central Indiana, a submarine channel fill: Lexington, Kentucky Geological Survey, Report of Investigations, Series 10, v. 11, 24 p.

GARRASINO, C. A., 1981, Late Paleozoic Tarija Formation, southern Bolivia and northern Argentina, *in* Hambrey, M. J. and Harland, W. B., eds., Earth's Pre-Pleistocene Glacial Record: Cambridge, Cambridge University Press, p. 853–854.

GOLDHAMMER, R. K., DUNN, P. A., AND HARDIE, L. A., 1990, Depositional cycles, composite sea-level changes, cycle stacking patterns, and the hierarchy of stratigraphic forcing: Examples from Alpine Triassic platform carbonates: Geological Society of America Bulletin, v. 102, p. 535–562.

GOLDHAMMER, R. K., OSWALD, E. J., AND DUNN, P. A., 1991, Hierarchy of stratigraphic forcing: Example from Middle Pennsylvanian shelf carbonates of the Paradox basin, *in* Franseen, E. K., Watney, W. L., Kendall, C. G. St. C., and Ross, W., eds., Sedimentary Modeling: Computer Simulations and Methods for Improved Parameter Definition: Lawrence, Kansas Geological Survey Bulletin 233, p. 361–414.

GOODWIN, P. W., AND ANDERSON, E. J., 1985, Punctuated aggradational cycles: A general hypothesis of episodic stratigraphic accumulation: Journal of Geology, v. 93, p. 515–533.

HAMBREY, M. J., AND HARLAND, W. B., 1981, Earth's Pre-Pleistocene glacial record: Cambridge, Cambridge University Press, 1004 p.

HARRISON, R. S., AND STEINEN, R. P., 1978, Subaerial crusts, caliche profiles and breccia horizons: Comparisons of some Holocene and Mississippian exposure surfaces, Barbados and Kentucky: Geological Society of America Bulletin, v. 89, p. 385–396.

HECKEL, P. H., 1986, Sea-level curve for Pennsylvanian eustatic marine transgressive-regressive depositional cycles along the Midcontinent outcrop belt, North America: Geology, v. 14, p. 330–334.

HUFF, B. G., 1993, Analysis of the Aux Vases (Mississippian) petroleum reservoirs of Energy field, Williamson County: Champaign, Illinois State Geological Survey, Illinois Petroleum No. 141, 40 p.

HUNTER, R. E., 1993, An eolian facies in the Ste. Genevieve Limestone of southern Indiana, *in* Keith, B. D. and Zuppann, C. W., eds., Mississippian Oolites and Modern Analogs: Tulsa, American Association of Petroleum Geologists Studies in Geology No. 35, p. 31–48.

KISSLING, D. L., 1967, Environmental history of lower Chesterian rocks in southwestern Indiana: Unpublished Ph.D. Thesis, Bloomington, Indiana University, 367 p.

KOERSCHNER, W. F., III, AND READ, J. F., 1989, Field and modeling studies of Cambrian carbonate cycles, Virginia Appalachians: Journal of Sedimentary Petrology, v. 59, p. 654–687.

LEETARU, H. E., 1997, Sequence stratigraphy and resource assessment of Aux Vases Sandstone in Illinois: Unpublished Ph. D dissertation, University of Illinois, Champaign-Urbana, 133 p.

LEHMANN, P. J., COOK, H. E., ZEMPOLICH, W. G., ZHEMCHUZNIKOV, V. G., ZHAIMINA, V. Y., ZORIN, A. Y., BOWMAN, M., GIOVANELLI, A., VIAGGI, M., HUNT, D., FRETWELL, N., AND CRAIK, D., 1996, Paleosols and carbonate sequence stratigraphy, Carboniferous, S. Kazakstan (abs.): Tulsa, American Association of Petroleum Geologists Official Program, v. 5, p. A82.

LIEBOLD, A. W., 1982, Stratigraphy, petrology and depositional environment of the Bryantsville Breccia (Meramecian) of south-central Indiana: Unpublished M.S. Thesis, Indiana University, Bloomington, 171 p.

MALLOT, C. A., 1952, Stratigraphy of the Ste. Genevieve and Chester formations of southern Indiana: The Edwards Letter Shop, Ann Arbor, 105 p.

MAPLES, C. G., AND WATERS, J. A., 1987, Redefinition of the Meramecian/Chesterian boundary (Mississippian): Geology, v. 15, p. 647–651.

MCFARLAN, A. C., SWANN, D. H., WALKER, F. H., AND NOSOW, E., 1955, Some old Chester problems: Correlations of lower and middle Chester formations of western Kentucky: Lexington, Kentucky Geological Survey Series IX, Bulletin 16, 37 p.

MERKLEY, P. A., 1991, Origin and distribution of carbonate eolianites in the Ste. Genevieve Limestone (Mississippian) of southern Indiana and northwestern Kentucky: Unpublished M.S. Thesis, Indiana University, Bloomington, 147 p.

MITCHUM, R. M., JR., AND VAN WAGONER, J. C., 1991, High frequency sequences and their stacking patterns: Sequence-stratigraphic evidence of high-frequency eustatic cycles: Sedimentary Geology, v. 70, p. 131–160.

Montanez, I. P., and Read, J. F., 1992, Eustatic control on dolomitization of cyclic peritidal carbonates: Evidence from the Early Ordovician Knox Group, Appalachians: Geological Society of America Bulletin, v. 104, p. 872–886.

Off, T., 1963, Rhythmic linear sand bodies caused by tidal currents: American Association of Petroleum Geologists Bulletin, v. 47, p. 324–341.

Potter, P E., 1962, Late Mississippian sandstones of Illinois: Champaign, Illinois State Geological Survey Circular 340, 36 p.

Pryor, W. A., Lamborg, A. D., Roberts, M. J., Tharp, T. C., and Wilsey, W. M., 1990, Geologic controls on porosity in Mississippian limestone reservoirs in the Illinois basin, *in* Leighton, M. W., Kolata, D. R., Oltz, D. F., and Eidel, J. J., eds., Interior Cratonic Basins: Tulsa, American Association of Petroleum Geologists Memoir 51, p. 329–359.

Read, J. F., 1974, Calcrete deposits and Quaternary sediments, Edel Province, Shark Bay, Western Australia, *in* Logan, B. W., Read, J. F., Hagan, G. M., Hoffman, P., Brown, R. G., Woods, P. J., and Gebelein, C. D., eds., Evolution and Diagenesis of Quaternary Carbonate Sequences, Shark Bay, Western Australia: Tulsa, American Association of Petroleum Geologists Memoir 22, p. 250–282.

Read, J. F., 1995, Overview of carbonate platform sequences, cycle stratigraphy and reservoirs in greenhouse and icehouse worlds, *in* Read, J. F., et al., eds., Milankovitch Sea-Level Changes, Cycles, and Reservoirs on Carbonate Platforms in Greenhouse and Icehouse Worlds: Tulsa, SEPM (Society for Sedimentary Geology) Short Course Notes 35, 102 p.

Read, J. F., and Horbury, A. D., 1993, Eustatic and tectonic controls on porosity evolution beneath sequence-bounding unconformities and parasequence disconformities on carbonate platforms, *in* Horbury, A. D. and Robinson, A. G., eds., Diagenesis and Basin Development: Tulsa, American Association of Petroleum Geologists Studies in Geology No. 36, p. 155–197.

Rexroad, C. B., and Fraunfelter, G. H., 1977, Upper Mississippian conodonts and boundary relations in southern Illinois: Guidebook for post-meeting field trips, 11th Annual Meeting, North-Central Section, Geological Society of America, v. 2, p. 81–103.

Roberts, J., Claoue-Long, J., Jones, P. J., and Foster, C. B., 1995, SHRIMP zircon age control of Gondwanan sequences in Late Carboniferous and Early Permian Australia, *in* Non-biostratigraphical Methods of Dating and Correlation: London, Geological Society of London Special Publication No. 89, p. 145–174.

Ross, C. A., and Ross, J. R. P., 1988, Late Paleozoic transgressive-regressive deposition, *in* Wilgus, C. W., Posamentier, H. W., Ross, C. A., and Kendall, C. G. St. C., eds., Sea-Level Changes: An Integrated Approach: Tulsa, SEPM (Society for Sedimentary Geology)Special Publication 42, p. 227–247.

Ruddiman, W. F., Raymo, M., and Macintyre, A., 1986, Matuyama 41,000-year cycles: North Atlantic Ocean and Northern Hemisphere ice-sheets: Earth and Planetary Science Letters, v. 80, p. 117–129.

Sable, E. G., and Dever, G. R., 1990, Mississippian Rocks in Kentucky: Washington, D.C., United States Geological Survey Professional Paper 1503, 125 p.

Schlager, W., 1981, The paradox of drowned reefs and carbonate platforms: Geological Society of America Bulletin, v. 92, p. 197–211.

Scotese, C. R., and McKerrow, W. S., 1990, Revised world maps and introduction, *in* McKerrow, W. S. and Scotese, C. R., eds., Paleozoic Paleogeography and Biogeography: London, Geological Society of London Memoir 12, p. 1–24.

Seyler, B., 1986, Aux Vases and Ste. Genevieve Formations—A Core Workshop and Field Trip Guidebook: Champaign, Illinois Geological Society and Illinois State Geological Survey, 34 p.

Smith, L. B., 1996, High-resolution sequence stratigraphy of Late Mississippian (Chesterian) mixed carbonates and siliciclastics, Illinois basin: Unpublished Ph.D. Dissertation, Virginia Tech, Blacksburg, 146 p.

Smith, L. B., and Read, J. F., 1995, Stratigraphic evidence for increasing amplitude 4th-order eustasy during Late Mississippian greenhouse to icehouse transitions, Illinois basin (abs.): Geological Society of America Abstracts with Programs, v. 27, no. 6, p. A-332.

Smith, L. B., Al-Tawil, A. A., and Read, J. F., 1995, Regional, interbasinal fourth-order sequence stratigraphy, Late Mississippian, Illinois to Appalachian basins (abs.): Tulsa, American Association of Petroleum Geologists Official Program, v. 4, p. 90A.

Swann, D. H., 1963, Classification of Genevievian and Chesterian (Late Mississippian) rocks of Illinois: Champaign, Illinois State Geological Survey Report of Investigations 216, 91 p.

Swann, D. H., 1964, Late Mississippian rhythmic sediments of Mississippi Valley: American Association of Petroleum Geologists Bulletin, v. 48, p. 637–658.

Swann, D. H., and Atherton, E., 1948, Subsurface correlations of lower Chester strata of the Eastern Interior basin: Journal of Geology, v. 56, p. 269–287.

Vaughn, R. D., and Adams, A. E., 1984, Chadian and Arundian sedimentation in the Furness and Millom areas, South Cumbria (abs.), *in* European Dinantian Environments, 1st Meeting: Manchester, Open University, p. 120–123.

Walkden, G. M., 1987, Sedimentary and diagenetic styles in late Dinantian carbonates of Britain, *in* Miller, J., Adams, A. E., and Wright, V. P., eds., European Dinantian Environments, Chichester, John Wiley and Sons, p. 131–155.

Walls, R. A., Harris, W. B., and Nunan, W. E., 1975, Calcareous crust (caliche) profiles and early subaerial exposure of Carboniferous carbonates, northeastern Kentucky: Sedimentology, v. 22, p. 417–440.

Weber, L. J., Sarg, J. F., and Wright, F. M., 1995, Sequence stratigraphy and reservoir delineation of the Middle Pennsylvanian (Desmoinesian), Paradox basin and Aneth field, southwestern USA, *in* Read, J. F., et al., eds., Milankovitch Sea-Level Changes, Cycles, and Reservoirs on Carbonate Platforms in Greenhouse and Icehouse Worlds: Tulsa, SEPM (Society for Sedimentary Geology) Short Course Notes No. 35, 81 p

Weller, J. M., and Sutton, A. H., 1940, Mississippian border of Eastern Interior basin: American Association of Petroleum Geologists Bulletin, v. 24, p. 765–858.

Witzke, B. J., 1990, Paleoclimatic constraints for Paleozoic paleolatitudes of Laurentia and Euramerica, *in* McKerrow, W. S. and Scotese, C. R., eds., Paleozoic Paleogeography and Biogeography: London, Geological Society of London Memoir No. 12, p. 57–74.

Zuppann, C. W., 1993, Complex oolite reservoirs in the Ste. Genevieve Limestone (Mississippian) at Folsomville field, Warrick Co., Indiana, *in* Keith, B. D. and Zuppann, C. W., eds., Mississippian Oolites and Modern Analogs: Tulsa, American Association of Petroleum Geologists Studies in Geology No. 35, 265 p.

Zuppann, C. W., and Keith, B. D., 1988, Geology and petroleum production of the Illinois basin, volume 2: Bloomington, Indiana Geological Survey, and Lexington, Kentucky Geological Survey, 272 p.

APPENDIX 1. OUTCROP AND CORE LOCATIONS

The following are the outcrop and core locations for the data in the cross sections in Figure 4. Abbreviations: Alt.—alternate route, Co.—county, Ill.—Illinois, Ind.—Indiana, Ky.—Kentucky, S.R.—state road.

1 Core, Madison Co., C-141, Illinois State Geological Survey Core Facility, Champaign, Ill.

2 Core, Madison Co., C-140, Illinois State Geological Survey Core Facility, Champaign, Ill.

4 Core, Union Co., C-13619, Illinois State Geological Survey Core Facility, Champaign, Ill.

5 Outcrop, Anna Quarry, East Side of Anna, Ill., northwest of junction of S.R. 51 and S.R. 146

6 Outcrop, I-57, milepost 27.4, south of Anna, Ill., exit

7 Outcrop, Cypress Quarry, Columbia Quarries Inc., S.R. 37, 3 mi (about 5 km) south of Cypress, Ill.

8 Core, Pope County, C-46, Illinois State Geological Survey Core Facility, Champaign, Ill.

10 Core, Hardin Co., C-12781, Illinois State Geological Survey Core Facility, Champaign, Ill.

12 Core, Hardin Co., C-12782, Illinois State Geological Survey Core Facility, Champaign, Ill.
13 Core, Hardin Co., from Ozark-Mahoning Fluorspar mine near Denton, Ill.
14 Outcrop, Hardin County Materials Company, Pit 2, junction of S.R. 1 and S.R.146, Cave-in-Rock, Ill.
15 Outcrop, Cave-in-Rock Quarry, Martin-Marietta, on county roads east-northeast of Cave-in-Rock, Ill.
16 Outcrop, Three Rivers Quarry, Martin-Marietta, U.S. 60, 6.25 mi (10 km) northeast of Smithland, Ky.
17 Outcrop, Fredonia Quarry, Denny and Simpson, U.S. 641, 2.5 mi (4 km) south of Fredonia, Ky.
18 Outcrop, Princeton Quarry, Kentucky Stone Company, S.R. 91, 2.75 mi. south of Princeton, Ky.
21 Core, Caldwell Co., C-220, Kentucky Geological Survey Core Facility, Lexington, Ky.
22 Outcrop, S. Hopkinsville Quarry, Rogers Group, U.S. Alt. 41, 4.5 mi (about 7 km) south of Hopkinsville, Ky.
23 Outcrop, Christian Quarries Quarry (abandoned), U.S. 68, 0.5 mi (0.8 km) east of Hopkinsville, Ky.
24 Outcrop, Todd Quarry, Kentucky Stone Company Quarry, U.S. 68, 7 mi (11 km) west of Elkton, Ky.
25 Outcrop, new road cut for U.S. 68 bypass, west of Russelville, Ky.
26 Outcrop, Russelville Quarry and Core, Kentucky Stone Company, S.R. 79, 1 mi (1.6 km) east of Russelville, Ky.
29 Outcrop, road cut at milepost 9.1 on Green River Parkway north of Bowling Green, Ky.
30 Outcrop, Rockfield Quarry, Kentucky Stone Company, U.S. 68, Rockfield, Kentucky
33 Outcrop, road cut on I-65 N at milepost 48 north of Park City, Ky., exit
34 Outcrop, E. Park City Quarry (abandoned), Old Bardstown Road, 2 mi (about 3 km) east of Park City, Ky.
35 Outcrop, Cave City Quarry, Scotty's Paving Company, Route 90, about 5 mi (8 km) east of Cave City, Ky.
36 Outcrop, road cut on I-65 south, 1 mi (1.6 km) south of Munfordville, Ky., exit
37 Outcrop, road cut on I-65 North, 0.5 mi (0.8 km) north of Munfordville, Ky., exit
38 Outcrop, Upton Quarry, Kentucky Stone Company, Quarry Road, Upton, Ky.
40 Core, Hardin Co., C-117, Kentucky Geological Survey Core Facility, Lexington, Ky.
41 Outcrop, Cecelia Quarry, Larry Glass Paving, S.R. 62, 3 mi (about 5 km) west of Cecelia, Ky.
42 Core, Breckenridge Co., C-118, Kentucky Geological Survey Core Facility, Lexington, Ky.
43 Outcrop, Stephensburg Quarry (abandoned), on dirt road 2 mi (about 3 km) north of Stephensburg, Ky.
44 Outcrop, Irvington Quarry, Kentucky Stone Company, S.R. 477, 2 mi (about 3 km) northwest of Irvington, Ky.
45 Outcrop, Battletown Quarry, Kosmos Cement Company, S.R. 228, 0.5 mi (0.8 km) northwest of Battletown, Ky.
46 Outcrop, Cape Sandy Quarry, Pit 1, Mulzer Crushed Stone Inc., 4 mi (about 6.5 km) east of Alton, Ind.
47 Core, Harrison Company, C-339, Indiana Geological Survey Core Facility, Bloomington, Ind.
48 Core, Perry Company, C-303, Indiana Geological Survey Core Facility, Bloomington, Ind.
49 Outcrop, Tower Quarry, Mulzer Crushed Stone Company, 2 mi (about 3 km) north of Leavenworth, Ind.
50 Outcrop, road cut at milepost 99 on I-64, 7 mi (about 11 km) west of Corydon, Ind., exit
51 Outcrop, Corydon Quarry, Corydon Crushed Stone Inc., 2.5 mi (4 km) northwest of Corydon, Ind.
52 Outcrop, Robertson Quarry, Robertson Crushed Stone Inc., S.R. 64, 1 mi (1.6 km) west of Depauw, Ind.
56 Core, Crawford Co., C-626, Indiana Geological Survey Core Facility, Bloomington, Ind.
57 Core, Orange Co., C-338, Indiana Geological Survey Core Facility, Bloomington, Ind.
60 Core, Orange Co., C-651, Indiana Geological Survey Core Facility, Bloomington, Ind.
62 Outcrop, W. Paoli Quarry, Cave Quarries Inc., U.S. 150, 3.5 mi (about 5.5 km) northwest of Paoli, Ind.
63 Core, Martin Co., C-214, Indiana Geological Survey Core Facility, Bloomington, Ind.
64 Outcrop, Mitchell Quarry, Rogers Group, S.R. 60, 5 mi (8 km) west of Mitchell, Ind.
65 Core, Lawrence Co., C-340, Indiana Geological Survey Core Facility, Bloomington, Ind.
67 Core, Martin Co., C-797, Indiana Geological Survey Core Facility, Bloomington, Ind.
68 Outcrop, Sieboldt Quarry, Rogers Group, county roads, 5 mi (8 km) north of Oolitic, Ind.
69 Core, Greene Co., C-487, Indiana Geological Survey Core Facility, Bloomington, Ind.
70 Core, Monroe Co., C-672, Indiana Geological Survey Core Facility, Bloomington, Ind.
71 Outcrop, Bloomington Quarry, Rogers Group, Oard Road, 4 mi (about 6.5 km) west of Bloomington, Ind.
73 Core, Clay Co., C-444, Indiana Geological Survey Core Facility, Bloomington, Ind.
74 Outcrop, Putnamville Quarry, Kentucky Stone Company, S.R. 343, 1 mi (1.6 km) west of Cloverdale, Ind.
75 Core, Putnam Co., C-105, Indiana Geological Survey Core Facility, Bloomington, Ind.

CONTROLS ON STRATIGRAPHIC ARCHITECTURE OF ICEHOUSE MIXED CARBONATE-SILICICLASTIC SYSTEMS: A CASE STUDY FROM THE HOLDER FORMATION (PENNSYLVANIAN, VIRGILIAN), SACRAMENTO MOUNTAINS, NEW MEXICO

EUGENE C. RANKEY, STEVEN L. BACHTEL, AND JONATHAN KAUFMAN*

Exxon Production Research Company, P.O.Box 2189, Houston, Texas 77252-2189

ABSTRACT: The stratal architecture of the Holder Formation (Pennsylvanian, Virgilian) in the Sacramento Mountains, New Mexico, reveals the complexities of mixed carbonate-siliciclastic depositional systems. Thirty-three measured sections, correlated by physical tracing, and three research cores and well logs provide an extensive database for examination of two- and three-dimensional stratigraphic architecture. In the Holder Formation, carbonate and siliciclastic facies are arranged into 22 shallowing-upward high-frequency sequences, each bounded by either a surface of subaerial exposure developed on subtidal strata (an abnormal subaerial-exposure surface) or its correlative conformity. Vertical facies successions within individual high-frequency sequences vary significantly across the narrow shelf, as shown in the late 1960s by James Lee Wilson. High-frequency sequence boundaries cross lateral facies transitions and extend down paleoslope; such boundaries are the highest-resolution, correlatable chronostratigraphic surfaces in this succession.

Subaerial exposure of subtidal strata, fluvial incision through marine strata, and major downward and basinward facies shifts indicate that relative changes in sea-level significantly influenced stratal architecture. Measured minimum amplitudes of relative falls in sea level are as great as 32 m and exceed measured thicknesses of high-frequency sequences, suggesting that accommodation space was not completely filled. Stratigraphic response to relative changes in sea level was controlled by spatial and temporal changes in sediment supply and production and shelf position. Relative lows in sea level and the early parts of relative rises in sea level were characterized by siliciclastic sediment bypass of the emergent shelf and deposition and onlap in more downdip positions. Flooding of the shelf is indicated by a vertical facies transition from alluvial siliciclastics to marine carbonates and shales. During relative highs in sea level, shelf-wide carbonate sedimentation dominated, although localized deltaic siliciclastics were deposited on the shelf during the early part of the relative high. Responses to relative changes in sea level were mediated by shelf position, and strata at different shelf positions reflect distinct parts of the sea-level curve.

Syndepositional structural deformation affected stratigraphic architecture at several scales through its influence on siliciclastic sediment supply, subsidence, and uplift; such deformation accounts for some of the considerable vertical and lateral complexity in the system as first described by J. L. Wilson. A relatively steep depositional gradient and depositional topography during deposition of the lower part of the Holder Formation (partly controlled by La Luz anticline) led to distinct lateral changes in facies and laterally compressed facies belts; the less pronounced topography and gradient that existed during deposition of the upper Holder Formation, because of in-filled topography and the waning influence of La Luz anticline, resulted in more laterally continuous sheetlike deposits.

INTRODUCTION

Recent advances in carbonate sequence stratigraphy have highlighted secular variations in characteristics of thickness patterns, facies-tract distribution, subaerial-exposure surfaces, and larger-scale depositional geometries within depositional sequences (Wilson, 1975; Handford and Loucks, 1993; Read et al., 1995; Lehrmann and Goldhammer, this volume). Recognition of these different styles of sequence architecture has led to enhanced understanding of the controls on the origin of stratigraphic sequences. In general, most studies recognize two end-member types of sequences, "icehouse" and "greenhouse," in which depositional style is interpreted to be related to the character of eustatic changes. Icehouse sequences are characterized by evidence for pronounced, high-amplitude relative changes in sea level, including facies offsets and subaerial exposure of subtidal strata ("abnormal subaerial exposure"—Van Wagoner, 1995) (Wilson, 1967; Goldstein, 1988a, b; Soreghan, 1994; Read et al., 1995; Rankey, 1997). In contrast, greenhouse sequences include well-developed parasequences and facies tracts and contain little evidence for large facies offsets, suggesting less pronounced, lower-amplitude relative changes in sea level (Koerschner and Read, 1989; Goldhammer et al., 1990, 1993; Read et al., 1995).

These broad characterizations successfully capture important contrasts in the architecture of stratigraphic sequences, yet techniques for evaluating icehouse and greenhouse sequence stratigraphy have not evolved to account for these contrasts. For example, icehouse sequences are commonly high-frequency sequences, bounded by subaerial-exposure surfaces developed on subtidal strata, suggesting unfilled accommodation. In these sequences, then, the thickness of a high-frequency sequence is not indicative of the water-depth changes during development of the sequence (e.g., Heckel, 1977; Giannini and Simo, 1996). In contrast, greenhouse sequences only rarely show evidence for subaerial exposure of subtidal strata, and most interpretations suggest that all available accommodation was filled (e.g., Koerschner and Read, 1989; Osleger and Read, 1993; Goldhammer et al., 1993). Because of this simple contrast, standard analytical techniques, such as the Fischer plot (Fischer, 1964; Read and Goldhammer, 1988; Goldhammer et al., 1990), developed for greenhouse systems, may provide ambiguous results if applied to icehouse settings (cf. Giannini and Simo, 1996; Grammer et al., 1996).

The purposes of this paper are (1) to describe and interpret the depositional sequence architecture of an icehouse system in which chronostratigraphically significant surfaces were physically traced, (2) to discuss controls on mixed carbonate-siliciclastic sedimentation in this system, and (3) to illustrate complexities of sequence stratigraphic interpretation in this system. Furthermore, this paper highlights techniques for develop-

* *Present address: Sonat Exploration Company, P.O. Box 4792, Houston, Texas 77210-4792*

Advances in Carbonate Sequence Stratigraphy: Application to Reservoirs, Outcrops and Models, SEPM Special Publication No. 63
 ISBN 1-56576-060-3, p. 127–150

ing high-resolution chronostratigraphy in this icehouse system, illustrates the potential complicating effects of syndepositional deformation on architecture of mixed carbonate-siliciclastic systems, and describes the significant influence of shelf position on stratal patterns in icehouse systems.

Background and Geologic Setting

In the Sacramento Mountains of New Mexico, the Holder Formation (Pennsylvanian, Virgilian) consists of marine siliciclastics, marine carbonates, and alluvial siliciclastics arranged into generally shoaling-upward successions (Fig. 1; Cline, 1959; Pray, 1959; Wilson, 1967, 1975; Goldstein, 1988a; Carr and Scott, 1990). Marine siliciclastics include calcareous, fossiliferous claystone to siltstone and sandstone. Marine carbonates are peloidal, skeletal, and oolitic wackestone, packstone, grainstone, and phylloid algal boundstone (Wilson, 1967; Toomey et al., 1977a, b; Goldstein, 1988a, b; Toomey, 1991). Alluvial siliciclastics include pedogenically altered claystone to siltstone and lenticular sandstone and conglomerate.

The Holder Formation is underlain by the Beeman Formation (Pennsylvanian, Missourian to Virgilian) (Fig. 1). The Beeman Formation is 0–150 m (0–500 ft) thick (Pray, 1961; Raatz, 1996), and the lower one-third includes interbedded sandstone and shale and the upper two-thirds consists of argillaceous, thin-bedded limestone and thick shale (Pray, 1961; Wilson, 1967). The top of the Beeman Formation is defined as the base of the massive phylloid algal bioherm horizon (Pray, 1961), and our study includes some upper Beeman strata. The Holder-Beeman Formation contact is conformable, but probably time-transgressive, in the southern Sacramento Mountains (Wilson, 1967; Raatz, 1996).

The Laborcita Formation (Pennsylvanian–Permian, uppermost Virgilian to lower Wolfcampian) overlies the Holder Formation (Fig. 1) and includes interbedded shallow-marine siliciclastics, marine carbonates, and alluvial siliciclastics (Otte, 1959; Delgado, 1977; Carr, 1983; Fly, 1985; Rankey, 1996). The Laborcita Formation contains stacked, shallowing-upward successions similar to those in the Holder Formation, but with more siliciclastics and fewer carbonates (Winchester, 1976; Rankey, 1996). In the study area, the Holder-Laborcita Formation contact is an angular unconformity, but to the north, the contact is conformable (Pray, 1959; Otte, 1959). As much as 30 m (100 ft) of the uppermost Holder Formation may have been eroded from the study area by pre-Laborcita erosion.

These strata were deposited on a narrow shelf just west of the Pedernal uplift, part of the Ancestral Rocky Mountains tectonic province (Cline, 1959; Pray, 1959; Wilson, 1967, 1975; Goldstein, 1988a) (Fig. 2). The Pedernal uplift was a high-angle, fault-bounded basement block active in the Late Pennsylvanian–Early Permian and was bordered to the west by the down-dropped Orogrande basin (Pray, 1959; Otte, 1959; Meyer, 1966; Wilson, 1967; Kluth, 1986; Algeo et al., 1991; Soreghan, 1994). Syndepositional deformation during Pennsylvanian–Permian time is indicated by depositional thinning and shallowing across La Luz anticline, an angular relationship between upper Holder strata and

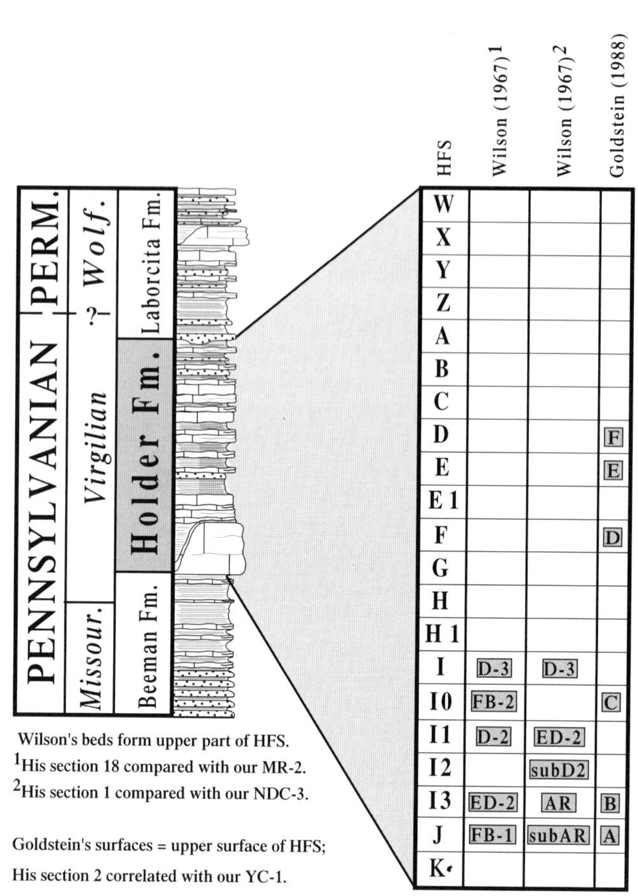

FIG. 1.—Stratigraphic setting and sequence stratigraphy of Holder Formation, Sacramento Mountains, New Mexico. Units in column labeled HFS (high-frequency sequence) are used in this paper. Sections from Wilson (1967) and Goldstein (1988a) are shown for comparison. Wilson's beds form the capping beds of our high-frequency sequences: (1) represents his section 18 compared with our MR-2 and (2) represents his section 1 compared with our NDC-3. Goldstein's surfaces are the upper boundaries of our high-frequency sequences; this figure compares the surfaces in his section 2 with our section YC-1. The Missourian/Virgilian boundary is as defined by Raatz (1996). The base of the Permian is defined by the first appearance of the "isolated-nodular" morphotype of the conodont *Strepognathus "wabaunsensis"* (Davydov et al., 1995), which probably occurs within the Laborcita Formation. Fm. = Formation, Perm. = Permian, Missour. = Missourian, Wolf. = Wolfcampian.

the overlying Laborcita Formation (Virgilian), and by an angular unconformity between the Laborcita Formation and the Abo Formation (Wolfcampian) (Otte, 1959; Pray, 1959; Wilson, 1967, 1972; Delgado, 1977). The study area was located near the equator during deposition of the Virgilian and Holder strata, which were deposited under seasonally wet-dry climatic conditions (Wilson, 1967; Winchester, 1976; Goldstein, 1988a).

Our study includes two transects roughly perpendicular to the Pennsylvanian–Permian shelf margin in Dry and Beeman Canyons (Fig. 2). Along these transects, the Orogrande basin and the Pedernal uplift lie to the west and the east, respectively, but subtle syndepositional structural elements also influenced depositional dip. In particular, because of the northwest plunge of La Luz anticline (Wilson, 1967, 1972), the Beeman Canyon transect (to the south) represents a slightly more updip setting than the Dry Canyon transect (Fig. 2B).

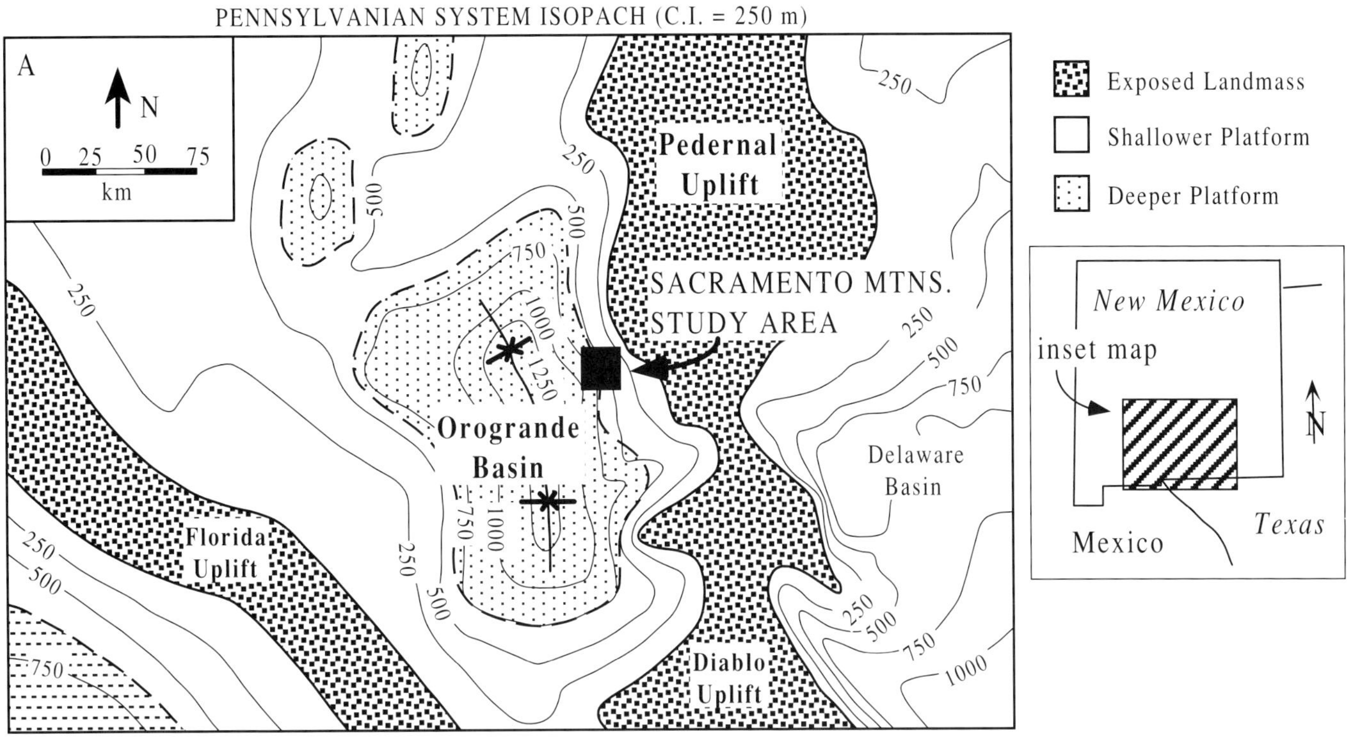

FIG. 2.—(A) Location map of Pennsylvanian Orogrande basin and Pedernal uplift, south-central, New Mexico. Isopach contour interval (C.I.) is 250 m. Modified from Algeo et al. (1991). (B) Location of study area, measured sections, and wells, Dry and Beeman Canyons, Sacramento Mountains, south-central New Mexico. Contour interval is 100 ft (= 30 m).

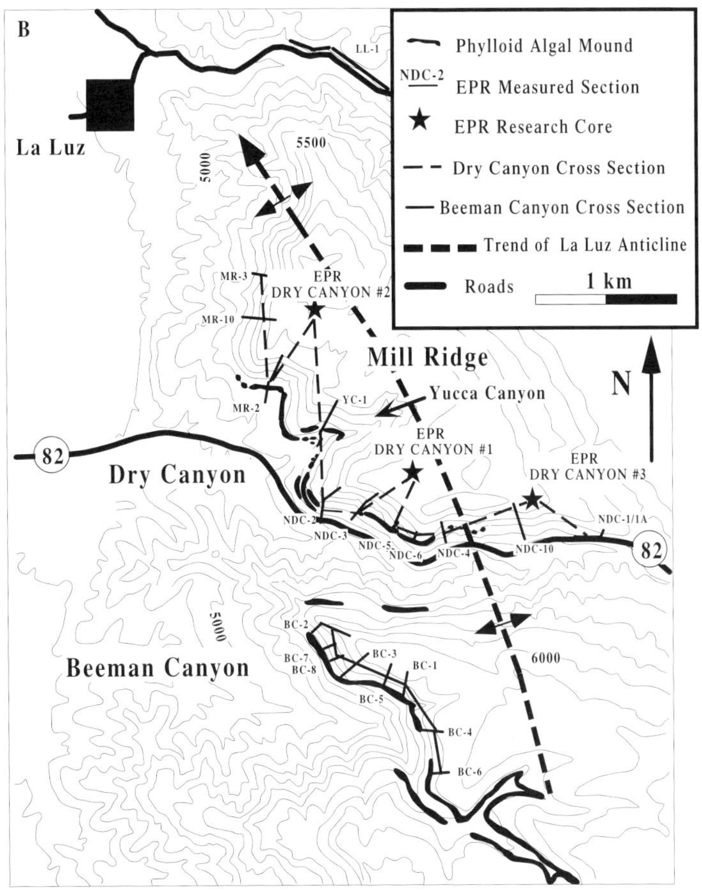

TABLE 1.—LITHOFACIES SUMMARY OF THE HOLDER FORMATION (VIRGILIAN, PENNSYLVANIAN), SACRAMENTO MOUNTAINS, NEW MEXICO.

Facies Association C

Name	Contacts and Associated Facies*	Component Grains†	Sedimentary Structures and Bedding‡	Interpretation
Phylloid algal-skeletal wackestone, packstone to boundstone	Gradational lower, commonly sharp upper contacts Nodular skeletal-peloid w-p-g leopard rock	**Major:** phylloid algae **Minor:** foraminifera, brachiopods, crinoids, bryozoans, rugose corals, echinoid spines, peloids, cement	Massive to wavy bedding Large-scale accretionary foresets as flank beds (R) Shelter pores (C), laminated internal sediment (C) Commonly grades upward to more diverse fauna with more bioclastic debris	Below wave base to wave-influenced; subtidal open marine
Leopard rock boundstone	Sharp lower, sharp upper contacts Nodular skeletal-peloid w-p-g Phylloid algal facies Fossiliferous gray claystone-siltstone	**Major:** encrusting foraminifera, cement, cyanobacteria **Minor:** brachiopods, crinoids, bryozoans	Bulbous hemispheroids, commonly with digitate morphology up to 3 m x 3 m Some layers continuous across several 100s of meters, one leopard rock horizon continuous across >5 km Skeletal-peloidal packstone may fill in between heads common in lows laterally equivalent to mounds	Below wave base; subtidal open marine Commonly flooding units
Oncoid-skeletal wackestone to packstone	Sharp to gradational lower and commonly gradational upper contacts Nodular skeletal w-p-g, phylloid w-p-g syringoporid coral	**Major:** oncoids, encrusting foraminifera **Minor:** foraminifera, brachiopods, crinoids, bryozoans, rugose corals, echinoid spines, peloids, cephalopods	Burrows Nodular bedded (A) Argillaceous seams	Deeper subtidal open marine
Fusulinid-peloid wackestone, packstone, to grainstone	Gradational lower from siltstone Gradational upper to massive contacts Peloid and skeletal w-p-g	**Major:** fusulinids, peloids **Minor:** brachiopods, bivalves, bryozoans, gastropods, phylloid algae, crinoids, quartz silt Thin, nodular (A) to thick bedded (C)	Argillaceous (C) or with thin clay to fine silt seams (C) Parallel to wavy laminated (C)	Subtidal open marine
Skeletal-peloid wackestone, packstone, to grainstone	Gradational to sharp lower and upper contacts to or from phylloid w-p-g, foraminiferid-fusulinid w-p-g, peloid-skeletal w-p-g	**Major:** diverse assemblage of skeletal debris, including brachiopods, fusulinids, gastropods, bivalves, trilobites, pin shells, foraminiferids, dasycladacean algae, phylloid algae, crinoids, echinoids, peloids, syringoporid corals **Minor:** quartz silt to sand, coated grains, fenestrate and ramose bryozoans, ostracodes, *Osagia*, encrusting foraminifera	Thin, nodular (C) to thick bedded (A) Planar-tabular cross-strata (15–80 cm) (R) Parallel and wavy laminations (C) Bellerophontid gastropods common at tops of some units (C)	Subtidal open marine
Peloid-skeletal wackestone, packstone, to grainstone	Gradational to sharp lower and upper contacts to or from phylloid w-p-b, foraminiferid-fusulinid w-p-g, skeletal-peloid w-p-g	**Major:** peloids, skeletal grains, including fusulinids, bryozoans, crinoids, echinoids, foraminiferids, gastropods, bivalves, dasycladacean algae, phylloid algae **Minor:** trilobites, ostracodes, quartz silt, brachiopods, ooids, echinoids	Thin, nodular (C) to thick bedded (A) parallel to wavy laminate (C) Distinct burrows to bioturbated (C) coarsening upward trend (R) Trough and planar-tabular cross-beds (R)	Subtidal open marine
Foraminiferal-peloid wackestone, packstone to grainstone	Gradational to sharp lower and upper contact to or from phylloid w-p-b, fusulinid w-p-g, skeletal-peloid w-p-g	**Major:** foraminiferids, peloids **Minor:** brachiopods, fusulinids, bivalves, bryozoans, gastropods, dasycladacean algae, phylloid algae, crinoids, quartz silt	Thin, nodular (R) to thick bedded (A) parallel to wavy laminate (A) Grainstone trough to low-angle cross-beds	Subtidal open marine
Ooid-skeletal packstone to grainstone	Gradational lower from siltstone Gradational to sharp upper contacts to phylloid w-p-b, fusulinid w-p-g, skeletal-peloid w-p-g	**Major**: ooids **Minor:** crinoids, brachiopods, peloids, foraminifera, quartz sand, bivalves, fusulinids	Thin to thick bedded Large-scale (up to 8 m) accretionary foresets (B) Coarse-fine laminations (C), fining upward (C) Trough to planar-tabular cross-beds (C)	Above wave base; subtidal open marine
Laminated crust	Sharp upper contact Sharp to gradational lower contact commonly developed on facies with open marine fauna	**Major:** enclosed skeletal grains	Wavy laminations (A), autoclastic breccia (C) Rhizoliths (C), root tubules (A) with drab halos Thin (<0.5 cm) to thick bedded Sheet cracks (C), circumgranular cracks (C)	Subaerial exposure; Paleosol

TABLE 1. — CONTINUED

	Name	Contacts and Associated Facies*	Component Grains†	Sedimentary Structures and Bedding‡	Interpretation
Facies Association B	Gray claystone-siltstone with fossils	Gradational lower contact from red siltstone Sharp lower from laminated crust Gradational to sharp upper to peloid-skeletal w-p-b, fusulinid w-p-g, skeletal-peloid w-p-g	Minor: brachiopods, fusulinids, gastropods, bivalves, ostracodes, fenestrate bryozoans, plant fragments, Orbiculoides	Platy to fissile bedding (A), thin-bedded parallel laminations (A), ripples (<5 cm amplitude) (R) Fewer fossils, more plant fragments upwards in many units Coarsening upward trend over 5-20 m (C)	Shallow to deeper marine; some delta front
	Sandstone to conglomerate	Sharp, commonly erosional base incising into gray clay-siltstone (with fossils) Sharp to gradational upper contact, into marine carbonate or fossiliferous claystone-siltstone	Matrix-sandy portions: arkosic arenite (poorly to moderately sorted subangular to subrounded) Fine silt to coarse sand; Q, B, M, C, F, fusulinids, brachiopods, bivalves Matrix may also be calcareous Clasts: well-rounded, poorly sorted; include, chert, quartzite, some carbonate	Trough to planar-tabular cross-strata (to 3 m) (A) Hummocky cross-strata (R) in finer grained, upper portions of conglomerates and in sandstone units Fining-upward trough sets (C) Wave and current ripple cross-laminae (<1 cm) (C) Gutter casts (R)	Delta stream Mouth bar
Facies Association A	Red claystone-siltstone nonfossiliferous	Sharp upper contact overlain by skeletal and peloid w-p-g or green shale Sharp to gradational lower contact	Well to poorly sorted clay to coarse silt Quartz, feldspar, micas	Platy (R) and subangular blocky structure (A), argillans (C), rhizotubules (A), rhizoliths, circumgranular cracks (C) carbonate glaebules (R to A)	Subaerial exposure; paleosol, alluvial
	Sandstone conglomerate	Sharp commonly erosional base incising to or through red claystone-siltstone or, less commonly, marine carbonate with subaerial exposure at top Sharp to gradational upper, possibly fining upward to red claystone-siltstone	Matrix: arkosic arenite (poorly sorted, subangular to subrounded) Fine silt to coarse sand; Q, B, M, C, F Clasts: well-rounded, poorly sorted; include carbonate, chert, quartzite; to 5 cm diameter	Trough cross-strata (to 3 m) (A) Fining-upward trough sets (C) Current-ripple cross-laminae (<1 cm) (C) Locally significant incision (to 52 m) (R) Gutter casts (R) Plant fragments (C)	Fluvial

*Abbreviations: m = mudstone, w = wackestone, p = packstone, g = grainstone; †Q = quartz, B = Biotite, M = Muscovite, C = Chlorite, F = Feldspar; ‡A = abundant, C = common, R = rare.

This study expands on the exceptional framework developed by previous workers, including J. L. Wilson, L. Pray, and their coworkers and students (e.g., Pray, 1959, 1961; Wilson, 1967, 1972, 1975; Winchester, 1976; Toomey et al., 1977a, b; Dunning, 1978; Goldstein, 1988a, b, 1991; see also Oppel, 1957; Reyer, 1958). James Lee Wilson (1967, 1972, 1975) recognized the cyclicity of the Holder Formation and refined the classic concept of cyclic-reciprocal sedimentation from observations of these strata. These concepts laid much of the conceptual framework for sequence stratigraphy. Many of our interpretations parallel those made by Wilson. The work of Bob Goldstein (Goldstein, 1988a, b, 1991; Goldstein et al., 1993) on the Holder Formation built upon that of Wilson and focused on early diagenesis and the character of the subaerial-exposure surfaces capping Wilson's cycles. These surfaces are central to our sequence stratigraphic correlations. Our study utilizes the insights from these previous studies, but builds upon the earlier work by interpreting the strata in a sequence stratigraphic framework. This study also included the entire Holder Formation, recognized more subaerial-exposure surfaces, described the lower-order stacking patterns, and included data from three research wells drilled adjacent to the outcrops.

In this study, we utilize the phrase *relative changes in sea level* to describe vertical changes in the elevation of shoreline relative to a datum, such as the top of the previous high-frequency sequence at a previously defined position (cf. Rankey and Lehrmann, 1996; Rankey, 1997). We do not utilize the position of the sediment-water interface as a datum because it changes through time. In this context, relative changes in sea level include components of eustatic change, subsidence, and uplift, but not sediment accumulation, which creates only a lateral shift in position of shoreline and changes in water depths. A *relative fall in sea level* thus requires a downward shift in the elevation of shoreline (cf. Wilson, 1967; Goldhammer and Elmore, 1984; Goldstein, 1988a; Christie-Blick, 1991; Van Wagoner, 1995; Goldstein and Franseen, 1995; Rankey and Lehrmann, 1996; Rankey, 1997).

The phrase *abnormal subaerial exposure* is an objective, rock-based term that describes subaerial-exposure features developed on subtidal facies (Van Wagoner, 1995), subaerial-exposure features extending down paleoslope in a marine succession, or vadose meteoric diagenetic features in subtidal strata. With this definition, abnormal subaerial exposure cannot result from vertical accretion to sea level ("normal" subaerial exposure; e.g., Shinn, 1973; Enos, 1977; Harris, 1979), but instead requires a relative fall in sea level (cf. Wilson, 1967; Goldhammer and Elmore, 1984; Hardie et al., 1986; Goldstein, 1988a; Goldstein et al., 1993; Rankey, 1997). As such, these surfaces contain important process information. That "abnormal" subaerial-exposure surfaces are common in the Holder Formation demonstrates that high-amplitude relative falls in sea level were common during deposition of these strata, unlike most greenhouse or transitional successions (Wilson, 1975; Read et al., 1995; Lehrmann and Goldhammer, this volume).

FIG. 3.—Marine and alluvial siliciclastic facies. (A) Sandstone; planar to low-angle cross-bedded laminations grading up into ripple cross-laminations, characteristic of waning-flow conditions. Some sandstone beds interpreted as marine also include hummocky cross-stratification and marine fossils. Hammer for scale is 28 cm long. (B) Claystone (recessive weathering) and marine carbonate (resistant weathering) interbedded at foot (approximately 0.3 m) scale. Note that from right to left, several carbonate beds thin as interbedded shales thicken, suggesting progradation of the carbonate system across areas that were previously characterized by deposition of shale. Distance between arrows (at right of photo) is roughly 3 m. (C) Fining-upward succession in sandstone interpreted as fluvial. Packages of similar fining-upward successions commonly become thinner and finer-grained upward. Pen for scale is 15 cm long. (D) Carbonate nodules (N) and prismatic structure (P) in interpreted paleosol. Pedogenic features not illustrated in this photograph include rhizoliths, root tubules, subangular blocky structure, and pedogenic slickensides. Hammer for scale is 28 cm long. (E) Coarse conglomerate, interpreted as fluvial, that fills a channel-form feature. Part of pencil in lower right corner for scale is 15 cm long. (F) Coarse conglomerate filling in channel form that cuts into marine carbonate below. Hammer for scale is 28 cm long.

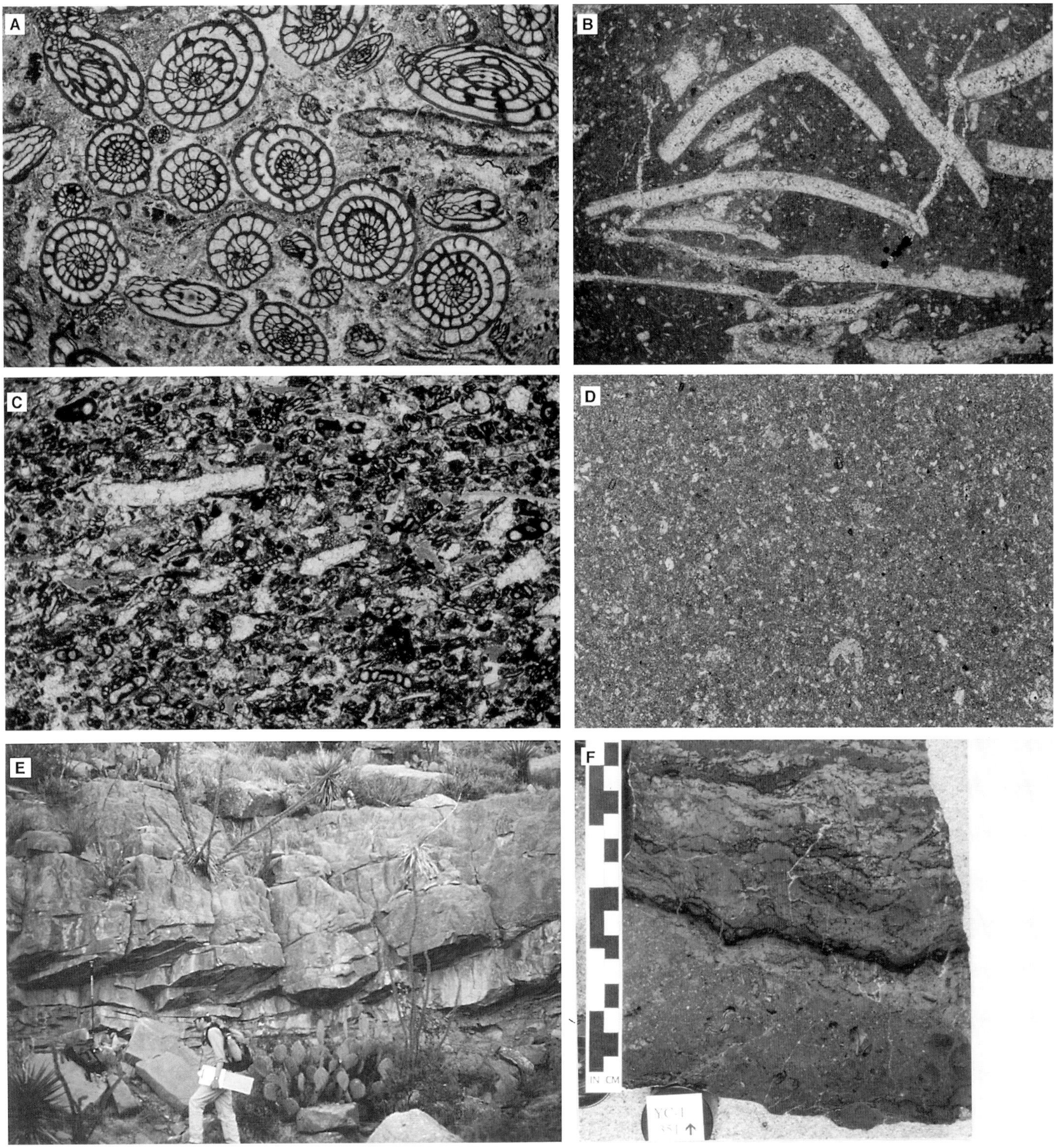

FIG. 4.—Carbonate facies. (A) Fusulinid-rich, skeletal, mud-lean packstone to grainstone. Phylloid algae, foraminifera, peloids, bryozoans, and small crinoids are also present. Long axis represents 11 mm. (B) Phylloid algal wackestone. Phylloids occur in more muddy facies, as in this photograph, or within more grainy, cement-rich facies. In the Holder Formation, very little depositional porosity remains owing to early meteoric diagenesis (see Goldstein, 1988b, 1991). Long axis represents 11 mm. (C) Foraminifera-rich, skeletal-peloid grainstone. Fossils also include an open-marine fauna such as phylloid algae, brachiopods, fusulinids, bryozoans, and crinoids. Other grains are peloids and minor quartz silt. There is no evidence for peritidal deposition (e.g., mudcracks, fenestrae, microbial laminites) in this facies. Long axis represents 11 mm. (D) Skeletal wackestone. Grains include crinoids, minor quartz silt, and bryozoans. Long axis represents 2.65 mm. (E) Shingles of ooid grainstone. The thick, cross-bedded accretionary foresets pass laterally into more nodular and slightly argillaceous packstone. Photograph is from the uppermost part of the Beeman Formation (cf. Fig. 5, section NDC-3, 8–23 ft interval). Person for scale. (F) Root tubules, rhizoliths, and laminated crusts on subaerial-exposure surface.

Methods

Thirty-three detailed measured sections and three research cores with complete wireline log suites make up the database for this study. Surfaces were correlated between measured sections by physical tracing where possible and by use of detailed photographic panoramas, with observations recorded directly on the photographs. The three research cores with 100% core recovery provide continuous information through the Holder Formation, which is not completely exposed in many outcrop areas.

FACIES AND FACIES ASSOCIATIONS

Holder strata are characterized by a diverse array of facies, summarized in Table 1 and illustrated in Figures 3 and 4. In general, most facies can be grouped into one of three facies associations, termed A, B, and C.

1. Facies association A includes fossiliferous gray to green claystone, siltstone, and fine sandstone with marine fossils and hummocky cross-stratification (Figs. 3A–B). Claystone and siltstone are parallel laminated (Fig. 3B), contain brachiopods (including *Orbiculoides*), fusulinids, crinoids, and other skeletal fragments, and are more fossiliferous near their base. These fine-grained siliciclastics contain more abundant plant debris upward and form a general coarsening-upward trend, from claystone to siltstone. Near the top of this coarsening-upward trend, some beds contain thin, sandy or silty layers with fining-upward, partial Bouma sequences (Tc, Td, Te). Siltstone and sandstone in this facies association contain marine fossils (fusulinids, foraminiferids, brachiopods, crinoids, bivalves), marine trace fossils (*Chondrites* and *Thallassinoides*?), trough or hummocky cross-stratification, and wave ripples (Fig. 3A). Sandstones are poorly to moderately sorted and subarkosic to arkosic arenite in composition. The upper contact of the siltstone and sandstone facies is in some places gradational with marine limestone, and lower contacts are sharp to gradational. On the basis of the open-marine fauna, marine trace fossils, hummocky cross-stratification, and gradational contacts with overlying marine carbonate, facies association A is interpreted to represent marine deposition (cf. Carr and Scott, 1990). Parallel-laminated, *Orbiculoides*-bearing, fossiliferous claystone units suggest deeper-water, offshore deposition, and the overlying fossiliferous siltstone or fine sandstone is interpreted to represent lower-shoreface deposits.
2. Facies association B includes red, nonfossiliferous claystone to siltstone and laterally continuous to lenticular sandstone and conglomerate (Fig. 3). Red claystone and siltstone commonly contain evidence for pedogenesis, including rhizoliths, root tubules, well-developed subangular blocky structure, argillans, carbonate glaebules with circumgranular cracks, and an absence of laminations, marine trace fossils, and primary depositional structures (e.g., Fig. 3D). Sandstone and conglomerate units are trough cross-stratified, poorly sorted arkosic to lithic arenite to wacke and consist of several thinning- and fining-upward packages (Fig. 3C). Conglomerate clasts are up to 75 cm (30 in.) long (most are 1–4 cm [0.3–1.5 in.] in diameter) and include limestone, chert, and quartzite (Fig. 3E). These units are either sheetlike or cut sharply through previously deposited strata, including red claystone to siltstone and, much less commonly, marine carbonate (Fig. 3F). Erosional relief at the base of conglomerate beds ranges between 1 and 20 m (3–66 ft). On the basis of the coarse-grained, channel-form deposits and intimately interbedded claystone and siltstone with pedogenic features, facies association B is interpreted to represent alluvial deposition on a coastal plain. The fine-grained components with subangular blocky structure, rhizoliths, and root tubules represent paleosols, whereas the coarser-grained sheetlike to channel-form strata are fluvial deposits.
3. Facies association C includes wackestone, packstone, and grainstone with peloids, skeletal fragments, oncoids, and ooids, and phylloid algal, coral, and foraminiferal-algal boundstone ("leopard rock" of Toomey et al., 1977a) (Fig. 4). Skeletal carbonate-grain types include foraminiferids, fusulinids, crinoids, brachiopods, bivalves, bryozoans, echinoid spines, corals, ostracodes, and dasycladacean algae (Figs. 4A–D). Nonskeletal grain types include peloids, ooids, and oncolites (Fig. 4E). Quartz silt is generally uncommon in carbonate units, although it is locally abundant. The most prominent boundstone features are the phylloid algal bioherms (some of which form cliffs up to 35 m [115 ft] tall) (internal mound facies are detailed in Plumley and Graves, 1953; Wilson, 1967, 1975; Parks, 1977; Toomey et al., 1977a, b; Toomey, 1991). Isolated syringoporid and encrusting foraminiferid boundstone and leopard rock form smaller bioherms as much as 1 m [3.3 ft] thick. Muddier, more nodular wackestone units commonly grade upward and generally coarsen upward into more diversely fossiliferous packstone containing some grainstone with subtle parallel or cross-laminations. The open-marine fauna, oolitic facies, and absence of peritidal features (mudcracks, fenestrae, microbial laminites) in these carbonates indicate deposition in a subtidal, open-marine setting. Oolitic units commonly are cross-laminated and contain accretionary foresets as thick as 2.5 m (8.2 ft) (Fig. 4E). In Beeman Canyon, several capping oolitic units contain low-angle cross-laminations, fine-graded laminations, and keystone vugs suggesting foreshore deposition. These types of features, as well as microbial laminites, laminar or irregular fenestrae, and mudcracks, are virtually absent in Dry Canyon. Fossiliferous marine carbonates at the top of high-frequency sequences are capped by a surface of laminated crusts containing rhizoliths and root tubules, interpreted as forming during subaerial exposure (see Goldstein, 1988a, b, 1991 for examples and detailed descriptions) (Fig. 4F). Our observations suggest that not all reddish zones are subaerial-exposure surfaces (cf. Goldstein, 1988a).

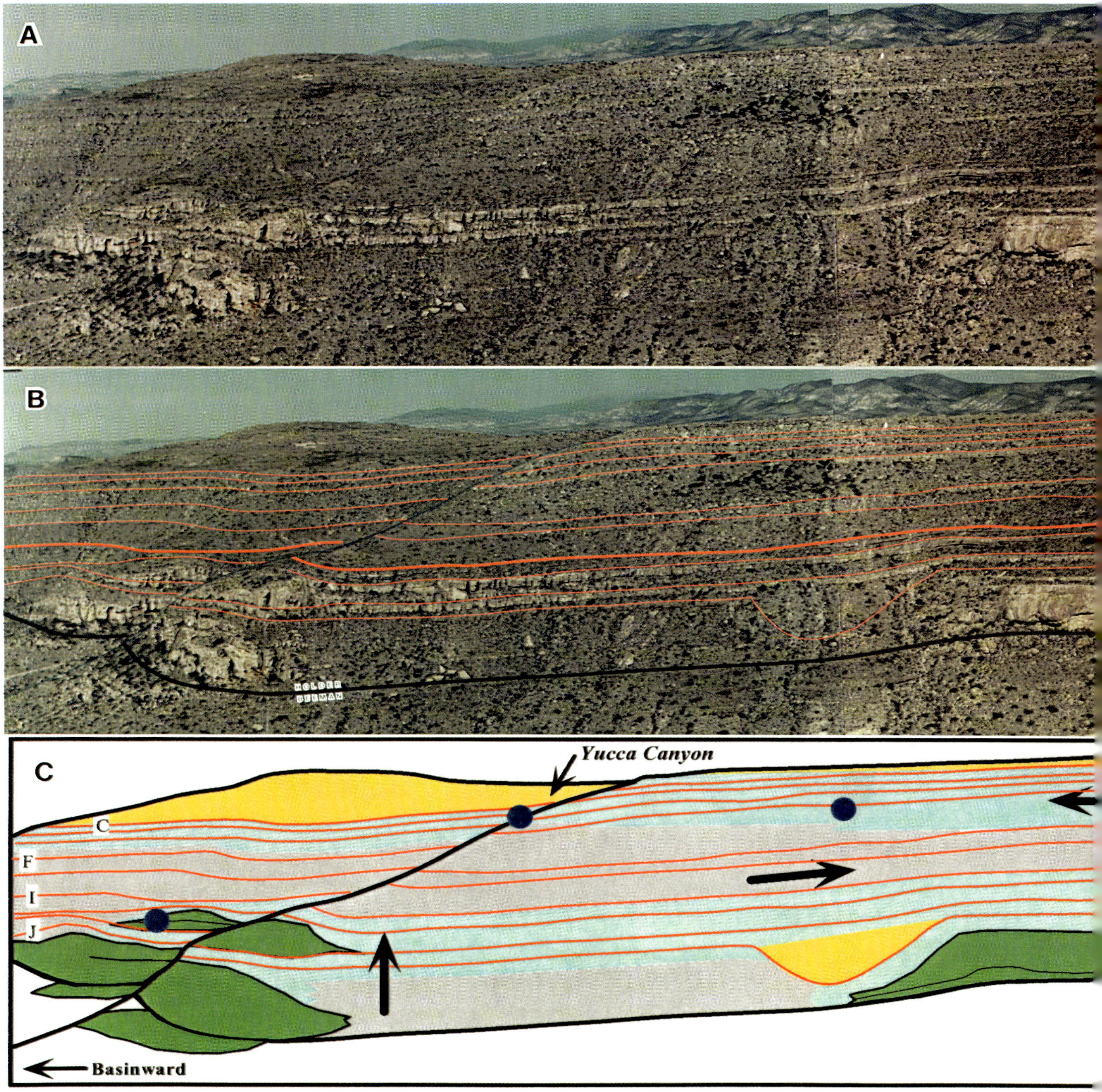

FIG. 6.—Photographic panorama of Dry Canyon study area. Basinward is to the left; landward is to the right. Width represented is approximately 2.5 km. (A) Uninterpreted photographic panorama. (B) Interpreted photographic panorama. Some high-frequency sequence boundaries distinguishable on this panorama are marked with red lines. (C) Schematic interpretive diagram in which some high-frequency sequence boundaries (C, F, I, and J) appear as red lines and are carried across the outcrop. Different colors schematically represent types of vertical successions: blue = carbonate dominated; gray = marine siliciclastic dominated; yellow = mixed non-marine-marine. Green areas are phylloid algal mounds. The large dots represent the most basinward occurrence (i.e., the facies turnaround) of thick, grainy marine car

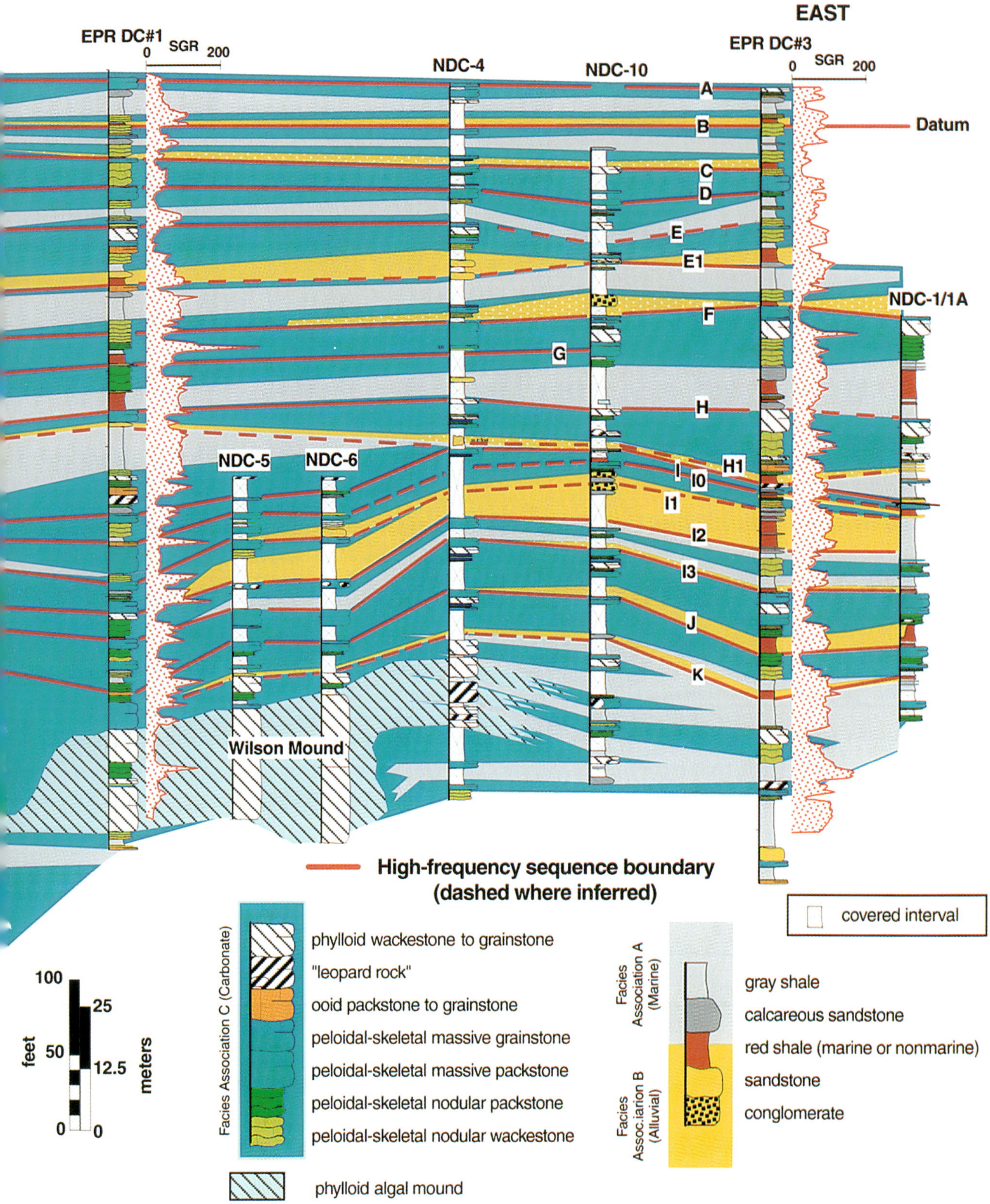
EAST
EPR DC#1
0 SGR 200
NDC-4
NDC-10
EPR DC#3
0 SGR 200
Datum
A
B
C
D
E
E1
F
G
H
H1
I
I0
I1
I2
I3
J
K
NDC-1/1A
NDC-5
NDC-6
Wilson Mound
High-frequency sequence boundary (dashed where inferred)
covered interval
Facies Association C (Carbonate)
phylloid wackestone to grainstone
"leopard rock"
ooid packstone to grainstone
peloidal-skeletal massive grainstone
peloidal-skeletal massive packstone
peloidal-skeletal nodular packstone
peloidal-skeletal nodular wackestone
phylloid algal mound
Facies Association A (Marine)
Facies Association B (Alluvial)
gray shale
calcareous sandstone
red shale (marine or nonmarine)
sandstone
conglomerate
feet
100
50
0
meters
25
12.5
0

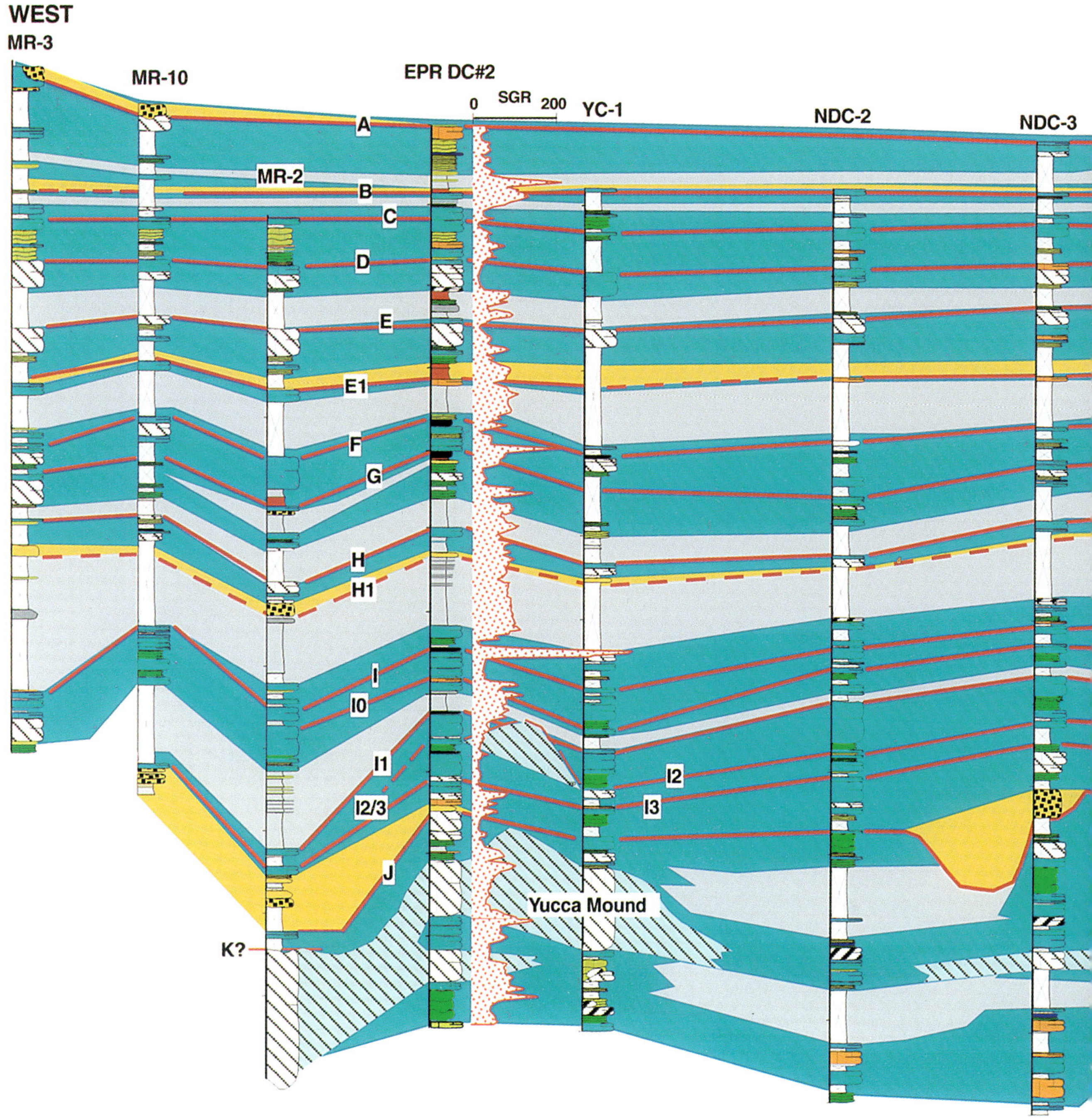

FIG. 5.—Stratigraphic cross section through the Pennsylvanian Holder Formation, Dry Canyon, New Mexico. The cross section covers approximately 3–4 km on a generally landward to basinward (east to west) transect. Data include both outcrop and core descriptions. La Luz anticline extends roughly perpendicular to this cross section at approximately the location of section NDC-10 (see Fig. 1). Gamma-ray logs are indicated at the positions of the wells.

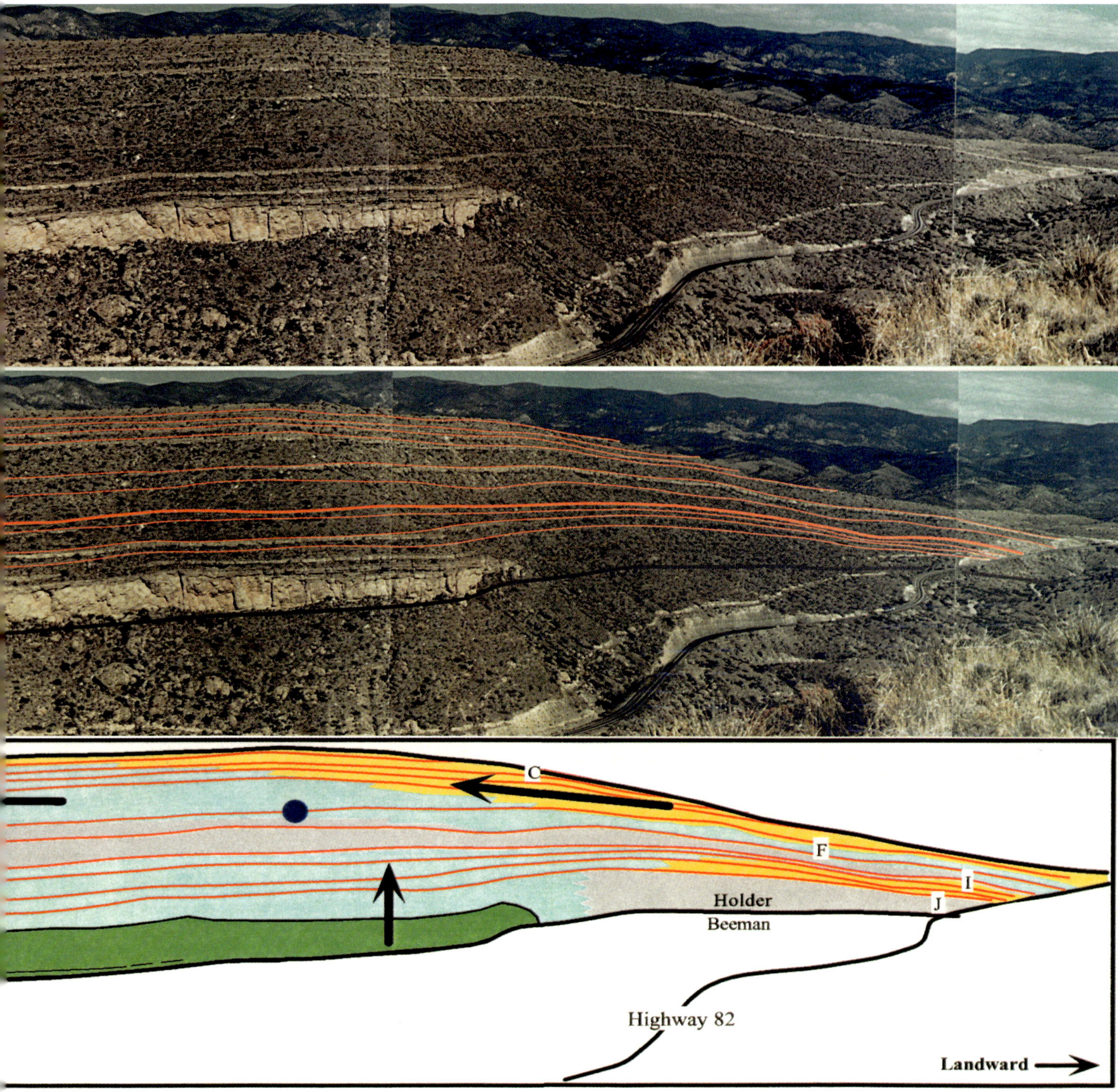

bonate within the individual high-frequency sequences. Basinward of these points, the thick-bedded carbonate passes laterally into more nodular, argillaceous, or muddy carbonate, commonly within several tens of meters. This transition is referred to in the text as the massive to nodular transition. Note the general aggradation (upward-pointing arrows) in the lower part of the section (below surface I), followed by the pronounced landward stepback of the carbonate-dominate facies (right-pointing arrow), then general progradation (left-pointing arrows). These trends are discussed in detail in the text.

HIGH-RESOLUTION SEQUENCE ARCHITECTURE

Facies of the Holder Formation are arranged into parasequences and high-frequency sequences (defined and described below). The Holder high-frequency sequences are comparable in scale to the classic "cyclothems" of many previous workers (e.g., Wanless and Shepard, 1936; Moore, 1964; Wilson, 1967; Heckel, 1977; Watney, 1980; Watney et al., 1989, 1996) and form the highest-resolution, correlatable, chronostratigraphic unit in our study (Figs. 5, 6, 7). We use an informal nomenclature to identify regionally or subregionally extensive high-frequency sequence boundaries on our cross sections, starting with A at the top, down to K near the base. At least four additional surfaces above surface A (named W, X, Y, Z) are present only in the single measured section (LL) that described the uppermost part of the Holder Formation. In this text, we name our sequences based on the designation of the immediately underlying sequence boundary. For example, high-frequency sequence C is underlain by high-frequency sequence boundary C and is capped by high-frequency sequence boundary B.

Parasequences

Individual facies in the Holder Formation stack into shoaling-upward parasequences bounded by a marine-flooding surface interpreted to represent an abrupt increase in water depth (cf. Van Wagoner et al., 1990). Abnormal subaerial-exposure surfaces are present only at the top of the uppermost parasequence in each high-frequency sequence.

Carbonate parasequences, bounded by marine-flooding surfaces, can be defined in all measured sections. Within parasequences, basal, nodular-bedded, argillaceous, fusulinid, oncoid, or phylloid algal packstone to wackestone grades upward into medium- to thick-bedded skeletal-peloid packstone (or less commonly grainstone). This vertical facies transition is interpreted to represent a shallowing-upward trend from lower-energy or deeper-water deposition to higher-energy or relatively shallow water deposition. A sharp contact with the next overlying nodular, argillaceous facies represents the parasequence boundary or marine-flooding surface. Figure 8 shows lateral facies relationships within Yucca Canyon (Fig. 1) and illustrates the apparent "facies mosiac" within high-frequency sequences dominated by carbonate parasequences. Between these closely spaced sections, parasequences cannot be confidently correlated, although high-frequency sequence boundaries can be easily traced. We should note that some of this facies complexity is due to depositional topography along this cross section (Fig. 8), although high lateral facies variability is a consistent theme in much of the Holder strata.

Marine siliciclastic parasequences contain a conformable coarsening-upward succession bounded by marine-flooding surfaces (Mitchum and Van Wagoner, 1991). The bottom part of these parasequences are platy-bedded, parallel-laminated, gray to greenish-gray claystone or siltstone that contains *Orbiculoides* or phosphatic fragments. Bed thickness and grain size increase upward, and beds include sandstone or siltstone with parallel laminations or wave ripples. Some parasequences are capped by fossiliferous, hummocky cross-stratified sandstone of facies association A. A sharp upward transition from the siltstone or sandstone to claystone represents an abrupt deepening and the marine-flooding surface (parasequence boundary). These siliciclastic parasequences may be correlative across the study area (e.g., marine siliciclastics of high-frequency sequences B, E, and F), although not all are continuous because of lapout or erosion (e.g., marine siliciclastics in high-frequency sequences I1 and I). Alluvial siliciclastic parasequences may be defined in Holder strata, although they were not subdivided in this study.

An autogenic origin for many carbonate and siliciclastic parasequences cannot be ruled out because of the absence of abnormal subaerial-exposure surfaces and the limited lateral continuity of some parasequences. The influence of progradation on parasequence development is illustrated in Figure 9. Alternatively, parasequences may reflect minor relative changes in sea level that did not subaerially expose any of the study area (Koerschner and Read, 1989; Watney et al., 1989; Goldhammer et al., 1990; Weber et al., 1995).

High-Frequency Sequences

High-frequency sequences of the Holder Formation form upward-shoaling successions like parasequences but are bounded by abnormal subaerial-exposure surfaces (including fluvial incision of marine strata) or their correlative conformities (Fig. 10) (Wilson, 1967). The sequence-bounding abnormal subaerial-exposure surfaces are laterally continuous and represent significant chronostratigraphic surfaces (Fig. 6), even though the facies that they bound are discontinuous. One means to describe lateral facies change within an individual high-frequency sequence is to characterize the vertical facies succession ("vertical successions" hereafter) at any one point along a shelf transect. We recognize three informal end-member, descriptive types of vertical successions: (1) marine siliciclastic dominated, (2) carbonate dominated, and (3) nonmarine-marine mixed (Fig. 10). Because of lateral changes, however, different vertical successions may be present at different positions along the shelf transect within the same high-frequency sequence. Lateral relationships among end members are discussed below. These vertical successions and lateral relationships are similar to those documented by Wilson (1967), except that our study did not include true basinal strata (e.g., Wilson's Jarilla and Franklin Mountains sections).

1. Marine siliciclastic-dominated vertical successions consist primarily of one or more parasequences with thick marine siliciclastics of facies association A, but in some cases becoming carbonate dominated near the top (Fig. 5; e.g., high-frequency sequence I1; MR-2, 43–66 m [140–216 ft] in height; Fig. 10). These successions generally coarsen upward, from marine claystone to siltstone to marine sandstone, and contain thicker beds upward. Subaerial exposure of subtidal facies is almost always developed on the marine carbonate strata rather than at the top of one of the siliciclastic facies. Marine siliciclastic-dominated vertical succes-

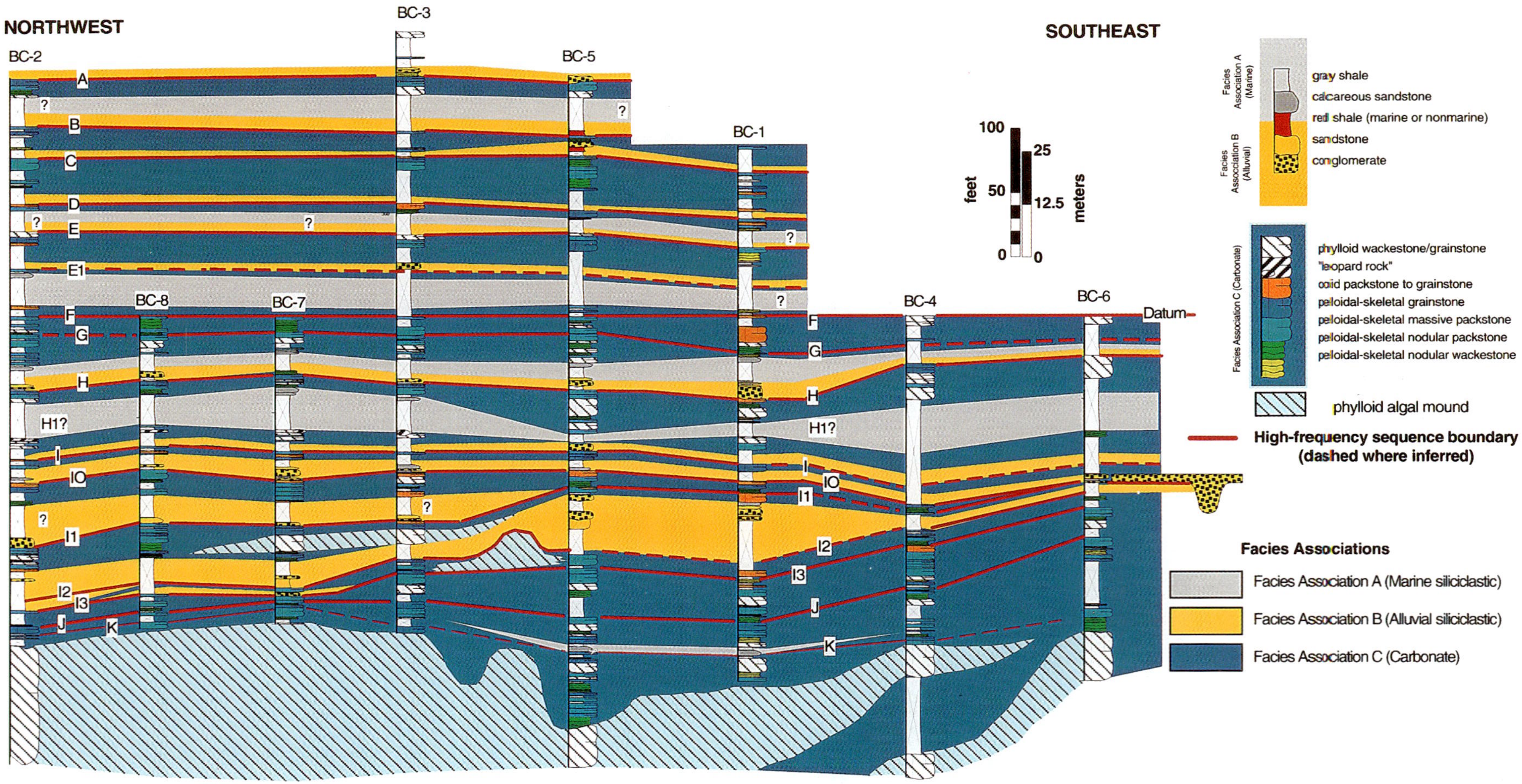

FIG. 7.—Stratigraphic cross section, Pennsylvanian Holder Formation, Beeman Canyon, New Mexico. Data include only outcrop descriptions. This cross section covers approximately 1.25 km in a slightly oblique landward to basinward (southeast to northwest) transect (see Fig. 1). Relative to the Dry Canyon cross section (Figs. 5, 6), this transect lies up structural dip, because of the northward plunge of La Luz anticline. The axis of the anticline lies just east of the cross section.

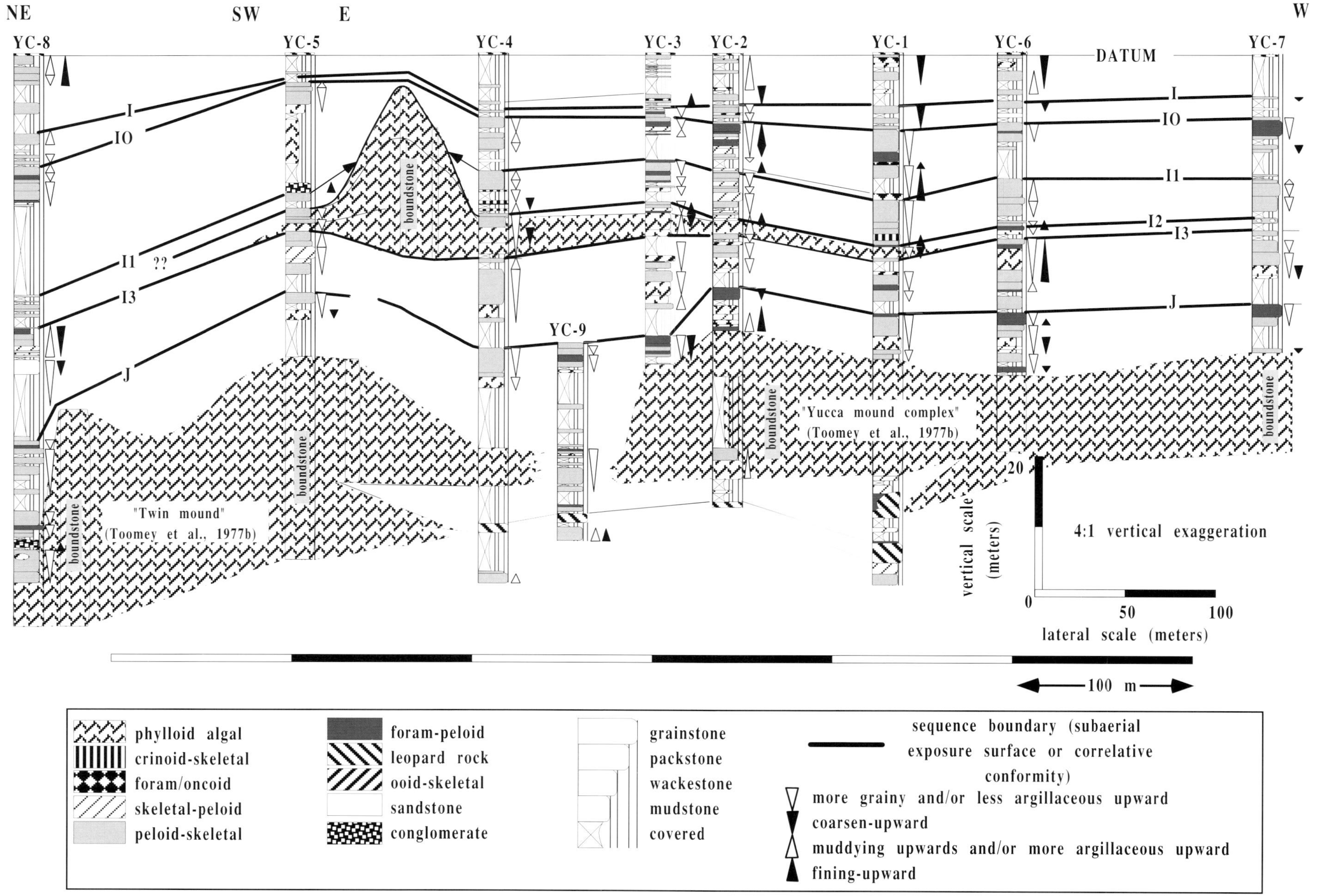

FIG. 8.—Yucca Canyon cross section (see Fig. 2B for location). Sections are closely spaced (average spacing is about 100 m), yet many individual facies and parasequences do not appear to correlate across the area. Correlations were made through physical tracing and from observations made in the field utilizing photographic panoramas. High-frequency sequence boundaries are marked by the labeled bold lines.

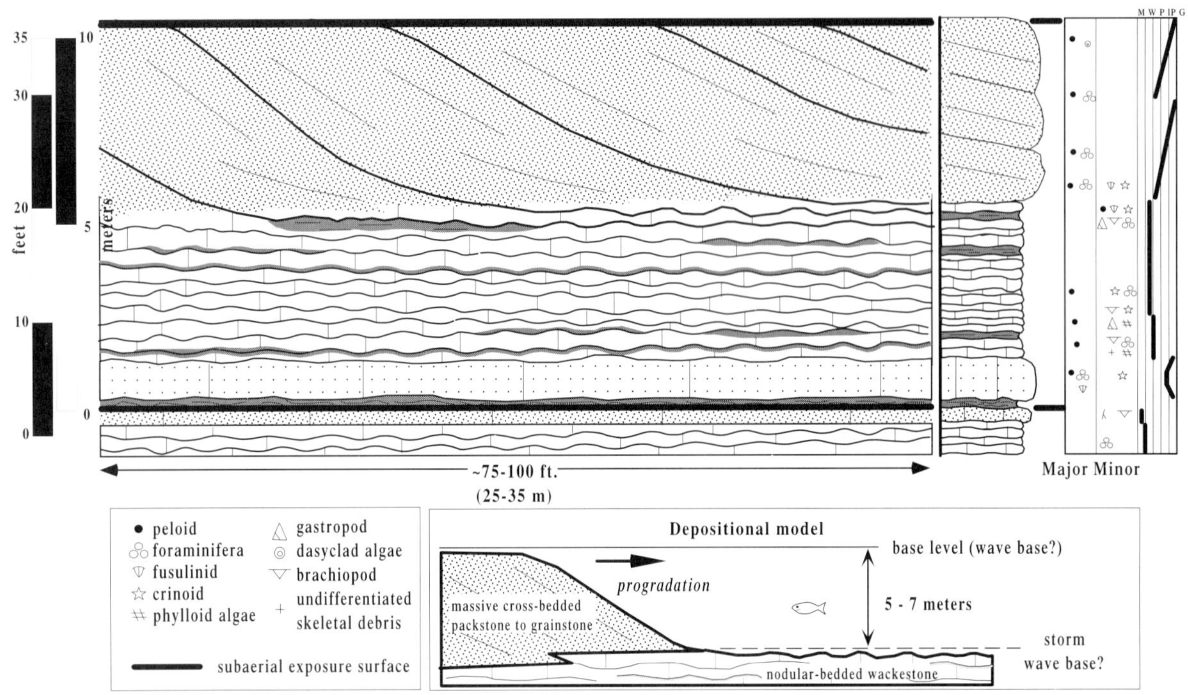

Fig. 9.—Measured section from approximately 0.4 km (0.25 mi) west of tunnel on Highway 82. In the column to the right of the schematic measured section, M = mudstone, W = wackestone, P = packstone, lP = mud-lean packstone, G = grainstone, "Major" refers to constituents that compose a majority of grains, and "Minor" refers to less abundant grain types. At this location, a lower subaerial-exposure surface is overlain by a condensed, hematitic zone and a fossiliferous, marine limestone. This thick-bedded carbonate is in turn overlain by a nodular-bedded, slightly argillaceous, skeletal wackestone that grades up into a cross-laminated thick-bedded carbonate with marine fossils (cf. Goldstein et al., 1991; J. A. French, 1987, unpublished document). Lateral relationships indicate that the nodular wackestone and the thick-bedded, cross-laminated grainstone represent time-equivalent strata formed by the progradation of the cross-laminated carbonate over the nodular-bedded carbonate. Compare this stacking pattern with that shown in Figure 4E.

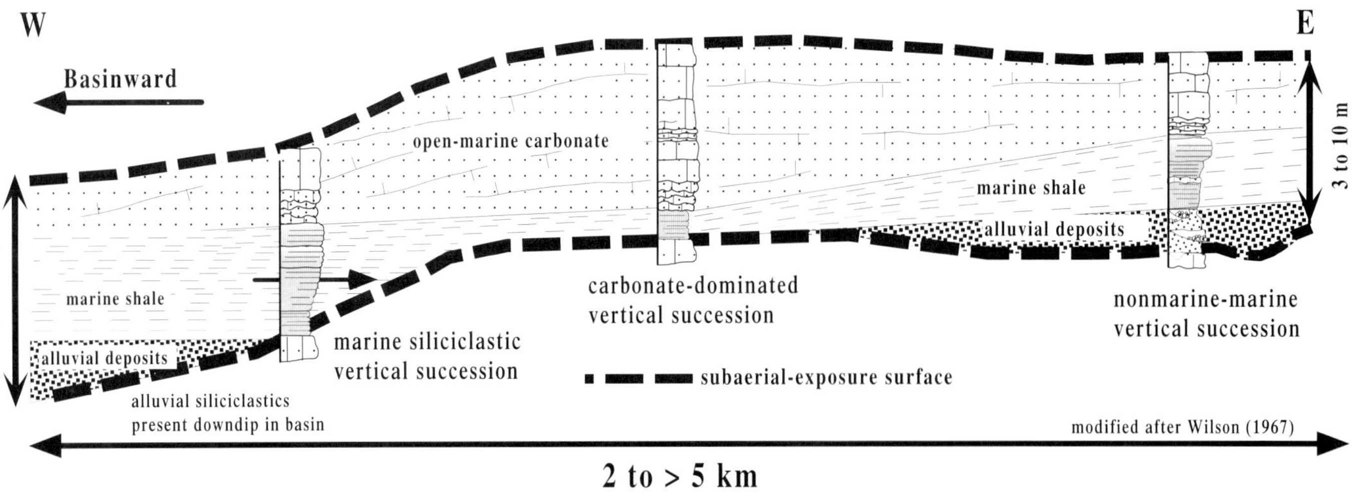

Fig. 10.—Schematic diagram illustrating types of vertical successions and their lateral relationships on a general landward to basinward transect (cf. Wilson, 1967, 1975; Heckel, 1977; Watney et al., 1989, 1996). Note that the internal components of the high-frequency sequences change significantly across the shelf (Wilson, 1967).

sions are common in basinward locations (Fig. 5) where bypassed coarse, alluvial siliciclastics are absent.

2. Carbonate-dominated vertical successions include thin to absent marine siliciclastics overlain by one or more parasequences with marine carbonates of facies association C (Fig. 5; e.g., high-frequency sequence I2, NDC-2, 63–74 m [207–242 ft] in height; Figs. 10, 11A). Flooding units at the base of the carbonate-dominated vertical successions are fossiliferous shales or nodular oncoidal, fusulinid-, or phylloid algal–rich carbonate. Upper units include a diverse open-marine fauna forming fossiliferous packstone, mud-lean packstone, or less commonly grainstone; ooid grainstone is also present locally, especially in Beeman Canyon. The upper surface of these vertical successions is a subaerial-exposure surface developed on subtidal marine carbonate facies. Carbonate-dominated vertical successions are best developed in the central part of the study area, basinward of the area of nonmarine-marine mixed vertical successions (Fig. 10).
3. Nonmarine-marine mixed vertical successions include marine siliciclastic, marine carbonate, and alluvial siliciclastic facies, each of which can be relatively thick (Fig. 5; e.g., high-frequency sequence J, section NDC-1, 15.5–27 m [51–90 ft] in height; Figs. 10, 11B). The base of this vertical succession commonly is formed of several meters of alluvial strata (facies association B) with strongly developed paleosols and fluvial channels. Alluvial facies of the nonmarine-marine mixed vertical succession may be confined to incised valleys that locally erode into the marine carbonates of the previous high-frequency sequence (Fig. 5, west of NDC-3). The alluvial strata are overlain by a flooding surface and fine-grained marine siliciclastics or nodular, muddy marine carbonates. As in the other vertical successions, some of the nonmarine-marine vertical successions include several parasequences, the uppermost of which is capped by an abnormal subaerial-exposure surface. In general, within each high-frequency sequence, the areas with a nonmarine-marine mixed vertical succession lie updip (e.g., Beeman Canyon) and landward (e.g., section NDC 1/1A) of the area in which vertical successions are carbonate dominated (Fig. 10).

Lateral changes in high-frequency sequences above surface G are quite subtle (Figs. 5, 7). In contrast, lateral changes are readily apparent in high-frequency sequences J, I3, I2, and I1 (Figs. 5, 7, 10, 11; Wilson, 1967, 1972). In high-frequency sequence J, basinward areas (MR-10, MR-2; Fig. 5) contain thick alluvial siliciclastics overlain by marine carbonates, representing the nonmarine-marine mixed vertical facies succession. These alluvial siliciclastics onlap the steep depositional topography created by the mounds. Other high-frequency sequences located in a more basinward position contain the marine-siliciclastic–dominated vertical facies succession. Slightly more landward positions (e.g., EPR DC#2, YC-1, NDC-2; Figs. 5, 8, 11A) include only thick carbonates of the carbonate-dominated vertical facies succession. Farther landward positions (EPR DC#3, NDC-1/1A; Figs. 5, 10, 11B) are characterized by alluvial, marine siliciclastic, and carbonate facies, composing the nonmarine-marine mixed vertical facies succession.

Vertical facies successions are distributed differently in high-frequency sequence I3. In this interval, basinward locations include only very thin marine siliciclastics and carbonates, and shelf-margin areas (e.g., EPR DC#2, NDC-2; Fig. 5) have carbonate-dominated vertical facies successions. Both landward (e.g., EPR DC#3, NDC-10; Fig. 5) and more structurally updip positions (e.g., Beeman Canyon; Fig. 7) contain the nonmarine-marine mixed vertical facies succession, with alluvial siliciclastics, marine siliciclastics, and marine carbonates. In contrast to high-frequency sequence J (previously described), the thickest siliciclastics in high-frequency sequence I3 are present in the landward and updip positions. This simple contrast illustrates the lateral and vertical variability in this system.

Interpretations—High-Frequency Sequences

Following Wilson (1967), high-frequency sequences of the Holder Formation can be interpreted in terms of high-frequency relative changes in sea level, which are indicated by the abnormal subaerial-exposure surfaces (some of which extend down paleoslope) and by incised valleys that cut through marine strata. The basal part of each high-frequency sequence overlies an abnormal subaerial-exposure surface or correlative conformity. In the Dry Canyon study area, downdip siliciclastics (alluvial and/or marine) onlap previous topography and the subaerial-exposure surface (Fig. 12A). These onlapping strata are limited to downdip locations and within incised valleys and are interpreted to reflect bypass of the shelf during a lowstand or the early phases of a relative rise in sea level (cyclic-reciprocal sedimentation of Wilson, 1967). Preservation of alluvial siliciclastics in the landward sections to the east in Figure 5 and within incised valleys may be related to alluvial aggradation or progradation during the early part of the relative rise in sea level (Wilson, 1967).

Alternatively, the landward alluvial strata may represent subaerial exposure of siliciclastics that were deposited during the latest highstand. Although we cannot rule out this possibility, it seems intuitive that the preservation of landward alluvial siliciclastics would be favored during periods of increasing accommodation, rather than decreasing accommodation. Similarly, there is commonly a sharp contact between the carbonates and pedogenically altered siliciclastics (Wilson, 1967), although many marine siliciclastic-carbonate contacts are gradational. Bob Goldstein (oral commun., 1997) has noted that subaerial-exposure surfaces are less well developed on carbonates landward of the axis of La Luz anticline, but that the overlying alluvial strata contain well-developed paleosols. This observation suggests that some siliciclastics were supplied to this area prior to subaerial exposure.

Some initial transgressive deposits include thin, muddying-upward wackestone to packstone (containing oncoids, or small syringoporid and leopard-rock bioherms). Most high-frequency sequences, however, coarsen upward locally and are characterized by a basal, shelf-wide, dark gray to green, fossil-bearing claystone and siltstone above the transgressive surface (Fig. 12B). These marine carbonate and siliciclastic deposits

A
J
13
12
B
J
13

are interpreted to represent continued relative rise in sea level to a stillstand in sea level (Wilson, 1967).

The upper part of most high-frequency sequences consists of one or more shallowing-upward carbonate parasequences (Fig. 8) in a succession that becomes less nodular, more diversely fossiliferous, and more grain-dominated upward (Fig. 12C). These subtidal carbonate units extend across the entire shelf, but some thin and become more nodular basinward and are capped by an abnormal subaerial-exposure surface, reflecting a relative fall in sea level. The carbonate deposits and the subaerial-exposure surfaces are interpreted to represent the late part of the relative stillstand to the early part of the relative fall in sea level (Wilson, 1967). Repeated abnormal subaerial-exposure surfaces indicate that available accommodation was not filled during deposition of each high-frequency sequence (cf. Gianinni and Simo, 1996).

One means to quantify the minimum amplitude of the relative falls in sea level is by measuring the vertical extent of subaerial-exposure surfaces down paleoslope and the depth of fluvial incision through marine strata (Wilson, 1967; Goldstein, 1988a; Goldstein and Franseen, 1995; Rankey and Lehrmann, 1996; Rankey, 1997). In this system, depth of incision is as much as 18 m, and the vertical change in the elevation of subaerial-exposure surfaces on marine strata is as much as 32 m (not accounting for compaction; cf. Wilson, 1967; Goldstein, 1988a). Other measurements are slightly less, but are still relatively high (Table 2). The high-amplitude relative changes in sea level are what would be expected during the Pennsylvanian global icehouse climate in which extensive continental glaciers were present (Wanless and Shepard, 1936; Van Siclen, 1958; Merriam, 1964; Wilson, 1967; Heckel, 1977; Veevers and Powell, 1987; Crowley and Baum, 1991), and many of these changes are probably eustatic. Nevertheless, the Pedernal shelf was tectonically active, and some of the relative changes in sea level might reflect syndepositional deformation (Wilson, 1967, 1972).

Fig. 11.—Core photographs of high-frequency sequence J in two cores. Scale on rulers is in inches; the values on top of the boxes are in depths in feet. Note the distinct differences in the facies, even though these cores were taken only approximately 2 km apart (see Fig. 2B). (A) Carbonate-dominated vertical succession from near the shelf margin (EPR Dry Canyon #2). Surface J (indicated by the arrows) overlies foraminiferal-peloid-skeletal packstone with vuggy porosity. The eroded and bored surface is overlain by an approximately 50-cm-thick, fining-upward succession of poorly sorted, matrix-supported conglomerate (clasts include chert and laminated crust) with a medium-silt to coarse-sand matrix, which grades up into a poorly sorted medium-grained siltstone to medium-grained sandstone with flasers and current ripples. This interval is overlain by an approximately 1.3-m-thick (depth, 421.8–418.0 ft) ooid-skeletal grainstone, probably reflecting initial marine flooding. A 15-cm-thick (depth, 418–417.5 ft) bioturbated, argillaceous skeletal wackestone is gradationally overlain by a phylloid-peloid-fusulinid packstone (depth, 417.5–415.1 ft). This unit is sharply overlain by a unit 120 cm (4 ft) thick of muddying-upward peloidal-skeletal packstone to wackestone with a diverse fauna. Above this interval is a peloidal-skeletal packstone to mud-lean packstone unit 1 m (3.3 ft) thick that contains irregular fenestrae and possible root tubules representing surface I3 (depth, 407.7 ft). A thin bivalve lag occurs above this surface. The next approximately 5.4 m (depth, 407.7.0–390.0 ft) includes primarily peloidal-skeletal packstone with abundant foraminifera, crinoids, brachiopods, corals, and phylloid algae. Surface I2 is characterized by a laminated crust and root tubules (cf. Goldstein, 1988a). (B) Nonmarine-marine mixed vertical succession from a landward position (EPR Dry Canyon #3). Surface J is a 0.6-m-thick alteration zone, with laminated crust, rhizoliths, circumgranular cracks, and alveolar (?) textures (not illustrated in this core photo). Sharply overlying this surface (depth, 382.0 ft; marked by an arrow) is red to gray-green, slightly calcareous shale (depth, 382.0–371.0 ft) with carbonate nodules, root tubules, and some parallel laminations. A flooding event is indicated by an approximately 2.4-m-thick interval of slightly to highly argillaceous, phosphatic skeletal packstone-wackestone with brachiopods, fusulinids, Osagia, and crinoids (depth, 371.0–363.3 ft). This argillaceous succession is overlain by burrowed phylloid algal, foraminiferid, crinoid packstone (depth, 363.3–355.0 ft) and phylloid algal and fossil boundstone (depth, 355.0–349.7 ft). The capping bed is fusulinid- and benthic foraminifera–bearing mud-lean packstone to grainstone, which is sharply overlain by a subaerial-exposure surface (depth, 340.3 ft) and a thin (<0.4 m) alluvial(?) red shale. The remainder of the illustrated core is calcareous, fossiliferous, glauconitic siltstone, with intraclasts in the lower half.

STACKING PATTERNS OF HIGH-FREQUENCY SEQUENCES

Stacking patterns can be defined by different stratal attributes, including one-dimensional thickness and facies patterns (e.g., Read and Goldhammer, 1988; Koerschner and Read, 1989; Osleger and Read, 1993; Goldhammer et al., 1993; Saller et al., 1994; Lehrmann and Goldhammer, this volume) and progradational, aggradational, and retrogradational patterns within parasequences or high-frequency sequences on cross sections (e.g., Mitchum and Van Wagoner, 1991; Weber et al., 1995; Watney et al., 1996). Trends in these attributes through time can then be used to define a lower-order stratigraphy (see references above).

One-dimensional stacking-pattern analysis utilizes attributes including thickness, facies, and presence or absence of subaerial exposure to define lower-order stacking patterns. In the Holder Formation, the thicknesses of individual high-frequency sequences vary significantly across the shelf (Fig. 13). This variability, and the presence of unused accommodation, suggests that thickness is not a consistent and diagnostic attribute related solely to long-term accommodation trends (Giannini and Simo, 1996; Grammer et al., 1996). Thickness is in part related to the distribution of facies across the shelf, which is not uniform or systematic (as illustrated below). These ambiguities suggest that one-dimensional analysis of facies or thickness trends in this system can be misleading. Likewise, all of the high-frequency sequences include abnormal subaerial-exposure surfaces, and many include fluvial incision; therefore, these characteristics are nondiagnostic for defining composite sequence boundaries.

There are many means by which two-dimensional stacking patterns can be defined, including characterizing geometric relationships (e.g., Vail et al., 1977; Sarg, 1988; Christie-Blick, 1991) and tracking the position of facies transitions or contacts (Mitchum and Van Wagoner, 1991). For example, Wilson (1967, 1972) noted the offlapping geometry present in the mound horizons of the lower part of the Holder Formation and used the geometry to describe the time-transgressive nature of the bioherms (see also Toomey et al., 1977a, b; Toomey, 1991). Most of the remainder of the Holder is more stratiform, however, and requires utilization of the position of facies transitions or contacts to define stacking patterns. Two such transitions are high-

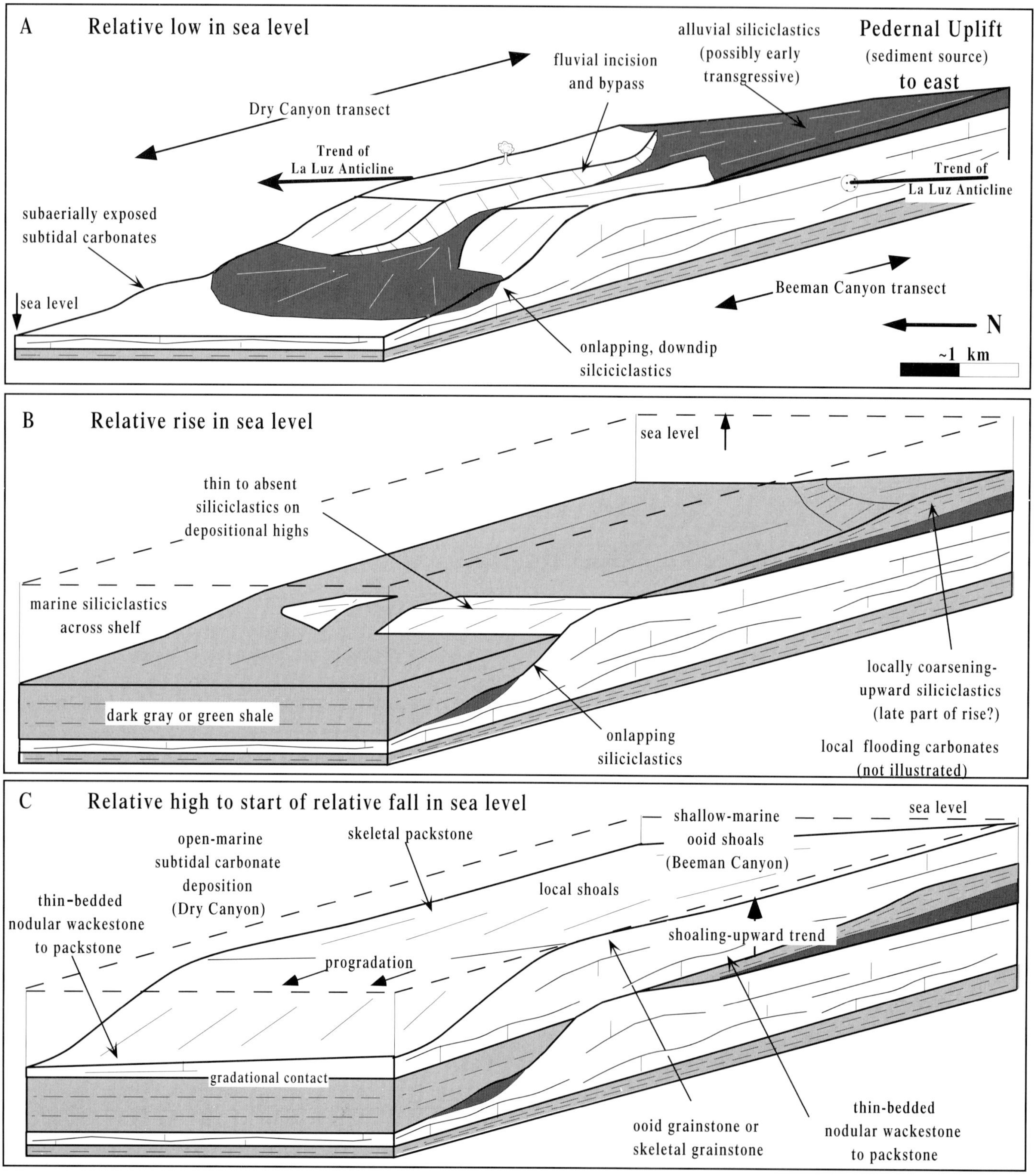

Fig. 12.—Schematic diagram illustrating interpreted evolution of an individual high-frequency sequence. Approximate positions of Dry and Beeman Canyon transects are indicated. La Luz anticline plunges to the north-northwest through an area near the center of the schematic profile. The Pedernal uplift is to the east (the right in this figure). The amplitudes of relative changes in sea level are 30 m or more, as discussed in the text. (A) Relative low in sea level, characterized by bypass and deposition of alluvial strata. (B) Relative rise in sea level, accompanied by significant marine flooding and deposition of dark gray or green shale. Thin, transgressive carbonates are present in many high-frequency sequences. (C) Relative high in sea level to start of relative fall in sea level, characterized by deposition of carbonate strata.

Table 2.—MEASURED MINIMUM AMPLITUDES OF RELATIVE CHANGE IN SEA LEVEL

Surface	Vertical extent of subaerial exposure features down slope[1]	Depth of fluvial incision[2]	Comments
A		2.5 m	Recognized in section MR-3.
I0		8.3 m	Valley located southeast of section BC-3. Incision cuts through I0, I1, and 1.3 m in I2.
I1	13.6 m ~10 m		Topography estimated from thickness changes between sections BC-1 and BC-2 (most downdip extent of recognized subaerial exposure features) (13.6 m). Topography estimated from thickness of onlapping beds draping subaerially exposed phylloid algal mound between sections YC-4 and YC-5 (~10 m).
I2	~17 m ~10 m 25 m		Topography estimated from thickness of subaerially exposed phylloid algal mound between sections YC-4 and YC-5 (~17 m). Estimated from topography on subaerially exposed phylloid algal mound between sections BC-3 and BC-5 (~10 m). Estimated from thickness changes between sections BC-7 and BC-9 (most downdip extent of recognized subaerial exposure features) (25 m). May have been enhanced due to pulift of the La Luz anticline.
I3	32 m		Calculated using the differences in elevation of surfaces I0 and I3 between sections BC-8 and BC-6. Used the base of alluvial conglomerate in BC-6, recognizing that this datum may result in a lower estimate of amplitude because of possible fluvial incision.
J	~30 m	18 m	Paleoslope estimated from thickness changes and thickness of onlapping beds between sections MR-2 and YC-8. Depth of fluvial incision measured just basinward of NDC-3.

[1]Measurements represent minimum amplitudes of change (cf. Wilson, 1967; Goldstein, 1988), as all represent a subaerial exposure of subtidal sediments in their most updip extent (reflecting unutilized accommodation). Possible errors of compaction, water loading, and sediment loading are not quantitatively considered (cf. Rankey and Lehrmann, 1996).

[2]Measurements represent maximum depth of fluvial incision through marine strata.

Note: cf. Wilson (1967), Goldstein (1998a), Goldstein and Franseen (1995), Rankey and Lehrmann (1996). Data are uncorrected for compaction, sediment and water loading, etc. (Wilson, 1967). Where the downdip areas are more argillaceous (e.g., basinward sections in **Fig. 5**), compaction would result in an overestimation of the amplitude of relative fall. In contrast, estimates of amplitude from Beeman Canyon (e.g., I3, I2, I1; **Fig. 7**) include thicker shaly intervals in updip (intermound) areas and so might underestimate amplitude of change.

lighted in Figure 14: (1) the basinward lateral transition from thicker, massive carbonate to thinner, nodular carbonate and (2) the basinward pinch-out of alluvial siliciclastics.

The geographic position of the turnaround from basinward-stepping to landward-stepping facies transitions in the two highlighted examples is different. For example, on the basis of the trends in the two-dimensional distribution of alluvial strata, a turnaround occurs above surface H1 (Fig. 14). Alluvial strata in high-frequency sequences K through H1 progressively step basinward. Above H1, however, no alluvial strata are observed in high-frequency sequences H and G, and in high-frequency sequence F, they are at least 2.2 km landward of those above H1 (Fig. 14), reflecting thelandward step. In the massive-to-nodular carbonate transition, however, a progradation-retrogradation turnaround is evident at surface I (Fig. 14), where the position of this transition steps landward approximately 2.5 km. The overall progradational pattern in the upper part of the Holder is apparent, but landward steps present in the alluvial strata (e.g., in high-frequency sequences D and C) are not reflected by similar steps in the carbonate strata. Similarly, the positions of the deepest incised valley (on surface J) or the thickest alluvial strata (on surface I3) do not appear to correspond with any of the other turnarounds.

Interpretation—Stacking Patterns of High-Frequency Sequences

The description above and Figure 14 illustrate that a longer-term basinward to landward turnaround of the facies transitions within the Holder Formation might be picked at different horizons, depending on the attributes by which the patterns are defined. Why is there this variability? To address this question, we should first consider the processes that might control these transitions that we are tracking.

The position of basinward pinch-out of alluvial siliciclastics might reflect several different processes. Deposition and preservation of alluvial siliciclastics are influenced by factors including shelf gradient, bypass, sediment supply, the duration and elevation of the relative low in base level (sea level), and the character of the relative rise in base level (sea level) (e.g., Schumm, 1993). Many alluvial siliciclastics below surface I thin and pinch out in a basinward direction near the axis of La Luz anticline and the mound horizons, suggesting that the depositional highs in this part of the section limited alluvial siliciclastics to these topographically lower, landward positions (cf. Wilson, 1967). The central, higher part of the shelf was bypassed by incised valleys, however, such that coarse alluvial strata were deposited in more basinward positions in some high-frequency sequences (e.g., high-frequency sequence J). The position at which bypass begins probably is influenced by gradient of the shelf also.

Likewise, the basinward extent of alluvial siliciclastics is influenced by siliciclastic sediment supply. For example, the paleosol at surface I2 is not exceptionally strongly developed, yet this high-frequency sequence includes the thickest alluvial section. These thick alluvial siliciclastics may reflect increased sediment influx related to uplift in the source area or climate change, rather than a long-term relative low in sea level. The Pedernal shelf was considerably closer to its source area than many other well-ordered icehouse systems (e.g., Midcontinent, Paradox basin), such that any climatic or tectonic changes in the uplands would probably have a large and immediate effect on sedimentation on the shelf.

Another control on the preservation of alluvial strata on the shelf might be the character of base-level change. For example, re-equilibration of stream gradients might cause aggradation on the shelf during a relative rise in base level (sea level). Alternatively, these alluvial deposits on the shelf may represent basinward progradation during a relative low in sea level, after the shoreline had migrated basinward. Rate of change might also be important. A rapid rate of rise, which might be expected with the relative changes in sea level of 30 m or more, might favor rapid migration of the shoreline, flooding the coastal plain before its slope re-equilibrated, resulting in thin alluvial deposits.

The position of the lateral transition from relatively shallow, massive carbonate to relatively deep, nodular carbonate (e.g., Figs. 6, 14) reflects the farthest basinward extent of this facies transition when sea level fell below this position on the shelf. The location of this massive-to-nodular carbonate transition is controlled by factors including depositional gradient, the rate of progradation, and the duration of submergence. The influence of depositional gradient and topography is illustrated by the aggradational to subtly progradational pattern in the lower part of the Holder Formation (Fig. 14). In these strata, the mounds created significant depositional topography that was not filled in for several high-frequency sequences, such that the shallow-marine carbonates could not prograde across the deeper area, and the position of the massive-to-nodular transition was fixed near the margins of the mounds. During deposition of the upper part of the Holder (above H1), when the depositional topography had decreased, the massive-to-nodular carbonate transition moved basinward several hundred meters in each successive high-frequency sequence. The rate of progradation and the duration

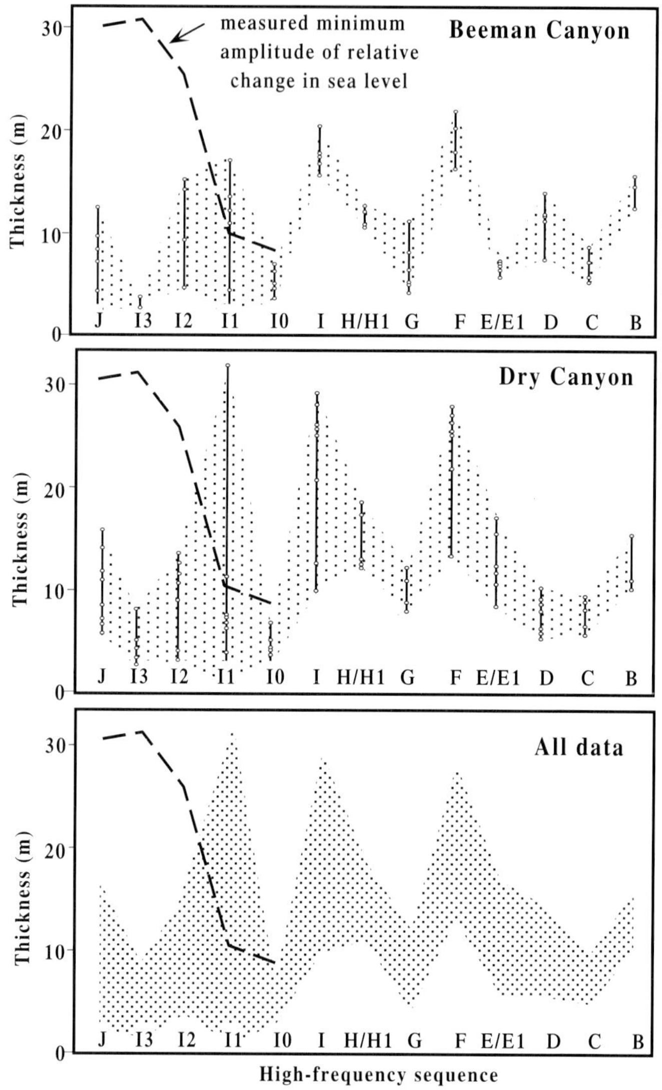

Fig. 13.—Thickness changes within and among individual high-frequency sequences. "Measured minimum amplitude" refers to the measured values (see Table 2). Note the broad range of thicknesses for each high-frequency sequence, the lack of systematic changes, and the fact that most thickness values are less than the measured minimum amplitude of change.

of submergence cannot be evaluated from our data, but a conceptual model for the character of carbonate strata is discussed below.

DISCUSSION

Periods with global icehouse climatic conditions are characterized by high amplitudes of relative change in sea level (e.g., Wanless and Shepard, 1936; Wilson, 1967; Heckel, 1977; Veevers and Powell, 1987; Crowley and Baum, 1991; Read et al., 1995; see Table 2). High amplitudes of change, and high rates of change, could explain some of the apparent variability in the stacking patterns of Holder strata.

As suggested by Wilson (1967) and described above, many alluvial siliciclastics on the shelf probably represent deposition during a relative low or rise in sea level, whereas the massive-

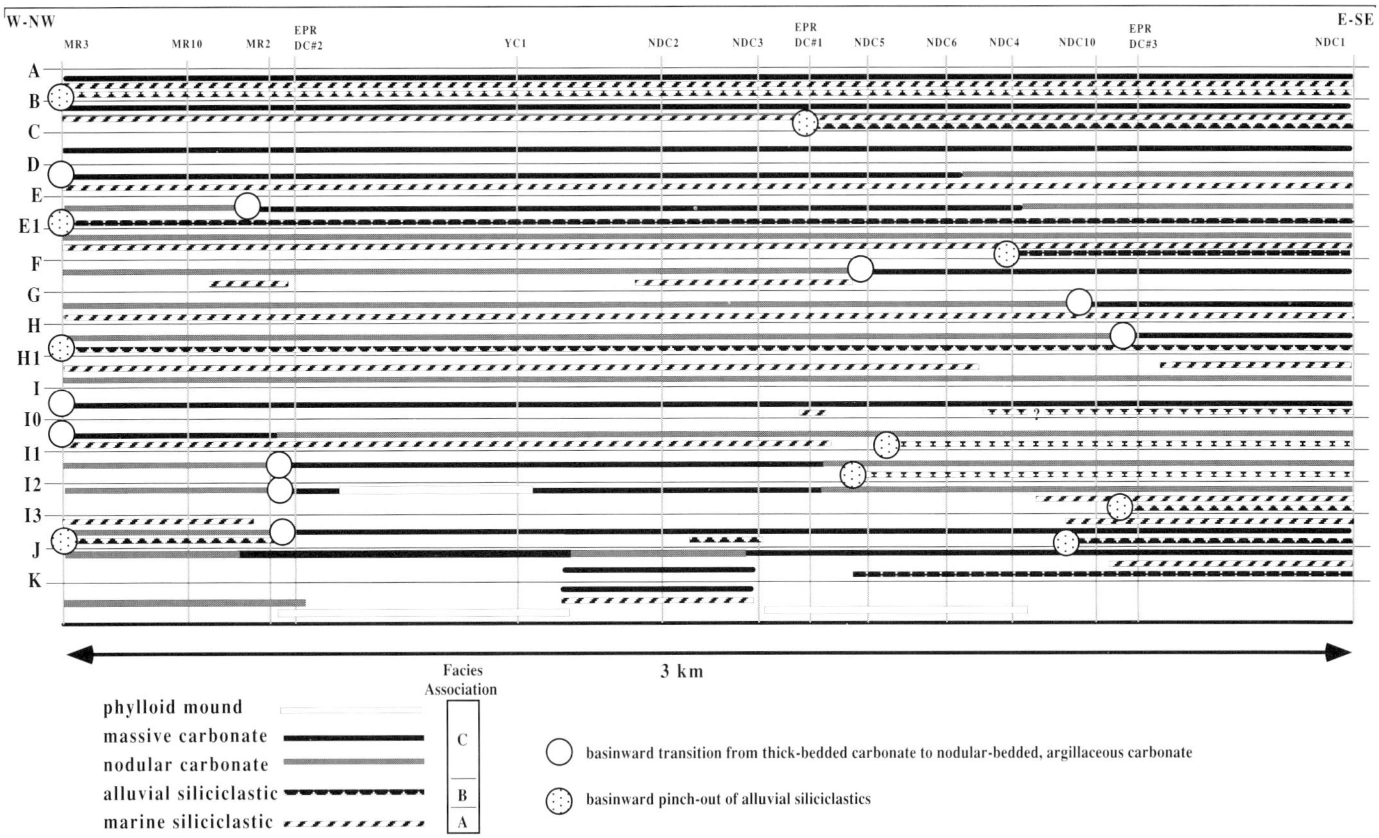

Fig. 14—Distribution of key facies and facies associations within high-frequency sequences from the base of the Holder Formation to surface A. The total length of the cross section is approximately 3 km. Because of projection from three-dimensional space onto the line of cross section, there are components of along-strike variability that may appear unusual, and the spacing is arbitrary. The white circles represent the positions of the basinward transitions from thick-bedded, shoal carbonate to nodular-bedded carbonate; patterned circles show the basinward positions of pinch-outs of alluvial facies. Note that (1) the trends in lateral position of the two transitions are not coincident and (2) the high-frequency sequences above surface H1 contain facies and facies associations that are more continuous than those below H1.

to-nodular carbonate transitions represent the farthest basinward progradation of that facies transition before relative fall in sea level exposed the shelf. In this case, these two facies, marine limestone and alluvial siliciclastics, reflect deposition during significantly different parts of the sea-level curve, thus representing nonlinked depositional systems. It would almost be surprising, then, if trends in the position of these two contacts (e.g., Fig. 14) were identical. This situation is quite different from that on many greenhouse carbonate platforms, where sequences are composed of simple shallowing-upward progressions (parasequences) and where component facies are depositionally linked and respond similarly to sea-level forcing mechanisms.

As noted by Wilson (1967) and Raatz (1996), shelf position and elevation provide significant controls on stratal architecture developed during global icehouse climatic periods that are characterized by high amplitudes and rapid rates of change in sea level (Fig. 15A). At low elevations (e.g., in the basin, Fig. 15A), facies representing the shallowest water depths are deposited during relative sea-level lowstands and may include sandstone, conglomerate, carbonates, or evaporites (Wilson, 1967; Hair, 1976; Jarilla, San Andres, and Franklin Mountains). In these areas, the relative rises and highs are reflected as black shales deposited in water depths greater than the amplitude of change in sea level (Wilson, 1967). At intermediate elevations (Fig. 15A), the shelf is subaerially exposed during the low, and the rate of relative rise in sea level is at a maximum when these positions are initially flooded. The high rate of rise results in carbonate strata that either actively aggrade or are flooded. The flooding carbonates at the base of some high-frequency sequences reflect this rapid rate of rise. The relative high in sea level is reflected by deposition of dark marine shales or a condensed carbonate section. A rapid rate of fall results in subaerial exposure of subtidal strata in these positions. Facies and stratigraphic architecture at these shelf positions are most sensitive to the rates of change of sea level during the rises and falls. At positions high on the shelf (Fig. 15A), most of the relative fall, the relative low, and the relative rise are represented as subaerial exposure or incised valleys. The relative rise may also result in alluvial aggradation on the shelf. This shelf position is inundated only during the relative highs of sea level, and maximum water depths are significantly less than the amplitude of change in sea level. While submerged, this shelf position remains in the photic zone, and carbonate-producing organisms are highly prolific, resulting in aggradational and progradational carbonates. Limited accommodation at times favors aggradation to sea level prior to the relative fall in sea level. In sum, the strata at positions high on the shelf most closely reflect the character of the relative highs in sea level, although alluvial facies (deposited during the

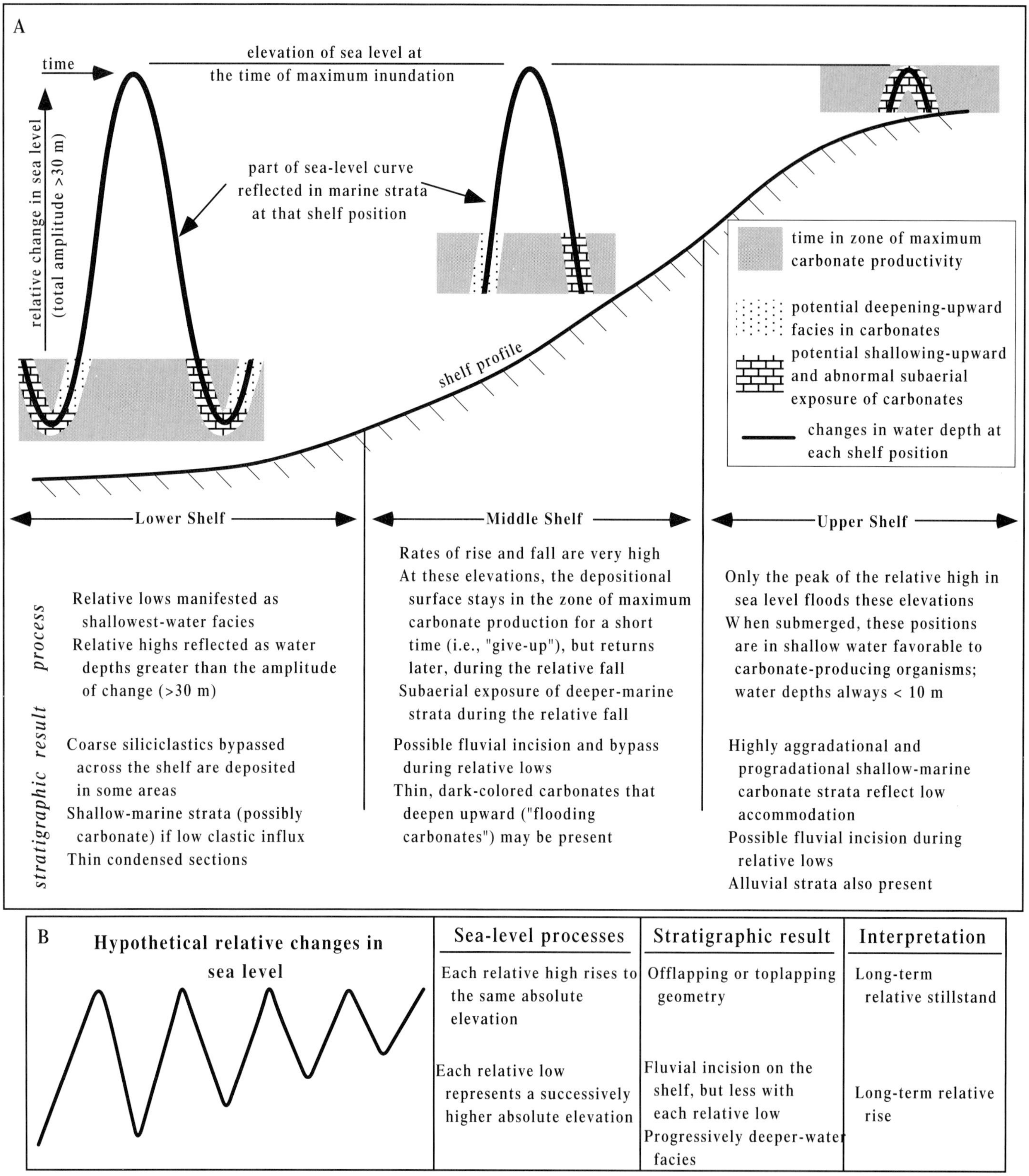

Fig. 15.—(A) Relationship between shelf position, relative changes in sea level, and shelf position or elevation. In icehouse systems, amplitudes of change are high, at least 30 m. Strata of different shelf positions reflect the character of different parts of the curve and must be considered in evaluating short-term variability in icehouse systems, as described in the text. (B) Hypothetical sea-level curve in which amplitudes of relative change in sea level became progressively less. As discussed in the text, changing amplitudes may lead to ambiguity in defining larger-scale sequence stratigraphy of the system, depending on which part of the curve the analyzed attributes were deposited. The differences between the long-term trends in alluvial siliciclastics and marine carbonates might be the result of variable amplitudes.

relative rise or perhaps the relative fall) may also be present.

In the Holder Formation, the characteristics of the lower part of the Dry Canyon cross section (below surface I; Fig. 5) in positions shelfward of the mound trend suggest "mid" to "high" shelf positions, such that subaerial exposure of subtidal facies is common and carbonates are more aggradational (Fig. 14). The upper part of the Holder contains more progradational deposits and fewer basal flooding carbonates, reflecting a relatively higher shelf position. In contrast, the Beeman Canyon high-frequency sequences reflect the "high" shelf position updip on La Luz anticline. Some high-frequency sequences are capped by carbonates with ooid shoal or foreshore deposits, which may reflect aggradation to sea level prior to the relative fall. Positions higher on the shelf (not present in Holder outcrops) would be expected to contain thinner marine carbonates within a thick alluvial succession. This pattern can be observed in the overlying Laborcita Formation (Otte, 1959; Winchester, 1976; Rankey, 1996).

If the basinward pinch-out of alluvial siliciclastics and the massive-to-nodular carbonate transition represent different parts of the sea-level curve, then they would be responsive to any variations in the amplitude of relative changes in sea level (cf. Rankey and Lehrmann, 1996; Felton and Heckel, 1996). For example, consider a system in which each of a series of short-term relative highs rises to the same elevation, yet because of changing amplitudes, the lows are at progressively higher elevations (Fig. 15B). The carbonate strata deposited on the shelf during the relative highs would be highly progradational, whereas deposits representing each of the relative lows or rises might contain progressively shallower incised valleys or thicker alluvial deposits (because of the decreased rates of rise) on the shelf and progressively deeper-water facies in the basin. Some of the variability in the long-term stacking patterns of the Holder Formation could be driven by changing amplitude of relative change in sea level.

In addition to reflecting processes inherent to icehouse systems, the variability in stacking patterns of Holder strata (Figs. 13, 14) may reflect a component of tectonic activity (cf. Wilson, 1972). Several previous studies of Pennsylvanian icehouse strata have documented a well-ordered stratal heirarchy (e.g., U.S. Midcontinent: Watney, 1985; Watney et al., 1989, 1996; Paradox basin: Goldhammer et al., 1993; Weber et al., 1995; Permian basin: Saller et al., 1994). These other well-ordered systems represent deposition on the distal margins of foreland basins or isolated platforms, however, rather than on a proximal, narrow, tectonically active attached shelf (Saller et al., 1994; Weber et al., 1995; Watney et al., 1996).

Several observations of large-scale features suggest that syndepositional deformation influenced the area (Wilson, 1972). The gross thickness and facies patterns reflect the position of La Luz anticline (Wilson, 1972). Overall thinning across the axis of the anticline is evident in Figure 5, and the composite sections in Dry Canyon (structurally downdip) are commonly 10–20 m thicker than those in Beeman Canyon (structurally updip) (Wilson, 1972). Stratal thinning into Beeman Canyon is also accompanied by a change to much shallower-water facies (oolitic grainstone commonly caps high-frequency sequences in Beeman Canyon) (Figs. 5, 7), and phylloid algal mounds are observed to shingle down the western flank of the anticline (illustrated in Wilson, 1967, 1972).

Some of the variability in longer-term stratal patterns might reflect tectonically induced variability in sediment supply and distribution or relative changes in sea level at short time scales. For example, the thickest alluvial package (e.g., high-frequency sequence I3) does not appear to correspond with a turnaround in stacking patterns, as would be expected if the character of alluvial siliciclastics reflected solely a lower-order sea-level signal. These thick alluvial strata may instead reflect a tectonically controlled influx of coarse siliciclastics out of phase with long-term changes in sea level, as described above. Similarly, surface H1 bevels the underlying dark, fossiliferous deep-marine shales across the axis of La Luz anticline, truncating approximately 18 m of section. This angular relationship probably reflects uplift of the structure at this time (Wilson, 1967, 1972). The overlying units are relatively sheetlike across the anticline, suggesting that its influence was less pronounced after this time. As suggested by the lack of any accompanying turnaround in stacking patterns (Fig. 14), rather than reflecting a long-term relative low, the pronounced incision of surface J or the thick siliciclastics of I3 may represent uplift independent of any lower-order sea-level trend. Thus, an interpretation of long-term stacking patterns that focused on these features (cf. Fig. 14) would illustrate considerable complexity in the system.

SUMMARY AND CONCLUSIONS

1. The Holder Formation consists of three general facies associations: marine siliciclastics, alluvial siliciclastics, and marine carbonates.
2. Facies are arranged into shoaling-upward parasequences, bounded by marine-flooding surfaces, interpreted to reflect an abrupt increase in water depth or decrease in energy. Many carbonate parasequences are not correlative across distances of more than several hundred meters; some siliciclastic parasequences are more extensive, although local lobes are less laterally extensive. Because of the lack of correlatability and the absence of abnormal subaerial exposure, an autogenic origin for carbonate and siliciclastic parasequences cannot be ruled out.
3. The 22 high-frequency sequences, each bounded by surfaces of subaerial exposure of subtidal strata (Goldstein, 1988a, 1991) or fluvial incision through marine strata, are laterally continuous across the shelf, but their component facies change laterally (Wilson, 1967). In landward positions, vertical facies successions include alluvial, marine siliciclastic, and marine carbonate strata. In middle-shelf locations, vertical successions are dominated by marine carbonates. In the more basinward locations, vertical facies successions include basal lowstand alluvial siliciclastics, thick marine siliciclastics, and marine carbonates. High-frequency sequence boundaries provide the highest-resolution, correlatable chronostratigraphic markers in this system.
4. As J. L. Wilson (1967) noted, the most pronounced control

on high-resolution stratigraphic architecture was repeated, high-amplitude relative changes in sea level. Measured minimum amplitudes of relative falls in sea level range up to approximately 30 m (Wilson, 1967; Goldstein, 1988a). The relative changes in sea level led to subaerial exposure of subtidal facies, facies offsets, and significant lateral shifts in the position of shoreline. Changes in sediment supply and production combined with the relative changes in sea level to create strata with significant lateral and vertical facies changes (Wilson, 1967). Relative lows and the early parts of relative rises in sea level were accompanied by abundant siliciclastic supply, favoring siliciclastic bypass of the emergent shelf but deposition in downdip areas. Relative highs in sea level were characterized by predominantly carbonate sedimentation, although some localized marine siliciclastics were present during the early part of the highstand.

5. Shelf position and elevation significantly influence facies distribution and stratal architecture. In downdip positions, facies representing the shallowest-water depths reflect the character of the relative lows in sea level; in more updip positions, facies reflecting the shallowest-water depths are influenced by the character of the relative highs in sea level.
6. Syndepositional structural deformation is suggested by low-angle unconformities, variable facies, and variable thicknesses across and up La Luz anticline (Wilson, 1967, 1972). Deformation of La Luz anticline or uplift in the Pedernal highlands influenced the complexities in the longer-term stacking patterns.
7. Longer-term stacking patterns of high-frequency sequences are ambiguous. One-dimensional thickness and facies patterns are highly variable, and the stratigraphic positions of the turnaround of significant facies contacts and pinch-outs are variable. Likewise, the positions of maximum fluvial incision and the thickest alluvial siliciclastics are not coincident with any of the facies turnarounds. In this system, stacking patterns do not solely reflect accommodation trends, but instead can be sensitive to the amplitudes and rates of change in sea level, shelf elevation, and the proximal source area.

ACKNOWLEDGMENTS

Thanks to John Mitchell, Steve Schutter, Gordon Start, Mike Porter, and Norm Corbett, who collected and interpreted data from the Sacramento Mountains (including core, well logs, and photographic panoramas) as part of the Seismic to Outcrop Calibration Study (SOCS) at Exxon Production Research Company in the late 1980s. The work done by these current and former EPR geoscientists advanced this project from the onset. Thanks also to Mike Kozar, who helped measure several sections (especially those shown in Figs. 5 and 7), and Jim Weber who participated in the planning and initial field reconnaissance. Rick Weber and Tom Frantes provided support and guidance during the planning, proposal, and completion of this research. Thanks to all the members of the Carbonate Reservoir Analysis Section at EPR who assisted in the editing of this manuscript and participated in many discussions on Holder Formation stratigraphy. Thanks to EPR reviewers K. Campion, J. C. Mitchell, L. J. Weber, and J. C. Van Wagoner for their helpful comments, discussion, and suggestions. Bob Goldstein, Bill Morgan, and Art Saller added many constructive comments for the SEPM volume that significantly enhanced the presentation of this paper. Thanks to Bob Goldstein and Lynn Watney for numerous insightful discussions on Pennsylvanian stratigraphy.

REFERENCES

ALGEO, T. J., WILSON, J. L., AND LOHMANN, K. C., 1991, Eustatic and tectonic controls on cyclic sediment accumulation patterns in Lower–Middle Pennsylvanian strata of the Orogrande basin, New Mexico, *in* Sierra Blanca, Sacramento, and Capitan Ranges: Socorro, New Mexico Geological Society, 42nd Field Conference Guidebook, p. 203–212.

BAARS, D. L., ROSS, C. A., RITTER, S. M., AND MAPLES, C. G., 1994, Proposed repositioning of the Pennsylvanian–Permian boundary in Kansas, *in* Baars, D. L., compiler, Revision of Stratigraphic Nomenclature in Kansas: Lawrence, Kansas Geological Survey Bulletin 230, p. 5–9.

CARR, D. L., 1983, Paleozoic siliciclastic shelf-bars, Sacramento Mountains, New Mexico: Unpublished M.S. Thesis, University of Texas, Austin, 127 p.

CARR, D. L., AND SCOTT, A. J., 1990, Late Pennsylvanian storm-dominated shelf-sand ridges, Sacramento Mountains, New Mexico: Journal Sedimentary Petrology, v. 60, p. 592–607.

CHRISTIE-BLICK, N., 1991, Onlap, offlap, and the origin of unconformity-bounded depositional sequences: Marine Geology, v. 97, p. 35–56.

CLINE, L. M., 1959, Preliminary studies of the cyclical sedimentation and paleontology of the upper Virgil strata of the La Luz area, Sacramento Mountains, New Mexico: Roswell, Roswell Geological Society and, Midland, Permian Basin Section–Society of Economic Paleontologists and Mineralogists Guidebook of the Sacramento Mountains, p. 172–185.

CROWLEY, T. J., AND BAUM, S. K., 1991, Estimating Carboniferous sea-level fluctuations from Gondwanan ice extent: Geology, v. 19, p. 975–977.

DAVYDOV, V. I., GLENISTER, B. F., SPINOSA, C., RITTER, S. M., CHERNYKH, V. V., WARDLAW, B. R., AND SNYDER, W. S., 1995, Proposal of Aidaralash as GSSP for the base of the Permian System: Permophiles: A Newsletter of SCPS (Subcommission on Permian Stratigraphy, International Commission on Stratigraphy of the International Union of Geological Sciences [IUGS]), no. 6, p. 1–9.

DELGADO, D. J., 1977, Conglomeratic lithofacies of the Laborcita and Abo Formations (Wolfcampian), north-central Sacramento Mountains: Sedimentologic and tectonic importance, in Geology of the Sacramento Mountains, Otero County, New Mexico: Midland, West Texas Geological Society Publication Number 1977-68, p. 90–101.

DUNNING, C. P., 1978, Provenance and paleotectonic setting of conglomerates in the Virgilian Holder Formation, northern Sacramento Mountains, New Mexico: Unpublished M.A. Thesis, Rice University, Houston, 91 p.

ENOS, P., 1977, Holocene sediment accumulation of south Florida shelf margin, *in* Enos, P. and Perkins, R. D., eds., Quaternary Sedimentation in South Florida: Boulder, Geological Society of America Memoir 147, p. 1–130.

FELTON, R. M., AND HECKEL, P. H., 1996, Small-scale cycles in Winterset Limestone Member (Dennis Formation, Pennsylvanian of northern Midcontinent) represent "phased regression," *in* Witzke, B. J., Ludvigson, G. A., and Day, J., eds., Paleozoic Sequence Stratigraphy: Views from the North American Craton: Boulder, Geological Society of America Special Paper 306, p. 389–397.

FISCHER, A. G., 1964, The Lofer cyclothems of the Alpine Triassic, *in* Merriam, D. F., ed., Symposium on cyclic sedimentation: Lawrence, Kansas Geological Survey Bulletin 169, p.

FLY, S. H.,1985, Shallow marine environments of the Laborcita Formation (Wolfcampian), Sacramento Mountains, New Mexico: Unpublished M.A. Thesis, University of Texas, Austin, 60 p.

GIANNINY, G. L., AND SIMO, J. A. T., 1996, Implications of unfilled accommodation space for sequence stratigraphy on mixed carbonate-siliciclastic platforms: An example from the lower Desmoinesian (Middle Pennsylvanian), southwestern Paradox basin, Utah, *in* Longman, M. W. and Sonnenfeld, M.

D., eds., Paleozoic Systems of the Rocky Mountain Region: Denver, Rocky Mountain Section–SEPM (Society for Sedimentary Geology), p. 213–234.

GOLDHAMMER, R. K., AND ELMORE, R. D., 1984, Paleosols capping regressive carbonate cycles in the Pennsylvanian Black Prince Limestone, Arizona: Journal of Sedimentary Petrology, v. 54, p. 1124–1137.

GOLDHAMMER, R. K., DUNN, P. A., AND HARDIE, L. A., 1990, Depositional cycles, composite sea-level changes, cycle stacking patterns, and the hierarchy of stratigraphic forcing—Examples from the platform carbonates of the Alpine Triassic: Geological Society of America Bulletin, v. 102, p. 535–562.

GOLDHAMMER, R. K., OSWALD, E. J., AND DUNN, P. A., 1993, Hierarchy of stratigraphic forcing: Example from Middle Pennsylvanian shelf carbonates of the Paradox basin, *in* Franseen, E. K., Watney, W. L., Kendall, C., and Ross, W., eds., Sedimentary Modeling: Computer Simulations and Methods for Improved Parameter Definition: Lawrence, Kansas Geological Survey Bulletin, v. 233, p. 361–415.

GOLDSTEIN, R. H., 1988a, Paleosols of Late Pennsylvanian cyclic strata, New Mexico: Sedimentology, v. 35, p. 777–803.

GOLDSTEIN, R. H., 1988b, Cement stratigraphy of Pennsylvanian Holder Formation, Sacramento Mountains, New Mexico: American Association of Petroleum Geologists Bulletin, v. 72, p. 425–438.

GOLDSTEIN, R. H., 1991, Stable isotope signatures associated with paleosols, Pennsylvanian Holder Formation, New Mexico: Sedimentology, v. 38, p. 67–77.

GOLDSTEIN, R. H., AND FRANSEEN, E. K., 1995, Pinning points: A method providing quantitative constraints on relative sea-level history: Sedimentary Geology, v. 95, p. 1–10.

GOLDSTEIN, R. H., HALLEY, R. B., AND SCHOLLE, P. A., 1991, Upper Paleozoic bioherms in the northern Sacramento Mountains, *in* Scholle, P. A. and Goldstein, R. H., eds., Classic Upper Paleozoic Reefs and Bioherms of West Texas and New Mexico: Tulsa, American Association of Petroleum Geologists, Field Guide, p. 147–246.

GOLDSTEIN, R. H., ANDERSON, J. E., AND BOWMAN, M. W., 1993, Diagenetic responses to sea-level change: Integration of field, stable isotope, paleosol, paleokarst, fluid inclusion, and cement stratigraphy to determine history and magnitude of sea-level fluctuation, *in* Franseen, E. K., Watney, W. L., Kendall, C., and Ross, W., eds., Sedimentary Modeling: Computer Simulations and Methods for Improved Parameter Definition: Lawrence, Kansas Geological Survey Bulletin 233, p. 163–176.

GRAMMER, G. M., EBERLI, G. P., VAN BUCHEM, F. S. P., STEVENSON, G. M., AND HOMEWOOD, P., 1996, Application of high-resolution sequence stratigraphy to evaluate lateral variability in outcrop and subsurface—Desert Creek and Ismay intervals, Paradox basin, *in* Longman, M. W. and Sonnenfeld, M. D., eds., Paleozoic Systems of the Rocky Mountain Region: Denver, Rocky Mountain Section–SEPM (Society for Sedimentary Geology), p. 235–266.

HAIR, G. L., 1976, The Upper Pennsylvanian Panther Seep Formation (Virgilian) of the Franklin Mountains, Doña Ana County, New Mexico and El Paso County, Texas, *in* LeMone, D. V. and Lovejoy, E. M. P., eds., Symposium on the Stratigraphy of the Franklin Mountains: El Paso, El Paso Geological Society, p. 61–66.

HANDFORD, C. R., AND LOUCKS, R. G., 1993, Carbonate depositional sequences and systems tracts—Responses of carbonate platform to relative sea-level changes: Loucks, R. G. and Sarg, J. F. eds., Carbonate Sequence Stratigraphy: Tulsa, American Association of Petroleum Geologists Memoir 57, p. 3–42.

HARDIE, L. A., BOSELLINI, A., AND GOLDHAMMER, R. K., 1986, Repeated subaerial exposure of subtidal carbonate platforms, Triassic, northern Italy: Evidence for high-frequency sea level oscillations on a 10 4 year scale: Paleoceanography, v. 1, p. 447–457.

HARRIS, P. M., 1979, Facies anatomy and diagenesis of a Bahamian ooid shoal: Miami, University of Miami, Comparative Sedimentology Laboratory, Sediments 7, 163 p.

HECKEL, P. H., 1977, Origin of phosphatic black shale facies in Pennsylvanian cyclothems of Mid-continent North America: American Association of Petroleum Geologists Bulletin, v. 61, p. 1045–1068.

KLUTH, C. F., 1986, Plate tectonics of the Ancestral Rocky Mountains, *in* Peterson, J. A., ed., Paleotectonics and Sedimentation in the Rocky Mountain Region, United States: Tulsa, American Association of Petroleum Geologists Memoir 41, p. 353–369.

KOERSCHNER, W. F., III, AND READ, J. F., 1989, Field and modeling studies of Cambrian carbonate cycles, Virginia Appalachians: Journal of Sedimentary Petrology, v. 59, p. 645–659.

MERRIAM, D. F., 1964, ed., Symposium on cyclic sedimentation: Lawrence, Kansas Geological Survey Bulletin 169, 640 p.

MEYER, R. F., 1966, Geology of Pennsylvanian and Wolfcampian rocks in southeast New Mexico: Socorro, New Mexico Bureau of Mines Memoir 17, 122 p.

MITCHUM, R. M., JR., AND VAN WAGONER, J. C., 1991, High frequency sequences and their stacking patterns: Sequence stratigraphic evidence of high-frequency eustatic cycles: Sedimentary Geology, v. 70, p. 131–160.

MOORE, R. C., 1964, Paleoecological analysis of Kansas Pennsylvanian and Permian cyclothems, *in* Merriam, D. F., ed., Symposium on Cyclic Sedimentation: Lawrence, Kansas Geological Survey Bulletin 169, p. 287–380.

OPPEL, T. W., 1957, The unconformity between the Fresnal Group and the Bursum Formation, Sacramento Mountains, New Mexico: Unpublished M.S. Thesis, University of Wisconsin, Madison, 66 p.

OSLEGER, D., AND READ, J. F., 1993, Comparative analysis of methods used to define eustatic variations in outcrop: Late Cambrian interbasinal sequence development: American Journal of Science, v. 293, p. 157–216.

OTTE, C., 1959, Late Pennsylvanian and Early Permian stratigraphy of the northern Sacramento Mountains, Otero County, New Mexico: Socorro, New Mexico Bureau Mines Bulletin 50, 111 p.

PARKS, J. M., 1977, Paleoecological evidence on the origin of the Dry Canyon Pennsylvanian bioherms, *in* Pray, L. C., Wilson, J. L., and Toomey, D. F., eds., Geology of the Sacramento Mountains, Otero County, New Mexico: Midland, West Texas Geological Society Guidebook, p. 25–38, 159–161.

PLUMLEY, W. L., AND GRAVES, R. W., 1953, Virgilian reefs of the Sacramento Mountains, New Mexico: Journal Geology, v. 61, p. 1–16.

PRAY, L. C., 1959, Stratigraphic and structural feature of the Sacramento Mountains escarpment, New Mexico: Roswell, Roswell Geological Society, and Midland, Permian Basin Section–Society of Economic Paleontologists and Mineralogists Guidebook, p. 87–130.

PRAY, L. C., 1961, Geology of the Sacramento Mountains escarpment, Otero County, New Mexico: Socorro, New Mexico Bureau of Mines Bulletin 35, 144 p.

RAATZ, W. D., 1996, Sequence stratigraphy of the Beeman Formation: Interaction of eustasy, tectonics, and climate on a mixed terrigenous clastic/carbonate shelf and basin, Upper Pennsylvanian, Sacramento Mountains, New Mexico: Unpublished Ph.D. Dissertation, University of Wisconsin, Madison, 404 p.

RANKEY, E. C., 1996, Isolating eustatic, climatic, tectonic, and autogenic controls on sedimentation: Spatial and temporal changes in upper Virgilian–lower Wolfcampian (Pennsylvanian–Permian) strata, western North America: Unpublished Ph.D. Dissertation, University of Kansas, Lawrence, 267 p.

RANKEY, E. C., 1997, Relations between relative changes in sea level and climate shifts: Pennsylvanian–Permian mixed carbonate-siliciclastic strata, western U.S.A.: Geological Society of America Bulletin, v. 109, p. 1089–1100.

RANKEY, E. C., AND LEHRMANN, D. J., 1996, Anatomy and origin of toplap in a mixed carbonate-clastic system, Seven Rivers Formation (Permian, Guadalupian), Guadalupe Mountains, New Mexico, USA: Sedimentology, v. 43, p. 807–826.

READ, J. F., AND GOLDHAMMER, R. K., 1988, Use of Fischer plots to define third-order sea-level curves in Ordovician peritidal cyclic carbonates, Appalachians: Geology, v. 16, p. 895–899.

READ, J. F., KERANS, C., WEBER, L. J., SARG, J. F., AND WRIGHT, F. M., eds., 1995, Milankovitch Sea-Level Changes, Cycles, and Reservoirs on Carbonate Platforms in Greenhouse and Ice-House Worlds: Tulsa, SEPM (Society for Sedimentary Geology) Short Course Notes No. 35, 81 p.

REYER, R. W., 1958, Virgil cyclothems of the Holder Formation in Beeman and Indian Wells Canyons in the Sacramento Mountains northeast of Alamogordo New Mexico: Unpublished M.S. Thesis, University of Wisconsin, Madison, 65 p.

SALLER, A. H., DICKSON, J. A. D., AND BOYD, S. A., 1994, Cycle stratigraphy and porosity in Pennsylvanian and Lower Permian shelf limestones, eastern Central Basin platform, Texas: American Association of Petroleum Geologists Bulletin, v. 78, p. 1820–1842.

SARG, J. F., 1988, Carbonate sequence stratigraphy, *in* Wilgus, C. K., Hastings,

B. S., Kendall, C. G. St. C., Posamentier, H. W., Ross, C. A., and Van Wagoner, J. C., ed., Sea-Level Changes: An Integrated Approach: Tulsa, Society of Economic Paleontologists and Mineralogists Special Publication 42, p. 155–182.

SCHUMM, S. A., 1993, River response to baselevel change: Implications for sequence stratigraphy: Journal of Geology, v. 101, p. 279–296.

SHINN, E. A., 1973, Carbonate coastal accretion in an area of longshore transport, northeast Qatar, Persian Gulf, *in* Purser, B. H., ed., The Persian Gulf—Holocene carbonate sedimentation in a shallow epicontinental sea: Berlin, Springer-Verlag, p. 179–192.

SOREGHAN, G. S., 1994, Stratigraphic responses to geologic processes: Late Pennsylvanian eustasy and tectonics in the Pedragosa and Orogrande basins, Ancestral Rocky Mountains: Geological Society of America Bulletin, v. 106, p. 1195–1211.

TOOMEY, D. F., 1991, Late Pennsylvanian phylloid-algal bioherms, Orogrande basin, south-Roswell, Roswell, New Mexico and west Texas, in Sierra Blanca, Sacramento, and Capitan Ranges: Socorro, New Mexico Geological Society Guidebook, 42nd Annual Field Conference, p. 213–220.

TOOMEY, D. F., WILSON, J. L., AND REZAK, R., 1977a, Evolution of Yucca Mound complex, Late Pennsylvanian phylloid algal buildup, Sacramento Mountains, New Mexico: American Association Petroleum Geologists Bulletin, v. 61, p. 2115–2133.

TOOMEY, D. F., WILSON, J. L., AND REZAK, R., 1977b, Growth history of a Late Pennsylvanian phylloid algal organic buildup, northern Sacramento Mountains, New Mexico, *in* Pray, L. C., Wilson, J. L., and Toomey, D. F., eds., Geology of the Sacramento Mountains, Otero County, New Mexico: Midland, West Texas Geological Society Guidebook, p. 8–24.

VAIL, P. R., MITCHUM, R. M., JR., AND THOMPSON, S., 1977, Seismic stratigraphy and global changes of sea-level: Part 4. Global cycles of relative changes of sea-level, *in* Payton, C. E., ed. Seismic Stratigraphy—Applications to Hydrocarbon Exploration: Tulsa, American Association of Petroleum Geologists Memoir 26, p. 83–97.

VAN SICLEN, D. C., 1958, Depositional topography—Examples and theory: American Association of Petroleum Geologists Bulletin, v. 42, p. 1897–1913.

VAN WAGONER, J. C., 1995, Overview of sequence stratigraphy of foreland basin deposits: Terminology, summary of papers, and glossary of sequence stratigraphy, *in* Van Wagoner, J. C. and Bertram, G. T., eds., Sequence Stratigraphy of Foreland Basin Deposits: Tulsa, American Association of Petroleum Geologists Memoir 64, p. ix–xxi.

VAN WAGONER, J. C., MITCHUM, R. M., CAMPION, K. M., AND RAHMANIAN, V. D., 1990, Siliciclastic sequence stratigraphy in well logs, cores, and outcrops: Concepts for high-resolution correlation of time and facies: Tulsa, American Association of Petroleum Geologists Methods in Exploration Series No. 7, 55 p.

VEEVERS, J. J., AND POWELL, C. MC. A., 1987, Late Paleozoic glacial episodes in Gondwanaland reflected in transgressive-regressive depositional sequences in Euramerica: Geological Society of America Bulletin, v. 98, p. 475–487.

WANLESS, H. R., AND SHEPARD, F. P., 1936, Sea level and climate related to late Paleozoic cycles: Geological Society of America Bulletin, v. 47, p. 1177–1206.

WATNEY, W. L., 1980, Cyclic sedimentation of the Lansing–Kansas City Groups in northwestern Kansas and southwestern Nebraska: Lawrence, Kansas Geological Survey Bulletin 220, 72 p.

WATNEY, W. L., 1985, Resolving controls on epeiric sedimentation using trend surface analysis: Mathematical Geology, v. 17, p. 427–454.

WATNEY, W. L., FRENCH, J. A., AND FRANSEEN, E. K., eds., 1989, Sequence stratigraphic interpretations and modeling of cyclothems in the Upper Pennsylvanian (Missourian) Lansing and Kansas City Groups in eastern Kansas: Lawrence, Kansas Geological Society, 41st Annual Field Trip Guidebook, 211 p.

WATNEY, W. L., FRENCH, J. A., DOVETON, J. H., YOULE, J. C., AND GUY, W. J., 1996, Cycle hierarchy and genetic stratigraphy of Middle and Upper Pennsylvanian strata in the upper Mid-Continent, *in* Hyne, N. J., ed., Sequence Stratigraphy of the Mid-Continent: Tulsa, Tulsa Geological Society Publication, p. 141–192.

WEBER, L. J., SARG, J. F., AND WRIGHT, F. M., 1995, Sequence stratigraphy and reservoir delineation of the Middle Pennsylvanian (Desmoinesian), Paradox basin and Aneth field, southwestern USA, *in* Read, J. F., Kerans, C., Weber, L. J., Sarg, J. F., and Wright, F. M., Milankovitch Sea-Level Changes, Cycles, and Reservoirs on Carbonate Platforms in Greenhouse and Ice-House Worlds: Tulsa, SEPM (Society for Sedimentary Geology) Short Course Notes No. 35, 81 p.

WILSON, J. L., 1967, Cyclic and reciprocal sedimentation in Virgilian strata of southern New Mexico: Geological Society of America Bulletin, v. 78, p. 805–818.

WILSON, J. L., 1972, Influence of local structure in sedimentary cycles of Beeman and Holder Formations, Sacramento Mountains, Otero County, New Mexico, *in* Elam, J. C. and Chuber, S., eds., Cyclic Sedimentation in the Permian Basin, A Symposium: Midland, West Texas Geological Society, p. 100–114.

WILSON, J. L., 1975, Carbonate facies in geologic history: New York, Springer-Verlag, 471 p.

WINCHESTER, P. D., 1976, Carbonate diagenesis and cyclic sedimentation in the Wolfcampian Laborcita Formation of the Sacramento Mountains, New Mexico: Unpublished Ph.D. Dissertation, Rice University, Houston, 155 p.

EUSTATIC AND TECTONIC CONTROLS ON CYCLIC DEPOSITION IN LOWER PERMIAN RAMP FACIES (CHASE GROUP AND BASAL WELLINGTON FORMATION) IN THE U.S. MIDCONTINENT

S. J. MAZZULLO

Department of Geology, Wichita State University, Wichita, Kansas 67260

ABSTRACT: Cyclic marine carbonate and marine to terrestrial siliciclastic strata in the upper Wolfcampian Chase Group and basal Wellington Formation were deposited on a low-relief, south-sloping ramp overlying the buried Nemaha ridge. The section comprises eight depositional sequences, each of which consists of two higher-frequency cyclothems of transgressive-regressive character. Each cyclothem includes a transgressive systems tract and ensuing highstand systems tract, and the maximum flooding surface of each sequence consistently occurs in the basal cyclothem of that sequence. Cyclothems are separated by relative lowstand systems tracts of dominantly marginal-marine facies, whereas sequence boundaries are either regionally extensive unconformities or lowstand systems tracts of paleosols. Thickness and facies-stacking patterns of the sequences are arranged so as to define three third-order cycles; component sequences and cyclothems are regarded as fourth- and fifth-order cycles, respectively. The regional occurrence of these cycles is interpreted to suggest glacio-eustatic forcing.

The complex internal architecture of the sequences reflects interplay among forcing parameters and periodic syndepositional tectonism along the Nemaha ridge. Although there is coincidence between maxima of tectonic activity and degree of architectural complexity of sequences within each third-order cycle, the episodic nature of tectonism does not appear to have been consistent enough to have forced cyclicity in this section. Rather, syndepositional tectonism merely overprinted thickness variations of systems tracts within sequences on the fundamentally glacio-eustatically forced sequence stratigraphic architecture of this section.

INTRODUCTION

Cyclic strata are the hallmark of the upper Paleozoic section in the U.S. Midcontinent. Proposed controls on cyclicity in this section include eustasy (Moore, 1936; Elias, 1937), principally glacio-eustasy (Wanless and Shepard, 1936; Heckel, 1986, 1994; Boardman et al., 1995), climatic changes (Brough, 1928; Wanless and Shepard, 1936; Wheeler and Murray, 1957; Cecil, 1990), and tectonics (Moore, 1936, 1964; Weller, 1956; Klein and Kupperman, 1992). Rocks of the Lower Permian Chase Group and basal Wellington Formation that crop out in Kansas and adjoining states record the final stage of cyclic marine carbonate and marine to terrestrial siliciclastic deposition in the Permian of the Midcontinent (Mazzullo et al., 1995). These rocks were deposited on a low-relief ramp and include proximal shoreline to distal-ramp, open-marine environments. Deposition occurred during the transition from an icehouse to a greenhouse world coincident with waning Gondwanan glaciation (Veevers, 1994). Hence, much of the cyclicity in this section likely was controlled largely by glacio-eustasy (Mazzullo et al., 1995, 1997). There is, however, evidence of recurrent tectonism at this time (DuBois and Wilson, 1978; Stander, 1989; Steeples, 1989; Marshak and Paulsen, 1996) that may have overprinted presumed glacio-eustatic cycles in these rocks.

This paper describes regional aspects of depositional environments, facies, and thickness relationships in the Chase Group and basal Wellington Formation. Following the initial description is an evaluation of the sequence stratigraphic architecture of the section, an attempt that is facilitated by the fact that exposures encompass the complete spectrum from shoreline proximal facies to more distal settings on the ramp. Distinctions among allocyclic versus autocyclic and tectonic controls on sequence development are considered. Finally, the paper compares the character of these cycles, deposited during the waning stages of glaciation and transition to a greenhouse world, to cycles in Pennsylvanian rocks deposited during the apex of Gondwanan glaciation in an icehouse world.

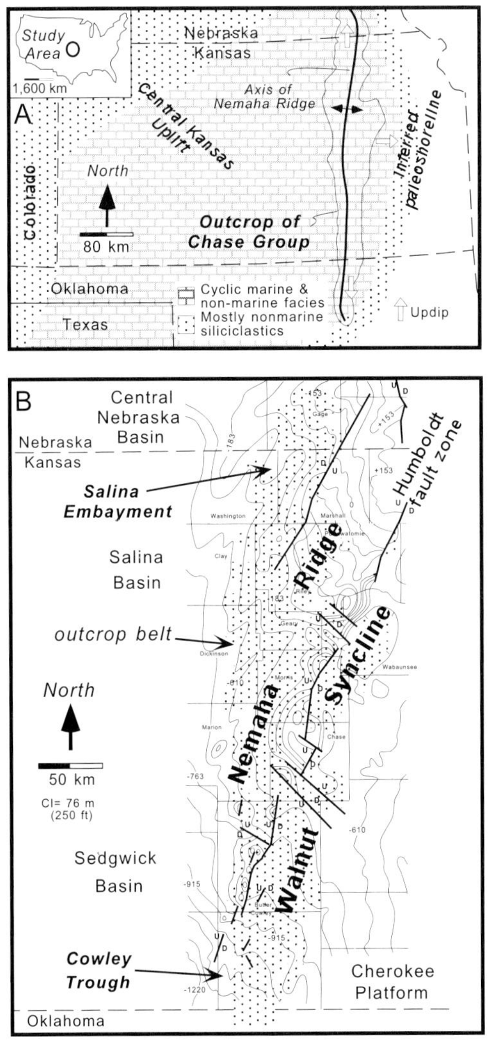

FIG. 1.—(A) Location of study area. (B) Paleogeographic and tectonic features referred to in the text and structure-contour map on the top of the Precambrian along and around the Nemaha ridge (after Burchett et al., 1985). CI = contour interval.

Advances in Carbonate Sequence Stratigraphy: Application to Reservoirs, Outcrops and Models, SEPM Special Publication No. 63

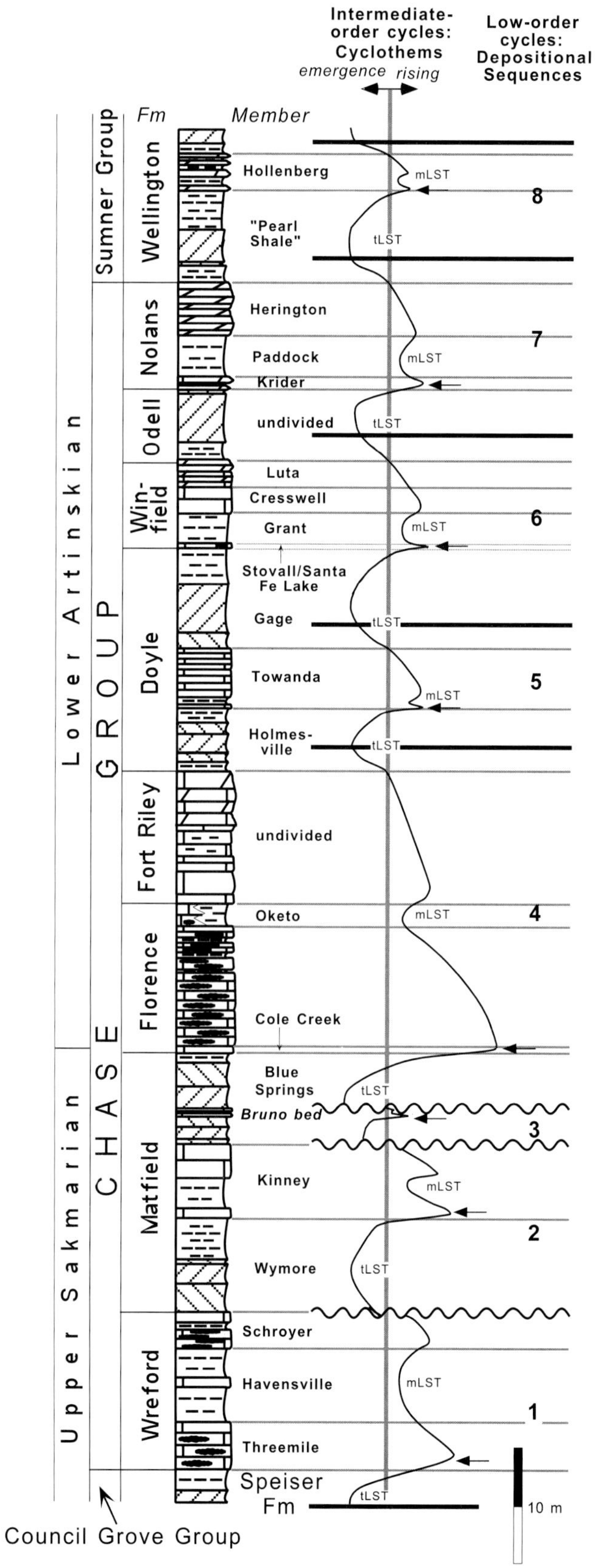

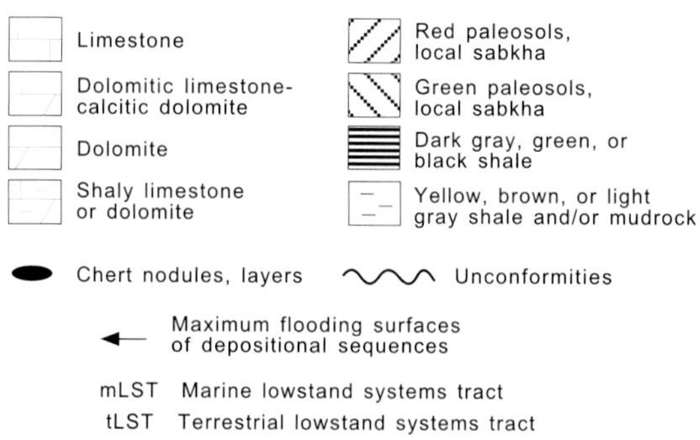

FIG. 2.—Stratigraphy (from Mazzullo et al., 1997) and principal aspects of cyclicity in the Chase Group and basal Wellington Formation (modified from Mazzullo et al., 1995, 1996).

STRATIGRAPHY AND LOCATION

The Chase Group and basal Wellington Formation are exposed from northern Oklahoma to southeastern Nebraska, and lithic correlatives are present in the subsurface to the west (Fig. 1A). This area was a shallow, epi-Pangean sea (Rascoe and Adler, 1983) located just north of the paleoequator in a warm, semiarid climate (Golonka et al., 1994; Parrish, 1995). The outcrop section is 87–136 m of cyclic carbonates and siliciclastics (Fig. 2) assigned to the upper Wolfcampian and basal lower Leonardian, and correlated to the upper Sakmarian to lower Artinskian of Russia (Ross, 1963; Baars et al., 1994a, b). Marine facies grade regionally into terrestrial siliciclastics south of the Kansas-Oklahoma border (Chaplin, 1988). The present study focuses on the cyclic facies north of this pinch-out; the database involved comprises the 200 measured outcrop sections described by Mazzullo et al. (1997). A similar facies transition presumably was present in Nebraska and to the east of the outcrop belt in Kansas, but the rocks have been eroded from these areas. Hence, the depositional setting was a low-relief ramp that became gradually shallower to the north, south, and east of the study area. Preliminary studies of the Chase Group appear in Mazzullo and Teal (1994) and Mazzullo et al. (1995, 1996).

The outcrop belt exposes regionally west-dipping strata overlying the south-plunging, buried Nemaha ridge and the margins of adjoining basins and platforms (Fig. 1B). The Nemaha ridge was uplifted during the Late Mississippian and Early Pennsylvanian and then buried beneath younger Pennsylvanian deposits. It remained as a subtle positive feature in northern Kansas and Nebraska during the Pennsylvanian (Heckel, 1994), and its south plunge served as the foundation of a ramp whereon facies in the Chase Group and basal Wellington Formation generally are of progressively shallower-water aspect northward from the Cowley trough (Fig. 1B). The Humboldt fault zone, a trend of down-to-the-east normal faults, marks the eastern side of the Nemaha ridge adjoining the Walnut syncline. A down-to-the-west normal fault is present along the western side of the Nemaha ridge in northern Kansas and southern Nebraska; the area to the immediate west of this fault is referred to as the Salina embayment. Farther south, the Nemaha ridge

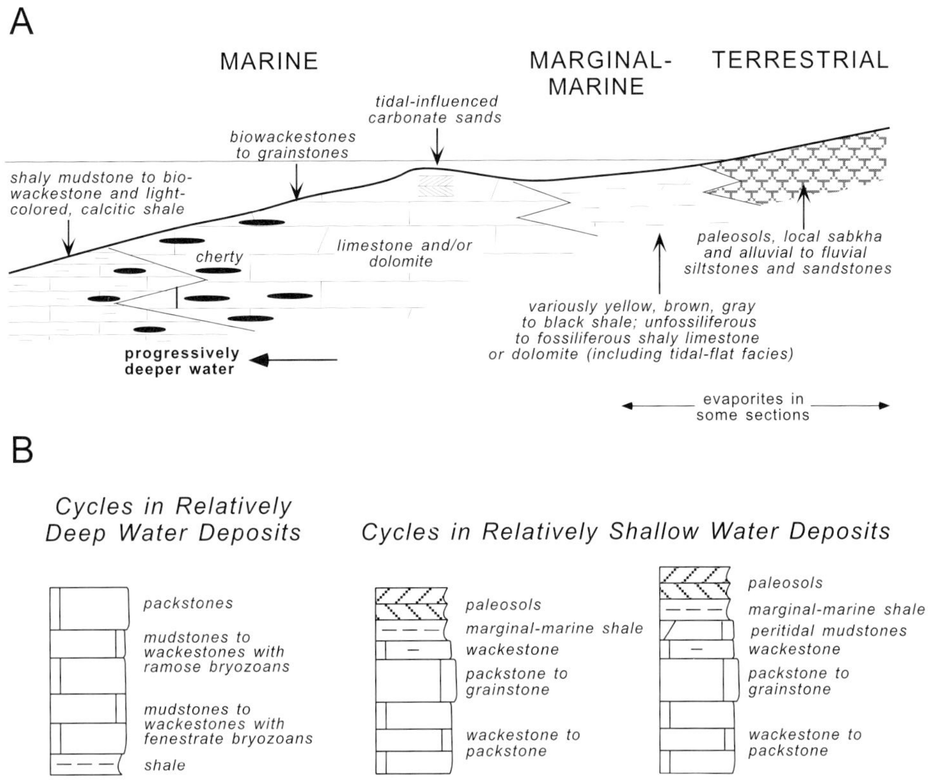

FIG. 3.—(A) Lithologies and inferred depositional environments recognized along the ramp in the study area. (B) Facies of highest-order shallowing-upward cycles.

passes indistinctly, seemingly without a bounding fault, into the Salina and Sedgwick basins. There has been recurrent, mild tectonism along the Nemaha ridge and adjoining areas since the Pennsylvanian (DuBois and Wilson, 1978; Stander, 1989; Steeples, 1989; Marshak and Paulsen, 1996).

LITHOLOGIES AND INFERRED DEPOSITIONAL FACIES

Five principal depositional environments, representing lateral gradation from terrestrial to distal marine facies, are inferred in the section (Fig. 3A). Terrestrial deposits are mostly unfossiliferous, red and green, silty mudrocks with pedogenic overprint, and caliches. Red or green mudrocks without obvious pedogenic overprint, but with relatively abundant geodes composed of mixtures of anhydrite, calcite, and quartz (replaced evaporites), are interpreted as terrestrial sabkha or salina deposits. Laminated to massive, red siltstones and sandstones, commonly with fossil root casts, are alluvial to fluvial deposits.

Marginal-marine facies include (1) lime mudstones, (2) unfossiliferous to sparsely fossiliferous shales, and (3) richly fossiliferous shales and shaly wackestones. The lime mudstones are thin bedded to laminated, variously with microbial laminites, desiccation cracks, fenestrae, intraclasts, lenticular to wavy bedding, paucity of fossils, and, locally, replaced evaporites. Presumed evaporite-dissolution collapse breccias are present in some units in the section. These rocks are interpreted as tidal-flat, including hypersaline sabkha, deposits. The unfossiliferous to sparsely fossiliferous (mainly pectinids and high-spired gastropods) shales are fissile and light gray, brown, dark green, or black. The commonly varved, wavy-laminated, or mottled black and dark green shales may be lagoonal and/or swamp deposits or, in some instances, lower-intertidal to shallow-subtidal deposits on the basis of their similarity to modern deposits (e.g., Reineck, 1967; Frey and Basan, 1978). Light colored shales and shaly wackestones contain pectinids, brachiopods, high-spired and bellerophontid gastropods, and some foraminifers and nautiloids.

Some units include cross-stratified carbonate grainstones with bimodal-bipolar dip directions, *Arenicolites*-type burrows, and some limestone-pebble intraclasts. Allochems are variously dominated by ooids, foraminifers, or oncolites with accessory high-spired and bellerophontid gastropods and pelecypods. These rocks are interpreted as tidal-influenced, nearshore, shallow-marine deposits. Progressively more open-marine, deeper-water environments are represented, in turn, by (1) locally cross-stratified packstones and grainstones on the distal inner ramp; (2) fossiliferous and commonly bioturbated (*Thalassinoides* and/or *Planolites*), fine-grained biopackstones and wackestones on the middle ramp; and (3) shaly mudstones, sparse biowackestones, and light colored shales of the outer ramp. Principal allochems variously include ramose and fenestrate bryozoans, brachiopods and, in some units, rugose and tabulate corals and fusulinids. Glauconite and/or skeletal phosphate, and the high concentration of conodonts in many of the outer ramp shales, identify them as partly condensed maximum flooding surfaces (Mazzullo et al., 1995; Boardman, 1997). Middle and outer ramp carbonates are cherty in the Wreford and Florence Formations.

SEQUENCE STRATIGRAPHY

Framework and Definitions

The Chase Group and basal Wellington Formation are divided into a hierarchy of low-order and intermediate-order cycles (Fig. 2). The eight low-order cycles are bounded by terrestrial deposits that compose terrestrial lowstand systems tracts (tLSTs). These cycles are considered depositional sequences, each of which in turn consists of two intermediate-order, transgressive-regressive packages of strata referred to as cyclothems. The cyclothems compose two sections of dominantly marine carbonates, near the bases and tops of sequences, separated by marginal-marine facies (Fig. 2). In a gross sense, the marine carbonates represent highstand conditions relative to the intervening marginal-marine facies that, therefore, are lowstand deposits. These lowstand tracts, however, are distinct from the tLSTs in the section; for this reason, they are not considered to enclose sequence boundaries. In a departure from established sequence stratigraphic terminology dictated by the nature of these cycles, these lowstand tracts are referred to as marine lowstand systems tracts (mLST). Each cyclothem includes a transgressive systems tract (TST), maximum-flooding surface (MFS), and an ensuing highstand systems tract (HST). The MFS of each sequence consistently occurs in the lower cyclothem of that sequence (Fig. 2); hence, inferred water depths of successive cyclothems within a sequence shallow upward (Mazzullo et al., 1997). Only the MFS of each sequence is shown in ensuing figures. Wherever possible, the HSTs of cyclothems are divided into an early HST, deposited when the rate of sea-level rise was declining, and a late HST that represents the regressive phase of cyclothem deposition. In standard terminology, the mLST would be included within the latest HST of the lower cyclothem, and the basal TST of the upper cyclothem within a sequence.

Higher-order, transgressive-regressive cycles of presumed lesser duration also are recognized in the section. Some of them are considered to be "minor cyclothems," but not in sense that this term was used by Heckel (1986, 1994). They are discussed in the following text. Meter and sub-meter scale, shallowing-upward facies tracts are the highest-order (i.e., highest frequency) cycles recognized (Fig. 3B).

Depositional Sequence 1

The thickness of this sequence is relatively uniform across the study area, although thickness relationships of the component cyclothems are complex (Fig. 4). The tLST is composed of stacked paleosols. Lenses of fluvial, cross-stratified sandstone in this tract are present only in southernmost Kansas and northern Oklahoma. Although not shown in Figure 4 because of space limitations, the tLST is thin within the Salina embayment and thickens to the south. The TST of the lower cyclothem deepens upward from marginal-marine shales and more open-marine limestones dominated by brachiopods (*Derbyia, Composita, Juresania*, and chonetids), to cherty biowackestones with bryozoans. The MFS is recognized throughout the study area as a partly condensed section of glauconitic, shaly biowackestones and fossiliferous shales. The early HST is mainly cherty, micritic limestones dominated by fenestrate and ramose bryozoans and accessory brachiopods (*Reticulatia* and *Composita*) and crinoids. There is no evidence of significant shallowing of depositional environments to the north within the TST, MFS, or the early HST. The entire cyclothem is very thin within the Salina embayment, and thickens southward into the Walnut syncline and Cowley trough. The late HST is composed of a wedge of shallow-water, noncherty foraminiferal grainstones that is present only within the central and southern parts of the Walnut syncline and in the Cowley trough. It overlies deeper-water cherty limestones with marked facies dislocation that suggests some degree of forced regression. There is no evidence, however, of an unconformity at the top of this section. To the north, this systems tract instead is a thin section of interbedded marginal-marine shales and shallower-water cherty biowackestones and packstones at the gradational top of the Threemile Member (Wreford Formation). The HST is anomalously thick where it overlies a graben along the edge of the Nemaha ridge (Fig. 4). The rocks are noncherty, bryozoan bafflestones interpreted as a bryozoan reef or a series of closely spaced, laterally coalesced reefs (Mazzullo et al., 1995, 1997).

The mLST is relatively thin and dominated by shallow-marine limestones, with a wedge of biograinstones at the top, in the Cowley trough. The environments of deposition generally shallowed and units thickened dramatically updip to the north, onto the Nemaha ridge and within the Salina embayment, where the rocks are mostly fossiliferous shales and some thin, shaly limestones. Figure 4 does not illustrate the regional lithologic continuity of this systems tract but, rather, shows important lithologic variations that are discussed below. Where the mLST is dominantly shale, for example, consistently near its middle is a thin limestone (horizontal arrows in Fig. 4) with open-marine fossils. It and immediately adjoining strata are interpreted as a minor cyclothem. Lenses of incipient paleosol occur locally above and below this limestone on the Nemaha ridge, but they are not regionally correlative. This limestone thickens abruptly into cross-stratified grainstones within an area that overlies a graben along the eastern side of the Nemaha ridge (southern Riley County section 1, Fig. 4). These grainstones are interpreted as barrier-beach deposits on the basis of geometry and facies relationships.

The upper cyclothem is anomalously thick within a linear trend that extends across the northern part of the Nemaha ridge and into the Salina embayment. Otherwise typically composed of marginal-marine shales, its TST in these areas is a thick section of tidal-flat lime mudstones overlain by tidal-influenced, cross-stratified biograinstones (Fig. 4). The remainder of the TST and the early HST along the central and northern parts of the ramp consists of cherty mudstones, wackestones, and some biograinstones dominated by fenestrate and ramose bryozoans.

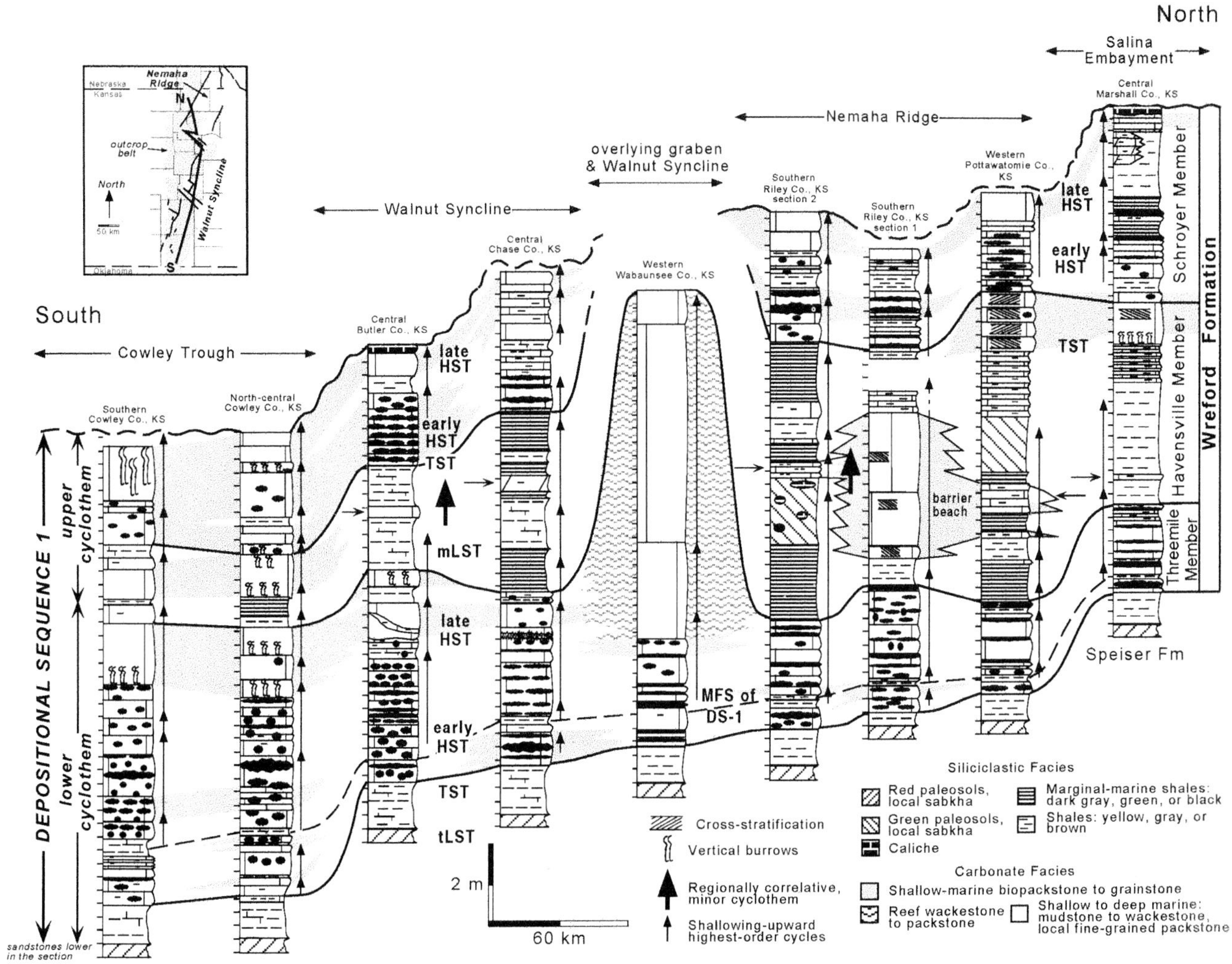

FIG. 4.—Sequence stratigraphy of depositional sequence 1 (DS 1) along a north-south section (see inset map for location) in Kansas (KS); counties (Co.) are identified in Figure 1B. Columnar sections in Figures 4–9 are divided into 0.3-m segments. LST = lowstands systems tract, tLST = terrestrial lowstand systems tract, mLST = marine lowstand systems tract, TST = transgressive systems tract, HST = highstand systems tract, MFS = maximum-flooding surface.

Southward shallowing into the southern Walnut syncline and Cowley trough is indicated, however, by the change in these areas to biopackstones and grainstones. The maximum water depth of this cyclothem is inferred to have been less than that in the lower cyclothem. Over most of the study area, the late HST is a section of shallow-water biopackstones and grainstones that thicken southward into the Walnut syncline and Cowley trough. The top of the sequence is a regional unconformity at the top of the Wreford Formation. It is identified by calichified horizons, eluviated green and red clay in secondary pores in the subjacent grainstones, and, locally, erosional thinning of these grainstones.

Depositional Sequence 2

The tLST of this sequence (Fig. 5A) is anomalously thick within the Salina embayment, overlying the similarly thick upper cyclothem of sequence 1. It consists of stacked paleosols that compose the entire Wymore Member (Matfield Formation); the TST of the lower cyclothem here is within the basal carbonates of the Kinney Member (Matfield Formation). The tLST generally thins southward toward the Cowley trough, but it is anomalously thin where it overlies a graben along the edge of the Nemaha ridge in central Kansas. Unfossiliferous marginal-marine shales overlie paleosols at the top of the Wymore Member in these areas and, together with basal carbonates in the Kinney Member, compose the TST of the lower cyclothem. The thickness of the overlying section appears to compensate for that of the tLST so that the thickness of the sequence is relatively uniform across the study area. However, within the Cowley trough and where the sequence overlies the graben along the edge of the Nemaha ridge in central Kansas, the sequence is anomalously thick and composed of only shallow- and marginal-marine deposits. The sequence is less extensive areally than sequence 1 in that it passes into terrestrial siliciclastics only a few kilometers south of the Kansas-Oklahoma border.

The TST and HST of the lower cyclothem include a complex mosaic of shallowing-upward cycles that compose facies tracts

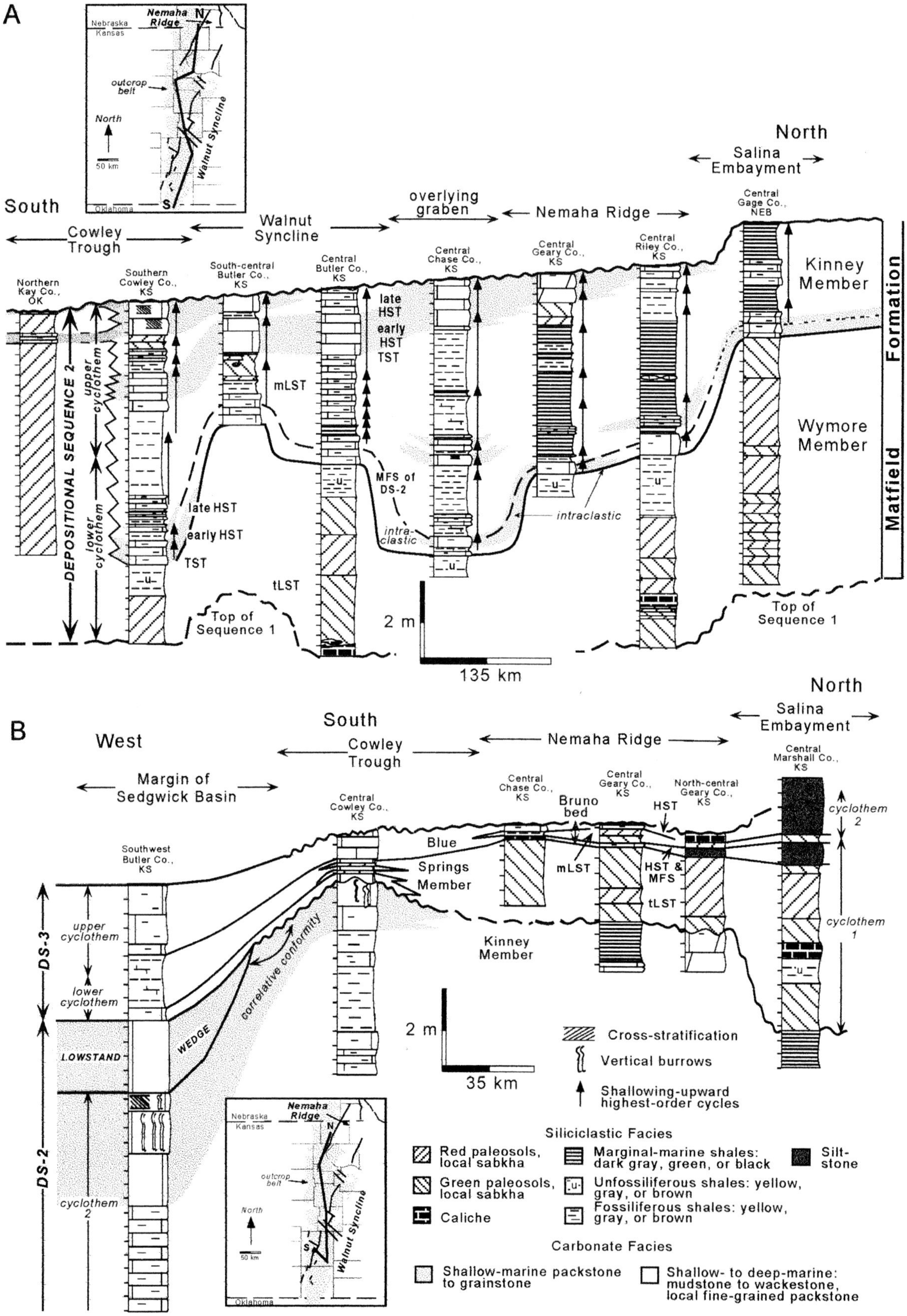

FIG. 5.—Sequence stratigraphy of depositional sequences 2 (A) and 3 (B). OK = Oklahoma, NEB = Nebraska; for explanation of other labels, see Figure 4.

wherein environments of deposition are of more open-marine aspect to the south. These facies are (1) pelecypod-rich (*Aviculopecten*) wackestones and packstones in updip locations, (2) intraclastic biowackestones with lenses of marginal-marine shale and/or bioclastic sandstone on the middle ramp, and (3) a thicker section of brachiopod-rich (mainly *Derbyia* and *Composita*) wackestones to packstones and fossiliferous shales downdip, in the Walnut syncline and Cowley trough. According to Boardman (1997), the MFS of this sequence is recognized within the latter facies tract only from southern to central Kansas. On the basis of lithology, biota, and the limited northward extent of the MFS, the maximum water depth of this cyclothem, and of the sequence, is inferred to have been considerably less than in the upper cyclothem of sequence 1. The HST grades upward into the mLST of the sequence which, over most of the study area, is dominated by *Derbyia*-rich shales.

The TST and HST of the upper cyclothem are composed of locally cross-stratified, foraminiferal packstones and grainstones that define sediment wedges that thicken into the Walnut syncline and Cowley trough (Fig. 5A). Farther updip, to the north, the depositional environments became progressively shallower, as shown by shaly, nearshore marine carbonates to marginal-marine shales. The top of the sequence is a regional unconformity at the top of the Kinney Member that is recognized by the same lines of evidence that identify the unconformity at the top of sequence 1. The cyclothem thickens westward into the southern Sedgwick basin, where the unconformity at its top passes downdip into a correlative conformity (Fig. 5B). The shallow-marine biopackstones and grainstones that overlie this correlative conformity are interpreted as a lowstand wedge deposited when most of the study area was subaerially exposed.

Depositional Sequence 3

This unconformity-bounded sequence is the thinnest and most areally restricted in the Chase Group (Fig. 5B). It is capped by the Bruno limestone bed, which generally is less than 0.3 m thick, and correlative strata. The Bruno actually is two beds of carbonate, separated by siliciclastics, that compose the two cyclothems in the sequence. The carbonates are present only in central and southern Kansas, and they pass updip to the north and south into terrestrial siliciclastics, the section of which progressively thickens onto the axis of the Nemaha ridge and into the Salina embayment. The lower cyclothem in distal-ramp locations, including the TST, MFS, and undivided HST, consists of foraminiferal sands deposited in shallow water. Farther north, this section is represented by locally intraclastic, micritic limestones with pelecypods. The inferred maximum water depth of this cyclothem and the sequence was much less than in sequence 2. The mLST is unfossiliferous shale that passes northward into paleosols. The upper cyclothem on the distal ramp consists mainly of shallow water, micritic limestones with pelecypods, and some interbeds of foraminiferal sands. At many localities, this cyclothem is either locally replaced by caliche or is absent as a result of postdepositional erosion. The sequence thickens to the west, and is of relatively more marine aspect, along the southern margin of the Sedgwick basin, where it overlies the lowstand wedge in sequence 2 (Fig. 5B).

Depositional Sequence 4

This sequence includes the thickest section of carbonates in the Chase Group, and the thickness relationships of component cyclothems and systems tracts are highly complex (Fig. 6). The thickness of the tLST, for example, varies considerably across the study area, although the entire lower cyclothem thins considerably into the Cowley trough. The TST grades upward from marginal-marine, pelecypod-rich shales to more open-marine shales, *Derbyia*-rich wackestones, and local foraminiferal biopackstones. The early HST is mainly cherty mudstones and biowackestones dominated by fenestrate and ramose bryozoans, fusulinids, and brachiopods (notably *Reticulatia*). Although the TST and early HST are pervasively replaced by dolomite to the north, there is no indication that depositional environments shallowed significantly in that direction. The MFS is a partly condensed section of light colored, fossiliferous shale, locally glauconitic and with skeletal phosphate, that is traced throughout the study area. The maximum water depth of this cyclothem and sequence appears to have been greater than in sequences 1–3. The late HST consists of interbedded cherty carbonates and marginal-marine shales in the north and a thinner section composed entirely of limestones of more open-marine aspect to the south. In middle and updip parts of the ramp, the mLST is a section of fossiliferous, and locally oncolitic, silty shales and shaly carbonates (Oketo Member of Florence Formation, Fig. 6). This systems tract appears to pass downdip into a correlative wedge of noncherty, relatively shallow water biowackestones and packstones—locally oncolitic and with small biostromes of rugose corals near the top—that thickens into the Cowley trough. A few centimeters of silty carbonate is present at the top of this section.

The upper cyclothem thickens southward into the Walnut syncline and Cowley trough and also into the northern Sedgwick basin. It includes depositional environments of significantly more shallow-water aspect than in the lower cyclothem, although facies are of progressively more marine aspect to the south and west. The TSTs and early HSTs, for example, pass in these directions from peritidal dolomites, to poorly fossiliferous nearshore facies, to a somewhat thicker wedge of shallow-water, oncolitic packstones and grainstones in the Cowley trough. What ostensibly would be considered the late HST of this cyclothem is architecturally complex. The middle of this section appears to be a minor cyclothem that thickens as a wedge into the Sedgwick basin and Walnut syncline and then thins southward into the Cowley trough. Its transgressive-regressive character is best observed in downdip locations, where the section consists of relatively open-marine facies that are overlain and underlain by marginal-marine deposits (e.g., central Butler County, Fig. 6). The unit consists regionally of a complex of highest-order cycles that include peritidal dolomites and marginal-marine shales on the north and marginal-marine shales, more fossiliferous nearshore carbonates (wackestones

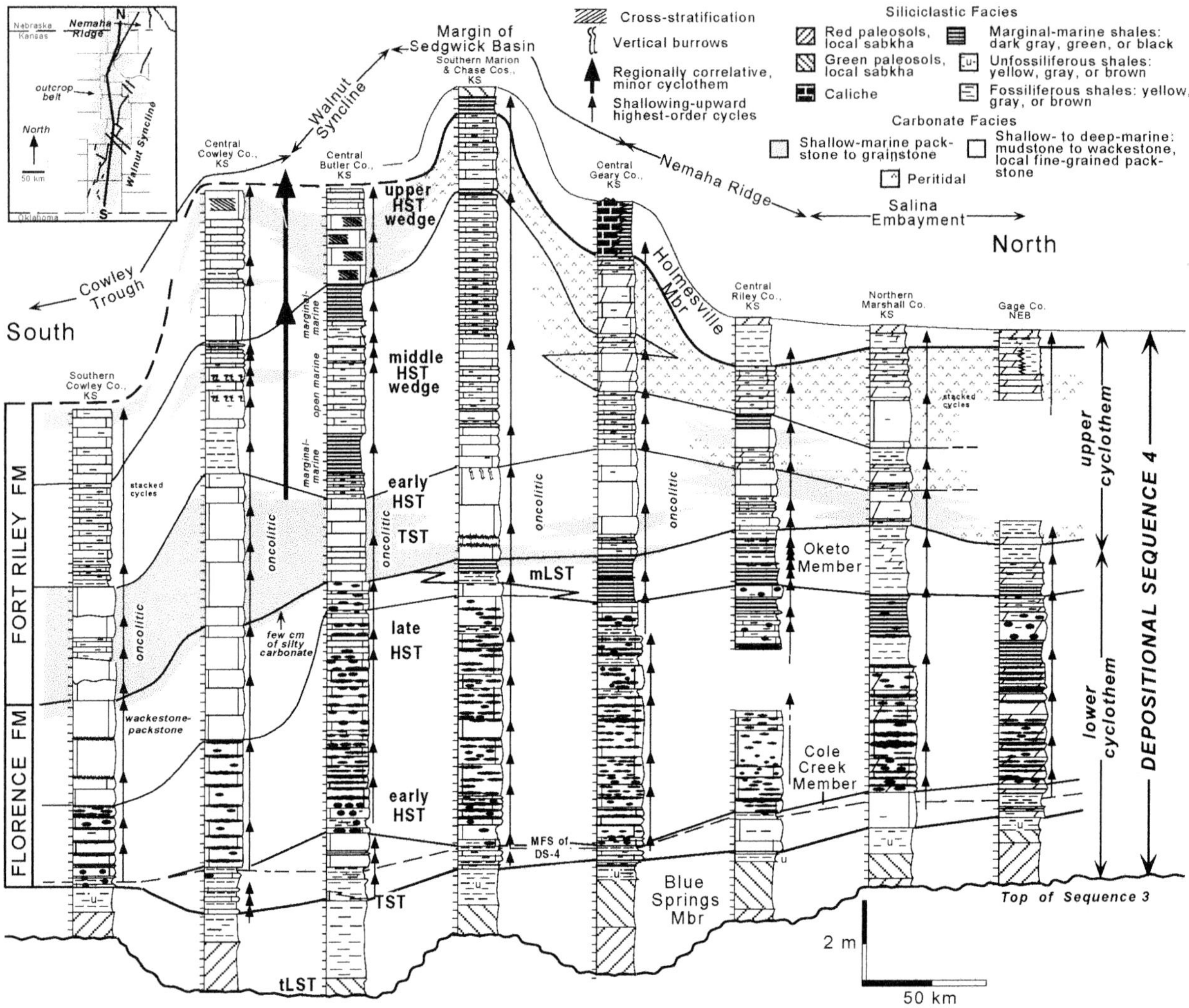

FIG. 6.—Sequence stratigraphy of depositional sequence 4. FM = formation, Mbr = member; for explanation of other labels, see Figure 4.

and some packstones) and, locally, grainstone units, to the south. It is recognized from northern Oklahoma into northernmost Kansas. It seemingly is overlain by another minor cyclothem that also thickens as a wedge into the Walnut syncline and northern Cowley trough and then thins farther to the south. Environments of deposition are represented by stacked peritidal deposits in the north and nearshore-marine and cross-stratified foraminiferal grainstones that shallow upward into lagoonal facies to the south. The unit is also recognized over much of the study area, although it loses its identity within stacked peritidal deposits in northern Kansas. The cyclothem grades upward into marginal-marine shales and, locally atop the Nemaha ridge, into caliche.

Depositional Sequence 5

This sequence (Fig. 7) is more restricted areally than sequence 4. The thickness is relatively uniform across the study area, although thickness relationships of the component systems tracts are moderately complex. The tLST, for example, is a wedge of stacked paleosols that thins northward onto the Nemaha ridge and into the Salina embayment. It includes a minor cyclothem of marginal-marine shales and, locally, micritic limestones, that is present at many localities in the Cowley trough, Walnut syncline, and the Sedgwick and Salina basins. The overlying marine and marginal-marine phases of the sequence thicken considerably into the Salina embayment, but they are anomalously thin at a locality along the transition from the Nemaha ridge into the Salina basin that appears to overlie a paleotopographic high (Morris County, Fig. 7).

Stacked peritidal cycles compose most of the TSTs and HSTs of the lower and upper cyclothems in updip-ramp locations to the north. To the south, these systems tracts are a complex mosaic of highest-order cycles that variously include peritidal limestones, marginal-marine shales, and shallow-marine, foraminiferal wackestones to locally cross-stratified grainstones. The MFS of the sequence is represented by a noncondensed section of oolite grainstones present only in downdip-ramp locations. The inferred maximum water depth is less than in sequences 1, 2, and 4. The mLST is dominantly

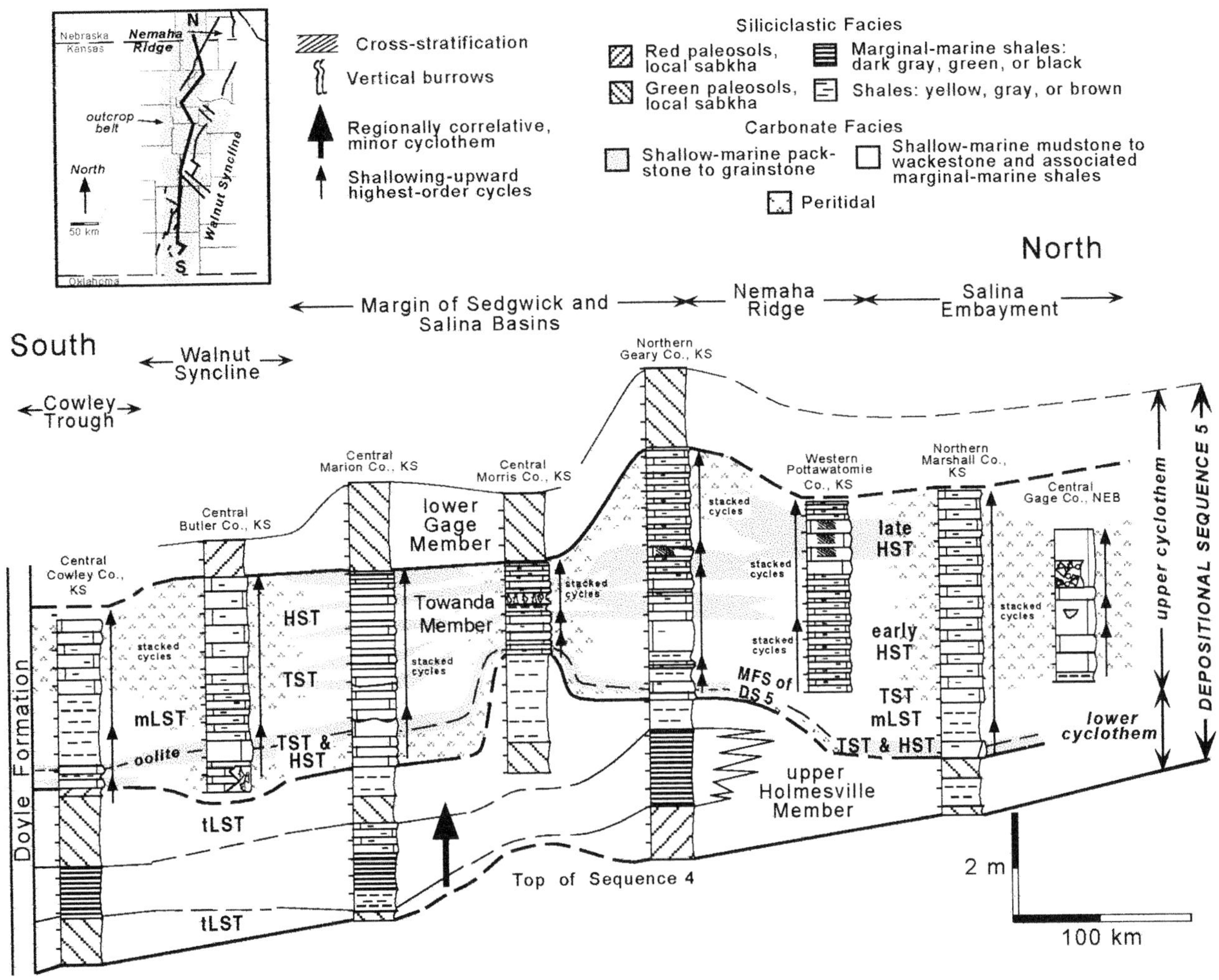

FIG. 7.—Sequence stratigraphy of depositional sequence 5. For explanation of labels, see Figure 4.

unfossiliferous shale that passes locally into peritidal or marginal-marine limestones within the Walnut syncline and Sedgwick and Salina basins. Locally rooted at the top, the cyclothem grades upward into incipient paleosols.

Depositional Sequence 6

The internal architecture of this sequence is highly complex (Fig. 8). It thickens within the Cowley trough, Sedgwick and Salina basins, and along the axis of the Nemaha ridge in northern Kansas. Stacked paleosols of the tLST, however, thin northward across the axis of the Nemaha ridge and into the Salina embayment, where most of the section is represented by two minor cyclothems of marginal-marine shale and/or micritic limestone. The lower minor cyclothem is restricted to the Salina embayment, whereas the upper one appears to be of more regional extent. The TST of marginal-marine shales passes upward into a thin, undivided HST and cryptic MFS of cherty biowackestone, with bryozoans and brachiopods (*Reticulatia* and *Derbyia*), on the middle and upper ramp (Stovall Member, Winfield Formation). The inferred maximum water depth of this sequence is greater than that in sequence 5. This unit passes southward, however, into a correlative section of shales and noncherty, shaly mudstones, locally with interbeds of oncolitic, foraminiferal wackestones (Santa Fe Lake Member). These rocks are interpreted as peritidal and nearshore-marine deposits. Facies relationships between the Santa Fe Lake and Stovall Members suggest regional shallowing to the south. This shallowing is not due to an increased thickness of subjacent tLST deposits for which overlying facies would have compensated.

The mLST is thin within the Cowley trough and thickens to the north. The change from highly fossiliferous shales and oncolitic limestones to less fossiliferous, marginal-marine shales indicates that the water depths once again shallowed northward. The upper cyclothem is relatively thin in the Salina embayment and thickens somewhat along the axis of the Nemaha ridge in northern Kansas. It thickens dramatically into the Salina and Sedgwick basins and then thins southward into the Cowley trough. The TST of fossiliferous shales passes upward into an early HST composed of stacked peritidal cycles in the north, and a wedge of shallow-marine, oncolitic biopack-

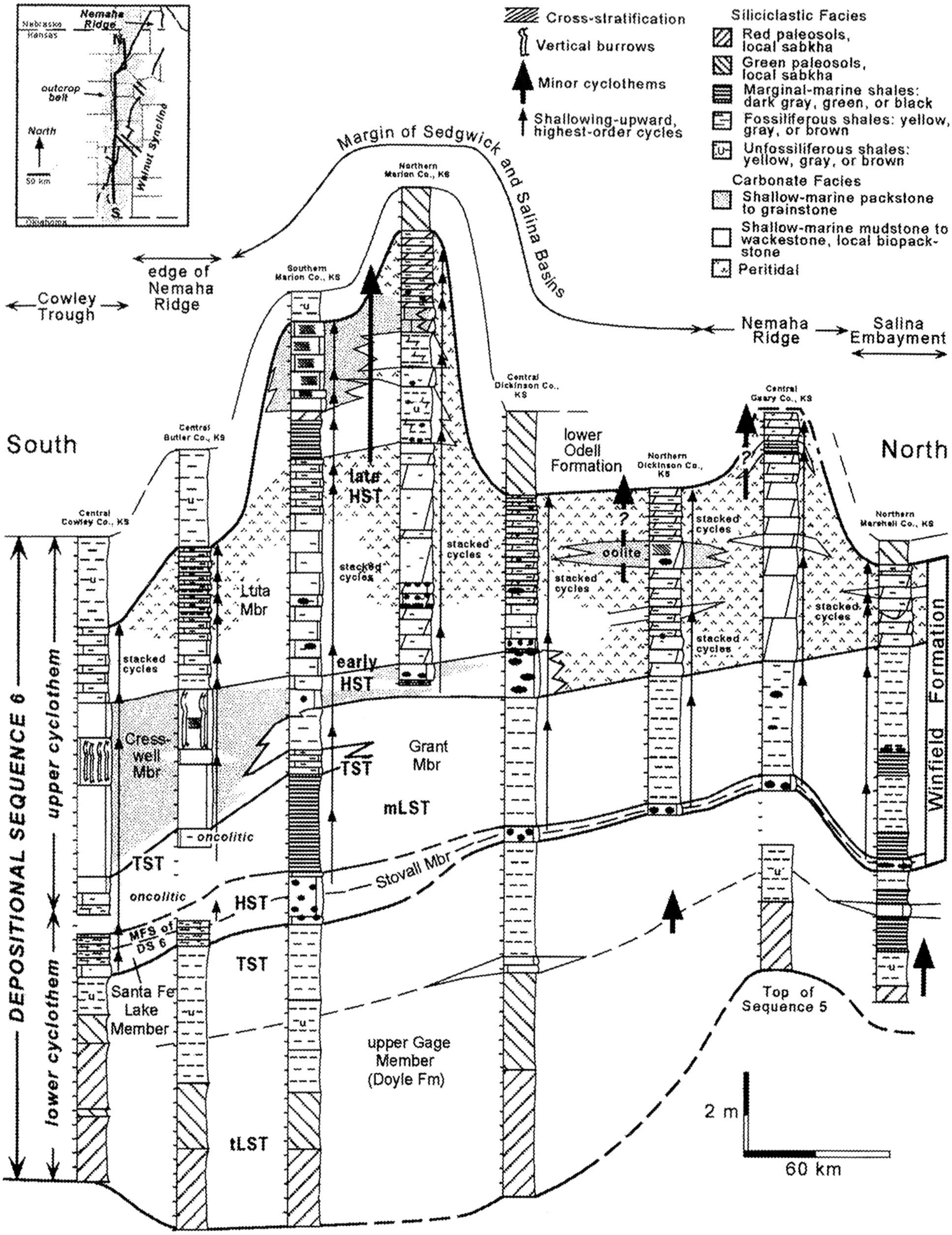

FIG. 8.—Sequence stratigraphy of depositional sequence 6. For explanation of labels, see Figure 4.

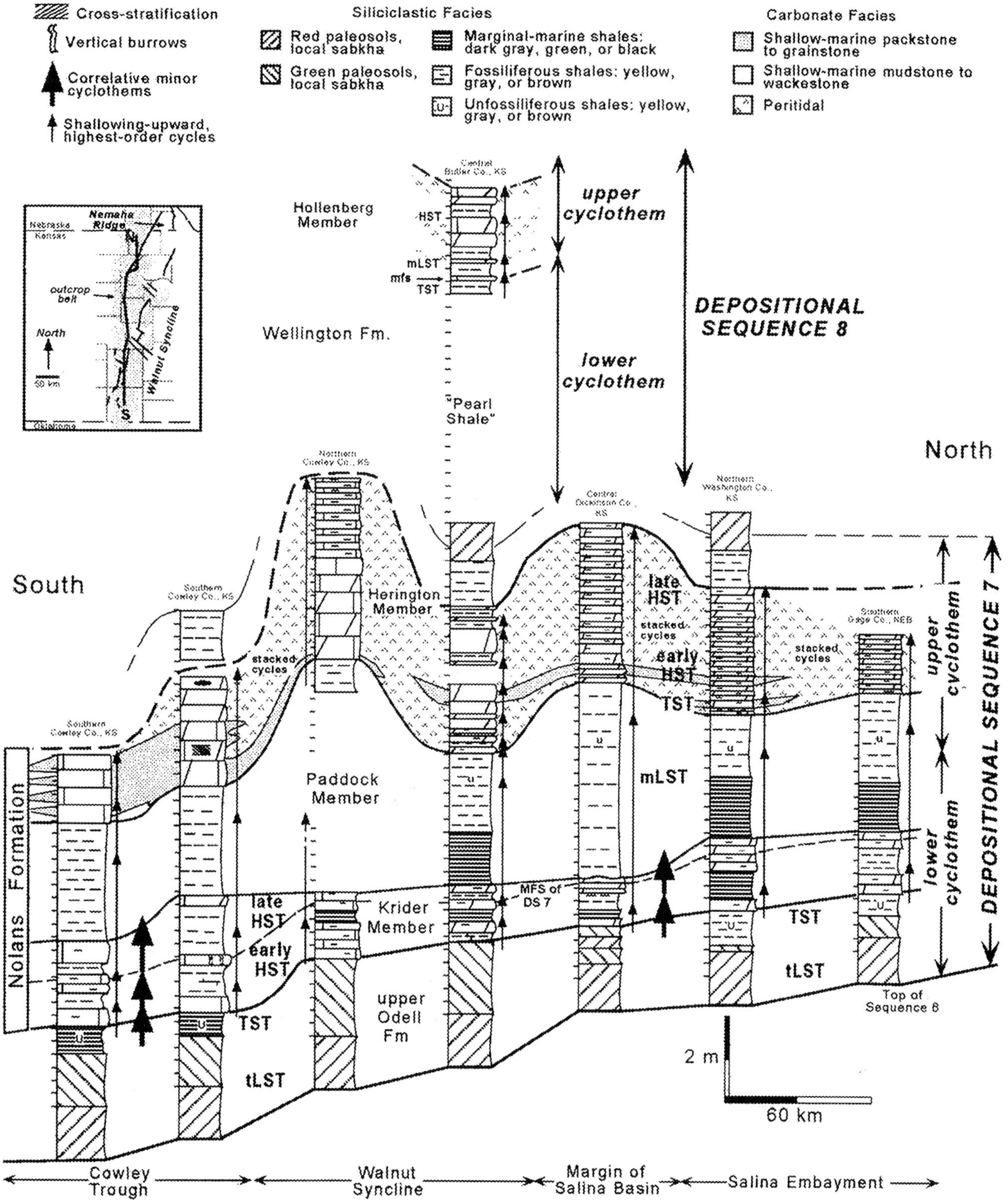

FIG. 9.—Sequence stratigraphy of depositional sequences 7 and 8. For explanation of labels, see Figure 4.

stones and grainstones that thickens southward into the Cowley trough. The late HST likewise consists dominantly of stacked peritidal cycles in the north, which prograded over nearshore-marine, generally sparsely fossiliferous (*Composita*, ramose bryozoans, crinoids) mudstones and wackestones to the south. The expanded section along the margin of the Salina and Sedgwick basins includes prominently cross-stratified, shallow-marine grainstones that pass updip into marginal-marine shales and peritidal dolomites. This section is considered a minor cyclothem that is readily recognized here and, possibly, at two other localities farther north (Fig. 8). The cyclothem grades upward into paleosols on the north and marginal-marine shales to the south.

Depositional Sequence 7

This sequence thickens within the Walnut syncline and along the margin of the Salina basin (Fig. 9). Stacked paleosols of the tLST are locally overlain by a basal TST of marginal-marine shales, both of which thicken slightly to the south. The TST occurs within basal beds of the Krider Member (Nolans Formation) in those areas where paleosols compose all of the upper Odell Formation. In the Cowley trough, the TST and HST of the lower cyclothem include three beds of wackestone to packstone with a fairly diverse biota (*Composita, Aviculopinna*, bryozoans, and crinoids) separated by marginal-marine, fossiliferous shales (*Composita* and oncolites). This stratigraphy also is recognized southward into Oklahoma. In contrast, only two beds of micritic carbonate separated by marginal-marine shales, both of which include a more restricted, shallower-water biota dominated by pectinids, are present to the north. This stratal architecture defines a sediment wedge of more open-marine facies that thickens into the Cowley trough. The regional extent of these limestone-shale beds suggests they are minor cyclothems. The MFS of the sequence appears to be in the middle limestone in southern Kansas and the correlative upper carbonate to the north. The inferred maximum water depth is less than that in sequence 6.

The mLST and basal TST of the upper cyclothem are poorly fossiliferous (mainly pectinids) shales that thicken into the Salina basin and Walnut syncline. The upper TST and the HST are peritidal dolomites over most of the ramp, and they pass downdip into a thinner section of shallow-marine biopackstones and grainstones within the Cowley trough. The cyclothem grades upward into marginal-marine shales, locally with abundant quartz-replaced evaporite nodules, in the basal Wellington Formation.

Depositional Sequence 8

This sequence (Fig. 9) is poorly exposed in the study area and is more restricted areally than sequence 7. The cyclothems compose highest-order cycles dominated by peritidal dolomites with replaced evaporites, unfossiliferous marginal-marine shales, and minor interbeds of mudstone to sparse biowackestone bearing fragments of thin-shelled pelecypods. The MFS is a bed of foraminiferal wackestone to packstone in the lower cyclothem, the inferred maximum water depth of which is less than that in sequence 7.

DISCUSSIONS

Allocyclicity and Cycle Hierarchy

The eight depositional sequences and component cyclothems in the Chase Group and basal Wellington Formation are recognized throughout the study area; in fact, they are correlated to equivalent cycles as far away as Texas (Boardman, 1997). They were deposited during a time of known Gondwanan glaciation (Veevers, 1994). Their deposition is therefore believed to have been driven by glacio-eustasy (Mazzullo et al., 1995, 1997). The sequences and cyclothems are stacked so as to define three lowest-order cycles (A, B, and C: Fig. 10), within each of which (1) the thickness and areal extent of successive sequences decrease upward, (2) the deepest-water sequence (i.e., the deepest-water MFS) occurs at the base of the cycle, and (3 depositional environments of successive cyclothems and sequences shallow upward. Such thickness and facies stacking patterns are consistent with those of eustatically forced, third-order sequences (e.g., Goldhammer et al., 1993). Accordingly, the component depositional sequences and cyclothems within each of these third-order cycles are regarded as fourth- and fifth-order cycles, respectively. The shapes of the curves of these cycles suggest rapid transgression followed by slow, prolonged regression that, according to Heckel (1994), is characteristic of glacio-eustatic forcing in Midcontinent Pennsylvanian strata.

Boardman (1997) regarded the underlying Council Grove Group, which likewise includes a hierarchy of depositional sequences and component cyclothems, as a single third-order sequence. A similar interpretation for the Chase Group and basal Wellington Formation is not considered tenable because the thickness and facies-stacking patterns of the sequences within cycle A thin and shallow upward rather than thicken and deepen upward, as might be expected if this cycle had been deposited during a time of increasing accommodation (overall transgression) that led to the maximum water depths (i.e., the maximum highstand) represented by depositional sequence 4 (cycle B). Thickness and inferred maximum-depth relationships among third-order cycles A, B, and C (Fig. 10) may suggest that the magnitudes of glacio-eustatic highstands increased and then decreased over time. Such a pattern of cyclicity may reflect relatively short-term changes in Gondwanan ice volumes during overall waning of glaciation in the Early Permian (Veevers, 1994). Sequence-bounding unconformities within cycle A truncate late-HST subtidal carbonate deposits, which suggests forced regression. Peritidal facies are not recognized in the sequences that compose this cycle. In contrast, sequence boundaries are of normal regressive character within cycles B and C and, with the exception of the lower cyclothem in sequence 4, involved shallowing upward of late HSTs of component cyclothems into peritidal facies. The presence or absence of sequence-bounding unconformities may be related to the timing and/or the availability of siliciclastic influx such that (1) unconformities develop on

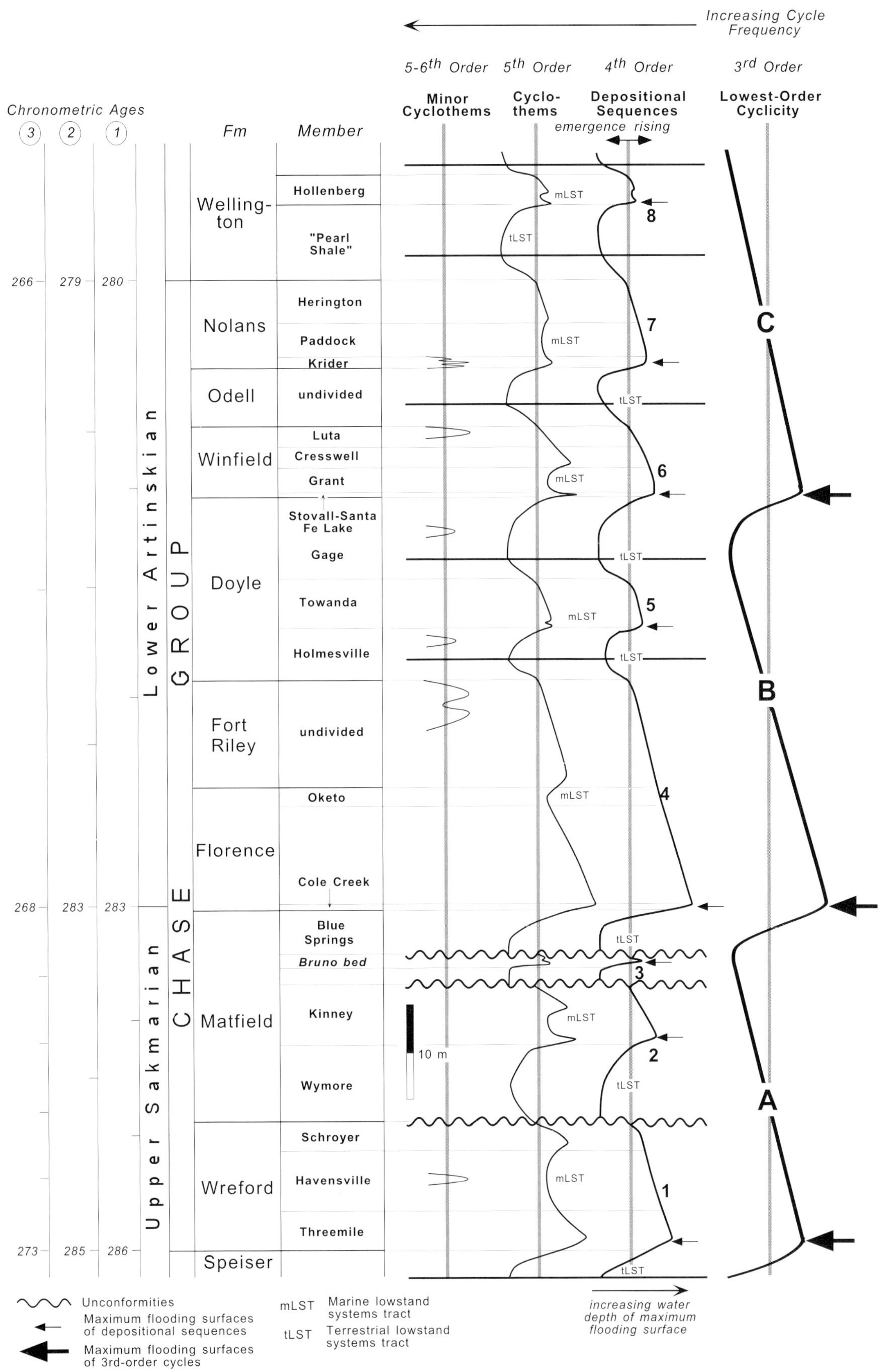

FIG. 10.—Detailed cyclostratigraphy of the Chase Group and basal Wellington Formation, and inferred relative water depths of the maximum flooding surfaces in depositional sequences. Chronometric scales (in millions of years) are from (1) Menning (1995), (2) Klein et al. (1994), and (3) Harland et al. (1982).

exposed carbonates when the siliciclastic influx initially lags behind regression; siliciclastics are then deposited on these surfaces during the later stages of lowstand; and (2) normal regressive sequence boundaries instead develop when regression is accompanied by rapid siliciclastic influx. Alternatively, the presence of sequence-bounding unconformities within cycle A may reflect pulses of tectonic uplift at these times.

Many of the minor cyclothems in the section appear to be of regional extent, and they also may be glacio-eustatically forced, fifth- or sixth-order cycles. Whereas the thicknesses of some of the highest-order, shallowing-upward cycles in the section display a stacking pattern that is consistent with eustasy (e.g., Goldhammer et al., 1993), many of them do not. Likewise, as much as 40% of these cycles are not regionally correlative (Figs. 4–9). It is therefore likely that they include both eustatically forced and Ginsburg-type autocycles, and it is possible that deposition of some of these autocycles was influenced by syndepositional tectonism (discussed below).

Cycle Periodicities

Cycle periodicities are impossible to estimate with any confidence because of the scant database of radiometric ages, disagreements concerning global correlations of Permian strata, and differences in reported dates (see Waterhouse, 1978). Compounding these problems is the currently untestable possibility that successively thinner fourth-order cycles within the third-order cycles are not of equal duration. With these reservations in mind, average durations of the eight sequences and component cyclothems (excluding minor cyclothems) were estimated on the basis of radiometric data from several sources (Table 1). These duration estimates ranged from 0.41 to 1.43 m.y. and from 0.2 to 0.72 m.y., respectively, and thus would have included Milankovitch eccentricity periodicities. If one assumes that all of the highest-order, shallowing-upward cycles were eustatically forced, then minimum periodicities of the average number of such cycles within sequences may be estimated to have ranged from 0.014 to 0.054 m.y., which seemingly would include the precession and obliquity bands.

TABLE 1.—ESTIMATED DURATIONS OF CYCLES IN THE CHASE GROUP AND BASAL WELLINGTON FORMATION (EXCLUDES MINOR CYCLOTHEMS)

	Duration (m.y.)		
	Menning (1995)	Klein et al. (1994)	Harland et al. (1982)
Cycle C	1.86	2.45	1.23
Each sequence (6, 7, and 8)	0.62	0.82	0.41
Each cyclothem*	0.31	0.41	0.2
Highest-order cycles**	0.02	0.027	0.014
Cycle B	2.04	2.58	1.82
Each sequence (4 and 5)	1.02	1.3	0.91
Each cyclothem*	0.51	0.65	0.45
Highest-order cycles**	0.032	0.04	0.028
Cycle A	2.6	1.72	4.3
Each sequence (1, 2, and 3)	0.87	0.57	1.43
Each cyclothem*	0.43	0.29	0.72
Highest-order cycles+	0.032	0.022	0.054

*Two cyclothems within each sequence.

**Average number within each cyclothem, assuming that all are of glacio-eustatic origin.

Sequence Development and Syndepositional Tectonic Overprint

The thicknesses of cyclothems and systems tracts in the section vary both temporally and spatially, but there is not a systematic relationship to the bathymetric profile of the low-relief ramp that could be allied to glacio-eustatically induced changes in accommodation (Figs. 4–9). That is, such thickness variations have not significantly altered basic depth-related facies patterns within cyclothems, and they do not appear to have been induced by differential compaction of shales (Mazzullo et al., 1997). Many areas of anomalous thickness directly overlie structural elements along and around the Nemaha ridge, which has remained active tectonically since the Pennsylvanian. For these reasons, the most prominent thickness variations in the section are considered to be proxies for tectonically induced changes in accommodation during deposition. Figure 11 equates thickness variations to what are considered tectonically induced changes in subsidence and accommodation. Construction of Figure 11 omitted likely compensatory thickness relationships between successive units, e.g., thick sections overlain by thin sections, and vice versa. Both regional and local-scale syndepositional tectonic effects on sequence stratigraphic architecture are recognized.

Regional Aspects.—

Facies and thickness patterns of systems tracts within cyclothems indicate that the regional southward dip of the ramp generally was maintained over time. Maintenance of such a bathymetry suggests that rates of subsidence must have increased from north to south otherwise recurring tLST deposition and ensuing highstand deposition quickly would have reduced the ramp to a relatively flat surface. The fact that not all cyclothems or sequences thicken to the south (Fig. 11) suggests, however, that subsidence was episodic rather than continual. In particular, the presence of sediment wedges composing late HSTs, that thicken into the Cowley trough and southern Walnut syncline in many of the sequences attests to the fundamental control of tectonically maintained bathymetry on development of the sequence stratigraphic architecture of the section. Increased rates of subsidence within the Walnut syncline likely resulted from down-dropping along the Humboldt fault zone; such tectonic activity continues along this fault zone today (Stander, 1989; Steeples, 1989). At times, regional rates of subsidence apparently also were greater toward the west, into the Sedgwick and Salina basins, as indicated by thickening of several sequences and/or cyclothems into these areas. Deposition of the minor cyclothems in sequences 5 and 6—which are variously recognized only in the Cowley trough, Walnut syncline, Sedgwick and Salina basins, and the Salina embayment—may have been in response to increased subsidence in these areas.

Reversal of regional bathymetry is indicated in the TST and HST of the lower cyclothem in sequence 6, which deepens to the north. Thickness relationships of these systems tracts and underlying deposits (Fig. 8) suggest that such reversal resulted from northward tilting of the ramp. Northward deepening of the early HST of the upper cyclothem in sequence 1 may similarly indicate northward tilting at this time. Alternatively, such facies relationships instead might reflect differences in the relative rates of sedimentation and accommodation change between these areas.

Local Depocenters and Arches.—

Periodic formation of local depocenters of increased marine accommodation, wherein systems tracts maintain their shallow-water depositional aspect, is evident in central and northern parts of the ramp. Some systems tracts in sequences 1 and 2 thicken abruptly, for example, where they overlie grabens along the eastern edge of the Nemaha ridge (Fig. 11). Likewise, many units also thicken abruptly into the half-graben of the Salina embayment to the immediate west of its bounding fault. Times of increased accommodation in these areas likely resulted from down-dropping along faults. Some systems tracts in sequences 1, 3, 5, and 6 are anomalously thick along or across the axis of the Nemaha ridge in northern Kansas; the mLST in sequence 1 locally consists of thick sections of peritidal or shallow-marine limestone instead of marginal-marine shales. These occurrences suggest episodic downwarping along the Nemaha ridge.

Areas of anomalously thin systems tracts are present in some parts of the section, and they are interpreted as local arches (Fig. 11) that likely resulted from uplift along the axis of the Nemaha ridge or along its bordering faults. Some examples include (1) the thin marine and marginal-marine phases of the cyclothems in sequences 2 and 3 that appear to overlie newly formed positive areas in south-central Butler County and central Cowley County, respectively (Fig. 5), and (2) regional northward thinning of the tLST in sequence 5 onto the Nemaha ridge and into the Salina embayment, where underlying sequence 4 is not anomalously thick (Fig. 7).

Tectonism and Sequence Development.—

The pulses of tectonism that apparently were coincident with deposition appear to have recurred with an episodic nature that overlapped that of glacio-eustasy at this time. The frequency of these pulses, however, cannot be estimated with any degree of confidence because of uncertainty as to the duration of deposi-

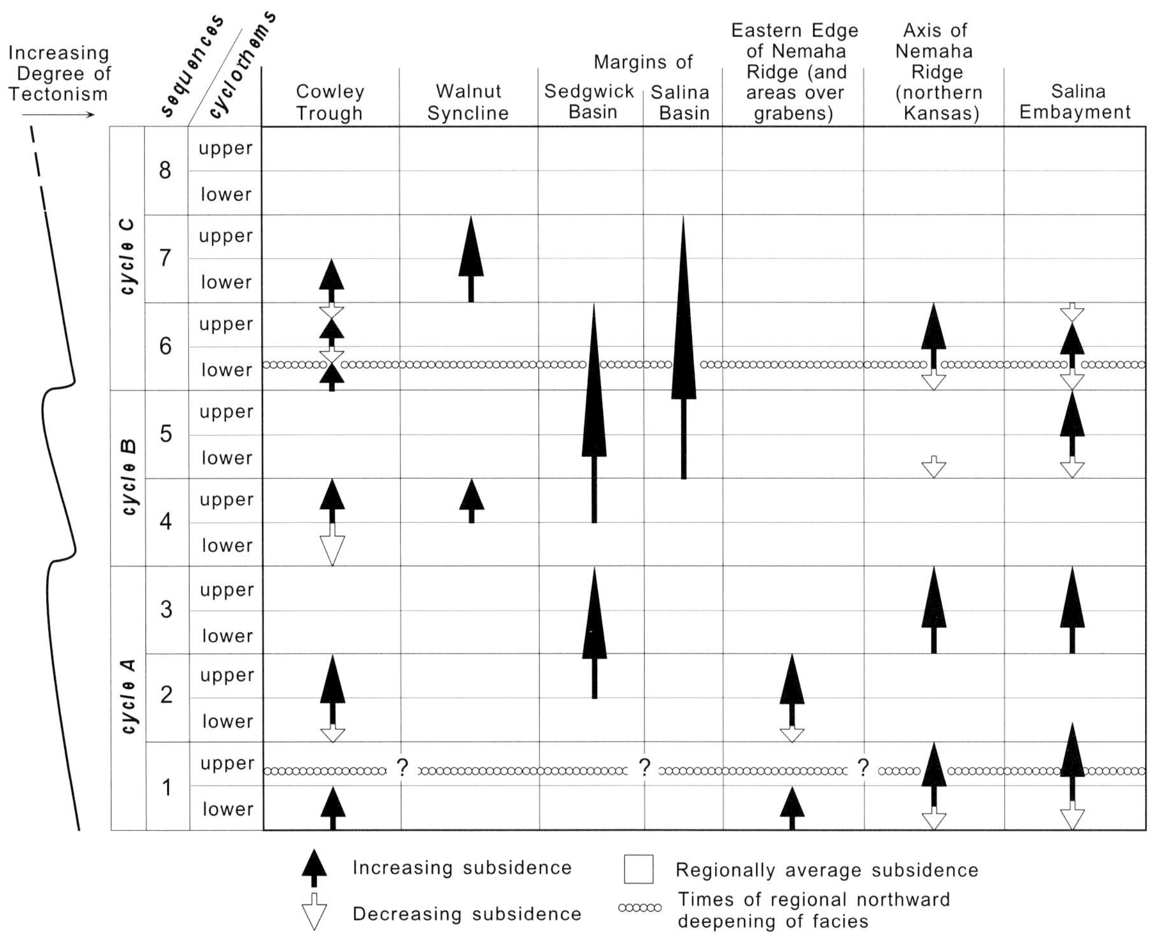

FIG. 11.—History of syndepositional tectonism inferred from anomalous thickness variations of sequences and component cyclothems in the study area.

tion of the Chase Group and basal Wellington Formation. If one assumes equal durations of cycles A, B, and C and a total duration of about 6 m.y. (Table 1) for deposition of these rocks, then the three major tectonic pulses indicated in Figure 11 recurred approximately every 2 m.y., with minor pulses therein recurring at shorter time intervals. Tectonism related to regional intraplate stress release, for example, is known to operate at a similar recurrence interval within 10^5–10^7 yr (Cloetingh, 1988; Cathles and Hallam, 1991; Klein, 1994). However, neither the apparent episodic nature due to intraplate stress release nor that indicated by various models of far-field deformation (e.g., Kluth and Coney, 1981; Klein and Willard, 1989; Goebel, 1991) seems applicable to the origin of the depositional sequences and component cyclothems in the section, which seemingly are of higher frequency. Furthermore, these sequences and cyclothems are developed across several different tectonic provinces as they are traced from Kansas to Texas.

The interpretation is that tectonic activity during deposition of third-order cycles A–C appears to have been at a maximum at the beginning of each cycle and then to have generally waned over time (Fig. 11). This tectonic framework is coincident with decreasing degrees of thickness variations and architectural complexity upward within each cycle (Figs. 4–9). However, each prominent tectonic pulse actually appears to be characterized by component pulses that are both temporally and spatially disparate in terms of their occurrence and intensity rather than having been regionally operative and of systematically decreasing intensity; furthermore, these tectonic pulses are focused specifically on major tectonic elements along and adjoining the Nemaha ridge (Fig. 11). Hence, it appears more likely that pulses of syndepositional tectonism resulted from differential movement along the Nemaha ridge. As shown in Dickinson et al. (1994, their Fig. 4), the recurrence of such local tectonism can operate at a quasi-periodicity of 10^4–10^6 yr, which is more consistent with the apparent timing of tectonic pulses inferred during deposition of the Chase Group and basal Wellington Formation (Fig. 11). Such rapid, relatively short-term tectonic pulses may be important in affecting local changes in accommodation space (e.g., Klein, 1994). Although the mechanisms that may have induced syndepositional tectonism along the Nemaha ridge are not known, it nevertheless appears that such tectonism had only two major effects on development of the section: (1) it overprinted regional and local thickness variations on the fundamental glacio-eustatically forced sequence stratigraphy of these cycles, and (2) it resulted in the geometrically complex internal architecture of sequences and cyclothems therein, including the development of sediment wedges that thicken into local depocenters.

Contrasting Styles of Cyclicity in Pennsylvanian and Lower Permian Strata

Comparison of cycle development in Pennsylvanian and Lower Permian Midcontinent strata (Fig. 12) offers some insight into the possible nature of the response of sequence stratigraphic development to the transition from an icehouse to

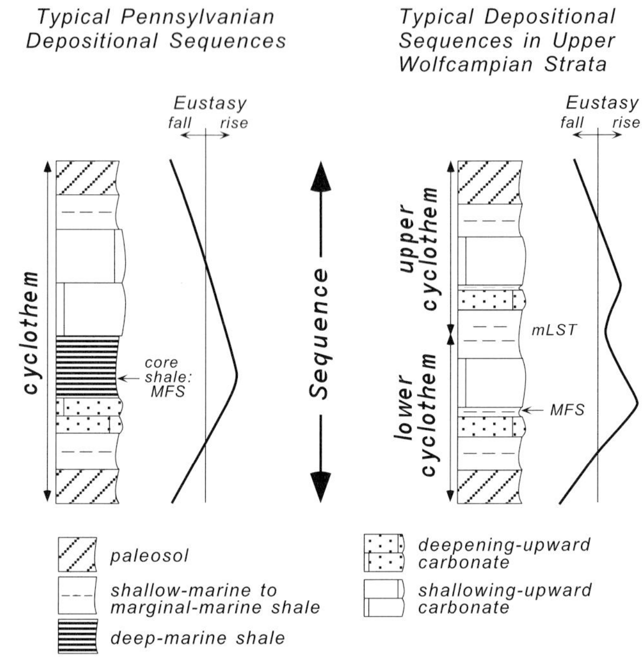

FIG. 12.—Comparison of sequence development in Pennsylvanian (after Heckel, 1986, 1994) and in Chase Group to basal Wellington strata.

a greenhouse world. According to Heckel (1986, 1994), for example, typical Pennsylvanian depositional sequences are single cyclothems deposited during major Gondwanan glaciation. The MFSs of most of these sequences are condensed sections of black, organic matter-rich, phosphatic shales ("core shales") interpreted as having been deposited in water depths of about 100 m. In contrast, the architecture of fourth-order depositional sequences in the Chase Group and basal Wellington Formation is much more complex. Their foremost characteristic is that they are composite in nature, being composed of two cyclothems and, in some sequences, one or more minor cyclothems of possible glacio-eustatic origin. Such an architecture is also characteristic of older Lower Permian sequences in the Midcontinent (Boardman et al., 1995; Boardman, 1997). Black core shales are present in some of the sequences in the lower Wolfcampian Admire and Council Grove Groups, but they are absent in the sequences in the Chase Group and basal Wellington Formation. If all of the sequences in upper Paleozoic strata were glacio-eustatically driven, then it appears that the magnitude of sea-level fluctuations, at least as suggested by the inferred depths of the MFSs of the depositional sequences, decreased into the Early Permian. Second, notwithstanding local thickness variations, average thickness relationships of successive cyclothems within each of the sequences in the Chase Group and basal Wellington section vary irregularly in a temporal sense. That is, those in sequence 1 thin upward, those in sequence 4 are of approximately equal thickness, and those in sequences 2 and 5–8 thicken upward (Fig. 10). Furthermore, there is temporal change from three sequences in cycles A and C to only two in intervening cycle B. These architectural relationships are regionally consistent, however, which suggests fundamental glacio-eustatic forcing of sequence development.

Temporal changes in the architectural styles of depositional sequences from Pennsylvanian into Lower Permian strata coincide with overall waning Gondwanan glaciation (Veevers, 1994) and likely reflect fundamental changes in the interplay among forcing periodicities. Single-cyclothemic sequences in Pennsylvanian strata presumably were forced by approximately 22 k.y., 120–220 k.y., and 235–400 k.y. Milankovitch orbital band periodicities (Heckel, 1994). Sequence-forcing periodicities in the Chase Group and Wellington Formation instead might have ranged from 0.62 to 1.43 m.y. (Table 1). Regardless of possible durations, however, the composite nature of these sequences, the irregular thickness relationships between successive cyclothems within sequences, and the changes from tripartite to bipartite sequences that variously compose cycles A–C suggest a much more complex interplay of forcing at this time. This complexity may be an inherent characteristic of Midcontinent upper Paleozoic sequences deposited during the transition from icehouse to greenhouse worlds, and the depositional pattern certainly was modified somewhat by syndepositional tectonism. Unfortunately, the specific attributes and temporal evolution of forcing periodicities, and how they might have been related to the history of late Paleozoic glaciation, cannot be determined at present because of the paucity of reliable radiometric ages for the upper Paleozoic section. Future studies should address these issues to evaluate possible relationships between sequence stratigraphy and global paleoclimates.

CONCLUSIONS

Upper Wolfcampian strata of the Chase Group and basal Wellington Formation in Kansas and Nebraska were deposited on a south-sloping ramp on which environments generally were of more open-marine aspect to the south. The section is 87–136 m of cyclically interbedded marine carbonate and marine to terrestrial siliciclastics within which eight depositional sequences are recognized. The sequences are separated either by regionally extensive unconformities developed on shallow-marine carbonates or by terrestrial lowstand systems tracts of stacked paleosols. Each sequence is composite in nature in that it comprises two transgressive-regressive cyclothems separated by relative lowstand systems tracts of dominantly marginal-marine shales. The composite nature of these sequences may be a characteristic attribute of Midcontinent upper Paleozoic cyclic strata deposited during the transition from icehouse to greenhouse worlds. Each cyclothem includes a transgressive systems tract and ensuing regressive highstand systems tract. The maximum flooding surface consistently occurs within the basal cyclothem of each sequence. The inferred maximum water depth of the deepest water of these sequences was less than that in Pennsylvanian sequences, which suggests overall decreasing magnitude of highstands during the late Paleozoic. Thickness and facies stacking patterns of the sequences and component cyclothems allow recognition of a lowest-order cyclicity, interpreted as third-order cycles, wherein the magnitude of eustatic fluctuations first increased and then decreased over time. This pattern may reflect perturbations in Gondwanan ice volumes over time. The included depositional sequences and cyclothems likely represent fourth- and fifth-order cycles, respectively.

The internal architecture of systems tracts within most of the sequences is highly complex. This inherent complexity resulted from regional and local variations in rates of subsidence and marine accommodation related largely to syndepositional tectonism, which (1) except for periodic reversals of bathymetry due to uplift, maintained the regional southward slope of the ramp; (2) resulted in the deposition of southward-thickening wedges of shallow-water sediment during the late-highstand phases in many cyclothems; and (3) affected periodic development of local arches and depocenters of increased shallow-marine accommodation along the ramp. Tectonism was related to movement along the bounding faults and to unknown processes that resulted in local uplift or downwarping along the buried Nemaha ridge. The episodic nature of tectonic activity appears to overlap that of glacio-eustasy at this time. However, syndepositional tectonism would have been too erratic to have forced cyclicity, especially considering that the cycles are correlated to equivalent cycles as far away as Texas and are present in several different tectonic provinces. Rather, tectonism merely caused these glacio-eustatically forced sequences to have an overprint of thickness variations in the component systems tracts.

REFERENCES

Baars, D. L., Ritter, S. M., Maples, C. G., and Ross, C. A., 1994a, Redefinition of the Upper Pennsylvanian Virgilian Series in Kansas: Lawrence, Kansas Geological Survey Bulletin 230, p. 11–16.

Baars, D. L., Ritter, S. M., Maples, C. G., and Ross, C. A., 1994b, Proposed repositioning of the Pennsylvanian–Permian boundary in Kansas: Lawrence, Kansas Geological Survey Bulletin 230, p. 5–10.

Boardman, D. R., 1997, Correlation of Upper Carboniferous and Lower Permian (Desmoinesian–Wolfcampian) 4th order (Milankovitch band) depositional sequences from the Midcontinent to the eastern shelf of the Midland basin: A new approach to high resolution stratigraphy: Tulsa, American Association of Petroleum Geologists, Annual Convention, Abstracts, v. 6, p. A-12.

Boardman, D. R., Nestell, M. K., and Knox, L. W., 1995, Depth-related microfaunal biofacies models for Late Carboniferous and Early Permian cyclothemic sedimentary sequences in mid-continent North America, in Hyne, N. J., ed., Sequence Stratigraphy of the Mid-Continent: Tulsa, Tulsa Geological Society Special Publication No. 4, p. 93–117.

Brough, J., 1928, On rhythmic deposition in the Yoredale series: University of Durham Philosophical Society Proceedings, v. 8, p. 118–126.

Burchett, R. R., Luza, K. V., Van Eck, O. J., and Wilson, F. W., 1985, Seismicity and tectonic relationships of the Nemaha uplift and Midcontinent geophysical anomaly (final project summary): Norman, Oklahoma Geological Survey Special Publication 85-2, 33 p.

Cathles, L. M., and Hallam, A., 1991, Stress-induced changes in plate density, Vail sequences, epeirogeny, and short-lived global sea-level fluctuations: Tectonics, v. 10, p. 659–671.

Cecil, C. B., 1990, Paleoclimatic controls on stratigraphic repetition of chemical and siliciclastic rocks: Geology, v. 18, p. 533–536.

Chaplin, J. R., 1988, Lithostratigraphy of Lower Permian rocks in Kay County, north-central Oklahoma, and their stratigraphic relationships to lithic correlatives in Kansas and Nebraska, *in* Morgan, W. A. and Babcock, J. A., eds., Permian Rocks in the Midcontinent: Tulsa, Midcontinent Section–SEPM (Society for Sedimentary Geology) Special Publication 1, p. 79–111.

Cloetingh, S., 1988, Intraplate stresses: A tectonic cause for third-order cycles in apparent sea level?, *in* Wilgus, C. K., Hastings, B. S., Kendall, C. G. St. C., Posamentier, H. W., Ross, C. A., and Van Wagoner, J. C., eds., Sea-Level Changes: An Integrated Approach: Tulsa, Society of Economic Paleontologists and Mineralogists Special Publication 42, p. 19–29.

DICKINSON, W. R., SOREGHAN, G. S., AND GILES, K. A., 1994, Glacio-eustatic origin of Permo-Carboniferous stratigraphic cycles: Evidence from the southern Cordilleran foreland region, *in* Dennison, J. M. and Ettensohn, F. R., eds., Tectonic and Eustatic Controls on Sedimentary Cycles: Tulsa, SEPM (Society for Sedimentary Geology) Concepts in Sedimentology and Paleontology No. 4, p. 25–34.

DUBOIS, S. M., AND WILSON, F. W., 1978, A revised and augmented list of earthquake intensities for Kansas, 1867–1977: Lawrence, Kansas Geological Survey Environmental Geology Series 2, 55 p.

ELIAS, M. K., 1937, Depth of deposition of the Big Blue (late Paleozoic) sediments in Kansas: Geological Society of America Bulletin, v. 48, p. 403–432.

FREY, R. W., AND BASAN, P. B., 1978, Coastal salt marshes, *in* Davis, R. A., ed., Coastal Sedimentary Environments: Berlin, Springer-Verlag, p. 101–169.

GOEBEL, K. A., 1991, Paleogeographic setting of Late Devonian to Early Mississippian transition from passive to collisional margin, Antler foreland, eastern Nevada and western Utah, *in* Cooper, J. D. and Stevens, C. H., eds., Paleozoic Paleogeography of the Western United States, II: Pacific Section SEPM, Book 67, p. 401–418.

GOLDHAMMER, R. K., LEHMANN, P. J., AND DUNN, P. A., 1993, The origin of high-frequency platform carbonate cycles and third-order sequences (Lower Ordovician, El Paso Gp., west Texas): Constraints from outcrop data and stratigraphic modeling: Journal of Sedimentary Petrology, v. 63, p. 318–359.

GOLONKA, J., ROSS, M. I., AND SCOTESE, C. R., 1994, Phanerozoic paleogeographic and paleoclimatic modeling maps, *in* Embry, A. F., Beauchamp, B., and Glass, D. J., eds., Pangea: Global Environments and Resources: Calgary, Canadian Society of Petroleum Geologists Memoir 7, p. 1–47.

HARLAND, W. B., COX, A. V., LLEWELLYN, P. G., PICKTON, C. A. G., SMITH, A. G., AND WALTERS, R., 1982, A Geologic Time Scale: Cambridge, Cambridge University Press, 131 p.

HECKEL, P. H., 1986, Sea-level curve for Pennsylvanian eustatic marine transgressive-regressive depositional cycles along Midcontinent outcrop belt: Geology, v. 14, p. 330–334.

HECKEL, P. H., 1994, Evaluation of evidence for glacio-eustatic control over marine Pennsylvanian cyclothems in North America and consideration of some possible tectonic effects, *in* Dennison, J. M. and Ettensohn, F. R., eds., Tectonic and Eustatic Controls on Sedimentary Cycles: Tulsa, SEPM (Society for Sedimentary Geology) Concepts in Sedimentology and Paleontology No. 4, p. 65–87.

KLEIN, G. D., 1994, Depth determination and quantitative distinction of the influence of tectonic subsidence and climate on changing sea level during deposition of Midcontinent Pennsylvanian cyclothems, *in* Dennison, J. M. and Ettensohn, F. R., eds., Tectonic and Eustatic Controls on Sedimentary Cycles: Tulsa, SEPM (Society for Sedimentary Geology) Concepts in Sedimentology and Paleontology No. 4, p. 35–50.

KLEIN, G. D., AND KUPPERMAN, J. B., 1992, Pennsylvanian cyclothems: Methods of distinguishing tectonically induced changes in sea level from climatically induced changes: Geological Society of America Bulletin, v. 104, p. 166–175.

KLEIN, G. D., AND WILLARD, D. A., 1989, Origin of the Pennsylvanian coal-bearing cyclothems of North America: Geology, v. 17, p. 152–155.

KLEIN, G. D., BEAUCHAMP, B., BAUD, A., CHUVASHOV, B. I., LOPEZ-GAMUNDI, O. R., PARRISH, J. T., ROSS, C. A., SCHOLLE, P. A., SCOTESE, C. R., AND WATNEY, W. L., 1994, Introduction: Project Pangea and workshop recommendations, *in* Klein, G. D., ed., Pangea: Paleoclimate, Tectonics, and Sedimentation During Accretion, Zenith, and Breakup of a Supercontinent: Boulder, Geological Society of America Special Paper 288, p. 1–12.

KLUTH, C. F., AND CONEY, P. J., 1981, Plate tectonics of the Ancestral Rocky Mountains: Geology, v. 9, p. 10–15.

MARSHAK, S., AND PAULSEN, T., 1996, Midcontinent U.S. fault and fold zones: A legacy of Proterozoic intracratonic extensional tectonism?: Geology, v. 24, p. 151–154.

MAZZULLO, S. J., AND TEAL, C. S., 1994, Road log and locality guide for day 1, a new look at old rocks: Stratigraphy, sedimentology, and structure of the Chase Group (Permian, Wolfcampian), south-central Kansas: Wichita, Kansas Geological Society Field Trip Guidebook, part 1, p. 1–60.

MAZZULLO, S. J., TEAL, C. S., AND BURTNETT, C. A., 1995, Facies and stratigraphic analysis of cyclothemic strata in the Chase Group (Permian, Wolfcampian), south-central Kansas, *in* Hyne, N. J., ed., Sequence Stratigraphy of the Mid-Continent: Tulsa, Tulsa Geological Society Special Publication No. 4, p. 217–248.

MAZZULLO, S. J., TEAL, C. S., AND BURTNETT, C. A., 1996, Chase Group strata in north-central Kansas and Nebraska: Outcrop analogs of stratigraphic-trap reservoirs, *in* Walters, R. F., ed., Field Trip Guidebook: Wichita, Kansas Geological Society, 92 p.

MAZZULLO, S. J., TEAL, C. S., AND BURTNETT, C. A., 1997, Outcrop Stratigraphy and Depositional Facies of the Chase Group (Permian, Wolfcampian) in Kansas and Southeastern Nebraska: Lawrence, Kansas Geological Survey Technical Series 6, 220 p.

MENNING, M., 1995, A numerical time scale for the Permian and Triassic Periods: An integrated time analysis, *in* Scholle, P. A., Peryt, T. M., and Ulmer-Scholle, D. S., eds., The Permian of Northern Pangea, Volume 1: Paleogeography, Paleoclimates, Stratigraphy: Berlin, Springer-Verlag, p. 77–97.

MOORE, R. C., 1936, Stratigraphic evidence bearing on problems of continental tectonics: Geological Society of America Bulletin, v. 47, p. 1785–1808.

MOORE, R. C., 1964, Paleoecological aspects of Kansas Pennsylvanian and Permian cyclothems, *in* Merriam, D. F., ed., Symposium on Cyclic Sedimentation: Lawrence, Kansas Geological Survey Bulletin 169, p. 87–380.

PARRISH, J. T., 1995, Geologic evidence of Permian climate, *in* Scholle, P. A., Peryt, T. M., and Ulmer-Scholle, D. S., eds., The Permian of Northern Pangea, Volume 1: Paleogeography, Paleoclimates, Stratigraphy: Berlin, Springer-Verlag, p. 53–61.

RASCOE, B., AND ADLER, F. J., 1983, Permo-Carboniferous hydrocarbon accumulations, Mid-Continent, U.S.A.: American Association of Petroleum Geologists Bulletin, v. 67, p. 979–1001.

REINECK, H. E., 1967, Layered sediments of tidal flats, beaches, and shelf bottoms of the North Sea, *in* Lauff, G. H., ed., Estuaries: Washington, D.C., American Association for the Advancement of Science Publication No. 83, p. 191–206.

ROSS, C. A., 1963, Standard Wolfcampian Series (Permian), Glass Mountains, Texas: Boulder, Geological Society of America Memoir 88, 205 p.

STANDER, T. W., 1989, Structural nature of the Humboldt fault zone in northeastern Nemaha County, Kansas, *in* Steeples, D. W., ed., Geophysics in Kansas: Lawrence, Kansas Geological Survey Bulletin 226, p. 117–128.

STEEPLES, D. W., 1989, Structure of the Salina–Forest City interbasin boundary from seismic studies, *in* Steeples, D. W., ed., Geophysics in Kansas: Lawrence, Kansas Geological Survey Bulletin 226, p. 31–52.

VEEVERS, J. J., 1994, Pangea: Evolution of a supercontinent and its consequences for Earth's paleoclimate and sedimentary environments, *in* Klein, G. D., ed., Pangea: Paleoclimate, Tectonics, and Sedimentation During Accretion, Zenith, and Breakup of a Supercontinent: Boulder, Geological Society of America Special Paper 288, p. 13–23.

WANLESS, H. R., AND SHEPARD, F. P., 1936, Sea level and climatic changes related to late Paleozoic cycles: Geological Society of America Bulletin, v. 47, p. 1177–1206.

WATERHOUSE, J. B., 1978, Chronostratigraphy for the world Permian, *in* Cohee, G. V., Glaessner, M. F., and Hedberg, H. D., eds., Contributions to the Geologic Time Scale: Tulsa, American Association of Petroleum Geologists Studies in Geology No. 6, p. 299–322.

WELLER, J. M., 1956, Argument for diastrophic control of late Paleozoic cyclothems: American Association of Petroleum Geologists Bulletin, v. 40, p. 17–50.

WHEELER, H. E., AND MURRAY, H. H., 1957, Base-level control patterns in cyclothemic sedimentation: American Association of Petroleum Geologists Bulletin, v. 41, p. 1985–2011.

THREE-DIMENSIONAL ARCHITECTURE OF UPPER PERMIAN HIGH-FREQUENCY SEQUENCES, YATES-CAPITAN SHELF MARGIN, PERMIAN BASIN, U.S.A.

DAVID A. OSLEGER

Department of Geology, University of California, Davis, California 95616

AND

SCOTT W. TINKER

Marathon Oil Company, Littleton, Colorado 80160

ABSTRACT: Dip-oriented profiles of the Yates-Capitan shelf margin were independently constructed from exceptional outcrops in McKittrick and Slaughter Canyons in the Guadalupe Mountains of west Texas and southern New Mexico. Comparison of the two profiles reveals significant similarities in the position and character of high-frequency sequence boundaries, the internal architecture of facies tracts and cycle-stacking patterns, and offlap angles of both the shelf and reef. The evolution of the Yates-Capitan shelf margin is recorded by systematic long-term trends in key depositional variables measured on individual high-frequency sequences in each canyon. This comparison of the two profiles, located along strike 25 km apart, provides a three-dimensional model of the extent and variability of genetic components of the Late Permian margin of the Northwest shelf of the Delaware basin.

Comparison of the sequence stratigraphic models for the Yates Formation in McKittrick and Slaughter Canyons indicates that four complete high-frequency sequences can be confidently correlated along strike. Fundamental architectural characteristics of the Yates-Capitan shelf margin are evident in the comparative profiles from each canyon. (1) The Yates Formation in both canyons exhibits initial aggradational geometries followed by strong progradational patterns. The volume of the shelf-crest pisolite complex in the Yates progressively expands through time in concert with a reciprocal contraction of the outer-shelf facies tract. The time-equivalent Capitan reef in both canyons exhibits remarkably similar patterns of stepwise alternations of aggradational and progradational growth that relate to changes in high-frequency sequence architecture through time. (2) The thickness of the Yates varies considerably both across the dip of the shelf margin and along strike between the two canyons. The 240%–600% increases in downdip thickness primarily reflect the inherited depositional topography. The thickness variation along the 25-km strike distance may indicate significant lateral variability in subsidence and/or accumulation rate along the Yates-Capitan shelf margin. (3) Several individual siliciclastic beds can be correlated between each canyon, and the distribution of siliciclastics within individual high-frequency sequences exhibits similar patterns of retrogradation, aggradation, and progradation. Correlative facies-stacking patterns integrated with long-term variations in progradation:aggradation ratio and derived offlap angle act as sensitive indicators of relative changes in base level.

Comparison of our integrated field observations with subsurface data from the northern Northwest shelf (65–90 km away), Central Basin platform (approximately 150 km away), and Eastern shelf (>300 km away) reveals a consistent, basin-wide pattern of the internal architecture of the high-frequency sequences that compose the Yates Formation. These widespread similarities provide compelling evidence for regional sea-level control on sequence development around the Permian basin.

INTRODUCTION

Considering the tremendous number of sequence stratigraphic studies that have been done over the past few decades, it is surprising that only a few high-resolution, two-dimensional profiles across carbonate platforms have been documented from continuous outcrops (e.g., Sonnenfeld and Cross, 1993; Franseen et al., 1993; Pomar, 1993; Kerans and Fitchen, 1995). Consequently, much of the lithologic and geometric detail regarding the internal architecture of depositional sequences that make up carbonate platforms remains vague. Laterally continuous chronostratigraphic data derived from "mapped" outcrop profiles are critical for providing the spatial information necessary for more accurate interwell correlations in subsurface hydrocarbon reservoirs. However, even two-dimensional profiles provide incomplete information about the distribution of facies tracts within genetic units because depositional processes operate in three dimensions. The near future of reservoir characterization in carbonates resides in three-dimensional modeling and visualization techniques (e.g., Tinker, 1996a; Kerans and Tinker, 1997) that are critically dependent on the veracity of the stratigraphic framework used to construct the three-dimensional model. Thus it is imperative that the underlying geologic data be laterally consistent, providing a ground-truth level of reliability that can be used to generate a three-dimensional reservoir model with predictive value.

The intent of this paper is to compare the stratigraphic architecture of two profiles, constructed independently of one another (Tinker, 1996b, 1998; Osleger, 1998), across the late Guadalupian Yates-Capitan shelf margin exposed in dip-oriented canyons in the Guadalupe Mountains of west Texas and southern New Mexico. The integration of digital photographs taken via helicopter, vertical measured sections, dip-oriented transects, and a depositional model that changes along dip and through time allowed us to "map" the facies distributions on continuous outcrop walls in McKittrick and Slaughter Canyons. The resulting two-dimensional profiles are essentially two-dimensional "measured sections." Characteristics shared by the two profiles along the strike-parallel distance of about 25 km provide a measure of three-dimensional control on the lateral extent and variability of genetic components and their bounding surfaces. Furthermore, comparison of our integrated field observations with subsurface data from the work of Borer and Harris (1991) on the northern Northwest shelf (65–90 km away), Central Basin platform (approximately 150 km away), and Eastern shelf (more than 300 km away) reveals a consistent, basin-wide pattern of the architecture of high-frequency sequences that form the Yates Formation. The integrated two-dimensional profiles of the

Advances in Carbonate Sequence Stratigraphy: Application to Reservoirs, Outcrops and Models, SEPM Special Publication No. 63
 ISBN 1-56576-060-3, p. 169–185

Yates-Capitan shelf margin resulting from this study provide a three-dimensional, regional, sequence stratigraphic model for comparable prograding shelf margins of other basins.

STUDY AREAS AND METHODS

Dip-oriented, cross-sectional "maps" of Capitan-equivalent shelf strata were constructed from exceptional exposures in McKittrick and Slaughter Canyons in the Guadalupe Mountains of west Texas and southern New Mexico (Fig. 1). In Slaughter Canyon, the upper Seven Rivers Formation through Yates Formation interval was mapped along a 3.2-km transect updip from the Capitan reef (Fig. 2A). The map of the outcrop wall in McKittrick Canyon extends for approximately 5 km behind the terminal reef margin and spans the entire Seven Rivers through Yates Formations as well as the Capitan reef and foreslope (Fig. 2B).

Numerous, closely spaced, vertical sections were measured in each canyon (Fig. 3), and a handheld scintillometer generated radioactivity (pseudo gamma-ray) logs for five of the sections in McKittrick Canyon. Correlations and stratal architecture were documented by physically walking out critical surfaces while simultaneously transcribing geometric and lithologic information to low-angle oblique photographs. Selected individual cycles were physically traced along dip to document facies changes and the nature of bounding discontinuities. Several hundred samples of representative lithofacies were collected in each canyon to provide petrographic detail to enhance the field descriptions. This paper focuses on the interval of stratigraphic overlap exposed in both canyons, specifically a 3.2-km transect spanning the Yates-Capitan shelf margin.

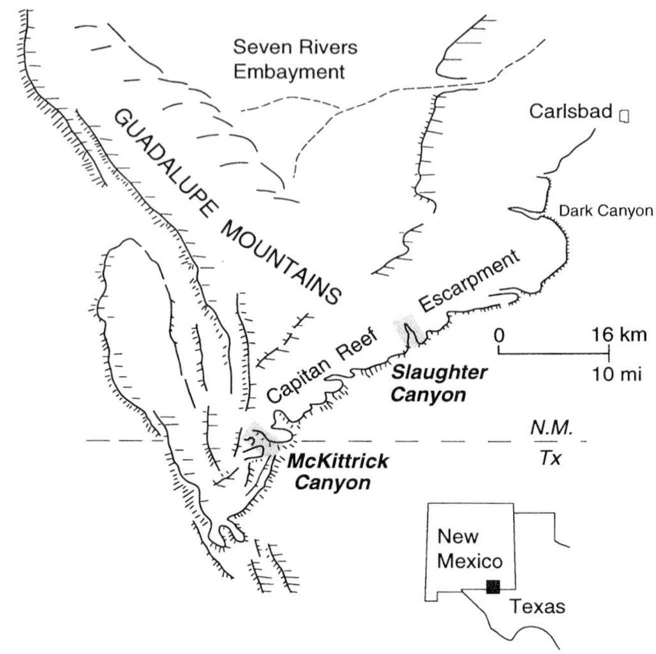

FIG. 1.—Location map of the Guadalupe Mountains (after King, 1948) showing the study areas (shaded) in McKittrick and Slaughter Canyons. Both canyons trend perpendicular to the Capitan reef escarpment, providing dip profiles across the shelf-to-basin transition.

STRATIGRAPHIC AND PALEOGEOGRAPHIC SETTING

Shallow-marine facies of the study interval were deposited on the Northwest shelf in the lee of the Capitan reef that rimmed the Delaware basin during the late Guadalupian (Fig. 4). Mixed carbonates and clastics of the Yates Formation are time equiva-

FIG. 2.—Panoramic photographs of the northeast walls of (A) Slaughter Canyon and (B) McKittrick Canyon, illustrating the seismic scale of the exposures (widths of the views are approximately 2 and 3 km, respectively). Most of the lower walls in both canyons are composed of Capitan reef and foreslope facies; the upper parts of the canyon walls are composed of back-reef shelf strata of the Seven Rivers and Yates Formations. Superimposed form lines mark the boundaries of high-frequency sequences through the Yates Formation. The transition between Yates Formation shelf strata and the Capitan reef and foreslope is shown in (A) by the subhorizontal dashed line against which shelf strata terminate (determined by mapping). Massive outcrops of Capitan reef and foreslope deposits are shown in (B) by the diagonal-line pattern.

lent to massive reef and fore-reef carbonates of the middle Capitan Formation and to basinal siliciclastics of the middle Bell Canyon Formation (Fig. 5). Shelf-carbonate accumulation occurred during high-accommodation phases when the carbonate factory was active, whereas basinal siliciclastic sediment accumulation occurred during low to negative accommodation phases.

Global climate models of Permian Pangea support the abundant lithologic evidence for widespread aridity throughout the region and further suggest monsoonal circulation patterns (Kutzbach and Gallimore, 1989). The Late Permian was a transitional phase between peak-icehouse conditions during the Pennsylvanian–Early Permian and peak-greenhouse conditions during the Jurassic–Cretaceous. Large continental ice sheets were gone by the late Guadalupian (Veevers and Powell, 1987), which limits glacio-eustasy as an exclusive explanation for the ubiquitous cyclicity recorded in the strata of the study interval.

FACIES TRACTS

The term *facies tract* (comparable to the facies belts of Wilson [1975]) is used in this paper as a genetically linked association of facies that records a discrete energy, water-depth, and sediment-supply setting (sensu Kerans and Fitchen, 1995). In McKittrick and Slaughter Canyons, the Yates-Capitan shelf margin is composed of three carbonate-dominated facies tracts (shelf crest, outer shelf, reef margin) and a continental siliciclastic facies tract. Lithofacies that make up each of these facies tracts define a dip-oriented succession of depositional environments that extends seaward from a topographically high, shelf-crest pisolite shoal complex, down a gently to steeply dipping outer shelf, then gradationally into the massive Capitan reef (marginal mound profile of Dunham, 1972). The lithofacies that are correlative with the Capitan reef and their interpreted depositional environments have been comprehensively documented in the literature (e.g., Kendall, 1969; Dunham, 1972; Smith, 1974; Esteban and Pray, 1977; Hurley, 1989; Garber et al., 1989; Borer and Harris, 1991; Kerans and Harris, 1993; Osleger, 1998; Tinker, 1998). Descriptions of the rock types composing the four dominant facies tracts are summarized below to aid in the subsequent discussion of meter-scale cyclicity.

The shelf-crest facies tract is primarily composed of pisolite dolorudstone and fenestral, peloidal dolowackestone-packstone. The presence of teepees, vadose caliche crusts, horizontal sheet cracks, and sandstone-filled vertical fissures all attest to intertidal to supratidal depositional environments. Seaward of the shelf crest, oolitic and coated-grain carbonate grainstones form a high-energy fringe separating the shelf crest from the gently to steeply dipping outer shelf where skeletal-peloidal packstones

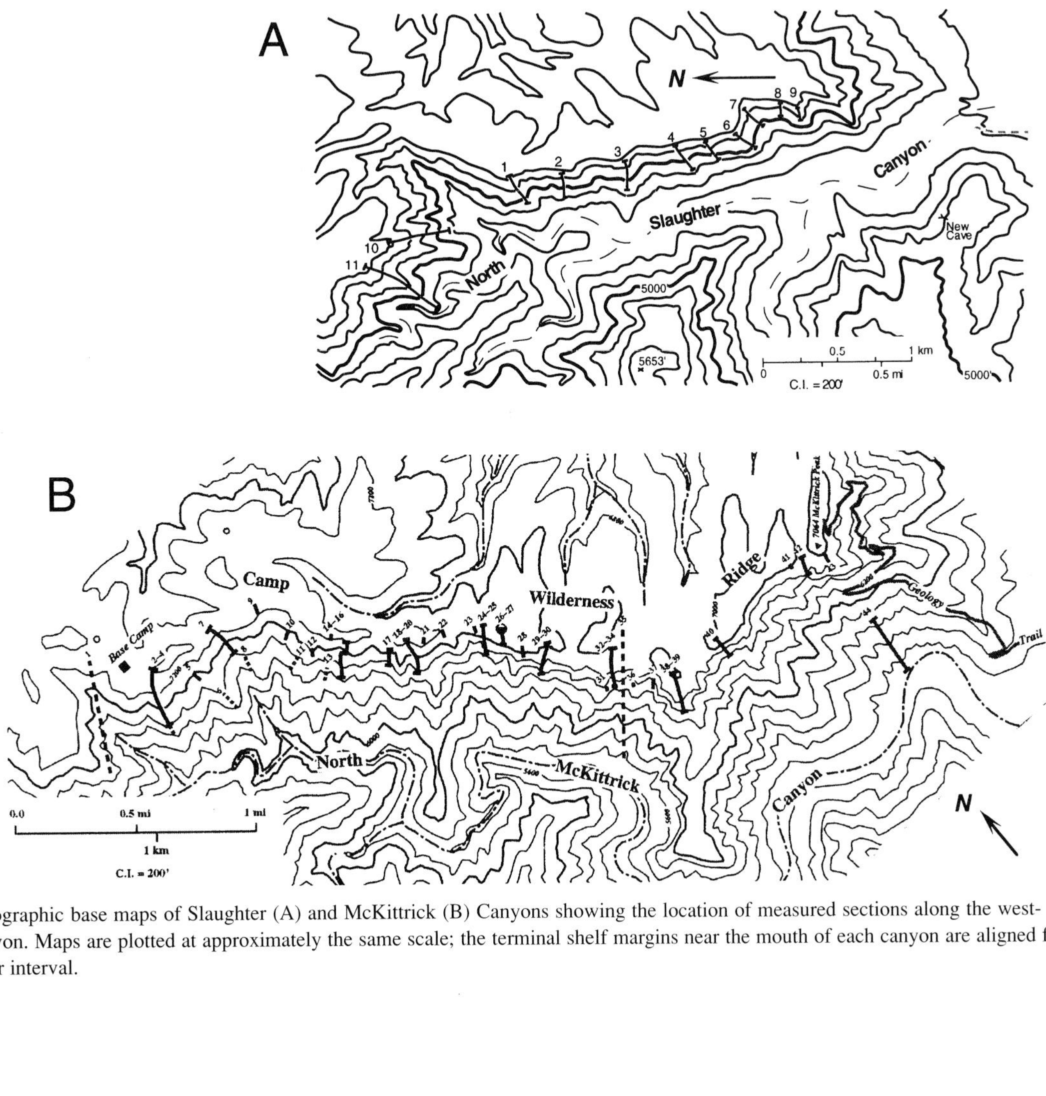

FIG. 3.—Topographic base maps of Slaughter (A) and McKittrick (B) Canyons showing the location of measured sections along the west- to southwest-facing walls of each canyon. Maps are plotted at approximately the same scale; the terminal shelf margins near the mouth of each canyon are aligned for ease of comparison. C.I. = contour interval.

and grainstones commonly form massive to thick, structureless beds. These deposits represent open-marine environments ranging in depth from near sea level to approximately 60 m (Osleger, 1998; Tinker, 1998). The gradational transition between outer-shelf environments and the submergent Capitan reef is recorded by intertonguing thick beds of fusulinid, crinoid, algal wackestone to grainstone and massive, muddy, reefal facies showing evidence of framework such as sponges, *Archaeolithoporella*, and botryoidal marine cements. Interbedded with the shelf-carbonate lithofacies are well-sorted fine-grained sandstones and siltstones of the continental siliciclastic facies tract that represent the seaward migration of coastal siliciclastic sabkhas and erg systems during episodes of lowstand exposure (Fischer and Sarnthein, 1988; Andreason, 1992; Mazzullo et al., 1996). Siliciclastic sediments that remained on the shelf were reworked in shallow-marine environments during subsequent transgressions (Borer and Harris, 1991).

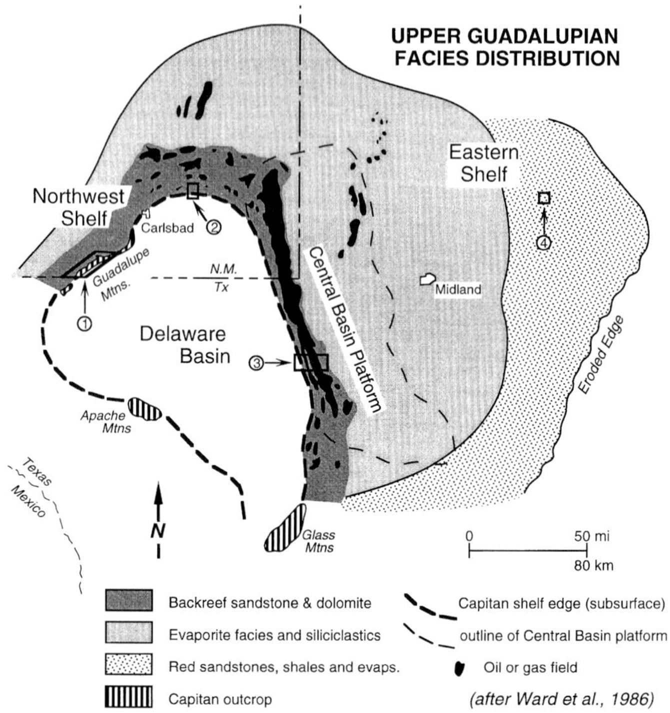

FIG. 4.—Distribution of shallow-marine to continental lithofacies around the perimeter of the deep-water Delaware basin, generalized for deposition of the Queen, Seven Rivers, and Yates Formations (Ward et al., 1986). Oil and gas fields producing from these formations are shown in black. Block outlines with arrows and numbers show the study areas of this project: (1) the Guadalupe Mountains, (2) the northern Northwest shelf (Borer and Harris, 1991), (3) Central Basin platform (Borer and Harris, 1991), and (4) an oil field from the Eastern shelf. N.M. = New Mexico, Tx = Texas.

Meter-Scale Cyclicity

Lateral migration of mixed carbonate-clastic shelf facies tracts in response to base-level changes resulted in the systematic arrangement of lithofacies into meter-scale cycles (Fig. 6). The architecture of individual cycles varies across the platform and also with position within high-frequency sequences (see Osleger [1998] and Tinker [1998] for specific examples of strata in McKittrick and Slaughter Canyons). Individual cycles thicken considerably in a seaward direction, mimicking larger-scale patterns evident in high-frequency sequences. Siliciclastics are interpreted to form the basal component of

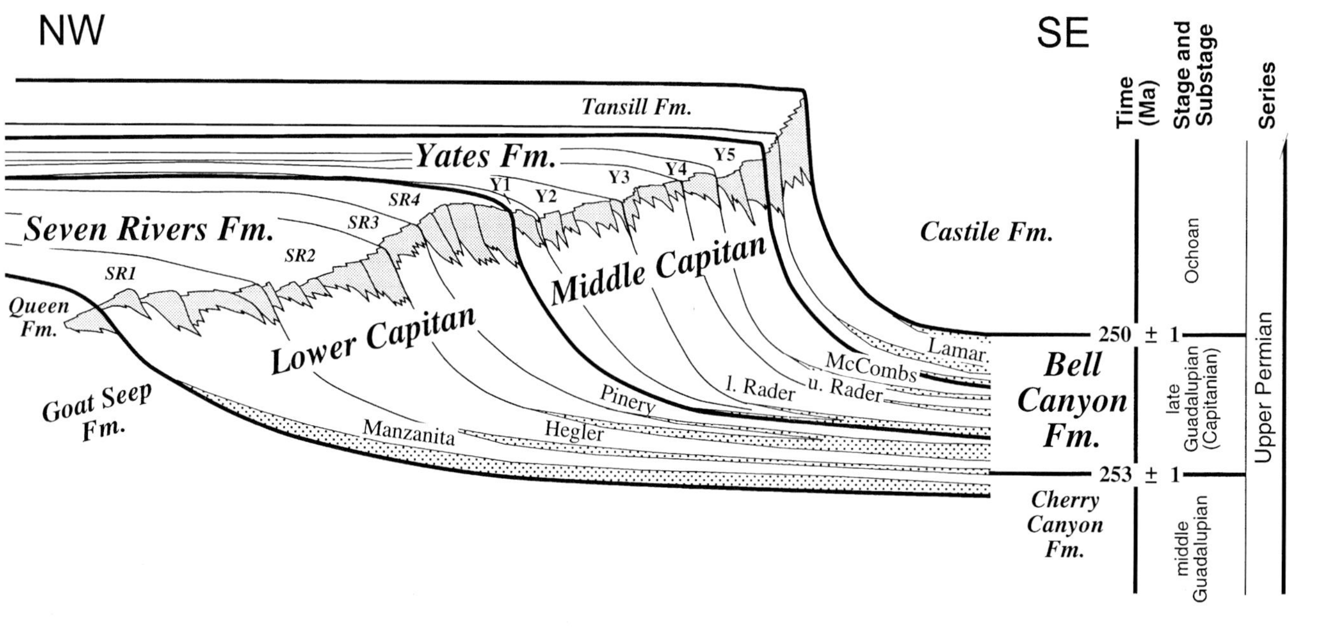

FIG. 5.—Simplified profile of the Late Permian shelf-to-basin transition exposed in the Guadalupe Mountains, illustrating temporal relationships and positions of high-frequency sequences (Tinker, 1996b). (HFS Y2 through Y5 in this paper correspond with HFS Y1 through Y4 in Osleger [1998]). Time estimates are from Ross and Ross (1987).

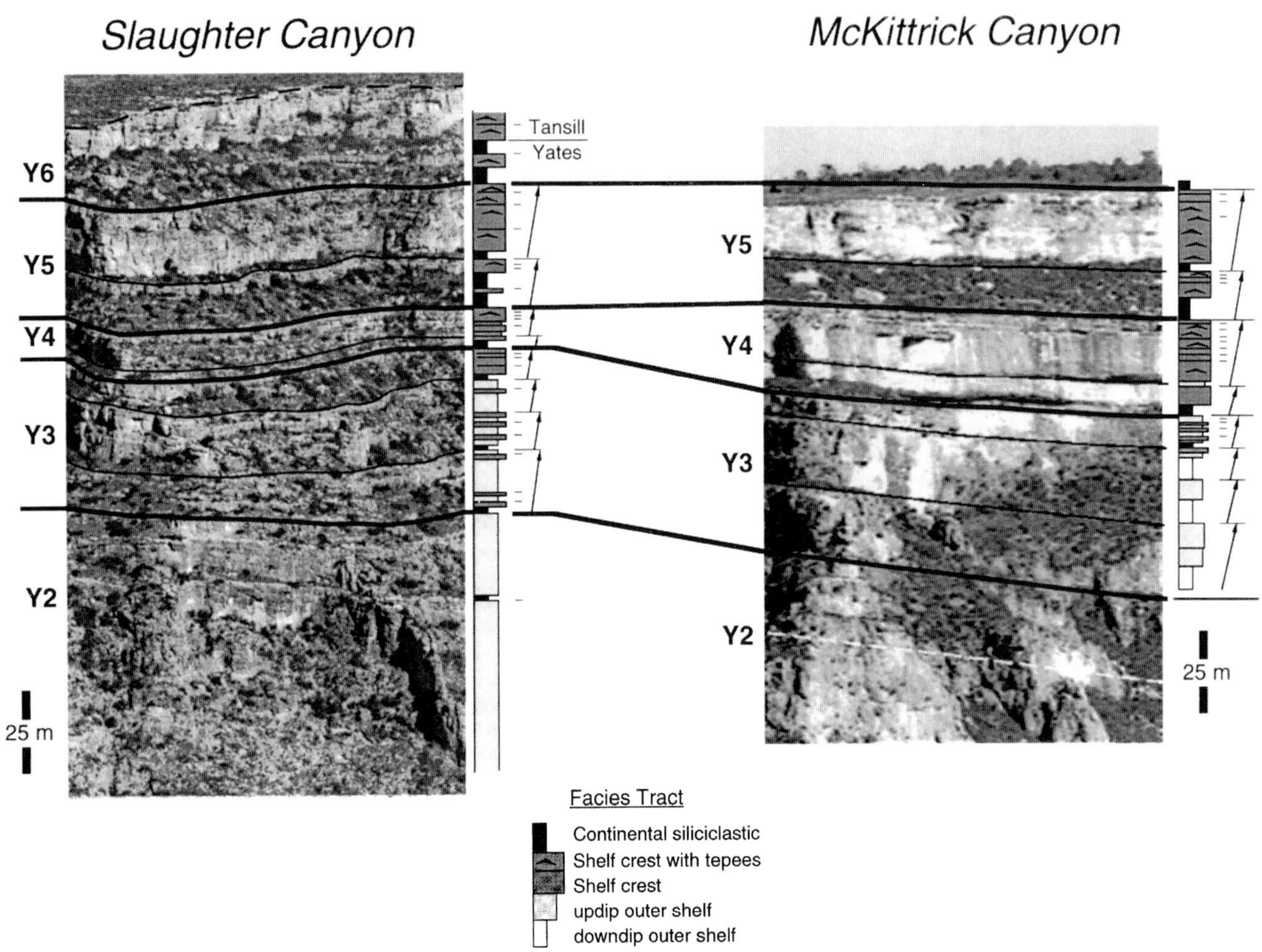

FIG. 6.—Comparison of vertical stacking patterns in a window located approximately 1.3 to 1.5 km landward of the terminal shelf margin in the Yates Formation in Slaughter and McKittrick Canyons. Each window spans approximately 100 to 150 m of lateral distance. The photographic interpretations have been scaled for ease of comparison. The simplified measured sections to the side of each photograph are composites of two or three sections merged to fit the location of the window. Tick marks denote meter-scale cycles determined in the field, and the inclined arrows separated by thin lines mark cycle sets. Thick lines on each photograph mark high-frequency sequence (HFS) boundaries. The McKittrick panel is positioned to the right of the Slaughter panel because it is located slightly more seaward, on the basis of the greater proportion of outer-shelf facies. The kinds of data represented here were collected for every measured section shown in Figure 3 as well as along physically traced lateral transects of selected cyclic intervals.

individual cycles. Siliciclastic beds are thick and amalgamated updip and progressively thin toward the outer shelf, eventually pinching out before reaching the reef. Basal siliciclastics typically exhibit an abrupt, erosional contact with carbonates of the underlying cycle and a gradational upper contact with overlying dolomitic subtidal to shelf-crest facies. Updip, Yates meter-scale cycles near the pisolitic shelf crest exhibit an asymmetric, upward-shallowing pattern. Downdip, Yates meter-scale cycles may become progressively more symmetric with basal siliciclastics grading upward through thin, tidal-flat dolomudstones into open-marine outer-shelf facies before progressively shallowing into shelf-crest facies. Most important for the purposes of this paper, meter-scale cycles of the Yates shelf exhibit systematic vertical and lateral stacking patterns that help to define larger-scale high-frequency sequences (Fig. 6), a developmental motif common to most shallow-marine carbonate platforms.

YATES SEQUENCE STRATIGRAPHY

The integrated sequence stratigraphic framework for the Yates described in this paper uses the terminology of Mitchum and Van Wagoner (1991). Large-scale "composite sequences" are internally composed of systematically stacked "high-frequency sequences" (HFSs) that in turn are composed of systematically stacked "meter-scale cycles." Each of these genetic units is bounded by a disconformable surface or its downdip conformable equivalent and, thus, has chronostratigraphic significance. A useful intermediate-scale unit is the cycle set, which is a bundle of meter-scale cycles that exhibit an overall upward-shallowing pattern bounded by marine-flooding surfaces (Kerans and Fitchen, 1995). The data for the Yates sequence stratigraphic model are derived from several thousand meters of measured section integrated with mapped photomosaics of each canyon; further documentation is provided in Tinker (1996b, 1998) and Osleger (1998).

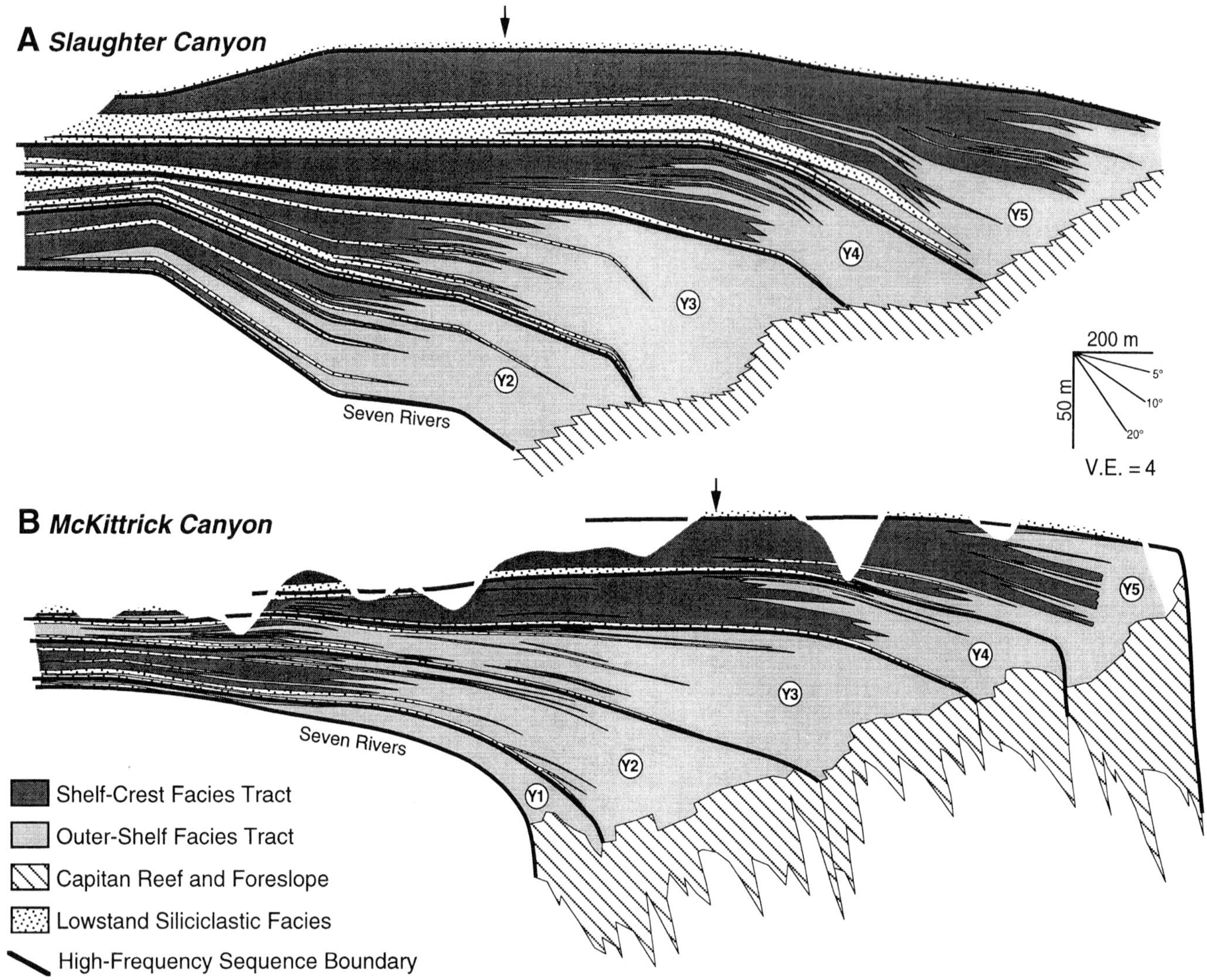

FIG. 7.—Two-dimensional cross-sectional maps of the Yates shelf margin, interpreted independently in (A) Slaughter and (B) McKittrick Canyons and showing similarities in high-frequency sequence architecture and facies-tract distribution. Each profile extends for approximately 3 km updip from the terminal Y5 shelf margin and is plotted to the same vertical scale for direct comparison. Facies tracts are generalized from assemblages of several distinct subfacies for purposes of display. Arrows indicate the positions of vertical sections shown in Figure 6. Data used to construct these profiles consist of thousands of meters of measured section integrated with physically mapped photomosaics, documented in Osleger (1998) and Tinker (1998).

Comparison of our sequence stratigraphic interpretations for the Yates Formation in McKittrick and Slaughter Canyons (Figs. 6, 7) indicates that four complete high-frequency sequences can be correlated along strike (Y2–Y5). Exposures in McKittrick Canyon reveal a basal HFS (Y1) that is not resolvable in Slaughter Canyon; this apparent discrepancy is evaluated in detail below because it concerns the placement of the contact between the Yates and Seven Rivers Formations, interpreted to be a major composite sequence boundary (Kerans et al., 1992). HFS boundaries on each profile appear to represent the same surface on the basis of detailed comparison of expanded-scale cross sections of the stratigraphic architecture in each canyon. From the perspective of predictability, it is exciting to emphasize that these genetic HFS interpretations were determined independently and have not been reinterpreted for purposes of this paper. This strike-parallel documentation of high-frequency sequences and bounding surfaces within the Yates offers a fourfold increase in chronostratigraphic resolution beyond that currently provided by the available biostratigraphic control.

Three fundamental characteristics of the Yates-Capitan shelf margin are evident in the comparative profiles from each canyon (Fig. 7). First, the Yates in both canyons exhibits initial aggradational architecture within Y2 and Y3 followed by strong progradation through Y4 and Y5. Within the 3-km-wide window, the volume of the shelf-crest facies tract expands from Y2 through Y5. The volume of the outer-shelf facies tract progressively contracts as it steps seaward in concert with the shelf crest. Moreover, the Capitan reef in both canyons exhibits similar patterns of stepwise alternations of aggradational and progradational growth that relate to HFS architectural changes through Yates time. The details of these variations in reef growth angles and their relationship to individual HFSs are discussed below.

Second, the thickness of the Yates varies considerably both across the dip of the shelf margin and along strike between the two canyons. Complete thickness data for the Yates cannot be determined precisely in McKittrick Canyon because the depth of erosional removal along Wilderness Ridge (Fig. 3B) extends progressively deeper into the Yates up-canyon, but total thicknesses can be estimated below a horizontal datum projected landward from the top of HFS Y5 (Fig. 7B). Complete thicknesses of the Yates can be determined in Slaughter Canyon, however, to a point about 3 km up-canyon from the terminal Yates shelf margin (Fig. 7A). In both profiles, the Yates is very thin updip, thickens to a maximum at a position around 1.5 km from the terminal shelf margin, and then thins seaward to zero near the Yates-Tansill-Capitan junction. Estimated updip thicknesses of the Yates (base Y2 to top Y5), measured at a common point approximately 3 km landward of the terminal shelf margin, vary from approximately 85 m in McKittrick Canyon to approximately 100 m in Slaughter Canyon. Maximum thicknesses of the Yates, estimated at a common point approximately 1.5 km landward of the terminal shelf margin, vary from approximately 165 m in McKittrick to approximately 220 m in Slaughter Canyon. These strike-parallel thickness increases of 18% (at 3 km) and 33% (at 1.5 km) over a 25-km distance between the two canyons are likely real (rather than measurement error) because comparable thickness differences are consistent through all four HFSs in each profile. Similar strike-parallel thickness differences were recognized by Borer and Harris (1991) in comparisons of Yates profiles between the northern Northwest shelf and Central Basin platform. These HFS thickness differences are discussed in detail below because they appear to indicate significant strike-parallel variability in subsidence and/or accumulation rates along the Yates-Capitan shelf margin.

The third primary characteristic of the Yates shelf margin revealed in the comparative profiles is the consistent strike-parallel distribution of fine siliciclastics. Several individual siliciclastic beds can be correlated between each canyon on the basis of stratigraphic position within individual HFSs. Furthermore, siliciclastic, shelf-crest, and outer-shelf units collectively exhibit similar patterns of retrogradation, aggradation, and progradation within individual HFSs, attesting to their value as sensitive indicators of relative changes in base level.

Vertical Cycle-Stacking Patterns

The component facies and thicknesses of individual meter-scale cycles vary systematically across the Yates shelf margin as a function of position within high-frequency sequences. Thus cycle-stacking patterns derived from individual measured sections (or drill holes) might be expected to have limited utility as predictive tools for recognizing larger-scale genetic units. However, this is not the case. If used in concert with a knowledge of relative position on a shelf to basin transect, a depositional model, and changes in facies proportions, vertical stacking patterns of cycle sets can be used judiciously to distinguish larger-scale genetic units and predict overall platform architecture. This point is illustrated by the reasonable degree of correlation of one-dimensional vertical stacking patterns of cycle sets exposed in the two canyons at comparable positions on the shelf margin (Fig. 6), as well as the two-dimensional stacking patterns shown in Figure 7. Meter-scale cycles may also be correlative, but this possibility is difficult to prove unequivocally.

Massive, ledge- and cliff-forming, outer-shelf facies characterize HFS Y2 in each window (Fig. 6). HFS Y3 in Slaughter Canyon is dominated by cyclic intertonguing of shelf-crest and outer-shelf facies, with siliciclastics marking the base of three of the four major cycles. Coeval Y3 strata in McKittrick Canyon are dominated by upward-coarsening outer-shelf cycles owing to the downdip position relative to the Slaughter window. Four cycle sets are interpreted for this window in McKittrick, however, and may be directly correlative with the four cycle sets in Y3 in Slaughter.

HFS Y4 is thicker in the McKittrick window than in the Slaughter window, but both exhibit the same basal siltstone and two cycle sets. A massive pisolitic cliff dominates Y4 in McKittrick, but coeval strata in the Slaughter window are characterized by intertonguing of outer-shelf and shelf-crest facies—hence, the ledgy, stair-step topography of the canyon wall.

In both windows, the Y5 HFS exhibits two cycle sets defined by upward-shallowing and upward-thinning bundles of meter-scale cycles. The thick siliciclastics overlying the basal Y5 HFS boundary and the ledge of pisolitic shelf-crest facies that defines the top of the lower cycle set are clearly correlative. The massive cliff of amalgamated shelf-crest facies exhibiting erosionally truncated tepees at the top of Y5 in each canyon also apparently represents the same event.

In sum, the overall upward progression from dominantly outer-shelf facies of Y2 to dominantly shelf-crest facies of Y5, punctuated by abrupt basinward offsets in facies above HFS boundaries, is clear in both windows (Fig. 6). These comparative intervals from each canyon illustrate how one-dimensional hierarchical patterns of cyclicity can be selectively used to reveal the overall seaward-stepping architecture of the Yates shelf margin or of any other highly progradational shelf margin. Specific positions on the shelf margin capture the fullest vertical extent of each high-frequency sequence and thus work best for this type of one-dimensional analysis (cf. Borer and Harris, 1991; Kerans and Harris, 1993). Stratigraphic intervals in extreme updip or downdip positions along the shelf profile would have limited utility, however, because of the likelihood of cycle amalgamation and "missing" cycles. Because no single one-dimensional interpretation can provide a unique solution, interpretations of platform architecture derived from vertical stacking patterns are greatly enhanced when placed in a two- or three-dimensional sequence stratigraphic framework where lateral variability in cycle-stacking patterns can be incorporated

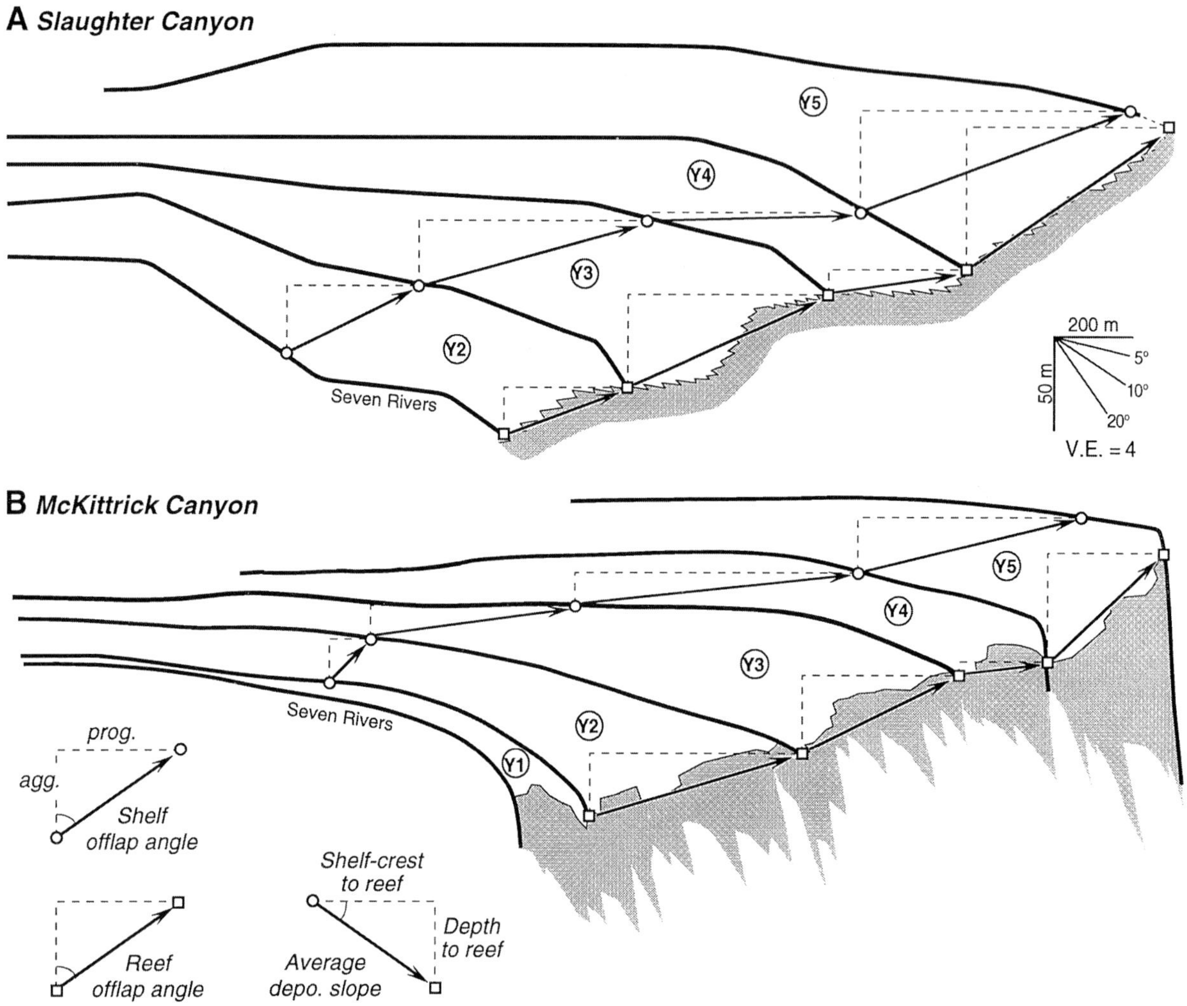

FIG. 8.—Simplified plot of Yates Formation profiles from each canyon, illustrating the method for determining depositional variables within each high-frequency sequence (HFS, plotted in Fig. 9). Open circles mark the seaward terminus of the shelf-crest facies tract just prior to exposure and formation of the HFS boundaries (assumed to represent a paleoshoreline and thus sea level). Squares mark the positions of the reef margins at the HFS boundaries. The offlap angles for both the shelf and reef are determined from the progradation:aggradation (prog.: agg.) ratio through HFS deposition. The average depositional (depo.) slope, distance from the shelf crest to the reef, and the depth of the reef along upper HFS boundaries are determined from simple measurements made between the terminal shelf crest and the time-equivalent reef margin. V.E. = vertical exaggeration.

into the interpretation. It is the two-dimensional "template" that provides a powerful tool when correlating one-dimensional well data in the subsurface.

Yates High-Frequency Sequence Comparison

The long-term evolution of the Yates-Capitan shelf margin from Y2 through Y5 is illustrated on plots of key depositional variables measured on individual high-frequency sequences in each canyon (Table 1; Figs. 8, 9). The bathymetric tie point used to calculate several of the depositional variables is the contact between the downdip limit of shelf-crest flanking facies (fenestral tidal-flat facies) and the updip, littoral limit of outer-shelf facies (oolites, coated grains, or skeletal grainstones). At any one time, this contact provides a reasonable approximation of the position of sea level. Systematic trends recorded in both two-dimensional profiles are evident in Figure 9, illustrating the strike-parallel development of the shelf margin through time. Furthermore, individual Yates HFSs exposed in the two canyons share several architectural characteristics that reveal a three-dimensional model of shelf-margin anatomy (Figs. 10, 11).

General characteristics.—
Some fundamental characteristics are shared by all HFSs in the Yates-Capitan shelf margin. In a vertical sense, the lower parts of Yates HFSs tend to be siliciclastic rich whereas the upper parts tend to be carbonate dominated. In a lateral sense, updip parts of Yates HFSs are characterized by abundant siliciclastics whereas downdip parts are dominated by massive outer-

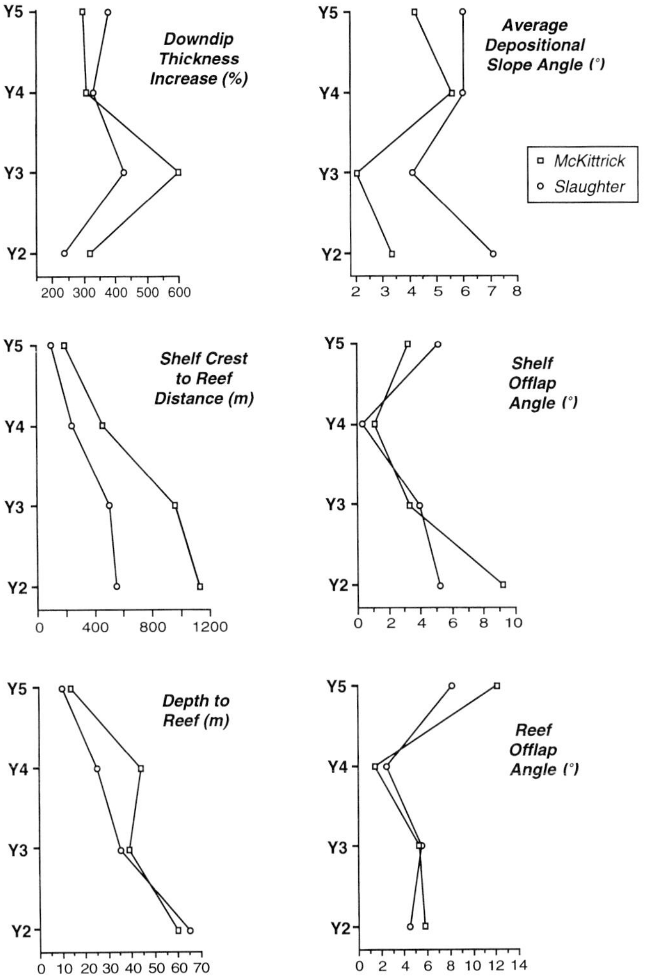

FIG. 9.—Trends in selected depositional variables during the deposition of the Yates Formation, plotted as stratigraphic changes. The methods used for measuring each variable are shown in Figure 8. The downdip thickness increase is determined from the difference between a minimum at a common position in both canyons about 3 km from the terminal Y5 reef margin and a maximum within each high-frequency sequence. Actual measurements were determined from expanded-scale cross sections calibrated against scaled photographic interpretations; regional structural dip was removed.

shelf carbonates. Yates HFSs also tend to thicken significantly downdip; increases range from 240% to 600% across the 3-km transect distances in each canyon (Table 1). The amount of thickness increase per HFS is very similar in both canyons and is discussed below.

Disconformable surfaces bounding Yates high-frequency sequences are distinguished by significant changes in vertical and lateral cycle-stacking patterns and relatively abrupt basinward shifts in facies. Thick siliciclastic beds commonly overlie HFS boundaries and extend far down onto the outer shelf, recording the seaward shift in the continental siliciclastic facies tract across the exposed carbonate platform. Cycles immediately overlying basal siliciclastics within individual HFSs typically exhibit mildly retrogradational or aggradational stacking patterns concurrent with significant accumulation of massive skeletal packstones and grainstones on the outer shelf. Uppermost cycles are strongly progradational; the shelf-crest facies tract thinned and expanded in width to within 1 km of or closer to the Capitan reef margin by the end of HFS deposition. These general characteristics shared by all HFSs in the Yates provide a framework for the HFS-specific variables discussed below.

High-frequency sequence Y2.—

The Seven Rivers Formation/Yates Formation boundary, defined by Hayes (1964) as the first significant siliciclastic bed above a thick stack of carbonates, was interpreted as a composite sequence boundary by Kerans et al. (1992). Depending upon the position along the shelf profile, this lithostratigraphic sandstone-to-carbonate contact could be picked incorrectly below Y5 (most seaward position), Y4, Y3, or correctly below Y2 (Slaughter) or Y1 (McKittrick). Therefore, given only vertical one-dimensional data, such as wireline logs, the Seven Rivers/Yates formational contact could be misinterpreted by as many as three or four HFSs.

The position of the genetic lower Yates composite sequence boundary below Y1 in McKittrick Canyon was documented by Tinker (1998) on the basis of negative progradation:aggradation ratios in underlying and overlying HFSs, stratigraphic fall of the Capitan reef above the Seven Rivers Formation (Fig. 10A), and sharp isotopic shifts in both $\delta^{18}O$ and $\delta^{13}C$ values across the boundary. These collective data are interpreted to reflect a relative fall in sea level below the shelf margin at the end of Seven Rivers deposition and a subsequent prolonged episode of subaerial exposure of the platform. The Y1 HFS of Tinker (1998) was not recognized in Slaughter, but an isolated sandstone bed overlying a disconformable surface occurs several meters below the basal Yates HFS boundary far updip in the Slaughter profile (base Y1, section 10 in Osleger, 1998); the possibility exists that this sandstone bed may overlie the actual composite sequence boundary.

Siliciclastic beds exhibit similar distribution patterns through Y2 (Fig. 10A). In both canyons, the basal siliciclastic bed above the lower Y2 boundary extends far seaward, reflecting the migration of continental siliciclastic environments across the exposed carbonate platform. Also in both canyons, overlying siliciclastic beds (including two siliciclastic beds that onlap the basal siliciclastic bed in McKittrick) step landward (backstep), accompanied by tongues of outer-shelf facies that extend far up onto the platform. Uppermost siliciclastic beds within Y2 prograde seaward in concert with the shelf-crest facies tract.

Near the end of Y2 deposition, the seaward front of the shelf crest reached to within approximately 1 km of the reef in the McKittrick area and to within approximately 0.5 km near Slaughter (Table 1), significantly narrowing the width of the outer shelf. The time-equivalent depth of the submerged Capitan reef was about 60–65 m (determined as the vertical distance between the terminal shelf crest and equivalent reef, Fig. 8). Average depositional slopes of the upper Y2 boundary surface calculated from these distances and depths range from 3.3° (McKittrick) to 7.1° (Slaughter). The apparent difference is simply due to the greater distance of progradation of the terminal

TABLE 1.—COMPARATIVE DEPOSITIONAL VARIABLES PER YATES HFS MEASURED IN McKITTRICK AND SLAUGHTER CANYONS

HFS	Downdip thickness increase (%)		Shelf crest to reef distance (m)		Depth to reef (m)		Average depositional slope angle (°)	
	McKittrick	Slaughter	McKittrick	Slaughter	McKittrick	Slaughter	McKittrick	Slaughter
Y5	300	380	195	~100	14	~10	4.1	6.0
Y4	310	330	446	250	44	25	5.6	6.0
Y3	600	430	1140	500	39	35	2.0	4.1
Y2	320	240	1035	550	60	65	3.3	7.1
average	383	345	704	350	39	34	3.8	5.8

HFS	Shelf Progradation:aggradation		Shelf Offlap angle (°)		Reef Progradation:aggradation		Reef Offlap angle (°)	
	McKittrick	Slaughter	McKittrick	Slaughter	McKittrick	Slaughter	McKittrick	Slaughter
Y5	18	11	3.2	5.1	5	7	12.2	8.2
Y4	51	170	1.1	0.3	37	23	1.5	2.5
Y3	15	15	3.8	3.9	11	10	5.3	5.5
Y2	6	11	9.2	5.2	10	13	5.8	4.5
average	22.5	52	4.3	3.6	16	13	6.2	5.2

Y2 shelf crest mapped in the Slaughter area. Thus the higher angle in Slaughter likely reflects steeper, downdip outer-shelf slopes and the lower angle in McKittrick likely reflects a broader distance and gentler average slopes. This relationship is consistent through all four HFSs in the Yates, with average depositional slope in the Slaughter profile roughly double that of the McKittrick profile (Fig. 9).

Relatively low progradation:aggradation ratios and consequent high offlap angles of the Y2 shelf (Table 1; Fig. 9) reflect the generally aggradational architecture of HFS Y2. The reasonably high growth angles of the time-equivalent Capitan reef (4.5°–5.8°) likewise reflect a dominantly upbuilding geometry, recording the biologic response to relatively high amounts of accommodation through Y2 deposition. The high accommodation is interpreted to be related to the position of Y2 near the beginning of the longer-term Yates composite sequence.

High-frequency sequence Y3.—

The most obvious characteristic of HFS Y2 in both profiles is the extreme downdip thickness increase, 430% to 600% beyond updip thicknesses (Table 1; Fig. 10B). The actual minimum and maximum thicknesses in Slaughter (22 m updip to 95 m downdip) are greater than those in McKittrick (12 and 72 m), but the percent increase is higher in the McKittrick profile. The greater overall thicknesses in Y3 in the Slaughter area are likely real; measured thicknesses through all four Yates HFSs are consistently higher overall in the Slaughter profile.

The distribution of siliciclastic beds is very consistent between the two profiles (Fig. 10B). The basal fine-grained sandstone overlying the lower HFS boundary extends to within or closer than 0.5 km of the Capitan reef. Physical tracing of this siliciclastic bed in both canyons reveals that it pinches out completely and is not evident in the time-equivalent reef, suggesting complete bypass of siliciclastics to the basin along the basal exposure surface. In the stratigraphic middle of HFS 3, a set of four siliciclastic beds located on the midshelf is likely the same set of beds in the two profiles. In Slaughter, they appear to be bifurcating tongues of single beds updip, whereas in McKittrick they appear to be discrete, isolated lenses. All four beds are stacked vertically, however, and record generally aggradational conditions during the middle phase of Y3 deposition. An upper siliciclastic bed located toward the top of Y3 is clearly correlative between both canyons on the basis of its stratigraphic position within the mildly progradational upper part of Y3 and its wide lateral extent across both profiles.

Near the end of Y3 deposition, just prior to the maximum sea-level fall along the exposed HFS boundary, the Capitan reef lay beneath 35 to 39 m of water (Table 1). Average depositional slopes reached their lowest values (2.0° to 4.1°) of all Yates HFS boundaries (Fig. 9). The shelf offlap angle for Y3 is somewhat higher in the Slaughter profile, probably because of the greater amount of shelf accumulation relative to McKittrick. The Capitan reef exhibits a significant phase of upbuilding during Y3 time, with average growth angles of 5.3° to 5.5°.

All of the depositional variables discussed above for HFS Y3 suggest that the creation of accommodation was at its highest during Y3 deposition. The downdip thickness increases for Y3 are the greatest of all four correlated Yates HFSs (Fig. 9), with thick piles of massively bedded, poor to moderately sorted, skeletal packstones and grainstones accumulating on the outer

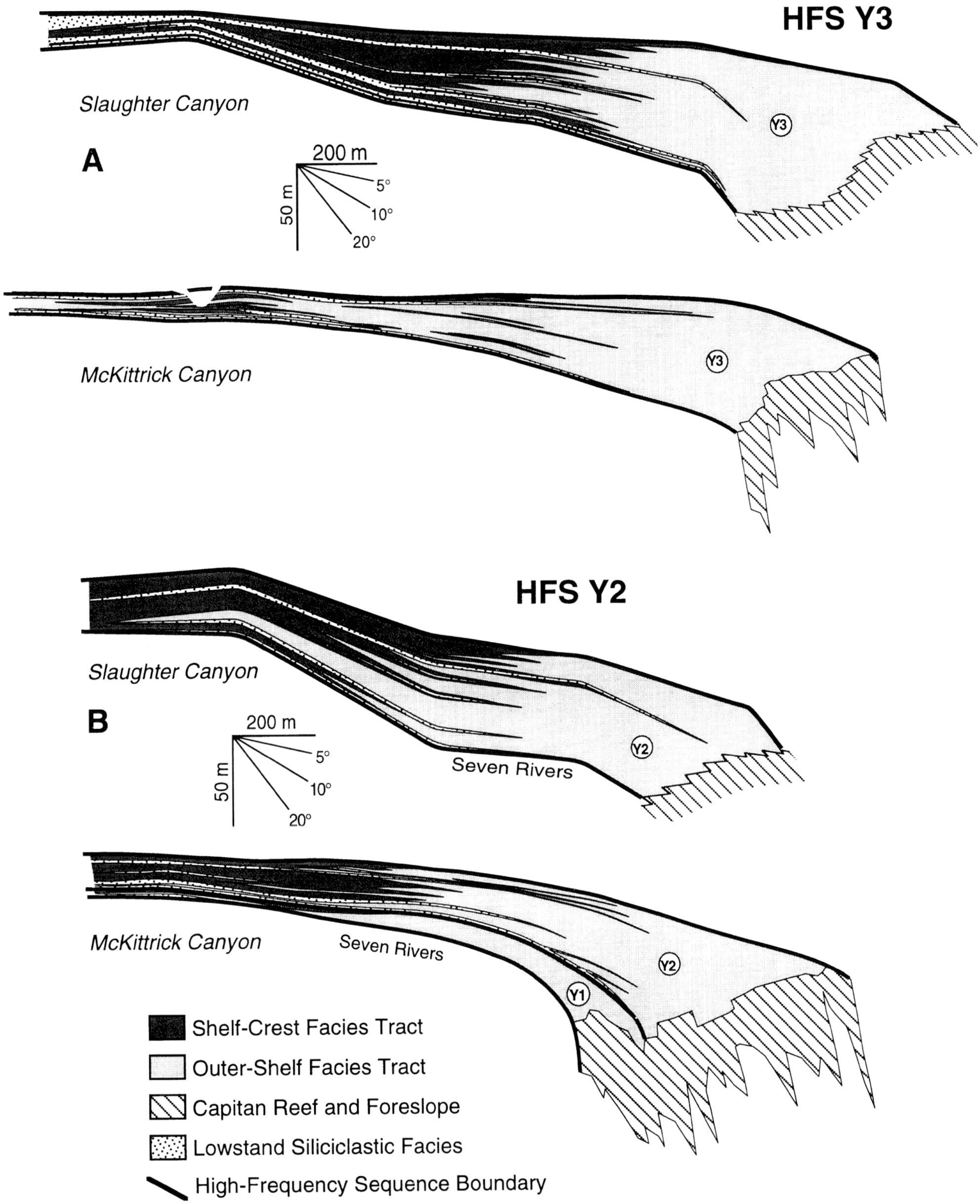

FIG. 10.—Comparison of internal anatomy of Yates Formation high-frequency sequences between Slaughter and McKittrick Canyons. (A) Y2 and (B) Y3 plotted as discrete wedges against the time-equivalent reef.

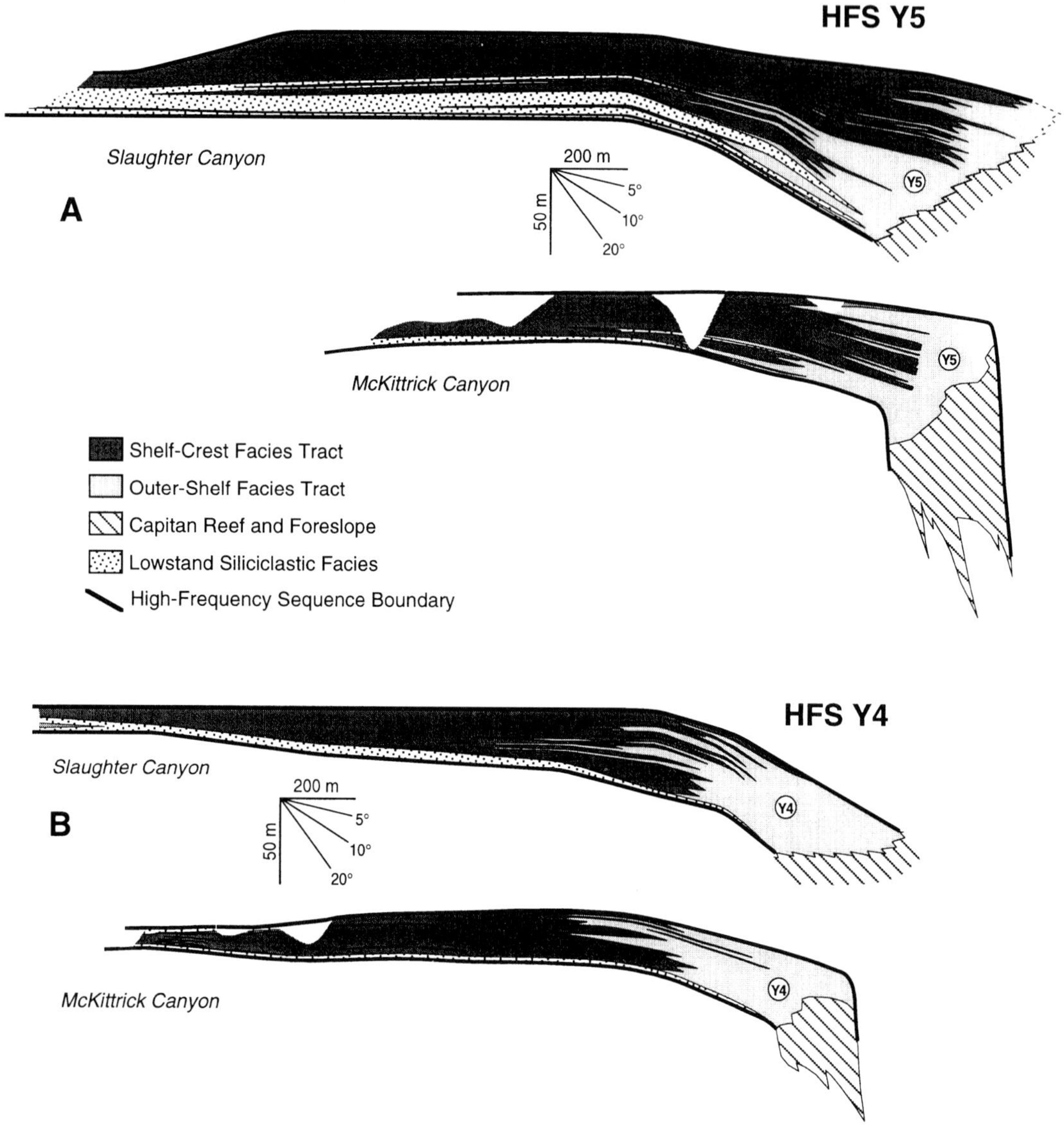

FIG. 11.—Comparison of internal anatomy of Yates Formation high-frequency sequences for (A) Y4 and (B) Y5 between Slaughter and McKittrick Canyons.

shelf. The vertical stacking of siliciclastics in the middle of Y3 also suggests dominantly aggradational deposition. Furthermore, the relatively flat depositional slopes on the outer shelf and the high reef-offlap angles attest to a major phase of upbuilding of the Yates-Capitan shelf margin during Y3 deposition, whereby accumulation rates in the outer shelf and across the margin kept up with increasing accommodation.

High-frequency sequence Y4.—

Extreme progradation characterizes HFS Y4 in both profiles, with the shelf-crest facies tract extending 400 to 800 m beyond the underlying Y3 terminal shelf crest (Figs. 7, 11A). This abrupt basinward shift is reflected in the high progradation:aggradation ratios of the shelf and reef in both canyons (Table 1). Y4 offlap angles of the shelf and reef derived from these ratios reached their lowest values ever on the Yates shelf margin (Fig. 9). Concurrent with the broad expansion of the shelf-crest pisolite complex, the outer-shelf facies tract underwent a significant contraction.

The base of Y4 is marked by a thick siliciclastic bed that blankets almost the entire shelf, reaching to closer than 100 m of the reef (Fig. 11A). Overlying the basal fine-grained sandstone, both profiles exhibit the abrupt basinward shift in the pisolite shoal complex to within 200 to 300 m of the reef margin; three discrete tongues are recognized at the seaward edge of the complex in both profiles. The remainder of Y4 is characterized by vertical upbuilding, with tongues of outer-shelf facies extending less than 1 km onto the shelf (vs. the 2 to 3 km of flooding distance in underlying Y2 and Y3).

All of these characteristics of Y4 indicate that deposition occurred during a phase of significantly reduced accommodation, forcing a seaward shift of the shelf crest and extensive outbuilding of the Capitan reef. On the basis of the extreme progradational geometry of Y4 and the dominantly aggradational geometry of overlying HFS Y5, we interpret the top of the Yates composite sequence to be along the upper Y4 HFS boundary.

High-frequency sequence Y5.—

Although HFS Y5 prograded farther basinward relative to HFS Y4, the internal architecture of HFS Y5 is characterized by dominantly aggradational stratal patterns, a widespread pisolite shoal complex, but only a minor progradational component (Fig. 11B). The limited progradation of the shelf crest is reflected in moderate shelf-offlap angles (3.2° in McKittrick, 5.1° in Slaughter). Profiles from both canyons reveal a remarkably similar disposition of carbonate facies tracts above the basal siliciclastic beds in Y5 (Fig. 11B). Tongues of outer-shelf facies in the lower part of Y5 extend less than 1 km landward and are overlain by a thick succession of aggradational shelf-crest facies in the middle of Y5. A second set of outer-shelf beds, reflecting renewed marine transgression, overlies the thick shelf-crest complex. The latest phase of Y5 is characterized by the seaward migration of the shelf crest to within 100 to 200 m of the reef. This complicated pattern of alternating episodes of progradation and retrogradation through Y5 may represent an intermediate-scale (cycle set) of cyclicity between the meter scale and HFS scale.

The growth angle of the Capitan reef reached its greatest value during Y5 time (12.2° in McKittrick, 8.2° in Slaughter), recording significant aggradation of the reef margin (Fig. 9). The depth of the reef by the end of Y5 deposition approached 10 to 14 m, providing further evidence for steepening and shallowing of the reef from Y2 through Y5. Kerans and Harris (1993) recognized an abrupt facies offset near the top Y5 surface in McKittrick Canyon (their Y4); the offset is marked by peritidal facies and exposure breccias juxtaposed above outer-shelf and reefal facies, suggesting a sea-level fall of approximately 12 m and near–sea-level conditions at this phase of Capitan reef development.

Shelf-Margin Evolution

A dynamic model for the evolution of the Yates-Capitan shelf margin from Y2 through Y5 can be constructed by integrating the long-term averages for each depositional variable (Table 1) with the graphs of changing variables through time (Fig. 9).

The downdip thickness increase is on average greater in the McKittrick area than near Slaughter (383% to 345%), even though total thickness is greater overall along the Slaughter profile (220 m maximum there vs. 165 m maximum near McKittrick). The thickness variation along the 25-km strike distance may indicate significant lateral variability in subsidence and/or accumulation rate along the Yates-Capitan shelf margin. The difference in downdip thickness increases suggests local variability in subsidence rates, with the cross-platform difference in subsidence rates near McKittrick being slightly greater than that near Slaughter. Over the long term, the thickness increases parallel one another (Fig. 9), reaching a maximum during Y3 deposition. The peak during Y3 is interpreted to reflect the greatest amounts of accommodation in the longer-term composite sequence at the Y3 phase of shelf-margin evolution.

The terminal shelf crest appears to have prograded consistently farther seaward in each HFS in the Slaughter area than near McKittrick (Table 1). This trend may be attributed to the smaller cross-platform difference in subsidence rates near Slaughter that would have permitted the greater seaward migration of the shelf crest in that area. Over the long term, however, the shelf crest-to-reef distance progressively decreases in both profiles from Y2 through Y5 (Fig. 9), reflecting the early aggradational, then strongly progradational architecture of the Yates shelf margin. By the end of Y5 time, the shelf crest extended to within less than 200 m of the reef edge.

The depth of the reef, measured from the terminal shelf crest below HFS boundaries, progressively decreased through time (Fig. 9). The actual depth of the reef likely varied episodically through Yates deposition, with shorter-term variations superimposed in response to cycle-scale and HFS-scale relative sea-level changes. For instance, the depth of the reef may have approached less than 10 or 20 m during subaerial exposure events at HFS boundaries, on the basis of the seaward extent of basal siliciclastic beds to within a few hundred meters of the reef margin in all four HFSs. The depth of the reef increased during subsequent marine transgression, potentially to depths well in excess of 60 m. These episodic changes in the depth of the reef are reflected in the geometric alternations between upbuilding during times of greater accommodation and outbuilding of the Capitan margin during times of lower accommodation (Fig. 7).

Both profiles show parallel long-term trends in average depositional slope, suggesting coincident episodes of outer-shelf flattening or steepening (Fig. 9). Average depositional slopes of the outer shelf were consistently higher near Slaughter, however, simply because the shelf crest appears to have prograded farther seaward (and downslope) at the very end of each HFS. Thus the average of 5.8° along the Slaughter outer shelf reflects the downdip part of the shelf, relative to the 3.8° average for McKittrick that includes the broad upper part of the outer shelf. These values corroborate the downdip increase in slope angle of the outer shelf measurable on outcrop in both canyon profiles.

Net offlap angles of the shelf-crest facies tract for the entire Yates Y2 through Y5 interval are reasonably similar (3.6° and 4.3°), with the difference primarily attributable to the much higher Y2 angle in McKittrick. Long-term changes track one another, progressively decreasing from Y2 through Y4 and then increasing through Y5 (Fig. 9). The decrease from Y2 through Y4 reflects the change from aggradation during times of greater accommodation in the longer-term Yates composite sequence (Y2 and Y3) to progradation during times of lower accommodation (Y4). The increase in offlap angle into Y5 records a renewed aggradational pulse of the shelf. This increase may have been in response to renewed accommodation driven by long-term sea-level rise, perhaps enhanced by the lack of a slope-sediment foundation over which to prograde in HFS Y5.

Somewhat similar trends are evident in the offlap angles of the Capitan reef. Minimum reef growth angles were attained during Y4 time, reflecting net outbuilding, followed by high angles of reef upbuilding during Y5. Net growth angles are very similar in both profiles from Y2 through Y4, but the Capitan reef appears to have grown more vertically in the McKittrick area during Y5 time (12.2° vs. 8.2°), resulting in the higher average growth angle for the reef near McKittrick (Table 1).

COMPARISON WITH CENTRAL BASIN PLATFORM AND EASTERN SHELF

A composite dip-oriented profile of the Yates shelf margin was constructed by Borer and Harris (1991) from subsurface data from the Central Basin platform and the PDB-04 research well from the Northwest shelf. A comparison of Borer and Harris's composite cross section with our two outcrop profiles from the Guadalupe canyons provides an interesting test of their model and permits an evaluation of the basin-wide character of the Yates-Capitan shelf margin rimming the Delaware basin (Figs. 4, 12). The model of Borer and Harris (1991) (reproduced in Fig. 12) has been rescaled for ease of comparison with the two Guadalupe canyon profiles. High-frequency sequence boundaries are superimposed on their cross section based on interpretations in Borer and Harris (1995). Carbonate facies tracts are simplified into one pattern on the composite cross section.

Several fundamental similarities can be discerned between the composite cross section and the two Guadalupe profiles (Fig. 12). The overall seaward progradation and thickening are clearly evident; siliciclastic beds extend progressively farther seaward in successive HFSs. The number and distribution of siliciclastic beds are very similar between the composite and Guadalupe profiles. Moreover, siliciclastic beds tend to amalgamate updip, and intervening carbonate beds thin dramatically.

Closer comparison reveals more specific similarities and some enlightening differences. Even though all of the shelf-margin profiles expand seaward, the composite profile constructed from data on the Central Basin platform and northern Northwest shelf appears to thicken the least. Updip thicknesses at a position about 4–5 km updip on the subsurface composite profile are substantially greater than projected thicknesses from the two outcrop profiles. Assuming accurate stratigraphic correlation, this difference is real, rather than apparent, because the subsurface composite section is constrained updip by abundant well control from the Central Basin platform where Yates siliciclastics dominate the middle shelf. Furthermore, a qualitative estimate suggests that the cumulative sandstone thickness appears to be volumetrically greatest on the northern Northwest shelf and Central Basin platform, slightly thinner near Slaughter, and thinnest near McKittrick. This apparent trend may reflect strike-parallel differences in subsidence rate, siliciclastic sediment supply, shelf gradient, transport distance, or perhaps even ambient wind orientations.

Overall thickness variations at similar positions on the platform may record strike-parallel differences in subsidence rate along the Yates-Capitan shelf margin. The thickness of the Yates in the PDB-04 well from the top of the Y5 HFS down to the estimated top of the reef is approximately 157 m. In McKittrick, shelf strata in this interval total approximately 150–160 m, comparable to the PDB-04 thickness. The projected position of this well in the Slaughter profile, based on distance from the terminal Y5 shelf margin and alignment of siliciclastic beds, suggests that the thickness of this same interval approaches approximately 175–180 m. The greater accumulation near Slaughter is likely real because similarly greater thicknesses are consistent between individual HFSs from separate profiles (Fig. 12). Subsidence and accumulation rates appear to have been highest in the Slaughter area relative to those elsewhere along the shelf margin.

Comparison of individual HFSs between the profiles illustrates the consistency of architectural patterns along the Yates-Capitan shelf margin. The Y2 HFS in the composite profile appears to include both the Y1 and Y2 HFSs of the McKittrick profile, based on total thicknesses and distribution of siliciclastic beds. Specifically, the two sandstones far updip at the base of the Yates on the composite profile may correlate with the two amalgamated sandstones at a similar platform position at the base of Y1 in McKittrick (Fig. 10A). Siliciclastic beds in Y2 (and Y1) in all three profiles exhibit similar retrogradational to aggradational to mildly progradational distribution, especially when compared to overlying HFSs. In HFS Y3, the number and distribution of siliciclastic beds are remarkably similar upon close inspection of all three profiles. HFS Y3 also exhibits comparable amounts of aggradational upbuilding on the composite profile relative to the two Guadalupe profiles. In HFS Y4, the extreme progradational anatomy and dominance of carbonate lithofacies are clear on all three profiles. Finally, HFS Y4 in all three profiles exhibits the same bifurcating siliciclastic units along the base and an apparent renewed aggradational component of the carbonate facies.

As discussed in the Vertical Cycle-Stacking Patterns section, a significant amount of interpretation can be made from one-dimensional stacking patterns alone. A cross section of total gamma-ray (GR) data from a three-dimensional geologic model of a field on the Eastern shelf (see Fig. 4 for approximate location) shows a cycle-stacking pattern of alternating feldspathic sandstones (high GR values owing to their potassium component) and evaporites (low GR values) (Fig. 13). Although the total thickness of the interval is only 30 m, the four HFSs that form the Yates composite sequence in outcrop are still apparent in the subsurface more than 300 km away. A similar distribution of siliciclastics and evaporites from subsurface data of the Central Basin platform (Dollarhide field log of Borer and Harris, 1991) corroborates the vertical stacking patterns on the Eastern shelf.

Collectively, the similarities described above indicate that a basin-wide driver must have governed the architecture of the Yates-Capitan shelf margin rimming the Delaware basin from the Guadalupe area to the Central Basin platform, as well as updip inner-shelf positions on the Eastern shelf. Relative sea-level changes likely played the critical role in determining the growth and development of the shelf margin. Variations in climate, tectonism, differential compaction, biologic productivity, and ambient oceanographic conditions likely influenced the evolution of the shelf margin and may have contributed to depositional variability along the margin. These background processes apparently exerted a subordinate overprint, however, compared

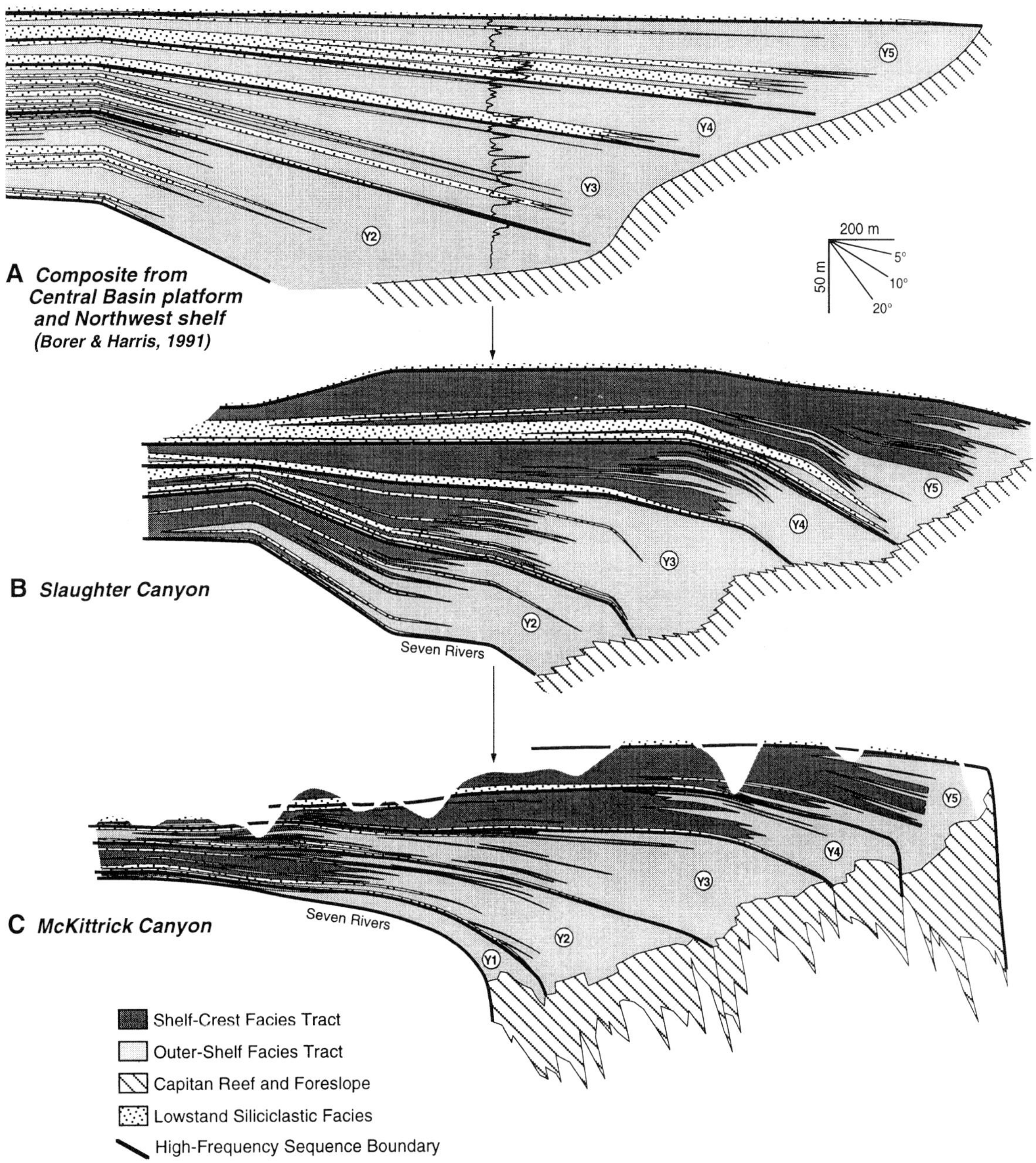

FIG. 12.—(A) Composite cross section constructed from subsurface data by Borer and Harris (1991), rescaled for comparison with the two profiles from Guadalupe outcrops; the interpreted boundary with the Capitan reef has been placed on the profile. Coarse-dot pattern represents siliciclastic facies; shaded pattern represents all carbonate facies (combined shelf crest and outer shelf). The gamma-ray log, from the PDB-04 research well (located along the Northern shelf margin [see Fig. 4]; Garber et al., 1989), is superimposed on the composite cross section. (B, C) The two Guadalupe profiles are aligned with the composite profile along the projected position of the PDB-04 well (arrows).

to the dominant role played by relative sea-level changes.

IMPLICATIONS

Integrating the independently generated two-dimensional profiles from the Guadalupe canyons significantly improves the three-dimensional visualization of the internal anatomy of genetic units that form the Yates-Capitan shelf margin. Our understanding of the evolution of the margin through time is further enhanced by comparison of key depositional variables such as progradation:aggradation ratios and derived offlap angles of the shelf and reef. These variables show systematic changes in

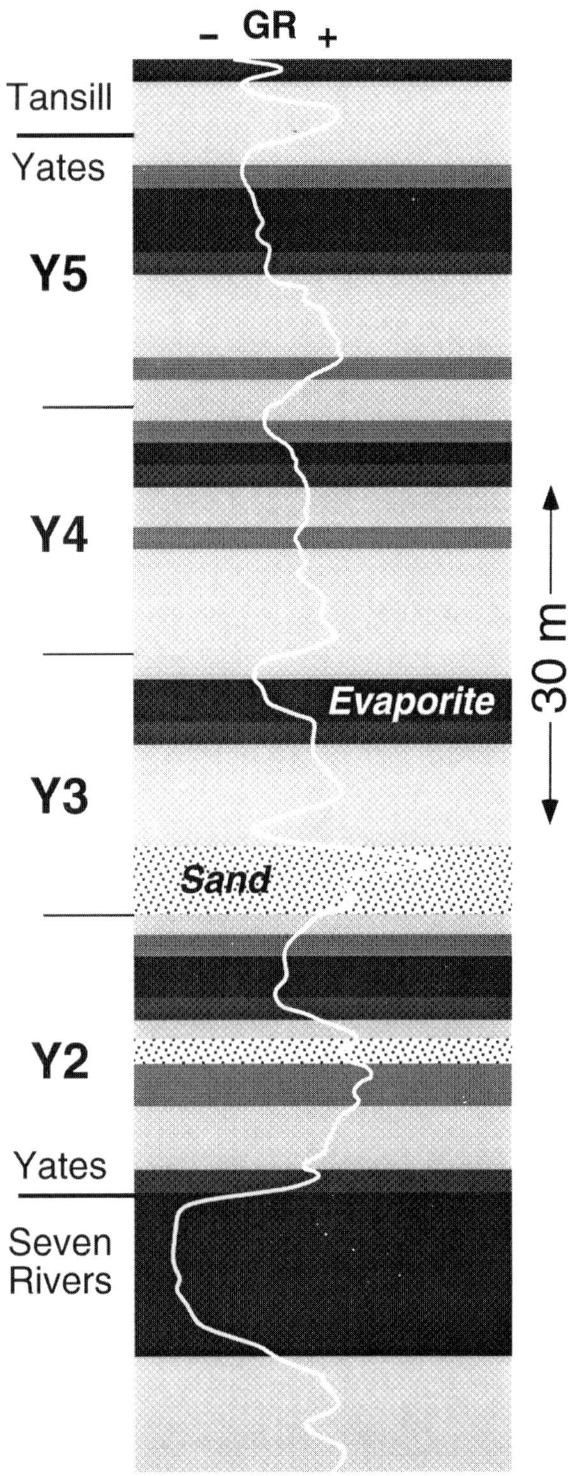

FIG. 13.—Cross section of total gamma-ray (GR) data from a three-dimensional geologic model of a field on the Eastern shelf showing four Yates Formation high-frequency sequences.

the architecture of Yates HFSs through time. Furthermore, comparison of our integrated field observations with subsurface data from the northern Northwest shelf, Central Basin platform, and Eastern shelf reveals a consistent, basin-wide pattern of the internal architecture of high-frequency sequences composing the Yates Formation. These widespread stratigraphic similarities provide compelling evidence for regional sea-level control on sequence development around the Permian basin.

High-resolution studies such as this provide a subseismic level of internal detail regarding how carbonate-platform margins are constructed and permit a direct comparison with published conceptual models. Moreover, the integrated two-dimensional profiles of the Yates-Capitan shelf margin provide a reliable geometric and chronostratigraphic "template" for subsurface correlation between one-dimensional well control points, as well as an architectural model for comparable prograding shelf-margin settings. The sequence stratigraphic framework established in this study provides a ground-truth level of reliability that can be used to increase confidence in the predictive value of a three-dimensional reservoir model in similar shelf-margin settings.

ACKNOWLEDGMENTS

Low-angle oblique photographs for both McKittrick and Slaughter Canyons were supplied by Charlie Kerans and Jerry Lucia of the University of Texas Bureau of Economic Geology. Hasselblad pans were supplied by Exxon Production Research. Permission to work in the two canyons was granted by the Guadalupe National Park Service and Carlsbad Caverns National Park Service. Constructive reviews by Tim McHargue, Jim Mazzullo, Mark Sonnenfeld, and Mitch Harris improved the paper. Charlie Kerans contributed significantly to the project through discussions regarding the regional stratigraphic framework of the Guadalupian system in the Permian basin. Support for this project was provided by Marathon Oil Company, Exxon Production Research Company, and the Petroleum Research Fund (administered by the American Chemical Society).

REFERENCES

ANDREASON, M. W., 1992, Coastal siliciclastic sabkhas and related evaporative environments of the Permian Yates Formation, North Ward-Estes field, Ward County, Texas: American Association of Petroleum Geologists Bulletin, v. 76, p. 1735–1759.

BORER, J. M., AND HARRIS, P. M., 1991, Lithofacies and cyclicity of the Yates Formation, Permian Basin: Implications for reservoir heterogeneity: American Association of Petroleum Geologists Bulletin, v. 75, p. 726–779.

BORER, J. M., AND HARRIS, P. M., 1995, Computer simulation of the Yates Formation (Permian, Delaware Basin)—Sequence stratigraphy and shelf-to-basin correlation implications, *in* Martin, R. L., ed., In Search of New Permian Basin Oil and Gas Fields: Using Today's Technologies and Tomorrow's Ideas for Exploration, Development, and 3D Seismic in a Mature Basin: Midland, West Texas Geological Society Symposium, Publication 95-98, p. 111–132.

DUNHAM, R. J., 1972, Capitan Reef, New Mexico and Texas: Facts and Questions to Aid Interpretation and Group Discussion: Midland, Permian Basin Section — Society of Economic Paleontologists and Mineralogists Publication No. 72-14, 297 p.

ESTEBAN, M., AND PRAY, L. C., 1977, Origin of the pisolite facies of the shelf crest, *in* Hileman, M. E. and Mazzullo, S. J., eds., Upper Guadalupian Facies, Permian Reef Complex, Guadalupe Mountains, New Mexico and West Texas: Midland, Permian Basin Section, Society of Economic Paleontologists and Mineralogists Publication No. 77-16, 1977 Field Conference Guidebook, v. 1, p. 479–486.

FISCHER, A. G., AND SARNTHEIN, M., 1988, Airborne silts and dune-derived sands from the Permian of the Delaware basin: Journal of Sedimentary Petrology, v. 58, p. 637–643.

FRANSEEN, E. K., GOLDSTEIN, R. H., AND WHITESELL, T. E., 1993, Sequence stratigraphy of Miocene carbonate complexes, Las Negras area, southeastern Spain: Implications for quantification of changes in relative sea level, *in* Loucks, R. G. and Sarg, J. F., eds., Carbonate Sequence Stratigraphy: Recent Developments and Applications: Tulsa, American Association of Petroleum Geologists Memoir 57, p. 409–434.

GARBER, R. A., GROVER, G. A., AND HARRIS, P. M., 1989, Geology of the Capitan shelf margin—Subsurface data from the northern Delaware basin, *in* Harris, P. M. and Grover, G. A., eds., Subsurface and Outcrop Examination of the Capitan Shelf Margin, Northern Delaware Basin: Tulsa, SEPM (Society for Sedimentary Geology) Core Workshop 13, p. 3–269.

HAYES, P. T., 1964, Geology of the Guadalupe Mountains, New Mexico: Washington, D.C., U.S. Geological Survey Professional Paper 446, 69 p.

HURLEY, N. F., 1989, Facies mosaic of the lower Seven Rivers Formation, McKittrick Canyon, New Mexico, *in* Harris, P. M. and Grover, G. A., eds., Subsurface and Outcrop Examination of the Capitan Shelf Margin, Northern Delaware Basin: Tulsa, SEPM (Society for Sedimentary Geology) Core Workshop 13, p. 325–346.

KENDALL, C. G. ST. C., 1969, An environmental reinterpretation of the Permian evaporite/carbonate shelf sediments of the Guadalupe Mountains: Geological Society of America Bulletin, v. 80, p. 2503–2526.

KERANS, C., AND FITCHEN, W. M., 1995, Sequence hierarchy and facies architecture of a carbonate-ramp system: San Andres Formation of Algerita Escarpment and western Guadalupe Mountains, west Texas and New Mexico: Austin, University of Texas, Bureau of Economic Geology, Report of Investigations No. 235, 86 p.

KERANS, C., AND HARRIS, P. M., 1993, Outer shelf and shelf crest, *in* Bebout, D. G. and Kerans, C., eds., Guide to the Permian Reef Geology Trail, McKittrick Canyon, Guadalupe Mountains National Park, West Texas: Austin, University of Texas, Bureau of Economic Geology, Guidebook 26, p. 32–43.

KERANS, C., AND TINKER, S. W., 1997, Sequence Stratigraphy and Characterization of Carbonate Reservoirs: Tulsa, SEPM (Society for Sedimentary Geology) Short Course Notes No. 40, 130 p.

KERANS, C., FITCHEN, W. M., GARDNER, M. H., SONNENFELD, M. D., TINKER, S. W., AND WARDLAW, B. R., 1992, Styles of sequence development within uppermost Leonardian through Guadalupian strata of the Guadalupe Mountains, Texas and New Mexico, *in* Mruk, D. H. and Curran, B. C., eds., Permian Basin Exploration and Production Strategies: Applications of Sequence Stratigraphic and Reservoir Characterization Concepts: Midland, West Texas Geological Society Symposium, Publication 92-91, p. 1–7.

KING, P. B., 1948, Geology of the southern Guadalupe Mountains, Texas: Washington, D.C., U.S. Geological Survey Professional Paper 215, 183 p.

KUTZBACH, J. E., AND GALLIMORE, R. G., 1989, Pangaean climates: Megamonsoons of the megacontinent: Journal of Geophysical Research, v. 94, p. 3341–3357.

MAZZULLO, J., DRONAMRAJU, S., JOHNSON, R., AND AHR, W., 1996, Facies and sequence stratigraphy of the Late Permian Yates Formation on the western margin of the Central Basin platform of the Permian basin, North Ward-Estes and South Ward fields, Ward County, Texas (abs.): Midland, West Texas Geological Society Symposium, Publication 96-101, p. 117–120.

MITCHUM, R. M., AND VAN WAGONER, J. C., 1991, High-frequency sequences and their stacking patterns: Sequence-stratigraphic evidence of high-frequency eustatic cycles: Sedimentary Geology, v. 70, p. 131–160.

OSLEGER, D. A., 1998, Sequence architecture and sea-level dynamics of Upper Permian shelfal facies, Guadalupe Mountains, southern New Mexico: Journal of Sedimentary Research, v. 68, p. 327–346.

POMAR, L., 1993, High-resolution sequence stratigraphy in prograding Miocene carbonates: Application to seismic interpretation, *in* Loucks, R. G. and Sarg, J. F., eds., Carbonate Sequence Stratigraphy: Recent Developments and Applications: Tulsa, American Association of Petroleum Geologists Memoir 57, p. 389–407.

ROSS, C. A., AND ROSS, J. R. P., 1987, Late Paleozoic sea levels and depositional sequences, *in* Ross, C. A. and Haman, D., eds., Timing and Depositional History of Eustatic Sequences, Constraints on Seismic Stratigraphy: Ithaca, Cushman Foundation for Foraminiferal Research, Special Publication 24, p. 137–149.

SMITH, D. B., 1974, Sedimentation of Upper Artesia (Guadalupian) cyclic shelf deposits of northern Guadalupe Mountains, New Mexico: American Association of Petroleum Geologists Bulletin, v. 58, p. 1699–1730.

SONNENFELD, M. D., AND CROSS, T. A., 1993, Volumetric partitioning and facies differentiation within the Permian upper San Andres Formation of Last Chance Canyon, Guadalupe Mountains, New Mexico, *in* Loucks, R. G. and Sarg, J. F., eds., Carbonate Sequence Stratigraphy: Recent Developments and Applications: Tulsa, American Association of Petroleum Geologists Memoir 57, p. 475–492.

TINKER, S. W., 1996a, Building the 3-D jigsaw puzzle: Applications of sequence stratigraphy to 3-D reservoir characterization, Permian basin: American Association of Petroleum Geologists Bulletin, v. 80, p. 460–485.

TINKER, S. W., 1996b, Reservoir-scale sequence stratigraphy: McKittrick Canyon and 3-D subsurface examples, west Texas and New Mexico: Unpublished Ph.D. Dissertation, University of Colorado, Boulder, 245 p.

TINKER, S. W., 1998, Shelf-to-basin facies distributions and sequence stratigraphy of a steep-rimmed carbonate margin: Capitan depositional system, McKittrick Canyon, New Mexico and Texas: Journal of Sedimentary Research, v. 68, No. 6, p. 1146-1174.

VEEVERS, J. J., AND POWELL, C. M., 1987, Late Paleozoic glacial episodes in Gondwanaland reflected in transgressive-regressive depositional sequences in Euramerica: Geological Society of America Bulletin, v. 98, p. 475–487.

WARD, R. F., KENDALL, C. G. ST. C., AND HARRIS, P. M., 1986, Upper Permian (Guadalupian) facies and their association with hydrocarbons—Permian basin, west Texas and New Mexico: American Association of Petroleum Geologists Bulletin, v. 70, p. 239–262.

WILSON, J. L., 1975, Carbonate Facies in Geologic History: New York, Springer-Verlag, 471 p.

SECULAR VARIATION IN PARASEQUENCE AND FACIES STACKING PATTERNS OF PLATFORM CARBONATES: A GUIDE TO APPLICATION OF STACKING-PATTERNS ANALYSIS IN STRATA OF DIVERSE AGES AND SETTINGS

DANIEL J. LEHRMANN

Department of Geology, Harrington Hall, University of Wisconsin, Oshkosh, Wisconsin 54901

AND

ROBERT K. GOLDHAMMER

4309 Yoakum Street, Houston, Texas 77006

ABSTRACT: Stacking-patterns analysis is defined as the identification, interpretation, and correlation of chronostratigraphic units from vertical patterns in facies, diagenetic attributes, and the thickness of high-resolution depositional cycles (parasequences). Such analysis has been widely used in outcrop and subsurface studies of shallow-marine carbonate strata to establish a stratigraphic hierarchy and erect a sequence stratigraphic framework. Recent debate about the levels of order and randomness preserved in the record of cyclic platform-carbonate strata, however, points to the likelihood that a broad spectrum of stratigraphic behavior exists, ranging from nearly random (stochastic) to highly ordered (deterministic). To evaluate the levels of order versus randomness inherent in different depositional systems and to detect secular variations in stratigraphic style, we have constructed a digital database of high-quality stratigraphic data from diverse settings and ages ranging across the Phanerozoic; we have analyzed these data (1) by employing statistical tests (Markov chain analysis, runs test, Durbin-Watson test, entropy analysis) and (2) through visual and analytical comparison of normalized stratigraphic plots.

The results of our analysis indicate that six empirical classes of stratigraphic style are recognizable. The classes are grouped by age and consist of (1) Proterozoic through Early Ordovician, (2) Silurian–Devonian, (3) Pennsylvanian, Early Permian, Late Triassic, and Neogene, (4) middle to Late Permian, (5) Early to Middle Triassic, and (6) Jurassic to Cretaceous. Differences in stratigraphic style between classes are interpreted to have resulted from differences in the amplitude of eustatic sea-level signatures between icehouse climates (with high-amplitude glacio-eustasy) and greenhouse climates (with low-amplitude eustasy) as well as differences in levels of autocyclicity dictated by tectonic setting and platform configuration. Secular variations in amplitudes of sea-level fluctuation have resulted in differences in the strength of allocyclic forcing and are reflected in the stratigraphic record by the degree to which sediment fill accurately recorded accommodation space.

Icehouse platforms (Pennsylvanian, Early Permian, and Neogene) contain stacks of fifth- or fourth-order high-frequency sequences capped with subaerial exposure surfaces, relatively chaotic vertical thickness distributions, and high lateral continuity. High lateral continuity of individual fifth- or fourth-order sequences is interpreted to result from the strong allocyclic forcing potential during icehouse periods. Third-order composite sequences are best identified from regional retrogradational to progradational patterns of platform-margin facies. Chaotic vertical thickness distributions and long-term stationarity (i.e., lacking in long-term directional change) in facies patterns of high-frequency sequences, resulting from incomplete recording of accommodation space, however, hamper the identification of the composite sequences on the basis of individual vertical sections in the platform interior.

Greenhouse platforms contain broad transgressive-regressive (or landward-stepping to seaward-stepping) sequences characterized by (1) long-term systematic change in parasequence thickness and facies proportions, (2) gross shallowing and deepening shifts in facies tracts defining maximum regressive and transgressive intervals, and (3) in some cases, a capping surface with extensive subaerial diagenetic modification (e.g., karst). Proterozoic through Ordovician sequences record long-term systematic changes in parasequence thickness and facies proportions. Jurassic and Cretaceous sequences tend to be represented as gross shallowing and deepening shifts in facies tracts with high levels of randomness at the parasequence scale due to greater levels of autocyclicity and the development of facies mosaics. Potential for greater discontinuity at the parasequence scale results from weaker allocyclic forcing by diminished fourth- and fifth-order amplitudes. The superposition of low-amplitude, high-frequency sea-level fluctuations on higher-amplitude fluctuations of lower frequencies, however, sets up the potential for the accurate recording of long-term accommodation and a full expression of the stratigraphic hierarchy (i.e., third-, fourth-, and fifth-order stratigraphic cycles).

INTRODUCTION

Stacking-patterns analysis can be considered a subset of sequence stratigraphy that focuses on the evaluation of the vertical arrangement of facies types, stratal thicknesses, and unconformities into cyclic stratal packages of various scales and character (e.g., parasequences, high-frequency sequences, depositional sequences, and composite sequences). The utility of stacking-patterns analysis for the definition and correlation of chronostratigraphic units stems from the recognition that in many settings, carbonate stratigraphy contains a hierarchy of finer-scale units nested within coarser-scale units, commonly interpreted to be driven primarily by hierarchical fluctuations in sea level in which lower-frequency sea-level cycles force the higher-frequency sea-level cycles through composite interaction (Goldhammer et al., 1990, 1993b; Tucker and Wright, 1990; Kerans and Fitchen, 1995; Read, 1995; Grammer et al., 1996; Kerans and Tinker, 1997; Fitchen, 1997). The concept of hierarchical stratigraphic forcing implies that the formation of fine-scale stratigraphic cycles (e.g., parasequences) is genetically linked to the development of coarser-scale stratigraphic cycles (e.g., depositional sequences). The lower-frequency sea-level forcing thus results in organized vertical and lateral distributions in parasequence thickness, facies character, lateral geometry, and early diagenetic attributes relative to the longer-term depositional sequences.

The whole issue of carbonate cyclicity and its relevance to the origin of shallow-marine carbonate successions of all ages is in an ongoing phase of development and controversy. Many authors, intent on understanding the processes by which high-

Advances in Carbonate Sequence Stratigraphy: Application to Reservoirs, Outcrops and Models, SEPM Special Publication No. 63

frequency stratigraphic cycles develop, have promoted a causal mechanism, be it allocyclic or autocyclic, driven by deterministic, chaotic, or stochastic means (e.g., Koerschner and Read, 1989; Osleger and Read, 1991; Goldhammer et al., 1993b; Read, 1995). Other authors, more concerned with the internal architecture and hierarchical arrangement of stratigraphic cycles and sequences, have therefore provided predictive models for facies distribution (e.g., Goldhammer et al., 1991a; Kerans and Fitchen, 1995; Grammer et al., 1996). This second group, although cognizant of causal mechanisms, has been less concerned with origin, but rather has capitalized upon the predictive nature of carbonate depositional systems to constrain distribution of hydrocarbon reservoirs (Tinker and Mruk, 1995; Tinker, 1996).

A third camp has questioned the fundamental concept of a hierarchical grouping of cyclicity in carbonates as part of an effort to demonstrate that parasequence stacking patterns are predominantly random (Drummond and Wilkinson, 1993; Wilkinson et al., 1997; Satterly, 1996). They maintain that parasequence patterns are erratic, bounced about by random unpredictable factors that wipe out whatever deterministic signature may exist. These authors are thus shifting attention away from the predictive capacity of cyclic stratigraphy and the utility of using parasequence stacking patterns as a tool for evaluating the origin of depositional sequences. This third camp has applied statistical analyses to cyclic carbonate successions and has concluded that the vertical arrangement of high-frequency stratigraphic cycles and subfacies is random. Furthermore, as the distribution is random, they have claimed that the process is more than likely random, and they have appealed to autocyclic phenomena in the spirit of complex Holocene facies models. In contrast, Sadler et al. (1993), using similar statistical tests, concluded that the vertical organization of long strings of high-frequency stratigraphic cycles is decidedly nonrandom.

The issue of the presence of order and hierarchical structure in cyclical carbonate successions is of prime importance for gauging the usefulness of the stacking-patterns technique in construction of high-resolution chronostratigraphic correlations and development of predictive models for facies distribution. Variation in opinion on this issue, as illustrated above, reflect the specific depositional systems in which the authors are working, for example, a low-angle ramp system developed in a period of greenhouse climate with low-amplitude sea-level fluctuations versus a reef-rimmed shelf developed in an icehouse climate with high-amplitude sea-level fluctuations. Recently, Goldhammer (1994), Goldhammer and Lehrmann (1996), and Lehrmann and Goldhammer (1996, 1997) have suggested that there may be a spectrum of stratigraphic behavior in carbonate depositional systems, ranging from nearly deterministic through chaotic to stochastic, as a function of secular variations in stratigraphic forcing (icehouse vs. greenhouse dynamics), platform geometry, tectonic setting, and local environmental conditions.

In this study we investigate stacking-pattern styles (both vertical and lateral) at all hierarchical scales from carbonate depositional systems of all stratigraphic ages (Proterozoic to Pleistocene) and platform morphologies (ramps, rimmed shelves, etc.). We have compiled high-quality stratigraphic data from more than 50 depositional systems in the context of a digital stratigraphic database. Digital capture of the data sets has allowed us to evaluate levels of order and hierarchical structure by using a variety of geostatistical analyses (runs tests, Durbin Watson, Markov chains, and entropy analysis) as well as to cast the data into normalized graphic plots for qualitative evaluation of stratigraphic styles.

On the basis of a comparison of these carbonate depositional systems, we recognize six empirical classes of architectectural organization grouped by age: (1) Proterozoic–Early Ordovician, (2) Silurian–Devonian, (3) Pennsylvanian, Early Permian, Late Triassic, and Neogene, (4) middle to Late Permian, (5) Early to Middle Triassic, and (6) Jurassic to Cretaceous. To account for differences in levels of order and hierarchical structure between classes, stacking-patterns techniques should be modified accordingly to produce accurate chronostratigraphic correlations and predictions of facies distribution. For example, accurate parasequence-scale correlations may not be expected in Jurassic depositional systems with high levels of autocyclicity, although larger-scale depositional sequences may be correlated on the basis of long-term changes in vertical facies proportions (e.g., Smackover and Haynesville sequences in the U.S. Gulf of Mexico). In a contrasting example, long-term systematic changes in facies proportions may not be expected in the platform-interior sections of icehouse depositional systems, although accurate high-resolution correlations of high-frequency sequences should be expected (e.g., the Pennsylvanian of the U.S. Midcontinent and Southwest).

Although of lesser importance than the recognition of secular classes, evaluation of a wide range of depositional systems has led us to the conclusion that the diverse sequence stratigraphic terminologies currently found in the literature have resulted from development of new models and terms to account for variations encountered in local examples (e.g., cf. Goldhammer et al., 1993a; Sarg et al., 1995; Kerans and Fitchen, 1995; Grammer et al., 1996; Kerans and Tinker, 1997; Fitchen, 1997). Evaluation of a wide range of data sets puts us in the position to develop a terminology that should be applicable to any carbonate depositional system regardless of age or setting.

We begin with a review of the fundamental concepts of carbonate sedimentation and the development of a sequence stratigraphic terminology that encompasses the full range of styles in shallow-marine carbonates. This review is followed by a presentation of the statistical and qualitative basis for an empirical classification as well as examples of the classes. Finally, we provide guidelines for the application of stacking-patterns techniques.

BASIC CONTROLS ON PLATFORM-CARBONATE SEDIMENTATION

Platform-carbonate sediments generally accumulate in shallow, tropical, marine environments away from centers of clastic influx (Wilson, 1975). One of the most important aspects of carbonate sedimentation that distinguishes it from siliciclastic deposition is that carbonate sediments are produced locally, in the "shallow-marine carbonate factory" of carbonate platforms and their seaward margins. A variety of factors control carbonate-sediment production including water depth (light), temperature,

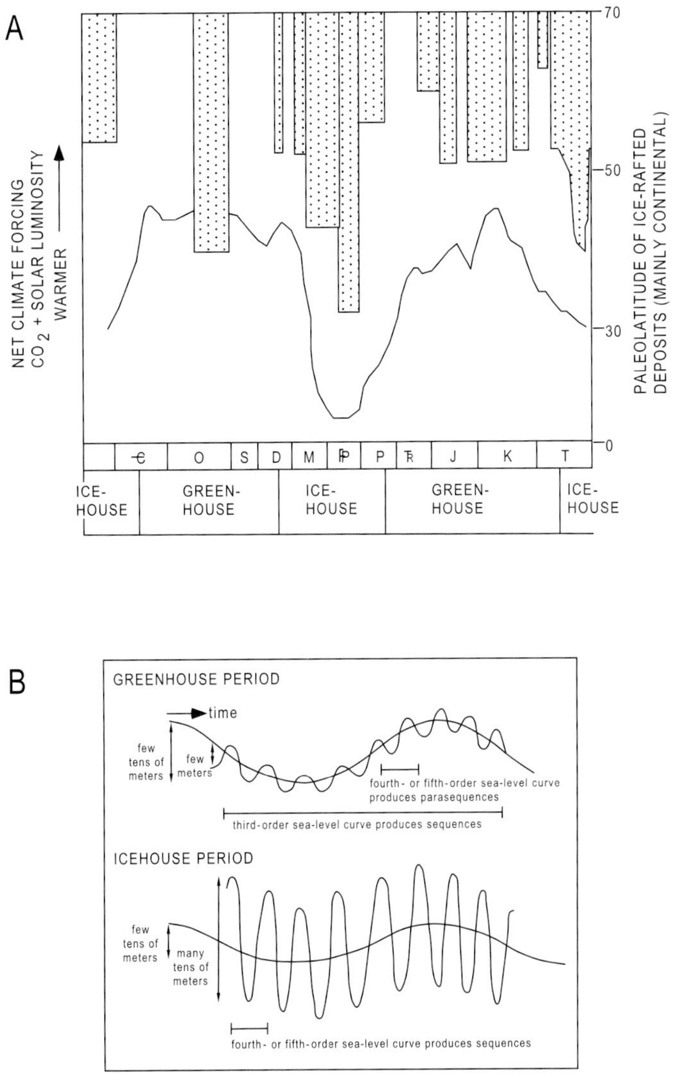

FIG. 1.—Climatic forcing and the nature of sea-level fluctuations from icehouse to greenhouse periods of Earth history. (A) Secular changes in climate from climate models and paleolatitude of ice-rafted deposits (modified from Read, 1995); original sources are Frakes and Francis (1988), Berner (1991), and Crowley and Baum (1991). (B) Schematic illustration of differences in signature of sea-level fluctuations, from Goldhammer (1994). Greenhouse periods are shown with low-amplitude, high-frequency sea-level fluctuations, and icehouse periods are shown with high-amplitude, high-frequency fluctuations.

salinity, nutrients, and siliciclastic turbidity (Bathurst, 1975; Lees and Buller, 1972; Hallock and Schlager, 1986; Schlager, 1992). Production of carbonate on tropical, marine platforms is strongly dependent on water depth, largely because of the volume of carbonate secretion by photosynthetic organisms. Carbonate production tends to be highest in the shallow-subtidal environments of platform tops and reef margins in water depths of less than 10 m.

Because carbonate sediment is produced locally and is strongly depth-dependent, carbonate-platform depositional systems are very sensitive to relative changes in sea level that result in fluctuations in accommodation space (space for sediment accumulation) on the platform and are recorded in the form of a hierarchy of sequence stratigraphic units such as parasequences and sequences (discussed below). Numerous other interrelated mechanisms influence the stratigraphy of carbonate sediments such as sedimentation rate, climate, water quality (oceanic circulation), tectonic setting, antecedent topography, compaction, and evolution of carbonate-secreting organisms. Of these, eustatic sea-level fluctuations and subsidence rates are commonly thought to have the strongest influence on the sequence stratigraphic record of carbonate platforms (e.g., Kendall and Schlager, 1981; Crevello et al., 1989; Loucks and Sarg, 1993).

Through geologic history, the signature of eustatic sea-level fluctuations has changed with changes in global climate, depending on whether the Earth's climate was generally cool and conducive to continental glaciation (icehouse periods) or whether the Earth's climate was warm and without continental glaciers (greenhouse periods) (Fig. 1). During icehouse periods, the large volumes of water removed from and released to oceans between glacial and interglacial periods, driven by climate changes, result in high-amplitude, high-frequency (fourth- and fifth-order) sea-level fluctuations superposed on lower-amplitude, low-frequency (third-order) sea-level fluctuations (Fig. 1). During greenhouse periods, the higher-frequency, climatically driven cycles (fourth and fifth order) have lower amplitude owing to fluctuations in alpine glaciers and in ground-water and lake storage (Tucker and Wright, 1990; Read, 1995; Jacobs and Sahagian, 1993). Thus, during greenhouse periods, the lower-frequency (third-order) sea-level changes should dominate the eustatic signature.

Sediment-production rates in shallow-marine carbonate platforms are extremely rapid, generally exceeding long-term subsidence rates and eustatic sea-level rises except in the case of high-amplitude glacio-eustasy (Schlager, 1981). Such high rates mean that platform-carbonate sediments generally fill to an equilibrium state, closely tracking relative changes in sea level especially during greenhouse and transitional periods. During icehouse periods, rates of high-frequency glacio-eustatic sea-level rises may outstrip carbonate-production rates, resulting in incipient drowning of carbonate platforms (cf. Schlager, 1992) and nonequilibrium fill whereby shallowing-upward carbonate sedimentation is not rapid enough to fill to sea level before the subsequent fall (Gianniny and Simo, 1996; Grammer et al., 1996; Rankey et al., this volume).

FUNDAMENTAL SEQUENCE STRATIGRAPHIC UNITS

Vertical stacking-patterns analysis is made possible by the fact that facies, stratal thicknesses, and unconformities in carbonate strata are systematically grouped within a hierarchy of chronostratigraphic units. The systematic packaging results because carbonate sedimentation (facies and stratal thickness) and erosion (unconformities) are strongly controlled by water depth or accommodation space on the platform, which is in turn largely driven by hierarchical fluctuations in sea level (e.g., Goldhammer et al., 1990; Kerans and Tinker, 1997). The hierarchy of cyclicity commonly observed in the stratigraphic record is most conveniently discussed in terms of orders of temporal duration: second (10–100 m.y.), third (1–10 m.y.), fourth

TABLE 1.–SEQUENCE STRATIGRAPHIC TERMINOLOGY

Stratal Unit	Definition	Characteristics	Temporal Order (and Duration)	Synonymous Terms
Parasequence	A relatively conformable succession of genetically related beds or bedsets bounded by marine flooding surfaces. May be subtidal or peritidal, symmetric or asymmetric.	Stratal packages with a shallowing-upward progression from deeper-marine facies, above the flooding surface, to shallow-subtidal to peritidal facies. Capped by subaerial exposure surfaces developed on peritidal facies, marine hardground, or shallow- to deeper-subtidal facies shift.	fifth (0.01–0.1 m.y.) fourth (0.1–1 m.y.)	Meter-scale shallowing-upward cycle (James, 1979; Goldhammer et al., 1987); punctuated aggradational cycle (PAC; Goodwin and Anderson, 1985)
Parasequence Set	A succession of genetically related parasequences that forms a progradational, aggradational, or retrogradational stacking pattern bounded by a marine flooding surface. May be coincident with systems tract in cases where a depositional sequence contains only two or three parasequence sets.	Capped by subaerial exposure features developed on peritidal facies or abrupt deepening from shallow- to deeper-subtidal facies. Parasequence set boundaries may display greater facies offset (landward stepping and deepening) and/or greater subaerial erosional alteration of bounding surfaces than parasequence boundaries within parasequence set.	fourth (0.1–1 m.y.) third (1–10 m.y.)	Megacycle (Goldhammer et al., 1990); cycle set (Kerans, 1995); bedding rhythm (Fischer, 1986)
Depositional Sequence, High-Frequency Sequence	A relatively conformable succession of genetically related strata bounded by unconformities or their correlative conformities.	Bounded by surfaces or facies shifts demonstrating relative fall in sea level such as (1) surface of abnormal subaerial exposure, (2) surface or deep subaerial erosional truncation, or (3) abrupt basinward facies offset superposing shallow-marine facies on deep-subtidal facies.	third (1–10 m.y.) (fourth [0.1–1 m.y.] or fifth [0.01–0.1 m.y.] for high-frequency sequences)	Synthem (North American Commission on Stratigraphic Nomenclature, 1983); transgressive-regressive cycle (Johnson et al., 1985); genetic stratigraphic sequence (Galloway, 1989)
Composite Sequence	A succession of genetically related sequences or high-frequency sequences in which the individual sequences are arranged into sequence sets with retrogradational, aggradational, and progradational stacking patterns (Mitchum and Van Wagoner, 1991).	Composite sequences composed of numerous high-frequency sequences are common in icehouse periods with high-amplitude glacio-eustasy. In some cases, composite sequences may contain a combination of subsidiary high-frequency sequences and parasequence sets.	third (1–10 m.y.) second (10–100 m.y.)	Supersequence
Abnormal Subaerial Exposure	Surface of subaerial exposure along which subtidal marine facies beneath the surface have been diagenetically altered by processes of subaerial exposure (e.g., paleosols, caliche).	Unambiguously demonstrates relative fall in sea level, found in depositional sequences and high-frequency sequences, unlike the subaerial exposure surfaces developed on peritidal facies (which commonly form bounding surfaces of parasequences) that may result from sediment accumulation to sea level.		
Systems Tract	Linkage of contemporaneous depositional systems (Brown and Fisher, 1977) represented in the rock record as three-dimensional facies assemblages. Defined on the basis of bounding surfaces, position within a sequence, and parasequence stacking patterns (Van Wagoner et al., 1988).	Include (1) transgressive systems tract with retrogradational parasequence and facies stacking patterns, (2) highstand systems tract with aggradational to progradational parasequence and facies stacking patterns, and (3) lowstand systems tract commonly represented on carbonate platforms as fluvial or eolian siliciclastic deposits or ponded lagoonal evaporites directly overlying sequence boundaries.	fourth (0.1–1 m.y.) third (1–10 m.y.)	Parasequence set, cycle set (in cases where a depositional sequence contains only two or three parasequence sets and/or cycle sets)
Greenhouse Sequence	A relatively conformable retrogradational to progradational succession of genetically related strata bounded by minimum-accommodation zones.	These stratal packages are common in carbonate-platform strata deposited during greenhouse periods with low-amplitude eustatic fluctuations. Previously they have been classified as depositional sequences, but with the recognition that in many cases a distinct bounding surface of abnormal exposure or subaerial erosional truncation is difficult to recognize.	third (1–10 m.y.) fourth (0.1–1 m.y.)	Previously classed as depositional sequences (Read and Goldhammer, 1988; Koerschner and Read, 1989; Goldhammer et al., 1990, 1993b; Montanez and Osleger, 1993, 1996)
Minimum-Accommodation Zone	Zone of thinnest parasequences with greatest proportion of peritidal facies or shallowest subtidal facies marking the progradational to retrogradational turn around or a greenhouse sequence.			Sequence boundary (Goldhammer et al., 1993a, b); sequence boundary zone (Montanez and Osleger, 1996); regional cryptic unconformity (Anderson and Goodwin, 1990)
Maximum-Accommodation Zone	Zone of thickest parasequences and deepest subtidal facies marking maximum deepening near the center of a greenhouse sequence.			Analogous with maximum flooding surface (Van Wagoner et al., 1988)

NOTE: Modified from Lehrmann (1995) and Fitchen (1997).

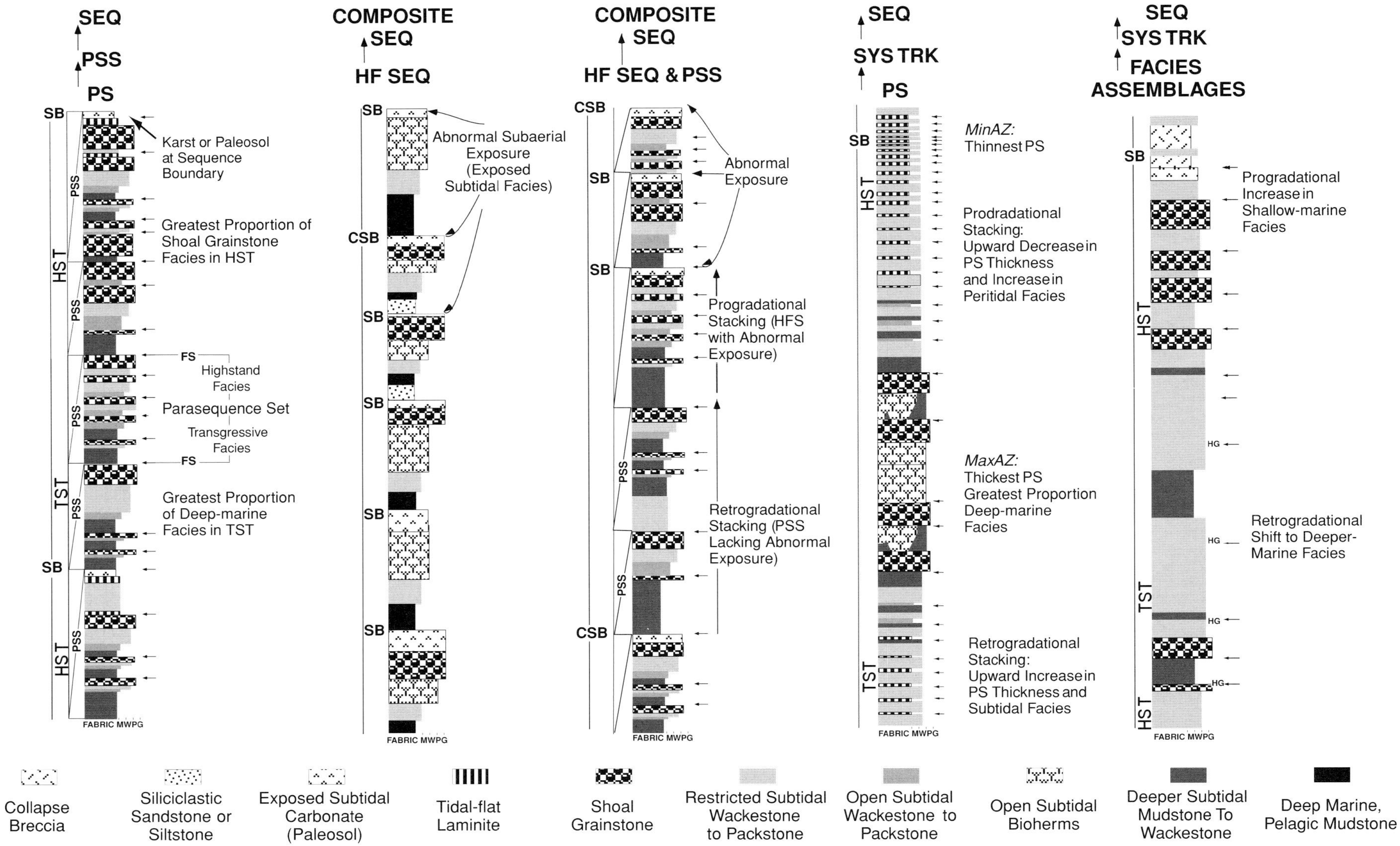

FIG. 2.—Different styles of hierarchical stacking patterns of platform carbonates recognized in strata from diverse ages and settings. SB—sequence boundary, CSB—composite sequence boundary, TST—transgressive systems tract, HST—highstand systems tract, MinAZ—minimum accommodation zone, MaxAZ—maximum accommodation zone, HFS—high-frequency sequence, PSS—parasequence set, PS—parasequence, SEQ—depositional sequence, HG—hardground, FS—flooding surface, M—mudstone, W—wackestone, P—packstone, G—grainstone, ◄— = Parasequence Boundary

(0.1–1 m.y.), and fifth (0.01–0.1 m.y.; e.g., Goldhammer et al., 1993b; Kerans and Tinker, 1997; Fitchen, 1997). The fourth and fifth orders are expressed as parasequences, parasequence sets, and high-frequency sequences (Table 1) recognized at the scale of outcrop sections, cores, and well logs, whereas the third and second orders are represented as sequences, composite sequences, and "greenhouse sequences" (Table 1) recognizable from large-scale outcrop patterns and conventional seismic data. Physical definitions and the driving mechanisms (primarily eustasy, subsidence, and sediment supply) of these sequence stratigraphic units are discussed below. In our survey of carbonate depositional systems through Earth history, we have encountered several different styles of hierarchical development in platform carbonates (Fig. 2):

1. Fifth-order parasequences grouped into fourth-order parasequence sets, which in turn form the systems tracts of third-order sequences.
2. Fourth- or fifth-order, high-frequency sequences grouped into sequence sets of third-order composite sequences.
3. Combination of fourth-order parasequence sets and high-frequency sequences forming systems tracts of third-order composite sequences.
4. Fifth-order parasequences forming the systems tracts of third-order sequences without development of distinct parasequence sets.
5. Facies assemblages, lacking cyclicity or with poorly developed cyclicity, forming the systems tracts of third-order sequences.

Parasequences

Parasequences are defined as a conformable succession of genetically related beds or bedsets bounded by marine-flooding surfaces (Table 1; Van Wagoner et al., 1988). Parasequences are recognized as stratal packages with a progression from deeper-marine facies types, above the flooding surface, that grade upward to shallow subtidal or peritidal facies. Parasequences are the fundamental, fifth-order building blocks of carbonate platforms except in cases where high-frequency sequences (described below) are the basic unit. Carbonate parasequences may be aggradational or progradational and therefore shallow upward. Parasequences can be most broadly grouped into (1) peritidal types capped by subaerial exposure features formed by progradation of tidal flats and (2) subtidal types formed by progradation of high-energy grainstone shoals over deeper-subtidal facies (Fig. 3).

Progradation of carbonate tidal flats or subtidal shoals, responsible for the formation of parasequences, can be driven by *allocyclic* or *autocyclic* mechanisms. Allocyclic mechanisms have external drivers such as sea-level fluctuations, whereas autocyclic mechanisms are driven internally by depositional processes and fluctuations in sediment supply or production (see Hardie and Shinn, 1986, for a review). Figure 4A illustrates an allocyclic model for formation of peritidal parasequences during fifth-order relative cycles of sea-level change. Tidal flats nucleate along the shoreline or off an island. Carbonate sediment produced on the shallow-subtidal platform accretes on the tidal flat, causing it to prograde basinward; eventually progradation is halted by sea-level fall. The subsequent rise in sea level floods the surface of the previous tidal flat and initiates another cycle of tidal-flat progradation. This process is repeated to form a stack of peritidal parasequences. Asymmetric parasequences are formed because carbonate production does not commence immediately after the initial flooding of the platform, but is initiated after a finite lag time resulting in deposition of the deepest-water facies at the base of the parasequence.

Ginsburg (1971) proposed an autocyclic model that suggested that peritidal parasequences could be generated under conditions of stationary sea level with constant subsidence (Fig. 4B). In this model, progradation of tidal flats is driven by changes in sediment supply. Tidal flats prograde across the platform as sediment produced in the shallow-subtidal carbonate factory is accreted to the front of the flats. Eventually, progradation causes the tidal flat to advance to the edge of the platform margin, reducing the area and thus the sediment-production rate of the subtidal factory; progradation is therefore halted. Continued subsidence floods the platform, and sediment accumulation commences after a finite lag time, initiating another prograding tidal flat. An alternative is the tidal-flat–island model that invokes the complex migration of shoals and tidal-flat "islands" through time, with stationary sea level and constant subsidence, that produces vertical shallowing-upward facies assemblages (parasequences) in any given vertical section (cf. Chow and James, 1987; Pratt and James, 1986). Allocycles are generally thought to have greater continuity than autocycles. At the scale of a producing hydrocarbon field or prospect, it is important to keep in mind that autocyclic parasequences generated by the Ginsburg model have the potential to extend at least from the platform interior out to near the platform margin; this mechanism has the potential to generate parasequences with high lateral continuity, especially on broad platforms with low wave energy (Dunn, 1991). In contrast, the migrating tidal-flat–island model of Pratt and James (1986) would produce highly discontinuous parasequences.

From the stratigraphic record, there is no simple key to determining whether parasequences were driven by autocyclicity or allocyclicity. Only in the case of high-frequency sequences (discussed below) can an allocyclic mechanism be unambiguously demonstrated. Attempts have been made to decipher the levels of autocyclicity vs. allocyclicity from platform successions of parasequences, by using statistical tools such as spectral analysis and forward modeling (e.g., Koerschner and Read, 1989; Goldhammer et al., 1991a, 1993b). Results indicated that a combination of both mechanisms operated within many depositional systems.

Parasequence Sets

A parasequence set is defined as a succession of genetically related parasequences that form a distinctive stacking pattern (progradational, aggradational, or retrogradational) that is bounded by a marine flooding surface (Table 1; Van Wagoner et al., 1988). In carbonate literature, fourth-order parasequence sets

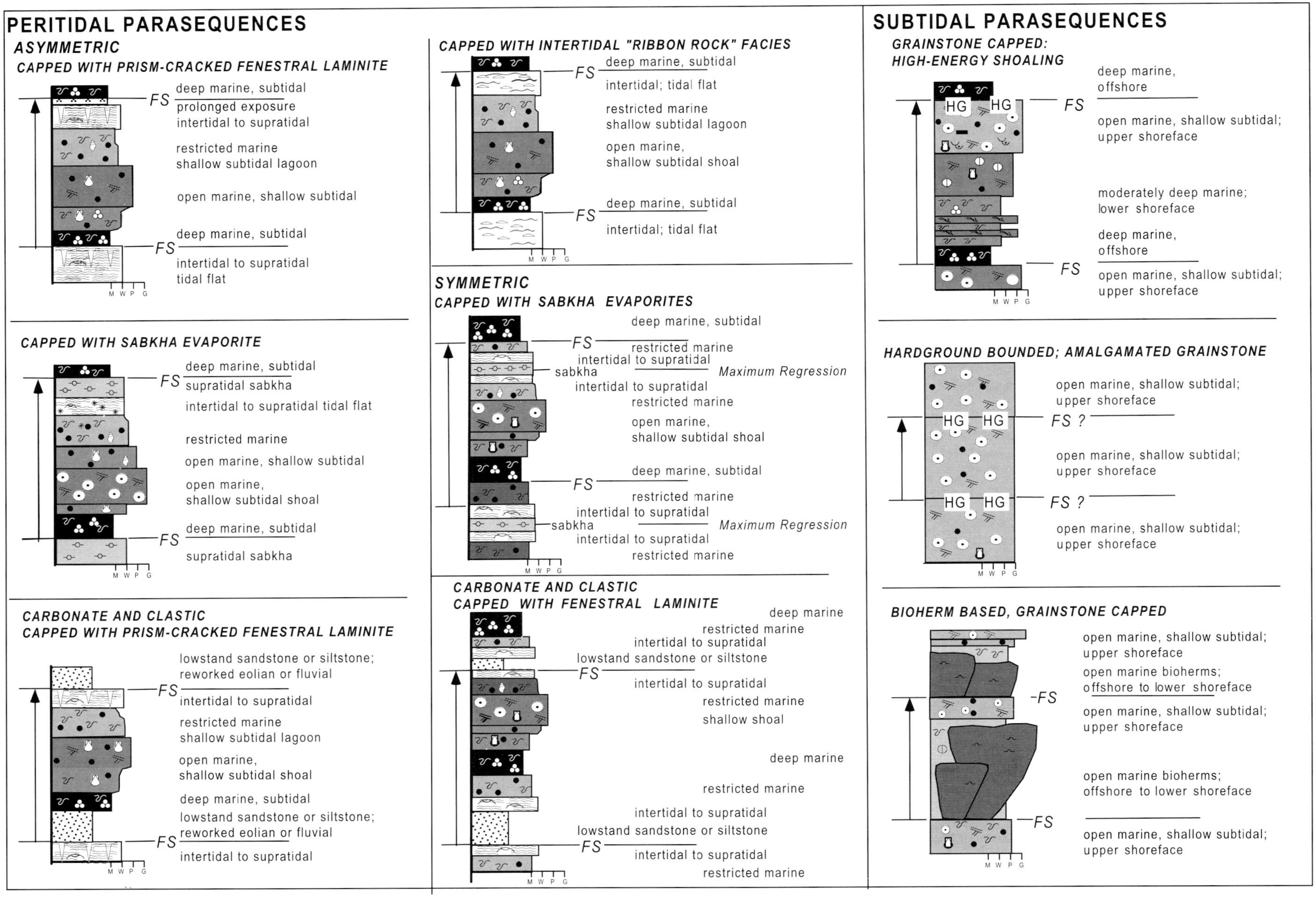

FIG. 3.—Types of parasequences found in platform carbonates (from Lehrmann, 1995). Peritidal parasequences are capped by intertidal and supratidal facies. Subtidal parasequences shallow upward from deeper-subtidal facies to shallow-subtidal or shoal facies. Labels as in Figure 2.

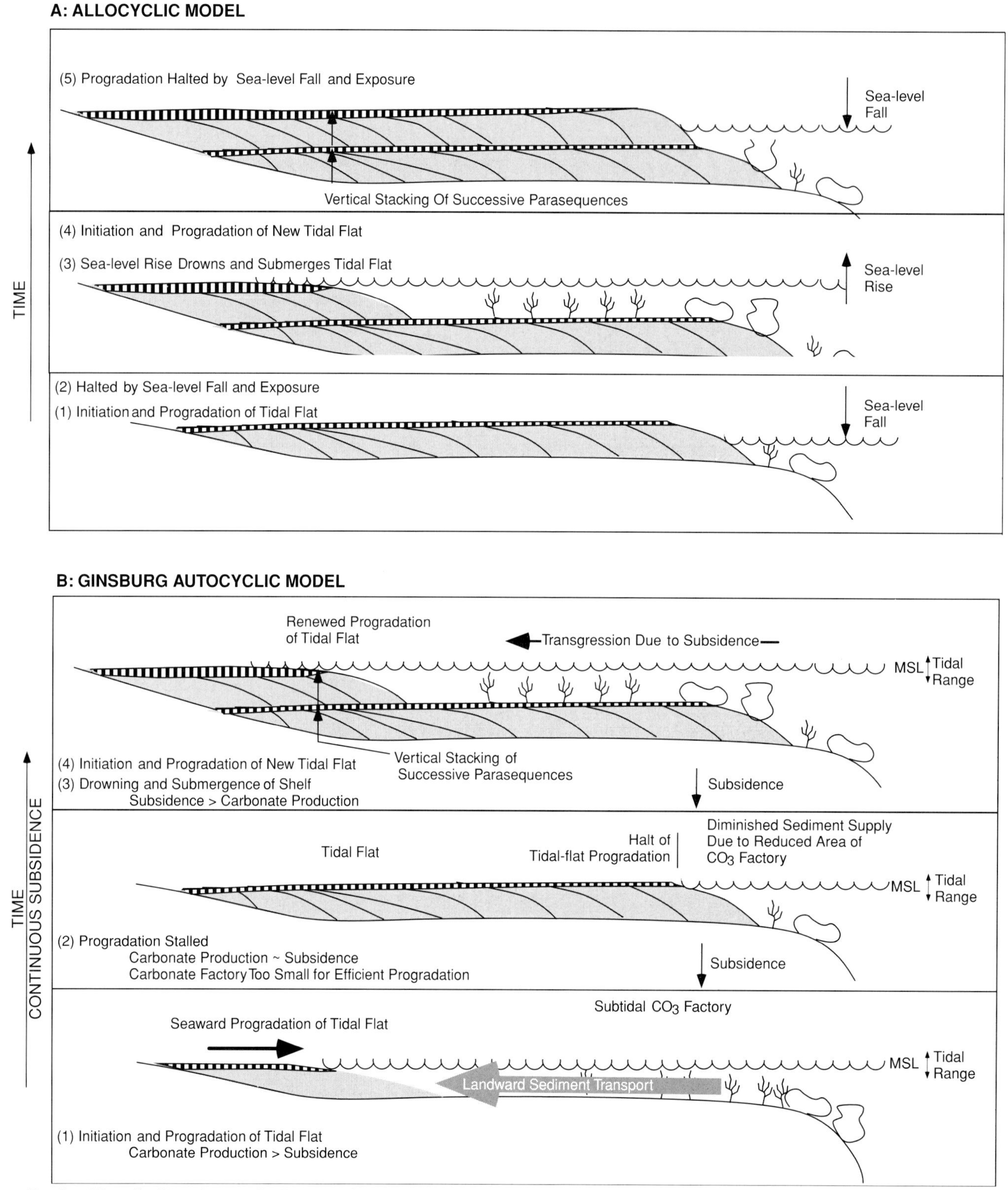

FIG. 4.—Allocyclic (A) and autocyclic (B) models for the genesis of parasequences (from Lehrmann, 1995). The allocyclic mechanism is externally controlled by relative sea-level changes. The autocyclic model is modified from Ginsburg (1971). Autocyclicity is an internally controlled mechanism for the generation of parasequences, driven primarily by repeated progradation and flooding of tidal flats due to fluctuations in sediment supply. MSL—mean sea level.

commonly have been referred to as cycle bundles or megacycles (Dunn, 1991; Goldhammer et al., 1993a). In carbonate sequences that contain several parasequence sets, the sets are most commonly progradational. Each parasequence set is recognized as a group of parasequences having an abrupt shift to deep-marine facies above a flooding surface at the base of the set followed by progradational parasequence stacking patterns to the top of the set (Fig. 2). The progradational stacking pattern is recognized by

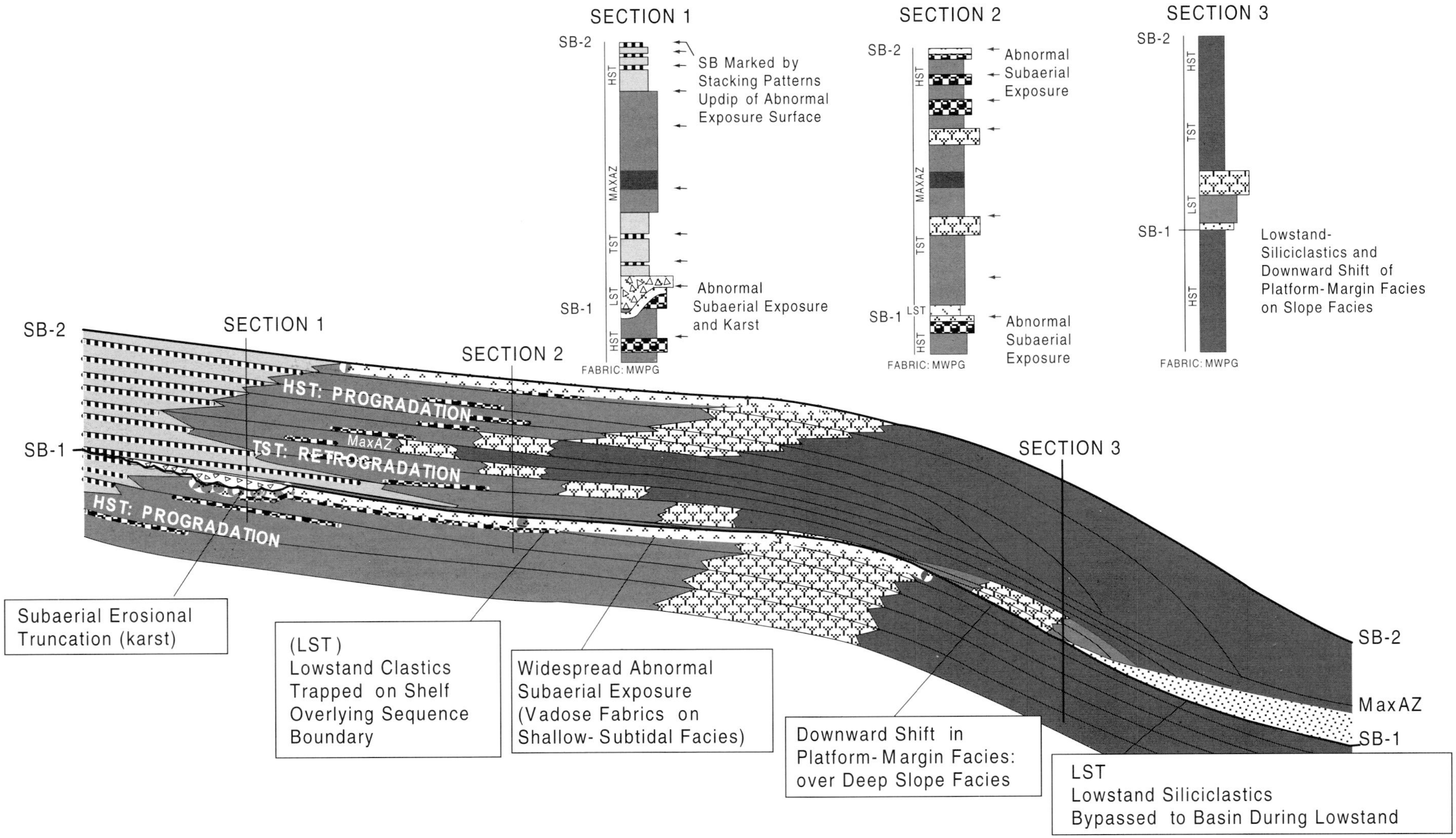

FIG. 5.—Stratigraphic features of depositional sequences formed during icehouse periods. Icehouse sequence boundaries contain evidence for abnormal subaerial exposure (subaerial exposure of subtidal facies), subaerial erosional truncation, and basinward facies shift that provide unambiguous evidence for relative sea-level fall. These features tend to be more common in icehouse sequences due to the high-amplitude sea-level fluctuations characteristic of icehouse periods. Labels as in Figure 2.

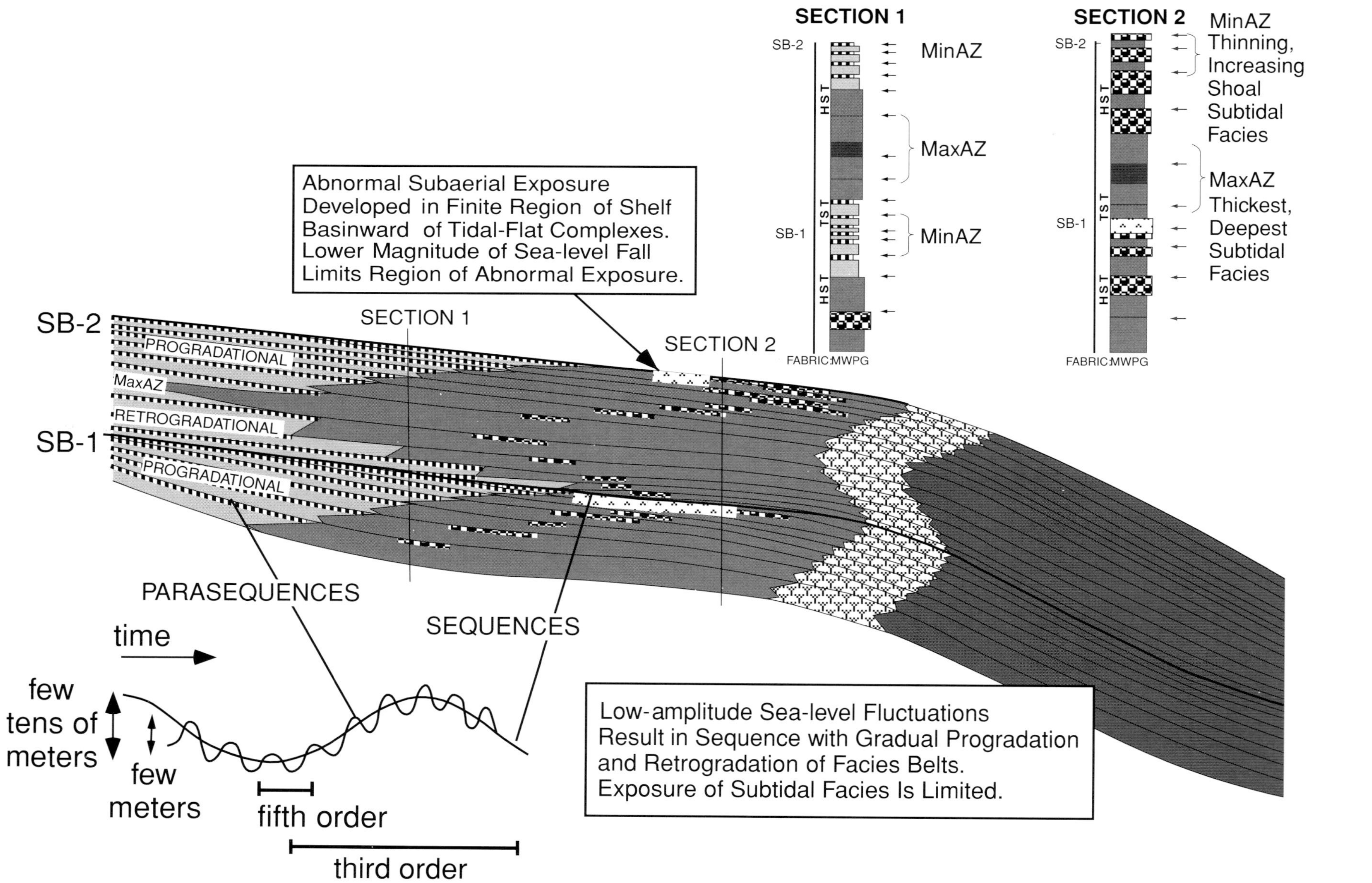

FIG. 6.—Stratigraphic features of depositional sequences formed during greenhouse periods. Greenhouse sequences tend to be represented as long-term progradation and retrogradation of platform facies. The progradation near sequence boundaries (minimum accommodation zones) is represented in the platform interior by thinning-upward parasequences and an upward change to parasequences dominated by peritidal facies (section 1). Retrogradation in the interior is represented by thickening-upward parasequences with a greater proportion of open-marine subtidal facies marking the maximum accommodation zone (section 1). In the outer platform, the progradational to retrogradational patterns are represented by thinning to thickening parasequence stacking patterns and the shift from shoal-subtidal to deeper-subtidal facies (section 2). In greenhouse periods, abnormal subaerial-exposure surfaces (exposed subtidal facies) are found only within a limited area seaward of the tidal flats because of the low-amplitude sea-level fluctuations in a greenhouse climate. Labels as in Figure 2.

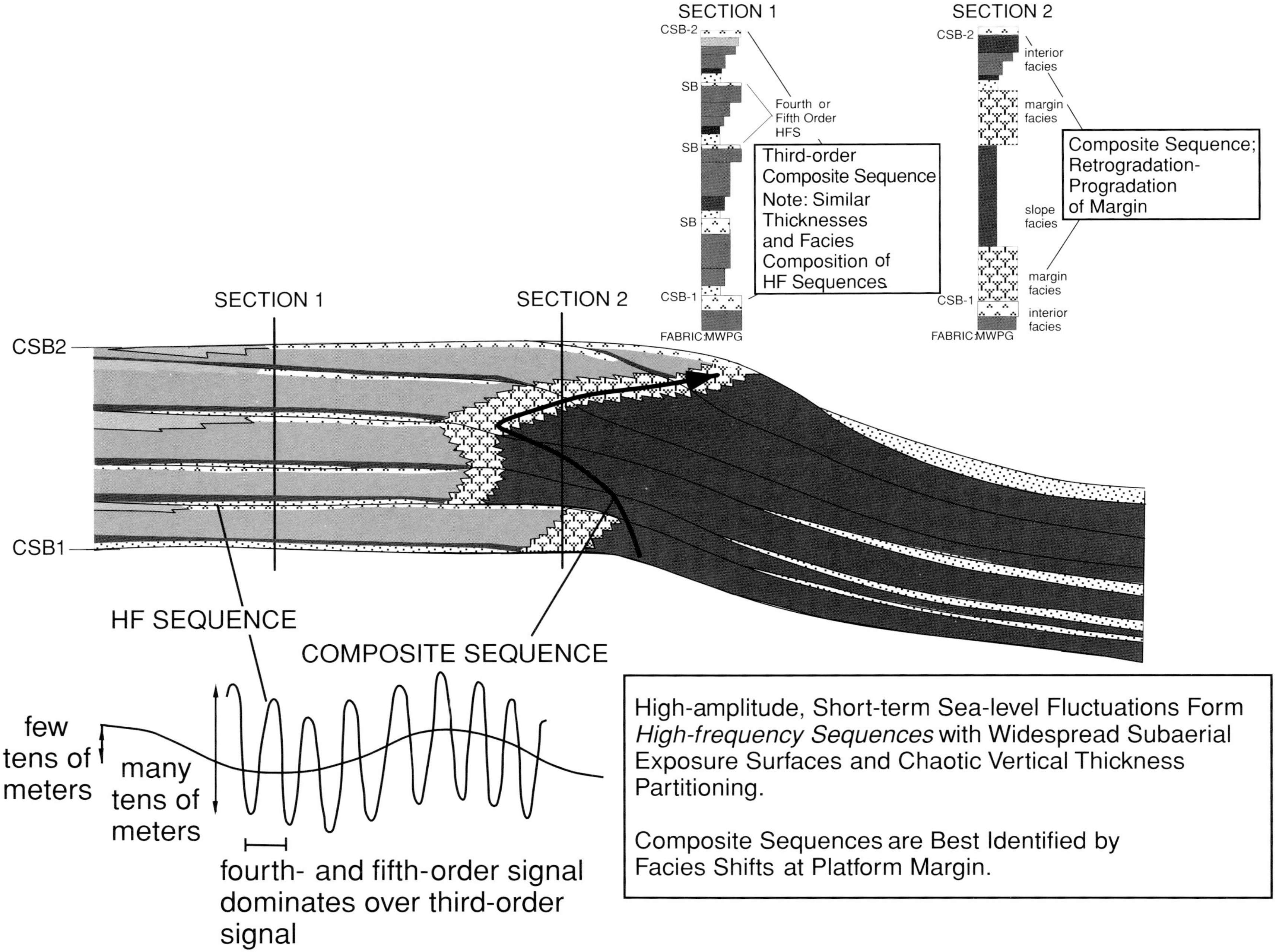

FIG. 7.—Stacking patterns and geometry of a composite sequence, composed of a stack of high-frequency sequences, each bounded by an abnormal subaerial exposure surface. Near the platform margin, the high-frequency sequences form a retrogradational to progradational pattern representing the larger composite sequence. Composite sequences are commonly formed during icehouse periods with high-amplitude, high-frequency sea-level fluctuations. Labels as in Figure 2.

a progressive upward thinning and an increase in the proportion of shoal subtidal and peritidal facies in the parasequences.

Sequences

Mitchum et al. (1977) defined a depositional sequence as a relatively conformable succession of genetically related strata bounded by unconformities or their correlative conformities (Table 1). Van Wagoner et al. (1988) narrowed the definition by redefining the concept of a sequence-bounding unconformity as a "surface separating younger from older strata along which there is evidence of subaerial erosional truncation (and, in some cases, correlative submarine erosion) or subaerial exposure, with a significant hiatus indicated." The definition essentially attempts to restrict the usage of the term *sequence* to genetically related stratal packages formed during a relative rise and fall of sea level. In carbonate strata, unambiguous relative falls in sea level are recorded as (1) *abnormal subaerial exposure,* defined as subaerial exposure of subtidal facies (Table 1), (2) abrupt basinward facies offset that superposes shallow-subtidal or peritidal facies on deep-marine facies, (3) down-stepping of shelf-edge deposits into the basin (basinally restricted sequences), or (4) subaerial erosional truncation of the inner shelf.

The identification of a sequence boundary on the basis of an abnormal subaerial exposure surface is a "best-case scenario" that unambiguously demonstrates a relative fall in sea level. In many cases, a given outcrop area or hydrocarbon field may be positioned too far downdip or updip of the depositional shoreline (depositional shoreline break of Van Wagoner et al., 1988) to record an abnormal subaerial exposure surface resulting from a relative fall in sea level. In carbonate platforms with extensive tidal flats, the depositional shoreline occurs at the seaward margin of tidal flats; relative fall in sea level will result in abnormal exposure of subtidal sediments outboard—and may result in erosional truncation of tidal-flat strata inboard—of the depositional shoreline or a downward facies shift near the shelf margin *if* the magnitude, rate, and duration of sea-level fall are great enough to form these features (Fig. 5). If the magnitude and duration of sea-level falls are not great enough, abnormal subaerial exposure will occur only within a finite region outboard of the depositional shoreline; updip and downdip of this region, sequences can be recognized by a symmetrical to semi-symmetrical turnaround from progradational to retrogradational parasequence stacking patterns (Fig. 6). Such transitional expressions of sequence boundaries are typical of greenhouse climatic periods when amplitudes of sea-level fluctuations are generally low enough to prevent widespread development of abnormal subaerial exposure.

Greenhouse sequences.—

Carbonate-platform strata deposited during greenhouse climates contain numerous examples of thick symmetrical or semisymmetrical, transgressive-regressive stratal packages (<30 to <300 m thick) characterized by (1) "long-term" systematic retrogradational to progradational parasequence stacking patterns (as deduced by variations in parasequence thickness and facies proportions) and/or (2) gross deepening to shallowing facies shifts defining maximum transgressive and regressive intervals. These stratigraphic packages have been classified as second- or third-order sequences, but with the recognition that in many cases a distinct surface of prolonged subaerial exposure is difficult to recognize. Instead, a transitional minimum-accommodation zone has been used to correlate the sequence boundary (e.g., Read and Goldhammer; 1988; Koerschner and Read, 1989; Goldhammer et al., 1990, 1993b; Montanez and Osleger, 1993). Because these packages do not contain unambiguous evidence for abnormal subaerial exposure, they are informally termed greenhouse sequences (1) to signify their transitional boundaries and their interpreted genesis as the result of long-term accommodation changes and (2) to emphasize the difference in stacking-patterns approach that is required to define and correlate these units (Fig. 6).

A greenhouse sequence is a relatively conformable, transgressive-regressive succession of genetically related strata bounded by minimum-accommodation zones (MinAZ) in the accommodation-limited settings of the platform and by their chronostratigraphically equivalent surfaces in the outer platform, slope, and basin (Table 1). The MinAZ is the symmetric or semisymmetric turnaround from progradational to retrogradational parasequence and facies stacking patterns of the greenhouse sequence (Table 1). The MinAZ is marked by the thinnest parasequences with greatest proportion of exposure-prone or shallowest-subtidal facies and a subtle toplap geometry (Fig. 6). Greenhouse sequences boundaries are most logically correlated by choosing a surface of maximum regression in the center of the MinAZ. This surface will occur at the top of the thinnest, most exposure-prone parasequence (Figs. 2, 6) and should be approximately coeval with a surface of abnormal subaerial exposure outboard or subaerial erosional truncation inboard of the depositional shoreline break. This surface should closely approximate a traditional sequence boundary and therefore has greater utility for regional correlation than an overlying flooding surface (see discussion in Van Wagoner et al., 1990). The maximum accommodation zone (MaxAZ) of a greenhouse sequence is analogous to the maximum flooding surface (Table 1).

In carbonate platforms, greenhouse sequences form by gradual, symmetric progradation and retrogradation of tidal-flat complexes and associated subtidal-facies belts across the shelf or ramp (e.g., Goldhammer et al., 1992, 1993b; Montanez and Osleger, 1993, 1996). Forward modeling of greenhouse sequences reveals that subtle toplapping and onlapping geometries occur along a distinct surface within the (Min AZ) (Dunn, 1991), but the geometries are extremely low angle (owing to the low gradient of shelves or ramps) and are therefore difficult to detect in outcrops without continuous exposure over vast distances.

Much of the current debate about the driving mechanisms of depositional sequences centers around the problem that sequences as defined by Mitchum et al. (1977) could be formed by variations in sediment supply (in addition to relative changes in sea level) in cases where the bounding surfaces lack subaerial erosional truncation and represent nondepositional hiatuses (i.e., toplap without subaerial erosional truncation or downward facies shift) (cf. Schlager, 1992; Christie-Blick, 1991; Christie-

TABLE 2.—DATA SETS USED IN STATISTICAL ANALYSES.

Age	Formation and Location	Code*	Reference
Proterozoic	Rocknest Fm., Northwest Territories	ROK	Grotzinger (1985)
	Katakturuk Fm., Alaska	KAT	Clough and Goldhammer (1992)
Cambrian	Bonanza King Fm., California	FUN, DES	Montanez and Osleger (1993)
Ordovician	Arbuckle Gp., Oklahoma	ARB	Goldhammer (unpub. data)
	El Paso Gp., Texas	EFR, EBE	Goldhammer et al. (1993b)
	Beekmantown Gp., Maryland	BEK	Goldhammer et al. (1992)
Silurian	Laketown Dolomite, Utah	BARN	Harris and Sheehan (1996)
Devonian	Cairn Fm., Alberta	CAR	McLean and Mountjoy (1994)
	Pillara Fm., Australia	PIL	Read (1973)
	Swan Hills Fm., Alberta	S-H	Wong and Oldershaw (1980)
Mississippian	Madison Fm., Wyoming	MAD	Sonnenfeld (1996)
	St. Louis and St. Genevieve Fms., Kansas	SLG	Hanford (1988)
Pennsylvanian	Hermosa Gp., Utah	HON	Goldhammer et al. (1991b)
		HNG	Gianniny (1995)
	Holder Fm., New Mexico	HOLD	Exxon Dry Canyon #1 (Rankey et al., this volume)
	Lansing–Kansas City Gp., Kansas	KSP	Watney et al. (1989)
Permian	Snake Canyon Fm., Idaho	SCAN	Rankey (1996)
	San Andres Fm., Texas	MEA	Exxon, Means field
		LAW	Kerans and Fitchen (1995)
	Grayburg Fm., Texas	GBPR	Barnaby and Ward (1995
	Queen Fm., Texas	DEV	Kerans and Nance (1991
	Seven Rivers and Tansill Fms., Texas	PDB04	Garber et al. (1989)
		SLAU	Rankey and Lehrmann (1996)
Triassic	Latemar Fm., Italy	LAT	Goldhammer et al. (1993a)
	Dawen Fm., China	DAWE	Lehrmann (1993)
	Dachsteinkalk, Austria		Goldhammer et al. (1990)
Jurassic	Arab Fm., Saudi Arabia	ARA	Mitchell and Lehmann (1988)
	Haynesville Fm., Texas	HANE	Goldhammer and McDonald (1996)
	Smackover Fm., Alabama	SMK	Prather (1992)
	Zuluoaga Fm., Mexico	ZULU	Goldhammer et al. (1991a), Johnson (1991)
Cretaceous	El Abra Fm., Mexico	ELAB	Minero (1983)
	Aurora Fm., Mexico	AUR	Lehmann et al. (1995)
	Cupido Fm., Mexico	CUP	Goldhammer et al. (1991a)
	Segovia Fm., Texas	SEG	Kerans (1997)
Tertiary–Quaternary	Bahamas (Pliocene–Pleistocene)	UNDA	Kievman (1996)
	Mallorca (Miocene), Spain	MALL	Oswald (1992)
	Key Largo Fm.–Miami Oolite, Florida	KLM	Enos and Perkins (1977)
	Suwanne River Fm., Florida	SUW	Hammes (1992)

NOTE: Fm. = formation, Fms. = formations, Gp. = group. *Codes are used in Table 3 and Figures 8–10.

Blick and Driscoll, 1995). With the refined terminology of Van Wagoner et al. (1988), such retrogradational to progradational packages bounded by marine-flooding surfaces are not considered as depositional sequences, but are more appropriately classified as parasequences or parasequence sets. Although greenhouse sequences commonly do not contain unambiguous evidence for a relative fall of sea level, such as abnormal subaerial exposure or abrupt basinward facies offset, regional correlation and forward stratigraphic modeling indicate that such sequences are formed largely by relative changes in sea level. Forward stratigraphic simulations (Dunn, 1991) show that, although "Ginsburg style" autocyclicity can develop thick stacks of peritidal parasequences, fluctuations in sediment supply that would be required for autocyclic development of the longer-duration greenhouse sequences fall far outside the ranges of carbonate sedimentation rates typical of "healthy" carbonate platforms (Dunn, 1991, p. 724; Enos, 1991). Processes that would decimate carbonate sediment production to a sufficient degree to force retrogradation, such as excessive nutrient or siliciclastic influx, most commonly result in asymmetric shifts to deep-marine facies or siliciclastics and often termination of the carbonate platform (Hallock and Schlager, 1986; Schlager, 1989). Furthermore, in forward simulations, it is difficult to form progradational sets of parasequences by simply increasing

TABLE 3.—STATISTICAL RESULTS.

Age	Data Set	Number of PS or HFS	Number of F	Number of Runs	Z-score	Longest Run*	D-W	Markov χ^2	Expected Value	E_{sys}
Proterozoic	ROK1	122	11	40	–3.60	19	1.24	264	112	0.71
	ROK2	161	11	56	–2.98	22	1.40	390	112	0.75
	ROK3	118	9	38	–2.92	19	1.26	301	73	0.77
	KAT	165	10	52	–3.66	21	1.23	295	92	0.82
Cambrian	FUN	239	10	72	–4.39	24	1.77	234	92	0.75
	DES	260	10	91	–3.31	40	1.37	639	92	0.70
Ordovician	ARB	269	9	101	–3.68	21	1.46	193	73	0.91
	EFR	116	7	29	–5.12	19	0.71	156	43	0.82
	EBE	82	8	30	–2.27	9	1.29	486	57	0.86
	BEK	80	7	28	–1.94	16	1.43	55	43	0.79
Silurian–Devonian	BARN	104	11	38	–1.84	13	1.14	150	112	0.77
	CAR	68	7	26	–1.16	11	1.92	44	43	0.85
	PIL	88	5	37	–1.14	8	1.66	178	20	0.79
	S-H	28	7	15	+0.59	8	1.28	168	43	0.70
Mississippian	MAD	94	15	31	–2.92	16	1.59	420	213	0.67
	SLG	13	5	5	–1.15	5	1.32	16	20	0.72
Pennsylvanian	HON	35	7	19	+0.55	5	1.85	260	43	0.83
	HNG	44	7	20	–0.36	5	1.70	29	43	0.79
	HOLD	28		12	+0.84	4	1.87	140	112	0.79
	KSP	7	14	5	+0.93	2	4.37	159	185	0.68
Permian	SCAN	49	11	22	+0.20	8	2.15	104	112	0.77
	MEA	61	8	27	–1.29	16	1.45	60	57	0.69
	LAW	71	13	31	–0.24	5	1.26	401	159	0.73
	GBPR	48	12	23	+0.01	8	1.55	138	134	0.75
	DEV	38	10	11	–2.52	15	1.03	94	92	0.79
	PDBO4	54	9	19	–1.96	11	1.43	87	73	0.84
	SLAU	22	8	9	–1.28	4	1.11	61	57	0.78
Triassic	LAT	499	2	143	–7.19	29	0.87	..	..	..
	DAWE	83	6	25	–1.63	9	1.37	76	30	0.69
Jurassic	ARAB	40	10	15	–1.74	8	1.15	370	92	0.67
	HANE	19	8	9	+0.13	5	1.98	47	57	0.81
	SMK	14	7	7	–0.27	5	1.88	22	43	0.55
	ZULU1	99	8	35	–1.92	13	1.24	128	57	0.82
	ZULU2	68	4	10	–1.23	5	1.73	123	11	0.73
	ZULU3	54	3	23	+0.22	12	1.50	6	4	0.93
Cretaceous	AUR	94	12	41	–1.059	6	1.68	592	134	0.60
	ELAB	28	9	11	–0.51	4	1.72	53	57	0.78
	CUP	92	7	42	–1.0	10	1.11	286	43	0.74
	SEG	36	14	13	–0.17	14	1.52	148	185	0.70
Tertiary–Quaternary	UNDA	12	10	10	+3.29	2	2.97	182	92	0.64
	MALL-1	24	9	9	–1.15	7	1.30	180	73	0.63
	MALL-2	15	5	4	–1.88	8	2.59	9	17	0.66
	KLM	9	9	3	–1.23	6	2.90	46	73	0.64
	SUW	25	6	6	–2.34	12	1.32	43	30	0.83

NOTE: Data-set codes are keyed to Table 2. PS = parasequence, HFS = high-frequency sequence, F = facies unit; D-W = Durbin-Watson value, χ^2 = the chi-squared statistic, expected value = that value that the chi-squared statistic must exceed to reject the null hypothesis of random facies transitions at the 95% confidence interval, E_{sys} = system entropy.

*Number of parasequences.

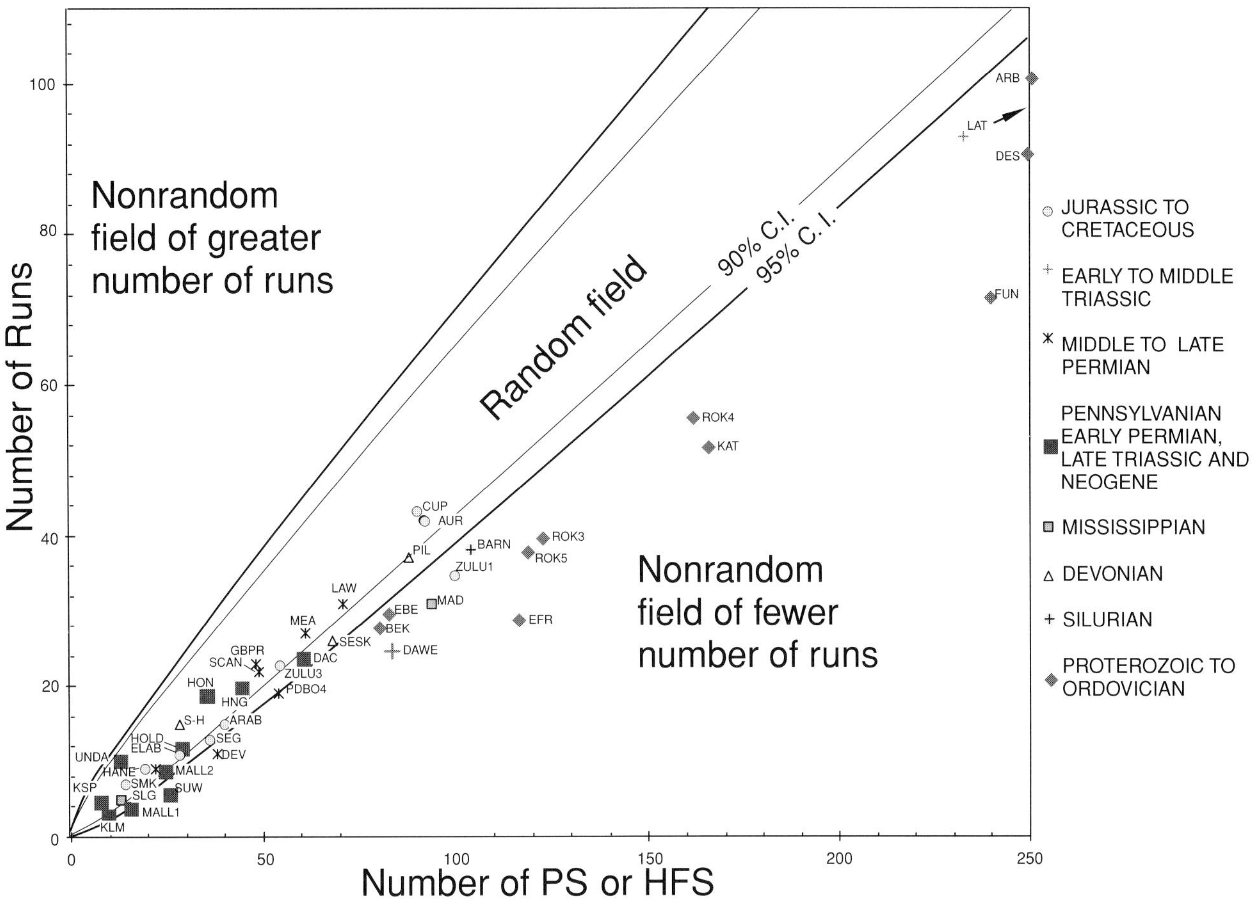

FIG. 8.—Plot of number of runs vs. number of parasequences or high-frequency sequences. Data sets are arranged by age (symbols on right). Abbreviations of stratigraphic sections (shown next to each data point) are from Tables 2 and 3. A run is a string of consecutive parasequences with thinner- or thicker-than-average thickness. Sections with random vertical parasequence patterns tend to have numbers of runs close to half the number of parasequences in the section (random field; cf. Drummond and Wilkinson, 1993). Sections that are nonrandom and have thinner and thicker parasequences are grouped together, have fewer numbers of runs, and plot in the nonrandom field (lower right).

carbonate-sediment supply because autocyclic-driven prograding tidal flats tend to stall when they reach the point of stepping seaward of the underlying tidal-flat surface (Dunn, 1991). Thus, the repeated symmetric parasequence and facies stacking patterns of greenhouse sequences are more likely to be driven by accommodation cycles than by changes in sediment supply. The origin of the accommodation cycles, however, remains equivocal (see Goldhammer et al., 1993b).

Icehouse sequences.—

Sequences developed during icehouse periods tend to be more clearly bounded by surfaces indicating relative fall in sea level (Fig. 5) and thus fit the classical definition of a depositional sequence (Van Wagoner et al., 1990) because high-amplitude glacio-eustasy results in widespread abnormal subaerial exposure and abrupt basinward facies shifts. This style of sequence development in carbonate strata is somewhat more analogous to the case in siliciclastic depositional systems in which sequences are clearly bounded by dramatic surfaces of unconformity marked by erosion of tens to hundreds of meters of section and an abrupt shift of fluvial deposits over marine facies within incised valleys (J. C. Van Wagoner, pers. commun., 1995). During icehouse periods, sequences tend to form at the high-frequency scale (fourth or fifth order) and may be grouped into larger composite sequences.

Composite sequences.—

A composite sequence is a succession of genetically related sequences in which individual sequences stack into sequence sets with progradational, aggradational, and retrogradational stacking patterns (Table 1; Fig. 7) (Mitchum and Van Wagoner, 1991). Composite sequences in carbonate strata are generally second- or third-order packages composed of high-frequency (fourth- and fifth-order) sequences (e.g., Kerans and Fitchen, 1995; Sarg et al., 1995). Composite sequences may also contain a combination of both subsidiary parasequence sets (without evidence of relative fall in sea level) and high-frequency sequences (e.g., Kerans and Fitchen, 1995). In this case, fourth-order parasequence sets occur in the retrogradational part, and fourth-order sequences occur in the progradational part, of the composite sequence. The parasequence sets and high-frequency sequences are genetically equivalent except that, in the composite sequence, the parase-

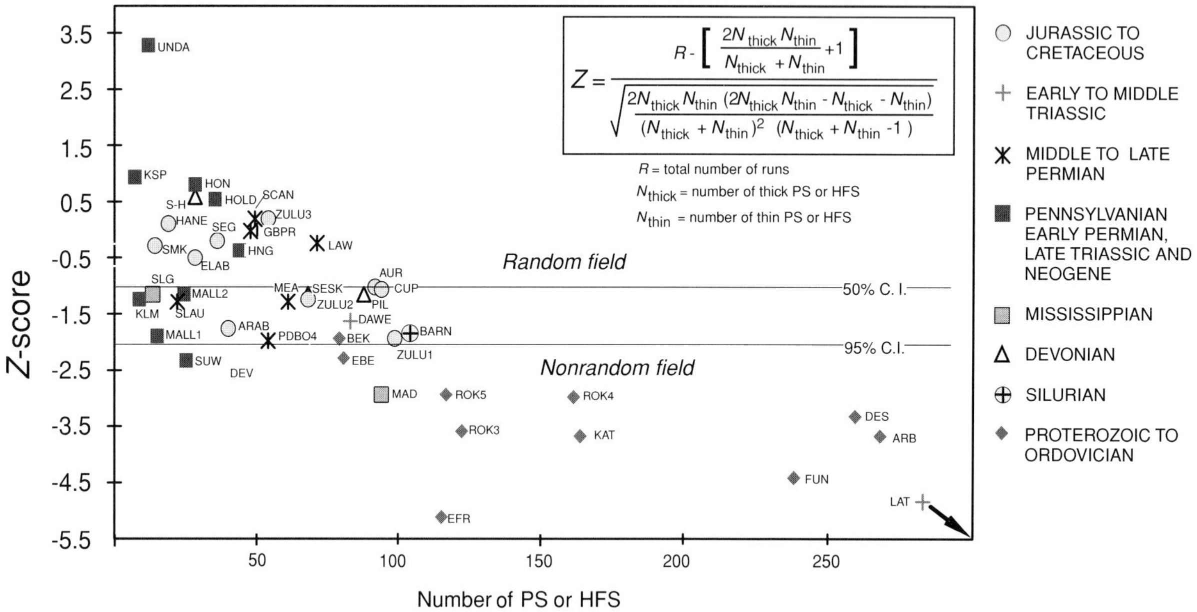

$$Z = \frac{R - \left[\frac{2N_{thick}N_{thin}}{N_{thick}+N_{thin}}+1\right]}{\sqrt{\frac{2N_{thick}N_{thin}(2N_{thick}N_{thin}-N_{thick}-N_{thin})}{(N_{thick}+N_{thin})^2(N_{thick}+N_{thin}-1)}}}$$

FIG. 9.—Plot of Z-scores vs. parasequence or high-frequency sequence number. Data sets are arranged by age (symbols on right). Abbreviations of stratigraphic sections (shown next to each data point) are from Table 3. The Z-score expresses the number of runs in a section as the number of standard deviations from the value expected from a random distribution. Z-scores less than –2.4 indicate nonrandom groupings of thin and thick parasequences at the 95% C.I. Note the clear separation of populations by age class. Proterozoic–Ordovician sections occur near or below the 95% C.I. cutoff and show ordered behavior. Jurassic–Cretaceous depositional systems tend to plot within the field of randomness and icehouse depositional systems (Pennsylvanian, Late Triassic, and Neogene) plot mainly in the field of randomness and show a spread toward the highest Z-scores. See text for details.

quence sets were formed during a long-term relative rise in sea level, whereas the high-frequency sequences were formed during a long-term relative fall in sea level.

Systems Tracts

Depositional sequences can be subdivided into component *systems tracts,* defined as linkages of contemporaneous depositional systems (Brown and Fisher, 1977) that are represented in the rock record as three-dimensional facies assemblages (Table 1). Systems tracts are defined objectively on the basis of their types of bounding surfaces, position within a sequence, and stacking patterns of parasequence and parasequence sets (Van Wagoner et al., 1988) (Figs. 5, 6). Transgressive and highstand systems tracts are the ones most commonly encountered in the interior of carbonate platforms. Transgressive systems tracts contain retrogradational parasequences and parasequence sets, which in vertical sections may be recognized as thickening-upward parasequences and parasequence sets and overall deepening-upward facies trends. Highstand systems tracts contain aggradational to progradational parasequences and parasequence sets, which in vertical sections may be recognized as thinning-upward and overall shallowing-upward facies patterns. Lowstand systems tracts are recognized in the interior of some carbonate platforms as siliciclastic deposits (fluvial or eolian) or ponded lagoonal evaporites directly overlying sequence boundaries.

LEVELS OF ORDER AND HIERARCHICAL DEVELOPMENT

To investigate the utility of vertical patterns in thickness and facies proportions of parasequences or high-frequency sequences for interpreting and correlating lower-order depositional sequences, we have employed a series of statistical tests (runs tests, Durbin-Watson, Markov chains, and entropy analysis) to evaluate levels of order in carbonate stratigraphic sections. In an effort to understand secular differences in behavior, the data sets have been organized by age (Tables 2, 3). The tests, discussed below, primarily evaluate vertical successions of stratal thicknesses and facies transitions. The basic foundation of stacking-patterns analysis relies on the recognition of ordered or quasi-ordered vertical trends in parasequence or high-frequency sequence thicknesses and facies composition within a nested hierarchy in which patterns of finer-scale units such as parasequences can be used to decipher and correlate large-scale units such as parasequence sets and depositional sequences. If these vertical patterns are nonrandom, then one can use the stacking-patterns technique for chronostratigraphic correlation. If randomness occurs at a certain scale or in a specific stratal attribute, then one should rely on other scales or attributes to constrain lateral stratigraphic architecture.

The statistical tests below comment on the level of order of vertical (one-dimensional) stratigraphic patterns. In this context, one should keep in mind that two-dimensional stratigraphic patterns documented in detailed outcrop and subsurface studies (e.g., Sonnenfeld and Cross, 1993; Kerans and Fitchen, 1995;

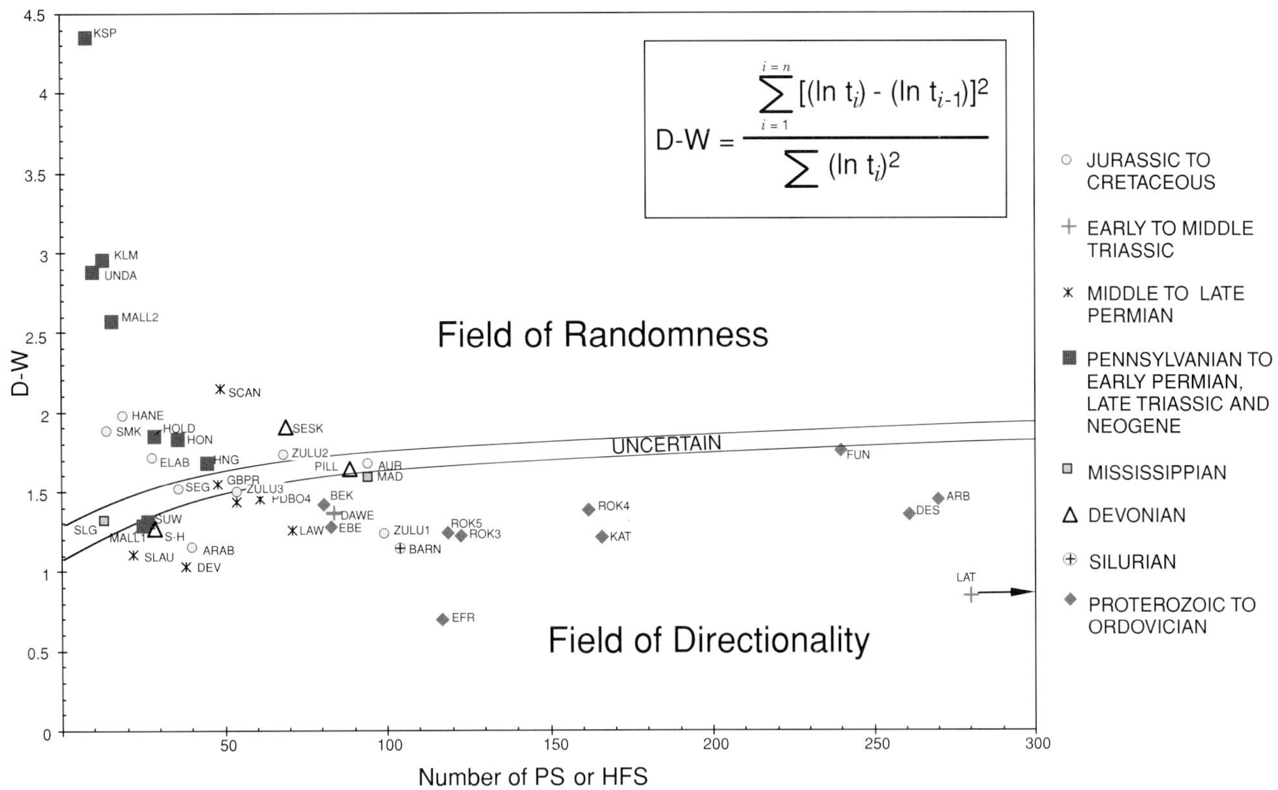

FIG. 10.—Plot of Durbin-Watson (D-W) values, calculated for entire sections, vs. number of parasequences or high-frequency sequences. Data sets are arranged by age (symbols on right). Abbreviations of stratigraphic sections (shown next to each data point) are from Table 3. Equation used to calculate Durbin-Watson values is shown in upper right (ln ti = natural log of thickness t of parasequence i, ln ti–1 = natural log of thickness of succeeding parasequence i – 1). High Durbin-Watson values indicate large differences in the thickness of consecutive parasequences, suggesting erratic or random patterns. Low Durbin-Watson values, in contrast, indicate small differences in consecutive thicknesses or smooth, ordered, thickness patterns. Note the separation of populations by age: Proterozoic–Ordovician sections all have smooth, ordered, thickness patterns, all Jurassic–Cretaceous sections but one plot within the field of randomness or uncertainty, and icehouse sections plot mainly in the random field and have the highest Durbin-Watson values (i.e., very "jumpy" or erratic thickness patterns).

Tinker and Mruk, 1995; Rankey and Lehrmann, 1996; Tinker, 1996) reveal order that is undetectable by these tests. A qualitative evaluation of two-dimensional order is integrated with our analysis in the following section.

Vertical Thickness Patterns

Runs tests.—

A runs test evaluates whether the long-term changes in parasequence or high-frequency sequence thickness are nonrandom. In other words, it addresses the question of whether thinner-than-average or thicker-than-average stratal units occur in nonrandom groupings. Sadler et al. (1993) applied runs tests to evaluate the long-term order of parasequence thicknesses expressed as the rising and falling portions of Fischer plots. Fischer plots can be used as a graphic aid for visualizing trends in parasequence thickness with the goal of identifying clusters of thin parasequences within the MinAZ of depositional sequences (e.g., Goldhammer et al., 1993b) or for identification of parasequence sets (e.g., Goldhammer et al., 1990). Rising limbs of Fischer plots occur where thicker-than-average parasequences are grouped together, and falling limbs occur where there are runs of thinner-than-average parasequences. Sadler et al. (1993) defined a run as a consecutive occurrence of like parasequences (either thick or thin). Runs begin and end where a thinner-than-average parasequence is followed by one that is thicker than average or vice versa.

As pointed out by Sadler et al. (1993) and Drummond and Wilkinson (1993), the number of runs in a given section are compared against those values expected from a random walk. A random walk generates a random string of parasequences by selecting thick or thin parasequences from the distribution present in the actual data set on the basis of a random process such as a coin toss. When the random-walk experiment is repeated multiple times, the number of runs should converge on a value of one-half the number of parasequences. From the experiment, it is possible to establish fields of randomness and order at the 90%–95% confidence intervals (C.I., Fig. 8). For nonrandom data sets, those in which thinner- or thicker-than-average parasequences are grouped together, the number of runs will be less than that expected from a random walk.

Application of the runs test to our data produced a broad spread in values (Fig. 8) similar to the results reported by Drummond and Wilkinson (1993). At the 95% C.I., 14 out of 40 and, at the 90% C.I., 19 out of 40 data sets fall outside the field

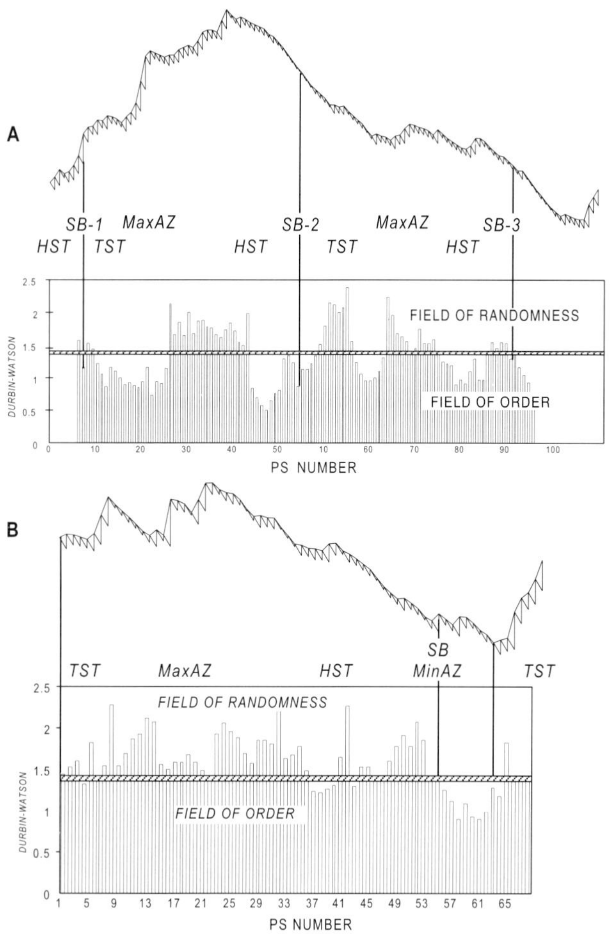

FIG. 11.—Plots from greenhouse sections showing Durbin-Watson values vs. parasequence number, calculated for a moving window of 15 parasequence widths. Results demarcate fields of order and randomness. Durbin-Watson values are plotted beneath Fischer plots to illustrate relationships between Durbin-Watson values and thickness patterns. (A) The Ordovician El Paso Group, Franklin Mountains, Texas, and (B) the Lower Triassic Dawen Formation, Nanpanjiang basin, south China, show long-term changes (nonstationarity) in the level of thickness ordering; greater order occurs near minimum-accommodation zones. Labels as in Figure 2.

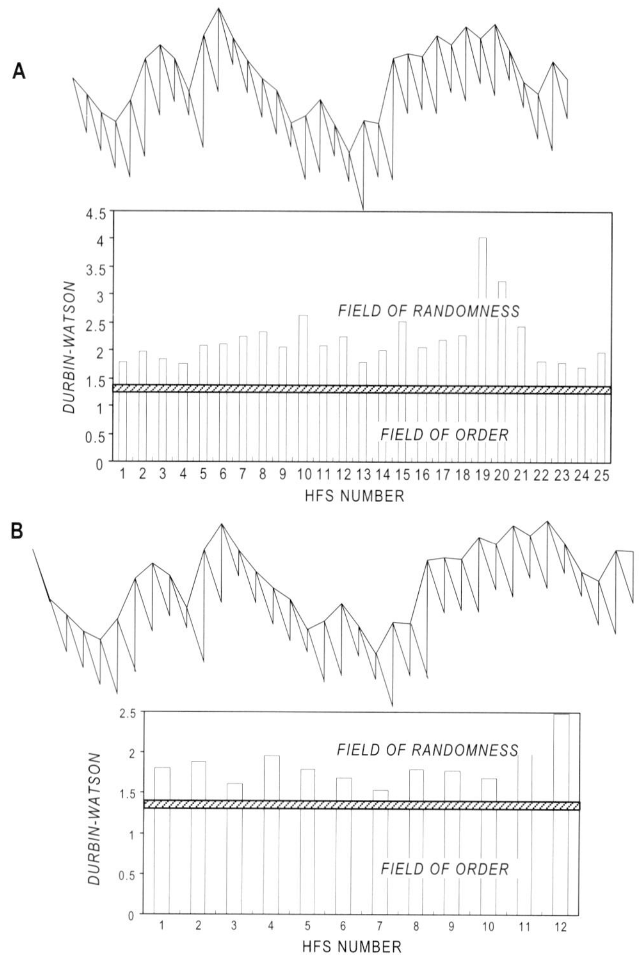

FIG. 12.— Plots from icehouse sections showing Durbin-Watson values vs. high-frequency sequence number, calculated for a moving window of 15 parasequence widths. Results demarcate fields of order and randomness. Durbin-Watson values are plotted beneath Fischer plots to illustrate relationships between Durbin-Watson values and thickness patterns. (A) The Pennsylvanian Hermosa Group, Paradox basin, Utah, and (B) Pennsylvanian Holder Formation, Orogrande basin, New Mexico, show random thickness patterns throughout the sections (stationary, random behavior). Labels as in Figure 2.

of randomness. If we evaluate the results in the context of secular classes, we find that all of the greenhouse depositional systems of Proterozoic–Ordovician age and Early to Middle Triassic age fall outside the random field at the 95% C.I.. In contrast, greenhouse depositional systems of Jurassic to Cretaceous age fall predominantly within the field of randomness at the 90% or 95% C.I. (Fig. 8). Icehouse depositional systems generally fall within the field of randomness, and transitional depositional systems occur in both fields (Fig. 8).

For greenhouse depositional systems of Proterozoic-Ordovician and Early to Middle Triassic age, the ordered groupings of thin and thick parasequences may be interpreted to result from long-term accommodation change driven by third- or second-order relative changes in sea level. An ice-free climate results in high-frequency (fifth-order) sea-level fluctuations with lower amplitude superposed on lower-frequency (third- or second-order) fluctuations with higher amplitudes. The superposition of low-amplitude fluctuations with high frequency on higher-amplitude fluctuations of lower frequency allows these carbonate depositional systems to record a hierarchy in which long-term accommodation changes are systematically recorded as long-term thickness changes of parasequences nested within the larger third- or second-order depositional sequences. Forward stratigraphic modeling of these depositional systems yields a similar conclusion (e.g., Goldhammer et al., 1993a; Osleger and Read, 1991). From these results, it is apparent that vertical thickness patterns as revealed in Fischer plots serve as a useful stacking-patterns analysis tool for identification and correlation of depositional sequences in Proterozoic–Ordovician and Lower to Middle Triassic strata.

For greenhouse depositional systems of Jurassic and Cretaceous age, the apparent lack of order in parasequence groupings is surprising given that sea-level fluctuations should exhibit a similar style of forcing as in other greenhouse depositional systems. This apparent lack of order may be interpreted to have resulted from autocyclic processes and facies amalgamation, noting that the lower amplitude of high-frequency sea-level fluctuations also represents weaker stratigraphic forcing as compared to icehouse periods. Weaker stratigraphic forcing may result in erratic parasequence thickness distributions, if the thicknesses are generated by autocyclic processes. This type of behavior would certainly be expected in Jurassic ramp settings characterized by extensive development of facies mosaics (e.g., the Smackover and Haynesville of the U.S. Gulf of Mexico). Another possibility is that the high-frequency sea-level changes forcing these depositional systems are random in period and amplitude; thus random thickness patterns result. Such a process might be expected in the Cretaceous Period in light of the extremely warm climate generally interpreted for this period (cf. Donn and Shaw, 1977; Frakes et al., 1992; Warrick et al., 1993).

Sadler et al. (1993) introduced the Z-score as a statistic of the level of order in thickness groupings, as generated by the runs test, for direct comparison between sections with different numbers of parasequences. In Figure 9 we have plotted the Z-scores for comparison of greenhouse depositional systems. All Proterozoic–Ordovician depositional systems have Z-scores less than or equal to –2. A Z-score of <–2 means an r value (where r is the total number of runs) two standard deviations below the expectation of random stacks. According to Sadler et al. (1993), a Z-score of <–2 indicates that there is less than a 2.3% chance that random stacking could produce fewer runs. Furthermore, several of these data sets have Z-scores of <–3.5, indicating that there is less than a 0.02% chance that the number of runs could have resulted from a random process. These depositional systems exhibit ordered, nonstationary thickness patterns (nonstationary patterns exhibit long-term directional change; Kruskkal and Tanur, 1978) in which there are significant vertical groupings of thinner- and thicker-than-average parasequences. Most Jurassic and Cretaceous depositional systems, in contrast, have Z-scores near zero, indicating random thickness arrangements. Some Jurassic–Cretaceous depositional systems, such as the Arab, Aurora, and Zuluoaga Formations have Z-scores between 1 and 2, indicating possible nonrandom groupings. For Z-scores between 1 and 2, there is a 2.3–16% chance that the thickness distributions resulted from a random process.

All but one icehouse depositional system has numbers of runs consistent with a random walk at the 95% C.I. (Fig. 9). This behavior may be interpreted to result from the incomplete recording of accommodation space owing to high-amplitude, high-frequency, glacio-eustasy. High-frequency sequences of icehouse depositional systems exhibit non-Waltherian facies offsets and subaerial exposures of subtidal facies. These relationships indicate that most high-frequency sequences do not fill to sea level, and thus their thicknesses generally should not be used to interpret a record of long-term accommodation changes (Rankey et al., this volume). The lack of systematic groupings of thinner- and thicker-than-average units can be considered a stationary behavior in which thicknesses fluctuated randomly about the average but lack "long-term" vertical groupings (cf. Kruskkal and Tanur, 1978).

Durbin-Watson analysis.—
Drummond and Wilkinson (1993) used the Durbin-Watson test to evaluate the "directionality" of parasequence thickness patterns. As pointed out by Drummond and Wilkinson (1993), the runs test does not consider differences in thickness between contiguous parasequences. If there is a significant "directionality" within a stack of parasequences, neighboring parasequences would display similar thicknesses, whereas if the stack is nondirectional, contiguous parasequences would exhibit highly variable thicknesses. The Durbin-Watson value is calculated as illustrated in the equation in Figure 10. Successions with gradually changing parasequence thickness will result in low Durbin-Watson values, whereas highly variable, "jumpy," thickness successions will result in high Durbin-Watson values.

From the analysis of our data sets, we found results that are generally consistent with those derived from the runs test (Fig. 10). Proterozoic–Ordovician sections and Lower to Middle Triassic sections tend to have the lowest values, indicating ordered, directional behavior in parasequence thicknesses. Jurassic–Cretaceous sections, with the exception of the Arab Formation and one of the Zuluoaga sections, fall within the field of randomness. Likewise, most icehouse sections fall within the field of randomness with the exception of the Suwanne Formation, the Dalksteinkalk, and one of the Miocene sections. Notably, the highest Durbin-Watson values (most erratic thickness patterns) were found in icehouse depositional systems (Fig. 10). Sections from transitional depositional systems fall in both fields.

The results may be interpreted in the same way that has been discussed for the runs test: (1) Proterozoic–Ordovician and Triassic depositional systems exhibit ordered long-term thickness patterns representing accommodation changes driven by a hierarchy of sea-level fluctuations in a greenhouse depositional system, (2) Jurassic–Cretaceous depositional systems exhibit "jumpy" random thickness changes due to autocyclicity, and (3) icehouse depositional systems exhibit highly variable "jumpy" thickness patterns owing to an incomplete recording of accommodation space caused by high-amplitude sea-level fluctuations.

To illustrate the differences in thickness stacking patterns between icehouse and greenhouse depositional systems, Durbin-Watson values were calculated for windows or subpopulations of parasequence stacks within the Ordovician El Paso Group, the Lower Triassic Dawen Formation, the Pennsylvanian Hermosa Group, and the Pennsylvanian Holder Formation (Figs. 11, 12). By designating a window of 15 parasequences, parts of the data sets with random and nonrandom behavior are discernible. According to Drummond and Wilkinson (1993), ordered behavior is demarcated by Durbin-Watson values of less than 1.4. In each of the data sets, the Durbin-Watson values are plotted beneath a Fischer plot (Figs. 11, 12).

In the greenhouse example from the Ordovician El Paso Group, a large part of the section exhibits ordered behavior (Fig. 11). Interestingly, the highstand systems tract (falling limb of the Fischer plot) is most directional, whereas parts of the transgressive systems tract are random (Fig. 11). The greenhouse example from the Lower Triassic Dawen Formation exhibits lesser order overall compared to the El Paso Group, but—as is typical of greenhouse depositional systems—it shows disorder in the parasequence thickness patterns of the TST and order in the HST and MinAZ (Fig. 11). These results may be interpreted in the context of long-term changes in accommoda-

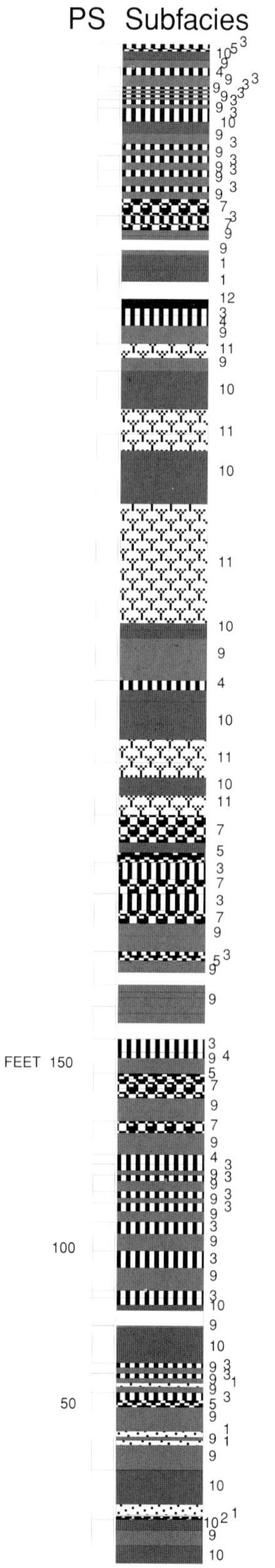

Observed Facies Transition Matrix

from \ to	1	2	3	4	5	7	9	10	11	12
1							3	1		
2	1									
3				1	1	2	18	1		1
4			2				3			
5			3			1	1			
7			3		1		2		1	
9	3		16	3	2	4		4	1	
10		1	2	1	1		3		3	
11							2	4		
12									1	

Expected Facies Transition Matrix (for Random Succession)

from \ to	1	2	3	4	5	7	9	10	11	12
1	0.128	0.031	0.97	0.161	0.161	0.229	1.847	0.374	0.195	0.031
2	0.031	0.008	0.237	0.039	0.039	0.056	0.451	0.091	0.048	0.008
3	0.97	0.237	7.376	1.223	1.223	1.742	14.04	2.844	1.48	0.237
4	0.161	0.039	1.223	0.203	0.203	0.289	2.329	0.472	0.245	0.039
5	0.161	0.039	1.223	0.203	0.203	0.289	2.329	0.472	0.245	0.039
7	0.229	0.056	1.742	0.289	0.289	0.412	3.317	0.672	0.35	0.056
9	1.847	0.451	14.04	2.329	2.329	3.317	26.74	5.414	2.818	0.451
10	0.374	0.091	2.844	0.472	0.472	0.672	5.414	1.096	0.571	0.091
11	0.195	0.048	1.48	0.245	0.245	0.35	2.818	0.571	0.297	0.048
12	0.031	0.008	0.237	0.039	0.039	0.056	0.451	0.091	0.048	0.008

χ^2 Matrix = (Observed - Expected)^2 / Expected

from \ to	1	2	3	4	5	7	9	10	11	12
1	0	0.031	0.97	0.161	0.161	0.229	0.719	1.047	0.195	0.031
2	30.12	0	0.237	0.039	0.039	0.056	0.451	0.091	0.048	0.008
3	0.97	0.237	0	0.041	0.041	0.038	1.115	1.195	1.48	2.459
4	0.161	0.039	0.493	0	0.203	0.289	0.194	0.472	0.245	0.039
5	0.161	0.039	2.581	0.203	0	1.75	0.758	0.472	0.245	0.039
7	0.229	0.056	0.908	0.289	1.75	0	0.523	0.672	1.209	0.056
9	0.719	0.451	0.273	0.194	0.046	0.141	0	0.369	1.173	0.451
10	0.374	9.042	0.25	0.592	0.592	0.672	1.076	0	10.34	0.091
11	0.195	0.048	1.48	0.245	0.245	0.35	0.238	20.61	0	0.048
12	0.031	0.008	0.237	0.039	0.039	0.056	0.451	0.091	19.08	0

Cumulative χ^2 (Sum of Matrix) = 126.6

Degrees of Freedom = (Number of Facies Units - 1)^2 - Number of Facies Units = 71

Minimum χ^2 to Reject Null Hypothesis of Random Facies Succession = 91.67

First-Order Markov Properties at 95% C.I.

FIG. 13.—Illustration of Markov chain analysis performed on Silurian strata of the Barn Hills Formation, Utah (Harris and Sheehan, 1996). In this example, the cumulative χ^2 statistic (126.6) exceeds the minimum χ^2 statistic (91.67) required to reject the null hypothesis of random facies transitions at the 95% C.I. The facies succession is therefore statistically nonrandom. Facies codes are as follows: 1—quartz sandstone, 2—karst, 3—cryptalgal, prism-cracked, and fenestral laminite, 4—mechanically laminated lime mudstone, 5—stromatolitic wackestone to packstone, 7—ripple cross-laminated wackestone to packstone, 9—bioturbated mudstone, 10—bioturbated skeletal-peloidal wackestone to packstone, 11—coral-stromatoporoid wackestone to packstone, 12—argillaceous wackestone. C.I.—confidence interval.

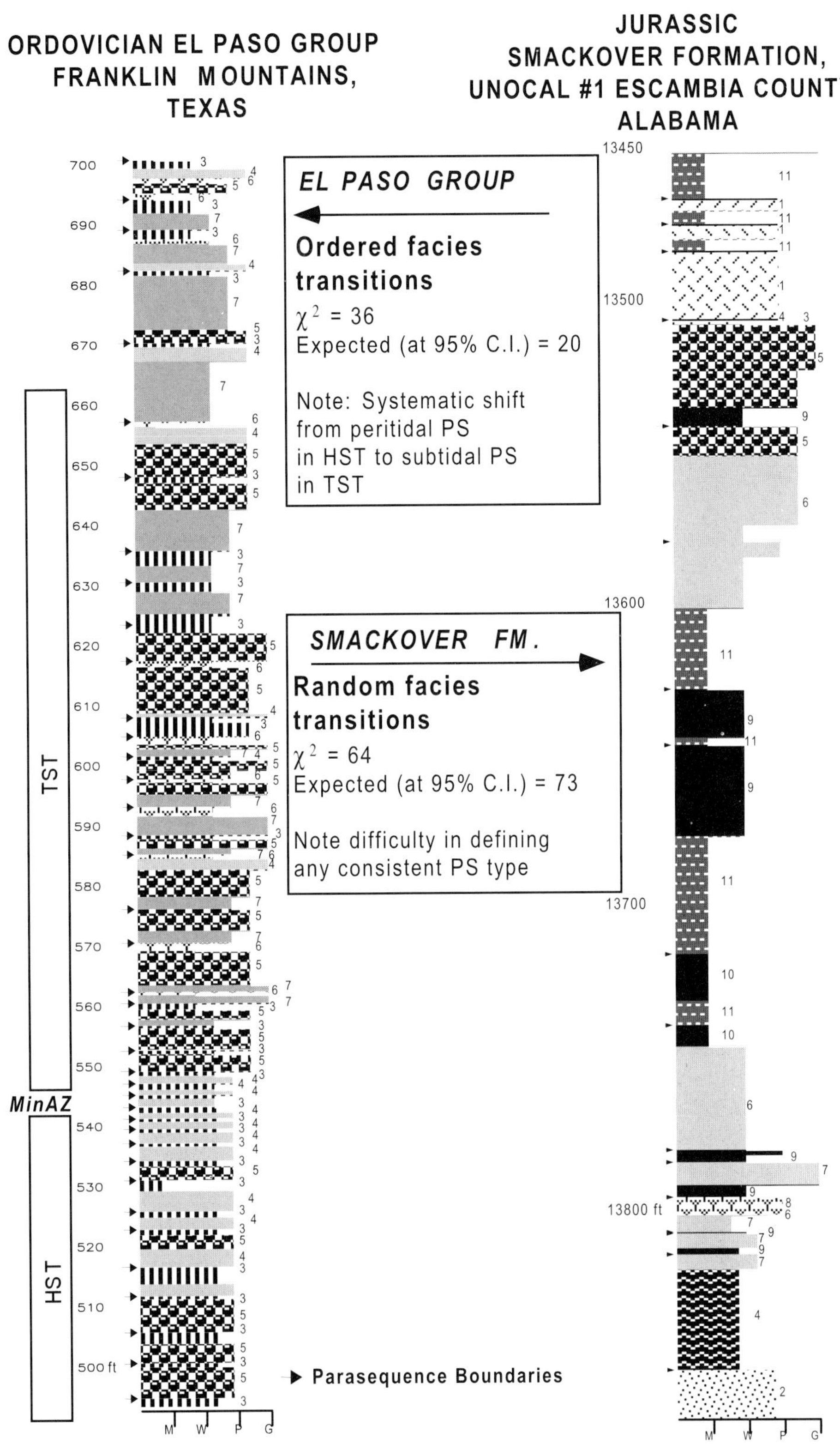

FIG. 14.—End-member greenhouse examples of vertical facies ordering as evaluated by using Markov chain analysis. Markov scores are plotted beside graphic sections of the Ordovician El Paso Group and the Jurassic Smackover Formation. Note that the El Paso Group, as in other Proterozoic–Ordovician sections (Table 3), has ordered facies transitions. The Smackover Formation fails to reject the null hypothesis of random facies transitions. Facies for the El Paso Group: 1—evaporite, 2—quartz sandstone, 3—prism-cracked, cryptalgal laminite, 4—ribbon rock, 5—cross-bedded ooid grainstone, 6—intercalated thin-bedded grainstones and packstones, 7—stromatolites, 8—burrowed skeletal-peloidal wackestone, 9—open-marine bioherms. Facies for the Smackover Formation: 1—sabkha evaporite, 2—eolian sandstone, 3—paleosol, 4—cryptalgal laminite, 5—oolite grainstone, 6—peloidal packstone, 7—oncolite peloidal grainstone to packstone, 8—hydrozoan-algal boundstone, 9—skeletal peloidal wackestone to mudstone, 10—offshore, dark laminated lime mudstone, 11—shale. Note that thickness in feet is given on left sides of sections. C.I.—confidence interval; other labels as in Figure 2

tion through systems tracts. Accommodation space is limited in the highstand systems tract because all parasequences fill to sea level; thus contiguous parasequences cannot exhibit extreme thickness changes. Accommodation space is typically much greater in the transgressive systems tract, and parasequences do not fill all of the available accommodation space—that is, they are subtidal parasequences. Contiguous parasequences in the transgressive systems tract, therefore, exhibit more "jumpy" thickness changes as the thickness of any given parasequence is dependent upon factors other than accommodation space. In other words, these depositional systems not only exhibit nonstationary groupings of thin and thick parasequences, but the *level* of thickness ordering also shows long-term vertical patterns.

In contrast to the greenhouse examples, Durbin-Watson values for the icehouse Pennsylvanian Hermosa Group and Holder Formation fall entirely in the field of randomness (Fig. 12). The consistently high Durbin-Watson values and the rapidly varying form of the Fischer plots suggest erratic "jumpy" thickness changes between contiguous high-frequency sequences in these icehouse depositional systems. This finding is interpreted to result from incomplete recording of accommodation space in high-frequency sequences driven by high-amplitude glacio-eustasy. High-frequency sea-level fluctuations with amplitudes of as much as or more than 50 m (cf. Mesolella et al., 1969; Ross and Ross, 1987; Goldstein, 1988; French and Heckel, 1994; Rankey et al., this volume) outstrip carbonate production during rises and fall before sedimentation fills accommodation space, resulting in subaerial exposure of subtidal facies and stationary, erratic changes in the thicknesses of individual high-frequency sequences.

Vertical Facies Succession

Embedded Markov chains.—

Markov chain analysis is a technique used to determine if the vertical succession of facies types within a given stratigraphic section exhibits nonrandom behavior (Gingerich, 1969; Krumbein and Dacey, 1969; Carr, 1982; Powers and Easterling, 1982; Davis, 1986). Specifically, Markov chain analysis addresses the question of whether the frequency distributions of facies transitions differs significantly from that of a strictly random distribution. Vertical facies transitions (facies 1 to 2, 1 to 3, 2 to 3, etc.) are tabulated in a facies-transition matrix (Fig. 13). Through a series of steps, the facies-transition matrix is compared against a matrix of expected facies transitions for a random succession in which the occurrence of each facies type is independent of the occurrence of underlying and overlying types (Davis, 1986). The null hypothesis of randomness or independence is evaluated by using a χ^2 test to compare the actual facies-transition matrix against the expected matrix for a random succession (Fig. 13). Embedded Markov chain analysis differs from conventional Markov analysis in that the facies transitions are tabulated as they occur without regard to thickness. As a result, transitions from a given facies to itself are not counted, and the diagonal values of the transition matrix are forced to zero. To account for this difference, diagonal transition frequencies in the expected transition matrix for independence are calculated through an iterative routine in which diagonal transition probabilities are derived from the marginal probabilities of the actual transition matrix (cf. Davis, 1986).

The results of our analyses are presented in Table 3. More than half of the stratigraphic sections (68%; 30 of 44) reject the null hypothesis and have statistically nonrandom facies successions. We also evaluated the results in the context of secular classes. In greenhouse periods, 100% (10 of 10) Proterozoic–Ordovician and (2 of 2) Triassic sections are nonrandom, whereas only 60% (6 of 10) Jurassic–Cretaceous sections are nonrandom. In transitional periods, 83% (5 of 6) Silurian–Mississippian and 57% (4 of 7) Permian depositional systems are nonrandom. Icehouse depositional systems exhibit the lowest levels of order; only 50% (4 of 8) of the results for those sections reject the null hypothesis of random facies successions.

Three different depositional factors may be called upon to interpret these differences in facies ordering: (1) stationary vs. nonstationary behavior, (2) completeness of the accommodation record, or (3) levels of autocyclicity. With regard to the first, *stationarity* is a stratigraphic behavior in which proportions of facies types (or thicknesses, as discussed earlier) fluctuate erratically about the mean through the section, whereas *nonstationary* behavior is reflected in long-term changes in facies proportions or thicknesses (cf. Kruskkal and Tanur, 1978). Greenhouse depositional systems typically exhibit strongly nonstationary behavior. They contain long-term changes in facies proportions, from subtidal-prone parasequences through peritidal-prone parasequences (Fig. 14). In such depositional systems, we interpret this nonstationarity to result from a long-term deterministic signature (i.e., change in accommodation space) from which the changes in facies proportion can be used in the interpretation of depositional systems tracts and the identification and correlation of second- or third-order depositional sequences. In Proterozoic–Ordovician and Triassic sections, the long-term shifts in facies proportions result in high Markov scores because peritidal and shoal-subtidal facies types are systematically grouped together in the HST and MinAZ, whereas deeper-subtidal facies units are grouped in the TST and MaxAZ (e.g., Fig. 14). Thus, even if randomness occurs within parts of the section (e.g., random facies intercalations induced by autocyclicity at the parasequence scale), the long-term accommodation driver imparts order in the overall facies succession.

In contrast, carbonate-platform strata of icehouse depositional systems tend to exhibit greater stationarity. They tend to lack the long-term changes in facies proportions exhibited by greenhouse depositional systems (e.g., Fig. 15). Thus, in such sections, a relatively high level of facies ordering (e.g., 1, 2, 3, 4, 1, 2, 3, 4) would be required at the high-frequency sequence scale in order to reject the null hypothesis of a random facies succession. In other words, because of lesser long-term facies change (stationarity), order in such depositional systems must reside in the smaller-scale high-frequency sequences. The result that 50% of the icehouse sections failed to reject the null hypothesis of random facies succession is probably best evaluated at the scale of high-frequency sequences.

These results bring us to the second process factor: completeness of the accommodation record. The low Markov scores

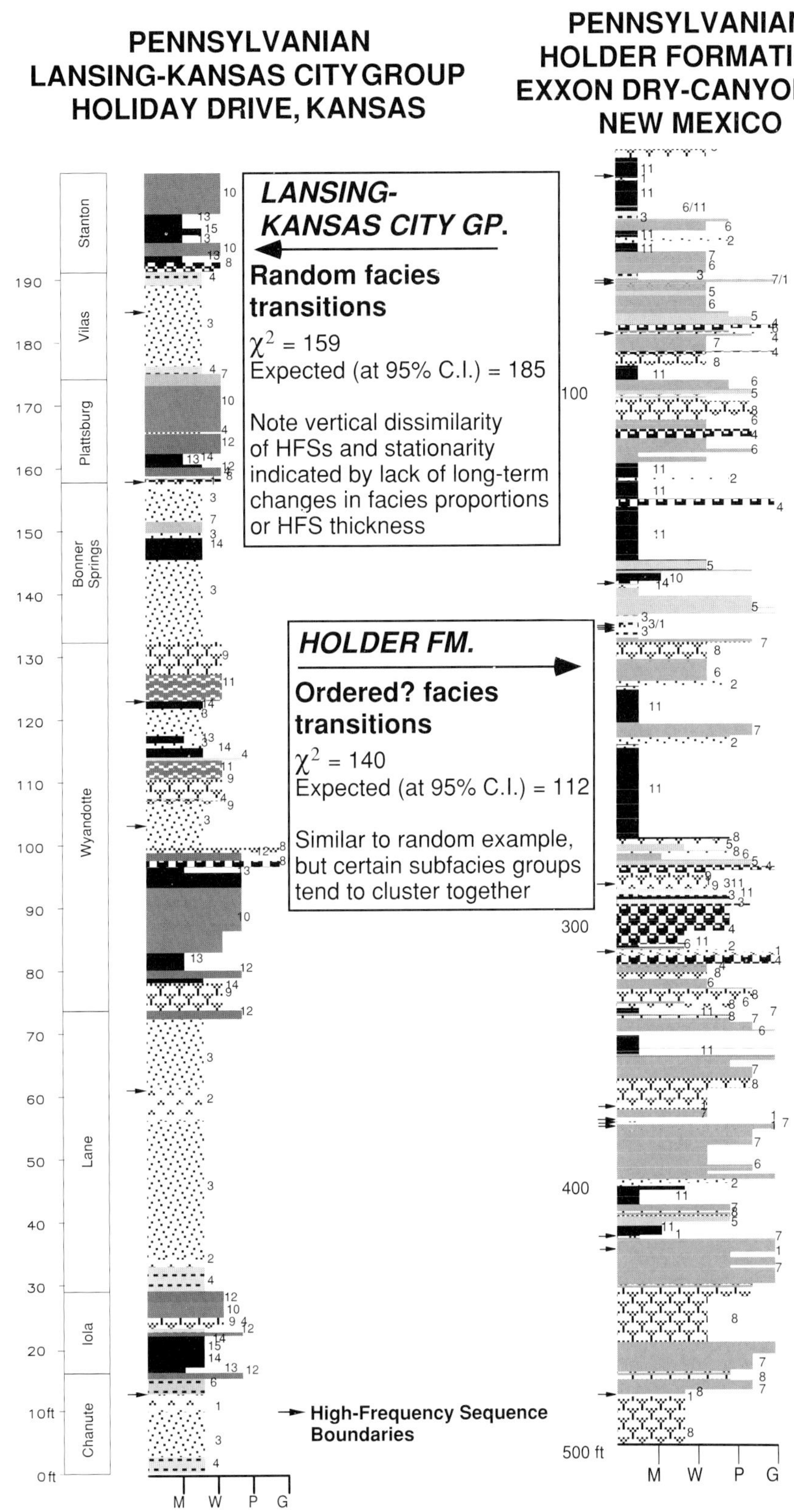

FIG. 15.— End-member icehouse examples of vertical facies ordering as evaluated by using Markov chain analysis. Markov results are plotted beside graphic sections of the Pennsylvanian Lansing–Kansas City Group and Pennsylvanian Holder Formation. Facies codes for the Lansing Kansas City Group: 1—paleosol, 2—terrestrial red shale, 3—shaly quartz sandstone to siltstone, 4—calcareous quartz sandstone to siltstone, 5—mudstone, 6—calcareous shale, 7—sandy molluscan packstone, 8—ooid packstone to grainstone, 9—phylloid algal wackestone to bafflestone, 10—diverse open-marine skeletal wackestone to packstone, 11—domal stromatolite wackestone, 12—undifferentiated skeletal wackestone, 13—lime mudstone, 14—gray core shale, 15—black core shale. Facies codes for the Holder Formation: 1—caliche, 2—quartz sandstone to conglomerate, 3—terrestrial red shale, 4—skeletal oolitic grainstone to packstone, 5—oncolitic packstone to wackestone, 6—normal-marine skeletal packstone to wackestone, 7—fusulinid, encrusting foraminiferal packstone, 8—phylloid algal bafflestone, 9—leopard rock boundstone, 10—lime mudstone, 11—black organic shale. Note that thickness in feet is given on left sides of sections. C.I.—confidence interval; other labels as in Figure 2.

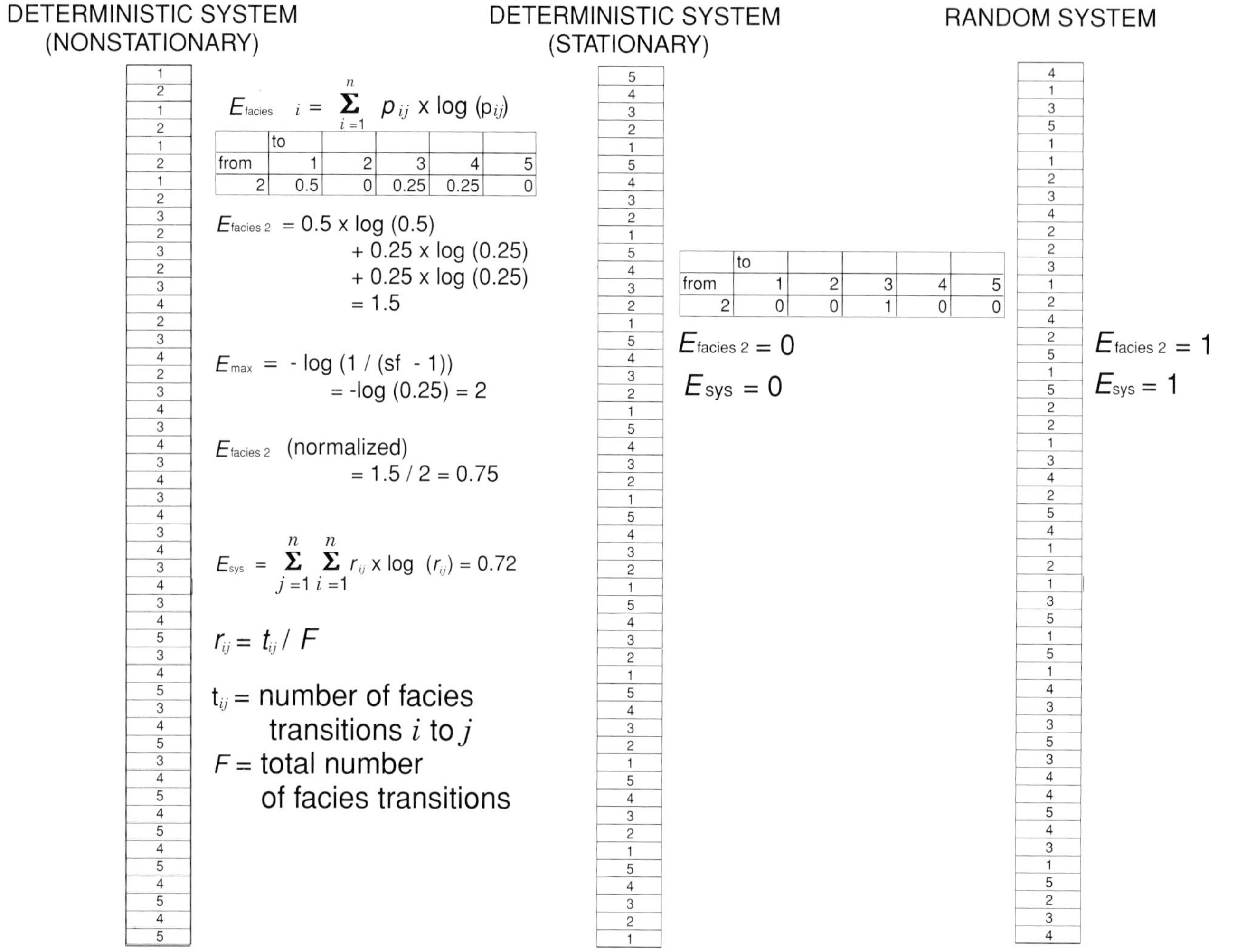

FIG. 16.—Illustration of the entropy technique. Facies entropy (E_{facies}) and system entropy (E_{sys}) are calculated for the synthetic data sets shown. The entropy statistic is normalized to range from 0 to 1 (ordered to random facies transitions). For the system entropy to be zero, the section must be stationary (identical cyclic facies patterns throughout), and the parasequences must be perfectly ordered and asymmetrical (i.e., 1, 2, 3, 1, 2, 3, . . .).

for icehouse sections are probably best interpreted to result from incomplete recording of accommodation space in high-frequency sequences because of high-amplitude glacio-eustasy. High-amplitude changes in sea level result in "upward truncation" of high-frequency sequences caused by subaerial exposure of subtidal facies. In other words, because the amplitude of sea-level rise is so high, sedimentation fails to fill accommodation space before being halted by subaerial exposure, and thus the facies succession fails to record the full accommodation cycle. "Upward cycle truncation" is frequently cited as a cause for randomness in stacking patterns (e.g., Soreghan and Dickinson, 1994; Gianniny and Simo, 1996).

Finally, the high level of randomness in facies successions of Jurassic–Cretaceous depositional systems may be explained by the third factor, dominance of autocyclic processes. Although, like the other greenhouse depositional systems, the Jurassic–Cretaceous sections exhibit nonstationarity in facies proportions, 4 out of 10 sections from these periods fail to reject the hypothesis of random facies successions as determined by Markov chain analysis. The indication of random facies successions in these depositional systems may be interpreted as the result of weak stratigraphic forcing (due to low-amplitude sea-level fluctuations) coupled with random, autocyclic migration of facies mosaics. The Jurassic Smackover Formation is a classic candidate for the interpretation of autocyclic migration of facies mosaics on a high-energy, low-gradient ramp. Although the Jurassic section exhibits a long-term nonstationary shift in facies proportions, due to progradation and shoaling of the ramp depositional system, the facies transitions appear random (Fig. 14). The autocyclic explanation is ambiguous, however, because out of the Jurassic–Cretaceous sections studied, only the sections with lesser numbers of parasequences or facies units failed the Markov test (Table 3). One might argue that the sections failed to pass the Markov tests because the sample size (number of facies units) is too small. Even though the number of facies transitions is smaller in these examples, the cumulative stratigraphic thickness matches or exceeds those of sections that passed the Markov test (e.g., Fig. 14). An alternative way to view this problem is that weaker levels of stratigraphic forcing may have *resulted* in substantial facies amalgamation and led to the lesser numbers of parasequences and subfacies units. Thus, the random facies successions could be viewed as the result of autocyclicity coupled with facies amalgamation.

TABLE 4.—EMPIRICAL CLASSES OF STACKING-PATTERN STYLE.

Secular Class	Depositional Setting	Vertical Stacking Patterns	Lateral Stacking Patterns	Database Examples	Additional Examples
Proterozoic–Ordovician	Greenhouse climate Low-gradient ramps and rimmed shelves Passive continental margins	Low-order, greenhouse SEQs composed of PSs Systematic, gradual change in PS thickness and facies proportions through MinAZ and MaxAZ	Greenhouse SEQs with regional extent PSs may extend several kilometers to 200 km or may be locally discontinuous	Rocknest Fm. Katakturuk Fm. Bonanza King Fm. Arbuckle Gp. El Paso Gp. Beekmantown Gp.	Elbrook, Nolichucky, Allentown, and Conococheague Fms. (Demicco, 1981; Koerschner and Read, 1989; Osleger and Read, 1991)
Silurian–Devonian	Transitional climate Broad ramps and rimmed shelves or isolated carbonate platforms Passive to convergent continental margins	Low-order depositional SEQs commonly marked by abrupt facies shifts Fourth-order PSSs may be identified by platform-margin step-backs PSs may exhibit quasi-systematic thinning and change in facies proportions at SEQ boundaries or may lack such changes with respect to SEQ boundaries or PSSs Similar facies motif repeated in PSs	Depositional SEQs with potential regional extent PSs locally discontinuous or may extend several kilometers	Barn Hills Fm. Southesk Fm. Pillara Fm. Swan Hills Fm. Laketown Dolomite	Helderberg Gp. (Anderson and Goodwin, 1990) Lost Burrow Fm. (Yang et al., 1995)
Pennsylvanian–Early Permian, Late Triassic, and Neogene	Icehouse climate Flat-topped shelves or isolated platforms with abrupt barrier-reef or discontinuous patch-reef margins Intracratonic compressional basins or passive continental margins	HFSs, capped with subaerial exposure surfaces, exhibit chaotic vertical thickness patterns and long-term stationarity in facies proportions HFSs grouped into composite SEQs Long-term changes in thickness and facies proportions usually not diagnostic of composite SEQs, which are best recognized from regional retrogradational to progradational trends in facies tracts and stratal geometries, especially at platform margin.	HFSs extend across platform; some HFSs are correlatable as far as 600 km	Hermosa Gp. Holder Fm. Lansing–Kansas City Gp. Dachsteinkalk Pliocene–Pleistocene Bahamas Miocene Mallorca Key Largo Fm.–Miami Oolite Suwanne Fm.	Midcontinent cyclothems (Heckel, 1986; Busch and Rollins, 1984; French and Heckel, 1994) Permian basin (Brown et al., 1990; Dickson and Saller, 1995) Miocene Spain (Franseen et al., 1993)
Middle Permian–Late Permian	Transitional climate Ramps with grainstone shoal margins or flat-topped shelves with steep reef-rimmed margins Intracratonic compressional basins	Middle Permian: PSs grouped into PSSs or HFSs with chaotic vertical thickness and relatively stationary facies proportions relative to low-order depositional SEQs or composite sequences Late Permian: PSs with systematic thickness patterns and facies proportions defining depositional SEQs	Continuity of PSs, PSSs, and HFSs is dependent on shelf profile. Continuous as far as 50 km in platform interior, but highly discontinuous at platform margin	Snake Canyon Fm. San Andres Fm. Grayburg Fm. Queen Fm. Seven Rivers–Tansill Fms.	San Andres Fm. (Sonnenfeld and Cross, 1993)
Early Triassic–Middle Triassic	Greenhouse climate Isolated carbonate platforms in rift basins	Highly ordered packaging of PSs into PSSs and depositional SEQs with Milankovitch-type hierarchy	Continuous across platform	Latemar Fm. Dawen Fm.	
Jurassic–Cretaceous	Greenhouse climate Broad, low-gradient ramps, shelves, and isolated platforms Passive continental margins	Depositional SEQs characterized by gross transgressive-regressive facies shifts Quasi-systematic grouping of thin peritidal prone PSs at SEQ boundaries in some systems PSs poorly developed in other systems	Depositional sequences with regional extent Continuous PSSs in some systems. Probable highly discontinuous PSs	Arab Fm. Haynesville Fm. Smackover Fm. Zuluoaga Fm.	Also from database: Aurora Fm. El Abra Fm. Cupido Fm. Segovia Fm.

NOTE: SEQ = depositional sequence, CSEQ = composite sequence, PS = parasequence, PSS = parasequence set, HFS = high-frequency sequence, MinAZ = minimum accommodation zone, MaxAZ = maximum accommodation zone, Fm. = formation, Fms. = formations, Gp. = group.

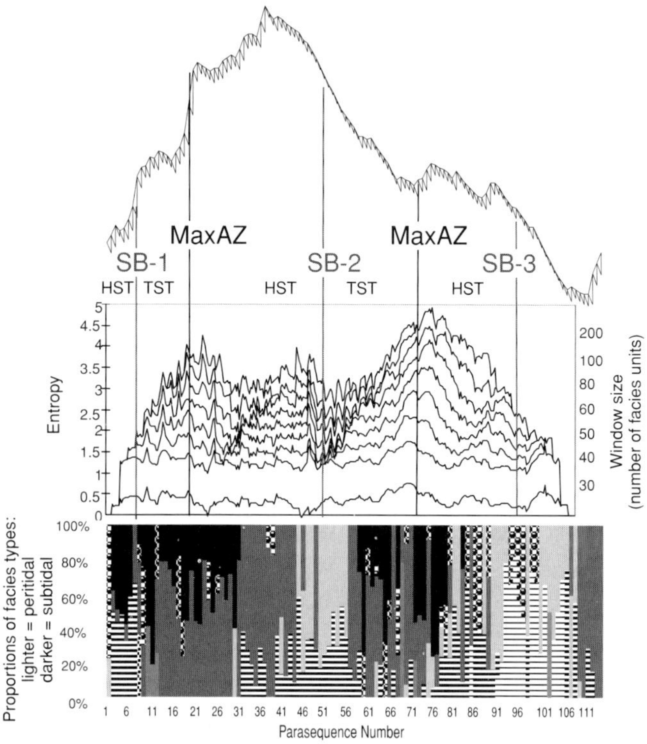

FIG. 17.—Interval entropy, calculated for moving windows (width increased from 10 to 200 facies units) stepped through the stratigraphic section of the Ordovician El Paso Group, Franklin Mountains, Texas, and plotted vs. parasequence number. The interval entropy values are compared against facies proportions (facies histogram) and thickness trends (Fischer plot) of systems tracts and depositional sequences. Note that the long-term variation in the level of entropy correlates with nonstationary trends in facies and thickness. The section exhibits long-term, nonstationary, patterns of parasequence thickness and facies proportions corresponding to system tracts, and the level of facies ordering shows a long-term modulation within the systems tracts. Higher entropy occurs in the TST and MaxAZ; lower entropy occurs in the HST and MinAZ. Labels as in Figure 2.

Entropy analysis.—

Hattori (1976) introduced facies and system entropy statistics as an extension of Markov chain analysis for evaluation of the levels of facies order in cyclic stratigraphic successions. Facies entropy is calculated from the transitions probability matrix through the use of the following formula: $E_{\text{facies } i} = \Sigma p_{ij}$ $(\log_2 (p_{ij}))$, where p_{ij} is the probability of transition from facies i to facies j. In a deterministic system where facies i is always followed by the same facies type (i.e., 1 always followed by 2) $p_i = 1$, and thus $E_{\text{facies } i} = 0$. If the transition probabilities for facies i are equally spread among all possible alternatives (i.e., there is no preferential transition from 1 to 2, 3, 4, etc.), $E_{\text{facies } i \text{ (normalized)}} = 1$. System entropy ($E_{\text{system}}$) is calculated by using a similar approach, essentially summing entropies of all possible transitions from the probability matrix (Fig. 16).

As noted by Hattori (1976), the entropy statistic is highly sensitive to parasequence symmetry. In other words, only a strictly deterministic and stationary system with asymmetric parasequences (e.g., 1, 2, 3, 4, 1, 2, 3, 4, . . . ; Fig. 16) will yield an $E_{\text{system}} = 0$. A strictly deterministic, stationary system with symmetrical parasequences (1, 2, 3, 4, 3, 2, 1, . . .) will yield a higher E_{system} because transition probabilities associated with a given facies will be split between two other facies. From our experiments, we have also found that E_{system} is highly sensitive to stratigraphic nonstationarity. As illustrated in Figure 16, a deterministic system in which parasequence types progressively change up section because of a long-term change in facies proportions (e.g., 4, 5, 4, 5, 4, 3, 4, 3, 2, 4, 3, 2, 3, 2, . . .) will yield a relatively high E_{system}. E_{system} can also be expected to change on the basis of randomness resulting from "cycle truncation" and autocyclicity. Because of all of these sources of variation in E_{system}, we have found it an unsatisfactory technique for between-system comparisons of order and randomness in the overall depositional process. Nevertheless, because E_{system} offers the possibility of calculation of *levels* of facies ordering, we were encouraged to use it as a technique for gauging vertical changes in the level of facies ordering within nonstationary systems.

Many greenhouse depositional systems exhibit different levels of facies ordering depending on which part of the section is evaluated. From the Ordovician example in Figure 14, the HST passes the Markov test ($\chi_2 = 21$; minimum $\chi_2 = 11$) with a relatively low entropy ($E_{\text{system}} = 0.74$). In contrast, the TST failed the Markov test ($\chi_2 = 13$; minimum $\chi_2 = 19$) and has a relatively high entropy ($E_{\text{system}} = 0.98$). These results suggest that *levels* of facies ordering may exhibit systematic up-section changes resulting from a long-term accommodation driver. In order to explore this possibility further, we calculated E_{system} for segments or windows of the stratigraphic section ranging from 10 to 200 facies units long. For each window size, E_{system} was repeatedly calculated (and normalized to the number of facies types present) as the window was stepped vertically upward through the stratigraphic section (Fig. 17). This technique reveals long-term vertical changes in the level of facies ordering through a given stratigraphic section.

The interval entropy values for successively higher stratigraphic horizons in the Ordovician El Paso Group resemble a long-term wave form (Fig. 17). Furthermore, the rise and fall of the entropy values may be interpreted to closely follow the systems tracts of the third-order sequences in the section, as indicated by parasequence thickness and facies proportions (e.g., greater parasequence thickness and proportion of deeper-subtidal facies in the TST). Entropy values are lowest in the HST and the MinAZ, whereas they are highest in the TST and MaxAZ. This analysis suggests the exciting possibility that not only does long-term third-order accommodation drive vertical changes in parasequence thicknesses and nonstationarity in facies proportions, but it also *modulates* the *levels* of facies ordering within the depositional system. In terms of process, these patterns may be interpreted to result from greater randomness introduced by autocyclicity in the deeper-subtidal parts of the depositional system, where facies distribution is not strictly controlled by accommodation, and greater order in the peritidal parts of the section resulting from allocyclic processes controlled by accommodation. From application of the interval entropy technique to numerous greenhouse examples, we consistently found patterns of substantial long-term fluctuations in entropy, compatible with the nonstationarity of the depositional systems. In many cases, the interval entropy reflects the systems tracts and third-order

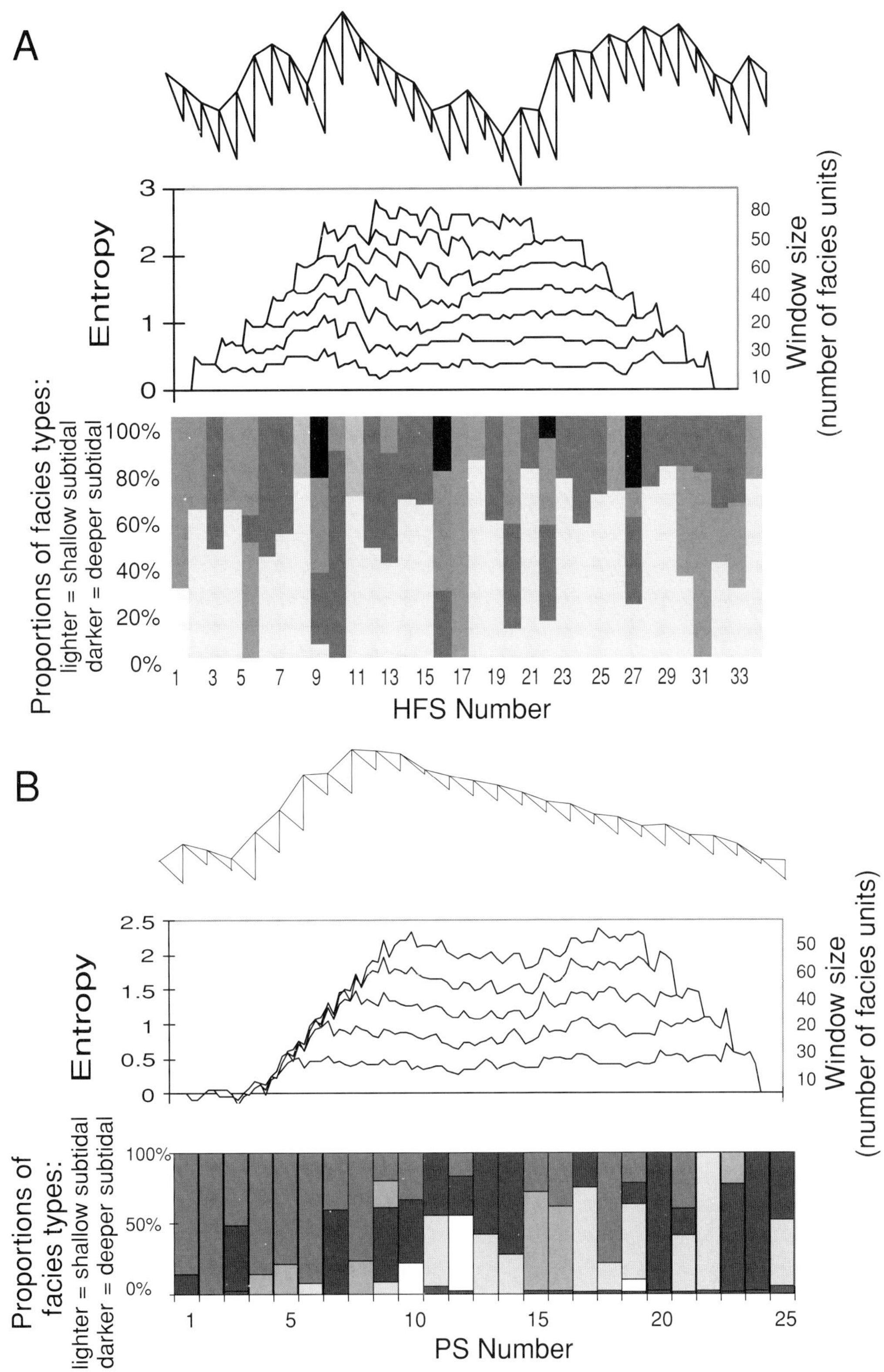

FIG. 18.—Interval entropy plotted vs. parasequence or high-frequency sequence number and compared against facies proportions (facies histogram) and thickness trends (Fischer plot). Icehouse examples are (A) from the Pennsylvanian Hermosa Group, Paradox basin, Utah, and (B) the Oligocene Suwanne Formation, Florida. Note that the entropy values, as well as the parasequence thickness and facies proportions, lack the long-term changes seen in greenhouse sections (e.g., Fig. 17). That is, the icehouse example exhibits stationary behavior not only in parasequence or high-frequency sequence thickness and facies proportions but also in the level of facies ordering through the section. Labels as in Figure 2.

sequences in a manner similar to that described above for the El Paso Group. In other examples, however, the pattern is almost opposite: higher entropy values tend to occur in the HST rather than the TST, or the entropy fluctuations do not exhibit a clear correspondence to systems tracts. These "reversed" or unpredictable interval entropy patterns can result from randomness introduced in the highstand systems tract by unconformities or facies with numerous hiatuses such as teepee complexes. Icehouse depositional systems, in contrast, tend *not* to exhibit such long-term variations in entropy. These depositional systems tend to exhibit uniform, relatively high levels of entropy consistent with their stationarity (Fig. 18).

QUALITATIVE SYNTHESIS OF SECULAR CLASSES

The following section presents a series of six secular classes in stacking-patterns style derived empirically from examination of carbonate strata from a wide range of ages and depositional settings. In construction of the secular classes, we have qualitatively integrated (1) statistical results from the preceding section, (2) grouping of stratigraphic styles based on visual comparisons of normalized stratigraphic sections (all plotted at the same scale with identical symbols for facies, texture, stratigraphic surfaces, sedimentary structures, etc.) cast from the digital data base, and (3) considerations of lateral stratigraphic behavior derived from detailed cross sections in the literature and panoramic outcrop photos.

Paleontologists have noted that the illustration drawn from the careful examination of many individuals is a far superior guide to identifying a given species than is a photograph of a single individual. It is in this spirit that we have attempted to distill the enormous diversity of carbonate stratigraphic styles into the following empirical classes and their accompanying figures. We believe that the following classes should serve as a useful guide to the major differences in stratigraphic styles that must be recognized in order to accurately employ the stacking-patterns technique for chronostratigraphic correlation.

Table 4 gives a summary of stacking-pattern style for the secular classes, which are (1) Proterozoic–Ordovician, (2) Silurian–Devonian, (3) Pennsylvanian, Early Permian, Late Triassic, and Neogene, (4) Middle to Late Permian, (5) Early to Middle Triassic, and (6) Jurassic–Cretaceous. In terms of genesis, the empirical classes can be interpreted as a first-order grouping into periods of Earth history with icehouse, greenhouse, and transitional climate dynamics; the classes may then be ranked into lower-order groupings based on platform geometry, tectonic setting, and local environmental conditions.

1. Platform carbonates in icehouse periods (Pennsylvanian, Early Permian, Late Triassic, and Neogene) are generally stacks of fifth- or fourth-order high-frequency sequences capped with subaerial exposure surfaces. The high-frequency sequences exhibit stationary, chaotic vertical thickness distributions (3–15 m thick), and high lateral continuity. Successive high-frequency sequences record a stationary, relatively self-similar or random facies pattern with non-Waltherian facies offsets. Successive stacks of high-frequency sequences generally lack long-term changes in facies proportions relative to "third-order" composite sequences. Where high-frequency sequences are composed of subsidiary parasequences, the parasequences lack systematic thickness trends or grouping of similar numbers relative to the high-frequency sequences (cf. Goldhammer et al., 1991b). High lateral continuity of high-frequency sequences results from the strong allocyclic forcing potential during icehouse periods. Chaotic vertical thickness distributions and the lack of long-term change in facies proportions result from incomplete recording of accommodation space, making identification of the composite sequences difficult from vertical patterns of individual sections in the platform interior. Third-order composite sequences are best identified from regional retrogradational to progradational patterns of platform-margin facies (e.g., Watney et al., 1996).
2. Platform carbonates in greenhouse periods contain broad transgressive-regressive sequences (<30–<300 m thick) characterized by (a) "long-term" systematic change in parasequence thickness and facies proportions, (b) gross shallowing and deepening shifts in facies tracts defining maximum regressive and transgressive intervals, and (c) in some cases, a sequence boundary capped with extensive subaerial diagenetic modification (e.g., karst), especially in proximal, platform-interior sections. In contrast to the icehouse setting, a greenhouse climate results in low-amplitude, high-frequency sea-level fluctuations superposed on lower-frequency fluctuations of higher amplitude. This climate sets up a situation whereby a full hierarchy of parasequences, parasequence sets, and third-order sequences is more likely to be deposited and preserved in the stratigraphic record. Proterozoic through Lower Ordovician strata, deposited on extensive flat-topped shelves or ramps, tend to contain laterally extensive parasequences that exhibit systematic long-term changes in thickness and facies proportions relative to second- or third-order depositional sequences. In rare cases, where autocyclicity is inhibited, on Early to Middle Triassic isolated platforms, a highly structured hierarchy with a Milankovitch signature is recorded (e.g., Goldhammer et al., 1990; Lehrmann and Goldhammer, 1998). Alternatively, the lesser forcing potential (lower amplitudes) of the high-frequency sea-level fluctuations in a greenhouse setting coupled with the development of facies mosaics leads to greater autocyclicity and facies amalgamation. These characteristics result in chaotic parasequence distributions and poorly developed parasequences in Jurassic and Cretaceous ramps or platforms.
3. During transitional climatic periods (Devonian–Mississippian and middle to Late Permian), platform carbonate style is transitional between icehouse and greenhouse styles. A good example is the style of Upper Permian stacking patterns in the Guadalupe Mountains of Texas and New Mexico. Leonardian–lower Guadalupian strata exhibit a hierarchy of parasequences and parasequence sets nested within third- and fourth-order sequences bounded by subaerial unconformities (e.g., Kerans et al., 1992; Kerans and Fitchen, 1995). Similar to sequences developed in icehouse periods, these depositional systems commonly contain sequences bounded by subaeri-

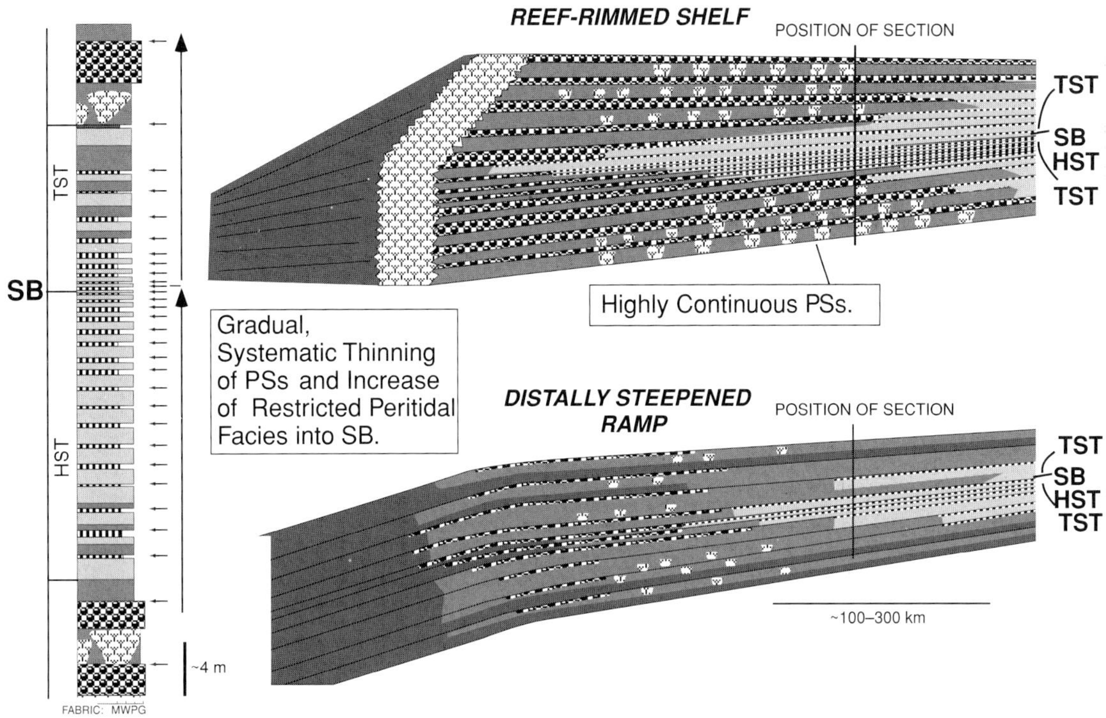

FIG. 19.—Stacking-pattern style of Proterozoic–Ordovician carbonate platforms. The long-term nonstationary stacking patterns of Proterozoic–Ordovician depositional systems are reflected in systematic progradation and retrogradation of tidal-flat–capped parasequences on the interior of broad, flat-topped platforms or distally steepened ramps. Parasequences show systematic thickening and thinning patterns and changes in facies proportions relative to low-order depositional sequences bounded by minimum-accommodation zones. Parasequences tend to have high lateral continuity. Labels as in Figure 2.

al unconformities. Similar to the stacking patterns of greenhouse periods, they contain a well-developed stratigraphic hierarchy, but the stacking patterns lack systematic long-term thickness trends relative to the sequence boundaries and Milankovitch-type grouping of similar numbers of parasequences into parasequence sets. As continental glaciation collapsed and amplitudes of high-frequency sea-level fluctuation declined in the late Guadalupian, stacking patterns returned to a style more typical of greenhouse periods whereby systematic parasequence thickness and facies-stacking trends became important in defining low-order (third-order) sequence boundaries (e.g., Borer and Harris, 1995; Lehrmann et al., 1997).

Proterozoic, Cambrian, and Early Ordovician

Proterozoic through Early Ordovician shallow-marine carbonate depositional systems formed in a greenhouse climate on broad, low-gradient, carbonate ramps or rimmed shelves that extended hundreds of kilometers in dip profile and thousands of kilometers along strike (Wilson, 1975). These vast platforms developed on passive continental margins after breakup of the Proterozoic supercontinent Rodinia. Proterozoic–Lower Ordovician strata contain second- or third-order depositional sequences as much as 300 m thick characterized by gradual, systematic thinning and thickening of parasequences relative to sequence boundaries (Fig. 19, Table 4). Facies are increasingly subtidal-prone away from sequence boundaries in the TST with the deepest-water facies occurring in the MaxAZ generally near the middle of sequences. Facies become progressively peritidal-prone near sequence boundaries (MinAZs). Parasequence types change from one systems tract to the next. The TST typically contains muddy, open-marine bioherm- or argillaceous wackestone-based parasequences that are capped by skeletal or nonskeletal grainstones and hardground surfaces. The HST contains tidal-flat parasequences with either restricted skeletal wackestone to packstone or stromatolitic bases and muddy, flaser-bedded, to mud-cracked, cryptalgal laminite caps. Near sequence boundaries, successive parasequences may contain increasing evidence of prolonged exposure such as tufa or small teepee structures and intercalated lowstand siliciclastics.

Low-order (third-order) sequences in these depositional systems have been correlated across the passive margins of the North American craton, indicating a high level of lateral continuity on an interbasinal scale (Osleger and Read, 1993; Goldhammer et al., 1993b). Individual parasequences have been traced for several kilometers (Montanez and Osleger, 1993) to more than 200 km (Grotzinger, 1985). Although some studies have recognized discontinuity of parasequences on a local scale

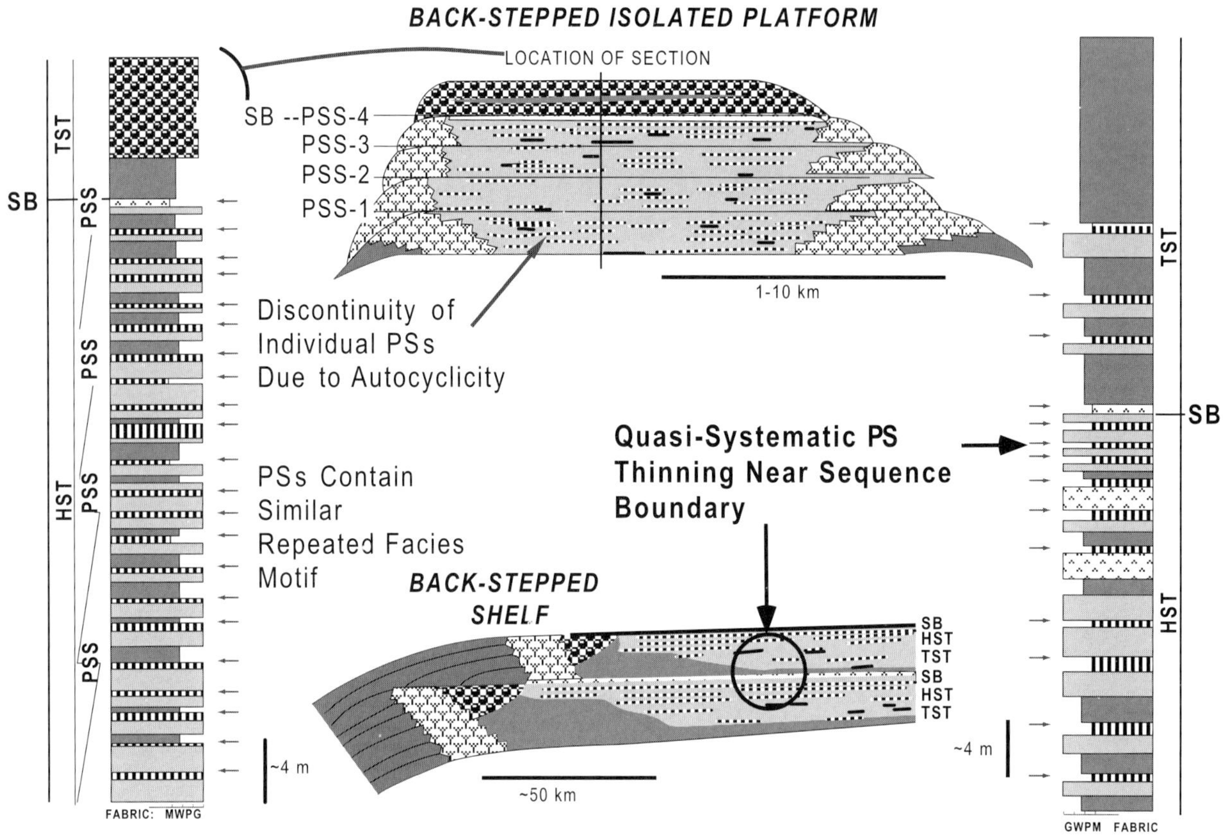

FIG. 20.—Stacking-pattern styles of Silurian and Devonian carbonate platforms. Stacking patterns are variable and tend to exhibit quasi-systematic groupings of thin, peritidal-prone parasequences near sequence boundaries or low-order sequences bounded by abrupt facies shifts. Individual parasequences in Devonian depositional systems tend to have self-similar vertical facies successions and may be highly discontinuous. Common themes include repeated, similar facies motif at parasequence scale and potential parasequence discontinuity at interwell scale. Labels as in Figure 2.

(Chow and James, 1987; Pratt and James, 1986) these depositional systems probably contain some of the most laterally continuous parasequences developed in geologic history. Facies bodies such as oolite shoals and cyanobacterial bioherms, however, are commonly discontinuous across a lateral distance of 1 m to several hundreds of meters.

The parasequence stacking styles described above suggest that the sequence stratigraphic units with the greatest degree of order are the second- or third-order sequences and system tracts. This impression is supported by the recognition that such low-order sequences can be correlated regionally and by systematic parasequence thickness and facies trends relative to sequence boundaries. Higher levels of randomness are found at the parasequence scale, especially in the TST, which commonly contains a greater diversity of parasequence types and greater levels of facies disorder as discussed above.

Silurian–Devonian

Silurian–Devonian carbonate platforms formed in a transitional climate on broad homoclinal ramps, flat-topped carbonate shelves, and isolated carbonate platforms in passive to convergent tectonic settings. Although there have been relatively few studies of the stratigraphic hierarchy of Silurian and Devonian carbonates, some generalizations can be made: (1) parasequence types are generally self-similar in facies succession, especially in Devonian depositional systems, (2) some depositional systems exhibit a quasi-systematic grouping of thin, peritidal-prone parasequences near sequence boundaries, and (3) some depositional systems lack such groupings and contain low-order sequences bounded by abrupt facies shifts (Fig. 20, Table 4).

The only detailed parasequence stacking-patterns analysis of Silurian carbonates comes from Harris and Sheehan's (1996) study of the Laketown Dolomite in Utah. These strata exhibit quasi-systematic changes in parasequence thickness and facies proportions that define low-order depositional sequences. Facies show nonstationary change in proportions, and their transitions are ordered as indicated by Markov scores. Thickness patterns occur in the field of randomness in the runs test (at the 95% C.I.) and in the ordered field of the Durbin-Watson test. Qualitatively, these relationships appear in stratigraphic sections as patterns of thickness and facies proportions that together define depositional sequences but tend to be more erratic than patterns of the Proterozoic–Lower Ordovician strata. Some of the depositional sequence boundaries are marked by an abrupt landward shift in facies.

Devonian platforms tend to contain shallowing-upward, tidal-flat parasequences that grade upward through the follow-

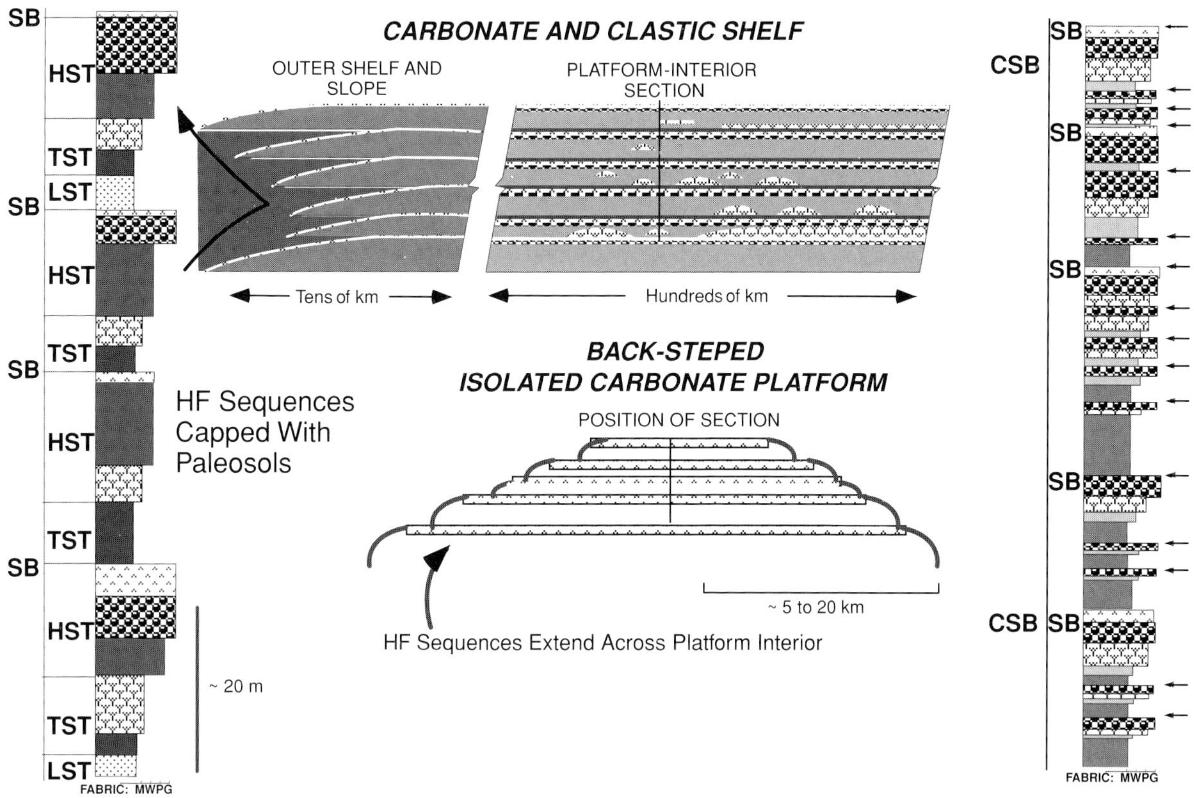

FIG. 21.—Stacking-pattern style of Pennsylvanian, Early Permian, Late Triassic, and Neogene carbonate platforms. These icehouse depositional systems contain high-frequency sequences that have high lateral continuity, are capped with subaerial exposure surfaces, and are bounded by abrupt non-Waltherian facies shifts. Vertical stacking patterns of high-frequency sequences (with high-amplitude, high-frequency, glacio-eustatic signatures) tend to be uniform (stationary) on the platform interior and lack long-term systematic changes in facies proportions and thickness. Labels as in Figure 2. The stacking pattern is interpreted to result from high-amplitude, high-frequency sea-level fluctuations masking the lower-frequency sea-level signature.

ing facies: (1) domal-stromatoporoid floatstone to rudstone, (2) *Amphipora* wackestone to grainstone, and (3) fenestral-laminite mudstone to wackestone. The style of parasequence development is surprisingly similar in vertical succession and among different localities and settings; however, subtidal coarsening-upward parasequences have also been described, and the caps of tidal-flat parasequences may contain a fenestral (or keystone vug) grainstone facies rather than a muddy fenestral laminite.

From a study of stacking patterns in the interior of the Judy Creek isolated buildup (Devonian of Canada), we found little facies or parasequence thickness differentiation relative to the fourth-order parasequence sets defined independently by step backs of the platform margins (Lehrmann, 1995). Facies proportions within the parasequences remained relatively constant, whereas parasequence thicknesses were relatively uniform or randomly distributed in relationship to the parasequence sets. Vertical stacking of parasequences with similar facies proportions and uniform to random thickness trends probably resulted from high levels of accommodation due to the Frasnian transgression and the highly aggradational to backstepping style of the Judy Creek buildup. The rate of sediment aggradation was high enough to allow accumulation of similar stacks of parasequences on the platform interior despite the transgressive pulses resulting in platform-margin step backs. An abrupt facies shift, just prior to drowning of the buildup, defines a regional third-order sequence boundary. Similar styles have been described elsewhere (Harvard and Oldershaw, 1976; Keith, 1970; Wong and Oldershaw, 1980).

Other studies of Devonian strata have found more systematic thickness and facies trends relative to low-order sequences (McLean and Mountjoy, 1994; Yang et al., 1995). The stacking patterns in the Lower Devonian Helderburg Group also contain somewhat systematic parasequence thinning and change in facies proportions relative to low-order sequence boundaries. Anderson and Goodwin (1990) referred to the parasequences as punctuated aggradational cycles and the low-order sequence boundaries as "regional cryptic unconformities." Yang et al. (1995) suggested the possibility of a Milankovitch-type signature in the Lost Burro Formation (Death Valley, California). Interwell discontinuity of individual parasequences in the Devonian strata of Judy Creek suggests that autocyclicity played a significant role in parasequence genesis (Lyndon Yose, pers. commun., 1994). McLean and Mountjoy (1994) traced individual parasequences along strike for 13 km in the Miette platform (Alberta, Canada).

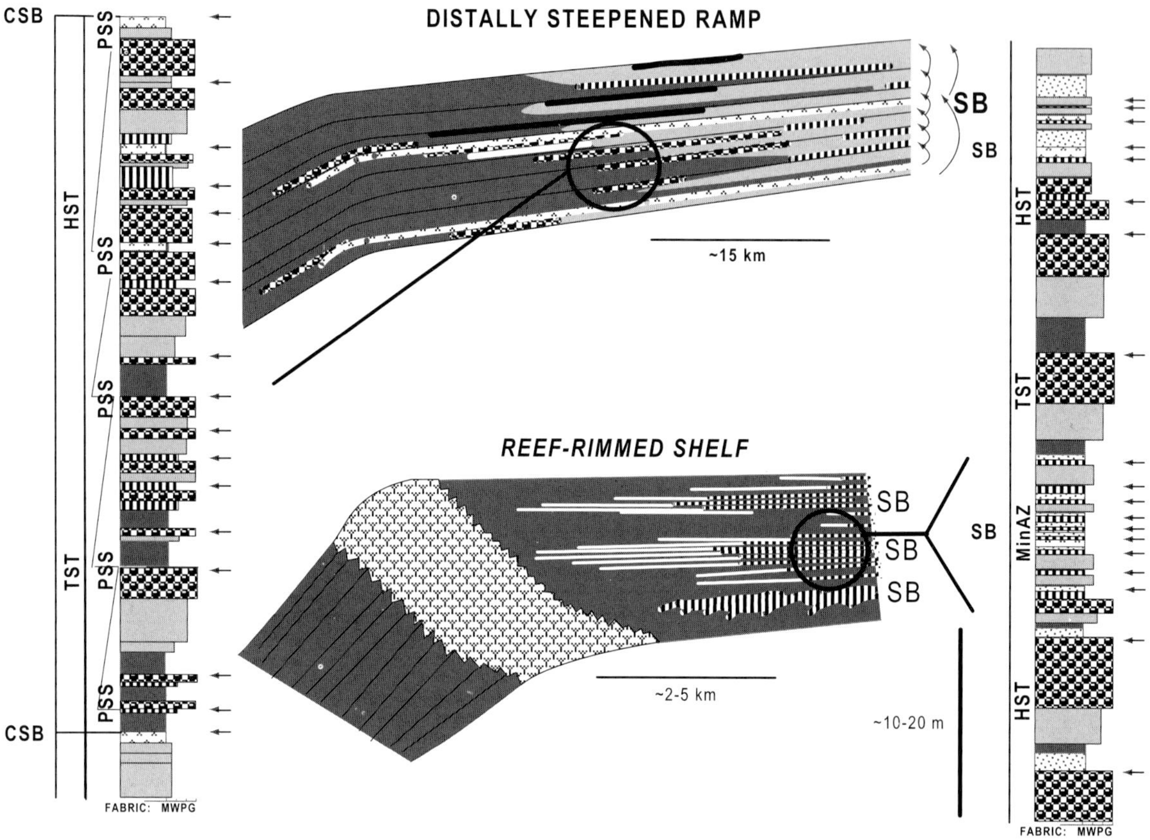

FIG. 22.—Stacking-pattern style of middle Permian–Late Permian carbonate platforms. These depositional systems tend to exhibit patterns that are transitional between the stationary icehouse style of the Pennsylvanian and the hierarchical greenhouse style of the Early to Middle Triassic. Middle Permian depositional systems may contain parasequence sets and high-frequency sequences with long-term nonstationary thickness and or facies patterns. Upper Permian depositional systems may contain quasi-systematic changes in parasequence thickness and facies proportions relative to lower-order sequences. Labels as in Figure 2

Pennsylvanian to Early Permian, Late Triassic, and Neogene

Pennsylvanian, Early Permian, and Neogene carbonate platforms formed during periods of Earth history with icehouse climates. These depositional systems formed on broad, flat-topped, rimmed shelves or isolated carbonate platforms in intracratonic, compressional basins, in vast epicontinental seas, or on passive continental margins. The stratigraphic style of these depositional systems is dominated by high-frequency sequences, generally ranging from 3 to 30 m in thickness, bounded at their tops by abrupt subaerial exposure of subtidal carbonate and commonly overlain by lowstand siliciclastics (Fig. 21, Table 4). The high-frequency sequences stack into composite sequences with chaotic vertical thicknesses and stationary facies proportions (cf. Grammer et al., 1996; Goldhammer et al., 1990, 1991b; Rankey et al., this volume).

The high-frequency sequences tend to be highly continuous, usually extending across the platform in which they occur (Wilson, 1967). Some of these sequences have been correlated farther than 600 km (Heckel, 1991, 1994). Although these sequences are laterally continuous, individual parasequences and facies within the sequences are commonly discontinuous (Watney et al., 1996; Rankey et al., this volume) owing to broad facies changes across the platform or to development of local bioherms or shoals (Goldhammer et al., 1991b; Watney et al., 1996). The great lateral continuity and high probability of glacio-eustatic forcing (Busch and Rollins, 1984; Heckel, 1986; Goldhammer et al., 1991b) suggest that the sequences have a relatively high level of order in a lateral sense. Low levels of stratigraphic order are exhibited in vertical thickness and facies ordering, however, primarily resulting from stratigraphic incompleteness or "upward cycle truncation" at sequence boundaries (Soreghan, 1994; Rankey et al., this volume). Facies arrangement within the sequences ranges from ordered, shallowing-upward patterns to nearly random distributions (Rankey et al., this volume). Randomness in vertical facies arrangement results from lateral development of facies mosaics and from facies truncation at sequence boundaries.

Pennsylvanian–Lower Permian strata tend to contain asymmetric to symmetric anoxic-shale–based or "–cored" high-frequency sequences (analogous to cyclothems) that shallow upward to open-marine boundstone or shoal-grainstone carbonate or, rarely, cryptalgal laminite. Subaerial exposure surfaces that cap the sequences tend to occur most frequently on subtidal shoal facies (e.g., Goldhammer et al., 1991b) but may occur on a wide variety of facies types (Gianniny and Simo, 1996). High-frequency sequences in Neogene strata contain diverse facies compositions including muddy, restricted carbonate; open-

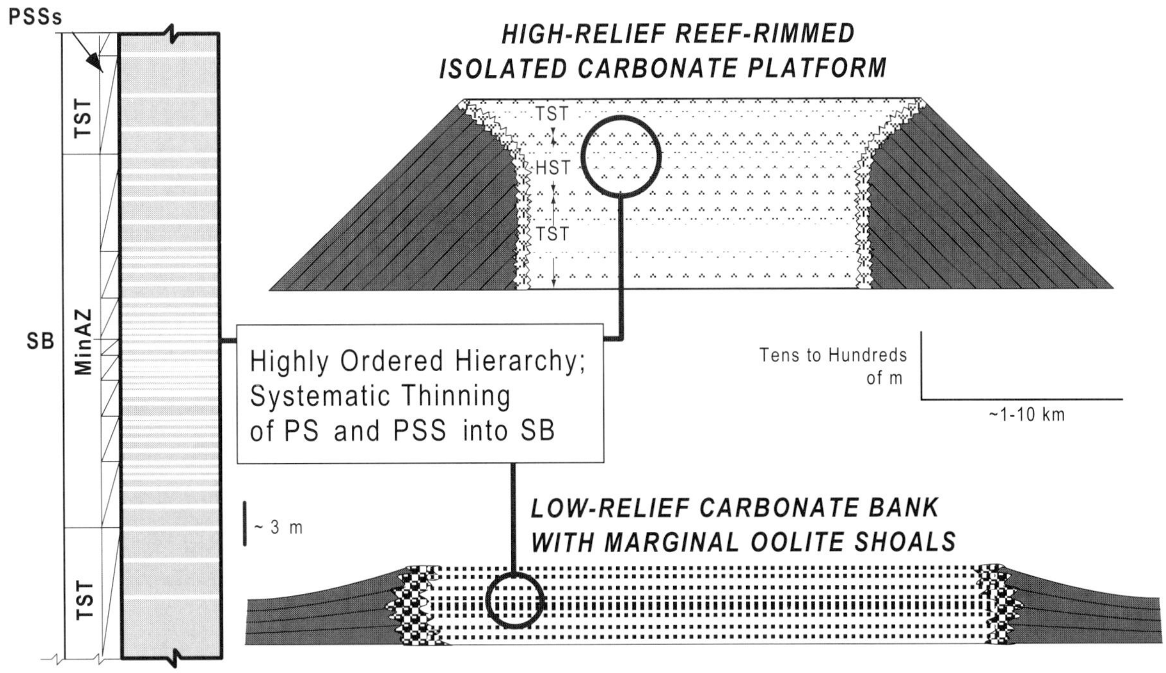

FIG. 23.—Schematic diagram illustrating stacking-pattern style of Early to Middle Triassic isolated carbonate platforms. These depositional systems contain systematic, nonstationary thickness and facies stacking patterns defining low-order sequences and may contain an extremely well developed stratigraphic hierarchy with a systematic bundling of parasequences into sets with a Milankovitch signature. The highly systematic style is interpreted to result from a greenhouse signature (low-amplitude sea-level fluctuations of high frequency superposed on higher-amplitude sea-level fluctuations of lower frequency), tuned subsidence, and inhibited autocyclicity. Labels as in Figure 2.

marine boundstone; and shoal grainstone with numerous subaerial exposure surfaces. Regional correlation indicates that each subaerial exposure surface is commonly underlain by a wide range in facies types resulting from across-platform facies changes (Enos and Perkins, 1977).

Although not strictly from an icehouse period (Fig. 1), we have included the Upper Triassic Dachsteinkalk Formation (Loferites) in this class because it has a stratigraphic style similar in several respects to the icehouse examples. Goldhammer et al. (1990) noted that the Dachsteinkalk Formation contains numerous soil horizons and a stratigraphic hierarchy less complete than other Triassic sections such as the Latemar Formation. Through forward stratigraphic modeling, they concluded that the lesser hierarchical development resulted from higher-amplitude sea-level fluctuations and thus a record of "missed beats" in sea-level changes. When the Dachsteinkalk section is plotted at the same scale and compared with other icehouse depositional systems, the "cycle bundles" or parasequence sets closely resemble high-frequency sequences in that they exhibit chaotic vertical thickness trends, long-term stationarity in facies proportions, and numerous subaerial exposure surfaces.

Middle to Late Permian

Stratigraphic depositional systems that make up this secular class are best studied in the Permian basin of Texas and New Mexico. These depositional systems formed in a transitional climate on ramps rimmed with grainstone shoals or on flat-topped shelves with steep reef margins ranging from 5 to more than 50 km in width. The stratigraphic style is transitional between the chaotic vertical partitioning of high-frequency sequences of the Pennsylvanian and the systematically stacked Milankovitch-type parasequences of the Early to Middle Triassic (Fig. 22, Table 4). Middle Permian examples usually contain either high-frequency sequences with stacking patterns similar to Pennsylvanian examples or a combination of high-frequency sequences and parasequence sets. These may be grouped into third-order sequences or composite sequences bounded by major subaerial unconformities. Parasequence sets, high-frequency sequences, and the larger composite sequences are defined dominantly by facies trends and unconformities; there generally is little trend in thickness stacking to aid in their definition. In the Upper Permian, systematic parasequence thickness trends become important in differentiating third-order depositional sequences. Parasequences systematically thin and become progressively peritidal-prone upward into sequence boundaries; parasequences progressively thicken and become progressively subtidal-prone upward away from the sequence boundaries (Lehrmann et al., 1997).

Cyclic Permian strata exhibit a wide range of levels of stratigraphic order. Lower Permian sequences with chaotic stratigraphic partitioning probably have levels of order similar to the Pennsylvanian examples. Middle Permian examples with parasequences grouped into parasequence sets and sequences based on transgressive-regressive facies trends, flooding surfaces, and unconformities appear to have a relatively low degree of stratigraphic order in the sense that there are no systematic trends of parasequence thickness or grouping of similar numbers of para-

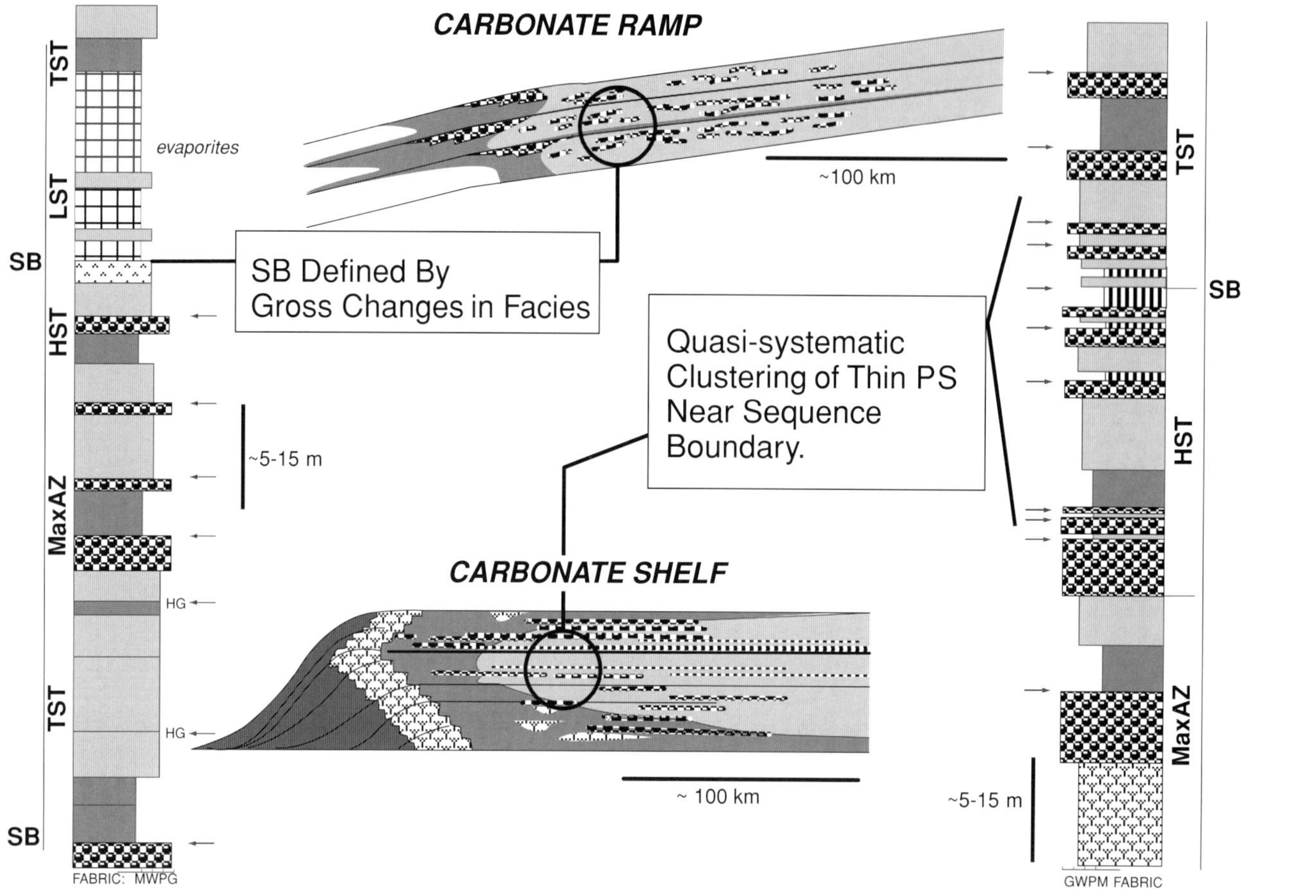

FIG. 24.—Stacking-pattern style of Jurassic and Cretaceous carbonate platforms. Jurassic–Cretaceous depositional systems exhibit nonstationary facies patterns in which changes in gross facies assemblages define depositional sequences. Thickness and facies patterns tend to exhibit high levels of randomness, and parasequences tend to be highly discontinuous because of autocyclicity and facies amalgamation. Some depositional systems exhibit quasi-systematic changes in parasequence thickness and facies proportions that define depositional sequences. Labels as in Figure 2.

sequences relative to parasequence sets or sequence boundaries. These successions are nonrandom as demonstrated by the two-dimensional geometries and the possibility of defining a stratal hierarchy based on facies patterns (e.g., Kerans and Fitchen, 1995). In practice, the lack of systematic parasequence thickness trends simply means that one must rely more on facies patterns for identification and correlation. For the Upper Permian, a fairly high degree of order based on systematic patterns in parasequence thickness and facies stacking is exhibited by the third-order depositional sequences (Lehrmann et al., 1997).

Lateral parasequence continuity in middle to Upper Permian strata is strongly dependent on the slope of the shelf profile and shelf position. Parasequence sets and sequences tend to be highly continuous (as far as 50 km in dip direction) on the interiors of broad ramp profiles and in the flat-topped interiors of rimmed shelves (Kerans and Fitchen, 1995). Individual parasequences are also generally continuous in the interiors, although occasionally subtle topography or facies amalgamation results in discontinuity of parasequences. Lateral facies continuity within parasequences and parasequence sets also depends largely on platform position. Facies bodies tend to have greatest lateral continuity in the inner shelf (commonly >17 km; Cowan and Harris, 1986) and the lowest continuity where shoal bodies are developed in the shelf crest and shelf margin (Harris et al., 1993). Near ramp or platform margins, parasequences commonly occur in discontinuous, shingled clinoform strata (Rankey and Lehrmann, 1996; Sonnenfeld and Cross, 1993).

Early to Middle Triassic

The Early to Middle Triassic depositional systems included in this class formed in a greenhouse climate on relatively small isolated carbonate platforms, 5–20 km across, developed on structural highs in extensional rift basins. In these systems, fifth-order parasequences are systematically stacked into thinning and thickening packages with similar numbers of parasequences forming fourth-order parasequence sets, and with long-term vertical thickness and facies trends defining third-order depositional sequences (Fig. 23, Table 4). Spectral analysis reveals thickness distributions consistent with a hierarchy of Milankovitch-type, eustatic sea-level fluctuations (Goldhammer et al., 1993a). Extremely high levels of vertical order occur at the parasequence and parasequence-set scale especially in the early HSTs where complete 5 to 1 or 10 to 1 Milankovitch-type rhythms are recorded in parasequence sets. Decreased levels of order at the parasequence-set scale occur in the TSTs and late HSTs in which the record of high-frequency sea-level fluctuations is incomplete (missed beats) owing to facies amalgamation or prolonged exposure. Like other greenhouse examples, the superposition of sea-level fluctuations with higher amplitudes at progressively lower frequencies sets up the potential for recording a full hierarchy. Autocyclicity was inhibited by the small size of the platforms, and the lack of sufficient calcareous mud required for the development of prograding tidal flats (Lehrmann et al., 1998; Lehrmann and Goldhammer, 1998). Facies, parasequence sets, and sequences are continuous across the interior of isolated platforms (5–20 km). Locally discontinuous bioherms developed near the bases of parasequences in the Dawen Formation (Lehrmann, 1993). It should be noted that these highly ordered records of sea-level oscillations may be extremely rare for shallow-marine carbonate platforms. The record of Milankovitch-type frequencies was permitted by a combination of ideal sea-level amplitudes, subsidence and accumulation "in tune" with the sea-level signature, and inhibited autocyclicity.

Jurassic and Cretaceous

Shallow-marine carbonates of Jurassic and Cretaceous age formed in a greenhouse climate on broad, low-gradient carbonate ramps, broad flat-topped rimmed shelves, and isolated platforms. These platforms formed in extensional and passive continental margins resulting from the breakup of Pangea. These depositional systems are characterized by second- or third-order depositional sequences 30 to 300 m thick differentiated mainly on the basis of gross changes in facies assemblages. Peritidal and subtidal parasequences (0.3 to 30 m thick) are well developed in some settings or may be poorly developed or absent because of facies amalgamation in other settings (Fig. 24). Low-order sequences contain gross facies shifts from open-marine shallow-subtidal skeletal-peloidal wackestone-packstones and bioherms in the TSTs to shallow-subtidal oolitic-peloid, skeletal grainstones in the HSTs and finally peritidal carbonates, paleosols, and/or sabkha evaporites in the late HSTs adjacent to sequence boundaries. In contrast to Proterozoic–Ordovician ramp and shelf deposits, Jurassic and Cretaceous deposits generally lack systematic trends in parasequence thickness and facies proportions relative to sequence boundaries. An overall tendency for parasequences to be thinner adjacent to sequence boundaries has been demonstrated in some cases (Goldhammer et al., 1991a), but most sections contain a sporadic distribution of thicknesses with both thin and thick parasequences occurring in the TSTs and HSTs. Parasequence types are also highly variable and commonly are sporadically distributed. Parasequence types typically include subtidal, skeletal-peloidal packstone to grainstone or hardground-capped subtidal parasequences and fenestral-laminite– or evaporite–capped peritidal parasequences. Some examples contain a more or less random intercalation of different subtidal facies without consistently repeated motifs probably resulting from facies amalgamation and autocyclic deposition (e.g., Prather, 1992).

Overall, these depositional systems appear to contain higher levels of randomness than the other stratigraphic classes. High levels of randomness occur at the facies and parasequence scale, as evidenced by random facies successions and sporadic distribution of parasequence thicknesses. Parasequences are probably highly discontinuous because of autocyclicity (e.g., see cross section in Johnson, 1991). The second- or third-order sequences with the greatest continuity are the units with the highest degree of order. The sequences can be correlated regionally, and in some depositional systems may be the only definable sequence stratigraphic unit.

GUIDELINES FOR STACKING-PATTERNS ANALYSIS

The goal of stacking-patterns analysis is to use vertical stratigraphic patterns from isolated outcrops or subsurface cores and logs to identify and correlate a hierarchy of sequence stratigraphic units for the construction of a high-resolution chronostratigraphic framework. From our discussions of differences in the levels of stratigraphic order and style, it is obvious that different stratal attributes must be employed and different scales of resolution will be possible depending on the depositional system studied. For example, if a particular depositional system lacks systematic vertical trends in parasequence thicknesses, then thickness trends should not be used to interpret parasequence sets and depositional sequences. Within the same depositional system, long-term systematic changes in facies proportions may prove more useful for identifying depositional sequences. In one depositional system (e.g., in the Pennsylvanian), a high-resolution framework may be obtained with a high level of confidence in the correlation of high-frequency sequences or parasequence sets. In another depositional system (e.g., in the Jurassic), attempts to correlate individual parasequences may be misleading owing to random facies patterns resulting from autocyclicity.

A practical approach to developing a high-resolution sequence stratigraphic framework is to (1) identify the stratigraphic style to determine the stratal units that are most likely to be correlated confidently, (2) determine the highest-resolution chronostratigraphic unit permitted by the stratal style and log response, (3) correlate units within a nested hierarchy, recognizing a higher level of confidence in larger-scale units such as sequences or composite sequences that have regional extent, and (4) utilize whatever two- or three-dimensional information is available such as outcrop mapping or seismic data to constrain discontinuities and shingled geometries.

CONCLUSIONS

Stacking-patterns analysis is a powerful tool for recognizing and correlating sequence stratigraphic units in the construction of high-resolution chronostratigraphic frameworks. The technique is especially important in hydrocarbon exploitation where the correlation of fine-scale chronostratigraphic units is needed to develop an accurate reservoir zonation and where vertical stratigraphic patterns in core and well logs are commonly the only available data. In outcrop studies, the technique has also facilitated increasingly finer-scale chronostratigraphic analysis yielding better understanding of the dynamics of platform sedimentation and the evolution of depositional systems (e.g.,

Goldhammer et al., 1993b; Kerans and Fitchen, 1995; Rankey and Lehrmann, 1996; Sonnenfeld, 1996).

Although the validity of the fundamental concepts of carbonate cyclicity and stratigraphic hierarchy have been challenged on the bases of randomness found in stratal thickness or facies succession (e.g., Drummond and Wilkinson, 1993; Wilkinson et al., 1997), all of the platform-carbonate depositional systems we have studied exhibit stratigraphic order at one scale or another. Previous studies of the levels of order in shallow-marine carbonates (e.g., Drummond and Wilkinson, 1993; Sadler et al., 1993; Wilkinson et al., 1997) have not attempted to identify populations of behavior on the basis of age. Shallow-marine carbonate depositional systems exhibit a spectrum of behaviors from nearly random to deterministic. Order and randomness reside in different stratal attributes and at different scales and can be interpreted in the context of secular classes of stratigraphic style. On the basis of an understanding of secular changes in stratigraphic style, the stacking-patterns approach should be modified accordingly to utilize the stratal attributes with the greatest order and at the appropriate scale so that a chronostratigraphic zonation may confidently be constructed.

Secular variations in stratigraphic order and hierarchy recognized in this study are interpreted to result from differences in the nature of sea-level fluctuations through icehouse and greenhouse climates, platform morphology, and tectonic settings. Sea-level fluctuations influence the level of order and hierarchical style in a delicate balance between lower amplitudes (superposed on higher amplitudes of lower frequency)—that, in greenhouse settings, permit the record of a full hierarchy, but also enhance autocyclicity—and high amplitudes (of high frequency)—that, in icehouse settings, impart a strong allocyclicity, but mask the overall hierarchy.

ACKNOWLEDGMENTS

This paper is an outgrowth of work begun at Exxon Production Research. We thank Exxon for sponsoring the data-compilation phase and early phases of qualitative stratigraphic analysis of this project. Statistical analysis and the final stages of synthesis were conducted at the University of Wisconsin–Oshkosh. We thank the University of Wisconsin–Oshkosh for continuing support of this project. B. H. Wilkinson and E. C. Rankey are thanked for reviewing early versions of this manuscript. We are grateful to P. M. Harris, R. G. Loucks, and D. A. Osleger for comments and suggestions that substantially improved the manuscript.

REFERENCES

ANDERSON, E. J., AND GOODWIN, P. W., 1990, The significance of metre scale allocycles in the quest for a fundamental stratigraphic unit: Journal of the Geological Society of London, v. 147, p. 507–518.

BARNABY, R. J., AND WARD, W. B., 1995, Sequence stratigraphic framework, high frequency cyclicity and three dimensional heterogeneity: Grayburg Formation, Brokeoff Mountains, New Mexico, *in* Pause, P. H. and Candelaria, M. P., eds., Carbonate Facies and Sequence Stratigraphy: Practical Applications of Carbonate Models: Midland, Permian Basin Section–SEPM (Society of Sedimentary Geology) Publication No. 95-5, p. 37–50.

BATHURST, R. G. C., 1975, Carbonate Sediments and Their Diagenesis: Amsterdam, Elsevier, 658 p.

BERNER, R. A., 1991, A model for atmospheric CO2 over Phanerozoic time: American Journal of Science, v. 291, p. 339–376.

BORER, J. M., AND HARRIS, P. M., 1995, Computer simulation of the Yates Formation (Permian, Delaware Basin)—Sequence stratigraphy and shelf-to-basin correlation implications, *in* Martin, R. L., ed., In Search of New Permian Basin Oil and Gas Fields: Using Today's Technologies and Tomorrow's Ideas for Exploration, Development, and 3D Seismic in a Mature Basin: Midland, West Texas Geological Society Symposium Publication No. 95-98, p. 111–132.

BROWN, L. F., AND FISHER, W. L., 1977, Seismic-stratigraphic interpretation of depositional systems: Examples from Brazil rift and pull-apart basins, *in* Payton, C. E., ed., Seismic Stratigraphy—Applications to Hydrocarbon Exploration: Tulsa, American Association of Petroleum Geologists Memoir 26, p. 213–248.

BROWN, L. F., SOLIS-IRIARTE, R. F., AND JOHNS, D. A., 1990, Regional systems tracts, paleogeography, and sequence stratigraphy, Upper Pennsylvanian and Lower Permian strata, north- and west-central Texas: Austin, University of Texas, Bureau of Economic Geology, Report of Investigations No. 197, 116 p.

BUSCH, R. M., AND ROLLINS, H. B., 1984, Correlation of Carboniferous strata using a hierarchy of transgressive-regressive units: Geology, v. 12, p. 471–474.

CARR, T. R., 1982, Log-linear models, Markov chains and cyclic sedimentation: Journal of Sedimentary Petrology, v. 52, p. 905–912.

CHOW, N., AND JAMES, N. P., 1987, Cambrian grand cycles: A northern Appalachian perspective: Geological Society of America Bulletin, v. 98, p. 418–429.

CHRISTIE-BLICK, N., 1991, Onlap, offlap, and the origin of unconformity-bounded depositional sequences: Marine Geology, v. 97, p. 35–56.

CHRISTIE-BLICK, N., AND DRISCOLL, N. W., 1995, Sequence stratigraphy: Annual Review of Earth and Planetary Sciences, v. 23, p. 451–478.

CLOUGH, J. G., AND GOLDHAMMER, R. K., 1992, Third order vertical variation in parasequence character of the lower Gray Craggy Member, Katakturuk Dolomite (Proterozoic), northeastern Brooks Range, Alaska: Tulsa, American Association of Petroleum Geologists, Annual Convention, Abstracts, p. 21.

COWAN, P. E., AND HARRIS, P. M., 1986, Porosity distribution in San Andres Formation (Permian), Cochran and Hockley Counties, Texas: American Association of Petroleum Geologists Bulletin, v. 70, p. 888–897.

CREVELLO, P. D., WILSON, J. L., SARG, J. F., AND READ, J. F., eds., 1989, Controls on Carbonate Platform and Basin Development: Tulsa, Society of Economic Paleontologists and Mineralogists, Special Publication No. 44, 405 p.

CROWLEY, T. J., AND BAUM, S. K., 1991, Toward reconciliation of Late Ordovician glaciation with very high CO2 levels: Journal of Geophysical Research, v. 96, p. 22597–22610.

DAVIS, J. C., 1986, Statistics and Data Analysis in Geology: New York, John Wiley and Sons, 646 p.

DEMICCO, R. V., 1981, Comparative sedimentology of an ancient carbonate platform: The Conococheague Limestone of the central Appalachians: Unpublished Ph.D. Dissertation, Johns Hopkins University, Baltimore, 333 p.

DICKSON, J. A. D., AND SALLER, A. H., 1995, Identification of subaerial exposure surfaces and porosity preservation in Pennsylvanian and Lower Permian shelf limestones, eastern Central Basin Platform, Texas, *in* Budd, D. A., Saller, A. H., and Harris, P. M., eds., Unconformities and Porosity in Carbonate Strata: Tulsa, American Association of Petroleum Geologists Memoir 63, p. 239–258.

DONN, W. L., AND SHAW, D. M., 1977, Model of climate evolution based on continental drift and polar wandering: Geological Society of America Bulletin, v. 88, p. 390–396.

DRUMMOND, C. N., AND WILKINSON, B. H., 1993, On the use of cycle thickness diagrams as records of long-term sea level change during accumulation of carbonate sequences: Journal of Geology, v. 101, p. 687–702.

DUNN, P. A., 1991, Diagenesis and cyclostratigraphy: An example from the Middle Triassic Latemar platform, Dolomites Mountains, northern Italy: Unpublished Ph.D. Dissertation, Johns Hopkins University, Baltimore, 865 p.

ENOS, P., 1991, Sedimentary parameters for computer modeling, *in* Franseen, E. K., Watney, W. L., Kendall, C. G. St. C., and Ross, W., eds., Sedimentary modeling: Computer simulations and methods for improved parameter definition: Lawrence, Kansas Geological Survey Bulletin, no. 233, p. 63–100.

ENOS, P., AND PERKINS, R. D., 1977, Quaternary sedimentation in South Florida:

Boulder, Geological Society of America Memoir 147, 198 p.

FISCHER, A. G., 1986, Climatic rhythms recorded in strata: Annual Review of Earth and Planetary Sciences, v. 14, p. 351–376.

FITCHEN, W. M., 1997, Carbonate sequence stratigraphy and its application to hydrocarbon exploration and reservoir development, *in* Ibrahim, P. and Marfurt, K. J., eds., Carbonate Seismology: Tulsa, Society of Exploration Geophysicists, Geophysical Developments Series No. 6, p. 121–178.

FRAKES, L. A., AND FRANCIS, J. E., 1988, A guide to Phanerozoic cold polar climates from high-latitude ice-rafting in the Cretaceous: Nature, v. 333, p. 547–549.

FRAKES, L. A., FRANCIS, J. E., AND SYKTUS, J. I., 1992, Climate models of the Phanerozoic: The history of Earth's climate over the past 600 million years: Cambridge, Cambridge University Press, 274 p.

FRANSEEN, E. K., GOLDSTEIN, R. H., AND WHITESEL, T. E., 1993, Sequence stratigraphy of Miocene carbonate complexes, Las Negras area, southeastern Spain: Implications for quantification of changes in relative sea-level, *in* Loucks, R. G. and Sarg, J. F., eds., Carbonate Sequence Stratigraphy: Tulsa, American Association of Petroleum Geologists Memoir 57, p. 409–434.

FRENCH, J. A., AND HECKEL, P. H., 1994, Probable high amplitude glacial-eustatic origin for Pennsylvanian cyclothems requires modification of applied sequence stratigraphic classification and terminology: Tulsa, American Association of Petroleum Geologists, Annual Convention, Abstracts, p. 152.

GALLOWAY, W. E., 1989, Genetic stratigraphic sequences in basin analysis: I. Architecture and genesis of flooding-surface bounded depositional units: American Association of Petroleum Geologists Bulletin, v. 73, p. 125–142.

GARBER, R. A., GROVER, G. A., AND HARRIS, P. M., 1989, Geology of the Capitan shelf margin—Subsurface data from the northern Delaware basin, *in* Harris, P. M. and Grover, G. A., eds., Subsurface and Outcrop Examination of the Capitan Shelf Margin, Northern Delaware Basin: Tulsa, Society of Economic Paleontologists and Mineralogists, Core Workshop No. 13, p. 3–372.

GIANNINY, G. L., 1995, Facies and sequence stratigraphic evolution of the mixed carbonate-siliciclastic strata, lower Desmoinesian, southwest Paradox basin: Unpublished Ph.D. Dissertation, University of Wisconsin, Madison, 368 p.

GIANNINY, G. L., AND SIMO, J. A., 1996, Implications of unfilled accommodation space for sequence stratigraphy on mixed carbonate-siliciclastic platforms—An example from the lower Desmoinesian (Middle Pennsylvanian), southwestern Paradox basin, *in* Longman, M. W. and Sonnenfeld, M. D., eds., Paleozoic systems of the Rocky Mountain region: Rocky Mountain Section–SEPM (Society for Sedimentary Geology), p. 213–234.

GINGERICH, P. D., 1969, Markov analysis of cyclic alluvial sediments: Journal of Sedimentary Petrology, v. 39, p. 330–332.

GINSBURG, R. N., 1971, Landward movement of carbonate mud: New model for regressive cycles in carbonates (abs.): American Association of Petroleum Geologists Bulletin, v. 55, p. 340.

GOLDHAMMER, R. K., 1994, Sequence stratigraphic architecture and cycle stacking patterns of platform carbonates—An evaluation of autocyclic, deterministic and stochastic controls: Tulsa, American Association of Petroleum Geologists 1994–1995 Distinguished Lecture Tours: American Association of Petroleum Geologists Bulletin, v. 78, p. 1808.

GOLDHAMMER, R. K., AND LEHRMANN, P. J., 1996, Chaos in El Paso?, *in* Stoudt, E. L., ed., Precambrian–Devonian Geology of the Franklin Mountains, West Texas—Analogues for Exploration and Production in Ordovician and Silurian Karsted Reservoirs in the Permian Basin: Midland, West Texas Geological Society Publication No. 96-100, Annual Field Trip Guide Book, p. 125–140.

GOLDHAMMER, R. K., AND MCDONALD, K. A., 1996, High resolution reservoir architecture of Late Jurassic Haynesville ramp carbonates in the Gladewater field, East Texas salt basin: Tulsa, American Association of Petroleum Geologists, Annual Convention, Abstracts, v. 5, p. 54.

GOLDHAMMER, R. K., DUNN, P. A., AND HARDIE, L. A., 1987, High-frequency glacio-eustatic sea-level oscillations with Milankovitch characteristics recorded in Middle Triassic platform carbonates in northern Italy: American Journal of Science, v. 287, p. 853–892.

GOLDHAMMER, R. K., DUNN, P. A., AND HARDIE, 1990, Depositional cycles, composite sea-level changes, cycle stacking patterns, and the hierarchy of stratigraphic forcing: Examples from Alpine Triassic platform carbonates: Geological Society of America Bulletin, v. 102, p. 535–562.

GOLDHAMMER R. K., LEHMANN, P. J., TODD, R. G., WILSON, J. L., WARD, W. C., AND JOHNSON, C. R., 1991a, Sequence stratigraphy and cyclostratigraphy of the Mesozoic of the Sierra Madre Oriental, northeast Mexico, a field guidebook: Austin, Gulf Coast Section, Society of Economic Paleontologists and Mineralogists, 85 p.

GOLDHAMMER, R. K., OSWALD, E. J., AND DUNN, P. A., 1991b, Hierarchy of stratigraphic forcing: Example from Middle Pennsylvanian shelf carbonates of the Paradox basin, *in* Franseen, E. K., Watney, W. L., Kendall, C. G. St. C., and Ross, W., eds., Sedimentary modeling: Computer simulations and methods for improved parameter definition: Lawrence, Kansas Geological Survey Bulletin, no. 233, p. 361–413.

GOLDHAMMER, R. K., DUNN, P. A., AND LEHMANN, P. J., 1992, Cycle and sequence dynamics of Lower Ordovician platform carbonates of the Great American Bank (USA): Constraints from inverse and two-dimensional forward modeling: Tulsa, American Association of Petroleum Geologists, Annual Convention, Abstracts, p. 46–47.

GOLDHAMMER, R. K., HARRIS, M. T., DUNN, P. A., AND HARDIE, L. A., 1993a, Sequence stratigraphy and systems tract development of the Latemar platform, Middle Triassic of the Dolomites (Northern Italy): Outcrop calibration keyed by cycle stacking patterns: Tulsa, American Association of Petroleum Geologists Memoir 57, p. 353–388.

GOLDHAMMER, R. K., LEHMANN, P. J., AND DUNN, P. A., 1993b, The origin of high frequency platform carbonate cycles and third-order sequences (Lower Ordovician El Paso Group, West Texas): Constraints from outcrop data and stratigraphic modeling: Journal of Sedimentary Research, v. 63, p. 318–359.

GOLDSTEIN, R. H., 1988, Paleosols of Late Pennsylvanian cyclic strata, New Mexico: Sedimentology, v. 35, p. 777–803.

GOODWIN, P. W., AND ANDERSON, E. J., 1985, Punctuated aggradational cycles: A general hypothesis of episodic stratigraphic accumulation: Journal of Geology, v. 93, p. 515–533.

GRAMMER, G. M., EBERLI, H. P., VAN BUCHEM, F. S., STEVENSON, G. M., AND HOMEWOOD, P., 1996, Application of high-resolution sequence stratigraphy to evaluate lateral variability in outcrop and subsurface—Desert Creek and Ismay intervals, Paradox basin, *in* Longman, M. W. and Sonnenfeld, M. D., eds., Paleozoic systems of the Rocky Mountain region: Tulsa, Rocky Mountain Section–SEPM (Society for Sedimentary Geology), p. 235–266.

GROTZINGER, J. P, 1985, Evolution of Early Proterozoic passive-margin carbonate platform, Rocknest Formation, Wompay Orogen, Northwest Territories, Canada: Unpublished Ph.D. Dissertation, Virginia Polytechnic Institute and State University, Blacksburg, 303 p.

HALLOCK, P., AND SCHLAGER, W., 1986, Nutrient excess and the demise of coral reefs and carbonate platforms: Palaios, v. 1, p. 1156–1166.

HAMMES, U., 1992, Sedimentation patterns, sequence stratigraphy, cyclicity, and diagenesis of early Oligocene Suwannee Formation, southwest Florida: Unpublished Ph.D. Dissertation, University of Colorado, Boulder, 393 p.

HANFORD, R. C., 1988, Review of carbonate sand-belt deposition of ooid grainstones and application to Mississippian reservoir, Damme Field, southwestern Kansas: American Association of Petroleum Geologists Bulletin, v. 72, p. 1184–1199.

HARDIE, L. A., AND SHINN, E. A., 1986, Carbonate depositional environments, modern and ancient: Part 3. Tidal flats: Colorado School of Mines Quarterly, v. 81, 74 p.

HARRIS, M. T., AND SHEEHAN, P. M., 1996, Upper Ordovician–Lower Silurian depositional sequences determined from middle shelf sections, Barn Hills and Lakeside Mountains, eastern Great Basin, *in* Witzke, B. J., Ludvigson, G. A., and Day, J., eds., Paleozoic Stratigraphy: Views from the North American Craton: Boulder, Geological Society of America Special Paper 306, p. 161–176.

HARRIS, P. M., KERANS, C., AND BEBOUT, D. G., 1993, Ancient outcrop and modern examples of carbonate platform cycles—Implications for subsurface correlation and understanding reservoir heterogeneity, *in* Loucks, R. G. and Sarg, J. F., eds., Carbonate Sequence Stratigraphy: Tulsa, American Association of Petroleum Geologists Memoir 57, p. 475–492.

HARVARD, C. J., AND OLDERSHAW, A. E., 1976, Early diagenesis and back reef sedimentary cycles, Snipe Lake reef complex, Alberta: Canadian Petroleum Geology Bulletin, v. 24, p. 27–69.

HATTORI, I., 1976, Entropy in Markov chains and discrimination of cyclical patterns in lithologic successions: Journal of Mathematical Geology, v. 8, p. 477–497.

HECKEL, P. H., 1986, Sea-level curve for Pennsylvanian eustatic marine transgressive-regressive cycles along the midcontinent outcrop belt, North

America: Geology, v. 14, p. 330–334.

HECKEL, P. H., 1991, Thin widespread Pennsylvanian black shales of Midcontinent North America: A record of cyclic succession of widespread pycnoclines in a fluctuating epieric sea, *in* Tyson, R. V. and Pearson, T. H., eds., Modern and Ancient Continental Shelf Anoxia: London, Geological Society of London Special Publication No. 58, p. 259–273.

HECKEL, P. H., 1994, Evaluation of evidence for glacio-eustatic control over marine Pennsylvanian cyclothems in North America and consideration of tectonic events, *in* Dennison, J. M. and Ettensohn, F. R., eds., Tectonic and eustatic controls on sedimentary cycles: Tulsa, SEPM (Society for Sedimentary Geology) Concepts in Sedimentology and Paleontology, v. 4, p. 65–87.

JACOBS, D. K., AND SAHAGIAN, D. L., 1993, Climate induced fluctuations in sea level during non-glacial times: Nature, v. 361, p. 710–712.

JAMES, N. P., 1979, Shallowing-upward sequences in carbonates, *in* Walker, R. G., ed., Facies Models: Ottawa, Geoscience Canada Reprint Series 1, p. 109–119.

JOHNSON, C. R., 1991, Depositional cycles of the Zuluoaga (Oxfordian to Kimmerigian ?) Formation, Sierra del Bunelos, Coahuila, Mexico: Unpublished Ph.D. Thesis, University of New Orleans, New Orleans, 242 p.

JOHNSON, J. G., KLAPPER, G., AND SANDBERG, C. A., 1985, Devonian eustatic fluctuations in Euramerica: Geological Society of America Bulletin, v. 96, p. 567-587.

KEITH, J. W., 1970, Tectonic control on Devonian reef sedimentation (abs.): American Association of Petroleum Geologists Bulletin, v. 54, p. 854.

KENDALL, C. G. ST. C., AND SCHLAGER, W., 1981, Carbonates and relative changes in sea level: Marine Geology, v. 44, p. 181–212.

KERANS, C., 1995, Use of one- and two-dimensional cycle analysis in establishing high-frequency sequence frameworks, *in* Read, J. F., Kerans, C., and Weber, L. J., sponsors, Milankovitchian sea level changes, cycles and reservoirs in green-house and ice-house worlds: Tulsa, SEPM (Society for Sedimentary Geology) Short Course Notes No. 35, part 2, p. 1–20.

KERANS, C., 1997, Stratigraphic architecture of a greenhouse carbonate ramp, Cretaceous (Albian–Cenomanian) of Pecos River canyon, Texas (abs.): Tulsa, American Association of Petroleum Geologists, 1997 Annual Meeting, Abstracts, v. 6, p. 59–60.

KERANS, C., AND FITCHEN, W. M., 1995, Sequence stratigraphy and facies architecture of a carbonate ramp system: San Andres Formation of Algerita Escarpment and western Guadalupe Mountains, West Texas and New Mexico: Austin, University of Texas, Bureau of Economic Geology, Report of Investigations No. 235, 86 p.

KERANS, C., AND NANCE, H. S., 1991, High frequency cyclicity and regional depositional patterns of the Grayburg Formation, Guadalupe Mountains, New Mexico: Midland, Permian Basin Section–SEPM (Society for Sedimentary Geology) Publication No. 91-32, p. 53–69.

KERANS, C., AND TINKER, S., 1997, Sequence stratigraphy and characterization of carbonate reservoirs: Tulsa, SEPM (Society for Sedimentary Geology) Short Course Notes No. 40, 129 p.

KERANS, C., FITCHEN, W. M., GARDNER, M. H., SONNENFELD, M. D., TINKER, S. W., AND WARDLAW, B. R., 1992, Styles of sequence development within uppermost Leonardian through Guadalupian strata of the Guadalupe Mountains, Texas and New Mexico, *in* Mruk, D. H. and Curran, B. C., eds., Permian Basin Exploration and Production Strategies: Midland, West Texas Geological Society Publication No. 92-91, p. 1–6.

KIEVMAN, C. M., 1996, Sea-level effects on carbonate platform evolution: Plio–Pleistocene, northwestern Great Bahama Bank: Unpublished M.S. Thesis, University of Miami, Miami, 246 p.

KOERSCHNER, W. F., III AND READ, J. F., 1989, Field and modeling studies of Cambrian carbonate cycles, Virginia Appalachians: Journal of Sedimentary Petrology, v. 59, p. 654–687.

KRUMBEIN, W. C., AND DACEY, M. F., 1969, Markov chains and embedded Markov chains in geology: International Association of Mathematical Geology Journal, v. 1, p. 79–96.

KRUSKKAL, W. H., AND TANUR, J. M., eds., 1978, International Encyclopedia of Statistics: The Free Press, New York, 1350 p.

LEES, A., AND BULLER, A. T., 1972, Modern temperate-water and warm-water shelf carbonate sediments contrasted: Marine Geology, v. 13, p. 67–73.

LEHMANN, C., OSLEGER, D. A., AND MONTANEZ, I. P., 1995, Depositional facies, sequence stratigraphy and drowning of the Coahuila carbonate ramp (Albian), northeastern Mexico: Boulder, Geological Society of America Abstracts with Programs, v. 27, no. 6, p. 333.

LEHRMANN, D. J., 1993, The Great Bank of Guizhou: Birth, evolution and death of Triassic isolated carbonate platform, Guizhou Province, South China: Unpublished Ph.D. dissertation, University of Kansas, Lawrence, 570 p.

LEHRMANN, D. J., 1995, Guidelines for facies and parasequence stacking patterns analysis in shallow-marine carbonate platforms: Subsurface and outcrop examples from a diverse range of ages and settings: Exxon Production Research Co., unpublished report.

LEHRMANN, D. J., AND GOLDHAMMER, R. K., 1996, Secular variations in stacking patterns style of platform carbonates from ice-house to green-house periods in Earth history: Boulder, Geological Society of America Abstracts with Programs, v. 28, no. 7, p. A237.

LEHRMANN, D. J., AND GOLDHAMMER, R. K., 1997, Guidelines for application of stacking patterns techniques in platform carbonates from ice-house to greenhouse periods of Earth history: Tulsa, American Association of Petroleum Geologists, Annual Convention, Abstracts, p. A69.

LEHRMANN, D. J., AND GOLDHAMMER, R.K., 1998, Platform size as a fundamental control on architecture and sequence style: comparison of Triassic carbonate platforms of northern Italy and south China (abs): American Association of Petroleum Geologists, 1998 Annual Meeting, Extended Abstracts, p. 395.

LEHRMANN, D. J., RANKEY, E. C., AND GOLDHAMMER, R. K., 1997, Relationships between parasequence stacking patterns and depositional sequence geometries: Upper Guadalupian (Permian) strata of Slaughter Canyon, New Mexico: Tulsa, American Association of Petroleum Geologists, Annual Convention, Abstracts, v. 6, p. A68–A69.

LEHRMANN, D. J., WEI JIAYONG, AND ENOS, P., 1998, Controls on facies architecture of a large Triassic carbonate platform: The Great Bank of Guizhou, Nanpanjiang Basin, South China: Journal of Sedimentary Research, v. 68, p. 311–326.

LOUCKS, R. G., AND SARG, J. F., eds., 1993, Carbonate Sequence Stratigraphy: Recent Developments and Applications: Tulsa, American Association of Petroleum Geologists Memoir 57, 545 p.

MCLEAN, D. J., AND MOUNTJOY, E. W., 1994, Allocyclic control on Late Devonian buildup development, southern Canadian Rocky Mountains: Journal of Sedimentary Research, v. B64, p. 326–340.

MESOLELLA, K. J., MATTHEWS, R. K., BROECKER, W. S., AND THURBER, D. L., 1969, The astronomical theory of climate change: Barbados data: Journal of Geology, v. 77, p. 250–274.

MINERO, C. J., 1983, Discussion of back-reef and platform-interior sedimentation in the El Abra Formation, *in* Enos, P., Minero, C. J., Aguayo, C., Elby, D. E., and Clarke, R. T., eds., Sedimentation and Diagenesis of Mid-Cretaceous Platform Margin, East-Central Mexico, with Accompanying Field Guide: Dallas, Dallas Geological Society, p. 110–124.

MITCHELL, J. C., AND LEHMANN, P. J., 1988, Lithofacies, diagenesis and depositional sequence: Arab-D Member, Ghawar Field, Saudi Arabia, *in* Lomando, A. J. and Harris, P. M., eds., Giant Oil and Gas Fields: Tulsa, Society of Economic Paleontologists and Mineralogists Core Workshop No. 12, p. 459–514.

MITCHUM, R. M., AND VAN WAGONER, J. C., 1991, High-frequency sequences and their stacking patterns: Sequence-stratigraphic evidence of high-frequency eustatic cycles: Sedimentary Geology, v. 70, p. 131–160.

MITCHUM, R. M., VAIL, P. R., AND THOMPSON, S, 1977, The depositional sequence as the basic unit for stratigraphic analysis, *in* Payton, C. E., ed., Seismic Stratigraphy—Applications to Hydrocarbon Exploration: Tulsa, American Association of Petroleum Geologists Memoir 26, p. 53–62.

MONTANEZ, I. P., AND OSLEGER, D. A., 1993, Parasequence stacking patterns, third order accommodation events, and sequence stratigraphy of Middle to Upper Cambrian platform carbonates, Bonanza King Formation, southern Great Basin, *in* Loucks, R. G. and Sarg, J. F., Carbonate Sequence Stratigraphy: Recent Developments and Applications: Tulsa, American Association of Petroleum Geologists Memoir 57, p. 305–326.

MONTANEZ, I. P., AND OSLEGER, D. A., 1996, Contrasting sequence boundary zones developed within cyclic carbonates of the Bonanza King Formation,

Middle to Late Cambrian, southern Great Basin, *in* Witzke, B. J., Ludvigson, G. A., and Day, J., eds., Paleozoic Sequence Stratigraphy, Views from the North American Craton: Boulder, Geological Society of America Special Paper 306, p. 7–21.

North American Commission on Stratigraphic Nomenclature, 1983, North American Stratigraphic Code: American Association of Petroleum Geologists, Bulletin, v. 67, n. 5, p. 841-875.

Osleger, D., and Read, J. F., 1991, Relation of eustasy to stacking patterns of meter-scale carbonate cycles, Late Cambrian, USA: Journal of Sedimentary Petrology, v. 61, p. 1225–1252.

Osleger, D., and Read, J. F., 1993, Comparative analysis of methods used to define eustatic variations in outcrop: Late Cambrian interbasinal sequence development: American Journal of Science, v. 293, p. 157–216.

Oswald, E. J., 1992, Dolomitization of a Miocene reef complex, Mallorca, Spain: Unpublished Ph.D. dissertation, State University of New York, Stony Brook, 424 p.

Powers, D. W., and Easterling, R. G., 1982, Improved methodology for using embedded Markov chains to describe cyclical sediments: Journal of Sedimentary Petrology, v. 52, p. 913–923.

Prather, B. E., 1992, Evolution of the Late Jurassic carbonate/evaporite platform, Conecuh Embayment, northwestern Gulf Coast, USA: American Association of Petroleum Geologists Bulletin, v. 76, p. 164–190.

Pratt, B. R., and James, N. P., 1986, The St. George Group (Lower Ordovician) of western Newfoundland: Tidal flat island model for carbonate sedimentation in shallow epeiric seas: Sedimentology, v. 33, p. 313–343.

Rankey, E. C., 1996, Isolating eustatic, climatic, tectonic, and autogenic controls on sedimentation: Spatial and temporal changes in upper Virgilian–lower Wolfcampian (Pennsylvanian–Permian) strata, western North America: Unpublished Ph.D. dissertation, University of Kansas, Lawrence, 267 p.

Rankey, E. C., and Lehrmann, D. J., 1996, Anatomy and origin of toplap in a mixed carbonate-clastic system, Seven Rivers Formation (Permian, Guadalupian), Guadalupe Mountains, New Mexico: Sedimentology, v. 43, p. 807–826.

Read, J. F., 1973, Carbonate cycles, Pillara Formation (Devonian), Canning Basin, Western Australia: Canadian Petroleum Geology Bulletin, v. 21, p. 38–51.

Read, J. F., 1995, Overview of carbonate platform stratigraphy and reservoirs in green-house and ice-house worlds, *in* Read, J. F., Kerans, C., and Weber, L. J., sponsors, Milankovitchian sea level changes, cycles and reservoirs in green-house and ice-house worlds: Tulsa, SEPM (Society for Sedimentary Geology) Short Course Notes No. 35, part 1, p. 1–102.

Read, J. F., and Goldhammer, R.K., 1988, Use of Fischer plots to define third-order sea-level curves in Ordovician peritidal carbonates, Appalachians: Geology, v. 16, p. 895–899.

Ross, C. A., and Ross, J. R. P., 1987, Late Paleozoic sea levels and depositional sequences: Ithaca, Cushman Foundation for Foraminiferal Research Special Publication 24, p. 137–153.

Sadler, P. M., Osleger, D. A., and Montanez, I. P., 1993, On the labeling, length, and objective basis of Fischer plots: Journal of Sedimentary Petrology, v. 63, p. 360–368.

Sarg, J. F., Weber, L. J., and Markello, J. R., 1995, Carbonate sequence stratigraphy—A summary and perspectives, *in* Pause, P. H. and Candelaria, M. P., eds., Carbonate Facies and Sequence Stratigraphy: Practical Applications of Carbonate Models: Midland, Permian Basin Section–SEPM (Society for Sedimentary Geology) Publication No. 5-95, p. 181–200.

Satterley, A. K., 1996, Cyclic carbonate sedimentation in the Upper Triassic Dachstein Limestone, Austria—The role of patterns of sediment supply and tectonics in a platform-reef-basin system: Journal of Sedimentary Research, v. 66, p. 307–323.

Schlager, W., 1981, The paradox of drowned reefs and carbonate platforms: Geological Society of America Bulletin, v. 92, p. 197–211.

Schlager, W., 1989, Drowning unconformities on carbonate platforms, *in* Crevello, P. D., Wilson, J. L., Sarg, J. F., and Read, J. F., eds., Controls on Carbonate Platform and Basin Development: Tulsa, Society of Economic Paleontologists and Mineralogists, Special Publication 44, p. 15–26.

Schlager, W., 1992, Sedimentology and sequence stratigraphy of reefs and carbonate platforms: Tulsa, American Association of Petroleum Geologists, Continuing Education Course Notes Series No. 34, 71 p.

Sonnenfeld, M. D., 1996, Sequence evolution and hierarchy within the Lower Mississippian Madison Limestone of Wyoming, *in* Longman, M. W. and Sonnenfeld, M. D., eds., Paleozoic systems of the Rocky Mountain region: Tulsa, SEPM)Society for Sedimentary Geology), p. 165–192.

Sonnenfeld, M. D., and Cross, T. A., 1993, Volumetric partitioning and facies partitioning within the Permian Upper San Andres Formation of Last Chance Canyon, Guadalupe Mountains, New Mexico, *in* Loucks, R. G. and Sarg, J. F., eds., Carbonate Sequence Stratigraphy: Tulsa, American Association of Petroleum Geologists Memoir 57, p. 434–474.

Soreghan, G. S., 1994, Upper Pennsylvanian facies and cyclostratigraphy in Rhodes and Hembrillo Canyons, San Andres Mountains, *in* Garber, R. A. and Keller, D. R., eds., Field guide to the Paleozoic section of the San Andres Mountains: Midland, Permian Basin Section–SEPM (Society for Sedimentary Geology) Publication No. 94-35, p. 71–85.

Soreghan, G. S., and Dickinson, W. R., 1994, Generic types of stratigraphic cycles controlled by eustasy: Geology, v. 22, p. 759–761.

Tinker, S. W., 1996, Building the 3-D jigsaw puzzle: Applications of sequence stratigraphy to 3-D reservoir characterization, Permian basin: American Association of Petroleum Geologists Bulletin, v. 80, p. 460–485.

Tinker, S. W., and Mruk, D. H., 1995, Reservoir characterization of a Permian giant: Yates field, west Texas, *in* Stoudt, E. and Harris, P. M., eds., Hydrocarbon Reservoir Characterization, Geologic Framework and Flow-Unit Modeling: Tulsa, SEPM (Society for Sedimentary Geology) Short Course Notes No. 34, p. 51–128.

Tucker, M. E., and Wright, V. P., 1990, Carbonate Sedimentology: Oxford, Blackwell Scientific Publications, 367 p.

Van Wagoner, J. C., Posamentier, H. W., Mitchum, R. M., Vail, P. R., Sarg, J. F., Loutit, T. S., and Hardenbol, J., 1988, An overview of the fundamentals of sequence stratigraphy and key definitions, *in* Wilgus, C. K., Hastings, B. S., Kendall, C. G. St. C., Posamentier, H. W., Ross, C. A., and Van Wagoner, J. C., eds., Sea-Level Changes: An Integrated Approach: Tulsa, Society of Economic Paleontologists and Mineralogists, Special Publication 42, p. 38–45.

Van Wagoner, J. C., Mitchum, R. M., Campion, K. M., and Rahmanian, V. D., 1990, Siliciclastic sequence stratigraphy in well-logs, cores, and outcrops: Tulsa, American Association of Petroleum Geologists Methods in Exploration Series No. 7, 55 p.

Warrick, R. A., Barrow, E. M., and Wigley, T. M. L., 1993, Climate and Sea Level Change: Observations, Projections and Implications: Cambridge, Cambridge University Press, 424 p.

Watney, W. L., French, J. A., and Franseen, E. K., 1989, Sequence stratigraphic interpretations and modeling of cyclothems in the Upper Pennsylvanian (Missourian) Lansing and Kansas City Groups in eastern Kansas, *in* Watney, W. L. and Franseen, E. K., eds., First Annual Field Trip Guidebook: Lawrence, Kansas Geological Society, p. 108–114.

Watney, W. L., French, J. A., and Guy, W. J., 1996, Modeling petroleum reservoirs in Pennsylvanian strata of the Midcontinent, *in* Merriam, D. F., ed., Computer modeling and mapping: New York, Plenum Press, p. 43–77.

Wilkinson, B. H., Drummond, C. N., Rothman, E. D., and Diedrich, N. W., 1997, Stratal order on carbonate platforms: Journal of Sedimentary Research, v. 6, p. 1068–1082.

Wilson, J. L., 1967, Cyclic and reciprocal sedimentation in Virgilian strata of southern New Mexico: Geological Society of America Bulletin, v. 78, p. 805–818.

Wilson, J. L., 1975, Carbonate Facies in Geologic History: New York, Springer-Verlag, 471 p.

Wong, P. K., and Oldershaw, A. E., 1980, Causes of cyclicity in reef interior sediments, Kaybob reef, Alberta: Canadian Petroleum Geology Bulletin, v. 28, p. 411–424.

Yang, W., Harmsen, F., and Kominz, M. A., 1995, Quantitative analysis of a cyclic peritidal sequence, the Middle and Upper Devonian Lost Burro Formation, Death Valley, California—A possible record of Milankovitch climatic cycles: Journal of Sedimentary Research, v. B65, p. 306–322.

EFFECTS OF LONG-TERM ACCOMMODATION CHANGE ON SHORT-TERM CYCLES, UPPER PALEOZOIC PLATFORM LIMESTONES, WEST TEXAS

ARTHUR H. SALLER

Unocal, 14141 Southwest Freeway, Sugar Land, Texas 77478

J. A. D. DICKSON

Department of Earth Science, Cambridge University, Cambridge CB2 3EQ, United Kingdom

E. T. RASBURY

Department of Geosciences, State University of New York at Stony Brook, Stony Brook, New York 11794-2100

AND

TOSHIKAZU EBATO*

Japan National Oil Company, Technical Research Center, Chiba, Japan

ABSTRACT: Upper Paleozoic limestones were studied in the subsurface on the eastern side of the Central Basin platform, west Texas, to determine the effect of long-term change in accommodation on short-term cycles. Approximately 87 cycles are present in five stratigraphic intervals that are (from bottom to top) the Strawn, Canyon, Cisco, Wolfcamp detrital interval, and Wolfcamp "reef" interval. The lower three intervals are Middle to Upper Pennsylvanian, and the two Wolfcamp intervals are Lower Permian. The average cycle thickness is 3.7 m. Most cycles are bounded by subaerial-exposure surfaces, and each cycle apparently represents a glacio-eustatic fluctuation of sea level at an average frequency of approximately 163,000 yr per cycle. Characteristics of cycles change upward from the base of the Strawn (well depth of 2950 m [9700 ft]) to the top of the Cisco (depth of 2730 m [8950 ft]), and those changes include a general shallowing of depositional facies (more grainstones), a decrease in average cycle thickness (an average of 5.2 decreasing to 1.9 m), an increase in amount of shale, and a decrease in $\delta^{13}C$ and $\delta^{18}O$ values of bulk-rock samples. These patterns are interpreted to be a response to a long-term decrease in rate of accommodation. The changes upward through the Wolfcampian (between depths of 2620 and 2730 m [8600 and 8950 ft]), however, include a general thickening of the depositional cycles (an average of 2 increasing to 4.5 m), an increase in $\delta^{13}C$ and $\delta^{18}O$ values, and more deep-water facies (more wackestones and mudstones), apparently in response to a long-term increase in accommodation. A major drowning and backstepping of the carbonate-platform margin occurred at the top of the Wolfcamp reef interval.

The time represented by each cycle includes (1) time when the sediment surface was below sea level accumulating carbonate sediment and (2) time when the sediment surface was subjected to meteoric diagenesis above sea level and no carbonate sediment accumulated. Meteoric diagenesis apparently caused the stable carbon and oxygen isotope compositions of entire limestone cycles to become lighter with more prolonged subaerial exposure. Low $\delta^{13}C$ and $\delta^{18}O$ values in thin cycles suggest that the cycles were subjected to progressively longer subaerial exposures as the rate of accommodation decreased; finally, a major unconformity was created at the top of the Pennsylvanian. This scenario supports the idea that, compared to thin cycles, thick cycles spent more time below sea level accumulating sediment and less time above sea level being subjected to meteoric diagenesis. The most widespread porosity occurs in cycles of intermediate or greater thickness (>2 m) that were subaerially exposed for brief to moderately long periods (estimated at 5,000–50,000 yr) of freshwater diagenesis (upper part of the Strawn, lower and middle parts of the Canyon, and Wolfcamp reef intervals). Thin cycles (<2 m thick; upper part of the Cisco and lower part of the Wolfcamp detrital interval) have little porosity because (1) longer subaerial exposure resulted in more cements filling most matrix pores and (2) vugs, fractures, fissures, and caverns that formed during prolonged subaerial exposure were filled by shale.

INTRODUCTION

Early sequence stratigraphic research (Vail et al., 1977) concentrated on large-scale stratigraphic patterns observed in reflection seismic data. During the middle and late 1980s, sequence stratigraphic studies examined outcrops and wells, relating smaller scale changes in depositional facies patterns to larger scale sequence and seismic stratigraphic features (Sarg, 1988). Changes in cycle thickness and facies within those cycles have been related to long-term changes in sea level and/or accommodation (Goldhammer et al., 1990, 1991; Borer and Harris, 1991; Elrick and Read, 1991; Montanez and Read, 1992; Montanez and Osleger, 1993). Cycle stacking patterns have been used to delineate short- and long-term variations in eustatic sea level, subsidence, facies distribution, and/or major subaerial-exposure events (Sarg, 1988; Goldhammer et al., 1990, 1991; Borer and Harris, 1991; Elrick and Read, 1991; Montanez and Read, 1992; Montanez and Osleger, 1993).

Goldhammer et al. (1987, 1990) proposed a model for cyclic sedimentation that involved subtidal carbonate deposition alternating with hiatuses. In their model, cycle thickness and duration of hiatus are systematically affected by longer-term variation in rate of accommodation. When rates of long-term accommodation are high, periods of carbonate sedimentation should be long, cycles should be thick, and hiatuses between cycles should be short (Fig. 1). In contrast, periods of carbonate sedimentation should be short, cycles should be thin, and depositional hiatuses should be long during times when long-term rates of accommodation are low (Fig. 1). The purpose of this study is to test that part of the Goldhammer et al. model and show how long-term changes in accommodation alter the deposition, diagenesis, and porosity in cyclic Upper Pennsylvanian and Lower Permian limestones of the Central Basin platform, west Texas.

Most cycles examined in this study contain subtidal carbonates that were subaerially exposed. During subaerial exposure, the stable carbon and oxygen isotope compositions of subtidal carbonates are commonly altered (Gross and Tracy, 1966; Allan and Matthews, 1977, 1982; Hudson, 1977). Stable carbon and

Present address: Teikoku Oil de Venezuela, Caracas, Venezuela

Advances in Carbonate Sequence Stratigraphy: Application to Reservoirs, Outcrops and Models, SEPM Special Publication No. 63

oxygen isotope compositions of marine limestones become lighter because of recrystallization in meteoric water, which has generally lighter oxygen isotope compositions compared to those of seawater (Gross and Tracy, 1966; Allan and Matthews, 1977, 1982; Hudson, 1977; Hanor, 1978). Stable carbon and oxygen isotope compositions are useful indicators of relative length of subaerial exposure. If thin cycles (<2 m) were subaerially exposed for longer periods of time than thick cycles, as predicted by Goldhammer et al. (1987, 1990), then those cycles should have lighter carbon and oxygen isotope compositions compared to those of thick cycles.

The Southwest Andrews Area of west Texas (Fig. 2) is an excellent location to study cyclostratigraphy and porosity associated with subaerial exposure in carbonates because the Pennsylvanian and Lower Permian reservoir section contains limestones with numerous subaerial-exposure surfaces and variable porosity. The preliminary results of an earlier study in this area were presented in Saller et al. (1994). That study was based on approximately 350 m of widely scattered core from 10 wells and 10 two-dimensional seismic lines. More than half the core control was concentrated in a single 40-m stratigraphic thickness (the Wolfcamp "reef" interval). Thick nonporous beds in the Strawn interval and the Wolfcamp detrital interval were not cored, and cycle boundaries were inferred from well-log characteristics. This study is based on three new wells that completely cored the Pennsylvanian–Lower Permian reservoir section (>300 m) including the entire Strawn, Canyon, and Cisco, and much of the Wolfcamp intervals. Detailed biostratigraphic analyses of two wells allow much higher biostratigraphic resolution than was previously possible. Stable isotope profiles were constructed for each well; average sampling intervals were 1 m. Stable carbon and oxygen isotope analyses were performed by using methods described in Quinn (1991) with precision of better than ±0.2‰; values are reported relative to the PDB (Peedee belemnite) standard. A three-dimensional seismic survey over the entire field was also integrated into this study.

GENERAL SETTING

The 3000-m thickness of Pennsylvanian and Permian platform carbonates and evaporites (Galley, 1958) in the Permian basin indicates substantial regional subsidence during the late Paleozoic. The eastern part of the Central Basin platform and adjacent Midland basin subsided rapidly during the Pennsylvanian and Early Permian, resulting in deposition of approximately 500 m of Pennsylvanian and Wolfcampian strata that thin onto structural highs to the west (Figs. 2, 3). A major transgression occurred during middle to late Wolfcampian time during which the carbonate-platform margin stepped back landward (to the west) approximately 18 km (11 mi) between the time of deposition of the middle Wolfcampian strata and that of the Abo Formation (and equivalent Wichita Formation) (upper Wolfcampian–lower Leonardian; Fig. 3). The Clear Fork (Leonardian) and San Andres–Grayburg (Guadalupian) platform-margin deposits then prograded eastward over the Southwest Andrews Area into the Midland basin (Fig. 3).

The Southwest Andrews Area (which encompasses the

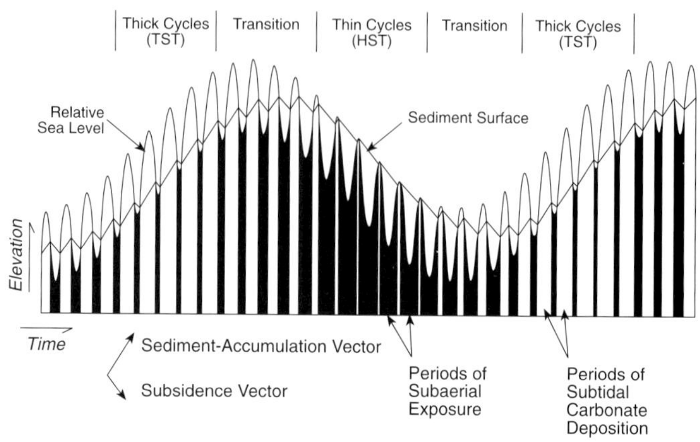

FIG. 1.—Model for cyclic carbonate sedimentation involving composite eustasy; a short-term sinusoidal sea-level fluctuation is superimposed on a long-term sinusoidal sea-level fluctuation. Rate of sediment accumulation is assumed to be constant when depositional surface is below sea level. Sediment-accumulation vector is (carbonate accumulation – subsidence)/time. No carbonate sedimentation occurs when depositional surface is above sea level. Rate of subsidence (subsidence vector) is assumed to be constant. Long-term rises of sea level have long periods of carbonate deposition and short subaerial exposures (black). Long-term falls of sea level have short periods of carbonate deposition and long subaerial exposures (black). Horizontal and vertical scales vary according to geologic time and location. Locations of highstand systems tract (HST) and transgressive systems tracts (TSTs) are approximate. See Goldhammer et al. (1987, 1990) for more detailed explanation of this type of modeling.

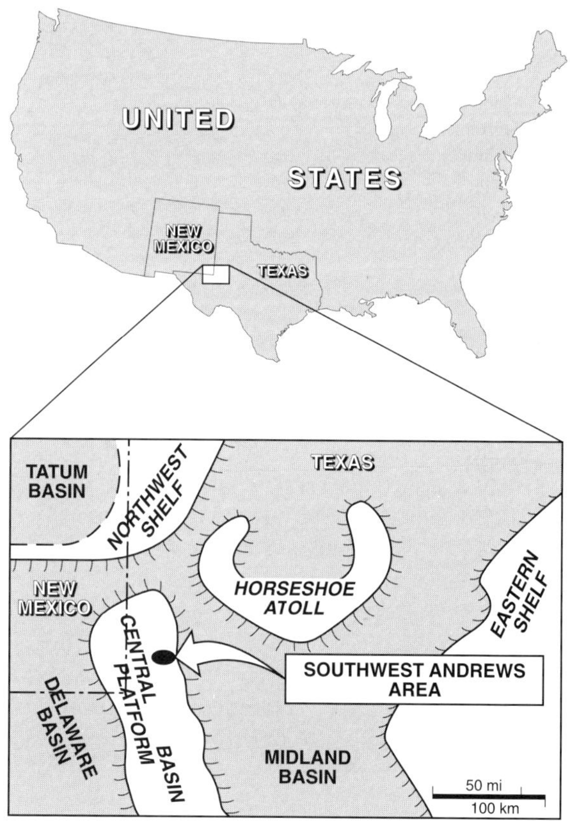

FIG. 2.—Map of west Texas and southeast New Mexico showing the location of the Southwest Andrews Area and paleogeography during Late Pennsylvanian time. White areas are shallow shelves, and shaded areas are basins.

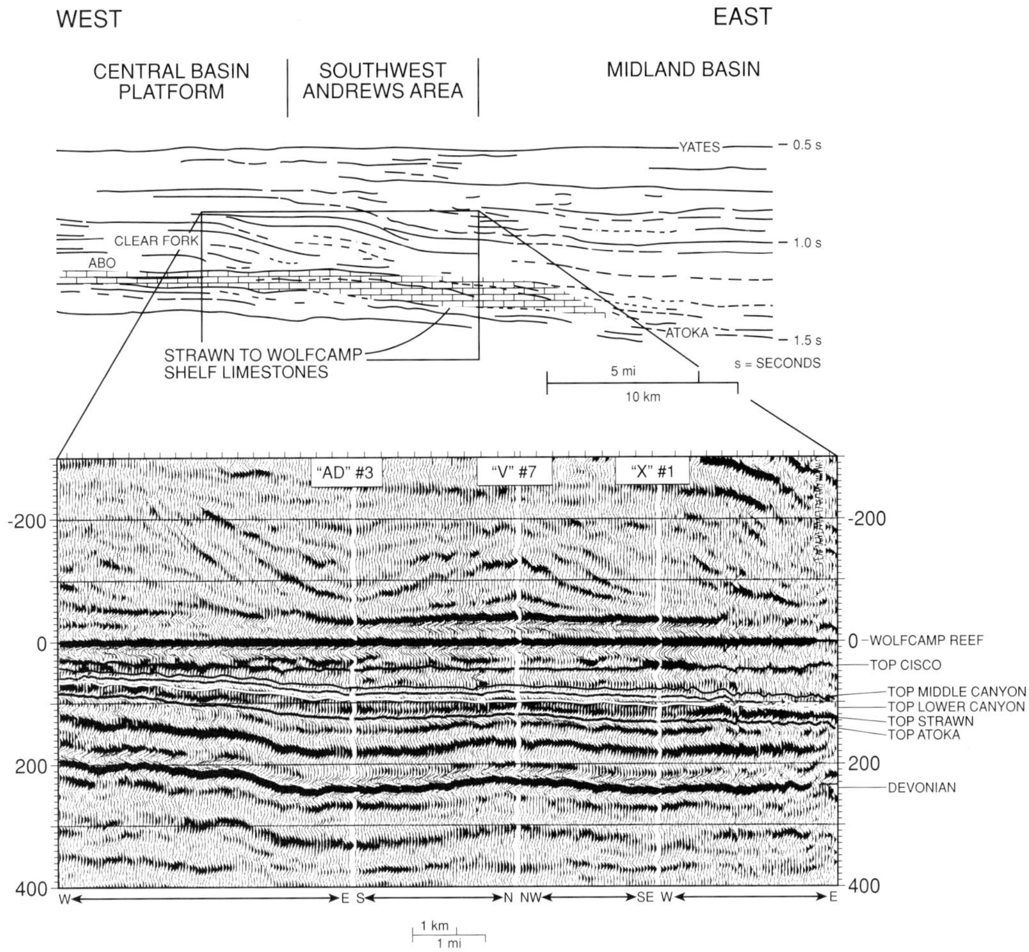

FIG. 3.—Line drawing of regional stratigraphy (above) and composite seismic line in the Southwest Andrews Area (below). Line drawing of regional stratigraphy is derived from an east-west reflection seismic line in the vicinity of the Southwest Andrews Area (modified from T. L. Elliott, pers. commun.). Abo carbonates are upper Wolfcampian and lower Leonardian. Clear Fork Formation is Leonardian (lower part of middle Permian). The shelf prograded eastward toward the Midland basin during deposition of Clear Fork and overlying strata. "Yates" is Guadalupian (upper part of middle Permian). The seismic line (below) is from a three-dimensional seismic survey and has been "flattened" on the top of the Wolfcamp reef interval. Other horizons shown are the top of the Cisco interval, a marker in the Canyon interval, the top of the lower part of the Canyon interval, a marker near the top of the Strawn interval, the top of the Atoka, and the top of the Devonian. Note that (1) the Strawn to Wolfcamp limestone section thins onto a depositional high to the west and (2) the outer Wolfcamp shelf drowned and the shelf margin stepped back landward approximately 17 km to its location in the Abo.

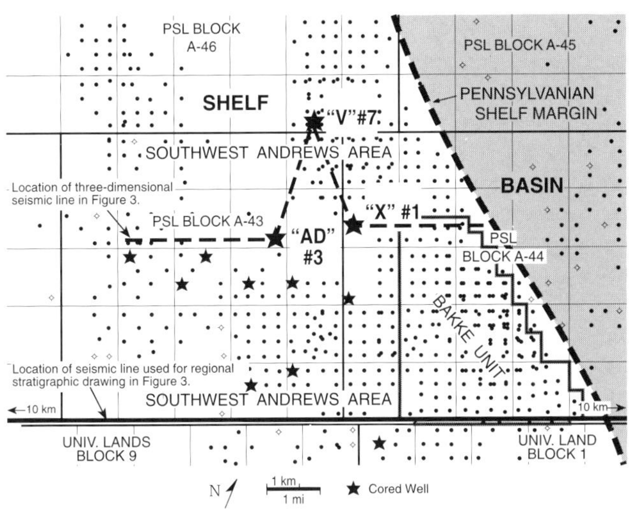

FIG. 4.—Map of the Southwest Andrews Area showing wells penetrating the Wolfcamp, the location of cored wells (stars), and the approximate location of the Pennsylvanian carbonate shelf (heavy dashed line). The shelf prograded farther eastward during Late Pennsylvanian and Wolfcampian time. PSL = Public School Land.

Parker, Andrews, and Deep Rock oil fields) is located on the eastern side of the Central Basin platform in Andrews County, Texas (Figs. 2, 4). Moderate to rapid subsidence and eustatic sea-level fluctuations affected stratigraphic patterns in the Southwest Andrews Area. The Middle to Late Pennsylvanian and Early Permian were times of high-amplitude (50–200 m) and high-frequency eustatic sea-level fluctuations that were probably caused by repeated building and melting of Pangean continental glaciers in the Southern Hemisphere (Heckel, 1986; Ross and Ross, 1987, 1988; Veevers and Powell, 1987; Crowley and Baum, 1991; Gonzalez-Bonorino and Eyles, 1995). The Central Basin platform was within a few degrees of the equator during the Pennsylvanian (Walker et al., 1991). The abundance of clay-rich shales and lack of evaporites suggest that the Pennsylvanian and Wolfcampian were times dominated by a humid to subhumid climate in the Permian basin. Algeo et al. (1992) proposed a similar humid tropical climate for the nearby Sacramento Mountains during the Middle Pennsylvanian. Depositional and diagenetic patterns observed in the strata of

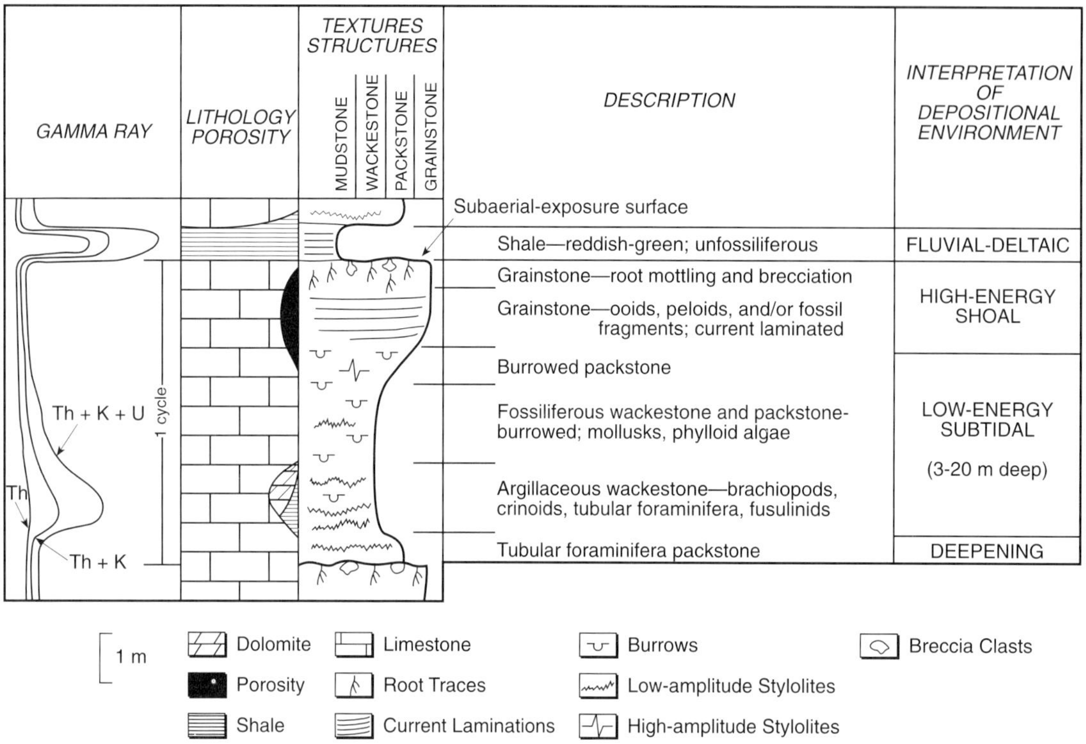

FIG. 5.—Idealized shallowing-upward cycle in Upper Pennsylvanian and Lower Permian carbonates in the Southwest Andrews Area. This idealized cycle shows depositional lithologies, interpreted depositional environments, and gamma-ray responses.

the Southwest Andrews Area are similar to those described by Wright (1992) for icehouse conditions during the Carboniferous.

DEPOSITIONAL FACIES AND THEIR DISTRIBUTION

Throughout this body of strata, short-term cyclic deposition is superimposed on systematic long-term changes in rate of accommodation. Most of the strata can be placed into one of four main facies groups: (1) fossiliferous mudstones, wackestones, and packstones, (2) grainstones, (3) phylloid-algal boundstones, and (4) terrigenous clastics. These depositional lithofacies generally occur in cycles bounded by subaerial-exposure surfaces (Fig. 5). A general depositional and sequence stratigraphic model for the Pennsylvanian and Wolfcampian at the Southwest Andrews Area is shown in Figure 6. Fossiliferous mudstones, wackestones, and packstones represent open-marine, subtidal deposition in a variety of water depths ranging from shallow (a few meters deep; some fusulinid and phylloid wackestones and packstones) to fairly deep (tens to possibly hundreds of meters; spiculitic lime mudstones). Grainstones (bioclastic, peloidal, oolitic) generally represent deposition on high-energy shoals, in water less than a few meters deep. Phylloid-algal boundstones accumulated as low-relief organic buildups. Thicknesses of phylloid-algal boundstones rarely exceed 8 m, suggesting maximum relief of 8 m. Analyses of well logs suggest that well-developed phylloid-algal boundstones are rarely present at the same level in adjacent wells at a 40-acre spacing, so they are probably less than 300 m across. Terrigenous clastics (mainly shales with a few sandstones and conglomerates) also represent deposition in a wide variety of environments ranging from fluvial (red shale and sandstones with caliche nodules) to deeper marine settings (dark gray shales with sponge spicules and sometimes crinoid fragments), in which the terrigenous clastics apparently accumulated below the depth of prolific carbonate production (>10–20 m).

The strata under study in the Southwest Andrews Area have been divided into five intervals on the basis of biostratigraphic and lithologic characteristics (from bottom to top): (1) Strawn, (2) Canyon, (3) Cisco, (4) Wolfcamp detrital, and (5) Wolfcamp reef (Fig. 7). We are using these informal subsurface names because of a lack of formal subsurface stratigraphic nomenclature in the immediate area. Although the Wolfcamp reef interval contains some boundstone, it was generally not deposited as a true reef. These strata are mainly limestone, except for the Wolfcamp detrital interval that contains approximately equal amounts of shale and limestone. Fusulinids in the studied strata have been analyzed biostratigraphically by Garner Wilde (pers. commun., 1994) and are reported here relative to the zonation of Wilde (1990). Mainly on the basis of fusulinids, Wilde (pers. commun., 1994) found four Desmoinesian biostratigraphic zones in the Strawn, five Missourian zones in the Canyon, three Virgilian zones in the Cisco, and four zones in the Wolfcampian section (Figs. 7, 8; our Wolfcamp reef and Wolfcamp detrital intervals represent approximately the first half of the Wolfcampian). The intervals discussed in this paper are based

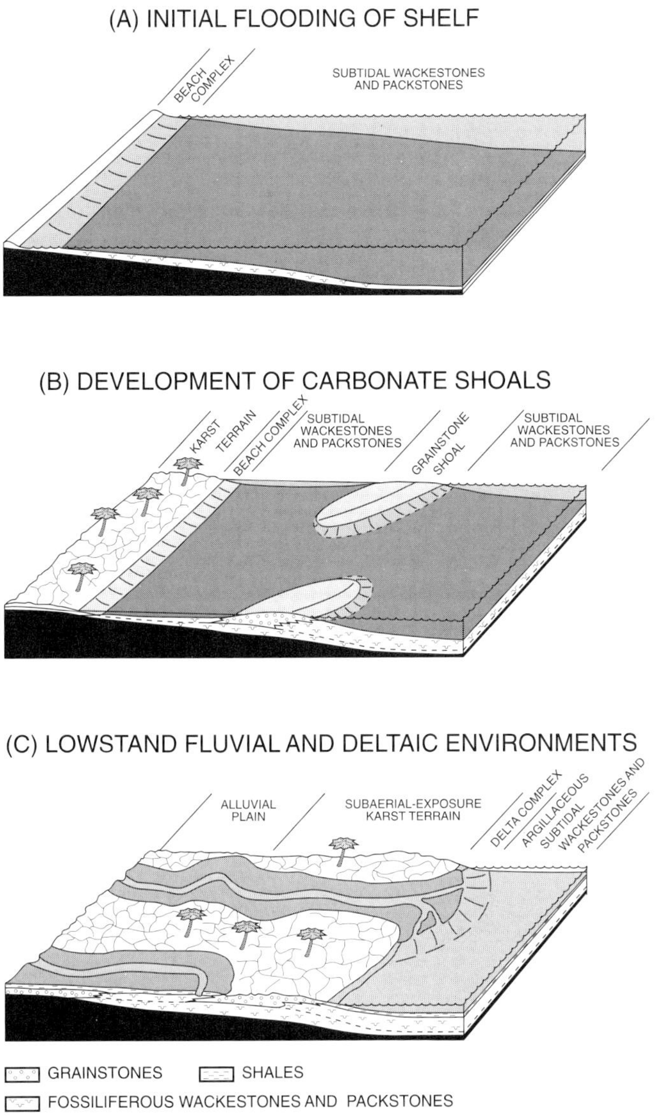

FIG. 6.—Model for deposition of a typical cycle. (A) Initial flooding and transgression of the platform occurred during a rapid sea-level rise. (B) When sea level fell slightly or carbonates built up near sea level, grainstone shoals developed. (C) At the end of depositional cycles, sea level fell, and the shelf was exposed. Terrigenous material was carried across the carbonate shelf in channels. Some of the channels were back-filled with sand and clay, probably as sea level began to rise again. This lowstand of sea level is represented by a subaerial-exposure surface at most locations on the carbonate shelf. Sea-level fluctuations were probably about 50 m (Crowley and Baum, 1991).

on lithology and depositional facies as well as biostratigraphy. Although our stratigraphic intervals are parallel to biostratigraphic zones, they do not necessarily coincide with them. Most intervals thin to the west, indicating a persistent depositional high to the west during the Late Pennsylvanian and Early Permian (Fig. 3).

Strawn Interval (Desmoinesian)

The Strawn interval in the Southwest Andrews Area consists of approximately 100 m of limestone with 16 correlated cycles. Dark gray Atokan shales underlie Strawn limestones. The top of the Strawn is a subaerial unconformity at the top of the Desmoinesian. The Strawn displays a general upward shallowing of depositional facies. We have divided the Strawn into four main parts based on depositional facies (from bottom to top): (1) lower Strawn, (2) *Komia*-rich middle Strawn, (3) phylloid-algal–rich middle Strawn, and (4) upper Strawn (Fig. 7). The boundaries between these parts of the Strawn follow biostratigraphic zones, but are not defined by them (Fig. 7).

1. The lower part of the Strawn interval (8–15 m thick) contains fossiliferous packstones and grainstones interbedded with sparsely fossiliferous wackestones and spiculitic mudstones (Fig. 7). Cycles are difficult to define in this interval because there are no distinct shallowing or deepening trends and no subaerial-exposure surfaces were identified. Two cycles were tentatively identified. Deposition is interpreted to have occurred in a relatively deep-water environment. Lime mud and sponge spicules were deposited slowly at depths of maybe 30–100 m, and the intercalated bioclastic sands were emplaced by episodic mass flows.
2. The *Komia*-rich middle part of the Strawn interval is approximately 20 m thick and characterized by the presence of the calcareous algae *Komia* in wackestones, packstones, and grainstones (Fig. 7). Three shallowing-upward cycles are defined by *Komia* wackestones grading upward into packstones and grainstones. The three cycles are capped by erosional surfaces interpreted as subaerial-exposure surfaces. Diagenesis associated with subaerial exposure is not intense and is confined to the top few centimeters of each cycle. The *Komia*-rich middle Strawn is interpreted as a time of generally shallow shelf deposition with three upward-shallowing cycles.
3. The phylloid-algal–rich middle part of the Strawn interval contains approximately 50 m of phylloid-algal wackestones, packstones, and boundstones (Fig. 7) with coral-rich, shaly, and cherty intervals. Nine depositional cycles are recognized. Seven are bounded by subaerial-exposure surfaces. Diagenesis associated with subaerial exposure is minor and includes thin caliche crusts, fractures, and root traces. Other cycle tops are identified by shallow-water facies overlain by distinctly deeper-water facies (spiculitic, crinoidal wackestones). Some cycles have wackestones and packstones overlain by grainstones indicating an upward shallowing. The phylloid-algal–rich middle Strawn is interpreted as having been deposited on a shallow-marine shelf with scattered phylloid-algal mounds separated by micritic, phylloid-algal–rich areas. Periodically, abrupt sea-level rises caused deepening of depositional environments; subsequent sea-level falls led to subaerial exposure.
4. The upper part of the Strawn interval (10–15 m thick) contains two depositional cycles with cross-bedded oolitic grainstones overlying cherty mudstones (Fig. 7). Both cycles are capped by subaerial-exposure surfaces. Alteration associated with exposure includes vugs and mottling that extend 1–2 m into cycles. Oolitic grainstones were probably deposited at or within 1–3 m of sea level, whereas cherty, spiculitic mudstones were deposited in much deeper water

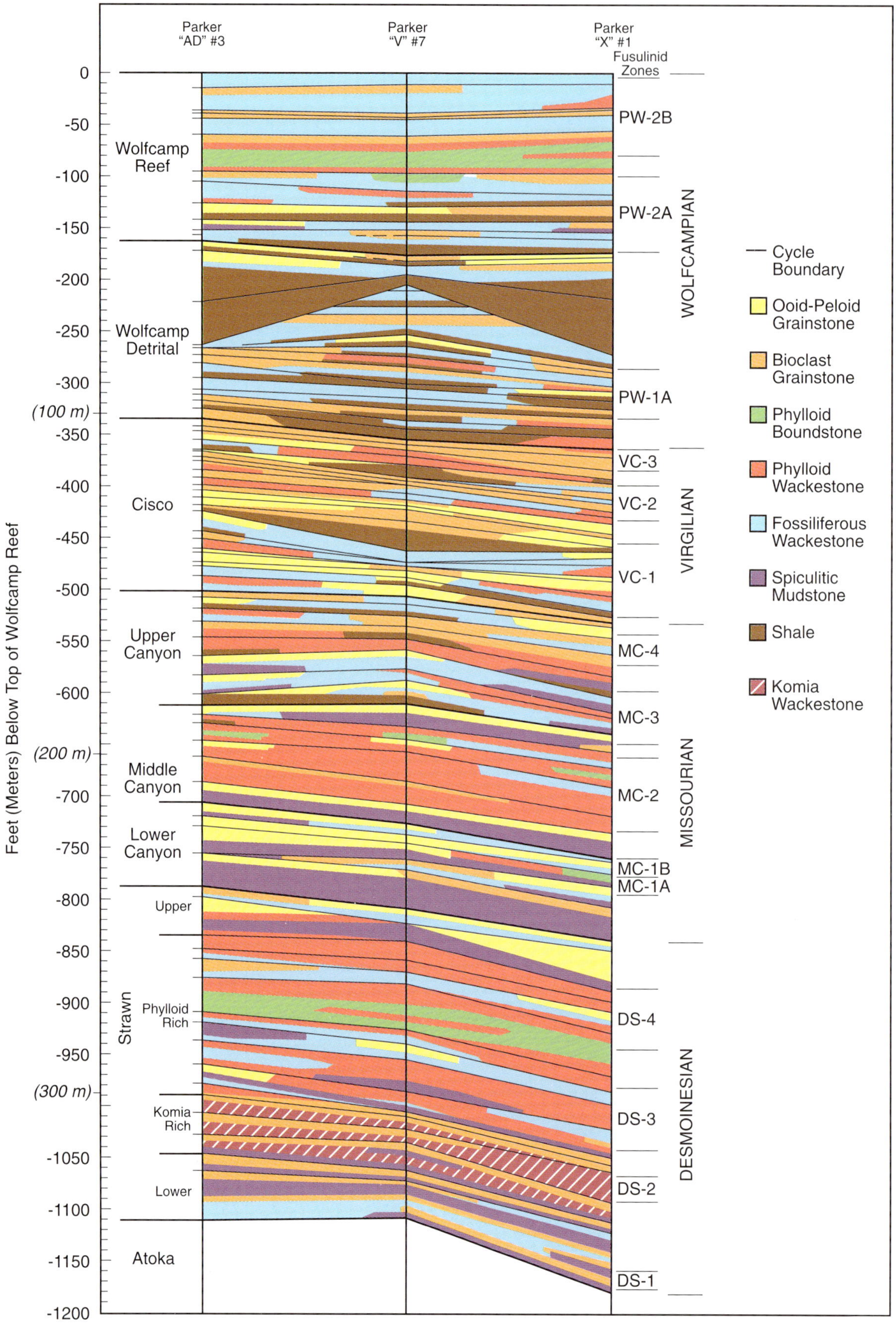

FIG. 7.—Correlation of cycles and facies between three project wells; fusulinid zones are shown at right. Fusulinid zones were determined by G. Wilde (pers. commun., 1994) and relate to the zonation of Wilde (1990)

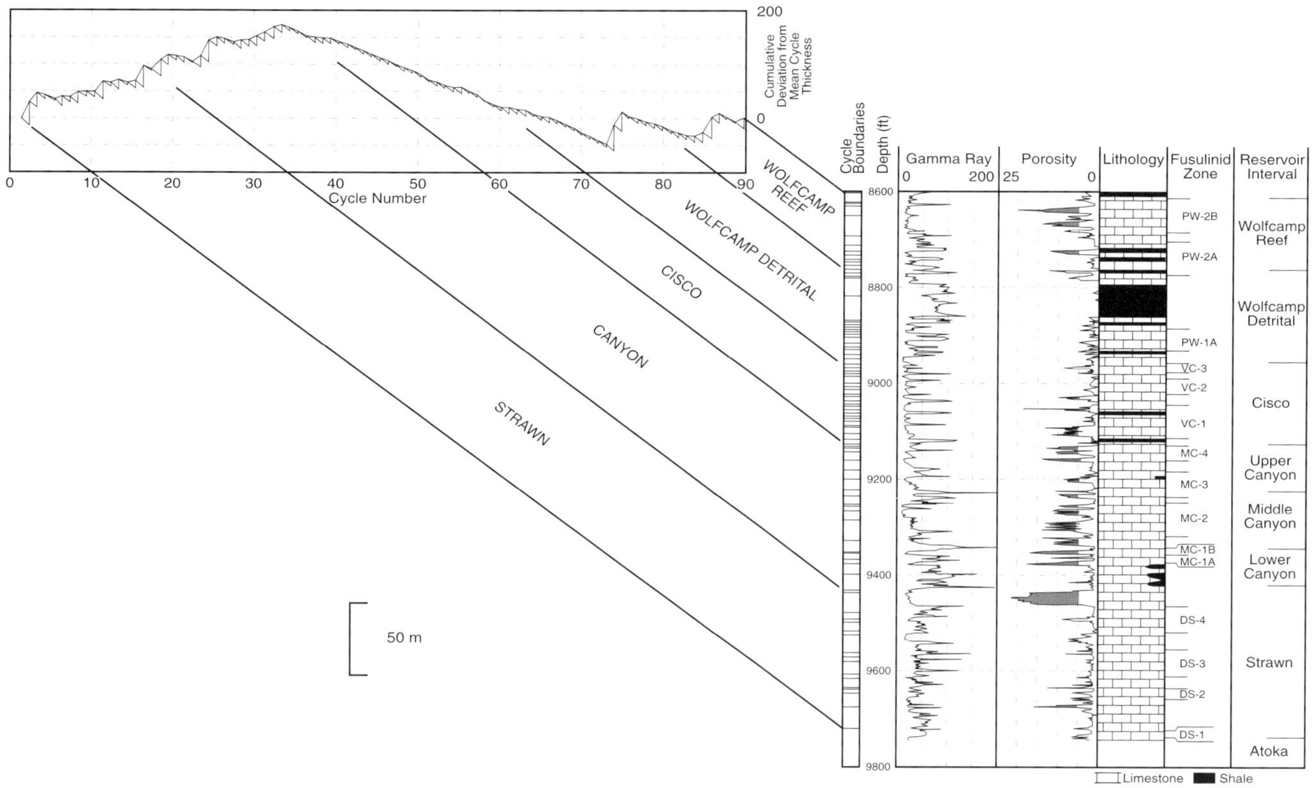

FIG. 8.—Conventional Fischer plot for the Parker "X" #1 (at left), and gamma-ray log, core-measured porosity, lithologies, and fusulinid zonation according to Wilde (1990). On the Fischer plot, horizontal scale is cycle number, and vertical scale is cumulative deviation from mean cycle thickness. Each vertical line is thickness of an individual cycle. Each inclined line is average accommodation per cycle. See Sadler et al. (1993) for explanation of Fischer plots. Note that the Strawn interval and lower to middle parts of the Canyon interval have thicker cycles, whereas the Cisco interval and lower part of the Wolfcamp detrital interval have generally thinner cycles.

(>20 m deep?). These differences suggest abrupt, large-amplitude sea-level fluctuations during the deposition of each cycle.

Porosity in the Strawn is generally low. Reservoir porosity (defined as 4% or greater) is confined to a few cycles that were subaerially exposed. Furthermore, reservoir porosity is facies selective; it occurs in a few *Komia* grainstones and phylloid-algal boundstones in the middle part and in oolitic grainstones in the upper part of the Strawn.

Canyon Interval (Missourian)

The Canyon interval is approximately 100 m thick in the Southwest Andrews Area and is divided into three parts (lower, middle, and upper) on the basis of depositional characteristics. The boundaries between those three parts parallel the five Missourian biostratigraphic zones of G. Wilde (pers. commun., 1994). The Canyon interval contains 19 correlated cycles that are all capped by subaerial-exposure surfaces. Cycles are generally thinner in the upper part of the Canyon and to the west. Lower Canyon cycles typically contain spiculitic wackestones or mudstones in the lower part and ooid or bioclast grainstones in the upper part (Fig. 7). The middle part of the Canyon interval is dominated by burrowed wackestones and packstones with phylloid-algal plates, tubular foraminifer fragments, and fusulinids. Grainstones are thin (<1 m) and preferentially occur in the uppermost part of cycles. Cycles in the upper part of the Canyon interval have fossiliferous wackestones passing up to bioclast-peloid grainstones with grainstones generally increasing in abundance upward in successive cycles. Subaerial-exposure surfaces occur at almost all cycle tops, and the intensity of exposure-related alteration (vugs, fissures, mottling) generally increases upward within the Canyon. Where present, porosity generally occurs in the upper part of cycles. Porosity preferentially occurs in thick grainstone, but some phylloid-algal wackestones are also porous. Cyclic porosity also tends to decrease upward in the Canyon in all three study wells (Figs. 9–11).

Shales vary in thickness and character vertically and laterally within the Canyon interval. Dark gray shales occur in the lower part of some cycles in the lower Canyon. The middle part of the Canyon contains almost no shale. Red and green shales increase in thickness upward and to the west (landward) in the upper part of the Canyon, with five shale beds more than 1 m thick in the "AD" #3 well, only three in the "V" #7 well, and none in the "X" #1 well. Depositional topography apparently influenced the color of the shale because much of the shale in the most landward, depositionally highest well ("AD" #3) is red

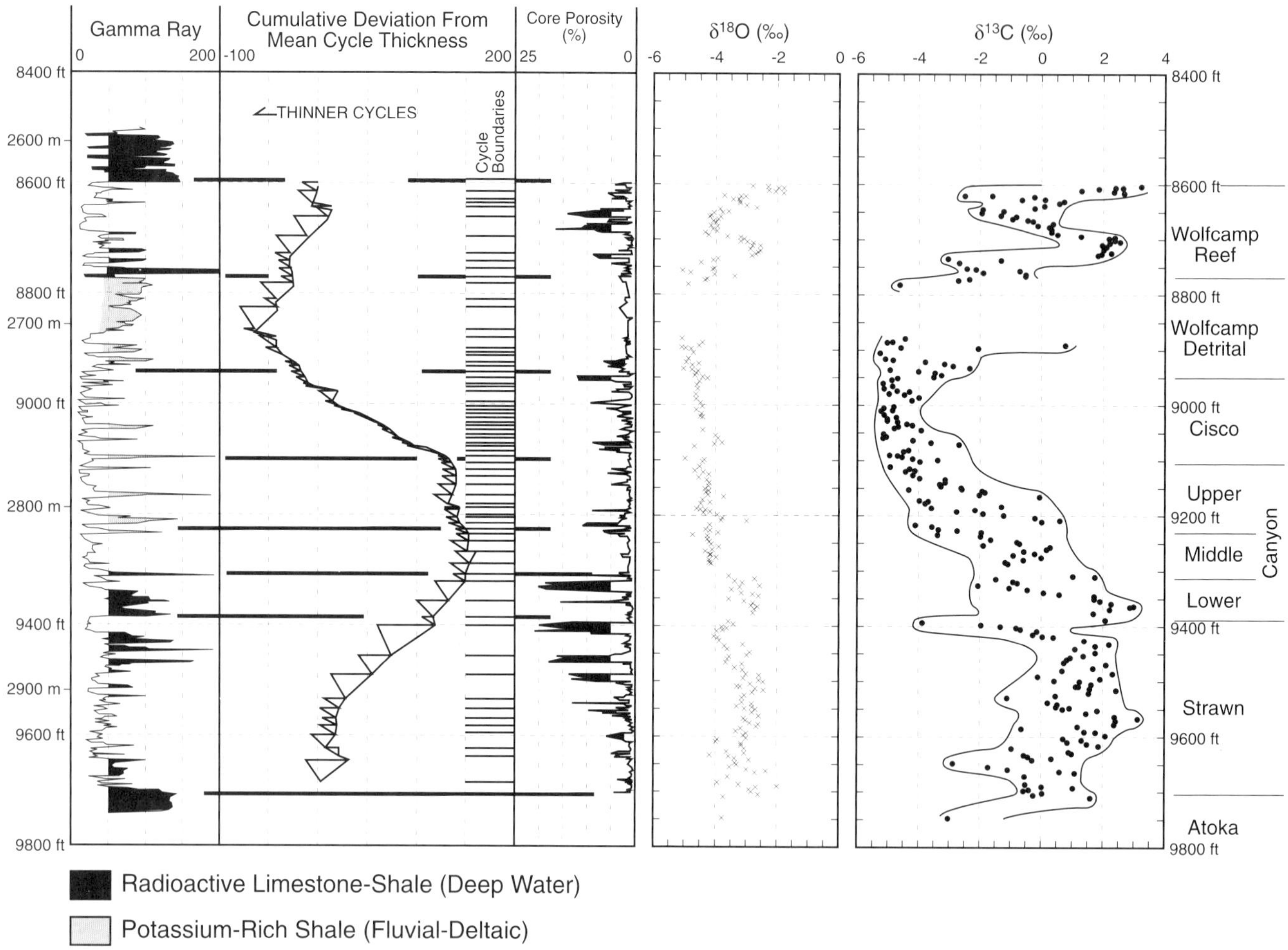

FIG. 9.—Plots of (from left to right) gamma ray, depth-based Fischer diagram, core-measured porosity, and stable isotope compositions for the Parker "AD" #3 well. The depth-based Fischer plot of cycle thickness indicates cumulative deviation of cycles from average cycle thickness. As the curve moves upward, deflections to the left indicate thinner-than-average cycles, and deflections to the right indicate thicker-than-average cycles. The Cisco interval and lower part of the Wolfcamp detrital interval have thinner-than-average cycles, whereas the Strawn interval and lower part of the Canyon interval have thicker-than-average cycles. Porosity >5% is shaded. Stable isotope analyses were of bulk limestone samples. Note the correlation of trends in stable isotope composition, porosity, and cycle thickness.

(indicating thorough oxidization), whereas shales in the "V" #7 well are green (indicating incomplete oxidization). Three-dimensional seismic data show a few isolated sinkholes in middle to upper Canyon strata that were filled with shale. One uncored well penetrated one of these sinkholes and has a 30-m-thick shale in the middle and upper parts of the Canyon interval that is absent from all adjacent wells.

Deposition in the Canyon interval occurred in a variety of environments that generally became shallower upward. As in the upper part of the Strawn interval, oolitic grainstones of the lower Canyon are interpreted as having been deposited very near sea level, whereas cherty, spiculitic mudstones were deposited in deeper water. Middle and upper Canyon cycles of fossiliferous wackestones that pass up into bioclastic grainstones are interpreted as having commenced deposition in the shallow-subtidal environment; then the depositional environment became shallower, approached sea level, and finally became subaerial during a sea-level drop. Shales were deposited during the early stages of some transgressions following subaerial exposure (Fig. 6).

Cisco Interval (Virgilian)

The Cisco interval (Virgilian) is the stratigraphically youngest part of the Pennsylvanian section. Depositional cycles in this interval are thin; there are 22 cycles in the 50-m-thick Cisco section, as compared to 19 cycles in the 100-m-thick Canyon section (Figs. 7, 8). Cycles in more basinward wells (to the east and south) are generally thicker. Cisco limestones are dominated by fossiliferous packstones and grainstones containing fusulinids, crinoids, brachiopods, and phylloid algae. All cycles in the Cisco are capped with subaerial-exposure surfaces. Soil-zone diagenesis is widespread including root traces, caliche crusts, *in situ* brecciation, vugs, and fissures. Shales are generally red and occur immediately above exposure surfaces. Many of the packstone- and grainstone-rich limestones in the central and western part of the field (around the "AD" #3 well and to the west) have shale that filled large pores in breccias, vugs, and fissures. Porosity in the Cisco is distinctly lower than in the Canyon, and where developed, occurs in grainstones.

Most Cisco limestones are interpreted as accumulating in shallow, episodically agitated open-marine water during marine

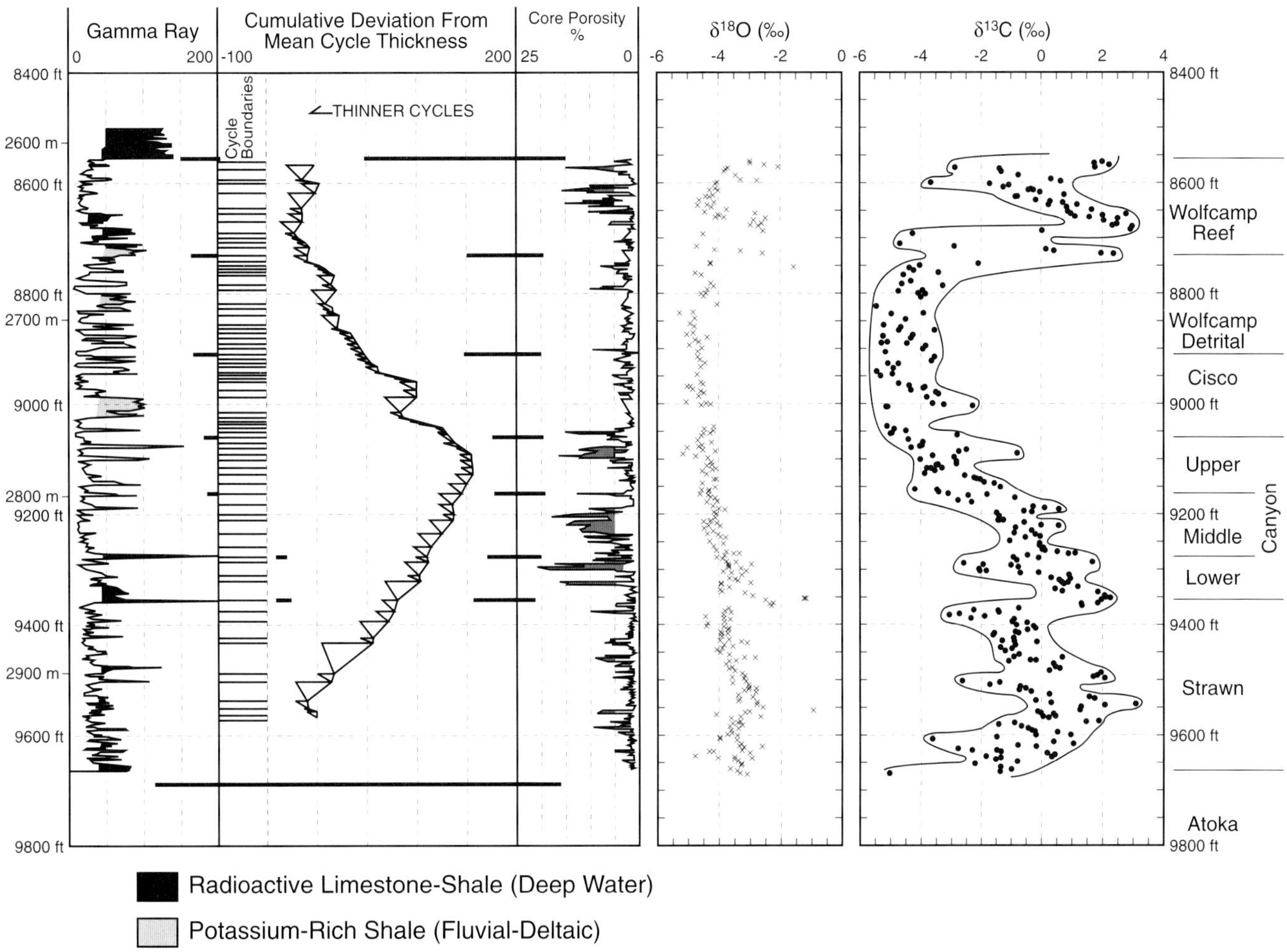

FIG. 10.—Plots of (from left to right) gamma ray, depth-based Fischer diagram, core-measured porosity, and stable isotope compositions for the Parker "V" #7 well. See Figure 9 for explanation.

flooding of the platform. Prolonged periods of subaerial exposure often caused intense alteration at exposure surfaces. Because the shales fill large soil-zone pores, they had to be emplaced after the subaerial exposure when the pores formed. However, neither the large soil-zone pores nor the shale is present in overlying, more pure limestones, indicating formation of the pores and emplacement of the shales before deposition of the next cycle. Shale thicknesses are highly variable, and imaging by three-dimensional seismic methods suggests that the shales filled irregular karst topography and channels incised during lowstands of sea level. For example, a 7-m-thick shale in the "V" #7 well is interpreted to fill a Cisco channel, but that shale is absent in adjacent wells (Fig. 7).

Wolfcamp Detrital Interval

The Wolfcamp detrital interval is a mixture of discontinuous limestones and shales with 18 cycles in a 50-m-thick unit. Limestones, 0.3–6 m thick, are mainly fossiliferous wackestones and packstones with phylloid algae and fusulinids. Shale bed thicknesses vary from less than 1 m to more than 25 m (Fig. 7). Almost all cycles in the Wolfcamp detrital interval are capped with subaerial-exposure surfaces. Limestones commonly have vugs and fissures filled with shale. Significant porosity (>4%) is rarely present in limestones of the Wolfcamp detrital interval.

Wolfcamp detrital interval limestones are interpreted to have accumulated in shallow, open-marine water during brief periods when sea-level rises flooded the platform. Large vugs and fissures formed in limestones and are associated with intense weathering during cyclic subaerial exposure. Larger shale bodies have morphologies that suggest they filled stream channels cut during lowstands of sea level.

Wolfcamp Reef Interval

The Wolfcamp reef interval contains approximately 12 depositional cycles in a section that is 50–55 m thick and dominantly limestone (Figs. 7, 8). Many of the cycles in the lower part of the Wolfcamp reef interval are relatively thin and shaly, and many of the cycles in the upper part are thick and consist of pure limestones. Most of the Wolfcamp reef interval is com-

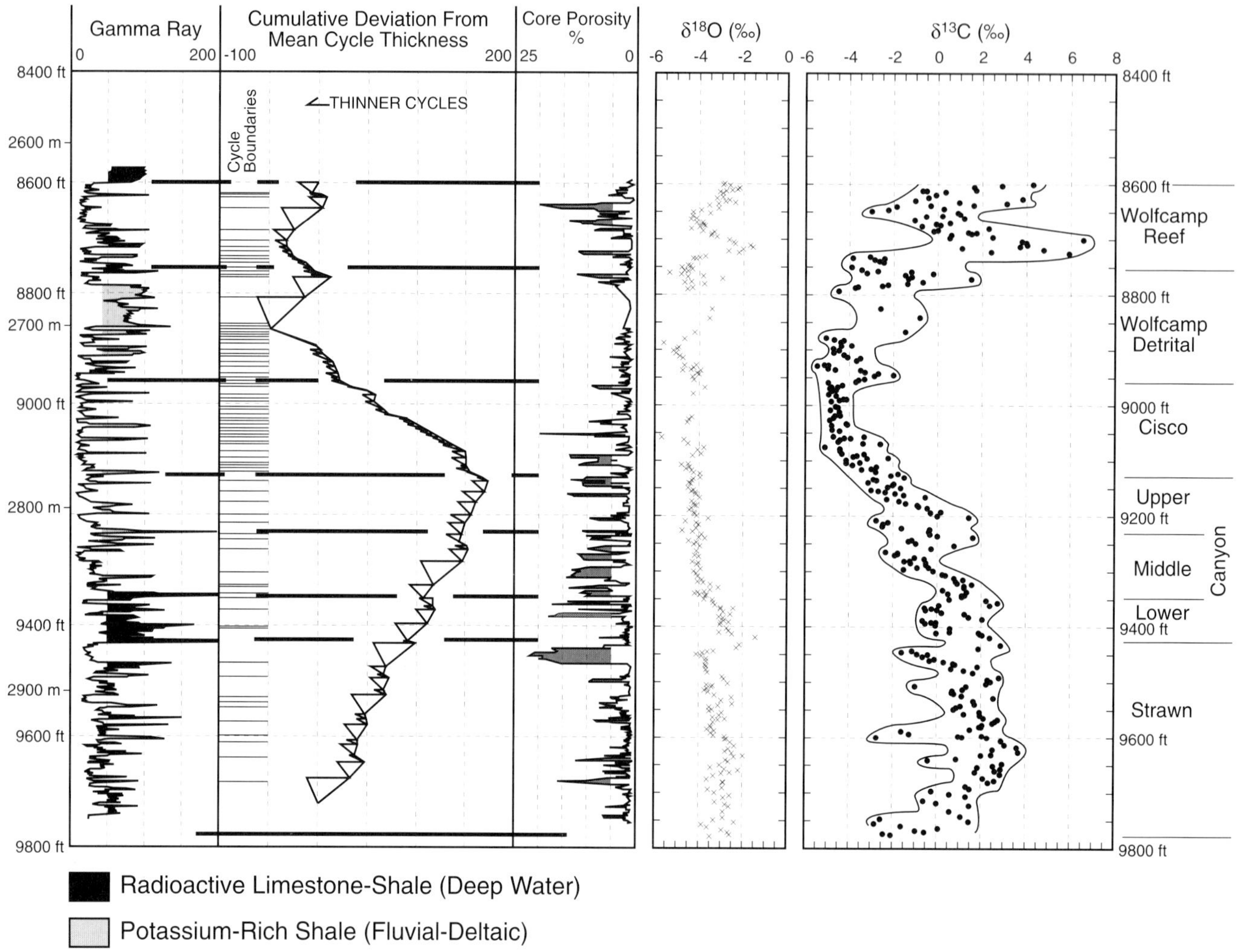

FIG. 11.—Plots of (from left to right) gamma ray, depth-based Fischer diagram, core-measured porosity, and stable isotope compositions for the Parker "X" #1 well. See Figure 9 for explanation.

posed of fossiliferous wackestones and packstones. Some lower cycles are mainly dark wackestone with coral, brachiopods, and crinoids, whereas cycles in the middle to upper parts of the Wolfcamp reef interval are composed largely of wackestones and packstones with phylloid algae, tubular foraminifers, and fusulinids. One thick cycle in the middle part of the Wolfcamp reef interval contains discontinuous phylloid boundstone up to 6 m (20 ft) thick. Most cycles are capped with distinct subaerial-exposure surfaces, though a few cycle tops in the lower part and at the top of the Wolfcamp reef interval do not have evidence of subaerial exposure. Substantial porosity is developed 0.5 to 5 m below subaerial-exposure surfaces in the middle to upper parts of the Wolfcamp reef interval. Porosity occurs in many different facies including wackestones, packstones, grainstones, and phylloid-algal boundstones.

Deposition of the Wolfcamp reef interval occurred mainly in open-marine, subtidal environments. Coral, brachiopods, and crinoids in dark wackestones of the lower part of the Wolfcamp reef interval suggest deposition in deeper water compared to the wackestones and packstones with phylloid algae, tubular foraminifers, and fusulinids in the middle part of the Wolfcamp reef interval. Phylloid-algal boundstones in the middle part of the Wolfcamp reef interval represent development of discontinuous mounds, perhaps 5–8 m high.

CYCLE STRATIGRAPHY

Use of cycle stacking patterns (Fischer plots) is somewhat controversial (Drummond and Wilkinson, 1993). Gianniny and Simo (1996) showed some shortcomings in using cycle stacking patterns in icehouse settings in which uneven sedimentation occurred and unfilled accommodation space was present at the end of cycles; these characteristics caused problems in interpreting cycle stacking patterns. Clearly our data set contains some of those problems. Phylloid-algal mounds are present in the Strawn and the Wolfcamp reef intervals, indicating uneven sedimentation. Valleys were cut by streams during deposition of the Cisco and the Wolfcamp detrital intervals. These features do cause one or two cycle anomalies in the cycle stacking patterns; however, through time and during deposition of 87 cycles, the carbonate depositional system has generally compensated for

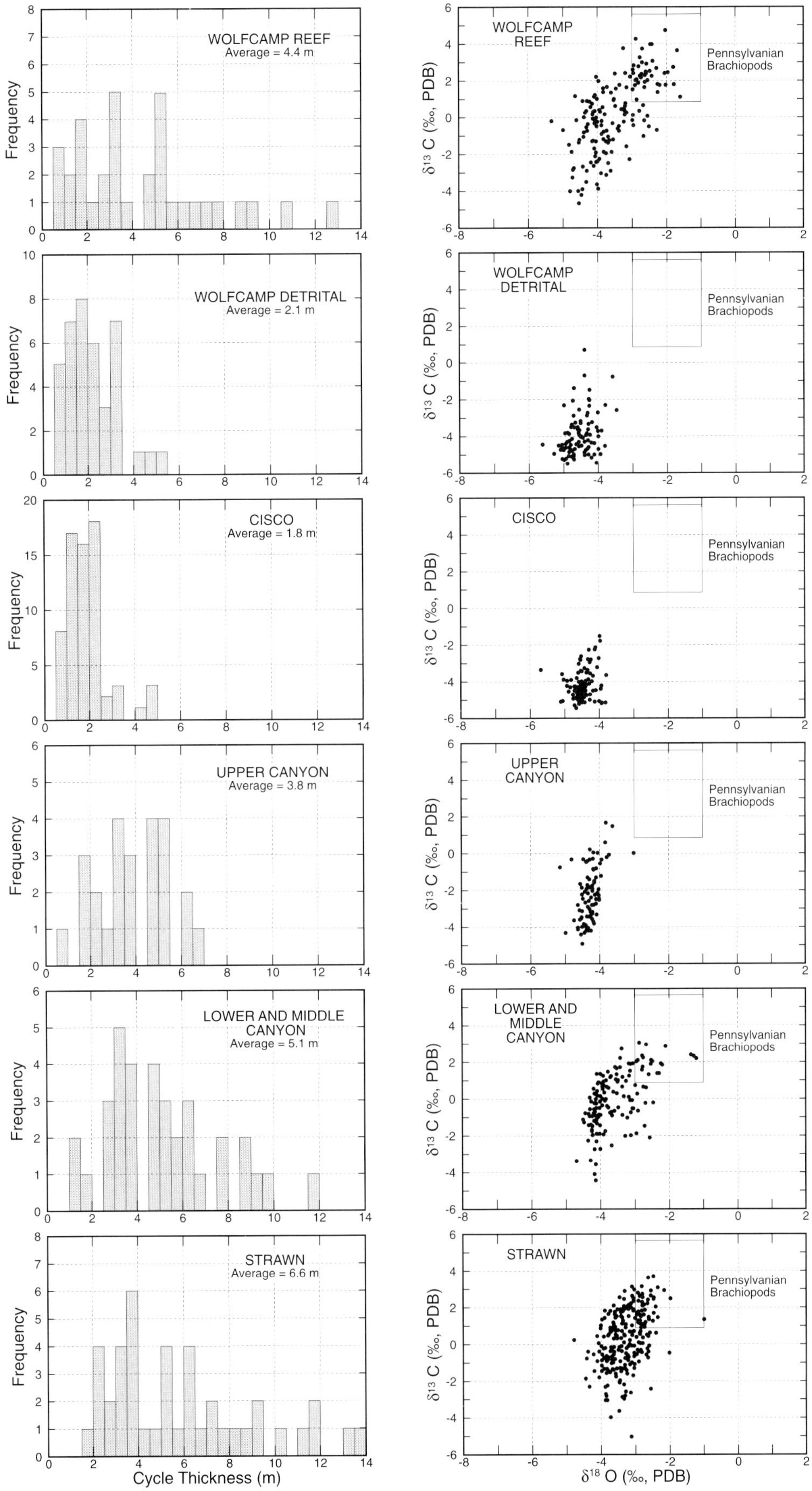

FIG. 12. Histograms of cycle thickness and $\delta^{13}C$ vs. $\delta^{18}O$ by stratigraphic interval. PDB is Peedee belemnite standard.

those local anomalies by filling the lows during subsequent cycles. The general cycle stacking patterns at Southwest Andrews are repeated in three separate locations, Suggesting that cycle stacking patterns indicate some real trends that are logical and follow the sequence stratigraphic models of Goldhammer et al. (1987, 1990) and Montanez and Osleger (1993).

The Strawn to Wolfcamp reef section (300–350 m thick) in the Southwest Andrews Area contains approximately 87 short-term (meter-scale) cycles—16 in the Strawn interval, 19 in the Canyon interval, 22 in the Cisco interval, 18 in the Wolfcamp detrital interval, and 12 in the Wolfcamp reef interval. Variations in cycle thickness are illustrated on a conventional Fischer plot for the Parker "X" #1 well in Figure 8. (See Sadler et al. [1993] for explanation of conventional Fischer plots.) Fischer diagrams modified to plot by depth are shown in Figures 9–11 with gamma-ray, porosity, and stable isotope data. (See Day [1997] for an explanation of depth-based Fischer plots.)

Variations in cycle thickness are fairly consistent within the three continuously cored wells (Table 1). The Strawn interval, lower part of the Canyon interval, and the Wolfcamp reef interval have many thick cycles (3–9 m; Fig. 12), more spiculitic mudstones to wackestones, and more radioactive limestone. These features suggest deeper-water deposition and greater accommodation (Figs. 7–11). Cycles in the middle and upper parts of the Canyon interval generally have an intermediate thickness (1.5–5.5 m; Fig. 12) and shale beds that increase in abundance up section and to the west (toward the "AD" #3 well; Fig. 7). The Cisco and the lower part of the Wolfcamp detrital intervals are dominated by thinner cycles (0.3–3 m; Figs. 8–12) and have few radioactive or spiculitic limestones but more potassium-rich shales that are discontinuous and interpreted to be of fluvial or deltaic origin (Fig. 7). The start of intermediate-thickness cycles in the Canyon interval is lower in the most landward well ("AD" #3) and higher in the most basinward well ("X" #1) (Fig. 13). The start of thin cycles is at about the Canyon/Cisco boundary in all three project wells (Fig. 13).

Relationship of Short-Term Cycles to Long-Term Stratigraphic Patterns

Sedimentary and stratigraphic features suggest that the Strawn to Wolfcamp reef section contains one long-term regression (Strawn to Cisco) and one long-term transgression (Wolfcamp detrital and Wolfcamp reef intervals). Grainstones interpreted as shallow-water deposits increase and deep-water spiculitic wackestone and mudstones decrease upward from the base of the Strawn to the top of the Cisco, suggesting a general regression of depositional environments in that part of the section. Candelaria et al. (1992) recognized a major unconformity (sequence boundary) at the top of the Cisco (Pennsylvanian) on the east side of the Central Basin platform. Depositional cycles generally thin upward in the Canyon and Cisco intervals as they approach that major unconformity (sequence boundary) at the top of the Cisco (Figs. 8–11).

In the Wolfcampian, depositional facies show a general deepening trend from the base of the Wolfcamp detrital interval to the top of the Wolfcamp reef interval. The base of the

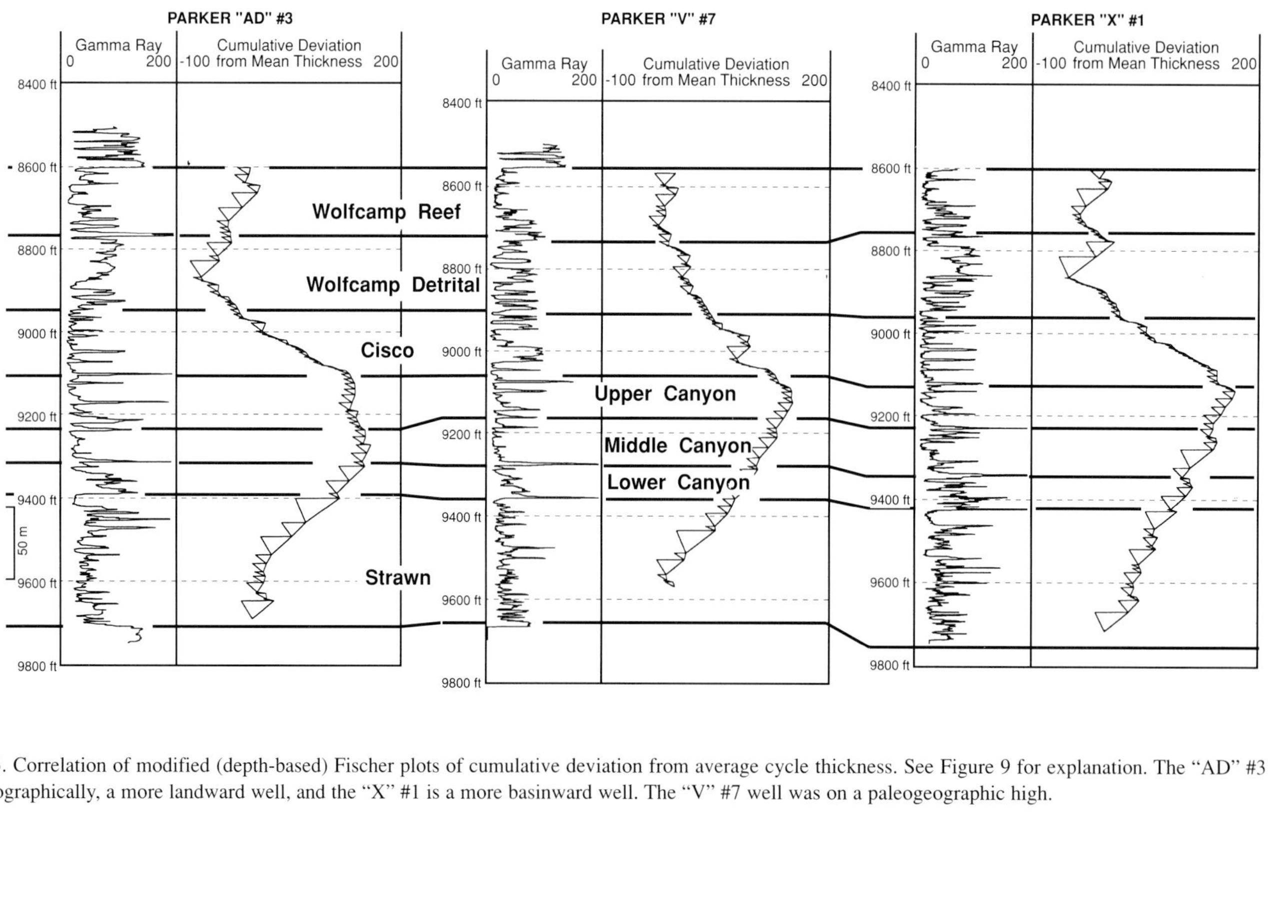

FIG. 13. Correlation of modified (depth-based) Fischer plots of cumulative deviation from average cycle thickness. See Figure 9 for explanation. The "AD" #3 is, paleogeographically, a more landward well, and the "X" #1 is a more basinward well. The "V" #7 well was on a paleogeographic high.

Wolfcamp detrital interval is dominated by fluvial-to-deltaic shales and shallow-marine packstone-grainstones. In contrast, the top of the Wolfcamp reef is dominated by wackestones with open-marine fossils. At the end of Wolfcamp reef deposition (fusulinid zone PW-2; Wilde, 1990; Fig. 7), the platform margin apparently drowned and stepped back landward approximately 18 km (11 miles) to where the Abo (and the stratigraphically equivalent Wichita) platform margin began to aggrade (Fig. 3). Therefore, we interpret the top of the Wolfcamp reef interval as a maximum-flooding surface as defined by Van Wagoner et al. (1988). Cycles generally thicken upward as the maximum-flooding surface at the top of the Wolfcamp reef interval is approached. Therefore, this data set supports sequence stratigraphic models that predict (1) a thinning of depositional cycles as the rate of accommodation decreases where a major unconformity (sequence boundary) is approached and (2) a thickening of cycles as the rate of accommodation increases where a maximum-flooding surface is approached (Sarg, 1988; Goldhammer et al., 1990; Montanez and Oslcger, 1993).

Comparison with Other Areas

The Southwest Andrews Area has a similar number of cycles as time-equivalent sections in other basins of North America, supporting a glacio-eustatic control on those Pennsylvanian and Early Permian cycles. Heckel (1986) reported 17 Missourian cycles in eastern Kansas, Nebraska, and Iowa. This result compares favorably with 19 Missourian cycles in the Southwest Andrews Area. Wilson (1967) reported approximately 20 cycles in the Virgilian Holder Formation in outcrops in the Sacramento Mountains of southeast New Mexico. The Virgilian may extend a few cycles (two?) higher in the Laborcita Formation (Otte, 1959) and possibly one cycle below the Holder. Therefore, approximately 23 cycles occur in Virgilian rocks of the tectonically active Orogrande basin. That number of cycles in the Orogrande basin compares favorably to the number of cycles (22) that we observed in the Virgilian strata at the western margin of the Midland basin. Similar numbers of short-term cycles in different basins suggest that those short-term cycles are correlative and eustatic and support conclusions of Heckel (1986, 1994) and Dickinson et al. (1994) that Pennsylvanian and Lower Permian cycles are related to glacio-eustasy.

Duration of Depositional Cycles

Previous workers have made estimates of Middle to Late Pennsylvanian cycle duration, including 235,000–400,000 yr by Heckel (1986), 330,000–370,000 yr by Algeo (1991), 230,000–385,000 yr by Goldhammer et al. (1991), and 352,000 yr by Connolly and Stanton (1992). The Cisco and Canyon intervals in the Southwest Andrews Area show distinct shallow-marine cycles that are suitable for determining duration of cycles. The Canyon (Missourian) and Cisco (Virgilian) correlate to the European Stephanian Stage (Heckel, 1986; Harland et al., 1990). The Stephanian in Europe was thought to represent approximately 13 m.y. (290–303 Ma; Harland et al., 1990). The Canyon and Cisco in the Southwest Andrews Area have approximately 41 cycles, i.e., an average of one cycle per 317,000 yr. However, the above calculations are dependent on dating boundaries of geologic intervals in absolute time (years). These dates are variable and potentially have large errors. For example, many authors (including us) use absolute ages for boundaries from Harland et al. (1982, 1990); these boundary ages have errors (1σ) of 5 m.y. to 10 m.y. for intervals with estimated durations of 13 m.y. Therefore, the above calculations represent only very crude estimates of cycle duration.

Recent work on radiometric dating of caliches in core and outcrops has allowed much higher precision in estimating duration of cycles (Rasbury et al., 1997, 1998). A caliche one cycle below the top Wolfcamp cycle in the Parker "X" #1 was dated as 298 (±1.4 Ma; 2σ; Rasbury et al., 1997). A caliche approximately two cycles above the Missourian/Virgilian contact in the Holder Formation of the Sacramento Mountains has a radiometric age of 306 ±2.6 Ma (2σ; Rasbury et al., 1998). In the Southwest Andrews Area, we have 29 Wolfcampian cycles and approximately 20 Virgilian cycles in the strata between the two paleosols with radiometric ages. Therefore, 49 cycles were deposited in approximately 8 m.y. (±4 m.y; 2σ), i.e., an average of 163,000 yr per cycle. Additional Virgilian cycles are probably missing because of cycle amalgamation; hence, the true average cycle duration was probably less than 163,000 yr.

If the cycle duration is assumed to be approximately 163,000 yr, approximate long-term accommodation rates can be calculated for the Southwest Andrews Area. In the Cisco interval and lower part of the Wolfcamp detrital interval, 30 cycles have a total thickness of approximately 72 m. If cycles average 163,000 yr in length, the Cisco and the lower part of the Wolfcamp detrital intervals represent approximately 5 m.y. Deposition of the cycles at the base of the strata under study started at sea level, and cycles repeatedly built to near sea level during that period of time. Hence, accommodation was 72 m in 5 m.y. or approximately 14 m/m.y. The lower and middle parts of the Canyon interval are approximately 60 m (200 ft) thick and contain 11 cycles. Deposition during the lower and middle Canyon started and ended at sea level. If 163,000 yr per cycle is assumed, lower and middle Canyon deposition spanned 3.1 m.y. Approximately 60 m of accommodation in 1.8 m.y. gives a rate of accommodation of 33 m/m.y. for the lower and middle Canyon section. Strawn deposition started in deep water; hence, the Strawn is not suitable for calculating longer-term accommodation.

STABLE ISOTOPE TRENDS

Stable carbon and oxygen isotope compositions give insight into the diagenetic history of carbonates (Moore, 1989). A large amount of oxygen is present in water, so when carbonates precipitate or recrystallize in meteoric water, the ratio of heavy (^{18}O) to light (^{16}O) oxygen isotopes of the resulting carbonate can change (decrease) relatively quickly to be in equilibrium with the isotopically lighter meteoric waters (Gross and Tracy, 1966; Hudson, 1977). In contrast, the amount of carbon that can be dissolved in water is relatively low (Garrels and Christ, 1965,

p. 76), and, hence, large volumes of water carrying dissolved carbonate with distinctly different carbon isotope compositions are required to change the carbon isotope composition of a limestone, even during complete recrystallization.

The stable carbon isotope compositions of many limestones subjected to subaerial exposure become lighter because of preferential incorporation of light carbon (^{12}C, as opposed to heavy carbon, ^{13}C) from soil-borne (soil-gas) carbon dioxide and decay of terrestrial organic material (Allan and Matthews, 1977, 1982; Hudson, 1977; Hanor, 1978). Although incorporation of light carbon isotopes is concentrated in calcite immediately below subaerial-exposure surfaces, cements and recrystallized fossils can also incorporate substantial amounts of light carbon isotopes substantially below the exposure surface (Gross and Tracy, 1966; Saller and Moore, 1991; Saller, 1992). Because light oxygen and carbon isotopes are preferentially added to limestones during subaerial exposure, it is logical to conclude that longer subaerial exposure will result in limestones with progressively lighter carbon and oxygen isotope compositions. Therefore, carbon and oxygen isotope compositions of limestones should be useful as indicators of relative length of subaerial exposure. Stable isotope data and their relationships to subaerial exposure were previously discussed by Dickson and Saller (1995) for the lower part of the Canyon interval and middle to upper parts of the Wolfcamp reef interval in the Southwest Andrews Area. Our new data set covers a much thicker section that allows correlation of stable isotope trends with the Strawn to Cisco regression and the Wolfcamp detrital to Wolfcamp reef transgression. Figures 9–11 and Table 1 show the relationships between patterns in cycle thickness, porosity, and bulk-rock isotope compositions for the three wells in the study area.

Several lines of evidence suggest that most of the stable isotope variations in bulk-rock samples in this data set are diagenetic, not depositional. First, the range of stable isotope compositions observed in the Southwest Andrews Area extends from an isotopically heavy end member ($\delta^{13}C_{PDB}$ = +2.5‰ to +4.90‰; $\delta^{18}O_{PDB}$ = −1.5‰ to −3‰) similar to Late Pennsylvanian marine calcite (Grossman et al., 1991) to lighter values ($\delta^{13}C_{PDB}$ as low as −6‰ and $\delta^{18}O_{PDB}$ as low as −5.5‰; Fig. 12). Second, there is no correspondence of $\delta^{13}C$ and $\delta^{18}O$ variations observed in our data set with known secular marine trends (Grossman et al., 1991). Third, limestones with similar depositional compositions (similar types of grains and amounts of micrite) have substantially different isotope compositions in different parts of the reservoir section. For example, phylloid-algal wackestones in the Strawn have $\delta^{13}C$ values of +2.2‰ and $\delta^{18}O$ values of −3.0‰ (PDB), whereas phylloid-algal wackestones in the lower part of the Wolfcamp detrital interval have $\delta^{13}C$ PDB values of −4.6‰ and $\delta^{18}O_{PDB}$ values of −4.3‰. Fourth, large isotope variations within similar facies within the same cycle also suggest that isotope variations are mainly diagenetic.

Three fields of stable isotope composition are present in bulk-rock limestones from the Southwest Andrews Area: (1) consistent, relatively heavy carbon isotopes and relatively heavy oxygen isotopes, (2) variable carbon and consistent, moderately light oxygen isotopes, and (3) relatively light carbon and relatively light oxygen isotopes.

1. Relatively heavy carbon ($\delta^{13}C_{PDB}$ of 0 to +2‰) and oxygen isotopes ($\delta^{18}C_{PDB}$ of approximately −3‰) are found in the lower and middle part of many cycles in the Strawn interval, the lower part of cycles in the lower Canyon interval, the lower part of cycles in the lower part of the Wolfcamp reef interval, and the uppermost Wolfcamp reef cycle (Figs. 9–12, 14). Those strata are characterized by deeper-water facies (spiculitic mudstones), radioactive limestones, relatively thick cycles, low porosity, and little diagenesis related to subaerial exposure (Figs. 7–12).
2. Cycles with variable carbon ($\delta^{13}C_{PDB}$ of −4‰ to 0‰) and intermediate to light oxygen ($\delta^{18}O_{PDB}$ of −4‰ to −4.5‰) isotope compositions are found in the uppermost Strawn, uppermost part of the lower Canyon, middle part of the Canyon, and middle to upper parts of the Wolfcamp reef intervals (Figs. 7–12). Light carbon isotopes generally predominate in the uppermost parts of these cycles, and heavier isotopes predominate in the lower parts of the cycles (similar to cycles of Allan and Matthews, 1977, 1982). Strata with variable carbon and moderately light oxygen isotope values are characterized by cycles having intermediate thickness and porosity in their upper parts (Fig. 14).
3. Strata with light carbon ($\delta^{13}C_{PDB}$ of −4‰ to −5.5‰) and light oxygen isotopes ($\delta^{18}O_{PDB}$ of −4‰ to −5‰) include the uppermost Canyon rocks, the Cisco interval, and most of the Wolfcamp detrital interval. These have generally thin cycles (<2 m) and low porosity (Figs. 7 12, 14).

Substantial variation in isotope compositions is present within cycles (Fig. 14). In thick cycles with distinct subaerial exposure (in the upper part of the Strawn, the Canyon, and the Wolfcamp reef intervals), the $\delta^{13}C$ values are generally low immediately below the subaerial surfaces and increase downward within each cycle (Figs. 14, 15). This pattern is attributed to soil-gas CO_2, which primarily contains light carbon and is most abundant immediately below the exposure surface, but decreases with depth (similar to profiles shown by Allan and Matthews [1977, 1982]). $\delta^{18}O_{PDB}$ values immediately below subaerial-exposure surfaces are relatively low (−4‰ to −5‰), but the lower part of some thick cycles (>5 m) in the Strawn, the lower part of the Canyon, and the Wolfcamp reef intervals have bulk-rock $\delta^{18}O$ values of −2‰ to −3‰, which are similar to calcite precipitated from Late Pennsylvanian seawater (Grossman et al., 1991; Grossman, 1994). These findings suggest that meteoric diagenesis did not penetrate deeply into those cycles, and, hence, the lower parts of those cycles did not recrystallize in meteoric water.

Large-scale trends in $\delta^{13}C$ and $\delta^{18}O$ are superimposed on the meter-scale $\delta^{13}C$ and $\delta^{18}O$ variations within cycles. The large-scale variations in isotope composition follow cycle stacking patterns. As cycles generally decrease in thickness from the upper part of the Strawn to the top of the Cisco, the $\delta^{13}C$ and $\delta^{18}O$ compositions of bulk-rock samples generally decrease in all three study wells (Fig. 15). Because lower $\delta^{13}C$ and $\delta^{18}O$ values indicate greater interaction with meteoric water containing isotopically light carbon and oxygen from soil gas and/or terrestrial organic material (Hudson, 1977), lighter $\delta^{13}C$ and $\delta^{18}O$ compositions of whole cycles should indicate that subaerial

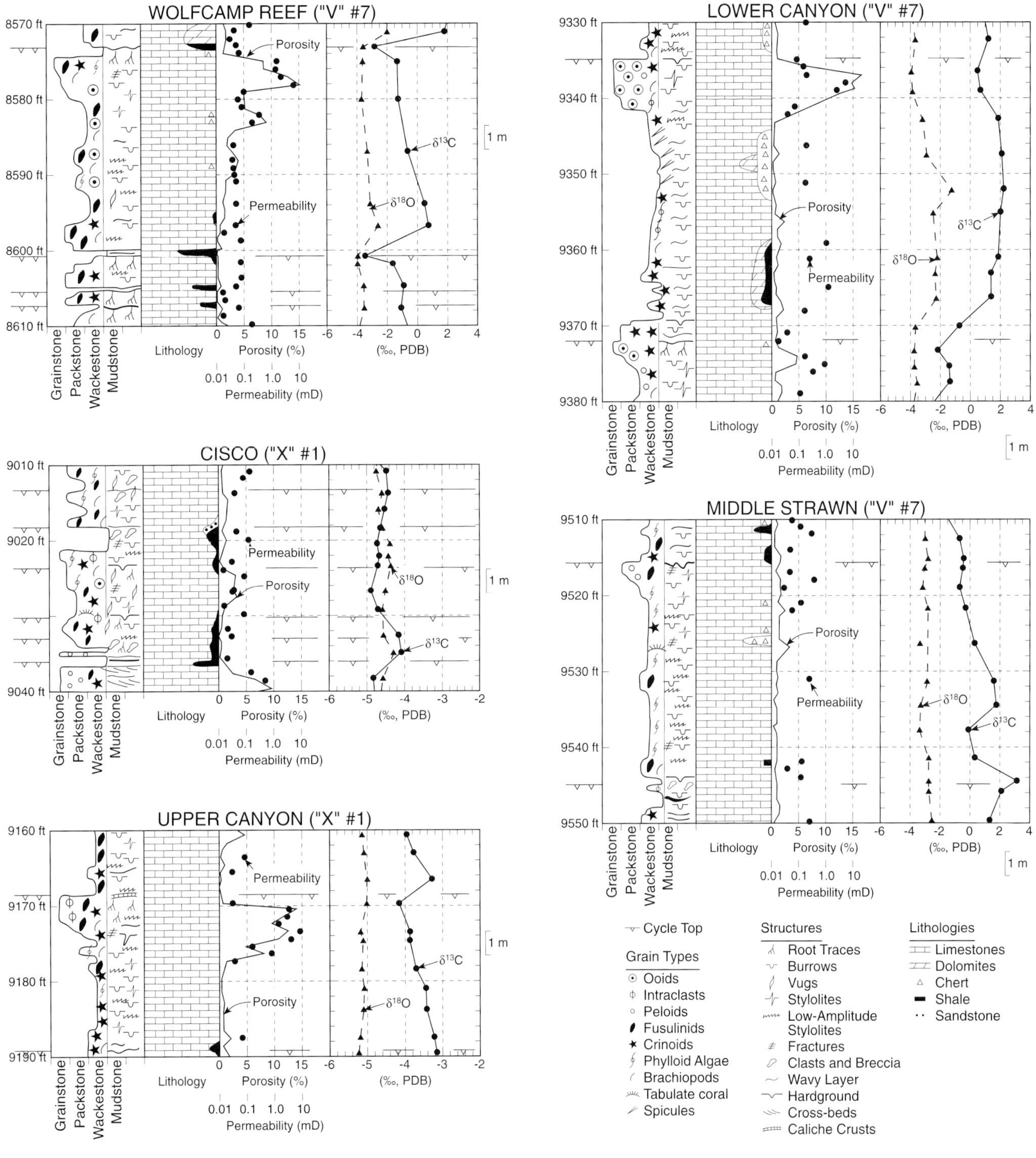

FIG. 14.—Characteristics of "typical cycles" in the Strawn, the lower part of the Canyon, the upper part of the Canyon, the Cisco, and the Wolfcamp reef intervals. PDB is Peedee belemnite standard.

exposure and associated meteoric diagenesis lasted longer. Low $\delta^{13}C$ and $\delta^{18}O$ values ($\delta^{13}C_{PDB}$ of −4‰ to −5.5‰ and $\delta^{18}O_{PDB}$ of −4‰ to −5‰), which are pervasive in Cisco and lower Wolfcamp detrital cycles, indicate that much isotopically light carbon and oxygen have been incorporated into these otherwise normal-marine limestones (Fig. 14). A prolonged period of subaerial exposure is required to allow a sufficient volume of water to transport enough ^{12}C-rich terrestrial organic carbon and soil-gas CO_2 into a cycle to decrease the $\delta^{13}C_{PDB}$ value of the entire cycle from a depositional composition of approximately +2‰ to the current composition of −4‰ to −5.5‰. The general large-scale decrease in $\delta^{13}C$ and $\delta^{18}O$ values from the upper part of the Strawn to the Cisco is, therefore, thought to be due to a general increase in duration of cyclic subaerial exposure.

TABLE 1.—AVERAGE VALUES FOR CYCLE THICKNESS, STABLE ISOTOPE COMPOSITION, AND POROSITY BY WELL.

Interval	"AD" #3				"V" #7				"X" #1			
	Cycle Thickness (m)	$\delta^{13}C$ (‰)	$\delta^{18}O$ (‰)	Porosity (%)	Cycle Thickness (m)	$\delta^{13}C$ (‰)	$\delta^{18}O$ (‰)	Porosity (%)	Cycle Thickness (m)	$\delta^{13}C$ (‰)	$\delta^{18}O$ (‰)	Porosity (%)
Wolfcamp Reef	5.1	+0.3	–3.4	3.2	4.3	+0.4	–3.6	3.6	4.1	+0.2	–3.5	3.5
Wolfcamp Detrital	2.1	–3.7	–4.7	2.2	2.7	–4.2	–4.5	2.4	1.7	–3.8	–4.4	2.6
Cisco	2.3	–4.5	–4.5	2.4	2.1	–4.3	–4.6	2.6	2.5	–4.0	–4.4	3.7
Upper Canyon	4.1	–2.7	–4.2	1.5	3.5	–2.8	–4.3	1.1	4.0	–1.2	–4.2	3.9
Middle Canyon	4.0	–1.2	–4.0	2.2	5.8	–0.4	–4.0	5.5	5.2	0.0	–3.9	5.7
Lower Canyon	6.1	+0.8	–2.6	4.5	5.1	+0.3	–3.2	4.0	4.8	+0.9	–2.7	4.1
Strawn	7.5	+0.5	–3.2	2.8	6.3	–0.5	–3.5	1.9	6.0	+1.1	–3.1	3.3

Note: Stable isotope compositions are reported relative to the PDB (Peedee belemnite) standard

These isotope data are independent support for the Goldhammer et al. (1987, 1990) model. That model assumes cycles of equal duration. The model predicts that a thicker cycle is deposited during times of high long-term rates of accommodation when the area spends most of the time represented by the cycle in a subtidal environment accumulating carbonate sediment. In contrast, a thinner cycle is predicted to be deposited during times of low long-term rates of accommodation when the area spends only a short amount of the time represented by the cycle in a subtidal environment accumulating sediment; because most of the time represented by the cycle is spent above sea level, the cycle is characterized by subaerial exposure, meteoric diagenesis, and little carbonate accumulation. The stable isotope compositions of the Pennsylvanian and Lower Permian cycles support the idea that each thin cycle (<2 m; Cisco and lower part of the Wolfcamp detrital intervals) spent a substantial length of time being subaerially exposed. In contrast, generally heavier isotopes in thicker cycles (Strawn, lower part of the Canyon, and Wolfcamp reef intervals) suggest only a brief period of subaerial exposure.

The stable isotope patterns in the three project wells are distinct enough that they could be used for correlation of thick (hundreds of meters) and thin stratigraphic units (1–5 m). A large-scale trend from heavy isotopes in the Strawn interval to light isotopes in the Cisco interval is present in all three wells (Fig. 15; Table 1). The light isotope compositions of the Cisco and Wolfcamp detrital intervals occur in all three wells (Figs. 9–11). On a smaller scale, cycles with heavier stable isotope compositions found in the middle part of the Strawn, lower part of the Canyon, and lower part of the Wolfcamp reef intervals are particularly distinctive and correlative (correlation horizons "B" and "D" in Fig. 15). Some isotope signatures associated with subaerial exposure in single cycles in the uppermost Strawn, uppermost lower part of the Canyon, middle to upper parts of the Canyon, and upper part of the Wolfcamp reef intervals are distinct and correlative (correlation horizons "A," "C," and "E" in Fig. 15).

DISTRIBUTION OF POROSITY

Porosity is highly variable and shows relationships to stratigraphic position and depositional facies in the Southwest Andrews Area. Only 10% to 25% of the Strawn through Wolfcamp reef section has reservoir-grade porosity (4% or more). Reservoir-grade porosity is generally found in cycles that are thicker than 2 m and that were subjected to subaerial exposure for a brief to moderate amount of time. Many >2-m-thick cycles in the lower part of the Strawn interval (Fig. 16) were

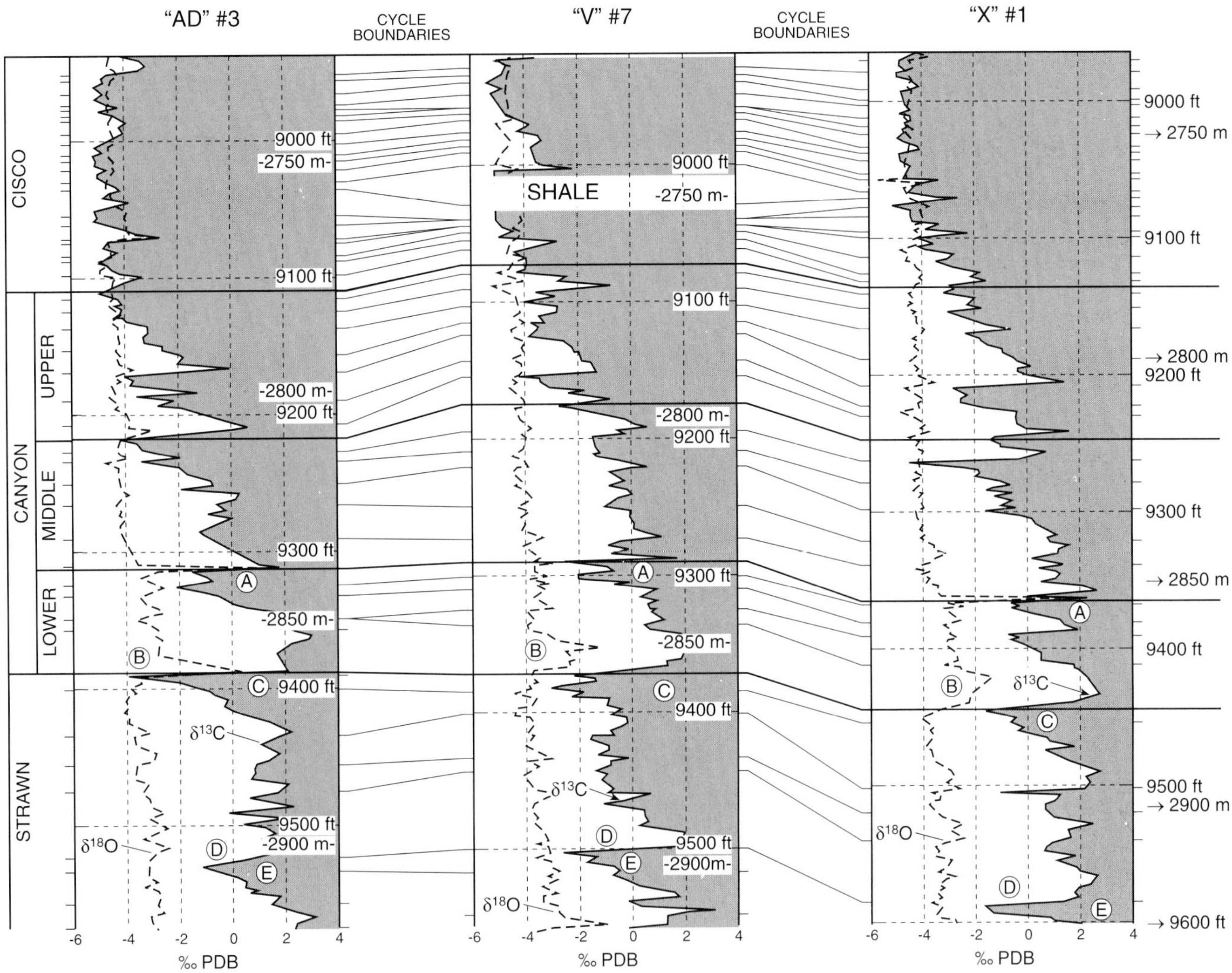

FIG. 15.—Comparison and correlation of stable carbon and oxygen isotope compositions of the upper part of the Strawn, the Canyon, and the Cisco intervals of the three cored wells. PDB is Peedee belemnite standard.

subjected to little or no subaerial exposure and have no significant porosity. Thin cycles (<2 m thick), which apparently underwent prolonged subaerial exposure (possibly >80% of the time represented by a cycle), generally have little or no porosity.

Depositional facies were important in determining a limestone's porosity and permeability in the subsurface. Phylloid boundstones are rare, but where present, have fair to good porosity (4%–17%) with variable and often high permeability (1–1000 mD). Phylloid-rich wackestones and packstones have generally low porosity and permeability except in the Strawn interval and middle part of the Canyon interval in the eastern part of the field, where some phylloid wackestones and packstones are porous and permeable. Other types of fossiliferous wackestones and packstones are generally not porous, except for two 5–10-m-thick cycles in the middle part of the Wolfcamp reef interval that have fair to good porosity (4%–14%) and permeability (0.07–10 mD). Bioclastic grainstones commonly have porosity greater than 10% where they are immediately below subaerial-exposure surfaces and in packages greater than 1 m thick; however, bioclastic grainstones in the lower (transgressive) parts of cycles are generally not porous. Where porous, bioclastic grainstones have 4%–15% porosity with moderate permeability (0.1–10 mD). Many ooid-peloid grainstones have fair to excellent porosity (4%–20%) with low to moderate permeability (0.1–10 mD). Ooid-peloid grainstone samples with permeability greater than 10 mD generally have significant intergranular and/or vuggy porosity.

The distribution of porosity in the three project wells is shown in Figure 17 and Table 1. The Parker "AD" #3 well is in a more landward paleogeographic location and has less porosity in the Cisco interval and the middle to upper parts of the Canyon interval than the Parker "V" #7 and Parker "X" #1 wells. Lower porosity in the Cisco interval and the middle to upper parts of the Canyon interval of the "AD" #3 is related to several factors: (1) thinner grainstones, (2) less phylloid boundstone, (3) thinner cycles, (4) more prolonged subaerial exposure between deposition of cycles, and (5) more interbedded shale. The "AD" #3 well has more porosity than the "V" #7 and "X" #1 in the middle to upper parts of the Strawn. This is apparently related to more grainstones and boundstones and longer, though still brief, subaerial exposure (perhaps 10% of the time represented by the cycles) during deposition of the middle to upper parts of the

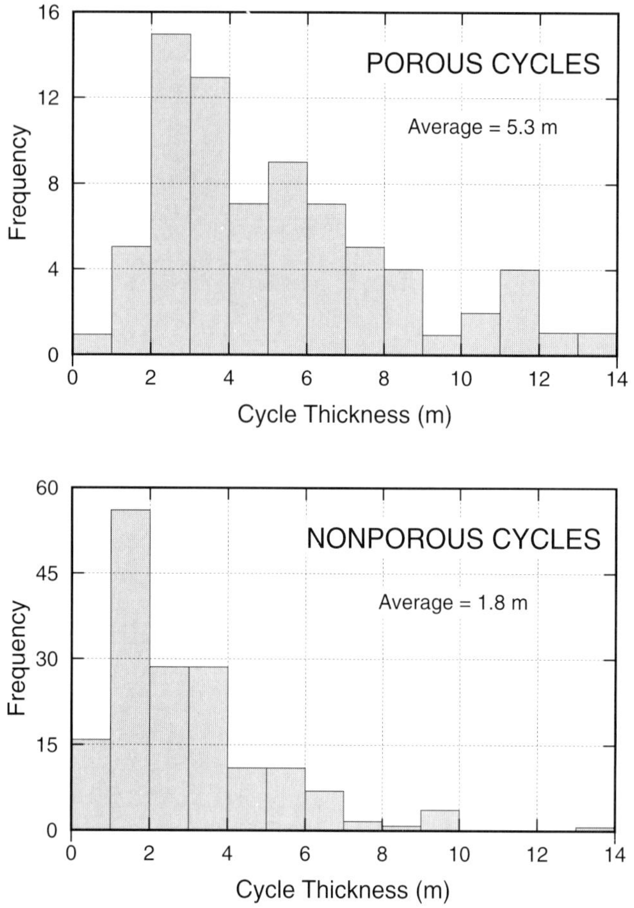

FIG. 16.—Histograms of thickness of porous cycles and nonporous cycles. Porous cycles are defined as cycles with >1 m of >5% porosity.

Strawn in the Parker "AD" #3 well. In summary, a number of different factors have determined the porosity and permeability of these limestones. Although much of the porosity occlusion in these rocks occurred during moderate to deep burial (1000–2500 m), depositional processes and early burial diagenesis largely determined how a given rock responded to burial stresses (Dickson and Saller, 1995). Furthermore, porosity shows distinct relationships to patterns of cycle thickness and duration of subaerial exposure.

CONCLUSIONS

Long-term variations in accommodation affect many important attributes of individual depositional cycles including cycle thickness, depositional facies, amount of interbedded clastics, duration of subaerial exposure, and amount of porosity. Several features are present during more rapid long-term increases in accommodation (more than 33 m/m.y.?) including (1) 3–15-m-thick cycles, (2) deposition dominated by mudstones and wackestones, (3) brief subaerial exposure (less than 20% of the time represented by a cycle; i.e., less than 30,000 yr?) with minor to moderate diagenetic alteration, (4) rare interbedded clastics, and (5) porosity in the upper part of cycles that were subjected to subaerial exposure (Fig. 14). Cycles deposited during slow long-term accommodation (approximately 14 m/m.y.) are thin, contain many packstone and grainstone beds, and were subjected to prolonged subaerial exposure (more than 80% of the time represented by a cycle; i.e., greater than 130,000 yr?) resulting in intense diagenetic alteration (Fig. 14). Those thin cycles have many interbedded shales and little porosity. This data set supports the idea that many depositional cycles have a similar duration (approximately 163,000 yr), but thickness is determined largely by the amount of time the sediment surface spent below sea level, and intensity of subaerial exposure is determined by the amount of time the sediment surface spent above sea level (Goldhammer et al., 1987, 1990).

ACKNOWLEDGMENTS

Results reported in this paper are part of a joint research project involving Unocal and the Japan National Oil Company–Technical Research Center (JNOC/TRC). Many people from both companies helped in this study including H. Matsuda, N. Shimada, T. Kozawa, H. Ishida, and Y. Tsumuraya of JNOC/TRC, and Skip Walden, P. Johnston, S. Boyd, G. Fitzgerald, S. Frost, S. Smith, F. Martens, and R. Todor of Unocal. We thank P. Day (Unocal) for making our depth-based Fischer plots. A. Walden and F. Martens helped us in interpreting and plotting the seismic information. Stable carbon and oxygen isotope analyses were performed by Robert Fifer at the Benedum Stable Isotope Laboratory, Brown University. G. Wilde analyzed samples for biostratigraphic correlations. We thank Clint Cowan, Jim Markello, and Toni Simo for thoughtful reviews of this manuscript. We also thank Mike Kirby for drafting all figures.

REFERENCES

ALGEO, T. J., 1991, Lower–Middle Pennsylvanian Gobbler Formation: Eustatic and tectonic controls on carbonate shelf cyclicity: Geological Society of America Abstracts with Programs, v. 23, no. 4, p. 2.

ALGEO, T. J., WILKINSON, B. H., AND LOHMANN, K. C., 1992, Meteoric-burial diagenesis of Middle Pennsylvanian limestones in the Orogrande basin, New Mexico: Water/rock interactions and basin geothermics: Journal of Sedimentary Petrology, v. 62, p. 652–670.

ALLAN, J. R., AND MATTHEWS, R. K., 1977, Carbon and oxygen isotopes as diagenetic and stratigraphic tools: Surface and subsurface data, Barbados, West Indies: Geology, v. 5, p. 16–20.

ALLAN, J. R., AND MATTHEWS, R. K., 1982, Isotope signatures associated with meteoric diagenesis: Sedimentology, v. 29, p. 797–817.

BORER, J. M., AND HARRIS, P. M., 1991, Lithofacies and cyclicity of the Yates Formation, Permian basin: Implications for reservoir heterogeneity: American Association of Petroleum Geologists Bulletin, v. 75, p. 726–779.

CANDELARIA, M. P., SARG, J. F., AND WILDE, G. L., 1992, Wolfcamp sequence stratigraphy of the eastern Central Basin platform, *in* Mruk, D. H. and Curran, C., eds., Permian Basin Exploration and Production Strategies: Application of Sequence Stratigraphic and Reservoir Characterization Concepts: Midland, West Texas Geological Society Publication 92-91, p. 27–44.

CONNOLLY, W. M., AND STANTON, R. J., 1992, Interbasinal cyclostratigraphic correlation of Milankovitch band transgressive-regressive cycles: Correlation of Desmoinesian-Missourian strata between southeastern Arizona and the Midcontinent of North America: Geology, v. 20, p. 999–1002.

CROWLEY, T. J., AND BAUM, S. K., 1991, Estimating Carboniferous sea-level fluctuations from Gondwanan ice extent: Geology, v. 19, p. 975–977.

DAY, P. I., 1997, The Fischer diagram in the depth domain: A tool for sequence stratigraphy: Journal of Sedimentary Research, v. 67, p. 982–984.

DICKINSON, W. R., SOREGHAN, G. S., AND GILES, K. A., 1994, Glacio-eustatic ori-

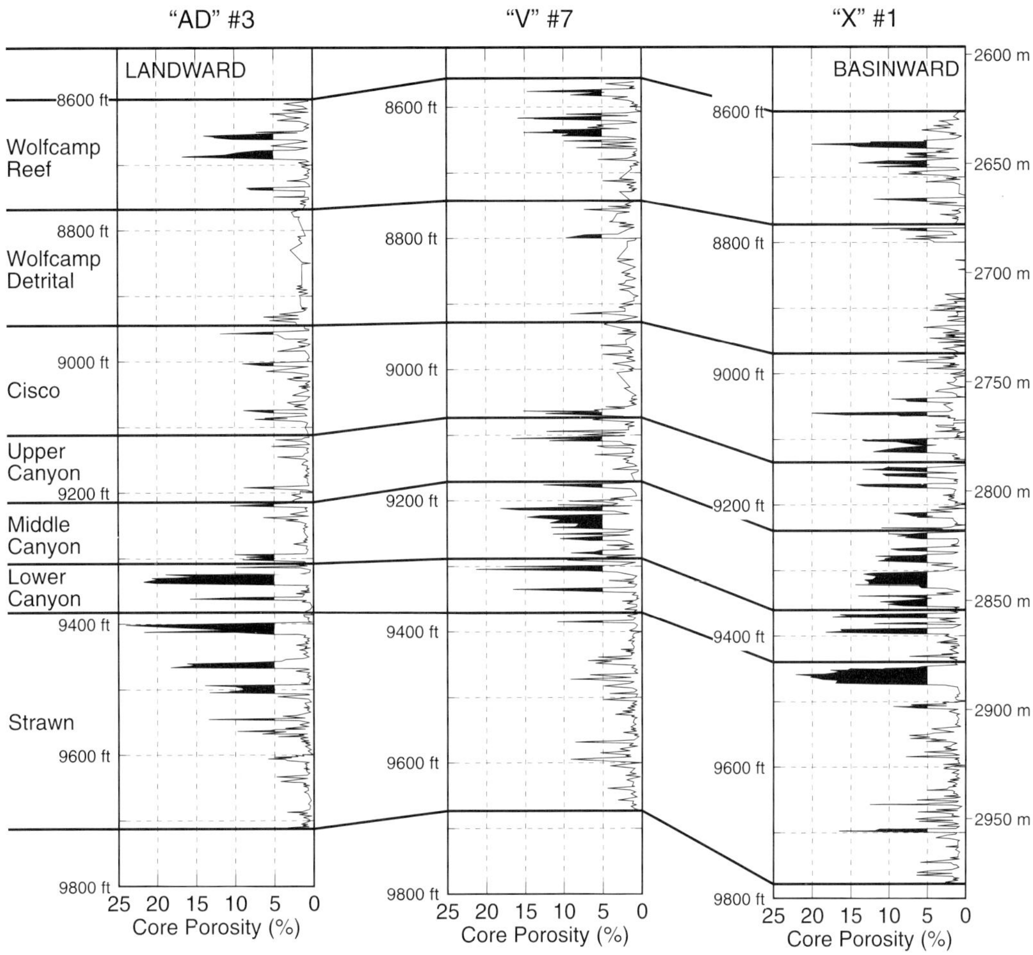

FIG. 17.—Correlation of core-measured porosity between wells.

gin of Permo-Carboniferous stratigraphic cycles: Evidence from the southern Cordilleran foreland region, *in* Dennison, J. M. and Ettensohn, F. R., eds., Tectonic and Eustatic Controls on Sedimentary Cycles: Tulsa, SEPM (Society for Sedimentary Geology) Concepts in Sedimentology and Paleontology No. 4, p. 25–34.

DICKSON, J. A. D., AND SALLER, A. H., 1995, Identification of subaerial exposure surfaces and porosity preservation in Pennsylvanian and Lower Permian shelf limestones, eastern Central Basin platform, Texas, *in* Budd, D. A., Saller, A. H., and Harris, P. M., eds., Unconformities in Carbonate Strata—Their Recognition and the Significance of Associated Porosity: Tulsa, American Association of Petroleum Geologists Memoir 57, p. 239–257.

DRUMMOND, C. N., AND WILKINSON, B. H., 1993, Carbonate cycle stacking patterns and hierarchies of orbitally forced eustatic sea-level change: Journal of Sedimentary Petrology, v. 63, p. 369–377.

ELRICK, M., AND READ, J. F., 1991, Cyclic ramp-to-basin carbonate deposits, Lower Mississippian, Wyoming and Montana: A combined field and computer modeling study: Journal of Sedimentary Petrology, v. 61, p. 1194–1224.

GALLEY, J. E., 1958, Oil and geology in the Permian basin of Texas and New Mexico, *in* Weeks, L. G., ed., Habitat of Oil—A Symposium: Tulsa, American Association of Petroleum Geologists, p. 395–446.

GARRELS, R. M., AND CHRIST, C. L., 1965, Solutions, Minerals, and Equilibria: New York, Harper and Row, 450 p.

GIANNINY, G. L., AND SIMO, J. A. T., 1996, Implications of unfilled accommodation space for sequence stratigraphy on mixed carbonate-siliciclastic platforms: An example from the lower Desmoinesian (Middle Pennsylvanian), southwest Paradox basin, Utah, *in* Longman, M. W. and Sonnenfeld, M. D., eds., Paleozoic Systems of the Rocky Mountain Region: Denver, Rocky Mountain Section–SEPM (Society of Sedimentary Geology), p. 213–234.

GOLDHAMMER, R. K., DUNN, P. A., AND HARDIE, L. A., 1987, High frequency glacio-eustatic sea-level oscillations with Milankovitch characteristics recorded in Middle Triassic platform carbonates in northern Italy: American Journal of Science, v. 287, p. 853–892.

GOLDHAMMER, R. K., DUNN, P. A., AND HARDIE, L. A., 1990, Depositional cycles, composite sea-level changes, cycle stacking patterns, and the hierarchy of stratigraphic forcing: Examples from Alpine Triassic platform carbonates: Geological Society of America Bulletin, v. 102, p. 535–562.

GOLDHAMMER, R. K., OSWALD, E. J., AND DUNN, P. A., 1991, Hierarchy of stratigraphic forcing: Example from Middle Pennsylvanian shelf carbonates of the Paradox basin, *in* Franseen, E. K., Watney, W. L., Kendall, C. G. S. G., and Ross, W., eds., Sedimentary modelling: Computer simulations and methods for improved parameter definition: Lawrence, Kansas Geological Survey Bulletin, v. 233, p. 361–413.

GONZALEZ-BONORINO, G., AND EYLES, N., 1995, Inverse relation between ice extent and the late Paleozoic glacial record of Gondwana: Geology, v. 23, p. 1015–1018.

GROSS, M. G., AND TRACEY, J. I., Jr., 1966, Oxygen and carbon isotope compositions of limestones and dolomites, Bikini and Eniwetok Atolls: Science, v. 151, p. 1082–1084.

GROSSMAN, E. L., 1994, The carbon and oxygen isotope record during the evolution of Pangea: Carboniferous to Triassic, *in* Klein, G. D., ed., Pangea: Paleoclimate, Tectonics, and Sedimentation during Accretion, Zenith, and Breakup of a Supercontinent: Boulder, Geological Society of America Special Paper 288, p. 207–228.

GROSSMAN, E. L., ZHANG, C., AND YANCEY, T. E., 1991, Stable-isotope stratigraphy of brachiopods from Pennsylvanian shales in Texas: Geological Society of America Bulletin, v. 103, p. 953–965.

HANOR, J. S., 1978, Precipitation of beachrock cements: mixing marine and meteoric waters vs. CO_2-degassing: Journal of Sedimentary Petrology, v. 48, p. 489–502.

HARLAND, W. B., COX, A. V., LLEWELLYN, P. G., PICKTON, C. A. G., SMITH, A. G., AND WALTERS, R., 1982, A geological time scale: Cambridge, Cambridge University Press, 128 p.

HARLAND, W. B., ARMSTRONG, R. L., COX, A. V., CRAIG, L. E., SMITH, A. G., AND SMITH, D. G., 1990, A geological time scale 1989: Cambridge, Cambridge University Press, 263 p.

HECKEL, P. H., 1986, Sea-level curve for Pennsylvanian eustatic marine transgressive-regressive depositional cycles along the Midcontinent outcrop belt, North America: Geology, v. 14, p. 330–334.

HECKEL, P. H., 1994, Evaluation of evidence for glacio-eustatic control over marine Pennsylvanian cyclothems in North America and consideration of possible tectonic effects, *in* Dennison, J. M. and Ettensohn, F. R., eds., Tectonic and Eustatic Controls on Sedimentary Cycles: Tulsa, SEPM (Society for Sedimentary Geology) Concepts in Sedimentology and Paleontology No. 4, p. 65–87.

HUDSON, J. D., 1977, Stable isotopes and limestone lithification: Journal of the Geological Society of London, v. 133, p. 637–660.

MONTANEZ, I. P., AND OSLEGER, D. A., 1993, Parasequence stacking patterns, third-order accommodation events, and sequence stratigraphy of Middle to Upper Cambrian platform carbonates, Bonanza King Formation, southern Great Basin, *in* Loucks, R. G. and Sarg, J. F., eds., Carbonate sequence stratigraphy: Recent developments and advancements: Tulsa, American Association of Petroleum Geologists Memoir 57, p. 305–326.

MONTANEZ, I. P., AND READ, J. F., 1992, Eustatic control on dolomitization of peritidal carbonates: Evidence from the Early Ordovician Knox Group, Appalachians: Geological Society of America Bulletin, v. 104, p. 872–886.

MOORE, C. H., 1989, Carbonate diagenesis and porosity: New York, Elsevier, 338 p.

OTTE, C., 1959, Late Pennsylvanian and Early Permian stratigraphy of the northern Sacramento Mountains, Otero County, New Mexico: Socorro, New Mexico Bureau of Mines and Mineral Resources Bulletin 50, 111 p.

QUINN, T. M., 1991, Meteoric diagenesis of post-Miocene limestones on Enewetak Atoll: Journal of Sedimentary Petrology, v. 61, p. 681–703.

RASBURY, E. T., HANSON, G. N., MEYER, W. J., AND SALLER, A. H., 1997, Dating of the time of sedimentation using U-Pb ages for paleosol calcite: Geochimica et Cosmochimica Acta, v. 61, p. 1525–1529.

RASBURY, E. T., HANSON, G. N., MEYER, W. J., HOLT, W. E., GOLDSTEIN, R. H., AND SALLER, A. H., 1998, U-Pb dates of paleosols: Constraints on late Paleozoic cycle durations and boundary ages: Geology, v. 26, p. 403–406.

ROSS, C. A., AND ROSS, J. R. P., 1987, Late Paleozoic sea levels and depositional sequences: Ithaca, Cushman Foundation for Foraminiferal Research, Special Publication 24, p. 137–153.

ROSS, C. A., AND ROSS, J. R. P., 1988, Late Paleozoic transgressive-regressive deposition, *in* Wilgus, C. K., Hastings, B. S., Posamentier, H., Ross, C. A., and Kendall, C. G. S. C., eds., Sea-Level Changes: An Integrated Approach: Tulsa, Society of Economic Paleontologists and Mineralogists, Special Publication 42, p. 227–247.

SADLER, P. M., OSLEGER, D. A., AND MONTANEZ, I. P., 1993, On the labeling, length and objective basis of Fischer plots: Journal of Sedimentary Petrology, v. 63, p. 360–368.

SALLER, A. H., 1992, Calcitization of aragonite in Pleistocene limestones of Enewetak Atoll, Bahamas, and Yucatan—An alternative to thin-film neomorphism: Carbonates and Evaporites, v. 7, p. 56–73.

SALLER, A. H., AND MOORE, C. H., 1991, Geochemistry of meteoric calcite cements in some Pleistocene limestones: Sedimentology, v. 38, p. 601–621.

SALLER, A. H., DICKSON, J. A. D., AND BOYD, S. A., 1994, Cycle stratigraphy and porosity in Pennsylvanian and Lower Permian shelf limestones, eastern Central Basin platform, Texas: American Association of Petroleum Geologists Bulletin, p. 1820–1842.

SARG, J. F., 1988, Carbonate sequence stratigraphy, *in* Wilgus, C. K., Hastings, B. S., Posamentier, H., Ross, C. A., and Kendall, C. G. S. C., eds., Sea-Level Changes: An Integrated Approach: Tulsa, Society of Economic Paleontologists and Mineralogists, Special Publication 42, p. 156–181.

VAIL, P. R., MITCHUM, R. M., JR., AND THOMPSON, S., III, 1977, Seismic stratigraphy and global changes of sea level, *in* Payton, C. E., ed., Seismic Stratigraphy—Applications to Hydrocarbon Exploration: Tulsa, American Association of Petroleum Geologists Memoir 26, p. 83–97.

VAN WAGONER, J. C., POSAMENTIER, H. W., MITCHUM, R. M., VAIL, P. R., SARG, J. F., LOUTIT, T. S., AND HARDENBOL, J., 1988, An overview of the fundamentals of sequence stratigraphy and key definitions, *in* Wilgus, C. K., Hastings, B. S., Posamentier, H., Ross, C. A., and Kendall, C. G. S. C., eds., Sea-Level Changes: An Integrated Approach: Tulsa, Society of Economic Paleontologists and Mineralogists, Special Publication 42, p. 39–45.

VEEVERS, J. J., AND POWELL, C. MCA., 1987, Late Paleozoic glacial episodes in Gondwanaland reflected in transgressive-regressive depositional systems in Euramerica: Geological Society of America Bulletin, v. 98, p. 475–487.

WALKER, D. A., GOLONKA, J., REID, A. M., AND REID, S. A. T., 1991, The effects of late Paleozoic paleolatitude and paleogeography in the Midland basin, Texas, *in* Candelaria, M. P., ed., Permian Basin Plays—Tomorrow's Technology Today: Midland, West Texas Geological Society Publication 91-89, p. 141–162.

WILDE, G. L., 1990, Practical fusulinid zonation: The species concept; with Permian basin emphasis: West Texas Geological Society Bulletin, v. 29, no. 7, p. 5–34.

WILSON, J. L., 1967, Cyclic and reciprocal sedimentation in Virgilian strata of southern New Mexico: Geological Society of America Bulletin, v. 78, p. 805–818.

WRIGHT, V. P., 1992, Speculations on the controls on cyclic peritidal carbonate; icehouse versus green-house eustatic controls: Sedimentary Geology, v. 76, p. 1–5.

SEDIMENTARY INPUT AND DIAGENESIS ON A CARBONATE SLOPE (BAHAMAS): RESPONSE TO MORPHOLOGIC EVOLUTION OF THE CARBONATE PLATFORM AND SEA-LEVEL FLUCTUATIONS

HILDEGARD WESTPHAL*

Geomar Research Center for Marine Geosciences, Wischhofstrasse 1-3, 24148 Kiel, Germany

JOHN J. G. REIJMER

Geomar Research Center for Marine Geosciences, Wischhofstrasse 1-3, 24148 Kiel, Germany

AND

MARTIN J. HEAD

Department of Geology, University of Toronto, 22 Russell Street, Toronto, Ontario M5S 3B1, Canada

ABSTRACT: Periplatform carbonates of flat-topped, steep-sided carbonate platforms are well known to record the varying environmental conditions that prevail on the platform top. The depositional record of slope sediments on carbonate ramps sensu lato is less understood. This study describes and interprets the textural and compositional signatures of Pliocene periplatform sediments from the core CLINO, located on the leeward paleoslope of the Great Bahama Bank (i.e., the present-day margin). During the Pliocene, the morphology of the Great Bahama Bank evolved from a distally steepened ramp into a steep-sided, flat-topped platform. Hence, the Pliocene deposits offer an important opportunity to compare slope sediments of the different morphology types.

Compositional analyses reveal that the upper Pliocene periplatform sediments, deposited on the slope of a steep-sided platform, clearly record the repetitive submergence and subaerial exposure of the platform top. Highstand deposits are characterized by a predominance of fine-grained sedimentary matrix exported from the platform interior (mainly aragonite needles). The lowstand deposits, in contrast, typically consist of coarse-grained skeletal material and cortoids, with fine-grained material from the platform interior being rare. The interpretation of sea-level highstand and lowstand deposits is substantiated by palynomorph assemblages that offer a direct proof of the conditions that prevailed on the platform top. Sea-level lowstand assemblages contain high proportions of terrestrial pollen, whereas the highstand assemblages are dominated by dinoflagellate cysts. The selected upper Pliocene section thus represents a straightforward image of highstand shedding.

The selected lower Pliocene section was deposited when the platform exhibited a distally steepened ramp morphology and shows a more complex picture. Composition of the sediments is rather uniform and does not reveal the clear cyclic nature seen in the upper Pliocene strata. This difference is partly attributed to the ramp morphology's being less sensitive to sea-level fluctuations, because facies belts are able to migrate upslope and downslope, following the changing sea level. Additionally, however, sea-level fluctuations in the early Pliocene were overlain by an overall sea-level rise that could have obscured the signatures of short-term sea-level drops.

Diagenesis of the periplatform sediments was found to be strongly influenced by sea-level–induced compositional changes in the upper Pliocene section. As seen in scanning electron microscopic examinations, the coarse-grained lowstand deposits are diagenetically more mature than the fine-grained highstand intervals. This pattern is thought to reflect the high primary permeability of the coarse-grained lowstand layers. Fine-grained highstand deposits from the upper Pliocene section and the overall fine-grained lower Pliocene sediments, in contrast, appear much less mature. The low initial permeability seems to have restricted fluid flow that could have driven diagenetic alterations.

INTRODUCTION

The dependence on light of most carbonate-secreting organisms requires that they build up close to sea level where they frequently form flat-topped carbonate platforms (Schlager, 1981). On these flat-topped platforms, even low-amplitude sea-level fluctuations commonly result in repeated exposure and flooding of the platform top. Subaerial exposure of the platform top drastically cuts down the production of sediment in the shallow-water carbonate factory (Schlager, 1981; Droxler and Schlager, 1985).

Flooded carbonate platforms, like the present Little Bahama Bank, produce up to three times the amount of carbonate mud as can be stored within their shallow-water area, the excess sediment being exported to the periplatform realm (Neumann and Land, 1975). The periplatform sediments are predominantly composed of mud-sized and larger grains produced on and exported from the platform top, rim, and foreslope; these grains are mixed with planktic input (Schlager and James, 1978). Because sediment export during sea-level highstands—when the shallow-water carbonate factory is submerged and hence operational—by far exceeds the export to the slopes during lowstands, periplatform records are usually characterized by higher sedimentation rates during highstands (highstand shedding *sensu* Droxler and Schlager, 1985).

Under optimal conditions (zero to moderate winnowing), the mixture of pelagic sediment and components from the platform results in a more continuous sedimentation on the slope than on the periodically exposed platform top. Slope sediments can thus be employed to study environmental changes affecting the shallow-water carbonate factory by deciphering their compositional and textural signatures (e.g., Droxler and Schlager, 1985; Schlager and Camber, 1986; Goldhammer and Harris, 1989; Haak and Schlager, 1989; Brachert and Dullo, 1991, 1994; Grammer et al., 1991; Grammer and Ginsburg, 1992; Harris, 1994).

The concept of highstand shedding, however, applies mainly to flat-topped carbonate platforms with high-angle slopes (Schlager et al., 1994). On many carbonate ramps having low-angle slopes and no pronounced edge, deposition in the various facies belts shifts downslope during minor sea-level falls

*Present address: RSMAS, University of Miami, 4600 Rickenbacher Causeway, Miami, Florida 33149-1098

Advances in Carbonate Sequence Stratigraphy: Application to Reservoirs, Outcrops and Models, SEPM Special Publication No. 63
 ISBN 1-56576-060-3, p. 247–274

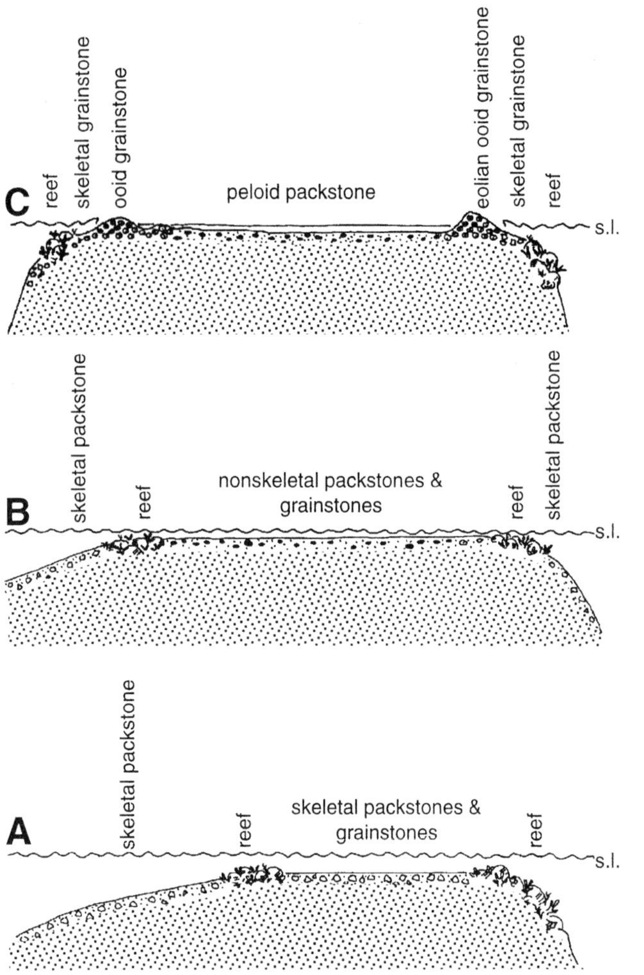

FIG. 1.—Schematic morphologic evolution of the Great Bahama Bank from the Pliocene to the late Pleistocene. During the early Pliocene (A), the Great Bahama Bank was characterized by a distally steepened ramp morphology. Steepening of the leeward flank (left side) due to progradation resulted in the flat-topped, steep-sided morphology of the late Pliocene platform (B). Further steepening in the Pleistocene (C) finally resulted in present-day morphology (not shown) that is typified by a virtually vertical wall down to depths of 140 m below sea level (after Reijmer et al., 1992).

("forced regression" of Posamentier et al., 1992) rather than being "switched off" as in the shallow-water factory of a flat-topped platform (Burchette and Wright, 1992; Bachmann and Willems, 1996). Such downslope facies shifts result in smaller changes in composition and texture between lowstand and highstand sediments deposited on the slope. Burchette and Wright (1992) noted that the periplatform record of carbonate ramps, in comparison to flat-topped platforms, is largely uninvestigated. Since this statement, research on periplatform facies of fossil ramps has increased (e.g., Elrick and Read, 1991; Bachmann et al., 1996; Bachmann and Willems, 1996), but many questions remain subject to discussion. Among those are many aspects of the sequence stratigraphic expression on a small scale (microfacies successions) and facets of diagenetic alteration in slope sediments of a carbonate ramp.

To contribute to the discussion, this study focuses on the signatures of sea-level fluctuations of the order of 100 k.y. (fourth order of Vail et al., 1991) in periplatform sediments of two contrasting settings: a flat-topped platform and a ramp-type configuration. During the Pliocene, the Great Bahama Bank evolved from a ramp sensu lato featuring atoll-like bioherms to a flat-topped, rimmed platform (Fig. 1; Beach and Ginsburg, 1980; Schlager and Ginsburg, 1981; Beach, 1982; McNeill et al., 1988; Reijmer et al., 1992). Therefore, the periplatform sediments of the Great Bahama Bank offer an opportunity to investigate slope sediments of both types of morphology at the same locality.

This study addresses two major questions: First, what are the compositional and textural input patterns observed in periplatform sediments and their relationship to sea-level fluctuations and the evolution of platform morphology? Second, how does the early diagenesis of these periplatform sediments vary with respect to their particular compositions and textures?

STUDY AREA AND MATERIAL EXAMINED

The western, leeward slope of the Great Bahama Bank is characterized by low-angle prograding clinoforms of Tertiary to Quaternary age (Eberli and Ginsburg, 1987, 1989). Seismic sequences are stacked (Fig. 2), a pattern that is thought to be induced by changes in sea level because down-stepping of the margins has been identified (Eberli and Ginsburg, 1987, 1989;

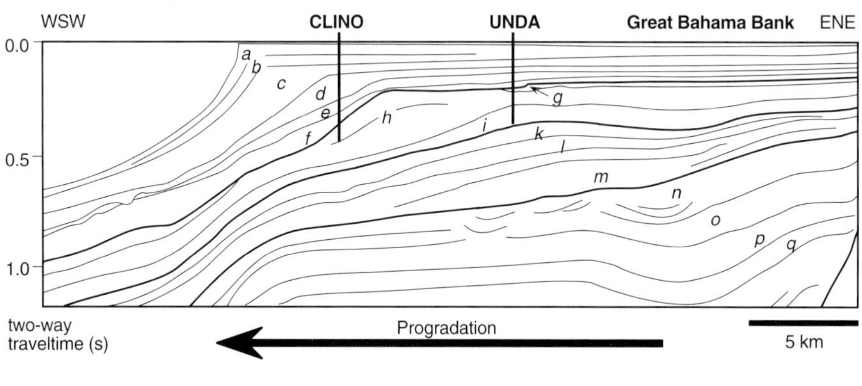

FIG. 2.—Traces of seismic reflections (letters identify the seismic sequences; there is no sequence *j* in this section) of part of the Western Geophysical seismic line showing the leeward clinoforms that prograde westward. The boreholes CLINO and UNDA of the Bahamas Drilling Project penetrate the Miocene *(i* and *h)* to present *(a)* prograding succession of slope sediments. Intervals selected from CLINO for this study are in seismic sequences *f* and *d* (from Eberli et al., 1997).

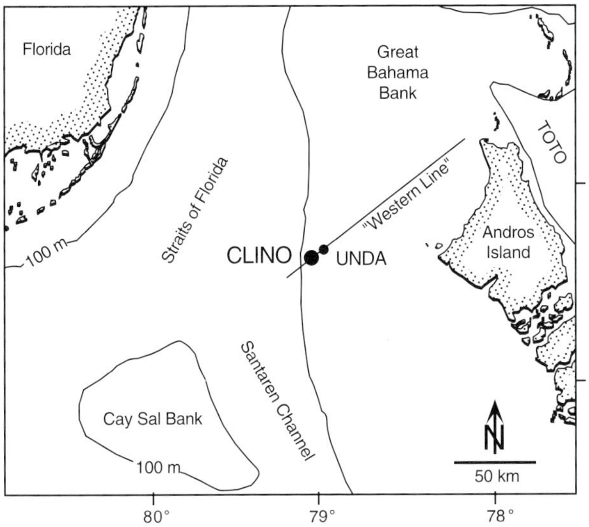

FIG. 3.—Localities of boreholes CLINO and UNDA of the Bahamas Drilling Project and Western Geophysical seismic line, Great Bahama Bank. Borehole CLINO, the subject of this study, is located on the leeward present-day platform margin. TOTO = Tongue of the Ocean (from Eberli et al., 1997).

Eberli et al., 1997). To gain a better understanding of seismic sequences and their correlation to lithologic variations, two boreholes, CLINO and UNDA were drilled in 1990 by the Comparative Sedimentology Laboratory of the University of Miami (Bahamas Drilling Project; R. N. Ginsburg, pers. commun., 1997). The Bahamas Drilling Project boreholes penetrated the proximal parts of the prograding seismic sequences seen on the Western Geophysical line (Fig. 2; Eberli et al., 1997). UNDA is located on the present platform top in a water depth of 6.70 m and aimed to penetrate an upper-slope to shallow-water succession within platform-margin deposits. CLINO is positioned 8.5 km farther to the southwest in a water depth of 7.60 m at the present margin of the platform (Fig. 3). It was designed to penetrate a series of inclined seismic reflectors that are thought to image slope deposits underneath a thin, upper Pleistocene and Holocene platform succession (Eberli et al., 1997). CLINO recovered a succession of Neogene slope sediments and reached the upper Miocene at a total depth of 677.27 m below mud pit (mbmp; mud pit was 7.30 m above sea level).

Lithologic examinations of CLINO led to the recognition of three large-scale depositional sequences in the slope facies that are separated by distinct unconformities (Fig. 4; J. A. M. Kenter, R. N. Ginsburg, and S. R. Troelstra, pers. commun., 1997). Dating of the cores was done by integrating biostratigraphic, magnetostratigraphic, and strontium isotope data (Lidz and McNeill, 1995a, 1995b; P. K. Swart, H. Elderfield, and K. Beets, pers. commun., 1997; for a summary see Eberli et al., 1997). The lower Pliocene deposits (depositional sequence III) are dominated by fine-grained skeletal and nonskeletal sediments. The lower (depositional sequence III) and upper Pliocene (depositional sequence II) deposits are separated by an unconformity at 367.03 mbmp. This boundary correlates with the onset of the first post-Miocene cooling (J. A. M. Kenter, R. N. Ginsburg, and S. R. Troelstra, pers. commun., 1997). The upper Pliocene (depositional sequence II) to lowermost Pleistocene (depositional sequence I) sediments are characterized by fine-grained components. Between the lowermost Pleistocene and the remaining Quaternary deposits, an unconformity is observed at 197.44 mbmp. Above this unconformity, shallow-water and reefal deposits dominate. Overall, the Pliocene to Holocene lithology of CLINO reflects a shallowing-upward trend that corresponds to the progradation of the slope across the locality of the drilling site.

Internally the large-scale depositional sequences are composed of subordinate cycles in the order of tens of meters (J. A. M. Kenter, R. N. Ginsburg, and S. R. Troelstra, pers. commun., 1997). The nature of the smaller-scale variations, as represented by changes in grain sizes and degree of cementation, is investigated in this study. To assess the effects of changes in platform morphology (ramp vs. steep-sided) on the composition and texture of periplatform carbonates, a section from the upper Pliocene (seismic sequence *d*) was selected to represent a steep-sided carbonate platform, whereas a lower Pliocene section (seismic sequence *f*) was chosen to represent a distally steepened ramp (Fig. 2). The parts of the core sampled are macroscopically rather uniform, fine-grained carbonates, but some intervals are conspicuous in having a coarser, biodetrital composition. Both sections appeared suitable for a study of the subordinate cycles, as they exhibit macroscopically visible variations in grain size, color, and porosity. Also, the gamma-ray log displays one conspicuous cycle in the selected upper Pliocene section and three well-developed saw-tooth–shaped cycles in the selected lower Pliocene section.

The upper Pliocene section (in foraminiferal zone N22, 217.04–264.11 mbmp, 182 samples; Fig. 4) was sampled at distances of 30 cm between samples. Parts that appeared interesting in having a more variable composition and texture were sampled at closer spacing. This selected section represents slope sediments from a rimmed, flat-topped platform (J. A. M. Kenter, R. N. Ginsburg, and S. R. Troelstra, pers. commun., 1997). It consists of one macroscopically visible cycle with two thinner beds of coarse-grained material bounding a thicker bed of fine-grained material.

The lower Pliocene section (in foraminiferal zone N19, 451.71–509.47 mbmp, 287 samples; Fig. 4) was sampled at distances of 15 cm. Locally, denser or wider sampling was applied. The upper 15 m of this section were only sampled selectively. The lower Pliocene section investigated was deposited during times when a ramp morphology prevailed (J. A. M. Kenter, R. N. Ginsburg, and S. R. Troelstra, pers. commun., 1997). It is composed of three subtle, macroscopically distinguishable cycles with thicknesses on the order of tens of meters. They are defined by wackestone to packstone alternations in which grain sizes do not differ significantly. At the base of the selected section, coarser-grained deposits were sampled that characterize the lowermost Pliocene.

Below 145 mbmp (thereby including the two sections examined), the core CLINO offers an opportunity to investigate the sedimentary record of slope carbonates beneath the reach of meteoric diagenetic fluids that are assumed to be typical for periplatform carbonates (Melim et al., 1995; Melim, 1996).

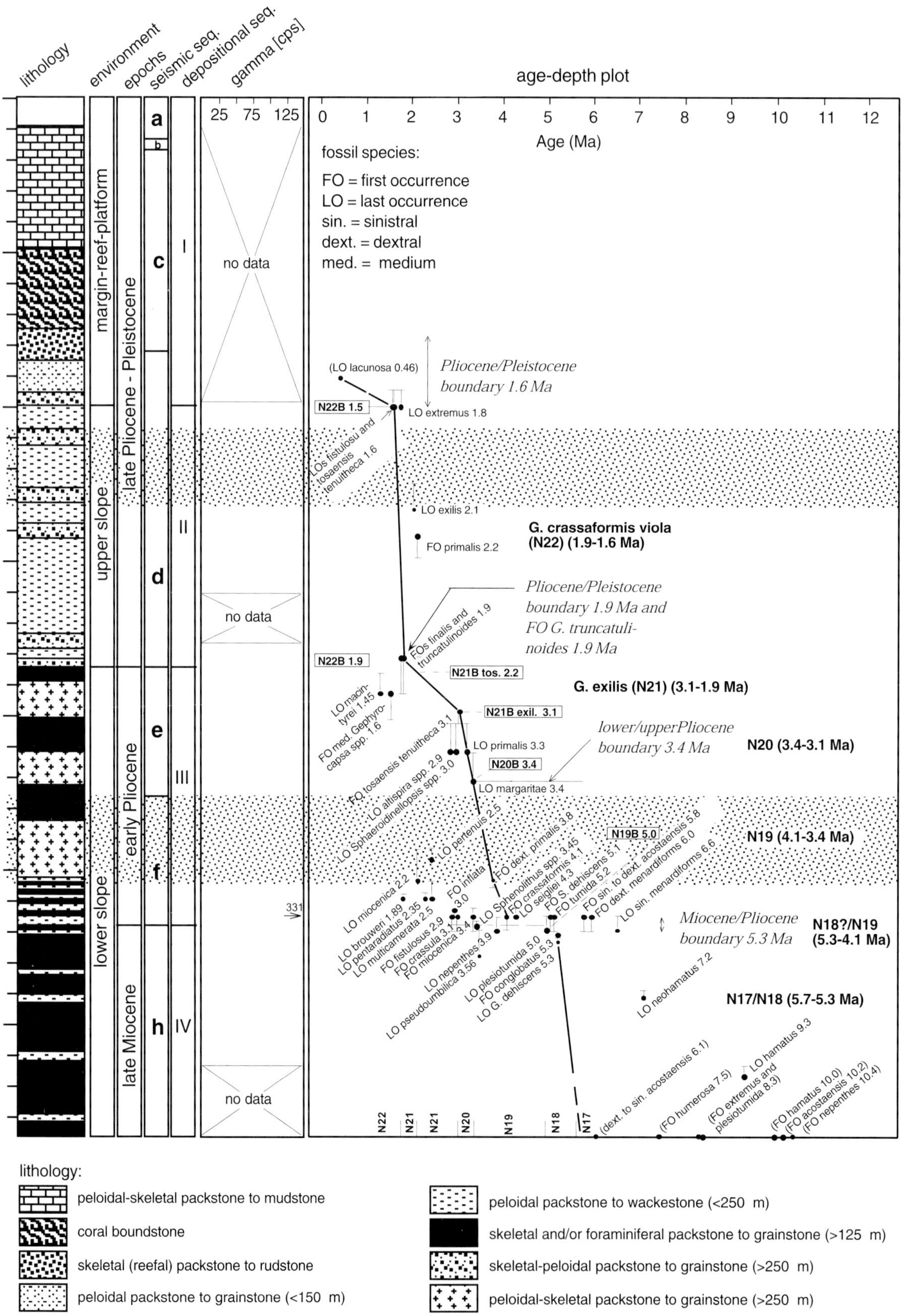

FIG. 4.—Data obtained from borehole CLINO. The two sections examined in this study cover part of the upper Pliocene planktic foraminiferal zone N22 and part of the lower Pliocene biozone N19 (shaded intervals). Lithological logs (mbmp = meters below mud pit), interpretation of depositional environments, and gamma-ray curve (cps = counts per second) are from J. A. M. Kenter, R. N. Ginsburg, and S. R. Troelstra (pers. commun., 1997), seismic sequences (seq.) and (simplified) depositional sequences are from Eberli et al. (1997), and age vs. depth relationship and dating are from D. F. McNeill, G. P. Eberli, B. H. Lidz, P. K. Swart, and J. A. M. Kenter (pers. commun., 1997) who have used the time scale of Berggren et al. (1985).

METHODS

Point Counting

A quantitative component analysis has been undertaken by point counting 369 thin sections (200 points each) using the method of Chayes (1956). The samples were counted volumetrically (Flügel, 1978), i.e., individual grains are counted several times if they occupy several points of the grid.

The quantitative component analysis aims at differentiating the various sediment sources of the periplatform sediments (platform interior, rim, slope, open marine). This method has been shown to be efficient for toe-of-slope and slope sediments where the specific signatures of different production zones were recognized (Haak and Schlager, 1989; Everts, 1991; Reijmer and Everaars, 1991; Reijmer et al., 1991; Everts and Reijmer, 1995). Groups of components have been distinguished that are thought to represent specific paleoenvironments (Fig. 5; based on, e.g., Flügel, 1978; Enos, 1983; Haak and Schlager, 1989; Ginsburg et al., 1991; Burchette and Wright, 1992; Everts and Reijmer, 1995). The same groups have been considered for both flat-topped platform and ramp morphologies. The distribution and widths of the production zones, however, differ slightly between the different morphologies (Fig. 5). Also, on ramps, the transitions between the different production belts are broader and less well defined. Classification after Dunham (1962) was based on point-counting results. The following 14 groups were distinguished (Fig. 5):

A. *Skeletal grains:* (1) *Halimeda* plates that originate predominantly from the rim and upper slope. (2) Red algae that are represented by nodules and branches, also derived from rim to upper-slope environments. (3) Mollusks that are dominantly shed from the platform top. (4) Bryozoans that are rim to slope inhabitants. (5) Benthic foraminifers, mainly shallow-water groups such as amphisteginids, miliolids, and specific rotaliids, and deeper-water forms such as some textulariids. (6) Planktic foraminifers, mainly globigerinids, but rarely globorotaliids, that represent the open-ocean source environment; and pteropods, that also originate from the open ocean but which are extremely rare. (7) Ostracodes that could not be further differentiated are nondiagnostic with respect to their source environment. (8) Echinoderms that are nondiagnostic, because in the samples examined, forms that occur on the platform top and rim usually cannot be differentiated from forms occurring in deeper water. (9) Nondeterminable grains. The objective was to quantify the composition of the predominantly allochthonous sediment at the time of deposition on the slope. Therefore, diagenetically altered grains (molds, neomorphic grains, small-sized biodetritus) have been counted as their primary precursors when recognizable. Otherwise, they have been incorporated into the category of nondeterminable grains.

B. *Nonskeletal components:* (10) Cortoids (*sensu* Flügel, 1982) that originate mainly from carbonate sandy shoals. (11) Intraclasts derived from a muddy environment and redeposited when still unlithified. This is indicated by incorporated fine grains that are indented rather than cut by the boundaries of the intraclasts. (12) Fine-grained material, constituting the matrix, that is thought to be derived originally from the platform interior. (13) Peloids that originate mainly from the platform interior. The predominantly fecal origin of the pellets is indicated by tiny burrows, which are characteristic for soft-material burrowers (coprophages), in the pellets themselves. In thin sections of strongly recrystallized samples, however, homogenized grains could not always be distinguished unequivocally from peloids of presumably fecal origin. (14) Primary interparticle pore space (generally preserved as sparry cement).

Numerical Analyses

Numerical analyses have been applied to quantify the compositional variations in the sections examined. Frequency distribution and basic summary statistics were calculated to estimate the significance of point-count groups in characterizing specific deposits and the variance of each point-count group within the two sections. The statistical similarity between the individual groups was assessed by using Pearson's correlation coefficient analysis. Principal components were calculated from the raw data to determine which groups steer the characterization of the point-count data set. Factors as linear combinations of related variables (dimensionality reduction) are based on eigenvalue-eigenvector analyses. Numerical methods have been applied to similar data sets by Reijmer and Everaars (1991), Reijmer et al. (1991, 1992), and Bachmann and Willems (1996). (For a detailed description of these methods, see Davis [1973] and Rock [1988].)

For statistical analyses, four duplicate samples in the selected lower Pliocene section were excluded, leaving a data set of 238 samples. In the upper Pliocene section, in two intervals of special interest, significantly denser sampling was performed (5–10-cm distance, in contrast to the usual 30 cm). To obtain an approximately equidistant data set, i.e., not to overemphasize the more densely sampled intervals, these additional 24 samples were excluded, leaving a data set of 155 samples.

Quantitative Analysis of Foraminiferal Associations

The abundance of benthic and planktic foraminifers within 69 thin sections was determined by counting the individuals present in an area of 1–9 cm^2, depending on the homogeneity of the thin section. Thus, statistically significant numbers were achieved that could not otherwise be obtained by point-counting analysis because of the small size of most individuals (and thus their small areal percentage in the thin sections). Foraminiferal assemblages were determined in order to extract environmental information on the source area with respect to restriction and salinity and on the facies belts productive during the time of deposition.

In contrast to unconsolidated samples, foraminifers in thin sections of carbonate rocks are more difficult to examine (see Rose and Lidz, 1977). Therefore, only clearly morphologically separable groups are distinguished (Fig. 6). The following envi-

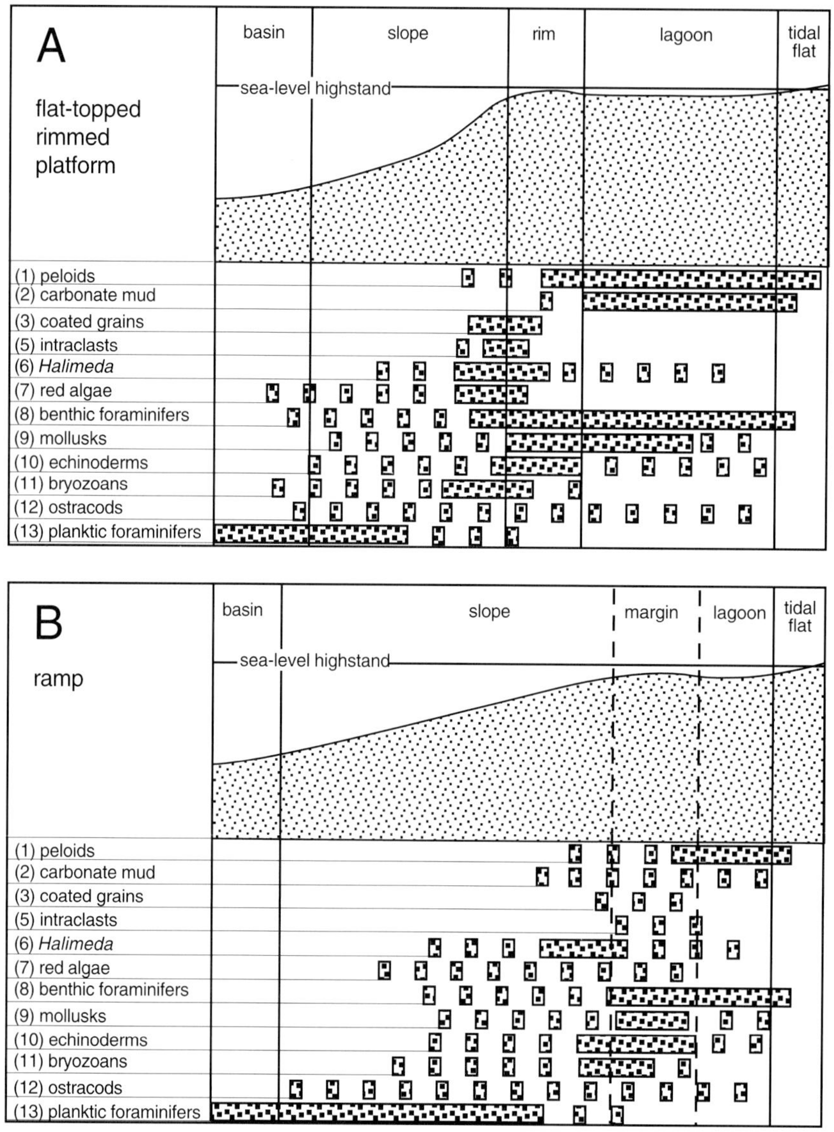

FIG. 5.—Distribution of the facies belts that yielded the constituents distinguished in point counting. (A) Source areas on a steep-sided, flat-topped platform. (B) Source areas on a distally steepened ramp. Based on Flügel (1978), Enos (1983), Haak and Schlager (1989), Ginsburg et al. (1991), Burchette and Wright (1992), and Everts and Reijmer (1995).

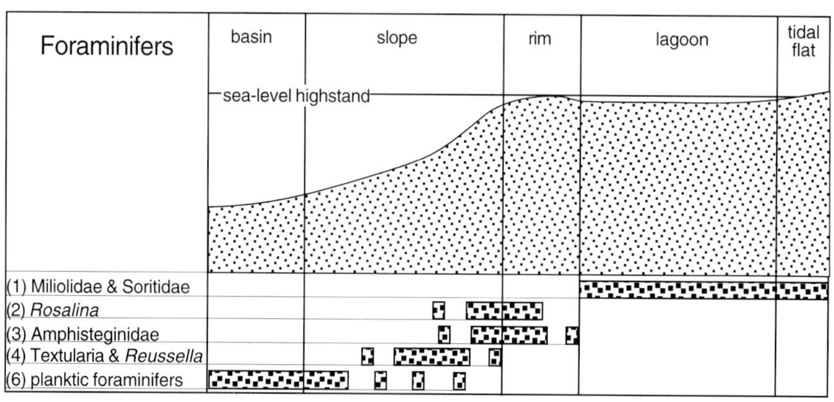

FIG. 6.—Foraminiferal groups typical of specific environments. The distribution on the present-day steep-sided, flat-topped platform is shown. Compiled from Murray (1973), Rose and Lidz (1977), and Lidz and McNeill (1995a).

ronmental interpretation of these groups is based on studies by Murray (1973), Rose and Lidz (1977), and Lidz and McNeill (1995a) (numbers as in Fig. 6):

A. *Platform top:* (1) Miliolids are interpreted as indicators of shallow water. Rotund forms are indicative of extremely shallow water. Soritids are similarly indicative of shallow water.
B. *Margin:* (2) *Rosalina* is found in outer-shelf sediments and is closely related to the presence of reefs. (3) Amphisteginids have a very limited distribution and flourish only in outer-margin and shallow-slope environments.
C. *Slope:* (4) Textulariids and *Reussella* are typical of deeper settings and are easily distinguishable even in thin section.
D. *Other benthic foraminifers:* Benthic groups with insufficient individuals, nondiagnostic groups, and indeterminable specimens are included in group (5) representing "other benthic foraminifers." They mainly comprise various rotaliids, but also include elphiids, eponids, and some others.
E. *Open ocean:* (6) Pelagic forms like globigerinids and globorotaliids occur in slope to basinal sediments and are rare in outer-margin deposits.

Palynologic Examinations

Palynologic determinations have been carried out for 14 samples (eight upper Pliocene and six lower Pliocene) by using standard palynological methods. The samples were treated with cold HCl and HF. Alkalis and hot acids were avoided since these can damage certain organic-walled microfossils, especially cysts of certain heterotrophic dinoflagellates. *Lycopodium* spore tablets were added to facilitate the determination of absolute abundances. Slides were made by using residues sieved at (1) >10 µm for the quantitative study of dinoflagellate cysts and pollen and for the search for acritarchs and (2) >20 µm to search for rare dinoflagellate species.

Scanning Electron Microscopy (SEM)

To examine diagenetic alterations of the carbonate samples, a bulk-rock preparation method was adopted similar to that of Lasemi and Sandberg (1984) as modified by Munnecke (1997). Bulk-rock samples were cut perpendicular to the bedding, mounted on SEM stubs, and polished with corundum powder 2000. Subsequently, they were etched in 0.1M hydrochloric acid for an optimal relief for SEM examinations and a better recognition of the boundaries of constituents. Also, dolomite is recognized more easily in etched samples by its elevated relief compared to calcite and aragonite. The samples were finally coated with gold and palladium. Examinations of 84 samples were performed on a CamScan Cs 44 scanning electron microscope of the Department of Geology and Paleontology at University of Kiel, Germany. Information on the chemical composition of constituents (low-Mg calcite, high-Mg calcite, dolomite) was gained by an EDX (energy-dispersive X-ray) device integrated with the SEM. As aragonite and low-Mg calcite show the same elementary composition, their identification was based on morphologic characteristics. Additionally, aragonite was identified by using selective staining of aragonitic constituents with diluted Feigel's solution (Schneidermann and Sandberg, 1971). Aragonite, being slightly more soluble in acids, reacts faster with Feigel's solution. The solution leaves traces of Ag precipitates on the aragonitic constituents. Using the EDX device, the stained aragonitic components are marked by small Ag peaks, whereas calcitic constituents lack an Ag peak.

RESULTS

Sediment Input

Distribution of components within the selected upper Pliocene section.—

The selected upper Pliocene section in foraminiferal zone N22 serves as an example of periplatform sediments on a steep-sided, flat-topped carbonate platform. It exhibits two distinct types of sediments: thick, fine-grained beds and thin, coarse-grained beds (Fig. 7). The gamma-ray log shows high values in the lower coarse-grained beds (about 256 mbmp), whereas the upper coarse-grained beds are indicated by only a slight increase in gamma-ray values (about 222 mbmp; Figs. 3 and 6).

The fine-grained beds (217.0–219.5 mbmp, 256.0–231.5 mbmp) are characterized by a predominance of sedimentary matrix averaging about 70% and in some samples exceeding 90%. Cortoids and intraclasts, characteristic of rim environments, are virtually absent from these fine-grained samples. Mollusks, echinoderms, and ostracodes occur subordinately. On the level of identification performed here, these biota are non-diagnostic with respect to distinct facies belts. Fine-grained biodetritus and other nondeterminable grains account for about 15%. Planktic foraminifers are common. Benthic foraminifers increase toward the top of the fine-grained intervals, showing maxima around the base of the overlying coarse-grained deposits. Parallel to the benthic foraminifers, peloids gradually increase up-core toward the top of the fine-grained intervals, resulting in a coarsening-upward trend. Their maxima, exceeding 40% of the constituents, are located slightly below those of the benthic foraminifers. In some strongly compacted samples (e.g., 261.7–261.9 mbmp), the true amount of peloids might be higher, as homogenized peloids can appear as a pseudomatrix. Both benthic foraminifers and peloids exhibit a subordinate increase in abundance about 244 mbmp.

Above the peloidal packstones at the top of the fine-grained intervals, an increase in cortoids and intraclasts marks a dramatic change in sediment composition (at about 261.5 and 224.0 mbmp). Cortoids, although virtually absent from the fine-grained layers, suddenly exceed abundances of 50%. Parallel to this trend, the amount of sparry cement increases, i.e., the amount of sedimentary matrix decreases. Besides cortoids, the coarse-grained intervals are mainly composed of intraclasts, benthic foraminifers, and fragments of red and green algae. These intervals are thus mainly composed of biota that are characteristic of rim to upper-slope facies belts. Bryozoans occur exclusively in the coarse-grained intervals. The top of the

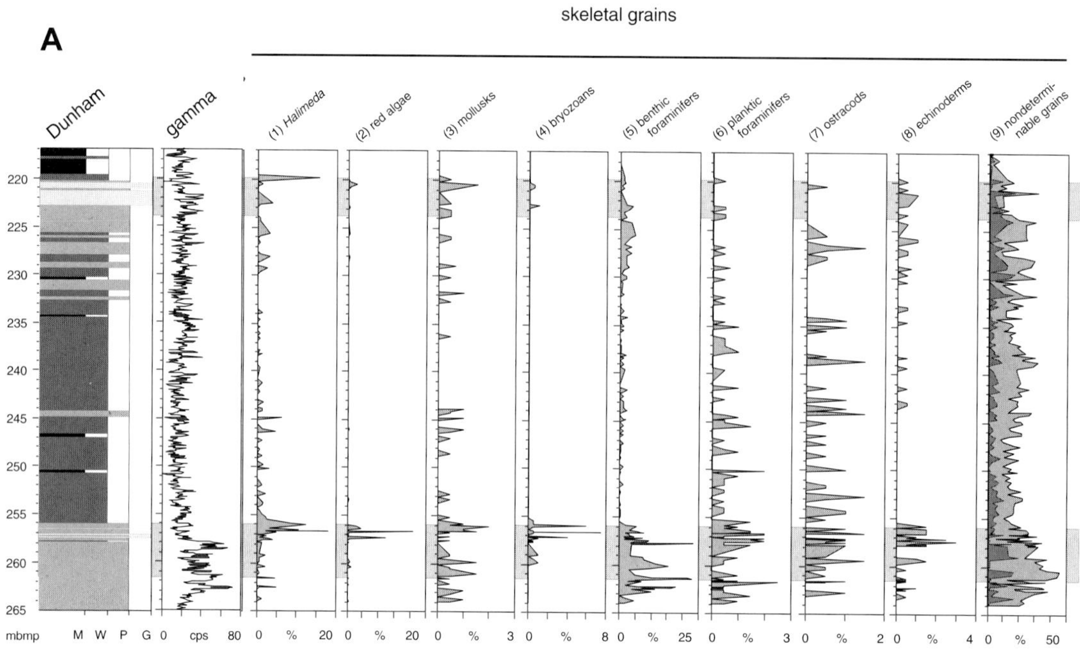

FIG. 7.—Results of components analysis of the selected upper Pliocene section. The abundance of the different skeletal (A) and nonskeletal (B) components is shown vs. depth. Darker gray curve overlain on the nondeterminable grains (constituent 9, rightmost curve in A) shows the fraction of molds that contribute to this group. Note the strong variations in the percentage of fine-grained matrix and the antithetic variations in the amounts of cortoids and sparry cement. The two coarse-grained intervals are underlain by gray bands. The left column shows the classification after Dunham (1962; G = grainstones, P = packstones, W = wackestones, M = mudstones; mbmp = meters below mud pit). The second column shows the gamma-ray curve (cps = counts per second) of J. A. M. Kenter, R. N. Ginsburg, and S. R. Troelstra (pers. commun., 1997).

coarse-grained layers is defined by an abrupt decrease in cortoids and an equally abrupt increase in fine-grained matrix.

Distribution of components within the selected lower Pliocene section.—

The lower Pliocene section from foraminiferal zone N19 represents slope sediments from a distally steepened ramp. The point-counting results reveal a rather uniform composition (Fig. 8). The fine-grained matrix averages about 50% throughout the entire succession. Sparry cement is absent. A large fraction of the components distinguished consists of nondeterminable fine-grained biodetritus. Peloids show maximum amounts of 15%. Three subtle cycles in the abundance of peloids are observed that show a rapid increase in quantity at their bases (501.50, 491.50, and 483.50 mbmp) and a more gradual decrease upward. The benthic foraminifers may parallel the peloids in the up-core trend by increases in abundance at 502.50, 491.50, and 481.50 mbmp. The saw-tooth shape of the gamma-ray curve with peaks at 477 and 498 mbmp is not reflected in the cycles of abundance of benthic foraminifers or peloids. Bryozoans are extremely rare. In analogy to the upper Pliocene section, their occurrence at 576.5 mbmp could indicate platform-rim input. Planktic foraminifers generally are more abundant than in the N22 succession. A strong peak with up to 20% of planktic foraminifers characterizes the base of the section examined. A similar maximum at the base of that section is exhibited by the echinoderms.

Components that in the upper Pliocene section show a characteristic distribution and are indicative of rim and upper-slope environments, such as cortoids, intraclasts, *Halimeda*, and red algae, are virtually absent.

Numerical Analyses of the Point-Counting Results

Statistics of the upper Pliocene composition.—

Numerical analysis of the data set of 155, roughly equidistant, upper Pliocene samples has led to the following observations. In the selected section in foraminiferal zone N22, the frequency distribution shows specific patterns for the different point-count groups (Fig. 9). The histogram of the fine-grained matrix is characterized by a bimodal distribution. One peak plots at about 25% and the other at about 85% of the total sediment volume. This finding indicates that there is a fundamental difference between the coarse-grained and the fine-grained upper Pliocene deposits. Other histograms (e.g., those for cortoids and sparry cement) exhibit an asymptotic shape (high skewness), reflecting the fact that in most samples these point-count groups are rare to absent, but in some samples they are common to abundant (Table 1).

For most point-count groups, the skewness value is greater than 1.5, indicating a strongly asymmetric distribution. The category of nondeterminable grains exhibits a normal distribution with a lower skewness of 0.67. This group is not a genetic one and partly depends on factors (mainly diagenesis) that could differ from those of the other point-count groups that directly reflect sediment input.

The sensitivity of a point-count group with respect to changes in the input is expressed by its coefficient of variation, i.e., the standard deviation over the mean. Point-count groups that are spatially clearly restricted to a definite facies belt are characterized by high variation coefficients (Table 1). In the selected succession of zone N22, the highest coefficients are

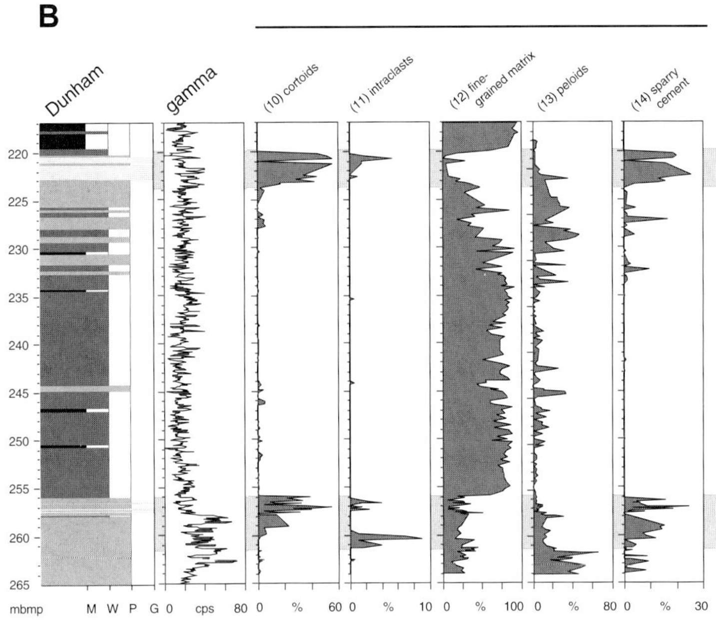

FIG. 7.— Continued.

found for red algae, bryozoans, intraclasts, *Halimeda,* and cortoids in descending order. For most of these groups, an origin from rather well-defined facies zones is assumed, as in the case of the red algae that come from the high-energy foreslope.

Correlation coefficients characterize the statistical dependence between different point-count groups. The nondeterminable grains have been included in statistical analyses as they probably represent aragonitic skeletal grains such as coral fragments. The highest positive correlation is observed between sparry cement and cortoids (0.80; Table 2). Cortoids also correlate positively with red algae and bryozoans (0.36 and 0.37, respectively) that are thought to originate from a similar high-energy environment. As expected, cortoids as well as sparry cement (primary interparticle space) correlate strongly negatively with the fine-grained matrix (–0.64 and –0.72, respectively). The negative correlation between fine-grained matrix and peloids (–0.61) and between fine-grained matrix and benthic foraminifers (–0.59) reflects the high amounts of peloids and foraminifers in the samples at the base of the coarse-grained intervals. The correlation matrix thus describes the compositional differences between coarse-grained and fine-grained sediments, and the transitions, that were observed in the upper Pliocene section.

In the selected upper Pliocene section, the first three unrotated factors calculated by principal-components analysis (Table 3) explain 52.7% of the total variation. This result is characteristic for complex systems. The first factor (28.0%) is dominantly controlled by sparry cement, cortoids, benthic foraminifers, and intraclasts (positive loadings) and by fine-grained matrix (negative loading). This factor characterizes coarse-grained, matrix-poor deposits derived from platform-margin environments. The second factor (13.4%) is positively loaded with nondeterminable components (that are thought to represent skeletal grains), peloids, and benthic foraminifers and is negatively loaded with cortoids and red algae. This factor represents the peloid-rich sediments at the top of the coarsening-upward trends below the coarse-grained intervals. The third factor (11.3%) is positively loaded with *Halimeda* and bryozoans and negatively loaded with sparry cement, cortoids, and peloids. This factor might represent the *Halimeda*-rich packstones at the top of the coarse-grained intervals.

Statistics of the lower Pliocene composition.—

In the selected lower Pliocene section, numerical analyses of the 238 samples revealed the following statistical behavior. Of the frequency histograms, only a few show an asymptotic shape (e.g., planktic foraminifers), whereas others exhibit a bell curve (e.g., fine-grained matrix; Fig. 9) and are characterized by a low skewness (0.11, –0.42, respectively; Table 4). This result indicates that these point-count groups have a similar frequency in most samples, and the Gaussian tails are statistical deviations of the normal distribution rather than indicative of systematic differences in the input. Components found to be most indicative in the selected upper Pliocene section (intraclasts, cortoids, and sparry cement) are absent from the lower Pliocene section examined.

The coefficient of variation, typifying the spatial restriction of the facies belt from which a particular point-count group originates, shows highest values for red algae, bryozoans, and *Halimeda.* These groups are present in the selected lower Pliocene section in amounts too low to be statistically significant. The point-count groups present in significant amounts are characterized by low coefficients. They seem to be derived from facies belts that are spatially not very well defined, such as, for example, a broad platform margin that interfingers with a gentle foreslope or with a backward lagoon.

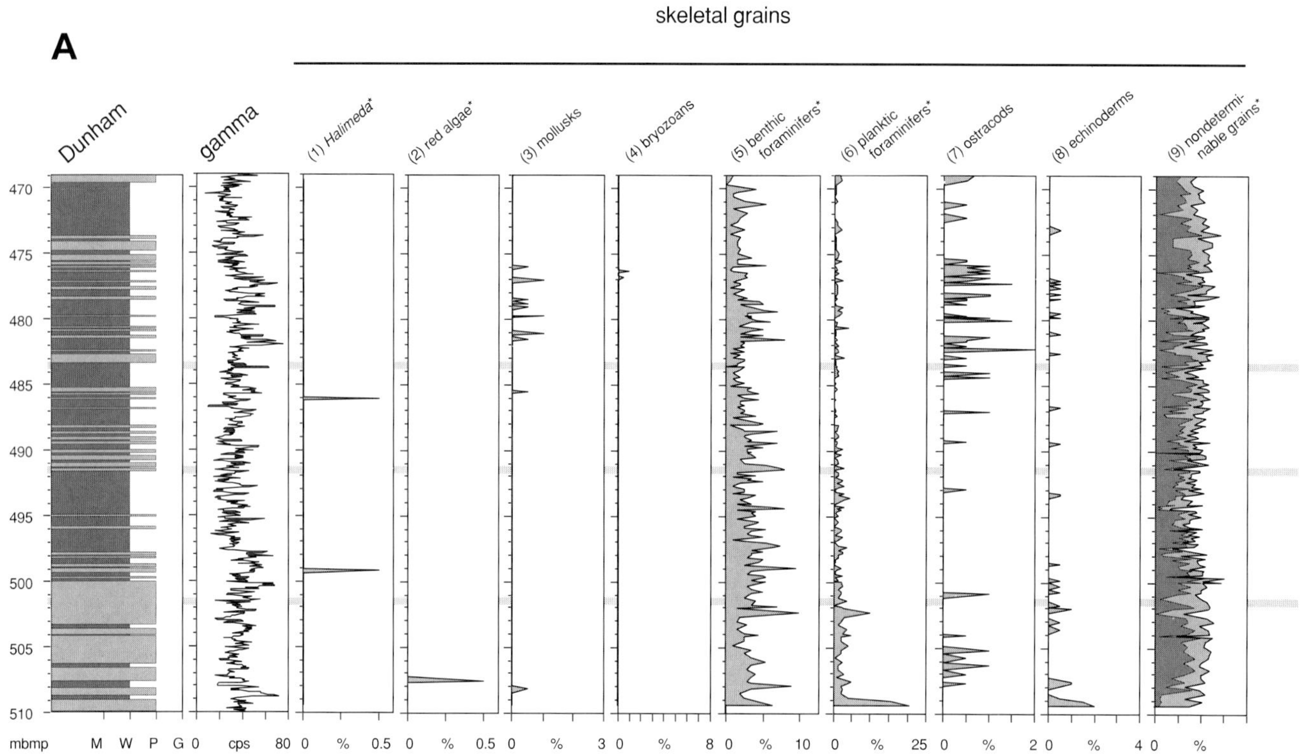

FIG. 8.—Results of components analysis of the selected lower Pliocene section. The densely sampled interval between 469 and 510 mbmp (meters below mud pit) is shown. The abundance of the different skeletal (A) and nonskeletal (B) components is shown vs. depth. Darker gray curve overlain on the nondeterminable grains (constituent 9, rightmost curve in A) shows the fraction of molds that contribute to this group. Variations in the composition are less pronounced than in the upper Pliocene section. Three subtle cycles are observed in the abundances of benthic foraminifers and peloids. Their bases are marked by gray lines. *X*-axis scale is the same as in Figure 7, unless the component is marked by an asterisk. The left column shows the classification after Dunham (1962; G = grainstones, P = packstones, W = wackestones, M = mudstones). The second column shows the gamma-ray curve (cps = counts per second) of J. A. M. Kenter, R. N. Ginsburg, and S. R. Troelstra (pers. commun., 1997).

TABLE 1.—SUMMARY STATISTICS OF THE 14 POINT-COUNT GROUPS COUNTED IN THE SELECTED 155 SAMPLES FROM THE UPPER PLIOCENE INTERVAL.

	Mean	Standard Deviation	Standard Error	Minimum	Maximum	Variance	Coefficient of Variation	Skewness
(1) Halimeda	0.78	2.03	0.16	0.0	15.5	4.10	2.61	4,827
(2) Red algae	0.33	2.01	0.16	0.0	21.0	4.06	6.06	8,563
(3) Molluscs	0.20	0.37	0.03	0.0	2.0	0.14	1.86	2,128
(4) Bryozoans	0.09	0.51	0.04	0.0	6.0	0.26	5.89	10,124
(5) Benthic foraminifera	1.59	2.45	0.20	0.0	15.5	5.99	1.54	2,762
(6) Planktic foraminifera	0.31	0.48	0.04	0.0	2.5	0.23	1.51	1,894
(7) Ostracodes	0.22	0.39	0.03	0.0	1.5	0.15	1.78	1,757
(8) Echinoderms	0.15	0.33	0.03	0.0	1.5	0.11	2.21	2,479
(9) Nondeterminable	18.36	8.79	0.71	1.5	46.0	77.22	0.48	0.669
(10) Cortoids	5.16	12.60	1.01	0.0	56.5	158.64	2.44	2,807
(11) Intraclasts	0.28	1.12	0.09	0.0	9.0	1.24	3.97	5,565
(12) Fine-grained matrix	59.23	26.96	2.17	1.0	96.5	726.82	0.46	–0.603
(13) Peloids	10.65	13.02	1.05	0.0	63.5	169.41	1.22	1,670
(14) Sparry cement	2.65	5.68	0.46	0.0	26.0	32.28	2.15	2,320

NOTE: In each thin section, 200 points were counted.

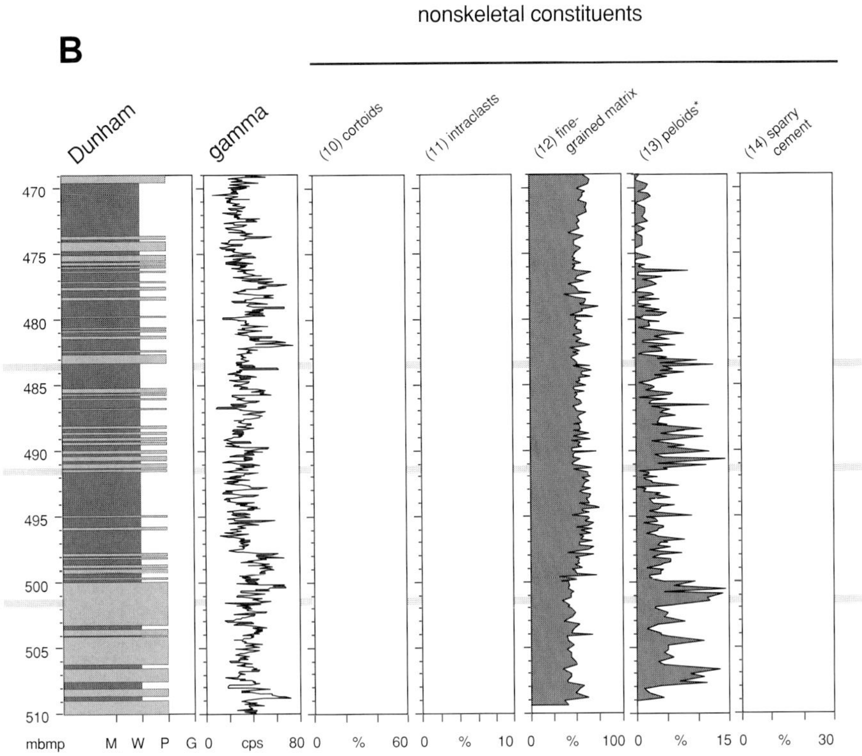

FIG. 8.—Continued.

In the selected lower Pliocene section, correlations between the point-count groups are generally very weak (Table 5). The highest correlation observed is the high negative correlation between fine-grained matrix and the nondeterminable grains (–0.86). As the fine-grained matrix and the nondeterminable, diagenetically altered and/or small-sized components are by far the dominant constituents of the samples from the examined lower Pliocene section, their quantity is expected to correlate negatively. A second strong correlation is the positive correlation between planktic foraminifers and echinoderms (0.63) that reflects the co-occurrence of these two groups in the lowermost samples of the section examined. All other groups are neither strongly positively nor strongly negatively correlated and thus show a rather neutral behavior.

TABLE 2.—CORRELATION MATRIX DESCRIBING THE INTERRELATIONSHIP OF THE COMPONENTS PRESENT IN THE SELECTED UPPER PLIOCENE INTERVAL.

	(1)	(2)	(3)	(4)	(5)	(6)	(7)	(8)	(9)	(10)	(11)	(12)	(13)	(14)
(1) *Halimeda*	1.00	0.10	0.36	0.49	0.13	0.16	0.02	0.31	0.07	0.16	–0.00	–0.15	–0.13	–0.02
(2) Red algae	0.10	1.00	0.08	0.15	0.06	0.37	0.06	0.15	–0.03	0.36	0.23	–0.28	–0.07	0.24
(3) Molluscs	0.36	0.08	1.00	0.17	0.33	0.10	0.05	0.31	0.30	0.26	0.24	–0.41	0.10	0.25
(4) Bryozoans	0.49	0.15	0.17	1.00	0.21	0.18	0.00	0.03	–0.04	0.37	0.06	–0.24	–0.07	0.12
(5) Benthic foraminifera	0.13	0.06	0.33	0.21	1.00	0.18	0.19	0.26	0.50	0.13	0.48	*–0.59*	0.31	0.35
(6) Planktic foraminifera	0.16	0.37	0.10	0.18	0.18	1.00	0.03	0.12	0.10	0.10	0.01	–0.27	0.20	0.03
(7) Ostracodes	0.02	0.06	0.05	0.00	0.19	0.03	1.00	0.05	0.21	0.00	0.07	–0.10	–0.04	0.05
(8) Echinoderms	0.31	0.15	0.31	0.03	0.26	0.12	0.05	1.00	0.15	0.27	0.12	–0.36	0.08	0.30
(9) Nondeterminable	0.07	–0.03	0.30	–0.04	0.50	0.10	0.21	0.15	1.00	–0.22	0.23	–0.39	0.23	–0.05
(10) Cortoids	0.16	0.36	0.26	0.37	0.13	0.10	0.00	0.27	–0.22	1.00	0.29	*–0.64*	1.40	*0.80*
(11) Intraclasts	–0.00	0.23	0.24	0.06	0.48	0.01	0.07	0.12	0.23	0.29	1.00	–0.41	0.00	0.40
(12) Fine-grained matrix	–0.15	–0.28	–0.41	–0.24	*–0.59*	–0.27	–0.10	–0.36	–0.39	*–0.64*	–0.41	1.00	*–0.61*	*–0.72*
(13) Peloids	–0.13	–0.07	0.10	–0.07	0.31	0.20	–0.04	0.08	0.23	1.40	0.00	*–0.61*	1.00	0.17
(14) Sparry cement	–0.02	0.24	0.25	0.12	0.35	0.03	0.05	0.30	–0.05	*0.80*	0.40	*–0.72*	0.17	1.00

NOTE: Nontransformed point-count values of 155 samples are included. Correlations that exceed 0.50 are in italics.

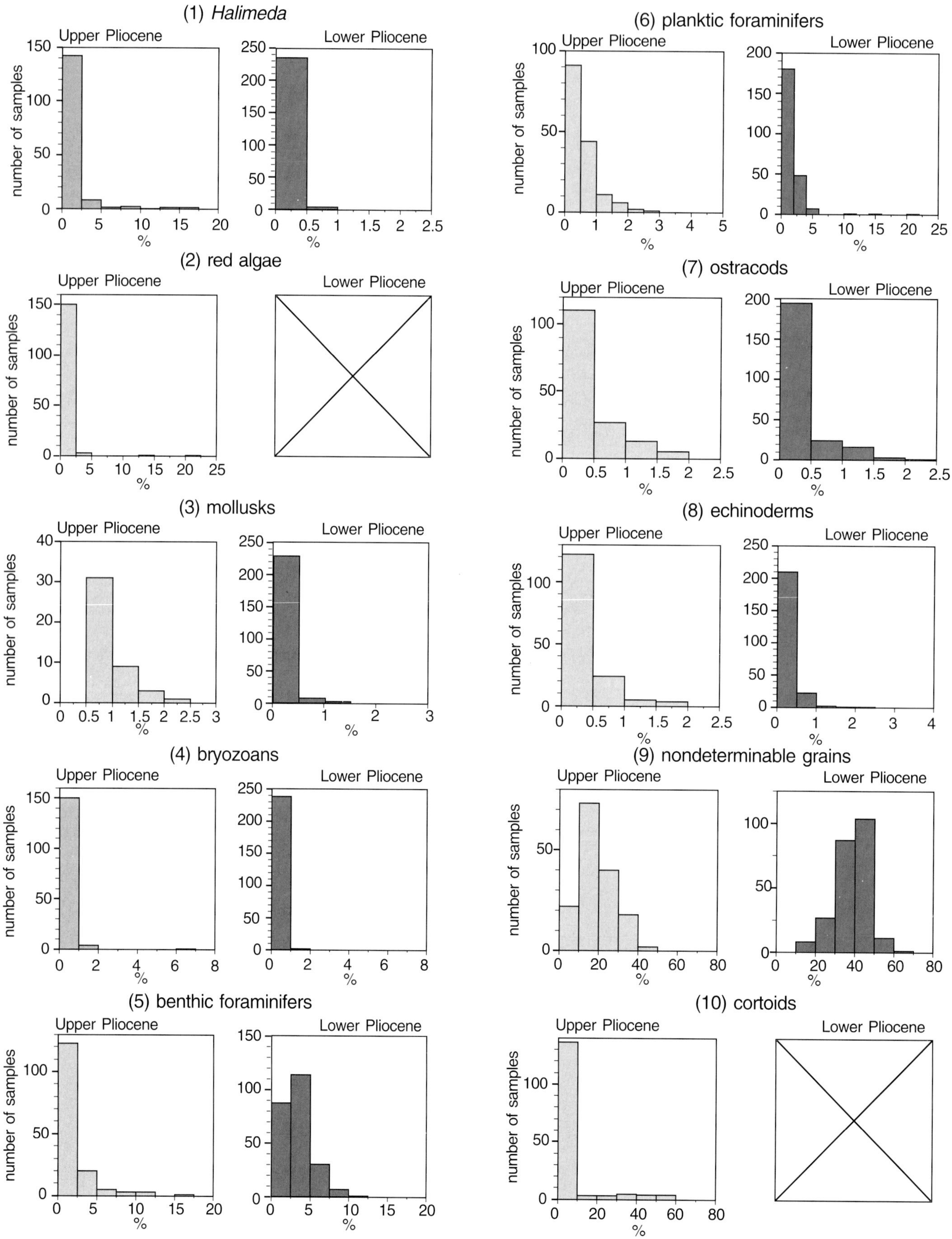

FIG. 9.—Frequency distribution of constituents from the selected upper (left) and lower Pliocene (right) sections. Samples shown are those included in statistical analyses (upper Pliocene: $n = 155$; lower Pliocene: $n = 225$). Note the bimodal distribution of fine-grained matrix (constituent 12) in the upper Pliocene section, implying that there were two distinct modes of sedimentation from particular sources.

TABLE 3.—FIRST THREE UNROTATED FACTORS DESCRIBING 52.7% OF THE VARIATIONS IN THE COMPOSITION OF THE SELECTED UPPER PLIOCENE INTERVAL

	Factor 1	Factor 2	Factor 3
(1) *Halimeda*	0.32	–0.28	*0.76*
(2) Red algae	0.39	–0.37	–0.02
(3) Molluscs	0.56	0.09	0.34
(4) Bryozoans	0.37	–0.41	0.48
(5) Benthic foraminifera	0.67	0.45	0.08
(6) Planktic foraminifera	0.33	–0.02	0.26
(7) Ostracodes	0.15	0.24	0.12
(8) Echinoderms	0.50	–0.03	0.19
(9) Nondeterminable	0.35	*0.73*	0.26
(10) Cortoids	0.68	*–0.59*	–0.29
(11) Intraclasts	0.55	0.10	–0.21
(12) Fine-grained matrix	*–0.92*	–0.14	0.20
(13) Peloids	0.35	0.50	–0.26
(14) Sparry cement	*0.73*	–0.27	*–0.49*

NOTE: Highest positive and negative values for each factor are in italics.

Of the total variation, 46.5% are explained by the first three factors (Table 6). The dominance of the fine-grained matrix and nondeterminable components is also reflected in the factor analysis. The first factor (18.5%) is loaded negatively with nondeterminable components and positively with the fine-grained matrix. It represents the packstone-wackestone alternations that characterize the lower Pliocene section. The second factor (16.6%) is positively loaded with planktic foraminifers and echinoderms and is negatively loaded with fine-grained matrix and subordinately with ostracodes and bryozoans. It is thought to characterize the varying influence of the platform top (various skeletal grains) and the open-marine environment (planktic foraminifers, echinoderms). The third factor (11.4 %) is typified by positive loadings of peloids and benthic foraminifers, and negative loadings of mollusks and nondeterminable grains. It reflects the slight input cycles observed in peloids and benthic foraminifers and thus possibly the productivity and/or the export from the platform top.

Foraminiferal Assemblages

Foraminiferal assemblages exhibit the mixed shallow-water and open-marine characteristics typical of periplatform sediments. As determined by point counting, benthic foraminifers account for an average of 2.7% in the selected upper Pliocene section (with two maxima below the bases of the two coarse-grained layers, a very prominent one of 23.5% at 257.81 mbmp and a subordinate one of 4.5% at 225.70 mbmp). In the selected lower Pliocene section, an average of 3.0% per sediment volume is composed of benthic foraminifers. A maximum of 10% is found at 502.46 mbmp, but small-scale variations exceed the large-scale trends. Planktic foraminifers are present with abundances of up to 2.5% in the upper Pliocene section (average 0.4%). The fraction of planktic foraminifers is higher in the lower Pliocene section examined (1.3% average), and at the base of that section (509.47 mbmp), a distinct peak of 20.5% is observed. Quantitative analysis of foraminifers, by determining the abundance of specimens per square centimeter, complements the point-counting analysis.

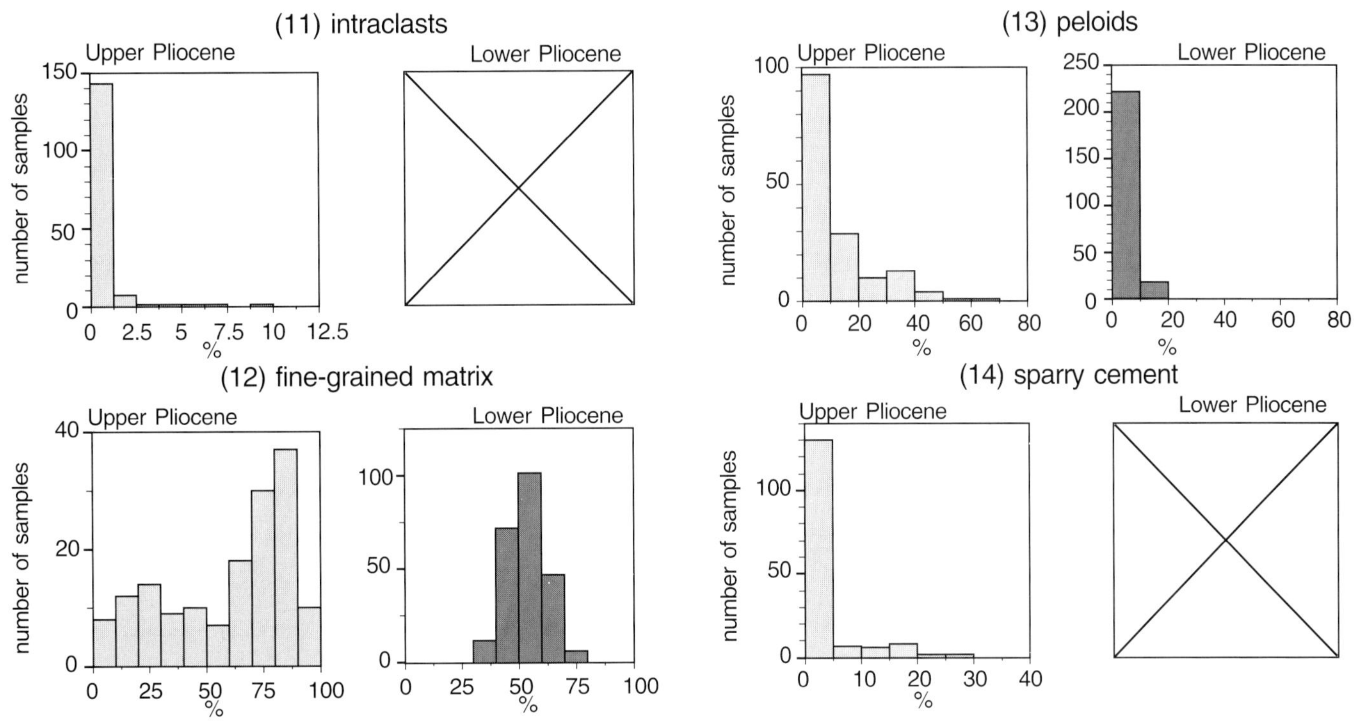

FIG. 9.—Continued.

TABLE 4.—SUMMARY STATISTICS OF THE 11 POINT-COUNT GROUPS COUNTED IN THE SELECTED 238 SAMPLES FROM THE LOWER PLIOCENE INTERVAL.

	Mean	Standard Deviation	Standard Error	Minimum	Maximum	Variance	Coefficient of Variation	Skewness
(1) *Halimeda*	0.01	0.06	0.00	0.0	0.5	0.00	8.87	8.74
(2) Red algae	0.00	0.03	0.00	0.0	0.5	0.00	15.43	15.33
(3) Molluscs	0.03	0.14	0.01	0.0	1.0	0.02	4.84	5.25
(4) Bryozoans	0.01	0.07	0.00	0.0	1.0	0.01	11.48	12.30
(5) Benthic foraminifera	2.99	1.81	0.12	0.0	10.0	3.29	0.61	1.09
(6) Planktic foraminifera	1.30	1.96	0.13	0.0	20.5	3.84	1.50	6.06
(7) Ostracodes	0.15	0.34	0.02	0.0	2.0	0.12	2.36	2.57
(8) Echinoderms	0.08	0.24	0.02	0.0	2.0	0.06	3.17	4.30
(9) Nondeterminable	38.47	8.60	0.56	12.5	60.5	73.95	0.22	–0.40
(12) Fine-grained matrix	53.13	8.50	0.55	31.5	74.5	72.21	0.16	0.10
(13) Peloids	3.85	3.12	0.20	0.0	14.5	9.75	0.81	1.29

NOTE: The three additional point-count groups that are shown for the upper Pliocene interval, but are missing in this table (cortoids, intraclasts, sparry cement), are absent from the lower Pliocene or were only found once. In each thin section, 200 points were counted.

TABLE 5.—CORRELATION MATRIX DESCRIBING THE INTERRELATIONSHIP OF THE COMPONENTS PRESENT IN THE SELECTED LOWER PLIOCENE INTERVAL

	(1)	(2)	(3)	(4)	(5)	(6)	(7)	(8)	(9)	(12)	(13)
(1) *Halimeda*	1.00	–0.01	–0.02	–0.01	–0.01	–0.01	–0.05	–0.04	0.10	–0.11	0.02
(2) Red algae	–0.01	1.00	–0.01	–0.01	0.48	0.12	–0.03	0.25	0.05	–0.11	0.07
(3) Molluscs	–0.02	–0.01	1.00	–0.02	–0.02	–0.02	0.04	0.06	–0.09	0.11	–0.08
(4) Bryozoans	–0.01	–0.01	–0.02	1.00	–0.06	–0.06	0.13	–0.03	–0.01	–0.02	0.09
(5) Benthic foraminifera	–0.01	0.48	–0.02	–0.06	1.00	0.23	–0.12	0.06	–0.33	0.03	0.11
(6) Planktic foraminifera	–0.01	0.12	–0.02	–0.06	0.23	1.00	–0.08	*0.63*	–0.15	–0.14	–0.00
(7) Ostracodes	–0.05	–0.03	0.04	0.13	–0.12	–0.08	1.00	0.03	0.10	–0.09	–0.03
(8) Echinoderms	–0.04	0.25	0.06	–0.03	0.06	*0.63*	0.03	1.00	0.03	–0.21	–0.04
(9) Nondeterminable	0.10	0.05	–0.09	–0.01	–0.33	–0.15	0.10	0.03	1.00	*–0.86*	–0.14
(12) Fine-grained matrix	–0.11	–0.11	0.11	–0.02	0.03	–0.14	–0.09	–0.21	*–0.86*	1.00	–0.25
(13) Peloids	0.02	0.07	–0.08	0.09	0.11	0.00	–0.03	–0.04	–0.14	–0.25	1.00

NOTE: Untransformed point-count values of 238 samples are included. Correlations that exceed 0.50 are in italics.

Distribution of foraminifers in the upper Pliocene section.— The benthic foraminifers exhibit specific vertical variations in distribution (Fig. 10). Miliolids are dominated by very small specimens (about 0.2 mm in diameter) with mostly rotund, thin tests that occur throughout the entire succession. Although numbers of about 10 specimens per square centimeter are common, miliolids are most abundant in the uppermost part of the fine-grained deposits where they reach quantities exceeding 200 specimens per square centimeter. Large tests of soritids, as for rotund miliolids that are typical of very shallow water, are absent from most samples. Nevertheless, they are observed in and below a few coarse-grained deposits. Large amphisteginid tests are usually absent from the fine-grained deposits. They are almost exclusively found in the coarse-grained beds and show maximum quantities of 20 specimens per square centimeter at the top of these beds. *Rosalina* exceeds 100 specimens per square centimeter in some samples, but more commonly, 5–10 specimens per square centimeter are found. Peaks in their distribution occur at the top of the fine-grained sediments and in the coarse-grained beds. Textulariids and *Reussella* are typical of slope environments and are observed in low amounts in the coarse-grained beds. Textulariids also occur sporadically in fine-grained samples. As with *Rosalina*, the group of other benthic foraminifers that is dominated by rotaliids shows low numbers in the fine-grained deposits. Toward the top of the fine-grained deposits, this group increases in numbers and reaches maxima

TABLE 6.—FIRST THREE UNROTATED FACTORS DESCRIBING 46.5% OF THE VARIATIONS IN THE COMPOSITION OF THE SELECTED LOWER PLIOCENE INTERVAL.

	Factor 1	Factor 2	Factor 3
(1) *Halimeda*	–0.18	–0.08	0.26
(2) Red algae	–0.25	0.35	–0.04
(3) Molluscs	0.16	0.04	*–0.46*
(4) Bryozoans	–0.04	–0.14	0.05
(5) Benthic foraminifera	0.26	0.48	0.43
(6) Planktic foraminifera	–0.16	*0.85*	–0.09
(7) Ostracodes	–0.17	–0.19	–0.36
(8) Echinoderms	–0.35	0.76	–0.31
(9) Nondeterminable	*–0.89*	*–0.35*	–0.09
(12) Fine-grained matrix	*0.94*	0.01	–0.21
(13) Peloids	–0.10	0.10	*0.71*

NOTE: Highest positive and negative values for each factor are in italics.

of about 50 specimens per square centimeter.

Planktic foraminifers (mostly globigerinids, rarely globorotaliids) are constantly present in the fine-grained deposits of the selected upper Pliocene section with two to five specimens per square centimeter. In the uppermost part of the fine-grained interval and in the coarse-grained sediments, they are most abundant, reaching up to 40 specimens per square centimeter.

Distribution of foraminifers in the lower Pliocene section.— In the section examined from the lower Pliocene, the quantities of some benthic foraminiferal groups are distinctly lower than in the selected upper Pliocene section (Fig. 11). Miliolids and soritids do not exceed 15 specimens per square centimeter and show no clear trend. Amphisteginids are extremely rare. *Rosalina* averages about 15 specimens per square centimeter and exhibits a maximum of 60 specimens per square centimeter in the lower part of the section examined (504 mbmp). Other benthic foraminifers, mainly rotaliids, show a peak exceeding 150 specimens per square centimeter, in parallel with the maximum of *Rosalina*. Above this peak, varying abundances of this group between 10–120 specimens per square centimeter lack a clear trend. Benthic foraminifers typical of slope environments (*Reussella*, textulariids) repeatedly occur in low numbers throughout the succession.

In the selected lower Pliocene section, planktic foraminifers are more abundant than in the selected upper Pliocene section and average about 50 specimens per square centimeter. At the bottom of the section examined (below 509 mbmp), numbers of planktic foraminifers increase to a maximum of 275 specimens per square centimeter.

Overall, the distribution patterns observed in the lower Pliocene section are less clear than the patterns in the upper Pliocene section. An obvious trend is the exceptionally high peak of planktic foraminifers at the base of the latter section.

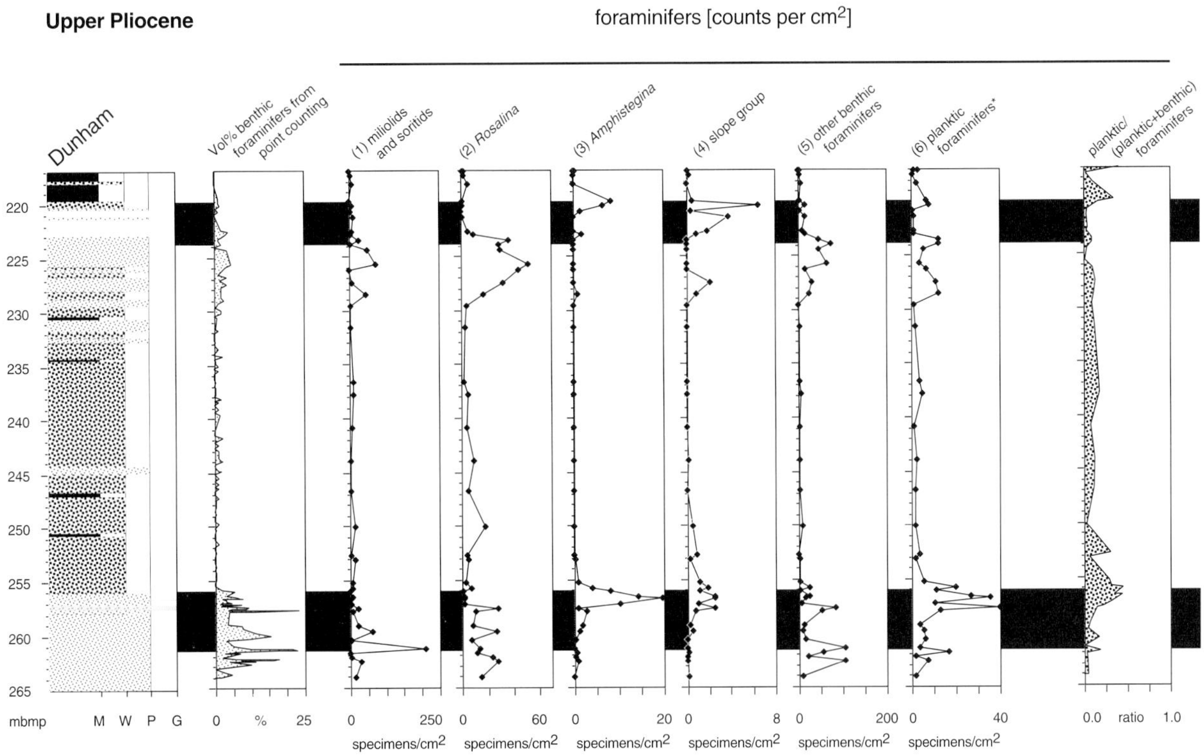

FIG. 10.—Abundance of foraminifers in the selected upper Pliocene section (in specimens per square centimeter). Note the distinctly higher numbers in benthic foraminifers in the coarse-grained intervals. Left column shows classification of the samples after Dunham (1962; G = grainstones, P = packstones, W = wackestones, M = mudstones; mbmp = meters below mud pit). Second column is the volumetric proportion of benthic foraminifers, derived from point counting (see Fig. 7). Right column shows the relative amounts of planktic foraminifers per square centimeter (coarse pattern) vs. benthic foraminifers per square centimeter. Columns with scales differing from those of the corresponding column in Figure 11 are marked with an asterisk. The two shaded bands represent the two coarse-grained beds.

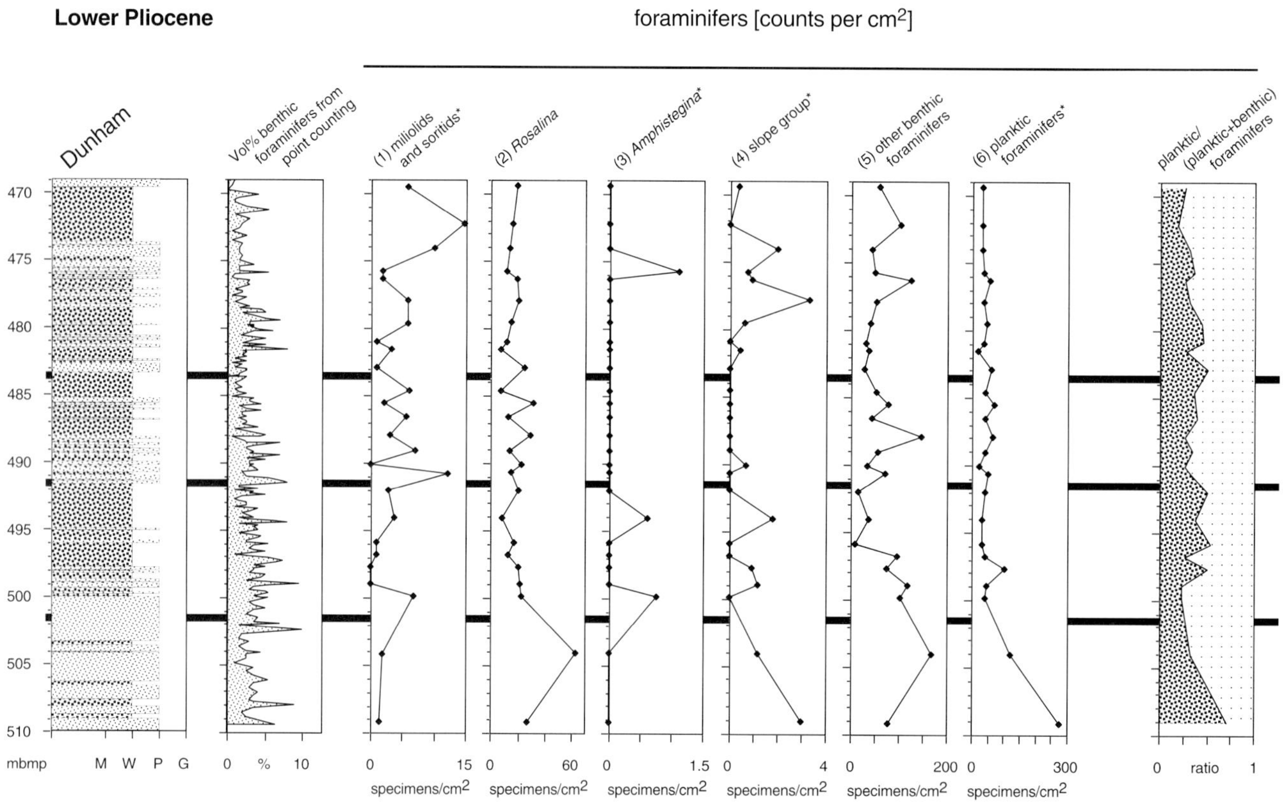

FIG. 11.—Abundance of foraminifers in the selected lower Pliocene section (in specimens per square centimeter). Left column shows classification of the samples after Dunham (1962; G = grainstones, P = packstones, W = wackestones, M = mudstones; mbmp = meters below mud pit). Second column is the volumetric proportion of benthic foraminifers, derived from point counting (see Fig. 8). Right column shows the relative amounts of planktic (coarse pattern) vs. benthic foraminifers (fine pattern). Columns with scales differing from those of the corresponding column in Figure 10 are marked with an asterisk. The bases of the three subtle cycles are marked by gray lines (cf. Fig. 8).

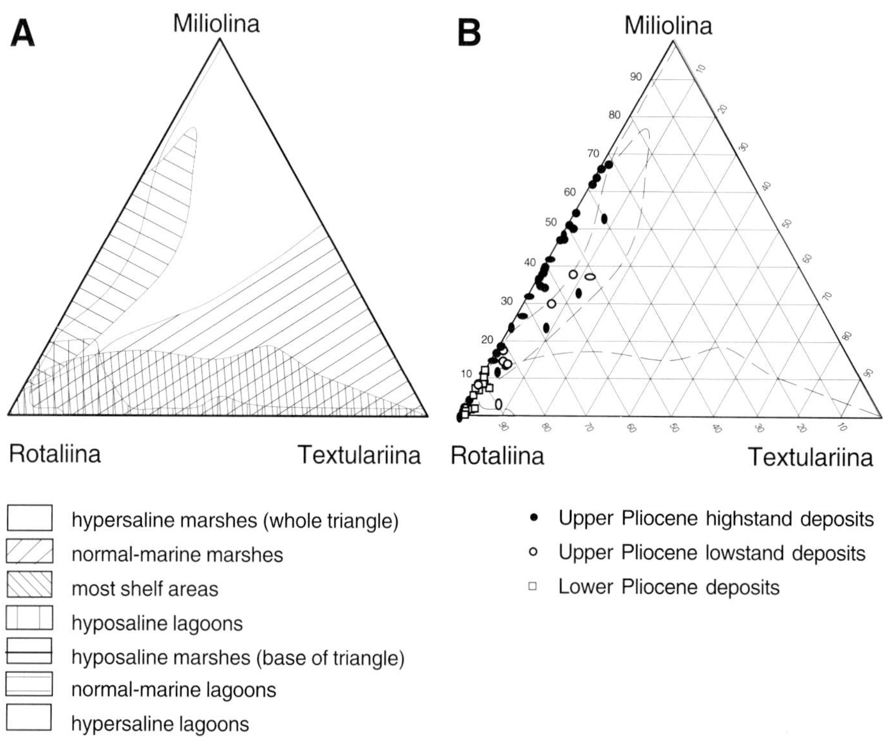

FIG. 12.—(A) Environmental information drawn from *in situ* foraminiferal assemblages (Murray, 1973). The relative abundances of the three large groups of present-day benthic foraminifers (Rotaliina, Miliolina, Textulariina) reflect the environmental conditions with respect to salinity, restriction, and water depth. (B) Allochthonous foraminiferal assemblage in samples from CLINO plotted in the triangle of Murray (1973). These are not *in situ* assemblages, but they yield information on the environmental conditions of the source area.

The variations in the abundances of the different benthic groups, however, show no systematic correlations and do not seem to correlate to the ambiguous cycles of peloids and total benthic foraminifers as observed in component analysis (point counting; Fig. 8 and second column in Fig. 11).

Environmental significance of the foraminiferal assemblages.— Murray (1973) devised a practical method to deduce environmental information from foraminiferal assemblages. The method is based on the fact that all modern benthic foraminifera belong to one of the three suborders: Miliolina, Rotaliina, and Textulariina (Loeblich and Tappan, 1964). According to Murray (1973), environmental properties are expressed in the relative abundances of these three large suborders (Fig. 12). This method has been developed for *in situ* assemblages. Nevertheless, the separation of the allochthonous upper and lower Pliocene assemblages, and the trends observed between fine-grained and coarse-grained deposits within the upper Pliocene samples, suggests that a bank-top signature has been preserved.

The samples from the selected Pliocene sections plot in different clusters (Fig. 12). The upper Pliocene samples are concentrated at the leg of the triangle that connects the Miliolina with the Rotaliina. Foraminiferal assemblages of the fine-grained deposits plot predominantly in the field that characterizes hypersaline lagoons. Coarse-grained deposits, in contrast, are found in the field of normal-marine lagoons. Lower Pliocene samples are concentrated near the corner that represents 100% Rotaliina. This area is less clearly defined, because here the fields of hypersaline lagoons, normal saline lagoons, and marine shelves overlap.

Palynological Examinations

The palynological assemblages (Table 7) contain abundant dinoflagellate cysts, microforaminiferal linings, and debris of predominantly marine origin. The dinoflagellate assemblages are of low species diversity and dominated by autotrophs. Heterotrophic taxa are persistent but scarce and are represented by only a few species. Terrestrial material is rare in most samples and consists mainly of bisaccate pollen of *Pinus* (Fig. 13A), which is readily transported in the water column (Traverse and Ginsburg, 1966), and rare angiosperm pollen, fern and bryophyte spores, and fungal remains, that were not identified to species level. No conspicuous reworking was identified.

Samples from the lower Pliocene section contain high proportions of *Polysphaeridium zoharyi* (Rossignol, 1962; Fig. 13B), an extant tropical to subtropical cyst that is adapted to restricted-marine environments and very high salinities. It currently is found off the west coast of Andros Island (Wall et al., 1977) and is abundant over much of the Persian Gulf where salinities typically exceed 40‰ (Bradford and Wall, 1984). Highest occurrences of *P. zoharyi* today are in depths of less than 22 m (Edwards and Andrle, 1992). Other neritic, thermophilic species in the lower Pliocene section include *Capisocysta lata* (Head, 1998), *Tuberculodinium vancampoae* (Rossignol, 1962), *Lingulodinium machaerophorum* (Deflandre and Cookson, 1955), and *Selenopemphix quanta* (Bradford, 1975). Neritic taxa dominate, and outer neritic to oceanic taxa such as *Impagidinium* are rare or absent from samples.

The upper Pliocene section continues to be dominated by neritic taxa. Two types of assemblages occur: one in fine-grained samples thought to represent highstand events, the other in coarse-grained samples of presumed lowstand events. Assemblages from the fine-grained samples, when compared with the lower Pliocene section, show a conspicuous reduction in *P. zoharyi* and a corresponding increase in *L. machaerophorum*. Today, the replacement of *P. zoharyi* by *L. machaerophorum* can occur where initially strongly hypersaline conditions become somewhat less restricted (Bradford and Wall, 1984). *Lingulodinium machaerophorum* is a temperate to tropical euryhaline species that can also thrive in areas of diffuse upwelling (i.e., slight increase in nutrients; Dale, 1996). Terrestrial material remains scarce in these fine-grained samples.

Assemblages from the coarse-grained samples of the upper Pliocene section show a return of abundant *P. zoharyi*. Although these samples have low concentrations of palynomorphs, there is a significant increase in the proportion of terrestrial palynomorphs (Table 7, Fig. 14).

Diagenetic Features

SEM observations have led to the recognition of distinct types of textures present in the Pliocene periplatform carbonates. In the following, first the diagenetic patterns of the upper Pliocene fine-grained and coarse-grained sediments are described, and then the lower Pliocene deposits are treated.

In the upper Pliocene section, coarse-grained and fine-grained deposits show different types of cementation. Coarse-grained deposits are tightly cemented (Fig. 15E). Most coarse-grained samples show blocky low-Mg calcite cements (10–50 µm). *Halimeda* plates are frequently found to be cemented by low-Mg calcite crystals that enclose the skeletal aragonite needles. Epitaxial overgrowth is common both on planktic and benthic foraminifers. Internal as well as external cements are observed to follow the crystallographic direction of the test crystallites. A few samples in the lower coarse-grained deposits of the upper Pliocene (below 256 mbmp) are strongly dolomitized. In these dolomitized samples, the matrix consists of fine-grained dolomite crystals (<4 µm).

The fine-grained deposits of the upper Pliocene section show a different picture. They appear to be immature and contain high amounts of (mineralogically metastable) aragonite needles. Beds alternate with varying degrees of cementation (Figs. 15A–B). Compaction of the uncemented to partially cemented layers is indicated, e.g., by deformed peloids and flattened organic microfossils, as seen in the SEM (Westphal and Munnecke, 1997). Cemented layers are characterized by blocky low-Mg calcite cements with grain sizes on the order of 10 µm (Westphal, 1998; Munnecke et al., 1997). These cement crystals engulf smaller constituents such as aragonite needles and border larger components. In the cemented layers, compaction-sensitive components

TABLE 7.—PALYNOMORPH ASSEMBLAGES DETERMINED IN 14 PLIOCENE SAMPLES FROM THE CORE CLINO.

Age:	Upper Upper Pliocene (ca. 2.0–2.2 Ma)								Upper Lower Pliocene (3.4–3.9 Ma)					
Chron:	C2r (lower Matuyama)								C2Ar (upper Gilbert)					
Depositional environment:	Upper Slope								Lower Slope					
Depth (feet below mud pit):	723.38	741.5	763.5	768.13	839.5	842.5	858.5	860.17	1,563	1,563.6	1,565.0	1,567.5	1,621.5	1,622
Depth (meters below mud pit):	220.48	226.01	232.71	234.12	255.88	256.79	261.67	262.18	476.4	476.61	477.04	477.77	494.23	494.39
Sample number:	CGS-1	CUS-K10	CUS-M7	CUS-K11	CUS-M8	CGS-2	CUS-K12	CUS-M9	CLS-K4	CLS-M1	CLS-K5	CLS-M2	CLS-K6	CLS-M3
Dinoflagellates (Raw Counts)														
Achomosphaera/Spiniferites spp. (total)	57	159	173	151	136	42	137	162	86	139	131	175	42	109
Capisocysta lata	1	29	15	35	1	2	11	13	5	2	8	3	+	+
Dapsilidinium pseudocolligerum	1	10	4	6	3	1	3	1		+		1	2	2
Kallosphaeridium sp.	*	*	*	*	*	*	*	*	*	*	*	*	*	*
Lingulodinium machaerophorum	7	25	33	28	77	2	56	42	13	15	18	5	33	16
Operculodinium n. sp.	+	5	1	+	3		3	+	+	+	+			+
Operculodinium spp. including O. *israelianum*	3	16	11	9	9	3	33	29	24	42	32	24	9	12
Polysphaeridium zoharyi	27				4	44	+	+	116	49	47	33	160	104
Round brown cysts	2	2	6	3	+	1	3	1	4		12	6	3	4
Spiniferites mirabilis	*		*	*	*		*	*	*	*	*	*		*
Spiniferites n. sp.	*	*	*	*	*		*	*	*	*		*	*	*
Lejeunecysta sp. cf. *L. marieae*	2													
Selenopemphix quanta		+	1	+	1	1	1	2		+	+	+	+	2
Selenopemphix nephroides		+		+	+						+			+
Tuberculodinium vancampoae	3		1						1	2	1	2	+	+
Ataxiodinium zevenboomii		(+)?		+	+	+	+	+				1		
Melitasphaeridium choanophorum		1	6	17	10									
Operculodinium psilatum			+										+	+
Bitectatodinium raedwaldii			+		+		3	+	+	+	+			
Bitectatodinium cf. *raedwaldii*				+	+									
Operculodinium? n. sp.					2	1								
Lejeunecysta marieae					+									
Nematosphaeropsis rigida					*4*									
Protoperidinioid cyst sp. A					+									
Impagidinium paradoxum						3	+		+					
Hystrichokolpoma rigaudiae									+	+		+	1	+
Operculodinium? *longispinigerum*									+	1	1	+		1
Bitectatodinium tepikiense									1					
Impagidinium sp. A												+		
Operculodinium janduchenei												+		
Total *in situ* dinoflagellates	100	250	250	250	250	100	250	250	250	250	250	250	250	250
Marine algae incertae sedis (raw counts)														
Cyclopsiella sp.	1				1									
Nannobarbophora walldalei	3	19	9	44	1		29	3	14	51	52	32	5	18
Small spiny acritarchs	16	23	51	63	68	25	108	43	93	112	131	158	11	43
Leiospheres		*	*	*	*	*	*	*	*	*	*	*	*	*
Acritarch sp. 1	*	*				*		*						
Incertae sedis sp. A											+	1		
Incertae sedis sp. B									+					
Other marine palynomorphs (raw counts)														
Foraminiferal linings (six or more chambers)	6	10	6	9	12		15	9	34	31	19	31	22	11
Trochospiral microforaminiferal lining B				*										
Scolecodonts	1					1			4		2	1		1
Copepod fragments			+		+								1	
Terrestrial palynomorphs (raw counts)														
Bisaccate pollen	88	3	2	2	6	28	4				1	1	4	1
Angiosperm pollen	9	+			3	6			1		1		1	
Fern and bryophyte spores			+		1	+	1			1				
Fungal spores and hyphae					1	2								
Total terrestrial palynomorphs	97	3	2	2	11	36	5	0	1	1	2	1	5	1
Terrestrial palynomorphs/dinoflagellates ratio	0.970	0.012	0.008	0.008	0.044	0.360	0.020	0.000	0.004	0.004	0.008	0.004	0.020	0.004
Sample data														
Sample dry weight (in grams)	6.3	3.2	2.2	18.6	5.9	6.6	3.9	8	10.4	14.3	13	3.5	2.1	8.6
Quantity of *Lycopodium clavatum* tablets	1	1	1	3	1	1	1	1	1	2	2	1	1	1
Quantity of *Lycopodium clavatum* spores counted	1,933	605	117	176	241	3,663	278	95	173	35	90	94	325	27
Palynomorphs per gram (dry wt concentrations)														
Estimated total dinoflagellates	114	1,796	13,511	3,187	2,446	58	3,208	4,576	1,933	13,897	5,945	10,571	5,096	14,977
(estimated standard error)	(±13)	(±162)	(±1,655)	(±327)	(±252)	(±7)	(±322)	(±596)	(±214)	(±2,555)	(±760)	(±1,382)	(±497)	(±3,123)
Estimated total terrestrial palynomorphs	111	22	108	25	108	21	64	0	8	56	48	42	102	60
(estimated standard error)	(±13)	(±13)	(±77)	(±18)	(±34)	(±4)	(±29)	(±0)	(±8)	(±57)	(±34)	(±42)	(±46)	(±61)

NOTE: The lower Pliocene assemblages, with their neritic character and tropical to subtropical, restricted-marine signatures, are rather similar to those found on the present-day Great Bahama Bank. The coarse-grained upper Pliocene samples (CGS-1 and CGS-2), interpreted to represent highstand deposits, have distinctly higher proportions of terrestrial palynomorphs. These two samples also show a temporary return of abundant P. zoharyi, indicative of restricted-marine conditions on the platform top. Symbols: +, an enumerated taxon recorded only outside of counts; *, a qualitative record of a taxon not specifically included during counting. Numerical ages are based on the Berggren et al. (1985) time scale.

like peloids and thin-walled organic microfossils (mainly dinoflagellate cysts) show no deformation (Westphal and Munnecke, 1997). Small dolomite rhomboids are common.

The lower Pliocene section, being rather uniform with respect to its components, shows different degrees of cementation similar to the fine-grained deposits in the upper Pliocene, but aragonite needles are less abundant (Figs. 15C–D). Cemented layers show a tight mosaic of low-Mg calcite cement crystals that engulf small components and show sharp boundaries along larger components. Unlithified layers in the lower Pliocene consist largely of loose, small, low-Mg calcite crystallites (about 2 μm) and small dolomite rhomboids.

In both sections, microstructures of components initially consisting of low-Mg calcite (e.g., ostracodes, planktic foraminifers) are well preserved. These shells are sharply bounded by cement crystals. Molds are common especially in the lower Pliocene section and are thought to originate from dissolution of initially metastable bioclasts. Most molds are partly filled by secondary calcite and dolomite cements.

DISCUSSION

Sediment Input

Sediment input characteristics of the upper Pliocene and lower Pliocene sections differ remarkably. The clear differences in the upper Pliocene section of lowstand and highstand deposits are absent in the lower Pliocene section. The latter displays a rather uniform distribution of constituents with only obscured indications of cyclic changes in quantity. The cyclic vs. noncyclic characteristics might indicate that there is an underlying difference between the two sections.

Sedimentation of the upper Pliocene section.—

The selected upper Pliocene section represents upper-slope sediments deposited in a water depth of roughly 200 m. The distance to the margin, as marked by reefal deposits penetrated by the UNDA borehole, was about 8.5 km (Eberli et al., 1997; G. P. Eberli, J. A. M. Kenter, F. McNeill, R. N. Ginsburg, P. K. Swart, and L. A. Melim, pers. commun., 1997).

The Pliocene periplatform carbonates from the CLINO borehole show a variety of compositional signatures. In most samples, large amounts of bank-derived constituents could be distinguished, pointing to the allochthonous nature of these slope sediments. Typical for periplatform deposits, the shallow-water–derived constituents occur together with planktic foraminifers.

Although the majority of the upper Pliocene sediments are characterized by small-sized constituents and a fine-grained matrix, two thin, coarser-grained intervals are conspicuous. These two coarse-grained intervals in the selected section (about 220.0–224.0 mbmp and 256.0–261.5 mbmp) are characterized by poorly sorted packstones to grainstones composed largely of rim to upper-slope skeletal and nonskeletal grains. Among these are predominantly cortoids, *Halimeda* fragments, smaller amounts of benthic foraminifers, red algae, echinoderms, and mollusks. On the present-day Bahamas, photic organisms such as *Halimeda* and specific benthic foraminifers like amphiste-

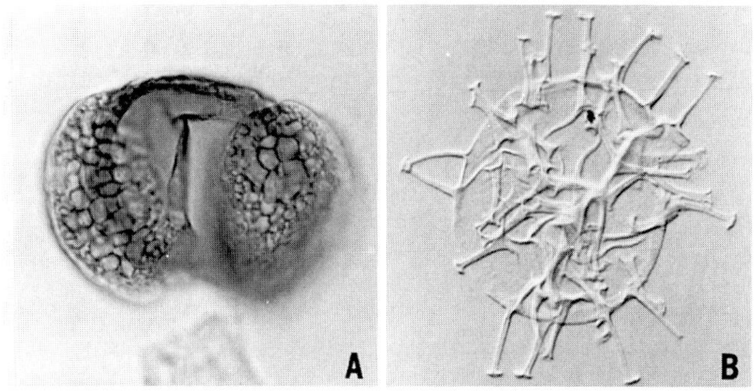

FIG. 13.—Examples of palynomorphs from the Pliocene periplatform carbonates. (A) Bisaccate pollen of the genus *Pinus;* length is 49 μm (from upper Pliocene; 220.48 mbmp [meters below mud pit]). (B) *Polysphaeridium zoharyi* (Rossignol, 1962); maximum diameter of central body is 58 μm (from lower Pliocene; 476.40 mbmp).

ginids are known to prosper in rim to upper-slope regions (Rose and Lidz, 1977; Ginsburg et al., 1991). These organisms are able to continue production and maintain export of calcareous material at least during minor sea-level lowstands by slightly shifting their facies belt downslope. Cortoids are almost exclusively present in the coarse-grained deposits. They are interpreted as allochthonous components whose source was the margin and upper slope of the carbonate platform. If it is assumed that the majority of the micrite envelopes are of algal origin, the large numbers of cortoids probably originated from high-energy

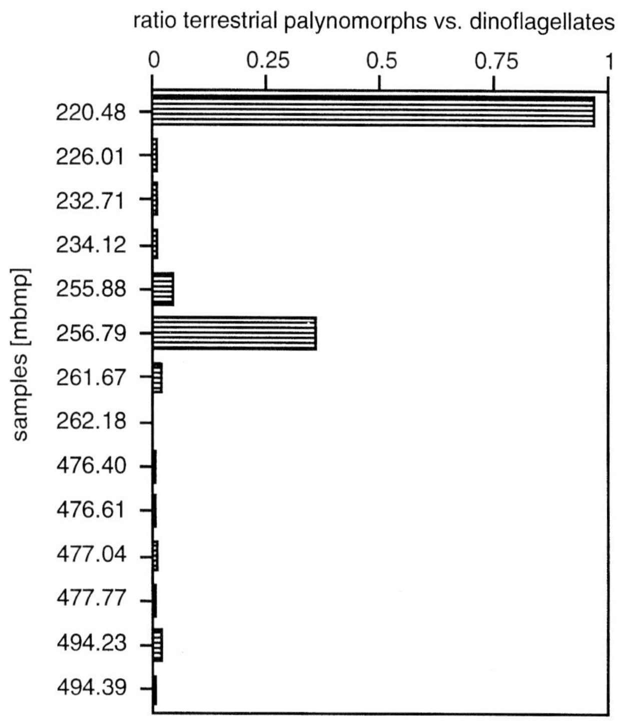

FIG. 14.—Ratio of terrestrial palynomorphs vs. dinoflagellates. In the diagram, eight upper Pliocene and six lower Pliocene samples are shown. Note the strikingly higher proportion of terrestrial palynomorphs in the two coarse-grained upper Pliocene samples at 220.48 and 256.79 mbmp (meters below mud pit).

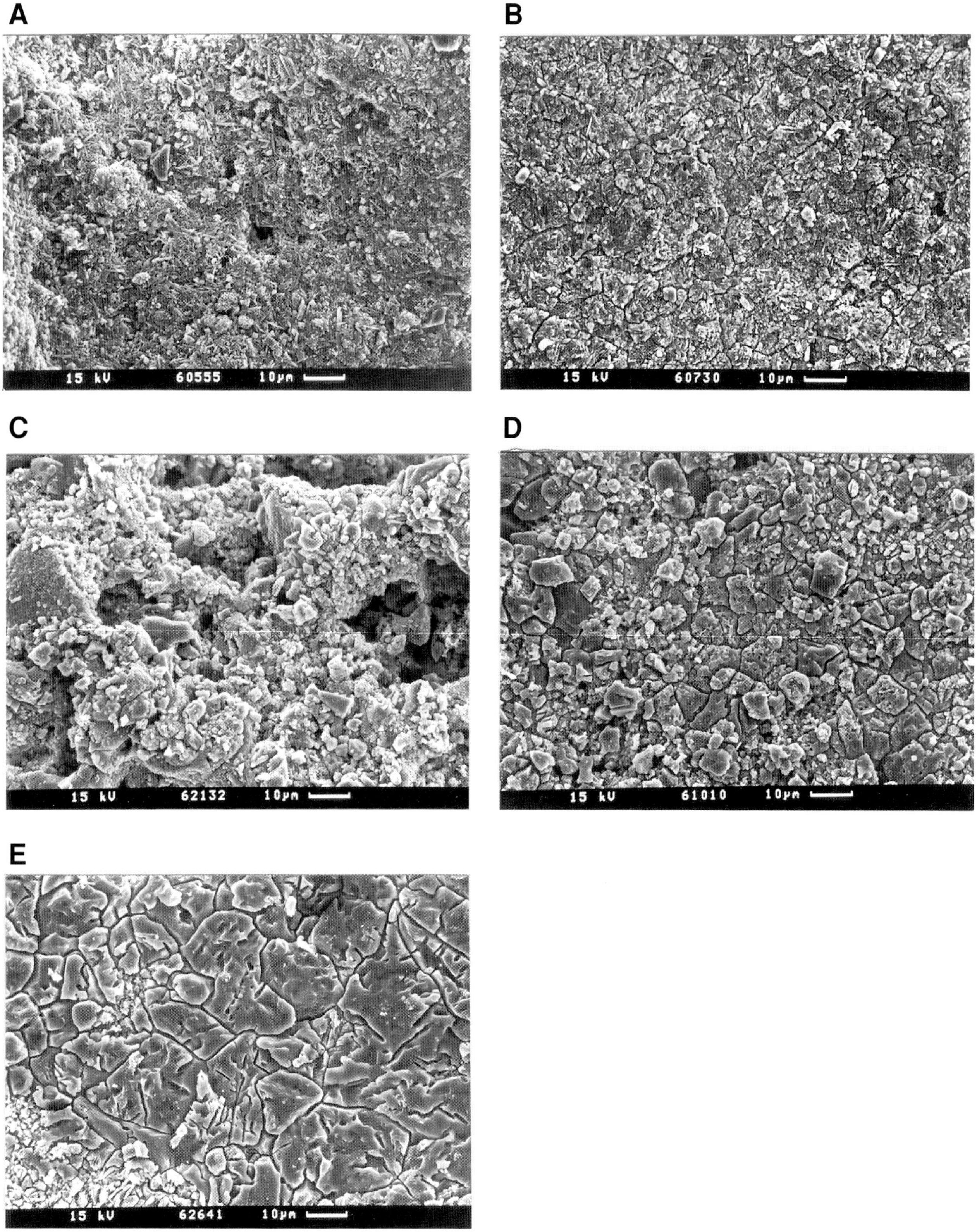

FIG. 15.—Scanning electron micrographs of polished, etched samples. All micrographs are shown at the same magnification; scale bar represents 10 μm. (A) Uncemented sample from the upper Pliocene highstand deposits (219.94 mbmp [meters below mud pit]); aragonite needles are well preserved. (B) Cemented sample from the upper Pliocene highstand deposits (217.17 mbmp); aragonite needles are enclosed in low-Mg calcite cement crystals. (C) Uncemented sample from the lower Pliocene section (479.82 mbmp); aragonite needles are rare, whereas calcite crystallites and some dolomite rhombs are present; high intercrystalline porosity characterizes these sediments. (D) Tightly cemented sample from the lower Pliocene (497.74 mbmp). (E) Tightly cemented lowstand deposit from the upper Pliocene (220.83 mbmp); only a few fine-grained constituents are visible in the cement.

shoals (Gunatilaka, 1976; Flügel, 1982). As algal activity is also well known from more protected environments (Flügel, 1982), a contribution from lower-energy lagoons along a more articulated shoreline is possible. The large amounts of photic organisms and the poor sorting (including very coarse to very fine sand-sized grains and mud-sized matrix constituents) indicate that the material was shed from the platform rim or upper slope probably as episodic mass flows during lowstand conditions.

Platform mud is rare to absent in the lowstand sediments, reflecting the fact that sediment production within the platform interior was sharply reduced by subaerial exposure of the bank top. Selective bypassing of carbonate mud is considered inappropriate as an explanation for the virtual absence of fine-grained matrix in many of these sediments. During such a situation, density cascading (Wilson and Roberts, 1992, 1995) would still transport fine-grained material to the slope environment. Winnowing of the fine-grained material by currents, leaving a coarse-grained lag deposit, also seems unlikely, as then a better sorting and more breakage of thin-shelled planktic organisms would be expected. These phenomena, however, are not observed.

On the basis of lithologic examinations and seismic geometries, J. A. M. Kenter, R. N. Ginsburg, and S. R. Troelstra (pers. commun., 1997) and Eberli et al. (1997) interpreted the three major coarse-grained slope intervals from core CLINO (at about 540, 370, and 200 mbmp; Fig. 4) as lowstand deposits of larger-scale sea-level cycles. For the smaller-scale variations examined in this study, an analogous interpretation appears reasonable. The associations of palynomorphs offer direct proof for the steering role of sea level on the fine-grained vs. coarse-grained sediment composition in the upper Pliocene section. The fine-grained intervals are strongly dominated by dinoflagellate cysts and other marine palynomorphs and have only trace amounts of terrestrial palynomorphs (Table 7). The coarse-grained intervals, in contrast, show a significant increase in the proportion of terrestrial palynomorphs, most of which are pollen of Pinus (Fig. 14). These pollen, once deposited in the water, can remain in the water column for many months (Traverse and Ginsburg, 1966). Assuming that the circulation patterns during the Pliocene were similar to the present situation, the sediment transport was dominantly from the east to the west. Westward transport during the Pliocene is also indicated by seismic geometries that reveal thick, prograding clinoforms to the west of the Great Bahama Bank. This pattern indicates that the pine pollen as well as the dinoflagellate cysts found in the periplatform sediments were probably transported from the platform top of the Great Bahama Bank to the western, leeward slope. The production of pine pollen requires subaerial exposure of at least of parts of the platform top. Numerous exposure surfaces have been described from the upper Pliocene platform-top section (Beach, 1982). Nevertheless, lithoclasts that would indicate erosion of an emerged platform top are not observed in these lowstand deposits. The occurrence of lithoclasts could be restricted to turbidite gullies that were not penetrated, so that such gully in-fills were simply not recovered. Dravis (1996), however, has shown that emerged platforms potentially are lithified extremely rapidly by meteoric diagenesis ("freezing"), thus inhibiting considerable erosion. Similarly, Schlager et al. (1994) have shown that early marine and meteoric diagenesis will lead to a rapid lithification of the platform top that hinders mechanical erosion.

The locally higher, but strongly variable amounts of planktic foraminifers in the lowstand deposits of the upper Pliocene section could reflect a lowered sedimentation rate during sea-level lowstands, when the voluminous export from the platform interior ceased and episodic mass flows might have dominated. A hardground at 256.20 mbmp mentioned by L. A. Melim, P. K. Swart, and R. G. Maliva (pers. commun., 1997) and J. A. M. Kenter, R. N. Ginsburg, and S. R. Troelstra (pers. commun., 1997) that would be located at the top of a coarse-grained interval examined in this study, however, could not unequivocally be confirmed.

The micritic sediments at the top of the coarse-grained intervals are typified by the third factor of the factor analysis, i.e., they are dominated by *Halimeda* fragments and bryozoans, whereas sparite, cortoids, and peloids are absent. These sediments could be the result of the early flooding event when most organisms were drowned. *Halimeda* seems to have given up shortly thereafter, when mud was deposited on top. The cause of this pulse of exported fine-grained material can only be speculated on. In previous studies, aragonite percentages have been documented to show similar, sudden, saw-tooth–shaped increases in younger intervals of Bahamian toe-of-slope sediments, although in the aragonite record, dissolution cycles could have modified the image (e.g., Kier and Pilkey, 1971; Droxler et al., 1983; Boardman et al., 1986; Reijmer et al., 1988). An unequivocal explanation for this pattern, however, has not yet been proposed. Below the "wall" of the present-day Great Bahama Bank, Wilber et al. (1990) observed Holocene subsurface sediments that are composed of 96% aragonite mud; 70% of the bulk sediment is of clay size. These fine-grained, muddy deposits are thought to represent the Holocene flooding event. Possibly the micritic pulse observed in the selected upper Pliocene section from CLINO also reflects the flooding event that led to the geologically instantaneous production of fine-grained material (chemical precipitation of aragonite needles and/or algal skeletal needles) whereas most organisms lagged behind. For the Holocene flooding event, lag times between submergence and colonization by reef organisms ranged from 500 to 2,000 yr on different Indo-Pacific islands (Montaggioni, 1988). Also, higher tidal energy in the deeper water of early flooding could have led to an enforced export of fine-grained material.

The fine-grained highstand intervals in the selected upper Pliocene section are characterized by bioclastic to peloidal packstones, wackestones, and mudstones. They are rich in material exported from the platform interior, namely, aragonite needles. The extensive shallow-water carbonate factory was submerged at the time of deposition and thus able to produce and export great amounts of calcareous material. As Neumann and Land (1975) have shown, on the present-day Great Bahama Bank, vast volumes of excess material are exported from the bank top. During the late Pliocene, a similarly voluminous fine-grained platform input could have diluted the skeletal material shed from the rim and upper slope. The varying amounts of peloids, which show a minor increase in abundance at 244 mbmp, could indicate subordinate changes in the conditions on the platform top or in the transport processes. The relatively good sorting and

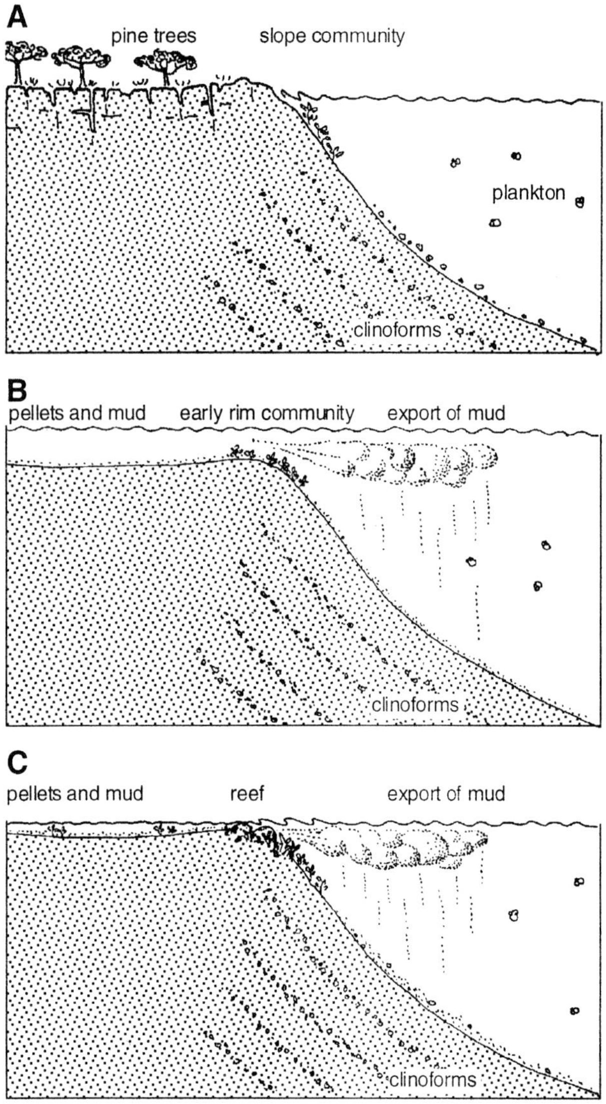

FIG. 16.— The influence of sea-level fluctuations on the late Pliocene flat-topped platform. (A) During sea-level lowstands, the platform top is exposed and colonized by terrestrial plants. Sedimentation on the slope is dominated by slope biota. (B) Flooding of the platform leads to lag-time deposition of muddy sediment. (C) During sea-level highstands, mud and pellets are exported from the platform interior to the slopes where they mix with margin-derived biota and planktic foraminifers.

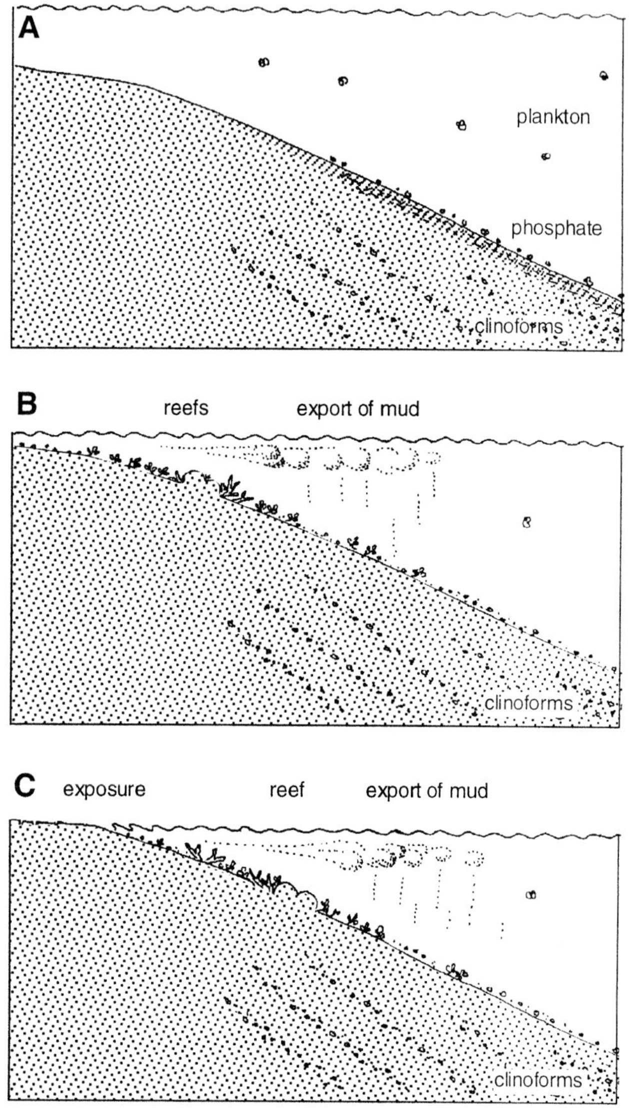

FIG. 17.— The influence of sea-level fluctuations on the early Pliocene ramp-shaped platform. (A) The earliest Pliocene is characterized by a lower-order drowning event, including pelagic sedimentation and phosphatization of hardgrounds. (B) When carbonate production on the platform recovered, platform-top–derived input reached the periplatform realm. (C) During higher-order sea-level drops, facies belts shifted downslope, and the productivity of the platform continued without drastic change.

the dominantly small grain sizes indicate that the material settled from suspension. The absence of clear graded bedding, elsewhere commonly associated with settling from suspension, is probably a result of bioturbation.

In the selected upper Pliocene section from CLINO, the highstand deposits coarsen up-core, and packstones become more abundant. Late highstand deposits above 233 mbmp show a gradual increase in the amount of peloids. At the base of the section examined (below 262 mbmp), large amounts of peloids are similarly observed that culminate in peloid packstones. These peloid packstones precede the base of the overlying lowstand as marked by the appearance of cortoids. Peloids are a dominant constituent in surface sediments of present-day Bahamian platforms (Purdy, 1963). The increase of peloids associated with a coarsening-upward trend might reflect the progressive decrease in the space for accommodation that is typical of late highstands when platforms are forced to build outward. It could be associated with an increased export of coarser-grained material from the platform interior such as fecal pellets. In parallel to this, there is an increase in the export of some groups of benthic foraminifers. Small, rotund miliolids occur in large numbers in the late highstand deposits. These miliolids are evidence for elevated saline conditions in the platform interior, coupled with the late sea-level highstand conditions. Miliolid packstones are restricted to the upper part of the fine-grained interval (260 mbmp) below the base of the coarse-grained interval.

The foraminiferal assemblages of the highstand sediments plot predominantly in the field of hypersaline lagoons, whereas the lowstand samples plot preferentially in the field of normal-marine lagoons. During highstand conditions, large areas of the platform top were covered by shallow water, which could have had elevated salinities. During lowstands, the shallow-water areas were not available, so that mostly more open-marine organisms were shed downslope from the marginal areas of the platform.

Dinoflagellate assemblages from two separate lowstand deposits (at 256.79 and 220.48 mbmp) give a somewhat different picture, having high proportions of *P. zoharyi* and low proportions of *L. machaerophorum,* in contrast to the lowstand assemblages of the upper Pliocene section where this pattern is reversed (Table 7). *Polysphaeridium zoharyi* is adapted to strongly hypersaline conditions today, and these conditions possibly prevailed in at least some shallow embayments around the margins of the exposed platform top, even though the foraminifers indicate the predominance of normal-marine lagoons.

Overall, the upper Pliocene microfacies succession draws a picture typical of a flat-topped, steep-sided carbonate platform subjected to successive flooding, sea-level highstand, and subaerial exposure (Fig. 16). The numerical analyses of the point-count results support the idea that two different modes of sediment production and transport prevailed during sea-level highstands and lowstands, respectively. This interpretation is best illustrated by the bimodal distribution of the percentages of fine-grained matrix (Fig. 9) in which the peak at lower values represents the matrix-poor lowstand samples, whereas the peak at higher values corresponds to the matrix-rich highstand deposits. The facies patterns observed in the upper Pliocene periplatform sediments thus represent a straightforward picture of the principle of highstand shedding *sensu* Droxler and Schlager (1985). The sea-level–induced cycles are a smaller-scale repetition of the patterns observed by J. A. M. Kenter, R. N. Ginsburg, and S. R. Troelstra (pers. commun., 1997) in the large-scale cycles, which these workers describe as being composed of the same elements, coarse-grained lowstand deposits and fine-grained highstand deposits. In contrast to these larger-scale cycles, the smaller-scale cycles examined within this study seem to lack the clear drowning signature on top of the lowstand sediments (e.g., hardgrounds and phosphatization of the lowstand deposits). This difference possibly reflects the shorter time spans involved in the high-frequency sea-level fluctuations. A stacking of small-scale cycles seems to form the large-scale cycles.

Sedimentation of the lower Pliocene section.—

The selected lower Pliocene section is composed of lower-slope sediments deposited at a depth of the order of 300–400 m. The distance to the shallow-water carbonate factory is estimated to have been 16 km (Eberli et al., 1997; G. P. Eberli, J. A. M. Kenter, F. McNeill, R. N. Ginsburg, P. K. Swart, and L. A. Melim, pers. commun., 1997).

The lower Pliocene section has been selected in spite of its rather uniform macroscopic appearance, because the gamma-ray log seems to indicate the presence of higher-order cycles. Sea-level fluctuations are known from the entire Pliocene. On the basis of the correlation of isotope events and sequence stratigraphic events, V. S. Abreu and G. A. Haddad (in press) and V. S. Abreu and J. B. Anderson (1998) have argued that by the early Pliocene, high-frequency sea-level fluctuations (including 100 k.y. cycles) had already occurred. Also, sedimentary cycles in the platformward borehole UNDA (J. A. M. Kenter, R. N. Ginsburg, and S. R. Troelstra, pers. commun., 1997), and discontinuity surfaces on the platform top at that time (Beach and Ginsburg, 1980), indicate sea-level fluctuations throughout the Pliocene. On the basis of sedimentation rates (Lidz and McNeill, 1995a), the upper Pliocene section examined here is calculated to have covered a time span of about 100 k.y., whereas the selected lower Pliocene section is thought to represent about 300 k.y. Thus both sections are thought to have been subjected to environmental changes resulting from higher-frequency sea-level fluctuations.

In contrast to the clear patterns observed in the selected upper Pliocene section, drastic variations are absent from the selected lower Pliocene section, and more subtle changes in composition are observed. The palynomorph associations of the selected lower Pliocene section appear rather uniform, and terrestrial palynomorphs are rare in the samples examined. Therefore, a clear interpretation of environmental changes based on palynomorphs is not possible. Bioclast packstones and bioclast wackestones contain the same components (although large amounts are nondeterminable biodetritus) and are mainly distinguished by differences in the amount of sedimentary matrix. Both packstones and wackestones are moderately to well sorted and fine-grained and suggest that the material settled from suspension. Gradual shifts in the abundances of peloids and benthic foraminifers mark a slight cyclic tendency. These two groups of components could represent sea-level fluctuations by reflecting the changing distance of the locality from the shallow-water carbonate factory. During sea-level falls, the facies belts that were the source areas of the peloids and benthic foraminifers migrated down the gently dipping slope and thereby moved closer to the site of deposition; the belts retreated upslope during sea-level rises. A parallel increase in both groups would thus be expected during sea-level lowstands.

Globigerinid packstones represent a striking signal at the base of the lower Pliocene section examined (about 509 mbmp). The samples from the lowermost few meters of the section are dominated by planktic foraminifers and relatively high amounts of echinoderm fragments. The proportion of planktic foraminifers decreases rapidly upward. The lowermost Pliocene strata in CLINO, below the selected lower Pliocene section, extend over 30 m (536.33–507.70 mbmp) and has been described by J. A. M. Kenter, R. N. Ginsburg, and S. R. Troelstra (pers. commun., 1997) to show a similar predominance of planktic foraminifers and echinoderm fragments. These authors interpreted this pattern as an expression of a temporary drowning event in the earliest Pliocene. The Miocene platform was exposed by a pronounced sea-level drop at the

Miocene-Pliocene transition, and subsequent flooding led to temporary drowning of the platform and the development of a marine hardground. Deposition above the hardground described by J. A. M. Kenter, R. N. Ginsburg, and S. R. Troelstra (pers. commun., 1997) was pelagic until shallow-water–production material was restored on the retreated platform, which by then showed the gently dipping slope of a distally steepened ramp. The fine-grained interval of the selected lower Pliocene section examined in this study represents slope sediments of the recovered carbonate platform (Fig. 17).

In the fine-grained sediments that make up the majority of the selected lower Pliocene section, the fraction of skeletal grains (including the fine biodetritus and other grains not further determinable) is about 50% of the sediment. The large amounts of fine-grained biodetritus and the strong diagenetic alteration of some components might be in part responsible for the monotonous appearance of the lower Pliocene section. The smaller grain sizes of the biodetritus (compared to the upper Pliocene section) reflect the lower-slope position of CLINO and its greater distance from the platform margin during the early Pliocene. Throughout the selected lower Pliocene section, higher amounts of planktic foraminifers are observed than in the selected part of the upper Pliocene section (Fig. 11). The higher relative and absolute abundance of planktic foraminifers indicates a stronger open-marine influence in the lower Pliocene that at least partly is attributed to the lower-slope setting. Additionally the low numbers of miliolids either point to a greater distance from the platform interior or might result from the deeper and less protected environment on the top of the ramp in comparison with the flat-topped platform.

The relatively high amount of skeletal material in the selected lower Pliocene section from CLINO (approximately 50%) corresponds to the high abundance of skeletal material observed on the platform top in approximately contemporaneous Bahamian platform-top sediments of the so-called pre-Lucayan Formation (to which foraminiferal zone N19 as a whole corresponds; Beach and Ginsburg, 1980). Reijmer et al. (1992) also observed a skeletal-rich composition in toe-of-slope turbidites in the Exuma Sound that are time equivalent to the pre-Lucayan Formation. Components deposited on the platform interior during deposition of the pre-Lucayan sediments are small miliolids, agglutinating foraminifers, amphisteginids, Halimeda, red algae, echinoids, and corals (Beach, 1982). With the exception of corals (which might be represented by the diagenetically altered biodetritus), this association corresponds to that present in the CLINO samples from the studied lower Pliocene section.

Beach and Ginsburg (1980) concluded from the composition that open circulation and a relatively great water depth on the bank itself, in combination with the absence of distinct marginal barriers, steered the development of these facies belts. Although the sediment composition observed in the selected lower Pliocene section is in accordance with the findings of Beach and Ginsburg (1980), the dinoflagellate associations point to more restricted conditions on the bank top. Large amounts of *Polysphaeridium zoharyi* suggest that elevated salinity prevailed in the source area. Laterally extensive banks are prone to restriction even if devoid of margins, just because their large size hinders rapid water exchange (Simms, 1984). The foraminiferal assemblages in the selected lower Pliocene section, in contrast, exhibit no clear signature with respect to their source environment (Fig. 12). In the triangle, the samples plot close to the corner that represents 100% Rotaliina. This area is an overlap of different possible environments—hypersaline lagoons, normal-marine lagoons, and normal open shelves. Although an interpretation can only be tentative, the possible shelf signature could reflect the absence of a strictly defined rim and the gently dipping slope morphology. Thus, the signature of dinoflagellates and foraminifers could represent different facies belts of the platform.

Numerical analyses emphasize that there are no clearly separable statistical clusters that might represent typical highstand vs. lowstand deposits. The slight fluctuations in the amount of peloids and benthic foraminifers show traces of cycles but no significant shifts in constituents (Fig. 11). The more subtle variations in the lower Pliocene section examined (except the aforementioned lowermost samples) are interpreted to reflect the gently dipping morphology of the platform that prevailed during the lower Pliocene (Beach and Ginsburg, 1980; Beach, 1982; Eberli and Ginsburg, 1987, 1989; McNeill, 1989). Slope angles between the platform margin and CLINO are about 3.5° (the slope is thus steeper than the slope of a ramp in the strict sense of Ahr, 1973). The absence of well-marked highstand-lowstand differences could thus result from this more open, gently dipping morphology. The effect of the distally steepened ramp morphology, however, in the example presented here, is overlain by an overall long-term sea-level rise. The small-scale sea-level fluctuations that have been examined are superimposed by a long-term sea-level rise and could have resulted in a stepwise sea-level rise rather than a succession of symmetric rises and falls. Sea-level fluctuations in the lower Pliocene are recorded as infrequent exposure surfaces and facies shifts in sediments of the platform interior (Beach, 1982). A direct correlation of these observations to the slope sediments examined here, however, is not possible.

In summary, the apparent absence of clear compositional cycles can be attributed to a number of causes. First, the assumed high-frequency eustatic sea-level fluctuations are overlain by a long-term sea-level rise that caused the backstepping or retreat of the platform. Second, the section sampled represents lower-slope facies deposited at some distance from the shallow-water factory, where compositional variations are possibly less pronounced. It is true that compositional signatures of various platforms are known to be traceable far out into the basin (open shelf, Cretaceous: Everts, 1994; Everts and Reijmer, 1995; ramp, Mississippian: Elrick and Read, 1991). Nevertheless, it is probable that no clear signatures are present in the lower Pliocene slope sediments of CLINO. It should also be noted that slight faunal changes described as a response to sea-level fluctuations in the Cretaceous (Burchette and Wright, 1992; Bachmann et al., 1996; Bachmann and Willems, 1996) seem to be less pronounced or absent in the lower Pliocene of CLINO. The slightly cyclic nature of the abundance of benthic foraminifers and peloids could possibly represent shifts of facies belts without faunal changes.

Clearly, more examples of ramps have to be investigated before a general rule can be extracted. Nevertheless, the distributions of peloids and foraminifers indicate that smaller-scale sea-level changes resulted in a shift of the facies belts, not in changes in the environments that were the source areas for the sediments.

Diagenesis

Sea-level position controls the texture and composition of the slope sediments, and texture and composition influence later diagenetic alterations.

Diagenetic changes in selected upper Pliocene section.—
As seen in the upper Pliocene section, the compositional and textural differences were a cause of differences in the cementation and the diagenetic maturity of a sediment. Light-microscope and SEM analyses have shown that highstand deposits are represented by fine-grained carbonates with high amounts of aragonite needles, whereas lowstand deposits consist mainly of grainstones with coarser components such as *Halimeda* plates and cortoids. Highstand deposits are diagenetically immature and contain large amounts of aragonite. The low initial permeability of these sediments precluded rapid fluid migration and protected these sediments from fluid-driven diagenetic alteration. These fine-grained deposits are partially cemented by presumably locally derived calcium carbonate. The diagenetic fluid could have been derived from normal-marine water. The high strontium contents in pore-water samples from CLINO indicate that diagenetic processes are still proceeding (P. K. Swart, H. Elderfield, and G. Ostlund, pers. commun., 1997). Unfortunately, the CLINO water samples are from intervals different from those selected for this study; therefore, no direct comparison was possible. In the fine-grained highstand slope deposits, considerable amounts of metastable aragonite survive early diagenesis. Thus, these sediments would be capable of forming considerable secondary porosity by later diagenetic alteration.

The coarse-grained lowstand deposits of the selected upper Pliocene section, in contrast, are tightly cemented. As seen under the SEM and confirmed with the EDX, dolomite occurs in strongly varying amounts and in some samples makes up most of the matrix. Typically, the coarse-grained lowstand deposits have higher initial porosities and permeabilities that could have focused early postdepositional fluid flow and thus could have resulted in an accelerated diagenesis. Halos of alteration to low-Mg calcite around grainstone beds in CLINO that are described by Melim et al. (L. A. Melim, P. K. Swart, and R. G. Maliva, pers. commun., 1997) emphasize the influence of high porosity and permeability on cementation.

Diagenetic changes in selected lower Pliocene section.—
The lower Pliocene section, typified by a rather uniform compositional and textural signature, shows an alternation of different states of cementation that is similar to the upper Pliocene highstand deposits, but aragonite needles are less common, and dolomite rhomboids are more abundant in the matrix. Therefore, the lower Pliocene sediments appear mineralogically more mature, but, in contrast to the coarse-grained beds of the upper Pliocene, cementation has not lithified the entire succession. The lower Pliocene sediments are as fine grained as the upper Pliocene highstand deposits, but the proportion of grains (most presumably of skeletal origin) is higher, and the proportion of sedimentary matrix is lower, resulting in a higher initial permeability. Therefore, the lower Pliocene sediments are more mature and contain more dolomite rhombs. Also, molds are more common in the lower Pliocene sediments, possibly implying somewhat higher fluid flow. Nevertheless, diagenesis was not fluid driven and, as indicated by compactional features, was not as rapid as that in the coarse-grained upper Pliocene lowstand beds.

Reasons for contrasts in diagenetic processes.—
The pronounced variability in diagenesis in the upper Pliocene sediments and their pronounced compositional and textural differences contrast with the uniformity of the lower Pliocene section.

The reason for the complex behavior of the fine-grained lower Pliocene sediments as well as the upper Pliocene highstand deposits showing small-scale alternations between varying degrees of cementation is still subject to discussion (Westphal, 1998). Slight changes in the carbonate input (production cycles or sea-level changes) might be reinforced by diagenesis (Ricken, 1986, 1987), or slight shifts in organic carbon input (climate?) might result in differences in dissolution of the aragonite constituents during early diagenesis (Munnecke and Samtleben, 1996).

Sequence Stratigraphic Implications

Sequence stratigraphic interpretations of sedimentary records of flat-topped carbonate platforms based on compositional and textural analyses show the significance of sediment composition and texture with respect to sea-level fluctuations (e.g., Haak and Schlager, 1989; Reijmer, 1991; Everts, 1994). Burchette and Wright (1992) proposed the application of stratigraphic analysis, including facies successions, to supplement sequence stratigraphic interpretations of classical ramps that, because of their large lateral extension and comparably low relief, frequently appear rather featureless. In some cases, clinoforms and sequence geometries of ramps have been successfully mapped by using careful correlations of wireline, biostratigraphic, and sedimentary logs (Stoakes, 1980; Cutler, 1983; Burchette and Britton, 1985; Chatellier, 1988; Dix, 1989). Bachmann et al. (1996) and Bachmann and Willems (1996) observed upward and downward shifts of the facies belts and could thereby prove the applicability of compositional and textural analyses of ramp-slope sediments.

Nevertheless, this study demonstrates that facies analysis of the periplatform sediments of a distally steepened ramp might not yield a clear record of sea-level fluctuations. In response to a small relative sea-level fall, facies belts even on a distally steepened carbonate ramp are likely to shift basinward. Whereas parts of the bank top will become exposed, production of carbonate sediment can continue on the former deeper slope. Therefore, only slight changes in composition and texture of the periplatform sediments of a ramp are likely to result from sea-level fluctuations.

For seismic sequence stratigraphy, the diagenetic behavior of the slope sediments is significant. Different diagenetic patterns are produced in periplatform sediments of different morphologies. On steep-sided, flat-topped carbonate platforms with clear compositional and textural signatures, differences in diagenesis will result in differences in the physical properties. Where pronounced differences in the primary permeability are absent, as in the distally steepened ramp deposits examined, diagenesis is less variable. As cementation is an important steering factor for acoustic impedance of carbonates (F. S. Anselmetti and G. P. Eberli, pers. commun., 1997), differences in the degree of cementation could also have implications for the interpretation of seismic profiles.

CONCLUSIONS

1. This study has demonstrated that recording of higher-order sea-level cycles in periplatform sediments depends on platform morphology. Sea-level fluctuations that drastically affect the shallow-water factory of the interior of flat-topped platforms are clearly recorded in abrupt textural changes. Lowstand deposits are characterized by coarse-grained sediments that lack significant amounts of fine-grained sedimentary matrix. Highstand deposits, in contrast, are characterized by a predominance of fine-grained material derived from the platform top. These observations, based on the upper Pliocene of the CLINO borehole, once more verify the concept of highstand shedding (Droxler and Schlager, 1985). In contrast, periplatform sediments of distally steepened carbonate ramps—where production zones migrate with the higher-order sea level rather than being switched on and off—are relatively monotonous with only slight shifts in relative amounts of constituents. In the lower Pliocene section examined, however, an overall sea-level rise is superimposed on the high-frequency sea-level fluctuations, thereby possibly rendering the sea-level falls less pronounced.
2. The general compositional changes from skeletal to dominantly nonskeletal components as a result of morphologic evolution of a carbonate platform have been observed in the lower- to upper-slope sediments penetrated by CLINO. This observation is consistent with the data from the lower Pliocene to upper Pliocene sediments in the platform interior (Beach and Ginsburg, 1980) and from toe-of-slope turbidites (Reijmer et al., 1992).
3. Palynologic information has become a valuable tool in interpreting sea-level history of a carbonate platform. Dinoflagellate cyst associations as well as terrestrial palynoflora supplement petrographic observations by indicating environmental conditions such as salinity and temperature of the bank-top waters. Also the existence of subaerially exposed land can be corroborated by a strong rise in the proportion of terrestrial palynomorphs.
4. The textural signature of sea-level fluctuations influences the early diagenetic patterns via initial porosity and permeability. The dependence of diagenesis on primary permeability is a link between diagenesis on one side and sea-level fluctuations and platform-morphology changes on the other. In the upper Pliocene section, the fine-grained highstand deposits had a low initial permeability that precluded rapid fluid migration and protected these sediments from fluid-driven diagenetic alteration. Therefore, they are diagenetically immature in that they have uncemented, partly cemented, and completely cemented layers and have retained large amounts of aragonite needles. Coarse-grained lowstand deposits, in contrast, had a high primary permeability. They are usually tightly cemented and uncompacted and can contain large amounts of dolomite. They could have had high initial permeabilities, making them subject to greater fluid flux. The fine-grained lower Pliocene lowstand deposits had somewhat higher initial permeabilities than the upper Pliocene highstand deposits, because of the lower amounts of sedimentary matrix in the former that result from the higher-energy wave action and less restricted conditions on the platform interior. The lower Pliocene lowstand deposits appear more mature than the fine-grained upper Pliocene highstand deposits, but less mature than the upper Pliocene lowstand deposits. This finding implies that the permeability was not high enough to allow for a rapid fluid-driven lithification.

ACKNOWLEDGMENTS

We are grateful to Robert N. Ginsburg and Gregor P. Eberli (Comparative Sedimentology Laboratory, University of Miami) for giving us the opportunity to study the CLINO core material. Many discussions with Florian Böhm were helpful and inspiring. Thanks are also due to Christian Dullo for discussions and suggestions. An earlier version of the manuscript was kindly reviewed by Adam Vecsei, Florian Böhm, and Rebecca Rendle. The manuscript also greatly benefited from peer reviews by Leslie Melim, Mark Harris, and Toni Simo.

The investigations were supported by a grant of the German Science Foundation (DFG Re1051/3 to H. W. and J. R.). M. J. H. is grateful to G. Norris (University of Toronto) for support from his Natural Sciences and Engineering Research Council of Canada operating grant. The core CLINO was collected with funding from the U.S. National Science Foundation under Grants OCE 891-7295 and OCE 910-4294, from the Industrial Associates of the Comparative Sedimentology Laboratory of the Rosenstiel School of Marine and Atmospheric Sciences, and from the Swiss National Science Foundation.

REFERENCES

Abreu, V. S., and Anderson, J.B., 1998, Glacial eustacy during the Cenozoic: Sequence stratigraphic implications: American Association of Petroleum Geologists Bulletin, v. 82, p. 1385–1400.

Abreu, V. S., and Haddad, G. A., in press, Glacioeustatic fluctuations: The mechanism linking isotope stratigraphy and sequence stratigraphy from the

Oligocene to middle Miocene, *in* De Graciansky, P.-C., Hardenbol, J., Jacquin, T., Vail, P. R., and Farley, M. B., eds., Sequence Stratigraphy of European Basins: Tulsa, SEPM (Society for Sedimentary Geology) Special Publication 60.

AHR, W. M., 1973, The carbonate ramp: An alternative to the shelf model, *in* Today's New Technology, Tomorrow's New Targets: Transactions of the Gulf Coast Association of Geological Societies, v. 23, p. 221–225.

BACHMANN, M., AND WILLEMS, H., 1996, High-frequency cycles in the upper Aptian carbonates of the Organyà basin, NE Spain: Geologische Rundschau, v. 85, p. 586–605.

BACHMANN, M., BANDEL, K., KUSS, J., AND WILLEMS, H., 1996, Sedimentary processes and Intertethyal comparisons of two Early/Late Cretaceous carbonate ramp systems (NE-Africa and Spain), *in* Reitner, J., Neuweiler, F., and Gunkel, F., eds., Global and Regional Controls on Biogenic Sedimentation: II. Cretaceous Sedimentation: Göttingen, Göttinger Arbeiten in Geologie und Paläontologie, v. Sb2, p. 151–163.

BEACH, D. K., 1982, Depositional and diagenetic history of Pliocene-Pleistocene carbonates of northwestern Great Bahama Bank: Evolution of a carbonate platform: Unpublished Ph.D. Dissertation, University of Miami, Miami, 447 p.

BEACH, D. K., AND GINSBURG, R. N., 1980, Facies succession of Pliocene-Pleistocene carbonates, northwestern Great Bahama Bank: American Association of Petroleum Geologists Bulletin, v. 64, p. 1634–1642.

BERGGREN, W. A., KENT, D. V., FLYNN, J. J., AND VAN COUVERING, J. A., 1985, Cenozoic geochronology: Geological Society of America Bulletin, v. 96, p. 1407–1418.

BOARDMAN, M. R., NEUMANN, C. A., BAKER, P. A., DULIN, L. A., KENTER, R. J., HUNTER, G. E., AND KIEFER, K. B., 1986, Banktop responses to Quaternary fluctuations in sea level recorded in periplatform sediments: Geology, v. 14, p. 28–31.

BRACHERT, T. C., AND DULLO, W.-C., 1991, Laminar micrite crusts and associated foreslope processes, Red Sea: Journal of Sedimentary Petrology, v. 61, p. 354–363.

BRACHERT, T. C., AND DULLO, W.-C., 1994, Micrite crusts on Ladinian foreslopes of the Dolomites seen in the light of a modern scenario from the Red Sea: Geologische Bundesanstalt, Abhandlungen, v. 50, p. 57–68.

BRADFORD, M. R., AND WALL, D. A., 1984, The distribution of Recent organic-walled dinoflagellate cysts in the Persian Gulf, Gulf of Oman and Northwestern Arabian Sea: Palaeontographica, ser. B, v. 192, p. 16–84.

BURCHETTE, T. P., AND BRITTON, S. R., 1985, Carbonate facies analysis in the exploration for hydrocarbons: A case-study from the Cretaceous of the Middle East, *in* Brenchley, P. J. and Williams, B. P. J., eds., Sedimentology—Recent developments and applied aspects: Oxford, Blackwell Scientific Publications, p. 311–338.

BURCHETTE, T. P., AND WRIGHT, V. P., 1992, Carbonate ramp depositional systems: Sedimentary Geology, v. 79, p. 3–57.

CHATELLIER, J.-Y., 1988, Carboniferous carbonate ramp, the Banff Formation, Alberta, Canada: Bulletin des Centres de Recherches Exploration-Production Elf-Aquitaine, v. 12, p. 569–599.

CHAYES, F., 1956, Petrographic modal analysis: New York, John Wiley & Sons, 113 p.

CUTLER, W. G., 1983, Stratigraphy and sedimentology of the Upper Devonian Grosmont Formation, northern Alberta: Canadian Petroleum Geology Bulletin, v. 31, p. 282–235.

DALE, B., 1996, Dinoflagellate cyst ecology: Modelling and geological applications, *in* Jansonius, J. and McGregor, D. C., eds., Palynology: Principles and Applications, v. 3: Dallas, American Association of Stratigraphic Palynologists Foundation, p. 1249–1275.

DAVIS, J. C., 1973, Statistics and Data Analysis in Geology: New York, John Wiley & Sons, 550 p.

DIX, G. R., 1989, High-energy, inner shelf carbonate facies along a tide-dominated nonrimmed margin, northwestern Australia: Marine Geology, v. 89, p. 347–362.

DRAVIS, J. J., 1996, Rapidity of freshwater calcite cementation—Implications for carbonate diagenesis and sequence stratigraphy: Sedimentary Geology, v. 107, p. 1–10.

DROXLER, A. W., AND SCHLAGER, W., 1985, Glacial versus interglacial sedimentation rates and turbidite frequency in the Bahamas: Geology, v. 13, p. 799–802.

DROXLER, A. W., SCHLAGER, W., AND WHALLON, C. C., 1983, Quaternary aragonite cycles and oxygen-isotope record in Bahamian carbonate ooze: Geology, v. 11, p. 235–239.

DUNHAM, R. J., 1962, Classification of carbonate rocks according to depositional texture, *in* Ham, W. E., ed., Classification of Carbonate Rocks: Tulsa, American Association of Petroleum Geologists Memoir 1, p. 108–121.

EBERLI, G. P., AND GINSBURG, R. N., 1987, Segmentation and coalescence of Cenozoic carbonate platforms, northwestern Great Bahama Bank: Geology, v. 15, p. 75–79.

EBERLI, G. P., AND GINSBURG, R. N., 1989, Cenozoic progradation of northwestern Great Bahama Bank, a record of lateral platform growth and sea-level fluctuations, *in* Crevello, P. D., Wilson, J. L., Sarg, J. F., and Read, J. F., eds., Controls on carbonate platform and basin development: Tulsa, Society of Economic Paleontologists and Mineralogists, Special Publication 44, p. 339–351.

EBERLI, G. P., SWART, P. K., MCNEILL, D. F., KENTER, J. A. M., ANSELMETTI, F. S., MELIM, L. A., AND GINSBURG, R. N., 1997, A synopsis of the Bahama Drilling Project: Results from two deep core borings drilled on the Great Bahama Bank, *in* Eberli, G. P., Swart, P. K., and Malone, M. J., et al., Proceedings of the Ocean Drilling Program, Initial Reports: College Station, Ocean Drilling Program, v. 166, p. 23–41.

EDWARDS, L. E.,, AND ANDRLE, V. A. S., 1992, Distribution of selected dinoflagellate cysts in modern marine sediments, *in* Head, M. J. and Wrenn, J. H., eds., Neogene and Quaternary Dinoflagellate Cysts and Acritarchs: College Station, American Association of Stratigraphic Palynologists Foundation, p. 259–288.

ELRICK, M., AND READ, J. F., 1991, Cyclic ramp-to-basin carbonate deposits, Lower Mississippian, Wyoming and Montana: A combined field and computer modeling study: Journal of Sedimentary Petrology, v. 61, p. 1194–1224.

ENOS, P., 1983, Shelf environment, *in* Scholle, P. A., Bebout, D. G., and Moore, C. H., eds., Carbonate depositional environments: Tulsa, American Association of Petroleum Geologists Memoir 33, p. 267–295.

EVERTS, A. J. W., 1991, Interpreting compositional variations of calciturbidites in relation to platform-stratigraphy: An example from the Paleogene of SE Spain: Sedimentary Geology, v. 71, p. 231–242.

EVERTS, A. J. W., 1994, Carbonate sequence stratigraphy of the Vercors (French Alps) and its bearing on the Cretaceous: Unpublished Ph.D. Dissertation, Vrije Universiteit, Amsterdam, 176 p.

EVERTS, A. J. W., AND REIJMER, J. J. G., 1995, Clinoform composition and margin geometries of a Lower Cretaceous carbonate platform (Vercors, SE France): Palaeogeography, Palaeoclimatology, Palaeoecology, v. 119, p. 19–33.

FLÜGEL, E., 1978, Mikrofazielle Untersuchungsmethoden von Kalken (1st edition): Berlin, Springer-Verlag, 454 p.

FLÜGEL, E., 1982, Microfacies Analysis of Limestones: Berlin, Springer-Verlag, 633 p.

GINSBURG, R. N., HARRIS, P. M., EBERLI, G. P., AND SWART, P. K., 1991, The growth potential of a bypass margin, Great Bahama Bank: Journal of Sedimentary Petrology, v. 61, p. 976–987.

GOLDHAMMER, R. K., AND HARRIS, M. T., 1989, Eustatic controls on the stratigraphy and geometry of the Latemar Buildup (Middle Triassic), the Dolomites of northern Italy, *in* Crevello, P. D., Wilson, J. L., Sarg, J. F., and Read, J. F., eds., Controls on carbonate platform and basin development: Tulsa, Society of Economic Paleontologists and Mineralogists, Special Publication 44, p. 323–338.

GRAMMER, G. M., AND GINSBURG, R. N., 1992, Highstand versus lowstand deposition on carbonate platform margins: Insight from Quaternary foreslopes in the Bahamas: Marine Geology, v. 103, p. 125–136.

GRAMMER, G. M., GINSBURG, R. N., AND MCNEILL, D. F., 1991, Morphology and development of modern carbonate foreslopes, Tongue of the Ocean, Bahamas *in* Larue, D. K. and Draper, G., eds., 12th Caribbean Geological Conference, St. Croix, U.S. Virgin Islands: Miami, Miami Geological Society, p. 27–32.

GUNATILAKA, A., 1976, Thallophyte boring and micritization within skeletal sands from Connemara, Western Ireland: Journal of Sedimentary Petrology, v. 46, p. 548–554.

HAAK, A. B., AND SCHLAGER, W., 1989, Compositional variations in calciturbidites due to sea-level fluctuations, late Quaternary, Bahamas: Geologische Rundschau, v. 78, p. 477–486.

HARRIS, M. T., 1994, The foreslope and toe-of-slope facies of the Middle Triassic Latemar buildup (Dolomites, Northern Italy): Journal of Sedimentary

Research, v. B64, p. 132–145.

KIER, J. S., AND PILKEY, O. H., 1971, The influence of sea level changes on sediment carbonate mineralogy, Tongue of the Ocean, Bahamas: Marine Geology, v. 11, p. 189–200.

LASEMI, Z., AND SANDBERG, P. A., 1984, Transformation of aragonite-dominated lime muds to microcrystalline limestones: Geology, v. 12, p. 420–423.

LIDZ, B. H., AND MCNEILL, D. F., 1995a, Deep-sea biostratigraphy of prograding platform margins (Neogene, Bahamas): Key evidence linked to depositional rhythm: Marine Micropaleontology, v. 25, p. 87–125.

LIDZ, B. H., AND MCNEILL, D. F., 1995b, Reworked Paleogene to early Neogene planktic foraminifera: Implications of an intriguing distribution at a late Neogene prograding margin, Bahamas: Marine Micropaleontology, v. 25, p. 221–268.

LOEBLICH, A. R., AND TAPPAN, H., 1964, Sarcodina, chiefly "thecameobians" and Foraminiferida, in Moore, A. R., ed., Treatise of Invertebrate Paleontology, Part C, Protista 2: Lawrence, Kansas, University of Kansas and Geological Society of America, 510 p.

MCNEILL, D. F., 1989, Magnetostratigraphic dating and magnetization of Cenozoic platform carbonates from the Bahamas: Unpublished Ph.D. Dissertation, University of Miami, Miami, 210 p.

MCNEILL, D., GINSBURG, R. N., CHANG, S.-B. R., AND KIRSCHVINK, J. L., 1988, Magnetostratigraphic dating of shallow-water carbonates from San Salvador, Bahamas: Geology, v. 16, p. 8–12.

MELIM, L. A., 1996, Limitations on lowstand meteoric diagenesis in the Pliocene–Pleistocene of Florida and Great Bahama Bank: Implications for eustatic sea-level models: Geology, v. 24, p. 893–896.

MELIM, L. A., SWART, P. K., AND MALIVA, R. G., 1995, Meteoric-like fabrics forming in marine waters: Implications for the use of petrography to identify diagenetic environments: Geology, v. 23, p. 755–758.

MONTAGGIONI, L. F., 1988, Holocene reef growth history in mid-plate high volcanic islands, *in* Choat, J. H., Barnes, D., Boronitzka, M. A., et al., eds., Proceedings of the 6th International Coral Reef Symposium, v. 3: Townsville, Australia, University of Townsville, p. 455–460.

MUNNECKE, A., 1997, Bildung mikritischer Kalke im Silur auf Gotland: Courier Forschungsinstitut Senckenberg, v. 198, p. 1–71.

MUNNECKE, A., AND SAMTLEBEN, C., 1996, The formation of micritic limestones and the development of limestone-marl alternations in the Silurian of Gotland, Sweden: Facies, v. 34, p. 159–176.

MUNNECKE, A., WESTPHAL, H., REIJMER, J. J. G., AND SAMTLEBEN, C., 1997, Microspar development during early marine burial diagenesis: A comparison of Pliocene carbonates from the Bahamas with Silurian limestones from Gotland (Sweden): Sedimentology, v. 44, p. 977–990.

MURRAY, J. W., 1973, Distribution and Ecology of Living Benthic Foraminiferids: London, Heinemann Educational Books, 274 p.

NEUMANN, A. C., AND LAND, L. S., 1975, Lime mud deposition and calcareous algae in the Bight of Abaco, Bahamas: A budget: Journal of Sedimentary Petrology, v. 59, p. 147–161.

POSAMENTIER, H. W., ALLEN, G. P., JAMES, D. P., AND TESSON, M., 1992, Forced regressions in a sequence stratigraphic framework: Concepts, examples and exploration significance: American Association of Petroleum Geologists Bulletin, v. 76, p. 1687–1709.

PURDY, E. G., 1963, Recent calcium carbonate facies of the Great Bahama Bank: Journal of Geology, v. 71, p. 334–355.

REIJMER, J. J. G., 1991, Sea level and sedimentation on the flanks of carbonate platforms: Unpublished Ph.D. Dissertation, Vrije Universiteit, Amsterdam, 162 p.

REIJMER, J. J. G., AND EVERAARS, J. S. L., 1991, Carbonate platform facies reflected in carbonate basin facies (Triassic, Northern Calcareous Alps, Austria): Facies, v. 25, p. 253–278.

REIJMER, J. J. G., SCHLAGER, W., AND DROXLER, A. W., 1988, Site 632: Pliocene-Pleistocene sedimentation cycles in a Bahamian basin, *in* Austin, J. A., Schlager, W., Comet, P. A., et al., Proceedings of the Ocean Drilling Program, Scientific Results: College Station, Ocean Drilling Program, v. 101, p. 213–220.

REIJMER, J. J. G., TEN KATE, W. G. H.Z., SPRENGER, A., AND SCHLAGER, W., 1991, Calciturbidite composition related to exposure and flooding of a carbonate platform (Triassic, Eastern Alps): Sedimentology, v. 38, p. 1059–1074.

REIJMER, J. J. G., SCHLAGER, W., BOSSCHER, H., BEETS, C. J., AND MCNEILL, D. F., 1992, Pliocene/Pleistocene platform facies transition recorded in calciturbidites (Exuma Sound, Bahamas): Sedimentary Geology, v. 78, p. 171–179.

RICKEN, W., 1986, Diagenetic bedding: A model for limestone-marl alternations: Berlin, Springer-Verlag, 210 p.

RICKEN, W., 1987, The carbonate compaction law: A new tool: Sedimentology, v. 34, p. 1–14.

ROCK, N. M. S., 1988, Numerical geology: Berlin, Springer-Verlag, 427 p.

ROSE, P. R., AND LIDZ, B., 1977, Diagnostic foraminiferal assemblages of shallow-water modern environments: South Florida and the Bahamas: Sedimenta, Miami, University of Miami, Comparative Sedimentology Laboratory, v. VI, 55 p.

SCHLAGER, W., 1981, The paradox of drowned reefs and carbonate platforms: Geological Society of America Bulletin, v. 92, p. 197–211.

SCHLAGER, W., AND CAMBER, O., 1986, Submarine slope angles, drowning unconformities, and self-erosion of limestone escarpments: Geology, v. 14, p. 762–765.

SCHLAGER, W., AND GINSBURG, R. N., 1981, Bahama carbonate platforms—The deep and the past: Marine Geology, v. 44/45, p. 1–24.

SCHLAGER, W., AND JAMES, N. P., 1978, Low-magnesian calcite limestones forming at the deep-sea floor, Tongue of the Ocean, Bahamas: Sedimentology, v. 25, p. 675–702.

SCHLAGER, W., REIJMER, J. J. G., AND DROXLER, A. W., 1994, Highstand shedding of carbonate platforms: Journal of Sedimentary Petrology, v. B64, p. 270–281.

SCHNEIDERMANN, N., AND SANDBERG, P. A., 1971, Calcite-aragonite differentiation by selective staining and scanning electron microscopy: Transactions of the Gulf Coast Association of Geological Societies, v. 21, p. 349–352.

SIMMS, M., 1984, Dolomitization by groundwater flow systems in carbonate platforms: Transactions of the Gulf Coast Association of Geological Societies, v. 34, p. 411–420.

STOAKES, F. A., 1980, Nature and control of shale basin fill and its effect on reef growth and termination: Upper Devonian Duvernay and Ireton Formations of Alberta, Canada: Canadian Petroleum Geology Bulletin, v. 28, p. 345–410.

TRAVERSE, A., AND GINSBURG, R. N., 1966, Palynology of the surface sediments of Great Bahama Bank, as related to water movement and sedimentation: Marine Geology, v. 4, p. 417–459.

VAIL, P. R., AUDEMARD, F., BOWMAN, S. A., EISNER, P. N., AND PEREZ-CRUZ, C., 1991, The stratigraphic signatures of tectonics, eustasy and sedimentology—An overview, *in* Einsele, G., Ricken, W., and Seilacher, A., eds., Cycles and Events in Stratigraphy: Berlin, Springer-Verlag, p. 617–659.

WALL, D., DALE, B., LOHMANN, G. P., AND SMITH, W. K., 1977, The environmental and climatic distribution of dinoflagellate cysts in modern marine sediments from regions in the north and south Atlantic Oceans and adjacent seas: Marine Micropaleontology, v. 2, p. 121–200.

WESTPHAL, H., 1998, Carbonate platform slopes—A record of changing conditions: Springer, Heidelberg, New York, Lecture Notes in Earth Sciences, v. 75, 196 p.

WESTPHAL, H., AND MUNNECKE, A., 1997, Mechanical compaction versus early cementation in fine-grained limestones: Differentiation by the preservation of organic microfossils: Sedimentary Geology, v. 112, p. 33–42.

WILBER, R. J., MILLIMAN, J. D., AND HALLEY, R. B., 1990, Accumulation of bank-top sediment on the western slope of Great Bahama Bank: Rapid progradation of a carbonate megabank: Geology, v. 18, p. 970–974.

WILSON, P. A., AND ROBERTS, H. H., 1992, Carbonate-periplatform sedimentation by density flows: A mechanism for rapid off-bank and vertical transport of shallow-water fines: Geology, v. 20, p. 713–716.

WILSON, P. A., AND ROBERTS, H. H., 1995, Density cascading: Off-shelf sediment transport, evidence and implications, Bahama Banks: Journal of Sedimentary Research, v. A65, p. 45–56.

SEDIMENTOLOGIC AND SEQUENCE STRATIGRAPHIC INTERPRETATION OF A MIXED CARBONATE-SILICICLASTIC RAMP, MIDCONTINENT EPEIRIC SEA, MIDDLE TO UPPER ORDOVICIAN DECORAH AND GALENA FORMATIONS, WISCONSIN

YONG SEOK CHOI*, J. A. (Toni) SIMO, AND BEVERLY Z. SAYLOR

Department of Geology and Geophysics, University of Wisconsin, Madison, Wisconsin 53706

ABSTRACT: The Middle to Upper Ordovician Decorah and Galena Formations in Wisconsin are composed of stratigraphically condensed sequences (net accumulation rate: 0.35–0.9 cm/k.y., characterized by extreme lateral continuity of facies and dominated by fine-grained carbonate with numerous omission surfaces. Depositional sequences are defined by relatively conformable successions bounded by identified or interpreted exposure surfaces across which siliciclastics and skeletal grains increase abruptly. Depositional sequences are subdivided in a nested hierarchy into cycle sets, cycles, and omission-surface–bounded packages.

The lower sequence, sequence D, corresponds to the Decorah Formation and has a siliciclastic-rich fining-upward succession followed by a coarsening-upward carbonate succession. The phosphate-enriched condensed section above the fining-upward succession represents maximum transgression. Cycles in sequence D are represented by the upward increases in skeletal-grain and tempestite abundance. The middle sequence, sequence G1, corresponds to the Galena Formation (lower to middle Dunleith Member) and shows an overall transition from a shale- and skeletal-grain–rich base to a carbonate-mud–dominated top. Both shale content and the abundance of skeletal-grain–supported rocks increase northward in eastern Wisconsin and westward in southern Wisconsin toward the inferred paleoshoreline. The upper sequence, sequence G2, corresponds to the Galena Formation (upper part of Dunleith, Wise Lake, and Dubuque Members) and consists of a skeletal-grain–rich base with minor shale content grading upward into mudstones and into a skeletal-grain–rich top.

Cycle sets and cycles in the sequences G1 and G2 are characterized by relatively sharp boundaries across which there is an abrupt increase in shale content and skeletal-grain–rich facies grading upward to a carbonate-mud–dominated facies. The asymmetry of cycles—represented by the abrupt increase followed by gradual decrease in shale content and skeletal grains—is inconsistent with progradation of high-energy shale-rich facies over low-energy mudstone. This facies stacking pattern is better explained by reciprocal sedimentation in which skeletal-grain–rich shale deposition and carbonate-mud deposition are disassociated from each other both in time and space. The omission-surface–bounded packages form the smallest recognized sedimentary units within the sequences G1 and G2 and represent alternations of periods of sediment deposition and starvation.

The Decorah and Galena Formations are dominated by mudstone-rich lithofacies and abundant omission surfaces, and their contained fauna consists predominantly of heterotrophs. These characteristics are in contrast with those of the Lower Ordovician and Silurian "Bahama-like" carbonate strata. Apparently, in addition to tectonics, antecedent topography, and changes in eustasy, siliciclastic influx, and storm transport, the paleoclimatic and paleoceanographic conditions in the Late Ordovician epeiric sea had important controls on depositional models, sequence development, and facies stacking patterns.

INTRODUCTION

The Middle to Upper Ordovician Decorah and Galena Formations of the Sinnipee Group (Fig. 1) were deposited across much of the Midcontinent in a vast epeiric sea (Figs. 2, 3). These strata are characterized by exceptional lateral continuity of facies such that lithostratigraphic members only 5 m thick extend without change for hundreds of kilometers (Willman and Kolata, 1978; Delgado, 1983). The "layer-cake" stratigraphy of the Middle to Upper Ordovician carbonate strata in the Midcontinent is well known (Agnew et al., 1956; Templeton and Willman, 1963; Ostrom, 1969; Willman and Kolata, 1978; Witzke, 1980; Delgado, 1983; Bakush, 1985; Witzke and Kolata, 1988; Choi, 1995; Choi et al., 1997), and the great extent of lateral continuity is well constrained by K-bentonite correlation (Kolata et al., 1983, 1986, 1987, 1996). However, the sedimentology, depositional settings, sequence stratigraphy, and paleoceanographic interpretations are far from being understood. Previous geologic studies have concentrated along the Upper Mississippi Valley where facies are homogeneous; those studies did not include areas with rapid facies changes such as those observed in northeastern Wisconsin.

The Decorah and Galena Formations differ substantially from the Lower Ordovician and Silurian successions, which are "Bahama-like" (Ginsburg, 1982; Mitchell, 1985), consisting of rapidly accumulated micrite, coated grains, peloids, and calcareous algae and stromatolites. The Decorah and Galena Formations are stratigraphically condensed (net accumulation rate: 0.35–0.9 cm/k.y., Delgado, 1983), dominated by fine-grained carbonate with numerous iron-mineralized omission surfaces (Byers, 1996). The omission surfaces mark pauses in deposition with little or no erosion (Bromley, 1975; Kennedy and Garrison, 1975).

The main purposes of this paper are to describe the Upper Ordovician mixed carbonate-siliciclastic Decorah and Galena Formations in Wisconsin and to interpret the depositional history of these strata in relationship to relative sea-level variations, changes in siliciclastic supply, antecedent topography, and climate. This work has been accomplished by interpretation of depositional facies and facies stacking patterns within a hierarchically ordered sequence framework.

GEOLOGIC SETTING

During the Middle and Late Ordovician, Wisconsin was situated between 10° and 20° south latitude and was part of an epeiric sea south of the Transcontinental arch (Witzke, 1980; Fig. 2). After a prolonged Early Ordovician global fall in sea

* *Present address: Korea National Oil Corp., 1588-14, Kwanyan-dong, Anyang City, Kyungki-do, Korea 431-711*

Advances in Carbonate Sequence Stratigraphy: Application to Reservoirs, Outcrops and Models, SEPM Special Publication No. 63

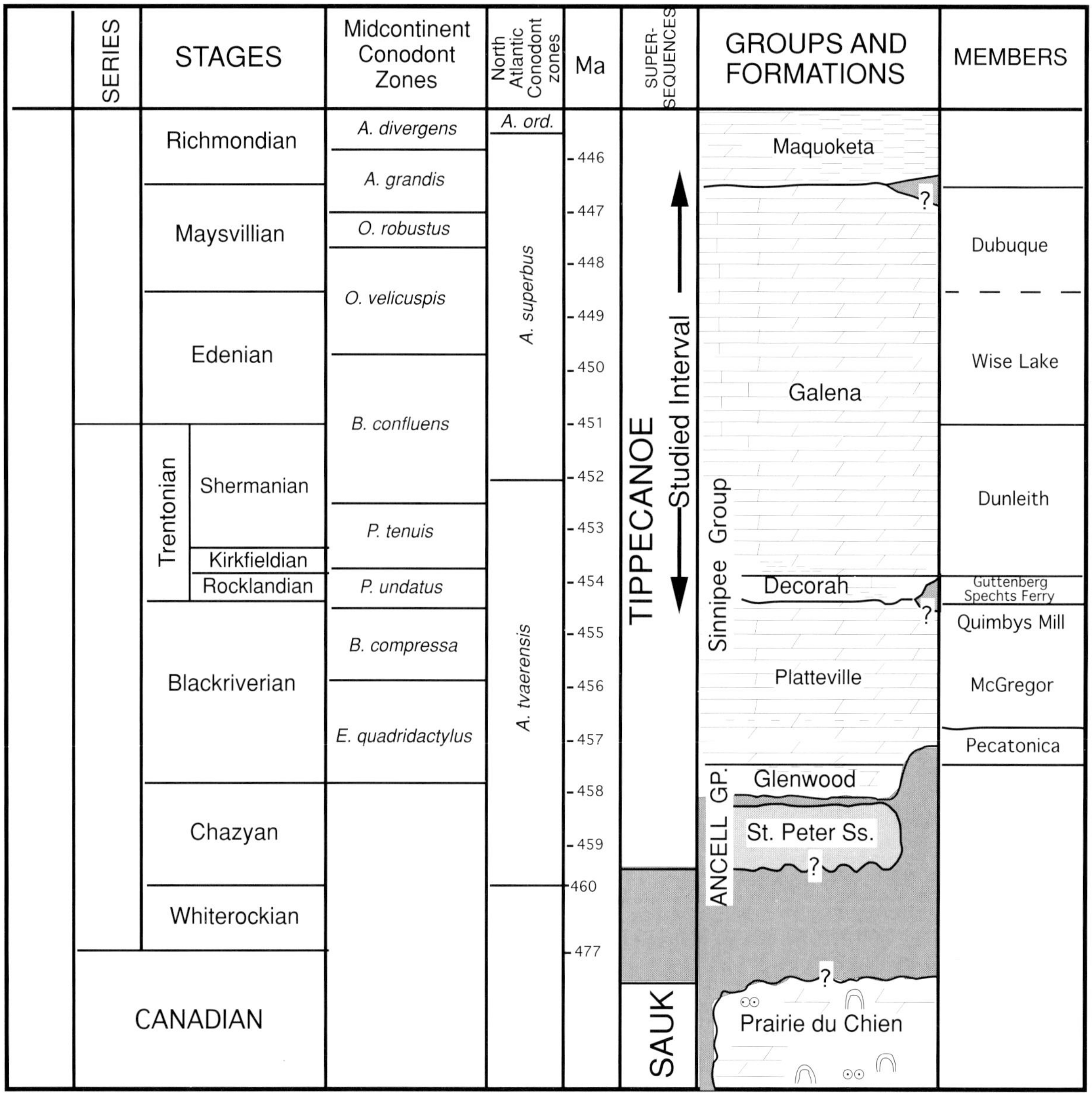

FIG. 1.—Chronostratigraphy (Bergström and Mitchell, 1992; Goldman et al., 1994) and lithostratigraphy (Templeton and Willman, 1963; Ostrom, 1969) of the Ordovician rocks in Wisconsin. The age column is calculated by a linear conversion, assuming an age of 454.2 Ma for the Deicke K-bentonite and an age of 443.5 Ma for the Ordovician/Silurian boundary (after Fordham, 1992).

level, the sea gradually transgressed over the North American continent during Middle Ordovician time, and a thick sequence of Middle to Upper Ordovician strata blanketed much of the Midwest (Tippecanoe A Supersequence; Sloss, 1963; Fig. 1). The Transcontinental arch, Wisconsin dome, Ozark dome, and Taconic orogenic belt were the siliciclastic sediment sources in the Upper Midwest during the Middle Ordovician through Devonian (Templeton and Willman, 1963; Witzke, 1980).

The Middle to Upper Ordovician stratigraphy of the Upper Mississippi Valley region has been the object of intensive study since the 1950s (Agnew et al., 1956; Templeton and Willman, 1963; Ostrom, 1969; Willman and Kolata, 1978; Sloan, 1987; Delgado, 1983; Bakush and Carozzi, 1986; Witzke and Kolata, 1988; Ludvigson et al., 1996). Stratigraphic work has concentrated on lithologic descriptions and correlation of mixed siliciclastic and carbonate facies, omission surfaces, bioturbation, terrigenous siltstones, and packages with increased storm-bed abundance. Lithostratigraphic classification in this study follows Ostrom (1969) with adoption of lithostratigraphic names from Templeton and Willman (1963), in which group, formations, and members are regarded as formations, members, and beds, respectively, as depicted in Figure 1.

The Decorah Formation consists of shale and argillaceous carbonate and unconformably overlies the Platteville Formation with a low angular discordance. The Decorah Formation is subdivided into two members (Fig. 1). The Spechts Ferry Shale Member is a green shale with thin beds of calcareous mudstone to wackestone, skeletal grainstone, and K-bentonite beds. The

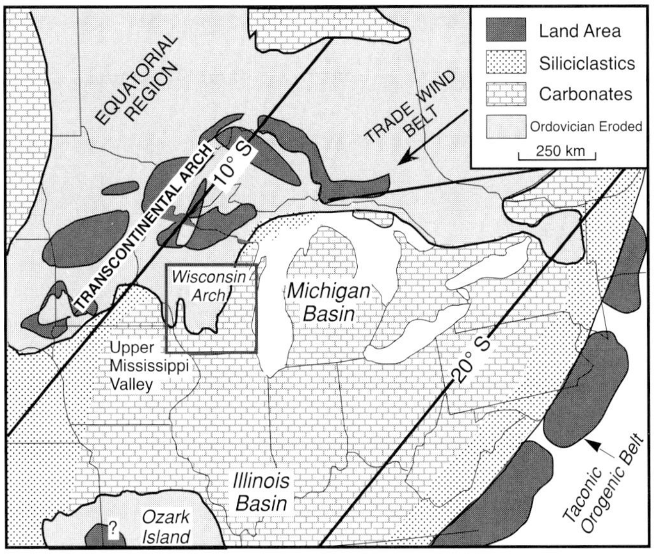

FIG. 2.—Late Middle to early Late Ordovician paleogeographic reconstruction of the eastern half of the United States and Canada showing major land areas, mountains, and basins. Inset box represents study area (see Fig. 3). Modified from Witzke (1980).

overlying Guttenberg Limestone Member consists of light pinkish-gray carbonate mudstone to wackestone with thin, dark brown, organic-rich shale partings and thick-bedded grainstones. The Decorah Formation is limited in exposure to southern Wisconsin, thins to the east and northeast, and grades into shale to the west (Witzke, 1980; Choi, 1995; Ludvigson et al., 1996).

The Galena Formation in Wisconsin conformably overlies the Decorah Formation or unconformably overlies the Platteville Formation with a very sharp boundary. Three members are distinguished in the Galena Formation: Dunleith, Wise Lake, and Dubuque (Ostrom, 1969; Fig. 1). The Dunleith Member is carbonate mudstone to packstone with intercalated, thin, grainstone beds and green shale. Mineralized omission surfaces are abundant in the Dunleith Member, particularly in carbonate strata. The lower Dunleith Member is more argillaceous. The Wise Lake Member overlies the Dunleith Formation and is a fine- to medium-grained crystalline dolostone with mudstone textures. The uppermost unit, the Dubuque Member, is characterized by abundant disseminated crinoid fragments and common brown organic-rich shale partings that gradually increase in abundance upward. The Dubuque Member is capped by a prominent hardground and is overlain by the Maquoketa Shale.

Witzke and Kolata (1988) and Witzke and Bunker (1996) subdivided the Decorah and Galena Formations in Iowa and Illinois into several "transgressive-regressive cycles." Noting that shale abundance generally increases toward the Transcontinental arch, they interpreted the large-scale alternations of green shale and carbonate in terms of progradation of terrestrial material during times of falling sea level, high fluvial runoff, and expansion of carbonate deposition during transgression and landward stepping of the siliciclastic shoreline. Organic-rich facies have generally been interpreted as condensed intervals associated with maximum flooding (Witzke and Kolata, 1988; Witzke and Bunker, 1996; Ludvigson et al., 1996).

LITHOFACIES AND DEPOSITIONAL ENVIRONMENTS

The carbonate-ramp model (Ahr, 1973) is particularly appropriate to interpret carbonate depositional settings in ancient epeiric seas (e.g., Shaw, 1964; Irwin, 1965; Laporte, 1969). In this model, fair-weather wave base and storm wave base subdivide the ramp into three distinct energy zones: the inner-, mid-, and outer-ramp environments (Burchette and Wright, 1992). The water depths that these two bases represent vary depending on physiography, wind direction, and paleoclimate. In this paper, the three environments are used as relative bathymetric zones in relationship to fair-weather wave base and storm wave base.

Eight lithofacies are defined in the Decorah and Galena Formations on the basis of detailed sedimentologic logging of sedimentary structures, shale content, and Dunham's (1962) textural classification (Table 1). Textural information and sedimentary structures are regarded as the most important criteria for paleo–water-depth interpretation. Bioturbation, biotic composition, and taphonomy provide additional information. Grain-supported textures with trough and ripple cross-bedding are interpreted to have been deposited above the fair-weather wave base in a high-energy ramp setting. Mud-rich textures and uncommon, thin, graded beds are interpreted to represent deposition below the storm wave base or above the fair-weather wave base in a low-energy ramp setting. Grain-size trends, intensity in bioturbation and cross-bedding preservation, and abundance of graded beds provide clues to the relative proximity to the fair-weather and storm wave bases (Aigner, 1985; Burchette and Wright, 1992). On the basis of these interpretations, the eight lithofacies are grouped into two facies associations. The first one, facies association A, includes the following lithofacies: thin-bedded mudstone, thick-bedded grainstone, interbedded grainstone and shale, and skeletal wackestone to packstone. This association is interpreted to have been deposited in upper-mid-ramp to inner-ramp environments and is characterized by high skeletal-grain content except for the thin-bedded mudstone lithofacies. A particular aspect of this association is the presence of thin shale beds and partings interbedded with cross-bedded and bioturbated skeletal grainstones and wackestone to packstones. The abundance of the shale beds and partings increases landward, in the same direction in which the abundance of grain-supported textures and high-energy facies increases. The mixture of the low-energy shale and high-energy grainstone facies is interpreted to represent (1) high terrigenous input due to the proximity to the source area and (2) the inability of the system to completely winnow the shales prior to deposition of grain-supported sediments.

The second association, facies association B, includes the following lithofacies: interbedded mudstone and organic-rich shale, thick-bedded mudstone to wackestone, crinoidal wackestone, and shale. This association is interpreted to have been deposited at deeper-water depths, largely in lower-mid-ramp to outer-ramp environments. In particular, the interbedded mudstone and organic-rich shale lithofacies is interpreted as the deepest-water facies and indicates times of low supply of both carbonate and siliciclastic sediments.

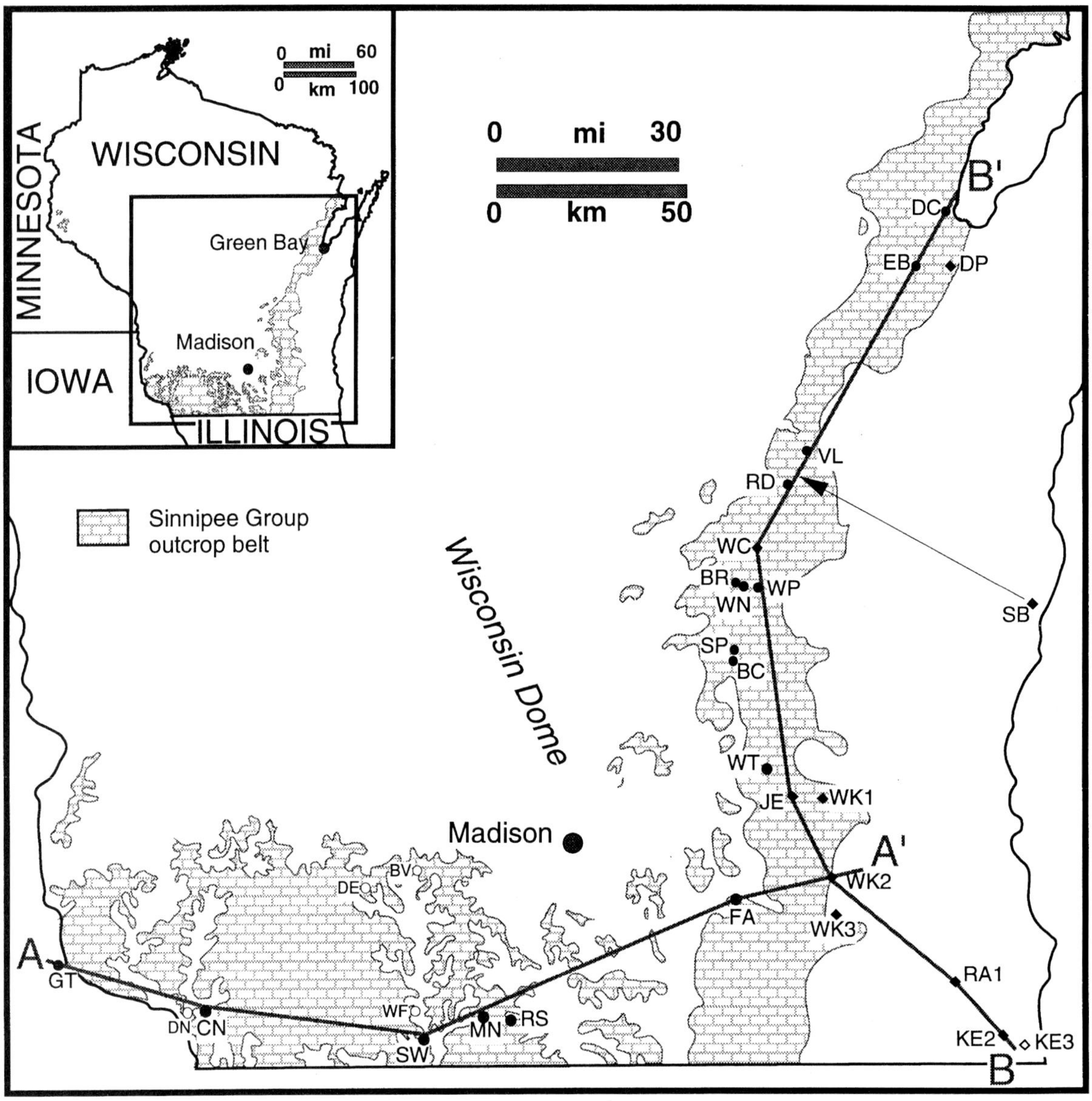

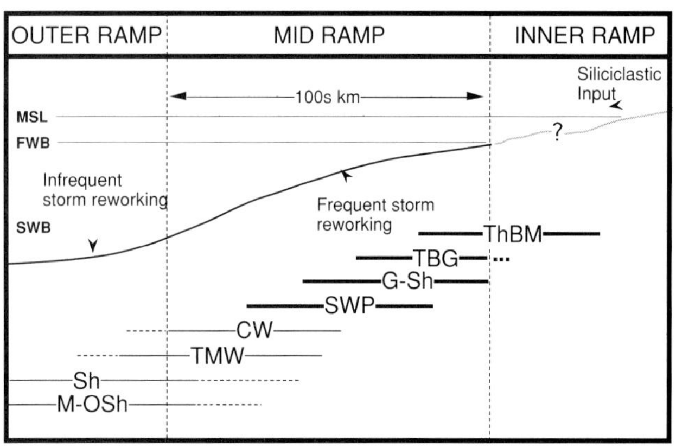

FIG. 4.— Depositional model and distribution of lithofacies along a ramp profile. Horizontal lines below the profile represent the range of depositional environments for each lithofacies; thick lines indicate the lithofacies in facies association A, and thin lines indicate the lithofacies in facies association B. See Table 1 for lithofacies abbreviations and descriptions. MSL = mean sea level, FWB = fair-weather wave base, SWB = storm wave base.

Figure 4 shows the interpreted depositional model and distribution of lithofacies along a ramp profile. The profile shows deposition of high-energy, grainier facies near the fair-weather wave base and low-energy, muddier facies near the storm wave base. The two exceptions are the thin-bedded mudstone and interbedded grainstone and shale lithofacies that are interpreted to have been deposited in shallow-water environments. The thin-bedded mudstone lithofacies is characterized by weak bioturbation, open-marine fauna, and restriction to the proximal areas from which it grades basinward into the upper-mid-ramp skeletal wackestone to packstone lithofacies. Thus, the thin-bedded mudstone lithofacies is interpreted as having been deposited on the upper-mid ramp to inner ramp during times of low siliciclastic influx. As discussed above, the association of thin shale beds and partings with grainstones supports an interpretation of shallow-water conditions during times of high terrigenous clay supply. The interpreted depositional profile is a high-energy ramp for the Decorah Formation and a relatively lower energy ramp for the Galena Formation. The Decorah

FIG. 3.—Middle to Upper Ordovician Sinnipee Group outcrop belt in Wisconsin. Closed circles and diamonds represent outcrops and cores, respectively, that were measured in detail. Open circles and diamonds represent outcrops and cores, respectively, in which lithologies along sequence boundaries were studied. Two cross sections along A–A′ and B–B′ are shown in Figures 7 and 8. The locations of outcrops and cores follow. DC = Daanen and Janssens Duck Creek Quarry: Glendale Road, Town of Howard, Brown County; SW/sec. 10, T. 24 N., R. 20 E. EB = Daanen and Janssens Ebben Quarry: South side of Highway EE, 6.4 km (4 mi) west of Highway 41, Brown County; NE/sec. 27, T. 23 N., R. 19 E. VL = Vulcan Materials Oshkosh Quarry: West side of Knapp Street, 0.2 km (0.13 mi) north of Highway 44, Oshkosh, Winnebago County; NE/sec. 27, T. 18 N., R. 16 E. RD = Michels Quarry: West side of Highway 26, just on the border of Fond du Lac and Winnebago Counties, 12.9 km (8 mi) southwest of Oshkosh, Fond du Lac County; SE/sec. 31, T. 17 N., R. 15 E. BR = Michels Waupun Quarry: West side of Highway 49, 2.4 km (1.5 mi) west of Waupun, Fond du Lac County; NW/sec. 36, T. 14 N., R. 14 E. WN = C. C. Linck Quarry: East side of Highway 49, 0.8 km (0.5 mi) west of Waupun, Fond du Lac County; SE/sec. 36, T. 14 N., R. 14 E. WP = Navis Quarry: East side of Highway 26 (Business 151), northeast corner of Waupun, Fond du Lac County; SW/sec. 33, T. 14 N., R. 15 E. BC = Quarry on south side of Breezy Point Road, just west of Highway A, 6.4 km (4 mi) north of Beaver Dam, Dodge County; NW/sec. 9, T. 12 N., R. 14 E. SP = Quarry on north side of Spruce Road, 2.4 km (1.5 mi) east of Highway 33, 0.8 km (0.5 mi) west of Highway W, 4.8 km (3 mi) north of Beaver Dam, Dodge County; SW/sec. 9, T. 12 N., R. 14 E. WT = Koplin Excavation Quarry: East side of Highway 16, 4 km (2.5 mi) north of Watertown, Dodge County; NW/sec. 21, T. 9 N., R. 15 E. FA = Quarry on north side of Highway M, 0.8 km (0.5 mi) east of Highway 12, 1.6 km (1 mi) southeast of Fort Atkinson, Jefferson County; SE/sec. 10, T. 5 N., R. 14 E. GT = Road cut on east side of U.S. Highway 52, 2.4 km (1.5 mi) south of Guttenberg, Clayton County, Iowa; SW/sec. 29, T. 92 N., R. 2 W. DN = Road cut on north side of U.S. Highway 61, 6.4 km (4 mi) northwest of Dickeyville, Grant County; NW/sec. 7, T. 2 N., R. 2 W. CN = Quarry on west side of Church Road, 0.8 km (0.5 mi) north of U.S. Highway 151, 3.2 km (2 mi) south of Cornelia, Grant County; NE/sec. 11, T. 2 N., R. 2 W. RS = Quarry on south side of Highway 59, 4 km (2.5 mi) northeast of Monroe, Green County; NE/sec. 30, T. 2 N., R. 8 E. MN = Quarry on north side of Highway Y, 1.6 km (1 mi) west of Highway 81, 3.2 km (2 mi) northwest of Monroe, Green County; SW/sec. 20, T. 2 N., R. 7 E. WF = Quarry on west side of North Road, 8.8 km (5.5 mi) west of Woodford, Lafayette County; SW/sec. 6, T. 2 N., R. 5 E. SW = Quarry on east side of Highway 176, 0.8 km (0.5 mi) south of South Wayne, Lafayette County; NE/sec. 15, T. 1 N., R. 5 E. DP = De Pere drill core: Drilled by U.S. Geological Survey, 1993, at West De Pere High School, De Pere, Brown County; NW/sec. 28, T. 23 N., R. 20 E. WC = Waupun drill core: Drilled by U.S. Geological Survey, 1995, at the intersection of Highway MM and M, Fond du Lac County; SE/sec. 8, T. 14 N., R. 15 E. SB = Sheboygan drill core: Texas-Wisconsin Explore Van Driest No. 1, Sheboygan County; sec. 12, T. 13 N., R. 22 E. JE = Jefferson drill core: Mobil Oil Corporation No. 8-16-21-Z1, Jefferson County; sec. 21, T. 8 N., R. 16 E. WK1 = Waukesha drill core: Mobil Oil Corporation No. 8-17-18-Z1, Waukesha County; sec. 18, T. 8 N., R. 17 E. WK2 = Waukesha drill core: Mobil Oil Corporation No. 6-17-22-Z1, Waukesha County; sec. 22, T. 6 N., R. 17 E. WK3 = Waukesha drill core: Mobil Oil Corporation No. 5-17-33-Z1, Waukesha County; sec. 33, T. 5 N., R. 17 E. RA1 = Racine drill core: Mobil Oil Corporation No. 3-20-35-Z1, Racine County; sec. 35, T. 3 N., R. 20 E. KE2 = Kenosha drill core: Mobil Oil Corporation No. 1-21-2-Z1, Kenosha County; sec. 2, T. 1 N., R. 21 E. KE3 = Kenosha drill core: Mobil Oil Corporation No. 1-22-6-Z1, Kenosha County; sec. 6, T. 1 N., R. 22 E.

Formation contains the thick-bedded grainstone lithofacies at its top that indicates a high-energy setting near a shoal. The Galena Formation does not contain high-energy facies. Instead, it contains alternations of grainstone and thin shale, and thin-bedded mudstone as the shallowest-water facies exposed (Figs. 5, 6). The low-energy ramp existed under the hydraulic conditions of a broad epeiric shelf with very low gradient on which incoming oceanic waves lost their energy very gradually over wide areas (e.g., Irwin, 1965).

SEQUENCE STRATIGRAPHIC INTERPRETATION

A nested hierarchy of sequences, cycle sets, cycles, and omission-surface–bounded packages are recognized in the Decorah and Galena Formations. Depositional sequences are relatively conformable successions bounded by surfaces across which there is an abrupt increase in siliciclastics and skeletal grains. These surfaces—interpreted as sequence boundaries—may show indication of truncation, and they always show abrupt, basinward, facies shifts. Individual sequences are correlated laterally for several hundreds of kilometers and vary in thickness and lithology regionally (Figs. 7, 8). Facies are arranged in repetitive patterns that define cycles within cycle sets. Generally, cycles consist of several decimeter- to meter-scale asymmetric facies successions with sharp bases across which there is abrupt increase in shale content and skeletal grains. Cycle sets are bundles of cycles that show a consistent trend in cycle stacking pattern. The consistent trend is commonly represented as either progradational, aggradational, or retrogradational (e.g., Kerans and Tinker, 1997). Cycle sets are meters to tens of meters thick, and individual cycle sets are correlated laterally for tens to hundreds of kilometers (Figs. 7, 8).

Numerous iron-mineralized omission surfaces punctuate the succession and define omission-surface–bounded packages, which are commonly less than 0.5 m thick and which form the smallest recognized sedimentary unit. Omission-surface–bounded packages are continuous for hundreds of meters, and some appear to be continuous for tens of kilometers.

Sequences

Three depositional sequences are identified in the Decorah and Galena Formations (Fig. 5). Figures 7 and 8 show the lateral facies variability and continuity of individual sequences and cycle sets. The lower sequence, sequence D, corresponds to the Decorah Formation and to the T-R (transgressive-regressive) cycle 4 of Witzke and Bunker (1996). The middle sequence, sequence G1, corresponds to the lower to middle parts of the Dunleith Member of the Galena Formation and to the T-R cycle 5A of Witzke and Bunker (1996). The upper sequence, sequence G2, corresponds to the upper part of the Dunleith, Wise Lake, and Dubuque Members of the Galena Formation and to the T-R cycles 5B, 5C, and 6A of Witzke and Bunker (1996). All sequences thin to the north in eastern Wisconsin (Fig. 8), and sequence D and sequence G2 thin to the east (Fig. 7). The fact that shale in sequence D becomes more dominant to the west suggests that the Transcontinental arch was the major source area for siliciclastics (Fig. 2). However, in sequences G1 and G2, shale becomes more dominant not only to the west but also to the north, and this fact suggests supply from both the Transcontinental arch to the west and Wisconsin dome to the north (Fig. 2). The regional thickness changes in the sequences (Figs. 7, 8) are attributed to depositional controls rather than erosion of the strata.

Table 1.—LITHOFACIES SUMMARY OF THE DECORAH AND GALENA FORMATIONS

Lithofacies and Abbreviations (See Figures)	Lithofacies Description	Sedimentary Structures and Type of Bedding	Biota (Ordered by Relative Abundance)	Bioturbation and Trace Fossils (If Any)	Diagenetic Features and Omission Surfaces	Interpretation
Interbedded mudstone and organic-rich shale (M-OSh)	Interbedded carbonate mudstone to wackestone (light orange-brown noncarbonate grains) and organic-rich shale (dark brown)	Thin and wavy bedded; fine lamination in shale beds	Strophomenid brachiopods, pelecypods, trilobites, and the organic-walled microfossil *G. prisca* in shale beds	Rare horizontal burrows; some *Chondrites*	Early cementation in marine environments; compaction in shale beds	Lower-mid ramp to outer ramp
Shale (Sh)	Fossiliferous; terrigenous siltstone common; green	Fissile, very thin bedded	Orthid brachiopods, bryozoans, and crinoids	Weak	Compaction	Outer ramp
Thick-bedded mudstone to wackestone (TMW)	Light brown to tan fossiliferous mudstone to wackestone	Medium to thick bedded and some very thin grainstone beds	Brachiopods, crinoids, ostracodes, gastropods, rugose corals, bryozoans, trilobites, pelecypods, and *Receptaculites*	Moderate to pervasive; *Thalassinoides*	Some calcite spar nodules; omission surfaces common to abundant	Mid to outer ramp
Crinoidal wackestone (CW)	Crinoidal wackestone with dark brown organic-rich shale partings; light pinkish brown noncarbonate grains	Thin to medium bedded	Crinoids, brachiopods, and the organic-walled microfossil *G. prisca*	Weak	Compaction	Lower-mid ramp
Skeletal wackestone to packstone (SWP)	Slightly argillaceous; greenish gray to medium gray noncarbonate grains; some floating quartz grains	Medium to thick bedded	Brachiopods, crinoids, bryozoans, trilobites, and rugose corals	Strong to pervasive; *Chondrites*, *Thalassinoides*	Omission surfaces common	Mid ramp
Interbedded grainstone and shale (G-Sh)	Skeletal grainstone to packstone interbedded with thin green shale; some intraclasts in grainstone beds; terrigenous siltstone beds common; some floating quartz grains	Thin bedded with erosive lower contacts; normal grading	Bryozoans, brachiopods, crinoids, and trilobites	Weak to strong; pervasive; *Chondrites* bioturbation leads to skeletal packstone in a clay matrix	Omission surface rare	Upper-mid ramp
Thick-bedded grainstone (TBG)	Skeletal packstone to grainstone; light orange-brown noncarbonate grains	Thick bedded with some thin-bedded tempestites	Brachiopods, crinoids, gastropods, bryozoans, and trilobites	Weak to strong; *Chondrites*	Omission surface rare	Upper-mid ramp
Thin-bedded mudstone (ThBM)	Light brown to tan with green to brown shale partings	Thin and wavy bedded with some very thin grainstone beds	Brachiopods, gastropods, crinoids, bryozoans, and *Receptaculites*	Commonly weak, rarely strong; *Chondrites*	Compaction; omission surfaces rare to common	Inner ramp

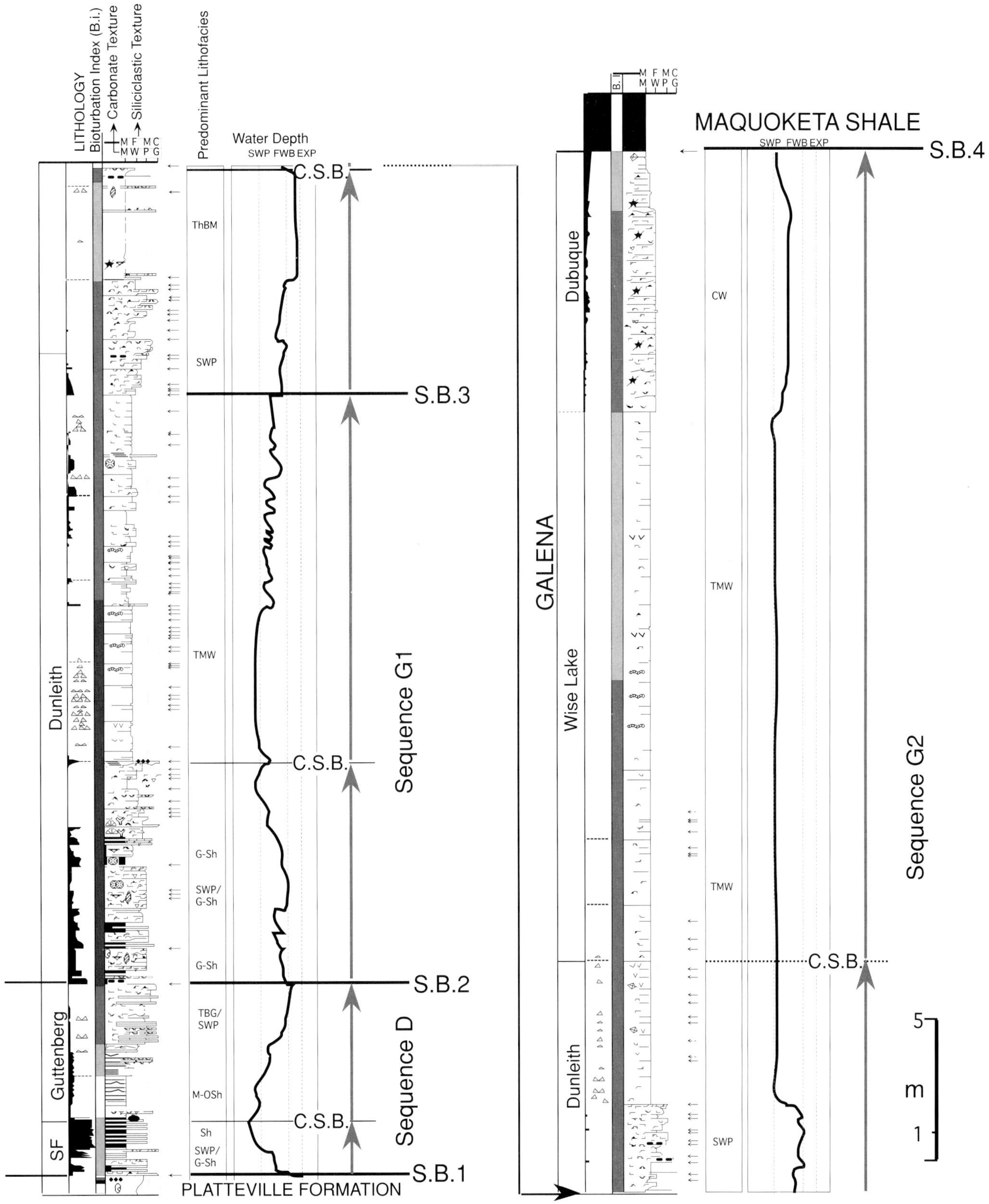

FIG. 5.—Composite section of the Decorah and Galena Formations in Wisconsin. The section does not reflect local variations in unit thickness and facies changes. See Table 1 for lithofacies abbreviations and descriptions. Siliciclastic textures include grain sizes (M = mud, F = fine-grained, second M = medium-grained, and C = coarse-grained). Carbonate textures follow Dunham's (1962) classification: M = mudstone, W = wackestone, P = packstone, G = grainstone. SF = Spechts Ferry Member, FWB = fair-weather wave base, SWB = storm wave base, EXP = exposure surface, S.B. = sequence boundary, C.S.B. = cycle set boundary. Vertical arrows represent cycle sets. Key is shown in Figure 6.

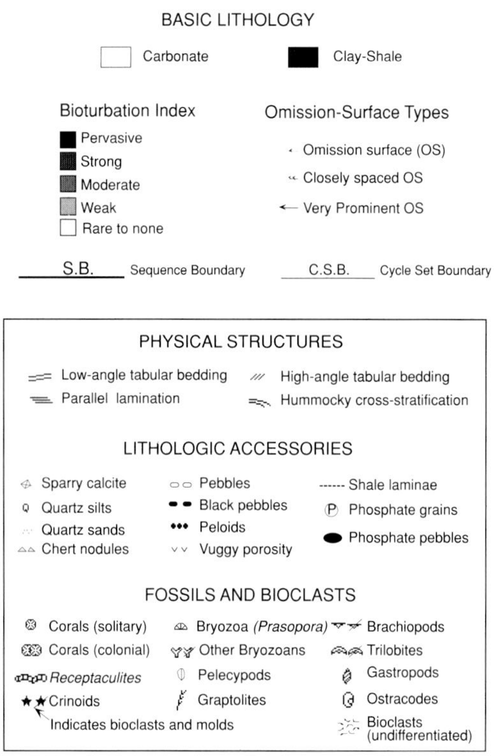

FIG. 6.—Key to stratigraphic sections.

Sequence D.—

Sequence D is bounded below by sequence boundary 1 (Figs. 5, 7, 8) on which there is low erosional relief; sequence D shows only slight angular discordance with the underlying Platteville Formation (Fig. 9). Sequence boundary 2 is the upper boundary of sequence D. Sequence boundary 1 is probably a submarine-modified subaerial-exposure surface truncating the underlying Platteville strata (Fig. 9). Petrographically, this surface is characterized by features such as nontectonic fractures and brecciation, and the underlying strata show leached and vuggy porosity (Fig. 9). Platteville lithoclasts suggest early cementation. Medium to coarse quartz grains are present above sequence boundary 1. Sequence D is thickest (7 m) in southwestern Wisconsin and thins to the east and north. Ludvigson et al. (1996) interpreted the thinning to the east as the result of downlap onto the top of the Platteville Formation. However, a similar thinning can result if sequence D onlapped a topographic high. The last interpretation is favored because sequence D also thins to the northeast across an area that was a depositional high throughout most of the early Paleozoic.

Sequence D consists of a basal, fining-upward, facies succession followed by a coarsening-upward succession. The base consists of coarse-grained carbonates overlain by alternating beds of fine-grained mudstone and green shale capped by fissile green shale punctuated by very thin bedded (<1 cm) brachiopod coquina. Ludvigson et al. (1996) reported a gradual upward increase in skeletal packstone beds in this interval (Spechts Ferry Shale Member) in eastern Iowa, however. The strata at the transition from the fining-upward to the coarsening-upward succession contain a phosphate-enriched bed (<1–3 cm thick) with 1–4-mm-sized, well-rounded phosphate grains. Above the phosphate-rich bed, the succession consists of interbedded mudstone and organic-rich shale overlain by skeletal wackestone to packstone and next by thick-bedded grainstone. The gradual upward increase in the thin-bedded grainstone lithofacies indicates depositional shallowing.

The interpreted depositional history of sequence D in the Upper Mississippi Valley area is summarized as follows. Skeletal-grain–rich carbonate sediments were deposited during periods of initial transgression, incorporating lithoclasts derived from the Platteville Formation below. Siliciclastics, supplied from the west, were deposited to form the eastward-thinning Spechts Ferry Shale Member. A rapid increase in accommodation drowned topographic highs, shut off the siliciclastic supply, and allowed deposition of the phosphate-rich condensed section. As the rate of accommodation decreased, carbonate sediments gradually prograded from depositional highs, a process that resulted in a shallowing-upward facies succession.

Sequence G1.—

Sequence G1 is bounded below by a very sharp and widespread discontinuity surface (sequence boundary 2) on top of the Guttenberg Member of the Decorah Formation. Sequence boundary 2 merges with sequence boundary 1 in much of northeastern Wisconsin where sequence G1 directly overlies the Platteville Formation without intervening sequence D (Figs. 7, 8). Sequence boundary 2 is a very prominent omission surface and is characterized by an abrupt increase in siliciclastics above it. Whether this surface was a subaerially exposed surface modified by a submarine omission surface is unclear, but the erosional nature (Fig. 10) suggests that significant truncation must have taken place after early cementation.

Sequence G1 shows an overall transition from a base rich in shale and skeletal grains to a fine-grained, carbonate-dominated top (Fig. 5). The lithofacies association A, upper-mid-ramp facies, sharply overlies sequence boundary 2 and shows gradual landward stepping to the north and west followed by seaward stepping (Figs. 7, 8). Shale intercalations are abundant at the base and diminish toward the top (Figs. 5, 7, 8). Northward and westward, shale content increases (Witzke, 1980; Choi et al., 1997). The lithofacies association B—the lower-mid-ramp to outer-ramp, mud-dominated facies—is thickest in southwestern Wisconsin and thins toward northeastern Wisconsin (Figs. 7, 8). Iron-mineralized omission surfaces are abundant in sequence G1; they occur at an average interval of 30 cm. Some of them are interpreted as submarine hardgrounds (cf. Delgado, 1983). The close association of thin shale beds and partings with the coarse-grained facies (Figs. 7, 8) suggests that most of the shale in sequence G1 was derived from nearby exposed terranes and that the depositional environments were shallower. Thus, sequence G1 shows a deepening-upward trend with a statewide, shallower-water facies at the base and a progressive encroachment of both grainy facies and shale to the north. This overall trend is punctuated by several cycles and cycle sets described below.

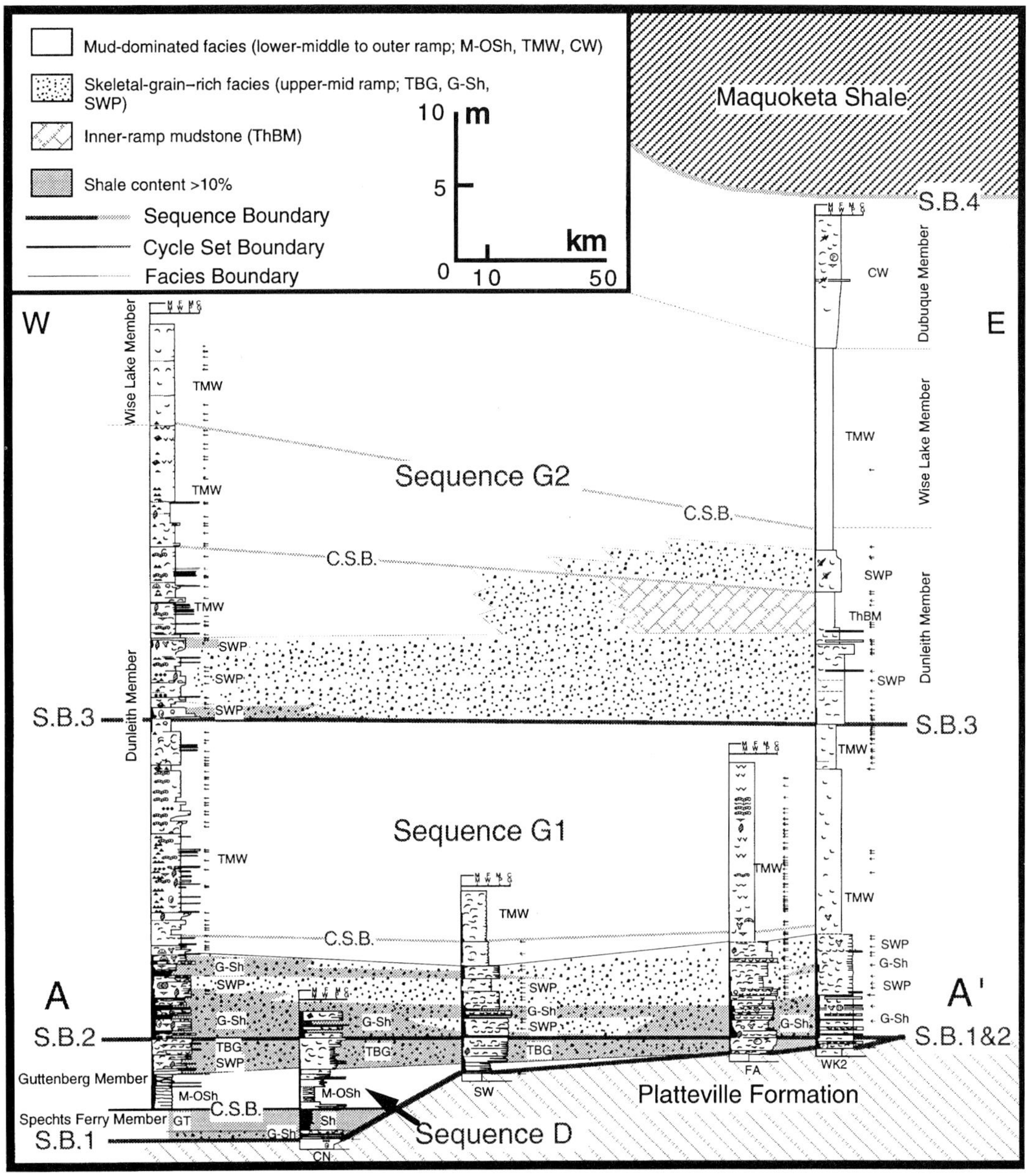

FIG. 7.—East-west cross section along the correlation line A–A′ and representative stratigraphic sections for which locations are shown in Figure 3. Correlation of sequence boundaries 1 and 4 to the west is based on thickness information of the Decorah and Galena Formations obtained from the composite section in the Guttenberg area by Witzke and Glenister (1987). The datum is the base of the Galena Formation (sequence boundary 2). See Table 1 for lithofacies abbreviations, Figure 5 caption for textural abbreviations, and Figure 6 for key.

Sequence G2.—

Sequence G2 is bounded above and below by very sharp discontinuity surfaces (sequence boundaries 3 and 4). Sequence boundary 3 is a prominent iron-mineralized omission surface (base of the Rivoli Bed, after Templeton and Willman, 1963), commonly recognized by an abrupt increase in shale and grain-supported rocks across it. The discontinuity surface (sequence boundary 4) along the contact between the Galena Formation and the Maquoketa Shale is widely recognized all over the eastern United States (Bergström and Mitchell, 1992). Sequence boundary 4 is characterized by nontectonic brecciation with clay matrix, infiltrated from above, and dissolution vugs, and it is interpreted as an exposure surface (Fig. 11). Lithoclasts are also present above sequence boundary 4.

Sequence G2 consists of a skeletal-grain– and shale-rich base and carbonate-mud–dominated top (Figs. 7, 8). Thin shale beds and partings are sparse in southeastern Wisconsin but increase to the west and north (Figs. 7, 8). The upper-mid-ramp, skeletal-grain–rich facies is thickest in northeastern Wisconsin and thins to the south and west (Figs. 7, 8). The lower-mid-ramp to outer-ramp, mud-dominated facies is thickest in southwestern Wisconsin where the total thickness of sequence G2 is the greatest. The inner-ramp mudstone facies is thickest in northeastern Wisconsin, thins to the south (Fig. 8), and is absent in southwestern Wisconsin (Fig. 7). Iron-mineralized omission surfaces are also abundant in sequence G2, but their number decreases upward. The uppermost part of sequence G2 (Dubuque Member) shows a slight increase in skeletal grains (mostly

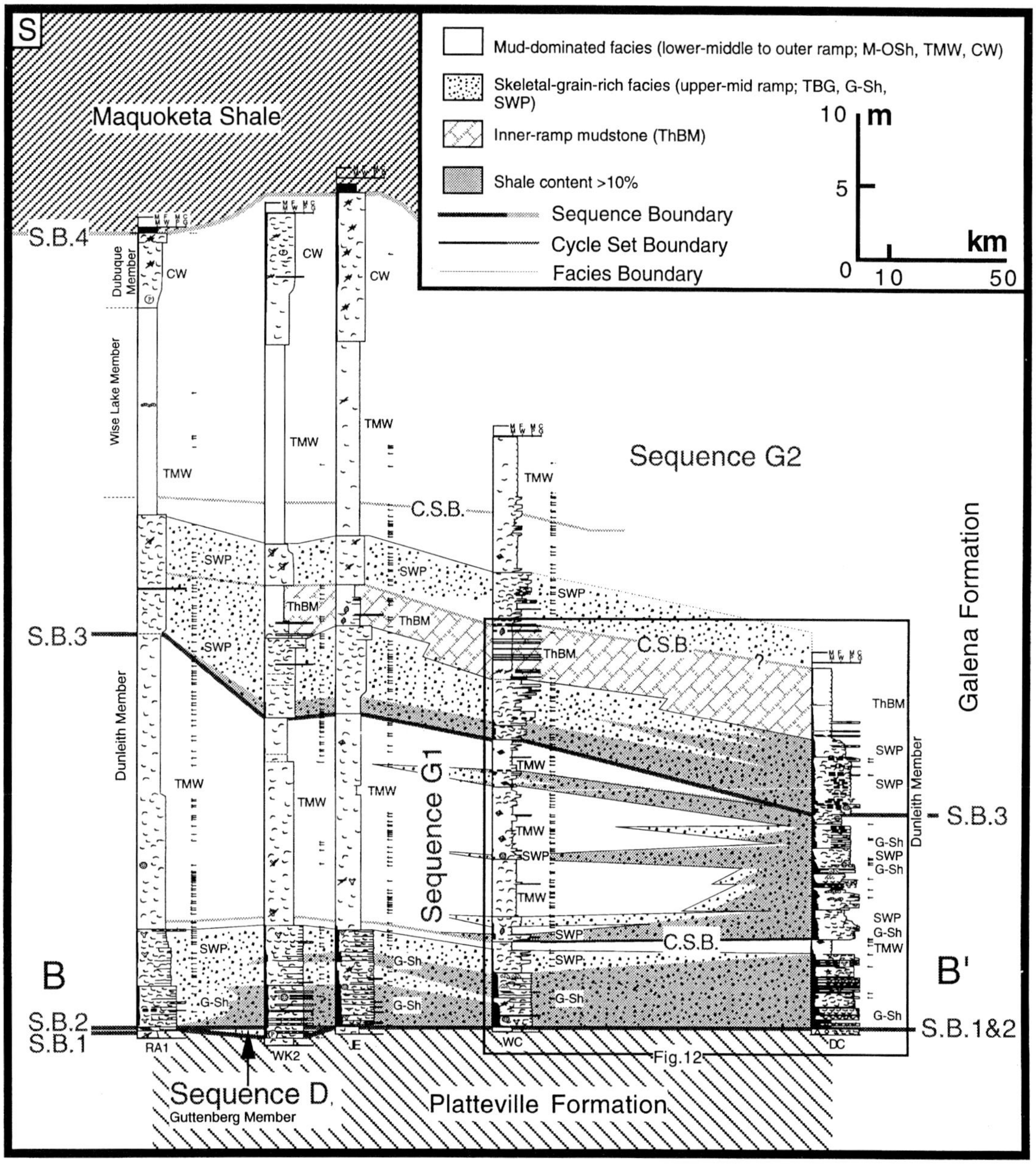

FIG. 8.—North-south cross section along the correlation line B–B' and representative stratigraphic sections for which locations are shown in Figure 3. The datum is the base of the Galena Formation (sequence boundary 2). See Table 1 for lithofacies abbreviations, Figure 5 caption for textural abbreviations, and Figure 6 for key.

crinoids) and organic-rich shale partings upward. The uppermost part is only exposed in southern Wisconsin. The interpretation of the depositional history of sequence G2 is very similar to that of sequence G1, with an overall deepening-upward trend. However, the uppermost part may represent a shallowing, but outcrops in a landward direction are absent, and this interpretation cannot be corroborated.

Cycles and Cycle Sets

Meter-scale repetition of facies successions defines cycles. Individual cycles are not well represented in sequence D, but alternations of skeletal-grain– and shale-rich packages and carbonate mudstone packages are well recognized in sequences G1 and G2. Figures 12 and 13 show details of facies variability and traceability of cycles and cycle sets in sequences G1 and G2 in northeastern Wisconsin. Individual cycles have a skeletal-grain– and shale-rich base that grades upward into a mud-dominated top (Fig. 13). The transition from the mud-dominated facies to the skeletal-grain– and shale-rich facies is sharp (Fig. 12), often marked by an iron-mineralized omission surface. These characteristics contrast with typical subtidal cycles in which deeper-water, fine-grained facies grade upward into shallower-water, coarser-grained facies (Aigner, 1985; Burchette and Wright, 1992). Some cycles may be correlated regionally more than 100 km. Cyclicity is best recognized in northeastern Wisconsin where the two lithologies interdigitate and is poorly defined to the south where cycles are characterized by mud-dominated

FIG. 9.—Field photograph showing bed truncation below sequence boundary 1 that is highlighted by a prominent hardground (lower arrow). Another prominent hardground (upper arrow) separates sequence D (Decorah Formation) from sequence G (Galena Formation). Hammer handle is approximately 30 cm long for scale (location FA, Fig. 3).

FIG. 10.—Field photographs of sequence boundary 2 showing erosional relief (dashed) on a laminated grainstone bed in the uppermost Decorah Formation. Note the dark impregnation at the discontinuity surface. Pen is 3 cm long for scale (location RS, Fig. 3).

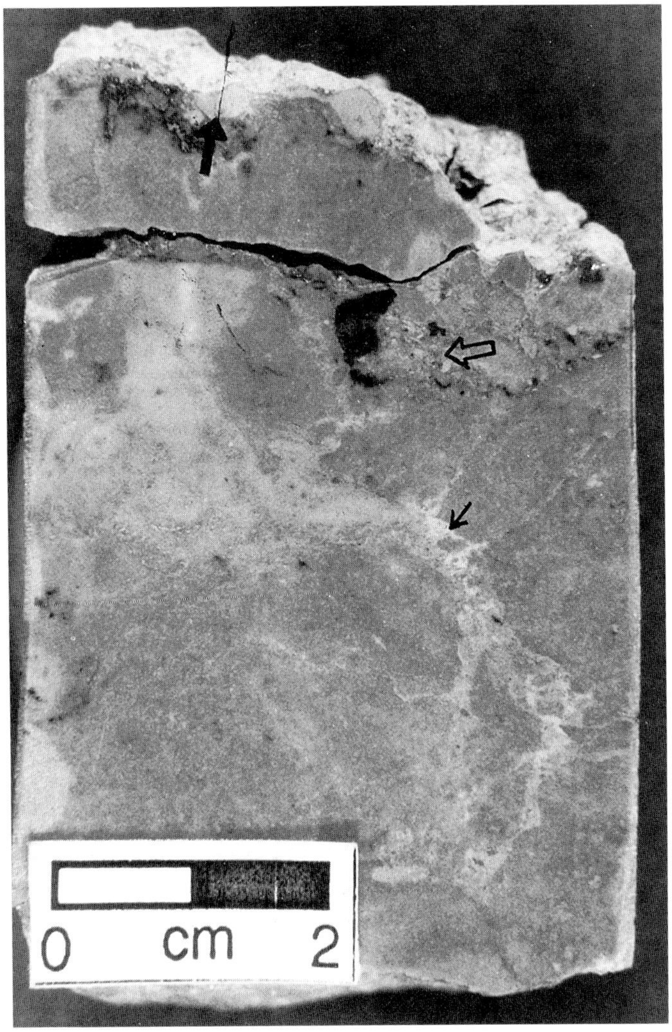

FIG. 11.—Core-slab photograph showing erosional relief and non-tectonic brecciation on top of the Galena Formation. Both features are indicative of subaerial exposure. Clays from the Maquoketa Shale above percolated down through the cracks (small arrow). Note also the round pebbles (large black arrow) above the discontinuity surface (sequence boundary 4) and within cracks (open arrow) (location KE3, Fig. 3).

lithologies (Figs. 12, 13). Cycle thickness ranges between 0.5 and 2.0 m and averages 1.1 m.

Bundles of cycles showing a consistent trend compose a cycle set. Two cycle sets are recognized in sequence D (Figs. 5, 7). The lower cycle set shows a retrogradational trend, and the upper cycle set shows a progradational trend. These two cycle sets are separated by a phosphate-rich condensed section deposited during a depositional hiatus (e.g., Loutit et al., 1988), which is interpreted to represent the maximum-flooding surface in sequence D.

Sequence G1 consists of a retrogradational cycle set showing an overall decrease in shale and skeletal grains followed by a progradational cycle set showing an overall increase in shale and skeletal grains (Figs. 5, 7, 8). The lower-mid-ramp to outer-ramp, mud-dominated facies between the two sets is interpreted to represent a maximum flooding episode in sequence G1 (Figs. 7, 8, 12). The base of this flooding interval is rich in stony bryozoans (*Prasopora*), representing "Epiboles" (Brett et al., 1990), and is widely correlated across several hundred kilometers in Wisconsin, Iowa (Witzke and Bunker, 1996), and Illinois (Willman and Kolata, 1978).

Three cycle sets are recognized in sequence G2 mostly on the basis of overall upward facies transitions (Figs. 5, 7, 8). Cycles in the upper two cycle sets are dominated by outer-ramp mudstones, and identification of cycles is difficult. The lower cycle set shows a progradational trend with the upper-mid-ramp skeletal-grain–rich facies at the base overlain by the inner-ramp mudstone facies (Figs. 7, 8). The middle cycle set shows a retrogradational trend in which the upper-mid-ramp skeletal-grain–rich facies at the base passes up into the lower-mid-ramp to outer-ramp, mud-dominated facies (Figs. 7, 8). The upper cycle set is dominated by lower-mid-ramp to outer-ramp mud-dominated facies, but the gradual increase in skeletal grains in the uppermost part (Dubuque Member) may represent a progra-

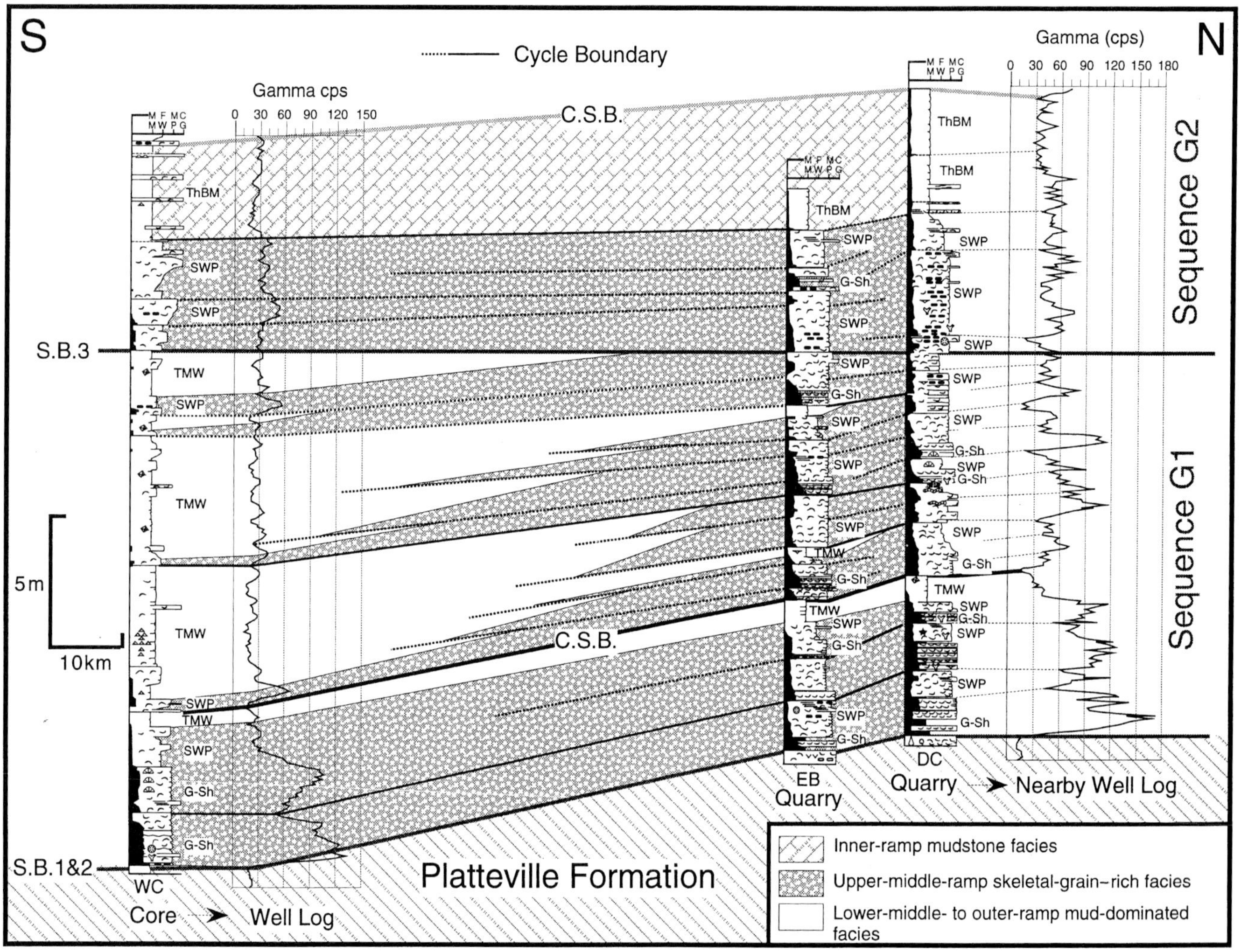

FIG. 12.—Detailed correlation of cycles in northeastern Wisconsin. (See Fig. 3 for location of sections.) Stratigraphic sections are simplified. Relative shale content is represented in the left parts of the stratigraphic columns by the width of the black portion. Two gamma-ray well logs (cps = counts per second) are provided, one from the same well as core WC and the other from a well 200 m west of the quarry section DC. Meter-scale cycles represented by changing shale content and skeletal grains are clearly seen in the well logs. In the north, several cycles are amalgamated with only shale-rich and skeletal-grain–rich facies whereas in the south they are amalgamated with only stacked mud-dominated facies. Recognition of cycles is difficult in the south where there is only the mudstone-dominated association (facies association B). The datum is sequence boundary 3. See Table 1 for lithofacies abbreviations, Figure 5 caption for textural abbreviations, and Figure 6 for key.

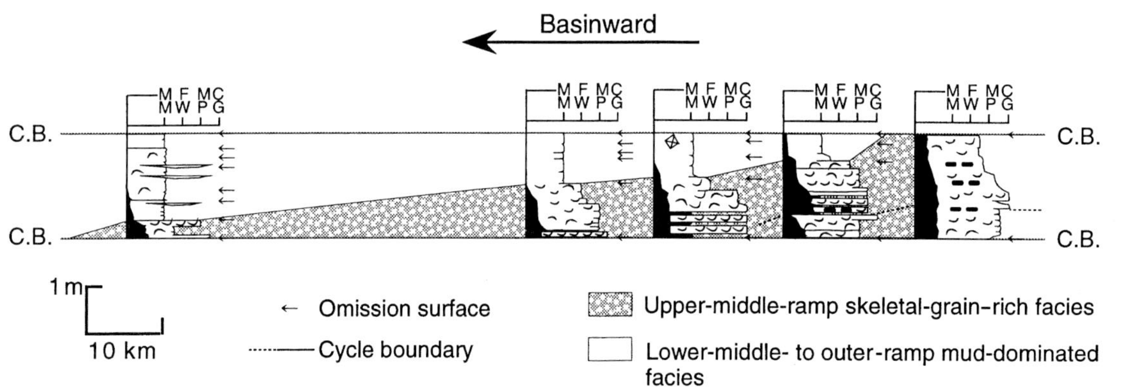

FIG. 13.—Summary of the lateral and vertical facies variation within cycles in sequence G as observed in the field. The omission surfaces at the contact between the shale- and skeletal-grain–rich association (facies association A) below and the mud-dominated association above (facies association B) are commonly coated by phosphate, indicating little or no sedimentation. See Figure 5 caption for textural abbreviations and Figure 6 for key.

dation. The maximum flooding episode in sequence G2 is represented at the base of the upper cycle set (which approximately corresponds to the base of Wise Lake Member).

Omission-Surface–Bounded Packages

Mineralized omission surfaces are particularly abundant in sequences G1 and G2 (Fig. 5). The omission surfaces mark pauses in deposition with little or no erosion (Bromley, 1975; Kennedy and Garrison, 1975). Some of the omission surfaces in the Decorah and Galena Formations are early lithified submarine hardgrounds (cf. Delgado, 1983), which are evidenced in most places by borings and truncation of grains. Other omission surfaces are recognized only by faint traces of iron mineralization, and evidence for early cementation is absent. Phosphate-laminated crusts less than 1 mm thick coat the tops of some omission surfaces. In some cases, terrigenous silt is concentrated at the omission surfaces. Omission surfaces are associated with nearly every lithofacies and, hence, with a wide range of inferred energy regimes. Omission surfaces form the boundaries of recognized sequences and cycles, but also are present within cycles and within otherwise homogeneous facies successions. Omission-surface morphologies range from flat-topped and planar to hummocky with overhangs and cavities formed by large *Thalassinoides* burrows (Fig. 14). These variations are interpreted to reflect differing degrees of erosional planation (Saylor et al., 1997).

Omission-surface–bounded packages reflect alternation between times of deposition and no deposition. Possible interpretations of no deposition include winnowing during low sea level (Byers and Geary, 1993) and alternation between times of carbonate production and starvation. These possibilities are discussed below.

DISCUSSION

The sudden increase in thin shale beds and partings associated with skeletal-grain–rich facies is observed in many orders of cyclicity: the sequence, the cycles, and some cycle sets. Shale abundance increases to the north and west, the direction of the emergent land (Figs. 2, 7, 8), and often is associated with skeletal-grain–supported rocks. These facies changes are indicative of proximity to the source as well as shallower water conditions. Thus, at the scales of sequences, cycle sets, and cycles, the shale- and skeletal-grain–rich facies is interpreted to have accumulated during periods of low accommodation space and to reflect an expansion of the shale domain and a concomitant increase in energy associated with lowering of relative sea level. It is important to note that characteristic cycles are not recognized in a basinward direction where the succession consists predominantly of mudstone punctuated by closely spaced omission surfaces.

FIG. 14.—Field photograph of a skeletal wackestone bed showing two omission surfaces (at the base and the top of the bed) and associated *Thalassinoides* burrows. Pen is 12 cm long for scale (location SP, Fig. 3).

The upward decrease in shale and skeletal grainstones in the cycles of sequences G1 and G2 can be interpreted as a retrogradational succession. A scenario that can account for these relationships is to invoke reciprocal sedimentation (e.g., Wilson, 1975) such that the shale- and skeletal-grain–rich facies and carbonate mud–dominated facies do not grade laterally into each other. Instead, deposition of these facies may have been disassociated in both time and space (Fig. 13). The deposition of skeletal-grain–rich facies and shale is interpreted as a combination of high siliciclastic influx and frequent storm reworking in shallow-water conditions during periods of low sea level. The gradational upward decrease in shale and skeletal-grain content may reflect a decrease in siliciclastic influx and a return to the "normal" mud-dominated mode of deposition associated with an upward decrease in depositional energy. This relationship explains the lateral facies distribution and vertical facies stacking patterns.

The mud-dominated and omission-surface–dominated lithofacies in sequences G1 and G2 is the trademark of the Middle to Upper Ordovician rocks in the Midcontinent (i.e., this lithofacies forms up to 90% of the total volume of rock). These strata differ substantially from the heterogeneous, stromatolitic and oolitic carbonate facies that constitute the underlying Lower Ordovician and overlying Silurian carbonates. Faunal assemblages in the Decorah and Galena Formations are dominated by benthic filter-feeding organisms such as brachiopods, bryozoans, crinoids, mollusks, and rugose corals. Stromatolitic and oolitic facies as well as carbonate buildup facies are absent in spite of the common reef-forming organisms in the Middle to Upper Ordovician (James and Bourque, 1992). The dominance of heterotrophs and the paucity of photoautotrophs are reminiscent of modern cool-water and, probably, of deep-water carbonate environments, which are characterized by overall slower sediment accumulation rates compared to those of their tropical equivalents. In addition, the abundance of omission surfaces suggests that the carbonate factory was near a threshold, prone to frequent and extended periods of nondeposition. However, the latitude reconstructions based on paleomagnetic data (Van der Voo, 1988) and the abundance of mud suggest that temperature alone is not the only process affecting deposition.

Omission surfaces form the boundaries of recognized sequences and cycles, but also are present within cycles and within otherwise homogeneous facies successions. Omission surfaces occur in all the lithologies as well as in the proximal and distal parts of the ramp. Byers and Geary (1993) suggested that the repetitive occurrences of omission surfaces in the Galena Formation in the Upper Mississippi Valley resulted from shoaling events that brought the sea floor within reach of the fair-weather wave base, preventing further accumulation of carbonate mud and cementing the sediment-water interface. However, (1) there is no preferential relationship between omission surfaces and substrate lithology, and (2) individual omission-surface–bounded units do not show a shallowing upward trend; these observations suggest that alternative mechanisms—such as a decrease in carbonate production during periods of inundation by waters with different temperature and/or chemistry (salinity, saturation state)—were important controls. Episodic and prolonged shutdown of carbonate mud production may have been linked to paleoceanographic changes that led to shale progradation, perhaps as a result of decrease in carbonate saturation state during times of increased fluvial runoff during relative fall in sea level. Such a link between shale progradation and carbonate mud production would result in reciprocal sedimentation and the development of the asymmetric, cleaning-upward, and fining-upward cycles characteristic of the Galena Formation.

CONCLUSIONS

1. Three depositional sequences are identified in the Middle to Upper Ordovician Decorah and Galena Formations in Wisconsin. Depositional sequences are relatively conformable successions bounded by identified or interpreted exposure surfaces across which there are abrupt increases in siliciclastics grading upward to carbonates.
2. Sequence D corresponds to the Decorah Formation. Sequence D consists of siliciclastic-dominated fining-upward successions (retrogradational cycle set) followed by coarsening-upward carbonate successions (progradational cycle set). The phosphate-enriched condensed section above the fining-upward succession represents maximum transgression. Deposition of the fine-grained siliciclastics in sequence D resulted from an abundant siliciclastic supply from the west.
3. Sequences G1 and G2 correspond to the Galena Formation. The boundary between these two sequences (sequence boundary 3) correlates to the base of the Rivoli Bed of the Dunleith Member. Sequence G1 consists of a retrogradational cycle set showing an upward decrease in shale- and skeletal-grain–rich facies and an upward increase in the mudstone facies with abundant omission surfaces followed by a progradational cycle set showing the opposite changes. Sequence G2 consists of three cycle sets: the lower progradational, middle retrogradational, and upper progradational cycle sets.
4. In contrast to sequence D, the shale in the sequences G1 and G2 thickens to the west and north, indicating two source areas in the direction of the emergent land. The shale is indicative of proximity to the source as well as shallower-water conditions.
5. Cycles of sequence G1 and the lower part of sequence G2 are characterized by lower shale- and skeletal-grain–rich facies and upper carbonate mud–dominated facies. Amalgamation of shale- and skeletal-grain–rich facies dominates cycles landward, and amalgamation of carbonate mud–dominated facies dominates cycles basinward.
6. The asymmetry of the cycles in sequences G1 and G2, represented by the abrupt increase followed by gradual decrease in shale and skeletal grains, is explained by reciprocal sedimentation, in which deposition of skeletal grains and siliciclastics is disassociated from deposition of carbonate mud both in time and space.
7. Iron-mineralized omission surfaces are abundant in the sequences G1 and G2. The observations that these omission surfaces developed widely across facies transitions suggest that decrease in productivity during periods of inundation by waters with different temperature and/or chemistry (salinity, saturation state) was an important control for the development of the omission surfaces.
8. The characteristic features of the Middle to Upper Ordovician Decorah and Galena Formations—such as laterally continuous subtidal facies, absence of records of shoaling to peritidal facies and "temperate-type" carbonate lithology in which abundant mudstones are punctuated by numerous omission surfaces—probably reflect the combined effects of paleoceanography and paleoclimate in the Late Ordovician epeiric sea in the Midcontinent.

ACKNOWLEDGMENTS

The authors wish to thank Steve Bachtel, Dennis Kolata, and Art Saller for their invaluable assistance in reviewing this paper.

REFERENCES

AGNEW, A. F., HEYL, A. V., BEHRE, C. H., AND LYONS, E. J., 1956, Stratigraphy of Middle Ordovician rocks in the zinc-lead district of Wisconsin, Illinois, and Iowa: Washington, D.C., United States Geological Survey Professional Paper 274-K, p. 248–312.

AHR, W. M., 1973, The carbonate ramp—An alternative to the shelf model: Transactions of the Gulf Coast Association of Geological Societies, v. 23, p. 221–225.

AIGNER, T., 1985, Storm depositional systems: Berlin, Springer-Verlag, 174 p.

BAKUSH, S. H., 1985, Carbonate microfacies, depositional environments and diagenesis of the Galena Group (Middle Ordovician) along the Mississippi River (Iowa, Wisconsin, Illinois, and Missouri), U.S.A.: Unpublished Ph.D. thesis, University of Illinois, Urbana-Champaign, 223 p.

BAKUSH, S. H., AND CAROZZI, A. V., 1986, Subtidal storm-influenced carbonate ramp model: Galena Group (Middle Ordovician) along the Mississippi River (Iowa, Wisconsin, Illinois, and Missouri), U.S.A.: Genève, Archives des Sciences, v. 39, p. 141–183.

BERGSTRÖM, S. M., AND MITCHELL, C. E., 1992, The Ordovician Utica Shale in the eastern Midcontinent region: Age, lithofacies, and regional relationships, *in* Chaplin, J. R. and Barrick, J. E., eds., Special paper in paleontology and stratigraphy: A tribute to Thomas W. Amsden: Norman, Oklahoma Geological Survey Bulletin 145, p. 67–86.

BRETT, C. W., MILLER, K. B., AND BAIRD, G. C., 1990, A temporal hierarchy of paleoecologic processes within a Middle Devonian epeiric sea, *in* Miller, W., III, ed., Paleocommunity temporal dynamics: The long-term development of multispecies assemblies: Knoxville, Paleontological Society Special Publication No. 5, p. 178–209.

BROMLEY, R. G., 1975, Trace fossils at omission surfaces, *in* Frey, R. W., ed., The study of trace fossils: A synthesis of principles, problems, and procedures in sedimentology: New York, Springer-Verlag, p. 399–428.

BURCHETTE, T. P., AND WRIGHT, V. P., 1992, Carbonate ramp depositional systems: Sedimentary Geology, v. 79, p. 3–57.

BYERS, C. W., 1996, The Cretaceous is the key to the Ordovician, *in* Carbonates and Global Change: An Interdisciplinary Approach: SEPM/IAS Research Conference, p. 34.

BYERS, C. W., AND GEARY, D. H., 1993, Milankovitch cyclicity in the Ordovician: Periodic omission surface generation in the Galena Group, northern Mississippi Valley: Geological Society of America Abstracts with Programs, v. 25, no. 6, p. 338.

CHOI, Y. S., 1995, Stratigraphy and sedimentology of the Middle Ordovician Sinnipee Group, eastern Wisconsin: Unpublished M.S. thesis, University of Wisconsin, Madison, 299 p.

CHOI, Y. S., SIMO, J. A., AND FREIBERG, P. G., 1997, Geology of the Middle and Upper Ordovician carbonates in eastern Wisconsin, *in* Mudrey, M. G., ed., Guide to Field Trips in Wisconsin and Adjacent Areas of Minnesota, 31st Annual Meeting of North Central Section, Geological Society of America: Madison, Wisconsin State Geologic and Natural History Survey, p. 97–114.

DELGADO, D. J., 1983, Deposition and diagenesis of the Galena Group in the Upper Mississippi Valley, *in* Delgado, D. J., ed., Ordovician Galena Group of the Upper Mississippi Valley—Deposition, diagenesis, and paleoecology: Guidebook for the 13th Annual Field Conference: Madison, Great Lakes Section — Society of Economic Paleontologists and Mineralogists, p. A1–A17.

DUNHAM, R. J., 1962, Classification of carbonate rocks according to depositional texture, *in* Ham, W. E., ed., Classification of Carbonate Rocks: Tulsa, American Association of Petroleum Geologists Memoir 1, p. 108–121.

FORDHAM, B. G., 1992, Chronometric calibration of mid-Ordovician to Tournaisian conodont zones: A compilation from recent graphic correlation and isotope studies: Geological Magazine, v. 129, p. 709–721.

GINSBURG, R. N., 1982, Actualistic depositional model for the Great American bank—Cambro-Ordovician: International Congress on Sedimentology, v. 11, p. 114.

GOLDMAN, D., MITCHELL, C. E., BERGSTRÖM, S. M., DELANO, J. W., AND TICE, S., 1994, K-bentonites and graptolite biostratigraphy in the Middle Ordovician of New York State and Quebec: A new chronostratigraphic model: Palaios, v. 9, p. 124–143.

IRWIN, M. L., 1965, General theory of epeiric clear water sedimentation: American Association of Petroleum Geologists Bulletin, v. 49, p. 445–459.

JAMES, N. P., AND BOURQUE, P. A., 1992, Reefs and mounds, *in* Walker, R. G. and James, N. P., eds., Facies models-response to sea level change: Calgary, Canadian Society of Petroleum Geologists, p. 323–347.

KENNEDY, W. J., AND GARRISON, R. E., 1975, Morphology and genesis of nodular chalks and hardgrounds in the Upper Cretaceous of Southern England: Sedimentology, v. 22, p. 311–386.

KERANS, C., AND TINKER, S. W., 1997, Sequence Stratigraphy and Characterization of Carbonate Reservoirs: Tulsa, SEPM (Society for Sedimentary Geology) Short Course No. 40, 130 p.

KOLATA, D. R., HUFF, W. D., AND FROST, J. K., 1983, Correlation of K-bentonites in the Decorah Subgroup of the Upper Mississippi Valley by chemical fingerprinting, *in* Delgado, D. J., ed., Ordovician Galena Group of the Upper Mississippi Valley—Deposition, diagenesis, and paleoecology: Guidebook for the 13th Annual Field Conference, p. F1–F15.

KOLATA, D. R., FROST, J. K., AND HUFF, W. D., 1986, K-bentonites of the Ordovician Decorah Subgroup, Upper Mississippi Valley: Correlation by chemical fingerprinting: Urbana, Illinois State Geological Survey Circular 537, 30 p.

KOLATA, D. R., FROST, J. K., AND HUFF, W. D., 1987, Chemical correlation of K-bentonite beds in the Middle Ordovician Decorah Subgroup, Upper Mississippi Valley, *in* Sloan, R. E., ed., Middle and Late Ordovician lithostratigraphy and biostratigraphy of the Upper Mississippi Valley: Minneapolis, Minnesota Geological Survey Report of Investigations 35, p. 52–56.

KOLATA, D. R., HUFF, W. D., AND BERGSTRÖM, S. M., 1996, Ordovician K-bentonites of eastern North America: Boulder, Geological Society of America Special Paper 313, 84 p.

LAPORTE, L. F., 1969, Recognition of a transgressive carbonate sequence with an epeiric sea: Helderverg Group (Lower Devonian) of New York State, *in* Friedman, G. M., ed., Depositional Environments in Carbonate Rocks: Tulsa, Society of Economic Paleontologists and Mineralogists Special Publication No. 14, p. 98–119.

LOUTIT, T. S., HARDENBOL, J., VAIL, P. R., AND BAUM, G. R., 1988, Condensed sections: The key to age determination and correlation of continental margin sequences, *in* Wilgus, C. K., Hastings, B. S., Posamentier, H. W., Van Wagoner, J., Ross, C. A., and Kendell, C. G. St. C., eds., Sea Level Changes—An Integrated Approach: Tulsa, Society of Economic Paleontologists and Mineralogists, Special Publication 42, p. 183–213.

LUDVIGSON, G. A., JACOBSON, S. R., WITZKE, B. J., AND GONZÁLEZ, L. A., 1996, Carbonate component chemostratigraphy and depositional history of the Ordovician Decorah Formation, Upper Mississippi Valley, *in* Witzke, B. J., Ludvigson, G. A., and Day, J., Paleozoic Sequence Stratigraphy, Views from the North American Craton: Boulder, Geological Society of America Special Paper 306, p. 67–86.

MITCHELL, R. W., 1985, Comparative sedimentology of shelf carbonates of the Middle Ordovician St. Paul Group, central Appalachians: Sedimentary Geology, v. 43, p. 1–41.

OSTROM, M. E., 1969, Champlainian Series (Middle Ordovician) in Wisconsin: American Association of Petroleum Geologists Bulletin, v. 53, p. 671–678.

SAYLOR, B. Z., CHOI, Y. S., AND SIMO, J. A., 1997, Waupun core, Platteville and Galena Formations, Fond du Lac County, Wisconsin, *in* Mudrey, M. G., ed., Guide to Field Trips in Wisconsin and Adjacent Areas of Minnesota, 31st Annual Meeting of North Central Section, Geological Society of America: Madison, Wisconsin State Geologic and Natural History Survey, p. 1–19.

SHAW, A. B., 1964, Time in Stratigraphy: New York, McGraw-Hill, 365 p.

SLOAN, R. E., 1987, Tectonics, biostratigraphy and lithostratigraphy of the Middle and Late Ordovician of the Upper Mississippi Valley, *in* Sloan, R. E., ed., Middle and Late Ordovician lithostratigraphy and biostratigraphy of the Upper Mississippi Valley: Minneapolis, Minnesota Geological Survey Report of Investigations 35, p. 7–20.

SLOSS, L. L., 1963, Sequences in the cratonic interior of North America: Geological Society of America Bulletin, v. 74, p. 93–114.

TEMPLETON, J. S., AND WILLMAN, H. B., 1963, Champlainian Series (Middle Ordovician) in Illinois: Urbana, Illinois State Geological Survey Bulletin 89, 260 p.

VAN DER VOO, R., 1988, Paleozoic paleogeography of North America, Gondwana, and intervening displaced terranes: Comparisons of paleomagnetism with paleoclimatology and biogeographical patterns: Geological Society of America Bulletin, v. 100, p. 311–324.

WILLMAN, H. B., AND KOLATA, D. R., 1978, The Galena and Galena Groups in northern Illinois: Urbana, Illinois State Geological Survey Circular 502, 75 p.

WILSON, J. L., 1975, Carbonate Facies in Geologic History: Berlin, Springer-Verlag, 471 p.

WITZKE, B. J., 1980, Middle and Upper Ordovician paleogeography of the region bordering the transcontinental arch, *in* Fouch, T. D. and Magathan, E. R., eds., Paleozoic paleogeography of the west-central United States: Denver, Rocky Mountain Section — Society of Economic Paleontologists and Mineralogists, p. 1–18.

WITZKE, B. J., AND BUNKER, B. J., 1996, Relative sea-level changes during Middle Ordovician through Mississippian deposition in the Iowa area, North American craton, *in* Witzke, B. J., Ludvigson, G. A., and Day, J., eds., Paleozoic Sequence Stratigraphy: Views from the North American Craton: Boulder, Colorado, Geological Society of America Special Paper 306, p. 307–330.

WITZKE, B. J., AND GLENISTER, B. F., 1987, The Ordovician sequence in the Guttenberg area, Iowa: Boulder, Geological Society of America, Centennial Field Guide—North-Central Section, p. 93–96.

WITZKE, B. J., AND KOLATA, D. R., 1988, Changing structural and depositional patterns, Ordovician Champlainian and Cincinnatian Series of Iowa-Illinois, *in* Ludvigson, G. A. and Bunker, B. J., eds., New Perspectives on the Paleozoic History of the Upper Mississippi Valley—An Examination of the Plum River Fault Zone: Tulsa, Society of Economic Paleontologists and Mineralogists, Guidebook for the 18th Annual Field Conference of the Great Lakes Section, p. 55–77.

STRATIGRAPHIC ARCHITECTURE OF A LOWER CRETACEOUS–LOWER TERTIARY CARBONATE BASE-OF-SLOPE SUCCESSION: GRAN SASSO D'ITALIA (CENTRAL APENNINES, ITALY)

JAN-HENK VAN KONIJNENBURG*, DANIEL BERNOULLI, AND MARIA MUTTI*

Geological Institute, ETH-Zentrum, CH-8092 Zürich, Switzerland

ABSTRACT: Carbonate slopes are key areas for correlating shelf and basinal sediments. They provide important clues to the problem of correlation of sequential units on platforms with unconformity-bounded and conformable units on the slope and in the deep basin. This study aims to define packages of genetically related sedimentary units in a base-of-slope succession and to compare them to second-order sequences defined on a nearby carbonate-platform margin.

The Lower Cretaceous to Oligocene pelagic and redeposited carbonate sediments, now lithified and exposed in the Gran Sasso d'Italia area, were deposited along the base of slope separating the Lazio-Abruzzi carbonate platform to the southwest and the Maiella platform margin of the Apulia carbonate platform to the southeast from the Umbria-Marche basin to the north. Three major facies types can be distinguished: (1) breccias and megabreccias, containing a large variety of platform- and slope-derived lithoclasts, originating from the collapse of platform-margin areas, and minor skeletal fragments; (2) turbiditic packstone and grainstone beds, with mainly skeletal fragments and minor amounts of smaller lithoclasts shed from the nearby platforms; and (3) pelagic foraminiferal lime mudstones and wackestones. The interplay of the different sedimentary processes caused changes in facies associations, not only downslope but also along depositional strike. Although slope deposits were reworked, most of the redeposited material was derived from the adjacent platforms and, therefore, provides additional evidence for platform evolution.

The sedimentary succession of the Gran Sasso area is subdivided into six formations, separated from each other by unconformities or rapid vertical changes in facies associations. Each of these formations is characterized by a distinct depositional system, which changed drastically across formational boundaries. These changes in depositional system were largely controlled by changes in sedimentary regime that took place on the adjacent platforms and that were related to second-order (10–20 m.y.) platform cycles. The platform cycles were in turn controlled by relative changes in sea level, which in the case of the middle Cretaceous unconformity were related to tectonic events. A comparison of geohistory (i.e., depth vs. time) curves of the Maiella platform-margin, the Gran Sasso base of slope, and the Umbria-Marche basin deposits indicates that platform evolution also appears to have controlled the amount of sediment that accumulated on the base of slope. Along the base of slope, higher sedimentation rates during times of massive redeposition were compensated by periods of submarine erosion and nondeposition, resulting in bulk sedimentation rates comparable to those in the Umbria-Marche basin where pelagic sedimentation was essentially continuous.

INTRODUCTION

In recent years, sequence stratigraphic concepts (Vail et al., 1977; Posamentier and Vail, 1988) have been applied to carbonate successions, both in outcrop and reflection seismic studies (Sarg, 1988; Handford and Loucks, 1993). Although the basic concepts are the same, there are important differences between carbonate systems, in which most if not all of the sediment is produced within the system, and siliciclastic systems in which sediment is ultimately derived from a continent. As pointed out by Schlager (1993), changes in sediment supply induced by environmental changes other than sea level may generate sequence boundaries (e.g., drowning unconformities) and influence the internal stacking pattern of sequences. In particular, the mechanisms responsible for the generation of surfaces bounding carbonate depositional sequences and the way in which these surfaces are expressed in reflection seismic profiles, as well as in outcrop, still remain under debate (Schlager, 1989, 1993; Everts et al., 1995; Mutti et al., 1996).

Carbonate slope and base-of-slope successions form an integral part of the platform-to-basin transect and are therefore a key component for basin-wide sequence stratigraphic interpretations. Many studies have been conducted with respect to sedimentary processes on modern carbonate slopes in the Bahamas (Grammer and Ginsburg, 1992; Grammer et al., 1993) and the Florida Straits (Mullins et al., 1984, 1986; Mullins and Hine, 1989), and in the subsurface of these areas (Mullins and Cook, 1986; Harwood and Towers, 1988; Mullins et al., 1988; Eberli and Ginsburg, 1989). Furthermore, various studies have documented the sedimentology and stratigraphy of ancient carbonate slope deposits in outcrop (Cook and Mullins, 1983; Enos and Moore, 1983; Coniglio, 1986; Colacicchi and Baldanza, 1986; Watts and Garrison, 1986; Kenter, 1990; Watts and Blome, 1990; Enos and Stephens, 1993; Colacicchi and Monaco, 1994). However, so far, relatively few studies have been published that apply sequence stratigraphy to outcrop studies of ancient carbonate slope and base-of-slope successions (Bosellini et al., 1993a; Eberli et al., 1993; Handford and Loucks, 1993; Hunt and Tucker, 1993; Tucker et al., 1993; Everts et al., 1995; Vecsei et al., 1998).

In the Gran Sasso d'Italia area, large seismic-scale outcrops offer an excellent opportunity to study Lower Cretaceous to Oligocene base-of-slope deposits in an almost three-dimensional framework. The succession was deposited on the slope and base-of-slope adjacent to Bahamian-type carbonate platforms to the southwest (Lazio-Abruzzi) and southeast (Apulia) and merges with the pelagic successions of the Umbria-Marche basin to the north (Fig. 1). This particular situation has allowed us to study the stratigraphic architecture of these deposits and to contrast it to the sequence architecture identified in the shallow-water and upper-slope areas investigated by previous studies (Eberli et al., 1993; Mutti et al., 1996; Vecsei et al., 1998). Our study aims to (1) determine the stratigraphy, sedimentology, and depositional geome-

*Present addresses: (van Konijnenburg) SIEP-RTS, Volmerlaan 8, P.O. Box 60, NL-2280 AB Rijswijk, The Netherlands; (Mutti) Department of Earth Sciences, University of Southern California, Los Angeles, California 90089-0740

Advances in Carbonate Sequence Stratigraphy: Application to Reservoirs, Outcrops and Models, SEPM Special Publication No. 63

tries of the carbonate slope deposits through time; (2) determine how sequences and sequence boundaries defined on the adjacent carbonate platforms are expressed in the slope deposits; and (3) evaluate how changes in the sedimentary regime of the adjacent carbonate platforms influenced deposition on the base of slope and over what time scale these changes took place. To this end, we compare the lithostratigraphic units distinguished in the Gran Sasso area with the second-order sequences (supersequences) defined on the adjacent platforms. Comparison with the adjacent Lazio-Abruzzi platform (Praturlon and Sirna, 1976; Accordi and Carbone, 1988) is somewhat hampered by the poor outcrop conditions that hardly allow the reconstruction of depositional geometries; however, in the magnificently exposed Montagna della Maiella, the sequence stratigraphic framework of a nearby platform margin has been well established (Eberli et al., 1993; Mutti et al., 1996; Vecsei et al., 1998). Likewise, the stratigraphic and sedimentary evolution of the Umbria-Marche basin to the north is well known and allows for comparison with a deep pelagic setting (Arthur et al., 1977; Montanari et al., 1989).

GEOLOGIC FRAMEWORK

The Gran Sasso area is part of the east-vergent late Tertiary fold-and-thrust belt of the external Apennines. This fold-and-thrust belt was detached along Upper Triassic evaporites (Burano Anhydrite) and gradually thrust onto the upper Miocene to Quaternary foreland deposits, which are also involved in thrusts in marginal areas (Ghisetti and Vezzani, 1983, 1991; Bally et al., 1986). Prior to the Alpine orogeny, the area was part of the southern continental margin of the Alpine segment of the Tethys ocean. This southern continental margin of Tethys—Adria—extended over what are now the external fold belts of the Hellenides, Dinarides, Southern Alps, and Apennines and their Adriatic foreland. Where the basement of these units is exposed or has been drilled, the Mesozoic limestone sequences of Adria are underlain by upper Paleozoic and Triassic clastic rocks and evaporites and continental crust of Variscan age. Flooding by shallow seas in the Late Triassic built up a carbonate megabank extending over much of the central Mediterranean area (Hauptdolomit–Dolomia Principale–Dachstein megabank). In the central and southern Apennines, this megabank persisted until the early Liassic (Calcare Massiccio Formation), when it disintegrated during rifting and a number of isolated, Bahamian-type, carbonate platforms—the so-called peri-Adriatic platforms—came into existence (Bernoulli et al., 1979). A strong similarity exists between these platforms and the still active ones of the Bahamas, not only in terms of facies, size, shape, time-space relationships, and subsidence rates (Bernoulli, 1972; D'Argenio et al., 1975), but also in terms of internal architecture (Eberli and Ginsburg, 1989).

The long-lived peri-Adriatic platforms were separated from each other by deeper basins, comparable to the present-day deep troughs of the Bahamian archipelago. These troughs were remote from terrigenous input and thus were the site of pelagic deposition, while the platform slopes accumulated turbiditic and other gravitational carbonates. In most places, the platform margins were the site of tectonic decoupling during Alpine thrusting; however, where the platform margins are preserved, the platforms were bordered by steep, nondepositional escarpments (de Alteriis and Aiello, 1993; Bosellini et al., 1993a; Eberli et al., 1993) that were inherited from Liassic fault scarps (cf. Bice and Stewart, 1990) and that were subsequently shaped by gravitational collapse and submarine erosion (Bosellini et al., 1993a; Eberli et al., 1993). Windward margins to the northeast remained steep and in a sedimentary bypass mode (cf. de Alteriis and Aiello, 1993), whereas leeward or oblique margins locally prograded over the slope sediments shed from the margin (Mutti et al., 1996). This situation is conspicuously similar to the leeward progradation of Andros Bank and Bimini Bank in the Bahamas described by Eberli and Ginsburg (1989; cf. Fig. 3 in Bernoulli et al., 1996).

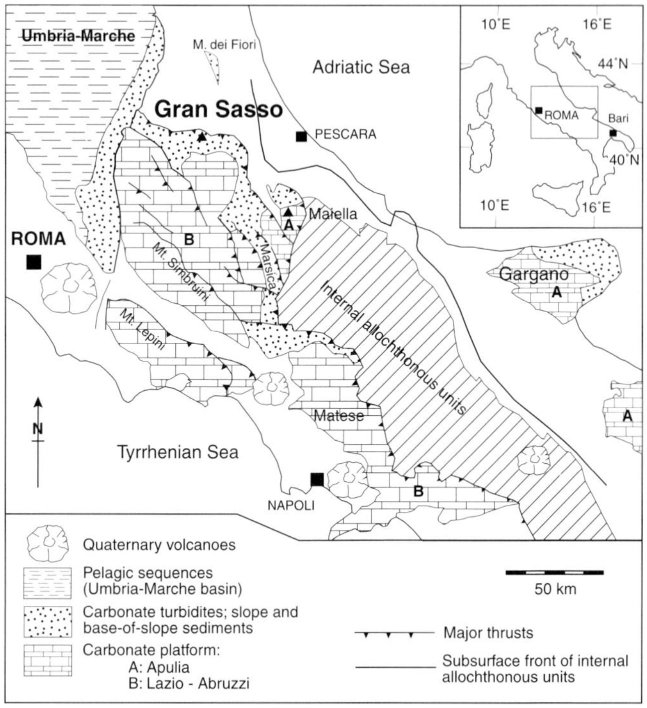

FIG. 1.—Location of the Gran Sasso d'Italia and its present-day position with respect to (A) Apulia and (B) Lazio-Abruzzi carbonate platforms to the southeast and southwest, respectively, and the Umbria-Marche basin to the north. The Lazio-Abruzzi platform and its slope in the Gran Sasso area were thrust onto basinal Mesozoic–Tertiary sequences and Tertiary flysch, whereas the Apulia platform was only marginally involved in Tertiary thrusting in the Montagna della Maiella area. Facies distribution refers to the Upper Cretaceous strata (after Eberli et al., 1993).

The pelagic and redeposited limestones of the Gran Sasso area were deposited along the base of slope adjacent to one of the peri-Adriatic platforms—the Lazio-Abruzzi platform—and not far from the margin of the southerly Apulia platform. This margin is magnificently exposed in Montagna della Maiella (Fig. 1; referred to as the Maiella platform margin) and will serve for comparison. In the Gran Sasso area, disintegration and drowning of the Early Jurassic Calcare Massiccio platform are recorded by deposition of the pelagic and turbiditic limestones of the Corniola Formation on top of the Calcare Massiccio Formation (Bernoulli, 1967; Bernoulli et al., 1979). The

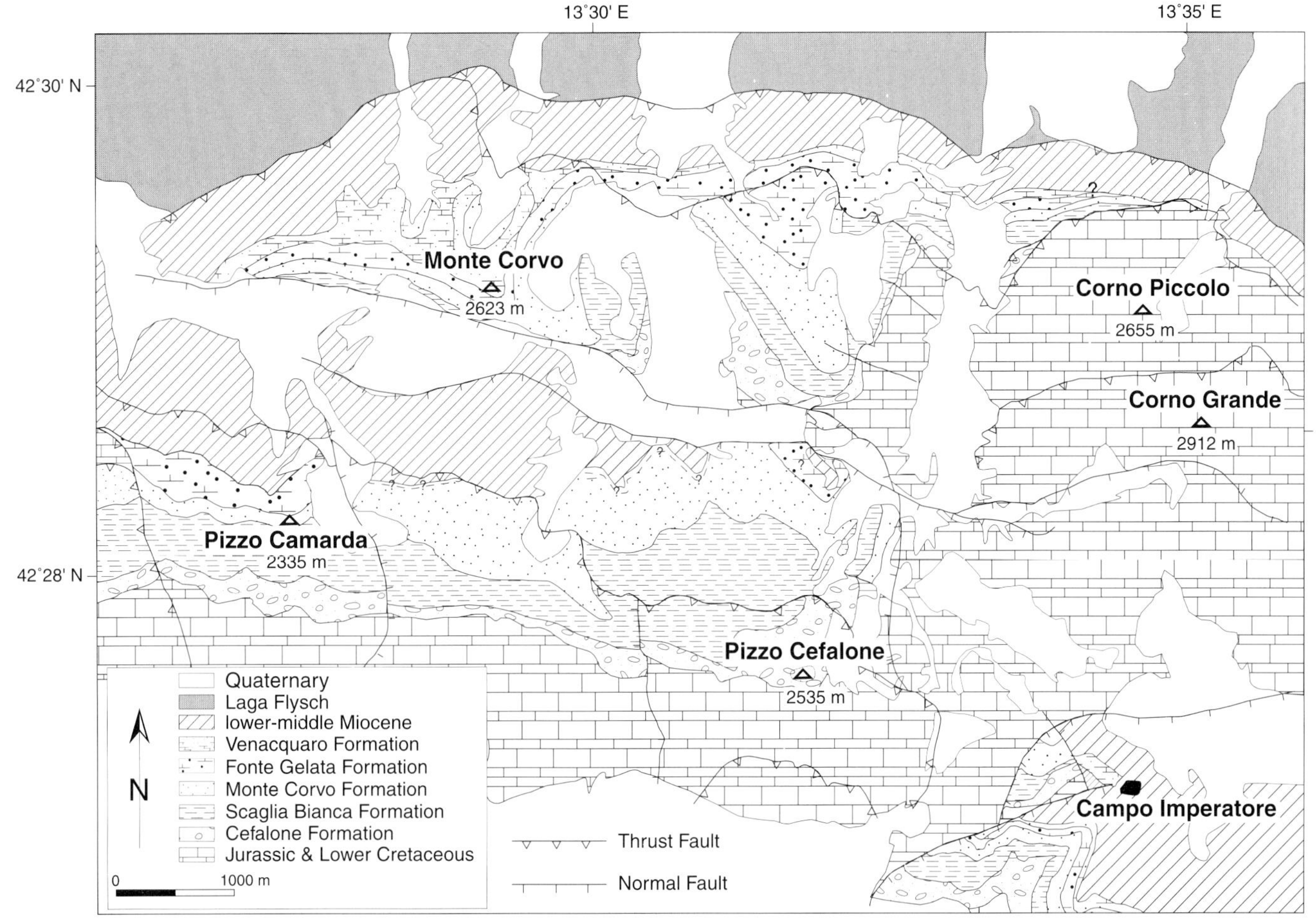

FIG. 2.—Geologic map of the central Gran Sasso d'Italia. Modified after Ghisetti and Vezzani (1990a, b, 1991).

Corniola Formation also marks the beginning of basinal (and seamount) deposition in the Umbria-Marche basin to the north (Santantonio, 1993; Fig. 1). During its Jurassic to Oligocene history, the Gran Sasso area was the site of redeposition from the adjacent platform(s), culminating in the deposition of huge megabreccias during the Late Jurassic (Bernoulli et al., 1979), the middle Cretaceous, and the early Tertiary (this paper).

Although the stratigraphy of the Lower Cretaceous to Oligocene succession in the study area is by now relatively well known (Zamparelli, 1966; Crescenti et al., 1969; Dela Pierre and Bruzzone, 1991; Dela Pierre, 1992; Dela Pierre and Clari, 1994), correlation with the adjacent platform deposits and interpretation in a sequence stratigraphic context have never been established. This situation is due to lack of preservation of the northern margin of the Lazio-Abruzzi platform. However, it appears that the history of the Lazio-Abruzzi platform margin largely parallels that of the Apulia margin that crops out in Montagna della Maiella to the south, and we assume that both platforms underwent the same type of thermal subsidence and sea-level changes after Early Jurassic rifting and prior to the onset of Alpine orogeny. We therefore compare the history of Gran Sasso with the well-documented sequence stratigraphic history of the Maiella-Apulia platform margin (Accarie et al., 1986; Accarie, 1988; Vecsei, 1991; Eberli et al., 1993; Sanders, 1994; Mutti, 1995; Mutti et al., 1996; Vecsei et al., 1998). However, no physical continuation between the platforms and the base-of-slope deposits of the Gran Sasso area is preserved.

The Upper Triassic to Miocene deposits of the Gran Sasso d'Italia are stacked in a set of thrust sheets that form one of the frontal thrust zones of the central Apennines (Bally et al., 1986; Ghisetti and Vezzani, 1990a, 1990b, 1991; Figs. 1, 2). After deposition of deeper-marine limestones and marlstones in the early Miocene to Tortonian, the onset of thrusting in the Gran Sasso area was heralded by the Messinian siliciclastic deposits of the Laga Flysch (Bally et al., 1986; Ghisetti and Vezzani, 1991; Ghisetti et al., 1993). Thrusting in the Gran Sasso mainly occurred between Messinian and middle Pliocene time (Ghisetti and Vezzani, 1991). Subsequent erosion of this stack of thrust sheets exposed spectacular kilometer-wide outcrops of the base-of-slope deposits. Although the original geometry of the sedimentary pile was disrupted by the thrusts, the position of the profiles relative to each other was not fundamentally altered (Figs. 2, 3).

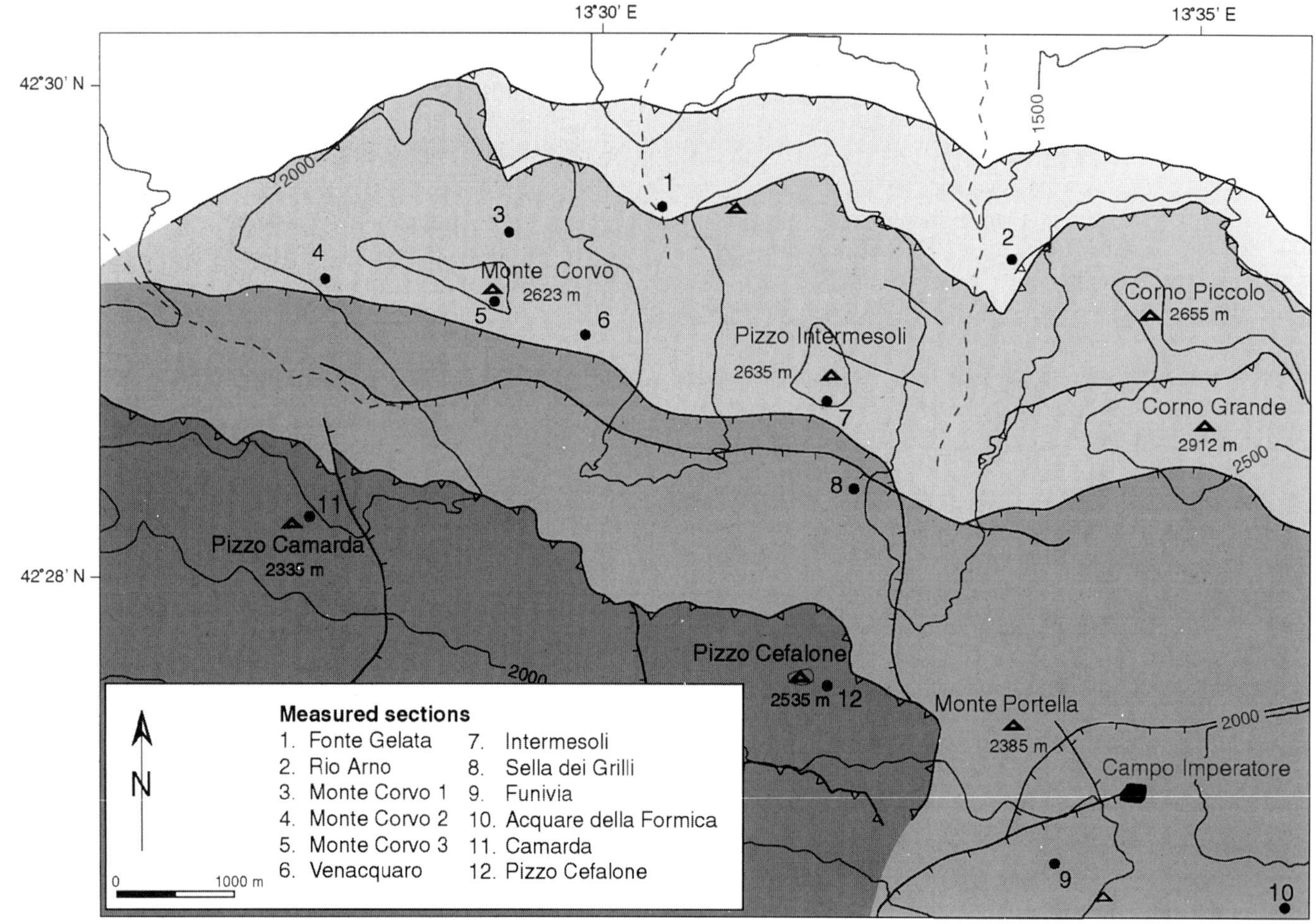

FIG. 3.—Simplified tectonic map of the central Gran Sasso d'Italia and the locations of the measured sections in Figure 4. The various shaded areas are the most important fault panels within the stack of thrust sheets. Modified after Ghisetti and Vezzani (1990a, b, 1991).

METHODS AND DEFINITIONS

Outcrops of the base-of-slope deposits in the Gran Sasso d'Italia reach several kilometers in length and up to 1 km in height and, thus, allow for the examination of large-scale depositional geometries (cf. Figs. 6, 8, 9, 14). These geometries are comparable to those observed in reflection seismic sections and were analyzed in an analogous way on panoramic photomosaics of mountain sides. By this procedure, the successions were subdivided into genetically related sedimentary packages bounded by unconformities or conformable surfaces according to the method described by Mitchum et al. (1977) and Schlager (1992). Twelve measured stratigraphic sections (Figs. 3, 4) served to calibrate the distinguished units with respect to facies and biostratigraphy.

In recent years, considerable debate has been raised over the term *depositional sequence* and the mechanisms responsible for the formation of depositional sequences and of their bounding surfaces. Vail et al. (1977) initially defined a sequence as "a stratigraphic unit composed of genetically related strata bounded at its top and base by unconformities or their correlative surfaces." In their concept, the sequence-bounding unconformities are related to changes in sea level. Posamentier et al. (1988, Table 1, p. 110) interpreted a depositional sequence "to be deposited between *eustatic* fall inflection points." *If* unconformities are related to relative changes in sea level, then a depositional sequence in a given area can be assigned to a chronostratigraphic interval—i.e., a sechron—that is defined by the age of the conformities correlative to the sequence boundaries (Vail et al., 1977). However, as Schlager (1992) pointed out, some unconformities identified in seismic sections may simply represent changes in sediment input and dispersal in the basin. On isolated carbonate platforms, subaerial exposure and down-stepping of shelf breaks are therefore the only unambiguous criteria for recognizing relative sea-level cycles as opposed to sediment-supply cycles (Schlager, 1993; Mutti et al., 1996). In addition, unconformities in deep water may be unrelated to changes in sea level (cf. Austin et al., 1988, for the example of the Blake Plateau), and in these cases unconformity-bounded packages may not correlate chronologically with sequences defined on the adjacent shelf or platform.

However, the extent of resedimented material derived from the nearby carbonate-platform margins suggests that the sedimentary evolution and, thus, the stratigraphic architecture of the base-of-slope deposits are, at least to some extent, determined by the evolution of the platforms. One of our goals is therefore to compare the chronological extent of unconformity-bounded packages of the base-of-slope section with (second-order) sequences defined along and on the adjacent platform margins in order to evaluate the influence of platform evolution on the sedimentary evolution of the base-of-slope deposits. In the following, we shall distinguish lithostratigraphic units that are separated from each other by regional unconformities associated with a depositional hiatus and/or by rapid vertical changes in facies associations, reflecting a manifest change in sediment input and dispersal. In the study area, the boundaries of these units coincide with formation boundaries defined either by previous authors or during this study. The geometric relationships between the distinguished formations and their chronostratigraphy are shown in Figures 4 and 5. These units will then be compared with the second-order depositional sequences distinguished along the Maiella platform margin (Eberli et al., 1993; Vecsei et al., 1998). Within the distinguished formations are minor unconformities; however, they appear to have been caused by local events and cannot be followed across the Gran Sasso area, let alone correlated to third-order sequence boundaries recognized on the platform or its upper slope (Vecsei, 1991; Mutti et al., 1996).

Since no physical connection between the platform sequences and the base-of-slope deposits exists, our comparisons are largely based on biostratigraphic correlation and facies analysis. Biostratigraphic correlation also provides a link to the strata of the Umbria-Marche basin to the north (Arthur et al., 1977). Biostratigraphy of the base-of-slope sedimentary units is dominantly based on planktic organisms in pelagic limestones, determined in thin sections (M. Caron, pers. commun., 1996; van Konijnenburg et al., 1998). Larger benthic foraminifera present in redeposited limestones are of limited use; however, they provide oldest possible ages in the absence of autochthonous fossils. Our biozone subdivision (Fig. 5) follows, with modifications, the zonal scheme for the Gran Sasso by Dela Pierre (1992) and Dela Pierre and Clari (1994), which is correlated with the chronostratigraphic scheme of Haq et al. (1988).

LITHOSTRATIGRAPHY AND SEDIMENTARY EVOLUTION OF THE GRAN SASSO BASE-OF-SLOPE DEPOSITS

Facies types

The base-of-slope deposits of Gran Sasso are characterized by three major facies types: (1) breccias and megabreccias, with a large variety of platform- and slope-derived lithoclasts and minor skeletal fragments; (2) turbidite beds, with mainly shallow-water skeletal fragments and minor amounts of smaller lithoclasts; and (3) pelagic-periplatform limestones, i.e., foraminiferal lime mudstones and wackestones. Within these pelagic deposits, winnowing by contour currents shaped residual deposits consisting of a lag of planktic foraminifera. These individual facies types form different facies associations.

Breccias and turbidite beds generally have a limited lateral extent, and individual beds often cannot be followed from one section to the other, despite continuity of the outcrops. Facies associations change not only downslope, but also laterally along depositional strike. Slope erosion is indicated by relief along the base of the gravity-driven mass flows that truncates the underlying beds and by slumping in the pelagic sedimentary units.

Maiolica Formation (Tithonian to Upper Barremian)

Upper Tithonian to upper Barremian pelagic limestones form the base of the succession considered in this paper. These rocks are white to light gray lime mudstones and wackestones consisting of beds a few to 10s of centimeters thick; the limestones were derived from the lithification of coccolith oozes (Bernoulli, 1972) and include lenses and bands of diagenetic replacement chert. Remnants of radiolarians occur as calcite-filled molds, and in the lower part of the formation, calpionellids are common. Deposits of this facies are typically found in the Lower Cretaceous section along a wide tropical belt extending from the Caribbean across the central Atlantic and through the Mesozoic Tethys as far east as Oman (Bernoulli, 1972; Bernoulli and Jenkyns, 1974). In the Gran Sasso area, the Maiolica Formation lies conformably on Upper Jurassic turbiditic bioclastic calcarenites of the Calcare ad Entrochi Formation (Adamoli et al., 1982).

The limestones of the upper part of the Maiolica Formation contain thin (<30 cm), discontinuous (1–5 m wide) beds of pebbly mudstones composed of lime mudstone pebbles similar to the surrounding pelagic limestones. Less commonly, 5–10-m-thick, sheetlike, clast-supported lithoclastic breccias with a limited lateral extent (50–500 m wide) and abrupt lateral terminations are intercalated with the pelagic lime mudstones (Fig. 6). These breccias contain cemented, platform-derived lithoclasts and slope-derived pelagic mudstone clasts, as well as rudist and other mollusk fragments.

The boundary between the Maiolica Formation and the overlying Cefalone Formation is an erosional unconformity. Pelagic limestones of the Maiolica Formation are truncated by the breccia beds of the Cefalone Formation, the deposition of which eroded as deep as 4 m into the underlying limestones (Figs. 7, 8). The uppermost preserved beds in the Maiolica Formation are late Barremian in age: these beds contain nannoconids typical of the late Barremian "nannoconid bloom" including *Nannoconus bucheri*, *N. globulus*, *N. steinmanni*, and *N. steinmanni minor* (K. von Salis Perch-Nielsen, pers. commun., 1996). Although an earliest Aptian age cannot entirely be excluded, this association suggests a late Barremian age, prior to the "nannoconid crisis" of the early Aptian (Perch-Nielsen, 1985). As the positive $\delta^{13}C$ excursion typically observed in lower Aptian deposits (Weissert and Lini, 1991) was not encountered in our sections (Pizzo Cefalone, Rio Arno, van Konijnenburg, 1997), a late Barremian age for the top of the Maiolica Formation is suggested.

Cefalone Formation (Middle Albian to Upper Cenomanian)

The Cefalone Formation consists of redeposited beds, mainly lithoclastic breccias and turbiditic lithoclastic to bioclastic packstones, interlayered with discontinuous pelagic (marly) lime wackestones. The formation (Figs. 2, 4, 5) was previously described by Crescenti (1969) and Crescenti et al. (1969) as the lower part of their Monte Acquaviva Formation and was correlated chronologically with the Aptian–Albian Marne ad Fucoidi of the Umbria-Marche basin (Bernoulli, 1967).

In the central part of the area, coarse-grained lithoclastic

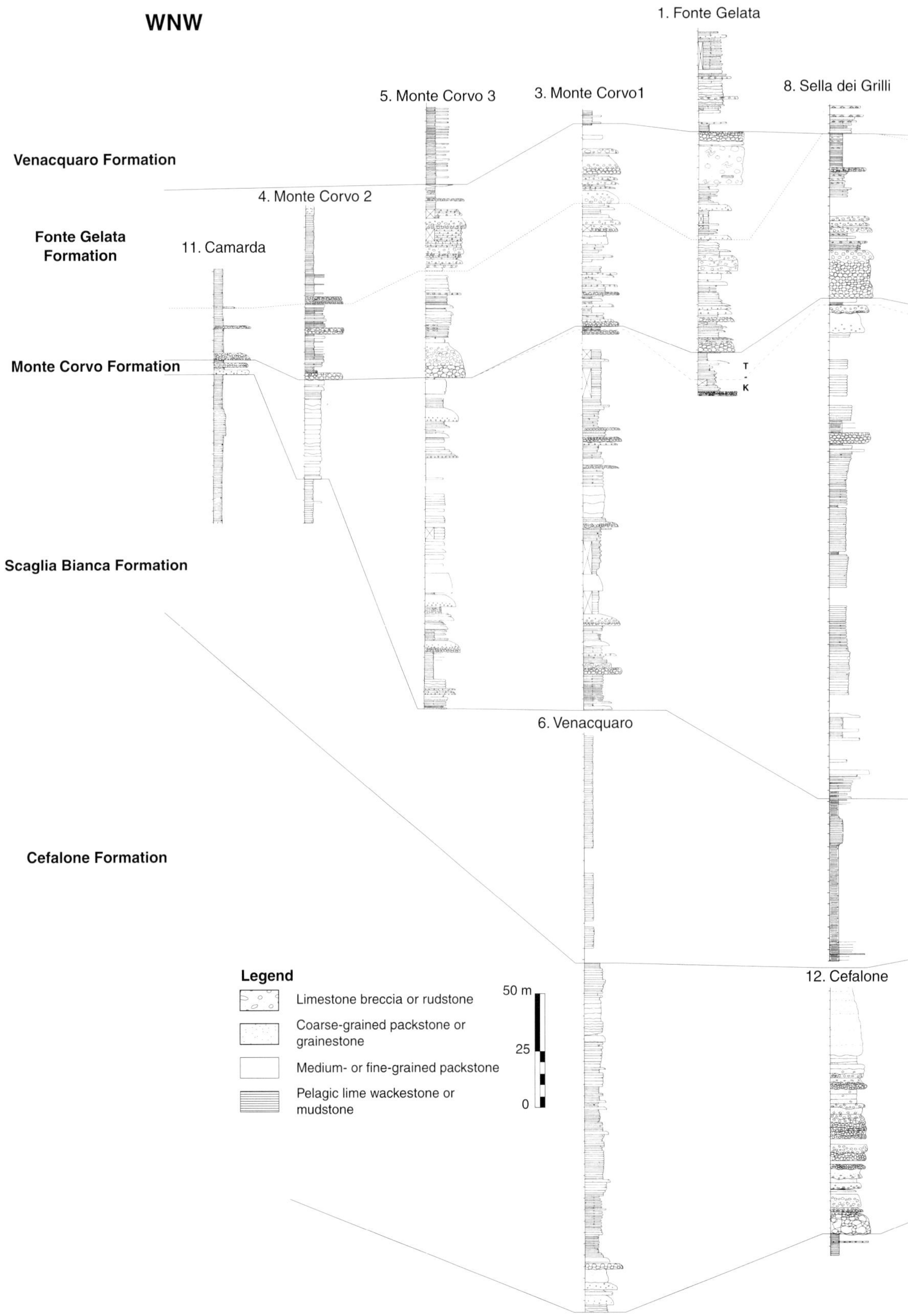

FIG. 4.—Correlation of the measured stratigraphic sections (section locations are shown in Fig. 3). Solid lines represent formation boundaries; the dashed line shows the Cretaceous/Tertiary boundary where it is observed. The dotted line represents the hiatus within the Fonte Gelata Formation.

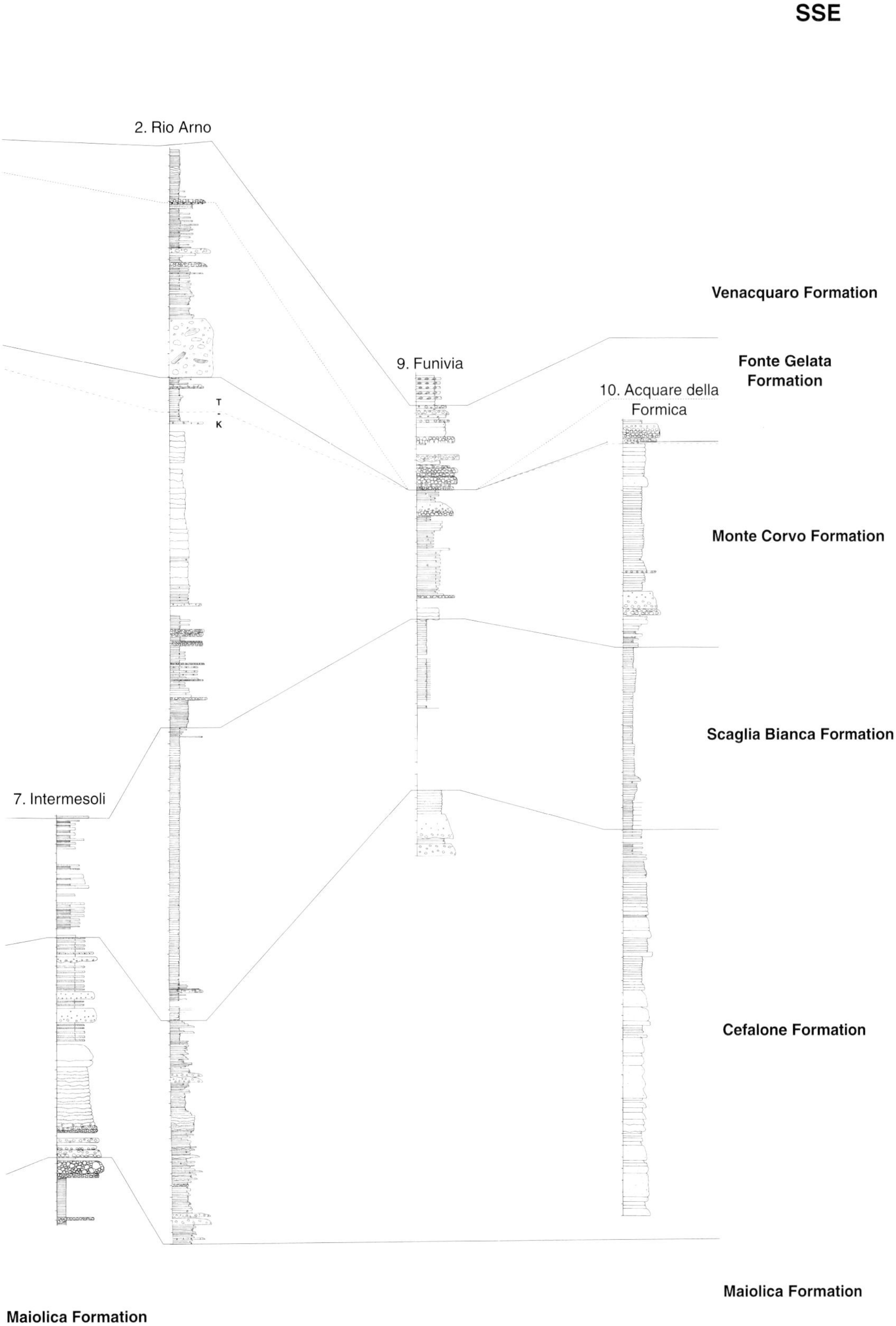

FIG. 4.—Continued.

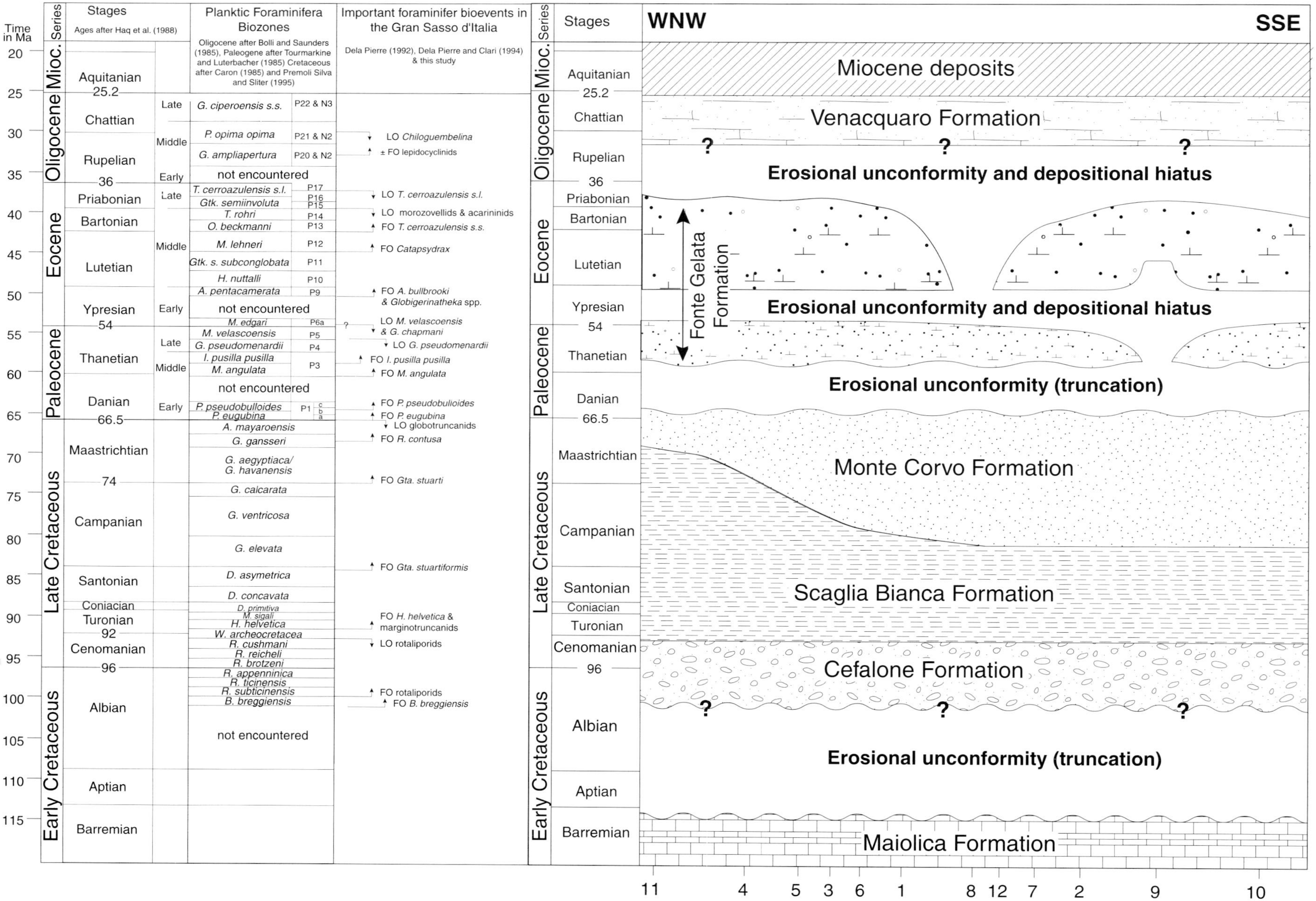

FIG. 5.—Stratigraphic correlation diagram, Albian to Chattian (left), and chronostratigraphic diagram of the Gran Sasso base of slope (right). Chronology and numerical ages are after Haq et al. (1988). Horizontal axis in the chronostratigraphic diagram shows lateral variations from west-northwest to south-southeast, based on the sections in Figure 4. The relative positions of sections 1–12 (for section names, see Figs. 3 and 4) are indicated on the lower horizontal axis. FO = first occurrences, LO = last occurrences of taxa.

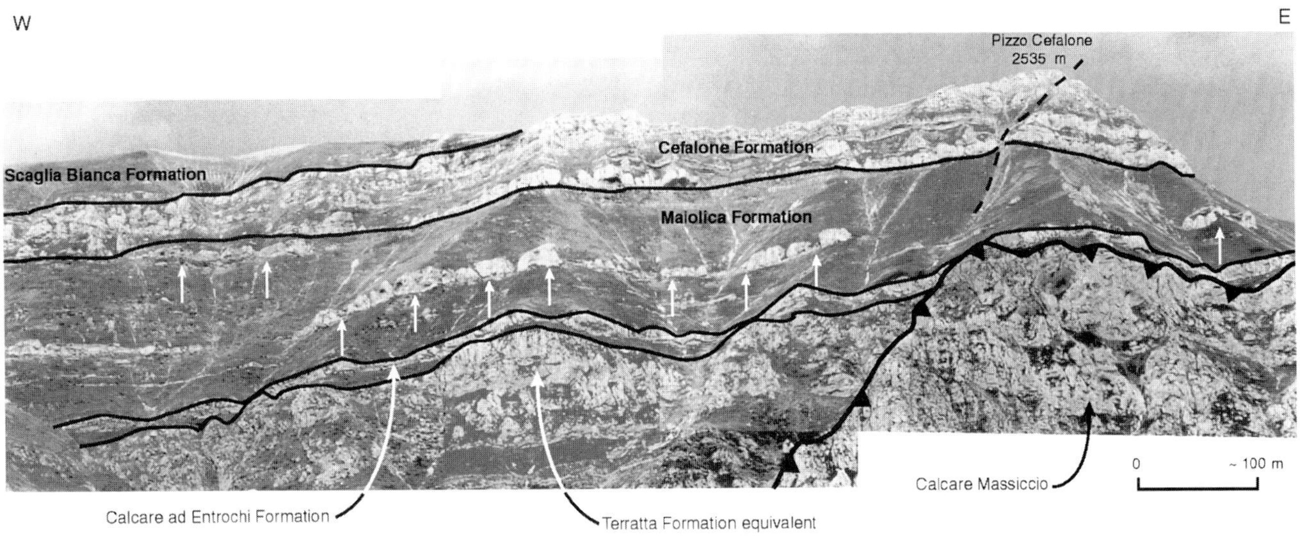

FIG. 6.—Panoramic view of the southern flank of Pizzo Cefalone (location shown in Fig. 2). The massive beds in the lower part of the section are Upper Jurassic megabreccias that are time equivalent to the Terratta Formation, interbedded with and overlain by thinner-bedded, backweathering bioclastic (crinoidal) turbidites (Calcare ad Entrochi Formation, Upper Jurassic) and the pelagic limestones of the Maiolica Formation (upper Tithonian–upper Barremian). Lenticular bodies of lithoclastic breccias in the Maiolica Formation are indicated by white arrows. Also note the erosive nature of the contact along the lower boundary of the first breccia bed of the middle Cretaceous Cefalone Formation.

breccias and very coarse grained lithoclastic to bioclastic packstones dominate the lower 30–40 m of the formation; laterally to the west and east and distally to the north, they become thinner and less common (Figs. 4, 9). Although the individual breccia layers are as thick as 15 m, their lateral extent is limited (≤700 m), and correlation of a single breccia bed from one section to another is often impossible (Figs. 6, 9). The breccia beds are dominantly clast supported, but relatively matrix rich; they have sheetlike bed geometries, and their bases are either flat or undulating, but always erosive. Their internal fabric is usually chaotic, but flow banding is observed in some beds. Usually the top 0.5–1.5 m of a bed is graded (Bouma T_a or $T_{a\text{-}b}$ interval). Matrix-supported breccias are thinner and finer grained, and their deposition did not visibly erode into the underlying strata. The breccias contain a variety of platform-derived lithoclasts, including coral boundstones, bioclastic grainstones, cemented micritized fragments, and isolated bioclasts. A 2-m-long block containing part of a rudist biostrome was observed in one of the breccia beds. Interlayered with the breccias are graded, 1–3-m-thick bioclastic packstones with both sheetlike and broad (>1000 m wide) channelized-bed geometries. Bouma T_a and T_b intervals are commonly observed in these packstones, which contain bioclasts and minor amounts of small lithoclasts (<2 cm). Isolated bioclasts in the redeposited beds include fragments of rudists, echinoderms, corals, bryozoans, *Bacinella*, and orbitolinids.

Between 30 and 40 m above the base of the formation, the breccia beds gradually give way to graded, commonly amalgamated bioclastic packstones, several meters thick (Fig. 10). Such amalgamation, together with the observed grading and Bouma intervals, suggests that these packstones were deposited by turbidity currents. These turbidites show fining-upward (from very coarse to medium- or fine-grained arenite) and thinning-upward (several meters to ≤1 m) trend in the top 30 m of the section (Fig. 7), which contains increasing numbers of thin pelagic interbeds. This trend is accompanied by an increase in bioclastic material with respect to lithoclasts.

The hiatus represented by the unconformity at the base of the Cefalone Formation is bracketed by the late Barremian age of the topmost strata of the Maiolica Formation and the age of the overlying breccia beds. The age of the lowermost redeposited lithologies is weakly constrained. Displaced fossils (*Palorbitolina lenticularis*, *Orbitolina* sp., and *Bacinella*) in the lower 10–20 m of the Cefalone Formation indicate a maximum age of early Aptian; however, as most of them occur within lithic fragments showing the imprint of meteoric diagenesis, we consider them to be reworked from older strata. The oldest planktic foraminifera, found approximately 10–15 m above the base of the formation, belong to the late Albian *Biticinella breggiensis–Rotalipora subticinensis* biozone. We therefore consider the unconformity to span the late Barremian to middle Albian. In view of the high sedimentation rates of the megabreccias and breccias dominating the lower part of the Cefalone Formation, we think that this age assignment is reasonable.

The depositional system clearly changed in the middle Cretaceous. Large amounts of displaced, platform-derived carbonate debris entered the Gran Sasso base-of-slope area. The observed lateral facies change from breccias in the center to turbiditic packstones in the east, west, and north could suggest the former existence of a broad feeder channel through which the debris flows that deposited the breccias traveled. During the Cenomanian, dominantly bioclastic turbidites were deposited on the base of slope. The disappearance of these bioclastic turbidites marks the boundary with the overlying Scaglia Bianca Formation.

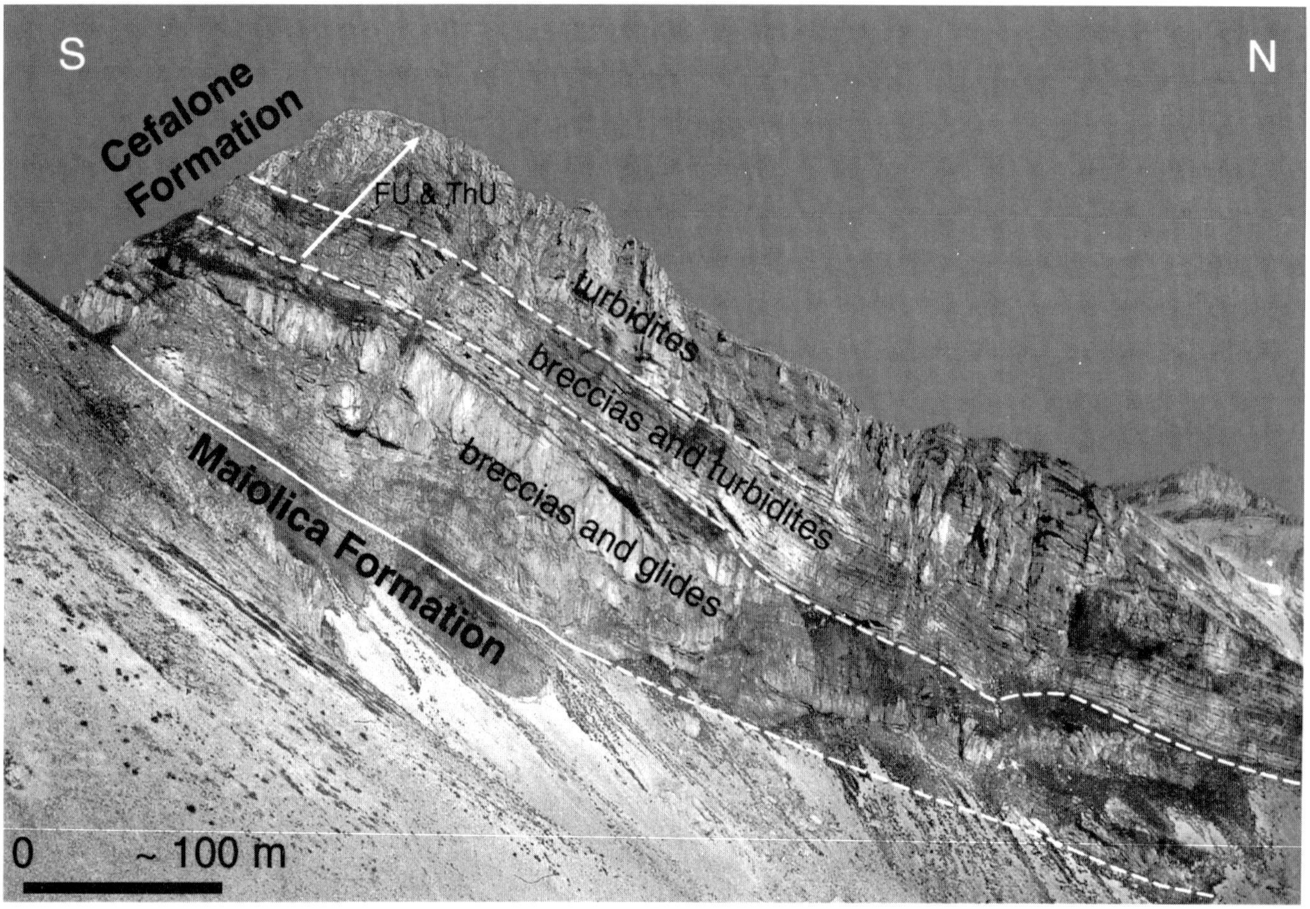

FIG. 7.—View of the eastern side of Pizzo Cefalone. Note the unconformity at the base of the Cefalone Formation, the large sediment glides and the penecontemporaneous deformation of the debris-flow deposits (basal breccias) in the lower part of the formation, and the well-bedded turbidites in the upper part. A thickness of approximately 120 m has been measured for the Cefalone Formation in this outcrop. FU and ThU = fining- and thinning-upward calcirudites and calcarenites.

FIG. 8.—Panoramic view of the profile from Pizzo Cefalone to Sella dei Grilli (sections 8 and 12 [shown in Fig. 4] are labeled), as seen from the east. The contact between the Cefalone Formation (middle Albian–upper Cenomanian) and the Scaglia Bianca Formation (here uppermost Cenomanian–lower Campanian) is a conformable but rapid transition. Likewise, the pelagic limestones of the Scaglia Bianca Formation are conformably overlain by the calcareous turbidites of the Monte Corvo Formation (here middle Campanian–lowermost Paleocene). The Fonte Gelata Formation (here upper Paleocene) overlies the Monte Corvo unconformably and has a 25-m-thick breccia at its base (see Fig. 17). The Oligocene Venacquaro Formation overlies the Fonte Gelata Formation unconformably; no Eocene strata are present in this section.

FIG. 9.—Panoramic view of the eastern side of the Rio Arno valley (section 2; see Fig. 4). The breccia beds in the Cefalone Formation are distinctly thinner, finer, and less common than in the more proximal Cefalone section (see Figs. 7, 8). The Scaglia Bianca Formation consists of well-bedded pelagic limestones. In the Monte Corvo Formation, an overall coarsening- and thickening-upward trend is visible, reflecting the progradation of a turbidite fan. The top of the Monte Corvo Formation, above the Cretaceous/Tertiary boundary, consists of thin-bedded pelagic limestones, which are unconformably overlain by the Fonte Gelata Formation. The measured section runs along the base of the cliffs. F = thrust fault.

Scaglia Bianca Formation (Uppermost Cenomanian to Lower Campanian–Middle Maastrichtian)

At the base of the Scaglia Bianca Formation, there is a gradational transition (within 10 m) from turbiditic packstones interlayered with pelagic lime wackestones in the topmost part of the Cefalone Formation to white to yellowish, thin-bedded pelagic lime wackestones with planktic foraminifera (Fig. 8). In the field, the lower boundary of the Scaglia Bianca Formation is placed at the topmost significant (i.e., ≥1 m thick) turbidite bed. This transition takes place within the *R. cushmani* biozone and is not visibly associated with a hiatus.

The name *Scaglia* is widely used throughout Italy and refers to biomicritic, pelagic limestones or marlstones with a scaly fracture and characterized by thin tabular bedding (e.g., Montanari et al., 1989). Different names (Scaglia Bianca ["white"], Scaglia Rossa ["red"], and Scaglia Cinerea ["ash gray"]) have been used to distinguish between different subfacies, and often an age connotation has been attributed to these different types of "Scaglia" (Arthur et al., 1977; Colacichi and Baldanza, 1986; Montanari et al., 1989; Dela Pierre, 1994). Ghisetti and Vezzani (1990a) used the term *Scaglia Rossa* on their map of the Gran Sasso to include all pelagic facies of our Scaglia Bianca, Monte Corvo, and Fonte Gelata Formations. In this study, the term *Scaglia Bianca Formation* refers to pelagic limestones deposited from latest Cenomanian to early Campanian–middle Maastrichtian, which was the only period during which the entire Gran Sasso base-of-slope site was dominated by pelagic facies.

The pelagic limestone beds of the Scaglia Bianca Formation have flat to slightly undulating surfaces and are commonly intensely bioturbated. Nodules of diagenetic replacement chert either occur parallel to bedding in the center of the beds or replace burrows ("burrow flint"). Despite the facies change, the lower 10 m of the Scaglia Bianca Formation still contain intercalated 15–50-cm-thick beds of graded and laminated calcarenites (Fig. 11). Throughout the Scaglia Bianca Formation, thin packstones to grainstones are commonly observed, which we attribute to the activity of contour currents on the basis of their bed forms, composition, and texture. Two types of contourites are distinguished: The first type is a packstone to grainstone forming 0.5–10 cm thick beds that can be traced over tens of meters. These beds have undulating or flat bases, sharp tops, and a near-homogeneous grain-size distribution (0.3–0.5 mm). The grains are well-sorted echinoderm and mollusk fragments together with planktic foraminiferal tests. Interstitial mud is rare or absent. Stow and Lovell (1979) described texturally similar deposits in terrigenous deep-water successions, which they termed *sandy contourites*. The second type is a packstone consisting almost entirely of planktic foraminiferal tests. These beds are considerably thinner (only a few millimeters thick), have a lenticular shape, and are commonly 5–10 cm wide. They have a sharp, flat base, are medium sorted and nongraded, and contain interstitial mud (Fig. 12). These contourites are textu-

FIG. 10.—Incomplete Bouma cycle (T_a–T_c) in fine-grained, bioclastic calcirudite and calcarenite, eroded and overlain by another calcirudite. Cefalone Formation, Pizzo Cefalone.

FIG. 11.—Calcareous turbidite in the Scaglia Bianca Formation, showing complete Bouma cycle (T_a–T_e). The pebbles at the base are reworked pelagic lime mudstone clasts, and the components of the arenitic part are mainly bioclasts. Lowermost Turonian, Rio Arno (section 2; see Fig. 4). Scale in centimeters.

rally equivalent to the muddy contourites of Stow and Lovell (1979).

Local, minor discontinuities are present throughout the Scaglia Bianca Formation; they include slumped intervals (Fig. 13) and low-angle angular unconformities. No hiatus of biostratigraphic resolution is associated with these surfaces. On the other hand, no obvious discontinuity surfaces were observed where local biostratigraphic hiatuses are documented within the Scaglia Bianca Formation. Therefore, we interpret these latter

FIG. 12.—Thin-section photomicrograph of pelagic lime mudstone containing fine packstone laminae consisting entirely of planktic foraminiferal tests and fine-grained echinoderm fragments. These laminae, interpreted as contourites, have been disturbed by burrowing. Scaglia Bianca Formation, lowermost Campanian, Rio Arno (section 2; see Fig. 4).

hiatuses to have most likely formed by the winnowing activity of contour currents that removed sediment without forming obvious erosional features (cf. Stow and Lovell, 1979). In contrast, the minor discontinuity surfaces probably resulted from local slope instabilities, causing breaks in the sedimentary record too small to be resolved by biostratigraphy. Lateral and downslope displacement of fine sediment is also suggested by the large variations in the thickness of the purely pelagic, uppermost Cenomanian to lower Campanian interval (55–160 m; Fig. 4). A transition into the skeletal packstones of the overlying Monte Corvo Formation forms the diachronous upper boundary of the Scaglia Bianca Formation.

Monte Corvo Formation (Lower Campanian [or Maastrichtian]–lowermost Danian)

The lower boundary of the Monte Corvo Formation is placed at the facies change from the pelagic limestones of the Scaglia Bianca Formation to overlying skeletal packstones to grainstones. At places, this change forms a sharp contact, where-

Fig. 13.—Slumped pelagic limestones in the Scaglia Bianca Formation (Campanian), section 11 (see Fig. 4), below Pizzo Camarda. The hammer handle for scale is 35 cm long.

as elsewhere, a gradual transition over an interval of 2 m is observed. In the central part of Gran Sasso, the lower beds are in the *G. elevata* biozone (i.e., early Campanian), whereas to the west, at Pizzo Camarda, they are in the *G. gansseri* biozone (i.e., middle Maastrichtian; Figs. 4, 5). The thickness of the Monte Corvo Formation is highly variable: it ranges from a few meters of breccias (Pizzo Camarda) to more than 220 m (Sella dei Grilli, Figs. 4, 8).

Crescenti (1969) described the Monte Corvo Formation as part of the "Formazione Monte Acquaviva, Membro Orfento," which he recognized throughout the central Apennines (Crescenti, 1969; Crescenti et al., 1969). In the type area (Montagna della Maiella), the "Membro Orfento" has also been referred to as the "Orfento Formation" (Vecsei, 1991) or the "Calcari Cristallini" (Catenacci, 1965). The Orfento Formation is time equivalent to our Monte Corvo Formation, but lithologically distinct.

The Monte Corvo Formation is characterized by fine- to medium-grained skeletal packstone or grainstone beds, intercalated with thin (2–5 cm thick) layers of pelagic lime mudstones and wackestones. The packstone and grainstone beds were deposited in broad, flat, sheetlike or slightly channelized beds, hundreds to more than 1000 m wide (Fig. 14). Bed surfaces are flat or slightly undulating and virtually always erosive; amalgamation of beds is common and obscures some of the individual layers. In places, minor internal unconformities are observed to be associated with these channelized deposits. On the other hand, beds with a positive relief and associated onlap and offlap geometries are observed, even along the same stratigraphic level (Fig. 14). The packstones and grainstones show graded bedding and are interpreted to result from deposition by turbidity currents (Figs. 15A–B). They consist of fine to medium skeletal fragments, dominantly of rudists, inoceramids, echinoderms, and larger benthic foraminifera (*Orbitoides media, Siderolites*).

Clast-supported, intraclastic breccia beds are found mainly in the upper part of the formation; in the lower part, they are uncommon. The breccia beds fill broad, 100–200-m-wide channels (Fig. 14), and their overbank deposits extend laterally for several hundred meters. They contain intraclasts of fine- to medium-grained packstone and grainstone and of pelagic lime mudstone, which are from 2 cm to more than 1 m in diameter. Many of these intraclastic breccias are graded, and all have an erosive base (Figs. 14, 16). In the upper part of the formation, the frequency and thickness of the breccias increase considerably in the center of the Gran Sasso (around Sella dei Grilli and Monte Corvo; Figs. 4, 14). This increase in breccia deposition started during the late Maastrichtian. Laterally, to the west and east, this trend is not observed, and breccia deposits remain uncommon. At Pizzo Camarda, resedimented deposits corresponding to the lower part of the Monte Corvo Formation are absent; here, the age-equivalent limestones are purely pelagic and therefore included within the Scaglia Bianca Formation (Fig. 4).

Channel-levee complexes typical of basin-floor submarine fans are characterized by the association of (1) channels filled by debris-flow or turbidity-current deposits whose bases indicate that the depositing currents eroded into the substrate, (2) flat-based turbiditic calcarenites with a positive relief, and (3) onlap and offlap geometries (Mullins and Cook, 1986; Harwood and Towers, 1988; Coniglio and Dix, 1992; Walker, 1992). Such facies models also provide an explanation for the considerable thickness variation in the Monte Corvo Formation. The increased number and thickness of channelized debris-flow deposits in the central part of the Gran Sasso area is explained by progradation of such a channel-levee complex.

The youngest sedimentary rocks of the Monte Corvo Formation are pelagic limestones of earliest Danian age (planktic foraminifera biozone P1a–P1c). They overlie resedimented deposits or, at one location (Rio Arno), pelagic limestones of latest Maastrichtian age. Whereas fine-grained carbonate detritus is ubiquitous in the uppermost Maastrichtian, the pelagic limestones above the K/T boundary are exclusively composed of small planktic foraminifera, accompanied up section by calcispheres (calcareous dinocysts). Thin (20–50 cm) bioclastic packstone beds occur between the pelagic limestones only in the Rio Arno

FIG. 14.—Panoramic view of the eastern side of Monte Corvo (section 5; see Fig. 4). In the lower part of the Monte Corvo Formation, thick-bedded bioclastic turbidites having a low but positive relief and associated onlap geometries (white arrows) are overlain by sheetlike intraclastic breccias (b). The upper part of the Monte Corvo Formation consists of stacked, channelized intraclastic breccias (outlined in white) and calcarenite turbidites. The boundary between the Monte Corvo Formation (here lowermost Danian) and the overlying Fonte Gelata Formation shows no major erosional relief, although it is associated with a major hiatus. The same is the case between the Fonte Gelata Formation (here upper Paleocene–middle Eocene) and the Venacquaro Formation (middle to upper Oligocene).

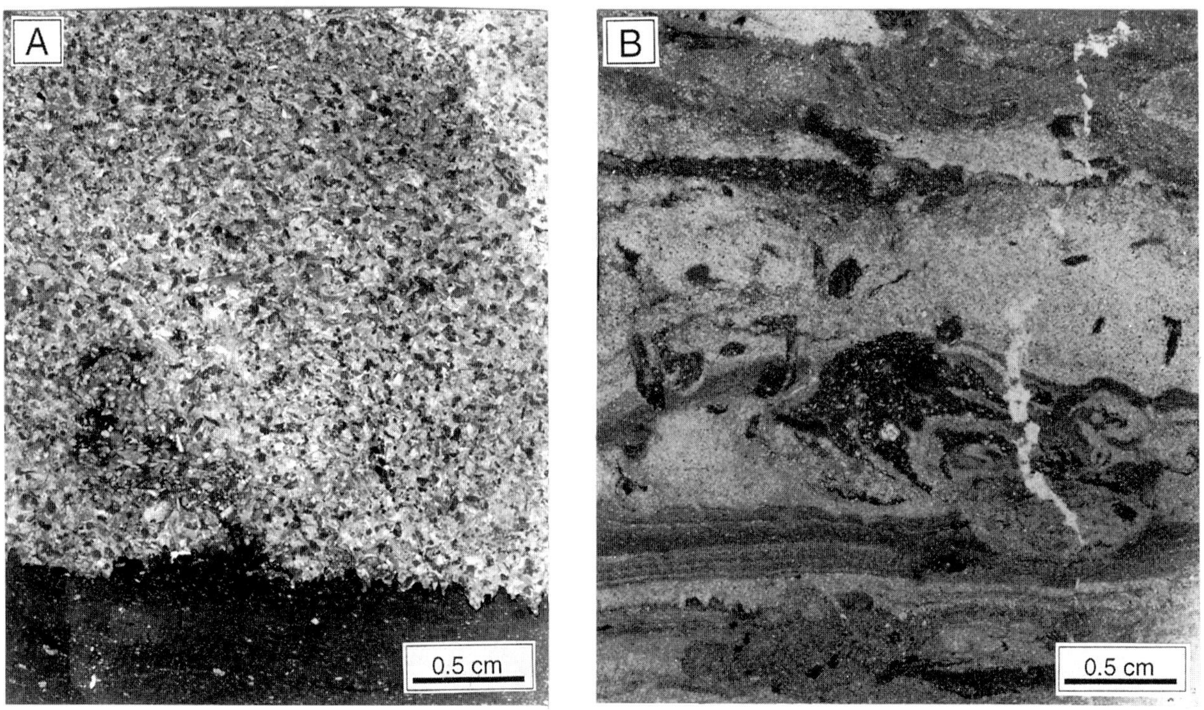

FIG. 15.—Thin-section photomicrographs of calcareous turbidites of the Monte Corvo Formation. (A) Graded skeletal calcarenite with a sharp base overlying a pelagic-periplatform limestone. Lower Campanian, Sella dei Grilli (section 8; see Fig. 4). (B) Graded laminae of redeposited calcisiltite, churned by intense bioturbation. Middle Campanian, Sella dei Grilli. The microfacies of these redeposited layers is conspicuously similar to that of Upper Cretaceous calcareous turbidites from the Northeast Providence Channel, Bahamas (see Bernoulli, 1972, Plate 13).

FIG. 16.—Close-up of stacked, channelized intraclastic breccias in the upper part of the Monte Corvo Formation. The channel in the center is approximately 50 m wide. East side of Monte Corvo (see Fig. 14).

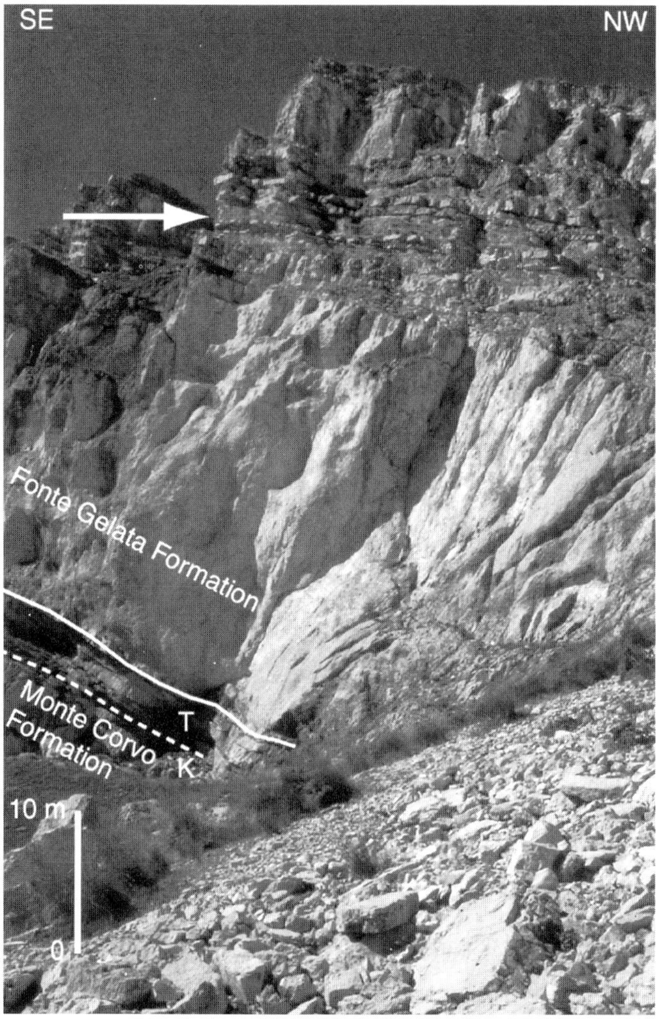

Fig. 17.—An approximately 25-m-thick breccia at the unconformable base of the Fonte Gelata Formation. The associated hiatus spans the interval between the earliest Paleocene (early Danian) and the early Thanetian. Note the low-angle unconformity within the overlying turbidites (arrow). Sella dei Grilli (see Fig. 8).

FIG. 18.—Pebbly calcarenite that displays a plastically deformed matrix, which was partly silicified after deformation. The hammer handle for scale is 33 cm long. Fonte Gelata Formation, upper part, middle Eocene, Rio Arno (section 2; see Fig. 4).

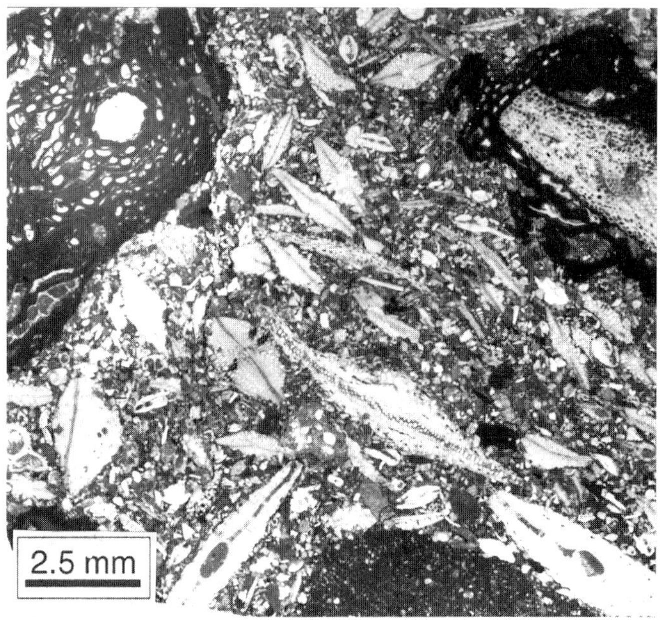

FIG. 19.—Thin-section photomicrograph of a bioclastic turbidite containing tests of *Lepidocyclina* and bioclasts encrusted by red algae. Venacquaro Formation, middle Oligocene, Fonte Gelata (section 1; see Fig. 4).

and Fonte Gelata sections (biozone P1b–P1c to P2?). However, more commonly, these pelagic limestones were eroded before or during the deposition of the overlying breccias of the Fonte Gelata Formation. This erosional unconformity is the upper boundary of the Monte Corvo Formation.

Fonte Gelata Formation (Lower Thanetian–Bartonian [Priabonian])

The Fonte Gelata Formation is dominated by coarse-grained, lithoclastic to lithoclastic and bioclastic breccias and bioclastic packstones, interbedded with minor, thin, pelagic lime wackestone beds (Fig. 14); the formation includes all Paleocene-Eocene deposits in the Gran Sasso area. The hiatus associated with the unconformity at the base of the formation spans an interval from the late Maastrichtian or early Danian (P1c, locally P2?) to the early Thanetian (P3a). In the lower (Paleocene) part of the formation, planktic foraminifera represent biozones P3a to P5 and possibly P6a.

Within the Fonte Gelata Formation, a major hiatus ranges from biozones P6b to P8. Dela Pierre (1992) and Dela Pierre and Clari (1994) also claimed the absence of biozones P5 and P6a. However, we observed the occurrence of *Morozovella velascoensis*, together with *Globanomalina chapmani*, after the last appearance of *Globanomalina pseudomenardii*, indicating the presence of biozone P5 or P6a? below the hiatus. Although this hiatus was clearly identified biostratigraphically over the whole area, it has no sedimentological nor lithological expression, and there are no major erosional features associated with it. Both the upper and lower units are therefore included in one formation. Even though the lithologies of the lower and upper part of the formation look very similar in the field, the difference in constituents and three-dimensional facies organization requires a separate description of the lower and the upper parts of the formation. Locally, the lower (Paleocene) part of the Fonte Gelata Formation is missing, and deposits of the Monte Corvo Formation are directly overlain by the middle Eocene strata (Funivia section; Fig. 4). In other places (Sella dei Grilli; Fig. 4), the lower (Paleocene) part of the Fonte Gelata Formation is directly overlain by the Oligocene Venacquaro Formation.

Lower part (Lower Thanetian–lowermost Ypresian?).—
The lower part of the Fonte Gelata Formation is characterized by coarse-grained lithoclastic breccias and bioclastic packstones interlayered with minor pelagic lime wackestones. The lowermost bed of the formation is always a very coarse grained lithoclastic breccia overlying a submarine erosion surface cutting as much as 3 m into the underlying deposits. Its thickness varies from 2 m at Pizzo Camarda to more than 25 m at Sella dei Grilli (Fig. 17), but on average it is 5 m thick. Pebble size ranges from 2 to 15 cm; the pebbles consist of coral boundstones, cemented bioclastic platform material comprising corals, bryozoans, red algae, clasts containing *Microcodium*, platform- and slope-derived packstone and grainstone fragments, and slope-derived wackestone clasts. Many clasts are reworked from Cretaceous deposits and contain rudist fragments and/or Cretaceous planktic and benthic foraminifera. The matrix of the breccias consists of a medium- to coarse-grained packstone made up of Paleocene planktic and larger benthic foraminifera, skeletal fragments, red algae, bryozoans, and little interstitial lime mud.

The basal breccia is overlain by lithoclastic breccias with channelized bed geometries, 200–500 m wide. These breccias are finer grained, but contain similar components. Up section, the reworked Cretaceous material disappears as penecontemporaneously displaced bioclastic material increases. The breccias are dominantly clast supported and contain little matrix; they are occasionally graded near the top of the beds. They are interpreted to result from grain or debris flows with relatively little matrix (cf. Nardin et al., 1979; Cook and Mullins, 1983).

Coarse bioclastic packstones are interbedded with the breccias. They have sheetlike bed geometries (500–800 m wide) and minor, thin-bedded, pelagic lime wackestone interbeds. The breccias disappear 15 to 20 m above the base of the formation. In several places, low-angle unconformities were observed within these bioclastic packstone beds (Fig. 17); however, no biostratigraphically significant hiatuses are associated with these unconformities. The packstone beds show graded bedding and are interpreted to have been deposited by turbidity currents. They are dominantly composed of echinoid, red algal, and bryozoan fragments, larger benthic foraminifera (*Discocyclina* sp., *Alveolina* spp., *Operculina,* and Rotalidae; in the uppermost part, also *Nummulites* spp. and *Heterostegina* sp.), peloids, small (≤3 mm) fragments of platform-derived grainstones, and pelagic lime wackestone clasts. To the west, most of the lower part of the formation consists of pelagic limestones (Pizzo Camarda; Fig. 4).

Upper part (uppermost Ypresian–uppermost Bartonian or Priabonian).—
The upper part of the Fonte Gelata Formation is composed of interbedded clast- to matrix-supported lithoclastic limestone breccias and conglomerates, bioclastic packstones, and minor pelagic lime wackestone beds. The breccia and conglomerate beds have slightly channelized bed forms (300–500 m wide and 2–4 m thick) and bases indicating that the depositing currents eroded into the substrate. The beds commonly show normal size grading, from pebble to granule size, and are interpreted to have been deposited by debris flows. Lithoclasts range from 2 to 20 cm in size (Fig. 18) and are composed of cemented platform material (which includes Miliolidae and red algae skeletons), platform-derived bioclastic packstones, slope-derived pelagic wackestones, and bioclastic material such as echinoderm, bryozoan, coral, red algae, and mollusk fragments and a diverse association of larger benthic foraminifera. *Microcodium* in clasts at the base of the succession again documents subaerial diagenesis in the source area.

Medium- to coarse-grained bioclastic packstones are interlayered with these breccias. They are commonly graded, have sheetlike bed geometries (300–700 m wide, 0.5–2 m thick), and flat bases showing no or only slight incision. These beds were deposited by turbidity currents. The grain size of the bioclasts typically ranges from 0.5 to 4 mm. Bioclasts are dominantly larger benthic foraminifera (mostly *Discocyclina* sp., *Nummulites*

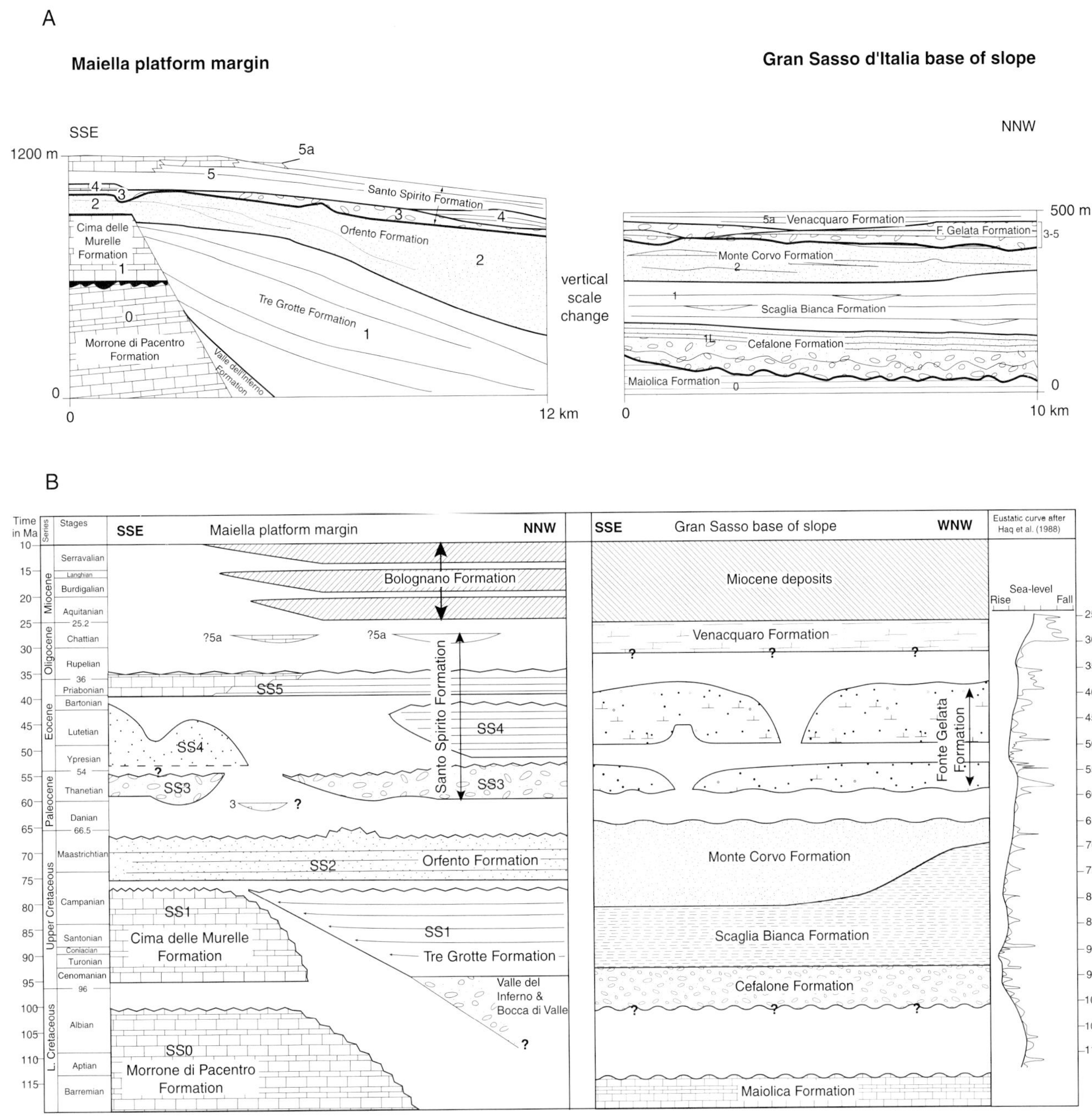

FIG. 20.—Comparison between the stratigraphic architecture of the Maiella platform margin (after Eberli et al., 1993; Mutti et al., 1996; Vecsei et al., 1998) and the Gran Sasso base of slope. (A) Depositional geometries. (B) Chronostratigraphic relationships. The two areas are physically not connected (see Fig. 1), and correlation is based on biostratigraphy and facies evolution. Numbers refer to supersequences as defined along the Maiella carbonate-platform margin and the correlative units on the base-of-slope site as defined in this paper. The middle Cretaceous emergence of the platforms (SS 0/SS 1 supersequence boundary on the Maiella platform margin) is reflected on the base of slope by the deposition of the megabreccias and lithic breccias of the Valle dell'Inferno and the Cefalone Formations, resulting from platform-margin collapse during sea-level lowstand. Aggradation of the platforms from the middle Cenomanian (Maiella: Cima delle Murelle Formation, SS 1) or the Turonian (Monte Gargano; Bosellini et al., 1993b) to the middle Campanian is reflected at the foot of the escarpments by deposition of breccias during third-order lowstands and skeletal turbidites during transgression and highstand (Tre Grotte Formation, SS 1; Eberli et al., 1993; Lampert et al., 1997; Vecsei, 1991; Vecsei et al., 1998); however, such resediments did not reach the base of slope of the Gran Sasso area where the pelagic-periplatform limestones of the Scaglia Bianca Formation were deposited. After the late Campanian, the erosional escarpment bordering the Maiella platform margin was eventually buried, and the rapid platform progradation during long-term sea-level lowering (Orfento Formation, SS 2) is reflected by the progradation of a basin-floor fan on the base-of-slope site (Monte Corvo Formation). During platform emergence and formation of an unconformity along the upper slope in the latest Maastrichtian–early Danian, pelagic limestones were deposited on the base of slope (uppermost Monte Corvo Formation). Platform-margin erosion and backstepping in the late Paleocene–middle Eocene interval (Santo Spirito Formation, lower part, SS 3 and SS 4) is reflected on the base of slope by lithoclastic breccias yielding fragments of Upper Cretaceous to lower Tertiary limestones (Fonte Gelata Formation). This interval can only be correlated by and large with the evolution of the Maiella margin; in particular the different episodes of platform-margin erosion and aggradation cannot be correlated in detail with the base-of-slope evolution, which seemed to have been dominated by slope processes. It is surprising that the late Eocene–earliest Oligocene platform progradation (Santo Spirito Formation, upper part, SS 5) is only locally recorded on the base-of-slope site. The pelagic limestones and skeletal turbidites of the middle to upper Oligocene Venacquaro Formation were deposited during low relative sea level, which is suggested by the downward shift of the middle Oligocene reefs on the Maiella margin (SS 5a).

spp., *Assilina* sp., *Operculina* sp., *Heterostegina* sp., etc.) associated with red and green algae, bryozoan, echinoderm, and mollusk fragments. Thin-bedded (3–30 cm), pelagic lime wackestones are intercalated between the redeposited limestones.

In the central and northern part of the Gran Sasso (Monte Corvo), the upper part of the formation comprises an irregularly interlayered association of amalgamated turbidity and debris-flow deposits. To the west (west side of Monte Corvo, Pizzo Camarda) and east (Rio Arno), however, only the beds immediately overlying the hiatus are mass-flow deposits (Fig. 4). In these areas, the remaining two thirds of the Eocene deposits consist of hemipelagic marly lime wackestones, reaching up to P16–P17 (Dela Pierre, 1992; Dela Pierre and Clari, 1994). At Sella dei Grilli, no Eocene strata are present. South of Campo Imperatore (Funivia), the upper part of the Fonte Gelata Formation starts only with Bartonian breccias and conglomerates and continues into the middle to upper Priabonian (Figs. 4, 5). At Fonte Gelata, a very thick (23 m) chaotic debris-flow deposit forms the top of the Eocene. This deposit contains large blocks, as much as 2 m across, of reworked Maastrichtian, Paleocene, and middle Eocene sediments derived from the platform margin and/or the uppermost slope. Everywhere, a hiatus separates the Fonte Gelata Formation from the overlying pelagic marly limestones of the Venacquaro Formation.

Venacquaro Formation (Upper Rupelian–Chattian)

The lower boundary of the Venacquaro Formation is a paraconformity with a depositional hiatus; no obvious erosional features are observed along the contact. The boundary is easily recognized by the marked change from resedimented limestones to marly, hemipelagic lime wackestones. The associated hiatus ranges from latest Bartonian or latest Priabonian to the "middle" Oligocene. Genera and groups of species of planktic foraminifera date the Venacquaro Formation as "middle" to late Oligocene (P20–P22, van Konijnenburg et al., 1998), which is in line with the age of the redeposited larger foraminifera (*Lepidocyclina [Eulepidina]* sp., *Lepidocyclina [Nephrolepina]* sp.). In the westernmost part of the Gran Sasso (at Passo delle Capannelle), the lower boundary of the Venacquaro Formation is a hardground, accentuated by ferromanganese oxihydrate impregnations and borings. Dela Pierre and Clari (1994) interpreted this omission surface to have formed by the winnowing activity of contour currents.

Dela Pierre (1992) and Dela Pierre and Clari (1994) used the name *Scaglia Cinerea con Torbiditi* for these Oligocene sedimentary rocks. However, this name is rather generic and, to avoid confusion with the Scaglia Bianca Formation, we propose the name *Venacquaro Formation*. The Venacquaro Formation is characterized by greenish or reddish, marly hemipelagic lime wackestones with planktic foraminifera and commonly intercalated bioclastic packstones to grainstones dominated by the displaced tests of *Lepidocyclina* spp. together with minor *Amphistegina* sp., bryozoans, red algae, and sponge spicules (Fig. 19). The hemipelagic limestones are also the oldest units in the Gran Sasso to contain minor amounts of very fine siliciclastic debris (quartz, feldspar, and chlorite) and show burrows common in pelagic sediments (*Thalassinoides, Zoophycos, Chondrites*). The bioclastic packstones to grainstones occur as individual beds and have broad (1000–1500 m wide, 30–150 cm thick) sheetlike geometries with no obvious erosional features at the base; in places, they overlie slumped strata. As these rocks are composed almost entirely of uniformly sized tests of *Lepidocyclina,* only a faint grading is observed at the top of these beds. Because of the association with slump features, the obviously displaced shallow-water biota, and the faint grading, we interpret the rocks to have been deposited by turbidity currents.

The Venacquaro Formation has a thickness of 45–65 m all over the Gran Sasso area, and this absence of significant differences in the thickness of the turbidite beds seems to indicate that they were deposited as distal turbidites. The upper boundary of the Venacquaro Formation is defined by a marked change to coarse-grained, white, redeposited skeletal packstones bearing glauconite, and virtually no hemipelagic marly limestones are found. This boundary coincides approximately with the Oligocene/Miocene boundary.

COMPARISON WITH ADJACENT CARBONATE-PLATFORM AND BASINAL AREAS

Correlation from Platform to Basin

Along the Maiella platform margin, a number of second-order sequences (supersequences, SS) have been defined (Figs. 20A–B; Vecsei, 1991; Eberli et al., 1993; Vecsei et al, 1998). They are bounded by truncation surfaces, associated with a stratigraphic hiatus and with a change in depositional system. On the platform top, the truncation surfaces are generally characterized by subaerial exposure (Accarie, 1988; Accarie and Delamette, 1991; Mutti, 1995), whereas on the upper slope of the platform margin, they are associated with submarine erosion or nondeposition. Where the unconformities are associated with prolonged subaerial exposure and can be traced from the flat platform top to the slope and into the basin, they are sequence boundaries in the sense of Vail et al. (1977) and can be linked to relative changes in sea level. However, where there is no physical connection between platform and base-of-slope deposits, a direct connection to sea-level changes cannot be established for slope unconformities, because submarine erosion and mass wasting are not necessarily related to sea-level change. In the case of Gran Sasso, the major unconformities and rapid vertical facies changes are associated with a change in depositional system and thus correspond to sequence boundaries as redefined by Schlager (1993); however, a connection to sea-level change remains to be proven.

In the following section we show that during the Cretaceous and early Tertiary, the evolution of the platform(s) determined sedimentation on the base of slope to a large extent. Each major change in depositional system on the base of slope appears to be related to a major change in sedimentary regime on the adjacent platform(s) accompanying the formation of the supersequence boundaries. However, the third-order sequences recognized along the platform margin could not be traced into the base-of-slope area, where intraformational unconformities such as in the Scaglia Bianca and Fonte Gelata Formations appear to be related to sedimentary processes that were not directly related to sea-level changes.

Maiolica Formation: Basinal Sedimentation During Platform Aggradation: SS 0

The nannofossil oozes of the Maiolica Formation were deposited under pelagic conditions. Thin, slope-derived pebbly mudstones suggest slope instabilities on a gently inclined slope with fine-grained sediments (Kenter, 1990). Basinward, in the Umbria-Marche area, the Maiolica sediments covered the fault blocks resulting from Jurassic rifting and smoothed the pre-existing sea-floor topography (Bernoulli, 1972; Bernoulli et al., 1979; Koopman, 1983). In the Gran Sasso area, quiet pelagic sedimentation was occasionally interrupted by debris flows carrying shallow-water carbonate debris onto the base-of-slope site. Compared with the underlying Upper Jurassic limestones (Calcare ad Entrochi Formation and Terratta Formation equivalent), such redeposited beds reflecting episodes of platform-margin collapse are relatively scarce; apparently the platforms were mainly aggrading during long-term sea-level rise (Morrone di Pacentro Formation, SS 0 in Montagna della Maiella, Figs. 20A–B).

Cefalone Formation and Middle Cretaceous Unconformity: SS 1 Lowstand

In the Albian, large parts if not all of the Lazio-Abruzzi and the Apulia platform margins became subaerially exposed. The platform tops underwent karstification, and karstic bauxites accumulated in depressions on the platform surface (Carannante et al., 1988). The duration of this middle Cretaceous exposure may vary from place to place, as it cannot be established in all locations because of the lack of good biomarkers. Subaerial exposure in the middle Cretaceous is documented in Marsica (Colacicchi and Praturlon, 1965a, b), the central Lazio-Abruzzi platform (Praturlon and Sirna, 1976), and to the south in the mountains of Campania (Carannante et al., 1988). In the Maiella platform margin, the time of subaerial exposure is well constrained: platform deposition was interrupted in the middle Albian (unconformity between SS 0 and SS 1 in Figs. 20A–B; Eberli et al., 1993), and the platform was reflooded in the middle Cenomanian, as documented by ammonites in the strata above the unconformity (Accarie and Delamette, 1991). This middle Cretaceous platform exposure is in sharp contrast to the worldwide drowning and retreat of carbonate platforms during the same time interval (Schlager, 1981). Subaerial exposure of the peri-Adriatic platforms may therefore have been caused by intraplate stresses in connection with early Alpine movements (D'Argenio and Mindszenty, 1991; Eberli et al., 1993).

Deposition of megabreccias and lithoclastic debris-flow deposits of the Cefalone Formation in the Gran Sasso coincided, at least partly, with the exposure of the platforms. Although the lowermost part of the Cefalone Formation is not directly dated, our arguments for an approximately middle Albian age for the lower boundary of the formation imply a long hiatus in deposition and/or submarine erosion between the Barremian and the middle Albian. During the time of exposure, no coeval carbonate material was exported from the platform, but platform escarpments were shaped by submarine erosion. This scenario is documented by our observations of (1) Lower Cretaceous platform-interior limestones that are exposed in the escarpment along the Maiella platform margin, implying collapse and/or erosion of the platform rim, and (2) the presence of middle Cretaceous megabreccias in the adjacent basin. Collapse of a platform margin during the middle Cretaceous relative sea-level lowstand is well documented in the eastern part of the Gargano promontory (Fig. 1) where the deposition of megabreccias also coincided with emergence of the platform (Bosellini et al., 1993a, b). The formation of extensive cave porosity in the platform-margin limestones could have led to partial collapse of the margin and deposition of the megabreccias and breccias in the lower part of the Cefalone Formation. In the Umbria-Marche basin, pelagic and hemipelagic limestones, marly limestones, and marlstones were deposited (Marne a Fucoidi) that are devoid of redeposited material except for areas adjacent to the Lazio-Abruzzi platform. They record the anoxic events of the Aptian (Livello Selli, a pronounced interval of black shales; Coccioni et al., 1987) and the Albian (Arthur and Premoli Silva, 1982), which were not recorded in the base-of-slope area.

Reflooding of the platforms apparently did not take place everywhere at the same time. In places, e.g., in the Gargano promontory, the oldest sediments overlying the middle Cretaceous bauxites on the platform top are reported to be Turonian in age (Bosellini et al., 1993b; Neri, 1993), but in other areas, the upper Cenomanian interval (above the main bauxite horizon) is characterized by extensive karstification indicating a second major episode of platform emergence (Carannante et al., 1988). Unfortunately, we cannot constrain the duration of the middle Cretaceous exposure event(s) on the Lazio-Abruzzi carbonate platform. In any case, the gradual transition from lithoclastic breccias and turbidites to skeletal turbidites in the upper part of the Cefalone Formation might reflect reflooding of the platform(s).

Scaglia Bianca Formation and Platform Aggradation: SS 1

With the onset of deposition of the Scaglia Bianca Formation, the Gran Sasso base-of-slope site rapidly became starved of shelf-derived sediments. Sedimentation of pelagic-periplatform ooze prevailed up to the early Campanian, and locally into the late Maastrichtian, and the resulting ooze deposits were continuously shaped by contour currents, causing hiatuses in some places and depositing contourites in others. Sediment apparently was deposited on a gentle slope, as documented by the common presence of slumped strata.

During the middle Cenomanian or Turonian to middle Campanian, the peri-Adriatic platforms aggraded (SS 1, Figs. 20A–B), and only minor platform-derived material, except

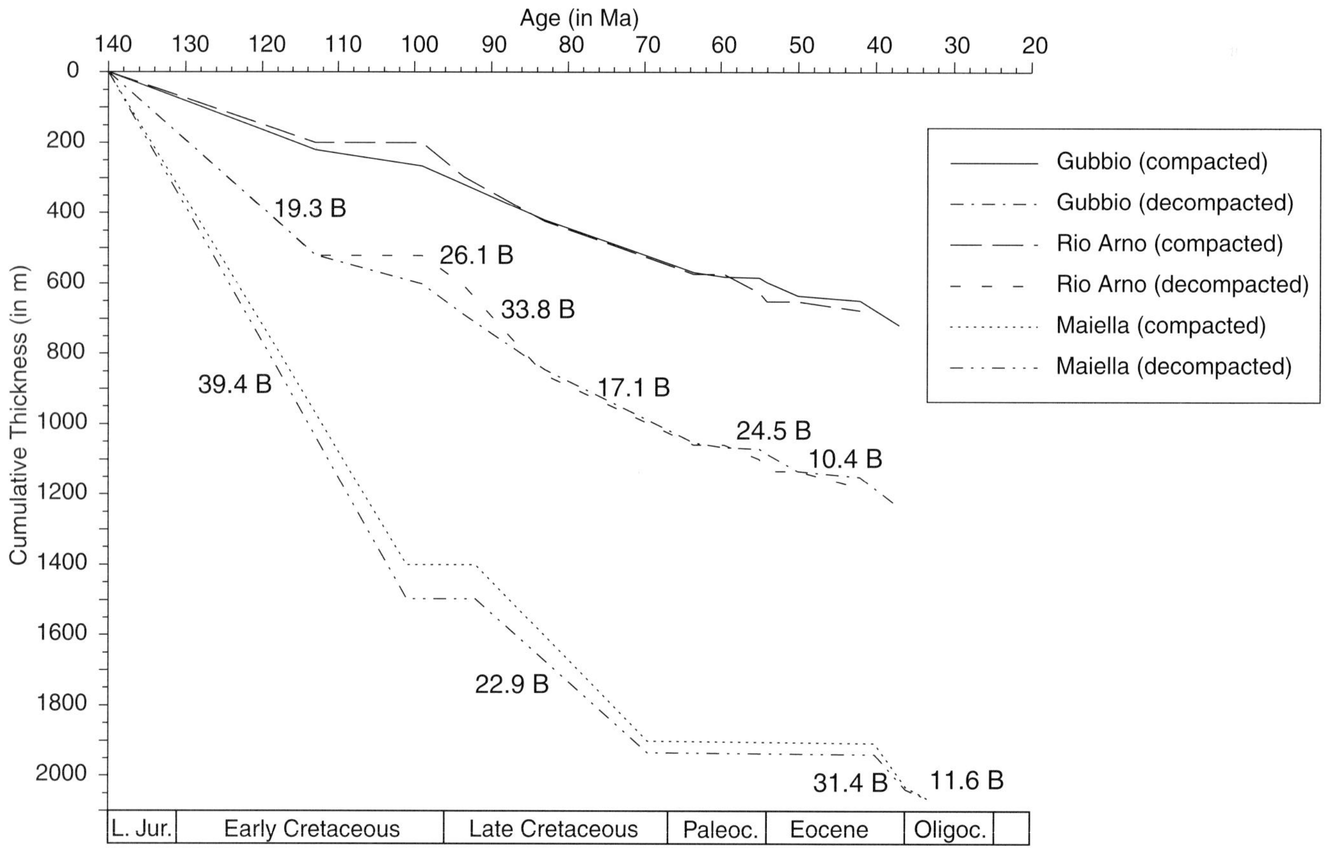

Fig. 21.—Accumulation curves for the Maiella platform margin (Vecsei, 1991), the Gran Sasso base-of-slope area (the Rio Arno section; see Fig. 4), and the Umbria-Marche basin (Gubbio section, Arthur et al., 1977). Curves uncorrected (i.e., compacted) and corrected (i.e., decompacted) for compaction are given. Sedimentation rates in Bubnoff (B) units (m/m.y.) are given for the decompacted curves of Rio Arno (Gran Sasso) and Maiella. Numerical ages are after Haq et al. (1988).

fine carbonate mud, reached the base-of-slope site in the Gran Sasso during this interval. Much material was, however, exported from the platforms during this time. In Montagna della Maiella, nearly 1000 m of pelagic-periplatform ooze, skeletal turbidites, breccias, and megabreccias were deposited at the foot of the escarpment (Tre Grotte Formation, SS 1, Fig. 20A; Eberli et al., 1993; Vecsei, 1991; Vecsei et al., 1998). To the north, within 14 km, these deposits pass into pelagic limestones with only rare turbiditic intercalations (Crescenti, 1969). A similar facies association is recorded in the eastern part of the Gargano promontory, but passes farther eastward into pelagic limestones (Bernoulli, 1972; Bosellini et al., 1993a, b). In the Gran Sasso, the pure pelagic-periplatform ooze sedimentation of the Scaglia Bianca Formation may be related to a retreat of the Lazio-Abruzzi platform, an interpretation supported by the presence of Upper Cretaceous pelagic limestones with skeletal turbidites that overlie platform limestones west of Aquila (Dela Pierre, 1994). We speculate that, by analogy to the Maiella platform margin (Eberli et al., 1993) and the eastern part of the Gargano promontory (Bosellini et al., 1993a, b), these pelagic and turbiditic limestones may onlap one or more middle Cretaceous scars of platform collapse in the Lazio-Abruzzi platform.

In the Gran Sasso area, the anoxic event of the Cenomanian/Turonian boundary (bituminous shales of the Livello Bonarelli in the Umbria-Marche basin; Arthur and Premoli-Silva, 1982) is not recorded. Although there are internal stratigraphic gaps, caused by erosion, within the Scaglia Bianca Formation, the accumulation rates are comparable with those in the corresponding interval of the Gubbio section in the Umbria-Marche basin (Fig. 21). This fact suggests that the hiatuses are compensated by fine sediment derived from the platforms, similar to what can be observed today along the Bahamian carbonate platforms, particularly along their leeward side (Boardman and Neumann, 1984; Mullins et al., 1985; Glaser and Droxler, 1993).

Monte Corvo Formation and Late Cretaceous Sea-Level Fall: SS 2

During the late Campanian to early Maastrichtian, the escarpment of the Maiella platform margin was buried, and a distally steepened ramp was established during progradation of the platform (Eberli et al., 1993). Mutti et al. (1996) documented repetitive downward shifts of the shoreline, indicating long-term lowering of sea level during the late stages of progradation. Mass-flow deposits and skeletal turbidites interfingering with pelagic limestones were deposited at the foot of the distally steepened ramp; 14 km north of the escarpment, however, only rare intercalations of calcareous turbidites are found (Crescenti,

1969). The turbidity currents carrying the arenitic material into the basin appear to have been of relatively low transport capacity (cf. Mutti, 1979); in the Umbria-Marche basin, turbidity currents derived from carbonate-platform margins also rarely traveled more than 30 km (Montanari et al., 1989).

Along the Gran Sasso base of slope, no unconformity is present at the base of the Monte Corvo Formation, and the passage from pelagic-periplatform lime mudstones of the Scaglia Bianca Formation to the limestone turbidites and associated breccias of the Monte Corvo Formation is diachronous. A progradational pattern is recognized in the resedimented units of the Monte Corvo Formation, which are basin-floor submarine-fan deposits in which the increase of channelized debris-flow deposits in the central part of the Gran Sasso area is explained by the progradation of a channel-levee complex. This fan was established in the middle Campanian and prograded basinward during the Maastrichtian. The generally low transport capacity of the mud-poor turbidity currents would imply that the platform source of the calcarenites was probably within 20 km. Other margins such as the platform margin in the eastern part of the Gargano promontory (upper part of Monte Acuto Formation, Bosellini et al., 1993a, b; Neri, 1993) show a similar progradation of turbiditic successions in the Campanian–Maastrichtian. Although a northwest-directed progradation of shallow-water facies over basinal deposits is observed in the Maiella platform margin, other margins seem to have retained their steep slopes (e.g., northeast of the Apulia carbonate platform; de Alteriis and Aiello, 1993). This dissimilarity may be explained by the orientation of the different margins: the northeastern margin of the Apulia carbonate platform appears to have been a windward margin, whereas the northwestern margin of that platform—i.e., part of the Montagna della Maiella platform margin—was a leeward one (cf. Eberli and Ginsburg, 1989; Schlager, 1993).

By the end of the Maastrichtian, the platform tops of both the Apulia platform and the Lazio-Abruzzi platform became subaerially exposed again (Praturlon and Sirna, 1976; Accordi and Carbone, 1988; Mutti, 1995). The base of the Tertiary on the platform top in Montagna della Maiella and along its proximal margin is a pronounced unconformity associated with subaerial exposure reaching the shelf margin (supersequence boundary SS2/SS3, Figs. 20A–B; e.g., Mutti, 1995; Lampert et al., 1997). However, a thin package of pelagic limestones of early Danian age is present at the top of the Monte Corvo Formation below the unconformity at the base of the Thanetian. During the early Danian, when the platforms most probably were exposed, only thin pelagic limestones were deposited in the basin. In Umbria-Marche, the Cretaceous/Tertiary boundary is often associated with a hiatus (Luterbacher and Premoli Silva, 1964). Younger pelagic limestones (*S. trinidadensis* zone) occur above the turbiditic units of the Monte Acuto Formation along the east Gargano base-of-slope area below an unconformity at the base of the Eocene deposits (Neri, 1993).

The entire succession from the Cefalone Formation to the Monte Corvo Formation can be interpreted as a major cycle, composed of at least two second-order cycles (supersequences 1 and 2 of the Maiella platform margin, Figs. 20A–B; Eberli et al., 1993; Vecsei et al., 1998), governed by long-term sea-level change. The lower part of the Cefalone Formation coincides with subaerial exposure of the platform, and the Scaglia Bianca Formation is coeval with the aggradational phase of the platform, whereas the Monte Corvo Formation reflects platform progradation during long-term sea-level fall. By and large, this cycle corresponds to the Monte Sant'Angelo sequence of the eastern part of the Gargano promontory (Bosellini et al., 1993a).

Fonte Gelata Formation: Slope Erosion and Redeposition: SS 3 to SS 5

During the late Paleocene and the early to middle Eocene, the platform margin was mostly in a deeper-water environment and underwent extensive submarine erosion. Platform flooding in the late Danian(?) to early Thanetian allowed growth of reefs dominated by corals and red algae, which, however, are exclusively preserved as blocks in megabreccias present on the slope of the platform (Sanders, 1994; Moussavian and Vecsei, 1995). Extensive erosion and deepening during the late Thanetian is documented by large incised channels filled with megaconglomerates, interbedded with bioclastic turbidites and pelagic limestones. Overall, during the Thanetian to Bartonian interval, the Maiella-Apulia platform margin was shaped by subaerial exposure, reflooding, and backstepping of the platform margin, submarine erosion, and renewed progradation, induced by third-order changes in relative sea level (Vecsei, 1991; Vecsei et al., 1998). Because of low overall aggradation and submarine erosion on the slope, deposits are often confined to erosional remnants that are separated by unconformities (Figs. 20A–B; supersequences 3 and 4 of Vecsei, 1991; Eberli et al., 1993; Sanders, 1994; Vecsei et al., 1998).

After the deposition of the basin-floor turbidite fan of the Monte Corvo Formation, the Gran Sasso base-of-slope area underwent submarine erosion in the early Paleocene. In the late Paleocene to earliest Eocene (Fonte Gelata Formation, lower part), debris flows, carrying platform-derived, shallow-water carbonate debris, recorded the erosion of a Cretaceous platform margin and of younger, Paleocene, deposits. The occurrence of *Microcodium* in limestone clasts indicates that the source area was emergent and that platform-margin collapse affected rocks previously exposed to subaerial alteration (cf. Esteban and Klappa, 1983). In turn, the associated bioclastic turbidites reflect reflooding and reestablishment of reefs and carbonate production on the platform(s); however, the biostratigraphic resolution available does not allow for a correlation of the lithic breccias or skeletal turbidites with the third-order sequences defined along the Maiella platform margin (Vecsei, 1991; Sanders, 1994; Vecsei et al., 1998). The depositional hiatus within the Fonte Gelata Formation partly coincides with the deposition of megabreccias, lithoclastic calcirudites, bioclastic calcarenites, and pelagic limestones on the slope of the Maiella margin (Fig. 20B). Dela Pierre and Clari (1994) explained the hiatus by winnowing activity of contour currents; however, pelagic sediment could also have been eroded by bypassing turbidity currents. During the late Ypresian to Bartonian, a thin and incomplete sedimentary record with several internal hiatuses characterizes the Maiella platform margin. Its equivalent on the base-of-slope site occurs in the upper part of the Fonte Gelata Formation, which consists of an irregular alternation of debris-flow deposits, turbidites, and pelagic limestones. Episodes of deposition and submarine erosion by mass wasting alternated closely in space and time. This upper part of the Fonte Gelata Formation is age equivalent to the Monte Saraceno sequence (middle Eocene) on the eastern part of the Gargano promontory (Bosellini et al., 1993a, b).

In Montagna della Maiella, a pronounced truncation surface separates the upper Bartonian or lower Priabonian to lower Rupelian deposits (supersequence 5) from the underlying Maastrichtian (SS 2) or Paleocene–Eocene (SS 3 and SS 4) strata. During the Priabonian to earliest Oligocene, skeletal calcarenites prograded over the former slope environment, interspersed with patch reefs. During the early Oligocene, coral and red algal reefs developed, probably on a distally steepened ramp. This platform progradation is only locally recorded on the base of slope. Thin Priabonian debris-flow deposits are locally present, and Dela Pierre (1992) recorded Priabonian pelagic deposits from the northern and western part of Gran Sasso. Deposits of early Oligocene age seem to be entirely lacking. Obviously, submarine erosion prevailed at this time over deposition on the base-of-slope site of the Gran Sasso.

Venacquaro Formation: Base-of-Slope Deposition During Lowstand SS 5a

After a long interval of nondeposition and/or submarine erosion in the latest Eocene–early Oligocene, the base-of-slope site in the Gran Sasso was dominated by hemipelagic sedimentation and frequent deposition of bioclastic turbidites derived from shallow-water areas. This interval is more or less coeval with the occurrence of patch reefs with lepidocyclinids on the Maiella platform margin; Vecsei (1991) and Vecsei et al. (1998) described the remnants of these patch reefs as "middle Oligocene limestones of uncertain supersequence affinity" (SS 5a in Fig. 20). The patch-reef remnants are isolated on the slope and apparently reflect a downward shift of facies during sea-level lowstand. In the Gran Sasso area, the boundary with the overlying Miocene deposits (SS 6) appears to be conformable.

Sedimentation Rates

The sedimentation rates on the Maiella platform margin (based on Vecsei, 1991), in the Umbria-Marche basin (based on Arthur et al., 1977; Cresta et al., 1989) and on the base of slope of the Gran Sasso area (Rio Arno section) are compared in Figure 21. The depth to "basement" curves show depths to the top of the Upper Jurassic. The Maiella curve documents the aggradation on the shallow-water platform and can be regarded as a first approximation of subsidence. The curve shows a clear exponential decrease in subsidence from the Early Cretaceous to the Miocene, reflecting decaying thermal subsidence of the continental margin. This decrease in rate of subsidence might primarily be responsible for the generally lower sedimentation rates during the early to middle Tertiary and the longer duration of hiatuses on the platform (Vecsei et al., 1998). Considerably less sediment accumulated on the base-of-slope site and in the Umbria-Marche basin than on the Maiella platform margin. Bulk sedimentation rates in these two areas are remarkably similar, although the base-of-slope succession includes important hiatuses. These hiatuses were partly caused by submarine erosion, preceding or associated with events of mass wasting along the slope (base of Cefalone Formation, base of upper Paleocene, and lower Eocene unconformity), and by slope processes such as winnowing by contour currents (Scaglia Bianca Formation and possibly the base of the Venacquaro Formation). Obviously, these hiatuses compensate for the extra sediment introduced to the base-of-slope site from the platforms. The base of slope alternated among accretionary, neutral (bypass), and erosional depositional regimes over short time intervals.

CONCLUSIONS

From the middle Cretaceous until the Oligocene, the area that now forms the Gran Sasso d'Italia was situated along the base of the slope of the adjacent Lazio-Abruzzi carbonate platform. This base-of-slope site was characterized by pelagic-periplatform sedimentation, by deposition from mass flows originating from the collapse of platform-margin areas, and by shedding of bioclastic material from the nearby platform(s). The sedimentary record is punctuated by episodes of submarine erosion and nondeposition, leading to unconformities and depositional hiatuses. Contour currents swept the base-of-slope site from time to time, depositing winnowed lag deposits and causing local stratigraphic gaps in the pelagic successions. Such contour currents might also be responsible in part for longer periods of nondeposition during the Eocene and the Oligocene (Dela Pierre and Clari, 1994). The interplay of these sedimentary processes caused changes in facies associations, not only downslope but also along depositional strike.

The Lower Cretaceous to Oligocene base-of-slope deposits of the Gran Sasso area are subdivided into six formations, separated from each other by unconformities or rapid vertical changes in facies associations. Each of these formations is characterized by a distinct association of depositional facies, which changed drastically across formational boundaries. These changes in depositional system were related to changes in sedimentary regime due to second-order (10–20 m.y.) platform cycles on the adjacent platforms. These platform cycles were largely controlled by relative changes in sea level, which in the case of the middle Cretaceous unconformity were related to tectonic events (Eberli et al., 1993; Vecsei et al., 1998). Third-order cycles recognized on the nearby Maiella platform margin cannot be recognized in the base-of-slope section with the available biostratigraphic resolution. Unconformities of higher order within the base-of-slope units seem to be related to slope processes, in particular to submarine erosion or mass wasting.

A comparison of geohistory curves of the Maiella platform margin, the Gran Sasso base-of-slope site, and the Umbria-Marche basin indicates that platform evolution also seems to have controlled the amount of sediment accumulating on the base of slope. Along the base of slope, higher sedimentation rates during times of massive redeposition were compensated by submarine erosion and nondeposition, resulting in bulk sedimentation rates comparable to those in the Umbria-Marche basin where pelagic sedimentation was essentially continuous. Although sedimentation along the base of slope closely reflects evolution of the carbonate-platform margin and upper slope, a correlation with the platform has to be established for each depositional system, because erosional and depositional processes not directly related to sea-level changes operated along the base of slope.

ACKNOWLEDGMENTS

We would like to acknowledge the help of M. Caron, K. von Salis, and R. Wernli in the determination of microfossil assemblages. W. Schlager, F. Ghisetti, L. Vezzani, P. Clari, and F. Dela Pierre are thanked for inspiring discussions. We are grateful to I. Stössel and D. W. Schmid for their support in the field and many discussions and to P. M. Harris, S. Mazzullo, and J. F. Read for constructive reviews. We thank U. Gerber for his photographic work. Financial support was provided by the Swiss National Science Foundation (grants 20-35907.92 and 20-45131.95).

REFERENCES

ACCARIE, H., 1988, Dynamique sédimentaire et structurale au passage plate-forme/bassin. Les faciès carbonatés Crétacés et Tertiaires: Massif de la Maiella (Abruzzes, Italie): Paris, École des Mines de Paris, Mémoires des Science de la Terre, v. 5, 162 p.

ACCARIE, H., AND DELAMETTE, M., 1991, Découverte d'ammonites dans le massif Apennin de la Maiella (Plate-forme Lazio-Abruzzaise, Italie centrale): Précision sur la durée du hiatus bauxitique médio-Crétacé: Cretaceous Research, v. 12, p. 81–90.

ACCARIE, H., BEAUDOIN, B., CUSSEY, R., JOSEPH, P., AND TRIBOULET, S., 1986, Dynamique sédimentaire et structurale au passage plate-forme/bassin, les faciès carbonatés Cretacés du massif de la Maiella (Abruzzes, Italie): Rome, Memorie della Società Geologica Italiana, v. 36, p. 217–231.

ACCORDI, G., AND CARBONE, F., 1988, Sequenze carbonatiche meso-cenozoiche: Rome, Quaderni della Ricerca Scientifica, v. 114, no. 5, p. 11–92.

ADAMOLI, L., MANGANELLI, V., PIERUCCINI, U., AND ROMANA, A., 1982, Ricerche geologiche sul Gran Sasso d'Italia (Abruzzo): VII. La zona tra Valle Fredda e Valle dell'Inferno (Provv. di l'Aquila e Teramo): Camerino, Studi Geologici Camerti, v. 7, p. 105–114.

ARTHUR, M. A., AND PREMOLI SILVA, I., 1982, Development of widespread organic carbon–rich strata in the Mediterranean Tethys, *in* Schlanger, S. O. and Cita, M. B., eds., Nature and Origin of Cretaceous Carbon-Rich Facies: London, Academic Press, p. 7–54.

ARTHUR, M. A., FISCHER, A. G., PREMOLI SILVA, I., LOWRIE, W., ALVAREZ, W., ROGGENTHEN, W. M., AND NAPOLEONE, G., 1977, Upper Cretaceous–Paleocene magnetic stratigraphy at Gubbio, Italy: Geological Society of America Bulletin, v. 88, p. 367–389.

AUSTIN, J. A., JR., SCHLAGER, W., et al., 1988, Leg 101—An overview, in Austin, J. A., Jr., Schlager, W., et al., eds., Proceedings of the Ocean Drilling Project: Scientific Results, v. 101: College Station, Ocean Drilling Program, p. 455–472.

BALLY, A. W., BURBI, L., COOPER, C., AND GHELARDONI, R., 1986, Balanced sections and seismic reflection profiles across the Central Apennines: Rome, Memorie della Società Geologica Italiana, v. 35, p. 257–310.

BERNOULLI, D., 1967, Probleme der Sedimentation im Jura Westgriechenlands und des zentralen Apennins: Basel, Verhandlungen der Naturforschenden Gesellschaft, v. 78, p. 35–54.

BERNOULLI, D., 1972, North Atlantic and Mediterranean Mesozoic facies: A comparison, *in* Hollister, C. D., Ewing, J. J., et al., eds., Initial Reports of the Deep Sea Drilling Project, v. 11: Washington, D.C., United States Government Printing Office, p. 801–871.

BERNOULLI, D., AND JENKYNS, H. C., 1974, Alpine, Mediterranean and central Atlantic Mesozoic facies in relation to the early evolution of the Tethys, *in* Dott, R. H. and Shaver, R. H., eds., Modern and Ancient Geosynclinal Sedimentation: Tulsa, Society of Economic Paleontologists and Mineralogists, Special Publication 19, p. 129–160.

BERNOULLI, D., KÄLIN, O., AND PATACCA, E., 1979, A sunken continental margin of the Mesozoic Tethys: The northern and central Apennines, *in* Purser, B. and Beaudoin, B., eds., Sédimentation Jurassique W. Européen: Paris, Association Sédimentologique de France, Publication Spéciale No. 1, p. 197–201.

BERNOULLI, D., ANSELMETTI, F. S., EBERLI, G. P., MUTTI, M., PIGNATTI, J. S., SANDERS, D. G. K., AND VECSEI, A., 1996, Montagna della Maiella: The sedimentary and sequential evolution of a Bahamian-type carbonate platform of the South-Tethyan continental margin: Rome, Memorie della Società Geologica Italiana, v. 51, p. 7–12.

BICE, D. M., AND STEWART, K. G., 1990, The formation and drowning of isolated carbonate seamounts: Tectonic and ecological controls in the northern Apennines, in Tucker, M. E., Wilson, J. L., Crevello, P. D., Sarg, J. F., and Read, J. F., eds., Carbonate Platforms: Facies, Sequences and Evolution: Oxford, Blackwell Scientific Publications, International Association of Sedimentologists Special Publication 9, p. 145–168.

BOARDMAN, M. R., AND NEUMANN, A. C., 1984, Sources of periplatform carbonates, Northwest providence channel, Bahamas: Journal of Sedimentary Petrology, v. 54, p. 1110–1123.

BOLLI, H. M., AND SAUNDERS, J. B., 1985, Oligocene to Holocene low latitude planktic foraminifera, *in* Bolli, H. M., Saunder, J. B., and Perch Nielsen, K., eds., Plankton Stratigraphy, v. 1: Cambridge, Cambridge University Press, p. 155–262.

BOSELLINI, A., NERI, C., AND LUCIANI, V., 1993a, Platform margin collapses and sequence stratigraphic organization of carbonate slopes: Cretaceous–Eocene Gargano Promontory, southern Italy: Terra Nova, v. 5, p. 282–297.

BOSELLINI, A., NERI, C., AND LUCIANI, V., 1993b, Guida ai carbonatici cretaceo–eocenici di scarpata e bacino del Gargano (Italia Meridionale): Ferrara, Annali dell'Università di Ferrara, v. 4 supplemento, 77 p.

CARANNANTE, G., D'ARGENIO, B., DELLO IACOVO, B., FERRERI, V., MINDSZENTY, A., AND SIMONE, L., 1988, Studi sul carsismo Cretacico dell'Appennino Campano: Rome, Memorie della Società Geologica Italiano, v. 41, p. 733–759.

CARON, M., 1985, Cretaceous planktic foraminifera, *in* Bolli, H. M., Saunders, J. B., and Perch-Nielsen, K., eds., Plankton Stratigraphy, v. 1: Cambridge, Cambridge University Press, p. 17–86.

CATENACCI, E., 1965, Resoconto sommario delle osservazioni stratigrafiche compiute sulla Maiella (Appennino abruzzese): Rome, Servizio Geologico d'Italia, Bollettino, v. 86, p. 17–25.

COCCIONI, R., NESCI, O., TRAMONTANA, M., WEZEL, F. C., AND MORETTI, E., 1987, Descrizione di un livello-guida "Radiolaritico-bituminoso-ittiolitico" alla base delle Marne a Fucoidi nell'Appennino umbro-marchigiano: Bollettino della Società Geologica Italiana, v. 106, p. 183–192.

COLACICCHI, R., AND BALDANZA, A., 1986, Carbonate turbidites in a Mesozoic pelagic basin: (Scaglia Formation, Apennines)—Comparison with siliciclastic depositional models: Sedimentary Geology, v. 48, p. 81–105.

COLACICCHI, R., AND MONACO, P., 1994, Pure carbonate gravity flow deposits of the Scaglia basin compared with central Apennine siliciclastics (Marnoso-Arenacea and Laga): Analogies and differences: Memorie Padova, di Scienze Geologiche, v. 46, p. 23–41.

COLACICCHI, R., AND PRATURLON, A., 1965a, Il problema delle facies nel Guirese della Marsica Nord-orientale: Bollettino della Società Geologica Italiana, v. 84, p. 55–66.

COLACICCHI, R., AND PRATURLON, A., 1965b, Stratigraphical and paleogeographical investigations on the Mesozoic shelf-edge facies in eastern Marsica (Central Apennines, Italy): Geologica Romana, v. 5, p. 89–118.

CONIGLIO, M., 1986, Synsedimentary submarine slope failure and tectonic deformation in deep-water carbonates, Cow Head Group, western Newfoundland: Canadian Journal of Earth Sciences, v. 23, p. 476–490.

CONIGLIO, M., AND DIX, G. R., 1992, Carbonate slopes, *in* Walker, R. G. and James, N. P., eds., Facies Models: Response to Sea-Level Change: St. John's, Geological Association of Canada, p. 349–374.

COOK, H. E., AND MULLINS, H. T., 1983, Basin margin, *in* Scholle, P. A., Bebout, D. G., and Moore, C. H., eds., Carbonate Depositional Environments: Tulsa, American Association of Petroleum Geologists Memoir 33, p. 539–618.

CRESCENTI, U., 1969, Stratigrafia della serie calcarea del Lias al Miocene nella regione marchigiano-abruzzese (parte I—Descrizione delle serie stratigrafiche): Rome, Memorie della Società Geologica Italiana, v. 8, p. 155–204.

CRESCENTI, U., CROSTELLA, A., DONZELLI, G., AND RAFFI, G., 1969, Stratigrafia della serie calcarea dal Lias al Miocene nella regione marchigiano-abruzzese (Parte II—Litostratigrafia, Biostratigrafia, Paleogeografia): Rome, Memorie della Società Geologica Italiana, v. 8, p. 343–420.

CRESTA, S., MONECHI, S., AND PARISI, G., 1989, Mesozoic-Cenozoic stratigraphy in the Umbria-Marche area: Geological fieldtrips in the Umbria-Marche Apennines (Italy), Servizio Geologico d'Italia, Memorie Descrittive della Carta Geologica d'Italia: Rome, Istituto Poligrafico e Zecca dello Stato, v. 39, p. 185.

D'ARGENIO, B., AND MINDSZENTY, A., 1991, Karst bauxites at regional unconformities and geotectonic correlation in the Cretaceous of the Mediterranean: Bollettino della Società Geologica Italiana, v. 110, p. 85–92.

D'ARGENIO, B., DE CASTRO, P., EMILIANI, C., AND SIMONE, L., 1975, Bahamian and Apenninic limestones of identical lithofacies and age: American Association of Petroleum Geologists Bulletin, v. 59, p. 524–533.

DE ALTERIIS, G., AND AIELLO, G., 1993, Stratigraphy and tectonics offshore of Puglia (Italy, southern Adriatic Sea): Marine Geology, v. 113, p. 233–253.

DELA PIERRE, F., 1992, Stratigrafia, sedimentologia e discontinuità stratigrafiche nella successione paleogenica dell'Abruzzo Settentrionale (Zona di transizione tra la piattaforma carbonatica laziale-abruzzese ed il bacino marchigiano): Unpublished Ph.D. Dissertation, Università di Torino, Torino, 147 p.

DELA PIERRE, F., 1994, Stratigrafia della successione Cretacico–Paleogenica affiorante tra la valle del Velino e la valle dell'Aterno (Appennino centrale): Pavia, Atti Ticinensi di Scienze della Terra, v. 2 (Serie Speciale), p. 119–130.

DELA PIERRE, F., AND BRUZZONE, B., 1991, Analisi sedimentologica e biostratigrafica di alcune sezioni paleogeniche della catena del Gran Sasso d'Italia: Bollettino della Società Geologica Italiana, v. 110, p. 25–34.

DELA PIERRE, F., AND CLARI, P., 1994, Discontinuità stratigrafiche ed evoluzione tettonico-sedimentaria paleogenica della catena del Gran Sasso d'Italia: Pavia, Atti Ticinensi di Scienze della Terra, v. 2 (Serie Speciale), p. 97–117.

EBERLI, G. P., AND GINSBURG, R. N., 1989, Cenozoic progradation of northwestern Great Bahama bank, a record of lateral platform growth and sea-level fluctuation, *in* Crevello, P. D., Wilson, J. L., Sarg, J. F., and Read, J. F., eds., Controls on Carbonate Platform and Basin Development: Tulsa, Society of Economic Paleontologists and Mineralogists, Special Publication 44, p. 339–352.

EBERLI, G. P., BERNOULLI, D., SANDERS, D., AND VECSEI, A., 1993, From aggradation to progradation: The Maiella platform (Abruzzi, Italy), *in* Simo, J. A., Scott, R. W., and Masse, J. P., eds., Cretaceous Carbonate Platforms: Tulsa, American Association of Petroleum Geologists Memoir 56, p. 213–232.

ENOS, P., AND MOORE, C. H., 1983, Fore-reef slope, *in* Scholle, P. A., Bebout, D. G., and Moore, C. H., eds., Carbonate Depositional Environments: Tulsa, American Association of Petroleum Geologists Memoir 33, p. 507–538.

ENOS, P., AND STEPHENS, B. P., 1993, Mid-Cretaceous basin margin carbonates, east-central Mexico: Sedimentology, v. 40, p. 539–556.

ESTEBAN, M., AND KLAPPA, C. F., 1983, Subaerial exposure, *in* Scholle, P. A., Bebout, D. G., and Moore, C. H., eds., Carbonate Depositional Environments: Tulsa, American Association of Petroleum Geologists Memoir 33, p. 1–54.

EVERTS, A. J. W., STAFLEU, J., AND SCHLAGER, W., 1995, Stratal patterns and sequence stratigraphy at the margin of the Vercors carbonate platform (Lower Cretaceous, SE France): Journal of Sedimentary Research, v. 65B, p. 119–131.

GHISETTI, F., AND VEZZANI, L., 1983, Deformazioni pellicorali mioceniche e plioceniche nei domini strutturali esterni dell'Appennino centro-meridionale (Maiella ed Arco Morrone—Gran Sasso): Rome, Memorie della Società Geologica Italiana, v. 26, p. 563–577.

GHISETTI, F., AND VEZZANI, L., 1990a, Carta Geologica del Gran Sasso d'Italia (da Vado di Corno a Passo Capannelle): Firenze, Società Elaborazioni Cartografiche s.r.l. (S. El. Ca.).

GHISETTI, F., AND VEZZANI, L., 1990b, Stili strutturali nei sistemi di sovrascorrimento della Catena del Gran Sasso (Appennino Centrale): Camerino, Studi Geologici Camerti, Volume Speciale 1990, p. 37–50.

GHISETTI, F., AND VEZZANI, L., 1991, Thrust geometries and sequence of imbrication in the Gran Sasso chain (Central Apennines, Italy): Bollettino della Società Geologica Italiana, v. 110, p. 427–440.

GHISETTI, F., BARCHI, M., BALLY, A. W., MORETTI, I., AND VEZZANI, L., 1993, Conflicting balanced structural sections across the central Apennines (Italy): Problems and implications, *in* Spencer, A. M., ed., Generation, Accumulation and Production of Europe's Hydrocarbons: Berlin, Springer-Verlag, Special Publication of the European Association of Petroleum Geoscientists, v. 3, p. 219–231.

GLASER, K. S., AND DROXLER, A. W., 1993, Controls and development of late Quaternary peri-platform carbonate stratigraphy in Walton basin (Northeastern Nicaragua Rise, Caribbean Sea): Paleoceanography, v. 8, p. 243–274.

GRAMMER, G. M., AND GINSBURG, R. N., 1992, Highstand versus lowstand deposition on carbonate platform margins: Insight from Quaternary foreslope in the Bahamas: Marine Geology, v. 103, p. 125–136.

GRAMMER, G. M., GINSBURG, R. N., SWART, P. K., MCNEILL, D. F., JULL, A. J. T., AND PREZBINDOWSKY, D. R., 1993, Rapid growth rates of syndepositional marine aragonite cements in steep marginal slope deposits, Bahama and Belize: Journal of Sedimentary Petrology, v. 63, p. 983–989.

HANDFORD, C. R., AND LOUCKS, R., 1993, Carbonate depositional sequences and system tracts—Response of carbonate platforms to relative sea-level changes, *in* Loucks, R. L. and Sarg, J. F., eds., Carbonate Sequence Stratigraphy: Recent Developments and Applications: Tulsa, American Association of Petroleum Geologists Memoir 57, p. 3–39.

HAQ, B. U., HARDENBOL, J., AND VAIL, P., 1988, Mesozoic and Cenozoic chronostratigraphy and cycles of sea-level change, *in* Wilgus, C. K., Hastings, B. S., Kendall, C. G. S. C., Posamentier, H. W., Ross, C. A., and van Wagoner, J. C., eds., Sea-Level Changes—An Integrated Approach: Tulsa, Society of Economic Paleontologists and Mineralogists, Special Publication 42, p. 72–108.

HARWOOD, G. M., AND TOWERS, P. A., 1988, Seismic sedimentologic interpretation of a carbonate slope, north margin of Little Bahama Bank, *in* Austin, J. A., Schlager, W., et al., eds., Proceedings of the Ocean Drilling Project: Scientific Results: College Station, Ocean Drilling Program, v. 101, p. 263–277.

HUNT, D., AND TUCKER, M. E., 1993, Sequence stratigraphy of carbonate shelves with an example from the mid-Cretaceous (Urgonian) of southeast France, *in* Posamentier, H. W., Summerhayes, C. P., Haq, B. U., and Allen, G. P., eds., Sequence Stratigraphy and Facies Associations: Oxford, International Association of Sedimentologists Special Publication 18, p. 307–341.

KENTER, J. A. M., 1990, Carbonate platform flanks: Slope angle and sediment fabric: Sedimentology, v. 37, p. 777–794.

KOOPMAN, A., 1983, Detachment tectonics in the Central Apennines, Italy: Utrecht, Geologica Ultraiectina, v. 30, 155 p.

LAMPERT, S. A., LOWRIE, W., HIRT, A. M., BERNOULLI, D., AND MUTTI, M., 1997, Magnetic and sequence stratigraphy of redeposited Upper Cretaceous limestones in the Montagna della Maiella, Abruzzi, Italy: Earth and Planetary Science Letters, v. 150, p. 79–93.

LUTERBACHER, H. P., AND PREMOLI SILVA, I., 1964, Biostratigrafia del limite Cretaceo–Terziario nell'Appennino centrale: Rivista Italiana di Paleontologia e Stratigrafia, v. 70, p. 67–128.

MITCHUM, R. M., VAIL, P. R., AND THOMPSON, S. I., 1977, Seismic stratigraphy and global changes of sealevel, part 2: The depositional sequence as a basic unit for stratigraphic analysis, *in* Payton, C. E., ed., Seismic Stratigraphy—Applications to Hydrocarbon Exploration: Tulsa, American Association of Petroleum Geologists Memoir 26, p. 53–62.

MONTANARI, A., CHAN, L. S., AND ALVAREZ, W., 1989, Synsedimentary tectonics in the Late Cretaceous–Early Tertiary pelagic basin of the northern Apennines, Italy, *in* Crevello, P. D., Wilson, J. L., Sarg, J. F., and Read, J. F., eds., Controls on Carbonate Platform and Basin Development: Tulsa, Society of Economic Paleontologists and Mineralogists, Special Publication 44, p. 379–399.

MOUSSAVIAN, E., AND VECSEI, A., 1995, Paleocene reef sediments from the Maiella carbonate platform, Italy: Facies, v. 32, p. 213–222.

MULLINS, H. T., AND COOK, H. E., 1986, Carbonate apron models: Alternatives to the submarine fan model for paleoenvironmental analysis and hydrocarbon exploration: Sedimentary Geology, v. 48, p. 37–79.

MULLINS, H. T., AND HINE, A. C., 1989, Scalloped bank margins: Beginning of the end for carbonate platforms: Geology, v. 17, p. 30–33.

MULLINS, H. T., HEATH, K. C., VAN BUREN, H. M., AND NEWTON, C. R., 1984, Anatomy of a modern open-ocean carbonate slope: Northern Little Bahama Bank: Sedimentology, v. 31, p. 141–168.

MULLINS, H. T., WISE, S. W., GARDULSKI, A. F., HINCHEY, E. J., MASTERS, P. M.,

AND SIEGEL, D. I., 1985, Shallow subsurface diagenesis of Pleistocene periplatform ooze: Northern Bahamas: Sedimentology, v. 32, p. 473–494.
MULLINS, H. T., GARDULSKI, A. F., AND HINE, A., 1986, Catastrophic collapse of the west Florida carbonate platform margin: Geology, v. 14, p. 167–170.
MULLINS, H. T., GARDULSKI, A. F., HINE, A. C., SHERWOOD, W. W., AND APPLEGATE, J., 1988, Three-dimensional sedimentary framework of the carbonate ramp slope of central west Florida: A sequential seismic stratigraphic perspective: Geological Society of America Bulletin, v. 100, p. 514–533.
MUTTI, E., 1979, Turbidites et cônes sous-marins profonds, *in* Homewood, P., ed., Sédimentation Détritique (fluviatile, littorale et marine): Fribourg, Troisième Cycle Romand en Sciences de la Terre, Institut de Géologie de l'Université de Fribourg (CH), p. 353–419.
MUTTI, M., 1995, Porosity development and diagenesis in the Orfento supersequence and its bounding unconformities (Upper Cretaceous, Montagna della Maiella, Italy), *in* Budd, D., Harris, P. M., and Saller, A., eds., Porosity in Carbonate Strata: Tulsa, American Association of Petroleum Geologists Memoir 61, p. 141–158.
MUTTI, M., BERNOULLI, D., EBERLI, G. P., AND VECSEI, A., 1996, Depositional geometries and facies associations in an Upper Cretaceous prograding carbonate platform margin (Orfento supersequence, Maiella, Italy): Journal of Sedimentary Research, v. 66B, p. 749–765.
NARDIN, T. R., HEIN, F. J., GORSLINE, D. S., AND EDWARDS, B. D., 1979, A review of mass movement processes, sediment and acoustic characteristics, and contrasts in slope and base-of-slope systems versus canyon-fan-basin floor systems, *in* Pilkey, O. H. and Doyle, L. S., eds., Geology of Continental Slopes: Tulsa, Society of Economic Paleontologists and Mineralogists, Special Publication 27, p. 61–73.
NERI, C., 1993, Stratigraphy and sedimentology of the Monte Acuto Formation (Upper Cretaceous–lower Paleocene, Gargano Promotory, southern Italy): Ferrara, Annali dell'Università di Ferrara, v. 4, p. 13–44.
PERCH-NIELSEN, K., 1985, Mesozoic calcareous nannofossils, *in* Bolli, H. M., Saunders, J. B., and Perch-Nielsen, K., eds., Plankton Stratigraphy, v. 1: Cambridge, Cambridge University Press, p. 329–554.
POSAMENTIER, H. W., AND VAIL, P. R., 1988, Eustatic controls on clastic deposition II—Sequences and system tract models, *in* Wilgus, C. K., Hastings, B. S., Kendall, C. G. S. C., Posamentier, H. W., Ross, C. A., and van Wagoner, J. C., eds., Sea-Level Changes—An Integrated Approach: Tulsa, Society of Economic Paleontologists and Mineralogists, Special Publication 42, p. 125–154.
POSAMENTIER, H. W., JERVEY, M. T., AND VAIL, P. R., 1988, Eustatic control on clastic deposition I—Conceptual framework, *in* Wilgus, C. K., Hastings, B. S., Kendall, C. G. S. C., Posamentier, H. W., Ross, C. A., and van Wagoner, J. C., eds., Sea-Level Changes—An Integrated Approach: Tulsa, Society of Economic Paleontologists and Mineralogists, Special Publication 42, p. 109–123.
PRATURLON, A., AND SIRNA, G., 1976, Ulteriori dati sul margine cenomaniano della piattaforma carbonatica laziale-abruzzese: Geologica Romana, v. 15, p. 83–111.
PREMOLI SILVA, I., AND SLITER, W. V., 1995, Cretaceous planktonic foraminiferal biostratigraphy and evolutionary trends from the Bottaccione section, Gubbio, Italy: Palaeontographica Italica, v. 82, p. 1–89.
SANDERS, D. G. K., 1994, Carbonate platform growth and erosion: The Cretaceous–Tertiary of Montagna della Maiella, Central Italy: Unpublished Ph.D. Thesis Nr. 10800, Eidgenössische Technische Hochschule, Zürich, 122 p.
SANTANTONIO, M., 1993, Facies associations and evolution of pelagic carbonate platform/basin systems: Examples from the Italian Jurassic: Sedimentology, v. 40, p. 1039–1067.
SARG, J. F., 1988, Carbonate sequence stratigraphy, *in* Wilgus, C. K., Hastings, B. S., Kendall, C. G. S. C., Posamentier, H. W., Ross, C. A., and van Wagoner, J. C., eds., Sea-Level Changes: An Integrated Approach: Tulsa, Society of Economic Paleontologists and Mineralogists, Special Publication 42, p. 155–181.
SCHLAGER, W., 1981, The paradox of drowned reefs and carbonate platforms: Geological Society of America Bulletin, v. 92, p. 197–211.
SCHLAGER, W., 1989, Drowning unconformities on carbonate platforms, *in* Crevello, P. D., Wilson, J. L., Sarg, J. F., and Read, J. F., eds., Controls on Carbonate Platform and Basin Development: Tulsa, Society of Economic Paleontologists and Mineralogists, Special Publication 44, p. 15–26.
SCHLAGER, W., 1992, Sedimentology and Sequence Stratigraphy of Reefs and Carbonate Platforms: Tulsa, American Association of Petroleum Geologists Continuing Education Course Note Series, v. 34, 71 p.
SCHLAGER, W., 1993, Accommodation and supply—A dual control on stratigraphic sequences: Sedimentary Geology, v. 86, p. 111–136.
STOW, D. A. V., AND LOVELL, J. P. B., 1979, Contourites: Their recognition in modern and ancient sediments: Earth-Science Reviews, v. 14, p. 251–291.
TOUMARKINE, M., AND LUTERBACHER, H. P., 1985, Paleocene and Eocene planktic foraminifera, *in* Bolli, H. M., Saunders, J. B., and Perch-Nielsen, K., eds., Plankton Stratigraphy, v. 1: Cambridge, Cambridge University Press, p. 87–154.
TUCKER, M. E., CALVET, F., AND HUNT, D., 1993, Sequence stratigraphy of carbonate ramps: System tracts, models and application to the Muschelkalk carbonate platform of eastern Spain, *in* Posamentier, H. W., Summerhayes, C. P., Haq, B. U., and Allen, G. P., eds., Sequence Stratigraphy and Facies Associations: Oxford, International Association of Sedimentologists Special Publication 18, p. 397–418.
VAIL, P. R., TODD, R. G., AND SANGREE, J. B., 1977, Seismic stratigraphy and global changes of sea level, Part five: Chronostratigraphic significance of seismic reflections, *in* Payton, C. E., ed., Seismic Stratigraphy—Application to Hydrocarbon Exploration: Tulsa, American Association of Petroleum Geologists Memoir 26, p. 99–116.
VAN KONIJNENBURG, J.-H., 1997, Sedimentology and stratigraphic architecture of a Cretaceous to lower Tertiary base-of-slope succession, Gran Sasso d'Italia, Central Apennines, Italy: Unpublished Ph.D. Thesis No. 12361, Eidgenössische Technische Hochschule Zürich, 147 p.
VAN KONIJNENBURG, J.-H., WERNLI, R., AND BERNOULLI, D., 1998, Tentative biostratigraphy of Paleogene planktic foraminifera in thin-section, an example from the Gran Sasso d'Italia (central Apennines, Italy): Eclogae Geologicae Helvetiae, v. 91, p. 203–216.
VECSEI, A., 1991, Aggradation und Progradation eines Karbonatplattform-Randes: Kreide bis mittleres Tertiär der Montagna della Maiella, Abruzzen: Mitteilungen aus dem Geologischen Institut der Eidgenössischen Technischen Hochschule und der Universität Zürich, Neue Folge 294, 169 p.
VECSEI, A., SANDERS, D. G. K., BERNOULLI, D., EBERLI, G. P., AND PIGNATTI, J. S., 1998, Sequence stratigraphy and evolution of the Maiella carbonate platform margin, Cretaceous to Miocene, Italy, *in* de Graciansky, P. Ch., Jacquin, T., and Vail, P. R., eds., Mesozoic and Cenozoic Sequence Stratigraphy of European Basins: Tulsa, Society of Economic Paleontologists and Mineralogists, Special Publication 60, p. 53–74.
WALKER, R. G., 1992, Turbidites and submarine fans, *in* Walker, R. G. and James, N. P., eds., Facies Models: Response to Sea-level Change: St. John's, Geological Association of Canada, p. 239–264.
WATTS, K. F., AND BLOME, C. D., 1990, Evolution of the Arabian carbonate platform margin slope and its response to orogenic closing of a Cretaceous ocean basin, Oman, *in* Tucker, M. E., Wilson, J. L., Crevello, P. D., Sarg, J. R., and Read, J. F., eds., Carbonate Platforms: Facies, Sequences and Evolution: Oxford, Blackwell Scientific Publications, International Association of Sedimentologists Special Publication 9, p. 291–324.
WATTS, K. F., AND GARRISON, R. E., 1986, Sumeini Group, Oman—Evolution of a Mesozoic carbonate slope on a South Tethyan continental margin: Sedimentary Geology, v. 48, p. 107–168.
WEISSERT, H., AND LINI, A., 1991, Ice age interlude during the time of Cretaceous greenhouse climate?, *in* Müller, D. W., McKenzie, J. A., and Weissert, H., eds., Controversies in Modern Geology: London, Academic Press, p. 173–192.
ZAMPARELLI, V., 1966, Le microfacies cretaceo–eoceniche nella serie di Rio Arno (Gran Sasso d'Italia): Bollettino della Società dei Naturalisti in Napoli, v. 75, p. 552–570.

SEDIMENTOLOGY AND ACCOMMODATION CYCLES OF PARIS BASIN CAMPANIAN CHALK: THE KEY TO HIGH-RESOLUTION STRATIGRAPHY AND SEISMIC SIGNATURE

MONIQUE METTRAUX

9, rue Maubec 64290 Gan, France

PETER HOMEWOOD

Elf Exploration Production, CSTJF, Avenue Larribau, 64018 Pau, France

ANNE SCHWAB*

Elf Exploration Production, CSTJF, Avenue Larribau, 64018 Pau, France

AND

FRANCOIS GUILLOCHEAU

Géosciences Rennes, Campus de Beaulieu, 35042 Rennes, France

ABSTRACT: Facies of Campanian chalk in the Paris basin are shown to change as a function of accommodation cycles. Fourteen facies—identified in the field at the Chaux-de-Boran and Précy-sur-Oise quarries at Précy-sur-Oise near the Oise River, north of Paris, and studied by thin-section and SEM (scanning electron microscope) analysis—make up four well-defined facies associations. Three nannofacies compose the background chalk sedimentation. Cherts are more abundant in the progradational phases of 5–10-m-thick, higher-frequency cycles. Depositional environments were generally below fair-weather wave base and built up from below storm wave base. The tops of some shallowing-upward cycles were reworked by current activity, perhaps near to fair-weather wave base.

Patterns of chert distribution and abundance, early-cemented firmgrounds, and evidence of storm-current reworking allow interpretation in terms of progradation, aggradation, and sediment bypass. Two sedimentological models illustrate the contrasting conditions under increasing and decreasing accommodation. Sedimentation took place over a low-relief profile far from any shoreline, with low-angle clinoforms prograding from storm-winnowed platform areas that lay above storm wave base into slightly deeper lows.

Symmetrical to asymmetrical cycles of facies associations, corresponding to progradational-aggradational cycles, define genetic stratigraphic units. These stratigraphic units resulted from cycles of increasing and decreasing accommodation of at least two frequencies. Major hardgrounds record sediment bypass at the turnarounds from longer-term decreasing to increasing accommodation.

A simple seismic model shows that the abundance and distribution of chert may explain amplitude changes observed on regional seismic data. Progradational intervals with higher concentrations of chert and capped by hardgrounds should create continuous reflections on a seismic line. These reflections should be of variable amplitude, depending upon the spacing between individual chert layers.

INTRODUCTION

Upper Cretaceous sedimentary rocks of the Paris basin are mainly composed of chalk. The three main lithological units, bounded by key markers (Plenus and Lewes Marls), are the "Craie Glauconieuse" or glauconitic chalk, the "Craie Argileuse" or argillaceous chalk, and the "Craie Blanche" or white chalk (Fig. 1).

Previous work on Paris basin chalk has provided an extensive regional stratigraphic database, together with a reliable biostratigraphic framework, whereas sedimentological interpretations have traditionally emphasized the pelagic nature of the chalk.

In one of these earlier studies, Juignet (1980) identified a type sequence ("séquence virtuelle"). This is a sort of facies model in which an ideal vertical section shows all the facies observed in a stratigraphic unit. They are drawn in a succession representing a proximal to distal trend (Lombard, 1956). Incompleteness of facies in a vertical section at any particular locality was explained by lack of preservation or a limited extent to transgressive-regressive cycles. In their analysis of the regional stratigraphy, Pomerol and Mortimore (1990) and Mortimore and Pomerol (1991) based their correlations on the detailed identification of trace fossils.

Quine and Bosence (1991) studied the Turonian to Coniacian chalk (Hancock's [1976] "white chalk facies") in Normandy at the Etretat Cliffs and distinguished 13 facies from thin section and SEM (scanning electron microscope) analysis. According to their observations, the large-scale geometries of middle Turonian to latest Coniacian erosive submarine channels (seen on the coastal cliffs, Quine, 1988) are the result of several phases of scouring. These channels were localized where submarine morphology was enhanced by local tectonic activity. Channel widths vary from 70 to 900 m, and their depths range from 10 to 60 m.

Previous work on seismic data within the often-called "homogeneous" Upper Cretaceous Chalk Group has been limited to observations of large-scale geometries, such as the erosional channels of Quine and Bosence (1991). These features were also documented in the subsurface (Hanot and Renoux, 1991). Regional onlaps observed on seismic sections have been associated with tectonic events (Nygaard et al., 1990). As far as we know, the effect of the chert layers on the seismic signal has never been investigated, although previous work (Kristensen et al., 1995) has shown that a hardground within the chalk produces a continuous, high-amplitude seismic reflection.

The aim of this study is to provide coherent sedimentological and genetic stratigraphic models for the white chalk of Campanian age in the central part of the Paris basin. These models combine the data from the literature with our own facies analysis in the field, under the SEM, and from thin-section analyses. These models are

Present address: Elf Exploration UK PLC, 1 Claymore Drive, Bridge of Don, Aberdeen AB23 8GB, United Kingdom

Advances in Carbonate Sequence Stratigraphy: Application to Reservoirs, Outcrops and Models, SEPM Special Publication No. 63
 ISBN 1-56576-060-3, p. 317–334

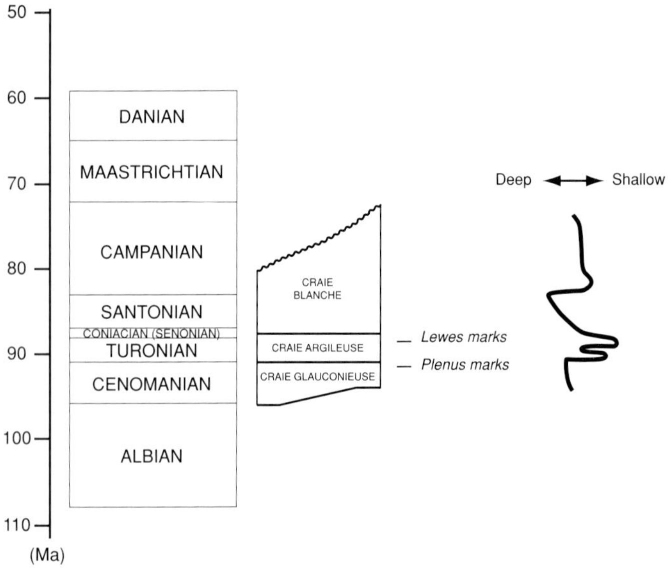

FIG. 1.—Stratigraphy of the Paris basin chalk. Sea-level curve was deduced from facies in the Paris basin.

the first step toward other studies under way that will yield basinwide high-resolution stratigraphic correlation, study of the postchalk erosion related to Tertiary inversion tectonics, and the better understanding of seismic images from the chalk. With regard to the latter, the distribution of chert as scattered or concentrated layers within the chalk is shown to depend upon genetic stratigraphic units and their architecture over longer-term periods of change in accommodation. We have therefore carried out preliminary modeling of the seismic response to compare with signatures known from proprietary data.

Two quarries (the Chaux-de-Boran and Précy-sur-Oise quarries at Précy-sur-Oise near the Oise River) in particular provided the field data. They are located on the vertical axis of the Bray anticline, near Chantilly, 60 km northeast of Paris (Fig. 2).

GEOLOGIC SETTING

The Paris basin lies in an intracratonic setting in which sedimentation, spanning Early Triassic to middle Miocene, continued over 230 m.y. (Mégnien, 1980a, b, c; Perrodon and Zabek, 1990; Dercourt, 1997). The basin fill records successive deformations of the western European lithosphere, responding to the evolution of the Atlantic and Tethys oceans (Robin, 1997). The

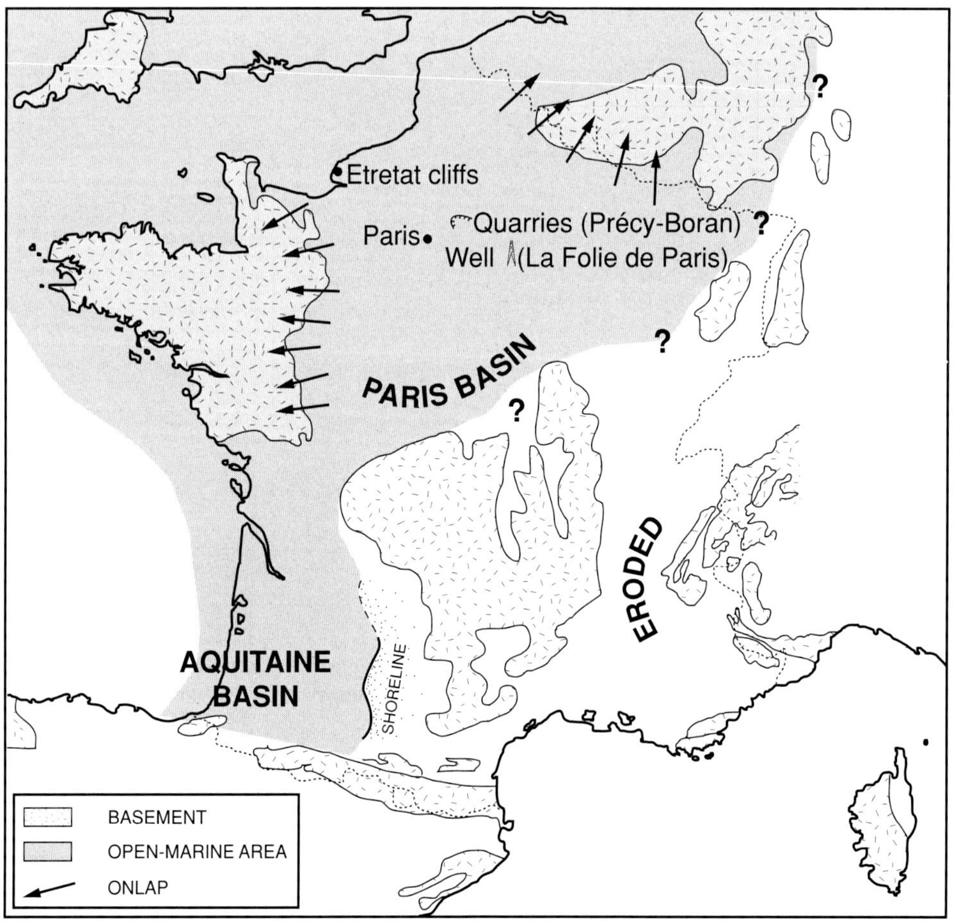

FIG. 2.—Location map showing known paleogeographical features at Campanian time. The quarries studied (Chaux-de-Boran and Précy-sur-Oise) and the well designated "La Folie de Paris" are located northeast and east of Paris. The Normandy cliffs and locality of Etretat are located to the northwest of Paris. The shoreline in the Aquitaine basin is based on Platel (1996).

sedimentary history of the basin may be subdivided into four periods: Triassic–Jurassic, Cretaceous, Paleogene, and Neogene.

A geohistory plot (tectonic subsidence vs. time curve, Fig. 3) documents a high rate of subsidence under an extensional regime from Triassic to Late Jurassic time. Early deposits were continental clastics, followed by four successive carbonate platforms. With the opening of the Bay of Biscay at the Jurassic/Cretaceous boundary, cessation of subsidence caused the base Cretaceous unconformity. The overlying Lower Cretaceous continental clastics were derived from weathering of the Armorican Massif. Renewed subsidence in the Late Cretaceous saw deposition of the Upper Cretaceous chalk (Fig. 4).

The Paris basin acquired its present morphology during medium-wavelength compressive deformation (150 km) at the Cretaceous/Tertiary boundary. The uppermost Cretaceous (Maastrichtian) section was eroded at that time, followed by compression recording Alpine deformation. Low rates of subsidence continued until the early Oligocene (or the middle Miocene in the Loire area) before the general uplift of western Europe.

The three Cretaceous chalk units are separated by the Plenus and the Lewes Marls (Fig. 1). The shallowest chalk facies was deposited during late Turonian time, and the deepest one was laid down during the Santonian. The Cenomanian/Turonian boundary coincided with a marked basin-wide shallowing. The detailed paleogeography during Campanian time (Fig. 2) is poorly known because of the broad onlap of the chalk over the surrounding basement. It is clear that true shorelines formed far south in the Aquitaine area (Platel, 1996), where emergence was caused by significant uplift due to the convergence between Europe and Iberia. During Campanian time, chalk deposits were laid down under small fluctuations in bathymetry, and they record a generally aggradational trend. No Campanian deposits are preserved in the southwestern part of the Paris basin.

METHODS

Field work comprised observations of facies, facies successions, and bedding geometries at scales from the hand lens to quarry faces (several hundred meters wide by 100 m high). Vertical sections were measured at 60–80-m spacing, and beds and chert layers were followed between measured sections. Sections were correlated between quarries on the basis of a well-identified hardground and several depositional sequences below the hardground.

Sample analysis at the macroscale, microscale, and nannoscale confirmed the identification of facies in the field. In

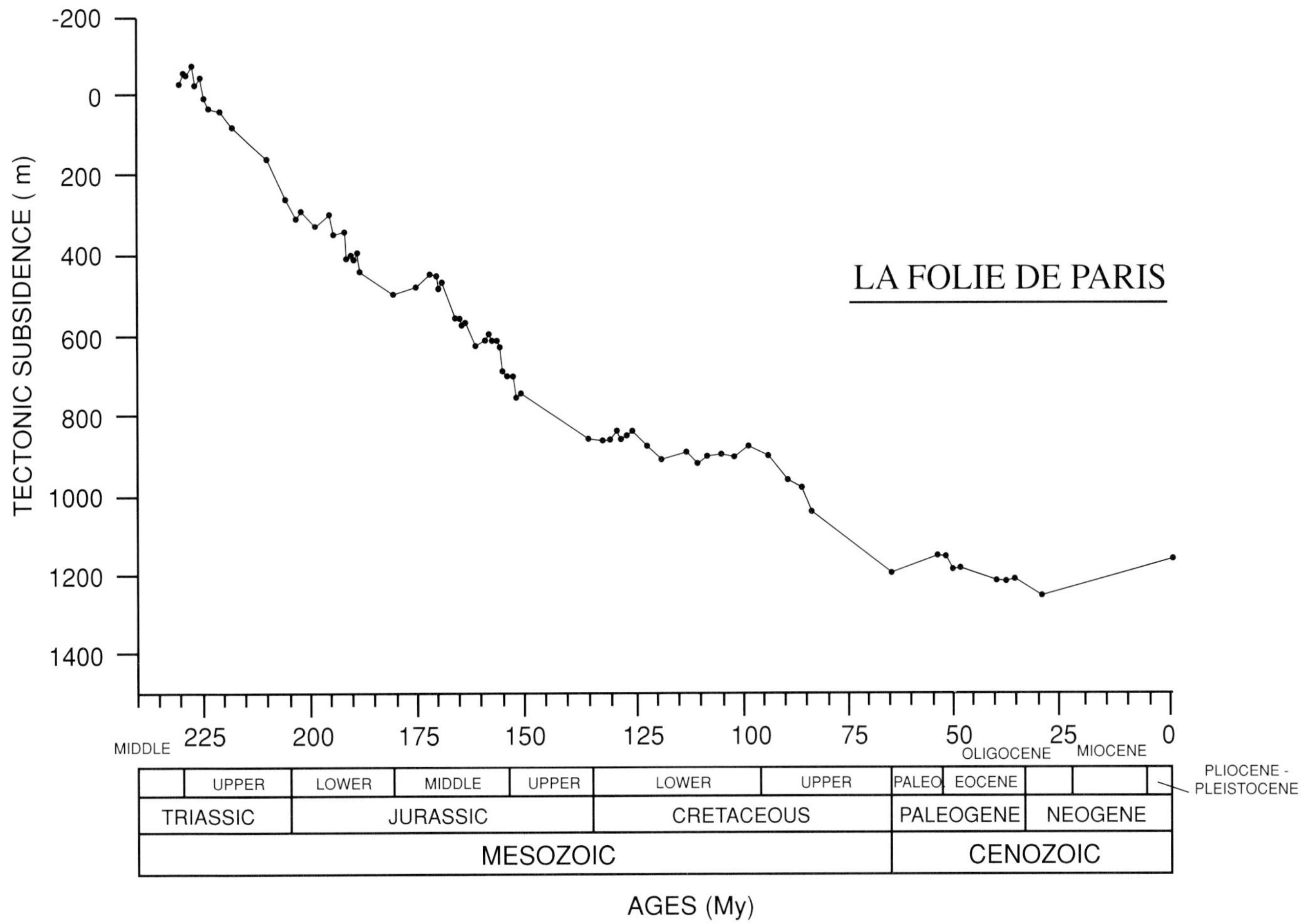

FIG. 3.—Geohistory of the region as shown by the tectonic subsidence documented in the well La Folie de Paris (see Fig. 2 for location). Data from Rennes University, GEOSCIENCES-RENNES database.

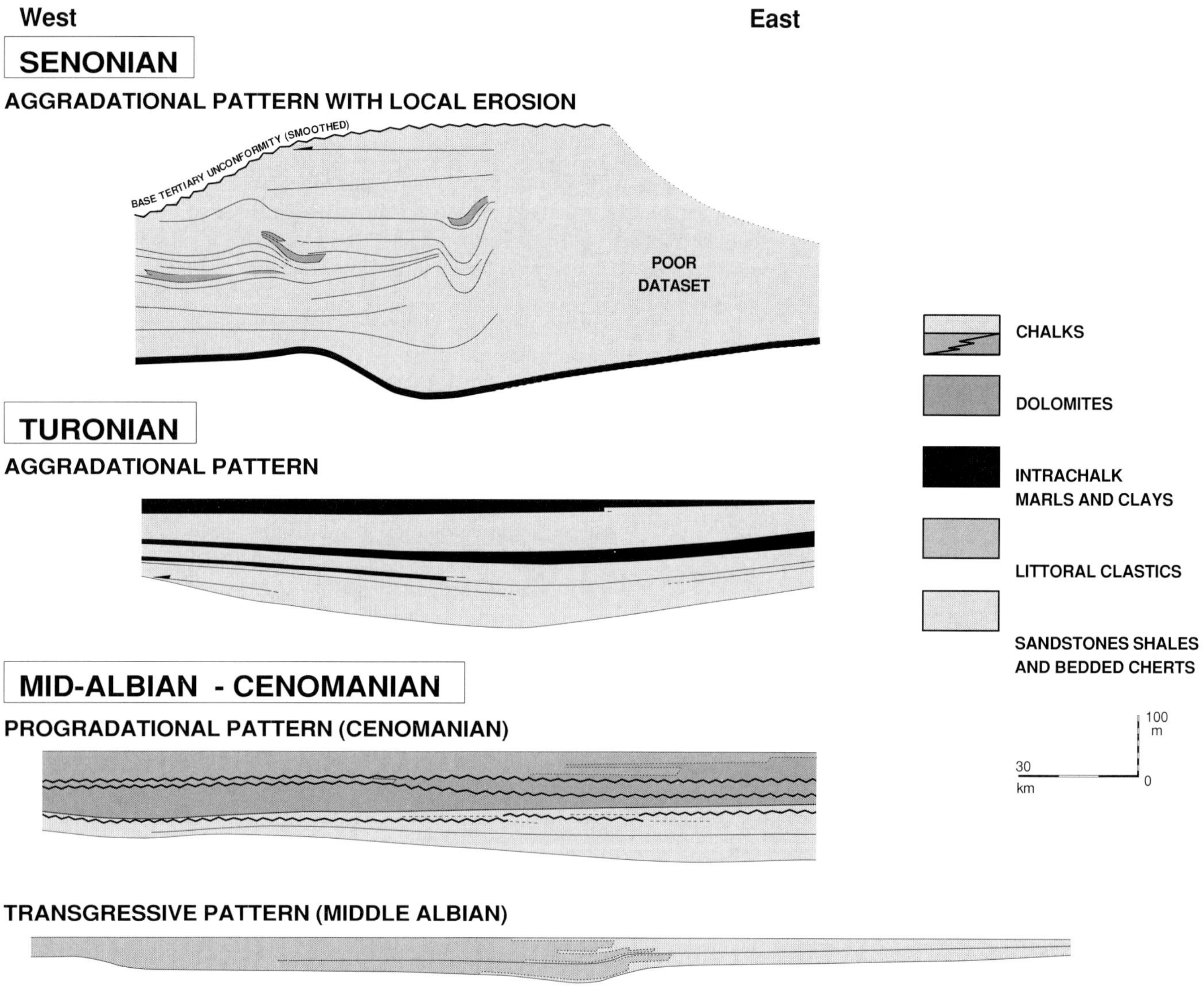

FIG. 4.—Paris basin west-east regional cross section (Rambouillet-Nancy), from mid-Albian to Senonian. Yellow designates the white and argillaceous chalks; oranges designate the glauconitic chalk. Wavy lines indicate extensive hardgrounds.

the quarries, 80 samples were collected to be slabbed in order to observe sedimentary structures after impregnation with oil (Bromley, 1981). Of these 80 samples, 41 were then chosen for thin sections, which were stained with Alizarin-red and Potassium Ferricyanide. Seven microfacies were identified by using a polarizing optical microscope. Ten samples, corresponding to the facies below, within and above the hardground, were analyzed under the SEM. Three different nannofacies compose the various chalk types.

Facies and facies associations were interpreted in terms of depositional processes and relative bathymetry, in order to establish a sedimentological model. Facies associations and sequences were interpreted in terms of increasing and decreasing accommodation from the combination of sedimentary petrography, facies interpretation in terms of depositional processes and bathymetry, and degrees of amalgamation and ratios of progradation to aggradation as deduced from bedding geometries. The cycles of increasing and decreasing accommodation define genetic stratigraphic units (Jervey, 1988; Cross, 1988; Homewood et al., 1992; Cross et al., 1993). The facies partitioning identified between times of increasing and decreasing accommodation dictated a separate facies model for each of these two modes.

Simple one-dimensional seismic modeling was carried out in order to identify the effects of the distribution of the chert layers within the progradational and aggradational cycles. This modeling was done with a zero-phase Ricker wavelet of SEG (Society of Exploration Geophysicists) reverse polarity.

SEDIMENTOLOGY

The white chalk in the Chaux-de-Boran and Précy-sur-Oise quarries is mainly composed of pelagic organisms (coccoliths, planktic crinoids, foraminifera, calcispheres) together with fragments of echinoderms, sponges, pelecypods, and some bryozoans. The chalk shows different degrees of bioturbation, firmground (Quine and Bosence, 1991) to hardground early lithification with

FIG. 5.— Chaux-de-Boran quarry outcrop. Chert forms stringers, dense layers, and dispersed nodules. A1 and A2: continuous layers of dense nodular chert. B: Continuous thin layer of nodular chert. C: Scattered nodular chert in the chalk. D: Swaley stringers of nodular chert. The section measures 6 m between A1 and A2.

indurated sediment partially reworked into microbreccias, extensive development of cherts (Fig. 5, Chaux-de-Boran quarry), and 1–10-m-scale horizontal to vertical fracture patterns. This white chalk is composed of 95% $CaCO_3$ with up to 40% porosity. Textures are frequently wackestone to packstone, rarely mudstone. In the field, 14 facies were identified and are presented here on a composite measured section (Fig. 6; note five chert layers).

Field Observations

Chert nodules in the Chaux-de-Boran quarry provide the most prominent features seen at the outcrop. The cherts show variations in their pattern of vertical distribution as well as in their lateral continuity. Six easily recognized chert types occur. Five are defined by the patterns described (Fig. 5), which seem to follow primary sedimentary features, whereas the sixth type is of later, diagenetic origin:

1. dispersed nodular chert,
2. oblique (slightly inclined) clinoformlike stringers,
3. festoons of concave-up, meter-scale stringers,
4. undulatory stringers, extensive flat stringers,
5. thin but laterally continuous dense layers of nodular chert, and
6. later, diagenetic chert, related to horizontal or vertical fractures.

The laterally continuous layers of chert are numbered upward from the Chaux-de-Boran quarry floor as it was in 1995 (Fig. 6). A conspicuous, 10-cm-thick hardground that is yellow when weathered occurs between dense nodular chert layers 4 and 5. This hardground is continuous throughout the whole quarry and provides a marker bed that is also present in the Précy-sur-Oise quarry 2 km away. The hardground is enriched in phosphates and contains glauconite. Thin gray or white beds, commonly associated with fine breccias of reworked chalk intraclasts are formed by firmgrounds or minor hardgrounds.

Macrofacies

Ten different lithologies are identified by the presence or absence of chert, presence or absence of bioturbation, chert den-

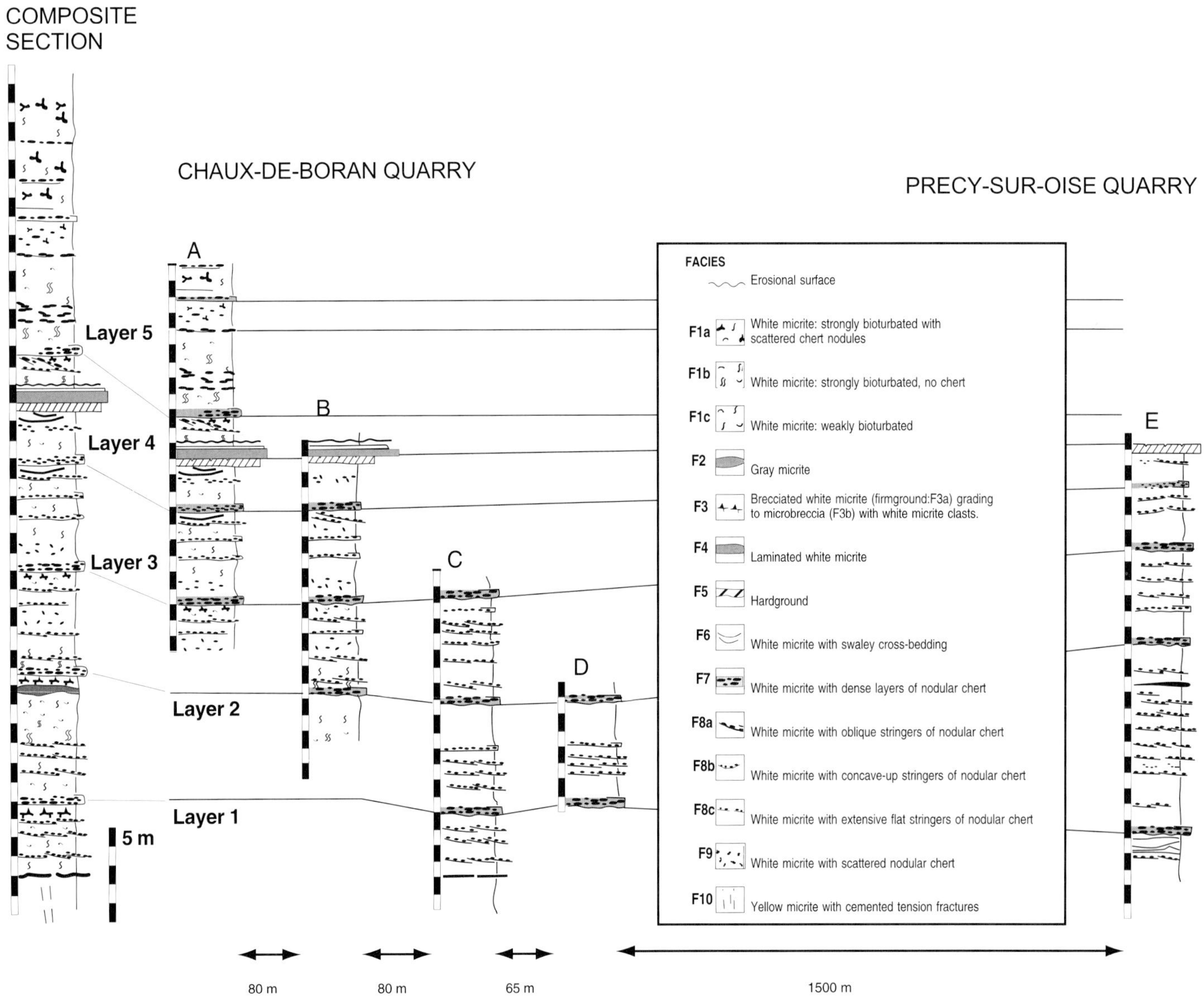

FIG. 6.—Measured sections. (A) Facies types located on a composite section from the two quarries. Layers 1, 2, 3, and 4 are continuous layers of dense nodular chert below the hardground. Layer 5 is a similar layer located above the hardground. (B) Five sections measured at the two outcrops. Sections A, B, C, and D are from the Chaux-de-Boran quarry, and section E is from the Précy-sur-Oise quarry. The two quarries are 1500 m apart.

sity, and the firmgrounds and the hardground. The patterns of the chert observed at the field scale (allowing it to be described as flat, oblique, or undulatory stringers, as festoons, or as laterally continuous dense layers) plus the degree of bioturbation and chert types within the "background" white chalk then allow us to distinguish 14 facies altogether (Table 1).

The 14 facies initially identified in outcrops (Fig. 6) were studied in further detail. Impregnation of slabbed samples with oil (Bromley, 1981) improved the visibility of numerous sedimentary features (Figs. 7, 8), but few new features were observed through this approach, when compared with those seen under fresh, damp conditions at the quarry faces. The facies are summarized in Table 1.

Microfacies

Microfacies (Fig. 9) are predominantly wackestones and packstones, but rarely are mudstones (Table 2). Allochems are bioclasts, comprising echinoderm fragments, sea-urchin spines, planktic crinoids (*Saccocoma*), benthic and planktic foraminifera, calcispheres, pelagic bivalves, siliceous sponges, and rare bryozoans.

Average porosity is 25%, and highest porosity is about 40%. Bioturbation, emphasized by differential cementation, is better seen on oil-impregnated surfaces and stained thin sections. The hardground has a microgranular siliceous cement, and the moldic porosity is partially filled with chalcedony. The firmgrounds have nonferroan calcite cements, and dolomite only occurs in cemented vertical tension fractures.

TABLE 1.—CHALK FACIES

Facies	Color	Primary Sedimentary Features	Lithification	Cherts
F1				
a	white	strongly bioturbated		scattered, chert nodules
b	white	strongly bioturbated		no chert
c	white	weakly bioturbated		rare chert
F2	gray	decimeter-scale wavelength, centimeter-scale amplitude, wavy laminae	early indurated	
F3	white	brecciated, grading to microbreccia with chalk clasts	early-indurated	angular, millimeter-sized chert clasts (rare)
F4	white like snow	thin layer; low-amplitude to undulatory laminae (wave ripples); sharp flat base and top; only seen directly above the hardground		
F5	weathers yellow	irregular upper surface, bored bioclasts	indurated hardground	
F6	white	low-angle oblique and swaley cross-stratification		
F7	white			dense thick or thin layers of nodular chert that are continuous throughout both quarries; occur at 6-8 m intervals
F8				
a	white			meter-scale stringers of nodular chert (i.e. in festoons)
b	white			clinoform, meter-scale stringers
c	white			undulatory, meter-scale stringers
F9	white			scattered nodular chert
F10	yellow			late-diagenetic, dolomite-cemented, vertical tension fractures

FIG. 7.—An oil-impregnated slab of laminated white micrite (facies F4). The slabbed sample (collected above the hardground-facies F5) shows a succession of light and darker layers containing scattered intraclasts. Cross-laminae and graded chalk intraclasts (microbreccias) are evidence for storm reworking. Surfaces of lighter-colored bands (firmgrounds) show erosional truncation (Sample BOH1).

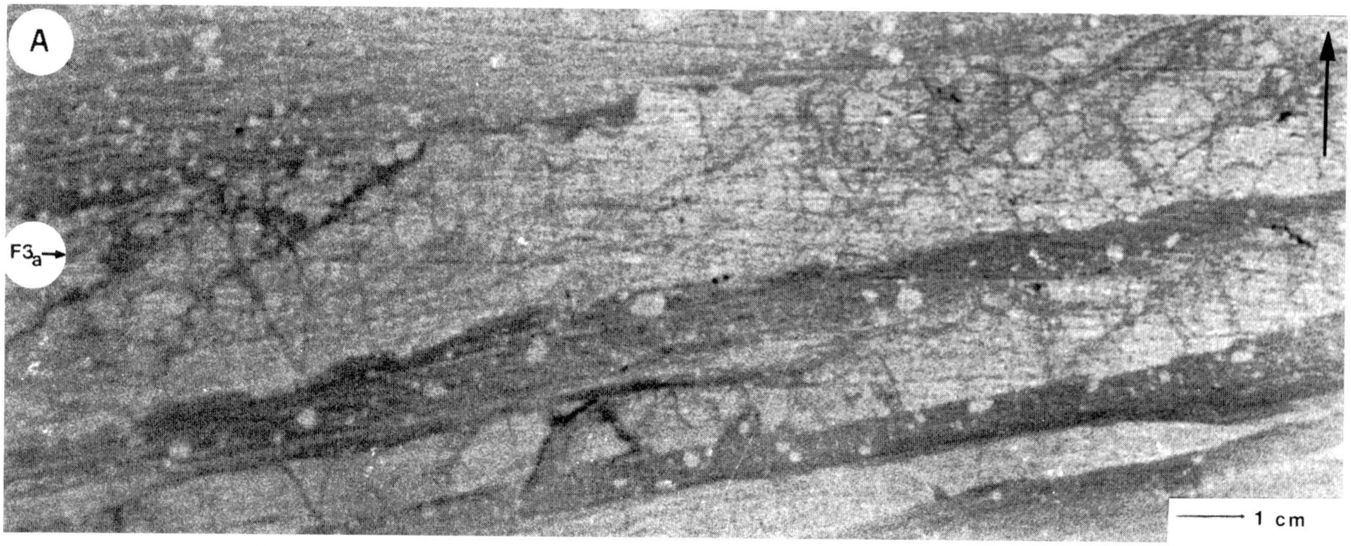

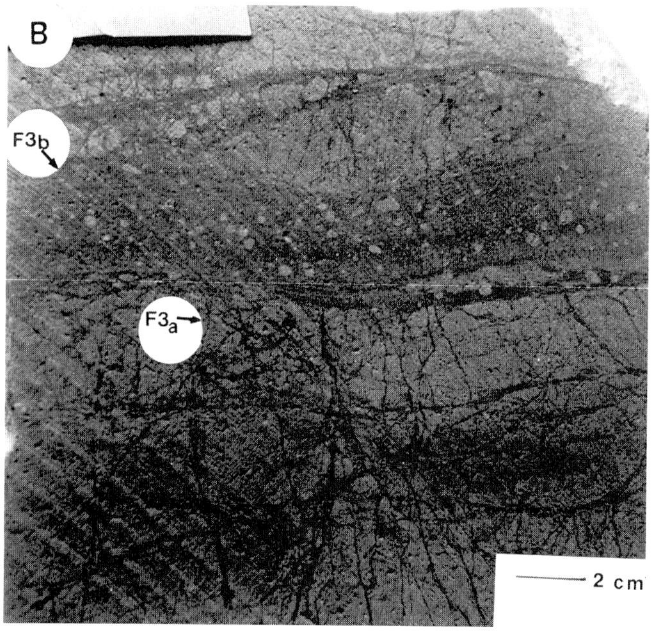

FIG. 8.—Facies F3 illustrated on oil-impregnated slabs of brecciated white micrite. (A) Sample BOH2. (B) Sample BOH3B. The white micrite (facies F3$_a$) is more cemented than the dark micrite and shows synsedimentary fracturing penetrating from the top of the layer. The darker micrite (facies F3$_b$) is less cemented and contains numerous reworked intraclasts. These samples show synsedimentary erosion and reworking of firmgrounds. The oil stain enhanced the sedimentary structures.

Nannofossils and Nannofacies

Smear slide samples (species identification by Dr. K. von Salis Perch-Nielsen, ETH Zürich, pers. commun., 1995) show coccolith assemblages (Fig. 10) that are typically Campanian in age. *Watznaueria barnesae* (Fig. 10C), *Micula decussata*, *Calculites obscrurus*, *Lucianorhabdus cayeuxii*, and *Prediscosphaera cretacea* are the most common coccoliths (Fig. 10A). *Broinsonia parca parca* (Fig. 10C) is the marker for the Campanian, together with *Rheinardites anthophorus* and *Eiffellithus eximius*.

Nannofacies show coccoliths and fine carbonate debris. Nannofossils show better preservation (Fig. 10A) above the hardground (Fig. 10B). Intra- and intergranular porosity is also higher above the hardground.

Facies Associations A, B, C, and D

Facies association A comprises facies F1a, F1b, and F1c. This association is the most commonplace white chalk, a moderately bioturbated wackestone, rarely a mudstone.

Facies association B comprises facies F1c, F2, F6, F8a, F8b, F8c, and F9. These are predominantly the white chalks containing different types of chert stringers; some of the more commonplace white chalks have little bioturbation or few cherts, and the facies with primary sedimentary structures formed under higher-energy conditions. In short, the association is characterized by slight bioturbation, structures such as swaley cross-bedding, and a wackestone to packstone texture.

Facies association C includes facies F3 and F7. The more highly cemented, commonly brecciated gray or white micrite (F3: "firmground") is always overlain by continuous dense layers of chert nodules (F7).

Facies association D comprises facies F4 and F5. These are the conspicuous hardground and the overlying thin layer of white chalk that shows fine grains of chalk reworked under high-energy conditions.

The four facies associations show vertical repetition in symmetrical to asymmetrical cycles (ABCBA, ABCABC, Fig. 11), and these cycles form seven successive units measuring 4–10 m in thickness (measured between recurring facies in each facies association A).

Facies Partitioning, Facies Interpretation, and Facies-Substitution Diagram

The observations at different scales (field, optical microscope, SEM) have thus been used to define 14 macrofacies (Table 1), 7 microfacies (Table 2), 3 nannofacies (Fig. 10), and 4 facies associations (Fig. 11). The same four facies associations are found both below and above the hardground. However, as

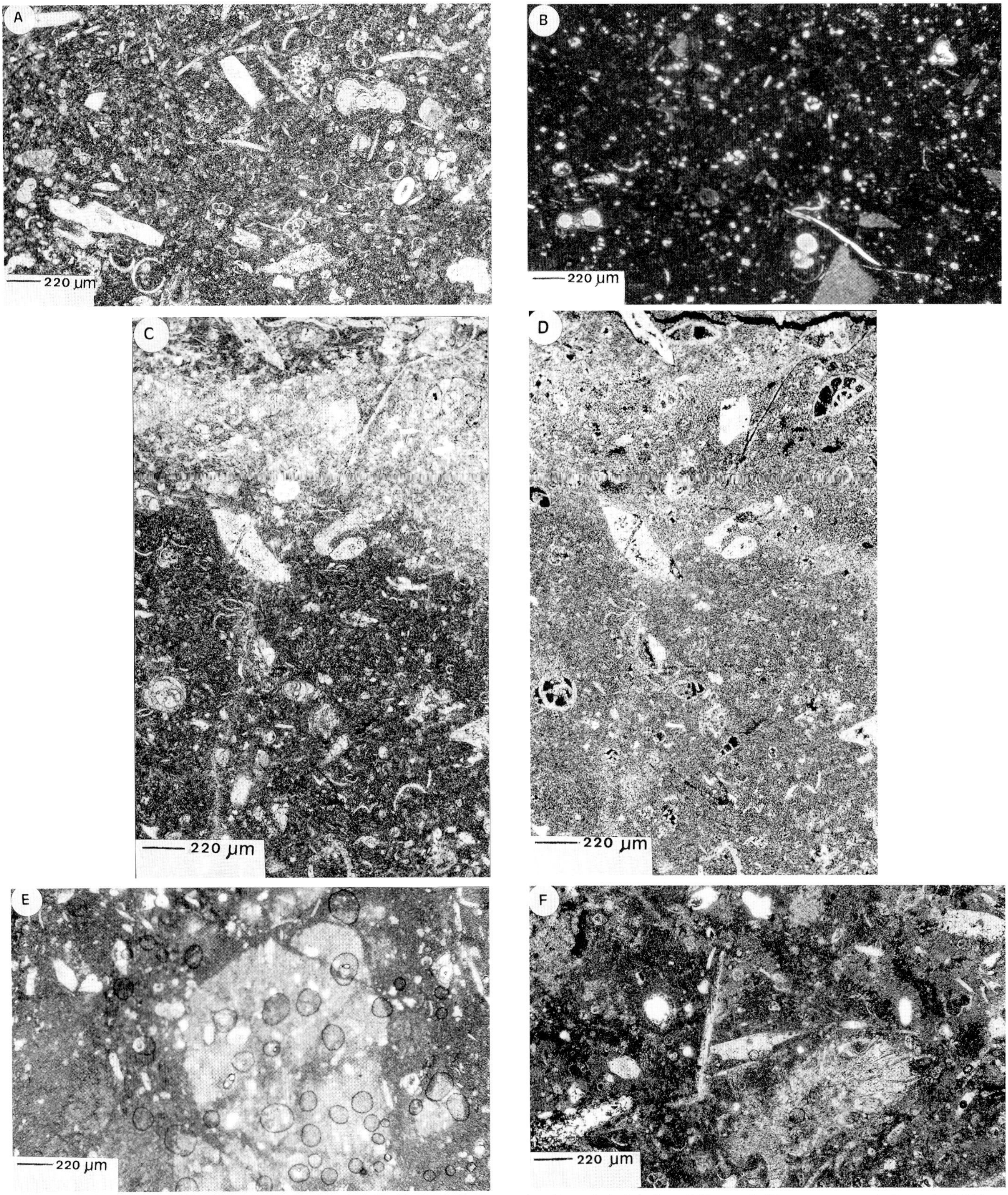

FIG. 9.—Microfacies. (A) Wackestone to packstone with echinoderm fragments, benthic and planktic foraminifera, and siliceous sponge spicules and exhibiting intergranular and moldic porosity (sample B1, microfacies MF4). (B) Wackestone with echinoderm fragments, benthic and planktic foraminifera, and calcispheres (sample BO5, microfacies MF2). (C) Wackestone to packstone with bored echinoderm fragments, benthic foraminifera, cemented by silica (hardground: sample BOHG, microfacies MF5). (D) View of C under crossed polarizers. (E) Microbreccia consisting of fragments of wackestone to packstone with echinoderm fragments and benthic and planktic foraminifera (sample BOH1, microfacies MF5). (F) Wackestone with echinoderm fragments, benthic and planktic foraminifera, and rare bryozoans (sample BOH14, microfacies MF6).

TABLE 2.—CHALK MICROFACIES

Micro-facies	Texture (Dunham, 1962)	Fauna: Benthic Foraminifera	Planktic Foraminifer	Echino-derms	Siliceous Sponges	Thin Pelecy-pods	Bryo-zoans	Calci-spheres	Cements†	Porosity‡	Facies
MF1	wackestone	+	+	+		+				m, i	F1c
MF3	wackestone-packstone	+	+	++						m, i	F2, F6
MF3	wackestone-packstone	+	+	++*	++	+			s	m	F5
MF4	wackestone-packstone	+	+	++	+	+			c	i, low	F4, F3
MF5	wackestone-packstone	+	+	++			+		ch	m	F7–F9
MF6	mudstone-wackestone	+	+					+	c	u	F1b
MF7	mudstone-		+		+	+	+			m, i	F1a

* Bored
† Siliceous cement = s, carbonate cement = c.
‡ Moldic porosity = m, intergranular = i.

may be seen on the composite section (Fig. 11), the units defined by the cyclic arrangement of facies associations become asymmetrical going up section toward the hardground (from ABCBA to ABCABC), whereas the cyclic arrangement of facies associations reverts to symmetry above the hardground.

The change in cycle symmetry from below to above the hardground is accompanied by a marked change in the abundance and distribution of facies (facies partitioning). Whereas the firmgrounds and dense layers of nodular chert are common and thicker below the hardground, they decrease in abundance over the first 10 m above the hardground, and the more commonplace white chalks then predominate. Some facies are different in the units below and above the hardground, notably in facies association B. The undulatory chert stringers (facies F8b) appear only in the upper part (Figs. 6, 11) where the other types of chert stringers are rare. Conversely, the swaley cross-bedding (facies F6; Figs. 6, 11) only occurs in the two units immediately below the hardground. These differences in types and abundance of facies are accompanied by the change in preservation noted above for the nannofacies (nannofossils are better preserved above the hardground). The origin of the diagenetic chert stringers that mimic primary sedimentary features (swales, festoons, clinoforms) might be related to the porosity and permeability patterns that resulted from primary sedimentary features such as troughs, swales, and pause planes along depositional clinoforms.

Within each of the seven units identified by the motif of facies associations, the energy of the depositional environment increased from bottom to top (Juignet, 1980). The highest-energy facies lie directly below and just on top of the hardground. This pattern suggests that the hardground represents deposition in the shallowest water (Jarvis, 1980). The cycles below the hardground are therefore shoaling cycles (terminating at the firmgrounds), and the cycles above the hardground, richer in bioturbated white chalk, indicate deepening. At the scale of individual cycles, the bioturbated white chalk, accumulated under low-energy conditions, is the deeper-water facies whereas shoaling is recorded by the chalks with chert stringers. Swaley cross-bedding and undulatory laminae suggest storm-wave influence in these chalks. Firmgrounds and hardgrounds (Quine and Bosence, 1991), documented here by better-lithified white chalks and the microbreccias lying immediately above, show scouring and winnowing, which suggest sediment bypass in an environment at or above storm wave base, possibly reaching fair-weather wave base for the hardground.

The facies partitioning documented at the longer-term scale is seen to exist also for the smaller-scale cycles. The different chert stringers are developed mostly within the progradational parts of cycles, whereas the bioturbated chalks, with or without scattered chert, were deposited during aggradation before each following progradational event.

The absolute water depth may be estimated with only little accuracy, but the shallower depths were presumably well within storm wave base in order to create swaley cross-bedding and wave-ripple lamination. The deeper-water facies were laid down below the reach of storm-wave activity. The size of the open-marine basin suggests a range for deposition of the deeper-water facies of 50–250 m at the most.

Since the facies types and proportions differ during general decrease or increase of water depth (i.e., facies partitioning commented on above), the facies-substitution diagram that illustrates the distribution of facies with regard to relative bathymetry (Fig. 12) shows increasing depth both below and above the shallowest position.

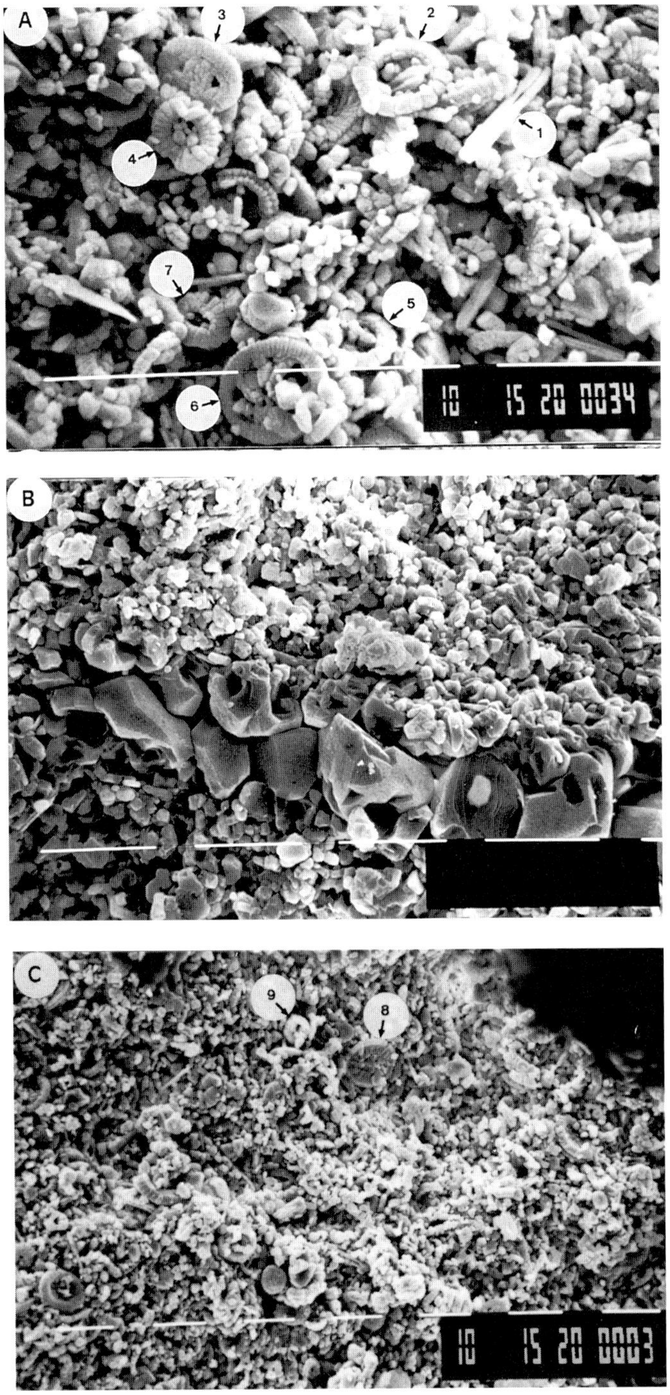

FIG. 10.—SEM images of nannofacies. (A) sample BOH 11; nannofossils are (1) *Microrhabdulus* sp., (2) *Cribosphaera ehrenbergii*, (3) *Rheinhardites* sp.?, (4) *Cretahabdus* sp., (5) *Prediscosphaera spinosa*, (6) *A. octoradiata*, and (7) *Biscutum* sp. (B) Sample BOHG; no determinable coccoliths are present in this example of the hardground. (C) Sample PO16; nannofossils are (8) *Broinsonia parca* (Campanian) and (9) *Watznaueria barnesae*. Each scale bar represents 10 µm.

Depositional Profile, Progradational-Aggradational Units, and Accommodation Cycles

The vertical succession of facies and facies associations within each unit provides the basis for describing the depositional profile in this part of the Paris basin. Far from any shoreline or isolated shallow platforms, the sedimentary surface sloped gently down from broad, shallower, platform areas that were winnowed and scoured by storm waves and currents. The deeper sounds lay no more than tens of meters lower than the shallower areas surrounding them.

Progradational units in the white chalk (ABC half-cycles) are characterized by a decrease upward in bioturbation and an increase of energy during deposition (from mudstone or wackestone to packstone); these units are capped by firmgrounds or hardgrounds that correspond to sediment bypass. The inclined chert stringers of facies F8 are suggestive of clinoform geometries along which progradation below wave base would have been caused by decreasing accommodation (facies association B, Fig. 11). Minimum accommodation, recorded by early lithification of firmgrounds and hardgrounds, is pinpointed at the outcrop by the dense layers of chert that overlie them. The dense layers of nodular chert thus mark bypass surfaces at the turnaround (change in trend) from decreasing to increasing accommodation, whereas the hardground records a major bypass surface (Jarvis, 1980). Increasing accommodation, allowing vertical aggradation of carbonate silts and muds in slightly deeper water with few high-energy sedimentary structures, resulted in more bioturbated chalks with scattered nodular cherts. Conditions of maximum accommodation are recorded by the homogeneous, bioturbated white chalk with or without scattered nodular chert (facies association A, Fig. 11).

Cycles of decrease and increase in accommodation may therefore be deduced from the progradational-aggradational cycles, as we have found no evidence for major change in rates of sediment supply at the scale in question. Moreover, the increase in water depth, when passing from storm-wave reworking to low-energy conditions below storm wave base, required an increase in accommodation even if sediment supply waned. Symmetrical cycles of facies associations ABCBA are full progradational-aggradational cycles and record decreasing and increasing accommodation. In the case of the asymmetrical facies cycles ABCABC, the time of increasing accommodation was recorded mostly by the firmground or hardground (in this case representing bypass, condensation, or ravinement), and maximum accommodation was recorded either by the firmground or by the thin, bioturbated overlying chalk. The pairs of black and white triangles in Figure 11 show the accommodation cycles defining the genetic units at a scale of 5–10-m thickness. They show an organization into one stratigraphic unit on a longer-term scale.

Sedimentological Models

In this study, the facies observed in the outcrops were identified and interpreted not only in terms of depositional processes (storm activity with local resedimentation, winnowing, accumulation of carbonate *in situ* under low-energy conditions through growth and decay of carbonate-producing planktic and benthic fauna and flora) but also in terms of progradational and aggradational cycles and longer-term increasing or decreasing accom-

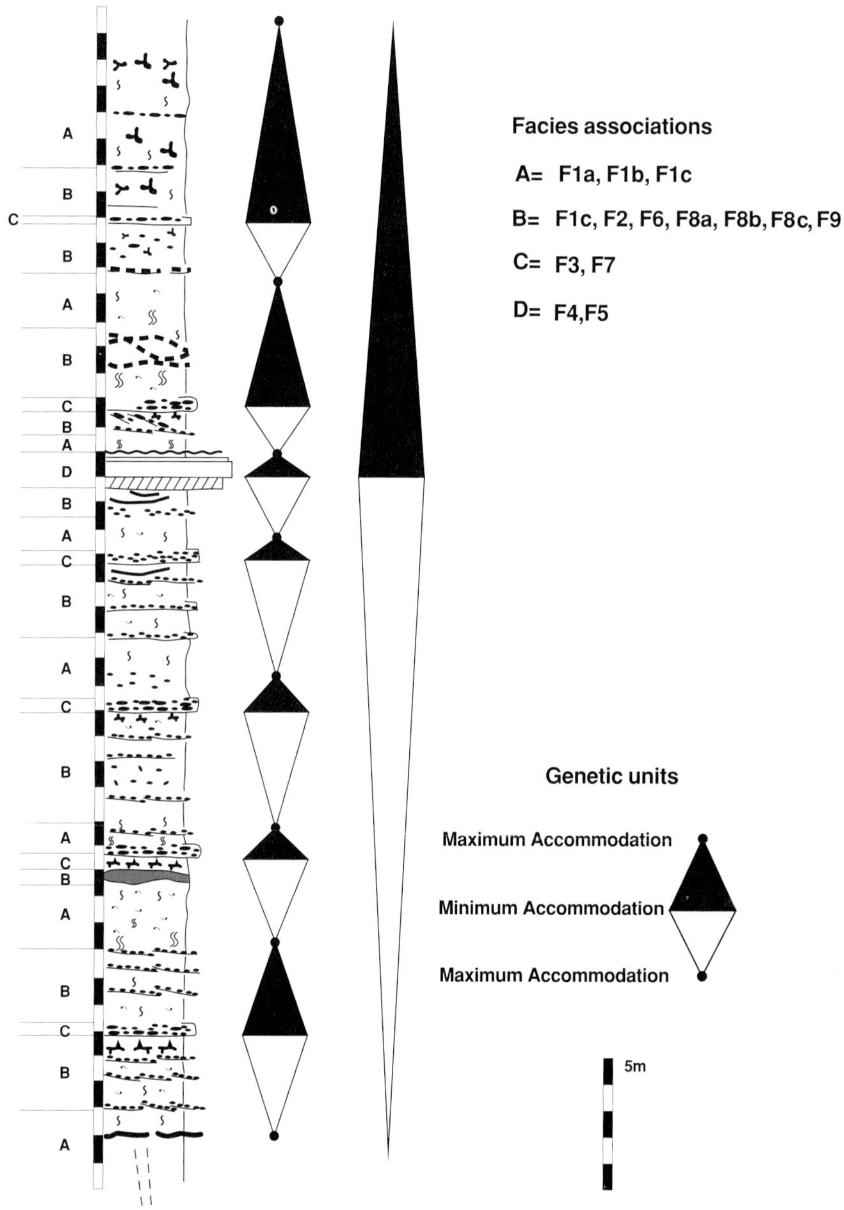

FIG. 11.—Stratigraphic model showing asymmetrical pattern of facies association repetition below the prominent hardground and symmetrical pattern above it. Facies listed are the same as in Figure 6.

modation. As a result, sedimentological models need to be established for at least the two contrasting sets of conditions that are given by increasing and decreasing accommodation (Fig. 13).

The high-resolution stratigraphic units described here correspond in scale and in character to genetic units from progradational shoreface environments, both from siliciclastic systems (e.g., McDonough and Cross, 1991; Homewood et al., 1992; Cross et al., 1993) and from carbonate systems (e.g., Sonnenfeld and Cross, 1993; Gianniny and Simo, 1996; Grammer et al., 1996). However, in the present case, the contemporary shorelines lay at a considerable distance, and the progradational-aggradational units, well below sea level, show that the base level for stratigraphic accommodation was not sea level but must have been lower (Barrell, 1917). The sediment input to the clinoforms that prograded under storm influence was not provided by a littoral carbonate factory; instead, the sediment was produced part in the water column and in part on the storm-winnowed platform itself. The sedimentological models are therefore described as "shore detached" in that the progradation of shoreface deposits made no contribution to sediment accumulation in this part of the Paris basin.

The sedimentological models are drawn for two settings (Fig. 13). First, during times of decreasing accommodation, the depositional environment (which is set far away from the shoreline and from shoreline progradation) would have sediments building up to wave base by progradation. Sediment type is characterized by poorly preserved nannofossils, more abundant echinoderms, and siliceous sponges. Hardground formation occurred during a longer period of sediment bypass (during a

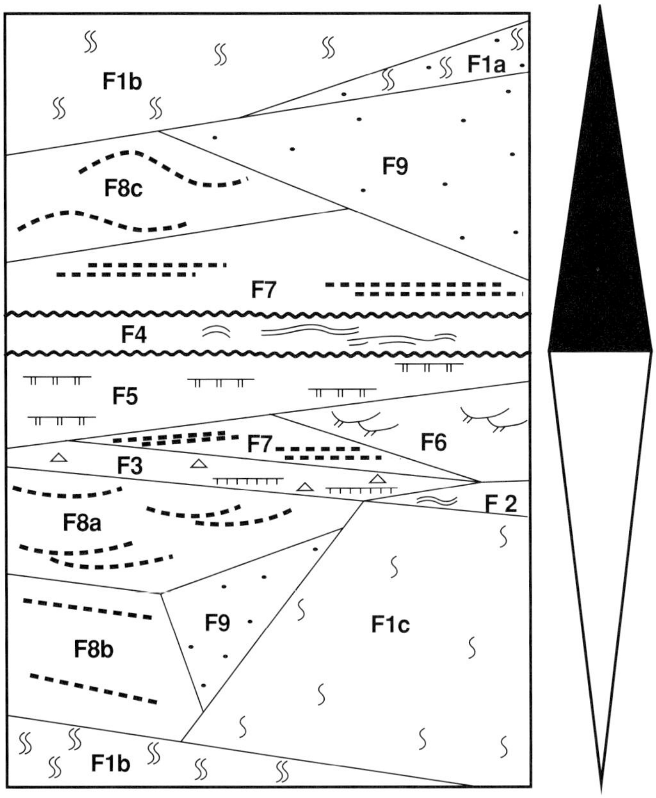

FIG. 12.—Facies substitution diagram showing distribution of facies as a function of water depth (no absolute depths can be given) and for both decreasing and increasing accommodation. Facies numbered as in the text. The lower half of the diagram shows decreasing water depth from the bottom to the wavy line between facies F5 and F4. This part of the diagram indicates possible lateral facies substitutions and vertical facies successions during *decreasing* accommodation. The upper part of the diagram shows increasing water depth from F4 to F1b. This part of the diagram indicates possible lateral facies substitutions and vertical facies successions during *increasing* accommodation. Facies shown are the same as in Figure 6.

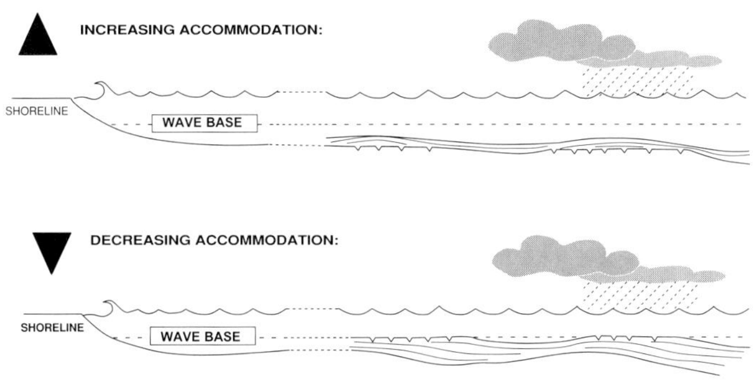

FIG. 13.—Sedimentological model showing that with decreasing accommodation in a depositional environment far from shoreline, sediment builds up to wave base, filling in hollows by progradation. Hardgrounds and erosion occur at or near storm wave base. With increasing accommodation and a slow rate of aggradation, the sediment surface sinks below storm wave base, a situation that may enhance residual sea-floor topography.

turnaround of longer-term decreasing to increasing accommodation). During this time, wave activity shaved sediment from the sea floor and transported it to deeper areas ("wave-base razor"; McDonough et al., 1991; Sonnenfeld and Cross, 1993). Second, during times of increasing accommodation (Fig. 13), the depositional environment is still "detached" from the shoreline, but the sediment surface deepens below wave base with aggradation of sediment and enhancement of residual sea-floor topography. Sediments are characterized by increased bioturbation, and nannofossils dominate and are better preserved. Bryozoans appear in aggradational phases of cycles during longer-term increasing accommodation (above the hardground).

STRATIGRAPHY

The basic motif of repetition of facies, together with the cycles of progradation and aggradation that led to the sedimentological models discussed above, are also the basis for a high-resolution (genetic) stratigraphic model.

At the scale of stacked genetic units (some 40 m thick), the turnaround from longer-term decreasing to increasing accommodation is marked by the hardground (facies association D, Fig. 11). Accommodation cycles at the genetic unit scale show a marked variation from less symmetrical during long-term decreasing accommodation and increasing accommodation (middle part of the composite section), to more symmetrical close to the longer-term, greatest accommodation (base of the composite section). These two scales of accommodation cycle provide the basic stratigraphic units that have been shown in other cases to correlate at kilometer- and basin-wide scales (e.g., Sonnenfeld, 1996; Robin, 1997).

It is important to notice that microfacies are not a simple key to stratigraphic units as they do not suffice alone to indicate increase, decrease, or turnaround in accommodation. However, degrees of preservation of nannofossils, and the appearance or relative abundance of some fossils (e.g., bryozoans, echinoderms) seem directly linked to longer-term decrease or increase in accommodation. Bryozoans and belemnites are present and coccoliths are also better preserved in aggradational units. Coccoliths are poorly preserved in progradational units, whereas sea urchins are common. Bioclasts are broken and frequently bored at the tops of progradational units (in particular in the hardground).

SEISMIC SIGNATURE

Previous work on seismic data within the "homogeneous" Upper Cretaceous Chalk Group has been limited to observations of large-scale geometries, such as erosional channels (Quine and Bosence, 1991; Hanot and Renoux, 1991) or regional onlaps associated with tectonic events (Nygaard et al., 1990). Recently, inversion of three-dimensional seismic data using amplitude to predict porosity and production in the chalk has been done for the producing fields in the Central Danish graben (Rasmussen and Maver, 1996; Campbell and Gravdal, 1995; Megson, 1992).

Lateral changes in the amplitude and continuity of seismic

reflections within the Upper Cretaceous chalk have been observed on regional proprietary Elf seismic data from the Paris basin and the Central Danish graben. The seismic data in the Paris basin is close to the surface (100–300 ms), so data quality is poor; nonetheless, amplitude and continuity reflection changes were observed in the Santonian interval (see Figs. 9, 11, 15, and 16 of Barchi, 1991). Core control showed that these discontinuous seismic reflections tied to a zone with abundant chert layers. These changes are similar to those seen in the seismic reflections of the low-velocity chalk in Figure 16 of Hanot and Renoux (1991). The Campanian interval we have studied falls within their zone of low-velocity chalk, which has numerous hardgrounds and bands of flint in the outcrops they studied (Hanot and Renoux, 1991). Amplitude and continuity reflection changes were also observed on Elf proprietary regional seismic lines in the Central Danish graben (similar to Fig. 3 [sequences 5 and 6] of Nygaard et al. (1990). Discontinuous seismic reflections corresponded to zones with chert layers in the Maastrichtian on well logs.

Since the genetic stratigraphic units in this study were defined by patterns of chert distribution and abundance, we studied the effect of the chert layers on the seismic signal. As far as we know, this effect has never been reported elsewhere. However, previous work on the Maastrichtian Dan field in the Central Danish graben has shown that a hardground produced a continuous, high-amplitude seismic reflection (Kristensen et al., 1995).

A simple one-dimensional seismic modeling based on the outcrop (Figs. 5 and 6), a long-term accommodation sequence, was done to identify the effects of the distribution of the chert layers within the progradational and aggradational cycles. The premise of this modeling was that a long-term turnaround hardground and underlying progradational cycle with numerous chert layers would produce a reflection, whereas an aggradational cycle with fewer chert layers would be reflection-free or only produce a weak reflection.

The synthetic trace of this 50-m-thick outcrop model (Fig. 14) is displayed at three different frequencies (30, 60, 90 Hz) to show the chert effects on seismic data gathered at different frequencies. The limit of resolution (i.e., ¼ wavelength, Sheriff, 1977) is 20–25 m for these chalks at 30 Hz. The modeling was done with a zero-phase Ricker wavelet of SEG reverse polarity. Constant interval velocities of 2700 m/s for the chalk (Hanot and Renoux, 1991) and 4240 m/s for the hardground and chert layers were used; these values were based on regional observations of velocities. Densities were computed from the interval velocities through the use of Gardner's equation.

Model A (Fig. 14) shows the synthetic seismic response caused by the hardground alone at the turnaround from long-term sea-level fall to rise. This model calibrates the seismic response of the hardground. Model B (Fig. 14) has four chert layers (1–4) in the progradational cycle (representing a decrease in accommodation), but only two in the aggradational cycle (layers 5–6; deposited during an increase in accommodation). This model illustrates the seismic response to the presence of chert layers around the hardground.

Comparison of models A and B (Fig. 14) shows that the chert layers have an effect on the seismic signal. Comparison of the higher-frequency (60 and 90 Hz) displays for Model B shows that the numerous chert layers in the progradational cycle (layer 1 to the hardground) diminish the amplitude of the reflection produced in this interval because of the combined destructive interference of the numerous chert beds, whereas the reflection in the aggradational cycle (from the hardground to layer 6) is higher amplitude because there are fewer thin beds to interfere with the reflection produced from the hardground. These amplitude changes are caused by the varying degrees of constructive and destructive interference due to the thickness and spacing of the chert beds. This modeling suggests that amplitude changes seen on the proprietary regional seismic data, and those cases discussed above, could be due to higher concentrations of chert layers in the progradational cycle of the chalk. This hypothesis needs to be tested further on a detailed data set near the outcrop and in other areas with well control.

DISCUSSION

Previous contributions concerning the white chalk of the Paris basin (e.g., Hancock, 1976) have addressed numerous aspects of sedimentology, sedimentary geometries, biostratigraphy, and seismic geometries. In addition to these aspects, this paper proposes a sedimentological model that differentiates between times of increasing and of decreasing accommodation and provides a basis for a basin-scale high-resolution stratigraphic model.

Sedimentology

Juignet (1980) reconstructed a "virtual sequence" that provides the various features required to construct a classical facies model, but in an idealized, complete vertical sequence (an approach pioneered by Lombard, 1956). As in the construction of facies models (e.g., Walker, 1979), this approach tends to combine elements that did not coexist. The resulting interpretation is inevitably generalized and precludes detailed correlation either of genetic stratigraphic units or of their stacking patterns. The sedimentological models given here, although based on essentially the same interpretations in terms of process and depositional environment as those of Juignet (1980), attempt to avoid overgeneralization by separating features observed according to whether they were produced during times of increasing or of decreasing accommodation.

The understanding of facies successions from the white chalk of the Paris basin in terms of varying accommodation leads to interpretations of the depositional environment that place more emphasis on rates of increase or decrease of relative sea level or accommodation, and less emphasis on absolute water depths. Although the sediment surface must have remained below storm wave base for much of the time during increasing accommodation, it must have reached above storm wave base during decreasing accommodation. Under this hypothesis, bathymetry may never have changed more than a few meters to tens of meters during a cycle of increasing and

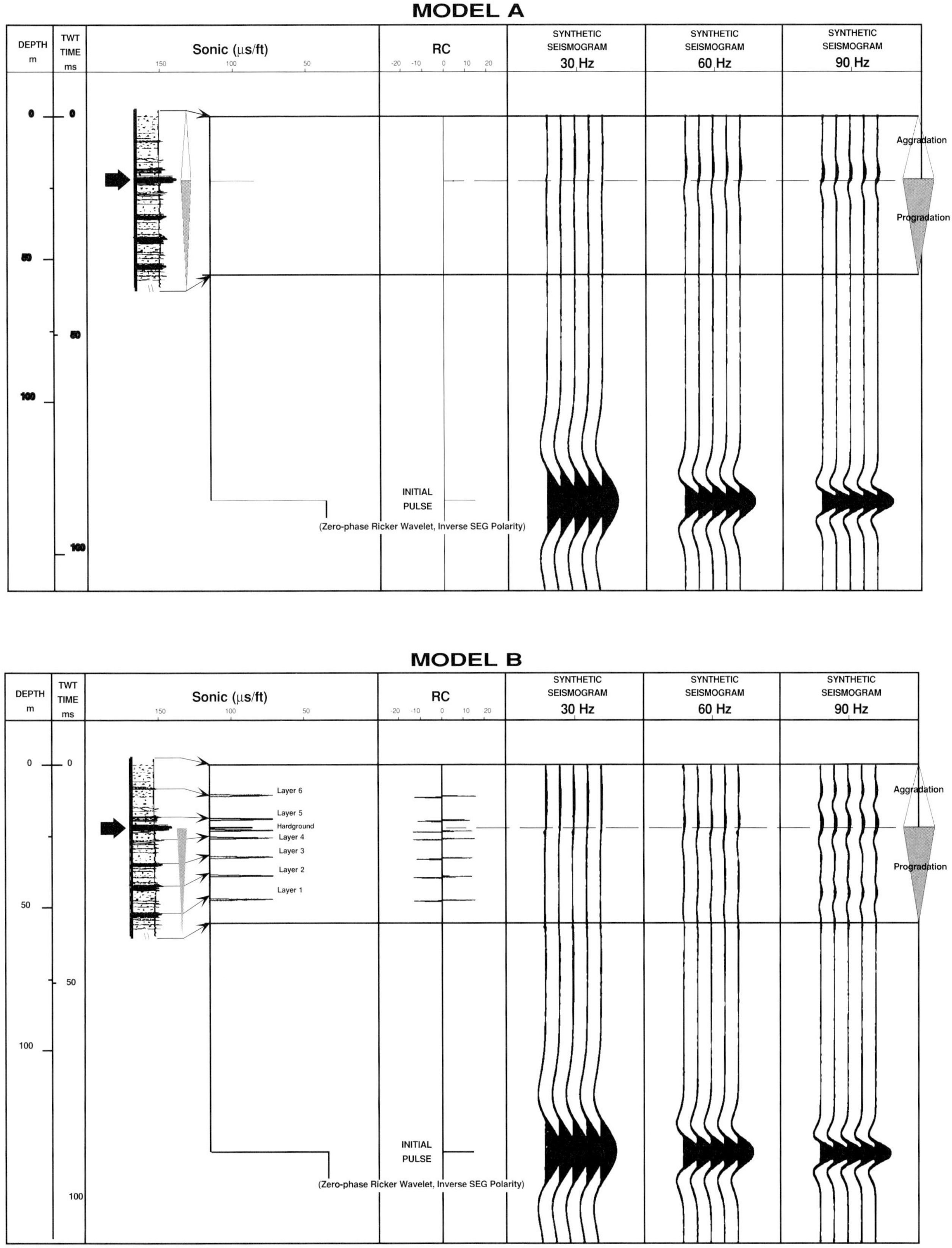

FIG. 14.—One-dimensional seismic modeling based on the vertical section from the outcrop (Fig. 5). The initial pulse, a zero-phase Ricker wavelet of SEG reverse polarity, is seen at the base of each synthetic trace. The synthetic seismograms are displayed at 30, 60, and 90 Hz frequencies. Model A models only the hardground. Model B models the chert layers and the hardground. TWT = two-way traveltime, RC = reflection coefficient.

then decreasing accommodation.

Slight changes in bathymetry are in sharp contrast with those proposed by interpretation under a classical approach, be it via a "virtual" sequence or via a facies model. When using the classical approach, in which facies are seen to reflect process alone, changes in water depths receive more emphasis. For example, changes in depth might have ranged over several tens to hundreds of meters during chalk deposition according to Juignet (1980). Major fluctuations in sea level are indeed required, in this approach, to explain the changes in facies that occur through a 1–10-m-thick cycle.

An approach that interprets changes in facies to result from variations in accommodation as well as from changes in depositional process, as used in the present paper, answers the common objection to the interpretation of cycles of shelf deposits in terms of sea-level fluctuations. A change in bathymetry of ±5 m where depth is about 50 m, or of ±10 m where depth is near 100 m, can hardly explain a sufficient change in depositional processes to be the cause of such cycles. However, changes in the rate of sediment supply or flux, and in the rate of increase or of decrease in accommodation space, would lead to significantly different deposits even where bathymetry changes less than a few tens of meters. This process would result in the formation of firmgrounds or hardgrounds during sediment bypass as accommodation decreases and aggradation below storm wave base as accommodation increases.

Quine and Bosence (1991) provided detailed descriptions of microfacies and nannofacies for the older (Cenomanian-Coniacian) and slightly shallower chalk (in terms of depositional environments) composing the coastal cliffs of Normandy (Etretat). They documented and discussed the firmgrounds and hardgrounds that are placed in the context of large-scale geometries. Scouring and aggradation obviously played a considerable role in generating these broad and deep submarine-channel morphologies, but Quine and Bosence (1991), although they explained the localization of scouring by substrate morphology due to regional fault activity, did not provide a sedimentological model combining facies and processes to link their observations, which were made at vastly different scales.

The sedimentological and stratigraphic observations made in the present paper, and the models proposed, do provide reasoning at an intermediate scale. The phases of major scouring, at the end of longer-term decrease in accommodation, could correspond to the sediment bypass events causing major hardgrounds such as those documented here. This interpretation follows the reasoning of Jarvis (1980), who associated the hardgrounds with strong current activity and scouring or sediment bypass. Subsequent increase in accommodation, at similar time scales, would then explain the aggradation over previously erosive-channel morphologies.

Stratigraphy

In much stratigraphic work, hardgrounds have been used as event horizons to assist construction of a regional stratigraphic framework. Biozonation, using the regionally correlated "events," has largely been based on echinoderms and trace fossils. The high-resolution stratigraphy presented here lies at a scale within the biostratigraphic framework previously established (Pomerol and Mortimore, 1990; Mortimore and Pomerol, 1991), although the correlation of hardgrounds from one location to the next might be called into question at the scale of genetic units and their stacking patterns.

Seismic Signature

Previous to this study, no work had been published on the effects of chert abundance and distribution on the seismic signature. Regional proprietary Elf seismic lines show lateral variations in amplitude and continuity of reflections in the Upper Cretaceous chalk. Two outcrop-based seismic models were run to compare the effects that the chert layers might have on the seismic signature. Results show that amplitude changes are due to the thickness and spacing of the thin chert beds. Numerous chert layers (i.e., in progradational intervals) diminish the amplitude of a reflection, whereas a reflection in an aggradational interval shows a higher, more continuous amplitude, because there are fewer thin chert beds to interfere with the reflection. Reflections are produced by the hardgrounds at long-term turnarounds of accommodation. This modeling suggests that reflection-amplitude and continuity changes seen on the regional seismic sections could be due to the higher concentrations of chert layers in the progradational interval of the chalk related to a decrease in accommodation. If this hypothesis holds true for other data sets, it could be a useful tool in the interpretation of regional seismic lines and seismic facies in terms of accommodation cycles and associated genetic stratigraphy.

CONCLUSIONS

This contribution adds a high-resolution stratigraphic analysis to previous regional and thematic sedimentological and stratigraphic studies of the Upper Cretaceous white chalk of the Paris basin. Genetic units measuring 5–8 m in thickness may be recognized on the basis of sedimentological models in which the depositional environment, just below and above storm wave base, evolves from a progradational mode under decreasing accommodation, to an aggradational mode under increasing accommodation.

The progradational-aggradational units of the Chaux-de-Boran and Précy-sur-Oise quarries, which are interpreted as high-resolution stratigraphic genetic units laid down during successive cycles of change in accommodation, must have compensated subsidence, sea-level change, and local topographic effects, as they all record similar changes in bathymetry near to storm and fair-weather wave base.

The seismic modeling of the pattern of distribution of chert beds and hardgrounds, as recognized in the field, offers a response that, together with the high-resolution stratigraphic model, may improve understanding of amplitude changes and geometries observed on regional seismic lines.

ACKNOWLEDGMENTS

The authors thank Elf Exploration Production and Elf Aquitaine Exploration Production France for permission to publish. G. Sambet, J. Blanco, K. von Salis Perch-Nielsen, M. Seigneur and M. Hirtig provided help and advice at different stages of this study. Very constructive reviews by D. Lehrmann, B. Saylor, J. A. Simo, and T. A. Cross were of help when rewriting the manuscript.

REFERENCES

BARCHI, P., 1991, Stratigraphie des marqueurs géophysiques dans la Craie: Unpublished M.S. Thesis, Université Pierre et Marie Curie, Paris, 100 p.

BARRELL, J., 1917, Rhythms and the measurement of geologic time: Geological Society of America Bulletin, v. 28, p. 745–904.

BROMLEY, R. G., 1981, Enhancement of visibility in marly chalk: Modification of the Buchinsky oil technique: Bulletin of the Geological Society of Denmark, v. 29, p. 111–118.

CAMPBELL, S. J. D., AND GRAVDAL, N., 1995, The prediction of high porosity chalks in the East Hod field: Petroleum Geoscience, v. 1, p. 57–69.

CROSS, T. A., 1988, Controls on coal distribution in transgressive-regressive cycles, Upper Cretaceous, Western Interior, USA, *in* Wilgus, C. K., Hastings, B. S., Kendall, C. G. St. C., Posamentier, H. W., Ross, C. A., and Van Wagoner, J. C., eds., 1988, Sea-Level Changes—An Integrated Approach: Tulsa, Society of Economic Paleontologists and Mineralogists, Special Publication 42, p. 371–380.

CROSS, T. A., BAKER, M. R., CHAPIN, M. A., CLARK, M. S., GARDNER, M. H., HANSON, M. H., LESSENGER, M. A., LITTLE, L. D., MCDONOUGH, K-J., SONNENFELD, M. D., VALASEK, D. W., WILLIAMS, M. R., AND WITTER, D. N., 1993, Applications of high-resolution sequence stratigraphy to reservoir analysis, *in* Eschard, R. and Doligez, B., eds., Proceedings of the 7th Exploration Production Research Conference: Paris, Editions Technip, p. 11–33.

DERCOURT, J., 1997, Géologie et géodynamique de la France, outre-mer et européenne: Paris, Dunod, 320 p.

DUNHAM, R. J., 1962, Classification of carbonate rocks according to depositional texture, *in* Ham, W. E., ed., Classification of Carbonate Rocks: Tulsa, American Association of Petroleum Geologists Memoir 1, p. 108–121.

GIANNINY, G. L., AND SIMO, J. A. T., 1996, Implications of unfilled accommodation space for sequence stratigraphy on mixed carbonate-siliciclastic platforms: An example from the lower Desmoinesian (Middle Pennsylvanian), southwestern Paradox basin, Utah, *in* Longmann, M. W. and Sonnenfeld, M. D., eds., Paleozoic Systems of the Rocky Mountain Region: Rocky Mountain Section–SEPM (Society for Sedimentary Geology), p. 213–234.

GRAMMER, G. M., EBERLI, G. P., VAN BUCHEM, F. S. P., STEVENSON, G. M., AND HOMEWOOD, P., 1996, Application of high-resolution sequence stratigraphy to evaluate lateral variability in outcrop and subsurface—Desert Creek and Ismay intervals, Paradox basin, *in* Longmann, M. W. and Sonnenfeld, M. D., eds., Paleozoic Systems of the Rocky Mountain Region: Rocky Mountain Section–SEPM (Society for Sedimentary Geology), p. 235–266.

HANCOCK, J. M., 1976, The petrology of chalk: London, Proceedings of the Geological Society of London, v. 86, p. 499–535.

HANOT, F., AND RENOUX, P., 1991, Petrophysical variations in the Senonian chalk of the Paris basin and their influence on static corrections: First Break, v. 9, p. 515–526.

HOMEWOOD, P., GUILLOCHEAU, F., ESCHARD, R., AND CROSS, T. A., 1992, Corrélations haute résolution et stratigraphie génétique: Une démarche intégrée: Bulletin des Centres de Recherches Exploration-Production Elf-Aquitaine, v. 16, p. 357–381.

JARVIS, I., 1980, The initiation of phosphatic chalk sedimentation—The Senonian (Cretaceous) of the Anglo-Paris basin, *in* Bentor, Y. K., ed., Marine Phosphorites: Tulsa, Society of Economic Paleontologists and Mineralogists, Special Publication 29, p. 167–192.

JERVEY, M. T., 1988, Quantitative geological modeling of siliciclastic rock sequences and their seismic expression, *in* Wilgus, C. K., Hastings, B. S., Kendall, C. G. St. C., Posamentier, H. W., Ross, C. A., and Van Wagoner, J. C., eds., 1988, Sea-Level Changes—An Integrated Approach: Tulsa, Society of Economic Paleontologists and Mineralogists, Special Publication 42, p. 47–69 .

JUIGNET, P., 1980, Transgressions-régressions, variations eustatiques et influences tectoniques de l'Aptien au Maastrichtien dans le Bassin de Paris occidental et sur la bordure du Massif Armoricain: Cretaceous Research, v. 1, p. 341–357.

KRISTENSEN, L., DONS, T., MAVER, K. G., AND SCHIOLER, P., 1995, A multidisciplinary approach to reservoir subdivision of the Maastrichtian chalk in the Dan field, Danish North Sea: American Association of Petroleum Geologists Bulletin, v. 79, p. 1650–1660.

LOMBARD, A., 1956, Géologie sédimentaire. Les séries marines: Paris, Masson, 560 p.

MCDONOUGH, K. J., AND CROSS, T. A., 1991, Cretaceous sea-level from a paleoshoreline: Journal of Geophysical Research, v. 96, p. 6591–6609.

MCDONOUGH, K. J., CROSS, T. A., HOMEWOOD, P. W., ASSOCIATED ELVES, ARNAUD, H., AND ARNAUD-VANNEAU, A., 1991, Sediments, geometries, and stratal architecture in the Combau Valley (S. Vercors, France): Facies partitioning by the carbonate factory (abs.): Brest, 3e Congrès Français de Sédimentologie, Publication de l'Association des Sédimentologues Français, v. 15, p. 211–212.

MÉGNIEN, C., coordinator, 1980a, Synthèse géologique du Bassin de Paris: Orléans, Bureau de Recherches Géologiques et Minières, Mémoire 101, 466 p.

MÉGNIEN, C., coordinator, 1980b, Synthèse géologique du Bassin de Paris: Orléans, Bureau de Recherches Géologiques et Minières, Mémoire 102, 54 p.

MÉGNIEN, C., coordinator, 1980c, Synthèse géologique du Bassin de Paris: Orléans, Bureau de Recherches Géologiques et Minières, Mémoire 103, 466 p.

MEGSON, J. B., 1992, The North Sea chalk play: Examples from the Danish Central graben, *in* Hardman, R. F. P., ed., Exploration Britain: Geological Insights for the Next Decade: London, Geological Society of London Special Publication 67, p. 247–282.

MORTIMORE, R. N., AND POMEROL, B., 1987, Correlation of the Upper Cretaceous white chalk (Turonian to Campanian) in the Anglo-Paris basin: London, Proceedings of the Geologists Association, v. 98, p. 97–143.

MORTIMORE, R. N., AND POMEROL, B., 1991, Upper Cretaceous tectonic disruptions in a placid Chalk sequence in the Anglo-Paris basin: Geological Society of London Journal, v. 148, p. 391–404.

NYGAARD, E., ANDERSEN, C., MOLLER, C., CLAUSEN, C. K., AND STOUGE, S., 1990, Integrated multidisciplinary stratigraphy of the Chalk Group: An example from the Danish Central trough, *in* Burland, J. B., ed., Chalk: Proceedings of the 1989 International Chalk Symposium: London, Imperial College, p. 195–201.

PERRODON, A., AND ZABECK, J., 1990, Paris basin, *in* Leighton, M. W., Kolata, D., Oltz, D. F., and Eidel, J. J., eds, Interior Cratonic Basins: Tulsa, American Association of Petroleum Geologists Memoir 51, p. 633–679.

PLATEL, J. P., 1996, Stratigraphie, sédimentologie et évolution géodynamique de la plate-forme carbonatée du Crétacé superieur du nord du Bassin d'Aquitaine: Géologie de la France, v. 4, p. 33–58.

POMEROL, B., AND MORTIMORE, C., 1990, Lithostratigraphie et cycles sédimentaires dans la craie du bassin anglo-parisien: Paris, Comptes Rendus de l'Académie des Sciences, ser. II, v. 310, p. 553–558.

QUINE, M., 1988, Sedimentology of the chalk of coastal Haute Normandie, France: Unpublished Ph.D. Thesis, University of London, London, 375 p.

QUINE, M., AND BOSENCE, D., 1991, Stratal geometries, facies and sea-floor erosion in Upper Cretaceous chalk, Normandy, France: Sedimentology, v. 38, p. 1113–1152.

RASMUSSEN, K. B., AND MAVER, K. G., 1996, Direct inversion for porosity of post stack seismic data, in Europe 3-D Reservoir Modelling Conference, Stavanger, Norway, April 16–17, 1996: Dallas, Society of Petroleum Engineering Report No. SPE 35509, p. 235–246.

ROBIN, C., 1997, Mesure stratigraphique de la déformation: Application à l'évolution jurassique du Bassin de Paris: Rennes, Mémoires Géosciences–Rennes, v. 77, 293 p.

SHERIFF, R. E., 1977, Limitations on resolution of seismic reflections and geologic detail derivable from them, *in* Payton, C. E., eds., Seismic stratigraphy-applications to hydrocarbon exploration, American Association of Petroleum Geologists, Memoir 26, p. 3–14.

SONNENFELD, M. D., 1996, Sequence Evolution and Hierarchy within the Lower Mississippian Madison Limestone of Wyoming, *in* Longmann, M. W. and Sonnenfeld, M. D., eds., Paleozoic Systems of the Rocky Mountain Region: Rocky Mountain Section–SEPM (Society for Sedimentary Geology), p. 165–192.

SONNENFELD, M. D., AND CROSS, T. A., 1993, Volumetric partitioning and facies differentiation within the Permian Upper San Andres Formation of Last Chance Canyon, Guadalupe Mountains, New Mexico, *in* Loucks, R. G., and Sarg, J. F., eds., Recent Advances and Applications of Carbonate Sequence Stratigraphy: Tulsa, American Association of Petroleum Geologists Memoir 57, p. 435–474.

WALKER, R. G., 1979, Facies models: St. John's, Geological Association of Canada, Geosciences Canada Reprint Series 1, 279 p.

SEDIMENTOLOGIC AND CHEMOSTRATIGRAPHIC RECOGNITION OF THIRD-ORDER SEQUENCES IN RESEDIMENTED CARBONATE: THE MIDDLE JURASSIC VAJONT LIMESTONE, VENETIAN ALPS, ITALY

WILLIAM G. ZEMPOLICH*

Department of Earth and Planetary Sciences, Johns Hopkins University, Charles and 34th Streets, Baltimore, Maryland 21218

AND

ELISABETTA ERBA

Dipartimento di Scienze della Terra, Università degli Studi di Milano, Via Mangiagalli, 34, I-20133, Milano, Italy

ABSTRACT: The Middle Jurassic Vajont Limestone is predominantly composed of oolite that was shed from the Friuli platform and resedimented by gravity flows into the Belluno basin. Detailed petrographic, geochemical, and biostratigraphic study of a 325-m-thick section reveals stratigraphic variations in nonskeletal-grain and skeletal-grain content and in $\delta^{13}C$, $\delta^{18}O$, and $^{87}Sr/^{86}Sr$ ratios measured in radial calcite ooids. New nannofossil and ammonite data provide tight age constraints on the stratigraphic section and suggest that deposition occurred during the interval between the late Aalenian/early Bajocian boundary and the late Bajocian—a period of approximately 4.5 m.y. Thus the grain contents and the isotope compositions of Vajont gravity-flow deposits are interpreted as a basinal record of the compositional, chemical, and sequence stratigraphic evolution of the Friuli platform during part of the Middle Jurassic.

Point-count analysis indicates that Vajont gravity-flow deposits are predominantly composed of nonskeletal grains (average = 89% by volume) with lesser amounts of skeletal grains (11%). Nonskeletal grain types of the Vajont Limestone include ooids (44%), peloids (30%), oncoids (7%), grain fragments (4%), and grapestone (3.75%). Overall, the Vajont exhibits an upward increase in skeletal, peloid, and grapestone contents and a concomitant decrease in ooid and oncoid content. Within the Vajont stratigraphic section, variations in the grain type forming the rock—either skeletal grains or nonskeletal grains—and the $\delta^{13}C$ values of radial calcite ooids define four decameter-scale sedimentologic associations. Each of two nonskeletal-grain–dominated associations is paired with an overlying skeletal-grain–rich association, and each of the couplets so formed is interpreted to represent an individual cycle of transgressive and highstand sedimentation (i.e., to be a third-order sequence). The absence of a definitive lowstand compositional signature—i.e., skeletal-grain—dominated gravity-flow deposits—between the two third-order sequences is attributed to the sustained flooding of the Friuli platform. Sustained flooding was most likely achieved by a combination of rapid platform subsidence and a long-term (second-order) rise in global sea level during middle Aalenian–late Bajocian time.

The ^{13}C fluctuations that are displayed by Vajont radial ooids are in agreement with ^{13}C profiles defined by study of time-equivalent pelagic marine calcite. Similarly, $^{87}Sr/^{86}Sr$ ratios measured in Vajont radial ooids and other calcitic components are consistent with a seawater Sr isotope curve defined by study of time-equivalent calcitic fossils. Positive excursions in $\delta^{13}C$ values (i.e., ^{13}C enrichments) correlate with increased skeletal-grain content in Vajont gravity-flow deposits and proposed sea-level highstands. Times of relative depletion in ^{13}C correlate with increases in the nonskeletal-grain content in Vajont gravity-flow deposits and proposed transgressions. Upward through the stratigraphic section, $^{87}Sr/^{86}Sr$ ratios exhibit an overall decrease on which is superimposed small increases that potentially correlate with maximum skeletal-grain contents within skeletal-grain–rich sedimentologic associations. These data suggest that third-order sea-level changes affected sediment production on the Friuli platform as well as the $\delta^{13}C$ values and perhaps the $^{87}Sr/^{86}Sr$ ratios of ambient seawater.

On the basis of the gradual upward enrichment in skeletal-grain content of Vajont gravity-flow deposits, the exposure of the Friuli platform following Vajont deposition, and the abrupt transition to skeletal-grain–dominated gravity-flow deposits in the overlying Fonzaso Formation (Callovian–Oxfordian), we conclude that the Vajont Limestone is a basinal record of a second-order depositional sequence. Within this second-order sequence, the decameter-scale couplets of nonskeletal-grain–dominated and skeletal-grain–rich sedimentologic associations are defined by upward increases in skeletal-grain and ^{13}C contents and are interpreted as the basinal record of two third-order depositional sequences. We use these results to support our proposal that third-order sequences in resedimented carbonates may be recognized through a combination of sedimentologic and chemostratigraphic studies

INTRODUCTION

Gravity-flow deposits composed of shallow-water carbonate detritus have been the subject of intense study in both ancient and modern environments (e.g., Cartwright, 1985; Cook, 1983; Cook et al., 1972; Crevello and Schlager, 1980; Droxler and Schlager, 1985; Eberli, 1987, 1988; Fulthorpe and Melillo, 1988; Haak and Schlager, 1989; Mullins et al., 1984; Mullins and Cook, 1986; Reijmer, 1991; Ruiz-Ortiz, 1983; Schlager, 1980; Schlager et al., 1994; Shanmugam and Moiola, 1984; Watts, 1987; Yose and Heller, 1989). Study of carbonate gravity-flow deposits is important because sediment composition, depositional geometry, and stacking patterns provide important clues as to relative changes in sea level, the in-filling history of the depositional basin, and the growth history of the neighboring platform (e.g., Boardman et al., 1986; Droxler and Schlager, 1985; Eberli, 1988, 1991; Eberli and Ginsburg, 1989; Glaser and Droxler, 1991; Haak and Schlager, 1989; Reijmer, 1991; Schlager et al., 1994; Yose and Heller, 1989). Although less understood, the diagenetic history of resedimented shallow-water carbonate may reveal important information with regard to the timing of porosity destruction and creation in slope and basinal settings (e.g., Enos, 1988; Halley et al., 1984; Melim et al., 1995; Schlager, 1980). Accordingly, a better understanding of the primary sedimentologic and diagenetic processes that control sediment composition, distribution, and alteration of resedimented carbonate is desirable not only for academic reasons but for economic reasons as well.

*Present address: Mobil (22), P.O. Box 139082, Dallas, Texas 75313

Advances in Carbonate Sequence Stratigraphy: Application to Reservoirs, Outcrops and Models, SEPM Special Publication No. 63
 ISBN 1-56576-060-3, p. 335–370

The Jurassic Vajont Limestone of the Venetian Prealps of northern Italy is an unusual formation because it is composed predominantly of oolite that was formed on a shallow-water platform and was subsequently deposited in deep water as a thick wedge in slope and basinal settings (Bosellini et al., 1981). The Vajont provides an unusual opportunity to examine both the deposition and diagenesis of resedimented carbonates as the Vajont is composed of calcitic oolite with varying amounts of aragonitic and calcitic skeletal-grain debris, oncoids, bedded hemipelagic mudstone, and platform and hemipelagic lithoclasts. This mixture of stable and metastable components allows for detailed diagenetic study, the results of which suggest that most Vajont sediments were not exposed to platform diagenetic processes such as meteoric stabilization and early dolomitization prior to being resedimented (Zempolich, 1995; Zempolich and Erba, 1997). The relative absence of early surface-related diagenesis allows for (1) the definition of primary sedimentologic and geochemical signatures that are present within the Vajont Limestone, (2) an assessment of the role that sea-level change may have played in the unusually abundant production of oolite, and (3) the potential use of these findings in the sequence stratigraphic interpretation of resedimented carbonate.

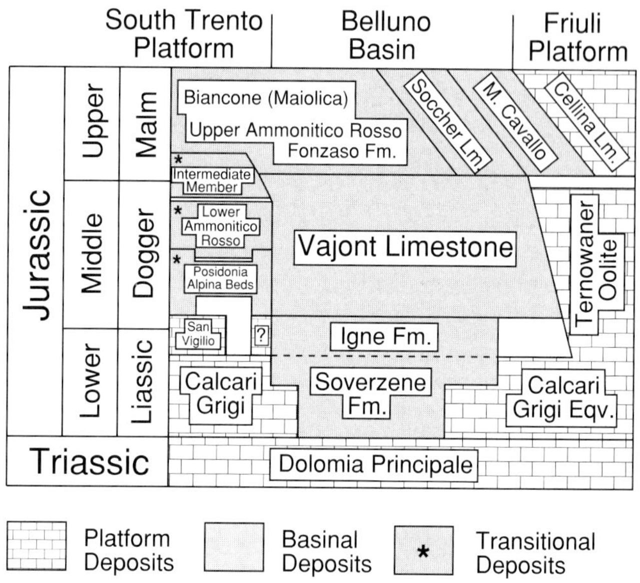

FIG. 1.—Stratigraphic-age relationships of the Friuli platform, Belluno basin, and South Trento platform (modified from Bosellini et al., 1981). The Vajont Limestone is the basinal equivalent of the Ternowaner Oolite, which was deposited on the Friuli platform (Bosellini et al., 1981). Ooids and skeletal-grain debris derived from the Friuli platform were deposited in the Belluno basin as a thick series of turbidites and debris flows.

THE VAJONT LIMESTONE AND ITS GEOLOGIC SETTING

Deposition of the Vajont Limestone (Middle Jurassic) is closely associated with the breakup of Pangea, during which time great volumes of oolite were deposited along the margins of the Tethys ocean in the present-day circum-Mediterranean region (Bosellini, 1989). Beginning in the Early Jurassic, Europe and northern Africa began to separate, and by Late Jurassic time, an extensive transform zone was created (Weissert and Bernoulli, 1985). This regional rifting event established a horst-and-graben tectonic setting along the southern Tethyan margin and led to the structural definition of local platforms and basins. The Trento platform—the most landward horst block of the Southern Alps—was bounded to the west by the Lombardy basin and to the east by the Belluno basin, which separated the Trento platform from the Friuli platform (i.e., the stable foreland). Thick sequences of Mesozoic platform and basin carbonates are found in these domains (Fig. 1). The Vajont Limestone records Middle Jurassic (Dogger) sedimentation in the Belluno basin.

The Vajont Limestone crops out extensively in the Venetian Alps of northeastern Italy (Fig. 2; Doglioni, 1990). Paleogeographic reconstructions suggest a depositional area in excess of 100 km along strike and 50 km across strike (Bosellini et al., 1981). Well penetrations in the Po Plain and northern Adriatic area suggest that the Vajont is also found to the south in the subsurface and that shallow-water platform equivalents are found to the east (Bosellini et al., 1981; Cati et al., 1987). The Vajont is a thick sequence (up to 600 m along the platform margins) predominantly composed of shallow-water ooids and skeletal grains that were redeposited by gravity-flow processes in a deep-marine environment (Masetti, 1971; Bosellini and Masetti, 1972; Bosellini et al., 1981). Ooids were derived from the western edge of the Friuli platform and deposited as a series of deep-water, slope fans in the Belluno basin (Figs. 2, 3). These submarine ooid fans thin out basinward and onlap parts of the submerged Trento platform toward the west. Overall, the Vajont carbonates are predominantly limestone, but in places large masses (up to 400 m thick) have been dolomitized (Zempolich, 1995; Zempolich and Hardie, 1997). The transition to the skeletal-grain–rich turbidites and bedded chert of the overlying Fonzaso Formation (Figs. 1–3) records a rather abrupt shutdown of ooid production on the Friuli platform and the initiation of reefal sedimentation, probably during the Callovian–Oxfordian (Bosellini et al., 1981).

Age dating of the Vajont is problematic. Stratigraphic age constraints from formations located below and above the Vajont suggest a general age range of Bajocian to Callovian (Bosellini et al., 1981; Casati and Tomai, 1969). Age constraints within the Vajont are provided by the overlapping ranges of the *Protopeneropolis striata* and *Trocholina* foraminiferal zones, which suggest that some deposition may have occurred during the late Bajocian to early Bathonian (Casati and Tomai, 1969).

METHODS

Both limestone and dolomite outcrops were measured, described, and sampled over a wide paleogeographic area (Zempolich, 1995; Zempolich and Hardie, 1997). The limestone section at Col Visentin (Fig. 4) was chosen for detailed sedimentologic and diagenetic study because of its proximal location to the platform margin, accessibility, and thick continuous exposure (Figs. 2, 3). The 325 m of limestone present at Col Visentin was measured bed-by-bed and sampled at approximately 1-m intervals for sedimentologic, geochemical, and biostratigraphic study (303 total samples). In the laboratory,

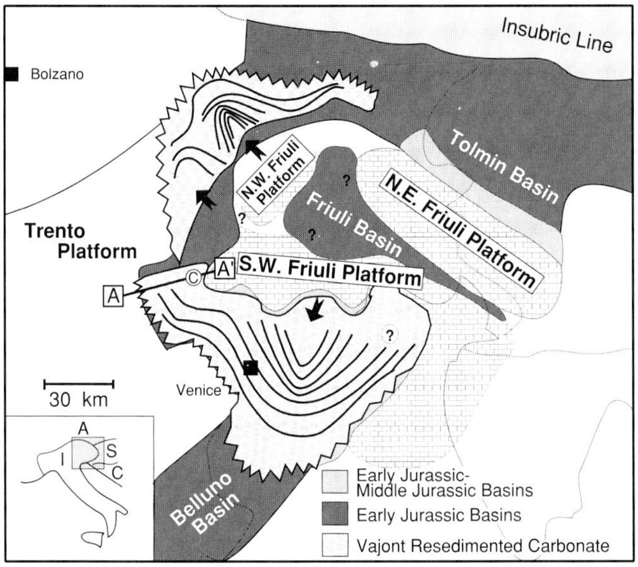

FIG. 2.—Early and Middle Jurassic paleogeography of the Venetian Alps (modified from Bosellini et al., 1981, and Cati et al., 1987). The study area is divided into several platform and basin domains including the Trento platform, Belluno basin, Friuli platform, and Tolmin basin. The subdivision of the Friuli platform into several platforms and basins is based on the interpretation of seismic data from the Friuli Plain and Po basin (Cati et al., 1987). Distribution of Vajont resedimented carbonate is based on both outcrop and subsurface data (modified from Bosellini et al., 1981, and Cati et al., 1987). The Vajont thickens toward the Friuli platform and attains a maximum thickness of 600 m near the platform margin (see Fig. 5). Toward the west, the Vajont thins and onlaps parts of the Trento platform. This sedimentary geometry suggests that the Vajont Limestone was deposited on the leeward side of the Friuli platform(s). A–A′ marks the location of the regional cross section depicted in Figure 3, and C marks the location of the Col Visentin section (Figs. 3, 4). Inset: I = Italy, A = Austria, S = Slovenia, C = Croatia.

detailed petrographic and geochemical study determined the primary mineralogy and chemistry of Vajont limestone components. A study of cement and replacement fabrics evaluated more extensively the diagenetic history of the limestone components (Zempolich, 1995).

To quantitatively examine the variation in sediment composition at Col Visentin, the medium- to coarse-size matrix of both turbidites and debris-flow deposits was sampled for point-count analysis. In all, 74 thin sections over the 325-m interval were point-counted by using the method described by Chayes (1956). For most thin sections, more than 200 points were volumetrically counted, thereby keeping statistical errors within acceptable limits (±3.5%; Van der Plas and Tobi, 1965).

To assess secular geochemical trends and diagenesis at Col Visentin, isotope and trace element data were gathered from the same samples used in the sediment compositional study and from additional samples over the stratigraphic interval. Samples of bedded hemipelagic mudstone were processed and examined for nannofossil content at the University of Milan.

Thin sections were examined by plane-light, cathodoluminescence light, and scanning, backscattered, and transmission electron microscopy (SEM–BSE–TEM) to identify and characterize primary and diagenetic components. Microdrilling techniques were employed to sample identified components for isotope and trace element analysis. C and O isotope analyses were completed at the University of Michigan Stable Isotope Laboratory under the direction of Dr. K. C Lohmann. Isotopic analyses were made by using an automated Kiel reaction device coupled with a Finnigan-MAT 251 mass spectrometer. Calcite and dolomite samples were reacted on-line at 73 °C with three drops of anhydrous phosphoric acid in separate reaction vessels. Trace element analyses via inductively coupled plasma (ICP) mass spectroscopy were completed at the University of Michigan with the help of Ted Huston under the direction of Dr. Lynn Walter. Samples were dissolved in 1 mL of dilute HCl-HNO_3 and simultaneously analyzed for Ca, Mg, Sr, Fe, and Mn by ICP-AES (atomic emission spectroscopy) by using a Leeman Labs Plasma-Spec III. Analytical precision of calcite analyses was ±1.9% for Ca, ±3.2% for Mg, ±2.5% for Sr, ±3.6% for Fe, and ±2.6% for Mn. Sr isotope analyses were completed at the University of Texas at Dallas by Dr. Tim Denison. Microdrilled ooid and whole-rock samples were separated by dissolution in 1*N* acetic acid, filtered, and then recovered from the filtrate by using a Sr-specific resin. After desiccation, Sr extracts were diluted in ultrapure water, and 2-μg aliquots were loaded onto zone-refined Ta filaments for analysis. Sr ratios for each unknown sample are reported as an absolute difference relative to NBS SRM 987, which is assumed to have a ratio of $^{87}Sr/^{86}Sr$ equal to 0.710240.

BIOSTRATIGRAPHY

New Age Assignments for the Vajont Limestone

Hemipelagic mudstone samples gathered from the Col Visentin yielded nannofossil assemblages, which provided tight age constraint on the stratigraphic section. In addition to nannofossils, a single ammonite was discovered within the lower Vajont Limestone at the nearby Col dei Moi section on the top of a bedding plane. The ammonite is prudently identified as *Docidoceras* (sp. ind.) of earliest Bajocian age (Zempolich, 1993).

Age Constraints

Diagnostic nannofossils were obtained from three samples from the Col Visentin section (Fig. 4; Table 1). When compared with nannofossil assemblages described from other Jurassic sections (Cobianchi, 1992; Cobianchi et al., 1992; Erba, 1990), the age ranges of these samples suggest that the Vajont Limestone was deposited between the late Aalenian/early Bajocian boundary and early to late Bajocian time. A basal age range of late Aalenian (Concavum ammonite biozone) to early Bajocian (Discites ammonite biozone) is established by the presence of *Watznaueria britannica* and *Cyclagelosphaera margerelii* in sample CVX1 and the concomitant absence of *W. communis*, *Watznaueria* aff. *W. manivitae*, and *W. manivitae*. A general age range of early Aalenian (Opalinum ammonite biozone) to late Bajocian (Parkinsoni ammonite biozone) is suggested by the presence of *Hexalithus magharensis* in sample CV230. An

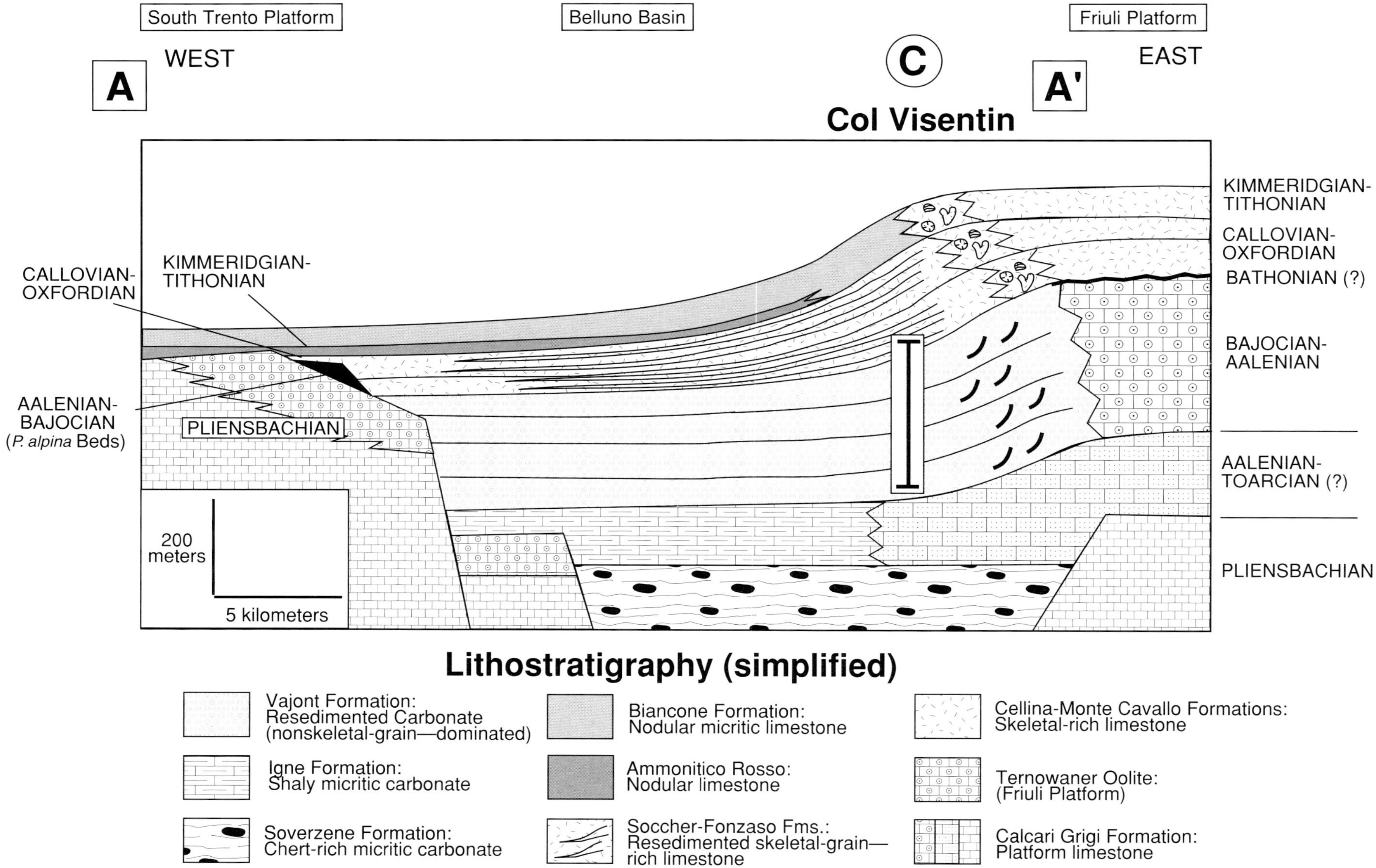

FIG. 3.—Schematic cross section of Mesozoic platform and basin facies in the Venetian Alps. Stratigraphic sections were sampled and measured along a transect that starts at the Friuli platform margin, crosses the Belluno basin, and overlaps the eastern edge of the South Trento platform (A–A´, Fig. 2). The Vajont Limestone was extensively sampled at the Col Visentin locality (labeled C) for detailed sedimentologic and stratigraphic study (Fig. 4). The position of the Col Visentin stratigraphic section shown as vertical bar within cross section.

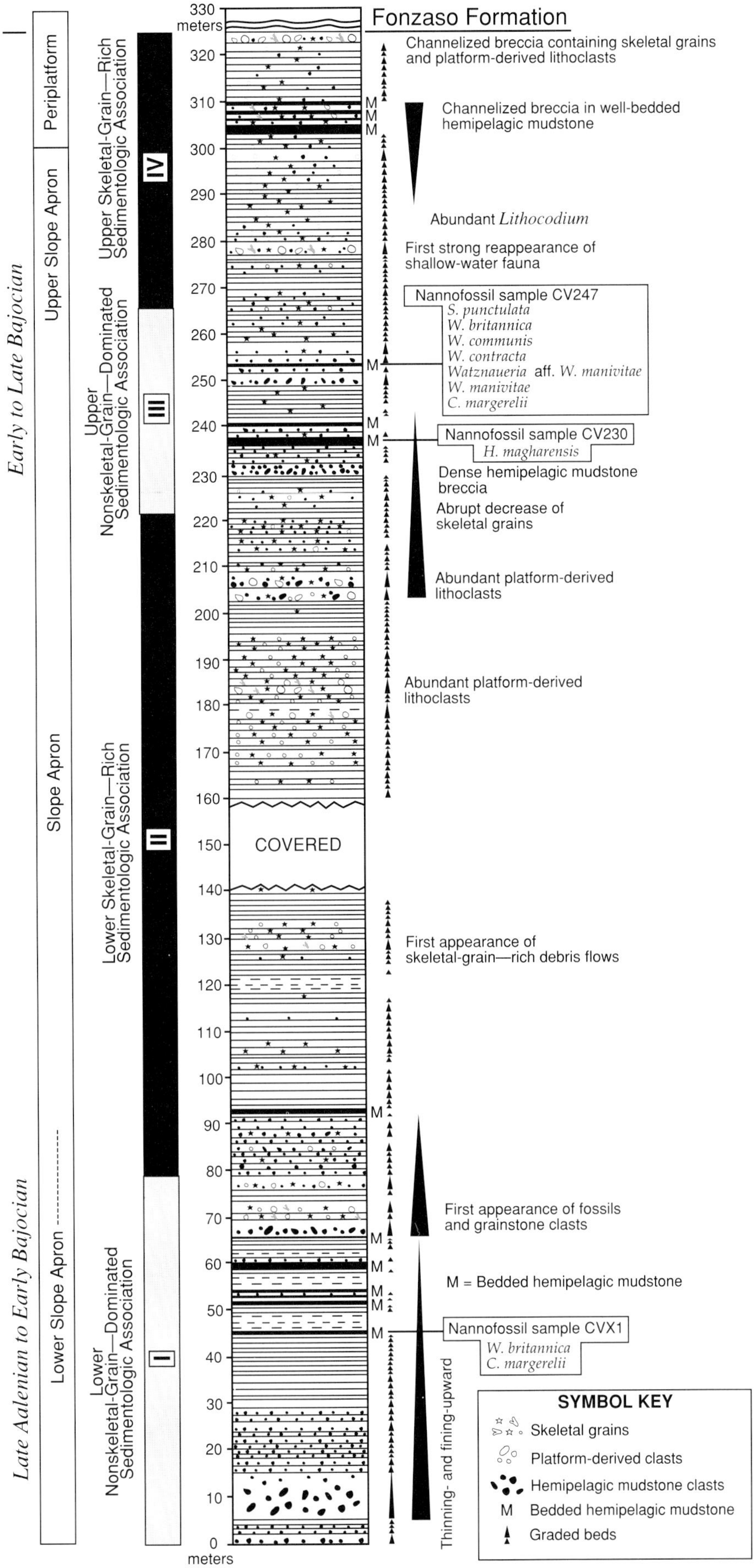

FIG. 4.—Measured stratigraphic section of the Vajont Limestone at Col Visentin. The section is composed of approximately 325 m of oolitic-rich gravity-flow deposits that are subdivided into two nonskeletal-grain–dominated and two skeletal-grain–rich sedimentologic associations: (I) a lower nonskeletal-grain–dominated sedimentologic association, (II) a lower skeletal-grain–rich sedimentologic association, (III) an upper nonskeletal-grain–dominated sedimentologic association, and (IV) an upper skeletal-grain–rich sedimentologic association. Sedimentary structures and facies succession suggest that the section can be divided into slope apron and periplatform facies. Thinning-upward and fining-upward units are distributed within an otherwise random stacking of debris flows and turbidites in slope deposits. A coarsening-upward unit composed of gravity-flow deposits, interbedded hemipelagic mudstone, and channelized breccia marks the transition to periplatform deposits. Nannofossil assemblages have been described from bedded hemipelagic mudstone in the lower and upper parts of the section. Field observations indicate that oncoids and periplatform-derived hemipelagic mudstone clasts are abundant components of oolitic gravity-flow deposits in both the lower and upper nonskeletal-grain–dominated sedimentologic association (I and III). Similarly, field observations indicate that platform-derived lithoclasts and skeletal grains are abundant components of skeletal-grain–rich gravity-flow deposits in both the lower and upper skeletal-grain–rich sedimentologic association (II and IV).

TABLE 1.—NEW AGE ASSIGNMENTS FOR THE VAJONT LIMESTONE BASED ON NANNOFOSSIL ASSEMBLAGES COLLECTED AT COL VISENTIN

Sample	Interval (m)	Nannofossils present	Age	Ammonite Biozone
CVX1	45.3	*Watznaueria britannica*	Late Aalenian–early Bajocian boundary	Concavum to Discites
		Cyclagelosphaera margerelii		
CV230	237	*Hexalithus magharensis*	Early Aalenian–late Bajocian	Opalinum to Parkinsoni
CV247	253	*W. manivitae*	Early Bajocian–late Bajocian	Laeviuscula to Garantiana
		Watznaueria aff. *W. manivitae*		
		W. communis		
		W. britannica		
		C. margerelii		
		W. contracta		
		Schizospherella punctulata		

upper age range of early Bajocian (Laeviuscula ammonite biozone) to late Bajocian (Garantiana ammonite biozone) is established by the presence of *W. communis, W. britannica, W. contracta, C. margerelii, Watznaueria* aff. *W. manivitae, W. manivitae,* and *Schizospherella punctulata* in sample CV247 and the absence of *W. barnesae*. These nannofossil assemblages are consistent with the early Bajocian age provided by the lone ammonite specimen and indicate that (1) Vajont sedimentation in the Belluno basin began by the time of the late Aalenian/early Bajocian boundary and (2) most sedimentation was completed by late Bajocian time. Furthermore, Sr isotope data gathered from radial calcite ooids, hemipelagic micrites, and a single belemnite yield ratios that trend upward through the stratigraphic section from $^{87}Sr/^{86}Sr = 0.707200$ to 0.707041. Both the values and the apparent decrease in $^{87}Sr/^{86}Sr$ over time are in agreement with the early to late Bajocian $^{87}Sr/^{86}Sr$ marine values reported by Jones et al. (1994a). Collectively, these biostratigraphic and geochemical data suggest that the Vajont accumulated over a relatively short time span (approximately 4.5 m.y.; cf. Haq et al., 1988, with reference to Harland et al., 1982, and Cope et al., 1980) and at a high rate of sedimentation (approximately 325 m over 4.5 m.y. = approximately 72 mm/k.y.).

SEDIMENTOLOGY

Facies Succession and Depositional Model

Vajont gravity-flow deposits and bedded hemipelagic mudstone comprise slope and periplatform facies that are arranged as an overall coarsening-upward sequence (Fig. 4). Within this sequence, several decameter-scale, thinning-upward and fining-upward units are contained within an otherwise unorganized succession of slope sediments. These data suggest that the Vajont Limestone was deposited as a slope apron (e.g., Mullins and Cook, 1986) adjacent to the Friuli platform.

Meter-scale debris flows are composed of boulder- to cobble-sized lithoclasts and oolitic matrix and display both normal and inverse grading (Figs. 5, 6). Vajont turbidites exhibit T_a, T_a and T_b, or base-cut-out T_b divisions of the Bouma turbidite sequence. Complete T_a–T_e turbidites and base-cut-out T_c–T_e structures are uncommon. T_a turbidites contain pebbly to coarse-size grainstone at the base and grade upward into fine-size oolitic and peloidal grainstone (Fig. 5). T_a–T_b turbidites display well-developed planar lamination toward the top (Fig. 5).

Bedded hemipelagic mudstone consists of dark to light brown carbonate mud that contains pelagic bivalves, pelagic and benthic foraminifers, poorly preserved radiolarian tests, and calcareous nannofossils (Fig. 7). Thin beds of hemipelagic mudstone are found toward the top of thinning- and fining-upward sequences in the lower and middle stratigraphic levels at Col Visentin (Fig. 4). Thick beds of hemipelagic mudstone in upper stratigraphic levels are incised by shallow, broad channels that are filled by carbonate breccia. Bedded hemipelagic mudstone displays a spectrum of soft-ground, firm-ground, and hardground bioturbation structures (Fig. 7). These features indicate that mudstone was continuously bioturbated and bored following deposition. Differing degrees of penecontemporaneous lithification shown by the bedded hemipelagic mudstone are also directly expressed in redeposited hemipelagic mudstone clasts. Clast morphologies vary from well-rounded softground lithoclasts to subangular firmground lithoclasts to scalloped hardground lithoclasts whose rims adjoin former excavation structures (Fig. 7). These features suggest that hemipelagic mudstone clasts were derived from periplatform locations and redeposited into slope and basin settings (Zempolich, 1995).

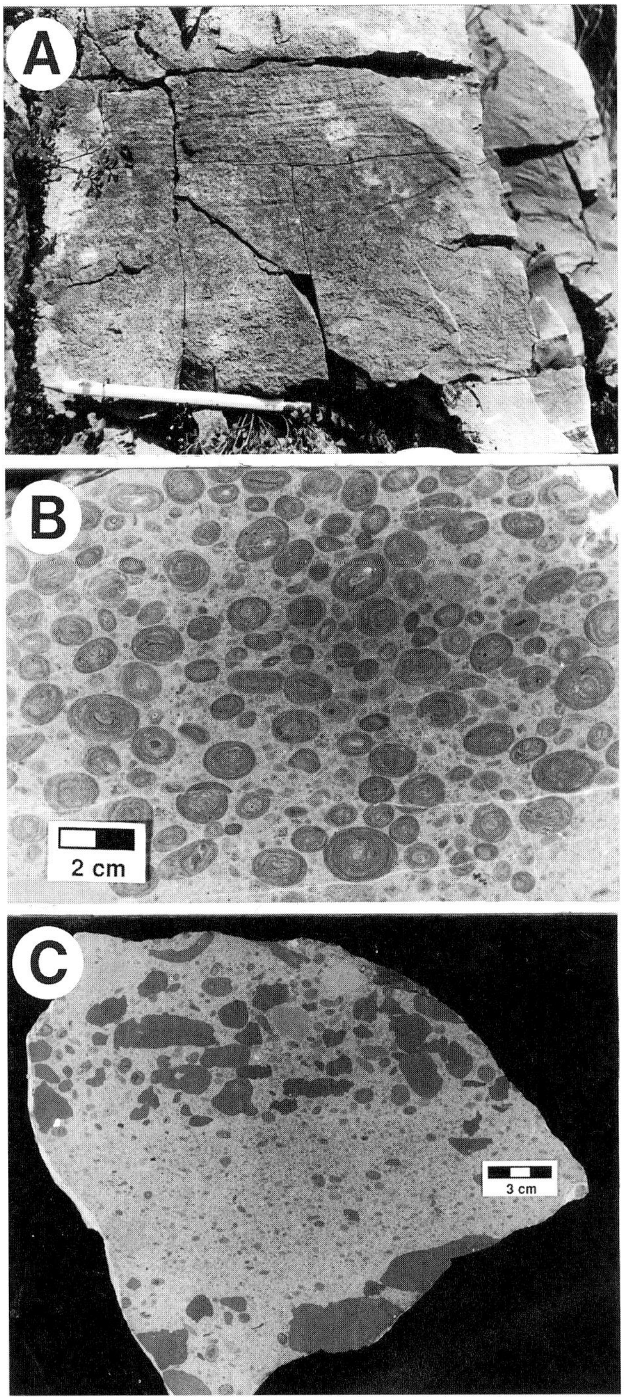

FIG. 5.—Nonskeletal-grain–dominated gravity-flow deposits composed of oolitic matrix, oncoids, and hemipelagic mudstone clasts (lower nonskeletal-grain–dominated sedimentologic association I, Col Visentin section). (A) Outcrop photograph of graded turbidite exhibiting transition from basal Ta unit composed of medium- to coarse-grained oolitic matrix to upper Tb unit composed of planar-laminated, medium- to fine-grained oolitic matrix. (B) Slab photograph of the base of an oncoid-rich turbidite. Vajont oncoids are composed of micritic laminations and are abundant components of Vajont gravity-flow deposits. Oncoid nuclei include peloids, aggregate grains, algal clasts, and shallow-water skeletal fragments. (C) Slab photograph of part of a large-scale debris flow (>4 m thick). Interlayered clast-rich and clast-poor horizons are part of overall graded-upward unit. Distribution of clasts indicates that oolitic matrix provided both dispersive pressure and buoyant lift during deposition.

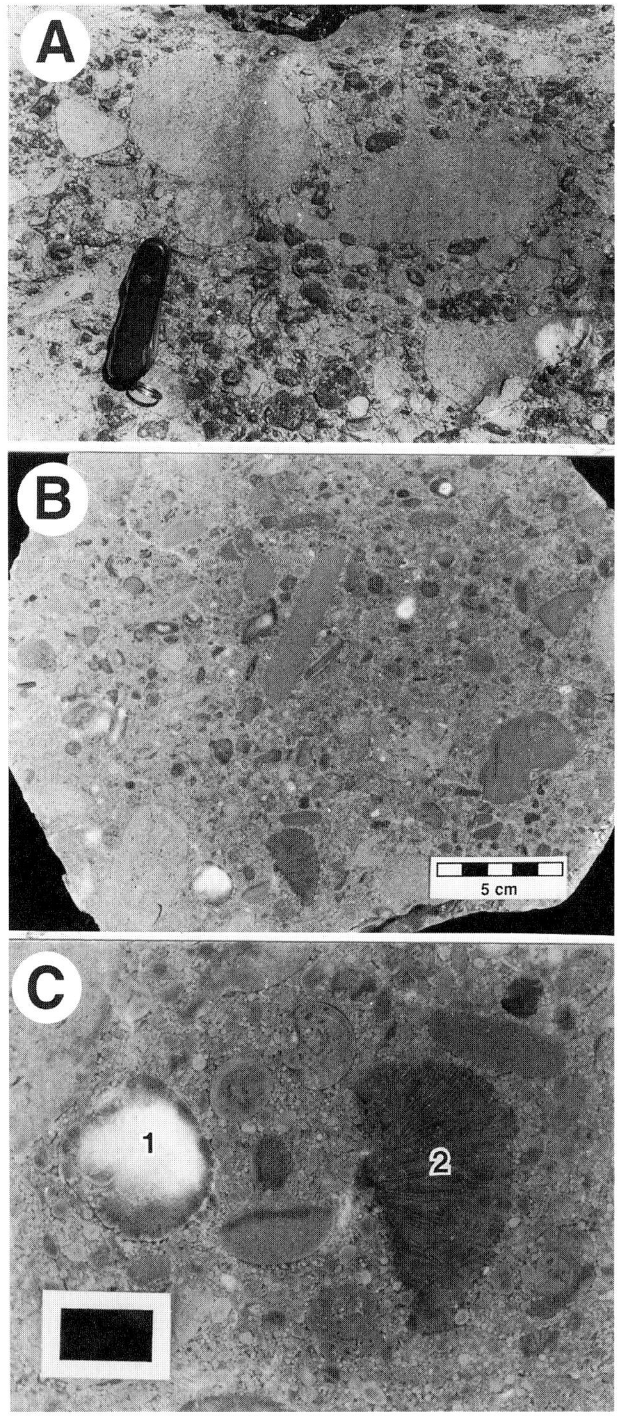

FIG. 6.—Skeletal-rich gravity-flow deposits composed of platform-derived lithoclasts, skeletal grains and fragments, and oolitic matrix (lower skeletal-grain–rich sedimentologic association II, Col Visentin section). (A) Inversely graded debris flow (1.3 m thick) contains oolitic lithoclasts, skeletal grains, and oolitic matrix. Platform-derived lithoclasts are well rounded and composed of oolitic grainstone, oolitic packstone, and skeletal-grain–rich wackestone. The presence of boulder-sized platform clasts in the middle parts of the flow indicate that oolitic matrix provided both dispersive pressure and buoyant lift during deposition. (B) Slab photograph of skeletal-grain–rich debris flow in (A). Shallow-water skeletal fragments include corals, calcareous algae, echinoderms, mollusks, gastropods, bryozoa, benthic foraminifers, etc. (C) Close-up slab photograph of skeletal grains in (B). Originally aragonitic skeletal grains (e.g., coral labeled 1) have been altered to low-Mg calcite. Originally calcitic skeletal grains have retained their primary skeletal microstructure (e.g., calcareous sponge labeled 2; e.g., Wendt, 1984). Scale bar represents 1 cm.

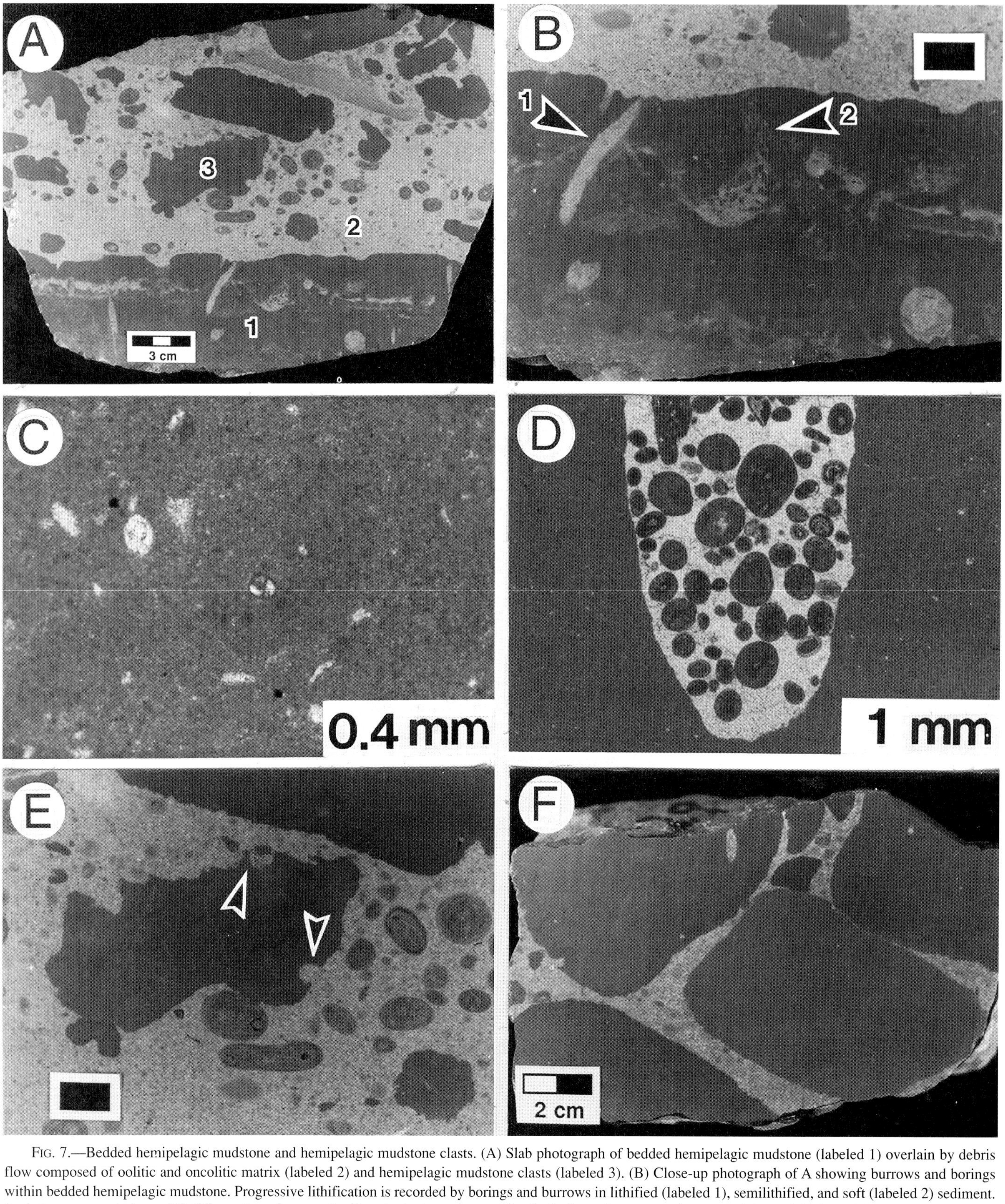

FIG. 7.—Bedded hemipelagic mudstone and hemipelagic mudstone clasts. (A) Slab photograph of bedded hemipelagic mudstone (labeled 1) overlain by debris flow composed of oolitic and oncolitic matrix (labeled 2) and hemipelagic mudstone clasts (labeled 3). (B) Close-up photograph of A showing burrows and borings within bedded hemipelagic mudstone. Progressive lithification is recorded by borings and burrows in lithified (labeled 1), semilithified, and soft (labeled 2) sediment. Scale bar represents 1 cm. (C) Thin-section photograph of hemipelagic mudstone. Hemipelagic mudstone contains protoglobergerinid foraminifers, calcareous nannofossils, poorly preserved radiolarian tests, and pelagic bivalves. (D) Thin-section photograph of boring 1 in B. Excavation of lithified sediment is noted by the formation of distinct walls within bedded mudstone and grainstone fill. (E) Close-up photograph of A showing the scalloped edges of hemipelagic mudstone clasts, which are thought to have been created by slope failure, and the preferential separation of semilithified mudstone along burrows and borings (arrows). Scale bar represents 1 cm. (F) Slab photograph of well-rounded hemipelagic mudstone clasts. Well-rounded texture indicates that clasts were eroded as soft sediment and redeposited prior to lithification.

TABLE 2.—SUMMARY OF VAJONT SEDIMENTOLOGIC ASSOCIATIONS AT COL VISENTIN

Sedimentologic Association (See Fig. 4)	Interval (m)	Number of Gravity Flows	Total Turbidite Thickness (m)	Number of Hemipelagic Mudstone Beds	Total Mudstone Thickness (m)	Number of Mixed Beds	Total Mixed Bed Thickness (m)
Upper skeletal-grain–rich (IV)	266.55–325	78	54.3	4	4.25	0	0
Upper nonskeletal-grain–dominated (III)	221.3–266.55	62	39.7	3	2.15	5	3.37
Lower skeletal-grain–rich (II)	79.0–221.3*	218	124.0	1	0.28	0	0
Lower nonskeletal-grain–dominated (I)	0–79.0	115	76.9	7	1.12	1	1
Totals	307	473	294.9	15	7.8	6	4.37
Calculated percentages			96.1%		2.5%		1.4%

*Includes 18 m covered interval.

Vajont slope facies are identified by the predominance of grain- and clast-supported conglomerates, T_a–T_b calcarenites, and T_a–T_e calcarenites (Mullins and Cook, 1986, Mutti and Ricci Lucchi, 1972). Upper-slope periplatform settings are indicated by the presence of channelized carbonate breccia in thick, well-bedded hemipelagic mudstone (Fig. 4). The absence of a steep sediment bypass facies in the Vajont, together with a gradual transition upward from slope and periplatform facies to coral-hydrozoan reef-margin facies (Fig. 3; Zempolich, 1995), argues against the presence of a high-relief platform and base-of-slope carbonate apron (e.g., Mullins and Cook, 1986; Schlager and Ginsburg, 1981; Schlager and Camber, 1986). Instead, Vajont facies distribution and depositional geometries collectively suggest that the Vajont Limestone was deposited as a low-angle (<4°) slope apron adjacent to, and transitional with, shallow-water oolite on the Friuli platform (Zempolich, 1995).

POINT-COUNT ANALYSIS

Two lithologic associations are apparent in field observations of the section at Col Visentin. The first is an association of hemipelagic mudstone clasts and oncoids within oolitic gravity-flow deposits (Figs. 5, 7). These nonskeletal-grain–dominated lithologies are prevalent within the lower 80 m of outcrop and within the interval from approximately 220 to 266 m (Table 2). The second association consists of skeletal grains and grain-rich lithoclasts within oolitic gravity-flow deposits (Fig. 6). These skeletal-grain–rich lithologies are prevalent within the interval from approximately 80 to 220 m and within the interval from approximately 266 to 325 m (Table 2). To quantify these observations, 74 thin sections of samples that span the 325-m section at Col Visentin were point-counted for their grain, carbonate mud, calcite cement, and dolomite content (Zempolich, 1995). Grains were subdivided into nonskeletal- and skeletal-grain types. These results provide a quantitative basis for separating the Vajont into nonskeletal-grain–dominated and skeletal-grain–rich sedimentologic associations (Table 3).

Grain Types

Nonskeletal grains.—

Nonskeletal grains include radial ooids, micritic ooids, oncoids, peloids, grapestone, and grain fragments. They are described as follows.

Ooids.—Ooid nuclei are composed of peloids, benthic foraminifers, echinoderms, and gastropods. Ooid cortical fabrics are nonluminescent and display a spectrum of depositional microfabrics that range from radial to micritic (Fig. 8). Fine-size ooids typically display cortices composed of *radially* oriented, small, subequant to bladed calcite crystals. Medium- to coarse-size ooids display fabrics that are composed of an inner radial cortex and outer micritized cortex; in the latter, the cortical fabrics are almost completely micritized but retain some relict radial fabric. Micritized textures are composed of unoriented equant calcitic nannograins. During point-count analysis, ooids were classified as either radial or micritic on the basis of the prevalent microfabric within the grain.

Oncoids.—Oncoids are round to ovoid in shape and range in size from several millimeters to several centimeters (Fig. 5B). Oncoids are composed of a nucleus and a dense micritic cortex. Nuclei consist of peloids, aggregate grains, reworked algal clasts, and shallow-water skeletal-grain fragments including pelecypods, small corals, and gastropods. Micritic cortices contain minute tubes and small tufts of the alga *Ortonella,* encrusting foraminifers, and the microproblematica *Lithocodium*(?), implying that some sediment was trapped and bound by organisms during growth.

Peloids, grapestone, and grain fragments.—The term "peloid" encompasses all those round to ovoid grains that lack identifiable internal microstructure. The shape and size of these grains suggest that many peloids were originally ooids that have been completely micritized. Grapestone includes aggregate

TABLE 3.—STATISTICAL SUMMARY OF VAJONT SEDIMENTOLOGIC ASSOCIATIONS

Sedimentologic Association (See Fig. 4)	$\delta^{18}O$ (PDB) (‰)	$\delta^{13}C$ (PDB) (‰)	Nonskeletal Grains (%)	Skeletal Grains (%)	Micrite (vol%)	Calcite Cement (vol%)	Dolomite (vol%)
Association IV	−2.87	2.38	78.56	21.44	0.71	10.51	0.00
Association III	−2.58	2.13	91.34	8.66	1.25	9.05	0.11
Association II	−3.48	2.14	88.80	11.20	0.04	10.73	0.00
Association I	−3.01	1.87	98.25	1.75	0.28	8.07	1.21
Totals	−3.12	2.13	88.88	11.12	0.39	9.85	0.31

Nonskeletal-Grain Type as a Percentage of Total Grains

	Ooids	Peloids	Oncoids	Grain Fragments	Grapestone
Association IV	32.81	33.27	4.29	2.86	5.33
Association III	43.85	36.74	4.18	2.16	4.36
Association II	41.65	30.06	8.12	5.30	3.67
Association I	59.13	24.01	9.28	3.81	2.01
Totals	43.99	30.09	7.04	4.01	3.75

Skeletal-Grain Type as a Percentage of Total Grains

	Benthic Foraminifers	Echinoderms	Corals	Bryozoans	Gastropods	Mollusks	Pelecypods	*Lithocodium*	Algal Grains	Encrusting Grains and Organisms	Unidentified Skeletal Grains (Former Aragonite)
Association IV	1.54	2.57	0.57	0.26	0.14	1.07	0.00	1.00	1.97	2.52	9.78
Association III	1.79	2.05	0.00	0.00	0.06	0.40	0.00	0.13	0.99	1.21	2.05
Association II	1.20	1.69	1.04	0.17	0.08	0.64	0.02	0.14	1.04	1.85	3.36
Association I	0.24	0.72	0.11	0.03	0.19	0.11	0.00	0.00	0.14	0.00	0.22
Totals	1.11	1.71	0.59	0.14	0.12	0.59	0.01	0.31	1.04	1.49	4.01

	Nonskeletal-Grain Type as a Percentage of Total Nonskeletal Grains						Ooid Type as a Percentage of Total Ooids	
	Radial Ooids	Micritic Ooids	Peloids	Oncoids	Grain Fragments	Grapestone	Radial Ooids	Micritic Ooids
Association IV	23.89	17.34	42.57	5.58	3.62	6.98	53.91	46.09
Association III	31.15	16.60	40.65	4.50	2.39	4.71	64.73	35.29
Association II	21.22	25.18	34.36	9.34	5.69	4.22	45.03	54.98
Association I	29.57	30.56	24.48	9.44	3.90	2.06	48.34	51.66
Totals	24.97	23.65	34.63	7.93	4.40	4.42	50.12	49.88

Skeletal-Grain Type as a Percentage of Total Skeletal Grains

	Benthic Foraminifers	Echinoderms	Corals	Bryozoans	Gastropods	Mollusks	Pelecypods	*Lithocodium*	Algal Grains	Encrusting Grains and Organisms	Unidentified Skeletal Grains (Former Aragonite)
Association IV	8.06	11.94	1.40	0.98	0.67	4.98	0.00	4.89	10.26	11.80	45.04
Association III	19.06	23.43	0.00	0.00	0.84	4.74	0.00	1.50	11.89	15.94	22.60
Association II	12.68	16.67	3.77	1.18	0.55	8.38	0.37	1.15	11.03	16.86	27.34
Association I	13.72	45.88	5.88	1.47	11.76	5.39	0.00	0.00	7.35	0.00	8.53
Totals	12.31	22.74	3.22	1.06	3.18	6.36	0.15	1.82	9.94	11.43	26.44

NOTE: PDB = Peedee belemnite standard.

grains composed of ooids and peloids that are bound by superficial oolitic laminations or meniscus cement. Grain fragments are composed of sharp angular fragments of various nonskeletal grains (i.e., broken ooids, peloids, and grapestone).

Skeletal grains.—
Skeletal grains within oolitic gravity-flow deposits include fine- to cobble-size fragments of shallow-water biota such as corals, mollusks, gastropods, calcareous sponges, echinoderms, bryozoans, benthic foraminifers, encrusting foraminifers, pelecypods, microproblematic algae, unidentified calcareous algal grains and clasts, encrusting organisms and clasts, and unidentified recrystallized and skeletal-moldic grains (Fig. 9). Belemnite rostrums and ammonites (rare) were the only pelagic grains found within the gravity-flow deposits. Calcareous algae include *Cayeuxia* and/or *Ortonella* and the problematic *Lithocodium* (Fig. 9). *Lithocodium* is particularly abundant in the upper skeletal-grain–rich sequence at Col Visentin. Unidentified calcareous algal grains and clasts contain clusters of *Ortonella* tubes in a nonlaminated micritic matrix.

Platform lithoclasts.—
Platform lithoclasts are composed of pebble- to boulder-size, well-rounded grainstone, packstone, and skeletal-grain–rich wackestone lithologies (Figs. 6, 10). Well-sorted grainstone clasts are cemented by nonluminescent, isopachous fibrous calcite cement (Fig. 10). Fibrous calcite cement and grains are truncated at clast peripheries, suggesting that these clasts were reworked in high-energy environments. Fibrous calcite cement was not observed as a primary precipitate in resedimented oolite. Other grainstone lithoclasts possess meniscus cement, which indicates that these clasts were cemented in vadose environments on the platform top (Fig. 10). Wackestone clasts contain a low-diversity fauna (e.g., pelecypods), suggesting deposition in somewhat restricted low-energy environments (e.g., lagoon). Some lithoclasts display well-developed micritized rinds (Zempolich, 1995).

Vajont Grain Content

Grains make up 89.5% of the total volume of rock whereas early and late calcite cement (9.85% by volume) occlude most primary depositional pore space (Table 3). In thin section, grains and calcite cement account for 99.35% of the total volume of gravity-flow deposits. Gravity-flow deposits account for 96.1% of the total measured thickness; bedded hemipelagic mudstone accounts for just 2.5% (Table 2). These field and quantitative petrographic data indicate that the Vajont was a grain-rich depositional system. Within the Vajont Limestone, 89% by volume of the grains are nonskeletal; the remainder are skeletal. Nonskeletal grains consist mainly of ooids (avg.= 44%) and peloids (avg.= 30%), but there are minor amounts of oncoids (avg.= 7%), grain fragments (avg.= 4%), and grapestone (avg.= 3.75%). The fact that radial and micritic ooids account for 44% of the total grains underscores the point that the Vajont Limestone was derived primarily from a "super-producing oolite factory" (Bosellini et al., 1981). Overall, the Vajont exhibits an upward increase in skeletal, peloid, and grapestone content and a concomitant decrease in ooid and oncoid content (Fig. 11).

Sedimentologic Associations

For each sequence of nonskeletal-grain–dominated and skeletal-grain–rich gravity-flow deposits observed in the field at Col Visentin, grain contents were averaged to delineate major sedimentologic associations (Table 3). Nonparametric correlation statistics were then used to identify significant relationships between the various grain types (Table 4). In this analysis, Spearman's correlation coefficient has been determined for pairs of data that are first ranked and then statistically compared (for further discussion, see Davis, 1986). The correlation coefficient expresses the similarity between the two sets of rankings and varies between +1.0 (perfect correlation) and –1.0 (perfect inverse correlation). A correlation coefficient of 0 indicates that the variables are independent. For the Vajont data set, correlation coefficients greater than +0.30 or less than –0.30 are statistically significant at the 99% confidence level (Table 4). Similarly, correlation coefficients greater than +0.23 or less than –0.23 are statistically significant at the 95% confidence level. These data point to several important relationships among Vajont sediment-grain and geochemical compositions and further support the division of the Vajont Limestone into nonskeletal-grain–dominated and skeletal-grain–rich sedimentologic associations (Fig. 4; Tables 2–4).

Nonskeletal-Grain–Dominated Sedimentologic Associations

Comparison of grain types reveal several important relationships. For example, ooid and skeletal-grain contents display a strong inverse relationship that can be observed in comparisons of both total-grain percentages (correlation coefficient ρ = –0.77) and radial and micritic ooid nonskeletal-grain contents with total skeletal-grain content (ρ = –0.49 and –0.36, respectively). Other grain contents, such as peloid and grapestone, exhibit a strong positive relationship with total skeletal-grain content (ρ = +0.63 and +0.56, respectively). Moreover, total ooid content exhibits an inverse relationship with peloid and grapestone contents (ρ = –0.63 and –0.41, respectively). These data suggest that increases in the sedimentation of ooids in shelf environments occurred at the expense of skeletal, peloidal, and grapestone sedimentation and vice versa. Accordingly, a predominance of ooids in Vajont grain counts is interpreted to reflect periods of platform sedimentation that were dominated by high-energy depositional environments.

Skeletal-Rich Sedimentologic Associations

The fact that individual skeletal-grain components represent only a small percentage of the total grain count coupled with the fact that several of these grain types can be found in both platform and upper-slope settings (e.g., benthic foraminifers, echin-

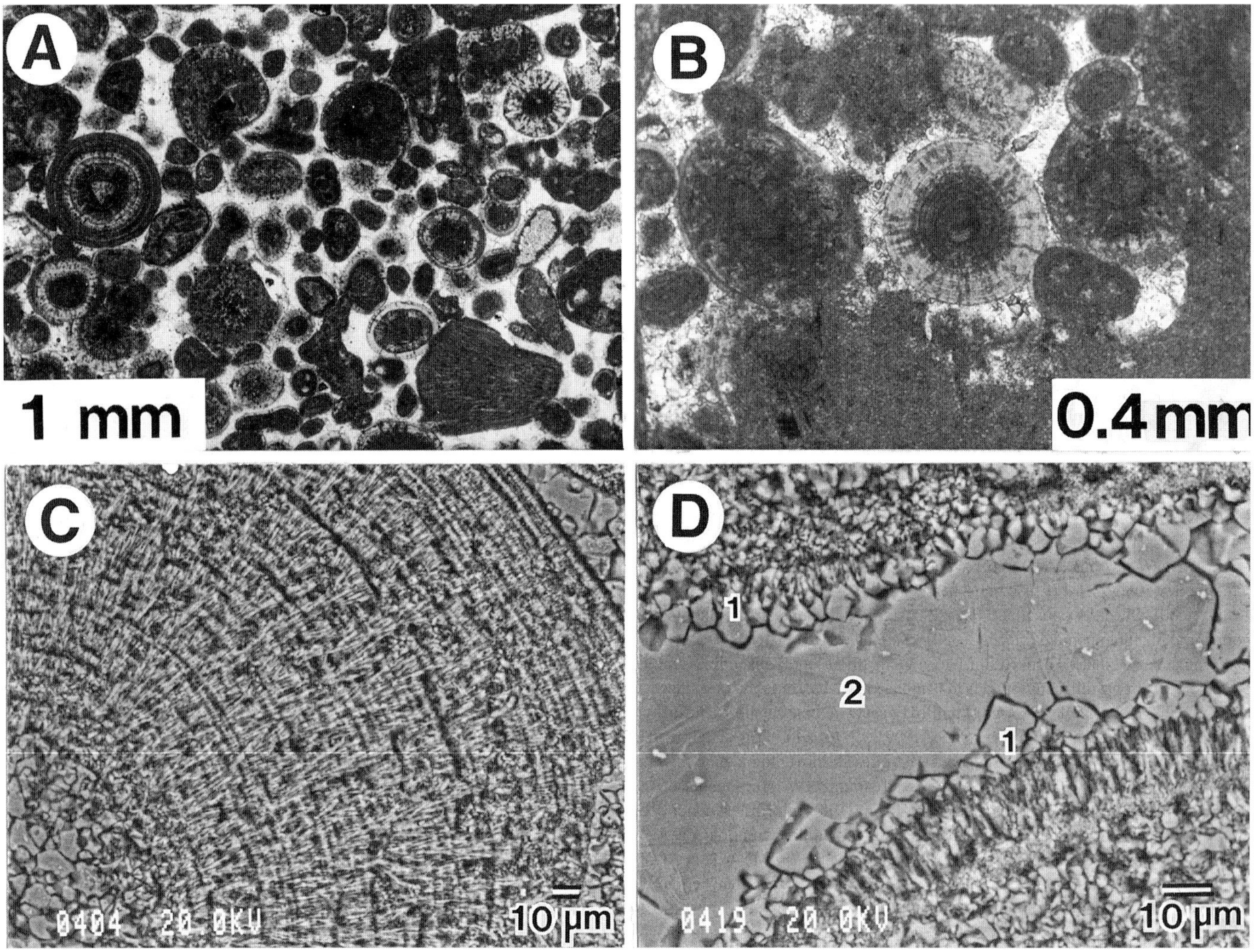

FIG. 8.—Petrography of ooids and oolitic grainstone in gravity-flow deposits of the lower and upper nonskeletal-grain–dominated sedimentologic associations (I and III) at Col Visentin. (A) Ooid-dominated grainstone is composed predominantly of radial and micritic ooids and is sparsely fossiliferous (cross-polarized light). (B) Radial ooid (center of photograph) and partly micritized radial ooid (right-center of photograph) fabrics (cross-polarized light). Radial ooids are composed predominantly of radially oriented, small, subequant to bladed calcite crystals. Micritized ooids are composed of unoriented, equant calcitic nannograins. (C) Radial ooid fabric (SEM photomicrographs; these and other samples used for SEM analysis were etched in 5% acetic acid for 30 s). Cortices of radial ooids are composed of thin laminae (3–10 µm in width) that contain radially oriented, small, subequant to bladed calcite crystals. (D) Calcite cement fabrics in oolitic matrix found in Vajont gravity-flow deposits. Oolitic grainstone was cemented first by isopachous coatings of nonluminescent, equant calcite (labeled 1) that was conformably overlain by banded, luminescent calcite (labeled 2) (cf. Fig. 18A). Early nonluminescent equant-calcite cement and banded, luminescent calcite cement occlude most primary depositional porosity within grainstone matrix.

oderms) makes interpretation of individual skeletal-grain contents difficult (see Reijmer, 1991, for discussion). However, several correlations may be noteworthy. Unidentified shallow-water skeletal-grain components (26.4% of total skeletal-grain content) composed of neomorphic, originally aragonite (corals?) and calcitic (sponges?) skeletal grains exhibit inverse correlations with total ooid contents and oncoid contents (ρ = –0.46 and –0.25, respectively) and positive correlations with peloid and grapestone contents (ρ = +0.23 and +0.26, respectively). Similarly, *Lithocodium* contents (0 to 4.9% of total skeletal-grain content) exhibit positive correlations with peloid and grapestone contents (ρ = +0.26 and +0.26, respectively) and weak inverse correlation with total ooid contents (ρ = –0.20). These correlations suggest that (1) skeletal grains derived from shallow-water skeletal fauna and *Lithocodium* were deposited at the same time as peloids and grapestone (representing deposition in only periodically active shoals) and ooids (representing deposition in constantly agitated subtidal shoals) and (2) during these periods, the platform was dominated by both moderate- and high-energy depositional environments.

SEDIMENTOLOGIC SEQUENCES AND MODELS OF CARBONATE-PLATFORM SHEDDING

As summarized by Schlager (1991), turbidite sediment compositions may provide information as to sediment composition and production on platforms, the export of this sediment to surrounding slopes and basins, and the dynamic interrelationship between sediment production and export with eustatic sea-level fluctuation. Comparison of Vajont sediment compositions is

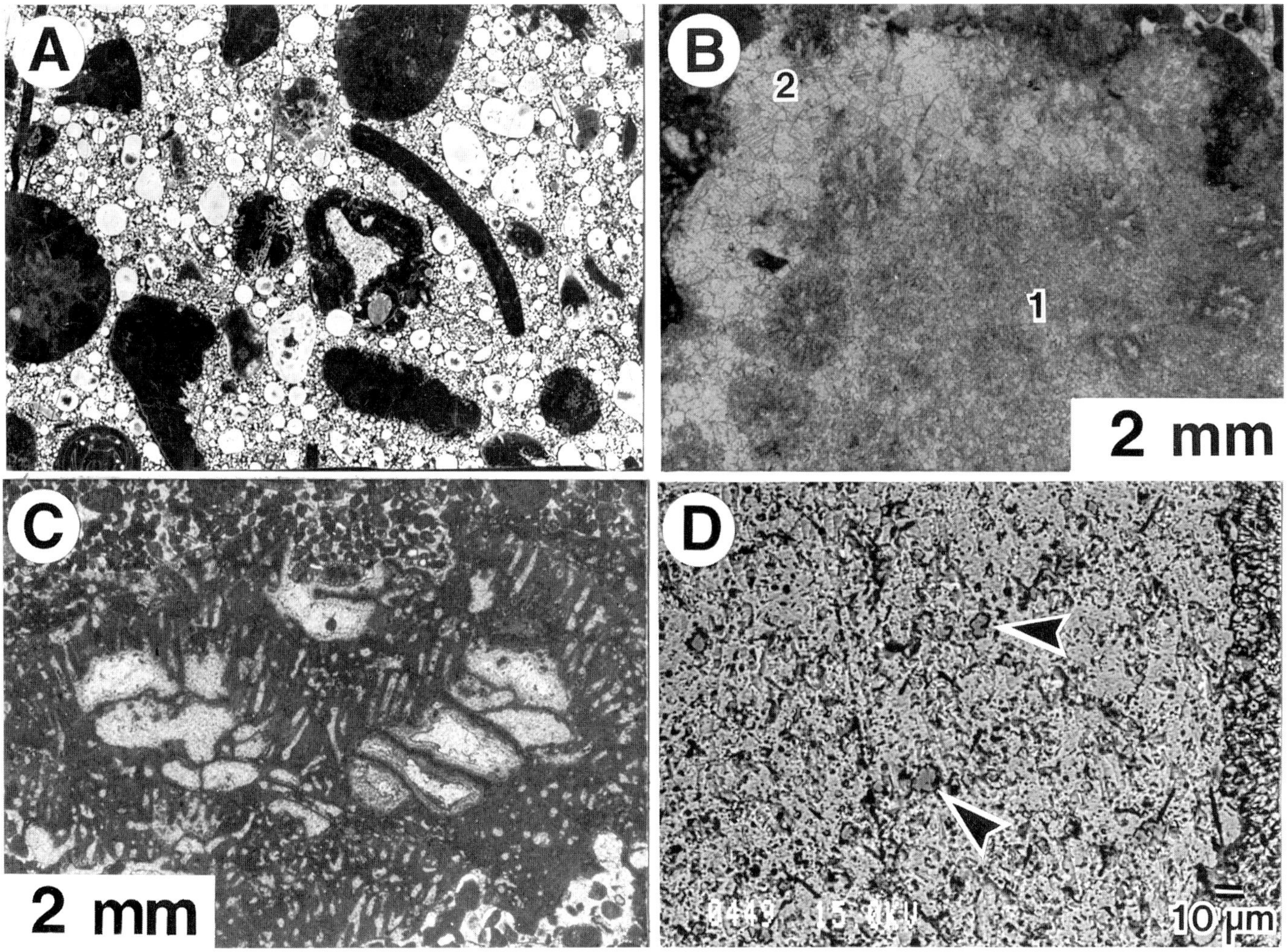

FIG. 9.—Petrography of skeletal grains deposited in skeletal-grain–rich gravity-flow deposits within the lower and upper skeletal-grain–rich sedimentologic associations (II and IV) at Col Visentin. (A). Negative thin-section print of skeletal-grain–rich matrix. Recrystallized corals and mollusks (coarse- to pebble-sized dark grains) are present in fine- to coarse-sized oolitic and oncolitic grainstone matrix. Width of photograph represents 4 cm. (B) Aggrading neomorphic calcite fabric after originally aragonitic coral (plane-polarized light). Relict polyp microstructure is preserved by fine-grained, nonluminescent neomorphic calcite (labeled 1). Primary skeletal-grain fabric is destroyed where replaced by luminescent, coarse-grained calcite (labeled 2). (C) Microproblematica *Lithocodium*. Pebble-sized grains composed of *Lithocodium* are a common skeletal-grain component of gravity-flow deposits in the upper skeletal-grain–rich sequence at Col Visentin. (D) Echinoderm fragment composed of low-Mg calcite and microdolomite (dark subhedral crystals noted by two arrows). The retainment of Mg in the form of microdolomites indicates that these grains were originally composed of high-Mg calcite. The absence of microdolomites in radial ooid cortices suggests that calcitic ooids were originally precipitated as low-Mg calcite.

made with Jurassic and Neogene carbonate systems to examine the potential relationship between compositional variation and sea-level fluctuation.

Jurassic Depositional Analogues

The grains and fauna that are represented in Vajont gravity-flow deposits are important components of platform-interior and platform-margin facies described in Upper Jurassic shelf sequences of the Northern Alps (Flügel, 1979) and Lower Jurassic shelf sequences of the South Trento platform (Beccarelli Bauck, 1986, 1988; Zempolich, 1993). In these Upper Jurassic examples, oncoids and calcareous algae such as *Cayeuxia* and *Lithocodium*(?) are common components of sparite facies belts located within open-shelf and restricted-platform settings immediately adjacent to oolitic margin facies (Flügel, 1979). Furthermore, peloids, micrite, and micritic intraclasts are common components of facies belts located within interior-shelf settings. Lower Jurassic (upper Pliensbachian) shelf-margin sequences of the South Trento platform are composed of small patch reefs (i.e., several meters in height and tens of meters in width) and oolitic grainstone, which are arranged in shallowing-upward cycles (Zempolich, 1993). These platform-margin facies include skeletal-grain–rich back-reef wackestone, skeletal-grain–rich boundstone, and cross-bedded oolitic, peloidal, and oncolitic grainstone. Skeletal grains include calcareous sponges, solitary and colonial corals, stromatoporoids, calcareous algae, gastropods, mollusks, and benthic foraminifers. This fauna is similar to that found in skeletal-grain–rich gravity-flow deposits of the Vajont and suggests that similar patch reefs may have been located just seaward of oolitic margin facies on the Friuli platform. On the basis of these Upper and Lower Jurassic depositional analogues, it is probable that the predominantly oolitic composition of Vajont gravity-flow deposits is due to initial sedimentation of mostly nonskeletal-grain types in open-shelf and margin environments of the Friuli platform, with a minor contribution of skeletal grains derived from small patch reefs. Although sediment compositions of the Vajont are consistent with Jurassic shallow-water facies models, the compositional variation exhibited by the alternation of nonskeletal-grain–dominated and skeletal-grain–rich sedimentologic associations remains unexplained.

Comparison of Vajont Gravity-Flow Deposits with Neogene Examples

Comparison of Vajont sediment compositions with Neogene resedimented carbonates suggests that the genesis of the Vajont is related to production of shallow-water sediment on the top of a flooded bank. Studies of Neogene carbonate gravity-flow deposits have shown a strong correlation between times of interglacial sea-level highstands and increased production and export of nonskeletal grains from the tops of carbonate banks to neighboring deeps (e.g., Cartwright, 1985; Haak and Schlager, 1989; Reijmer, 1991; Schlager, 1991). Conversely, these same studies show that during times of sea-level lowstand, a change to skeletal-grain–dominated compositions and a decrease in the total amount of sediment takes place.

In the Vajont carbonates, the nonskeletal-grain contents of both nonskeletal-grain–dominated sequences (avg. = 98.3% and 91.3%) and skeletal-grain–rich sequences (avg. = 88.8% and 78.6%) are much greater than the nonskeletal-grain contents (66.2% to 54.1%; Haak and Schlager, 1989) of Pleistocene interglacial "highstand" carbonate gravity-flow deposits (Fig. 12). It is important to note, however, that the skeletal-grain contents (21.4% and 11.2%) of the skeletal-grain–rich Vajont sequences do not approach those typical of glacial "lowstand" carbonate gravity-flow deposits (e.g., Haak and Schlager, 1989; avg. skeletal-grain content = 88.5% to 64.6% in Pleistocene glacial deposits). On the basis of this comparison, we conclude that both the nonskeletal-grain–dominated and skeletal-grain–rich sedimentologic associations in the Vajont Limestone are products of "highstand" shedding of shallow-water carbonate sediment.

To test this conclusion, the timing and frequency of Vajont deposition based on the new nannofossil ages allow for comparison of Vajont sedimentologic sequences with global sea-level curves and with Neogene resedimented carbonates. Comparison of the Vajont time-stratigraphic interval with the global sea-level curve of Haq et al. (1988) suggests that most Vajont deposition occurred over a period of approximately 4.5 m.y. (from the late Aalenian/early Bajocian boundary to the late Bajocian) during a long-term (second-order) rise in sea level (Fig. 13). Furthermore, this comparison suggests that the two couplets of nonskeletal-grain–dominated and skeletal-grain–rich sedimentologic associations (i.e., the couplet containing associations I and II and the couplet containing associations III and IV) were deposited during two short-term (third-order) fluctuations in sea level and prior to the beginning of a long-term (second-order) fall. Therefore, the scale, timing, and composition of the two couplets are consistent with sediment production on a persistently flooded platform during two third-order cycles of sea-level change.

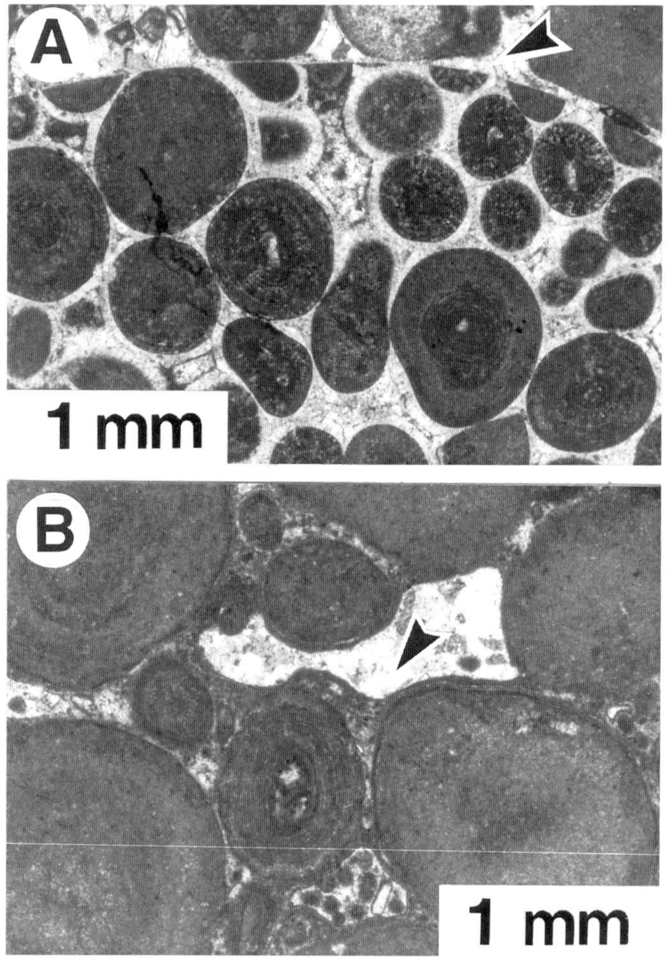

FIG. 10.—Platform-derived lithoclasts within the lower skeletal-grain–rich sedimentologic association II at Col Visentin. (A) Well-rounded lithoclast (most of photograph; note periphery at arrow) composed of oolitic grainstone that is cemented by marine fibrous calcite cement. Ooids and cement are truncated along the periphery of the grain (at arrow). Fibrous cement is only found in grainstone clasts, whereas isopachous nonluminescent, equant calcite is a pervasive intergranular cement in Vajont gravity-flow deposits. (B) Oolitic grainstone lithoclast cemented by meniscus calcite cement (arrow). The occurrence of meniscus cement implies that the grainstone was cemented in a vadose environment and was derived from a shallow-water platform. Collectively, well-rounded grainstone fabrics and meniscus cement suggest that grainstone lithoclasts were derived from shallow platform settings and reworked in high-energy environments prior to resedimentation into deep-water environments.

Turbidite frequency, turbidite accumulation rates, and bulk sedimentation rates were estimated from sedimentologic and biostratigraphic study of the section at Col Visentin (Tables 2 and 5). These values are comparable to highstand turbidite frequencies and accumulation rates deduced from study of Neogene carbonate gravity-flow deposits of the Bahamas and indicate that Middle Jurassic sediment production and export from the Friuli platform was similar to that of Neogene platforms. On the basis of these quantitative data, we conclude that the Vajont Limestone was the product of relatively continuous highstand shedding of shallow-water carbonate sediment from the Friuli platform during long-term sea-level rise. Furthermore, thick accumulation of Middle Jurassic shelf oolite on the Friuli platform is consistent with this interpretation and suggests that flooded conditions were also promoted by relatively rapid platform-subsidence rates (Ternowaner Oolite; 500–700 m; Cousin, 1980; Bosellini et al., 1981).

Generation of Nonskeletal-Grain–Dominated and Skeletal-Grain–Rich Sedimentologic Associations

Although relatively continuous sea-level rise and flooded platform conditions are in agreement with the overall nonskeletal-grain nature of the Vajont, subtle variations in platform and periplatform input are observed in field and petrographic data as an alternation between (1) nonskeletal-grain–dominated gravity-flow deposits that contain oncoids and hemipelagic mudstone clasts and exhibit an increase in radial ooid contents relative to a combination of micritic ooid and peloid contents and (2) skeletal-grain–rich gravity-flow deposits that contain an abundance of platform lithoclasts and exhibit a decrease in radial ooid contents relative to a combination of micritic ooid and peloid contents (Fig. 11; Tables 3, 4). Furthermore, each skeletal-grain–rich sedimentologic association exhibits an upward increase in skeletal-grain content. Because grain counts indicate that ooids and other nonskeletal grains were continuously being shed, the transition between skeletal-grain–rich associations and nonskeletal-grain–dominated associations must represent deposition during flooded-bank conditions without an intervening sea-level lowstand. As a result, the alternation between nonskeletal-grain–dominated and skeletal-grain–rich sedimentologic associations must reflect subtle changes in sea level and depositional environments of a *persistently* flooded platform.

Micritic oncoids and an abundance of hemipelagic mudstone lithoclasts in nonskeletal-grain–dominated sequences indicate that some carbonate mud was deposited in subtidal shoal settings and exported to the periplatform. Grain counts indicate that sediment on platform margins and inner-shelf areas was largely composed of active ooid and oncoid shoals. The lack of both skeletal-grain material and platform lithoclasts indicates that patch reefs and/or stabilized shoals were not pervasive along the platform margin. On the basis of these data, we conclude that nonskeletal-grain–dominated sedimentologic associations in the Vajont were formed during transgressions in which shallow-water sedimentation on the Friuli platform kept pace with relative sea-level rise (i.e., a transgressive systems tract [TST]). The efficiency of off-bank transport of nonskeletal grains and mud was probably at a maximum during deposition of these sequences (e.g., Hine, 1983) owing to a lack of wave-resistant margins (sediment catches) and island topography (sediment sieves). Rapid accumulation rates (Table 5) and deposition of thick gravity flows of oolitic debris and turbidites that are rich in oncoids and semiconsolidated hemipelagic mudstone clasts are consistent with this interpretation.

Platform lithoclasts represent both grain-rich shelf-margin and mud-rich inner-shelf facies. Large clast size, truncated cement fabrics, and well-rounded textures indicate that these clasts were reworked on the platform in high-energy environments prior to sedimentation in slope settings. Grain content, clast composition, and diagenetic fabrics suggest the presence of (1) an active margin environment including extensive ooid and oncoid grain shoals, small patch reefs with associated skeletal-grain sands (e.g., *Lithocodium*), and some stabilized and exposed islands and (2) a shelf-interior environment including semiactive shoals composed of ooids, peloids, and grapestone and lagoonal muds (e.g., Flügel, 1979). A gradual increase in the abundance of platform lithoclasts and skeletal grains is observed toward the tops of skeletal-grain–rich sedimentologic associations; an abrupt decrease is observed when passing upward into nonskeletal-grain–dominated sedimentologic associations. On the basis of these data, we conclude that (1) skeletal-grain–rich sedimentologic associations in the Vajont were formed during times of static sea level and regressions and (2) these associations are the basinal record of shallowing-upward platform sequences (i.e., highstand systems tracts [HST]) on the Friuli platform. Abrupt transition to nonskeletal-grain–dominated sedimentologic associations indicates that highstand sedimentation was followed by transgressive sedimentation (TST) without intervening lowstand sedimentation (i.e., a lowstand wedge [LSW]).

Analogous ooid-rich depositional environments, associated grain and clast types, and shallowing-upward sequences have been described from the Pleistocene and Holocene of Caicos platform, British West Indies (Lloyd et al., 1987; Waltz, 1988; Wanless et al., 1989). Caicos platform and its neighboring deeps are especially appealing analogues for the Friuli platform and Belluno basin because (1) the leeward margins and shelf interiors of Caicos are covered by widespread oolite and nonskeletal-grain shoals and (2) Vajont depositional geometry and facies distribution indicates that the redeposition of grains and micrite in periplatform and slope settings of the Belluno basin was the result of off-bank sediment transport along a leeward open margin (e.g., Hine et al., 1981a, b). Evolution of the leeward (western) margin of Caicos includes early Pleistocene highstand phase(s) during which off-bank transport of sediment effectively inhibited reef growth. Gradual stabilization of oolite during a middle Pleistocene highstand created islands and spits that fostered a shallowing-upward reef to oolite sequence. This sequence contains reef boundstone and detritus, skeletal-grain sand, oolitic and skeletal-grain rudite, and oolite (Waltz, 1988; Wanless et al., 1989). It is important to note that this mix of reefal material, skeletal- and nonskeletal-grain sand, and oolitic and skeletal-grain rudite was created during *highstands* of Pleistocene sea level. Similar lithologies are found accumulating in modern environments of Caicos during the present-day sea-level highstand and are actively transported to surrounding deeps. These data suggest, therefore, that exposure surfaces and platform-derived lithoclasts alone are not necessarily indicative of true lowstands, i.e., where sea level dropped below the platform margin and exposed carbonate-slope and carbonate-shelf

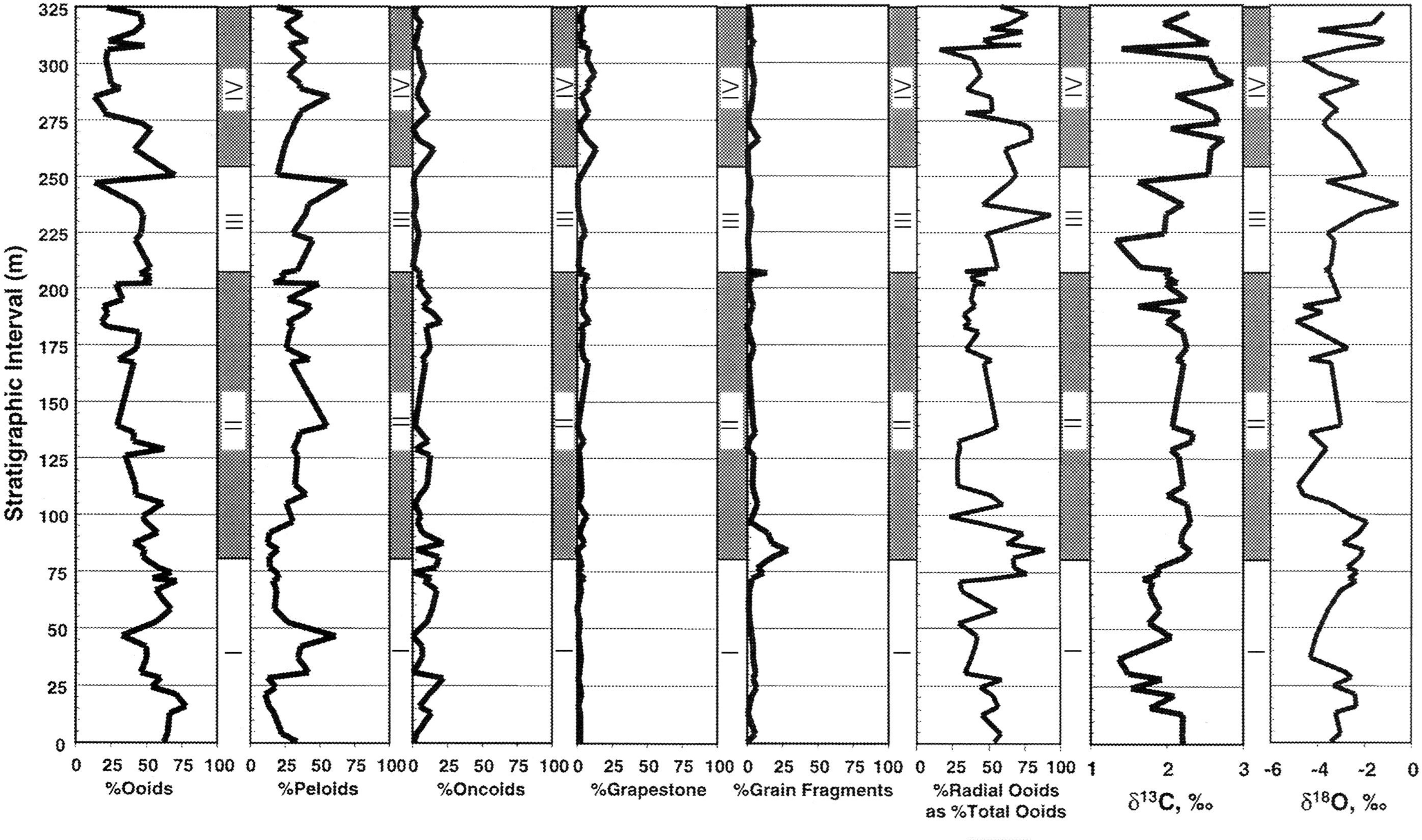

FIG. 11.—(A) Nonskeletal-grain and stable isotope profile, Col Visentin stratigraphic section. Total ooid and oncoid contents decrease upward whereas peloid, grapestone, and total skeletal-grain contents increase upward. These data suggest that nonskeletal-grain and skeletal-grain material were derived from an overall shallowing-upward platform. ^{13}C and total ooid content exhibit an inverse correlation (Spearman's correlation coefficient = –0.27; Table 4). $\delta^{13}C$ and oncoid content are apparently unrelated (Spearman's correlation coefficient = –0.01). $\delta^{13}C$ and grapestone nonskeletal-grain content exhibit a positive correlation (correlation coefficient = +0.35). These data suggest that sediment and geochemical composition were influenced by changes in relative sea level (for further discussion, see text).

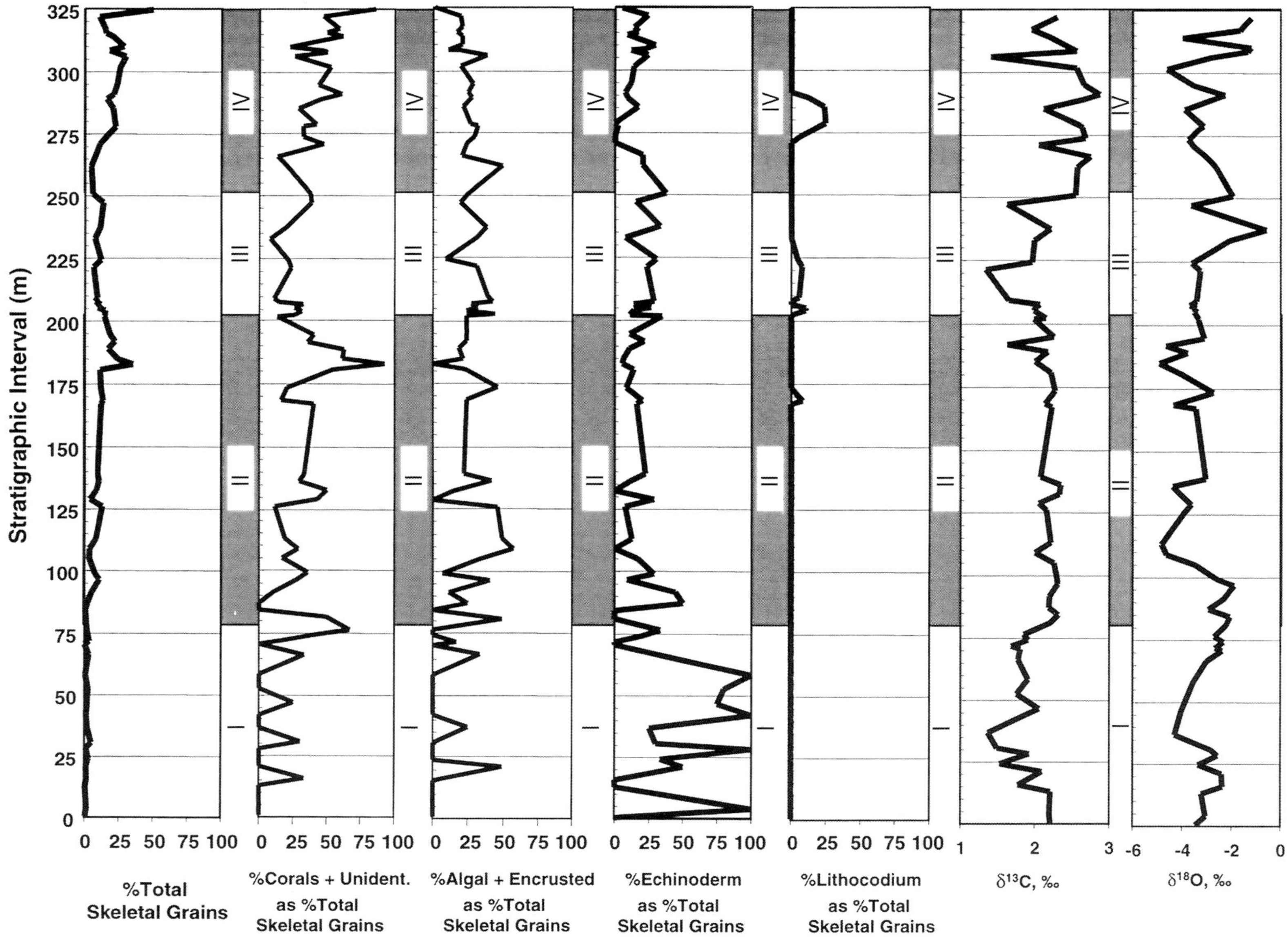

Fig. 11.—(B) Skeletal-grain and stable isotope profile, Col Visentin stratigraphic section. Total skeletal-grain content exhibits an overall increase upward. $\delta^{13}C$ and total skeletal-grain content exhibit a positive correlation (Spearman's correlation coefficient = +0.32). An upward increase in shallow-water skeletal-grain content (e.g., *Lithocodium*) in the Vajont suggests that nonskeletal-grain and skeletal-grain material was derived from an overall shallowing-upward platform. Total skeletal-grain content and peloid nonskeletal-grain content exhibit a strong positive correlation (Spearman's correlation coefficient = +0.63; Table 4). These data suggest that sediment and geochemical composition were influenced by changes in relative sea level (for further discussion, see text).

TABLE 4.—SPEARMAN CORRELATION COEFFICIENTS FOR NONSKELETAL-GRAIN AND SKELETAL-GRAIN COMPONENTS

	Component	As a Percentage of	1	2	3	4	5	6	7	8	9	10	11	12	13	14	15	16
1	Nonskeletal grains	total grains	1.00															
2	Skeletal grains	total grains	−1.00	1.00														
3	Ooids	total grains	*0.77*	*−0.77*	1.00													
4	Peloids	total grains	*−0.42*	0.42	*−0.63*	1.00												
5	Oncoids	total grains	0.12	−0.12	−0.03	*−0.43*	1.00											
6	Grain fragments	total grains	0.20	−0.20	0.03	−0.17	0.06	1.00										
7	Grapestone	total grains	*−0.48*	*0.48*	*−0.41*	0.00	*0.33*	−0.16	1.00									
8	Radial ooids	nonskeletal grains	*0.49*	*−0.49*	*0.71*	*−0.56*	*−0.33*	0.11	*−0.35*	1.00								
9	Micritic ooids	nonskeletal grains	0.36	−0.36	*0.50*	*−0.28*	*0.33*	−0.22	−0.01	−0.13	1.00							
10	Peloids	nonskeletal grains	*−0.63*	*0.63*	*−0.78*	*0.95*	*−0.40*	−0.20	0.09	*−0.61*	*−0.38*	1.00						
11	Oncoids	nonskeletal grains	0.02	−0.02	−0.11	*−0.40*	*0.99*	0.03	*0.36*	*−0.30*	0.19	*−0.33*	1.00					
12	Grain fragments	nonskeletal grains	0.09	−0.09	−0.05	−0.15	0.04	0.99	−0.11	0.07	−0.26	−0.15	0.03	1.00				
13	Grapestone	nonskeletal grains	*−0.56*	*0.56*	*−0.47*	0.03	*0.30*	−0.18	*0.99*	*−0.37*	−0.06	0.14	*0.34*	−0.12	1.00			
14	Radial ooids	total ooids	0.19	−0.19	0.25	−0.25	*−0.33*	0.20	−0.29	*0.80*	*−0.63*	−0.24	*−0.35*	0.20	−0.28	1.00		
15	Micritic ooids	total ooids	−0.19	0.19	−0.25	0.25	*0.33*	−0.20	0.29	*−0.80*	*0.63*	0.24	*0.35*	−0.20	0.28	−1.00	1.00	
16	$\delta^{18}O$ (PDB)		0.22	−0.22	*0.33*	*−0.36*	−0.12	0.17	−0.10	*0.42*	−0.02	*−0.35*	−0.13	0.17	−0.11	*0.33*	*−0.33*	1.0
17	$\delta^{13}C$ (PDB)		*−0.32*	*0.32*	−0.27	0.05	−0.01	0.00	*0.35*	−0.04	−0.23	0.10	0.02	0.05	*0.36*	0.05	−0.05	0.1

			Skeletal-Grain Type as a Percentage of Total Skeletal Grains					
	Component	As a Percentage of	Benthic Foraminifers	Echinoderms	*Lithocodium*	Algal Grains	Encrusting Grains	Unidentified Ske (Former Ara
3	Ooids	total grains	−0.05	0.22	−0.20	−0.14	*−0.47*	*−0.4*
4	Peloids	total grains	0.27	−0.03	0.26	0.08	*0.39*	0.2
5	Oncoids	total grains	*−0.37*	−0.06	−0.08	0.08	−0.07	−0.2
6	Grain fragments	total grains	−0.03	0.07	−0.20	0.06	−0.06	−0.1
7	Grapestone	total grains	−0.13	−0.15	0.26	0.06	*0.35*	0.2
8	Radial ooids	nonskeletal grains	0.07	0.05	−0.15	0.02	*−0.34*	−0.1
9	Micritic ooids	nonskeletal grains	−0.11	0.19	−0.11	−0.22	−0.07	*−0.3*
10	Peloids	nonskeletal grains	0.18	−0.10	0.27	0.08	*0.42*	*0.3*
11	Oncoids	nonskeletal grains	*−0.40*	−0.09	−0.08	0.08	−0.04	−0.1
12	Grain fragments	nonskeletal grains	−0.03	0.05	−0.19	0.06	−0.01	−0.0
13	Grapestone	nonskeletal grains	−0.14	−0.18	0.27	0.11	*0.36*	*0.3*
14	Radial ooids	total ooids	0.11	−0.02	−0.02	0.11	−0.24	0.0
15	Micritic ooids	total ooids	−0.11	0.02	0.02	0.15	0.24	−0.0
16	$\delta^{18}O$ (PDB)		−0.10	0.13	−0.18	0.02	−0.26	−0.0
17	$\delta^{13}C$ (PDB)		0.06	*−0.30*	0.07	0.11	0.27	*0.3*

NOTE: Correlation coefficients –≥+0.30 or ≤–0.30 (shown in italics) are significant at the 99% confidence level. PDB = Peedee belemnite standard.

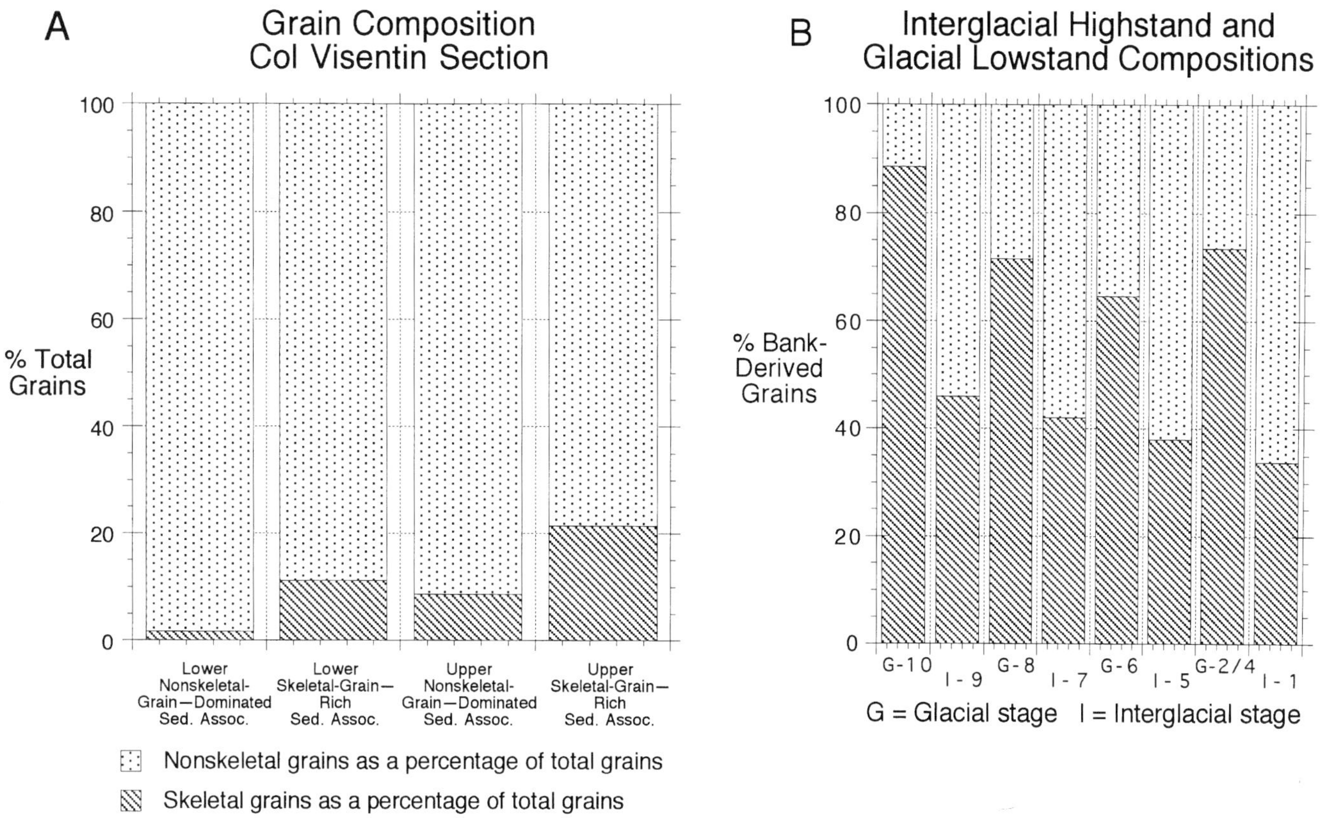

Fig. 12.—Quantitative comparison of Vajont (A) and Pliocene-Pleistocene (B) turbidite compositions. Nonskeletal-grain contents of both nonskeletal-grain–dominated sequences (avg. = 98.3% and 91.3%) and skeletal-grain–rich sequences (avg. = 88.8% and 78.6%) in the Vajont are much greater than nonskeletal-grain contents of interglacial highstand carbonate gravity-flow deposits (Haak and Schlager, 1989; avg. nonskeletal-grain content = 66.2% to 54.1%) described from the Pleistocene of the Bahamas. These data suggest that both nonskeletal-grain–dominated and skeletal-grain–rich sedimentologic associations in the Vajont Limestone resulted from the highstand shedding of shallow-water carbonate sediment.

TABLE 5.—COMPARISON OF VAJONT ACCUMULATION RATES WITH NEOGENE EXAMPLES FROM THE BAHAMAS

Study	Turbidite Frequency $[(k.y.)^{-1}]$	Turbidite Accumulation Rate (mm/k.y.)	Bulk Sedimentation Rate (mm/k.y.)
Vajont*	0.11	69	72
TOTO HS†	0.23–0.85 (0.56)	35–140 (70)	60–180 (120)
TOTO HS†	0.09–0.68 (0.34)	15–100 (51)	25–120 (80)
TOTO LS†	0–0.05 (0.02)	0–3 (1)	9–27 (18)
TOTO LS†	0–0.03 (0.01)	0–3 (1)	18–37 (25)
Exuma HS§		0–117 (21)	27–169 (63)
Exuma LS§		0–13 (5)	3–53 (24)

NOTE: Averages are shown in parentheses. HS = highstand, LS = lowstand.

* Assumes an age range of 171 to 166.5 Ma, estimated from Vajont nannofossil and $^{87}Sr/^{86}Sr$ data (this study) and the compiled ammonite-biozone time scale of Haq et al. (1988) (with reference to Cope et al., 1980, and Harland et al., 1982). Reported uncertainty of absolute age dates of Jurassic stage boundaries is ±3.5 m.y. (Haq et al., 1988).

† Pleistocene-Holocene, Tongue of the Ocean; Droxler and Schlager (1985).

§ Pliocene-Pleistocene, Exuma Sound; Reijmer (1991).

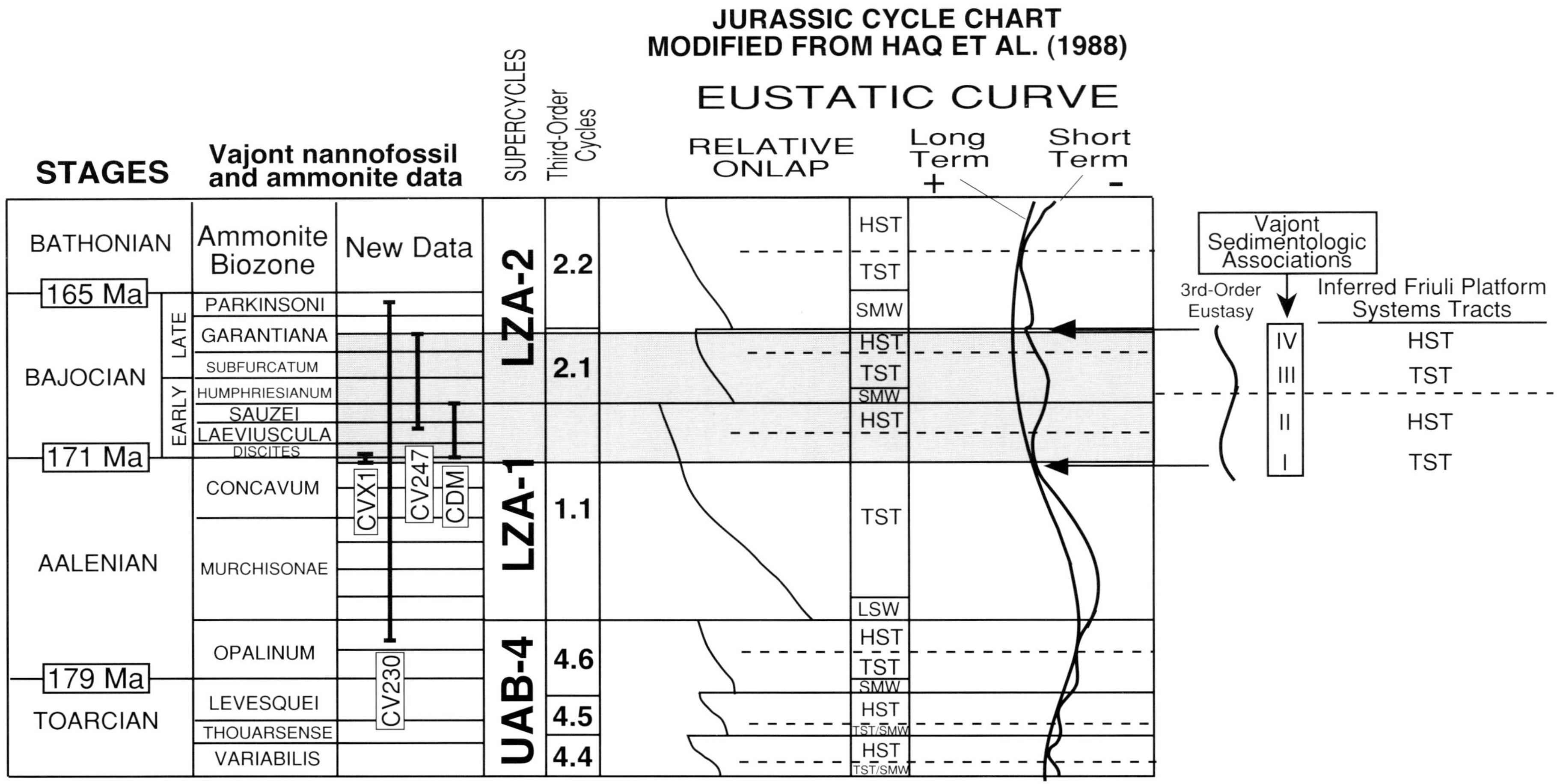

Fig. 13.—Comparison of new nannofossil and ammonite age dates of the Vajont Limestone with the Haq et al. (1988) sea-level curve. New age dates (Table 1) indicate that the Vajont Limestone was deposited over a period of approximately 4.5 m.y. during the late Aalenian to late Bajocian. Comparison with the global eustatic curve suggests that the Vajont was deposited during a sustained long-term (second-order) rise in sea level and prior to regression and development of a type 1(?) unconformity at the end of the late Bajocian. Moreover, the two pairs of nonskeletal-grain–dominated and skeletal-grain–rich sedimentologic associations (I and II; III and IV) correlate with two short-term (third-order) eustatic cycles. The absence of skeletal-grain–dominated gravity-flow deposits (i.e., lowstand gravity-flow deposits) between sedimentologic associations II and III suggests that sediment was derived from a persistently flooded platform through the early to late Bajocian. HST = highstand systems tract, TST = transgressive systems tract, SMW = shelf-margin wedge, LSW = lowstand wedge. CDM refers to an ammonite age assignment based on the presence of *Docidoceras* (sp. ind.) at the Col dei Moi section (Zempolich, 1993).

environments to significant erosion (Fouke et al., 1996; Sarg, 1988). Moreover, the occurrence of these lithologies on the leeward margin of Caicos, which is a wind-dominated platform affected by strong seasonal easterly trade winds (Wanless et al., 1989), suggest that these grain and clast types would be common components of carbonate gravity-flow deposits in periplatform and slope settings. Although different in scale, we interpret the sea-level history and depositional processes during highstand sedimentation at Caicos to be an analogue for similar lithologies inferred to have formed on the Friuli platform in response to short-term (third-order) transgression and regression during the Middle Jurassic.

On the basis of compositional study of Vajont gravity-flow deposits, we postulate transgressive and regressive depositional models (Figs. 14, 15) for the Friuli platform and Vajont Limestone; these models are consistent with Jurassic and Pleistocene-Holocene shallow-water depositional models (e.g., Flügel, 1979; Hine, 1983; Wanless et al., 1989; Zempolich, 1993), deep-water depositional models (e.g., Mullins et al., 1984; Mullins and Cook, 1986), study of Neogene carbonate gravity-flow deposits (e.g., Haak and Schlager, 1989; Cartwright, 1985), and postulated sea-level variations during the Middle Jurassic. These depositional models offer one explanation for the alternation of the nonskeletal-grain–dominated and skeletal-grain–rich sedimentologic associations that are observed in the Vajont Limestone.

GEOCHEMISTRY AND DIAGENESIS

Petrographic and geochemical data gathered from ooids, skeletal grains, and other diagenetic fabrics enable the delineation of primary marine and diagenetic compositions through the section at Col Visentin. Individual components were microsampled for C, O, and Sr isotope analyses and ICP trace element analysis. These data were then compared with sedimentologic compositions to further explore the apparent link between nonskeletal-grain–dominated and skeletal-grain–rich sedimentologic associations and sea-level change.

Grains and Platform Lithoclasts

Ooids.—
Only those ooids that displayed radial fabric were microsampled for geochemical analysis. A summary of these results is presented in Tables 6 and 7 (see also Zempolich, 1995). Isotope analyses are presented as a chemostratigraphic profile and cross-plot (Figs. 11, 16). Sr, Mg, Fe, and Mn contents of radial ooids are low (Table 6; Fig. 17; Zempolich, 1995). High Sr/Mn ratios of radial calcitic ooids (>4.3 to 46.4 [molar basis], avg. = 15.3, N = 38) suggest that these grains are likely composed of pristine marine carbonate (e.g., Derry et al., 1992; Kaufman et al., 1993; Denison et al., 1994; Jones et al., 1994a, b; Montanez et al., 1996).

Skeletal grains.—
Nonluminescent calcitic skeletal clasts (e.g., echinoderms, bryozoans, belemnites, calcareous sponges) that exhibit fabric-retentive or mimetic texture possess enrichments in ^{13}C and ^{18}O and low Fe and Mn contents (N = 18; $\delta^{18}O$ = –4.77‰ to –0.39‰, $\delta^{13}C$ = +0.52‰ to +2.7‰ ; Zempolich, 1995). The higher $\delta^{13}C$ and $\delta^{18}O$ values are comparable to those obtained from radial calcite ooids, allowing for small differences in $\delta^{13}C$ and $\delta^{18}O$ values caused by biological fractionation during skeleton precipitation (e.g., Gonzalez and Lohmann, 1985). Echinoderms possess high Mg contents and low Sr contents that reflect the retainment of some Mg in the form of microdolomite (Fig. 9D) and the loss of Sr during inversion of high-Mg calcite to low-Mg calcite (Fig. 17). Analysis of a belemnite yielded low Mg, Fe, and Mn content, high Sr content (707 to 980 ppm), high Sr/Mn (55.5 to 76.3 [molar basis]), and a characteristically marine Sr isotope composition ($^{87}Sr/^{86}Sr$ = 0.707070); these data suggest that this calcitic grain is pristine.

Originally aragonitic skeletal clasts (e.g., corals, gastropods, and mollusks) that are nonluminescent and exhibit preservation of fine-scale fabric possess ^{18}O and ^{13}C enrichments that fall within the field defined by the radial calcite ooids (Zempolich, 1995). Nonluminescent corals that exhibit aggrading neomorphic fabrics possess ^{13}C and ^{18}O enrichments (N = 8, $\delta^{18}O$ = –2.19‰, $\delta^{13}C$ = +1.73‰) similar to radial ooids but exhibit decreases in Sr contents from 1,460 to 70 ppm that correlate with increasing degrees of neomorphism (Fig. 17). The progressive loss of Sr and retainment of relatively invariant $\delta^{13}C$ and $\delta^{18}O$ values suggest that corals and other metastable skeletal clasts were transported directly from platform settings to deep-water slopes, where they stabilized in syndepositional marine or rock-dominated fluid (Zempolich, 1995; see also Melim et al., 1995).

Corals and mollusks that are not fabric retentive also exhibit aggrading neomorphic and cement-filled fabrics. Under cathodoluminescent light, these transitions exhibit a progression from nonluminescence to bright luminescence (Fig. 18). Early dissolution of some skeletal clasts is suggested by skeletal molds that possess an isopachous lining of early nonluminescent equant calcite cement and are filled by dull to brightly luminescent calcite cement. Brightly luminescent calcite is also observed to fill skeletal-moldic porosity without an early generation of nonluminescent equant calcite cement. These petrographic data and the depletion in ^{18}O imply that some dissolution of clasts and in-filling of molds occurred later in the diagenetic history under reduced conditions and at elevated temperature (Fig. 19; Zempolich, 1995).

Platform lithoclasts.—
Ooids and fibrous calcite cement in grain-rich clasts were sampled for isotope analysis to see whether a geochemical signature was present that might establish the origin of the clasts (e.g., subaerial exposure and incorporation of C depleted in ^{13}C and O depleted in ^{18}O). Of nine C and O analyses of ooids from grain-rich clasts, eight are indistinguishable from analyses of ooids sampled from gravity-flow matrix (N = 8, avg. $\delta^{13}C$ = +2.09‰ and $\delta^{18}O$ = –3.79‰ ; compare with Table 6). Only one analysis exhibited a minor depletion in ^{13}C accompanied by an

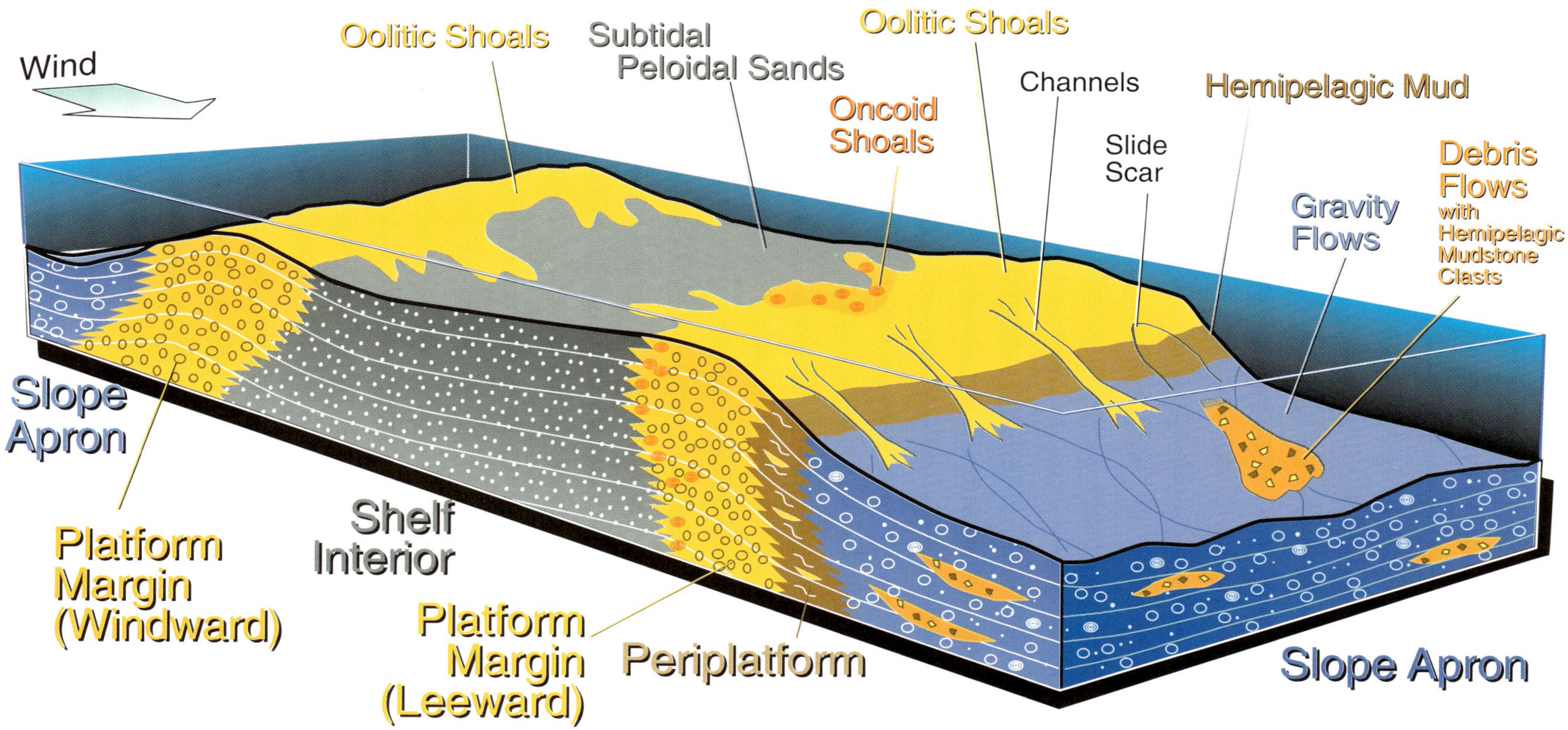

FIG. 14.—Transgressive depositional model of the submerged, isolated Friuli platform and Vajont slope system. In this hypothetical end-member model, nonskeletal-grain sedimentation in shallow-water platform settings and off-bank transport of this sediment to surrounding deeps are enhanced by deposition in a wind-dominated submerged setting and a lack of wave-resistant margins. The majority of shallow-water grain types exported to the Vajont slopes would include ooids, oncoids, peloids, and mud. Deep-water depositional facies would include (1) periplatform facies composed of bedded hemipelagic muds and channelized gravity-flow deposits and (2) slope-apron facies composed of nonskeletal-grain–dominated gravity-flow deposits that contain an abundance of periplatform-derived lithoclasts. Hemipelagic mudstone lithoclasts are thought to have formed through the failure of bedded hemipelagic mudstone in periplatform settings (Zempolich, 1995).

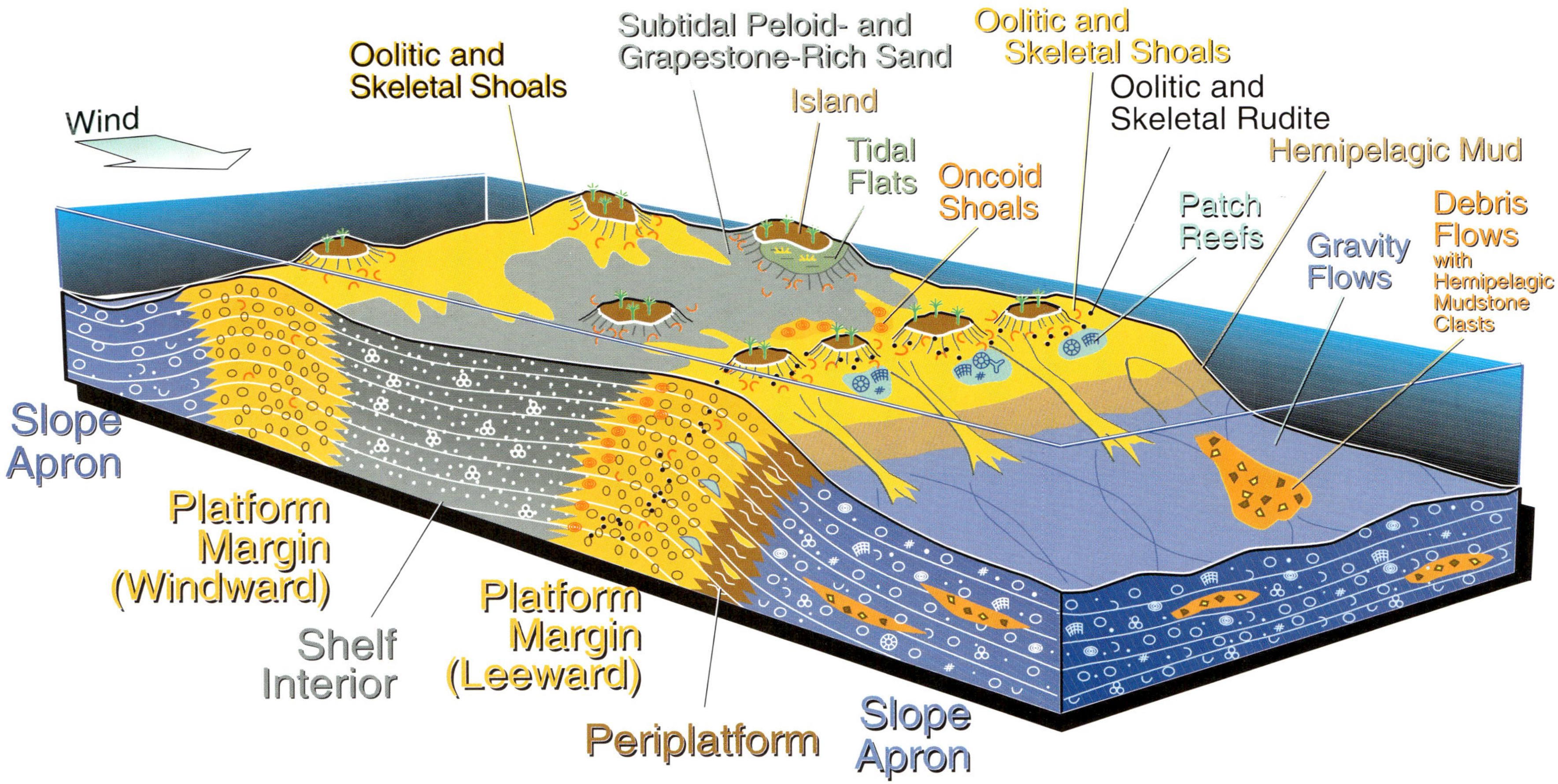

FIG. 15.—Regressive depositional model of the partly emergent, isolated Friuli platform and Vajont slope system. In this hypothetical end-member model, nonskeletal-grain sedimentation in shallow-water platform settings is enhanced by deposition in a wind-dominated submerged setting, but off-bank transport of this sediment to surrounding deeps is retarded by the wave-resistant margins of the exposed islands. By impeding the off-bank transport of nonskeletal-grain sediment, the presence of small islands is hypothesized to have (1) enhanced the abiotic and biotic alteration of primary nonskeletal grains (e.g., the micritization of radial ooids to form micritic ooids and peloids and the agglomeration and cementation of grains to form grapestone), (2) allowed for the sedimentation of small patch reefs and skeletal-grain sands, and (3) augmented the formation of nonskeletal-grain and skeletal-grain rudite. The majority of shallow-water grain types exported to the Vajont slopes would include ooids, oncoids, peloids, grapestone, mud, and skeletal grains. Deep-water depositional facies would include (1) periplatform facies composed of bedded hemipelagic muds and channelized gravity-flow deposits and (2) slope-apron facies composed of skeletal-grain–rich gravity-flow deposits that contain an abundance of platform-derived lithoclasts. As in the transgressive depositional model, hemipelagic mudstone lithoclasts are thought to have formed through the failure of bedded hemipelagic mudstone in periplatform settings.

TABLE 6. STABLE ISOTOPE AND MAJOR AND MINOR ELEMENT COMPOSITIONS OF RADIAL OOIDS FROM COL VISENTIN SECTION.

Sedimentologic Association (See Fig. 4)	$\delta^{13}C$ (‰) (Range)	$\delta^{18}O$ (‰) (Range)
Upper skeletal-grain rich association (IV) (N = 15)	+2.38 (1.41 to 2.87)	–2.87 (–4.61 to –1.19)
Upper nonskeletal-grain–dominated association (III) (N = 8)	+2.13 (1.34 to 2.75)	–2.58 (–3.64 to –0.57)
Lower skeletal-grain–rich association (II) (N = 28)	+2.14 (1.63 to 2.35)	–3.48 (–4.91 to –1.88)
Lower nonskeletal-grain–dominated association (I) (N = 16)	+1.87 (1.37 to 2.21)	–3.01 (–4.30 to –2.32)
Average (N=67)	+2.13 (1.34 to 2.87)	–3.12 (–4.91 to –0.57)

Major and Minor Element Compositions (N = 39)

	Ca (ppm)	Mg (ppm)	Sr (ppm)	Fe (ppm)	Mn (ppm)	Mg (mol%)
Mean	395,154	3,644	332	89	17	1.50
Median	395,000	3,630	332	73	15	1.49
Minimum	394,000	2,950	209	23	6	1.21
Maximum	396,000	4,540	505	350	45	1.86

enrichment in ^{18}O ($\delta^{13}C$ = –1.01‰ and $\delta^{18}O$= –0.56‰). These data suggest that the clasts were either geochemically stable during erosion in a subaerial platform setting or that the clasts were eroded in a predominantly marine environment.

Isotope analyses of isopachous fibrous cement contained within platform lithoclasts (N = 2, $\delta^{13}C$ = +1.91‰ and $\delta^{18}O$ = –2.00‰) yielded characteristically marine $\delta^{13}C$ and $\delta^{18}O$ values that are comparable with radial calcitic ooids of this study (Fig. 19). Similar fibrous calcite cement and $\delta^{13}C$ and $\delta^{18}O$ values have been described from time-equivalent shelf oolite in the United Kingdom by Marshall and Ashton (1980). These data suggest that the fibrous calcite cement was a marine precipitate and that cementation, erosion, and reworking of lithoclasts took place in a predominantly marine, shallow-water, high-energy environment.

Diagenetic Fabrics

Nonluminescent equant calcite.—

Nonluminescent equant-calcite cement occurs as thin isopachous linings in resedimented grainstone, as intraskeletal pore fillings, and as isopachous linings in skeletal molds (Figs. 8, 18). It is a common cement in coral skeletons that display well-preserved microfabrics. Nonluminescent equant cement is conformably overlain by banded luminescent equant-calcite cement in large pores (Fig. 18). Nonluminescent equant calcite also occurs as syntaxial, poikilotopic cement in association with echinoderm grains.

The $\delta^{13}C$ and $\delta^{18}O$ values of nonluminescent equant calcite are higher relative to those of other calcite cements and fall within the field defined by radial calcitic ooids (Fig. 19). Microprobe analysis indicates that these cements are composed of low-Mg calcite (N = 37, Mg = 0.5 to 2.0 mol%). Analogous equant–low-Mg calcite cement has been described in modern slope and basin settings by Schlager and James (1978). On the basis of these data, isopachous nonluminescent equant-calcite cement is interpreted as an early, characteristically marine precipitate in slope settings. Furthermore, its occurrence as a pore-lining phase within oolitic matrix, intraskeletal pore space, and skeletal molds indicates that precipitation began soon after deposition of carbonate in slope settings and continued during shallow-burial diagenesis (Zempolich, 1995). Similar fabrics and interpretation are reported by Melim et al. (1995) in a diagenetic study of Tertiary slope deposits of the Bahamas.

Late calcite cements.—

Late diagenetic calcite fabrics include banded, luminescent, equant calcite that overlies nonluminescent equant calcite and fills the remaining intergranular pore space, coarse luminescent calcite that fills molds of skeletal grains, and fracture-filling luminescent calcite that crosscuts all previously described fabrics. ^{18}O is progressively more depleted from banded, luminescent calcite (avg. $\delta^{18}O$ = –6.56‰) to mold-filling luminescent calcite (avg. = –8.68‰) and to fracture-filling luminescent calcite (avg. =–12.7‰); this pattern suggests progressive cementation in the burial environment (Fig. 19; Zempolich, 1995). It is important to note that late calcite cement occluded the majority of intergranular, intragranular, and moldic porosity that remained after early cementation and dissolution in slope and shallow burial environments. This process resulted in the formation of a relatively nonporous and impermeable limestone (Zempolich, 1995).

Primary Depositional Textures and Chemical Compositions

Radial ooids.—

Isotope, trace element, and fabric data suggest that Vajont ooid cortices were originally composed of radial low-Mg calcite. Radial calcite fabrics, nonluminescence, ^{13}C and ^{18}O

TABLE 7.— STRONTIUM ISOTOPE ANALYSES, COL VISENTIN SECTION, VAJONT LIMESTONE.

Sample	Stratigraphic Interval (m)	Component	$^{87}Sr/^{86}Sr$	Error ($\times 10^{-5}$)
CV280 A	324.70	radial ooids	0.707041	1.9
CV278	321.80	radial ooids	0.707510	1.2
CV275	315.30	radial ooids	0.707069	1.4
CV265	304.00	bedded hemipelagic micrite	0.707064	1.7
CV264	301.60	radial ooids	0.707083	4.0
CV258	278.80	radial ooids	0.707052	1.9
CV251	262.00	radial ooids	0.707050	1.8
CV247	253.10	bedded hemipelagic micrite	0.707069	1.2
CV241	243.00	peloidal hemipelagic micrite (bedded)	0.707063	1.9
CV224	224.40	radial ooids	0.707077	2.7
CV222	221.40	radial ooids	0.707114	1.7
CV222 B	221.40	belemnite	0.707070	0.7
CV216	206.60	radial ooids	0.707090	2.0
CV211	201.85	radial ooids	0.707057	1.7
CV209	197.40	peloidal hemipelagic micrite (bedded)	0.707079	1.2
CV207	194.90	radial ooids	0.707110	3.0
CV204	189.00	radial ooids	0.707059	1.4
CV183	183.00	peloidal micrite clast	0.707133	1.6
CV199	180.70	radial ooids	0.707158	1.2
CV195	173.50	radial ooids	0.707051	2.2
CV191	168.50	radial ooids	0.707176	1.4
CV190	167.20	radial ooids	0.707068	1.1
CV181	140.70	radial ooids	0.707086	0.6
CV178	136.30	radial ooids	0.707096	2.3
CV169	126.10	radial ooids	0.707074	1.6
CV164	109.00	radial ooids	0.707186	1.3
CV154	99.10	radial ooids	0.707084	2.3
CV150	93.00	bedded hemipelagic micrite	0.707096	1.6
CV148	91.60	radial ooids	0.707141	0.9
CV133	76.60	radial ooids	0.707173	1.9
CV126	70.60	radial ooids	0.707083	1.7
CV125	70.20	bedded hemipelagic micrite	0.707062	0.8
CV118	65.80	bedded hemipelagic micrite	0.707156	1.3
CV113	60.10	bedded hemipelagic micrite	0.707145	1.5
CV106	58.30	radial ooids	0.707128	1.6
CV90	52.60	radial ooids	0.707176	1.8
CV80	42.00	radial ooids	0.707180	1.8
CV68	37.00	radial ooids	0.707190	0.9
CV54	28.30	radial ooids	0.707160	1.9
CV44	24.20	radial ooids	0.707129	1.0
CV41	21.00	radial ooids	0.707140	1.9
CV35	15.80	radial ooids	0.707148	1.6
CV31	12.90	radial ooids	0.707186	1.2
CV29	10.90	hemipelagic micrite clast	0.707142	1.0
CV15	4.00	radial ooids	0.707195	1.7
CV1	0.45	radial ooids	0.707200	1.1

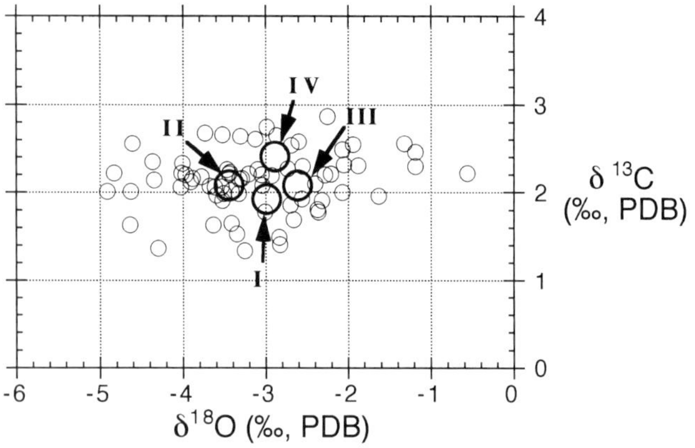

FIG. 16.—Stable isotope compositions of Vajont radial calcite ooids (N = 67). Average $\delta^{13}C$ and $\delta^{18}O$ values in the (I) lower nonskeletal-grain–dominated, (II) lower skeletal-grain–rich, (III) upper nonskeletal-grain–dominated, and (IV) upper skeletal-grain–rich sedimentologic associations are shown as bold circles (see Table 6).

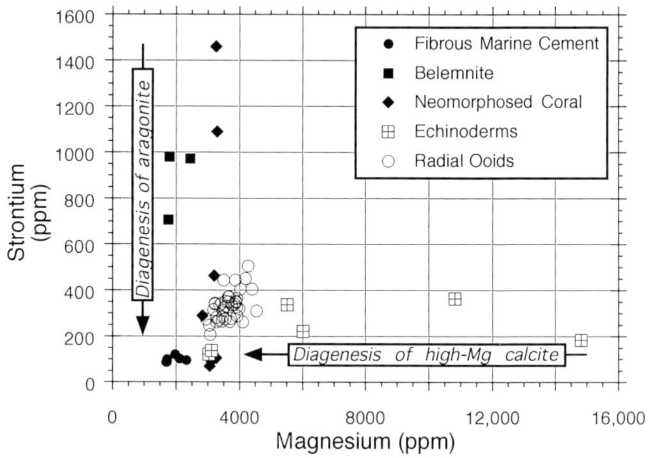

FIG. 17.—Cross-plot of the Sr and Mg contents of radial ooids and fibrous calcite cement (abiotic marine carbonate) and mimetic and fabric-retentive skeletal grains (biotic marine carbonate). A coral fragment that exhibits aggrading neomorphism and is nonluminescent contains variable Sr content that ranges from 1,460 to 70 ppm Sr. The decrease in Sr correlates with increasing degree of neomorphism, occurs over a relatively narrow range of Mg, and trends toward the low Sr values held by calcitic ooids and fibrous calcite cement. These data suggest that corals were originally composed of Sr-rich aragonite and were altered to low-Mg calcite early in the diagenetic history. Echinoderm fragments are nonluminescent and contain variable amounts of Mg and low Sr. These data are consistent with petrographic observations that indicate that stabilization of these originally high-Mg calcite grains produced low-Mg calcite and microdolomite inclusions (cf. Fig. 9D). Belemnites that are composed of radial low-Mg calcite possess high Sr contents relative to radial calcite ooids and fibrous calcite cement. This offset (500 to 900 ppm Sr) is probably due to nonequilibrium (biotic) and equilibrium (abiotic) precipitation from seawater (see Carpenter and Lohmann, 1992). Radial ooids and fibrous calcite cement contain low Sr and low Mg contents relative to all skeletal-grain components and define a positive covariant trend. These and other petrographic data suggest that radial ooids and fibrous cement (Zempolich, 1995) precipitated as primary low-Mg calcite in equilibrium with Middle Jurassic seawater.

enrichments, characteristically marine $^{87}Sr/^{86}Sr$ ratios, low and covariant Sr-Mg contents, low Mn-Fe contents, high Sr/Mn ratios, and an absence of neomorphic texture collectively suggest that Vajont ooids have undergone little diagenesis. Petrographic and geochemical comparison of Vajont ooid fabrics with those of modern and ancient ooids, and with diagenetic fabrics of coeval metastable skeletal grains, suggest that Vajont ooids were originally precipitated in shallow-marine environments as low-Mg calcite. These data suggest that Vajont ooids were redeposited in the Belluno basin as relatively pristine, mineralogically stable low-Mg calcite that underwent little diagenetic alteration.

Modern ooid cortical fabrics are composed of either low-Mg calcite (0–4% $MgCO_3$), high-Mg calcite (10–18% $MgCO_3$), aragonite or combinations of these minerals (Land et al., 1979; Medwedeff and Wilkinson, 1983; Popp and Wilkinson, 1983). Calcitic grains typically exhibit radial fabric composed of calcite crystals that are oriented normal to the direction of growth. Aragonite grains exhibit (1) tangential cortical fabric composed of micrometer to submicrometer-size aragonitic laths arranged parallel to the direction of growth and (2) rare radial cortical fabric composed of normal-oriented aragonite crystals. Diagenesis of these stable and metastable minerals can lead to very different results. For example, radial-calcite cortical fabrics are commonly found in ancient sequences. Although the original Mg content of these grains is unknown (Medwedeff and Wilkinson, 1983), these calcitic fabrics are attributed to the retainment of primary radial cortical texture. On the other hand, alteration of aragonite cortices to low-Mg calcite results in the formation of fabric-destructive, equant or bricklike, neomorphic calcite textures (e.g. Rich, 1982; Popp and Wilkinson, 1983; Tucker, 1984; Wilkinson et al., 1984). These fabric-destructive textures commonly contain elevated Sr contents (e.g., 500 to 3,000 ppm Sr; Zempolich et al., 1988), and together these features are considered reliable indicators of precursor aragonite (Sandberg, 1985).

Aragonitic skeletal grains in the Vajont were deposited along with radial ooids in gravity-flow deposits and are now replaced by low-Mg calcite. These grains exhibit a spectrum of fabric-retentive and fabric-destructive neomorphic fabrics and possess ^{13}C and ^{18}O enrichments. Sr contents within individual skeletal grains are variable and correlate with the degree of local fabric preservation—i.e., fabric-retentive regions possess high Sr contents (e.g., 1,460 ppm Sr) and fabric-destructive regions possess depleted Sr contents (e.g., 70 ppm Sr). These data indicate that original aragonite components were altered to low-Mg calcite early in the diagenetic history in either syndepositional marine or rock-dominated pore fluid. Vajont ooids do not display neomorphic cortical fabrics nor do they contain appreciable amounts of Sr (avg. = 332 ppm Sr). These data suggest that Vajont radial ooids were originally composed of either high- or low-Mg calcite. A similar argument is presented by Wilkinson and Landing (1978) in explanation of preferential diagenesis of the aragonite phase in bimineralic aragonite and calcite ooids in the Jurassic Twin Creek Formation and by Heydari and Moore (1994) in explanation of preferential diagenesis of aragonite ooids in the Upper Jurassic Smackover Formation.

Distinguishing whether ancient calcitic ooid fabrics were

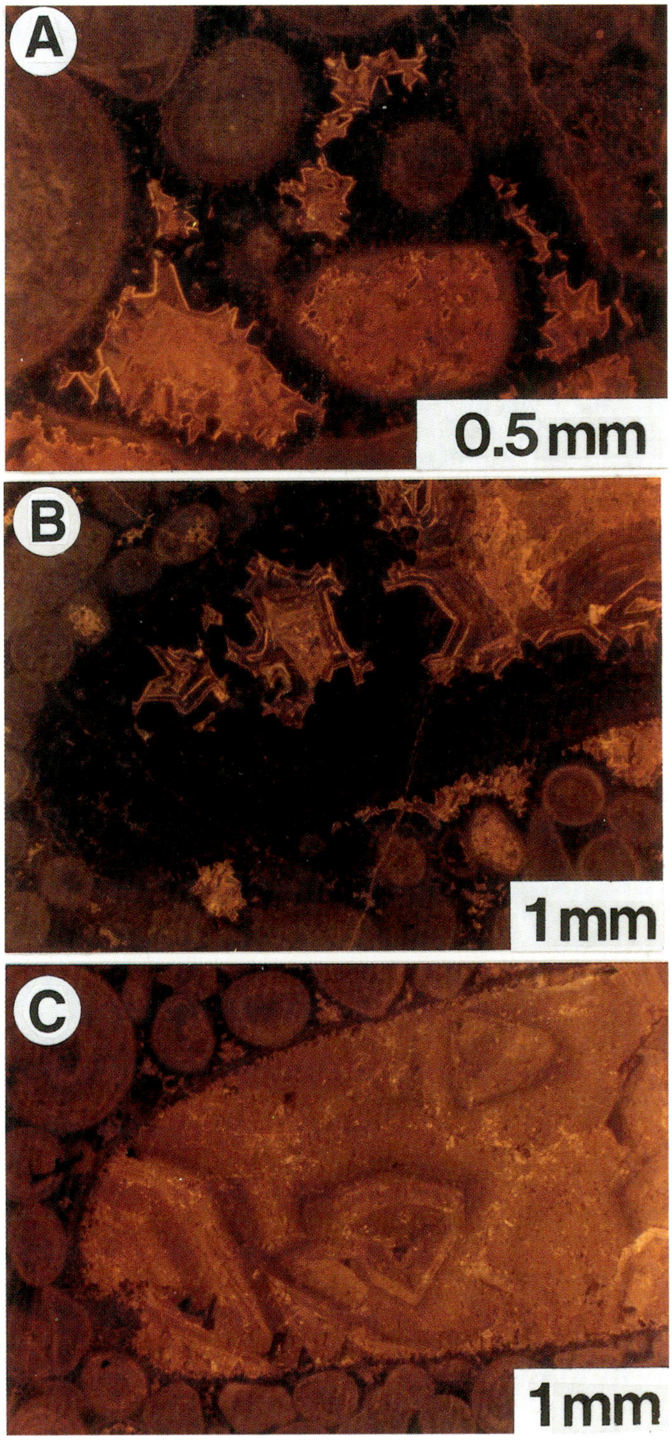

FIG. 18.—Early and late diagenetic fabrics (cathodoluminescent light). (A) Calcite cement fabrics in oolitic grainstone found in Vajont gravity-flow deposits. Oolitic grainstone was first cemented by an early generation of isopachous rims of nonluminescent equant calcite that were conformably overlain by banded, luminescent calcite cement (cf. Fig. 8D). (B) Aggrading neomorphic fabric after originally aragonitic mollusk grain. Transition from fine-grained neomorphic calcite at the grain rim to coarse-grained calcite in the grain interior is accompanied by an increase in luminescence. (C) Cement-filled fabric within originally aragonitic mollusk grain. The mollusk grain was first micritized prior to deposition and then cemented by early nonluminescent, equant calcite in a slope setting. The mollusk grain was then dissolved, and moldic pore space was filled by late-stage luminescent calcite. Transition from nonluminescence to bright, banded-luminescence in cement and skeletal-grain fabrics, together with increasingly depleted ^{18}O compositions, suggests that progressive replacement, dissolution, and cementation occurred in burial settings and reduced pore fluids.

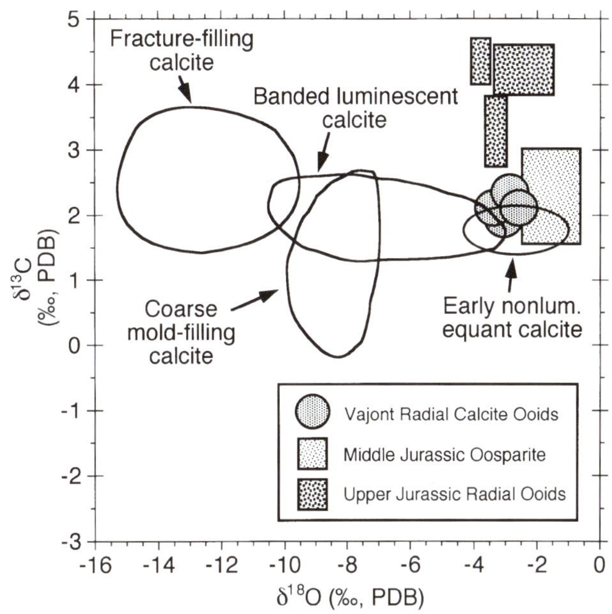

FIG. 19.— Summary of $\delta^{13}C$ and $\delta^{18}O$ values in Vajont radial ooids and cements with comparison to compositions of other Jurassic oolite. Average $\delta^{13}C$ and $\delta^{18}O$ values in the four sedimentologic associations—(I) lower nonskeletal-grain–dominated, (II) lower skeletal-grain–rich, (III) upper nonskeletal-grain–dominated, and (IV) upper skeletal-grain–rich—are shown as four small shaded circles (data from the Vajont radial calcite ooids; see Table 6 for isotope values). Vajont radial calcite ooids overlap the field (light shaded rectangle) defined by compositions of time-equivalent oosparite (Marshall and Ashton, 1980) and are similar to the compositions of Upper Jurassic radial ooids (dark shaded rectangles; Heydari and Moore, 1994). These and other petrographic data suggest that Vajont radial ooids record the primary composition of Middle Jurassic marine calcite. Compositions of early, nonluminescent, equant-calcite cement (small, labeled ellipse) overlap those of Vajont ooids and indicate that equant-calcite cement precipitated from marine-dominated fluids in slope settings. Banded luminescent calcite, mold-filling calcite, and fracture-filling calcite cements (three large, labeled ellipses) exhibit depleted ^{18}O compositions relative to Middle Jurassic marine values. These data indicate that late cementation and fracturing occurred during progressive burial and with increasing temperature (see Zempolich, 1995, for further detail).

originally composed of either high- or low-Mg calcite is difficult as both minerals display radial cortical fabrics (e.g., Land et al., 1979; Popp and Wilkinson, 1983). As discussed above, fabric-retentive radial texture, enrichments in ^{13}C and ^{18}O with respect to late diagenetic cement, nonluminescence, and high Sr/Mn ratios indicate that Vajont ooids underwent little to no diagenetic alteration prior to or after resedimentation. If a relatively closed system is assumed during diagenesis (on the basis of enriched isotopes and low Fe and Mn concentrations; Table 6), then alteration of high-Mg calcite to low-Mg calcite should result in the retainment of some Mg in the form of microdolomite inclusions (e.g., Lohmann and Meyers, 1977). In SEM and BSE examinations of radial ooids sampled throughout the 325-m section at Col Visentin, microdolomite was not observed within any of the ooid cortices. In contrast, microdolomite was a common component of echinoderm fragments (1.2 to 6.1 mol% Mg) examined over the same interval. These data suggest that radial ooids were originally composed of low-Mg calcite (1.2 to 1.9 mol% Mg).

Ooid isotope compositions.—

Ooid analyses display higher $\delta^{13}C$ and $\delta^{18}O$ values with respect to those of late diagenetic components and exhibit distinct shifts in $\delta^{13}C$ and $\delta^{18}O$ when plotted against stratigraphic interval in the section at Col Visentin (Fig. 11). $^{87}Sr/^{86}Sr$ ratios of radial ooids agree well with characteristic marine $^{87}Sr/^{86}Sr$ ratios and trends defined through analysis of other Jurassic calcitic fossils. Accordingly, Vajont radial ooids are interpreted to record the isotope composition of Middle Jurassic marine calcite.

$\delta^{13}C$ and $\delta^{18}O$ values of Vajont radial ooids (Fig. 19; average $\delta^{13}C$ = +2.13‰ and $\delta^{18}O$ = –3.12‰) overlap the marine values of other Middle Jurassic ooids derived by other workers (e.g., Marshall and Ashton, 1980; $\delta^{13}C$ = +2.24‰ and $\delta^{18}O$ = –1.53‰) and are also similar to the marine values of Upper Jurassic radial ooid isotope compositions (e.g., Heydari and Moore, 1994; $\delta^{13}C$ = +3.3‰ to +4.5‰ and $\delta^{18}O$ = –3.8‰ to –2.5‰). Furthermore, positive $\delta^{13}C$ excursions and progressive ^{13}C enrichment observed in the chemostratigraphic profile closely parallel a secular rise in $\delta^{13}C$ (+2.0‰ to +2.5‰ $\delta^{13}C$) from late Early Jurassic to Middle Jurassic time (Holser, 1984). Further evidence of the preservation of a marine signature is provided by the apparent match between the $\delta^{13}C$ profile defined by radial ooids of this study and a marine $\delta^{13}C$ curve presented by Bartolini et al. (1996), which was derived from time-equivalent pelagic successions of central Italy (Fig. 20).

$^{87}Sr/^{86}Sr$ ratios of Vajont radial calcite ooids, hemipelagic micrite, micritic clasts, and a belemnite decrease upward through the section from 0.707200 to 0.707041 (Table 7). These ratios fall within the range of other Bajocian marine $^{87}Sr/^{86}Sr$ ratios determined through the study of calcitic oysters and belemnites (Jones et al., 1994a; Fig. 21), which also show a decrease in $^{87}Sr/^{86}Sr$ from approximately 0.707284 (lowermost early Bajocian) to approximately 0.707034 (uppermost late Bajocian). Overall, the $^{87}Sr/^{86}Sr$ ratios of Vajont calcitic components appear to decrease rapidly from early Bajocian to late Bajocian time and improve upon the Bajocian seawater Sr isotope curve presented by Jones et al. (1994a).

Within the long-term trend of decreasing $^{87}Sr/^{86}Sr$ ratios, a scatter of data is apparent in several stratigraphic intervals (e.g., approximately 50–130 m and 160–225 m; Fig. 22). The scatter of data points within these intervals is problematic for the following reasons:

1. The range of scatter (as much as 0.000125) is greater than both the uncertainty due to analytical error (±0.000030) and the range of natural scatter of $^{87}Sr/^{86}Sr$ ratios displayed by well-preserved samples within discrete stratigraphic horizons (as much as approximately 0.00005; Jones et al., 1994b).
2. All values fall within the range of Bajocian marine values defined by Jones et al. (1994a) (cf. Fig. 21).
3. Analyses of radial ooids are in close agreement with analyses of stratigraphically related hemipelagic micrites, micritic clasts, and a belemnite (Fig. 22).

Scatter due to diagenetic alteration is considered unlikely because of the high Sr/Mn ratios, nonluminescence, and high $\delta^{13}C$ values displayed by the radial ooids. Therefore the scatter of $^{87}Sr/^{86}Sr$ ratios either reflects short-term fluctuations in marine $^{87}Sr/^{86}Sr$ ratios that are unresolvable with the current data set or is due to minor sedimentologic mixing associated with the gravity-flow redeposition of older, unlithified oolite derived from periplatform and upper-slope settings (e.g., Figs. 7, 10, 14, 15). The latter scenario is supported by the fact that some intervals that exhibit scatter in $^{87}Sr/^{86}Sr$ ratios are associated with an abundance of platform lithoclasts and hemipelagic micrite clasts that were redeposited in lower- to middle-slope settings (Fig. 4).

The apparent agreement of Vajont $\delta^{13}C$ and $^{87}Sr/^{86}Sr$ profiles with marine values determined from outside of the present study area strongly suggests that the isotope compositions of radial ooids have not been significantly altered during diagenesis. These data suggest that radial ooids record the isotope composition of Middle Jurassic marine calcite. Accordingly, isotope compositions of radial calcite ooids may be used as a chemostratigraphic "proxy" for secular trends in marine calcite compositions and in further analysis of Vajont short-term compositional trends, although with the caveat that some trends may be potentially modified by sedimentologic mixing associated with slope depositional processes.

Coupled Geochemical and Sedimentologic Variations

Profiles of the $\delta^{13}C$ values and $^{87}Sr/^{86}Sr$ ratios of radial ooids and the sedimentologic compositions at Col Visentin reveal several important trends. The first trend is an apparent depletion followed by a gradual enrichment in ^{13}C, coincident with a net decrease in $^{87}Sr/^{86}Sr$ ratio, upward through the first nonskeletal-grain–dominated sedimentologic association (I; Fig. 22). This trend is followed by rapid enrichment in ^{13}C, which correlates with sedimentation of the lowermost skeletal-grain–rich sedimentologic association (II, Fig. 22), and a scattered increase in the $^{87}Sr/^{86}Sr$ ratio, which appears to correlate with maximum skeletal content. A well-defined ^{13}C depletion and noisy but statistically verified $^{87}Sr/^{86}Sr$ decrease is then observed at the transition between the lower skeletal-grain–rich and upper nonskeletal-grain–dominated sedimentologic associations (II to III, Fig. 22). These changes are followed by punctuated enrichment in ^{13}C and a slight decrease in $^{87}Sr/^{86}Sr$ upward through the second nonskeletal-grain–dominated sedimentologic association (III, Fig. 22). ^{13}C enrichments then correlate with sedimentation of the upper skeletal-grain–rich sedimentologic association (IV, Fig. 22) and a slight increase in $^{87}Sr/^{86}Sr$ ratios that appears to correlate with a well-defined positive excursion in skeletal content. Similar to the point-count results, statistics are used to substantiate these and other important relationships.

Statistical comparison of isotope compositions and grain types is used to further define relationships between the two data sets (Table 4). These results include a positive correlation between $\delta^{13}C$ values and total skeletal-grain and grapestone contents (Fig. 11; ρ = +0.32 and +0.35, respectively), and an inverse correlation between $\delta^{13}C$ values and total ooid and

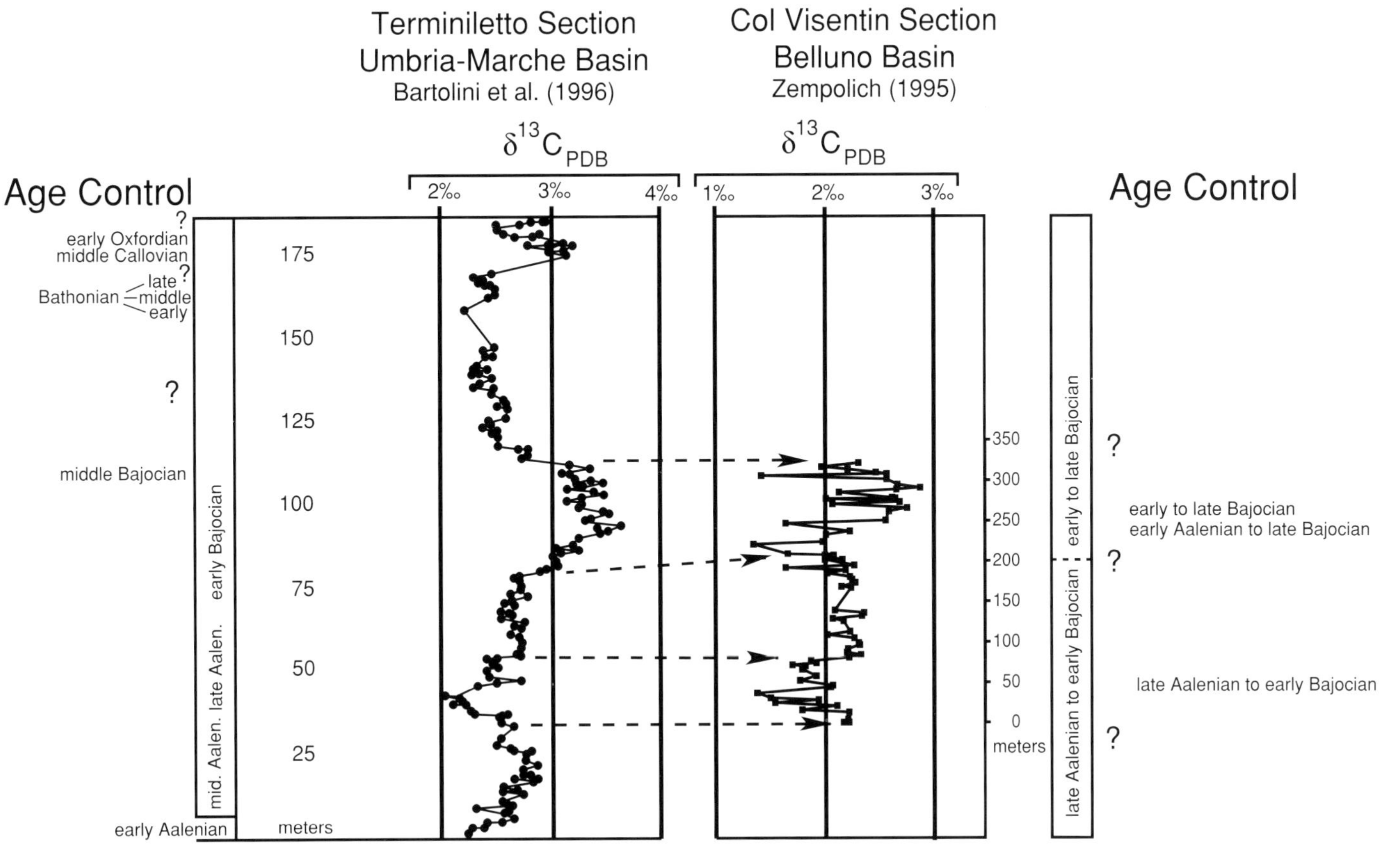

FIG. 20.—Comparison of the late Aalenian–late Bajocian $\delta^{13}C$ chemostratigraphic profile of the Col Visentin section (Zempolich, 1995) with that of the Terminiletto pelagic section, Umbro-Marche basin, Italy (Bartolini et al., 1996). Both profiles exhibit progressive enrichment in ^{13}C of approximately 1.5‰ through time, and correlative short-term excursions in $\delta^{13}C$ of 0.5‰ to 1.0‰. Correlation between such strikingly different lithologies and stratigraphic sections suggests that radial calcite ooids and pelagic calcite both reflect the primary composition of Middle Jurassic marine calcite. Active input of resedimented shelf carbonate is reflected in the difference in total thickness between the sections, i.e., the Vajont section accumulated at a rate roughly four times that of the pelagic Terminiletto section (note different thickness scales).

micritic ooid contents (Fig. 11; $\rho = -0.27$ at the 98% confidence level, $\rho = -0.23$ at the 95% confidence level, respectively). $\delta^{13}C$ values and peloid, oncoid, and radial ooid nonskeletal-grain contents appear to be unrelated (ρ = +0.05, –0.01, and –0.04, respectively). In addition, a positive correlation is displayed between $\delta^{18}O$ values and total ooid content and radial ooid nonskeletal-grain content (e.g., Fig. 11; ρ = +0.33 and +0.42, respectively). An inverse correlation is observed between $\delta^{18}O$ values and total peloid content ($\rho = -0.36$), and a weak inverse correlation is observed between $\delta^{18}O$ values and total skeletal-grain content ($\rho = -0.22$). These data suggest that correlations between grain contents and either $\delta^{13}C$ or $\delta^{18}O$ values are independent and that variations between $\delta^{13}C$ and $\delta^{18}O$ values are not directly related to each other (Fig. 13; $\rho = +0.13$). These correlations suggest that (1) $\delta^{13}C$ values are related to gross changes in sedimentologic composition, i.e., enrichment in ^{13}C corresponds with increases in total skeletal-grain and grapestone contents and depletion in ^{13}C corresponds with increases in total ooid and micritic ooid contents, and (2) $\delta^{18}O$ ratios are related to some variation in sedimentologic composition, where enrichment in ^{18}O corresponds with increases in total ooid and radial ooid contents and depletion in ^{18}O strongly corresponds with increases in total peloid content and weakly corresponds with increases in total skeletal-grain content.

Statistical comparison of radial ooid $^{87}Sr/^{86}Sr$ ratios with $\delta^{13}C$ values and total skeletal-grain contents yields correlation coefficients of –0.36 (N = 31, 95% confidence) and –0.66 (N = 34, 99% confidence), respectively. These negative correlations are interpreted to reflect the opposite directions in which long-term secular changes in these geochemical signatures occurred. For example, through the Middle Jurassic, ^{13}C increased (Fig. 20), and $^{87}Sr/^{86}Sr$ decreased (Fig. 21). Total skeletal-grain content exhibits an overall increase upward through the stratigraphic section and is negatively correlated with $^{87}Sr/^{86}Sr$ (ρ = –0.66). This finding is interpreted to reflect a general long-term shallowing of the platform and decreasing $^{87}Sr/^{86}Sr$ through the Bajocian. Within this decreasing $^{87}Sr/^{86}Sr$ trend, the Sr isotope compositions of Vajont radial ooids exhibit small, and sometimes scattered, increases in $^{87}Sr/^{86}Sr$ that are associated with maximum skeletal-grain contents (Fig. 22). The quantitative assessment of covariance between these short-term positive excursions awaits collection of a larger $^{87}Sr/^{86}Sr$ data set and more rigorous quantitative methods (e.g., statistical curve fitting).

Significantly, statistical and qualitative correlations of both sedimentologic features and $\delta^{13}C$ values through the stratigraphic section at Col Visentin support the division of the section into nonskeletal-grain–dominated and skeletal-grain–rich sedimentologic associations, as first observed in outcrop (Figs. 4, 11, 22; Tables 2–4).

DISCUSSION

Sequence Stratigraphic Interpretation

Second-order sequence.—

Overall, the Col Visentin section exhibits an upward increase in skeletal, peloid, and grapestone content and a concomitant decrease in ooid and oncoid content (e.g., Figs. 11, 22; Tables 2–4). As discussed, these compositional data suggest that lower-energy conditions and a general shallowing of depositional environments took place on the Friuli platform. The long-term compositional changes are consistent with second-order sea-level rise during the middle Aalenian–early Bajocian and with the initial relative fall in sea level during the late Bajocian (Fig. 13). Bosellini et al. (1981) reported that Middle Jurassic ooid sedimentation on the Friuli platform was abruptly terminated by a period of subaerial exposure that was followed in Callovian–Oxfordian time by sedimentation of skeletal-grain calcarenites and coral-hydrozoan boundstones at the platform margins. These stratigraphic relationships suggest the formation of a type 1 unconformity (e.g., Sarg, 1988) on the Friuli platform following ooid sedimentation, a conclusion that is in agreement with global data that suggest eustatic sea-level fall and formation of a type 1 or type 2 unconformity toward the end of the late Bajocian (e.g., Hallam, 1988; Haq et al., 1988; Fig. 13). Collectively, the gradual enrichment of skeletal-grain contents in Vajont gravity-flow deposits, the recognition of a significant unconformity on the Friuli platform, and the abrupt transition to the skeletal-grain–dominated gravity-flow deposits of the overlying Fonzaso Formation (Callovian–Oxfordian; Fig. 3) suggest that sedimentation of the Vajont Limestone reflects the *basinal* record of an adjacent, persistently flooded platform during long-term (second-order) sea-level rise (from the late Aalenian/early Bajocian boundary to the late Bajocian; Fig. 23).

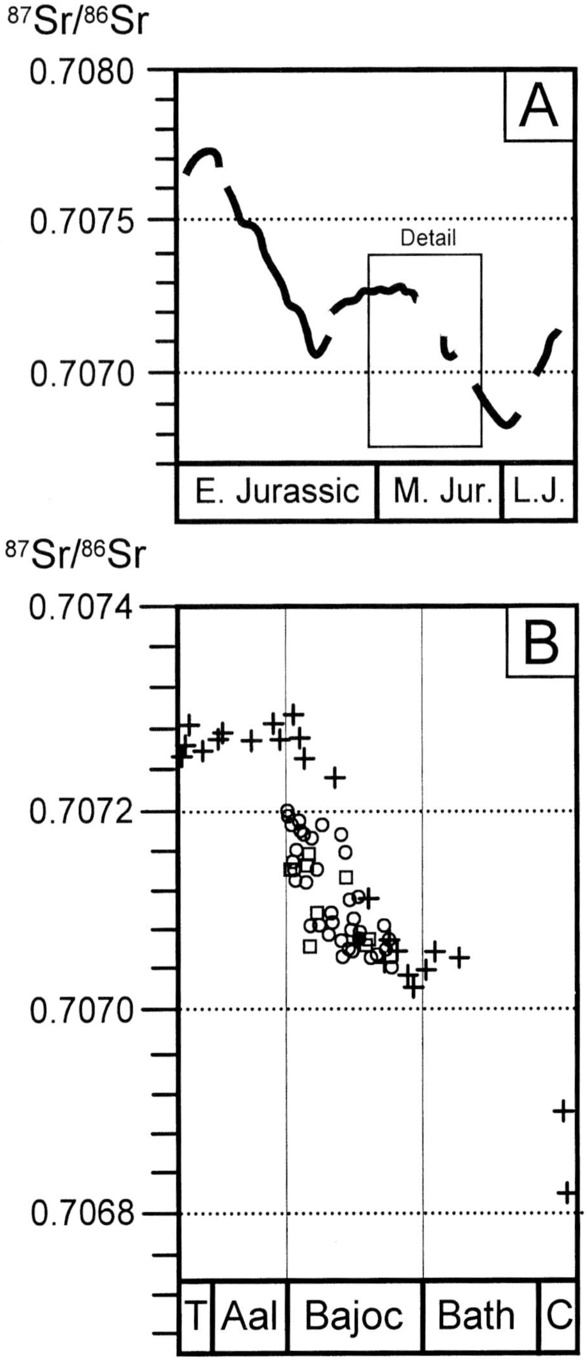

FIG. 21.—Comparison of Vajont $^{87}Sr/^{86}Sr$ analyses with best estimates of Jurassic seawater strontium isotope ratios. (A) Best-estimate seawater strontium isotope curve for the Jurassic (modified from Jones et al., 1994a). The curve was derived from $^{87}Sr/^{86}Sr$ analysis of well-preserved belemnite and oyster samples from the United Kingdom and well-documented ammonite biostratigraphy. Despite several gaps, the data clearly show an overall trend of decreasing $^{87}Sr/^{86}Sr$ ratios through the Early and Middle Jurassic. (B) Detail of the Middle Jurassic portion of the strontium isotope curve depicted in A. $^{87}Sr/^{86}Sr$ ratios of Vajont radial calcite ooids (open circles), hemipelagic carbonate mudstones (open squares), and a belemnite (filled circle) trend from 0.707200 to 0.707041 (Table 7) and define a decreasing trend through the Bajocian that is shared by analyses of well-preserved belemnite and oyster samples ("plus" symbols; Jones et al., 1994a). These and other petrographic and geochemical data suggest that Vajont radial calcite ooids were unaltered by diagenesis and thus record changes in seawater strontium isotopes through a portion of the Middle Jurassic. Note: Jurassic stages are as follows: T = Toarcian, Aal = Aalenian, Bajoc = Bajocian, Bath = Bathonian, C = Callovian. Vajont $^{87}Sr/^{86}Sr$ analyses are plotted for the age range—the late Aalenian/early Bajocian boundary to the late Bajocian—determined by nannofossil biostratigraphy (Tables 1, 7). Because of the potential error in comparison of different biostratigraphic data sets, Vajont $^{87}Sr/^{86}Sr$ analyses and $^{87}Sr/^{86}Sr$ analyses reported by Jones et al. (1994a) have not been adjusted for small differences (0.000014) in the laboratory measurement of NBS SRM 987 ($^{87}Sr/^{86}Sr$ = 0.710254; Jones et al., 1994b).

Third-order sequences.—

Lack of a definitive "lowstand" compositional signature (i.e., skeletal-grain–*dominated* gravity-flow deposits) between sedimentologic associations II and III may be attributed to rapid platform subsidence and sustained second-order relative rise in sea level. Accordingly, each pair of nonskeletal-grain–dominated

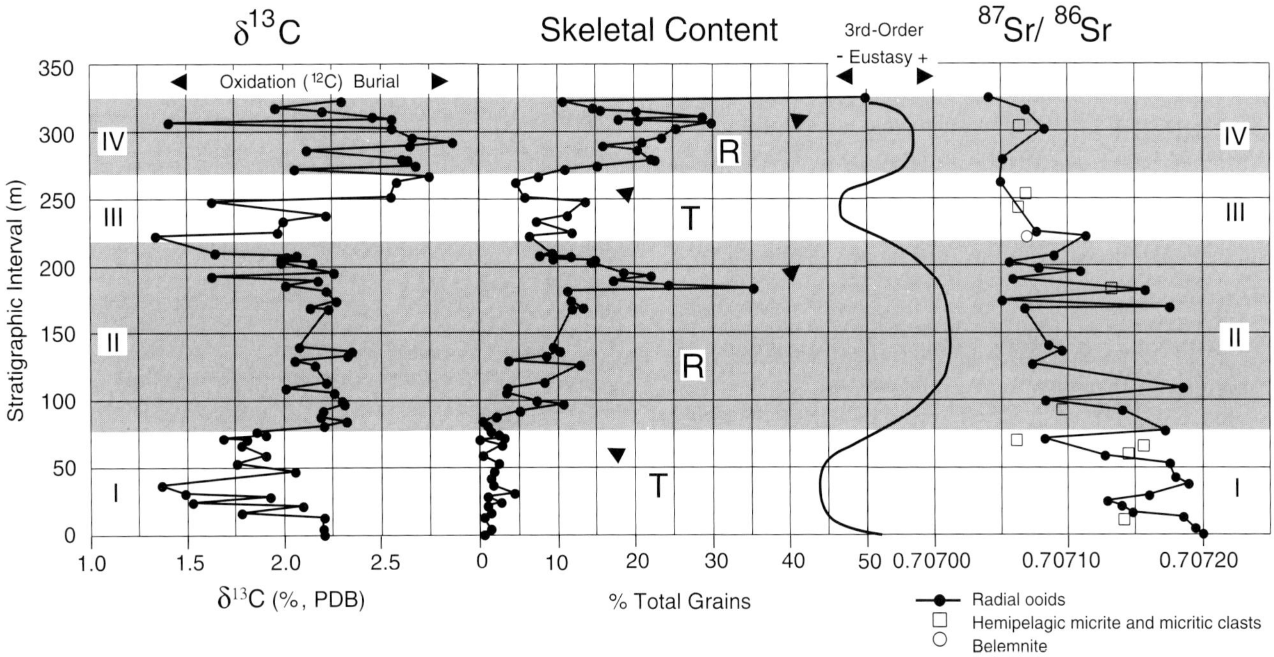

Fig. 22.—δ^{13}C and ^{87}Sr/^{86}Sr chemostratigraphic profiles and skeletal-grain profile, Col Visentin section. δ^{13}C and total skeletal-grain content exhibit short-term fluctuations upward through the stratigraphic section that are correlative (Spearman's correlation coefficient = +0.32). These relationships, together with other field and compositional data, suggest that alternation of nonskeletal-grain–dominated (intervals I and III) and skeletal-grain–rich (gray-shaded intervals II and IV) sedimentologic associations reflect transgressions (T) and regressions (R) that were controlled by short-term (third-order) eustasy. δ^{13}C and ^{87}Sr/^{86}Sr of radial ooids are negatively correlated (Spearman's correlation coefficient = –0.36). This result is interpreted to reflect the long-term secular variations in the δ^{13}C values (increasing) and the ^{87}Sr/^{86}Sr ratios (decreasing) of Middle Jurassic seawater. Similarly, total skeletal-grain content exhibits an overall increase upward through the stratigraphic section and is negatively correlated with ^{87}Sr/^{86}Sr (Spearman's correlation coefficient = –0.66). This result is interpreted to reflect a long-term shallowing of the platform and decreasing ^{87}Sr/^{86}Sr through the Bajocian. Small, and sometimes scattered, increases in ^{87}Sr/^{86}Sr appear to correlate with maximum skeletal-grain content in skeletal-grain–rich sedimentologic associations. If correlative, these positive excursions may reflect short-term relative falls in sea level (approximately 1–3 m.y.) and input of Sr with a high ^{87}Sr/^{86}Sr ratio (see text for further discussion).

and skeletal-grain–rich sedimentologic associations (I and II; III and IV) may reflect sedimentation of two third-order sequences on the Friuli platform during two short-term eustatic cycles that, when combined, compose the larger second-order sequence (Fig. 23).

Abrupt transition from the lower skeletal-grain–rich sedimentologic association (II) to the upper nonskeletal-grain–dominated sedimentologic association (III) in the Vajont, including the absence of an intervening lowstand sequence, suggests that transgression followed a sea-level highstand. Therefore each pair of nonskeletal-grain–dominated and skeletal-grain–rich sedimentologic associations may represent an individual cycle of transgressive and regressive deposition on the Friuli platform. If correct, Vajont nonskeletal-grain–dominated gravity-flow sequences may be interpreted as the basinal record of the flooding of the Friuli platform during short-term (third-order) sea-level rise and the development of transgressive systems tracts (TST; Fig. 23). Similarly, Vajont skeletal-grain–rich sedimentologic associations may be interpreted as the basinal record of regression on the Friuli platform following short-term (third-order) sea-level rise and development of highstand systems tracts (HST; Fig. 23). Because these cycles are bounded at the base by evidence of sustained platform flooding (i.e., nonskeletal-grain–dominated gravity-flow deposits) and lack evidence of lowstand sedimentation (i.e., there are no skeletal-grain–dominated gravity-flow deposits), these depositional units are interpreted to reflect the *basinal* record of the sedimentation of two third-order sequences on the Friuli platform (Fig. 23). These interpretations are further supported by relating sediment compositional variation with isotope compositional variation.

Geochemical Variation

The δ^{13}C values of radial ooids at Col Visentin exhibit ^{13}C depletions and enrichments that correlate with the alternation between nonskeletal-grain–dominated and skeletal-grain–rich sedimentologic associations (Fig. 4; Tables 2–4). Enrichments in ^{13}C and small and, in certain strata, scattered increases in ^{87}Sr/^{86}Sr ratios correlate with proposed relative falls in sea level as defined by sediment compositional changes; depletion in ^{13}C correlates with proposed relative rises in sea level as defined by sediment compositional changes. These data suggest that sediment content and δ^{13}C values, and possibly also ^{87}Sr/^{86}Sr ratios, were influenced by allocyclic processes.

Enrichments in ^{13}C in skeletal-grain–rich sedimentologic associations in the Vajont correlate with the long-term secular trend in ^{13}C (e.g., Holser, 1984) whereas nonskeletal-grain–dominated sedimentologic associations exhibit departures (i.e., depletions in ^{13}C) from the long-term secular trend. Because variation in $\delta^{13}C$ values is relatively insensitive to small-scale changes in temperature, these shifts probably reflect short-term variation in the $\delta^{13}C$ values of Middle Jurassic seawater (Zempolich, 1995). The correlation of short-term variations in $\delta^{13}C$ recorded by Vajont radial ooids and those reflected in basinal carbonate sequences (Bartolini et al., 1996) supports this conclusion (Fig. 20). Short-term shifts in oceanic $\delta^{13}C$ in conjunction with eustatic sea-level changes have been noted in many Phanerozoic sequences (e.g., Jenkyns, 1988). Possible causes of shifts in $\delta^{13}C$ include the mechanical storage and release of organic carbon (rich in ^{12}C) during sea-level retreats from land and continental shelves (e.g., Kennett, 1986) and/or the geochemical oxidation and release of organic carbon related to fluctuations in atmospheric and oceanic oxygen levels (e.g., Shackleton, 1987). More recently, Follmi et al. (1994) have proposed a more complicated system of causes and feedbacks that link positive excursions in $\delta^{13}C$ to rises in sea level and vice versa.

A "mechanical" explanation for the observed ^{13}C depletions in the Vajont is that short-term falls in relative sea level and/or forced regressions release significant amounts of oxidized organic material (enriched in ^{12}C) from continents to the oceanic reservoir. Such a scenario is supported by the fact that $\delta^{13}C$ values decrease in concert with transitions from regressive to transgressive depositional systems, but increase during transition to the following regressive depositional system (Fig. 22). Correlation of high $\delta^{13}C$ values with the long-term secular trend may suggest that the long-term balance between terrestrial, oceanic, and atmospheric carbon reservoirs is quickly reestablished following relative rises in sea level. Whatever the cause, the correlation between negative $\delta^{13}C$ excursions and the transition from regressive to transgressive Vajont sediment compositions—followed upward by positive $\delta^{13}C$ excursions that correlate with regressive Vajont sediment compositions—suggests that the Vajont $\delta^{13}C$ values and sediment compositions reflect sea-level eustasy and the development of transgressive and highstand systems tracts on the Friuli platform. Although no outcrop of time-equivalent shelf strata was available in the study area to corroborate this relationship, in a detailed study of Cretaceous shelf sedimentary deposits in the Arabian Gulf region, Vahrenkamp (1996) similarly related positive $\delta^{13}C$ excursions to highstand sedimentation and negative $\delta^{13}C$ excursions to forced regressions and lowstand sedimentation. The relationship between positive $\delta^{13}C$ excursions and regressive Vajont sediment compositions may be further supported by a link between positive excursions of $^{87}Sr/^{86}Sr$ ratios and total skeletal-grain content within the younger parts of the skeletal-grain–rich sedimentologic associations (Fig. 22), which is consistent with other studies (e.g., Oslick et al., 1994; Montanez et al., 1996; Ruppel et al., 1996) that suggest a link between short-term regressions (1–5 m.y.) and input of high $^{87}Sr/^{86}Sr$. Further interpretation of Vajont Sr isotope stratigraphy awaits collection and rigorous quantitative assessment of a larger data set.

CONCLUSIONS

The present study of the Vajont Limestone yields new information that aids in the interpretation of the deposition and diagenesis of shallow-water carbonate in deep-water settings. New ages based on ammonite and nannofossil data indicate that sedimentation of the Vajont Limestone in the Belluno basin was initiated at least by the time of the late Aalenian/early Bajocian boundary and that most, if not all, sedimentation was completed by late Bajocian time. Compositional data on resedimented gravity-flow deposits identify nonskeletal-grain–dominated and skeletal-grain–rich sedimentologic associations within the Vajont. These grain counts suggest that extensive mobile sand belts composed of ooids and oncoids were established on the Friuli platform during relative rises in sea level and that deposition of ooid sand belts, small patch reefs, and associated skeletal-grain sand, and shoaling and erosion of platform lithologies occurred during relative falls in sea level. Off-bank transport of these sediments from the platform interior to the western platform margins resulted in gravity-flow deposition and the asymmetric in-fill of the Belluno basin by Vajont slope apron deposits.

Transgressive and regressive depositional models for the Friuli platform and the Vajont slope system are supported by correlation of Vajont sedimentologic associations with proposed global sea-level curves, which indicate that most deposition occurred during long-term (second-order) sea-level rise. During the sea-level rise, the transport of nonskeletal-grain sediment directly from the Friuli platform to the Belluno basin was enhanced by a lack of extensive periods of platform-wide subaerial exposure (i.e., sea-level lowstands). Alternation of nonskeletal-grain–dominated and skeletal-grain–rich sedimentologic associations, together with the correlation of isotope data from the Vajont with secular variations in isotope compositions, suggests subtle changes in platform sediment input during relative sea-level fluctuations. These sediment and isotope variations are attributed to short-term (third-order) sea-level fluctuations.

With regard to sequence stratigraphy, study of Vajont gravity-flow deposits reveal several important relationships that may be useful in the sedimentologic and chemostratigraphic recognition of third-order sequences in resedimented carbonates. On the basis of petrographic, geochemical, and biostratigraphic study we conclude the following:

1. Vajont radial ooids were originally composed of low-Mg calcite and retain primary textures, ^{13}C and ^{18}O compositions, and $^{87}Sr/^{86}Sr$ ratios that were unaltered by either early (slope) or late (burial) diagenesis. Radial ooids therefore provide a proxy for Middle Jurassic marine calcite compositions in that the radial ooids record both long-term and short-term changes in seawater chemistry.
2. Grain contents, $\delta^{13}C$, and possibly $^{87}Sr/^{86}Sr$ ratios of radial ooids in Vajont gravity-flow deposits provide a basinal record of the sedimentary and eustatic history of the Friuli platform during the time from the late Aalenian/early Bajocian boundary to the late Bajocian.
3. An overall upward increase in skeletal, peloid, and grapestone contents and a concomitant decrease in ooid and

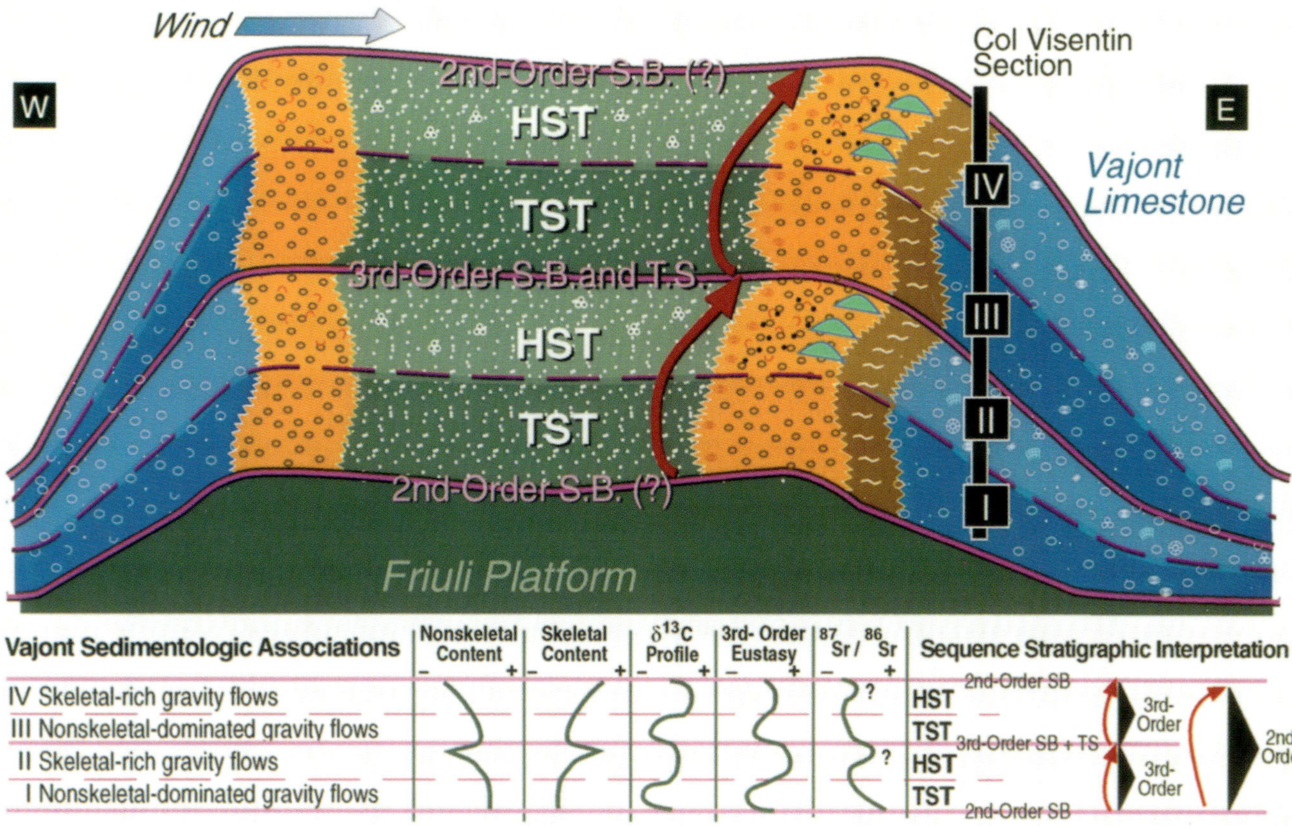

FIG. 23.—Sequence stratigraphic model of the Friuli platform and Vajont slope system. Sediment compositions of Vajont gravity-flow deposits indicate that Vajont sediment was derived from a persistently flooded platform. Lowstand (skeletal-grain–dominated) compositions are not recorded in Vajont gravity-flow deposits. These data are in agreement with a sustained rise (second-order) in global sea level during the late Aalenian to late Bajocian (Fig. 13). Exposure of the Friuli platform and abrupt change to skeletal-grain–dominated gravity-flow deposits of the overlying Fonzaso Formation (Bosellini et al., 1981) suggests the generation of a Type 1 unconformity on the platform following Vajont sedimentation. Thus the Vajont Limestone is interpreted as the basinal record of a second-order depositional sequence. Alternation between nonskeletal-grain–dominated sedimentologic associations that are relatively depleted in ^{13}C (I and III) and skeletal-grain–rich sedimentologic associations that are relatively enriched in ^{13}C (II and IV) are interpreted to reflect shallow-water sedimentation on the Friuli platform during two cycles (third-order), each consisting of a transgression and a highstand. Specifically, nonskeletal-grain–dominated sedimentologic associations in the Vajont are interpreted as the basinal record of deposition during relative rise in sea level and sedimentation of transgressive systems tracts (TST) on the Friuli platform (e.g., Fig. 14). Skeletal-rich sedimentologic associations in the Vajont are interpreted as the basinal record of deposition during static sea level and relative fall in sea level, and sedimentation of highstand systems tracts (HST) on the Friuli platform (e.g., Fig. 15). The absence of a lowstand wedge (LSW) between these two cycles may be explained by deposition during a sustained long-term (second-order) rise in relative sea level during the late Aalenian to late Bajocian.

oncoid contents of Vajont gravity-flow deposits reflects deposition during initial flooding and gradual shallowing of the Friuli platform during the time of the late Aalenian/early Bajocian boundary to the late Bajocian. Accordingly, the Vajont Limestone is interpreted as the basinal record of a second-order depositional sequence.

4. Within the Vajont, pairs of decameter-scale, nonskeletal-grain–dominated and skeletal-grain–rich sedimentologic associations are defined by covariation of grain contents and δ^{13}C values. Grain compositions of nonskeletal-grain–dominated and skeletal-grain–rich sedimentologic associations are consistent with persistent "highstand" shedding of carbonate, exhibit upward enrichment in skeletal-grain content and ^{13}C, display small increases in ^{87}Sr/^{86}Sr ratios, and are interpreted to reflect short-term (third-order) sea-level fluctuations and the development of transgressive and highstand systems tracts on the Friuli platform. As a result, each pair of nonskeletal-grain–dominated and skeletal-grain–rich sedimentologic associations is interpreted as the basinal record of one third-order depositional sequence.

If interpretation of sediment compositional variation and geochemical variation in the Vajont is correct, similar sedimentologic, petrographic, and chemostratigraphic study may provide an additional means by which third-order sequences may be identified in other resedimented carbonates.

ACKNOWLEDGMENTS

Danielle Masetti and Alfonso Bosellini of the University of Ferrara and Lawrie Hardie of Johns Hopkins University provided critical logistical support and guidance that made this study possible. Analyses used in this study were kindly provided by the following people and laboratories: K. C Lohmann, Jim Burdett, and Lora Wingate (University of Michigan) provided carbon and oxygen isotope analyses; Lynn Walters and Ted Huston (University of Michigan) provided trace element analyses, and Tim Denison (University of Texas at Dallas) provided Sr isotope analyses. Tim Denison and Nate Miller are thanked for helpful discussion and insight into Sr data interpretation and presentation. David Helber provided assistance with the preparation of color illustrations. We also thank Linda Hinnov and Jim Markello for reviewing early versions of this manuscript and Dan Lehrmann, Beverly Saylor, and Toni Simo for reviews of the final manuscript. A special thank you is extended to Beth Meany for her patience and support through the course of this study, which also benefited from the support and help of fellow students and friends including Michele Claps, Linda A. Hinnov, Joseph B. Paul, and Lyndon Yose. Grants from the following organizations enabled the completion of this study: the American Association of Petroleum Geologists, the Geological Society of America, Sigma Xi, Mobil Oil Italiana S.p.A., Johns Hopkins University Balk Fund, and the National Science Foundation (grant no. EAR910510).

REFERENCES

BARTOLINI, A., BAUMGARTNER, P. O., AND HUNZIKER, J., 1996, Middle and Late Jurassic carbon stable-isotope stratigraphy and radiolarite sedimentation of the Umbria-Marche basin (Central Italy): Eclogae Geologicae Helvetiae, v. 89, p. 811–844.

BECCARELLI BAUCK, L., 1986, Stylothalamien aus dem unterjurassischen Misone-kalk der Südalpen, Italien: Palaeontographica Abteilung A, v. 192, p. 1–13.

BECCARELLI BAUCK, L., 1988, Unter-bis mitteljurassische Karbonatformationen am Westrand der Trento-Plattform (Südalpen, Norditalien): München Geowissenschaften Abhandlungen, v. 13, p. 1–86.

BOARDMAN, M. R., NEUMANN, A. C., BAKER, P. A., DULIN, L. A., KENTER, R. J., HUNTER, G. E., AND KIEFER, K. B., 1986, Banktop responses to Quaternary fluctuations in sea level recorded in periplatform sediments: Geology, v. 14, p. 28–31.

BOSELLINI, A., 1989, Dynamics of Tethyan carbonate platforms, *in* Crevello, P. D., Wilson, J. L., Sarg, J. F., and Read, J. F., eds., Controls on carbonate platform and basin development: Tulsa, Society of Economic Paleontologists and Mineralogists, Special Publication 44, p. 3–13.

BOSELLINI, A., AND MASETTI, D., 1972, Ambiente e dinamica deposizionale del Calcare del Vajont (Giurassico medio, Prealpi Bellunesi e Friulane): Ferrara, Annali dell'Universita di Ferrara, v. 5, p. 87–100.

BOSELLINI, A., MASETTI, D., AND SARTI, M., 1981, A Jurassic "Tongue of the Ocean" infilled with oolitic sands: The Belluno trough, Venetian Alps, Italy: Marine Geology, v. 44, p. 59–95.

CARPENTER, S. J., AND LOHMANN, K. C, 1992, Sr/Mg ratios of modern marine calcite: Empirical indicators of ocean chemistry and precipitation rate: Geochimica et Cosmochimica Acta, v. 56, p. 1837–1849.

CARTWRIGHT, R. A., 1985, Provenance and sedimentology of carbonate turbidites from deep-sea fans, Bahamas: Unpublished Ph.D. dissertation, University of Miami, Coral Gables, Florida, 130 p.

CASATI, P., AND TOMAI, M., 1969, Il Giurassico ed il Cretacio del versante settentrionale del Vallone Bellunese e del Gruppo del M. Brandol: Rivista Italiana di Paleontologia e Stratigrafia, v. 75, p. 205–341.

CATI, A., SARTORIO, D., AND VENTURINI, S., 1987, Carbonate platforms in the subsurface of the northern Adriatic area: Memorie della Società Geologica Italiana, v. 40, p. 295–308.

CHAYES, F., 1956, Petrographic modal analysis: New York, John Wiley and Sons, 113 p.

COBIANCHI, M., 1992, Sinemurian–early Bajocian calcareous nannofossil biostratigraphy of the Lombardy basin (Southern Calcareous Alps; northern Italy): Atti Ticinensi di Scienze della Terra, v. 35, p. 61–106.

COBIANCHI, M., ERBA, E., AND RADRIZZANI, C. P., 1992, Evolutionary trends of calcareous nannofossil genera *Lotharingius* and *Watznaueria* during the Early and Middle Jurassic: Padova, Memorie di Scienze Geologiche, v. 43, p. 19–25.

COOK, H. E., 1983, Ancient carbonate platform margins, slopes, and basins, *in* Cook, H. E., Hine, A. C., and Mullins, H. T., eds., Platform Margin and Deep Water Carbonates: Tulsa, Society of Economic Paleontologists and Mineralogists, Short Course Notes No. 12, p. 5-1–5-189.

COOK, H. E., MCDANIEL, P. N., MOUNTJOY, E. W., AND PRAY, L. C., 1972, Allochthonous carbonate debris flows at Devonian bank ("reef") margins Alberta, Canada: Canadian Petroleum Geology Bulletin, v. 20, p. 439–497.

COPE, J. C., GETTY, T. A., HOWARTH, M. K., MORTON, N., TORRENS, H. S., DUFF, K. L., PARSONS, C. F., WIMBLEDON, W. A., AND WRIGHT, J. K., 1980, Jurassic, Parts 1 and 2: London, Geological Society of London, Special Reports, Nos. 14 and 15, 73 and 109 p.

COUSIN, M., 1980, Les rapports Alpes-Dinarides: Les confins de l'Italie et de la Yougoslavie: Unpublished Ph.D. dissertation, Université Pierre et Marie Curie, Paris, 517 p.

CREVELLO, P. D., AND SCHLAGER, W., 1980, Carbonate debris sheets and turbidites, Exuma Sound, Bahamas: Journal of Sedimentary Petrology, v. 50, p. 1121–1148.

DAVIS, J. C., 1986, Statistics and Data Analysis in Geology: New York, John Wiley and Sons, 646 p.

DENISON, R. E., KOEPNICK, R. B., FLETCHER, A., HOWELL, M. W., AND CALLOWAY, W. S., 1994, Criteria for the retention of original seawater $^{87}Sr/^{86}Sr$ in ancient shelf limestones: Chemical Geology, v. 112, p. 131–143.

DERRY, L. A., KAUFMAN, A. J., AND JACOBSEN, S. B., 1992, Sedimentary cycling and environmental change in the Late Proterozoic: Evidence from stable and radiogenic isotopes: Geochimica et Cosmochimica Acta, v. 56, p. 1317–1329.

DOGLIONI, C., 1990, The Venetian Alps thrust belt, in McClay, K. R., ed., Thrust Tectonics: London, Chapman and Hall, p. 319–324.

DROXLER, A. W., AND SCHLAGER, W., 1985, Glacial versus interglacial sedimentation rates and turbidity frequency in the Bahamas: Geology, v. 13, p. 799–802.

EBERLI, G. P., 1987, Carbonate turbidite sequences deposited in rift-basins of the Jurassic Tethys Ocean (Eastern Alps, Switzerland): Sedimentology, v. 34, p. 363–388.

EBERLI, G. P., 1988, The evolution of the southern continental margin of the Jurassic Tethys Ocean as recorded in the Allgau Formation of Austroalpine nappes of Graubunden (Switzerland): Eclogae Geologicae Helvetiae, v. 81, p. 175–214.

EBERLI, G. P., 1991, Calcareous turbidites and their relationship to sea-level fluctuations and tectonism, *in* Einsele, G., Ricken, W., and Seilacher, A., eds., Cycles and Events in Stratigraphy: New York, Springer-Verlag, p. 340–359.

EBERLI, G. P., AND GINSBURG, R. N., 1989, Cenozoic progradation of northwestern Great Bahama Bank, a record of lateral platform growth and sea-level fluctuations *in* Crevello, P. D., Wilson, J. L., Sarg, J. F., and Reid, J. F., eds., Carbonate Platform and Basin Development: Tulsa, Society of Economic Paleontologists and Mineralogists, Special Publication 44, p. 339–351.

ENOS, P., 1988, Evolution of pore space in the Poza Rica trend (Mid-Cretaceous), Mexico: Sedimentology, v. 35, p. 287–325.

ERBA, E., 1990, Calcareous nannofossils biostratigraphy of some Bajocian sections from the Digne area (SE France): Memorie Descriptive Carta Geologica d'Italia, v. 40, p. 237–256.

FLÜGEL, E., 1979, Paleoecology and microfacies of Permian, Triassic and Jurassic algal communities of platform and reef carbonates from the Alps: Bulletin de Centres Recherches Exploration-Production, Elf-Aquitaine, v. 3–2, p. 291–359.

FOLLMI, K. B., WEISSERT, H., BISPING, M., AND FUNK, H., 1994, Phosphogenesis, carbon-isotopic stratigraphy, and carbonate platform evolution along the Lower Cretaceous northern Tethyan margin: Geological Society of America Bulletin, v. 106, p. 729–746.

FOUKE, B. W., EVERTS, A. W., ZWART, E. W., SCHLAGER, W., SMALLEY, P. C., AND WEISSERT, H., 1996, Subaerial exposure unconformities on the Vercors carbonate platform (SE France) and their sequence stratigraphic significance, in Howell, J. A. and Aitken, J. F., eds., High Resolution Sequence Stratigraphy: Innovations and Applications: London, Geological Society of London Special Publication No. 104, p. 295–320.

FULTHORPE, C. S., AND MELILLO, A. J., 1988, Middle Miocene carbonate gravity flows in the Straits of Florida at site 626, *in* Austin, J. A., Schlager, W., et al., eds., Proceedings of the Ocean Drilling Program, Scientific Results, v. 101: College Station, Ocean Drilling Program, p. 179–191.

GLASER, K. S., AND DROXLER, A. W., 1991, High production and highstand shedding from deeply submerged carbonate banks, northern Nicaragua Rise: Journal of Sedimentary Petrology, v. 61, p. 128–142.

GONZALEZ, L. A., AND LOHMANN, K. C, 1985, Carbon and oxygen isotopic composition of Holocene reefal carbonates: Geology, v. 13, p. 811–814.

HAAK, A. B., AND SCHLAGER, W., 1989, Compositional variations in calciturbidites due to sea-level fluctuations, late Quaternary, Bahamas: Geologische Rundschau, v. 78, p. 477–486.

HALLAM, A., 1988, A reevaluation of Jurassic eustasy in the light of new data and the revised Exxon curve, *in* Wilgus, C. K., Hastings, B. S., Ross, C. A., Posamentier, H., Van Wagoner, J., and Kendall, C. G. St. C., eds., Sea-Level Changes: An Integrated Approach: Tulsa, Society of Economic Paleontologists and Mineralogists, Special Publication 42, p. 261–273.

HALLEY, R. B., PIERSON, B. J., AND SCHLAGER, W., 1984, Alternative diagenetic models for Cretaceous talus deposits, Deep Sea Drilling Project Site 536, Gulf of Mexico, *in* Buffler, R. T., Schlager, W., et al., eds., Initial Reports of the Deep Sea Drilling Project, v. 77: Washington, D.C., United States Government Printing Office, p. 397–408.

HAQ, B. U., HARDENBOL, J., AND VAIL, P. R., 1988, Mesozoic and Cenozoic chronostratigraphy and cycles of sea-level change, *in* Wilgus, C. K., Hastings, B. S., Ross, C. A., Posamentier, H., Van Wagoner, J., and Kendall, C. G. St. C., eds., Sea-Level Changes: An Integrated Approach: Tulsa, Society of Economic Paleontologists and Mineralogists, Special Publication 42, p. 71–108.

HARLAND, W. B., COX, A. V., LLEWELLYN, P. G., PICKTON, C. A. G., SMITH, D. G., AND WALTERS, R., 1982, A geologic time scale: Cambridge, Cambridge University Press, 131 p.

HEYDARI, E., AND MOORE, C. H., 1994, Paleoceanographic and paleoclimatic controls on ooid mineralogy of the Smackover Formation, Mississippi Salt basin: Implications for Late Jurassic seawater composition: Journal of Sedimentary Petrology, v. A64, p. 101–114.

HINE, A. C., 1983, Relict sand bodies and bedforms of the Northern Bahamas: Evidence of extensive early Holocene sand transport, *in* Peryt, T. M., ed., Coated Grains: New York, Springer-Verlag, p. 116–131.

HINE, A. C., WILBER, R. J., AND NEUMANN, A. C., 1981a, Carbonate sand bodies along contrasting shallow bank margins facing open seaways in northern Bahamas: American Association of Petroleum Geologists Bulletin, v. 65, p. 261–290.

HINE, A. C., WILBER, R. J., BANE, J. M., NEUMANN, A. C., AND LORENSON, K. R., 1981b, Offbank transport of carbonate sands along open, leeward bank margins: Northern Bahamas: Marine Geology, v. 42, p. 327–348.

HOLSER, W. T., 1984, Gradual and abrupt shifts in ocean chemistry during Phanerozoic time, *in* Holland, H. D. and Trendall, A. F., eds., Patterns of Change in Earth Evolution: New York, Springer-Verlag, p. 123–143.

JENKYNS, H. C., 1988, The early Toarcian (Jurassic) anoxic event: Stratigraphic, sedimentary, and geochemical evidence: American Journal of Science, v. 288, p. 101–151.

JONES, C. E., JENKYNS, H. C., COE, A. L., AND HESSELBO, S. P., 1994a, Strontium isotopic variations in Jurassic and Cretaceous seawater: Geochimica et Cosmochimica Acta, v. 58, p. 3061–3074.

JONES, C. E., JENKYNS, H. C., AND HESSELBO, S. P., 1994b, Strontium isotopes in Early Jurassic seawater: Geochimica et Cosmochimica Acta, v. 58, p. 1285–1301.

KAUFMAN, A. J., JACOBSEN, S. B., AND KNOLL, A. H., 1993, The Vendian record of Sr and C isotopic variations in seawater: Implications for tectonics and paleoclimate: Earth and Planetary Science Letters, v. 120, p. 409–430.

KENNETT, J. P., 1986, Miocene to early Pliocene oxygen and carbon isotope stratigraphy in the southwest Pacific, DSDP Leg 90, *in* Kennett, J. P., von der Borch, C. C., et al., Initial Reports of the Deep Sea Drilling Project, v. 90: Washington, D.C., United States Goverment Printing Office, p. 1388–1412.

LAND, L. S., BEHRENS, E. W., AND FRISCHMAN, S. A., 1979, The ooids of Baffin Bay, Texas: Journal of Sedimentary Petrology, v. 49, p. 1269–1278.

LLOYD, R. M., PERKINS, R. D., AND KERR, S. D., 1987, Beach and shoreface ooid deposition on shallow interior banks, Turks and Caicos Islands, British West Indies: Journal of Sedimentary Petrology, v. 52, p. 976–982.

LOHMANN, K. C., AND MEYERS, W. J., 1977, Microdolomite inclusions in cloudy prismatic calcites: A proposed criterion for former high-magnesium calcites: Journal of Sedimentary Petrology, v. 47, p. 1078–1088.

MARSHALL, J. D., AND ASHTON, M., 1980, Isotopic and trace element evidence for submarine lithification of hardgrounds in the Jurassic of eastern England: Sedimentology, v. 27, p. 271–291.

MASETTI, D., 1971, Sedimentologia e Paleogeografia del Giurassico tra Brenta e Piave: Unpublished thesis, University of Ferrara, Italy, 104 p.

MEDWEDEFF, D. A., AND WILKINSON, B. H., 1983, Cortical fabrics in calcite and aragonite ooids, *in* Peryt, T. M., ed., Coated Grains: New York, Springer-Verlag, p. 109–115.

MELIM, L. A., SWART, P. K., AND MALIVA, R. G., 1995, Meteoric-like fabrics forming in marine waters: Implications for the use of petrography to identify diagenetic environments: Geology, v. 23, p. 755–758.

MONTANEZ, I. P., BANNER, J. L., OSLEGER, D. A., BORG, L. E., AND BOSSERMAN, P. J., 1996, Integrated Sr isotope variations and sea-level history of Middle to Upper Cambrian platform carbonates: Implications for the evolution of Cambrian seawater $^{87}Sr/^{86}Sr$: Geology, v. 24, p. 917–920.

MULLINS, H. T., AND COOK, H. E., 1986, Carbonate apron models: Alternatives to the submarine fan model for paleoenvironmental analysis and hydrocarbon exploration: Sedimentary Geology, v. 48, p. 37–79.

MULLINS, H. T., HEATH, K. C., VAN BUREN, H. M., AND NEWTON, C. R., 1984, Anatomy of a modern open-ocean carbonate slope: Northern Little Bahama Bank: Sedimentology, v. 31, p. 141–168.

MUTTI, E., AND RICCI LUCCHI, F., 1972, Le torbiditi dell'Appenino settentrionale: Introduzione all, analisi di facies: Memorie della Società Geologica Italiana, v. 11, p. 161–199.

OSLICK, J. S., MILLER, K. G., AND FEIGENSON, M. D., 1994, Oligocene-Miocene strontium isotopes: Stratigraphic revisions and correlations to an inferred glacioeustatic record: Paleoceanography, v. 9, p. 427–443.

POPP, B. N., AND WILKINSON, B. H., 1983, Holocene lacustrine ooids from Pyramid Lake, Nevada, in Peryt, T. M., ed., Coated Grains: Berlin, Springer-Verlag, p. 143–153.

REIJMER, J. J. G., 1991, Sea level and sedimentation on the flanks of carbonate platforms: Unpublished Ph.D. dissertation, Vrije Universiteit, Amersterdam, 162 p.

RICH, M., 1982, Ooid cortices composed of neomorphic pseudospar: Possible evidence for ancient originally aragonitic ooids: Journal of Sedimentary Petrology, v. 52, p. 843–847.

RUIZ-ORTIZ, P. A., 1983, A carbonate submarine fan in a fault-controlled basin of the Upper Jurassic, Betic Cordillera, southern Spain: Sedimentology, v. 30, p. 33–48.

RUPPEL, S. C., JAMES, E. W., BARRICK, J. E., NOWLAN, G., AND UYENO, T. T., 1996, High-resolution $^{87}Sr/^{86}Sr$ chemostratigraphy of the Silurian: Implications for event correlation and strontium flux: Geology, v. 24, p. 831–834.

SANDBERG, P. A., 1985, Aragonite cements and their occurrence in ancient limestones, *in* Schneidermann, N. and Harris, P. M., eds., Carbonate Cements: Tulsa, Society of Economic Paleontologists and Mineralogists, Special Publication 36, p. 33–57.

SARG, J. F., 1988, Carbonate sequence stratigraphy, *in* Wilgus, C. K., Hastings, B. S., Ross, C. A., Posamentier, H., Van Wagoner, J., and Kendall, C. G. St. C., eds., Sea-Level Changes: An Integrated Approach: Tulsa, Society of Economic Paleontologists and Mineralogists, Special Publication 42, p. 155–181.

SCHLAGER, W., 1980, Mesozoic calciturbidites in DSDP hole 416A: recognition of a drowned carbonate platform, *in* Lancelot, Y. and Winterer, E. L., et al., eds., Initial Reports of the Deep Sea Drilling Project, v. 50: Washington, D.C., United States Government Printing Office, p. 733–749.

SCHLAGER, W., 1991, Depositional bias and environmental change—Important factors in sequence stratigraphy: Sedimentary Geology, v. 70, p. 109–130.

SCHLAGER, W., AND CAMBER, O., 1986, Submarine slope angles, drowning unconformities, and self-erosion of limestone escarpments: Geology, v. 14, p. 762–765.

SCHLAGER, W., AND GINSBURG, R. N., 1981, Bahama carbonate platforms—The deep and the past: Marine Geology, v. 44, p. 1–24.

SCHLAGER, W., AND JAMES, N. P., 1978, Low-magnesian calcite limestones form-

forming at the deep-sea floor, Tongue of the Ocean, Bahamas: Sedimentology, v. 25, p. 675–702.

SCHLAGER, W., REIJMER, J. J. G., AND DROXLER, A., 1994, Highstand shedding of carbonate platforms: Journal of Sedimentary Petrology, v. B64, p. 270–281.

SHACKLETON, N. J., 1987, The carbon isotope record of the Cenozoic: History of organic carbon burial in the ocean and atmosphere, *in* Brooks, J. and Fleet, A. J., eds., Marine Petroleum Source Rocks: London, Geological Society of London Special Publication No. 26, p. 319–322.

SHANMUGAM, G., AND MOIOLA, R. J., 1984, Eustatic control of calciclastic turbidites: Marine Geology, v. 56, p. 273–278.

TUCKER, M. E., 1984, Calcitic, aragonitic, and mixed calcitic-aragonitic ooids from the Belt Supergroup: Sedimentology, v. 31, p. 627–644.

VAHRENKAMP, V. C., 1996, Carbon isotope stratigraphy of the Upper Kharaib and Shuaiba Formations: Implications for the Early Cretaceous evolution of the Arabian Gulf region: American Association of Petroleum Geologists Bulletin, v. 80, p. 647–662.

VAN DER PLAS, L., AND TOBI, A. C., 1965, A chart for judging the reliability of point counting results: American Journal of Science, v. 263, p. 87–90.

WALTZ, M. D., 1988, The evolution of shallowing-upwards reef to oolite sequences at the leeward margin of Caicos platform, British West Indies: Unpublished M.Sc. thesis, University of Miami, Coral Gables, Florida, 95 p.

WANLESS, H. R., TEDESCO, L. P., ROSSINSKY, V., AND DRAVIS, J. J., 1989, Carbonate environments and sequences of Caicos platform: Washington, D.C., American Geophysical Union, International Geological Congress Field Trip Guidebook T374, 75 p.

WATTS, K. F., 1987, Triassic carbonate submarine fans along the Arabian platform margin, Sumeini Group, Oman: Sedimentology, v. 34, p. 43–71.

WEISSERT, H. J., AND BERNOULLI, D., 1985, A transform margin in the Mesozoic Tethys: Evidence from the Swiss Alps: Geologische Rundschau, v. 73, p. 665–679.

WENDT, J., 1984, Skeletal and spicular mineralogy, microstructure and diagenesis of coralline calcareous sponges: Paleontographica Americana, no. 54, p. 326–336.

WILKINSON, B. H., AND LANDING, E., 1978, "Eggshell diagenesis" and primary radial fabric in calcite ooids: Journal of Sedimentary Petrology, v. 48, p. 1129–1138.

WILKINSON, B. H., BUCZYNSKI, C., AND OWEN, R. M., 1984, Chemical control of carbonate phases: Implications from Upper Pennsylvanian calcite-aragonite ooids of southeastern Kansas: Journal of Sedimentary Petrology, v. 54, p. 932–947.

YOSE, L. A., AND HELLER, P. L., 1989, Sea-level control of mixed-carbonate-siliciclastic, gravity-flow deposition: Lower part of the Keeler Canyon Formation (Pennsylvanian), southeastern California: Geological Society of America Bulletin, v. 101, p. 427–439.

ZEMPOLICH, W. G., 1993, The drowning succession in Jurassic carbonates of the Venetian Alps, Italy: A record of supercontinent breakup, gradual eustatic rise, and eutrophication of shallow-water environments, *in* Loucks, R. G. and Sarg, J. F., eds., Carbonate Sequence Stratigraphy: Recent Developments and Applications: Tulsa, American Association of Petroleum Geologists Memoir 57, p. 63–105.

ZEMPOLICH, W. G., 1995, Deposition, early diagenesis, and late dolomitization of deepwater resedimented oolite: The Middle Jurassic Vajont Limestone of the Venetian Alps, Italy: Unpublished Ph.D. thesis, Johns Hopkins University, Baltimore, Maryland, 659 p.

ZEMPOLICH, W. G., AND ERBA, E., 1997, Sedimentologic and chemo-stratigraphic recognition of 3rd-order sequences in resedimented carbonate: The Middle Jurassic Vajont Limestone, Venetian Alps, Italy: Tulsa, American Association of Petroleum Geologists, Annual Convention, Abstracts, p. 130.

ZEMPOLICH, W. G., AND HARDIE, L. A., 1997, Geometry of dolomite bodies within deep-water resedimented oolite of the Middle Jurassic Vajont Limestone, Italy: Analogs for hydrocarbon reservoirs created through fault-related burial dolomitization, *in* Kupecz, J. A., Gluyas, J., and Bloch, S., eds., Reservoir quality prediction in sandstones and carbonates: Tulsa, American Association of Petroleum Geologists Memoir 69, p. 127–162.

ZEMPOLICH, W. G., WILKINSON, B. H., AND LOHMANN, K. C, 1988, Diagenesis of late Proterozoic carbonates: The Beck Spring Dolomite of eastern California: Journal of Sedimentary Petrology, v. 55, p.171–183.

LOWER AND MIDDLE MIOCENE CARBONATE FACIES IN THE CENTRAL MEDITERRANEAN: THE IMPACT OF PALEOCEANOGRAPHY ON SEQUENCE STRATIGRAPHY

MARIA MUTTI

Earth Science Department, University of Southern California, Los Angeles, California 90889-0740

DANIEL BERNOULLI

Geological Institute, ETH-Zentrum, CH-8092 Zürich, Switzerland

SILVIA SPEZZAFERRI

Geological Institute, ETH-Zentrum, CH-8092 Zürich, Switzerland

AND

PETER STILLE

Centre de la Géochimie de la Surface, CNRS, Rue Blessig 1, F-67084 Strasbourg, France

ABSTRACT: In the central Mediterranean area, a major, second-order transgressive event spanning from the Burdigalian to the Serravallian is recorded by the widespread deposition of deeper-water facies on carbonate shelves and is matched to a large extent by changes in carbonate facies, including the sudden disappearance of corals and coral reefs and the corresponding rise in the abundance of temperate and cool-water carbonate facies. Two sections from the Maiella platform margin in the southern Apennines and the Hyblean Plateau in Sicily provide us with time series based on oxygen and carbon isotopes and detailed biostratigraphic and strontium isotope ages of unprecedented resolution to compare these facies changes to the global Neogene cooling trends and to the paleoceanographic conditions existing at the time in the Mediterranean area.

An evaluation of the timing of facies change indicates that the major cooling interval in the Neogene, between 14 and 12 Ma, postdates the change from tropical to temperate carbonate facies in the Mediterranean, which began at approximately 20 Ma. Our oxygen isotope record from the Hyblean Plateau suggests relatively cool temperatures during the Serravallian (within biozones N10–N12) with progressive warming within biozone N11 (11.9–11.8 Ma). This signature is opposite to the one recorded in pelagic sections in other locations of the world and is interpreted to reflect regional conditions within the Mediterranean, as after 20 Ma, the connection to the Indian Ocean became closed or reduced to a very shallow sill, which prevented the outflow and inflow of intermediate and deeper waters.

The second-order transgressive event in the central Mediterranean coincides with an increase in the $\delta^{13}C$ values of skeletal carbonates. These isotope data, combined with biotic evidence, suggest increased productivity of surface waters during the transgressive event. The onset of this event predates by approximately 3 m.y. the major positive carbon isotope excursion recorded worldwide in pelagic sections (the Monterey event); its termination coincides with the end of the Monterey excursion.

Our data indicate how variations in water temperatures, coupled with drastic changes in water circulation, severely affected rates of carbonate production during the deposition of a second-order sequence and had the effect of accentuating the transgressive trend within the sequences. Rates of carbonate production reached a minimum during the transgressive and early-highstand systems tracts. This low production was not linked to a decrease in the size of the carbonate-producing area along the depositional ramp but rather to changes in production rates due to environmental change. A better understanding of the response of carbonate facies to paleoceanographic changes needs to be developed, so that these changes can be used to predict sequence stratigraphic architecture in a given time slice and to better estimate the reliability of the sea-level signature preserved in carbonate sequences.

INTRODUCTION

Sequence stratigraphy is considered a practical methodology for analyzing the development and history of carbonate platforms (Sarg, 1988; Eberli and Ginsburg, 1989; Handford and Loucks, 1993). Carbonate platforms are constructed and modified by depositional and erosional processes, acting under the controls exerted by eustasy, subsidence, sediment production, and climate (Sarg, 1988; Handford and Loucks, 1993). The first two variables interact to regulate relative sea-level changes, whereas climate mainly controls the type and amount of sediment that is being produced.

Intense debate has been generated in recent years regarding the accuracy and reliability of sea-level changes interpreted from sequence stratigraphic analysis in carbonate environments (e.g., Schlager, 1991, 1992). A fundamental control on sequence stratigraphic architecture is the relationship between accommodation space and rates of sediment production and accumulation. Rates of shallow-water carbonate production are dependent on a number of factors and change through time. Handford and Loucks (1993) have attempted to integrate into sequence stratigraphic models the variations in carbonate production and accumulation across the depositional shelf. However, the concept that variations in carbonate production can occur through time in similar shelf locations and the impact that such changes may have on sequence stratigraphic models have never been explored in detail.

In this paper we address how climatic and paleoceanographic conditions during the Miocene affected changes in rates of carbonate production, ultimately leaving an impact on the sequence stratigraphic packaging of carbonate strata. In the central and western Mediterranean area, a major, second-order transgressive event spanning from the Burdigalian to the Serravallian is recorded by the widespread deposition of deeper-water facies on carbonate shelves and has been identified at a regional scale (e.g., Pedley and Bennett, 1985; Simone and Carannante, 1988;

Advances in Carbonate Sequence Stratigraphy: Application to Reservoirs, Outcrops and Models, SEPM Special Publication No. 63

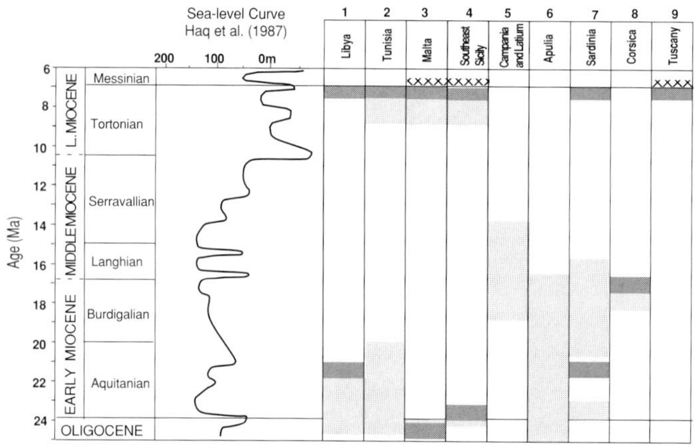

FIG. 1.—Stratigraphic distribution of specific carbonate-facies associations in the central Mediterranean region during the Miocene and their relationship to the sea-level curve of Haq et al. (1987, 1988). Light gray intervals represent red algal and bryozoan limestones; dark gray intervals represent *Porites*-dominated reefs, and crossed intervals represent stromatolites. Modified from Pedley (1996).

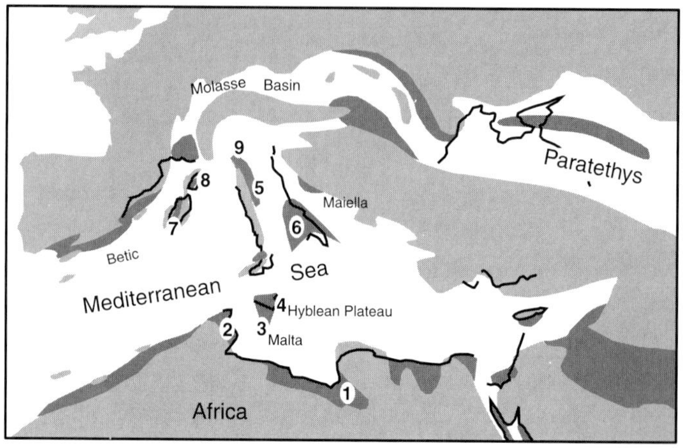

FIG. 2.—Paleogeographic map of the Mediterranean region during the early Miocene. Palinspastics are from Rögl and Steininger (1983). Continental areas are in gray, and shallow-water carbonate provinces are in dark gray (after Esteban, 1996). The numbers refer to the areas in Figure 1.

Grasso et al., 1994; Jacobs et al., 1996). We point out the association between this second-order relative sea-level event and carbonate-facies changes in the Mediterranean area during this time slice. This event is matched to a large extent by the sudden disappearance of corals and coral reefs and an increase in temperate and cool-water carbonate facies (Fig. 1).

The aim of this paper is to test the relationships of facies changes to global Neogene cooling trends and to the paleoceanographic conditions existing at that time in the Mediterranean area. We use two case studies of Miocene carbonates from the central Mediterranean area: the Maiella platform margin in the southern Apennines and the Hyblean Plateau in Sicily (Fig. 2). We use a multidisciplinary approach, integrating field-based stratigraphic data, biostratigraphic data, and carbon, oxygen, and strontium isotopes. In this paper, we show how changes in accumulation rates due to the switch from tropical to temperate carbonate assemblages might have contributed to amplify this second-order transgressive trend. This process is particularly important during formation of the transgressive systems tract when regionally decreased carbonate production rates combined with a rapid sea-level rise would result in a depositional architecture that mimics a sea-level rise of much larger extent.

MIOCENE CARBONATES IN THE CENTRAL MEDITERRANEAN AREA

Upper Miocene carbonate facies in the Mediterranean have attracted paleontologists and carbonate sedimentologists since the end of the last century. Starting in the 1970s, reef-building corals in upper Miocene facies were studied with emphasis on their significance in relationship to the Messinian evaporitic events. Work continued over the years, describing facies and depositional geometries (see Esteban, 1979, for a summary). Recently, the application of sequence stratigraphic concepts, modern petrographic techniques, and detailed diagenetic studies has significantly advanced our understanding of upper Miocene reefs in the western and central Mediterranean (Esteban, 1996; Franseen et al., 1996).

In contrast, lower and middle Miocene carbonate facies have received much less attention. An outstanding exception is the work by Carannante et al. (1988). The uppermost lower Miocene and the middle Miocene carbonate facies in the Mediterranean region comprise foraminiferal, red algal, and bryozoan limestones deposited in ramp settings, and coral-bearing associations are rare. Bryozoan-dominated ramps are typical of temperate to cool-water settings (James and Bone, 1991). Regional chronostratigraphic correlations are still uncertain, but available data indicate some temporal patterns in facies distribution (Fig. 1). Few coral reefs containing *Porites* are found in the upper Oligocene and in the lowermost Miocene (Fig. 1), and they seem to be geographically confined to the western central Mediterranean (Pedley, 1996). However, at ODP Sites 965 and 966 in the eastern Mediterranean, carbonate strata containing *Porites* were recovered and dated as earliest Miocene based on the occurrence of *Lepidocyclina* and *Miogypsina* (larger benthic foraminifers) (Emeis et al., 1996). Carbonate facies containing *Porites* or any other tropical assemblage have never been documented for the uppermost lower Miocene or the middle Miocene, whereas they again became common in upper Miocene strata (Esteban, 1996).

This temporal facies trend has been interpreted as a reflection of the global cooling trend and the northward displacement of the European plate. In particular, Esteban (1996) suggested that this facies trend reflects the transition between early Miocene open-oceanic, humid-tropical conditions and late Miocene landlocked, semiarid and subtropical environments. However, it is a paradox that global cooling alone caused this facies trend, as tropical facies are again present in the upper Miocene.

During the Miocene, the Mediterranean underwent a drastic change in paleogeography that may have had a major impact on climate, patterns of current circulation, and consequently on carbonate deposition. In fact, during the early Miocene, the connection to the Indian Ocean became closed, and the Mediterranean achieved a paleogeographic configuration similar to today's (Fig. 2).

METHODS

Samples for stable isotope analyses were generally taken at 1–2-m intervals, but sampling as close as 0.1-m intervals prevailed through some sections. Samples were reacted by using the common acid-bath method at 90 °C, and the CO_2 produced was analyzed by using a PRISM mass spectrometer. All samples analyzed for stable isotopes consisted of bulk carbonates, as the lithologies are too indurated and generally too fine grained to allow separation or sampling of different components through microdrilling.

Precise dating of the carbonate facies is essential to contrast timing of facies changes with the climate record inferred from deep-ocean strata. To overcome the poor biostratigraphic resolution typical of shallow-water carbonates, a carbon isotope stratigraphy has been generated and calibrated with the strontium isotope stratigraphy. The latest Oligocene–middle Miocene is a good interval to date marine sedimentary rocks by using strontium isotopes because the $^{87}Sr/^{86}Sr$ ratio in seawater increased steadily throughout this time (De Paolo and Finger, 1991; Hodell and Woodruff, 1994). Strontium isotopes were analyzed according to the procedures described in Stille et al. (1994) and Jacobs et al. (1996). In order to compare our data directly with the seawater curve of Hodell et al. (1991), we normalized our data to their measured $^{87}Sr/^{86}Sr$ ratio of 0.710235 for the NBS 987 standard. The corresponding ages were calculated by using their isotopic composition values and the regression equations developed by Hodell (1991) for the $^{87}Sr/^{86}Sr$ ratio of seawater as a function of age.

Total carbon content was determined according to standard procedures with a Carlo Erba elemental analyzer, and the total inorganic carbon (TIC) was measured with a Coulomat, both at EAWAG/ETH in Zurich. The total organic carbon (TOC) was determined by subtracting TIC from the total carbon.

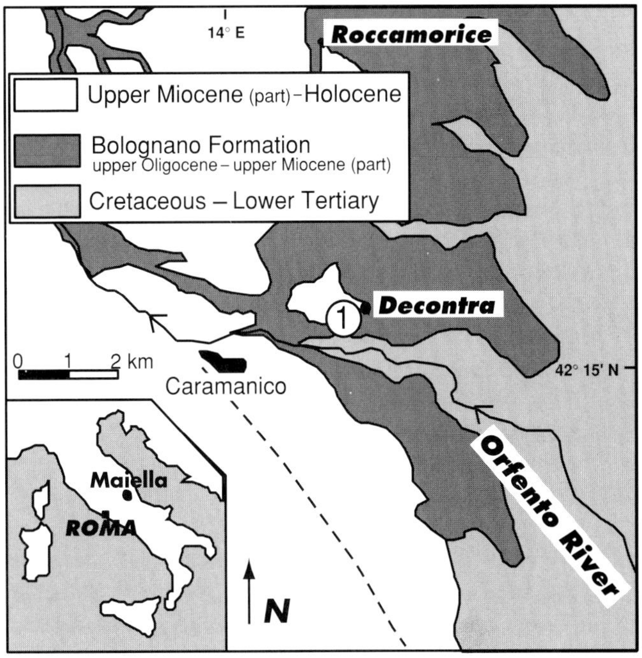

FIG. 3.—Geologic map of the northern Maiella area and location (labeled 1) of the section in the Bolognano Formation near Decontra.

THE MAIELLA PLATFORM MARGIN

In the southern Apennines, Miocene carbonate strata unconformably overlie tectonically tilted Cretaceous and lower Tertiary platform limestones, with the age of the transgression

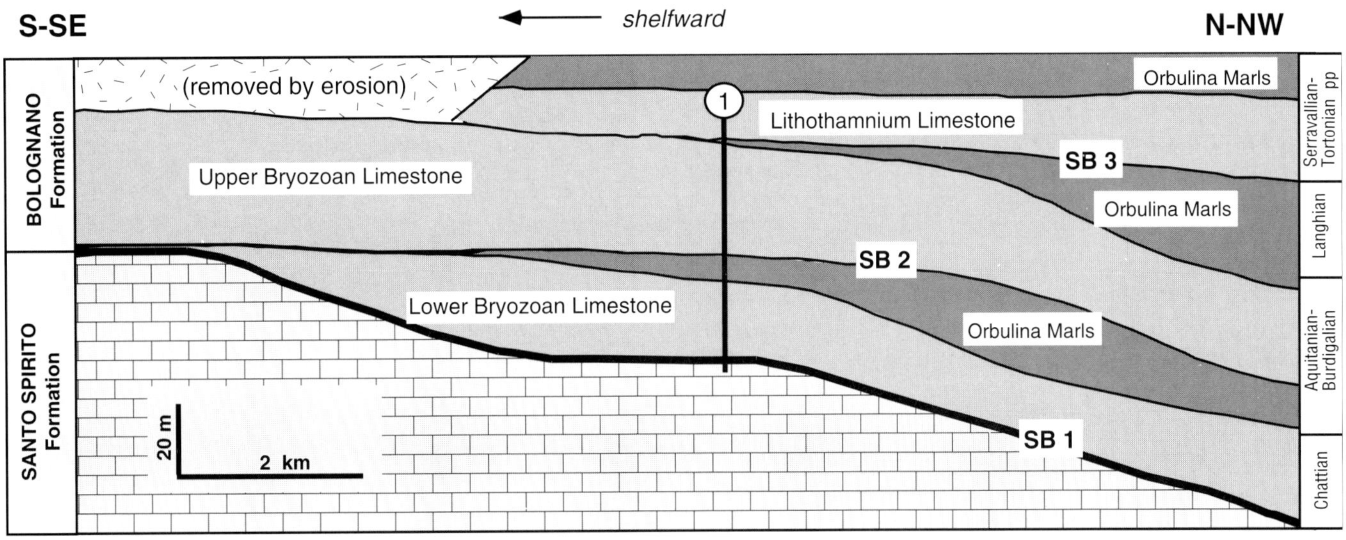

FIG. 4.—Shallow-water Miocene carbonates constitute the Bolognano Formation, which is subdivided into several informal members (Crescenti et al., 1969; Vecsei, 1991). The boundaries (SB = sequence boundaries) between these units are either erosional and associated with a hiatus or stratigraphically continuous and expressed by transitional facies changes. The bar labeled 1 indicates the location of the stratigraphic section illustrated in Figure 5. Only part of the Serravallian–Tortonian interval is represented in the cross section (as indicated by "pp"—per parte).

ranging between the latest Oligocene and the Serravallian (Carannante et al., 1988). Along the Maiella platform margin (Fig. 3), Miocene carbonate strata formed the last phase of deposition on a long-lived carbonate platform that persisted since the Jurassic. The platform was terminated by the late Miocene salinity crisis and ultimately became involved in the thrusting related to the Apennine orogeny during the Pliocene (Bernoulli et al., 1992).

Shallow-water Miocene carbonates constitute the Bolognano Formation (Crescenti et al., 1969; Vecsei, 1991), which is subdivided into several informal members (Fig. 4). Large-scale architecture of the sedimentary rocks reflects a second-order eustatic cycle. The Bolognano Formation contains four depositional sequences (Bernoulli et al., 1992). Only the first three sequences will be discussed in this paper. Each sequence is wedge shaped in cross section and consists of a lower, shallow-water unit with bioclastic grainstones and rudstones and an upper, deeper-water unit with hemipelagic wackestones-packstones (Vecsei, 1991).

Depositional facies associations and lithologies record three

TABLE 1.—STRONTIUM ISOTOPE DATA

Sample No.	Depth (m)	$^{87}Sr/^{86}Sr$ (Error $\times$ 10^{-6})	Age After Correction to NBS 987 Standard (Ma)
9439	14.5	0.708126 (6)	26.51 ± 0.74
9455	34.5	0.708142 (6)	26.24 ± 0.74
D11	48.6	0.708245 (8)	24.54 ± 0.74
D21 H	49.9	0.798457 (7)	21.04 ± 0.74
D21 L	50	0.708500 (6)	20.33 ± 0.74
9460	60	0.708639 (7)	18.03 ± 0.74
9476	76	0.708686 (7)	17.26 ± 0.74
9489	89	0.708719 (6)	16.71 ± 0.74

NOTE: The strontium isotope ages are reported in Figure 5; they were calculated by using the regression equations that were developed by Hodell (1991) for the $^{87}Sr/^{86}Sr$ isotope ratio of seawater as a function of age.

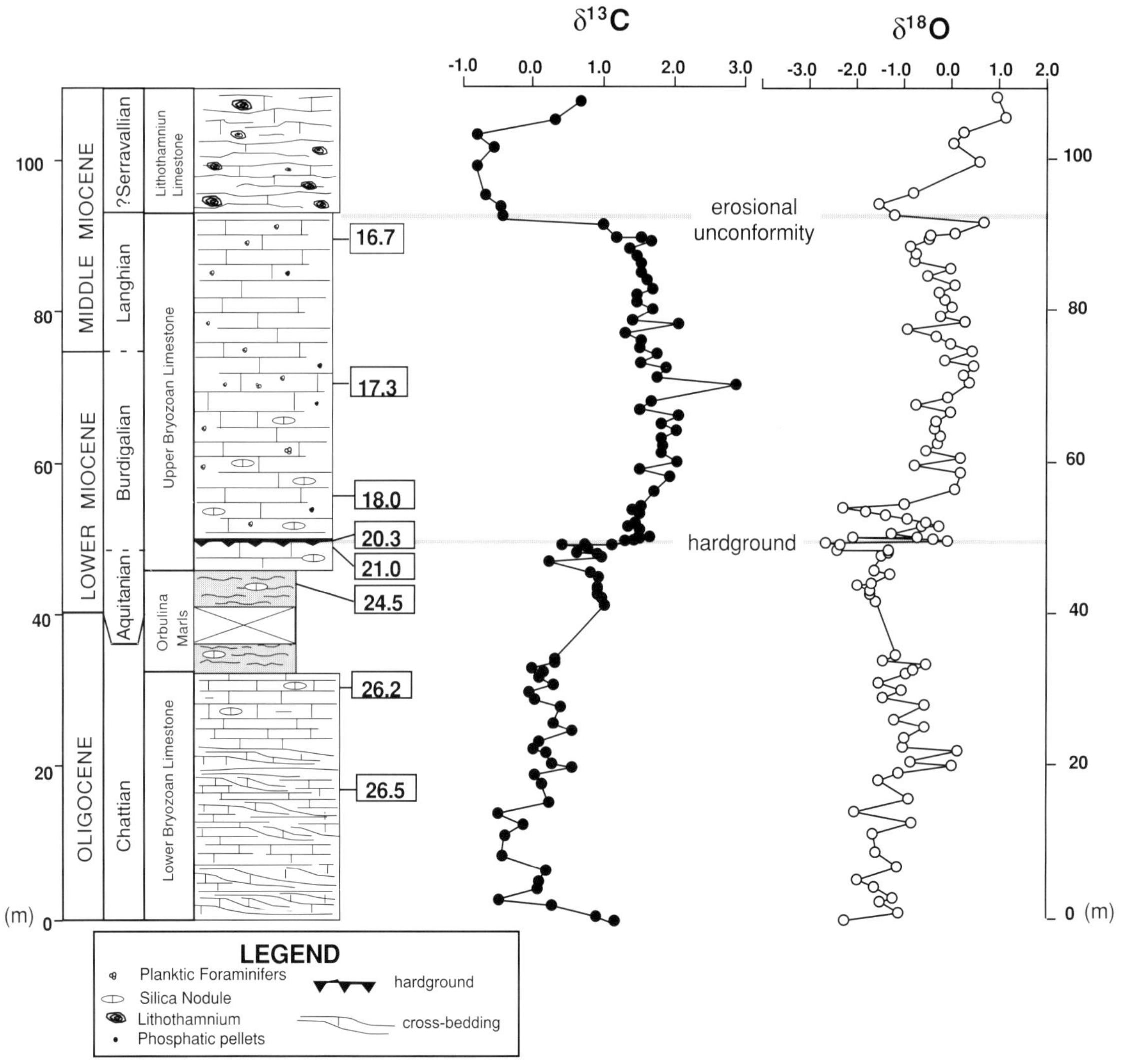

FIG. 5.—The Decontra section in the Bolognano Formation (see Fig. 4). The ages have been assigned on the basis of strontium isotope ratios and age-diagnostic microfossils; they are indicated by boxed numbers (in Ma) next to the stratigraphic profile. To the right are the $\delta^{13}C$ and $\delta^{18}O$ values from bulk samples. Note that the contact between the Upper Bryozoan Limestone and the Lithothamnium Limestone is an unconformity.

major phases of carbonate deposition on a temperate-water, low-angle ramp, interrupted by two episodes of deeper-water sedimentation. Strontium isotope data (Table 1) have allowed us to generate a stratigraphy of exceptionally high resolution for this type of setting (Fig. 5). The first shallow-water sequence, the Lower Bryozoan Limestone (uppermost Oligocene to Aquitanian), unconformably overlies deep-water deposits of Eocene age. The Lower Bryozoan Limestone is composed of bioclastic grainstone and rudstone dominated by larger benthic foraminifera, bryozoan fragments, red algae, and echinoid and mollusk fragments. Numerous and diversified larger benthic foraminifera indicate relatively warm-water conditions. This phase of deposition started before 26.5 Ma and lasted to 26.2 Ma (Fig. 5). Average accumulation rates calculated for this interval are approximately 6 cm/k.y. and fall into the normal range for temperate carbonates (Nelson, 1978; James and Bone, 1991). The Lower Bryozoan Limestone shifts relatively abruptly from neritic to the hemipelagic facies of the Orbulina Marls. The hemipelagic facies consists of strongly bioturbated, planktonic foraminifer-rich wackestones with minor radiolarians and siliceous sponge spicules. The abundance of biogenic silica in the strata and the presence of phosphate filling in foraminiferal chambers, as well as the predominantly small size of the planktonic foraminifera (Lipps, 1979), all point to increased surface-water productivity during the transgressive part of this first depositional sequence.

The second sequence (late Aquitanian to middle Langhian) consists of the Upper Bryozoan Limestone and a second unit of Orbulina Marls. The gradational base in the Decontra section (located in Fig. 3) can be traced updip along the depositional profile into an erosional unconformity, where the Upper Bryozoan Limestone unconformably overlies the Orbulina Marls (Fig. 4). A 1.5-m-thick, cross-bedded sand wave at the base of the sequence is capped by a 0.2-m-thick phosphatic hardground. The hardground has been dated at 20.3–21.0 Ma with strontium isotope stratigraphy. The rock is composed mainly of echinoid fragments, bryozoa, red algae, and subordinate benthic foraminifera. Reduction of both abundance and diversity of larger benthic foraminifera, as well as their replacement by smaller, cooler-water forms, suggests deposition under cooler-water conditions with respect to the underlying Lower Bryozoan Limestone. Above the hardground, a monotonous succession of decimeter-thick, parallel grainstone beds with low-angle cross-laminations forms most of the unit, suggesting the action of strong currents. These strata contain variable amounts of planktonic foraminifera and fragments of benthic biota, such as mollusks, bryozoa, and echinoids. Larger benthic foraminifera are very rare, and generally planktonic foraminifera, commonly filled by micritic cements, form more than 50% of the rock. In the strata of the Maiella platform margin, this unit corresponds to the most landward extent of deeper-shelf facies, at the transgressive peak of the second-order cycle.

Unconformably overlying the Upper Bryozoan Limestone is the third sequence: the Lithothamnium Limestone. The age of this sequence is still ambiguous, because biostratigraphic data are poor and strontium isotope ages are inconsistent, possibly because of reworking of older material. The Lithothamnium Limestone consists of floatstone to bindstone and contains coralline and other red algae, rhodolites, bivalves, and *Heterostegina*. Rhodolites are formed by the interlayering of coralline algae and bryozoan crusts with encrusting foraminifera. Deposition of this unit records renewed benthic carbonate production in a shallow-water depositional environment.

THE DECONTRA SECTION

A continuous, 110-m-thick section was measured through the Bolognano Formation at Decontra, a small village on the western side of the Maiella platform margin (Fig. 3). The Decontra section has been analyzed in detail for depositional and lithological facies variations, strontium isotopes, and stable isotope stratigraphy (Fig. 5).

The bulk-rock carbon isotope composition measured throughout the platform section varies significantly, from –0.9‰ to +2.7‰, with several notable excursions (Fig. 5). The first $\delta^{13}C$ shift is a gradual increase upward through the Orbulina Marls from 0.00 ± 0.40 to +0.70 ± 0.3‰. A second shift to +1.60 ± 0.3‰ occurs in the phosphatic hardground in the lower part of the Upper Bryozoan Limestone, and finally $\delta^{13}C$ decreases abruptly to (0.2‰ ± 0.7‰ at the unconformable boundary between the Upper Bryozoan Limestone and the Lithothamnium Limestone.

The bulk oxygen composition of the Lower Bryozoan Limestone ranges from –2.0‰ to +0.5‰, with a tendency toward higher values up section that ultimately peak 20 m above the base of the Decontra section. The interval spanning from the Orbulina Marls/Upper Bryozoan Limestone boundary to a few meters above the phosphatic hardground shows the widest range of $\delta^{18}O$ values, ranging from –3.0‰ to +0.5‰. The middle and upper part of the Upper Bryozoan Limestone has a fairly homogeneous oxygen isotope composition, ranging from –1.0‰ to +0.5‰. The Lithothamnium Limestone, resting above an unconformable boundary, shows a marked trend toward higher $\delta^{18}O$ values, which peak at +1.0‰.

Several lines of evidence suggest that the carbon isotope values do not reflect diagenetic modifications. The absence of subaerial-exposure horizons throughout the section reduces the probability of a postdepositional diagenetic alteration of the primary marine signature. In addition, sedimentological criteria indicate deposition on a relatively deep shelf, with a water depth of between 30 and 200 m. This setting rules out the possibility of an environment-related depletion in ^{13}C (Patterson and Walter, 1994). Therefore, we can assume that the $\delta^{13}C$ values reflect a primary, marine signature and changes in the $\delta^{13}C$ of the ambient dissolved inorganic carbon (DIC) at the time of carbonate precipitation. However, because of the presence in varying relative proportions of different biotic components with different isotopic fractionations, the observed $\delta^{13}C$ changes must be interpreted with caution. Within the Upper Bryozoan Limestone, characterized by a relatively uniform proportion of different biota, we observe a marked increase of $\delta^{13}C$, peaking at +2.5‰. The strontium isotope ages indicate that this shift occurs between approximately 21 and 16.5 Ma, when the record is interrupted by an unconformity.

The wide range of oxygen values in the lower part of the Upper Bryozoan Limestone (the interval between 45 and 55 m in stratigraphic depth, Fig. 5) reflects submarine diagenetic processes related to hardground formation and to later diagenetic events. In fact, as syndepositional cementation inhibited mechanical compaction, pore space was still available for further cementation (Mutti and Bernoulli, unpublished data). With the exception of this interval, syndepositional cements are absent, and cements related to postdepositional diagenetic processes are minimal, as mechanical compaction virtually eliminated any available pore space before cementation could take place. The measured oxygen values therefore reflect the composition of different proportions of bioclasts.

THE HYBLEAN PLATEAU

The eastern Hyblean Plateau of southeastern Sicily (see Fig. 6) is a wide upland area dominated by Miocene carbonates. The major part of the exposed succession in Sicily, the Palazzolo Formation, was dated as Tortonian in early studies (e.g., Cita, 1959), and later work has since subdivided the succession into the lithostratigraphic units shown in Figure 7 (Pedley, 1981, 1983; Grasso and Lentini, 1982).

Miocene carbonates in the eastern Hyblean Plateau preserve a nearly undeformed platform-to-basin transect approximately 40 km long. Overall, the large-scale architecture of the sequence reflects a second-order eustatic cycle. Limestones of the Ragusa Formation (upper Oligocene to lower Miocene) record deposition on a carbonate ramp. The Irminio Member (lower Miocene) consists of grainstones to packstones, deposited as cross-bedded sand waves that gradually pass up section into outer-shelf marls with rare hummocky beds (Mutti, unpublished data, 1995). The limestones contain bryozoa, larger foraminifera, and red algae (Grasso et al., 1979). A phosphatic hardground in the upper part of the Irminio Member has been dated as late Burdigalian on the basis of planktonic foraminifera in the upper part of the hardground (Grasso et al., 1994). A similar hardground is found in Malta in a comparable stratigraphic position and has been dated with strontium isotopes at 20 Ma (Jacobs et al., 1996). Jacobs et al. (1996) showed in their study in the Maltese Islands that the Malta-Ragusa platform underwent a "stepwise deepening" starting at 25 Ma and lasting until 15 Ma.

The Ragusa Formation is overlain by the Tellaro Formation, a transgressive package of packstone to lime mudstone strata that is rich in planktonic foraminifera. The Tellaro Formation is subdivided into the Giarratana Marl and Castelluccio Marl members. The Giarratana Marl Member is formed by plankton-rich packstone strata, whereas the Castelluccio Marl is a clay-rich, carbonate-poor hemipelagic unit, interpreted to reflect the maximum flooding within the transgressive basinal facies (Pedley, 1981). On the basis of field relationships, Pedley (1981) interpreted the Tellaro Formation as a wedge that terminated on the proximal shelf.

The proximal part of the succession consists of the coarse bioclastic strata of the Monti Climiti Formation, which includes the Melilli Member (Aquitanian) and the Siracusa Limestone Member (upper Serravallian–Tortonian). Burdigalian or Langhian facies have not been described in the literature, and this feature should be better investigated. A major subaerial-exposure surface is found at the base of the Siracusa Limestone Member (Mutti, unpublished data, 1995), but the duration of the hiatus is unknown. Facies associations and transitions along exposed valley walls suggest that the Siracusa Limestone Member is the shallow-water equivalent of, or at least part of, the outer-shelf carbonates of the Palazzolo Formation (Mutti, unpublished data, 1995). In addition, biostratigraphic data document a late Serravallian–Tortonian age for both units (Grasso and Lentini, 1982). The Siracusa Limestone Member contains abundant red algal facies and bears echinoids as well as isolated *Porites* corals in its uppermost 5 m (Pedley, 1981).

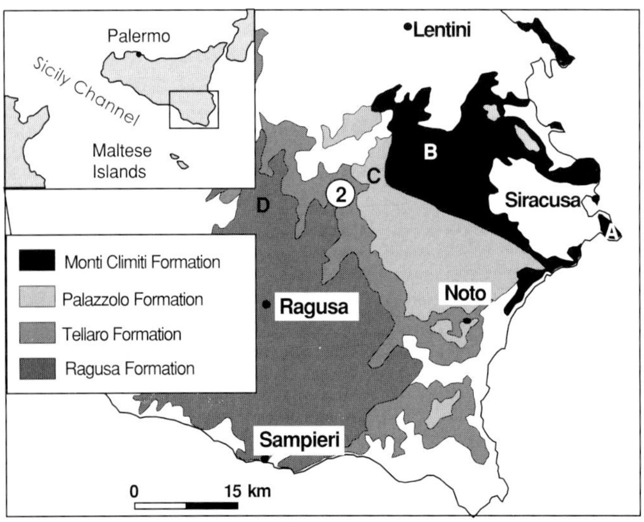

FIG. 6.—Geologic and location map of the Hyblean area, after Grasso and Lentini (1982). A–D and 2 are identified in the cross section in Figure 7.

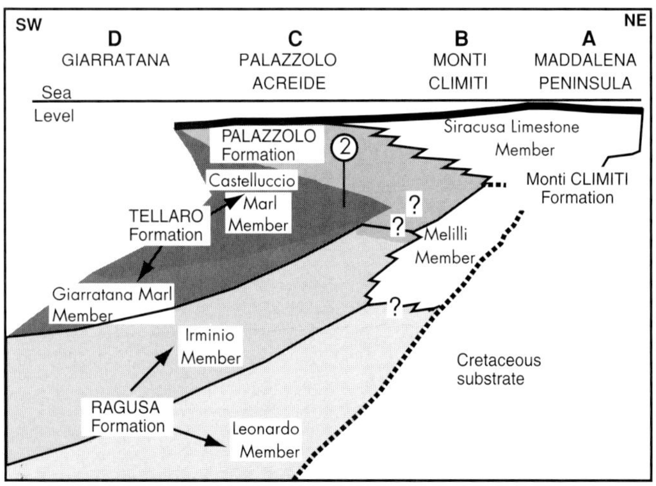

FIG. 7.—Schematic stratigraphic cross section of the Miocene formations in the Hyblean Plateau, in southeastern Sicily, modified from Pedley (1981). The horizontal distance is approximately 50 km. The thick black line indicates the reconstruction of the sea floor during the Tortonian, as suggested by Pedley (1981).

For the purpose of this paper, we focus on the younger part of the sequence, and in particular on the transition between the Tellaro Formation (Castelluccio Marl) and the Palazzolo Formation. The transition records the progressive increase in sediment export to the outer shelf from the shallow-water areas.

THE PALAZZOLO SECTION

A continuous section in the Palazzolo Acreide area, spanning 90 m through the upper part of the Tellaro Formation and the lower part of the Palazzolo Formation (Fig. 7), has been analyzed in detail for depositional and lithological facies variations and has been sampled at 1–2-m resolution for stable isotope stratigraphy (Fig. 8).

The Tellaro Formation is a clay-rich unit; its carbonate content averages 52.8% but ranges between 45% and 65%. The carbonate fraction consists mainly of planktonic foraminifers, but subordinate numbers of benthic foraminifers and *Bolboforma* (Protophyta) are also present. This latter microfossil is typical of middle and upper Miocene strata from temperate and cool regions in both the Southern and Northern Hemispheres, and it is commonly associated with siliceous assemblages consisting of diatoms and sponge spicules. In particular, *Bolboforma* generally occurs in cool-water masses at high latitudes, in cool surface-water currents, and on continental margins or submarine platforms (Qvale and Spiegler, 1989; Mackensen and Spiegler, 1992; Spiegler and von Daniels, 1991). Ostracodes are also present in the upper part of the Tellaro Formation, and their assemblages are interpreted to indicate a well-oxygenated upper-slope environment (Miculan, 1995). The unit contains organic carbon (TOC) in the range of 0.2% to 1.4% (Fig. 9).

The contact with the Palazzolo Formation is transitional and is expressed by the progressive upward increase of distinctive carbonate-rich beds, initially alternating with marly intervals that tend to decrease up section. This part of the Palazzolo Formation contains variable amounts of carbonate, ranging between 40% and 90% and averaging 62.1%.

The carbon and oxygen isotope compositions measured throughout the section vary significantly; $\delta^{13}C$ ranges from 0.2‰ to +1.0‰, and $\delta^{18}O$ ranges from –0.3‰ to +1.2‰ (Fig. 8). These values are from bulk samples, which include a mixture of several foraminiferal species. The good preservation of the foraminifers, as well as preliminary isotope data on separate benthic and planktonic foraminiferal species, suggests that these values do not reflect diagenetic modifications and can be interpreted in terms of marine environmental conditions. A $\delta^{13}C$ shift with a gradual increase from +0.2‰ to +1.0‰ in the lower part of the Tellaro Formation is followed upward by a rapid decrease to +0.2‰ in the upper part of the formation. The rapid decrease corresponds with the lithological change to the Palazzolo Formation, upward through which the $\delta^{13}C$ values show a gradual increase to +1.0‰, but do not define any significant shift at the resolution sampled.

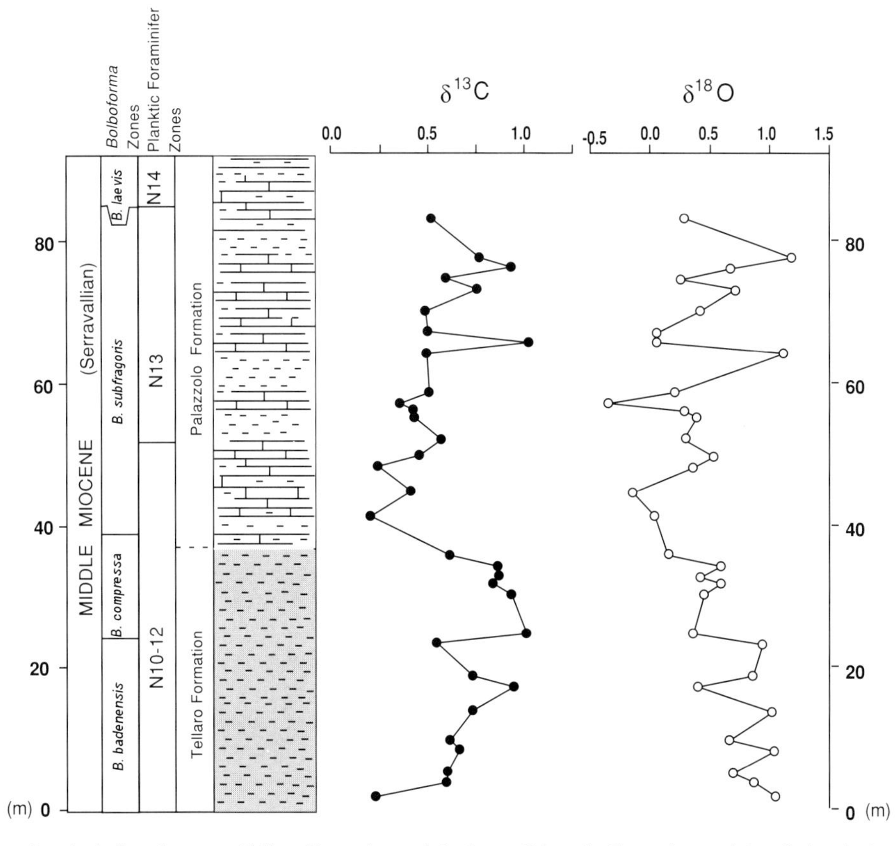

FIG. 8.—The Palazzolo section, including the upper Tellaro Formation and the lower Palazzolo Formation, and the planktonic foraminifera biozones of Blow (1969) and the *Bolboforma* zonation. To the right are the $\delta^{13}C$ and $\delta^{18}O$ values from bulk samples.

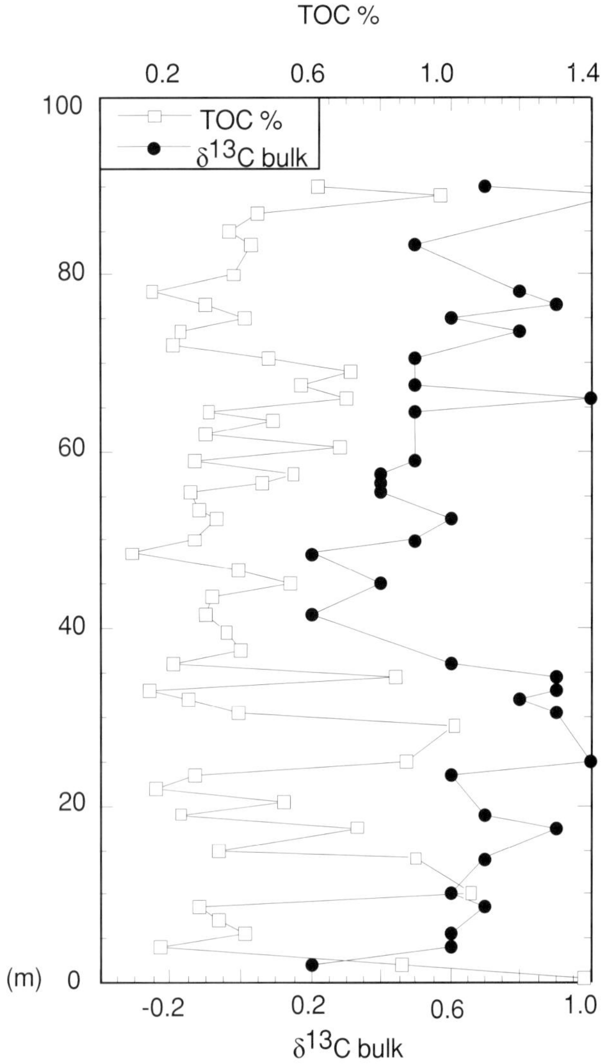

FIG. 9.—Total organic carbon (TOC) content and $\delta^{13}C$ versus depth in the Palazzolo section.

From the lower part of the section upward, the $\delta^{18}O$ values show a trend toward progressively more negative values, from +1.0‰ to –0.5‰; then the values increase through the upper part of the section. On the basis of our biostratigraphic data, the trend toward more negative values occurs within planktonic foraminifera biozones N10–N12 of Blow (1969) and terminates within biozone N13, when the trend is reversed toward heavier values. A correlation of Blow's biozones and numerical ages was established by the ODP Leg 166 Shipboard Party (Eberli et al., 1997). They identified the base of N10–N12 at 14.8 Ma, the base of N13 at 11.9 Ma, and the base of N14 at 11.8 Ma. These numerical ages are based on the geomagnetic polarity time scale of Berggren et al. (1995a, 1995b). On the basis of this correlation, the observed $\delta^{18}O$ trend in the Palazzolo section happened between 14.8 and 11.8 Ma.

CARBON ISOTOPES, ORGANIC MATTER BURIAL, AND PRODUCTIVITY

Carbon isotope data from marine carbonates are a powerful tool to evaluate the history of the global carbon system. As pointed out by Shackleton (1985), it is important to use $\delta^{13}C$ data from the appropriate part of the ocean. In fact, the history of the carbon isotope composition of ocean-dissolved CO_2 can be best obtained by analyzing benthic foraminifer carbonate because the average chemistry of the ocean is better represented at the sea floor than close to the surface. However, considering the 10^5-yr residence time of carbon in the ocean, the carbon isotope composition of the total carbonate may be also used to evaluate the long-term history of the global carbon system. Because bulk isotope data are discussed here, we look at the longer-term (>10^5-yr) variations of the global carbon system.

The $\delta^{13}C$ values measured in our two sections correlate well with the early to middle Miocene (Monterey) positive $\delta^{13}C$ excursion (Fig. 10) (Vincent and Berger, 1985; Wright and Miller, 1992; Wright et al., 1992; Hodell and Woodruff, 1994). Although the trends are similar, the Maiella record has a wider spread of values. However, geographical variations might be expected, as the $\delta^{13}C$ value of surface water is controlled by horizontal and vertical mixing of water masses and rate of input, degradation, and export of organic material.

The marine pelagic carbon isotope record is commonly interpreted to be an indicator of global-scale variations in burial rates of organic matter. Organic matter is enriched in ^{12}C; therefore any relative increase in organic carbon removal from the surface carbon reservoir will enrich the remaining dissolved inorganic carbon (DIC) in ^{13}C. Positive $\delta^{13}C$ excursions are commonly interpreted to reflect increased burial of organic carbon (Fischer and Arthur, 1977; Scholle and Arthur, 1980; Weissert, 1989; Compton et al., 1990, 1993). However, additional processes have to be considered. Lasaga et al. (1985) and Shackleton (1987) have suggested that the $\delta^{13}C$ decrease of marine carbonates throughout the Cenozoic reflects decreased burial of organic matter relative to input from weathering. In fact, because the formation of organic matter is accompanied by a 24–28‰ isotope fractionation relative to the inorganic carbon reservoir, any change in the relative amount of organic matter buried vs. weathered would be recorded by a change in the $\delta^{13}C$ of the inorganic carbon reservoir. An alternative view is the one of Derry and France-Lanord (1996), who have suggested, on the basis of isotopic mass-balance constraints, that a net addition of carbon to the sedimentary organic carbon reservoir occurred during the Neogene.

The increases in the $\delta^{13}C$ values of the carbonates deposited during the onset of the second-order transgressive event on the Maiella platform margin (Fig. 10) could be interpreted to reflect increased productivity of surface waters, organic matter exportation to the deep sea, and consequent enrichment in ^{13}C of shallow waters. Uppermost lower Miocene to lowermost middle Miocene strata with very high organic carbon burial rates were widely deposited around the Pacific coast. Higher accumulation rates of organic matter have further been related to trade-wind strength and coastal upwelling intensity (Vincent and Berger,

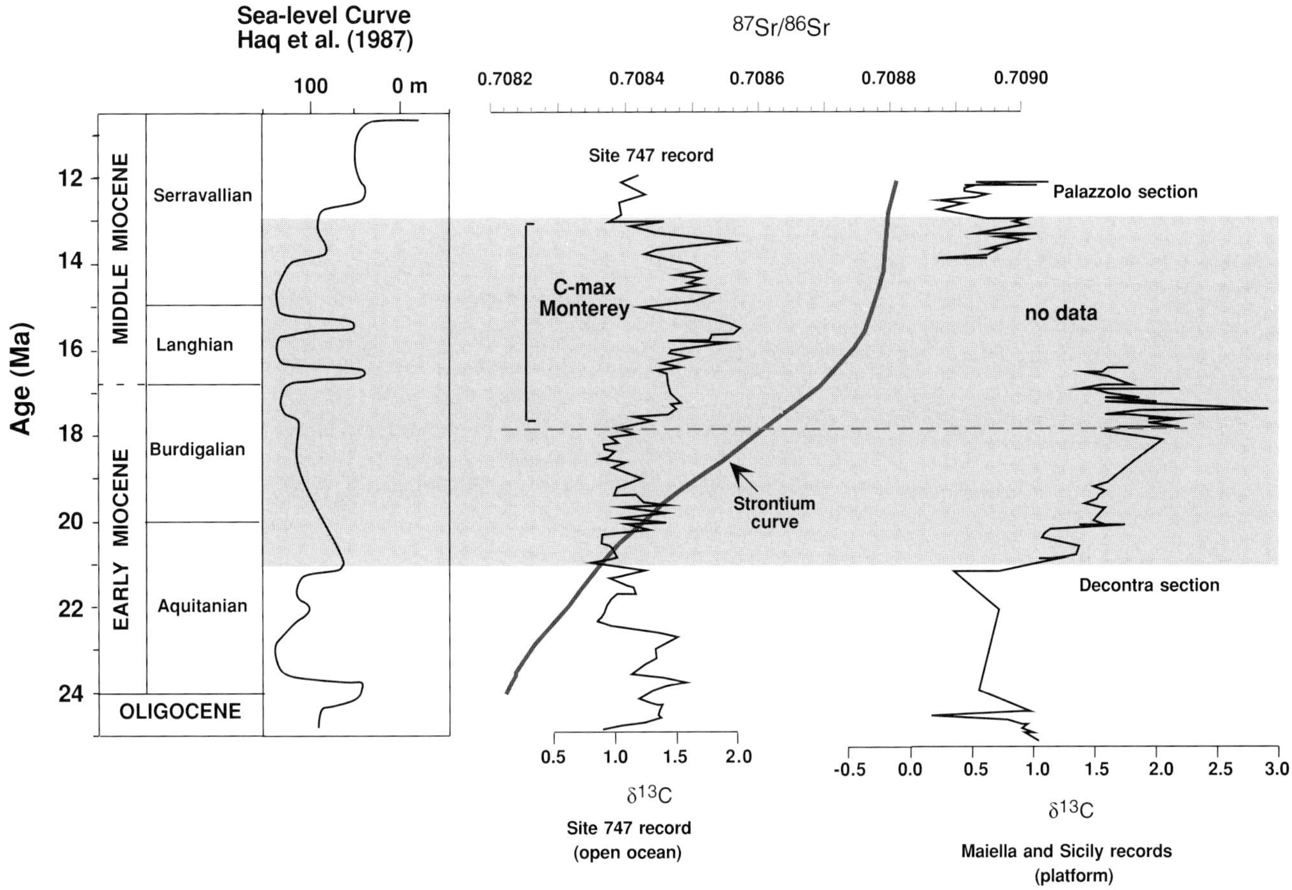

FIG. 10.—Comparison of $\delta^{13}C$ oceanographic records vs. age based on the $\delta^{13}C$ values from benthic foraminifera from the Indian Ocean (ODP Site 747; Wright and Miller, 1992), the $\delta^{13}C$ records of the Decontra section on the Maiella platform margin and the Palazzolo section, the global strontium isotope record (bold curve; data from ODP Site 747; Hodell and Woodruff, 1994), and the global sea-level curve (Haq et al., 1987). The gray band indicates the duration of the regional, second-order transgression and early highstand in the central Mediterranean. "C-max Monterey" refers to the global $\delta^{13}C$ maximum associated with the Monterey carbon isotope excursion.

1985), sea-level changes (Compton et al., 1990), and nutrient loading of upwelling zones (Raymo, 1994). Unfortunately, in the Maiella section, no significant amount of organic carbon is preserved in the strata (Mutti and Bernoulli, unpublished data). In addition, in the Decontra section the critical interval in which the $\delta^{13}C$ values decrease is cut out by the unconformity at the top of the Upper Bryozoan Limestone. However, phosphate-filled foraminifera and abundant chert nodules derived from biogenic silica suggest high primary productivity. In the Palazzolo section, the organic carbon content shows a significant range of variations with depth (Fig. 9), and, as a first approximation, increased burial rates of organic carbon can be correlated with increased $\delta^{13}C$ values. In addition, the shift of $\delta^{13}C$ values at the top of the Tellaro Formation in the Palazzolo section in correspondence with the lithological change to the Palazzolo Formation occurs within the upper part of Blow's (1969) biozones N10–N12; therefore, the formation boundary is somewhat older than 11.8 Ma. Both data sets seem to indicate that a renewal of benthic carbonate production was associated with termination of the $\delta^{13}C$ maximum associated with the Monterey event. However, the tropical facies containing abundant corals and coralline algae did not appear in the Mediterranean until late Tortonian time (see Fig. 1).

The $\delta^{13}C$ data from the two sections that we are discussing here, and comparison with other published records, suggest the existence of a link between regional changes in carbonate facies and the long-term carbon cycle. The regional demise of coral-reef facies (as summarized in Fig. 1) was much more extended in time than the Monterey positive $\delta^{13}C$ excursion (see Fig. 10). The global $\delta^{13}C$ maximum associated with the Monterey excursion occurred between approximately 18 and 13 Ma (Fig. 10), but its onset was predated by increasing values of $\delta^{13}C$ starting around 21 Ma (Vincent and Berger, 1985; Wright and Miller, 1992; Hodell and Woodruff, 1994). However, if we consider the extent of the transgressive event in the central Mediterranean, as dated in our two sections and in available literature data, its onset predated the major $\delta^{13}C$ excursion by approximately 3 m.y., whereas its termination coincided with the return to lower $\delta^{13}C$. The termination of the transgressive event coincided with renewed carbonate production in the area, but it was not until later, in the later Tortonian, that tropical facies were deposited again.

The link between drowning events on carbonate platforms, associated with phosphogenesis, and global positive $\delta^{13}C$ excursions has been documented in the Cretaceous (Föllmi et al., 1994). Jacobs et al. (1996) suggested also for the Miocene a link between carbonate burial rates and the carbon isotope record. A commonly observed pattern, yet to be explained, is that in general carbonate-facies changes predate the onset of positive $\delta^{13}C$ excursions by approximately 1–3 m.y., but coincide with the rising part of the $\delta^{13}C$ excursion. The return to normal carbonate facies generally coincides with the termination of the $\delta^{13}C$ excursion. The linking process between carbonate platform drowning and $\delta^{13}C$ excursions may be due to increased weathering rates, either as the consequence of reinforced greenhouse conditions (Föllmi et al., 1994) or because of the uplift of major mountain chains (Mutti et al., 1997). The latter process results in increased delivery of river nutrients to the ocean, which could have stimulated productivity in upwelling zones (e.g., Raymo, 1994). In the Miocene, the seawater strontium isotope curve can be used as a proxy for changes in weathering rates. This curve shows a rapid increase to more radiogenic values during the early Miocene between approximately 24 and 16 Ma (De Paolo and Finger, 1991; Richter et al., 1992; Hodell and Woodruff, 1994) (Fig. 9) as a consequence of the progressive uplift of the Tibetan-Himalayan region. The uplift of the Tibetan-Himalayan region coincides also with the collision between the Arabian and Turkish plates and must have resulted in increased weathering rates along the eastern and central Mediterranean coastlines. The increase in weathering rates, bringing more nutrients into the Mediterranean waters, might have had an impact on carbonate-facies development.

Several papers have suggested a link between carbonate-facies changes during the Miocene in the Mediterranean and nutrient availability; the differences mainly are in the mechanism suggested to produce these changes. Pedley (1996) noted that the new circulation pattern developed after the connection to the Indian Ocean became closed and the topography generated by the developing Apennine fold belt must have caused fundamental changes in nutrient supply and water flow. He added that there must have been a linkage between these processes and the middle Miocene reef retrenchment. Rouchy (1979) noticed the relationship between red algal facies, upwelling, and basinal diatomaceous deposits. Carannante and Simone (1996) identified the development of bryozoan and lithothamnian limestones in the southern Apennines as the result of upwelling of cold, nutrient-rich water on the shelves. Esteban (1996) finally proposed a sophisticated model to link carbonate-facies changes to changing circulation patterns, caused by variations in net precipitation and sea level.

OXYGEN ISOTOPES, COOLING TRENDS, AND PRECIPITATION PATTERNS

The most significant feature of the Miocene $\delta^{18}O$ benthic foraminifera curve (Fig. 11), for the purpose of this paper, is the fast, step-wise shift toward higher $\delta^{18}O$ values, which occurred at approximately 14 Ma and lasted until 12 Ma. This shift is commonly interpreted to reflect a major cooling and an ice-volume change, which was likely associated with Antarctic ice build-up (e.g., Savin et al., 1986). The Palazzolo section, which was deposited during the time interval over which this change occurred, recorded a shift toward lower $\delta^{18}O$ values throughout the planktonic foraminiferal biozones N10–N12 of Blow (1969), and this trend then reversed within N13 when values tended to shift toward higher $\delta^{18}O$ values (Fig. 8). The bulk samples measured in the Palazzolo section comprise carbonate from both planktonic and benthic foraminifera; therefore their composition reflects a mixture of seawater composition ranging from the shallow surface (plankton) to the sea floor (benthos). Ostracode data (Miculan, 1995) suggest an upper-slope environment, where the water depth probably did not exceed 500 m. If it is assumed that changes in temperature alone were the force driving this trend toward lower $\delta^{18}O$ values, then the trend indicates an increase in the average temperature at the end of the deposition of the Tellaro Formation. This interpretation of the data is in contrast to the trend documented in other areas of the world, where isotope records indicate a fast cooling trend between 14 and 12 Ma (e.g., Miller et al., 1987; Hodell and Woodruff, 1994).

Changes in precipitation patterns can also influence $\delta^{18}O$ values of marine carbonates. Increased precipitation will decrease the salinity of surface waters and concomitantly lower the $\delta^{18}O$ values in any carbonate that is formed. This process is observed today in areas of episodic rainfall such as the Tarawa Atoll in the western Pacific, where coral-skeletal carbonate precipitated from seawater carries the signature of rainfall variability over the past 100 years. The $\delta^{18}O$ values vary by ±1‰, and the associated salinity covaries from 35‰ to 31‰ (Dunbar and Cole, 1993). Episodic strong precipitation could have had an impact on the isotopic composition of Miocene Mediterranean surface waters. Existing clay mineralogy studies indicate that a humid climate developed in the central Mediterranean area (Visser, 1990). On the other hand, Rögl and Müller (1978) reported middle Miocene evaporite basins in the Paratethys (middle and eastern Europe), so distinctive that they even suggested the term "middle Miocene salinity crisis." During the Serravallian, several hundreds of meters of evaporite deposits accumulated in various areas of the Mediterranean region, including the Paratethys, the Red Sea, the Gulf of Suez, and the Sirte basin (Esteban, 1996). Unfortunately, the age determinations for these occurrences are controversial and complicated by the problem of correlating marine and continental sections.

On the basis of the existing data, i.e., the extent of the evaporite deposits, it seems that changes in precipitation during the Serravallian were not so extreme as to produce surface waters so depleted in ^{18}O that their oxygen isotope compositions simulate trends toward warm temperatures, at a time when the global trend was toward cooling. The question is, how is it possible that the Mediterranean water records this signature, which is so different from the global one? Several scenarios, summarized by Wright et al. (1992), have been proposed to explain the global middle Miocene $\delta^{18}O$ increase as linked to deep-water circulation patterns. Adding to the Miocene deep-water debate is the hypothesis that a warm, saline deep-water mass originated in the eastern Tethys during the early Miocene (Woodruff and Savin, 1989; Wright et al., 1992).

An answer to our question might lie in the degree to which

Mediterranean surface waters were stratified. The thermohaline circulation today in the Mediterranean is driven by evaporation, not by polar cooling, and it is strongly affected by precipitation related to continental meteorology (Hopkins, 1985). The mountain systems around the coasts of the Mediterranean impose prevailing wind directions (Flos, 1985), and strong winds and low pressures favor seasonal vertical mixing. In today's waters, vertical stratification is entirely lost during certain winters, when strong winds and low pressures favor seasonal vertical mixing. The paleogeographical configuration during the middle Miocene was not too different from today's, and we could argue that vertical stratification was reduced also then, resulting in an oxygen isotope composition controlled to a large extent by surface temperatures.

The Mediterranean change from tropical to temperate carbonate facies, which began at approximately 20 Ma, predated the major Neogene global cooling step between 14 and 12 Ma. Unfortunately, our oxygen isotope data, as recorded in the Maiella platform margin, cannot be used to constrain local temperature changes at the onset of this event. However, the oxygen isotope record from the Hyblean Plateau suggests relatively cool temperatures at the base of the section during the Serravallian (within Blow's [1969] biozones N10–N12) with progressive warming within biozone N11 (11.9–11.8 Ma). This signature is opposite to the one recorded in pelagic sections in other locations of the world (see Fig. 11). The mismatch between the global cooling history as recorded in other pelagic sequences and the data presented in this study needs to be tested more rigorously with similar studies in other locations, because it would provide important insights into the evolution of these Miocene carbonates. The oxygen isotope signature we have observed is indicative of the regional conditions within the Mediterranean, as after 20 Ma the connection to the Indian Ocean became closed or reduced to a very shallow sill, which prevented the outflow and inflow of intermediate and deeper waters. This trend is in agreement with earlier Sr and Nd isotope studies showing that the Indian-Tethys seaway started to close at approximately 25 Ma leading to the breakdown of the circum-equatorial circulation pattern of the world oceans at approximately 20 Ma (Stille et al., 1996).

PALEOCEANOGRAPHIC IMPACT ON SEQUENCE STRATIGRAPHY

We have discussed two stratigraphic sections that record the

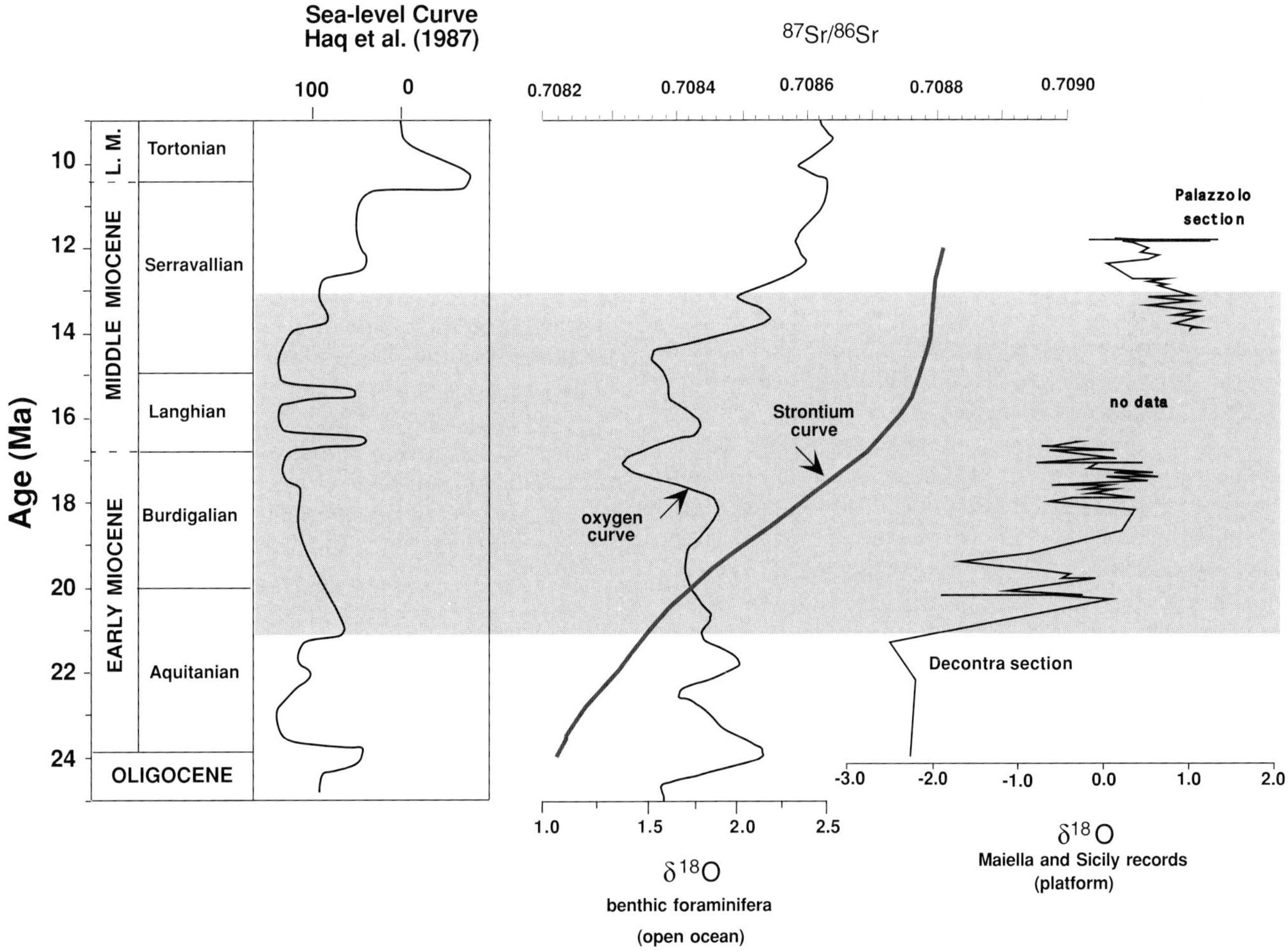

FIG. 11.—The eustatic sea-level record as established by Haq et al. (1987), the open-ocean $\delta^{18}O$ variations derived from *Cibicides* spp. (modified from Miller et al., 1996), and the Maiella and Sicily records plotted vs. age. Note how the Mediterranean Serravallian record indicates a shift toward lighter oxygen isotopes, in contrast with the open-ocean record. The gray band indicates the duration of the regional, second-order transgression and early highstand in the central Mediterranean.

onset and the termination of a second-order transgressive event in the central Mediterranean. Shallow-water carbonate facies were widespread in the Mediterranean before and after this event, but during this time interval, they were reduced and accumulated in only minor locations. By integrating the shallow-water and the pelagic records, through the use of higher-resolution dating than what was previously accomplished and then generating stable isotope time series, we have shown that facies changes in the central Mediterranean during the Miocene were linked to major changes in environmental parameters. Our data indicate that global cooling postdated the shift toward cooler-water facies in the central Mediterranean, but the return to subtropical facies in the late Miocene seems to be linked to regional warming. However, the specific role that regional temperature changes have played on facies changes remains controversial. Several lines of evidence—such as changes in biotic assemblages, abundance of phosphates and silica, and carbon isotopes—suggest that nutrient availability changed drastically during the Miocene in the central Mediterranean.

We have shown that, during the early and middle Miocene, carbonate accumulation rates reached a minimum while the transgressive systems tract was being deposited, as a consequence of environmental changes linked to regional circulation patterns and possibly climatic trends. The observed evolution of deepening-upward sequences in the more distal parts of the shelf and the general tendency to replace shallow-water by deeper-water facies up section are caused by changing rates of carbonate production and accumulation during deposition of the transgressive and highstand systems tracts. This depositional pattern is not linked to a change in the size of carbonate-producing area along the depositional ramp but rather to decreased amounts of carbonate produced as a consequence of environmental change.

Our results have major implications regarding the reliability of sea-level changes interpreted from sequence stratigraphic analysis in carbonate environments. When a regional decrease in carbonate production rates coincides with a rapid sea-level rise, the resulting architecture can mimic a sea-level rise of much more significant extent. Changing rates of carbonate production across the depositional shelf for a moment in time have been included in sequence stratigraphic models. However, we have shown that they also can change significantly through time in a certain region, as a response to local and more general processes occurring in the oceans. A better understanding has to be developed for the general response of carbonate facies to paleoceanographic changes, so that these changes can be used both to predict sequence stratigraphic architecture in a certain time slice and to better estimate the reliability of the sea-level signature preserved in carbonate sequences. In particular, our understanding is now limited to variations occurring at time scales of from 1 to 5 m.y., but changes at the orbital time scales are needed to integrate in a more predictive way the linked evolution of shallow-water and pelagic sequences.

ACKNOWLEDGMENTS

This research was supported by Swiss National Science Foundation grants 20-35907.92 and 20-45131.95. We greatly appreciate the support of the Stable Isotope Laboratory at ETH-Zürich (J. A. McKenzie and Stefano Bernasconi) and the EAWAG-Zurich for inorganic and organic carbon analyses. We thank M. Pika-Biolzi and D. Spiegler for biostratigraphic advice on the Palazzolo section, and Cliff Hammer and Daniel Stöckli for assistance in the field. We appreciated the stimulating discussions with Helmut Weissert and Karl Föllmi and the help of Mario Grasso with the local geology of Sicily. We thank Mitch Harris, Steve Dorobek, and Rick P. Major for their careful reviews and Mary C. Eberle for her helpful editing.

REFERENCES

Berggren, W. A., Hilgen, F. A. J., Langreis, C. G., Kent, D. V., Obradovich, J. D., Raffi, I., Raymo, M. E., and Shackleton, N. J., 1995a, Late Neogene chronology: New perspectives in high-resolution stratigraphy: Geological Society of America Bulletin, v. 107, p. 1272–1287.

Berggren, W. A., Kent, D. V., Swisher, C. C., III, and Aubry, M.-P., 1995b, A revised Cenozoic geochronology and chronostratigraphy, *in* Berggren, W. A., Kent, D. A., Aubry, M.-P., and Hardenbol, J., eds., Geochronology, Time Scales and Global Stratigraphic Correlation: Tulsa, SEPM (Society for Sedimentary Geology) Special Publication 54, p. 129–212.

Bernoulli, D., Eberli, G. P., Pignatti, J. S., Sanders, D., and Vecsei, A., 1992, Sequence Stratigraphy of Montagna della Maiella: Quinto Simposio di Ecologia e Paleoecologia delle Comunità Bentoniche, Paleobenthos: Roma, Libro-Guida delle Escursioni, p. 85–109.

Blow, W. H., 1969, Late middle Eocene to Recent planktonic foraminiferal biostratigraphy, *in* Brönnimann, P. and Renz, H. H., eds., Proceedings of the First International Conference on Planktonic Microfossils, Geneva, 1967, v. 1: Leiden, Brill, p. 199–422.

Carannante, G., and Simone, L., 1996, Rhodolith facies in the central-southern Apennine Mountains, Italy, *in* Franseen, E. K., Esteban, M., Ward, W. C., and Rouchy, J.-M., eds., Models for Carbonate Stratigraphy from Miocene Reef Complexes of Mediterranean Regions: Tulsa, SEPM (Society for Sedimentary Geology) Concepts in Sedimentology and Paleontology No. 5, p. 261–275.

Carannante, G., Esteban, M., Milliman, J. D., and Simone, L., 1988, Carbonate lithofacies as paleolatitude indicators: Problems and limitations: Sedimentary Geology, v. 60, p. 333–346.

Cita, M. B., 1959, Stratigrafia micropaleontologica del Miocene siracusano: Bollettino della Società Geologica Italiana, v. 77, p. 71–165.

Compton, J. S., Snyder, S. W., and Hodell, D. A., 1990, Phosphogenesis and weathering of shelf sediments from the southeastern United States: Implications for Miocene δ13C excursions and global cooling: Geology, v. 18, p. 1227–1230.

Compton, J. S., Hodell, D. A., Garrido, J. R., and Mallinson, D. J., 1993, Origin and age of phosphorite from the south-central Florida Platform: Relation of phosphogenesis to sea-level fluctuations and δ13C excursions: Geochimica et Cosmochimica Acta, v. 57, p. 131–146.

Crescenti, U., Crostella, A., Donzelli, G., and Raffi, G., 1969, Stratigrafia della serie calcarea dal Lias al Miocene nella regione Marchigiano-Abruzzese (Parte II—Litostratigrafia, biostratigrafia, paleogeografia): Memorie della Società Geologica Italiana, v. 8, p. 343–420.

De Paolo, D. J., and Finger, K. L., 1991, High-resolution strontium isotope stratigraphy and biostratigraphy of the Miocene Monterey Formation, central California: Geological Society of America Bulletin, v. 103, p. 112–124.

Derry, L. A., and France-Lanord, C., 1996, Neogene growth of the sedimentary organic carbon reservoir: Paleoceanography, v. 11, p. 267–275.

Dunbar, R. B., and Cole, J. E., 1993, Coral Records of Ocean-Atmosphere Variability: Washington, D.C., National Oceanic and Atmospheric Administration, Climate and Global Change Program, Special Report 10, 38 p.

EBERLI, G. P., AND GINSBURG, R. N., 1989, Cenozoic progradation of northwestern Great Bahama Bank, a record of lateral platform growth and sea level fluctuations, *in* Crevello, P. D., Wilson, J. L., Sarg, J. F., and Read, J. F., eds., Controls on Carbonate Platform and Basin Development: Tulsa, Society of Economic Paleontologists and Mineralogists, Special Publication 44, p. 339–351.

EBERLI, G. P., SWART, P. K., MALONE, M. J., et al., 1997, Proceedings of the Ocean Drilling Program, Initial Results, v. 166: College Station, Ocean Drilling Program, 850 p.

EMEIS, K.-C., ROBERTSON, A. H. F., RICHTER, C., et al., 1996, Proceedings of the Ocean Drilling Program, Initial Results, v. 160: College Station, Ocean Drilling Program, 972 p.

ESTEBAN, M., 1979, Significance of the upper Miocene reefs of the western Mediterranean: Palaeogeography, Palaeoclimatology, Palaeoecology, v. 29, p. 169–188.

ESTEBAN, M., 1996, An overview of Miocene reefs from Mediterranean areas: General trends and facies models, *in* Franseen, E. K., Esteban, M., Ward, W. C., and Rouchy, J.-M., eds., Models for Carbonate Stratigraphy from Miocene Reef Complexes of Mediterranean Regions: Tulsa, SEPM (Society for Sedimentary Geology) Concepts in Sedimentology and Paleontology No. 5, p. 3–53.

FISCHER, A. G., AND ARTHUR, M. A., 1977, Secular variations in the pelagic realm, *in* Cook, H. E. and Enos, P., eds., Deep Water Carbonate Environments: Tulsa, Society of Economic Paleontologists and Mineralogists, Special Publication 25, p. 19–50.

FLOS, J., 1985, The driving machine, *in* Marglef, R., ed., Western Mediterranean: Oxford, Pergamon Press, p. 60–99.

FÖLLMI, K. B., WEISSERT, H., BISPING, M., AND FUNK, H., 1994, Phosphogenesis, carbon isotope stratigraphy, and carbonate-platform evolution along the Lower Cretaceous northern Tethyan margin: Geological Society of America Bulletin, v. 106, p. 729–746.

FRANSEEN, E. K., ESTEBAN, M., WARD, W. C., AND ROUCHY, J.-M., eds., 1996, Models for Carbonate Stratigraphy from Miocene Reef Complexes of Mediterranean Regions: Tulsa, SEPM (Society for Sedimentary Geology) Concepts in Sedimentology and Paleontology No. 5, 391 p.

GRASSO, M., AND LENTINI, F., 1982, Sedimentary and tectonic evolution of the eastern Hyblean Plateau (S. E. Sicily) during Late Cretaceous to Quaternary time: Palaeogeography, Palaeoclimatology, Palaeoecology, v. 39, p. 261–280.

GRASSO, M., LENTINI, F., LOMBARDO, G., AND SCAMARDA, G., 1979, Distribuzione delle facies cretaceo-mioceniche lungo l'allineamento Augusta M. Lauro (Sicilia sud-orientale): Bollettino della Società Geologica Italiana, v. 98, p. 175–188.

GRASSO, M., PEDLEY, H. M., AND MANISCALCO, R., 1994, The application of a late Burdigalian–early Langhian high-stand event in correlating complex Tertiary orogenic carbonate successions within the central Mediterranean: Géologie Méditérranéenne, v. 21, p. 69–83.

HANDFORD, C. R., AND LOUCKS, R. G., 1993, Carbonate depositional sequences and systems tracts—Responses of carbonate platforms to relative sea-level changes, *in* Loucks, R. G. and Sarg, J. F., eds., Carbonate Sequence Stratigraphy: Tulsa, American Association of Petroleum Geologists Memoir 57, p. 3–41.

HAQ, B. U., HARDENBOL, J., AND VAIL, P. R., 1987, Chronology of fluctuating sea levels since the Triassic: Science, v. 235, p. 1156–1167.

HAQ, B. U., HARDENBOL, J., AND VAIL, P. R., 1988, Mesozoic and Cenozoic chronostratigraphy and eustatic cycles, *in* Wilgus, C. K., Hastings, B. S., Kendall, C. G. S. C., Posamentier, H. W., Ross, C. A., and Van Wagoner, J. C., eds., Sea-Level Changes: An Integrated Approach: Tulsa, Society of Economic Paleontologists and Mineralogists, Special Publication 42, p. 71–108.

HODELL, D. A., 1991, Variations in the Sr isotopic composition of seawater during the Neogene: Geology, v. 19, p. 24–27.

HODELL, D. A., AND WOODRUFF, F., 1994, Variations in the strontium isotopic ratio of seawater during the Miocene: Stratigraphic and geochemical implications: Paleoceanography, v. 9, p. 405–426.

HODELL, D. A., MUELLER, P. A., AND GARRIDO, J. R., 1991, Variations in the strontium isotopic composition of seawater during the Neogene: Geology, v. 19, p. 24–27.

HOPKINS, T. S., 1985, Physics of the sea, *in* Marglef, R., ed., Western Mediterranean: Oxford, Pergamon Press, p. 100–125.

JACOBS, E., WEISSERT, H., SHIELDS, G., AND STILLE, P., 1996, The Monterey event in the Mediterranean: A record from shelf sediments of Malta: Paleoceanography, v. 11, p. 717–728.

JAMES, N. P., AND BONE, Y., 1991, Origin of a cool-water, Oligo-Miocene deep shelf limestone, Eucla Platform, southern Australia: Sedimentology, v. 38, p. 323–341.

LASAGA, A. C., BERNER, R. A., AND GARRELS, R. M., 1985, An improved geochemical model of atmospheric CO_2 fluctuations over the past 100 million years, *in* Sundquist, E. T. and Broecker, W. S., eds., The Carbon Cycle and Atmospheric CO_2: Natural Variations Archean to Present: Washington, D.C., American Geophysical Union Geophysical Monograph 32, p. 455–468.

LIPPS, J. H., 1979, Ecology and paleoecology of planktic foraminifera, *in* Lipps, J. H., Berger, W. H., Buzas, M. A., Douglas, R. A., and Ross, C. A., eds., Foraminiferal Ecology and Paleoecology: Tulsa, Society of Economic Paleontologists and Mineralogists, Short Course No. 6, p. 67–74.

MACKENSEN, A., AND SPIEGLER, D., 1992, Middle Eocene to early Pliocene Bolboforma (Algae?) from the Kerguelen Plateau, southern Indian Ocean, *in* Wise, S. W., Jr., Schlich, R., et al., Proceedings of the Ocean Drilling Program, Initial Results, v. 120: College Station, Ocean Drilling Program, p. 675–682.

MICULAN, P., 1995, Middle Miocene ostracodes from the Hyblean Plateau, south eastern Sicily, Italy (abs.), *in* Riha, J., ed., Ostracoda and Biostratigraphy: Rotterdam, Balkema.

MILLER, K. G., FAIRBANKS, R. G., AND MOUNTAIN, G. S., 1987, Tertiary oxygen isotope synthesis, sea level history, and continental margin erosion: Paleoceanography, v. 2, p. 1–19.

MILLER, K. G., MOUNTAIN, G. S., et al. (Leg 150 Shipboard Scientific Party and members of the New Jersey Coastal Plain Drilling Project), 1996, Drilling and dating New Jersey Oligocene–Miocene sequences: Ice volumes, global sea level, and Exxon records: Science, v. 271, p. 1092–1094.

MUTTI, M., BERNOULLI, D., AND STILLE, P., 1997, Temperate carbonate platform drowning linked to Miocene oceanographic events: Maiella platform-margin, Italy: Terra Nova, v. 9, p. 122–125.

NELSON, C. S., 1978, Temperate shelf carbonate sediments in the Cenozoic of New Zealand: Sedimentology, v. 25, p. 737–771.

PATTERSON, P., AND WALTER, L. M., 1994, Depletion of ^{13}C in seawater ΣCO_2 on modern carbonate platforms: Significance for the carbon isotopic record of carbonates: Geology, v. 22, p. 885–888.

PEDLEY, H. M., 1981, Sedimentology and palaeoenvironment of the southeast Sicilian Tertiary platform carbonates: Sedimentary Geology, v. 28, p. 273–291.

PEDLEY, H. M., 1983, The petrology and paleoenvironment of the Sortino Group (Miocene) of S. E. Sicily: Evidence for periodic emergence: Journal of the Geological Society of London, v. 140, p. 335–350.

PEDLEY, H. M., 1996, Miocene reef distributions and their associations in the central Mediterranean region: An overview, *in* Franseen, E. K., Esteban, M., Ward, W. C., and Rouchy, J.-M., eds., Models for Carbonate Stratigraphy from Miocene Reef Complexes of Mediterranean Regions: Tulsa, SEPM (Society for Sedimentary Geology) Concepts in Sedimentology and Paleontology No. 5, p. 73–87.

PEDLEY, H. M., AND BENNETT, S., 1985, Phosphorites, hardgrounds and syndepositional subsidence structures in the Miocene Globigerina Limestone Formation, Maltese Islands: Sedimentary Geology, v. 45, p. 1–34.

QVALE, G., AND SPIEGLER, D., 1989, The stratigraphic significance of Bolboforma (Algae Chrysophyta) in Leg 104 samples from the Vøring Plateau, *in* Eldholm, O., Thiede, J., et al., Proceedings of the Ocean Drilling Program, Initial Results, v. 104: College Station, Ocean Drilling Program, p. 487–495.

RAYMO, M. E., 1994, The Himalayas, organic carbon burial, and climate in the Miocene: Paleoceanography, v. 9, p. 399–404.

RICHTER, F. M., ROWLEY, D. B., AND DE PAOLO, D. J., 1992, Sr isotope evolution of seawater: The role of tectonics: Earth and Planetary Science Letters, v. 109, p. 11–23.

RÖGL, F., AND MÜLLER, C., 1978, Middle Miocene salinity crisis and paleogeography of the Paratethys (middle and eastern Europe), *in* Montadert, L., Hsü, K. J., et al., Initial Reports of the Deep Sea Drilling Project, v. 42A: Washington, D.C., United States Government Printing Office, p. 985–990.

RÖGL, F., AND STEININGER, F., 1983 Vom Zerfall der Tethys zu Mediterran and Paratethys: Annalen des Naturhistorischen Museums Wien, v. 85/A, p. 135–163.

ROUCHY, J. M., 1979, La sédimentation évaporitique sur les marges méssiniennes: Athens, 7th International Congress on Mediterranean Neogene: Annales de Géologie des Pays Hélléniques, v. 3, p. 1051–1060.

SARG, J. F., 1988, Carbonate sequence stratigraphy, *in* Wilgus, C. K., Hastings, B. S., Kendall, C. G. S. C., Posamentier, H. W., Ross, C. A., and Van Wagoner, J. C., eds., Sea-Level Changes: An Integrated Approach: Tulsa, Society of Economic Paleontologists and Mineralogists, Special Publication 42, p. 155–182.

SAVIN, S. M., ABEL, L., BARRERA, E., HODELL, D., KELLER, G., KENNETT, J. P., KILLINGLEY, J., MURPHEY, M., AND VINCENT, E., 1986, The evolution of Miocene surface and near-surface marine temperatures: Oxygen isotopic evidence, *in* Kennett, J. P., ed., The Miocene Ocean: Paleoceanography and Biogeography: Boulder, Geological Society of America Memoir 163, p. 49–82.

SCHLAGER, W., 1991, Depositional bias and environmental change—Important factors in sequence stratigraphy: Sedimentary Geology, v. 70, p. 109–130.

SCHLAGER, W., 1992, Sedimentology and sequence stratigraphy of reefs and carbonate platforms: Tulsa, American Association of Petroleum Geologists, Continuing Education Course Note Series 34, 71 p.

SCHOLLE, P., AND ARTHUR, M. A., 1980, Carbon isotopic fluctuations in pelagic limestones: Potential stratigraphic and petroleum exploration tool: American Association of Petroleum Geologists Bulletin, v. 64, p. 67–87.

SHACKLETON, N. J., 1985, Oceanic carbon isotope constraints on oxygen and carbon dioxide in the Cenozoic atmosphere, *in* Sundquist, E. T. and Broecker, W. S., eds., The Carbon Cycle and Atmospheric CO_2: Natural Variations Archean to Present: Washington, D.C., American Geophysical Union Geophysical Monograph 32, p. 412–417.

SHACKLETON, N. J., 1987, The carbon oxygen record of the Cenozoic: History of organic carbon burial and oxygen in the ocean and atmosphere, *in* Brooks, J. and Fleet, A. J., eds., Marine Petroleum Source Rocks: London, Geological Society of London Special Publication 26, p. 423–434.

SIMONE, L., AND CARANNANTE, G., 1988, The fate of foramol ("temperate-type") carbonate platforms: Sedimentary Geology, v. 60, p. 347–354.

SPIEGLER, D., AND DANIELS, C. H. von, 1991, A stratigraphic and taxonomic atlas of *Bolboforma* (Protophytes, Incertae sedis, Tertiary): Journal of Foraminiferal Research, v. 21, p. 126–158.

STILLE, P., RIGGS, S., CLAUER, N., AMES, D., CROWSON, R., AND SNYDER, S., 1994, Sr and Nd isotopic analysis of phosphorite sedimentation through one Miocene high-frequency depositional cycle on the North Carolina continental shelf: Marine Geology, v. 117, p. 253–273.

STILLE, P., STEINMANN, M., AND RIGGS, S. R., 1996, Nd isotope evidence for the evolution of the paleocurrents in the Atlantic and Tethys oceans during the past 180 Ma: Earth and Planetary Science Letters, v. 144, p. 9–19.

VECSEI, A., 1991, Aggradation und Progradation eines Karbonatplattform-Randes: Kreide bis Mittleres Tertiär der Montagna della Maiella, Abruzzen: Mitteilungen aus dem Geologischen Institut der Eidgenössischen Technischen Hochschule und der Universität Zürich, Neue Folge, v. 294, 171 p.

VINCENT, E., AND BERGER, W. H., 1985, Carbon dioxide and polar cooling in the Miocene: The Monterey hypothesis, *in* Sundquist, E. T. and Broecker, W. S., eds., The Carbon Cycle and Atmospheric CO_2: Natural Variations Archean to Present: Washington, D.C., American Geophysical Union Geophysical Monograph 32, p. 455–468.

VISSER, J. P. DE, 1990, Clay mineral stratigraphy of Miocene to Recent marine sediments in the Central Mediterranean: Geologica Ultraiectina, v. 75, 243 p.

Weissert, H., 1989, C isotope stratigraphy, a monitor of paleoenvironmental change: A case study from the Early Cretaceous: Surveys in Geophysics, v. 10, p. 1–61.

WOODRUFF, F., AND SAVIN, S. M., 1989, Miocene deepwater oceanography: Paleoceanography, v. 4, p. 87–140.

WRIGHT, J. D., AND MILLER, K. G., 1992, Miocene stable isotope stratigraphy, Site 747, Kerguelen Plateau: Proceedings of the Ocean Drilling Program, Initial Results, v. 120: College Station, Ocean Drilling Program, p. 855–866.

WRIGHT, J. D., MILLER, K. G., AND FAIRBANKS, R. G., 1992, Miocene stable isotopes: implications for deepwater circulation and climate: Paleoceanography, v. 7, p. 357–389.

LATE QUATERNARY SHIFT OF MIXED SILICICLASTIC-CARBONATE ENVIRONMENTS INDUCED BY GLACIAL EUSTATIC SEA-LEVEL FLUCTUATIONS IN BELIZE

C. ELMER FERRO, ANDRÉ W. DROXLER, JOHN B. ANDERSON, AND DAVID MUCCIARONE

Department of Geology and Geophysics, MS-126, Rice University, Houston, Texas 77005-1892

ABSTRACT: The last glacial interval (oxygen isotope stages 4–2 or 75–12 ka) is considered overall as a sea-level lowstand relative to today (Holocene, i.e., since 12 ka) and the previous interglacial highstand stage 5 (130–75 ka). During this stage 4–2 interval, an earlier lagoon and barrier-reef system—established on the edge of the southern Belize shelf margin during the previous stage 5 interglacial highstand—was completely exposed. The exposed stage 5 lagoon area became a siliciclastic fluvial plain bounded on its eastern side by a ridge of karst topography created by the exposure of the stage 5 barrier reef.

A high-resolution single-channel seismic survey in the northern part of the present Southern Shelf Lagoon off the coast of Belize was used to image two late Quaternary lowstand fluvial drainage systems consisting of a series of well-developed incised valleys, trending parallel to the strike of the exposed stage 5 barrier reef. The divide between both drainage systems, at the latitude of Dangriga (just south of 17°N), appears to be rooted on a topographic high composed of upper Tertiary deposits. South of Dangriga, the system of buried incised valleys drained the area toward the south and merged into Victoria Channel and other channels separating modern rhomboid reefs. North of Dangriga, the system of buried incised valleys drained into the single, partially filled, and deeply incised valley of the English Cay Channel. This valley cut through the exposed stage 5 barrier-reef and back-barrier-reef system and linked the fluvial drainage system north of Dangriga to the Turneffe Basin.

The bathymetric contours at the mouth of the English Cay Channel protrude eastward across a thick, fanlike sedimentary body characterized by a series of stacked, prograding, and laterally shifting lobes. This sedimentary body is interpreted as a lowstand shelf-edge delta, similar in surface area to the modern highstand delta of the Belize River on which Belize City is located. Southerly directed currents partially reworked the lowstand shelf-edge delta and created an elongated slope fan at the toe of slope of the exposed barrier reef.

A model, tied to the established sea-level curve from 20 ka to the present, is developed to explain the lowstand geometry and morphology of the incised valleys and the sedimentary deposits in front of the modern Belize Barrier Reef. This model also describes, in a tentative time frame, the nature of the transgressive sedimentary fills in the valleys. Moreover, correlation of the single-channel high-resolution seismic grid along the western margin of the Turneffe Basin with a multichannel seismic dip line shows that the lowstand shelf-edge delta at the mouth of the English Cay Channel and the slope fan, developed on the toe of the barrier reef farther south, correspond to the latest lowstand prograding sedimentary bodies that are mostly observed underlying the modern barrier-reef and back-barrier-reef system. The establishment and growth of carbonates over siliciclastic deposits—expressed in the colonization of patch reefs over the lowstand shelf-edge delta, beach ridges, and flooded levees—appear to represent a possible mechanism by which the present barrier reef established itself. Results of our study suggest that carbonate deposition occurred during a very restricted time interval of the overall sea-level highstand and that the carbonates rest on a substratum of lowstand and transgressive siliciclastic deposits. However, the establishment of barrier reefs, stacked on top of one another during successive interglacial highstands, though exposed during intervening glacial lowstands, has influenced the orientation of the lowstand incised valleys on the exposed shelf. These valleys trend parallel to the strike of the exposed barrier reefs.

INTRODUCTION

Sequence stratigraphic models are built on a few basic principles. One of them expresses that, during sea-level lowstands, sediment in siliciclastic depositional systems is transported into the adjacent basins by rivers and deposited directly onto the upper slope, while previous highstand shelf deposits are eroded (Vail et al., 1977). This basic principle was tested in the past two decades in the context of pure-carbonate depositional systems by sequence stratigraphic studies of isolated carbonate platforms and shelves. These studies generated some controversies. On one hand, it is thought that carbonate systems respond to sea-level fluctuations as siliciclastic systems do, with only minor differences (Vail, 1987; Vail et al., 1991; Van Wagoner et al., 1992; Sarg, 1988; Jacquin et al., 1991). On the other hand, many studies have demonstrated that, in contrast to siliciclastic systems, carbonate production sites ("factories") export sediment to adjacent slopes and basins during sea-level highstands when carbonate bank tops are flooded but still situated within the photic zone (referred to as "highstand shedding"). Carbonate systems can be, therefore, completely out of phase in terms of sediment shedding relative to siliciclastic margins (Mullins, 1983; Droxler and Schlager, 1985; Schlager, 1992; Schlager et al., 1994; Handford and Loucks, 1993). In this context, some significant discrepancies are expected to arise in the construction of a eustatic sea-level curve that is based upon the seismic stratigraphic interpretation of contemporaneous siliciclastic and carbonate deposits of the shelf, slope, and rise.

Eustatic sea-level curves have been established, independently of sequence stratigraphic concepts, for the late Quaternary (Fig. 1; i.e., Lighty et al. [1982] for the past 10 k.y.; Fairbanks [1989] for the past 20 k.y.; Bard et al. [1990] for the past 140 k.y.; and Pirazzoli et al. [1993] for the past 500 k.y.). Because sea-level history is best known for the late Quaternary, sedimentary processes and primary sequence stratigraphic concepts, essentially tied to sea-level fluctuations, can be better tested in upper Quaternary sediments than in more ancient rocks. High-resolution sequence stratigraphic studies of Quaternary strata have been conducted along a number of different siliciclastic shelves and slopes (i.e., Colman and Mixon, 1988; Trincardi et al., 1994; Anderson and Thomas, 1991; Anderson et al., 1996), including studies of incised valleys (Thomas and Anderson, 1988, 1989, 1994; Carter et al., 1990; Anderson et al., 1990; Blum, 1993; Koss et al., 1994; Zaitlin et

Advances in Carbonate Sequence Stratigraphy: Application to Reservoirs, Outcrops and Models, SEPM Special Publication No. 63
 ISBN 1-56576-060-3, p. 385–411

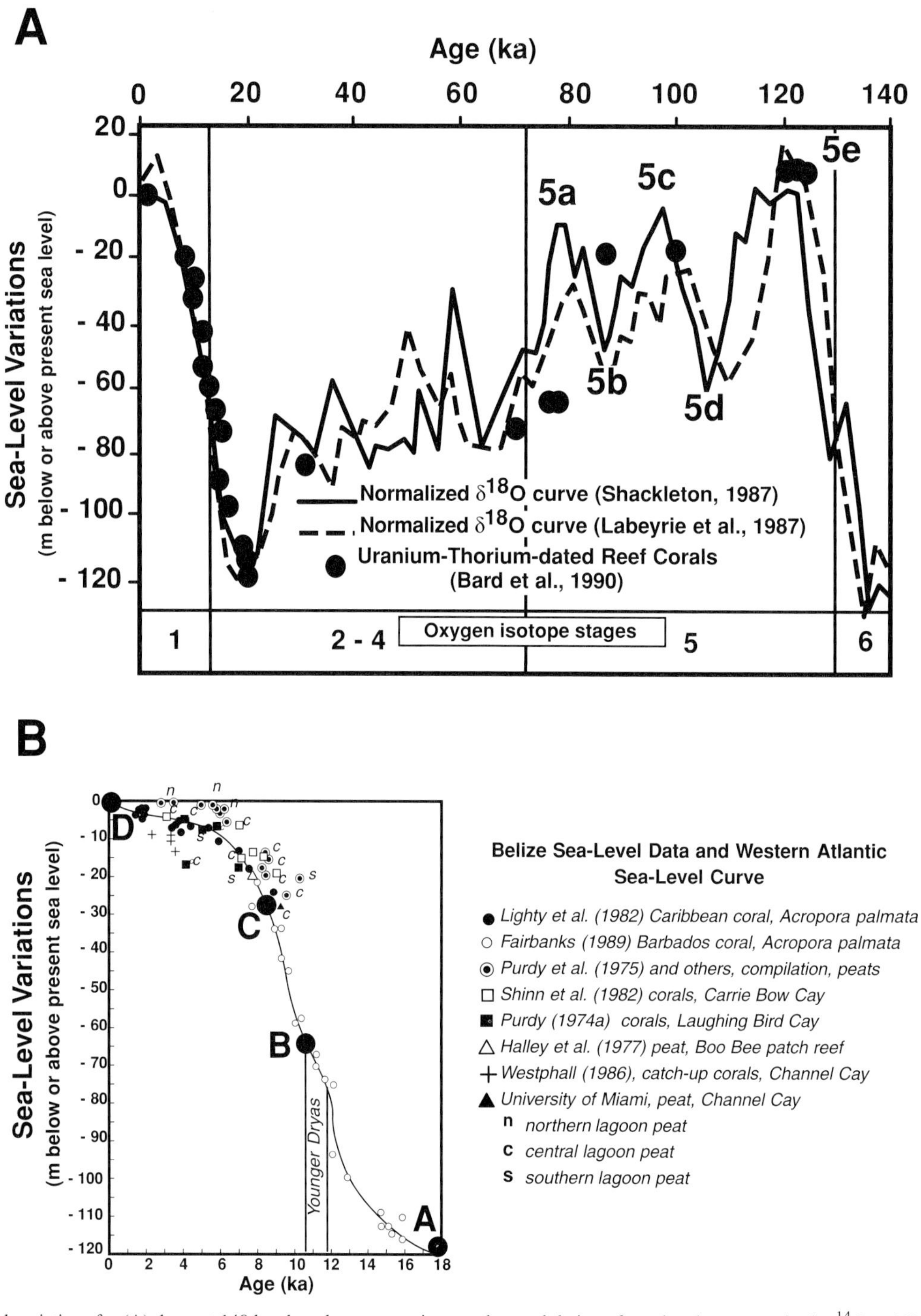

FIG. 1.—Sea-level variations for (A) the past 140 k.y. based on oxygen isotope data and dating of coral and peat samples by ^{14}C and U-Th methods and (B) the past 18 k.y. Letters A to D indicate the estimated levels of the ocean during four time intervals described in the model shown in Figure 18, which describes the evolution of the mixed siliciclastic-carbonate Belize system during the last sea-level cycle.

al., 1994; Kindinger et al., 1994; Harris, 1994; Allen and Posamentier, 1994; Nichol et al., 1994) and along carbonate-bank margins (Hine and Mullins, 1983; Droxler and Schlager, 1985; Wilber et al., 1990; Glaser and Droxler, 1991; Grammer and Ginsburg, 1992; Schlager, 1992; Schlager et al., 1994). However, studies testing sequence stratigraphic concepts either in late Quaternary or older mixed siliciclastic-carbonate sedimentary systems (i.e., Choi and Ginsburg, 1982; Belperio and Searle, 1988; Davies et al., 1989; Dolan, 1989; Peederman and Davies, 1993; Handford and Loucks, 1993; Brown and Loucks, 1993; Esker, 1998; Esker et al., 1995, Esker et al., 1998) are not as common.

Because modern siliciclastic and carbonate sediments coexist along the Belize continental margin (Purdy, 1974b; Choi and Ginsburg, 1982; Choi and Holmes, 1982) and because this margin is reasonably accessible, the Belize margin is an excellent area in which to study the association of siliciclastic and carbonate sediments within a sequence stratigraphic framework.

In a mixed sedimentary system such as the Belize margin, the migration and export of siliciclastic sediments toward the shelf margin during sea-level regressions and lowstands are expected to overwhelm and suffocate carbonate production. A prograding wedge of siliciclastic sediments is expected to be deposited by fluvial-deltaic processes below and on the seaward side of the exposed barrier reef during sea-level lowstands. This lowstand prograding wedge may provide a surface on which the new barrier reef becomes established sometime during the following sea-level transgression. Because the extent of the carbonate production is reduced to a minimum during sea-level lowstands, little carbonate resedimentation should occur along the slope and in adjacent basins. Conversely, as observed today, the Belize carbonate sediments are expected to thrive during sea-level transgressions and highstands, while siliciclastic sediments remain contained within the coastal zone.

We are reporting here on a high-resolution seismic survey of upper Quaternary strata of the Belize margin. This study was undertaken to understand how siliciclastic and carbonate systems have interacted in constructing the continental margin during the past few sea-level cycles. Another objective of this study was to investigate lateral and vertical shifts within the siliciclastic and carbonate environments caused by these sea-level fluctuations. This was accomplished by using two grids of high-resolution seismic lines within the northern part of the Southern Shelf Lagoon and the offshore Turneffe Basin.

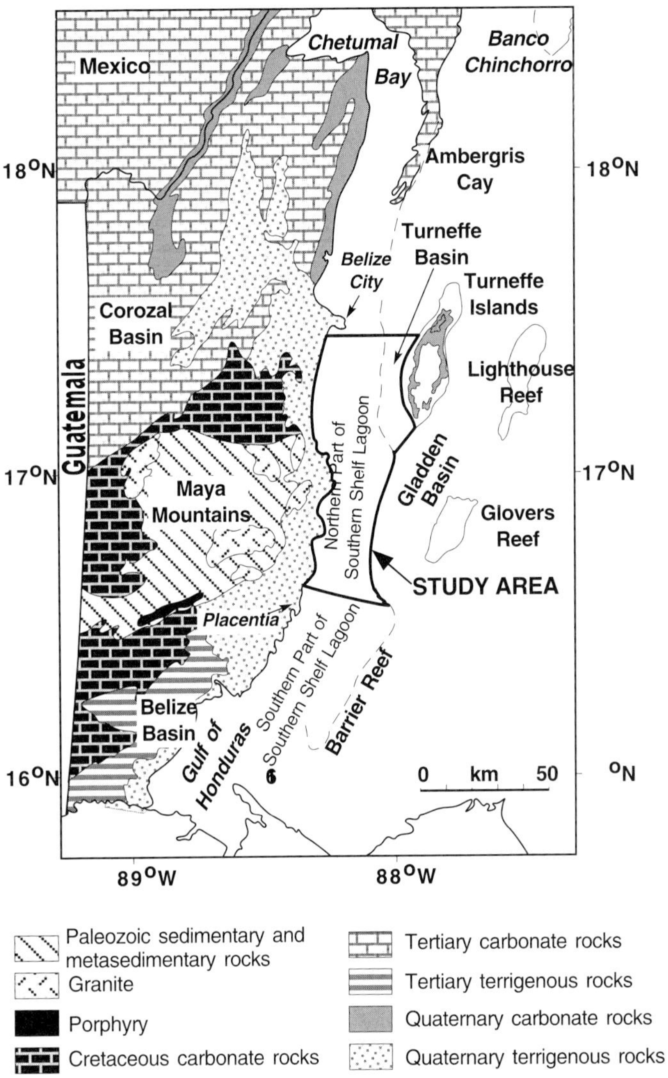

FIG. 2.—Belize geologic map modified from Wantland and Pusey (1971) and Purdy (1974a). The area where this study was conducted is also shown.

BACKGROUND

Belize exhibits two main geologic provinces linked to the Yucatan platform and the Sierras of northern Central America. Paleozoic metasedimentary rocks and granites, the oldest rock units in Belize, compose the core of the Maya Mountains. Relief in the central part of Belize reaches an elevation of more than 1,000 m at Victoria Peak (Bateson and Hall, 1977). Because of their relief and the intensity of chemical weathering, characteristic of the Belize tropical latitudes, the Maya Mountains have been a major and persistent source of siliciclastic sediments during the Cenozoic. Cretaceous carbonates, and Tertiary–Quaternary clastics and carbonates overlie the Paleozoic metasedimentary rocks and granites on the flanks of the Maya Mountains (Fig. 2).

The Belize shelf and the offshore basins have been studied to better understand their geology and structural evolution (Dillon and Vedder, 1973; Bateson and Hall, 1977; Morrice, 1992; Lara, 1993) as well as the sedimentological and paleontological aspects of their modern and past environments (Purdy et al., 1975; Wantland, 1975; James and Ginsburg, 1979; Choi and Holmes, 1982; Choi and Ginsburg, 1982; Shinn et al., 1982; Mazzullo et al., 1992; Precht, 1993, 1994). Our study area spans the northern part of the Belize Southern Shelf Lagoon and the Turneffe Basin (Figs. 3, 4). Recently, Esker (1998) and Esker et al., 1998, conducted a study similar and in parallel to ours by focusing on the southern part of the Belize Southern Shelf Lagoon. Esker emphasized the interplay between tectonics and eustasy in the Belize mixed carbonate-siliciclastic system. In Belize, pure siliciclastic muds, silts, and sands, originating from the Maya Mountains, are currently being deposited in the fluvial plain and along the coast. The shelf is covered with a mixture of siliciclastic clays and fine carbonate muds. The carbonate content of the shelf sediments systematically increases eastward (Pusey, 1975; Scott, 1975; Purdy et al., 1975). A well-established longshore current, flowing toward the south, confined most of the siliciclastic sediment to the coastal zone. Pure-carbonate sediment, composed of coral and coralline and green algae, occurs along the Belize barrier reef and in the back-barrier environment, as well as on the offshore atolls of Glovers, Turneffe, and Lighthouse (Gischler, 1994; Figs. 2, 3, 4). Periplatform oozes are being deposited in the basins adjacent to the barrier reef and the offshore atolls (James and Ginsburg, 1979), such as the Turneffe and Glovers Basins.

METHODS

Two grids of high-resolution single-channel seismic profiles were acquired in 1992–1993 onboard the R/V *Lone Star*

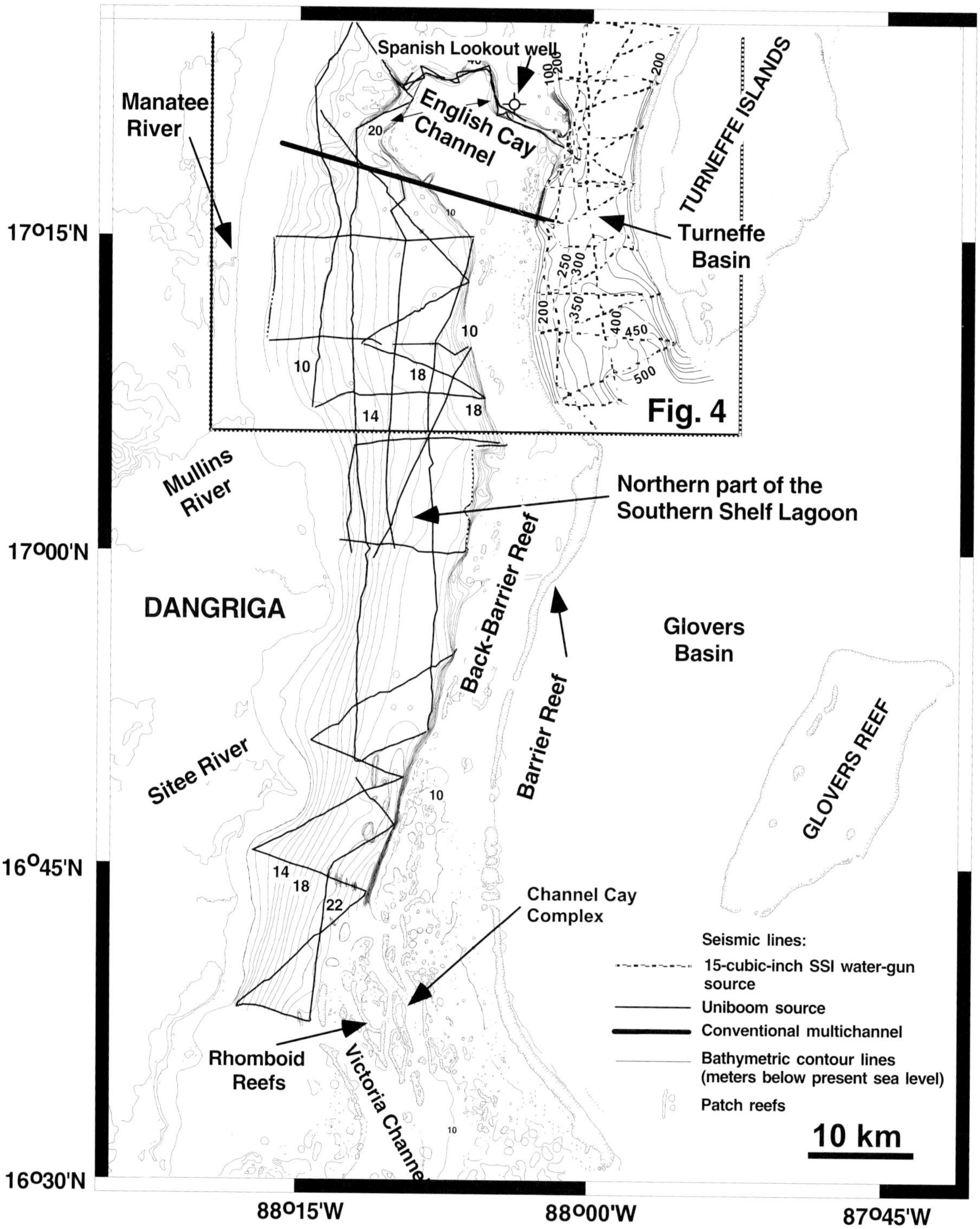

FIG. 3.—Modern bathymetry of the northern part of the Southern Shelf Lagoon and the Turneffe Basin modified on the basis of data sets newly collected during this study. These new data sets consist of two high-resolution single-channel seismic grids (data collected with the use of a Uniboom and a 15-in.^{3} SSI [Seismic Systems Incorporated] water gun), one conventional multichannel seismic line, and some lithologic and biostratigraphic information from the Spanish Lookout well. Note the overall southward dip of the modern sea floor from the entrance of the English Cay Channel (at the –20 m contour near the north edge of the map) to the southern part of the Southern Shelf Lagoon (just north of the Victoria Channel). Detailed modern bathymetry of the Turneffe Basin, the English Cay Channel, and the northern part of the Southern Shelf Lagoon at the western entrance of the English Cay Channel is shown in Figure 4.

(Fig. 3). The data sets were collected in the northern part of the Belize Southern Shelf Lagoon, the English Cay Channel, and in the basin—referred to here as the Turneffe Basin—located between the modern barrier reef and the atolls and islands of the Turneffe Islands (Fig. 3). A Uniboom (high-frequency seismic energy source) was used as the energy source to acquire about 500 km of data in the relatively shallow (<100 m) waters of the Southern Shelf Lagoon and the English Cay Channel, and a

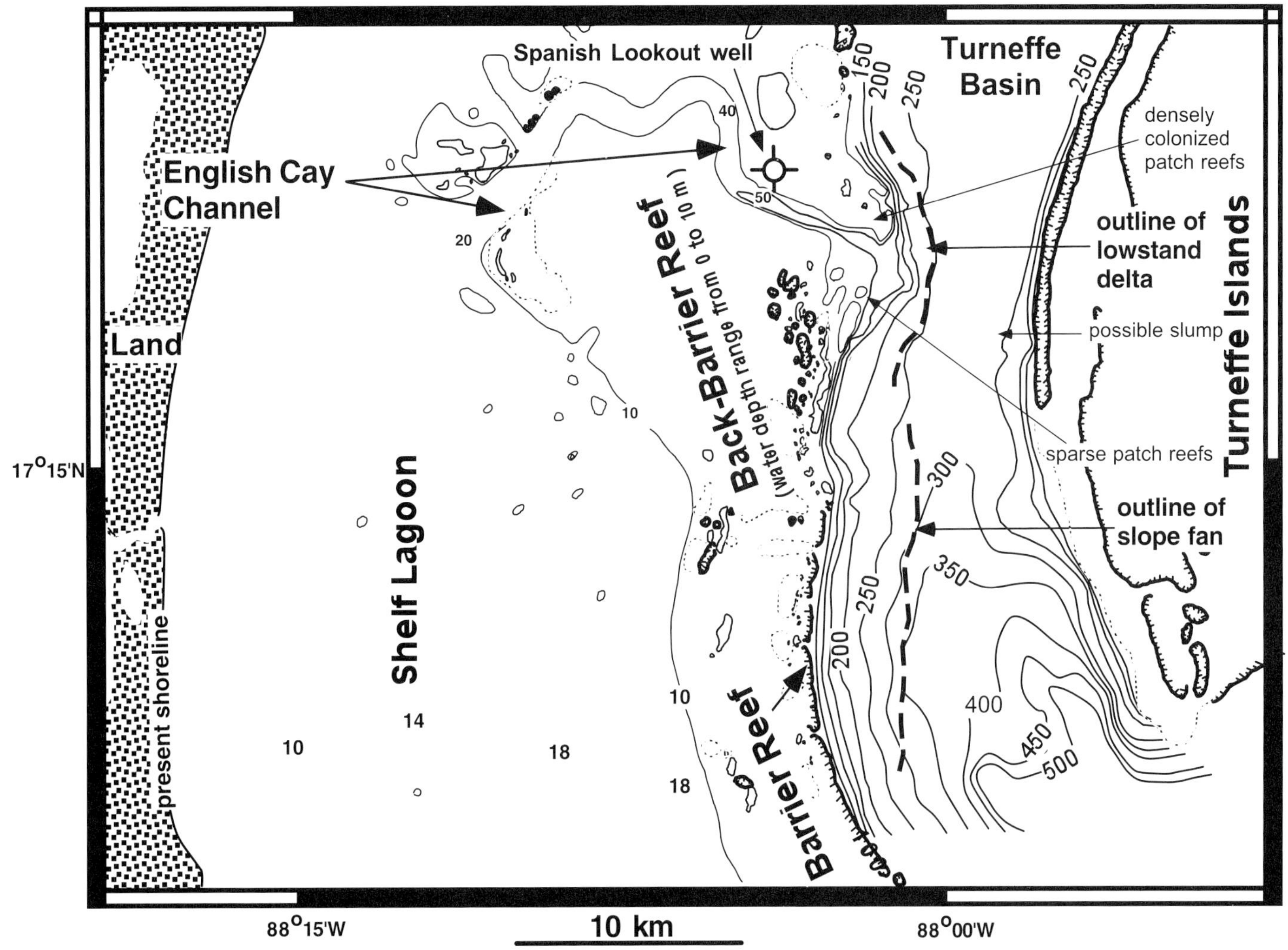

FIG. 4.—Detailed bathymetry of Turneffe Basin, the English Cay Channel, and the northern part of the Southern Shelf Lagoon at the western entrance of the English Cay Channel. Contour intervals are 10 m in Turneffe Basin and the English Cay Channel and 1 m in the Southern Shelf Lagoon. The overall southward-sloping floor of the Southern Shelf Lagoon is shown in Figure 3. From its entrance to its mouth, the English Cay Channel exhibits an eastward-sloping bathymetry. This map also shows a lowstand shelf-edge delta and a lowstand slope fan. Note also that the shallowest part of the Turneffe Basin sea floor is located in front (i.e., east) of the mouth of the English Cay Channel.

246-cm^3 (15-in.^{3}) water gun was used to acquire 800 km of data in the deep basins, including the Turneffe Basin. The seismic lines were digitally recorded in a pseudo SEG-Y format by using an Elics Delph2 system.

The seismic data were later processed at Rice University with PROMAX seismic processing software. Uniboom data were processed with the following sequence: single-filter Ormsby bandpass from 70 to 3200 Hz, time-offset variant gain, spiking-predictive deconvolution, automatic gain control (AGC), and trace muting. Water-gun data sets were processed twice with a single-filter Ormsby bandpass (from 80 to 900 Hz), AGC and trace muting. Seismic velocities of 1.5 km/s for the water and 1.7 km/s for the sediment were used to estimate the depth of sea floor and subsurface reflectors (Choi and Holmes, 1982). A 25-km-long multichannel seismic line (84–01) and well-log information and lithologic description of cuttings from the Spanish Lookout well were correlated with the high-resolution seismic grid (locations in Fig. 3).

RESULTS

Modern Bathymetry in the Northern Part of the Belize Southern Shelf Lagoon

The seismic data were used to map the bathymetry of the Belize Southern Shelf Lagoon. The lagoon floor slopes toward the south, from –11 m just north of the English Cay Channel to about –25 m near the northern boundary of Victoria Channel (Figs. 3, 4). The west-east bathymetric profile of the lagoon is asymmetric; the western floor slopes gradually toward the east, but the eastern floor of the lagoon is steep as it approaches the adjacent back-barrier-reef environment (Figs. 3, 4). Relatively flat areas along the axis of the shelf lagoon are located immediately south of the entrance to the English Cay Channel (Fig. 4) and in front of Dangriga (Fig. 3).

The overall southward slope of the lagoon contrasts with the English Cay Channel bathymetry, which slopes to the east. Water depths in the English Cay Channel range from about –20 m at the western head of the channel to about –60 m near its mouth (Fig. 4). The channel mouth expands onto a relatively flat surface that reaches a depth of about –90 m. The surface forms an eastward bulge in the bathymetric contours. Farther east, the floor slopes more steeply to about –250 m in Turneffe Basin. The northern part of the channel mouth is densely colonized by patch reefs, whereas the southern part has only sparse patch reefs (Fig. 4).

The shallowest depths (about 260 m) of the Turneffe Basin floor are found seaward of the English Cay Channel, where there are a few small, elongated, bathymetric ridges, interpreted as possible slumps (Fig. 4). The areas to the north and south of English Cay Channel reach maximum depths of about –280 m and –500 m, respectively. The western slope of Turneffe Basin is much gentler than its eastern slope. Comparisons of the maps shown in Figures 3 and 4 with previously published maps (Purdy et al., 1975; Defense Mapping Agency Hydrographic-Topographic Center, 1984; Morris, 1989) yield broadly similar features in the shelf lagoon, except that the areal extent of patch reefs is generally smaller in our maps (Fig. 3). Our Turneffe Basin bathymetric map (Figs. 3, 4) provides much greater detail than previously published maps.

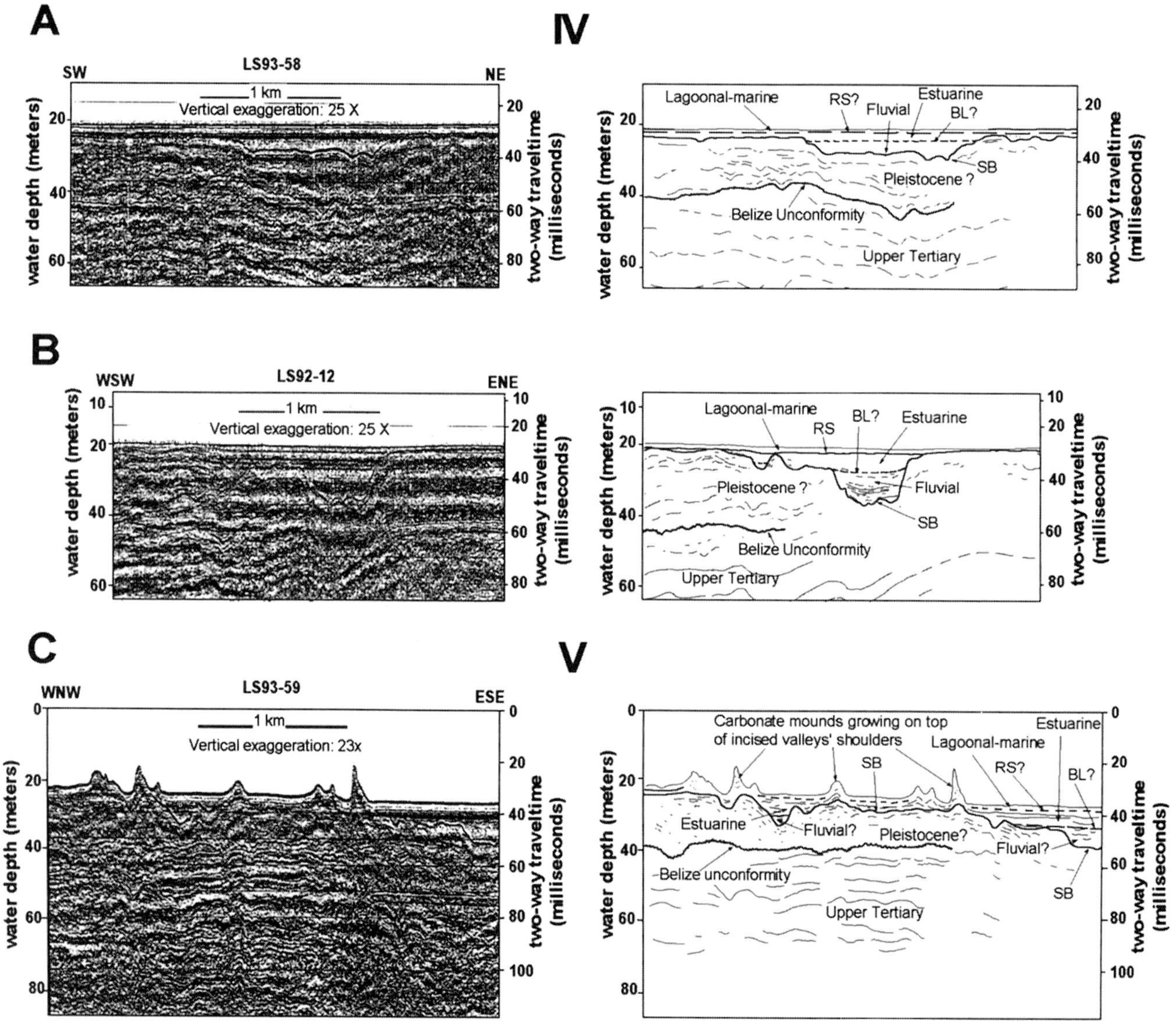

FIG. 5.—High-resolution single-channel seismic profiles (A–C), interpreted in adjacent line drawings, of the buried incised-valley system in the area of the Southern Shelf Lagoon south of Dangriga. These particular lines cross the complex drainage system of the lowstand fluvial plain between stations IV and V (located in Figs. 8, 10). In this area of the Southern Shelf Lagoon, the incised valleys are filled predominantly by seismically homogeneous deposits, which makes the separation of fluvial, estuarine, and lagoonal-marine seismic facies difficult. Patch reefs have developed over the levees of the incised valleys, as observed in C. RS = ravinement surface, SB = last glacial maximum sequence boundary, BL = bayline.

Incised Valleys

High-resolution seismic profiles from the shelf lagoon show a series of incised valleys (Figs. 5–7) that, when mapped, clearly delineate two paleodrainage systems separated by a drainage divide offshore Dangriga (Figs. 8, 9). South of Dangriga, individual valleys have U-shaped cross-sectional profiles (Figs. 5, 10A). Seismic lines display a complete cross-sectional profile of the main incised valley only in the northernmost part of the area. Farther to the south, it was not possible to acquire profiles across the valley because of shallow water depth. Hence, the deepest part of the incision was not always imaged. The maximum depth of the main channel ranges from –27 m in the northern part of the study area to –39 m in its southernmost part (Fig. 10B). A local gradient of 0.06° was calculated for the shallowest incision in the north to the deepest incision in the south (Fig. 10B). However, connection of the main incised valley in the north to the one reported by Choi and Holmes (1982) in the south yields an average gradient of 0.04° (Fig. 10B). The average subbottom depth of the highest interfluve areas near the back-barrier reef is –20 m. Patch reefs have commonly established themselves on top of the interfluves. In the south, the lowstand drainage configuration may have been influenced by local preexisting (substages 5e, 5c, and 5a?) topography consisting of exposed, elongate carbonate mounds. The occurrence of these older mounds resulted in the anastomosing character of the drainage pattern.

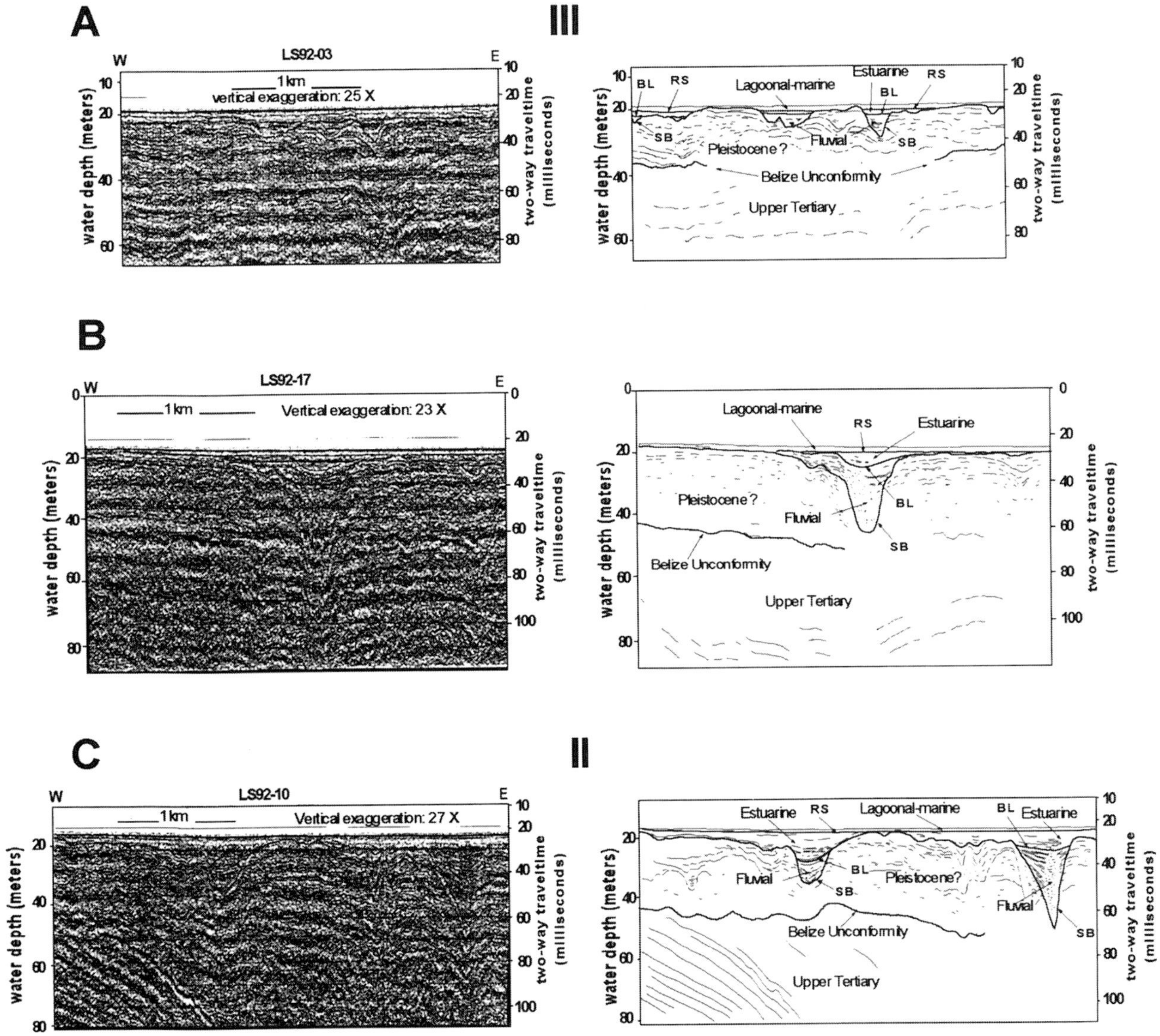

FIG. 6.—High-resolution single-channel seismic profiles (A–C), interpreted in adjacent line drawings, of the buried incised-valley system in the area of the Southern Shelf Lagoon north of Dangriga. These particular lines are located within the complex drainage system of the lowstand fluvial plain between stations II and III (located in Figs. 8, 10). In this area of the Southern Shelf Lagoon, incised valleys exhibit V-shaped cross-sectional profiles, and their fill is characterized by higher-amplitude seismic reflectors and more distinguishable seismic character than in the incised valleys south of Dangriga, allowing for better distinction between fluvial, estuarine, and lagoonal-marine seismic facies. The average depth of the incised valleys' shoulders is –18 m (Fig. 7A), increasing to about –23 m toward the entrance of the English Cay Channel (Fig. 7C). RS = ravinement surface, SB = sequence boundary formed at the last glacial maximum, BL = bayline.

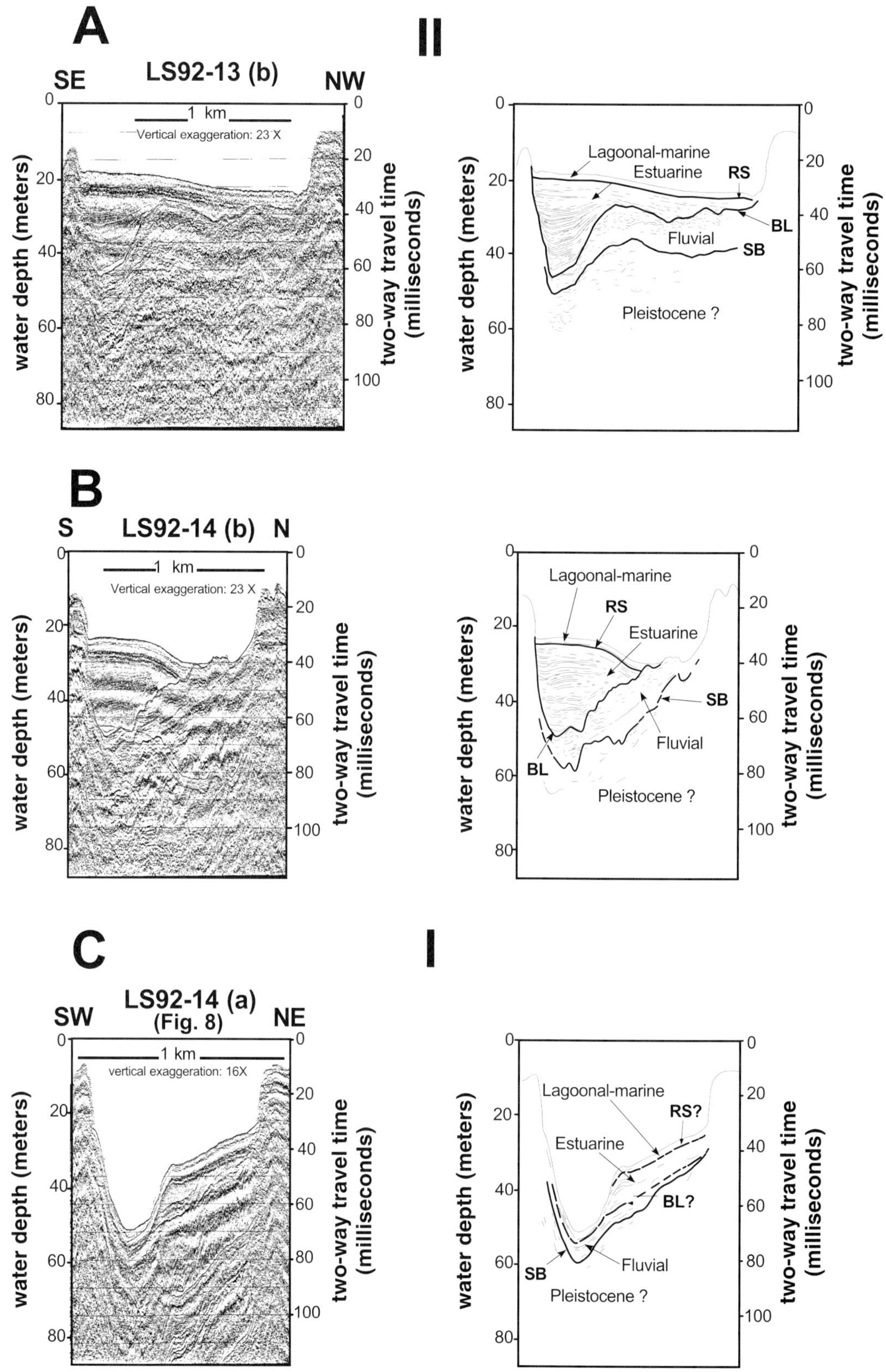

FIG. 7.—High-resolution single-channel seismic profiles (A–C), interpreted in adjacent line drawings, of the English Cay Channel incised valley. These particular lines are located within the complex drainage system of the lowstand fluvial plain between stations I and II (located in Figs. 8, 10). Sediment thickness of the valley fill increases dramatically upstream (from C to A). Note the difference in seismic character between the fluvial (chaotic) and the estuarine (laminar) seismic facies. As in the Southern Shelf Lagoon, the lagoonal-marine facies is very thin in the English Cay Channel. RS = ravinement surface, SB = sequence boundary formed at the last glacial maximum, BL = Bayline. Note the opposite asymmetries in the cross sections shown in A and B between the LGM deepest incision and the modern sea floor. This observation demonstrates that the modern sea-floor morphology of the English Cay Channel is mostly influenced by the westward flow of the incoming tide, whereas the LGM incision was cut by a river flowing from west to east.

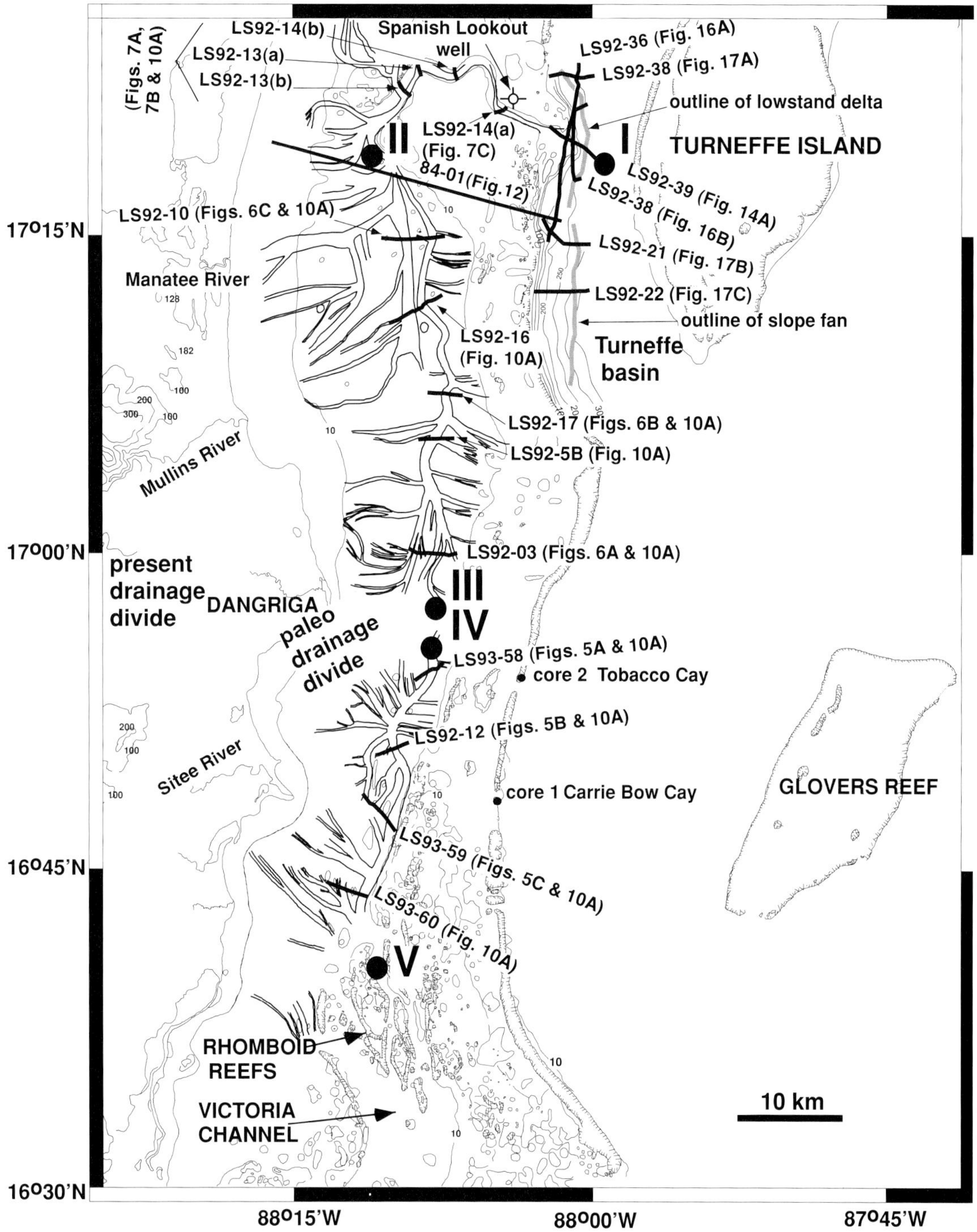

FIG. 8.—Paleodrainage during the last glacial maximum, 18–20 ka. Locations I to V are the positions of stations along the maximum incision of the main valley systems. Successive values of maximum incision were used to calculate the valley systems' gradient along the segment south of Dangriga (IV to V), the segment north of Dangriga (III to II), and the segment of English Cay Channel (II to I), shown in Figure 10B. Note how the complex drainage system north of Dangriga merges into the English Cay Channel and that the system south of Dangriga drains into the modern Victoria Channel complex. Compare the position of the paleo–drainage divide offshore Dangriga with the position of the dip divides of the upper Tertiary strata and the Belize unconformity (see Fig. 13).

A striking difference between the drainage system south and north of Dangriga is the occurrence of carbonate mounds in the southern area overlapping the shoulders of the incised valleys (Fig. 5). Although these features are local, they appear to be the northernmost expression of the well-developed, larger, rhomboid reefs in the south (Figs. 8, 9). These reefs are mainly growing on lowstand levee deposits (Choi and Ginsburg, 1982; Choi and Holmes, 1982) or on top of exposed stage 5(?) reefal limestone (Westphall, 1986).

North of Dangriga, the average subsea depth of the highest interfluve areas near the back-barrier reef is –19 m. The incised valleys show V-shaped cross-sectional profiles (Figs. 6, 10A).

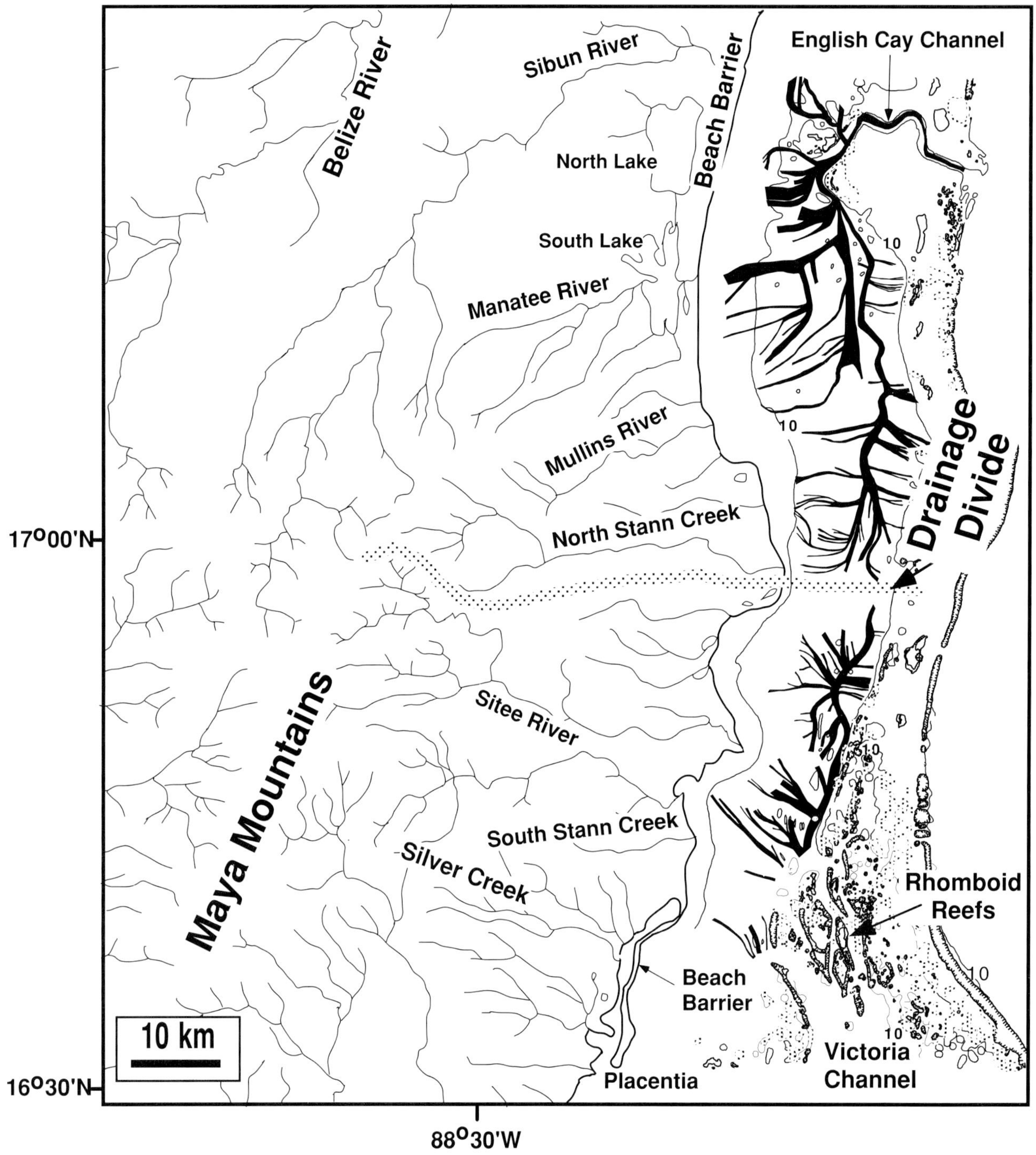

FIG. 9.—Comparison of the present onshore drainage with the paleodrainage during the last glacial maximum. The divides for both drainage systems at about lat 17°N appear to be connected.

In contrast with the area south of Dangriga, carbonate mounds are rare and small and appear to have a random distribution. The gradient of the main incised valley averages 0.032° (Fig. 10B), although gradients can be as steep as 0.22° (an order of magnitude steeper) in some segments of the main valley. In these segments, seismic facies of the valley fill are characterized by high-amplitude, chaotic reflectors, interpreted to correspond to coarse-grained sediment. A sharp increase in gradient occurs along an enclosed depression situated at approximately latitude 17°07′N (Fig. 10B). This depression has a depth of about 9 m below the local sea floor.

The English Cay Channel displays V-shaped, asymmetric cross sections, gradually deeper incisions toward the east, and a progressive decrease in thickness in its valley fill (Figs. 7, 10). In curved segments of the channel, the deepest fluvial incision is usually located at the inside bend of the channel (Fig. 7). This pattern contrasts with the overall asymmetric morphology of the modern sea floor that systematically displays the deepest part of

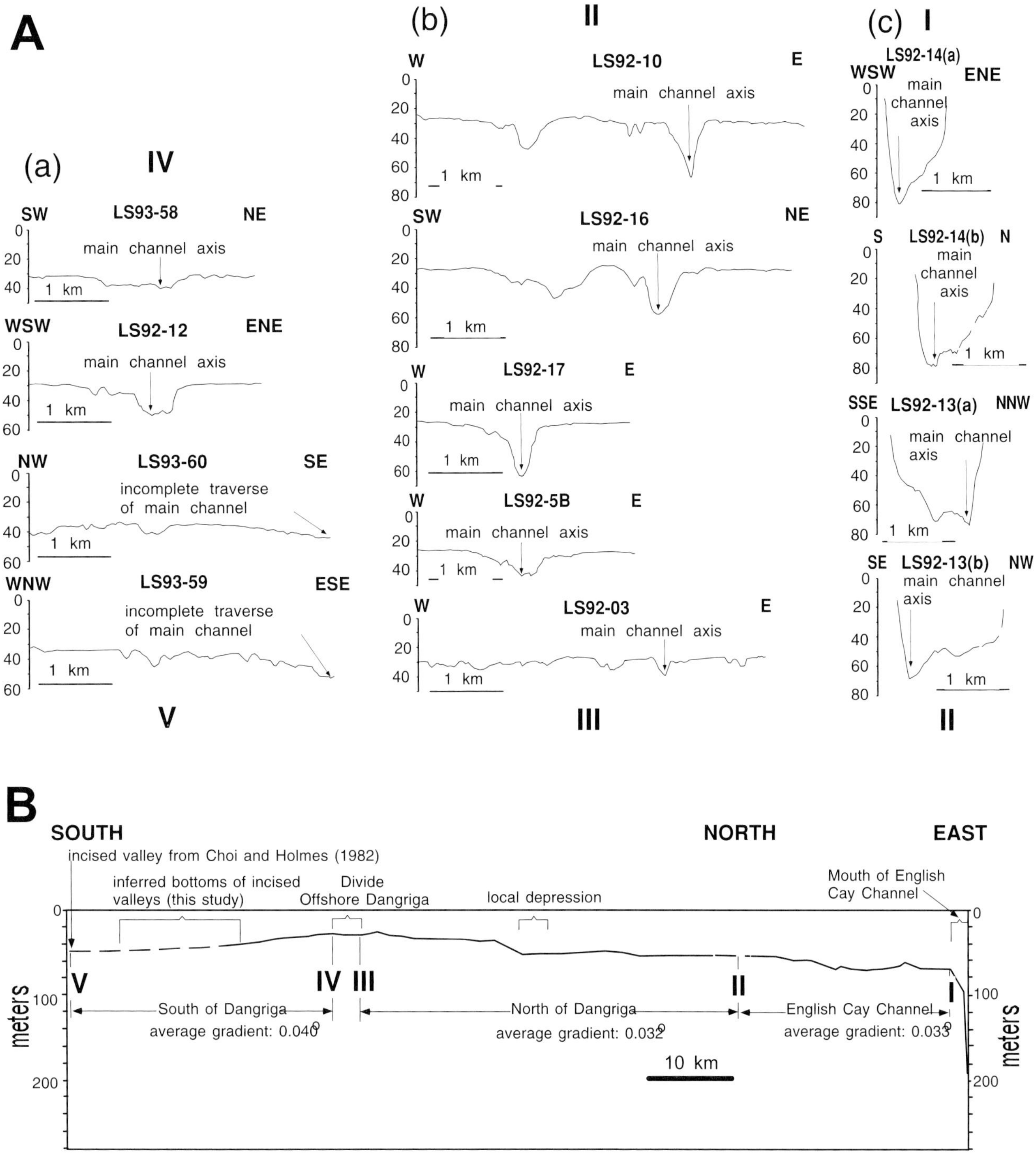

FIG. 10.—Cross sections of the valleys incised in the northern part of the Southern Shelf Lagoon and the English Cay Channel during the LGM, based on high-resolution seismic lines and the overall estimated sea-floor gradients. (A) Cross sections of incised valleys in three segments of the paleo–drainage system (located in Figs. 8, 9). (a) South of Dangriga, segment from IV to V. (b) North of Dangriga, segment from II to III. (c) English Cay Channel, segment from I to II. Vertical scale is two-way traveltime in milliseconds. (B) The longitudinal profiles of the main LGM incision, north and south of Dangriga, and along the English Cay Channel.

the channel floor on the opposite channel bank when compared with the last glacial maximum (LGM) incision (Fig. 7). This observation demonstrates that the modern sea-floor morphology of the English Cay Channel is mostly influenced by the westward flow of the incoming tide.

In summary, interpretation of the high-resolution seismic profiles from the Belize Southern Shelf Lagoon clearly shows two paleodrainage systems of incised valleys (Figs. 5–7), separated by a drainage divide offshore Dangriga (Figs. 8, 9). The valleys achieved maximum incision during the LGM (18–20 ka). During the last glacial interval, the fluvial plain north of Dangriga drained toward the north. When active, these valleys converged into the English Cay Channel, which drained across an exposed and karst-modified stage 5 barrier reef and into

Turneffe Basin (Figs. 8, 9). South of Dangriga, rivers flowed toward the south. The system of incised valleys, now buried in the shelf area between Dangriga and lat 16°40′N, merged into the modern channels that separate the rhomboid reefs (Figs. 8, 9).

Comparison Between Offshore Paleodrainage and Onshore Modern Drainage Systems

Two large dendritic rivers, the Belize and Sibun Rivers, and numerous streams (e.g., Mullins and Sitee Rivers, North Stann Creek, etc.) currently drain the eastern and northern flank of the Maya Mountains (Fig. 9). The Belize and Sitee Rivers carry relatively high sediment loads and form lobate deltas. Most of the streams carry less sediment and form small cuspate deltas (e.g., South Stann Creek). Southerly longshore currents partially rework these deltas and form a series of beach barriers composed of accreted beach ridges. Behind these barriers are small isolated lagoons such as North Lake and South Lake (Fig. 9). North of Dangriga, the modern river drainage is toward the east-northeast, whereas south of Dangriga, the drainage is toward the east-southeast. This modern drainage system between Belize City and Placentia contrasts with the overall north-south channel orientation of the paleodrainage system. The older north-south orientation appears to have been influenced by the overall north trends of the previous late Pleistocene interglacial stage 5 barrier reef.

The separation between the two different paleodrainage systems offshore Dangriga (Figs. 8, 9) appears to be structurally rooted on an upthrown block along a tectonic wrench zone in the central part of the Belize margin (Morrice, 1992). Although the structural high seems to have remained in the same location during most of the Quaternary, it is noteworthy that the paleodrainage system is in the opposite direction of the slope of the lagoon floor in the area between Dangriga and English Cay Channel (Figs. 3, 8, 9).

The paleodrainage system consists of numerous east-west tributaries feeding into the main north-south paleochannels. It is not possible to directly connect the paleodrainage system of the Southern Shelf Lagoon with the modern river system because the seismic profiles did not extend far enough landward. However, larger streams, like the Mullins River, North Stann Creek, and Sitee River, do not seem to be linked to any of the paleovalleys of the Southern Shelf Lagoon. The only exception is the Manatee River, which appears to be linked to an incised valley (Fig. 8). This apparent poor correspondence between modern and paleodrainage systems is in agreement with results of laboratory experiments that have demonstrated that, following an overall transgression, new drainage systems are usually established independently from paleodrainage systems (Koss et al., 1994).

The Belize Unconformity and Upper Tertiary Reflectors

Interpretation of the high-resolution seismic grids reveals that the paleodrainage was controlled by a topographic high in the upper Tertiary strata of the Southern Shelf Lagoon (Figs. 11–13). The dips of upper Tertiary main seismic reflectors define two sectors within the northern part of the Southern Shelf Lagoon (Figs. 11, 13). At the latitude of Dangriga, the reflectors are generally subhorizontal and extend farther south into smooth, concave, southward-dipping planes (Fig. 13). North of Dangriga, upper Tertiary strata dip toward the northeast, whereas farther to the north, they dip to the east. Closer to the coast, the upper Tertiary reflectors have dips ranging from 7° to 12° (Figs. 11–13). Farther offshore, these reflectors are characterized by gentler dips that range from 4° to 7° (Fig. 12). The variations of the upper Tertiary stratal geometries in the northern part of the Southern Lagoon define a topographic high, interpreted to have subsequently influenced the location of the late Quaternary drainage divide (Fig. 13A). A well-developed regional unconformity, observed within the entire Belize Southern Lagoon and referred to here as the "Belize unconformity," clearly truncates the upper Tertiary dipping reflectors (Figs. 6C, 11, 12). This unconformity appears to follow the physiography of the topographic high located offshore Dangriga (Fig. 13B).

Upper Quaternary Sedimentary Deposits

The incised valleys south of Dangriga seem to merge into the Victoria Channel system, whereas the incised valleys north of Dangriga are connected to the open sea through the English Cay Channel (Fig. 8). A lowstand shelf-edge delta and an overlying smaller delta, interpreted as a landward-stepped delta, occur at the mouth of the English Cay Channel. In Turneffe Basin, a slope fan occurs south of the shelf-edge delta and in front of the modern barrier reef. The two systems of incised valleys south and north of Dangriga show characteristic valley-fill seismic facies (Figs. 5–7). These seismic facies, which are better imaged in the incised valley north of Dangriga, are described here by area.

Incised valleys in the Belize Southern Shelf Lagoon, south of Dangriga.—

On the basis of studies of Quaternary incised valleys (i.e., Abdulah and Anderson, 1991; Thomas and Anderson, 1994; Kindinger et al., 1994; Nichol et al., 1994), parallel and laminar seismic reflectors are characteristic of estuarine and marine muds, whereas chaotic reflectors represent sandy fluvial sediment of the valley-fill succession. The incised valleys south of Dangriga are filled with low-amplitude, generally laminar reflectors (Fig. 5). Because of the subtle variations of seismic facies of the valley fill, it is difficult to divide it into fluvial, estuarine, and lagoonal-marine deposits. Faintly aligned reflectors, visible only locally, were selected as boundaries between the different seismic facies, although commonly they could only be inferred. On the basis of observations of sediment provenance, discussed later in this paper, the lower transparent seismic facies of these valley fills is inferred to represent fluvial facies consisting of fine-grained sands. This section is overlain by slightly more laminar reflectors interpreted to correspond to estuarine and marine muds.

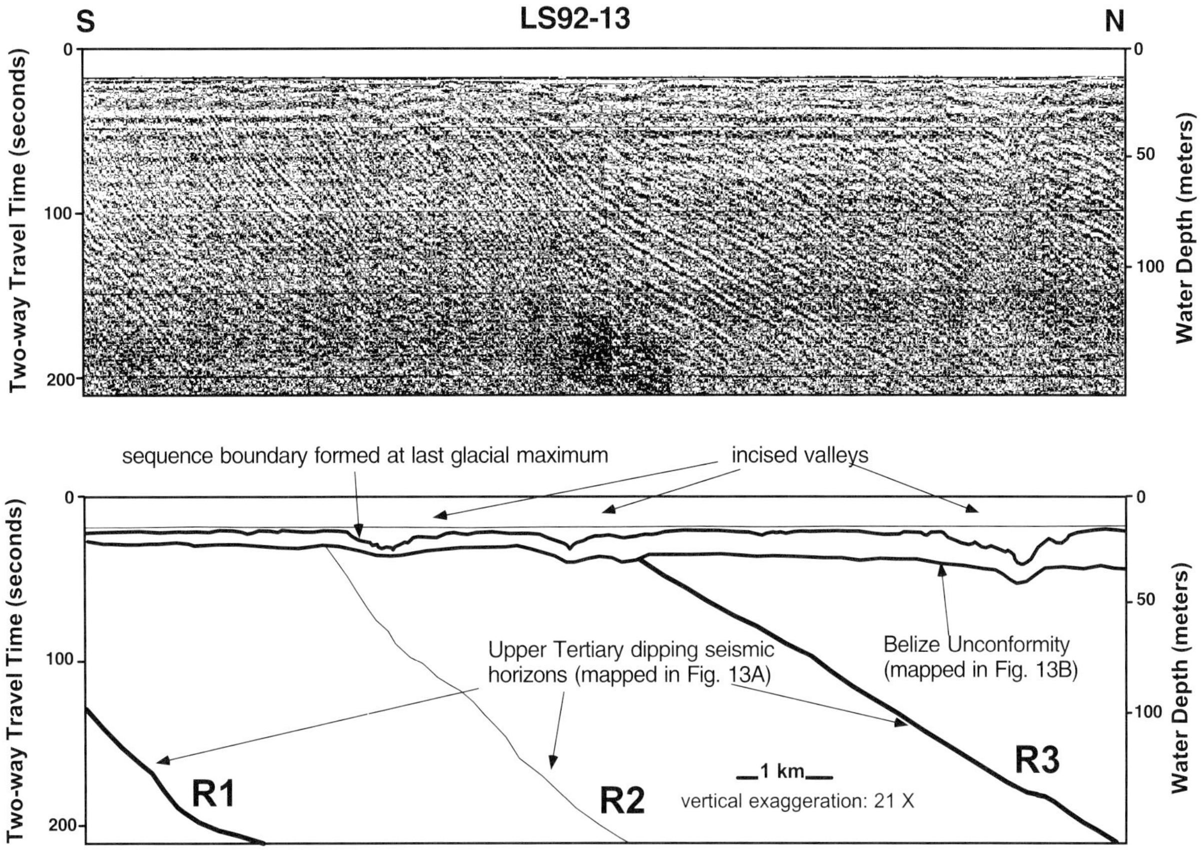

FIG. 11.—Segment of a south-to-north high-resolution single-channel seismic line located in the Southern Shelf Lagoon between Dangriga and the western entrance of the English Cay Channel located in Figure 13. Interpretation of the seismic line segment is shown in accompanying line drawing. Also shown are details of the relationship between the upper Tertiary dipping seismic reflectors, the Belize unconformity, and the incised valleys. The upper Tertiary seismic horizons R1, R2, and R3 as well as the surface representing the Belize unconformity are mapped in Figure 13.

Incised valleys in the Belize Southern Shelf Lagoon, north of Dangriga.—

In contrast to the seismic character of the valley fill south of Dangriga, the seismic character of the incised valley fill north of Dangriga can be more easily differentiated into three different seismic facies—(1) fluvial, (2) estuarine, and (3) lagoonal-marine sediments (Fig. 6)— presented below from bottom to top:

1. The fluvial facies is represented by high-amplitude and discontinuous reflectors, probably corresponding to coarse sediment. The fluvial sediments overly the surface of maximum incision, interpreted to be a sequence boundary cut during the LGM at 18–20 ka. The depth of maximum incision in the main channel ranges from –27 m offshore Dangriga to –53 m near the entrance to the English Cay Channel (Figs. 6–10).
2. The estuarine facies is characterized by laminar seismic reflectors representing alternating fine and coarse sediment (e.g., shale, silt, and fine sand). The estuarine sediment is separated from the underlying fluvial sediment by the bayline, a surface that formed as the sea transgressed within the incised valleys. The maximum depths of the bayline in the main channel range from –23 m offshore Dangriga to –35 m near the entrance of the English Cay Channel.
3. The lagoonal-marine facies is characterized by weak seismic reflectors, in some cases almost transparent, corresponding to fine homogeneous sediment, possibly a mixture of terrigenous and carbonate muds (Fig. 6). This seismic facies covers the entire shelf lagoon and overlies a widespread flat reflector, interpreted as a marine ravinement surface.

English Cay incised valley.—

Along the English Cay Channel, the vertical stacking of seismic facies is similar to that observed in the other incised valleys on the shelf, except that the estuarine deposits are thin in the eastern part of the channel and thicken abruptly toward its western part (Fig. 7). Estuarine sediment thickness ranges from about 4 m close to the mouth of the channel to 27 m in the western part of the channel. The maximum depth of the bayline surface is calculated to reach –61 m close to the mouth of the English Cay Channel and –47 m in the western part of the channel. The depth of the oldest fluvial facies, also corresponding to the level of maximum incision, ranges from –68 m in the eastern part of English Cay Channel to –53 m in the western part of the channel.

Elongated, commonly wedge-shaped sedimentary bodies

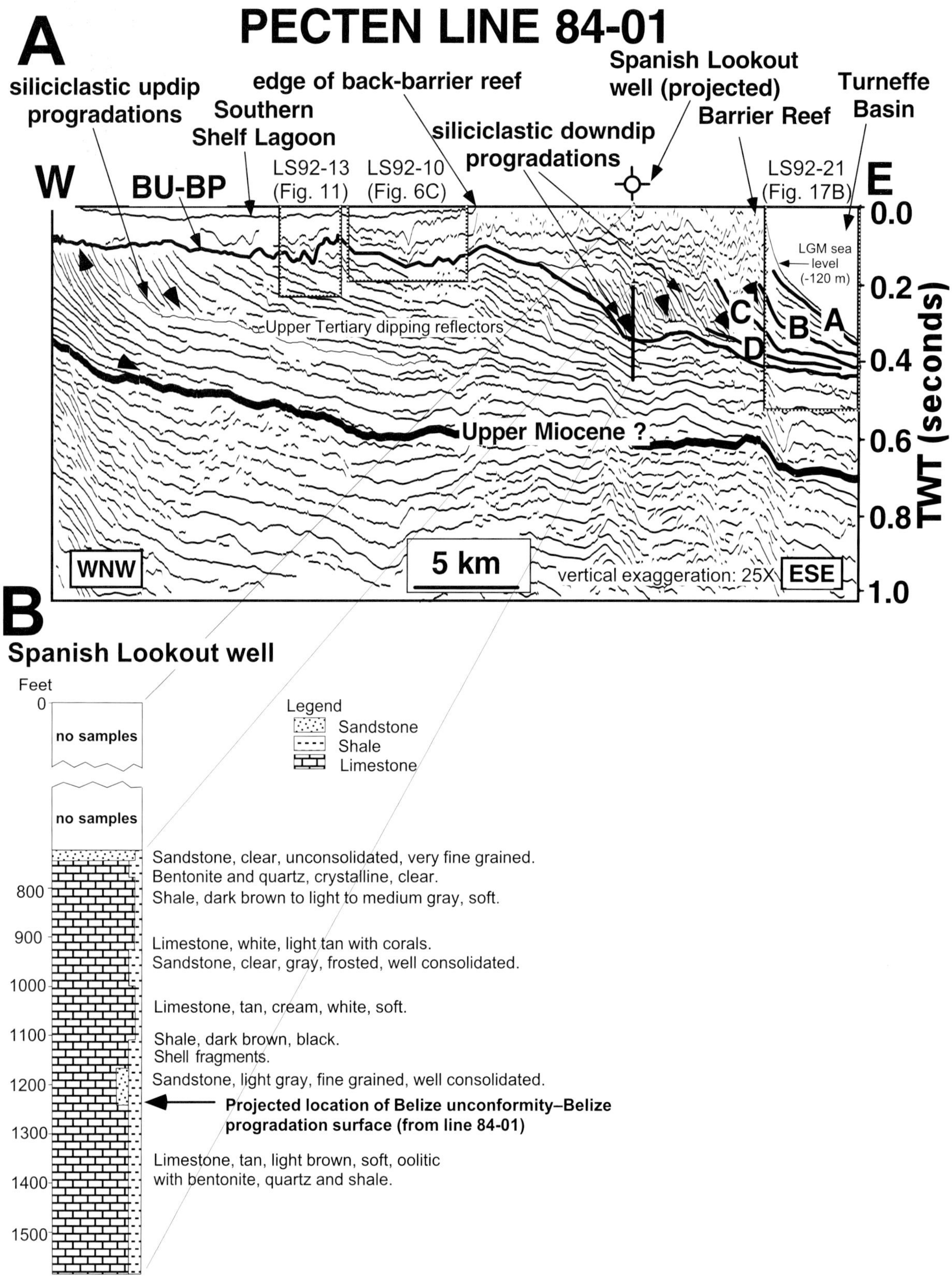

FIG. 12.—(A) Interpretation of multichannel seismic line 84-01 (located in Figs. 8, 13) and its correlation with the projected Spanish Lookout well. The outlined boxes indicate the projection of the high-resolution seismic lines shown in Figures 6C, 11, and 17B. The Belize unconformity truncates the top of the siliciclastic updip progradation complex in the west part of the seismic line and becomes a downlap surface for the siliciclastic downdip progradation complex observed beneath the barrier and back-barrier reef system in the east part of the seismic line. These progradations appear to be connected to wedges of sediment (LS92-21, Fig. 17B) associated with the lowstand shelf-edge delta and slope fan (LS92-39, Fig. 14A; LS92-22, Fig. 17C). Reflectors A to D can be correlated from this multichannel line to the high-resolution lines in Turneffe Basin (see Figs. 14A, 16A, 17B). The modern reef overlies wedges of downdip progradations. In Figure 12B, a schematic representation of the lithology at the Spanish Lookout well shows intercalations of siliciclastic and carbonate sediment in the interval corresponding to the downdip progradations. TWT = two-way traveltime.

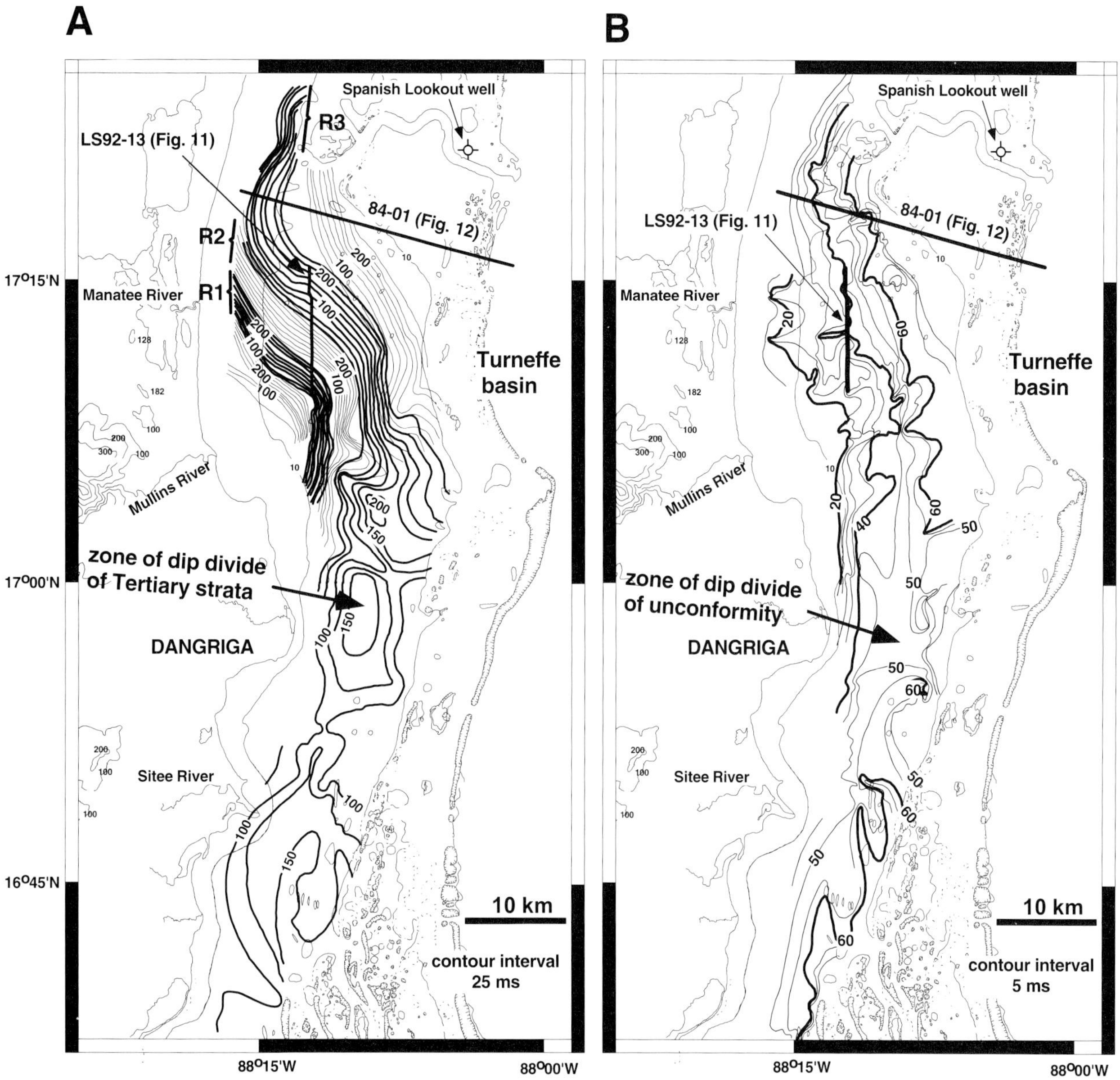

FIG. 13.—(A) Contour map of three upper Tertiary seismic horizons R1, R2, and R3 identified below the sea floor of the Southern Shelf Lagoon (Fig. 11). The R3 horizon displays a high, plateaulike area offshore Dangriga where the lowstand drainage system of valleys incised during the last glacial maximum shows a clear divide (see Figs. 8, 9). In the area north of the elevated plateau, the R1, R2, and R3 horizons dip toward the northeast, whereas in the area south of the plateau, the R3 horizon dips toward the south. (B) Contour map of the Belize unconformity identified below the sea floor of the Southern Shelf Lagoon (Figs. 5, 6, 11). The Belize unconformity displays a high, plateaulike elevated area offshore Dangriga where the lowstand drainage system of valleys incised during the last glacial maximum shows a clear divide (see Figs. 8, 9).

occur within the estuarine facies along English Cay Channel. These deposits generally show a laminar reflection character, and are locally chaotic. They are separated from the estuarine facies by an erosional surface and are interpreted to represent tidal creek bars.

Lowstand shelf-edge delta.—

A thick, eastward-bulging sedimentary deposit occurs at the mouth of the English Cay Channel (Figs. 3, 4). This sedimentary body is characterized in dip lines by a series of steep sigmoidal prograding reflectors with angles of about 10° (Figs. 14A, 15). This feature is interpreted as a lowstand shelf-edge delta. A strike section shows stacked delta lobes that display a clear shift of the depocenter toward the south (Fig. 16B). A high-amplitude seismic reflector cuts the upper part of the delta, clearly truncating the underlying dipping reflectors (Figs. 14A, 15). This particular reflector occupies the entire mouth of the English Cay Channel. The surface representation of this reflector reaches its

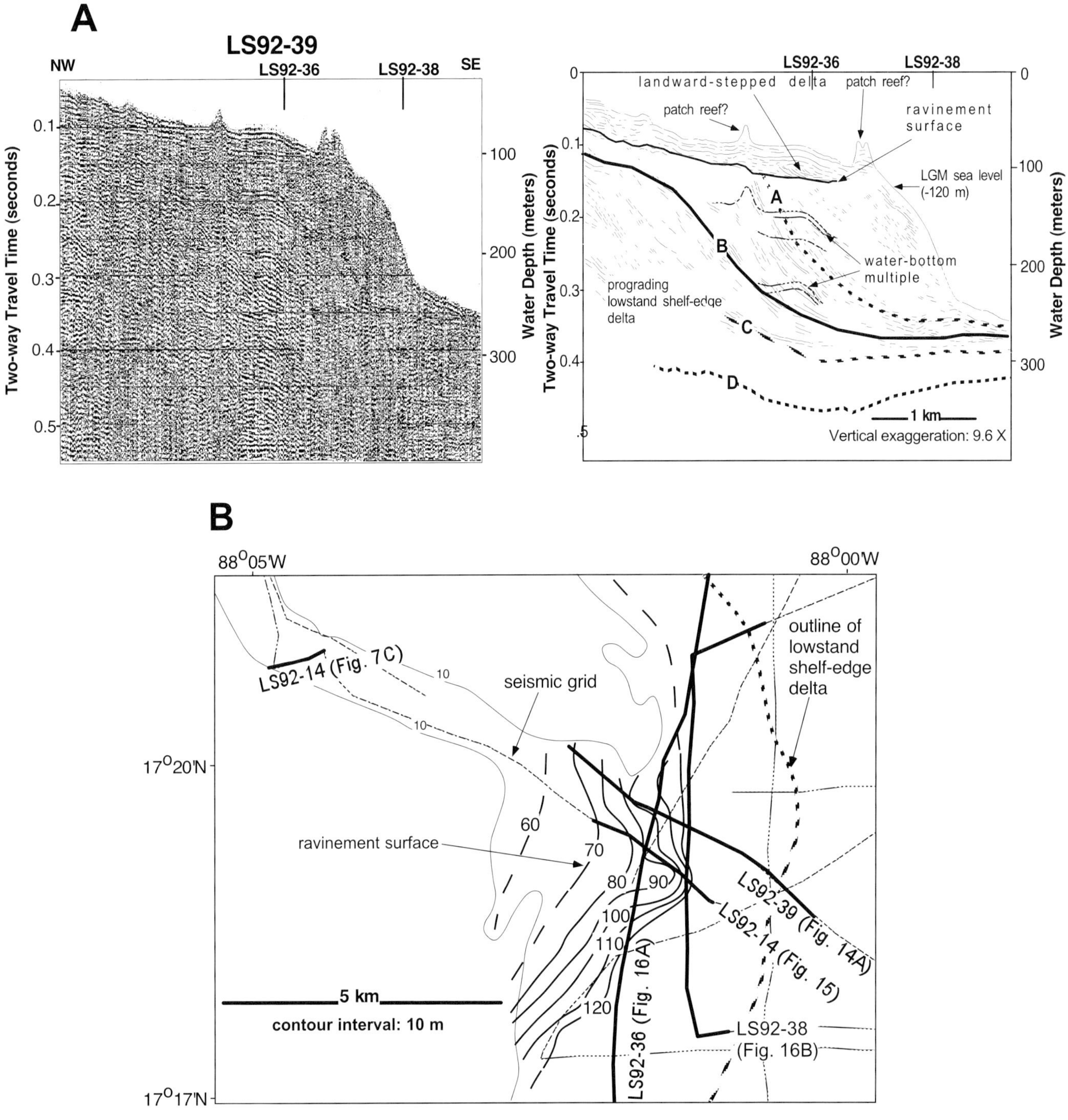

FIG. 14.—(A) High-resolution, single-channel, dip-oriented seismic profile, interpreted in the adjacent line drawing, showing a lowstand shelf-edge delta deposited at the mouth of the English Cay Channel (see location of line in B). A ravinement surface truncates the top of the prograding lowstand delta. A landward-stepped delta, deposited during the Younger Dryas, overlies the ravinement surface. Drowned patch reefs appear to have been established on top of the landward-stepped delta. Four seismic horizons (A to D) have been identified in the prograding lowstand shelf-edge delta. These horizons are correlated along dip and strike in single-channel and multichannel seismic lines in the Turneffe Basin (see Figs. 12, 16, 17). (B) Contour map of the ravinement surface on top of the lowstand shelf-edge delta. The ravinement surface reaches its lowest point at about 110 m below sea level in the most distal area of the channel mouth, which corresponds well to the LGM sea level (see Fig. 1).

lowest point at about –110 m in the most distal area of the channel mouth, rising upstream in a westerly direction to approximately –65 to –70 m (Figs. 14, 15). This reflector is interpreted to represent the ravinement surface generated during the initial and rapid sea-level transgression subsequent to the LGM lowstand.

A thin, delta-shaped sedimentary deposit overlies the ravinement surface (Fig. 14A). This body is characterized by lower-angle clinoforms relative to those in the underlying shelf-edge delta and extends up the English Cay Channel to –65 m (Figs. 14, 15). This sedimentary body is interpreted as a landward-stepped delta, resulting from fluviodeltaic sedimentation at the mouth of the English Cay Channel as sea level paused at ca. 11 ka during the Younger Dryas (Fig. 1; Fairbanks, 1989; Ruddiman and McIntire, 1981).

Slope fan.—

A strike line (Fig. 16A) that extends along the front of the barrier reef and across the mouth of the English Cay Channel shows that the lowstand shelf-edge delta ends abruptly to the north whereas it tapers more gradually to the south. About 15 km south of the lowstand delta, a morphologically distinctive, eastward-bulging sedimentary body is observed that is interpreted to correspond to a slope fan (Fig. 17C). This slope fan is connected to the lowstand delta by wedge-shaped slope deposits (Fig. 17B). The wedge-shaped deposits located north of the lowstand delta are much thinner than the ones south of the lowstand delta. This observation is illustrated in line LS92-38 (Fig. 17A), which exhibits a much thinner sediment wedge in the north than in the area south of the lowstand shelf-edge delta (Figs. 17B, 17C). The thickness of the sedimentary subunits defined by reflectors A, B, C, and D varies within the slope fan (i.e., interval between reflectors B and C, Fig. 17).

The slope-fan surface area is comparable to the surface area of the lowstand shelf-edge delta, about 8 km long and 4 km wide. A very steep slope connects the barrier reef to the top of the slope fan at about –200 m. From this depth to about –340 m, the slope is much gentler, forming an eastward-bulging surface, though the bathymetric expression of this bulge is not as accentuated as the bulge formed by the lowstand shelf-edge delta (Figs. 17C, 14A). In each subunit of the slope fan, laminar seismic reflectors downlap onto major reflectors A, B, C, and D, which define the top of a series of laminar concordant reflectors.

Correlation of Slope Deposits with Prograding Units Beneath the Barrier and Back-Barrier Reefs

Seismic line LS92-21, located south of English Cay Channel, shows wedgelike slope deposits in the basin (Fig. 17B). In strike lines, these deposits are shown to connect the lowstand shelf-edge delta at the mouth of English Cay Channel with the deposits in the slope fan (Fig. 16A). They are interpreted to result from the reworking of the lowstand shelf-edge delta by southward-directed currents.

Correlation of multichannel seismic dip line 84-01 (Fig. 12), acquired across the barrier reef, with line LS92-21 shows that the foreslope deposits in front of the barrier reef within Turneffe Basin represent the last depositional phase of a series of prograding clinoforms observed beneath the barrier reef itself (Figs. 12, 17B). The reef appears, therefore, to be resting on top of these prograding sedimentary bodies. The sediment underlying the back-barrier reef, partly recovered in Spanish Lookout well located 9 km to the north, consists of sandstones, shales, and limestone with corals and shell fragments (Fig. 12). These prograding units are overlain by high-amplitude discontinuous reflectors, interpreted to partially result from a series of collapse structures related to karst dissolution within the reefal cover. The base of this zone occurs at a mean depth of –110 m, calculated from a seismic velocity of 2.133 km/s (obtained from stacking velocities of nearby seismic lines and well information). The base of this zone is in good agreement with the lowest sea-level lowstands that occurred during several maximum glacial stages in the past 500 k.y. (Bard et al., 1990; Pirazzoli et al., 1993). This zone may, therefore, correspond to a succession of karst-modified surfaces included within the several generations of barrier reefs that grew during late Pleistocene highstands of sea level.

INTERPRETATION AND DISCUSSION

The pre-Holocene bathymetry of the Belize margin was characterized by a complex suite of incised valleys, organized in two distinct sea-level lowstand drainage systems. A high plateaulike area offshore Dangriga acted as a drainage divide. The main incised valleys within the lowstand fluvial plain did not run perpendicular to the basin axis, as is usually shown in the models for carbonate-siliciclastic margin evolution. Rather they trended parallel to the shoreline and the exposed, karst-modified paleo–barrier reef. The drainage system south of Dangriga merged farther south into different channels that currently separate the rhomboid reefs, whereas the drainage system north of Dangriga was linked to the English Cay Channel, currently a detached, single, large incised valley only partly in-filled with sediment. A well-developed lowstand shelf-edge siliciclastic delta formed at the mouth of the English Cay Channel in the Turneffe Basin and was partially reworked into a slope fan farther south along and in front of the modern barrier reef. This lowstand delta-slope fan system appears to be the last phase of a series of siliciclastic(?) progradations that are observed beneath the modern barrier and back-barrier reef system.

On the basis of these new findings, the following questions are posed. What controlled the distribution of the two paleodrainage systems and the observed differences between both systems in terms of (1) their general morphology and sedimentary infill and (2) the preferential occurrence of reefal mounds on top of channel levees developed along the drainage system south of Dangriga? How can the lowstand morphology of the Southern Shelf Lagoon be so different from its modern bathymetry? If, as postulated, the lowstand delta-slope fan system at the mouth of the English Cay Channel consists of the latest depositional phase in a series of lowstand coastal progradations, as observed beneath the barrier and back-barrier reef system, is it reasonable to think that the modern and late Pleistocene interglacial barrier and back-barrier reef systems were established on top of siliciclastic deposits? In this interpretation, did the carbonate sediment preferentially grow on top of, and subsequently cover, a series of lowstand siliciclastic paleocoastlines during successive late Quaternary sea-level highstands? In the following paragraphs, we will attempt to answer those questions.

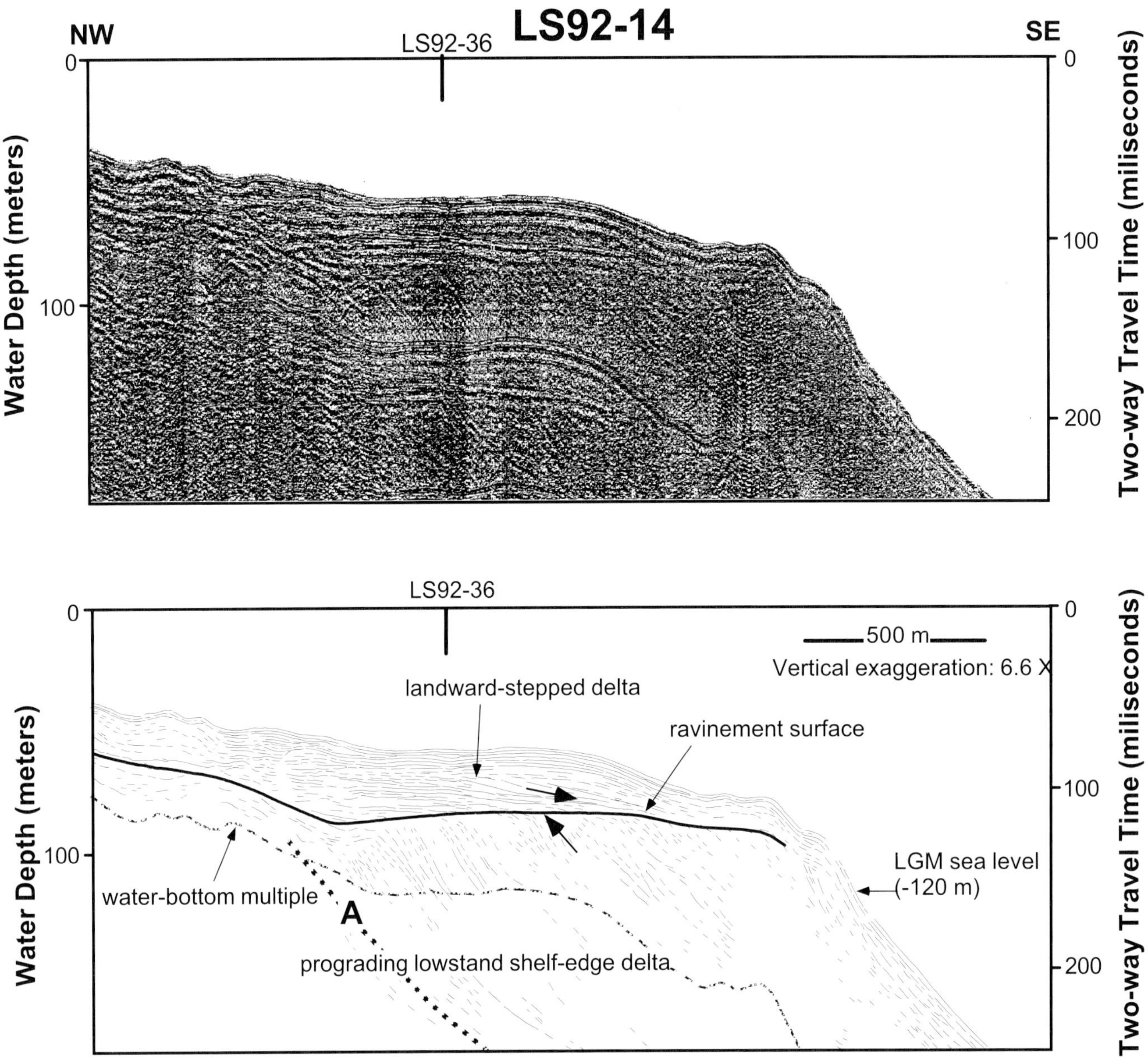

FIG. 15.—Interpreted high-resolution, single-channel, dip-oriented seismic profile (location in Fig. 14B; energy source was Uniboom), showing the upper part of a lowstand shelf-edge delta located at the mouth of the English Cay Channel. When compared with the water-gun seismic line (Fig. 14A), this Uniboom seismic line displays in greater detail the geometry of reflectors and relationship between the landward-stepped delta and the prograding lowstand shelf-edge delta.

1. Tectonic Control on the Organization of the Paleodrainage System

The organization of the incised valleys into two distinct drainage systems in the lowstand fluvial plain appears to have been dictated by a paleo–structural high at the latitude of Dangriga and by the karst-modified barrier reef running parallel to the paleocoastline. The two drainage systems of incised valleys follow a general trend already observed in the upper Tertiary reflectors and the upper Pliocene Belize unconformity (cf. Figs. 9, 13). Deposition of the upper Tertiary strata was probably influenced by a paleo–topographic high related to a narrow, upthrown, wrench-faulted zone reported by Morrice (1992). It appears, therefore, that tectonic elements have exerted an important control on sedimentary distribution patterns in the study area. Karst modification of limestones is also reported to have influenced fluvial drainage (Purdy, 1974a, 1974b). However, this study clearly demonstrates that the karst-modified barrier reef, established during the previous sea-level highstand(s), controlled the paleodrainage, mainly by causing the fluvial systems to run behind and parallel to the exposed barrier reef.

2. Differences Between Paleodrainage Systems

The sediment fill of the incised valleys north of Dangriga is characterized by high-amplitude, continuous to discontinuous, and chaotic reflectors (Fig. 6). This seismic character is inter-

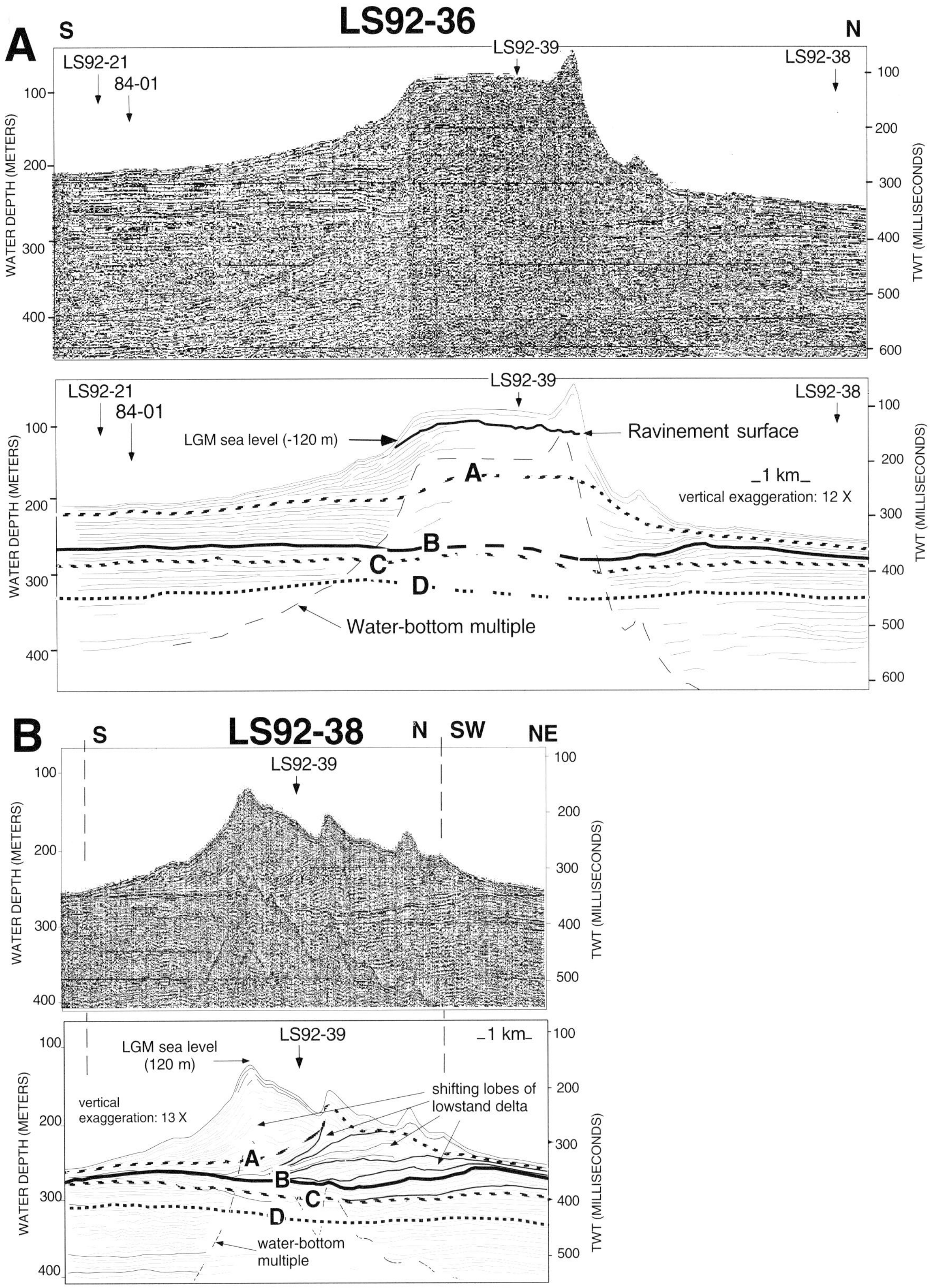

FIG. 16.—Interpreted high-resolution, single-channel, strike-oriented seismic profiles (location of lines in Fig. 14B; energy source was an SSI [Seismic System Inc.] 15-in.^{3} water gun), showing a lowstand shelf-edge delta located at the mouth of the English Cay Channel. Reflectors A to D were used for correlation between the different data sets and the multichannel seismic line (84-01) in the Turneffe Basin and beneath the barrier- and back-barrier-reef system (see Figs. 12, 14, 17). (A) Proximal, strike-oriented, seismic profile across the lowstand shelf-edge delta and along the slope of the barrier reef. Sediment thickness decreases gradually toward the south of the delta and abruptly toward the north, which reflects reworking of deltaic sediments by southward-flowing longshore currents during the last glacial maximum. Note that the reflectors A to D, directly connected to the lowstand delta, can be tied to the downdip progradations observed beneath the barrier and back-barrier reef system along the line 84-01 (Fig. 12). (B) Distal, strike-oriented, seismic profile across the lowstand shelf-edge delta. Note the southward shift of the delta lobes, also explained by southward-flowing longshore currents during the last glacial maximum. TWT = two-way traveltime.

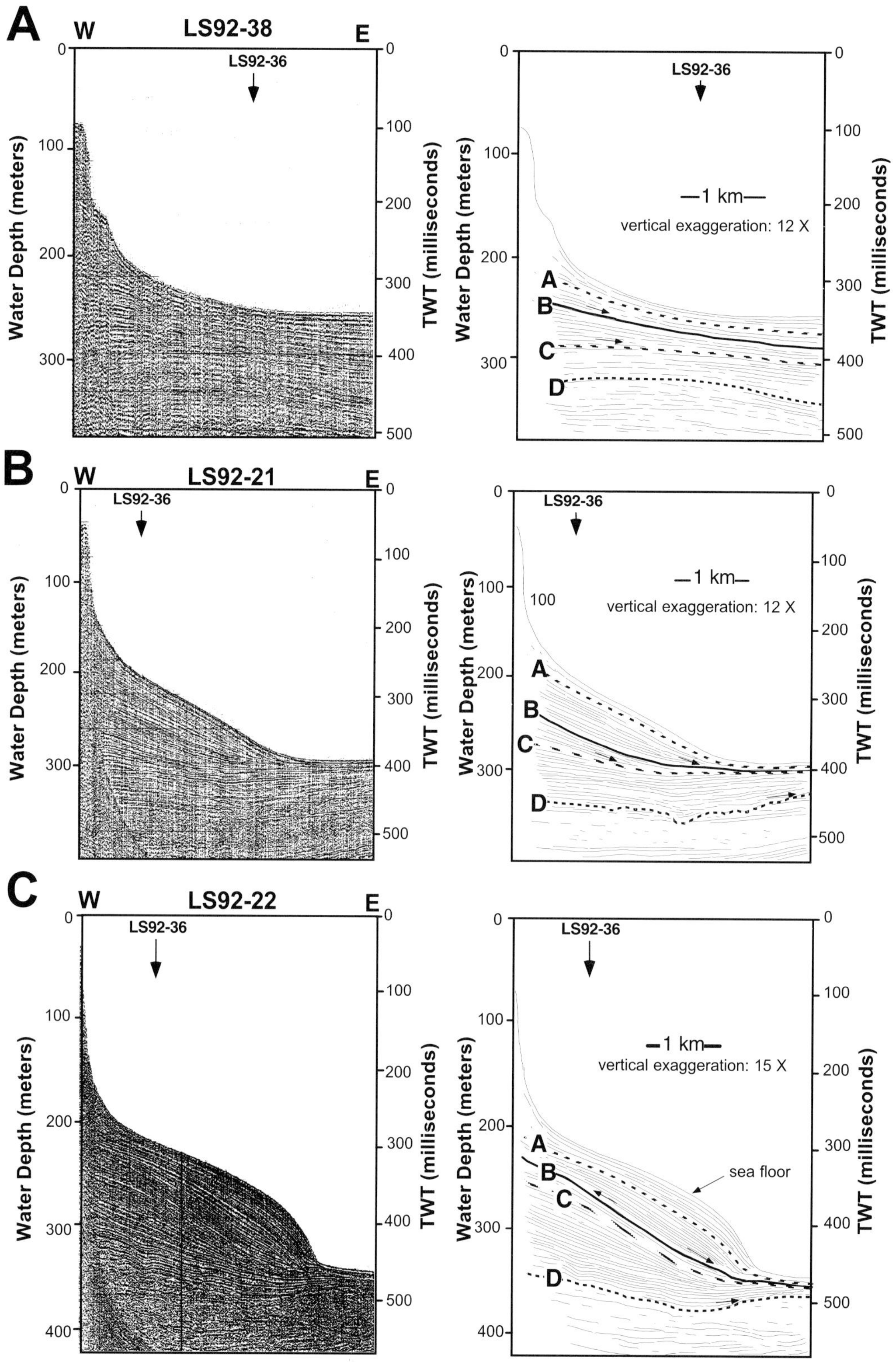

FIG. 17.—Interpreted high-resolution, single-channel, dip-oriented seismic profiles (location of lines in Fig. 8; energy source was an SSI [Seismic System Inc.] 15-in.^{3} water gun), showing the variable thickness of sediment along the slope of the western margin of the Turneffe Basin. The seismic lines are displayed from north (A) to south (C). Reflectors A to D have been identified and used for correlation in the seismic lines throughout the Turneffe Basin (see Figs. 12, 14, 16). (A) Dip-oriented seismic line north of the lowstand shelf-edge delta. The wedgelike slope deposit is very thin. (B) Dip-oriented seismic line between the lowstand shelf-edge delta and the slope fan. The slope sedimentary wedge displays a slight bulge. Reflectors B, C, and D identified in this wedge of slope sediment correlate with reflectors B, C, and D identified in the progradations that are observed beneath the barrier and back-barrier reef in the multichannel line (Fig. 12). (C) Dip-oriented seismic line across the lowstand slope fan. Note the slight bulge, mainly formed by reflectors that do not appear to connect with the barrier reef to the west. These reflectors are thought to represent an accumulation of reworked deltaic sediment by southward-flowing longshore currents transported from the lowstand shelf-edge delta. TWT = two-way traveltime.

preted to represent coarse sediment. In contrast, the equivalent fluvial seismic facies in the incised valleys south of Dangriga consists of low-amplitude and mostly transparent reflectors, interpreted to correspond to fine sediment (Fig. 5).

The Paleozoic Santa Rosa Group, which forms the bulk of the Maya Mountains, consists of coarse-grained arenaceous-rudaceous and fine-grained argillaceous rocks (Bateson and Hall, 1977). The contact between these arenaceous-rudaceous and argillaceous rocks strikes irregularly eastward at about lat 17°N. A close examination of the present Maya Mountains drainage systems shows that most of the area containing the coarse-grained Paleozoic rocks and granites is drained by rivers flowing to the north of Dangriga. Conversely, rivers reaching the modern shoreline south of Dangriga drain most of the terrane with fine-grained argillaceous rocks (Fig. 9). During sea-level lowstands, the drainage system north of Dangriga may have received even larger volumes of coarse sediment than the drainage system to the south. The coarse sediment carried through the lowstand incised valleys north of Dangriga, possibly bed-load–dominated streams, may also explain the formation of the characteristic V-shaped profiles of these valleys. The fine-sediment load of the lowstand rivers flowing south of Dangriga suggests that these were suspended-load–dominated rivers and streams and accounts for their U-shaped profiles (Fig. 10A).

The gradient in the paleodrainage system that extends north of Dangriga is similar to the gradient in the system that extends southward (Fig. 10B). This fact rules out gradient as a possible contributor to the morphological differences observed in the incised valley profiles and valley fill of the paleodrainage systems.

3. Mode and Timing of Shelf-Lagoon Flooding During the Last Glacial to Interglacial Transition

The occurrence of patch reefs on top of the incised valley shoulders south of Dangriga, their absence along the drainage system north of Dangriga, and the different depths of the interfluvial areas between both systems help us to understand the mode and timing of the most recent flooding of the Southern Shelf Lagoon. The current depths of the shallowest shoulders along the incised valleys offshore Dangriga and near the barrier reef average –18 m below sea level. Farther to the north, near the entrance of the English Cay Channel, the shoulders occur at about –23 m (Fig. 6). This increase in depth implies that, during the development of the fluvial system, the barrier reef was modified by karst processes and its top was at a depth shallower than –18 m. During the last transgression, long and narrow estuaries were formed by flooding the lowstand incised valleys. When sea level rose above –18 m, the karst-modified barrier reef still formed a protective barrier between the lagoon and the sea. The English Cay Channel was probably the only break within this barrier. The incised valley shoulders in the lagoonal area north of Dangriga were submerged by restricted, poorly ventilated, and brackish waters. These conditions, which are inimical to carbonate-mound establishment, would explain the absence of patch reefs in the shelf lagoon north of Dangriga.

The –18-m sea-level position corresponds to ca. 8 ka in the Fairbanks (1989) eustatic sea-level curve (Fig. 1). Existing data show that the reestablishment of corals over the karst-modified surface of the previous highstand barrier reef occurred more recently than 8 ka. By drilling on Carrie Bow Cay, Shinn et al. (1982) recovered, at a depth of –16.2 m, Pleistocene limestone with leaching and caliche staining, indicating subaerial exposure. This exposure surface is covered by corals in growth position, radiocarbon dated at ca. 7 ka (core 1 in Fig. 8). In an earlier study, Purdy (1974b) encountered, in a bore hole on Tobacco Cay (core 2 in Fig. 8), a leached and recrystallized coralline limestone surface at a subbottom depth of –14 to –16.5 m. On the basis of these two studies, the top of the karst-modified barrier reef in this central segment of the barrier is situated at about –16 m and was submerged at ca. 7 ka.

Sea-water circulation appears to have exerted an important influence on reef development in the shelf lagoon. In the lagoon south of Dangriga, patch reefs occur on channel levees (Fig. 5C). The same mode of growth has been found by Choi and Ginsburg (1982) and Choi and Holmes (1982) in the lagoon farther to the south. Choi and Holmes (1982) suggested that rhomboid and long sinuous reefs were established and grew on old river levees during successive late Pleistocene transgressions and highstands. Shinn et al. (1979) recovered coral-bearing Pleistocene (stage 5?) limestone beneath the outer rim at the Channel Cay Reef complex and nearby beneath the Boo Bee Patch Reef. On the basis of these findings, Shinn et al. (1979) and Westphall (1986) suggested that the substratum of the rhomboid reefs consists of a multistoried reef complex upon which the Holocene reefs became established. This interpretation has been strengthened by drilling (Robert Ginsburg, pers. commun., 1992) in the area of the rhomboid reefs, which recovered a 30-m-thick coral-bearing upper Pleistocene limestone. In several borings, the uppermost upper Pleistocene limestones overlie stiff clays, interpreted as Pleistocene levee and floodplain deposits that served as the substratum for the rhomboid reefs (Choi and Ginsburg, 1982; Choi and Holmes, 1982). Moreover, nonmarine quartz silts, underlying a peat layer radiocarbon dated at 9,200 yr B.P., were sampled in the enclosed lagoon of the Channel Cay Complex at –32 m (Westphall, 1986). By plotting radiocarbon ages of unaltered corals found in living position from the Channel Cay Complex and the barrier reef (Shinn et al., 1982) against a Belize sea-level curve, Westphall (1986) demonstrated that the Belize Barrier Reef was topographically higher, by as much as 13 m, than the Holocene reef at Channel Cay. According to their growth curve, the reefs at Channel Cay first grew in waters as deep as 15–12 m and subsequently grew to sea level ("catch-up" reefs). Westphall (1986) attributed this catch-up mode of Channel Cay reef growth to the semirestricted water circulation behind the barrier reef. Although the area of the rhomboid reefs remained restricted to the east by the barrier reef itself; it was open to normal-marine conditions from the south. In contrast, the shelf lagoon north of Dangriga, where the English Cay Channel was the only connection with the open sea during the initial flooding, a highly restricted marine environment remained. Conditions inimical to reef growth within this restricted-marine environment can explain the absence of patch reefs in this part of the shelf lagoon.

4. Paleobathymetry and Modern Bathymetry of the Southern Shelf Lagoon

It has long been thought that the modern bathymetry of the Southern Shelf Lagoon, characteristically sloping toward the south (Fig. 3), was inherited from the pre-Holocene bathymetry (Purdy, 1974a, 1974b; Purdy et al., 1975). In this context, the last marine transgression progressed from south to north across the shelf lagoon. The timing of Holocene reef establishment and growth behind the barrier reef was directly influenced by this south to north transgression across the Southern Shelf Lagoon (i.e., Precht, 1993, 1994). However, results of the present study demonstrate that the pre-Holocene bathymetry was characterized by two fluvial drainage systems separated by a drainage divide offshore Dangriga. In the area of the shelf lagoon north of Dangriga, the fluvial paleodrainage dips to the north whereas the modern lagoon floor slopes to the south (Figs. 3, 8). The gradient change from north during the LGM to south today may have its origin in the unusually large volume of lagoonal-marine sediment deposited during the early Holocene in the shelf lagoon north of Dangriga. During the last transgression, a large volume of sediment, comparable in size to the sediment volume that was funneled through the English Cay Channel into Turneffe Basin during the last sea-level lowstand, filled in, at least partially, the area adjacent to the western entrance of the English Cay Channel. Moreover, during the early highstand, sediment from the Belize City delta probably contributed to the sediment fill in the northern part of the lagoon and ultimately to the reversal in gradient.

5. Lowstand Delta and Slope-Fan Development in Front of the Exposed Reef: Ready-Made Substratum for Reef Development During the Following Transgression

Large rivers, such as the Belize River, North Stann Creek, and Sitee River, deliver large volumes of sediment to the modern Belize shoreline. At the mouth of these rivers, deltas are formed and then partially reworked by southward-flowing longshore currents. Reworked deltaic sediment is ultimately deposited along a series of well-developed prograding beach ridges on the south side of the main river mouths. The late Pleistocene lowstand shelf-edge delta at the mouth of the English Cay Channel appears to have been also influenced by southward-flowing longshore currents. Seismic profiles show that individual delta lobes shifted through time toward the south, and the overall deltaic package of strata thickens toward the south (Fig. 16). These observations indicate that the morphology of the lowstand shelf-edge delta was strongly influenced by longshore currents. The slope fan, located 10 to 15 km south of the English Cay Channel lowstand shelf-edge delta, was most likely formed as reworked deltaic sediment by southward-flowing longshore currents. Seismic lines in the lagoon show no incised valleys breaching the barrier reef at the location of the slope fan. Thus, the slope fan does not appear to have been built by sediment coming directly from the shelf lagoon, as was the lowstand shelf-edge delta. Dip sections across the slope fan show that some of the youngest sediment (represented by some reflectors overlying A and B in Figure 17 does not extend beneath the barrier reef. Instead, these younger strata appear to have been derived from the north and transported along strike by southward-flowing longshore currents that are acting today (Wüst, 1964).

In contrast, older slope deposits (sedimentary units between reflectors B, C, and D in seismic line LS92-21, Fig. 17) appear to be associated with prograding wedges, well developed beneath the modern barrier- and back-barrier-reef system (Fig. 12). Strike lines also show that the sedimentary units of the slope deposits (defined by reflectors B, C, and D in Fig. 17) are the condensed continuation of much thicker sedimentary units that form the lowstand shelf-edge delta, defined by reflectors B, C, and D (Figs. 14A, 16), and that the seismic facies between these reflectors do not change along strike (Fig. 16A). These correlations suggest that the progradation observed beneath the barrier reef was produced during a series of sea-level lowstands. The shelf-edge delta and slope fan observed at the mouth of the English Cay Channel, therefore, consist of the latest phase of a series of lowstand siliciclastic prograding wedges, which have served as a substratum for reef establishment during successive transgressions and highstands.

The Spanish Lookout well, located about 10 km to the north of this area, sampled a mixture of siliciclastic and carbonate sediment, which appears to correspond to the progradational section in line 84-01. The distance between the Spanish Lookout well and line 84-01 is too large for reliable correlation. Line 84-01 shows subhorizontal reflectors beneath the prograding complex (Fig. 12). Given these observations, it is feasible that the prograding unit imaged in line 84-01 consists, at least in part, of siliciclastic sediment. The correlation of lines 84-01, LS92-21, and LS92-36 (Figs. 12, 17B, and 16A, respectively) suggests that the prograding wedges beneath the barrier and back-barrier-reef complex and the shelf-edge delta-slope fan complex in front of the modern barrier reef were deposited during lowstands and are composed mostly of siliciclastic sediment. It is significant that the top of the prograding wedges beneath the barrier and back-barrier reef, in addition to the edge of the delta at the mouth of the English Cay Channel, range from about –100 and –120 m below sea level (Figs. 12, 14A, 15) and, therefore, correspond well to late Pleistocene lowstands (Fig. 1; Fairbanks, 1989; Bard et al., 1990; Pirazzoli et al., 1993).

6. Model for the Evolution of the Belize Siliciclastic-Carbonate System Tied to the Last Sea-Level Cycle

Despite the limited paleontologic age control for this study, we are confident that the high-resolution seismic data have imaged the youngest phase of the evolution of the Belize margin. The lowstand and transgressive deposits correlate with the well-established eustatic curves for the past 20 k.y. (Fig. 1) (Fairbanks, 1989; Bard et al., 1990).

During the LGM (18–20 ka), the shoreline was situated at what is now approximately –120-m depth (i.e., Fairbanks, 1989; Bard et al., 1990). In Belize, the shoreline shifted to the front of the previous interglacial stage 5 barrier reef, at about the location where the modern barrier reef exists today (Fig. 18A). A lowstand shelf-edge delta was deposited at the mouth of the English Cay Channel (Figs. 14, 15, 16, 18A). Longshore currents removed some sediment from this delta and redeposited it 15 km to the south into a slope fan. During the LGM, the Southern Shelf Lagoon was divided in two fluvial paleodrainage systems with a divide at the latitude of Dangriga (Figs. 9, 18A).

Following the LGM, the sea transgressed quickly across the top of the previously deposited shelf-edge delta and formed a clear ravinement surface (Figs. 14–15). The ravinement of the shelf-edge delta top reached its lowest level at about (110 m (relative to current sea level), corresponding well with the estimated LGM sea-level lowstand (Figs. 1, 14B). In the middle of this transgression, there was a pause or at least a significant decrease in the rate of sea-level rise for 1 k.y., at ca. 11 ka (Fig. 1B). This event is known as the Younger Dryas (Fig. 1B) and represents a time when the climate returned to almost full glacial (Fairbanks, 1989). During this pause, when sea level is estimated to have been at approximately –65 m, a new fluvio-deltaic system evolved, producing a landward-stepped delta on top of the ravinement surface (Figs. 14, 15, 18B).

After the Younger Dryas, sea level rose quickly and flooded the lower part of the English Cay Channel. The depth of the bayline in the lower part of the English Cay Channel is at about –61 m and marks the initial transgression along this segment of the valley. During this flooding episode, the river deposition was not able to keep up with the rapid sea-level rise and, as a result, only a thin (approximately 4 m) transgressive sedimentary section was deposited in the eastern segment of the English Cay Channel (Fig. 7C). As the rate of sea-level rise slowed, thicker transgressive sediment was deposited (Figs. 7A and 7B) within the upper reaches of the channel. In this part of the channel, the bayline was at a depth of –47 m and 27 m of estuarine sediment was deposited in the valley.

As the ocean flooded the incised valleys farther upstream, in the area of the lowstand fluvial plain that is today the Southern Shelf Lagoon, the river rapidly filled the valleys with estuarine deposits. Rates of sea-level rise had slowed during this time (Fig. 18C). Because the English Cay Channel was the only connection to the sea for the entire drainage area north of Dangriga, strong tidal currents swept along the English Cay Channel depositing tidal-creek bars, and the more brackish waters on the shelf prevented the growth of carbonate mounds. In contrast, South of Dangriga, the estuaries were connected to the open sea, promoting the growth of abundant patch reefs in that area of the Southern Shelf Lagoon.

At ca. 7 ka, the karst-modified surface of the previous (stage 5) barrier reef was finally submerged, and the modern barrier reef became established. Siliciclastic sediment became confined to the coastal area by the establishment of southward-flowing longshore currents. Thus, after a long interval (stages 4–2, or perhaps as long as 60 k.y.; Fig. 1) during which siliciclastic sediment was dominant along the Belize margin, carbonate production began again (Fig. 18D). The establishment of a pure carbonate barrier and back-barrier reef environment marks the beginning of the modern mixed siliciclastic-carbonate sedimentary system along the Belize margin.

7. Reservoir Implications

Understanding the origin of incised valleys and studying the sedimentary processes that are responsible for their development enhances oil exploration and production from these important reservoirs (Bowen et al., 1993). Models for mixed carbonate-siliciclastic margin evolution (i.e., Handford and Loucks, 1993) usually show the main incised valley system within a lowstand fluvial plain running perpendicular to the shoreline and a series of lowstand shelf-edge deltas consisting of several sedimentary bodies with point sources. In contrast to these models, the present study has demonstrated that the main lowstand fluvial-drainage systems of incised valleys on the Belize margin extended parallel to the shore and to the exposed barrier reef.

These results, among others of this study, could have some significant implications for hydrocarbon reservoir exploration and production along ancient mixed siliciclastic-carbonate margins. During sea-level lowstands, when the barrier and back-barrier reefs, deposited along the shelf edge during the previous highstand(s), became exposed, any rivers would tend to flow parallel to the reef. Thus the main reservoirs will trend parallel to the reef, not perpendicular. If sea level fell below the carbonate shelf edge, the fluvial system would dissect the shelf break. Massive coarse-grained lowstand sediment would subsequently be deposited in the basin at the mouth of the channel and along the trend of the exposed reef. Longshore currents would partially redistribute the sediments in front of the exposed barrier reef, forming a suite of subparallel beach ridges in the shallowest part of the lowstand coastline and a lowstand slope fan in deeper waters. However, if the subsequent sea-level rise was rapid, ravines would be cut into the upper parts of the lowstand delta, including the delta front, thus diminishing the quality of the reservoir.

CONCLUSIONS

1. Two lowstand fluvial drainage systems, consisting of a series of incised valleys, were observed in high-resolution seismic grids in the northern part of the Belize Southern Shelf Lagoon. These drainage systems, most likely developed during the LGM, are separated by a divide at the latitude of Dangriga. The drainage divide appears to be rooted on a topographic high in the upper Tertiary deposits of the Southern Shelf Lagoon offshore Dangriga.
2. South of Dangriga, a system of incised valleys drained the LGM fluvial plain toward the south and merged into Victoria Channel and other channels separating modern rhomboid reefs. North of Dangriga, a second system of incised valleys drained the area of the LGM fluvial plain toward the north into the incised English Cay Channel. These paleodrainage systems exhibit differences in the morphology of the incised valleys and the seismic character of their valley fill. The coarse-grained sediment carried through the incised valleys north of Dangriga may have contributed to the characteristic V-shaped profiles, whereas the fine-grained sediment load of the lowstand rivers flowing south of Dangriga could explain their characteristic U-shaped profiles.

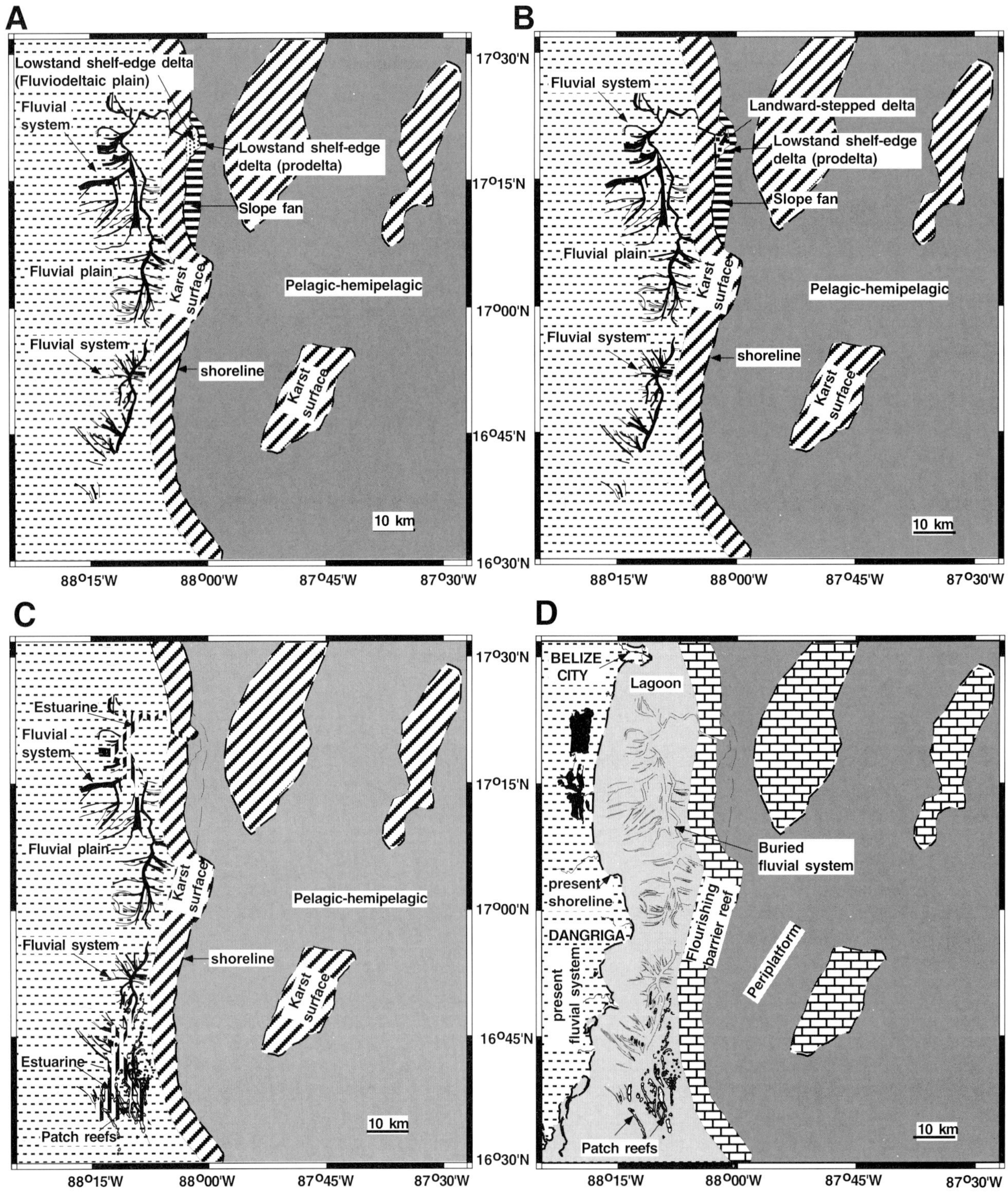

FIG. 18.—Model for the evolution of the mixed siliciclastic-carbonate system along the Belize margin during the last sea-level cycle. Letters A to D in Figure 1B indicate the estimated sea level at each of four time intervals described in the model. (A) During the last glacial maximum, 18–20 ka, the lowstand shoreline was located in front of (i.e., east of) the position of the modern barrier reef. A complex drainage system of incised valleys was developed in the lowstand fluvial plain parallel to and west of the exposed oxygen isotope stage 5 barrier reef. The drainage system south of Dangriga merged into the modern rhomboid reefs, whereas the drainage system north of Dangriga merged into a single incised valley of the English Cay Channel. A lowstand shelf-edge delta and associated slope fan were deposited at the mouth of the English Cay Channel along the western margin of the Turneffe Basin, located where the coastline was during the lowstand. (B) During the Younger Dryas, 11.8–10.6 ka, the rates of sea-level rise decreased, and sea level may have remained constant for about 1 k.y. As a result, a landward-stepped delta formed on top of the lowstand shelf-edge delta. (C) About 8.5 ka, the sea started to transgress across some parts of the lowstand fluvial plain, today the Southern Shelf Lagoon. Estuaries formed in the areas north and south of Dangriga. In the north, the estuaries were still funneled through the English Cay Channel. These environments therefore remained restricted. In the south, however, the estuaries were directly connected to the open sea, which probably favored the growth of patch reefs on river levees. Note that the dominant depositional environment in the shelf area was still siliciclastic. (D) During the past 7 ka, the sea flooded the top of the karst-modified barrier reef to establish the mixed siliciclastic-carbonate system that exists today.

3. A thick, fanlike sedimentary body, characterized by a series of stacked, prograding, and laterally shifting lobes, occurs at the mouth of the English Cay Channel. This sedimentary body is interpreted as a lowstand shelf-edge delta, similar in surface area to the modern highstand delta of the Belize River on which Belize City is located. Southerly longshore currents partially reworked the lowstand shelf-edge delta and created an elongated slope fan at the toe of the exposed barrier reef. The shelf-edge delta and slope fan, observed at the mouth of the English Cay Channel and in front of the barrier reef, served as a substratum for establishment of reefs during the last glacial to interglacial transgression.
4. The well-established sea-level fluctuations during the past 20 k.y. can explain the most recent sedimentary deposits observed in the high-resolution seismic grids across the Belize margin. The shelf-edge delta and slope-fan complex was deposited at the mouth of the English Cay Channel and in front of the barrier reef during the LGM –120 m sea-level lowstand. Subsequent rapid rise cut ravines into the top of the shelf-edge delta. A landward-stepped delta was deposited on top of the ravinement surface. This landward-stepped delta was probably deposited during the Younger Dryas in a 1-k.y.-long interval when rates of sea-level rise greatly diminished. Maximum rates of sea-level rise followed the Younger Dryas, resulting in a thin estuarine deposit in the lower part of the English Cay Channel. As the rate of sea-level rise slowed again, thicker estuarine sediment was deposited in the upstream segment of the English Cay Channel. Finally, as sea level reached the modern shelf lagoon (lowstand fluvial plain), sediment completely filled the incised valleys. During the last phase of the LGM to Holocene sea-level transgression, the entire area of the Southern Shelf Lagoon was covered by a thin blanket of lagoonal sediment.
5. Seawater circulation appears to have played a key role in reef development in the shelf lagoon and influenced the distribution of patch reefs. These reefs occur along the flanks of incised valleys south of Dangriga. Patch reefs are absent north of Dangriga,
6. For most of the past 30 k.y., and perhaps even the past 75 k.y., siliciclastic processes dominated sedimentation along the Belize margin. Carbonate environments were reestablished along the Belize margin only in the past 7 k.y. once the sea submerged the karst-modified barrier reef and flooded the LGM fluvial plain, creating the modern mixed siliciclastic-carbonate system.
7. The slope deposits in front of the modern barrier reef appear to be the continuation of well-developed prograding sedimentary units observed beneath the barrier and back-barrier reef complex. These findings suggest that the prograding strata beneath the barrier reef were produced during a series of sea-level lowstands. In this context, the shelf-edge delta and slope fan at the mouth of the English Cay Channel consist of the latest phase in a series of lowstand siliciclastic prograding wedges, which have served as a substratum for reef establishment during successive sea-level transgressions and highstands.
8. The mixed siliciclastic-carbonate system of the late Quaternary Belize margin is quite different from that of pure carbonate or siliciclastic systems. In the mixed system, the migration and export of siliciclastic sediments toward the shelf margin during sea-level regressions and lowstands overwhelms and suffocates carbonates. Prograding wedges of siliciclastic sediment are deposited below and on the seaward side of the exposed barrier reef by fluvial-deltaic processes during lowstands. These prograding wedges created a surface on top of which a new barrier reef was established during the following transgression. Because the extent of carbonate production is reduced to a minimum during lowstands, little carbonate resedimentation occurs along the slopes and in the adjacent basins. Belize carbonate sediment production was abundant during the succession of late Quaternary sea-level transgressions and highstands, whereas siliciclastic sediments were contained within the coastal zone and/or were transported along the shoreline by longshore currents. In Belize, both siliciclastic and carbonate systems have played an important role in building the continental margin, though at a different time of the sea-level cycle.
9. Results of this study could have significant implications for hydrocarbon reservoir exploration and production along ancient margins, where a siliciclastic shoreline coexisted with an offshore shelf-edge barrier and back-barrier reef system. Unlike most published models, where the lowstand incised valleys extend perpendicular to the reef, here the incised valleys run parallel to the strike of the reef and breach the exposed reef at only a few locations, where lowstand shelf-edge deltas are deposited. These deltas are partially reworked into slope fans in front of and in an orientation parallel to the exposed barrier reef.

ACKNOWLEDGMENTS

The data acquisition and the postcruise processing and interpretation of the seismic grids were partially funded by grants from Amoco, Conoco, Belize Natural Resources (BNR), Exxon (EPR), and Elf-Aquitaine. Special thanks are given to Adrian Pascouet and Jim Hedger at Seismic Systems Incorporated (SSI) in Houston, Robert Girault at Elics in Paris, (France), Susan Morrice at BNR, and Captain Mark Herring and the crew of the R/V *Lone Star*. We thank also Brenda Kirkland George, Mitch Harris, and a third anonymous reviewer for improving the first manuscript.

REFERENCES

Abdulah, K. C., and Anderson, J. B., 1991, Eustatic controls on the evolution of the Pleistocene Brazos-Colorado deltas, Texas: Gulf Coast Section, Society of Economic Paleontologists and Mineralogists Foundation, Twelfth Annual Research Conference, Programs and Abstracts, p. 1–7.

Allen, G. P., and Posamentier, H. W., 1994, Transgressive facies and sequence architecture in mixed tide- and wave-dominated incised valleys: Example from the Gironde estuary, France, *in* Dalrymple, R. W., Boyd, R., and Zaitlin, B. A., eds., Incised-Valley Systems: Origin and Sedimentary Sequences: Tulsa, SEPM (Society for Sedimentary Geology) Special Publication No. 51, p. 225–240.

ANDERSON, J. B., AND THOMAS, M. A., 1991, Marine ice-sheet decoupling as a mechanism for rapid, episodic sea-level change: The record of such events and their influence on sedimentation: Sedimentary Geology, v. 70, p. 87–104.

ANDERSON, J. B., SIRINGAN, F. P., AND THOMAS, M. A., 1990, Sequence stratigraphy of the late Pleistocene–Holocene Trinity/Sabine valley system: Relationship to the distribution of sand bodies within the transgressive systems tracts, in Sequence Stratigraphy as an Exploration Tool: Concepts and Practices in the Gulf Coast: Houston, Gulf Coast Section — Society of Economic Paleontologists and Mineralogists, Foundation, Eleventh Annual Research Conference, p. 15–20.

ANDERSON, J. B., ABDULAH, K., SARZALEJO, S., SIRINGAN, F., AND THOMAS, M. A., 1996, Late Quaternary sedimentation and high-resolution sequence stratigraphy of the east Texas shelf, *in* De Batist, M. and Jacobs, P., eds., Geology of siliciclastic shelf seas: London, Geological Society of London Special Publication 117, p. 95–124.

BARD, E., HAMELIN, B., AND FAIRBANKS, R. G., 1990, U-Th ages obtained by mass spectrometry in corals from Barbados: Sea level during the past 130,000 years: Nature, v. 346, p. 456–458.

BATESON, J. H., AND HALL, I. S. H., 1977, The Geology of the Maya Mountains, Belize: London, Institute of Geological Sciences, Natural Environment Research Council, Overseas Memoir 3, 44 p.

BELPERIO, A. P., AND SEARLE, E. D., 1988, Terrigenous and carbonate sedimentation in the Great Barrier Reef Province, *in* Doyle, L. J. and Roberts, H. H., eds., Carbonate Clastic Transitions: Amsterdam, Elsevier, v. 42, p. 143–174.

BLUM, M. D., 1993, Genesis and architecture of incised valley fill sequences: A late Quaternary example from the Colorado River, Gulf Coastal Plain, *in* Weimer, P. and Posamentier, H. eds., Siliciclastic sequence stratigraphy: Tulsa, American Association of Petroleum Geologists Memoir 58, p. 259–283.

BOWEN, D. W., WEIMER, P., AND SCOTT, A. J., 1993, The relative success of siliciclastic sequence stratigraphic concepts in exploration: Examples from incised valley fill and turbidite system reservoirs, *in* Weimer, P. and Posamentier, H., eds., Siliciclastic Sequence Stratigraphy, Recent Developments and Applications: Tulsa, American Association of Petroleum Geologists Memoir 58, p. 15–42.

BROWN, A. A., AND LOUCKS, R. G., 1993, Influence of sediment type and depositional processes on stratal patterns in the Permian basin-margin Lamar Limestone, McKittrick Canyon, Texas, *in* Loucks, R. G. and Sarg, J. F., eds., Carbonate Sequence Stratigraphy, Recent Developments and Applications: Tulsa, American Association of Petroleum Geologists Memoir 57, p. 133–156.

CARTER, L., CARTER, R. M., NELSON, C. S., FULTHORPE, C. S., AND NEIL, H. L., 1990, Evolution of Pliocene to Recent abyssal sediment waves on Bounty Channel levees, New Zealand: Marine Geology, v. 95, p. 97–109.

CHOI, D. R., AND GINSBURG, R. N., 1982, Siliciclastic foundations of Quaternary reefs in the southernmost Belize Lagoon, British Honduras: Geological Society of America Bulletin, v. 93, p. 116–126.

CHOI, D. R., AND HOLMES, C. W., 1982, Foundations of Quaternary reefs in South Central Belize Lagoon, Central America: American Association of Petroleum Geologists Bulletin, v. 66, p. 2663–2681.

COLMAN, S. M., AND MIXON, R. B., 1988, The record of major Quaternary sea level changes in a large coastal plain estuary, Chesapeake Bay, Eastern United States: Palaeogeography, Palaeoclimatology, Palaeoecology, v. 68, p. 99–116.

DAVIES, P. J., SYMONDS, P. A., FEARY, D. A., AND PIGRAM, C. J., 1989, The evolution of the carbonate platforms of northeast Australia, *in* Crevello, P. D., Wilson, J. L., Sarg, J. F., and Read, J. F., eds., Controls on Carbonate Platform and Basin Development: Tulsa, Society of Economic Paleontologists and Mineralogists, Special Publication 44, p. 233–258.

DEFENSE MAPPING AGENCY HYDROGRAPHIC/TOPOGRAPHIC CENTER, 1984, Loran C–Caribbean Sea–Belize: Ambergris Cay to Pelican Cays: Washington, D.C., Defense Mapping Agency Hydrographic/Topographic Center, 1:150,000.

DILLON, W. P., AND VEDDER, J. G., 1973, Structure and development of the continental margin of British Honduras: Geological Society of America Bulletin, v. 84, p. 2713–2732.

DOLAN, J. F., 1989, Eustatic and tectonic controls on deposition of hybrid siliciclastic/carbonate basinal cycles: Discussion with examples: American Association of Petroleum Geologists Bulletin, v. 73, p. 1233–1246.

DROXLER, A. W., AND SCHLAGER, W., 1985, Glacial versus interglacial sedimentation rates and turbidite frequency in the Bahamas: Geology, v. 13, p. 799–802.

ESKER, D., 1998, Interplay between tectonics and eustasy in the mixed carbonate-siliciclastic system of the southern part of the Belize Southern Shelf Lagoon: Unpublished Ph.D. Dissertation, University of Miami, Miami, 281 p.

ESKER, D., EBERLI, G. P., AND MCNEILL, D., 1995, The Neogene structural evolution of the southern Belize lagoon, in Belize Research Program, Industrial Sponsor's Report: Miami, University of Miami, Rosenstiel School of Marine and Atmospheric Sciences, p. 1–51.

ESKER, D., EBERLI, G. P., AND MCNEILL, D., 1998, The structural and sedimentological controls on the reoccupation of Quarternary incised valleys, Belize Southern Lagoon: American Association of Petroleum Geologists Bulletin, v. 82, p. 2075–2109.

FAIRBANKS, R. G., 1989, A 17,000-year glacio-eustatic sea level record: Influence of glacial melting rates on the Younger Dryas event and deep-ocean circulation: Nature, v. 342, p. 637–642.

GISCHLER, E., 1994, Sedimentation on three Caribbean atolls: Glovers reef, Lighthouse reef and Turneffe Islands, Belize: Facies, v. 31, p. 243–254.

GLASER, K. S., AND DROXLER, A. W., 1991, High production and highstand shedding from deeply submerged carbonate banks, northern Nicaragua Rise: Journal of Petrology, v. 61, p. 128–142.

GRAMMER, G. M., AND GINSBURG, R. N., 1992, Highstand versus lowstand deposition on carbonate platform margins: Insight from Quaternary foreslopes in the Bahamas: Marine Geology, v. 103, p. 125–136.

HALLEY, R. B., SHINN, E. A., HUDSON, J. H., AND LIDZ, B., 1977, Recent and relict topography of Boo Bee patch reef, Belize, *in* Taylor, D. L., ed., Proceedings, Third International Coral Reef Symposium: Miami, University of Miami, Rosenstiel School of Marine and Atmospheric Sciences, no. 2, p. 29–36.

HANDFORD, C. R., AND LOUCKS, R. G., 1993, Carbonate depositional sequences and systems tracts—Responses of carbonate platforms to relative sea-level changes, *in* Loucks, R. G. and Sarg, J. F., eds., Carbonate Sequence Stratigraphy, Recent Developments and Applications: Tulsa, American Association of Petroleum Geologists Memoir 57, p. 3–41.

HARRIS, P. T., 1994, Incised valleys and backstepping deltaic deposits in a foreland-basin setting, Torres Strait and Gulf of Papua, Australia, *in* Dalrymple, R. W., Boyd, R., and Zaitlin, B. A., eds., Incised-Valley Systems: Origin and Sedimentary Sequences: Tulsa, SEPM (Society for Sedimentary Geology) Special Publication No. 51, p. 98–108.

HINE, A. C., AND MULLINS, H. T., 1983, Modern carbonate shelf-slope breaks: Tulsa, Society of Economic Paleontologists and Mineralogists, Special Publication 33, p. 169–188.

JACQUIN, T., ARNAUD-VANNEAU, A., AND ARNAUD, H., 1991, Systems tracts and depositional sequences in a carbonate setting: A study of continuous outcrops from platform to basin at the scale of seismic lines: Marine and Petroleum Geology, v. 8, p. 122–139.

JAMES, N. P., AND GINSBURG, R. N., 1979, The Seaward Margin of Belize Barrier and Atoll Reefs: London, International Association of Sedimentologists Special Publication 3, 191 p.

KINDINGER, J. L., BALSON, P. S., AND FLOCKS, J. G., 1994, Stratigraphy of the Mississippi-Alabama shelf and the Mobile River incised-valley system, *in* Dalrymple, R. W., Boyd, R., and Zaitlin, B. A., eds., Incised-Valley Systems: Origin and Sedimentary Sequences: Tulsa, SEPM (Society for Sedimentary Geology) Special Publication 51, p. 83–95.

KOSS, J. E., ETHRIDGE, F. G., AND SCHUMM, S. A., 1994, An experimental study of the effects of base-level change on fluvial, coastal plain and shelf systems: Journal of Sedimentary Research, v. B64, p. 90–98.

LABEYRIE, L. D., DUPLESSY, J. C., AND BLANC, P. L., 1987, Variations in mode of formation and temperature of oceanic deep waters over the past 125,000 years: Nature, v. 327, p. 477–482.

LARA, M. E., 1993, Divergent wrench faulting in the Belize Southern Lagoon: Implications for Tertiary Caribbean plate movements and Quaternary reef distribution: American Association of Petroleum Geologists Bulletin, v. 77, p. 1041–1063.

LIGHTY, R. G., MACINTYRE, I. G., AND STUCKENRATH, R., 1982, Acropora palmata reef framework: A reliable indicator of sea level rise in the western Atlantic for the past 10,000 years: Coral Reefs, v. 1, p. 125–130.

MAZZULLO, S. J., ANDERSON-UNDERWOOD, K. E., BURKE, C. D., AND BISCHOFF,

W. D., 1992, Holocene coral patch reef ecology and sedimentary architecture, northern Belize, Central America: Palaios, v. 7, p. 591–601.

MORRICE, S., 1992, Gladden pull-apart basin, offshore Belize: Mesozoic and early Cenozoic development of the Gulf of Mexico and Caribbean region, a context for hydrocarbon exploration (abs.): Gulf Coast Section, Society of Economic Paleontologists and Mineralogists Foundation, Thirteenth Annual Research Conference, Programs and Abstracts, p. 50–51.

MORRIS, R. O., 1989, Colson Point to Belize City including Lighthouse Reef and Turneffe Islands—Monkey River to Colson Point: Washington, D.C., Defense Mapping Agency Hydrographic Office, 1:125,000.

MULLINS, H. T., 1983, Comment on "Eustatic control of turbidites and winnowed turbidites": Geology, v. 11, p. 57–58.

NICHOL, S. L., BOYD, R., AND PENLAND, S., 1994, Stratigraphic response of wave-dominated estuaries to different relative sea-level and sediment supply histories: Quaternary case studies from Nova Scotia, Louisiana and eastern Australia, *in* Dalrymple, R. W., Boyd, R., and Zaitlin, B. A., eds., Incised-Valley Systems: Origin and Sedimentary Sequences: Tulsa, SEPM (Society for Sedimentary Geology) Special Publication 51, p. 265–283.

PEEDERMAN, F. M., AND DAVIES, P. J., 1993, Sedimentological response of an outer-shelf, upper-slope sequence to rapid changes in Pleistocene eustatic sea level: Hole 820A, northeastern Australian margin, *in* McKenzie, J. A., Davies P. J., Palmer-Julson, A., et al., Proceedings of the Ocean Drilling Program, Scientific Results, v. 133: College Station, Texas, Ocean Drilling Program, p. 303–312.

PIRAZOLLI, P. A., RADTKE, U., HANTORO, W. S., JOUANNIC, C., HOANG, C. T., CAUSSE, C., AND BOREL BEST, M., 1993, A one million-year-long sequence of marine terraces on Sumba Island, Indonesia: Marine Geology, v. 109, p. 221–236.

PRECHT, W. F., 1993, Holocene coral patch reef ecology and sedimentary architecture, northern Belize, Central America: Palaios v. 8, p. 499–506.

PRECHT, W. F., 1994, Holocene reef development on the Belize Shelf (abs.): American Association of Petroleum Geologists Bulletin, v. 78, p. 1474.

PURDY, E. G., 1974a, Karst-determined facies patterns in British Honduras: Holocene carbonate sedimentation model: American Association of Petroleum Geologists Bulletin, v. 58, p. 825–855.

PURDY, E. G., 1974b, Reef configurations: Cause and effect, *in* Laporte, L. F., ed., Reefs in time and space, selected examples from the recent and ancient: Tulsa, Society of Economic Paleontologists and Mineralogists, Special Publication 18, p. 9–76.

PURDY, E. G., PUSEY, C. W., III, AND WANTLAND, K. F., 1975, Continental shelf of Belize—Regional shelf attributes, *in* Wantland, K. F. and Pusey, C. W., III, eds., Belize shelf-carbonate sediments, clastic sediments and ecology: Tulsa, American Association of Petroleum Geologists, Studies in Geology 2, p. 1–52.

PUSEY, W. C., III, 1975, Holocene carbonate sedimentation on northern Belize Shelf, *in* Wantland, K. F. and Pusey, W. C., III, eds., Belize Shelf—Carbonate sediments, clastic sediments and ecology: Tulsa, American Association of Petroleum Geologists, Studies in Geology 2, p. 131–233.

RUDDIMAN, W. F., AND MCINTYRE, A., 1981, The mode and mechanism of the last deglaciation: Oceanic evidence: Quaternary Research, v. 16, p. 125–134.

SARG, J. F., 1988, Carbonate Sequence Stratigraphy, in Wilgus et al., eds., Sea-Level Changes: An Integrated Approach: Tulsa, Society of Economic Paleontologists and Mineralogists Special Publication No. 42, p. 155–181.

SCHLAGER, W., 1992, Sedimentology and sequence stratigraphy of reefs and carbonate platforms: Tulsa, American Association of Petroleum Geologists, Continuing Education Course Note Series 34, 71 p.

SCHLAGER, W., REIJMER, J. J. G., AND DROXLER, A. W., 1994, Highstand shedding of carbonate platforms: Journal of Sedimentary Research, v. B64, p. 270–281.

SHACKLETON, N. J., 1987, Oxygen isotopes, ice volume and sea level: Quaternary Science Review, v. 6, p. 183–190.

SCOTT, M. R., 1975, Distribution of clay minerals on Belize shelf, *in* Wantland, K. F. and Pusey, W. C., III, eds., Belize shelf-carbonate sediments, clastic sediments and ecology: Tulsa, American Association of Petroleum Geologists, Studies in Geology 2, p. 97–130.

SHINN, E. A., HALLEY, R. B., HUDSON, J. H., LID, B., AND ROBBIN, D. M., 1979, Three dimensional aspects of Belize patch reefs (abs.): American Association of Petroleum Geologists Bulletin, v. 63, p. 528.

SHINN, G. A., HUDSON, H. J., HALLEY, R. B., LIDZ, B., ROBBIN, D. M., AND MACINTYRE, I. G., 1982, Geology and sediment accumulation rates at Carrie Bow Cay, Belize, *in* Rutzler, K. and Macintyre, I. G., eds., The Atlantic Barrier Reef Ecosystem at Carrie Bow Cay, Belize: I. Structure and Communities: Washington, D.C., Smithsonian Contributions to the Marine Sciences, no. 12, p. 63–75.

THOMAS, M. A., AND ANDERSON, J. B., 1988, The effect and mechanism of episodic sea level events: The record preserved within late Wisconsinan–Holocene incised valley-fill sequences: Transactions of the Gulf Coast Association of Geological Societies, v. 38, p. 399–406.

THOMAS, M. A., AND ANDERSON, J. B., 1989, Glacial eustatic controls on seismic sequences and parasequences of the Trinity/Sabine incised valley, Texas Continental Shelf: Transactions of the Gulf Coast Association of Geological Societies, v. 39, p. 563–569.

THOMAS, M. A., AND ANDERSON, J. B., 1994, Sea-level controls on the facies architecture of the Trinity/Sabine incised-valley system, Texas continental shelf, *in* Dalrymple, R. W., Boyd, R., and Zaitlin, B. A., eds., Incised-Valley Systems: Origin and Sedimentary Sequences: Tulsa, SEPM (Society for Sedimentary Geology) Special Publication No. 51, p. 63–82.

TRINCARDI, F., CORREGIARI, A., AND ROVERI, M., 1994, Late Quaternary transgressive erosion and deposition in a modern epicontinental shelf: The Adriatic semienclosed basin: Geo-Marine Letters, v. 14, p. 41–51.

VAIL, P. R., 1987, Seismic stratigraphy interpretation procedure, *in* Bally, A. W., ed., Atlas of Seismic Stratigraphy: Tulsa, American Association of Petroleum Geologists, Studies in Geology 27, v. 1, p. 1–10.

VAIL, P. R., MITCHUM, R. M., AND THOMPSON, 1977, Global cycles of relative changes of sea level, *in* Payton C. E., ed., Seismic Stratigraphy—Applications to hydrocarbon exploration: Tulsa, American Association of Petroleum Geologists Memoir 26, p. 83–97.

VAIL, P. R., AUDEMARD, F., BOWMAN, S. A., EISNER, P. N., AND PEREZ-CRUZ, G., 1991, The stratigraphic signatures of tectonics, eustasy and sedimentology—An overview, *in* Einsele, G., Ricken, W., and Seilacher, A., eds., Cycles and events in stratigraphy: Berlin, Springer-Verlag, p. 617–659.

VAN WAGONER, J. C., MITCHUM, R. M., CAMPION, K. M., AND RAHMANIAN, V. D., 1992, Siliciclastic sequence stratigraphy in well logs, cores, and outcrops: Concepts for high-resolution correlation of time and facies: Tulsa, American Association of Petroleum Geologists, Methods in Exploration Series 7, 55 p.

WANTLAND, K. F., 1975, Distribution of Holocene benthonic foraminifera on the Belize Shelf, *in* Wantland, K. F. and Pusey, W. C., III, eds., Belize shelf-carbonate sediments, clastic sediments and ecology: Tulsa, American Association of Petroleum Geologists, Studies in Geology 2, p. 332–399.

WANTLAND, K. F., AND PUSEY, W. C., III, 1971, A guidebook for the field trip to the Southern Shelf of British Honduras (October 10–13, 1971): 21st Annual Meeting, Gulf Coast Association of Geological Societies and Gulf Coast Section — Society of Economic Paleontologists and Mineralogists, p. 87.

WESTPHALL, M. J., 1986, Anatomy and history of a ringed-reef complex, Belize, Central America: Unpublished M.S. Thesis, University of Miami, Miami, 159 p.

WILBER, R. J., MILLIMAN, J. D., AND HALLEY, R. B., 1990, Accumulation of bank-top sediment on the western slope of Great Bahama Bank: Rapid progradation of a carbonate megabank: Geology, v. 18, p. 970–974.

WÜST, G., 1964, Stratification and circulation in the Antillean-Caribbean basins: Part 1. Spreading and mixing of the water types with an oceanographic atlas: New York, Columbia University Press, 201 p.

ZAITLIN, B. A., DALRYMPLE, R. W., AND BOYD, R.,1994, The stratigraphic organization of incised-valley systems associated with relative sea-level change, *in* Dalrymple, R. W., Boyd, R., and Zaitlin, B. A., eds., Incised-valley systems: Origin and sedimentary sequences: Tulsa, SEPM (Society for Sedimentary Geology) Special Publication No. 51, p. 45–60.

INDEX

D

T